The Genetic Code

First position	Second position U	C	A	G	Third position
U	Phe	Ser	Tyr	Cys	U
	Phe	Ser	Tyr	Cys	C
	Leu	Ser	Stop	Stop	A
	Leu	Ser	Stop	Trp	G
C	Leu	Pro	His	Arg	U
	Leu	Pro	His	Arg	C
	Leu	Pro	Gln	Arg	A
	Leu	Pro	Gln	Arg	G
A	Ile	Thr	Asn	Ser	U
	Ile	Thr	Asn	Ser	C
	Ile	Thr	Lys	Arg	A
	Met	Thr	Lys	Arg	G
G	Val	Ala	Asp	Gly	U
	Val	Ala	Asp	Gly	C
	Val	Ala	Glu	Gly	A
	Val	Ala	Glu	Gly	G

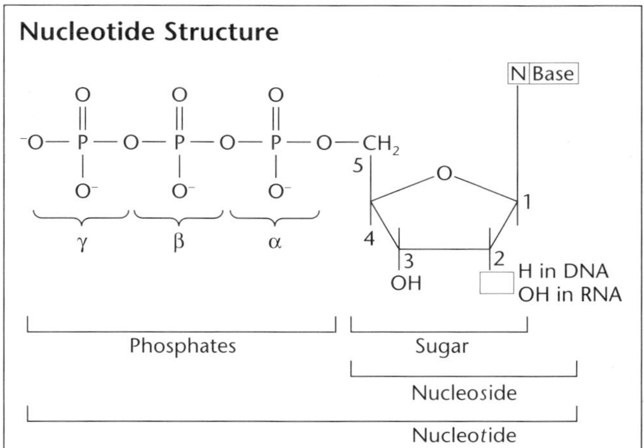

Names of Nucleic Acid Subunits

Base	Nucleoside	Nucleotide	Abbreviation RNA	DNA
Adenine	Adenosine	Adenosine triphosphate	ATP	dATP
Guanine	Guanosine	Guanosine triphosphate	GTP	dGTP
Cytosine	Cytidine	Cytidine triphosphate	CTP	dCTP
Thymine	Thymidine	Thymidine triphosphate		dTTP
Uracil	Uridine	Uridine triphosphate	UTP	

Snyder & Champness

Molecular Genetics of Bacteria

FIFTH EDITION

About the Companion Website

This book is accompanied by a companion website:

The URL is www.wiley.com/go/snyder/mgb5 

The website includes:

Figures from the book

Snyder & Champness

Molecular Genetics of Bacteria

FIFTH EDITION

Tina M. Henkin
Ohio State University, Columbus, Ohio

Joseph E. Peters
Cornell University, Ithaca, New York

Washington, DC

WILEY

Editorial Correspondence: ASM Press, 1752 N Street, NW, Washington, DC 20036-2904, USA

Registered Offices: John Wiley & Sons, Inc., 111 River Street, Hoboken, NJ 07030, USA

For details of our global editorial offices, customer services, and more information about Wiley products, visit us at www .wiley.com.
Wiley also publishes its books in a variety of electronic formats and by print-on-demand. Some content that appears in standard print versions of this book may not be available in other formats.

Library of Congress Cataloging-in-Publication Data
Names: Henkin, Tina M., author. | Peters, Joseph E., author. Molecular genetics of bacteria.
Title: Snyder and Champness molecular genetics of bacteria / Tina M. Henkin, Ohio State University, Columbus, Ohio, Joseph E. Peters, Cornell University, Ithaca, New York.
Other titles: Molecular genetics of bacteria
Description: Fifth edition. | Hoboken, NJ : John Wiley & Sons, Inc. ; Washington, D.C. : American Society for Microbiology, [2020] | Series: ASM books | Revised edition of: Molecular genetics of bacteria / Larry Snyder . . . [et al.]. 4th ed. c2013. | Includes bibliographical references and index.
Identifiers: LCCN 2020001847 (print) | LCCN 2020001848 (ebook) | ISBN 9781555819750 (hardback) | ISBN 9781555819767 (adobe pdf) | ISBN 9781683673576 (epub)
Subjects: LCSH: Bacterial genetics. | Bacteriophages—Genetics. | Molecular genetics.
Classification: LCC QH434 .S59 2020 (print) | LCC QH434 (ebook) | DDC 579.3/135—dc23
LC record available at https://lccn.loc.gov/2020001847
LC ebook record available at https://lccn.loc.gov/2020001848

Cover images:
Illustration: © 2017 Terese Winslow LLC
Micrograph: Scanning electron micrograph of adsorption of phage JWAlpha to *Achromobacter* DSM 11852 cells. From Wittmann J, Dreiseikelmann B, Rohde M, et al., Virology **11:**14 (2014), https://www.ncbi.nlm.nih .gov/pmc/articles/PMC3915230/, CC-BY 2.0
Authors: Henkin, courtesy Ohio State University; Peters, courtesy Jay N. Worley

Cover and interior design: Susan Brown Schmidler

Printed and bound in Singapore by Markono Print Media Pte Ltd

10 9 8 7 6 5 4 3 2 1

Contents

DNA Repair and Mutagenesis 385

Regulation of Gene Expression: Genes and Operons 425

Preface

Snyder and Champness Molecular Genetics of Bacteria is a new edition of a classic text updated to address the massive advances in the field of bacterial molecular genetics. We renamed the book as a tribute to the original authors, Larry Snyder and Wendy Champness, who welcomed us as coauthors for the 4th edition and trusted us to continue to build on the strong foundation of their multiple editions in carrying this important text forward. As with the previous editions, we have endeavored to keep the page length approximately the same. This meant making many hard choices of what to remove to make room for exciting new and important material. We are very happy that every illustration is now in full color, which offered us the opportunity to rethink each drawing and clarify and standardize features, which we believe will improve their use by instructors in classroom lectures.

Perhaps the most significant force in molecular genetics research since the last edition has been the plummeting cost of DNA sequencing. This factor has created an explosion of new sequence information of both independent genomes and microbial communities in the form of metagenomics, where DNA is extracted directly from all of the organisms in an environment. This information has vastly expanded our picture of the tree of life and the massive contribution of uncultured species. The broader availability of DNA sequencing at a reasonable price has also left its mark on genomic techniques. These new techniques and new information have had a considerable impact in every chapter and provided the impetus for a new chapter, "Genomes and Genomic Analysis" (chapter 13).

We expanded chapter 1, on DNA structure, DNA replication, and chromosome segregation, to include many advances in our understanding of how chromosomes are managed and the molecular machines that carry out these processes. Our understanding of the nature of FtsK and related DNA-pumping enzymes, the evolving role of SeqA, the mechanism of chromosome partitioning, and the domain structure of the chromosomes also benefited from multiple technological innovations. Chapter 2 focuses on mechanisms of gene expression, from transcription through mRNA turnover, translation, and post-translational effects, including protein targeting, which was moved into this chapter. We reduced the historical aspects of chapter 3, retaining key landmarks such as the important role of the F plasmid discovered by Esther Lederberg, so the chapter now focuses more on practical aspects of genetic analysis.

Newer molecular techniques that have replaced some of the classic approaches (e.g., for generation of targeted chromosomal mutations) are now discussed in the new chapter 13.

Chapter 4 presents a concise understanding of bacterial plasmids as important contributors to the genomic content in bacteria as well as essential tools in molecular biology. Significant additions to the chapter include an expanded discussion of the two major mechanisms of segregation and the ever-broadening view of toxin-antitoxin systems. Toxin-antitoxin systems were first discovered for their role in plasmid stabilization, but while the diversity of molecular mechanisms has expanded, important questions remain concerning the real function of these systems when situated in bacterial chromosomes. Chapter 5, which focuses on conjugation, continues to set its roots in the original conjugal plasmid, the fertility plasmid. We included considerable new information that relates to our recognition that conjugal systems appear to be as common in the form of integrating conjugative elements (ICEs) as they are in stand-alone plasmids. The diversity of ICEs is remarkable, and this chapter strives to provide a foundation for these dynamic elements, which are responsible for the largest known genomic islands transmitted between bacteria.

In chapter 6, we expanded the discussion of natural transformation and its regulation to include additional comparative information about how these systems vary in different groups of organisms. We consolidated the discussion of lytic and lysogenic bacteriophages and their roles in transduction of bacterial DNA as chapter 7. We organized the information on phage biology based on the different functions required for phage infection and replication, and followed this with a discussion of phage genetics, their use in bacterial genetic transfer, and their roles as tools for molecular biology.

We streamlined chapter 8, "Transposition, Site-Specific Recombination, and Families of Recombinases," to make room for additional families of elements including the exciting and still somewhat enigmatic HUH transposons, as well as group II mobile introns, and an advanced appreciation of the interrelationship between mobile elements and host DNA replication. Transposons continue to provide an important tool in genomics, and mobile genetic elements in general provide the most significant mechanisms for the transfer of antibiotic resistance. As the spread of antibiotic resistance is slowly nullifying the effectiveness of antibiotics worldwide, understanding the mechanisms of this spread is more important than ever. Chapter 9, "Molecular Mechanisms of Homologous Recombination," continues to be grounded in the central role that homologous recombination plays in the repair of DNA double-strand breaks. We expanded the chapter to include a better appreciation of the multiple pathways used to load the RecA recombinase onto different types of DNA substrates.

We broadly updated chapter 10, "DNA Repair and Mutagenesis," to reflect our increased mechanistic understanding across many DNA repair systems, as well as information on how mechanisms established in bacterial systems continue to contribute to our understanding of disease in humans. We extensively updated chapter 11, which focuses on mechanisms of gene regulation of individual genes and operons, to include new information as the field continues to advance. In chapter 12, we then applied the principles learned in chapter 11 to global regulatory systems that regulate multiple sets of genes and operons, often in response to multiple regulatory inputs. *Bacillus subtilis* sporulation, a complex developmental system, is presented in depth as a final example that integrates many of the different mechanisms that are introduced in chapters 11 and 12.

Chapter 13, "Genomes and Genomic Analysis," is a new chapter that consolidates relevant topics previously found elsewhere in the book and provides considerable new information on this topic. We provide background on the multiple mechanisms used for DNA sequencing, including the newest generations of high-throughput sequencing strategies. Having hundreds of thousands of bacterial genomes has allowed us to gain a better understanding of how genomes are organized as well as the relationship between core genes and genes acquired by horizontal gene transfer. The chapter also provides basic information on genome annotation and comparative genomics. Chapter 13 further presents an expanded picture of numerous systems that bacteria use to guard against horizontal gene transfer. Although horizontal gene transfer is by far the most important mechanism for evolution in bacteria and archaea, it also provides the greatest vulnerability, with the relentless onslaught of bacteriophages and mobile elements that can sap cellular resources or inactivate important or essential host genes. Significantly, host defense systems also provide the most important tools ever developed for molecular biology. The new chapter provides expanded background on diverse restriction endonucleases and the important roles they play in molecular biology. We cover the variety of tools that are available for cloning and gene assembly, as well as the advantages and disadvantages of these techniques to help guide the investigator. These techniques allow never-imagined possibilities for quickly and accurately constructing synthetic DNA fragments for testing ideas or allowing advances in engineering, including assembling entire bacterial genomes. We greatly expanded the section on CRISPR/Cas systems and chose the Cas9 system, important in many applications in a multitude of model systems and human genome engineering, to illustrate on the book's cover. CRISPR/Cas systems are very diverse, falling into six distinct types and tens of subtypes. We provide the reader with the background needed to understand how these fascinating systems evolved, the role they play in the natural environment, and the massive promise they hold in genome engineering.

Acknowledgments

We continue to be indebted to a large number of individuals for their help, talking through ideas, sharing insights from their area of expertise, or reading over sections for accuracy and clarity. Other individuals have alerted us to errors, suggested areas to include, or provided graphics. However, if errors remain in the text, they are our own.

We acknowledge those whose comments on earlier editions have carried over to the current edition. In addition, we thank Esther Angert, Melanie Berkmen, Briana Burton, Druba Chattoraj, Mick Chandler, Pete Christie, Alan Grossman, Alba Guarné, Claudia Guldimann (and the Zürich book club), John Helmann, Laura Hug, Ailong Ke, Bénédicte Michel, Kit Pogliano, Lise Raleigh, Phoebe Rice, Jim Samuelson, Mark Sutton, Anthony Vecchiarelli, Bob Wiess, Steve Winans, Wei Yang, and Steve Zinder.

It was a great pleasure to work with Director Christine Charlip and ASM Press, and we look forward to the new partnership with John Wiley & Sons, Inc. We enjoyed working first with Production Manager Larry Klein and then transitioning to Developmental Editor Ellie Tupper, who coordinated the project to the finish line. We continue to be grateful for the professional and patient work of Patrick Lane of ScEYEence Studios for making our visions come to life in these illustrations.

About the Authors

Tina M. Henkin is a Professor of Microbiology, Robert W. and Estelle S. Bingham Professor of Biological Sciences, and Distinguished University Professor at The Ohio State University, where she has been teaching microbiology and bacterial genetics since 1995. She received her B.A. in biology at Swarthmore College and her Ph.D. in genetics at the University of Wisconsin-Madison, and did postdoctoral work in molecular microbiology at Tufts University Medical School. Her research focuses on gene regulation in Gram-positive bacteria, primarily using *Bacillus subtilis* as a model. Her laboratory uncovered the T-box regulatory mechanism, in which the leader RNAs of bacterial genes bind a specific uncharged tRNA to modulate expression of the downstream genes. This work led to the discovery of riboswitch RNAs that bind cellular metabolites to mediate similar regulatory responses. Current work focuses on elucidating the basis for specific ligand recognition and molecular mechanisms for ligand-mediated changes in RNA structure in a variety of riboswitch classes. She is a Fellow of the American Academy of Microbiology, the American Association for the Advancement of Science, and the American Academy of Arts and Sciences, a member of the National Academy of Sciences, and co-winner of the National Academy of Sciences Pfizer Prize in Molecular Biology for her work on riboswitch RNAs.

Joseph E. Peters is a Professor of microbiology at Cornell University, where he has been teaching bacterial genetics and microbiology at the graduate and undergraduate level since 2002. He received his B.S. from Stony Brook University and his Ph.D. from the University of Maryland at College Park. He did postdoctoral work at the Johns Hopkins University School of Medicine, in part as an NSF-Alfred P. Sloan Foundation postdoctoral research fellow in molecular evolution. His research has focused on the intersection between DNA replication, recombination, and repair and how it relates to evolution, especially in the area of transposition. Most recently he has been interested in the evolution of defense systems like CRISPR/Cas systems and how they can be repurposed by mobile elements for new tasks. Research in his lab is funded by the National Science Foundation, the U.S. Department of Agriculture, and the National Institutes of Health. He is the director of graduate studies for the field of microbiology at Cornell.

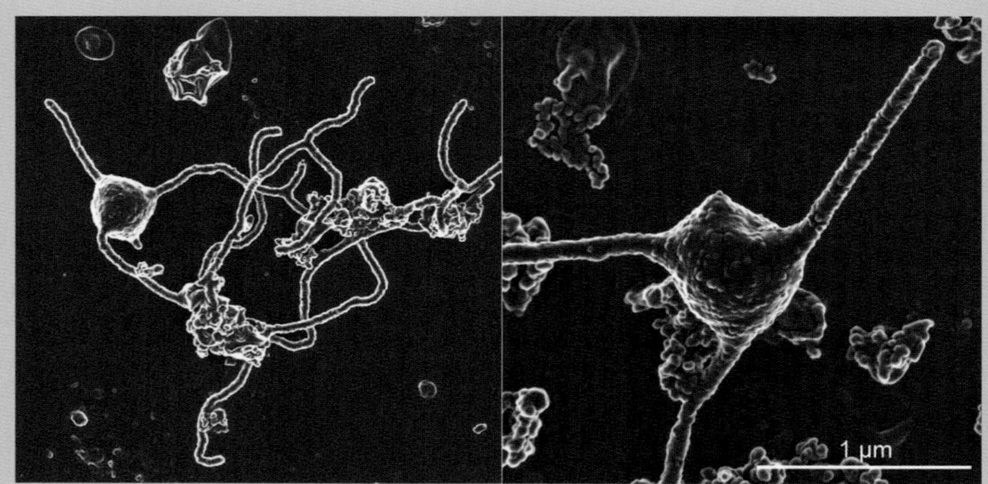

SEM images of the archaeon "*Candidatus* Prometheoarchaeum syntrophicum" strain MK-D1. Reprinted from Imachi H, et al, ©2020, Springer Nature, CC-BY 4.0, http://creativecommons.org/licenses/by/4.0/.

Introduction

THE GOAL OF THIS TEXTBOOK is to introduce the student to the field of bacterial molecular genetics. From the point of view of genetics and genetic manipulation, bacteria are relatively simple organisms. There also exist model bacterial organisms that are easy to grow and easy to manipulate in the laboratory. For these reasons, most methods in molecular biology and recombinant DNA technology that are essential for the study of all forms of life have been developed around bacteria. Bacteria also frequently serve as model systems for understanding cellular functions and developmental processes in more complex organisms. Much of what we know about the basic molecular mechanisms in cells, such as transcription, translation, and DNA replication, has originated with studies of bacteria. This is because such central cellular functions have remained largely unchanged throughout evolution. Core parts of RNA polymerase and many of the translation factors are conserved in all cells, and ribosomes have similar structures in all organisms. The DNA replication apparatuses of all organisms contain features in common, such as sliding clamps and editing functions, which were first described in bacteria and their viruses, called bacteriophages. Chaperones that help other proteins fold and topoisomerases that change the topology of DNA were first discovered in bacteria and their bacteriophages. Studies of repair of DNA damage and mutagenesis in bacteria have also led the way to an understanding of such pathways in eukaryotes. Excision repair systems, mutagenic polymerases, and mismatch repair systems are remarkably similar in all organisms, and defects in these systems are responsible for multiple types of human cancers.

In addition, as our understanding of the molecular biology of bacteria advances, we are finding a level of complexity that was not appreciated previously. Because of the small size of the vast majority of bacteria, it was impossible initially to recognize the high level of organization that exists in bacteria, leading to the misconception that bacteria were merely "bags of enzymes," where small size allowed passive diffusion to move cellular constituents around. However, it is now clear that movement and positioning within the bacterial cell are highly controlled processes. For example, despite the lack of a specialized membrane structure called the nucleus (the early defining feature of the "prokaryote" [see below]), the genome of bacteria is exquisitely organized to facilitate its repair and expression during DNA replication. In addition, advances facilitated by molecular genetics and microscopy have made it clear that many cellular processes occur in highly organized subregions within the cell. Once it was appreciated that bacteria evolved in the same basic way as all other living organisms, the relative simplicity of bacteria paved the way for some of the most important scientific advances in any field, ever. It is safe to say that a bright future awaits the fledgling bacterial geneticist, where studies of relatively simple bacteria, with their malleable genetic systems, promise to uncover basic principles of cell biology that are common to all organisms and that we can now only imagine.

However, bacteria are not just important as laboratory tools to understand other organisms; they also are important and interesting in their own right. For instance, they play essential roles in the ecology of Earth. They are the only organisms that can "fix" atmospheric nitrogen, that is, convert N_2 to ammonia, which can be used to make nitrogen-containing cellular constituents, such as proteins and nucleic acids. Without bacteria, the natural nitrogen cycle would be broken. Bacteria are also central to the carbon cycle because of their ability to degrade recalcitrant natural polymers, such as cellulose and lignin. Bacteria and some types of fungi thus prevent Earth from being buried in plant debris and other carbon-containing material. Toxic compounds, including petroleum, many of the chlorinated hydrocarbons, and other products of the chemical industry can also be degraded by bacteria. For this reason, these organisms are essential in water purification and toxic waste clean-up. Moreover, bacteria produce most of the naturally occurring so-called greenhouse gases, such as methane and carbon dioxide, which are in turn used by other types of bacteria. This cycle helps maintain climate equilibrium. Bacteria have even had a profound effect on the geology of Earth, being responsible for some of the major iron ore and other mineral deposits in Earth's crust.

Another unusual feature of bacteria and archaea (see below) is their ability to live in extremely inhospitable environments, many of which are devoid of life except for microbes. These are the only organisms living in the Dead Sea, where the salt concentration in the water is very high. Some types of bacteria and archaea live in hot springs at temperatures close to the boiling point of water (or above in the case of archaea), and others survive in atmospheres devoid of oxygen, such as eutrophic lakes and swamps.

Bacteria that live in inhospitable environments sometimes enable other organisms to survive in those environments through symbiotic relationships. For example, symbiotic bacteria make life possible for *Riftia* tubeworms next to hydrothermal vents on the ocean floor, where living systems must use hydrogen sulfide in place of organic carbon and energy sources. In this symbiosis, the bacteria obtain energy and fix carbon dioxide by using the reducing power of the hydrogen sulfide given off by the hydrothermal vents, thereby furnishing food in the form of high-energy carbon compounds for the worms, which lack a digestive tract. Symbiotic cyanobacteria allow fungi to live in the Arctic tundra in the form of lichens. The bacterial partners in the lichens fix atmospheric nitrogen and make carbon-containing molecules through photosynthesis to allow their fungal partners to grow on the tundra in the absence of nutrient-containing soil. Symbiotic nitrogen-fixing *Rhizobium* and *Azorhizobium* spp. in the nodules on the roots of legumes and some other types of higher plants allow the plants to grow in nitrogen-deficient soils. Other types of symbiotic bacteria digest cellulose to allow cows and other ruminant animals to live on a diet of grass. Bioluminescent bacteria even generate light for squid and other marine animals, allowing illumination, camouflage, and signaling in the darkness of the deep ocean.

Bacteria are also important to study because of their role in disease. They cause many human, plant, and animal diseases, and new diseases are continuously appearing. Knowledge gained from the molecular genetics of bacteria helps in the development of new ways to treat or otherwise control old diseases that can be resistant to older forms of treatment, as well as emerging diseases.

Some bacteria that live in and on our bodies also benefit us directly. The role of our commensal bacteria in human health is only beginning to be appreciated. It has been estimated that of the 10^{14} cells in a human body, only half are human! Of course, bacterial cells are much smaller than our cells, but this shows how our bodies are adapted to live with an extensive bacterial microbiome, which helps us digest food and avoid disease, among other roles, many of which are yet to be uncovered.

Bacteria have also long been used to make many useful compounds, such as antibiotics, and chemicals, such as benzene and citric acid. Bacteria and their bacteriophages are also the source of many of the useful enzymes used in molecular biology.

In spite of substantial progress, we have only begun to understand the bacterial world around us. Bacteria are the most physiologically diverse organisms on Earth, and the importance of bacteria to life on Earth and the potential uses to which bacteria can be put can only be guessed. Thousands of different types of bacteria are known, and new insights into their cellular mechanisms and their applications constantly emerge from research with bacteria. Moreover, it is estimated that less than 1% of the types of bacteria living in the soil and other environments have ever been isolated. Recent culture-independent mechanisms indicate that bacterial diversity is much greater than we ever imagined (see Hug et al., Suggested Reading). In this new picture, it seems that less than half of the major lineages of bacteria have representatives that have been cultured. Organisms in these uncharacterized groups of bacteria may have all manner of interesting and useful functions. Clearly, studies of bacteria will continue to be essential to our future efforts to understand, control, and benefit from the biological world around us, and bacterial molecular genetics will be an essential tool in these efforts. However, before discussing this field, we must first briefly discuss the evolutionary relationship of bacteria to other organisms.

The Biological Universe

The Bacteria

This textbook comes at a very exciting time in our understanding of the interrelationship of all living things on the planet. After the landmark work of Carl Woese, all organisms on Earth were assigned to three major groups called domains: the bacteria (formerly eubacteria), the archaea (formerly archaebacteria), and the eukaryotes (see Woese and Fox, Suggested Reading). However, it is now clear that two major divisions account for these three groups. Bacteria form one of these divisions, while eukaryotes are now believed to have diverged out of the archaea. Figure 1 shows the microbiologists' view of the living world, where microbes provide most of the diversity and eukaryotes occupy a relatively small niche. This is not a far-fetched concept. Sequence data show that we differ from chimpanzees by only 2% of our DNA sequence, while 25 to 50% of the genes in a typical bacterium are unique to the species. Furthermore, while mammals diverged from each other on the order of millions of years ago, the main bacterial lineages diverged billions of years ago.

Bacteria can differ greatly in their physical appearance under the microscope. Although most are single celled and rod shaped or spherical, some are multicellular and undergo complicated developmental cycles. The cyanobacteria (formerly called blue-green algae) are bacteria, but they have chlorophyll and can be filamentous, which is why they were originally mistaken for algae. The antibiotic-producing actinomycetes, which include *Streptomyces* spp., are also bacteria, but they form hyphae and stalks of spores, making them resemble fungi. Another bacterial group, the *Caulobacter* spp., have both free-swimming and sessile forms that attach to surfaces through a holdfast structure. Some of the most dramatic-appearing bacteria of all belong to the genus *Myxococcus*, members of which can exist as free-living single-celled organisms but can also aggregate to form fruiting bodies, much like slime molds. As mentioned above, bacterial cells are usually much smaller than the cells of higher organisms, but one very large bacterium, *Epulopiscium*, can be over half a millimeter long, longer than even most eukaryotic cells (see Angert, Suggested Reading). In addition, unlike most bacteria that multiply by simple division, *Epulopiscium* gives birth to multiple live progeny. Despite the fact that some bacteria are found in dramatically different shapes and sizes, they cannot be distinguished simply by their physical appearance; instead, it is necessary to use biochemical criteria, such as the sequences of their ribosomal proteins or RNAs (rRNAs), whose sequences are characteristic of the three domains of life.

GRAM-NEGATIVE AND GRAM-POSITIVE BACTERIA

Bacteria have historically been divided into two major subgroups, the **Gram-negative** and **Gram-positive** bacteria. This division was based on the response to a test called the Gram stain. "Gram-negative" bacteria retain little of the dye and are pink after this staining procedure, whereas "Gram-positive" bacteria retain more of the dye and turn deep blue. The difference in staining typically reflects the fact that Gram-negative bacteria are surrounded by a thinner structure composed of both an inner and an outer membrane, while the structure surrounding Gram-positive bacteria is much thicker, consisting of a single membrane surrounded by a thicker wall. However, this older form of classification is being replaced by talking about the phyla of bacteria as determined by the DNA sequence. The *Firmicutes* are a broad group containing *Bacillus*, clostridia, lactic acid bacteria, and the *Tenericutes*, including the mycoplasmas. Firmicutes have been referred to as low G+C Gram-positive bacteria based on the low percentage of guanine and cytosine (low G+C) compared to adenine and thymine often found in the genome sequence of members of this group (see chapter 1). However, having a low G+C genome is not a universal feature of the *Firmicutes*, which limits the utility of the designation. Another group of bacteria that were classically described as high G+C and Gram positive because they typically possess a higher percentage of guanine and cytosine includes the *Actinobacteria* (actinomycetes), such as *Streptomyces* and *Mycobacterium*.

The Gram designation system of classifying bacteria is particularly weak for capturing the diversity of the numerous phyla that stain Gram negative. While many bacteria historically referred to as Gram negative, such as *Escherichia coli*, *Pseudomonas*, and *Rhizobium*, fall within a broad group known as the *Proteobacteria*, many other characterized and uncharacterized groups also exist. It is also worth pointing out that relying on a staining form of classification is particularly contrived when talking about uncultured bacteria or those that are only capable of growth as symbionts in other organisms. Given all of these considerations, instead of using the Gram-positive and Gram-negative designations as a tool for

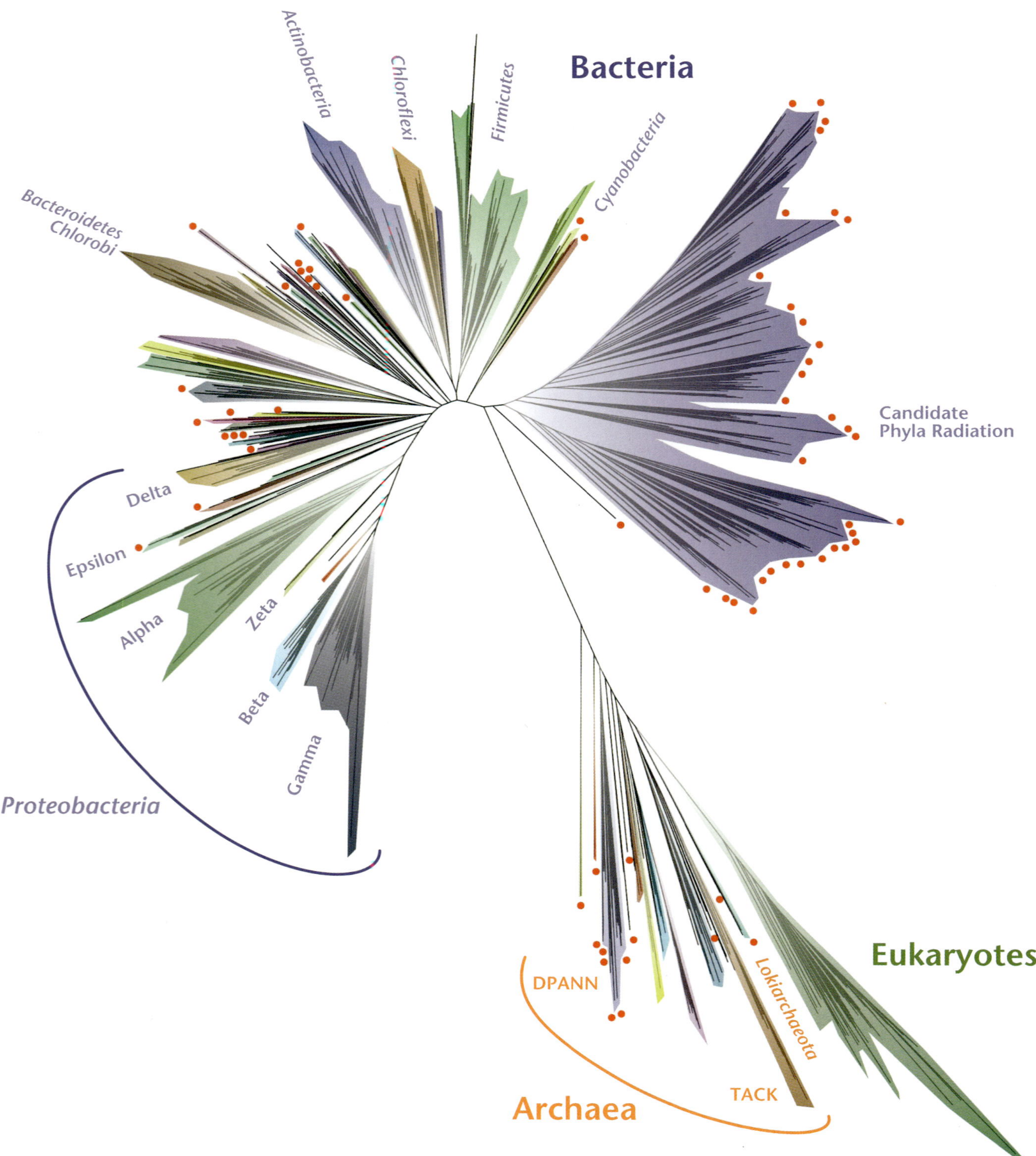

Figure 1 A molecular tree of life capturing diversity using ribosomal proteins from sequenced genomes (see Hug et al., Suggested Reading). Selected major linages within bacteria are indicated, including the *Proteobacteria* and the subgroups Alpha, Beta, Delta, Epsilon, Gamma, and Zeta, the *Firmicutes*, and the Candidate Phyla Radiation, which is almost completely devoid of cultured representatives. For the *Archaea*, two superphyla, TACK and DPANN, are indicated and described in the text. The position of the archaeal *Lokiarchaeota* lineage is indicated. Genome sequences from members of the *Lokiarchaeota* lineage indicate that they possess molecular systems previously believed to be found only in *Eukaryotes*. Red dots indicate lineages that have no cultured representatives. Adapted with permission from Hug L, et al, *Nat Microbiol* **1**:16048 (2016), https://doi.org/10.1038/nmicrobiol.2016.48.

describing the relatedness of groups of bacteria, we now opt for a more precise designation by referring to the phyla. In addition, while we will remind the reader of the older designations of Gram positive and Gram negative in the text when referring to the superstructure surrounding the cell, we have transitioned to describing the cell wall structure more directly.

The Archaea

The **archaea** (formerly called archaebacteria) are single-celled organisms that resemble bacteria. Bacteria and archaea were previously considered one group called the **prokaryotes**, which means "before the nucleus," and are still sometimes referred to by this name. However, there are hazards with this designation, as it erroneously implies that bacteria and archaea are more closely related to each other than they are to eukaryotes (see below). The term "prokaryote" also tends to obscure the idea that present-day bacteria and archaea have had as much time to evolve as eukaryotes since they separated and so did not "come before." Bacteria and archaea do lack a nucleus, the defined membrane structure found in eukaryotes that houses the vast majority of the genes of the organism. The presence or absence of a nuclear membrane greatly influences the mechanism used to manufacture proteins in the cell. The processes of messenger RNA (mRNA) synthesis and translation are coupled in bacteria and archaea, since no nuclear membrane separates the ribosomes (which synthesize proteins) from the DNA (see chapter 2). However, in most eukaryotes, mRNA made in the nucleus must be transported through the nuclear membrane before it can be translated into protein in the cytoplasm, where the ribosomes reside, and transcription and translation do not occur simultaneously.

Besides lacking a nucleus, bacterial and archaeal cells lack many other cellular constituents common to eukaryotes, including mitochondria and chloroplasts. They also lack such visible organelles as the Golgi apparatus and the endoplasmic reticulum. While bacteria can possess compartmentalized features for a wide variety of purposes such as the storage of carbon and nutrients and for specialized enzymatic processes, the absence of most organelles generally gives bacterial and archaeal cells a much simpler appearance under the microscope than eukaryotes.

Extremophiles (or "extreme-condition-loving" organisms), as their name implies, live under extreme conditions where other types of organisms cannot survive, such as at very high temperatures, in highly acidic environments, and at very high osmolality, such as in the Dead Sea. Most extremophiles are archaea. However, it is becoming clear that archaea also are important components of many less extreme environments; for example, archaea perform unique biochemical functions, such as making methane, and can be normal inhabitants of the human microbiome.

The archaea themselves are a very diverse group of organisms, and our understanding of the phylogeny of archaea is an area of intense research that is currently in flux. One exciting update to our understanding is that it is now clear that eukaryotes branch out of the archaea and, more specifically, out of the TACK superphylum (*Thaumarchaeota*, *Aigarchaeota*, *Crenarchaeota*, and *Korarchaeota*). Perhaps the most exciting area of research involves the discovery of uncultured archaeal lineages like the *Lokiarchaeota* that possess numerous molecular systems that were previously only associated with eukaryotes, providing a clear link between archaea and eukaryotes (see Spang et al., Suggested Reading). Excitingly, an archaeon that appears to be on the border between prokaryotes and eukaryotes has now been isolated and is pictured at the start of this chapter (see Imachi et al., Suggested Reading). The relationship between the superphylum TACK archaea and two other large divisions with the archaea, the *Euryarchaeota* and the superphylum DPANN (*Diapherotrites*, *Parvarchaeota*, *Aenigmarchaeota*, *Nanoarchaeota*, and *Nanohaloarchaeota*), is currently under investigation. While the basic molecular processes of archaea are an active area of investigation, much less is known about the archaea than about the bacteria.

The Eukaryotes

The **eukaryotes** are members of the third domain of organisms on Earth, which branches out of the archaea. This domain includes organisms as variable as plants, animals, fungi, and the highly diverse protists. The name "eukaryote" is derived from the presence of their nuclear membrane. Eukaryotic cells usually have a nucleus, and the word *karyon* in Greek means "nut," which is what the nucleus must have resembled to early cytologists. The eukaryotes can be unicellular, like yeasts, protozoans, and some types of algae, or they can be multicellular, like plants and animals. In spite of their widely diverse appearances, lifestyles, and relative complexity, however, all eukaryotes are remarkably similar at the biochemical level, particularly in their pathways for macromolecular synthesis.

MITOCHONDRIA AND CHLOROPLASTS AND THE ROLE OF ENDOSYMBIOSIS IN EVOLUTION

Essentially all eukaryotic cells contain something, the mitochondria, that ties them to the world of bacteria. The mitochondria of eukaryotic cells are the sites of efficient adenosine triphosphate (ATP) generation through respiration. Evidence, including the sequences of many genes in their rudimentary chromosomes, indicates that the mitochondria of eukaryotes are descended from free-living bacteria from the *Alphaproteobacteria* that formed a symbiosis with a primitive ancestor of eukaryotes. A specialized type of symbiosis, where the symbiont resides entirely within another organism, is called **endosymbiosis**.

Plant cells and some unicellular eukaryotic cells also contain chloroplasts, the site of photosynthesis. Like mitochondria, chloroplasts are also descended from free-living bacteria, in this case *Cyanobacteria*. Mitochondria and chloroplasts resemble bacteria in many ways. For instance, they contain DNA that encodes the components of oxidative phosphorylation and photosynthesis, as well as rRNAs and transfer RNAs (tRNAs). Even more striking, the mitochondrial and chloroplast rRNA and ribosomal proteins, as well as the membranes of the organelles, more closely resemble those of bacteria than they do those of eukaryotes. Comparisons of the sequences of highly conserved organelle genes, like the rRNAs, with those of bacteria also support this view of chloroplasts and mitochondria (see Yang et al., Suggested Reading).

Mitochondria and chloroplasts may have come to be associated with early eukaryotic/archaeal cells when these cells engulfed bacteria to take advantage of their superior energy-generating systems or their ability to obtain energy from light through photosynthesis. The engulfed bacteria eventually lost many of their own genes, which moved to the chromosome, from where they are expressed and their products are transported back into the organelle. The organelles had by then lost their autonomy and had become permanent endosymbionts of the eukaryotic cells. In one view, the role of endosymbiosis in the evolution of eukaryotes calls into question the use of phylogenetic trees as a tool to describe the interrelationship of entire organisms, given that many organisms represent a conglomeration of genomes (see Koonin, Suggested Reading).

Interestingly, members of a large newly identified group within bacteria with no cultured representatives, called the Candidate Phyla Radiation (Figure 1), have relatively small genomes and, to varying degrees, limited metabolic capacities. This has led to the idea that many of these lineages may be symbionts, something that has been shown in some cases.

What Is Genetics?

Genetics can be simply defined as the manipulation of DNA to study cellular and organismal functions. Since DNA encodes all of the information needed to make the cell and the complete organism, the effects of changes in DNA can give clues to the normal functions of the cell and organism.

Before the advent of methods for manipulating DNA in the test tube, the only genetic approaches available for studying cellular and organismal functions were those of **classical genetics**. In this type of analysis, mutants (i.e., individuals that differ from the normal, or wild-type, members of the species by a certain observable attribute, or **phenotype**) that have alterations in the function being studied are isolated. The changes in the DNA, or mutations, responsible for the altered function are then localized in the chromosome by genetic crosses. The mutations are then grouped into genes to determine how many different genes are involved. The functions of the genes can then sometimes be deduced from the specific effects of the mutations on the organism. The ways in which mutations in genes involved in a biological system can alter the biological system provide clues to the normal functioning of the system.

Classical genetic analyses continue to contribute greatly to our understanding of developmental and cellular biology. A major advantage of the classical genetic approach is that mutants with an altered function can be isolated and characterized without any *a priori* understanding of the molecular basis of the function. Classical genetic analysis also is often the only way to determine how many gene products are involved in a function and, through suppressor analysis, to find other genes whose products may interact either physically or functionally with the products of these genes.

The development of **molecular genetic techniques** has greatly expanded the range of methods available for studying genes and their functions. These techniques include methods for isolating DNA and identifying the regions of DNA that encode particular functions, as well as methods for altering or mutating DNA in the test tube and then returning the mutated DNA to cells to determine the effect of the mutation on the organism.

The approach of first cloning a gene and then altering it in the test tube before reintroducing it into the cells to determine the effects of the alterations is sometimes called **reverse genetics** and is essentially the reverse of a classical genetic analysis. In classical genetics, a gene is known to exist only because a mutation in it has caused an observable change in the organism. With the molecular genetic approach, a gene can be isolated and mutated in the test tube without any knowledge of its function. Only after the mutated gene has been returned to the organism does its function become apparent.

Rather than one approach supplanting the other, molecular genetics and classical genetics can be used to answer different types of questions, and the two approaches often complement each other. In fact, the most remarkable insights into biological functions have often come from a combination of classical and molecular genetic approaches.

Bacterial Genetics

In bacterial genetics, genetic techniques are used to study bacteria. Applying genetic analysis to bacteria is not different in principle from applying it to other organisms. However, the methods that are available differ greatly.

Some types of bacteria are relatively easy to manipulate genetically. As a consequence, more is known about some bacteria than is known about any other type of organism. Some of the properties of bacteria that facilitate genetic experiments are described below.

Bacteria Are Haploid

One of the major advantages of bacteria for genetic studies is that they are **haploid**. This means that they have only one copy, or **allele**, of each gene. This property makes it much easier to identify cells with a particular type of mutation.

In contrast, most eukaryotic organisms are **diploid**, with two alleles of each gene, one on each homologous chromosome. Most mutations are **recessive**, which means that they do not cause a phenotype in the presence of a normal copy of the gene. Therefore, in diploid organisms, most mutations have no effect unless both copies of the gene in the two homologous chromosomes have the mutation. Backcrosses between different organisms with the mutation are usually required to produce offspring with the mutant phenotype, and even then, only some of the progeny of the backcross have the mutated gene in both homologous chromosomes. With a haploid organism such as a bacterium, however, most mutations have an immediate effect and there is no need for backcrosses.

Short Generation Times

Another advantage of many bacteria for genetic studies is that they have very short generation times. The **generation time** is the length of time the organism takes to reach maturity and produce offspring. If the generation time of an organism is too long, it can limit the number of possible experiments. Some strains of the bacterium *E. coli* can reproduce every 20 minutes under ideal conditions. With such rapid multiplication, cultures of the bacteria can be started in the morning, and the progeny can be examined later in the day.

Asexual Reproduction

Another advantage of bacteria is that they multiply asexually, by cell division. Sexual reproduction, in which individuals of the same species must mate with each other to give rise to progeny, can complicate genetic experiments because the progeny are never identical to their parents. To achieve purebred lines of a sexually reproducing organism, a researcher must repeatedly cross the individuals with their relatives. However, if the organism multiplies asexually by cell division, all the progeny are genetically identical to their parent and to each other. Genetically identical organisms are called **clones**. Some simpler eukaryotes, such as yeasts, and some types of plants, such as water hyacinths, can also multiply asexually to form clones. Identical twins, formed from the products of the division of an egg after it has been fertilized, are clones of each other. While there are a few examples where mammals have been cloned by transplanting a somatic cell into the ovary, bacteria form clones of themselves every time they divide.

Colony Growth on Agar Plates

Genetic experiments often require that numerous individuals be screened for a particular property. Therefore, it helps if large numbers of individuals of the species being studied can be propagated in a small space.

With some types of bacteria, thousands, millions, or even billions of individuals can be screened on a single agar-containing petri plate. Once on an agar plate, these bacteria divide over and over again, with all the progeny remaining together on the plate until a visible lump, or **colony**, has formed. Each colony is composed of millions of bacteria, all derived from one original bacterium and hence all clones of the original bacterium

Colony Purification

The ability of some types of bacteria to form colonies through the multiplication of individual bacteria on plates allows colony purification of bacterial strains and mutants. If a mixture of bacteria containing different mutants or strains is placed on an agar plate, individual mutant bacteria or strains in the population each multiply to form colonies. However, these colonies may be too close together to be separable or may still contain a mixture of different strains of the bacterium. If the colonies are picked and the bacteria are diluted before replating, discrete colonies that result from the multiplication of individual bacteria may appear. No matter how crowded the bacteria were on the original plate, a pure strain of the bacterium can be isolated in one or a few steps of colony purification.

Serial Dilutions

To count the bacteria in a culture or to isolate a pure culture, it is often necessary to obtain discrete colonies of the bacteria. However, because bacteria are so small, a concentrated culture contains billions of bacteria per milliliter. If such a culture is plated directly on a petri plate, the bacteria all grow together, and discrete colonies do not form. **Serial dilutions** offer a practical method for diluting solutions of bacteria before plating to obtain a measurable number of discrete colonies. The principle is that if dilutions are repeated in succession, they can be multiplied to produce the total dilution. For example, if a solution is diluted in three steps by adding 1 ml of the solution to 99 ml of water (one in a hundred), followed by adding 1 ml of this dilution to another 99 ml of water, and finally, by adding 1 ml of the second dilution to another 99 ml of water, the final dilution is $10^{-2} \times 10^{-2} \times 10^{-2} = 10^{-6}$, or one in a million. To achieve the same dilution in a single step, 1 ml of the original solution would have to be added to 1,000 liters (about 250 gallons) of water. Obviously, it is more convenient to handle three solutions of 100 ml each than to handle a solution of 250 gallons, which weighs about 2,000 lb!

Selections

Probably the greatest advantage of bacterial genetics is the opportunity to do **selections**, by which very rare mutants and other types of strains can be isolated. To select a rare strain, billions of the bacteria are plated under conditions in which only the desired strain, not the bulk of the bacteria, can grow. In general, these conditions are called the **selective conditions**. For example, a nutrient may be required by most of the bacteria but not by the strain being selected. Agar plates lacking the nutrient then present selective conditions for the strain, since only the strain being selected multiplies to form a colony in the absence of the nutrient. In another example, the desired strain may be able to multiply at a temperature that would kill most of the bacteria. Incubating agar plates at that temperature would provide the selective condition. After the strain has been selected, a colony of the strain can be picked and the colony purified away from other contaminating bacteria under the same selective conditions.

The power of selection with bacterial populations is awesome. Using a properly designed selection, a single bacterium can be selected from among billions placed on an agar plate. If we could apply such selections to humans, we could find one of the few individuals in the entire human population of Earth with a particular trait.

Storing Stocks of Bacterial Strains

Most types of organisms must be continuously propagated; otherwise, they age and die. Propagating organisms requires continuous transfers and replenishing of the food supply, which can be very time-consuming. However, many types of bacteria can be stored in a dormant state and therefore do not need to be continuously propagated. The conditions used for storage depend on the type of bacteria. Some bacteria sporulate and so can be stored as dormant spores. Others can be stored by being frozen in glycerol or by being dried. Storing organisms in a dormant state is particularly convenient for genetic experiments, which often require the accumulation of large numbers of mutants and other strains. The strains remain dormant until the cells are needed, at which time they can be revived. Organisms that are continually propagated will also accumulate mutations, a problem that is eliminated with the proper storage of bacteria.

Genetic Exchange

Genetic experiments with an organism usually require some form of exchange of DNA or genes between members of the species. Most types of organisms on Earth are known to have some means of genetic exchange, which presumably accelerates evolution and increases the adaptability of a species.

Exchange of DNA from one bacterium to another can occur in one of three ways. In **transformation**, DNA released from one cell enters another cell of the same species (see chapter 6). In **conjugation**, discrete genetic elements transfer DNA from one cell to another (see chapter 5). Finally, in **transduction**, a bacterial virus accidentally picks up DNA from a cell it has infected and delivers this DNA into another cell (see chapter 7). The ability to exchange DNA between strains of a bacterium makes possible genetic crosses and complementation tests, as well as the tests essential to genetic analysis.

Phage Genetics

Some of the most important discoveries in genetics have come from studies with viruses that infect bacteria; these viruses are called **bacteriophages**, or **phages** for short (see chapter 7). Phages are not alive; instead, they are just genes wrapped in a protective coat of protein and/or membrane, as are all viruses. Because phages are not alive, they cannot multiply outside a bacterial cell. However, if a phage encounters a type of bacterial cell that is sensitive to that phage, the phage, or at least its DNA or RNA, enters the cell and directs it to make more phage.

Phages are usually identified by the holes, or **plaques**, they form in layers of sensitive bacteria. In fact, the name "phage" (Greek for "eat") derives from these plaques, which look like eaten-out areas. A plaque can form when a phage is mixed with large numbers of susceptible bacteria and the mixture is placed on an agar plate. As the bacteria multiply, one bacterial cell may be infected by the phage, which multiplies and eventually breaks open, or **lyses**, the bacterium, releasing more phage. As the surrounding bacteria are infected, the phage spread, even as the bacteria multiply to form an opaque layer called a **bacterial lawn**. Wherever the original phage infected the first bacterium, the plaque disrupts the lawn, forming a clear spot on the agar. Despite its empty appearance, this spot contains millions of the phage.

Phages offer many of the same advantages for genetics as bacteria. Thousands or even millions of phages can be put on a single plate. Also, like bacterial colonies, each plaque contains millions of genetically identical phage. By analogy to the colony purification of bacterial strains, individual phage mutants or strains can be isolated from other phages through plaque purification.

Phages Are Haploid

Phages are, in a sense, haploid, since they usually have only one copy of each gene. As with bacteria, this property makes isolation of phage mutants relatively easy, since all mutants immediately exhibit their phenotypes without the need for backcrosses.

Selections with Phages

Selection of rare strains of a phage is possible; as with bacteria, it requires conditions under which only the desired phage strain can multiply to form a plaque. For phages, these selective conditions may be a bacterial host in which only the desired strain can multiply or a temperature at which only the phage strain being selected can multiply. Note that the bacterial host must be able to multiply under the same selective conditions; otherwise, a plaque cannot form.

As with bacteria, selections allow the isolation of very rare strains or mutants. If selective conditions can be found for the strain, millions of phages can be mixed with the bacterial host, and only the desired strain multiplies to form a plaque. A pure strain can then be obtained by picking the phage from the plaque and purifying the phage strain under the same selective conditions.

Crosses with Phages

Phage strains can be crossed very easily. The same cells are infected with different mutants or strains of the phage. The DNA of the two phages is then in the same cell, where the molecules can interact genetically with each other, allowing genetic manipulations, such as gene-mapping and allelism tests.

A Brief History of Bacterial Molecular Genetics

Because of the ease with which they can be handled, bacteria and their phages have long been the organisms of choice for understanding basic cellular phenomena, and their contributions to this area of study are enormous. The following chronological list should give a feeling for the breadth of these contributions and the central position that bacteria have occupied in the development of modern molecular genetics. Some original references are given at the end of the chapter under Suggested Reading.

Inheritance in Bacteria

In the early part of the 1900s, biologists agreed that inheritance in higher organisms follows Darwinian principles. According to Charles Darwin, changes in the hereditary properties of organisms occur randomly and are passed on to the progeny. In general, the changes that happen to be beneficial to the organism are more apt to be passed on to subsequent generations.

With the discovery of the molecular basis for heredity, Darwinian evolution now has a strong theoretical foundation. The properties of organisms are determined by the sequence of their DNA, and as the organisms multiply, changes in this sequence sometimes occur randomly and without regard to the organism's environment. However, if a random change in the DNA happens to be beneficial in the situation in which the organism finds itself, the organism has an improved chance of surviving and reproducing.

As late as the 1940s, many bacteriologists thought that inheritance in bacteria was different from inheritance in other organisms. It was thought that rather than enduring random changes, bacteria could adapt to their environment by some sort of "directed" change and that the adapted organisms could then somehow pass on the change to their offspring. Such opinions were encouraged by the observations of bacteria growing under selective conditions. For example, in the presence of an antibiotic, all the bacteria in the culture soon become resistant to the antibiotic. It seemed as though the resistant bacterial mutants appeared in response to the antibiotic.

One of the first convincing demonstrations that inheritance in bacteria follows Darwinian principles was made in 1943 by Salvador Luria and Max Delbrück (see chapter 3 and Suggested Reading). Their work demonstrated that particular phenotypes, in their case resistance to a virus, occur randomly in a growing population, even in the absence of the virus. By the directed-change or adaptive-mutation hypothesis, the resistant mutants should have appeared only in the presence of the virus.

The demonstration that inheritance in bacteria follows the same principles as inheritance in eukaryotic organisms set the stage for the use of bacteria in studies of basic genetic principles common to all organisms.

Transformation

As discussed at the beginning of the Introduction, most organisms exhibit some mechanism for exchanging genes. The first demonstration of genetic exchange in bacteria was made by Fred Griffith in 1928. He was studying two variants of pneumococci, now called *Streptococcus pneumoniae*. One variant formed smooth-appearing colonies on plates and was pathogenic in mice. The other variant formed rough-appearing colonies on plates and did not kill mice. Only live, and not dead, smooth-colony-forming bacteria could cause disease, since the disease requires that the bacteria multiply in the infected mice. However, when Griffith mixed dead smooth-colony formers with live rough-colony formers and injected the mixture into mice, the mice became sick and died. Moreover, he isolated live smooth-colony formers from the dead mice. Apparently, the dead smooth-colony formers were "transforming" some of the live rough-colony formers into the pathogenic, smooth-colony-forming type. The "transforming principle" given off by the dead smooth-colony formers was later shown to be DNA, since addition of purified DNA from the dead smooth-colony formers to the live rough-colony formers in a test tube transformed some members of the rough type to the smooth type (see Avery et al., Suggested Reading). This method of exchange is called transformation (see chapter 6), and this experiment provided the first direct evidence that genes are made of DNA. Later experiments by Alfred Hershey and Martha Chase in 1952 (see Suggested Reading) showed that phage DNA alone is sufficient to direct the synthesis of more phages.

Conjugation

In 1946, Joshua Lederberg and Edward Tatum (see Suggested Reading) discovered a different type of gene exchange in bacteria. When they mixed some strains of *E. coli* with other strains, they observed the appearance of recombinant types that were unlike either parent. Unlike transformation, which requires only that DNA from one bacterium be added to the other bacterium, this means of gene exchange requires direct contact between two bacteria. It was later shown to be mediated by a genetic element that replicated separately from the chromosome, called a plasmid, in a process that was subsequently called conjugation (see chapter 5).

Transduction

In 1953, Norton Zinder and Joshua Lederberg (see Suggested Reading) discovered yet a third mechanism of gene transfer between bacteria. They showed that a phage of *Salmonella enterica* serovar Typhimurium could carry DNA from one bacterium to another. This means of gene exchange is called transduction and is now known to be quite widespread (see chapter 7).

Recombination within Genes

At the same time, experiments with bacteria and phages were also contributing to the view that genes were linear arrays of nucleotides in the DNA. By the early 1950s, recombination had been well demonstrated in higher organisms, including fruit flies. However, recombination was thought to occur only between mutations in different genes and not between mutations in the same gene. This led to the idea that genes were like "beads on a string" and that recombination is possible between the "beads," or genes, but not within a gene. In 1955, Seymour Benzer disproved this hypothesis by using the power of phage genetics to show that recombination is possible within the *r*II genes of phage T4. He mapped numerous mutations in the *r*II genes, thereby demonstrating that genes are linear arrays of mutable sites in the DNA. Later experiments with other phage and bacterial genes showed that the sequence of nucleotides in the DNA directly determines the sequence of amino acids in the protein product of the gene.

Semiconservative DNA Replication

In 1953, James Watson and Francis Crick published their structure of DNA. One of the predictions of this model is that DNA replicates by a semiconservative mechanism, in which specific pairing occurs between the bases in the old and the new DNA strands, thus essentially explaining heredity. In 1958, Matthew Meselson and Frank Stahl used bacteria to confirm that DNA replicates by this semiconservative mechanism.

mRNA

The existence of mRNA was also first indicated by experiments with bacteria and phages. In 1961, Sydney Brenner, François Jacob, and Matthew Meselson used phage-infected bacteria to show that ribosomes are the site of protein synthesis and confirmed the existence of a "messenger" RNA that carries information from the DNA to the ribosome.

The Genetic Code

Also in 1961, Francis Crick and his collaborators used phages and bacteria to show that the genetic code is unpunctuated, three lettered, and redundant. These researchers also showed that not all possible codons designate an amino acid and that some are "nonsense." These experiments laid the groundwork for Marshall Nirenberg and his collaborators to decipher the genetic code, in which a specific three-nucleotide set encodes one of 20 amino acids. The code was later verified by the examination of specific amino acid changes due to mutations in the lysozyme gene of phage T4.

The Operon Model

François Jacob and Jacques Monod published their operon model for the regulation of the lactose utilization genes of *E. coli* in 1961 as well. They proposed that a repressor blocks RNA synthesis on the *lac* genes unless the inducer, lactose, is bound to the repressor (see chapter 11). Their model has served to explain gene regulation in other systems, and the *lac* genes and regulatory system continue to be used in molecular genetic experiments, even in systems as far removed from bacteria as animal cells and viruses.

Enzymes for Molecular Biology

The early 1960s saw the start of the discovery of many interesting and useful bacterial and phage enzymes involved in DNA and RNA metabolism. In 1960, Arthur Kornberg demonstrated the synthesis of DNA in the test tube by an enzyme from *E. coli*. The next year, a number of groups independently demonstrated the synthesis of RNA in the test tube by RNA polymerases from bacteria. From that time on, other useful enzymes for molecular biology were isolated from bacteria and their phages, including additional RNA and DNA polymerases, polynucleotide kinase, DNA ligases, topoisomerases, and many phosphatases.

From these early observations, the knowledge and techniques of molecular genetics exploded. For example, in the early 1960s, techniques were developed for detecting the hybridization of RNA to DNA and DNA to DNA on nitrocellulose filters. These techniques were used to show that RNA is made on only one strand in specific regions of DNA, which later led to the discovery of promoters and other regulatory sequences. By the late 1960s, restriction endonucleases had been discovered in bacteria and shown to cut DNA in specific places (see Linn and Arber, Suggested Reading). By the early 1970s, these restriction endonucleases were being exploited to introduce foreign genes into *E. coli* (see Cohen et al., Suggested Reading), and by the late 1970s, the first human gene had been expressed in a bacterium. Also in the late 1970s, methods to sequence DNA by using enzymes from phages and bacteria were developed.

In 1988, a thermally stable DNA polymerase from a thermophilic bacterium was used to invent the technique called the polymerase chain reaction (PCR). This extremely sensitive technique allows the amplification of genes and other regions of DNA, facilitating their cloning and study. Thermally stable DNA polymerases are now an essential tool for genome sequencing.

Synthetic Genomics

More recently, advances in DNA synthesis and DNA recombination have been ushering in a new age of bacterial molecular genetics under the name of **synthetic genomics**, where massive strands of DNA large enough to comprise entire genomes can be made from the building blocks of DNA. In 2010, a significant milestone was reached with synthetic genomics when a derivative of the entire genome of *Mycoplasma mycoides* was synthesized from scratch, assembled by recombination, and used to replace the DNA in a related species (see Gibson et al., Suggested Reading). In one of the first demonstrations of the utility of this technique, in 2016, a minimal bacterial genome encoding only 473 genes was designed and synthesized, placing it as the smallest known genome in an autonomously replicating organism (see Hutchison et al., Suggested Reading). Amazingly and humbling, 149 of the 473 genes found necessary to support the growth of this organism are of unknown biological function, indicating that we still have much to learn about the molecular genetics of bacteria. While the ability to design bacteria *de novo* will likely have to include certain safeguards and greater public understanding, these types of experiments hold great promise for industrial use, as tools in medicine, and for addressing basic scientific questions, such as what is the minimum genetic requirement for life as a free-living organism.

These examples illustrate that bacteria and their phages have been central to the development of molecular genetics and recombinant DNA technology. Contrast the timing of these developments with the timing of comparable major developments in physics (early 1900s) and chemistry (1920s and 1930s), and you can see that molecular genetics is arguably the most recent major conceptual breakthrough in the history of science.

What Is Ahead

This textbook emphasizes how classical and molecular genetic approaches can be used to solve biological problems. As an educational experience, understanding the methods used and the interpretation of experiments is at least as important as the conclusions drawn. Therefore, whenever possible, the experiments that led to the conclusions are presented. The first two chapters, of necessity, review the concepts of macromolecular synthesis that are essential to understanding bacterial molecular genetics. However, they also introduce more current material, including interesting recent advances in bacterial cell biology. Chapter 1, besides reviewing the basics of DNA replication and the techniques of molecular biology, presents some recent advances in our understanding of how replication is coordinated with other cellular processes. Chapter 2, in addition to reviewing the basics of protein synthesis, presents current developments concerning protein folding, transport, and degradation. Chapter 3, similarly, reviews basic genetic principles, but with a special emphasis on bacterial genetics. Students are not likely to get some of this material in more general genetics courses, at least not in the same depth. This chapter also includes more current applications, such as gene knockouts, reverse genetics, and saturation genetics. Chapters 4 through 12 deal with more specific topics and the techniques that can be used to study them, with particular emphasis on recent evidence concerning the relatedness of seemingly disparate topics. The last chapter, chapter 13, focuses on how the genome is structured, the tools we use to analyze the genome, and even how to construct "new" genomes. We hope that this textbook will help put modern molecular genetics into a historical perspective, bring the reader up to date on current advances in bacterial molecular genetics, and position the reader to understand future developments in this exciting and rapidly progressing field of science.

SUGGESTED READING

Angert ER. 2012. DNA replication and genomic architecture of very large bacteria. *Annu Rev Microbiol* **66:**197–212.

Avery OT, Macleod CM, McCarty M. 1944. Studies on the chemical nature of the substance inducing transformation of pneumococcal types. I. Induction of transformation by a desoxyribonucleic acid fraction isolated from pneumococcus type III. *J Exp Med* **79:**137–158.

Brenner S, Jacob F, Meselson M. 1961. An unstable intermediate carrying information from genes to ribosomes for protein synthesis. *Nature* **190:**576–581.

Cairns J, Stent GS, Watson JD. 1966. *Phage and the Origins of Molecular Biology*. Cold Spring Harbor Laboratory Press, Cold Spring Harbor, NY.

Cohen SN, Chang ACY, Boyer HW, Helling RB. 1973. Construction of biologically functional bacterial plasmids *in vitro*. *Proc Natl Acad Sci USA* **70:**3240–3244.

Crick FHC, Barnett L, Brenner S, Watts-Tobin RJ. 1961. General nature of the genetic code for proteins. *Nature* **192:**1227–1232.

Gibson DG, Glass JI, Lartigue C, Noskov VN, Chuang R-Y, Algire MA, Benders GA, Montague MG, Ma L, Moodie MM, Merryman C, Vashee S, Krishnakumar R, Assad-Garcia N, Andrews-Pfannkoch C, Denisova EA, Young L, Qi Z-Q, Segall-Shapiro TH, Calvey CH, Parmar PP, Hutchison CA III, Smith HO, Venter JC. 2010. Creation of a bacterial cell controlled by a chemically synthesized genome. *Science* **329:**52–56.

Hershey AD, Chase M. 1952. Independent functions of viral protein and nucleic acid in growth of bacteriophage. *J Gen Physiol* **36:**39–56.

Hug LA, et al. 2016. A new view of the tree of life. *Nature Microbiol* **1:**16048. (Letter.) http://doi.org/10.1038/nmicrobiol.2016.48

Hutchison CA III, Chuang R-Y, Noskov VN, Assad-Garcia N, Deerinck TJ, Ellisman MH, Gill J, Kannan K, Karas BJ, Ma L, Pelletier JF, Qi Z-Q, Richter RA, Strychalski EA, Sun L, Suzuki Y, Tsvetanova B, Wise KS, Smith HO, Glass JI, Merryman C, Gibson DG, Venter JC. 2016. Design and synthesis of a minimal bacterial genome. *Science* **351:**aad6253.

Imachi H, et al. 2020. Isolation of an archaeon at the prokaryote-eukaryote interface. *Nature* **577:**519–525.

Jacob F, Monod J. 1961. Genetic regulatory mechanisms in the synthesis of proteins. *J Mol Biol* **3:**318–356.

Koonin EVE. 2015. Archaeal ancestors of eukaryotes: not so elusive any more. *BMC Biol* **13:**84.

Lederberg J, Tatum EL. 1946. Gene recombination in *Escherichia coli*. *Nature* **158:**558.

Leipe DD, Aravind L, Koonin EV. 1999. Did DNA replication evolve twice independently? *Nucleic Acids Res* **27:**3389–3401.

Linn S, Arber W. 1968. Host specificity of DNA produced by *Escherichia coli*. X. *In vitro* restriction of phage fd replicative form. *Proc Natl Acad Sci USA* **59:**1300–1306.

Luria SE, Delbrück M. 1943. Mutations of bacteria from virus sensitivity to virus resistance. *Genetics* **28:**491–511.

Meselson M, Stahl FW. 1958. The replication of DNA in *Escherichia coli*. *Proc Natl Acad Sci USA* **44:**671–682.

Nirenberg MW, Matthaei JH. 1961. The dependence of cell-free protein synthesis in *E. coli* upon naturally occurring or synthetic polyribonucleotides. *Proc Natl Acad Sci USA* **47:**1588–1602.

Olby R. 1974. *The Path to the Double Helix*. Macmillan Press, London, United Kingdom.

Olsen GJ, Woese CR, Overbeek R. 1994. The winds of (evolutionary) change: breathing new life into microbiology. *J Bacteriol* **176:**1–6.

Pace NR. 2009. Mapping the tree of life: progress and prospects. *Microbiol Mol Biol Rev* **73:**565–576.

Spang A, Saw JH, Jørgensen SL, Zaremba-Niedzwiedzka K, Martijn J, Lind AE, van Eijk R, Schleper C, Guy L, Ettema TJG. 2015. Complex archaea that bridge the gap between prokaryotes and eukaryotes. *Nature* **521:**173–179.

Schrodinger E. 1944. *What Is Life? The Physical Aspect of the Living Cell*. Cambridge University Press, Cambridge, United Kingdom.

Watson JD. 1968. *The Double Helix*. Atheneum, New York, NY.

Woese CR, Fox GE. 1977. Phylogenetic structure of the prokaryotic domain: the primary kingdoms. *Proc Natl Acad Sci USA* **74:**5088–5090.

Yang D, Oyaizu Y, Oyaizu H, Olsen GJ, Woese CR. 1985. Mitochondrial origins. *Proc Natl Acad Sci USA* **82:**4443–4447.

Zinder ND, Lederberg J. 1952. Genetic exchange in *Salmonella*. *J Bacteriol* **64:**679–699.

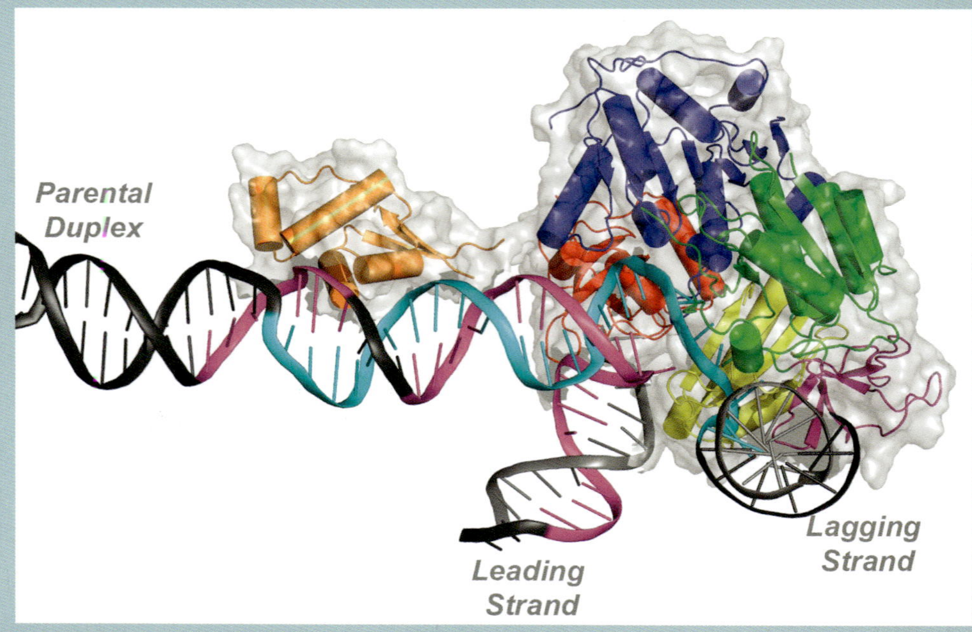

Model of the action of the PriA protein restarting a collapsed DNA replication fork. The DNA strands yet to be replicated (parental duplex) and the leading and lagging strands that have been replicated are labeled. The DNA strands shown in cyan and purple indicate regions that are believed to be bound by the PriA proteins based on biochemical experiments. The various subdomains of the PriA proteins are indicated in other colors. From Windgassen et al. (see Suggested Reading).

1 The Bacterial Chromosome: DNA Structure, Replication, and Segregation

DNA Structure

THE SCIENCE OF MOLECULAR GENETICS began with the determination of the structure of DNA. Experiments with bacteria and phages (i.e., viruses that infect bacteria) in the late 1940s and early 1950s, as well as the presence of DNA in chromosomes of higher organisms, had implicated this macromolecule as the hereditary material (see the introduction). In the 1930s, biochemical studies of the base composition of DNA by Erwin Chargaff established that the amount of guanine always equals the amount of cytosine and that the amount of adenine always equals the amount of thymine, independent of the total base composition of the DNA. In the early 1950s, X-ray diffraction studies by Rosalind Franklin and Maurice Wilkins showed that DNA is a double helix. Finally, in 1953, Francis Crick and James Watson put together the chemical and X-ray diffraction information in their famous model of the structure of DNA. This story is one of the most dramatic in the history of science and has been the subject of many historical treatments, some of which are listed at the end of this chapter.

Figure 1.1 illustrates the Watson-Crick structure of DNA, in which two strands wrap around each other to form a double helix. These strands can be extremely long, even in a simple bacterium, extending up to 1 mm—a thousand times longer than the bacterium itself. In a human cell, the strands that make up a single chromosome (which is one DNA molecule) are hundreds of millimeters, or many inches, long.

The Deoxyribonucleotides

If we think of DNA strands as chains, deoxyribonucleotides form the links. Figure 1.2 shows the basic structure of deoxyribonucleotides, called **deoxynucleotides** for short. Each is composed of a **base**, a **sugar**, and a **phosphate** group. The DNA bases are **adenine** (A), **cytosine** (C), **guanine** (G), and **thymine** (T), which have either one or two rings, as shown in Figure 1.2. The bases with two rings (A and G) are the **purines**, and those with only one ring (T and C) are **pyrimidines**. A third pyrimidine, uracil (U), replaces thymine in RNA. The carbons and nitrogens making up the rings of the bases are numbered sequentially, as shown in the figure. All four DNA bases are attached to the five-carbon sugar deoxyribose. This sugar is identical to ribose, which is found in RNA, except that it does not have an oxygen attached to the second

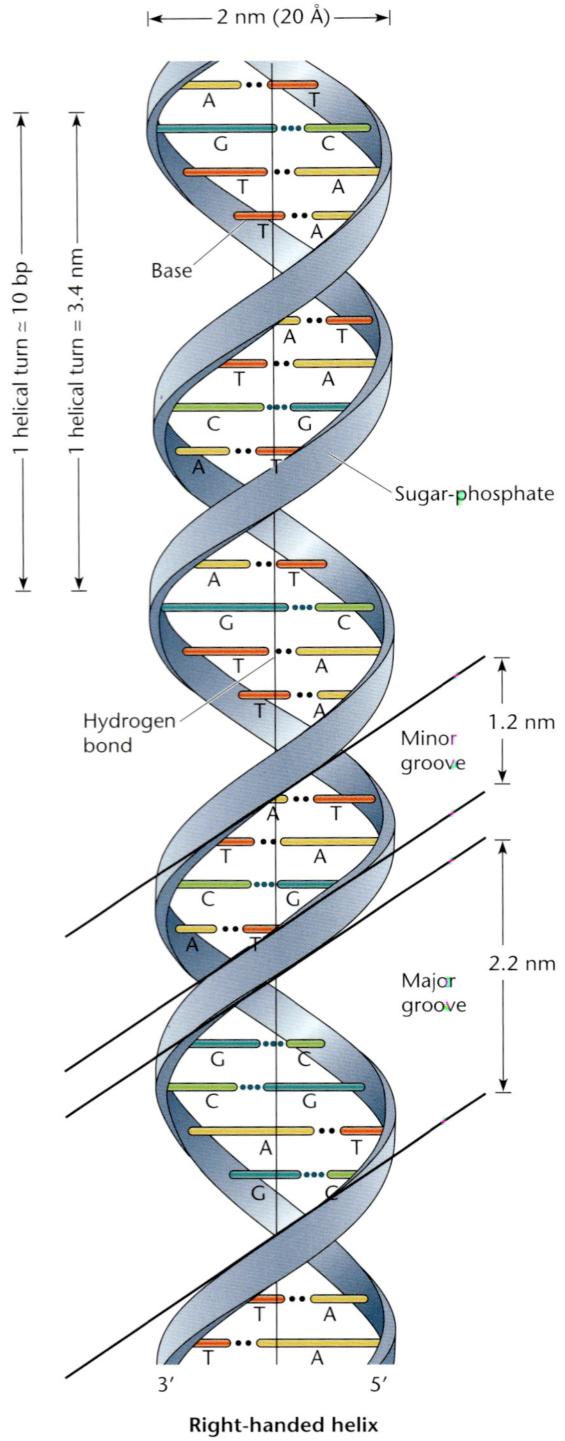

Figure 1.1 Schematic drawing of the Watson-Crick structure of DNA, showing the helical sugar-phosphate backbones of the two strands held together by hydrogen bonding between the bases. Also shown are the major and minor grooves and the dimensions of the helix.

carbon—hence the name deoxyribose. The carbons in the sugar of a nucleotide are also numbered 1, 2, 3, and so on, but they are labeled with "primes" to distinguish them from the carbons in the bases (Figure 1.2). The nucleotides also have one or more phosphate groups attached to a carbon of the deoxyribose sugar, as shown. The carbon to which the phosphate group is attached is indicated, although if the group is attached to the 5′ carbon (the usual situation), the carbon to which it is attached is often not stipulated.

The components of the deoxynucleotides have special names. A **deoxynucleoside** (rather than -tide) is a base attached to a sugar but lacking a phosphate. Without phosphates, the four deoxynucleosides are called **deoxyadenosine**, **deoxycytidine**, **deoxyguanosine**, and **deoxythymidine**. As shown in Figure 1.2, the deoxynucleotides have one, two, or three phosphates attached to the sugar and are known as deoxynucleoside monophosphates, diphosphates, or triphosphates, respectively. The individual deoxynucleoside monophosphates, called deoxyguanosine monophosphate etc., are often abbreviated dGMP, dAMP, dCMP, and dTMP, where the d stands for deoxy; the G, A, C, or T stands for the base; and the MP stands for monophosphate. In turn, the diphosphates are abbreviated dGDP, dADP, dCDP, and dTDP, and the triphosphates are abbreviated dGTP, dATP, dCTP, and dTTP. The phosphate attached to the sugar is called the α phosphate, while the next two are called the β and γ phosphates, respectively, as shown in the figure. Collectively, the four deoxynucleoside triphosphates are often referred to as dNTPs.

The DNA Chain

Phosphodiester bonds join each deoxynucleotide link in the DNA chain. As shown in Figure 1.3, the phosphate attached to the last (5′) carbon of the deoxyribose sugar of one nucleotide is attached to the third (3′) carbon of the sugar of the next nucleotide, thus forming one strand of nucleotides connected 5′ to 3′, 5′ to 3′, etc.

The 5′ and 3′ Ends

The nucleotides found at the ends of a linear piece of DNA have properties that are biochemically important and useful for orienting the DNA strand. At one end of the DNA chain, a nucleotide will have a phosphate attached to its 5′ carbon that does not connect it to another nucleotide. This end of the strand is called the 5′ end or the **5′ phosphate end** (Figure 1.3B). On the other end, the last nucleotide lacks a phosphate at its 3′ carbon. Because it has only a hydroxyl group (the OH in Figure 1.3B), this end is called the 3′ end or the **3′ hydroxyl end**.

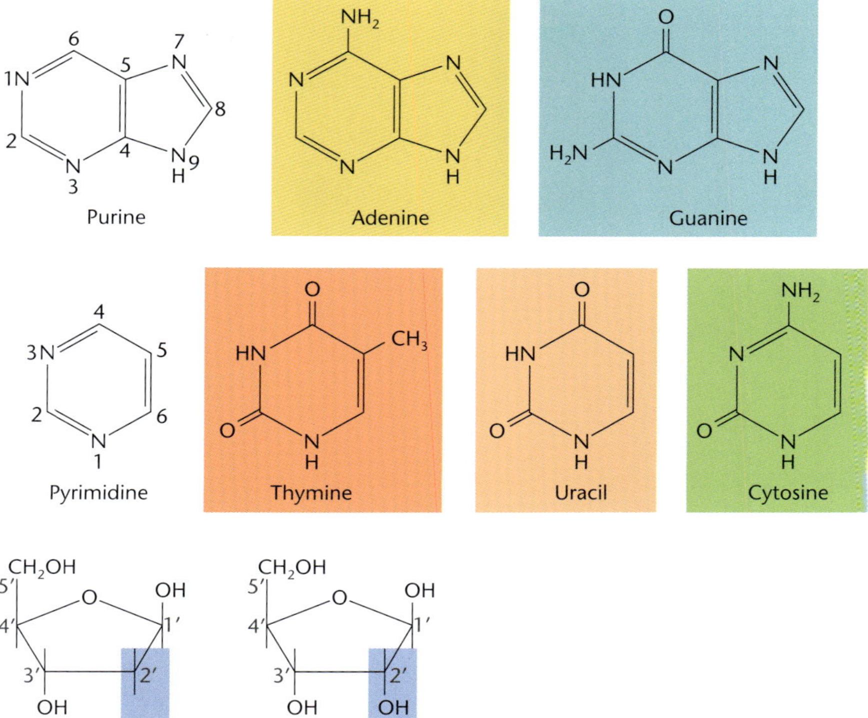

Sugars

2-Deoxyribose

Ribose

Nucleotides

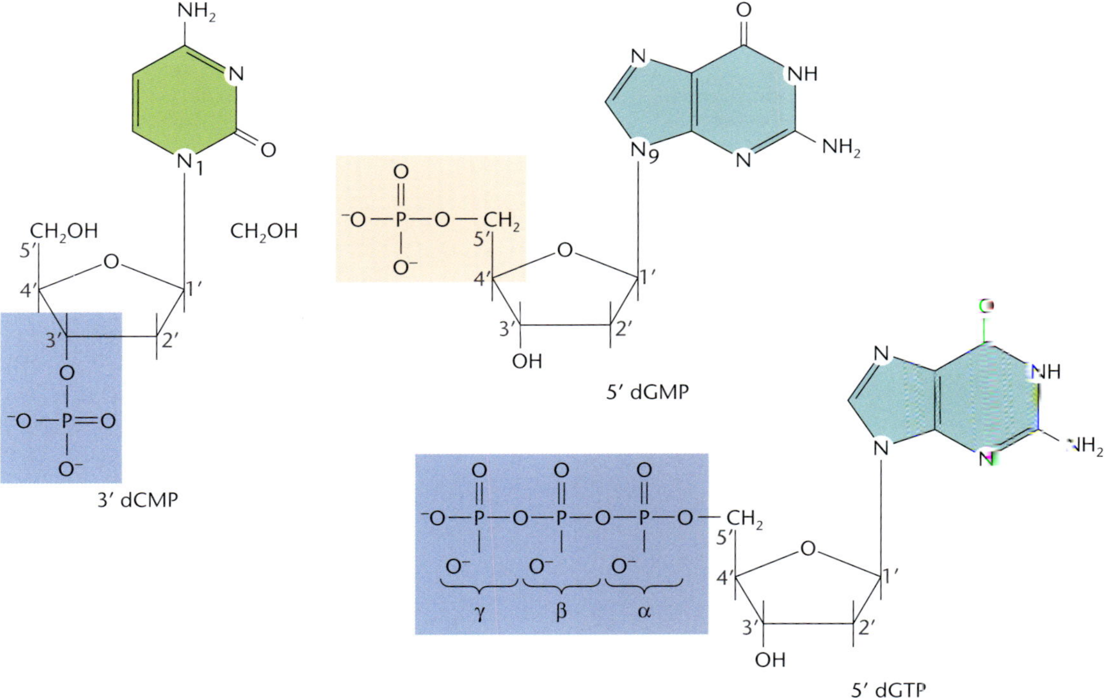

Figure 1.2 Chemical structures of deoxyribonucleotides, showing the bases and sugars and how they are assembled into a deoxyribonucleotide.

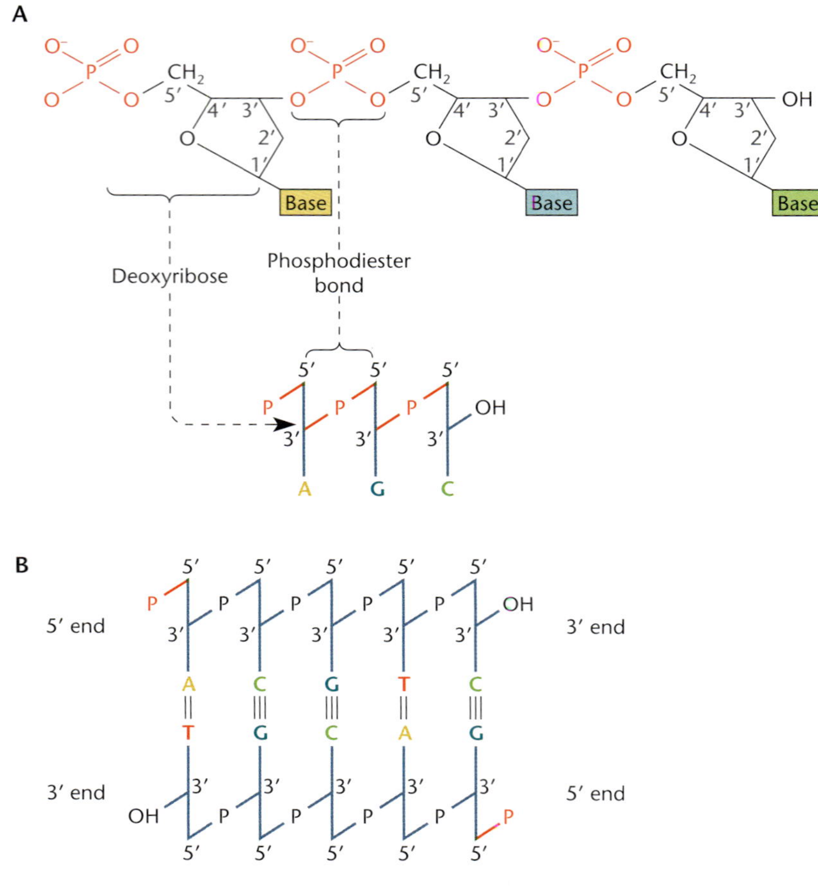

Figure 1.3 **(A)** Schematic drawing of a DNA chain, showing the 3′-to-5′ attachment of the phosphates to the sugars, forming phosphodiester bonds. **(B)** Two strands of DNA bind at the bases in an antiparallel arrangement of the phosphate-sugar backbones.

Base Pairing

The sugar and phosphate groups of DNA form what is often called a **backbone** to support the bases, which jut out from the chain. This structure allows the bases from one single strand of DNA to form hydrogen bonds with another strand of DNA, thereby holding together two separate nucleotide chains (Figure 1.3B). The first clue that pairing between specific bases could form the basis for the structure of DNA came from Erwin Chargaff's observation about the ratios of the bases; no matter the source of the DNA, the concentration of guanine (G) always equals the concentration of cytosine (C) and the concentration of adenine (A) always equals the concentration of thymine (T). These ratios, named Chargaff's rules, gave Watson and Crick one of the essential clues to the structure of DNA. They proposed that the two strands of the DNA are held together by specific hydrogen bonding between the bases in opposite strands, as shown in Figure 1.4. Thus, the amounts of A and T and of C and G are always the same because A's pair only with T's and G's pair only with C's to hold the DNA strands together. Each A-and-T pair and each G-and-C pair in DNA is called a **complementary base pair**, and the sequences of two strands of DNA are said to be complementary if one strand always has a T where there is an A in the other strand and a G where there is a C in the other strand.

It did not escape the attention of Watson and Crick that the complementary base-pairing rules essentially explain heredity. If A pairs only with T and G pairs only with C, then each strand of DNA can replicate to make a complementary copy, so that the two replicated DNAs will be exact copies of each other. Offspring containing the new DNAs would have the same sequence of nucleotides in their DNAs as their parents and thus would be exact copies of their parents.

Antiparallel Construction

As mentioned at the beginning of this section, the complete DNA molecule consists of two long chains wrapped around each other in a double helix (Figure 1.1). The

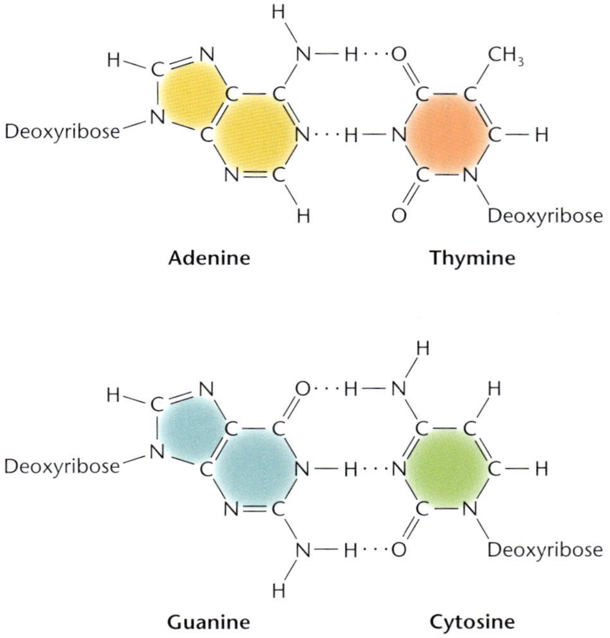

Figure 1.4 The two complementary base pairs found in DNA. Two hydrogen bonds form in adenine-thymine base pairs. Three hydrogen bonds form in guanine-cytosine base pairs.

double-stranded molecule can be thought of as being like a circular staircase, with the alternating phosphates and deoxyribose sugars forming the railings and the bases connected to each other forming the steps. However, the two chains run in opposite orientations, with the phosphates on one strand attached 5′ to 3′, 5′ to 3′, etc., to the sugars and those on the other strand attached 3′ to 5′, 3′ to 5′, etc. This arrangement is called **antiparallel**. In addition to phosphodiester bonds running in opposite directions, the antiparallel construction causes the 5′ phosphate end of one strand and the 3′ hydroxyl end of the other to be on the same end of the double-stranded DNA molecule (Figure 1.3B).

The Major and Minor Grooves

Because the two strands of DNA are wrapped around each other to form a double helix, the helix has two grooves between the two strands (Figure 1.1). One of these grooves is wider than the other, so it is called the **major groove**. The other, narrower groove is called the **minor groove**. Most of the modifications to DNA that are discussed in this and later chapters occur in the major groove of the helix.

The Mechanism of DNA Replication

The molecular details of DNA replication are probably similar in all organisms on Earth. The basic process of replication involves **polymerizing**, or linking, the nucleotides of DNA into long chains, or strands, using the sequence on the other strand as a guide. Because the nucleotides must be made before they can be put together into DNA, the nucleotides are an essential **precursor** of DNA synthesis.

Deoxyribonucleotide Precursor Synthesis

The precursors of DNA synthesis are the four deoxyribonucleoside triphosphates, dATP, dGTP, dCTP, and dTTP. The triphosphates are synthesized from the corresponding ribose nucleoside diphosphates by the pathway shown in Figure 1.5. In the first step, the enzyme **ribonucleotide reductase** reduces (i.e., removes an oxygen from) the ribose sugar to produce the deoxyribose sugar by changing the hydroxyl group at the 2′ position (the second carbon) of the sugar to a hydrogen. Then, an enzyme known as a **kinase** adds a phosphate to the deoxynucleoside diphosphate to make the deoxynucleoside triphosphate precursor.

The deoxynucleoside triphosphate dTTP is synthesized by a somewhat different pathway from the other three. The first step is the same. Ribonucleotide reductase synthesizes the nucleotide dUDP (deoxyuridine diphosphate) from the ribose UDP. However, from then on, the pathway differs. A phosphate is added to make dUTP, and the dUTP is converted to dUMP by a phosphatase that removes two of the phosphates. This molecule is then converted to dTMP by the enzyme thymidylate synthetase, using tetrahydrofolate to donate a methyl group. Kinases then add two phosphates to the dTMP to make the precursor dTTP.

Replication of the Bacterial Chromosome

Once the precursors of DNA replication are synthesized, they must be polymerized into long double-stranded DNA molecules. A very large complex of many enzymes

Figure 1.5 The pathways for synthesis of deoxynucleotides from ribonucleotides. Some of the enzymes referred to in the text are identified. THF, tetrahydrofolate; DHF, dihydrofolate.

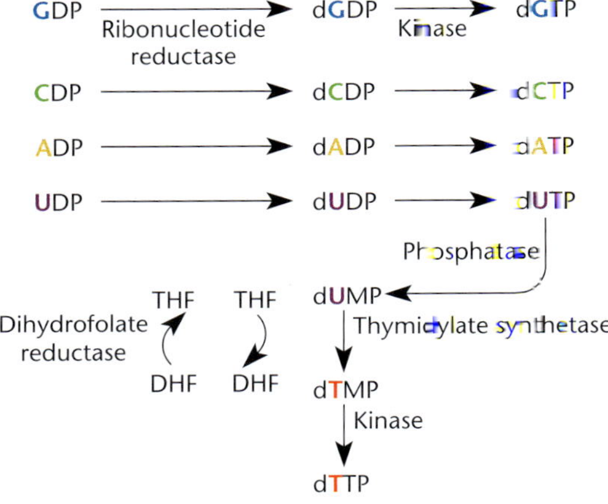

A Polymerization reaction

B Antiparallel strands

C Base flipping

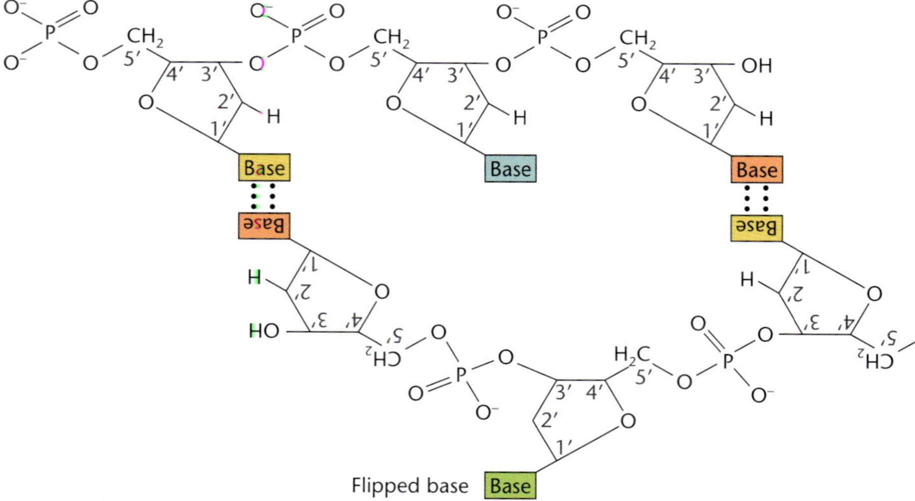

Figure 1.6 Features of DNA. **(A)** Polymerization of the deoxynucleotides during DNA synthesis. The β and γ phosphates of each deoxynucleoside triphosphate are cleaved off to provide energy for the polymerization reaction. **(B)** The strands of DNA are antiparallel. **(C)** A single base can be flipped out from the double helix, which could be important in recombination and repair.

assembles on the DNA and moves along the DNA, separating the strands and making a complementary copy of each of the strands. Thus, two strands of DNA enter the complex and four strands emerge on the other side, forming a branched structure. Each of the emerging branches contains one old "conserved" strand and one new strand, hence the name **semiconservation replication**. This branched structure where replication is occurring is called the **replication fork**. In this section, we discuss what is happening in the replication fork, including the mechanisms used to overcome obstacles to the progression of the replication fork on the chromosome.

DNA POLYMERASES

The properties of the DNA polymerases, the enzymes that join the deoxynucleotides together to make the long chains, are the best guides to an understanding of the replication of DNA. These enzymes make DNA by linking one deoxynucleotide to another to generate a long chain of DNA. This process is called DNA **polymerization,** hence the name DNA polymerases.

Figure 1.6 shows the basic process of DNA polymerization by DNA polymerase. The DNA polymerase attaches the first phosphate (the α phosphate) of one deoxynucleoside triphosphate to the 3′ carbon of the sugar of the next deoxynucleoside triphosphate, in the process releasing the last two phosphates (the β and γ phosphates) of the first deoxynucleoside triphosphate to produce energy for the reaction. Then the α phosphate of another deoxynucleoside triphosphate is attached to the 3′ carbon of this deoxynucleotide, and the process continues until a long chain is synthesized.

DNA polymerases also need a **template strand** to direct the synthesis of the new strand (Figure 1.7). As mentioned in "Base Pairing" above, complementary base pairing dictates that wherever there is a T in the template strand, an A is inserted in the strand being synthesized and so forth according to the base-pairing rules. The DNA polymerase can move only in the 3′-to-5′ direction on the template strand, linking deoxynucleotides in the new strand in the 5′-to-3′ direction. When replication is completed, the product is a new double-stranded DNA with antiparallel strands, one of which is the old template strand and one of which is the newly synthesized strand.

There are two DNA polymerases that participate in normal DNA replication in *Escherichia coli*; they are called DNA polymerase III and DNA polymerase I (Table 1.1). DNA polymerase III is a large protein complex in which the enzyme that polymerizes nucleotides works with numerous accessory proteins. In *E. coli* and other bacteria, DNA polymerase III is responsible for the bulk of DNA replication on both DNA strands. As discussed below, DNA polymerase I has a number of features that are important because replication is continually reinitiated on one of the DNA strands. It also plays a role in DNA repair, as discussed in chapter 10. Table 1.1 lists many of the DNA replication proteins, the genes encoding them, and their functions.

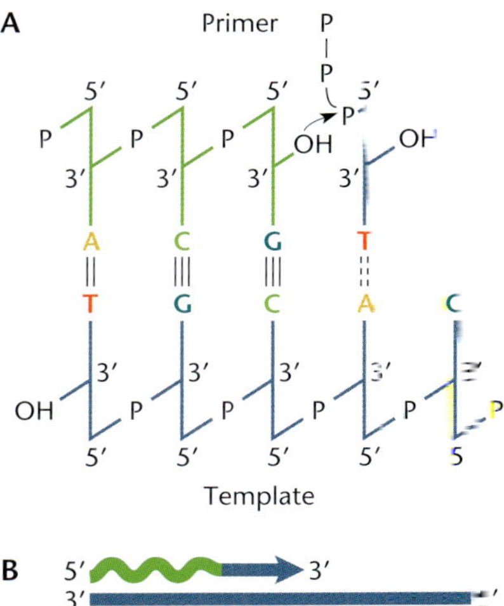

Figure 1.7 Functions of the primer and template in DNA replication. **(A)** The DNA polymerase adds deoxynucleotides to the 3′ end of the primer by using the template strand to direct the selection of each base. **(B)** Simple illustration of 5′-to-3′ DNA synthesis. The wavy green line represents the primer.

PRIMASES

One type of enzyme, called a **primase**, is required during DNA replication because DNA polymerases cannot start the synthesis of a new strand of DNA; they can only attach deoxynucleotides to a preexisting 3′ OH group. The 3′ OH group to which DNA polymerase adds a deoxynucleotide is called the **primer** (Figure 1.7). The requirement for a primer for DNA polymerase creates an apparent dilemma in DNA replication. When a new strand of DNA is synthesized, there is no DNA upstream (i.e., on the 5′ side) to act as a primer. Primase makes small stretches of RNA that are complementary to the template strand, which are in turn used to initiate, or prime, polymerization of a new strand of DNA (Figure 1.8). In some special situations, an RNA primer for DNA replication can be made by the RNA polymerase used for information processing (i.e., the RNA polymerase that makes all the other RNAs, including mRNA, tRNA, and rRNA; see chapter 2). Unlike DNA polymerase, primase and RNA polymerases do not require a primer to initiate the synthesis of new strands. During DNA replication, a special enzyme activity recognizes and removes the RNA primer (see below).

Table 1.1 Proteins involved in *Escherichia coli* DNA replication

Protein	Gene	Function
DnaA	*dnaA*	Initiator protein; primosome (priming complex) formation
DnaB	*dnaB*	DNA helicase
DnaC	*dnaC*	Delivers DnaB to replication complex
SSB	*ssb*	Binding to single-stranded DNA
Primase	*dnaG*	RNA primer synthesis
DNA ligase	*lig*	Sealing DNA nicks
DNA gyrase		Supercoiling
α	*gyrA*	Nick closing
β	*gyrB*	ATPase
DNA Pol I	*polA*	Primer removal; gap filling
DNA Pol III (holoenzyme)		
α	*dnaE*	Polymerization
ε	*dnaQ*	3′-to-5′ editing
RNase H	*rnhA*	Can aid in RNA primer removal
θ	*holE*	Present in core ($\alpha\varepsilon\theta$)
β	*dnaN*	Sliding clamp
τ[a]	*dnaX*	Organizes complex; joins leading and lagging DNA Pol III
γ[b]	*dnaX*	Binds clamp loaders and single-strand-binding protein
δ	*holA*	Clamp loading
δ'	*holB*	Clamp loading
χ	*holC*	Binds single-strand-binding protein
φ	*holD*	Holds χ to the clamp loader

[a]Full-length product of the *dnaX* gene.
[b]Shorter product of the *dnaX* gene produced by translational frameshifting (see chapter 2).

NUCLEASES

Enzymes that degrade DNA strands by breaking the phosphodiester bonds are just as important in replication as the enzymes that polymerize DNA by forming phosphodiester bonds between the nucleotides. These bond-breaking enzymes, called **nucleases**, can be grouped into two major categories. One type can initiate breaks in the middle of a DNA strand and so are called **endonucleases**, from a Greek word meaning "within," and the other type can remove nucleotides only from the ends of DNA strands and so are called exonucleases, from a Greek word meaning "outside." A special type of endonuclease activity, called a **flap endonuclease** activity, is involved in primer removal by DNA polymerase I. The flap endonuclease activity appears to be common to all organisms for removing RNA primers. In *E. coli*, DNA polymerase I displaces the RNA primer, making a flap-like structure, and then the flap endonuclease activity of Pol I cleaves away the oligonucleotide as indicated (Figure 1.9). The **exonucleases** can be subdivided into two groups. Some exonucleases can degrade only from the 3′ end of a DNA strand, degrading DNA in the 3′-to-5′ direction. These are called 3′ exonucleases; one example of their activity is their role in the editing function associated with DNA polymerases I and III, which is discussed below. Other exonucleases, called 5′ exonucleases, degrade DNA strands only from the 5′ end.

DNA LIGASES

DNA **ligases** are enzymes that form phosphodiester bonds between the ends of separate presynthesized chains of DNA. This important function cannot be performed by any of the known DNA polymerases. During replication, ligase joins the 5′ phosphate at the end of one DNA chain to the 3′ hydroxyl at the end of another chain to make a longer, continuous chain (Figure 1.8).

ACCESSORY PROTEINS

Replication of large DNAs requires many functions that reside in proteins separate from the subunit used for polymerizing the chain of nucleotides. These functions include the coordination of multiple DNA polymerases and tethering of these components to the template DNA strands as a moving production platform. DNA polymerase III is the major DNA replication protein in *E. coli* responsible for polymerizing the new complementary DNA strands, and it functions with multiple **DNA polymerase accessory proteins** that travel along the template strand with the molecule of DNA polymerase III. The term **DNA polymerase III holoenzyme** can be used to describe

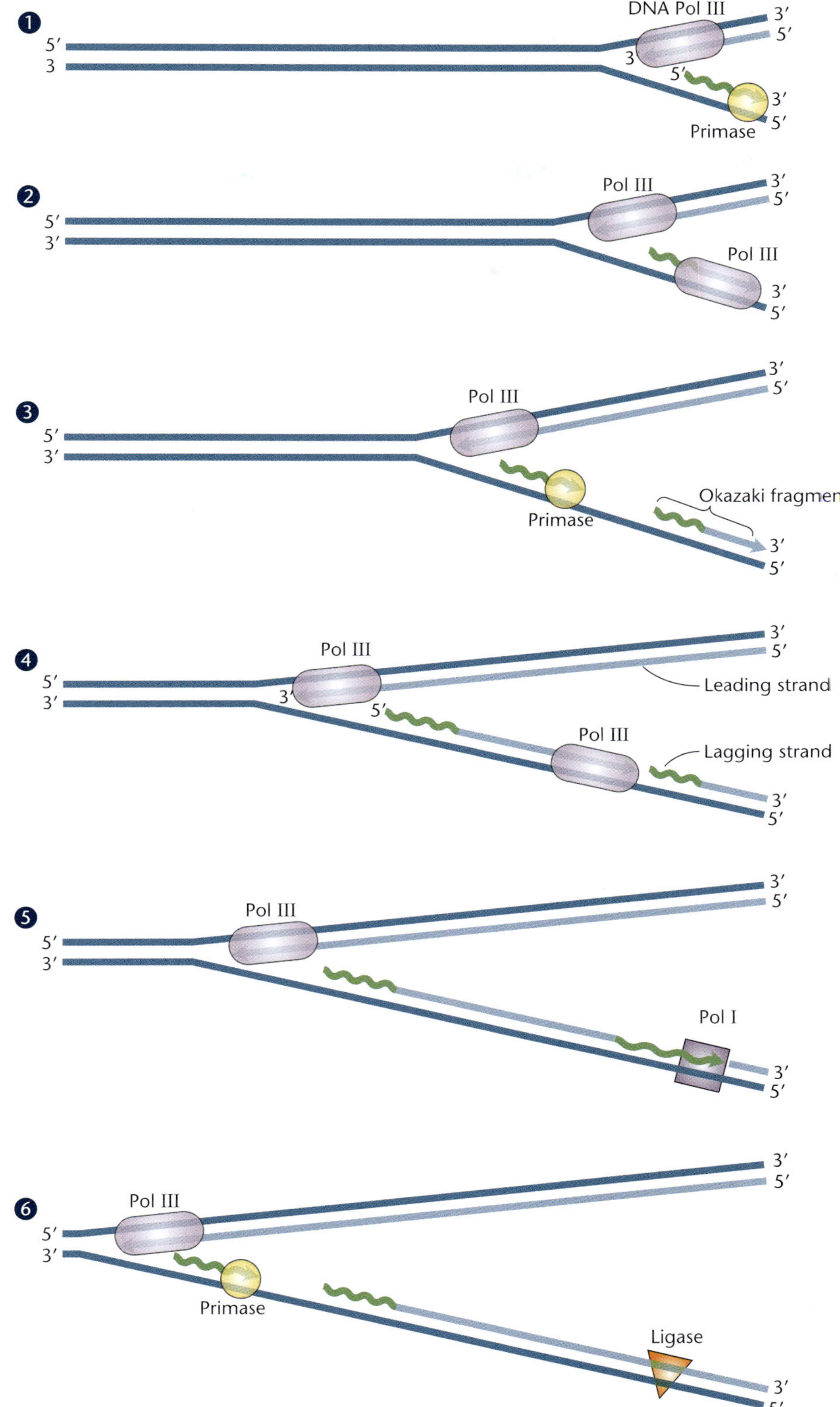

Figure 1.8 Discontinuous synthesis of one of the two strands of DNA during chromosome replication. **(1)** DNA polymerase (Pol) III replicates one strand, and the primase synthesizes RNA on the other strand in the opposite direction. **(2)** Pol III extends the RNA primer to synthesize an Okazaki fragment. **(3)** The primase synthesizes another RNA primer. **(4)** Pol III extends this primer until it reaches the previous primer. **(5)** Pol I removes the first RNA primer and replaces it with DNA. **(6)** DNA ligase seals the nick to make a continuous DNA strand, and the process continues. The strand that is synthesized continuously is the leading strand; the strand that is synthesized discontinuously is the lagging strand.

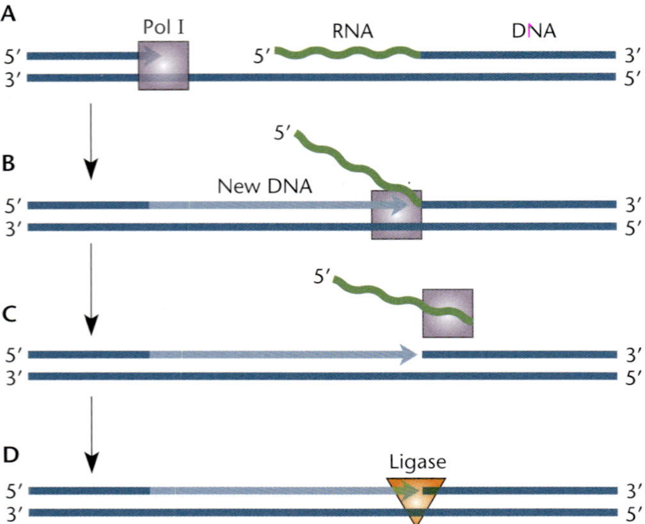

Figure 1.9 DNA polymerase I can remove an RNA primer by using strand displacement and endonuclease activity. **(A)** The DNA strand produced by DNA polymerase III holoenzyme is extended by DNA polymerase I until it encounters a previously synthesized RNA primer. **(B)** During the process of DNA replication, DNA polymerase I displaces the RNA primer. **(C)** An endonuclease activity in DNA polymerase I is used to cleave off the RNA primer. **(D)** Ligase joins the new Okazaki fragment to the previous Okazaki fragment to allow a contiguous DNA.

the entire complex of proteins. The various subunits and subassemblies of the DNA polymerase III holoenzyme were originally identified from fractionation procedures and were designated by Greek letters (Table 1.1).

One of the DNA polymerase accessory proteins forms a ring around the template DNA strand and is responsible for keeping DNA polymerase from falling off. Because this ring slides freely over double-stranded DNA and will not easily come off the DNA, it is also referred to as a sliding clamp, or β clamp. The β clamp provides the foundation of the mobile platform for DNA replication, allowing it to continue for long distances without being released. In bacteria, the β clamp is a product of the *dnaN* gene, where two head-to-toe molecules form the ring around the DNA. While it was first isolated as part of the DNA polymerase III holoenzyme, the β clamp protein is important for multiple DNA transactions.

A special subcomplex within DNA polymerase III is the **clamp loader**, which is responsible for loading β clamp proteins onto the DNA. The clamp-loading complex is also responsible for tethering proteins across the DNA replication fork; the clamp loader binds the DNA polymerases on both DNA template strands and the enzyme responsible for separating the DNA strands (see below). The clamp loader is a complicated structure that consists of one γ and two τ proteins and one each of δ and δ′, which form a five-sided structure, and two additional proteins, χ and ψ. The clamp loader complex is also responsible for removing β clamps. The rate of clamp removal allows many β clamps to reside on the DNA for a period of time after the replication fork passes (see Moolman et al., Suggested Reading). β clamps temporarily left behind on the newly replicated DNA play a role in helping to recruit other proteins responsible for various replication and repair functions described in this and other chapters.

Replication of Double-Stranded DNA

Additional complications of DNA replication come from the fact that the DNA is double stranded and the strands are antiparallel. The replication of all bacterial chromosomes begins at one point, called the **origin of replication**, with the replication enzymes moving in opposite directions from this point along the chromosome. In this process, both strands of DNA are replicated at the same time with a coordinated set of proteins. Replicating the antiparallel strands is further complicated by the above-mentioned fact that DNA polymerases can replicate only in the 5′-to-3′ direction. Therefore, one DNA strand is replicated in the same direction that the replication fork is moving, and in theory, replication of this strand could continue without the need for reinitiating in a process called **leading-strand** DNA synthesis. However, replication of the other DNA strand occurs in the opposite direction from the progression of the replication machinery. Replication of this strand must continually be reinitiated in a process known as **lagging-strand** DNA synthesis. Replication of double-stranded DNA requires coordination between multiple holoenzyme subunits and DNA polymerases, as well as a host of other replication proteins.

SEPARATING THE TWO TEMPLATE DNA STRANDS

To serve as templates for DNA replication, the two DNA strands must be separated, a task that DNA polymerase cannot perform on its own. The strands must be separated because the bases of the DNA are inside the double helix, where they are not available to pair with the incoming deoxynucleotides to direct which nucleotide will be inserted at each step. Proteins called DNA **helicases** separate the strands of DNA (see Singleton et al., Suggested Reading). Many of these proteins form a ring around one strand of DNA and propel the strand through the ring, acting as a mechanical wedge that strips the strands apart as it moves. It takes a lot of energy to separate the strands of DNA, and helicases cleave a lot of ATP for energy, forming ADP in the process. There are about 20 different helicases in *E. coli*, and each helicase works in only one direction, either the 3′-to-5′ or the 5′-to-3′ direction. The DnaB helicase that normally separates the strands of DNA ahead of the replication fork in *E. coli* is a large doughnut-shaped complex composed of six polypeptide products of the *dnaB* gene. It propels one strand, the template for lagging-strand DNA replication, through the center of the complex in the 5′-to-3′ direction, opening strands of DNA ahead of the replication fork (Figure 1.10). The DnaB ring cannot load onto single-stranded DNA on its own to start a DNA replication fork; it requires the loading protein DnaC. Other helicases are discussed in later chapters in connection with recombination and repair.

Once the strands of DNA have been separated, they also must be prevented from coming back together (or from annealing to themselves if they happen to be complementary over short regions). Separation of the strands is maintained by proteins called single-strand-binding (SSB) proteins or, less frequently, helix-destabilizing proteins. They are proteins that bind preferentially to single-stranded DNA and prevent double-stranded helical DNA from reforming prematurely. Interestingly, SSB activity goes beyond this passive role. SSB is also responsible for recruiting a number of replication and repair proteins through a specific set of amino acids encoded in the very C-terminal end of SSB, allowing it to serve as an organizational hub for other processes.

PROCESSING THE TWO TEMPLATE DNA STRANDS

As discussed above, the antiparallel configuration of DNA requires that the two DNA polymerases travel in two different directions while still allowing the larger replication machine to travel in one direction down the chromosome (Figure 1.8). This leads to fundamental differences in the natures of leading- and lagging-strand DNA replication. While replication of the leading-strand template can occur as soon as the strands are separated by the DnaB helicase, replication of the lagging-strand template is consistently reinitiated approximately every 1 to 2 kilobases (kb); this slows the process, hence the name lagging-strand synthesis. The short pieces of DNA produced from the lagging-strand template are called **Okazaki fragments**. Synthesis of each Okazaki fragment requires a new RNA primer about 10 to 12 nucleotides in length. In *E. coli*, these primers are synthesized by DnaG primase at the template sequence 3′-GTC-5′, beginning synthesis opposite the T. These RNA primers are then used to prime DNA synthesis by DNA polymerase III, which continues until it reaches the last RNA primer produced by DnaG (Figure 1.8). Before these short pieces of DNA that are annealed to the template can be joined to make a long, continuous strand of DNA, the short RNA primers must be removed. This process is carried out by DNA polymerase I using its flap exonuclease activity to displace and cleave the RNA strand (Figure 1.9). As DNA polymerase I displaces the RNA primer, it extends the upstream (i.e., 5′) DNA that was previously polymerized by DNA polymerase III (Figure 1.8). Ribonuclease (RNase) H may contribute to this process under some circumstances by using its ability to degrade the RNA strand of a DNA-RNA double helix (Table 1.1). The Okazaki fragments are then joined together by DNA ligase as the replication fork moves on, as shown in Figure 1.8. By using RNA rather than DNA to prime the synthesis of new Okazaki fragments, the cell likely lowers the mistake rate of DNA replication (see below).

What actually happens at the replication fork is more complicated than is suggested by the simple picture given so far. For one thing, this picture ignores the overall topological restraints on the replicating DNA. The **topology** of a molecule refers to its position in space. Because the circular DNA is very long and its strands are wrapped around each other, pulling the two strands apart introduces stress into other regions of the DNA in the form of **supercoiling**. If no mechanism existed to allow the two strands of DNA to rotate around each other, supercoiling would cause the chromosome to look like a telephone cord wound up on itself, an event that has been experimentally shown to eventually halt progression of the DNA replication fork. To relieve this stress, enzymes called **topoisomerases** work to help undo the supercoiling ahead of the replication fork. DNA supercoiling and topoisomerases are discussed below. The fork itself can also twist when the supercoiling that builds up ahead of the replication fork diffuses behind the replication fork, a process that twists the two new strands around one another and that is also sorted out by topoisomerases (see below).

COORDINATING REPLICATION OF THE TWO TEMPLATE STRANDS

The picture of the two strands of DNA replicating independently, as shown in Figure 1.8, does not take into consideration all of the coordination that must occur during DNA replication. The anatomy of the larger complex of replication factors remains unresolved; however, interactions among many of these components provide a hint as to how the larger complex functions (Figure 1.10). Rather than replicating independently, the DNA polymerases that produce the leading-strand and lagging-strand DNAs are joined to each other through the τ subunits of the holoenzyme (Table 1.1). In the holoenzyme there are two τ subunits and a derivative product called γ, which is incapable of interacting with DNA polymerase. The γ and τ subunits are encoded by the same gene, *dnaX*. Expression of the full gene results in production of the longer τ subunit, whereas a stutter in how the protein is produced from this gene, called a "frameshift," produces the shorter γ product. The configuration of having two τ and γ subunits may be important to ensure that only two DNA polymerase III molecules are at the replication fork, possibly facilitating the use of alternate polymerases for repair when needed (see below and Dohrmann et al., Suggested Reading).

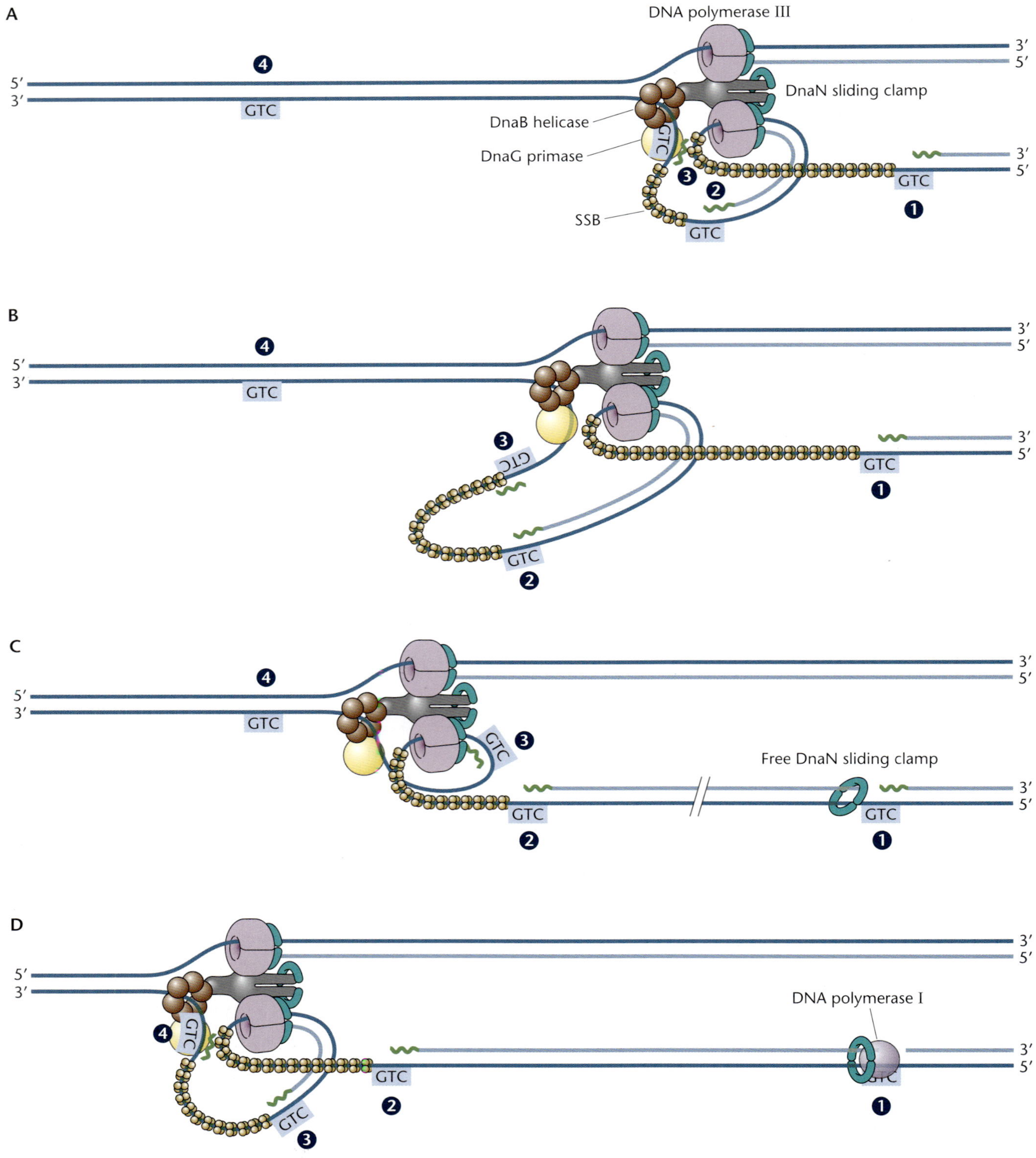
A
DNA polymerase III
5′
3′
GTC
3′
5′
DnaN sliding clamp
DnaB helicase
DnaG primase
SSB
GTC
GTC
3′
5′
B
5′
3′
GTC
3′
5′
GTC
GTC
GTC
3′
5′
C
5′
3′
GTC
3′
5′
GTC
Free DnaN sliding clamp
GTC
GTC
3′
5′
D
5′
3′
3′
5′
GTC
DNA polymerase I
GTC
GTC
3′
5′

To accommodate the fact that the two DNA polymerases must move in opposite directions and still remain tethered, the lagging-strand template probably loops out as an Okazaki fragment is synthesized. The loop is then relaxed as the polymerase on the lagging-strand template is released from the β clamp, allowing the DNA polymerase to rapidly and efficiently "hop" ahead to the next RNA primer to begin synthesizing the next Okazaki fragment (Figure 1.10). The polymerase associates with a new β clamp assembled by the clamp loader at the site of the new RNA primer, while the old β clamp is left behind. The β clamp left on the last Okazaki fragment plays important roles in finishing synthesis and joining the fragments of lagging-strand DNA via interactions with DNA polymerase I, ligase, and repair proteins. β clamps are eventually recycled, possibly through the removal function of the δ subunit of the clamp loader. This model involving the looping out of the lagging-strand template has been referred to as the "trombone" model of replication because the loops forming and contracting at the replication fork resemble the extension and return of the slide of the musical instrument. The situation is probably similar in all bacteria and even the other domains of life, although in some other bacteria, including *Bacillus subtilis*, and in eukaryotes, different combinations of DNA polymerases are used to polymerize the leading and lagging strands (see Sanders et al., Suggested Reading). In the case of *B. subtilis*, an additional DNA polymerase interacts with DnaG and the replicative helicase, extending the RNA primer with DNA before handing the template off to the DNA polymerase used for the majority of DNA replication on both strands.

In addition to its role in loading β clamps onto template DNA, the clamp loader also plays an important role in coordinating the various replication components. Not only does the τ subunit of the clamp loader interact with DNA polymerase on the leading-strand and lagging-strand templates, it also interacts with the DnaB helicase (Figure 1.10). Further coordination on the lagging-strand template is facilitated by interactions between the DnaG primase and DnaB helicase (Figure 1.10). Coordination through the clamp loader helps to focus the energy from DNA polymerization with the energy that powers the helicase, allowing a high rate of DNA replication. The interaction between DNA polymerase III and DnaB governs the speed of unwinding so that it matches the rate of DNA polymerization to prevent undue exposure of single-stranded DNA.

THE GENES FOR REPLICATION PROTEINS

Most of the genes for replication proteins have been found by isolating mutants defective in DNA replication, but not RNA or protein synthesis. Since a mutant cell that cannot replicate its DNA will die, any mutation (for definitions of mutants and mutations, see "Replication Errors" below and chapter 3) that inactivates a gene whose product is required for DNA replication will kill the cell. Therefore, for experimental purposes, only a type of mutant called a **temperature-sensitive mutant** can be usefully isolated with mutations in DNA replication genes. These are mutants in which the product of the gene is active at one temperature but inactive at another. The mutant cells can be propagated at the temperature at which the protein is active (the permissive temperature). However, shifting to the other (nonpermissive) temperature can test the effects of inactivating the protein. The molecular basis of temperature-sensitive mutants is discussed in more detail in chapter 3.

The immediate effect of a temperature shift on a mutant with a mutation in a DNA replication gene depends on whether the product of the gene is continuously required for replication at the replication forks or is involved only in the initiation of new rounds of replication. For example, if the mutation is in a gene for DNA polymerase III or in the gene for the DnaG primase, replication ceases immediately. However, if the temperature-sensitive mutation is in a gene whose product is required only for initiation of DNA replication, for example, the gene for DnaA or DnaC (see "Initiation of Chromosome Replication" below), the replication rate for the population will

Figure 1.10 "Trombone" model for how both the leading strand and lagging strand might be simultaneously replicated at the replication fork. RNA primers are shown in green, and their initiation sites are shown as the sequence 3′-GTC-5′ boxed in blue. **(A)** The Pol III holoenzyme synthesizes lagging-strand DNA initiated from priming site 2 and runs into the primer at site 1. **(B)** The DNA strand undergoing lagging-strand replication loops out of the replication complex as the leading-strand polymerase progresses and the lagging-strand polymerase replicates toward the last Okazaki fragment. **(C)** Pol III has been released from the lagging-strand template at priming site 1 and has hopped ahead, leaving the old β clamps behind, and has reassembled with a new β clamp on the DNA at primer site 3 to synthesize an Okazaki fragment. Both the leading-strand and lagging-strand Pol III enzymes remain bound to each other and the helicase through interactions with τ during the release and reassembly process. **(D)** Pol III continues synthesis of the lagging strand from priming site 3 while Pol I is removing the primer at site 1 and replacing it with DNA. The Pol III holoenzyme hops to the primer at site 4 after reaching the primer at site 2. The primers and Okazaki fragments are not drawn to scale.

slowly decline. Unless the cells have been somehow synchronized in their cell cycles, each cell is at a different stage of replication, with some cells having just finished a round of replication and other cells having just begun a new round. Cells in which rounds of chromosome replication were under way at the time of the temperature shift will complete their replication cycle but will not start a new round. Therefore, the rate of replication decreases until the rounds of replication in all the cells are completed.

Replication Errors

To maintain the stability of a species, replication of the DNA must be almost free of error. Changes in the DNA sequence that are passed on to subsequent generations are called **mutations** (see chapter 3). Depending on where these changes occur, they can severely alter the protein products of genes or other cellular functions. To avoid such instability, the cell has mechanisms that reduce the error rate.

As DNA replicates, the wrong base is sometimes inserted into the growing DNA chain. For example, Figure 1.11 shows the incorrect incorporation of a T opposite a G. Such a base pair in which the bases are paired wrongly is called a **mismatch**. Mismatches can occur when the bases take on forms called **tautomers**, which pair differently from the normal form of the base (see chapter 3). After the first replication shown in Figure 1.11, the mispaired T pairs with an A, causing a GC-to-AT change in the sequence of one of the two progeny DNAs and thus changing the base pair at that position on all subsequent copies of the mutated DNA molecule.

Editing

One way the cell reduces mistakes during replication is through editing functions. With some DNA polymerases the editing function resides in the same protein, while in other cases a separate protein performs the editing function. Editing proteins are aptly named because they go back over the newly replicated DNA looking for mistakes and then recognize and remove incorrectly inserted bases (Figure 1.12). If the last nucleotide inserted in the growing DNA chain creates a mismatch, the editing function stops replication until the offending nucleotide is removed. DNA replication then continues, inserting the correct nucleotide. Because the DNA chain grows in the 5′-to-3′ direction, the last nucleotide added is at the 3′ end. The enzyme activity found in a DNA polymerase or one of its accessory proteins that removes this nucleotide is therefore called a 3′ exonuclease. The editing proteins probably recognize a mismatch because the mispairing (between T and G in the example in Figure 1.11) causes a minor distortion in the structure of the double-stranded helix of the DNA.

DNA polymerase I is an example of a DNA polymerase in which the 3′ exonuclease editing activity is part of the DNA polymerase itself. However, in DNA polymerase III, which replicates the bacterial chromosome, the editing function resides in an accessory protein encoded by a separate gene whose product travels along the DNA with the DNA polymerase during replication. In *E. coli*, the 3′ exonuclease editing function is encoded by the *dnaQ* gene (Table 1.1), and *dnaQ* mutants, also called *mutD* mutants (i.e., cells with a mutation in this gene that inactivates the 3′ exonuclease function), show much higher rates of spontaneous mutagenesis than do cells containing the wild-type, or normally functioning, *dnaQ*

Figure 1.11 Mistakes in base pairing can lead to changes in the DNA sequence called mutations. If a T is mistakenly placed opposite a G during replication **(A)**, it can lead to an AT base pair replacing a GC base pair in the progeny DNA **(B to D)**.

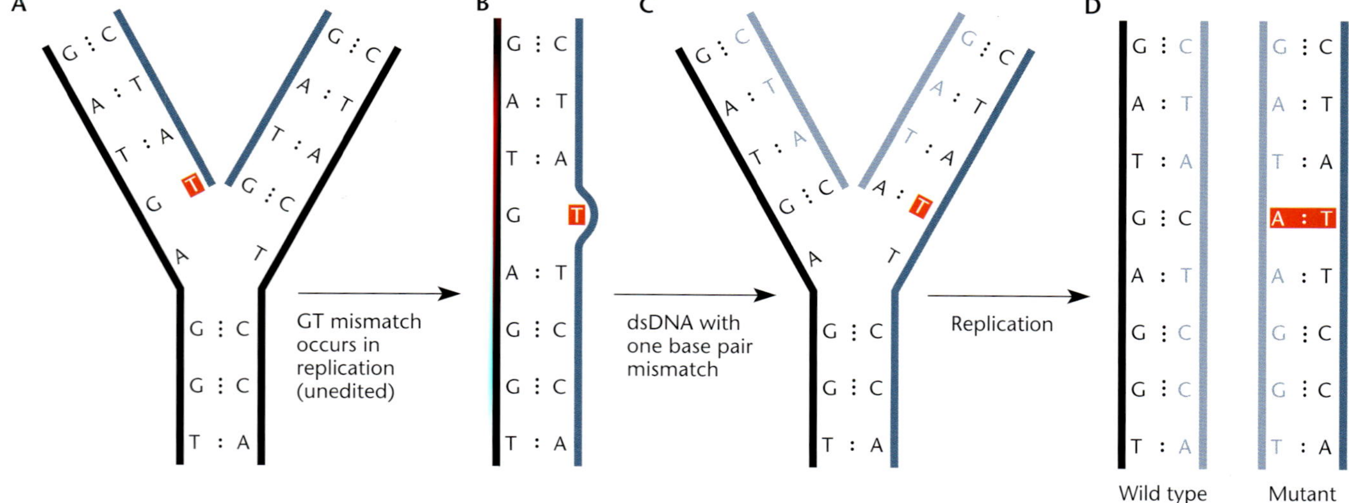

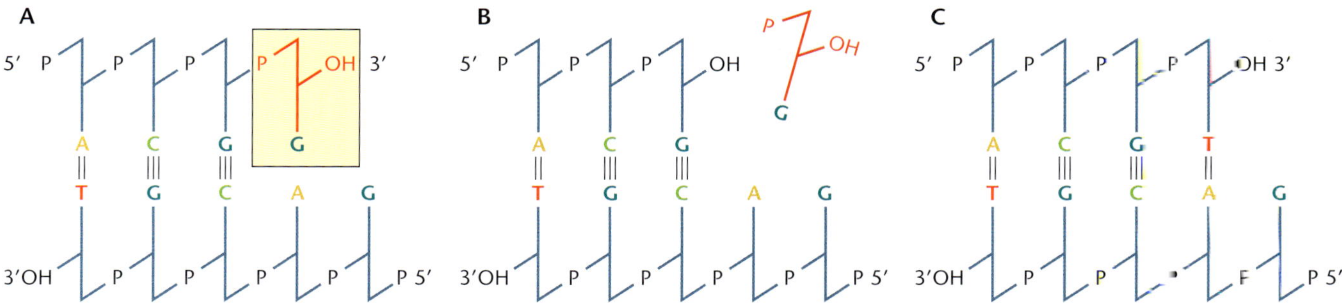

Figure 1.12 Editing function of DNA polymerase. **(A)** A G is mistakenly placed opposite an A while the DNA is replicating. **(B and C)** The DNA polymerase stops while the G is removed and replaced by a T before replication continues.

gene product. Because of their high spontaneous mutation rates, *mutD* mutants of *E. coli* can be used as a tool for mutagenesis, often combined with mutations in other genes whose products normally contribute to the correction of mismatches (see chapter 10).

RNA Primers and Editing

The importance of the editing functions in lowering the number of mistakes during replication may explain why DNA replication is primed by RNA rather than by DNA. When the replication of a DNA chain has just initiated, the helix may be too short for distortions in its structure to be easily recognized by the editing proteins. The mistakes may then go uncorrected. However, if the first nucleotides inserted in a growing chain are ribonucleotides rather than deoxynucleotides, an RNA primer is synthesized rather than a DNA primer. The RNA primer can be removed and resynthesized as DNA by using preexisting upstream DNA as a primer. Under these conditions, a distortion in the helix can be detected by the editing functions, and mistakes are avoided.

Another important system that safeguards the fidelity of the replication process is responsible for fixing mismatches after the growing DNA strand leaves the polymerase. In *E. coli* and its closest relatives, this process is guided by methylation and is termed **methyl-directed mismatch repair**. Related mismatch repair systems are used across all three domains of life, but the use of methylation signals is not widespread. The methyl-directed mismatch repair system is discussed in chapter 10.

Impediments to DNA Replication

While the process described above and diagrammed in Figure 1.10 would suffice for pristine DNA on a template that lacked any type of physical block to the progression of the DNA replication complex, in reality, the situation in the cell is rarely this tidy. DNA polymerases frequently encounter a number of different problems. Challenges to DNA polymerases include interruptions in the DNA template, bulky adducts that cannot be replicated by DNA polymerase III, and physical blocks mediated by supercoiling and proteins bound to, or acting on, the chromosome. One extreme form of impediment to a replication fork comes from **nicks** in either the leading strand or lagging-strand template DNAs in which the phosphate-deoxynucleotide chain is broken in one strand of the DNA. These nicks cause the destruction of the nick-containing arm of the replication fork, resulting in a broken chromosome and collapse of the DNA replication fork. Bacteria possess a highly efficient mechanism for priming repair of this broken end by using the broken DNA itself as a primer to reinitiate DNA replication. This process was likely the original driving force for the evolution of recombination and is described in chapter 9.

Damaged DNA and DNA Polymerase III

DNA polymerase III replicates DNA with incredibly high fidelity. Much of the fidelity of the enzyme comes from the structure of the catalytic pocket, where there is a **presynthetic** check for base pairing between the template strand and the incoming nucleotide. A side effect of this small binding pocket is the inability of the polymerase to tolerate **lesions** in which chemical changes have occurred in the base, the deoxyribose sugar, or even the phosphate on the DNA. There are many mechanisms for DNA replication to continue even when a cell is grown under conditions that result in highly damaged DNA. While early work suggested that the polymerization of the leading and lagging strand was so tightly coupled that a lesion on one strand would stop the entire DNA replication fork, more recent work indicates flexibility. It is now clear that although the polymerases producing the leading strand and lagging strand are physically coupled, the two complexes can be momentarily uncoupled by leaving a single-strand DNA gap at the point of the lesion that blocked one of the DNA polymerases. Other processes can repair these gaps, and in extreme cases, where there is extensive damage in the chromosome, these gaps initiate a DNA damage response called the SOS response (see chapter 10).

Mechanisms To Deal with Impediments on Template DNA Strands

The mechanisms used for momentarily functionally uncoupling synthesis of the two strands differ depending on whether the lesion occurs on the leading-strand or lagging-strand template. The discontinuous nature of replication on the lagging-strand template affords the opportunity to circumvent lesions that halt DNA polymerase III. Typically, DNA polymerase III is recycled onto a new DNA primer when a new RNA primer is deposited (Figure 1.10). However, a stalled lagging-strand DNA polymerase III can also be recycled by premature release when it stalls at DNA damage (Figure 1.13A). The single-strand DNA gap left behind is repaired by another mechanism.

Under the historical model of the function of DnaG primase, primers are placed only on the lagging-strand template. However, biochemical studies indicate that in cases where the leading-strand polymerase stalls, primase can also produce an RNA primer on the leading-strand template, allowing replication to continue but leaving a gap on this strand (Figure 1.13B). This process of lesion skipping on the leading-strand template and the ability to utilize alternative polymerases (described below) to copy over DNA damage provide complementary mechanisms to deal with damaged DNA template strands (see Gabbai et al., Suggested Reading).

While DNA polymerase III and DNA polymerase I are important for high-fidelity DNA replication, other DNA polymerases are found in *E. coli* that allow replication through damaged DNA in a process known as **translesion synthesis**. Most translesion polymerases appear to come with a trade-off in which the ability to copy damaged DNA results in a lower fidelity of DNA replication. As expected, the expression of these polymerases is induced as a response to DNA damage in the cell. In addition to controlling the amount of translesion polymerase present in the cell, access to the DNA replication fork by polymerases other than DNA polymerase III is regulated by a process called **polymerase switching**, a process by which one DNA polymerase replaces a polymerase already found at the 3′ OH end of a primed DNA template (Figure 1.13C). In *E. coli*, DNA polymerases II, IV, and V can be recruited to temporarily step in for DNA polymerase III at damaged DNA (more details of this system are described in chapter 10). Each of these polymerases has different attributes, ranging from fairly accurate and highly processive (DNA polymerase II) to very inaccurate and not very processive (DNA polymerases IV and V). Processivity refers to how far a DNA polymerase moves on the template before falling off.

Having multiple DNA polymerases with different properties appears to be common in all living organisms. How accurate or processive a given DNA polymerase is may also depend on the nature of the damage found in the template DNA and/or the availability of various accessory proteins. The regulation of the use of these DNA polymerases is still incompletely understood, but there appear to be highly evolved processes in which the system has been fine-tuned by the process of natural selection over a long time so that the most appropriate DNA-copying mechanism is used for each environmental challenge.

Physical Blocks to Replication Forks

Proteins bound to, or otherwise acting on, the chromosome can also stop the progression of DNA polymerase III. A programmed block to DNA replication occurs in some bacteria to terminate DNA replication within one region of the chromosome (see below). However, unintended blocks can occur in other situations, such as when DNA polymerase encounters RNA polymerase carrying out transcription (see chapter 2) or when RNA polymerase is stalled at sites of damage in the chromosome. Transcription appears to be the most significant impediment to replication, but a variety of protein complexes need to be displaced ahead of the DNA replication fork.

In addition to DnaB, other helicases help DNA replication to proceed through obstacles on template DNAs. Experimentally, it was shown that two other *E. coli* helicases, Rep and UvrD, allow replication to proceed through a protein block on template DNA (see Guy et al., Suggested Reading). These helicases travel in the 3′-to-5′ direction, and therefore, they are likely to travel on the leading-strand template while DnaB progresses forward on the lagging-strand template (i.e., the 5′-to-3′ direction). Rep interacts directly with the DnaB helicase, probably as a normal part of the **replisome**. The replisome is the collection of proteins that interact with one another (through DNA or other proteins) and that are involved in carrying out DNA replication. In *E. coli*, UvrD may be a helicase of general use for helping out when DNA replication is blocked by proteins or to remove recombination structures from the chromosome to allow DNA replication to proceed. As explained in more detail in chapter 10, UvrD allows strand displacement during mismatch repair and plays additional roles in nucleotide excision repair in *E. coli* in a repair process that is coupled with transcription.

Replication of the Bacterial Chromosome and Cell Division

So far, we have discussed the details of DNA replication, but we have not discussed how bacterial DNA as a whole is replicated, nor have we discussed how the replication process is coordinated with division of the bacterial cell. To simplify the discussion, we first consider only bacteria that grow as individual cells and divide by binary fission to form two cells of equal size, even though this is far

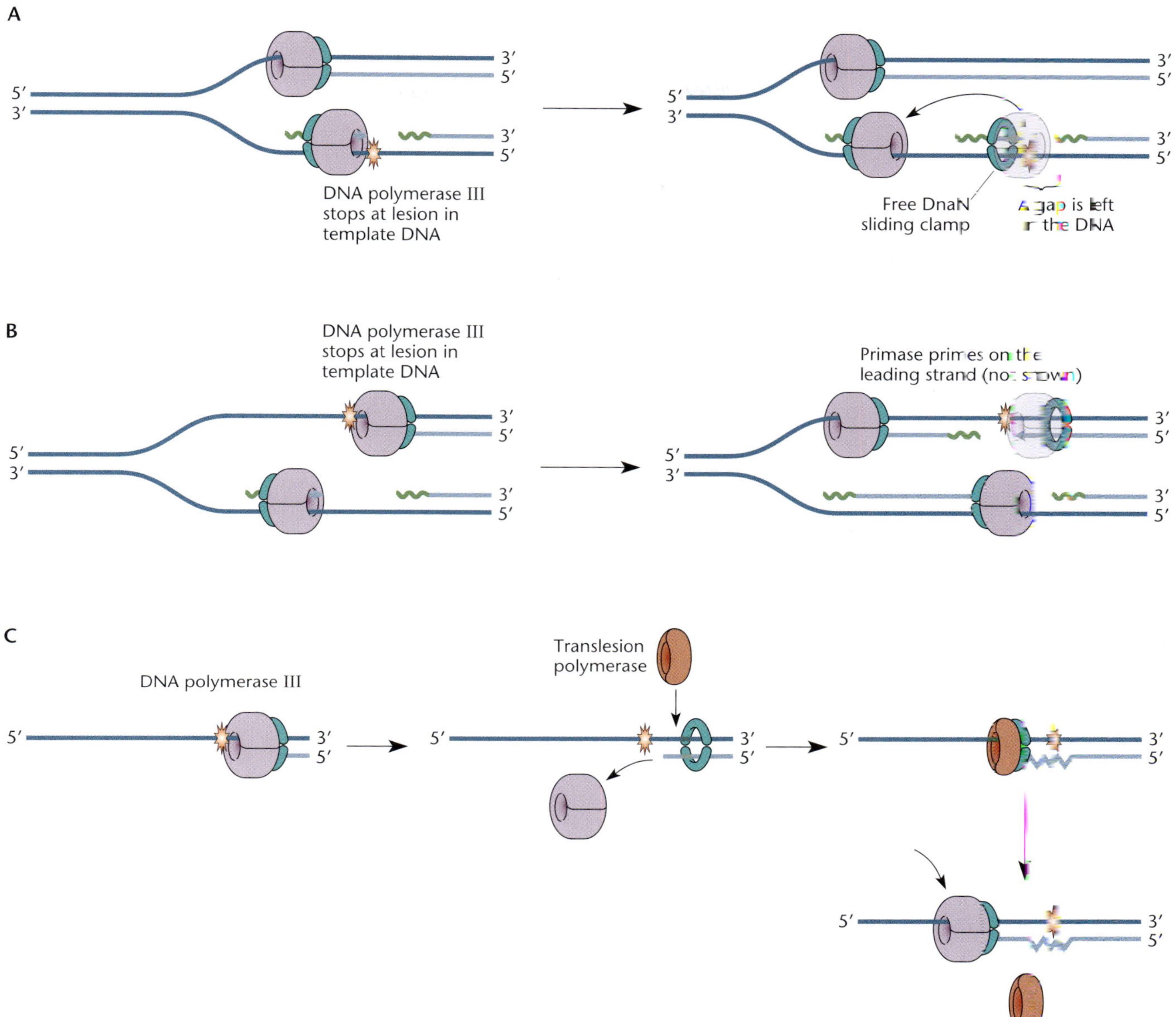

Figure 1.13 Physical blocks on template DNAs. **(A)** When DNA polymerase III stalls on the lagging-strand template strand, a new Okazaki fragment can be initiated and the stalled polymerase can be moved to the new RNA primer. The process leaves a gap that must be repaired by other means, probably involving RecFOR (see chapter 9). **(B)** When DNA polymerase III stalls on the leading strand, primase can restart DNA replication, leaving a gap that must be repaired by other means. **(C)** In some cases where DNA polymerase III stalls at damaged DNA, translesion polymerase can utilize the same sliding clamp to replicate through the lesion, often by error-prone DNA replication. The translesion polymerases have low processivity and fall off the template after a short distance, allowing the accurate DNA polymerase III replication to resume high-fidelity DNA replication.

from the only type of multiplication observed among bacteria.

The replication of the bacterial DNA occurs during the **cell division cycle**, which is the time during which a cell is born, grows larger, and divides into two progeny cells. Cell division is the process by which the larger cell splits into the two new cells. The **division time**, or **generation time**, is the time that elapses from the point when a cell is born until it divides. This time is usually approximately the same for all the individuals in the population under a given set of growth conditions. The original cell before cell division is called the mother cell, and the two progeny cells after division are called the daughter cells.

Structure of Bacterial Chromosomes

The DNA molecule of a bacterium that carries most of its normal genes is commonly referred to as its **chromosome**, by analogy to the chromosomes of higher organisms. This name distinguishes the molecule from plasmid DNA, which in some cases can be almost as large as chromosomal DNA but usually carries genes that are not always required for growth of the bacterium (see chapter 4). Most bacteria have only one chromosome; in other words, there is only one unique DNA molecule per cell that carries most of the normal genes. There are exceptions, and it is estimated that 10% of bacteria have more than one chromosome, including *Vibrio cholerae*, the bacterium responsible for the disease cholera. Even in bacteria that contain multiple chromosomes, the second chromosome shows more characteristics of a plasmid than of a chromosome, particularly in how it initiates replication. There appear to be special molecular systems for managing multiple chromosomes (see Fournes et al., Suggested Reading).

As discussed below, when bacteria, such as *E. coli*, are reproducing very rapidly, new rounds of replication initiate before others are completed, temporally increasing the DNA content of the cells until cellular division returns the number of copies to one unit chromosome per cell. It is important to note, however, that these individual chromosomal DNAs are not unique since they are directly derived from each other by replication.

The structure of bacterial DNA differs significantly from that of the chromosomes of higher organisms. One difference is that the DNA in the chromosomes of most bacteria is circular in the sense that the ends are joined to each other (for exceptions, see Box 4.1). In contrast, eukaryotic chromosomes are usually linear with free ends. As discussed in chapter 4, the circularity of bacterial chromosomal DNA allows it to replicate in its entirety without using telomeres, as eukaryotic chromosomes do, or terminally redundant ends, as some bacteriophages do (see chapter 7). Even in cases where bacterial chromosomes are linear, they do not use the same mechanism, involving telomerases to replicate their ends, that is used by eukaryotic chromosomes. Another difference between the DNA of bacteria and eukaryotes is that the DNA in eukaryotes is wrapped around proteins called histones to form nucleosomes. Bacteria have the proteins HU, HN-S, Fis, and IHF, around which DNA is often wrapped, and archaea do have rudimentary histones related to those of eukaryotes. However, in general, DNA is much less structured in bacteria than in eukaryotes.

Replication of the Bacterial Chromosome

Replication of a circular bacterial chromosome initiates at a unique origin of chromosomal replication, or ***oriC***, and proceeds in both directions around the circle. On the *E. coli* chromosome, *oriC* is located at 84.3 min. As mentioned above, the place in DNA at which replication occurs is known as the replication fork. Two replication forks start at *oriC* and proceed around the circle until they meet and **terminate** chromosomal replication. The DNA polymerases responsible for replicating the leading and lagging strands associate as a single holoenzyme. However, there is no association between the DNA polymerases at the two DNA replication forks to help drive the separation of chromosomes, and therefore, other force-generating mechanisms must be at play. As discussed in "Termination of Chromosome Replication" below, some bacteria actively terminate replication at a unique site in the DNA; however, these systems are not widespread, and most bacteria terminate replication using an unknown mechanism or simply terminate DNA replication where the two replication forks meet. Each time the two replication forks proceed around the circle and meet, a **round of replication** has been completed, and two new DNAs, called the **daughter DNAs**, are generated.

Initiation of Chromosome Replication

Much has been learned about the molecular events occurring during the initiation of replication. Some of this information has a bearing on how the initiation of chromosome replication is regulated and serves as a model for the interaction of proteins and DNA.

Two types of functions are involved in the initiation of chromosome replication. One consists of the sites or sequences on DNA at which proteins act to initiate replication. These are called ***cis*-acting sites**. The prefix *cis* means "on this side of," and these sites act only on the same DNA. The proteins involved in initiation of replication are examples of ***trans*-acting functions**. The prefix *trans* means "on the other side of," and these functions can act on any DNA in the same cell, not just the DNA from which they were made. The concepts of *cis*- and *trans*-acting properties are common in molecular genetics, and these references are used throughout this book.

ORIGIN OF CHROMOSOMAL REPLICATION

One *cis*-acting site involved in DNA replication is the *oriC* site, at which replication initiates. The sequence of *oriC* is well defined in *E. coli*, and the basic components that make up *oriC* are broadly similar in most bacteria. Figure 1.14 shows the structure of the origin of replication of *E. coli*. Less than 260 base pairs (bp) of DNA is required for initiation at this site. Within *oriC* are a series of binding sites for various proteins; the most important

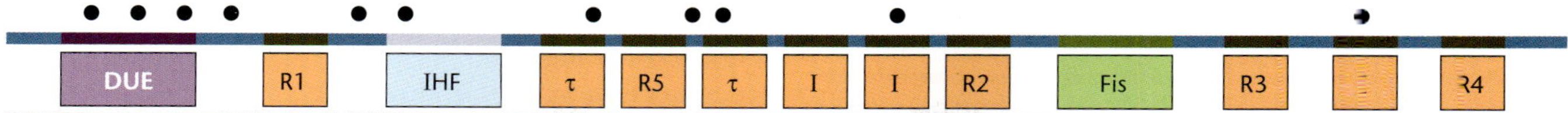

Figure 1.14 Structure of the origin of chromosomal replication (*oriC*) region of *Escherichia coli*. Shown are the positions of multiple types of DnaA-binding sequences, five DnaA boxes (R1 to R5) and other DnaA-binding sites (I and τ), and an AT-rich region that is unwound to allow loading of the replication apparatus, the DNA-unwinding element (DUE). Also shown are binding sites for the IHF and Fis proteins and a large number of GATC sites (black dots) that are important in regulating initiation by acting as sites of Dam methylation.

of these binding proteins is the master initiator protein in bacteria, called DnaA (see below). The **canonical** DnaA-binding sequences are 9 bp in length, and these sites are termed **DnaA boxes**. While five DnaA boxes exist within *oriC*, three of these sites bind DnaA particularly strongly, i.e., are of particularly high affinity, and are always bound by DnaA (Figure 1.14). The ability of DnaA to bind these high-affinity sites at all times can be considered analogous to the origin recognition complex associated with eukaryotes. Additional sites called "I" and "τ" sites, which differ from DnaA boxes, exist in the origin region but are occupied only by DnaA that is bound to ATP and not ADP (see below) (Figure 1.14). Finally, within an AT-rich region of DNA that is opened for initiation, called the DNA-unwinding element, are three additional sites for DnaA binding that are occupied only when DnaA is bound to ATP and not ADP. Binding sites for other DNA-binding proteins (IHF and Fis) are also found in this region.

INITIATION PROTEINS

Besides the *cis*-acting *oriC* site and DnaA, many *trans*-acting proteins are also required for the initiation of DNA replication, including the DnaB and DnaC proteins. DnaA is required only for initiation, allowing DnaC to load the DnaB helicase for establishing the DNA replication forks. Many proteins used in other cellular functions are also involved, such as the primase (DnaG), the normal RNA polymerase that makes most of the RNA in the cell, and the DNA-binding proteins IHF and Fis (Figure 1.14).

Figure 1.15 outlines how DnaA, DnaB, DnaC, and other proteins participate in the initiation of chromosome replication. As we will see throughout the remainder of this chapter, there are many points at which the initiation of DNA replication is controlled. One important regulatory consideration concerns the nucleotide-binding state of DnaA. While DnaA always binds some of the DnaA boxes, for initiation of DNA replication all of the boxes are bound, forming a special architecture that can open the DNA strands. This type of binding requires that DnaA be bound to ATP (DnaA-ATP). In biology, there are many examples where the nucleotide-bound state of a protein determines its activity. Proteins of this type have the capacity to hydrolyze nucleotides from the NTP to the NDP form, but the energy released is not directly used to actively carry out any particular task and instead allows the configuration of the proteins to change. In the case of DnaA, the ATP-bound form of the protein allows it to form a large multimer structure composed of many molecules of DnaA protein, where the DNA strands are opened through bending of the DNA by DnaA with the help of the IHF and Fis proteins. Within the special complex that productively opens the DNA strands in the origin, DnaA binding appears to take on a different form when it interacts with DUE (DNA unwinding element), preferentially engaging single-strand DNA in this region to facilitate strand opening (Figure 1.15). The binding and opening are also aided by supercoiling at the origin (see "Supercoiling" below) and by the SSB protein, which helps to keep the helix from reforming. DnaA binds directly to the helicase DnaB, and in a process involving DnaC, DnaB helicase is loaded onto *oriC*. Action of the DnaB helicase opens the strands further for priming and replication, and DnaC leaves the complex.

RNA Priming of Initiation

As described above, RNA primers are continuously needed during the DNA replication process. The complex that travels along the chromosome laying down RNA primers is called the **primosome** and contains DnaB and DnaG primase. The RNA polymerase that synthesizes most of the RNA molecules, including mRNA, in the cell (see chapter 2) is needed to initiate rounds of replication; however, transcription from RNA polymerase in this context from adjacent genes is involved in controlling the separation of the strands of DNA in the *oriC* region. DnaG primase is responsible for laying down RNA primers for DNA synthesis after replication is initiated at *oriC*.

Termination of Chromosome Replication

After replication of the chromosome initiates in the *oriC* region and proceeds around the circular chromosome in both directions, the two replication forks must meet somewhere on the chromosome and the two daughter chromosomes must separate. In some bacteria, including

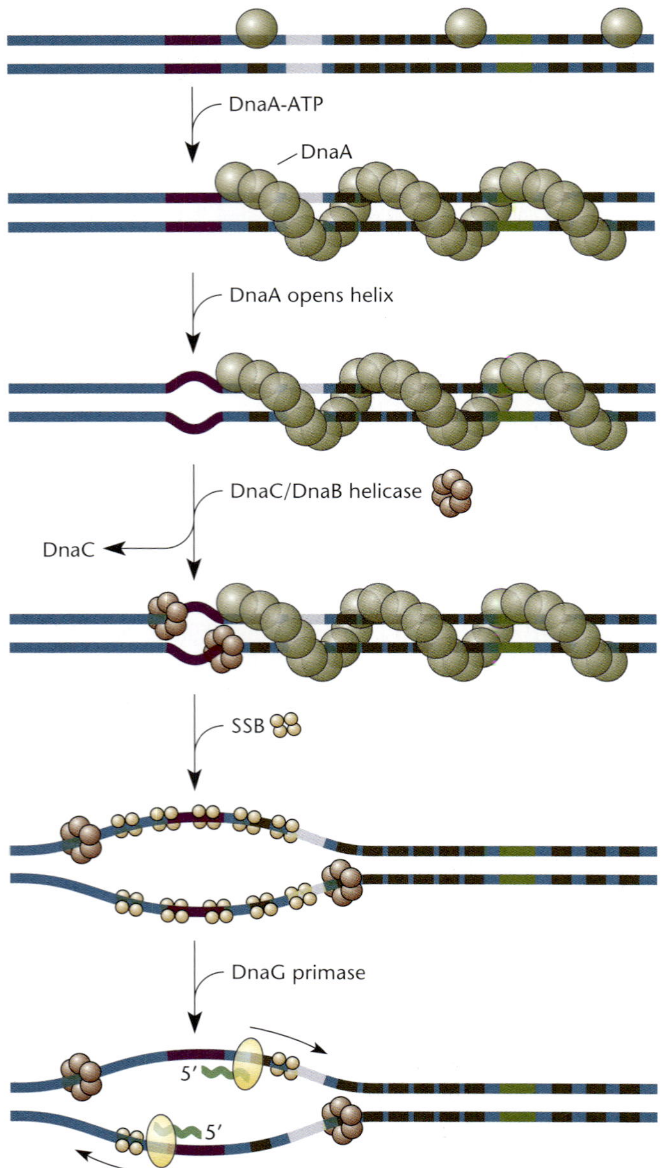

Figure 1.15 Initiation of replication at the *Escherichia coli* origin (*oriC*) region. DnaA is always bound to three DnaA boxes within *oriC*, even when DnaA is in its non-ATP-bound state acting as an origin recognition complex. About a dozen DnaA-ATP proteins bind to the origin, possibly by forming a type of helical filament and opening the helix and the DNA-unwinding element. DnaC helps the DnaB helicase to bind. The DnaG primase synthesizes RNA primers, initiating replication.

E. coli and *B. subtilis*, a specific system exists to control the region where replication forks meet. What happens in other organisms is less clear. As with most cellular processes, the process of termination of chromosome replication is especially well understood in *E. coli*. In this bacterium, termination is facilitated by DNA sequences, called *ter* sites, that are only ~22 bp long. These sites act somewhat like the one-way gates in an automobile parking lot, allowing the replication forks to pass through in one direction but not in the other.

Figure 1.16 shows how the one-way nature of *ter* sequences can cause replication to terminate in a specific region of the chromosome. In the illustration, two *ter* sites called *terA* and *terB* bracket the termination region. Replication forks are unaffected by the *terA* site in the clockwise direction but are terminated in the counterclockwise direction. The opposite is true for *terB*. Thus, the clockwise-moving replication fork progresses through *terA*, but if it gets to *terB* before it meets the counterclockwise-moving fork, it stalls, because it cannot move clockwise through *terB*. Similarly, the replication fork moving in the counterclockwise direction stalls at the *terA* site and waits for the clockwise-moving replication fork. When the counterclockwise and clockwise replication forks meet, at *terA*, *terB*, or somewhere between them, the two forks terminate replication, releasing the

Figure 1.16 Termination of chromosome replication in *Escherichia coli*. **(A)** The replication forks that start at the origin of chromosomal replication (*oriC*) can traverse *terA* and *terB* in only one direction, opposite that indicated by the black arrows. **(B)** When they meet, between or at one of the two clusters, chromosome replication terminates. f_L is the fork that initiated to the left and moved in a counterclockwise direction. f_R is the fork that initiated to the right and moved in a clockwise direction. Adapted from Camara JE, Crooke E, *in* Higgins MP, ed, *The Bacterial Chromosome* (ASM Press, Washington, DC, 2005), with permission.

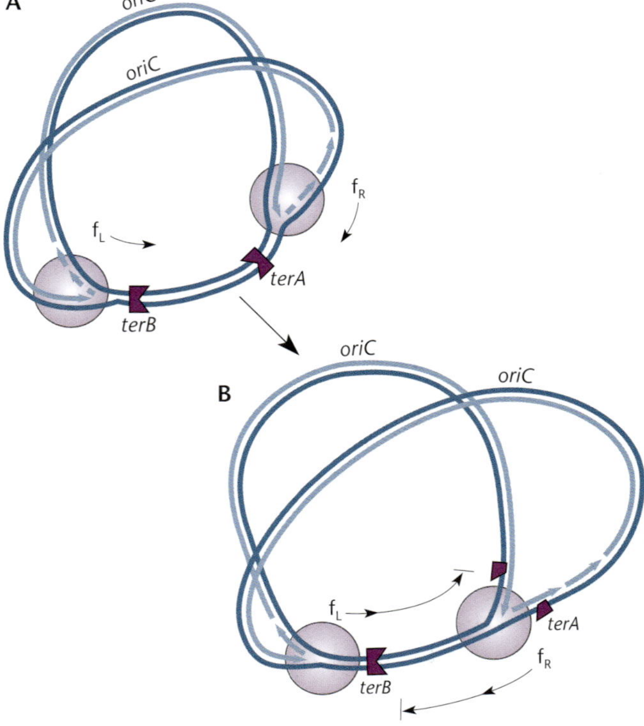

two daughter DNAs. In *E. coli*, it is known that most DNA replication termination occurs at one *ter* site, possibly because it is oriented to terminate replication forks traveling in the clockwise direction, which is shorter in most laboratory *E. coli* strains.

Encountering a *ter* DNA sequence, by itself, is not sufficient to stop the replication fork. A protein is also required to terminate replication at *ter* sites. The protein that works with *ter* sites, called the terminus utilization substance (Tus) in *E. coli* and the replication terminator protein (RTP) in *B. subtilis*, binds to the *ter* sites and stops the replicating helicase (DnaB in *E. coli*) that is separating the strands of DNA ahead of the replication fork. In both *E. coli* and *B. subtilis*, multiple *ter* sites bracket the terminus region, helping to ensure that replication proceeds from *oriC* to the terminus region (see Box 1.1). While the *ter* systems are not absolutely essential for *E. coli* and *B. subtilis* growing in the laboratory setting, they are important for other aspects of genome stability that are important for maintaining genome integrity over time in the natural environment (see Rudolph et al., Suggested Reading). It has been argued that the active termination systems involving *ter* sites and a *trans* acting protein originated in plasmids (chapter 5) and were domesticated in some branches of bacteria for use in the bacterial chromosome (see Galli et al., Suggested Reading).

Chromosome Segregation

While bacteria do not contain a special membrane compartment for chromosomal DNA like the nucleus of eukaryotes, even in bacteria the chromosome does not freely diffuse within the cytoplasm. In fact, as we learn more about bacterial chromosomes, we are realizing that they are maintained with an incredible amount of organization. Even with the aid of only a standard laboratory microscope and DNA stain, a mass of chromosomal DNA is easily observed in the center of the cell in a structure called the **nucleoid**, which is very compact, considering that the DNA is about a thousand times as long as the cell.

Because of the large size of the chromosome, the process of moving the replicating chromosomes to daughter cells, called **segregation**, is not trivial. Chromosome segregation encounters a number of obstacles. Obvious initial obstacles are viscous forces and torsional stress associated with unwinding the template strands of DNA. Advances in microscopy and techniques that allow the localization of certain regions of the chromosome are revealing the choreography involved in coordinating DNA

BOX 1.1

Structural Features of Bacterial Genomes

It is widely appreciated that the chromosomes are the information storehouse for an organism. What is less appreciated is that the chromosome as a structure has evolved sequence motifs that allow it to be efficiently replicated, repaired, and segregated into daughter cells. The distribution and orientation of these motifs are discussed here; the molecular biology of the systems that recognize these sequences is explained in greater detail in the text. The placement of these sequence motifs in the context of the chromosome is important for their function, as is the orientation of many of these sequences. Many of the motifs are oriented in one direction, which follows the direction of the DNA replication fork. DNA replication in *E. coli* and *B. subtilis* (and all bacteria studied to date) is initiated within a single *oriC* region and continues bidirectionally to a position on the chromosome equidistant from the origin (indicated by the long arrow-headed line in the figure). The *dif* site where the resolution of dimer chromosomes occurs is found near to where DNA replication normally terminates.

Certain DNA sequence motifs that guide DNA replication and DNA repair are polar in that they are not symmetrical and need to be in a specific 5′-to-3′ direction to carry out their functions. In other words, these sequences must be found in a certain orientation in the chromosome and will not work if they are flipped around. In *E. coli* and *B. subtilis*, DNA replication forks are actively terminated at specific sites called *ter* sites. The *ter* sites act as a trap for DNA replication forks, and these sites encompass a large portion of the chromosome, allowing DNA replication forks to pass when approaching from one direction but not the other. Multiple *ter* sites (10 in *E. coli* and 9 in *B. subtilis*) are found in the genome, and the redundancy of these sites may be important for catching replication events that get through the initial *ter* sites (indicated with a dashed line) or to stop replication forks that are initiated for DNA repair by recombination (see chapter 9). For unknown reasons, the central *ter* sites in *B. subtilis* are very close together, while the central sites in *E. coli* are separated by hundreds of thousands of base pairs. Along the path of DNA replication are sequence motifs involved in guiding the DNA translocase proteins involved in chromosome segregation: FtsK, found in *E. coli* (which recognizes motifs called KOPS), and SpoIIIE, found in *B. subtilis* (called SRS motifs). The StpA DNA translocase from *B. subtilis* may also recognize the SRS sites. Chromosomes also

(continued)

Structural Features of Bacterial Genomes

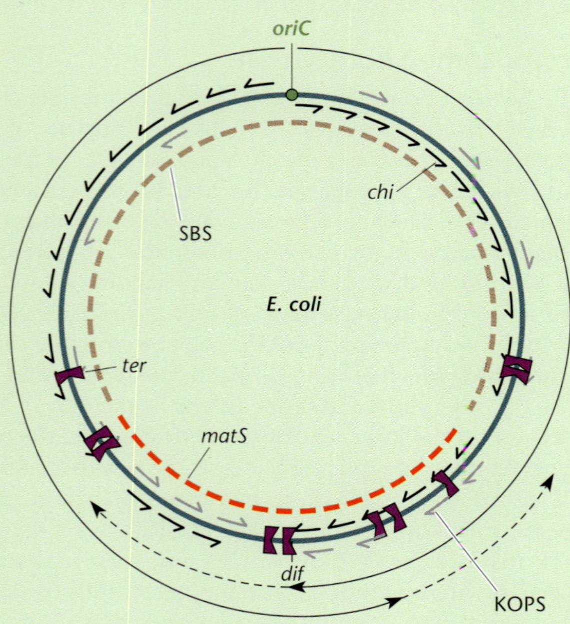

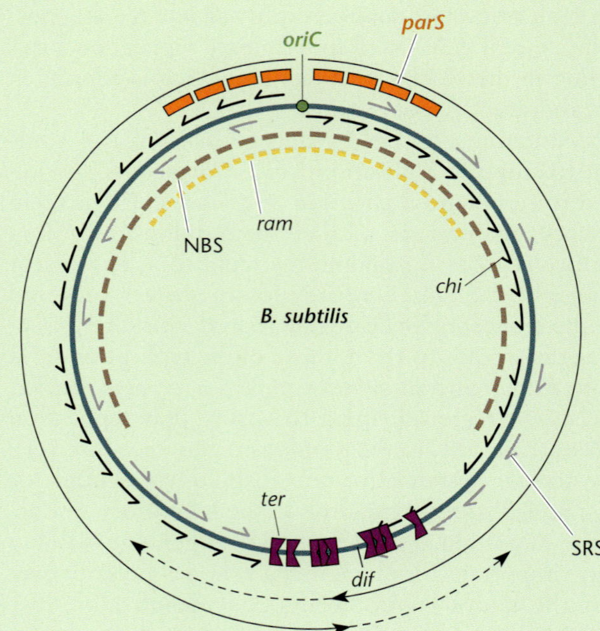

have polar DNA sequences that guide the recombination machinery (see chapter 9). DNA recombination is extremely important in bacteria as a way to repair DNA double-strand breaks that occur during DNA replication. Repair of these breaks occurs when recombination reestablishes a DNA replication fork using one broken end and the sister chromosome. Reestablishment of DNA replication involves an efficient processing event that utilizes polar sites called *chi* sites. The RecBCD complex in *E. coli* or the AddAB complex in *B. subtilis* carries out this processing activity using information found in the *chi* sites. *chi* sites are species specific and are common in genomes in one orientation from the origin to the terminus region on the leading strand (found about 1 every 5 kb in the *E. coli* chromosome).

Other polar sequence biases in the chromosome include an overrepresentation of the 5′-CTG-3′ sequence that primes lagging-strand DNA synthesis (not shown). Interestingly, the most common triplet codon is the CUG (5′-CTG-3′ in DNA) that codes for leucine, comprising almost 5% of all codons in *E. coli*. The 5′-CTG-3′ sequence is found in the *chi* sequence and all of the most frequent 8-bp sequences in the chromosome. It would be difficult to argue which came first, the use of this sequence by the primase or its frequency of use as a codon. Another type of sequence bias, but one that is not polar in nature, is a general sequence bias called the G/C skew, where G and C are overrepresented in the leading strand. The trend toward A and T in the lagging strand is believed not to have an adaptive value but to be a result of the way in which repair differs on the two strands.

There are other DNA sequences that are not polar but that show biases for regions of the chromosome. Around the origin of *B. subtilis*, an area recognized by the Spo0J protein for segregation of the origin region to daughter cells, the *parS* sequence, is found 8 times (the orange rectangular boxes in the figure). Also around the origin of *B. subtilis*, there is an enrichment of *ram* sites (short yellow dashes), a sequence recognized by the RacA protein for maintaining segregation during sporulation. Binding sites for the nucleoid occlusion proteins which prevent septum formation until division is nearly complete reside across the chromosome but are absent from the terminus region. SlmA-binding sites (SBS) in *E. coli* and Noc-binding sites (NBS) in *B. subtilis* (long brown dashes) are recognized by SlmA and Noc, respectively. The GATC sites (not shown), which are important for regulating the initiation of DNA replication at *oriC* in *E. coli*, also show enrichment in the *oriC* region, with a spacing that is important for SeqA binding. The organization of the large domain comprising the terminus region of the chromosome appears to be important in *E. coli*, where the MatP protein recognizes *matS* sites (red dashes in the *E. coli* diagram) found across this region.

References

Blattner FR, Plunkett G III, Bloch CA, Perna NT, Burland V, Riley M, Collado-Vides J, Glasner JD, Rode CK, Mayhew GF, Gregor J, Davis NW, Kirkpatrick HA, Goeden MA, Rose DJ, Mau B, Shao Y. 1997. The complete genome sequence of *Escherichia coli* K-12. *Science* **277**:1453–1462.

Touzain F, Petit M-A, Schbath S, El Karoui M. 2011. DNA motifs that sculpt the bacterial chromosome. *Nat Rev Microbiol* **9**:15–26.

replication and chromosome segregation. Microscopy experiments using green fluorescent protein (GFP) fused to replication proteins allow the localization of DNA replication forks, and GFP fused to proteins that bind to specific sites on DNA allows localization of the origin and terminus in the cell. These experiments show that soon after the initiation of DNA replication within the nucleoid, the origins start to move to the daughter cells. Once replication is complete, segregation would still be prevented if daughter chromosomes were joined by recombination, interlinked, or otherwise tangled during replication. Even if they were not physically joined, their separation would be very difficult if the two daughter chromosomes were randomly spread out throughout the cell. It is therefore not surprising that bacteria have a number of systems to ensure that their chromosomes segregate properly into the daughter cells during cell division. Molecular systems responsible for chromosome segregation are discussed separately below.

RESOLUTION OF DIMER CHROMOSOMES

During the process of DNA replication, two circular daughter chromosomes will periodically become joined, forming a chromosome **dimer**, in which they are joined end to end to form a double-length circle. Dimer chromosomes can result from recombination between the two replicating chromosomes and are fairly common. Recombination involved in restarting stalled DNA replication forks from a sister chromosome probably accounts for many of these events (see chapter 9). Such dimers prevent chromosome segregation because the two daughter chromosomes are part of the same large molecule.

If dimer chromosomes can be created by recombination, they can also be resolved into the individual chromosomes by a second recombination event. The general recombination system could in theory resolve the dimers between sister chromosomes by recombination; however, the general recombination system can both create and resolve dimers, depending on how many crossovers occur between the daughter DNAs. An odd number of crossovers occurring between any two sequences on the two daughter DNAs in the dimer will resolve the dimer, but an even number of crossovers will recreate a dimer.

All bacteria with a circular chromosome appear to have a system dedicated to dimer resolution. In *E. coli*, and in most bacteria, the so-called Xer recombination system is used to resolve chromosome dimers. Rather than using the general recombination system, the Xer systems involve a site-specific recombinase (see chapter 8) to resolve chromosome dimers. This system has evolved so that it resolves dimers into the individual chromosomes but does not create new dimers. Its action is also coordinated with division of the cell and other important chromosome-partitioning functions. The Xer recombination system consists of two proteins called XerC and XerD and a specific site in the chromosome called *dif*. If two copies of the *dif* site occur on the same DNA, as happens when the chromosome has formed a dimer, the Xer proteins promote recombination between the two *dif* sites, resolving the dimer into the individual chromosomes (Figure 1.17). The *dif* site is always found centrally located in the *ter* region in bacterial chromosomes. This is likely, in part, to help ensure that there is only one *dif* site in the cell until just before cell division so that it is not replicated until just before the chromosome has completed replication and just before the cell divides. As added insurance, the activity of the Xer site-specific recombination system is also dependent on the formation of the division septum through an interaction with the FtsK protein. As diagrammed in Figure 1.17, FtsK protein is localized to the region where the division septum pinches in during cell division, where it plays multiple roles, including facilitating dimer resolution. While the Xer proteins are needed for dimer resolution, they are actually active for full dimer resolution only when they interact with FtsK. As shown in Figure 1.17, the localization of FtsK at the septum limits the dimer resolution process temporally and spatially to when the daughter chromosomes are in the process of moving through the septum into daughter cells, a process that is facilitated and coordinated by FtsK itself (see below) (see Aussel et al., Suggested Reading).

The FtsK protein is a DNA translocase that can pump DNA through itself to help align the two *dif* sites in a dimer chromosomal DNA at the septum in the middle of the cell before they recombine, thereby facilitating segregation into daughter cells (Figure 1.17). See the reconstructed image of the structure of the motor domain of FtsK at the start of this chapter (see also Jean et al., Suggested Reading). An obvious question is how the FtsK protein "knows" which direction pumps to engage for transporting DNA in the correct direction to move a *dif* site to the septum. It does this by using sites on the DNA as an orientation cue. These sites are called the KOPS sites, for FtsK-orienting polar sequence. Polar means that they read differently in one direction on the DNA than in the other. These sequences are oriented in the DNA so that they will only be read progressing from the origin to the *dif* site in the chromosome; the FtsK protein, by translocating the DNA in the direction pointed to by the KOPS sites, will pump the DNA toward the *dif* sites close to the terminus of replication. Note in Figure 1.17 how the polar KOPS sites in the chromosome (shown as half-arrows) can be used as directional information by FtsK to mobilize the chromosome to the daughter cells while moving the *dif* sites toward the septum. This is just one example of how the chromosome, typically thought of as the informational storehouse of the cell, also contains structural information used by the many proteins

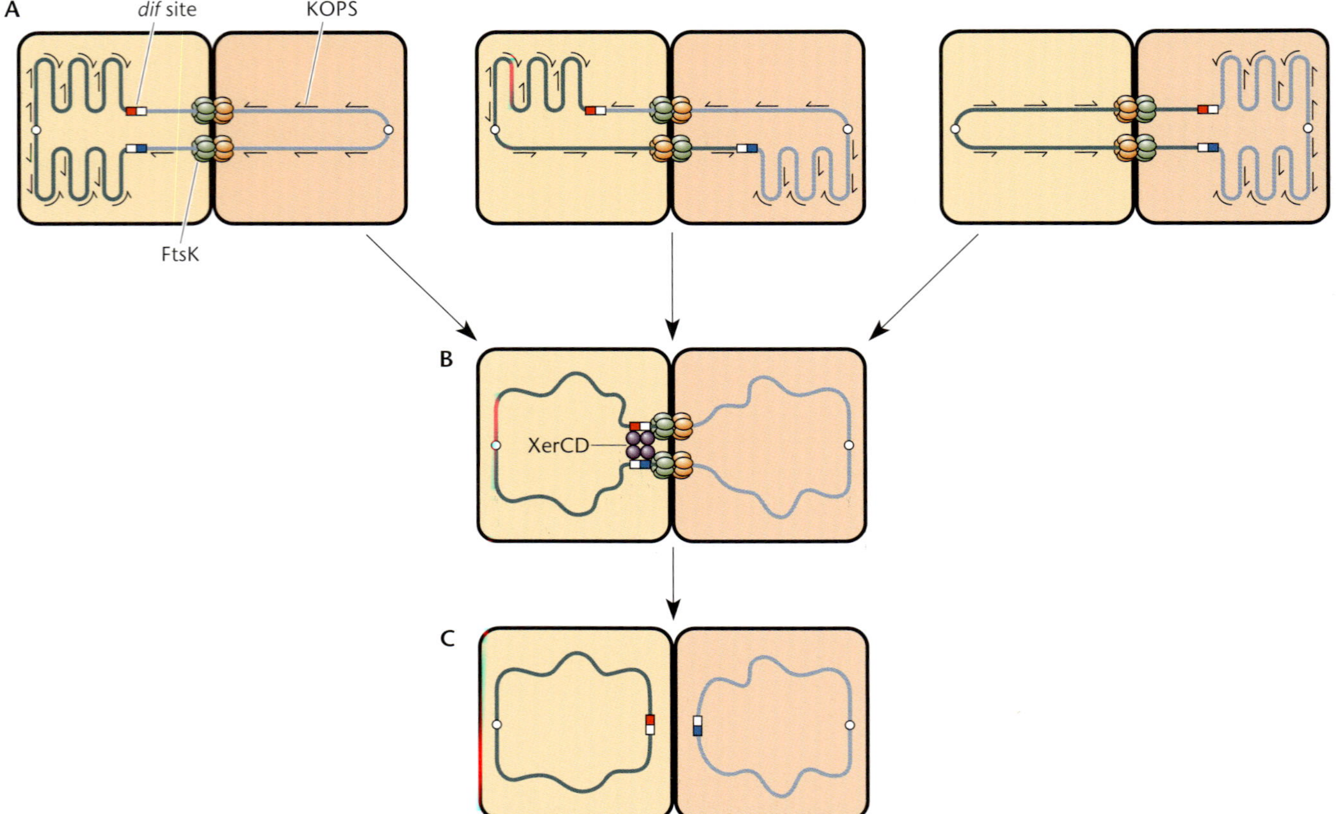

Figure 1.17 Model of the way in which chromosome translocation by FtsK coordinates chromosome segregation with dimer resolution. The two daughter chromosomes in a dimerized chromosome are shown in different shades of blue for emphasis. **(A)** Three possible distributions of the newly replicated dimer chromosome are shown following the start of cell septation. FtsK is a DNA translocase that can pump the chromosomes to the correct daughter cells using polar sequences in the chromosome called KOPS sites (shown as half arrows) while also moving the *dif* sites into alignment at the septum. Green indicates FtsK complexes that are actively pumping; orange indicates idle FtsK complexes where activity is set by the orientation of KOPS sites. **(B)** After aligning the *dif* sites, the FtsK protein also interacts with the XerCD enzyme, allowing it to resolve the dimer chromosomes at the *dif* sites. **(C)** The coordinated activities of the dimer resolution system and FtsK lead to monomer chromosomes that are capable of full segregation to daughter cells. From Camara JE, Crooke E, *in* Higgins NP (ed), *The Bacterial Chromosome* (ASM Press, Washington, DC, 2005).

that manage and repair the chromosome (Box 1.1). FtsK also has a domain that interacts with topoisomerase IV, a protein capable of untangling catenanes in the chromosomes (see below). Therefore, the FtsK protein coordinates a veritable clearinghouse of activities that help with chromosome management during chromosome replication and chromosome segregation.

Homologs of FtsK are widespread, and a homologous protein called SftA appears to carry out similar functions in *B. subtilis* (see Biller and Burkholder, Suggested Reading). Some bacteria have more than one FtsK-like protein, presumably for other specialized tasks. In the case of *B. subtilis*, another FtsK homolog, SpoIIIE, is responsible for translocating the final third of the chromosome into spores during spore development so the spore will get a complete copy of the chromosome (see chapter 12). How the roles of FtsK, SpoIIIE, and SftA differ and how the multiple members are involved in various DNA processing events in a single cell remain active areas of research with many questions still to be answered.

Interestingly, instead of XerC and XerD, *Streptococcus* and *Lactococcus* species have a system more closely related to bacteriophage integration systems that uses a single protein called XerS to carry out the same function (see Le Bourgeois et al., Suggested Reading). Archaea also seem to use a single protein (in this case called XerA) to resolve dimer chromosome at a *cis*-acting *dif* site (see Cortez et al., Suggested Reading). The regulation of dimer chromosome resolution in these systems and any involvement of FtsK-like proteins are unclear.

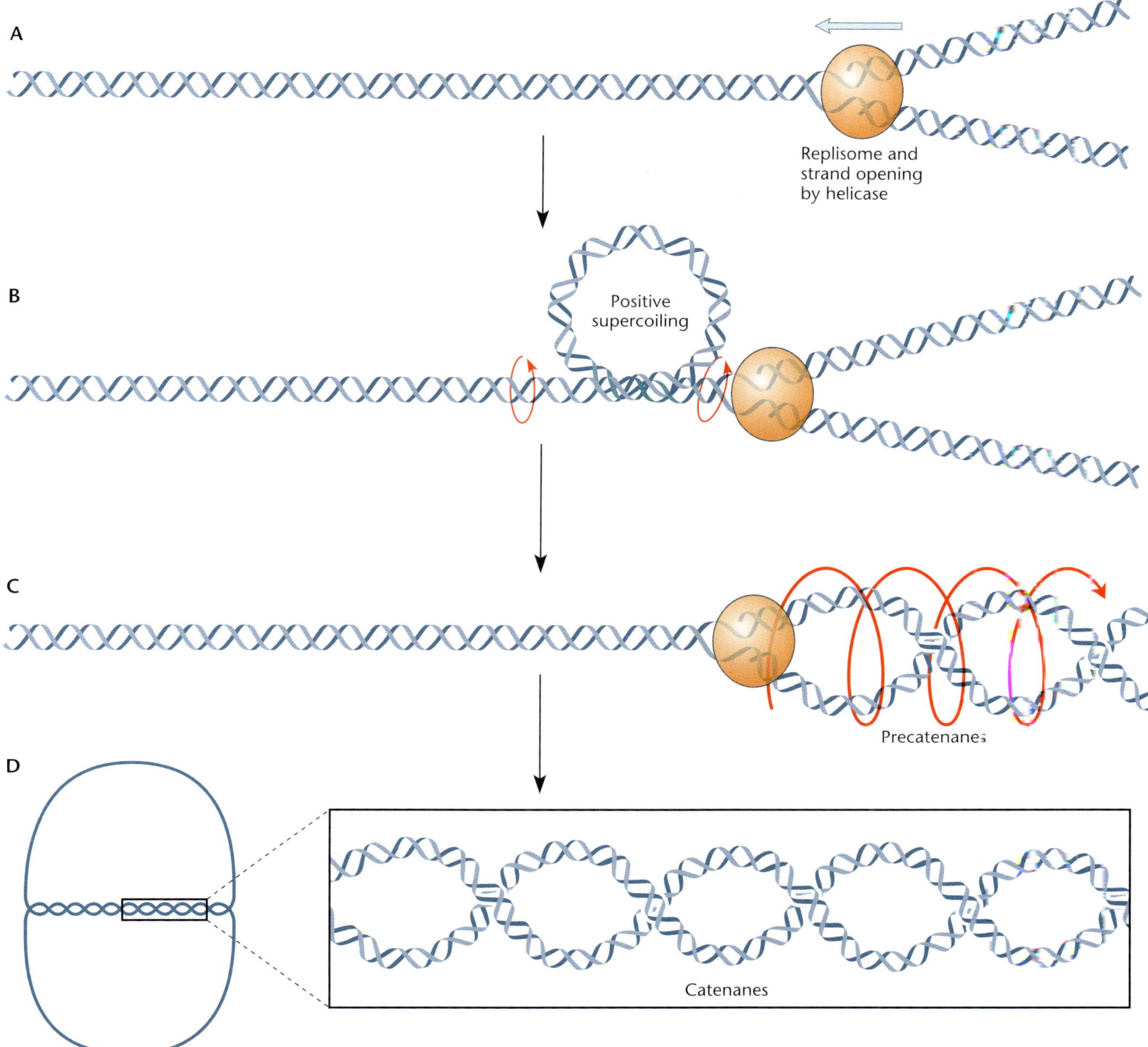

Figure 1.18 Model of the way in which unwinding of the template DNA strands can cause twists that can diffuse across the replication complex and twist the new DNA strands. **(A)** The replication machinery must open the double-stranded DNA to copy the template strands. **(B)** Unwinding the template DNA strands introduces twists (shown by thin red arrows) called positive supercoils ahead of the replication fork. **(C)** Some of the torsion that is generated ahead of the replication fork can be relieved by rotation of the replication complex itself. The torsional stress can spread behind the fork and intertwine the new copies of the chromosome (shown by a thick red arrow). The intertwined chromosomes are called precatenanes. **(D)** Precatenanes result in links in the daughter chromosomes called catenanes that must be unlinked for the chromosomes to separate into daughter cells.

DECATENATION

DNAs also become joined to each other through the formation of **catenanes**, where the daughter DNAs become interlinked like the links on a chain. These interlinks can form as a side effect of separating DNA strands during the process of DNA replication (Figure 1.18). As we discuss in "Supercoiling" below, DNA replication introduces a great deal of torsional stress ahead of the DNA replication fork (Figure 1.18A and B). This stress can be transferred across the DNA replication fork into the newly formed DNA strands in twists between the two new daughter strands called precatenanes (Figure 1.18C), because the twists will eventually form catenanes when replication is complete (Figure 1.18D). Once such interlinks are formed, the only way to unlink them is to break both strands of one of the two DNAs and pass the two strands of the other DNA through the break. The break must then be resealed. This double-strand passage, called **decatenation**, is one of the reactions performed by type II topoisomerases (see below). A type II topoisomerase called **topoisomerase IV** (Topo IV) plays a major role in removing most of the interlinks between the daughter DNAs in *E. coli*. The act of removing these links appears to remove one of the major cohesive forces between the daughter chromosomes prior to segregation. One of the major points of regulation appears to be spatial, where the two subunits that make Topo IV are most likely to interact through association with other proteins. In one case, this occurs following replication, when the chromosomes are being translocated across the division septum by FtsK (Figure 1.19). FtsK helps to regulate the decatenation process as it pulls the chromosome to the septum, because one of the subunits of Topo IV interacts with FtsK. Interaction between Topo IV and the FtsK protein puts the enzyme in a very appropriate position for the removal of catenanes just before chromosome segregation. Topo IV is also regulated through an interaction with a protein involved in the condensation of chromosomes following DNA replication (see "Condensation" below).

CONDENSATION

Bacterial cells have an important mechanism to help manage chromosomes, which is to condense them after DNA replication. If the daughter chromosomes are condensed, they do not overlap as much in the cell and so are less apt to become interlinked. Condensation of chromosomes prior to mitosis has been known for a long time to occur in eukaryotic cells, where the chromosomes are only clearly visible just before mitosis. We now know that bacteria also condense their daughter DNAs to make them easier to manage, even though it is more difficult to visualize the condensation of bacterial chromosomes because of their smaller size.

Condensins

Proteins called **condensins,** which help to condense DNA in the cell, were first discovered in eukaryotes. They are also known as **SMC proteins** for structural maintenance of chromosome. Condensins are long, dumbbell-shaped proteins with globular domains at the ends and a long coiled-coil region holding them together. The long coiled-coil region has a hinge so that it can fold back on itself and the two globular domains can bind together. Condensins work with partner proteins to help condense

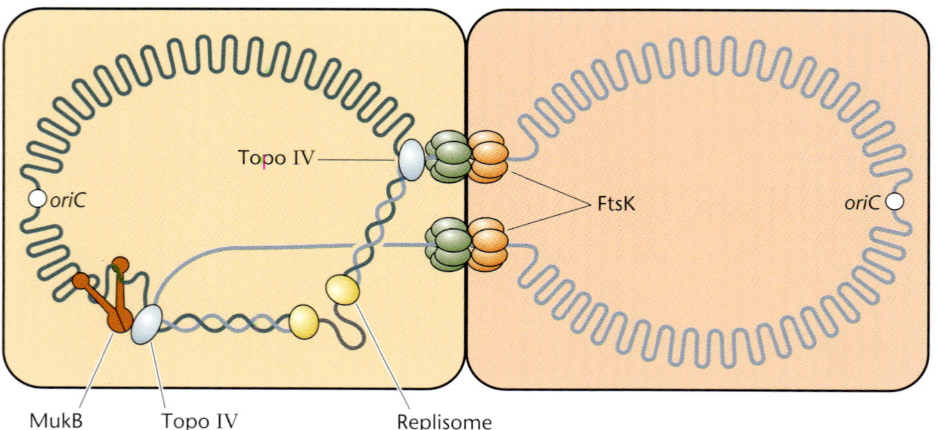

Figure 1.19 Model of the way in which chromosome decatenation by topoisomerase IV (Topo IV) is coordinated with chromosome condensation by MukB and chromosome translocation with FtsK. The two daughter chromosomes that have been replicated are indicated by light and dark blue lines, and the unreplicated portion of the chromosome is shown in gray. Unwinding of the template DNA strands that is associated with DNA replication twists the newly replicated strands of DNA, forming precatenanes that go on to become catenanes if not unlinked by the action of Topo IV. Topo IV can interact with the chromosome condensation protein MukB to remove catenanes before DNA is condensed. Topo IV also interacts with the FtsK translocase to coordinate decatenation with chromosome segregation. The replisome is shown as a yellow circle, and double-stranded DNA is shown as a single line for simplicity.

DNA into a higher-order structure. While still somewhat controversial, bacterial condensins may encircle DNA to gather regions on the DNA together that can subsequently be held together by a variety of other sequence-nonspecific DNA-binding proteins discussed elsewhere, such as HU, IHF, H-NS, and Fis.

The condensin of *E. coli*, called MukB, was found because mutations in its gene interfere with chromosome segregation. MukB was suspected of condensing the DNA because the protein and supercoiling of the DNA can compensate for each other in allowing proper segregation of the daughter chromosomes into daughter cells. As indicated above, supercoiling can lead to the formation of precatenanes and catenanes that are removed by Topo IV. The removal of precatenanes and catenanes also appears to be regulated with the condensation of chromosomes by an interaction between one of the Topo IV subunits and MukB (Figure 1.19) (see Hayama and Marians, Suggested Reading). MukB interacts with DNA through association with the partner proteins MukF and MukE. *B. subtilis* also has a condensin, which is more similar in amino acid sequence to the eukaryotic condensins and so was also named SMC protein (see Britton et al., Suggested Reading). It also has partner proteins named ScpA and ScpB. In *B. subtilis*, the link between chromosome condensation and partitioning is becoming clearer with the finding that proteins that directly recognize the region around the origin and are involved in partitioning are able to recruit the condensin SMC in this organism (see Thanbichler, Suggested Reading).

Supercoiling

Another way bacteria condense DNAs is through supercoiling. In bacteria, all DNAs are negatively supercoiled, which means that DNA is twisted in the opposite direction to the Watson-Crick helix, creating underwinds. As discussed in more detail below, the underwinds introduce stress into the DNA, causing it to wrap up on itself, much like a rope wraps up on itself if the two ends are rotated in opposite directions. This twisting occurs in loops in the DNA, causing the DNA to be condensed into a smaller space.

KEEPING NEW SISTER CHROMOSOMES SEPARATE INVOLVES COORDINATING MULTIPLE PROCESSES

Many processes ensure that the chromosomes are physically separate, yet are gathered together in a manageable way in the bacterial cell. With circular chromosomes, an uneven number of recombination events will regularly form dimer chromosomes, where the two circles join to form one large circle that cannot be subdivided into daughter cells. The highly controlled process of dimer resolution with XerC, XerD, and FtsK in *E. coli* ensures that dimer chromosomes are kept separate without accidently making dimers out of separate chromosomes before they can be passed to the daughter cells. Circular chromosomes can also become conjoined like the rings on a chain that cannot be passed on to daughter cells. Another highly regulated process by Topo IV removes these links between the chromosomes. Regulation of Topo IV activity with condensation and FtsK transport, as shown in Figure 1.19, plays an important role in making sure the enzyme separates interlinked chromosomes and does not link them together. A number of processes compact the chromosomes in bacteria. Condensins gather regions of the chromosome together that are, in turn, held together by a number of nonspecific DNA-binding proteins to allow compaction of the bacterial chromosome. This process is facilitated by supercoiling that allows the chromosome to twist around itself. In the next section, we learn how chromosomes are efficiently partitioned into each daughter cell during the process of cell division.

CHROMOSOME PARTITIONING

Not only must the two daughter chromosomes be segregated after replication, they also must be segregated in such a way that each daughter cell gets only one of the two copies of the chromosome. Otherwise one daughter cell would get two chromosomes and the other would be left with no chromosome and eventually would die. The apportionment of one daughter chromosome to each of the two daughter cells is called **partitioning**. Daughter cells that lack a chromosome after division are very rare, indicating that partitioning is a very efficient process in bacteria. Because of the importance of chromosome segregation, redundant mechanisms may have evolved to ensure that it occurs accurately. Indeed, many of the mechanisms that allow condensation of chromosomes can contribute to partitioning once the origin regions are located in the nascent daughter cells. While broad themes that describe partitioning across all bacteria have eluded our understanding, some important model systems are fairly well understood. In this section, we discuss what is known in the model bacteria.

The Par Proteins

Early work concentrated on the functions of the so-called partitioning proteins, the products of the *par* genes. The Par functions were first discovered in plasmids, which are small DNA molecules that are found in bacterial cells and that replicate independently of the chromosome (see chapter 4). Because they exist independently of the chromosome, plasmids usually must also have a system for partitioning; otherwise, they would often be lost from cells during the process of division. The Par systems of plasmids are known to fall into two families, one represented by the Par system of plasmid R1 and the other, much larger family represented by plasmids P1, F, and many others. It is the second of these families to which the known Par functions of chromosomes belong. The molecular details of Par systems are addressed in chapter 4, but some of the basics are described here as they apply to chromosome segregation systems.

The region of DNA that is to be partitioned, whether it is the origin region of the chromosome or a plasmid, contains a series of *cis*-acting sites called *parS* sites (Figure 1.20). One of two proteins in this system, ParB, binds to the *parS* sites. Partitioning of the ParB-bound *parS* DNA occurs because it is attracted to the other protein component of the system, active ParA*, which binds DNA. ParA* binds DNA nonspecifically and basically coats the entire nucleoid. The system works because ParB-*parS* is attracted to DNA-bound ParA*, but upon interacting with ParA* it helps to convert the ParA* into inactive ParA that does not bind DNA and is therefore released from the chromosome. As ParB-*parS* follows the gradient of ParA* across the nucleoid, it keeps the two *oriC* regions separate from one another. ParA eventually converts back to its active ParA* form and binds elsewhere on the nucleoid.

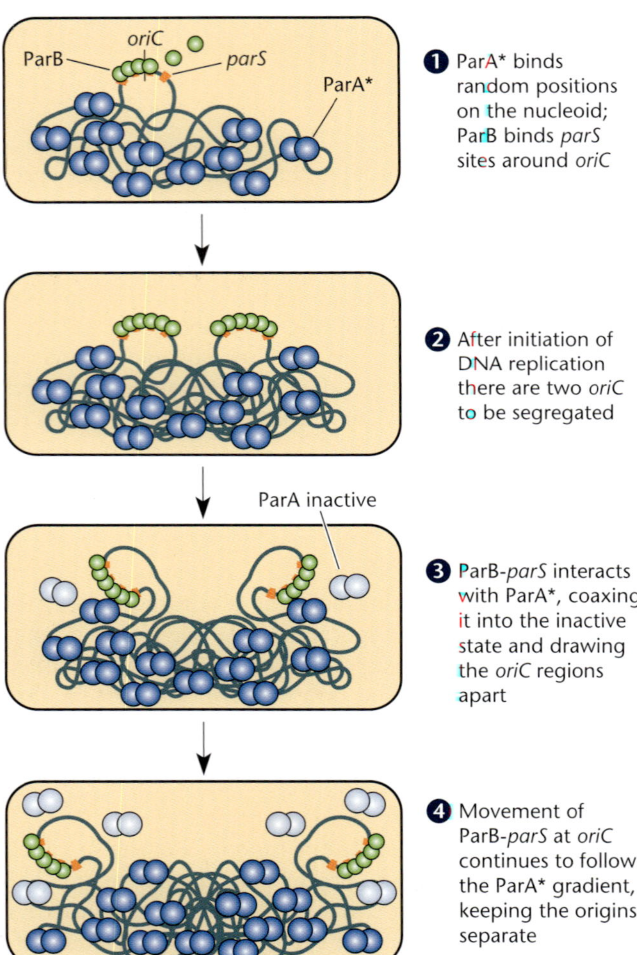
oriC
ParB
parS
ParA*
1 ParA* binds random positions on the nucleoid; ParB binds parS sites around oriC
2 After initiation of DNA replication there are two oriC to be segregated
ParA inactive
3 ParB-parS interacts with ParA*, coaxing it into the inactive state and drawing the oriC regions apart
4 Movement of ParB-parS at oriC continues to follow the ParA* gradient, keeping the origins separate

Figure 1.20 Model of how an origin region containing *parS* sites bound by the ParB protein is segregated by its attraction to DNA-bound ParA proteins (ParA*). Inactivation and displacement of ParA by the ParB-*parS* complex provide a mechanism to separate the origin region-containing ParB-*parS* complexes in the dividing cell.

Par functions in *B. subtilis*. The ParAB/*parS* system in *B. subtilis* provides some insight into how partitioning of *oriC* can work with condensation functions. In *B. subtilis*, the proteins analogous to the ParA and ParB proteins are called Soj and Spo0J, respectively. These names come from early genetic studies of *B. subtilis* sporulation, where *spo0J* was identified as a gene required for sporulation and *soj* was identified as a suppressor of *spo0J*. The *parS* sites close to the origin of chromosome replication (see Box 1.1) are bound by Spo0J(ParB), and these are pulled apart following the gradient to active Soj(ParA*), similar to what is shown in Figure 1.20. In *B. subtilis*, Spo0J(ParB) is capable of recruiting condensin protein SMC. Recruitment of SMC may play a critical role in organizing the chromosomes to help extrude the origin regions as they are gathered together with the SMC protein and other nonspecific DNA-binding proteins.

Macrodomains

Surprisingly, the chromosomes of *E. coli* and other *Enterobacteriaceae* do not have a recognizable Par-like system, despite the fact that plasmids with ParAB/*parS* systems are common in *Enterobacteriaceae*, and it seems likely that another system is responsible for the active partitioning of the chromosomes. Work in *E. coli* suggests that an incompletely understood system that hinges on large independently organizing regions called **macrodomains** functions in *Enterobacteriaceae*. Macrodomains (MD) were first established as large regions in the *E. coli* chromosome where recombination between two sites within a given macrodomain region occurs at a much greater frequency than outside the macrodomain region (see Valens et al. 2004, Suggested Reading). These regions localize to a specific subregion within the larger nucleoid and show individualized segregation properties. Four regions were identified as having these properties, one encompassing the origin region (Ori-MD), one encompassing the terminus region (Ter-MD), and two regions to the right (Right-MD) and left (Left-MD) of the terminus region (Figure 1.21). Two regions surrounding the origin macrodomain appear to not be as structured. Extensive molecular details exist for the Ter macrodomain, but little is known about the other macrodomain regions.

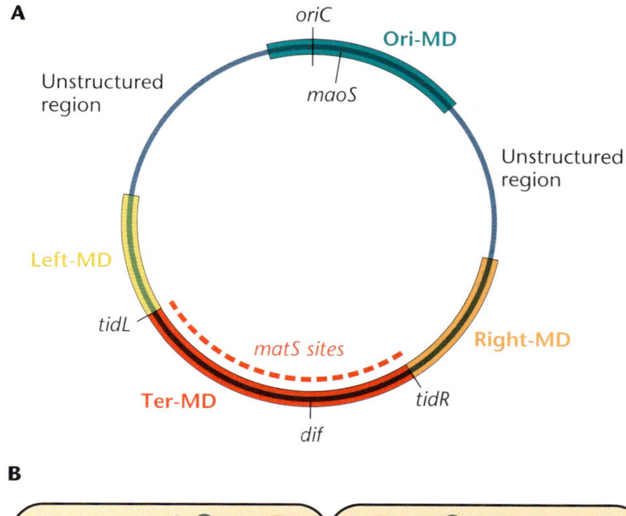

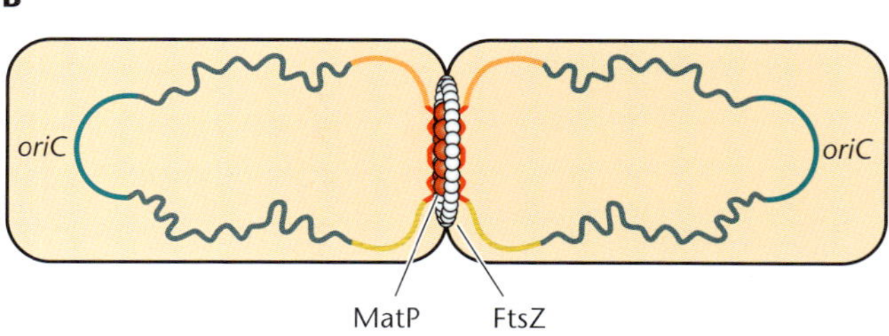

Figure 1.21 The *E. coli* chromosome has four structured regions called macrodomains (MDs) that aid in segregation of the chromosome. **(A)** Approximate genetic positions of the Ori-MD, the Right- and Left-MD, the Ter-MD, and the two unstructured regions of the *E. coli* chromosome. The Ter-MD is the best understood and is organized by the MatP protein via an association with its *matS* sites across the terminus region of the chromosome. The *tidL* and *tidR* sites, which are recognized by the YfbV protein (not shown), constrain the MatP protein from extending outside this region. **(B)** MatP compacts and localizes the Ter-MD through an association with ZapB (not shown), which is localized around the FtsZ ring region. The Ori-MD is organized via an association of the MaoP protein with a single site near the origin, *maoS*, via an unknown mechanism. The molecular mechanism responsible for organizing the Right- and Left-MDs and any associated proteins is unknown.

The Ter macrodomain: MatP and *matS*. The molecular basis for the Ter macrodomain involves a series of 23 *matS* sites found across an 800-kb region that are recognized by the protein MatP (see Mercier et al., Suggested Reading). MatP-*matS* complexes compact this region where they also associate with a specific set of proteins localized at the center, facilitating the process of orderly cell division (see below) (Figure 1.21B). The structure formed with MatP-*matS* appears to be constrained from spreading by two sites (*tidL* and *tidR*) which interact with a partner protein suggested to associate with the cell membrane (Figure 1.21A). Presumably, the MatP-*matS* system would coordinate the segregation of the terminus region as a late step in chromosome segregation. The molecular basis for the left and right macrodomains remains unknown.

The origin macrodomain: *maoS* and MaoP. Recombination studies were used to establish two important players in the origin macrodomain (see Valens et al. 2016, Suggested Reading). The explanation for the origin macrodomain is less clear, but appears to involve a single *cis*-acting site called *maoS*, found about 22 kb away from *oriC*, that is recognized by the MaoP protein (Figure 1.20). It remains unclear how a single site allows the formation of this large macrodomain or how it would function. The MaoP-*maoS* system is functionally distinct from the ParAB-*parS* partitioning systems found in plasmids and most other bacterial chromosomes, and placing the MaoP-*maoS* system onto a plasmid does not recapitulate the segregation found with the origin region, suggesting that important pieces are still missing from this story. A 25-bp site called *migS* was identified for its role in orienting the origin region within the larger nucleoid, but any role this site plays in segregation or any *trans*-acting factor that works with this site has yet to be identified (see Yamaichi and Niki, Suggested Reading).

Interestingly, a number of systems found in *E. coli* and other enteric bacteria appear to have coevolved: MatP/*matS*, MaoP/*maoS*, MukBEF, SeqA, and Dam methyltransferase along with a number of other proteins are only found in this subgroup of bacteria (see Brézellec et al., Suggested Reading). It is unclear how these systems functionally replace the ParAB-*parS* system found in most bacteria, and it will be interesting to discover how the systems relate and function in *Enterobacteriaceae*.

Coordinating Cell Division and Chromosome Partitioning in *E. coli* and *B. subtilis*

Much has also been learned about how the bacterial division septum forms. This process is called **cytokinesis**. A protein called FtsZ, which forms a ring around the midpoint of the cell, performs the primary step in this process (Figure 1.21b). This protein is related to tubulin of eukaryotes and forms filaments that grow and shorten by adding and removing shorter filaments, called protofilaments, to its ends in the presence of GTP. Before the cell is ready to divide, the FtsZ protein exists as helical filaments that traverse the cell. When the cell is about to divide, these filaments converge on the middle of the cell and form a ring at the site of the future septum. The FtsZ ring then attracts many other proteins, including the DNA translocase FtsK discussed above. FtsZ helps form the division septum, which eventually squeezes the mother cell at its center to allow the formation of the two daughter cells. The following major questions may be asked: why does the septum form only in the middle of the cell, and why does septum formation not occur over the nucleoid prior to chromosome segregation? The answers to these questions lie, at least in part, in two types of systems: the Min systems and the nucleoid occlusion systems.

The Min Proteins

In *E. coli*, three proteins called MinC, MinD, and MinE are known to be involved in localizing the division septum at the center of the cell. The *min* genes of *E. coli* were found because mutations in these genes can cause division septa to form in the wrong places, sometimes pinching off smaller cells called minicells. Apparently, in the absence of the Min proteins, division septa can form in places other than the middle of the cell. When this happens, smaller minicells that lack a chromosome are pinched off, hence the name Min proteins, for min cell-producing. It was predicted that the Min proteins would be localized in the ends of the *E. coli* cell, where they could prevent FtsZ from forming a division septum anywhere but the middle of the cell. However, when the localization of the Min proteins was studied using GFP fusions to the Min proteins, a very surprising result was revealed: the Min proteins oscillate from one pole of the cell to the other during the cell cycle. A model used to account for this finding held that oscillations of MinD and MinE drive the oscillation of MinC, which interacts with MinD (see MinCD in Figure 1.22) and is ultimately responsible for preventing FtsZ ring formation at the cell poles and enforcing the formation of a single FtsZ ring at mid-cell. The molecular mechanism that drives the redistribution of the proteins within the cell stems from the interaction of MinD (an ATPase) and MinE, which stimulates ATPase activity in MinD. MinD interacts with the membrane only in the ATP-bound state. More recent work with the system suggests that changes in the nature of MinD and MinE membrane-bound complexes and the states found in the cytoplasm are important for setting a distribution of these proteins (see Vecchiarelli et al., Suggested Reading). Ultimately, the concentration gradient set by the dynamic behavior of MinE and MinD sets a low concentration of MinCD at the center of the cell, allowing FtsZ to form a ring at the center of the cell.

Regulation of septum formation in *B. subtilis* differs from that found in *E. coli*. In *B. subtilis*, MinE is lacking and MinC and MinD do not oscillate. Instead, MinCD appears to tether directly to the cell poles by binding to another protein at the cell poles, called DivIVA. This binding creates a gradient of concentration of MinCD in the cell and similarly only allows formation of a single FtsZ ring at the center of the cell. Therefore, these two model bacteria use somewhat different mechanisms to establish a gradient of MinCD concentration and thereby restrict FtsZ ring formation to the center of the cell (Figure 1.22).

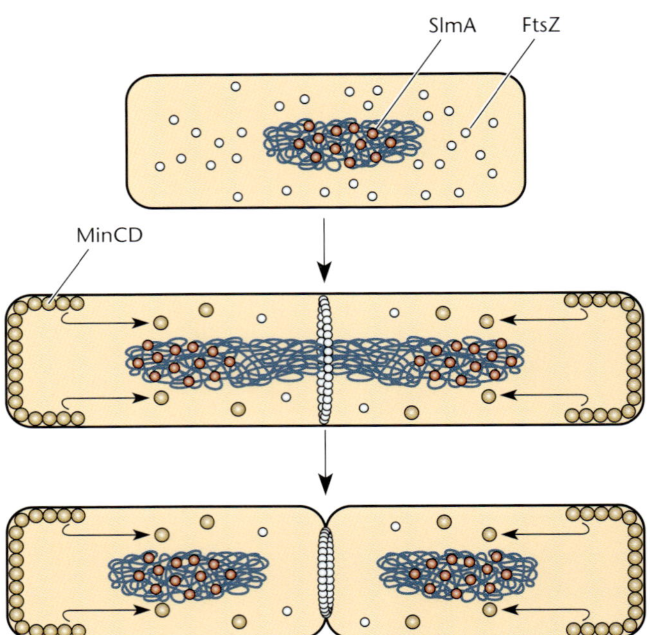

Figure 1.22 The MinCDE and nucleoid occlusion systems control placement of the FtsZ ring in *E. coli*. The FtsZ ring is an important marker of the central division site in a soon-to-divide cell which should not assemble until the chromosome is ready to segregate. MinC, which interacts with MinD, prevents a FtsZ ring from forming anywhere but at the center of the cell. Different mechanisms are used in bacteria to limit the concentration of MinCD at the center of the cell, as described in the text, to prevent FtsZ ring formation outside the center of the elongating cell. The nucleoid occlusion protein of *E. coli*, SlmA, is an important system that ensures that the FtsZ ring does not form over the nucleoid until the chromosomes are nearly completely replicated and segregated into daughter cells. Because the SlmA-binding sites are absent from the terminus region, the FtsZ ring can start to form once chromosomal replication is nearly complete and the origin region has progressed into the daughter cells away from the division site. See the text for details.

Nucleoid Occlusion

Not only should the FtsZ ring form only in the center of the cell, it also should not initiate the assembly of a division septum while the nucleoid is still occupying the center of the cell, or it might constrict the membrane around the chromosome. In fact, it was observed in *E. coli* that FtsZ rings never formed in the center of the cell when it was still occupied by the nucleoid, which had not yet segregated. Proteins that inhibit FtsZ ring formation in the presence of the nucleoid were discovered in both *E. coli* and *B. subtilis* at about the same time and were named **nucleoid occlusion proteins**. Both proteins were found because they are essential only if the Min system is inactivated. There is evidence that another process drives formation of the FtsZ at the midcell and that having both Min and nucleoid occlusion systems serves to increase the accuracy of the division process. The nucleoid exclusion protein in *B. subtilis*, named Noc, was found serendipitously, because its gene, *noc*, is adjacent to the genes for the Par functions, *soj* and *spo0J*, and it was observed that mutations in this gene could not be combined with mutations in the *minD* gene without making the cells very sick. The reason *noc* mutant strains were sick was that they were forming long filaments of cells because they were not dividing properly. The nucleoid occlusion protein in *E. coli*, named SlmA, was found directly by a synthetic lethal screen (see Bernhardt and de Boer, Suggested Reading). A synthetic lethal screen is a powerful genetic tool in which mutations in genes whose products are required only if another gene product is absent are isolated. The investigators looked for mutations in genes that were required only in the absence of the products of the *min* genes. They expressed the *min* genes from an inducible promoter and looked for mutants that were sick and failed to form colonies only in the absence of inducer. Some of these mutants had mutations in a gene that was named *slmA* by the investigators. While mutants deficient in Min proteins had more Z rings, they were never over the nucleoids (Figure 1.22). However, mutants that lacked both the Min proteins and SlmA often formed FtsZ rings over the nucleoids, as expected for a mutant deficient in nucleoid occlusion. The use of inducible promoters and other examples of synthetic phenotypes are discussed in more detail in later chapters.

The Noc and SlmA proteins seem to act by binding to DNA and then inhibiting FtsZ ring formation close to the DNA to which they are bound. There are known to be DNA sequences called Noc-binding sites (NBS) that are bound by Noc, and SlmA-binding sites (SBS) that are bound by SlmA; these sites are distributed across the chromosome and concentrated in the origin region but are absent from the terminus region (Box 1.1). This allows SlmA and Noc to help protect the nucleoid from the division septum until the final moments prior to the completion of DNA replication (see Wu et al., Suggested Reading) (Figure 1.22).

Note that this entire section has focused on simple division in rod-shaped cells. Fascinating adaptations to these simple ideas are known to occur in systems where the cells resulting from division are morphologically distinct. A particularly well-studied system in *Caulobacter crescentus* shows many adaptations when a mother cell gives rise to smaller motile daughter cells. Additionally, round or coccoid cells have distinct mechanisms that allow them to divide with clear cell poles as are found in rod-shaped cells. Given the extreme morphological variation known to occur across bacteria (see Kysela et al., Suggested Reading), interesting adaptations for division likely await discovery.

Coordination of Cell Division with Replication of the Chromosome

It is not sufficient to know how chromosomes replicate and then segregate into the daughter cells prior to division. Something must coordinate the replication of the chromosome with division of the cells. If the cells divided before the replication of the chromosome was completed, there would not be two complete chromosomes to segregate into the daughter cells, and one cell would end up without a complete chromosome. The mechanism by which cell division is coordinated with replication of the DNA is still not completely understood, but there is a lot of relevant information.

TIMING OF REPLICATION IN THE CELL CYCLE

It is important to know when replication occurs during the cell cycle. Experiments were designed to determine the relationship between the time of chromosome replication and the cell cycle in *E. coli* (see Helmstetter and Cooper, Suggested Reading). The conclusions are still generally accepted, so it is worth going over them in some detail. The investigators recognized that if the DNA content of cells at different stages in the cell cycle could be measured, it would be possible to determine how far chromosome replication had proceeded at that time in the cell cycle. Since bacterial cells are too small to allow measurements of DNA content in a single cell by the methods they had, it was necessary to measure the DNA content in a large number of pooled cells. However, cells growing in culture are all at different stages in their cell cycles. Therefore, to know how far replication had proceeded at a certain stage in the cell cycle, it was necessary to synchronize the cells in the population so that all were the same age or at the same point in their life cycles at the same time.

Helmstetter and Cooper accomplished this by using what they called a bacterial "baby machine." Their idea was to first label the DNA of a growing culture of bacterial cells by adding radioactively labeled nucleosides and then fix the bacterial cells on a membrane. When the cells on the filter divided, one of the two daughter cells would no longer be attached and would be released into the medium. All of the daughter cells released at a given time would be newborns and so would be the same age. This means that cells that divided to release the daughter cells at a given time would also be the same age and would have DNA in the same replication state. The amount of radioactivity in the released cells was then a measure of how much of the chromosome had replicated in cells of that age. This experiment was done under different growth conditions to show how the timing of replication and the timing of cell division are coordinated under different growth conditions.

Figure 1.23 shows the results of these experiments. For convenience, the following letters were assigned to each of the intervals during the cell cycle. The letter I denotes the time from when the last round of chromosome replication initiated until a new round begins. C is the time it takes to replicate the entire chromosome, and D is the time from when a round of chromosome replication is completed until cell division occurs. The top of the figure shows the relationship of I, C, and D when cells are growing slowly, with a generation time of 70 min. Under these conditions, I is 70 min, C is 40 min, and D is 20 min.

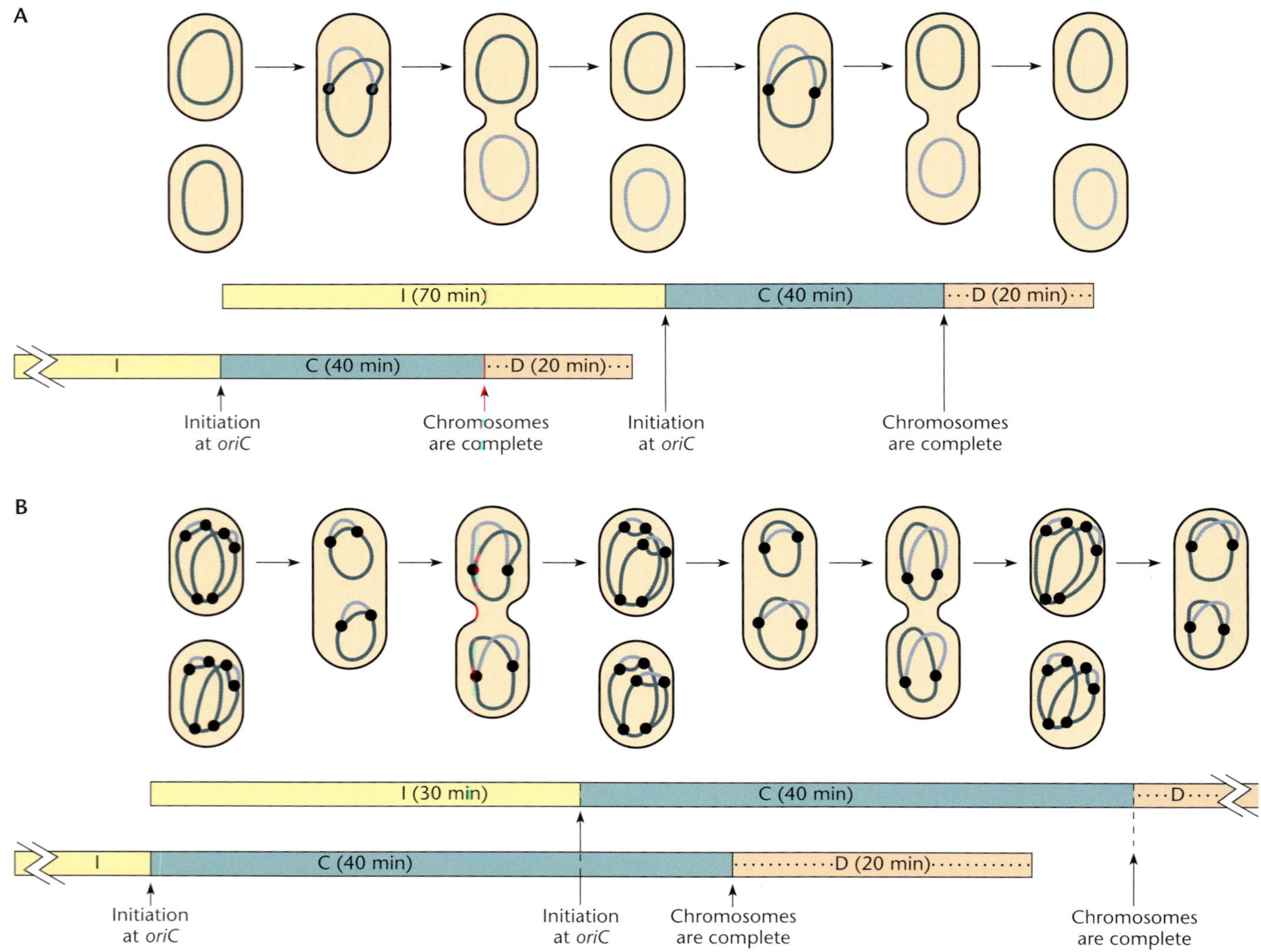

Figure 1.23 Timing of DNA replication during the cell cycle, with two different generation times **(A and B)**. The time between initiations (I) is the only time that changes. See the text for definitions of I, C, and D.

However, when the cells are growing in a richer medium and are dividing more rapidly, with a generation time of only 30 min, the pattern changes. The C and D intervals remain about the same, but the I interval is much shorter, only about 30 min.

Some conclusions may be drawn from these data. One conclusion is that the C and D intervals remain about the same independent of the growth rate. At 37°C, the time it takes the chromosome to replicate is always about 40 min, and it takes about 20 min from the time a round of replication terminates until the cell divides. However, the I interval gets shorter when the cells are growing faster and have shorter generation times. In fact, the I interval is approximately equal to the generation time—the time it takes a newborn cell to grow and divide. This makes sense because, as discussed below, initiation of chromosome replication appears to occur every time the cells reach a certain size. They reach this size once every generation time, independent of how fast they are growing.

Another point apparent from the data is that in cells growing rapidly with a short generation time, the I interval can be shorter than the C interval. If I is shorter than C, a new round of chromosomal DNA replication will begin before the old one is completed. This explains the higher DNA content of fast-growing cells as compared to slow-growing cells. It also explains the observation that genes closer to the origin of replication are present in more copies than are genes closer to the replication terminus.

Despite providing these important results, this elegant analysis does not allow us to tell whether division is coupled to initiation or termination of chromosomal DNA replication. The fact that the I interval always equals the generation time suggests that the events leading up to division are set in motion at the time a round of chromosome replication is initiated and are completed 60 min later independent of how fast the cells are growing. However, it is also possible that the act of termination of a round of chromosome replication sets in motion a cell division 20 min later. Multiple laboratories are working to resolve these issues.

Timing of Initiation of Replication

A new round of replication must be initiated each time the cell divides, or the amount of DNA in the cell would increase until the cells were stuffed full of it or decrease until no cell had a complete copy of the chromosome. Clearly, initiation of replication is exquisitely timed. In cells growing very rapidly, in which the next rounds of replication initiate before the last ones are completed, so that the cells contain a number of origins of replication, all of the origins in a cell "fire" simultaneously, indicating tight control.

A number of attempts have been made to correlate the timing of initiation of chromosome replication with other cellular parameters during the cell cycle. Most evidence from such attempts points to initiation of replication being tied to cell mass. After cells divide, their mass, or weight, continuously increases until they divide again. The initiation of chromosome replication occurs each time the cell achieves a certain mass, the initiation mass. If cells are growing faster in richer medium, they are larger and achieve the initiation mass sooner than do smaller, slower-growing cells, explaining why new rounds of chromosome replication occur before the termination of previous rounds in faster-growing cells but not in slower-growing cells. However, these experiments by themselves do not explain what it is about the cell mass that triggers initiation.

ROLE OF THE DnaA PROTEIN

DnaA is essential for initiating DNA replication at *oriC* and along with other components also helps to regulate the timing of DNA replication. DnaA primarily affects the frequency of initiation by changes in the ATP-binding versus ADP-binding state of DnaA and by access of DnaA to *oriC*. DnaA must be in its ATP-bound state to form the structure used to separate the DNA strands at *oriC* to initiate DNA replication. DnaA hydrolyzes ATP slowly; therefore, other factors are required to stimulate its ATPase activity to inactivate it until it is needed for the next round of DNA replication. The affinity of DnaA for the various binding sites within *oriC* (see "Origin of Chromosomal Replication" above) allows a mechanism for the origin to be identified but not immediately used as an origin until all conditions are in place for replication. A number of processes control the pool of DnaA found in the ATP-bound state. Reducing the pool of DnaA-ATP following initiation is an important mechanism for delaying reinitiation, thereby allowing replication to proceed in an orderly fashion. The production of new DnaA also plays a role in the ratio of ATP-bound versus ADP-bound DnaA, and because of the high ATP concentration in cells, newly made DnaA will be bound by ATP.

Inactivation of DnaA by hydrolysis to DnaA-ADP

Conversion to the DnaA-ADP state occurs primarily by two mechanisms following initiation of DNA replication. As mentioned above, only the ATP-bound form can bind all of the DnaA boxes in *oriC*. Only when all of the sites are bound can the structure form that will be needed to open the DUE to allow replication to initiate (Figure 1.15). This helps the DnaA protein to act as a switch that is largely independent of its concentration because other cellular inputs can control the ATP-bound versus ADP-bound state of DnaA. One cellular input that reduces the pool in the DnaA-ATP state involves the presence of the assembled replication fork at the origin region immediately after initiation of replication. The presence of the β sliding-clamp protein causes DnaA to hydrolyze ATP to ADP by interacting with another protein, a relative of DnaA called Hda (for homology to DnaA) (see Camara et al., Suggested Reading). This process is sometimes referred to as regulatory inactivation of DnaA, or RIDA. A second input that reduces the pool of DnaA-ATP involves a locus in the chromosome called *datA*. The *datA* locus, which is about 1 kb in length, has five DnaA-binding sites and a binding site for IHF. Binding of DnaA and IHF to this locus induces an endogenous ATPase activity of the DnaA protein, coaxing it into the inactive DnaA-ADP form in a process abbreviated as DDAH (*datA*-dependent DnaA-ATP hydrolysis). How specifically DDAH is controlled with the cell cycle remains unknown.

Reactivation of DnaA by nucleotide exchange to DnaA-ATP

Production of new DnaA in the ATP-bound form is one mechanism for increasing the DnaA-ATP/DnaA-ADP ratio in cells. There are also two mechanisms involved in "recycling" DnaA by encouraging it to exchange a nucleotide, putting it back into the DnaA-ATP form; one involves an exchange catalyzed by acidic phospholipids, and the other involves two sites found in the chromosome called DnaA-reactivating sequences (see Fujimitsu et al., Suggested Reading). The numerous inputs work together to limit the initiation of DNA replication to help keep the number of chromosomes consistent with the requirements for cell division.

SeqA-MEDIATED HEMIMETHYLATION AND SEQUESTRATION

In some types of bacteria, including *E. coli*, there is yet another means of delaying the initiation of new rounds of chromosome replication. As with the mechanisms described above, replication itself plays a role in this regulatory pathway where methylation of the DNA helps delay initiation. In *E. coli* and other enteric bacteria, the A's in the symmetric sequence GATC/CTAG are methylated at the N6 position. These methyl groups are added to the bases by the enzyme deoxyadenosine methylase (Dam or Dam methylase), but this occurs only after the nucleotides have been incorporated into the DNA. Since DNA replicates by a semiconservative mechanism, the A in the GATC/CTAG sequence in the newly synthesized strand remains temporarily unmethylated after replication of a region containing this sequence (Figure 1.24). The DNA at this site is said to be hemimethylated if the bases on only one strand are methylated.

The hemimethylated state is important in the context of regulation of initiation because a *trans*-acting protein called SeqA only binds with high affinity to hemimethylated GATC/CTAG sequences. SeqA is an essential facilitator of hemimethylation control of replication initiation and is found only in bacteria that have Dam. The sequence GATC/CTAG is found 11 times within *oriC*, much more often than would be expected by chance alone (Figure 1.15). GATC/CTAG sequences are also associated with the low-affinity DnaA-binding sites across the origin. Furthermore, the promoter region of the *dnaA* gene, the region in which mRNA synthesis initiates for the DnaA protein, also has GATC/CTAG sequences. SeqA is able to bind all of these strategically located GATC/CTAG sites after they are replicated (and therefore rendered hemimethylated), which has the effect of delaying the conversion to full methylation at these sites for about one-third of a cell cycle. SeqA also blocks binding of DnaA to the low-affinity sites and inhibits expression of the *dnaA* gene. SeqA bound to hemimethylated DNA may associate with the cell membrane to sequester the *oriC* region after initiation of DNA replication (Figure 1.25). Sequestration of the hemimethylated *oriC* region is predicted to render it temporarily nonfunctional for the initiation of a new round of replication and delays its methylation by the Dam methylase (see Slater et al., Suggested Reading).

SeqA activities outside of *oriC*

In addition to the GATC/CTAG sequences associated with *oriC* and the *dnaA* promoter, SeqA also interacts with the GATC/CTAG sequences as the replication forks progress around the chromosome, effectively marking the location of the replisome (Figure 1.25). SeqA bound to transiently hemimethylated GATC/CTAG sites immediately behind the DNA replication forks may be capable of bringing together the nascent sister chromosomes, and it also appears to negatively regulate the decatenation activity of topoisomerase (see Joshi et al., Suggested Reading). Both of these processes may mediate a form of sister chromosome cohesion that in turn may help to regulate processing of the new DNAs by positioning them for DNA repair and recombination. Interestingly, SeqA is essential in mutants that lack the major pathways of DNA recombination involving RecA, supporting an additional role in protecting genome integrity. Additionally, chromosomes appear to be vulnerable to many types of mobile DNA elements during DNA replication, especially when

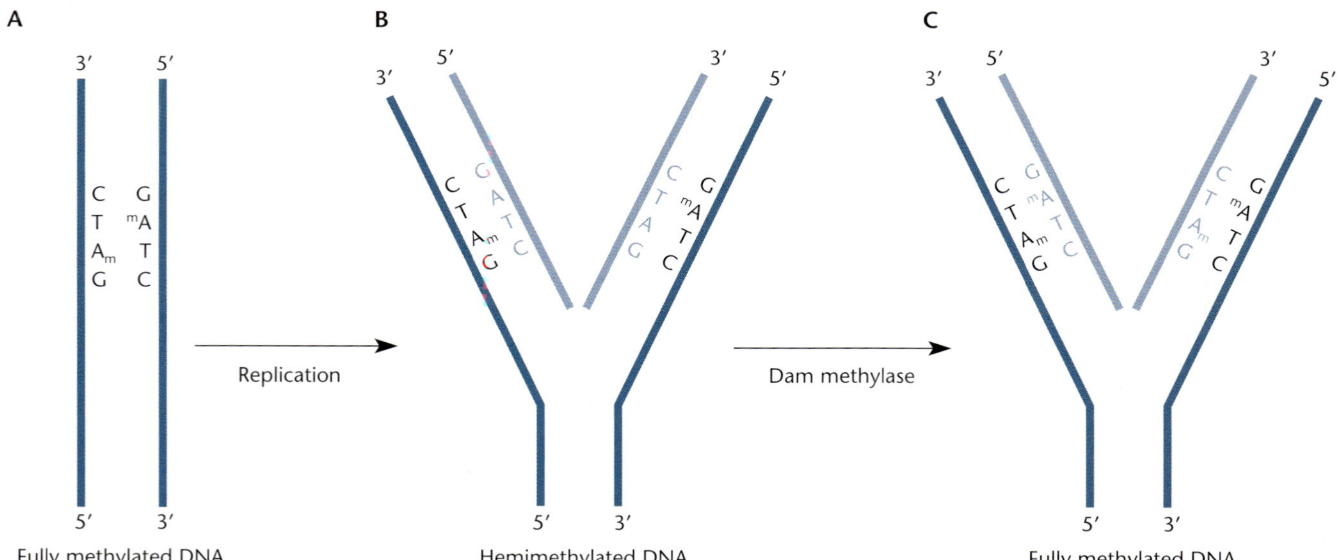

Figure 1.24 Replication creates hemimethylated DNA. **(A)** The A in the sequence GATC is methylated on both strands (A_m and ^mA). **(B)** After replication, the A in GATC in the new strand is not immediately methylated by the Dam methylase. **(C)** Eventually, GATC sites in the new strand are methylated, converting the DNA back to the fully methylated state.

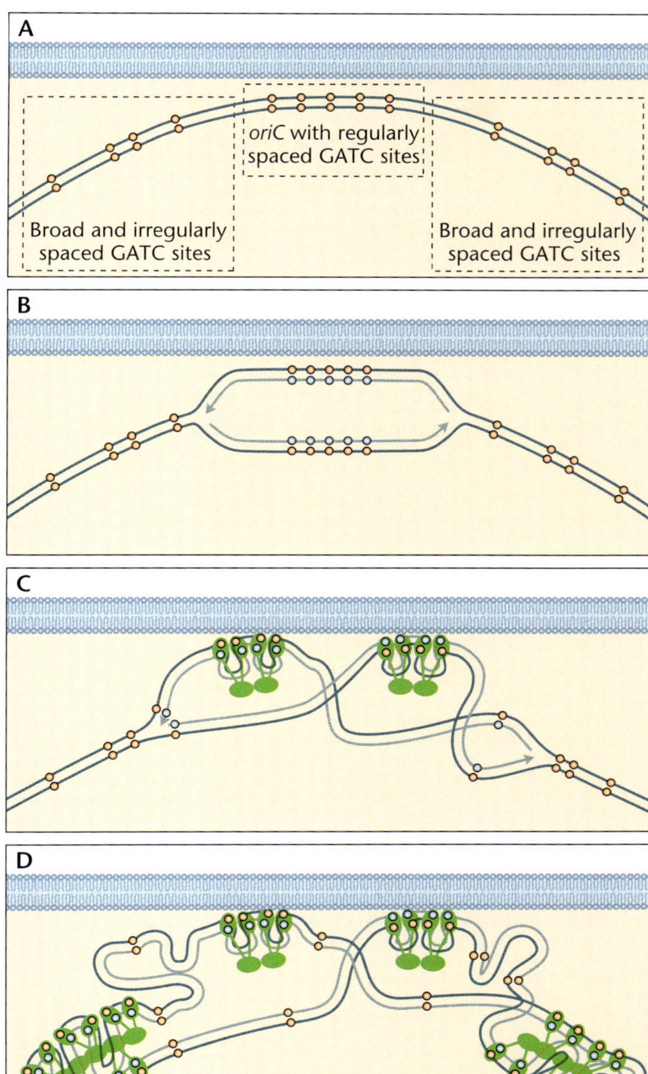

Figure 1.25 Model showing the possible functional consequences of SeqA binding with regularly and closely spaced GATC sites in the *oriC* region and broadly and irregularly spaced GATC sites outside the *oriC* region in the *E. coli* chromosome. Methylated GATC sites in a strand are represented by orange circles. Unmethylated GATC sites in a strand are represented by blue circles. **(A)** Before initiation, the *oriC* region is methylated in both strands. **(B)** After initiation, only one of the two strands is methylated; hence, the region is hemimethylated. The newly synthesized strand is shown in a lighter shade of blue. **(C)** SeqA binds the regularly spaced hemimethylated GATC sites in *oriC* and is able to interact with the membrane, thereby sequestering *oriC* and preventing further initiation and drastically slowing methylation. **(D)** Outside of the *oriC* region, where GATC sites are not regularly spaced and are situated farther apart, SeqA may be a bridge between the two hemimethylated strands to help coordinate processing events important for facilitating repair and recombination.

replication forks stall (see Fricker and Peters, Suggested Reading). SeqA-facilitated processes may help to protect DNA replication forks from mobile elements as one of multiple mechanisms of protecting genome integrity.

The Bacterial Nucleoid

The nucleoid was described with respect to chromosome segregation above. Indeed, experiments in many of the model systems indicate that bacteria carefully coordinate the position of the chromosome in the cell. Through techniques in which individual positions in the chromosome can be localized in whole cells, it has been shown that genes are located in the chromosome in roughly the same order as one would presume by looking at the DNA sequence. A variety of techniques are providing insight into how the structure of the chromosome is maintained in the cell. The molecular mechanisms that maintain the chromosome structure remain a mystery, but specific systems are likely to exist to ensure that the chromosome is available for transcription, recombination, and other functions.

Supercoiling in the Nucleoid

Supercoiling is one of the mechanisms that help compact and organize the chromosome. Supercoiling also affects the expression level of many genes. However, supercoiling will be lost if one of the strands of the DNA is cut, thereby allowing the strands to rotate around each other. The phosphodiester bond connecting the two deoxyribose sugars on the other strand serves as a swivel and rotates, resulting in relaxed (i.e., not supercoiled) DNA. A DNA with a phosphodiester bond broken in one of the two strands is said to be nicked. A variety of experiments suggest that the nucleoid is packaged in subregions that constrain supercoiling. Topological barriers would prevent the entire chromosome from losing supercoiling when there is a nick in the genome. Figure 1.26 illustrates supercoiling. In this example, the ends of a DNA molecule have been rotated in opposite directions, and the DNA has become twisted up on itself to relieve the stress. The DNA remains supercoiled as long as its ends are constrained and so cannot rotate, and a circular DNA has no free ends that can rotate. A linear DNA will not maintain supercoiling unless regions flanking the supercoiling are somehow otherwise constrained. A break or nick in a circular DNA should relax the whole DNA unless portions of the molecule are periodically attached to barriers that prevent rotation of the strands. Through the examination of the expression level of over 300 genes that are sensitive to supercoiling following the introduction of breaks in the chromosome, it has been estimated that the topologically isolated loops are about 10 kb in size. This is in good agreement with domain size as determined by directly measuring the lengths of loops under the microscope (see

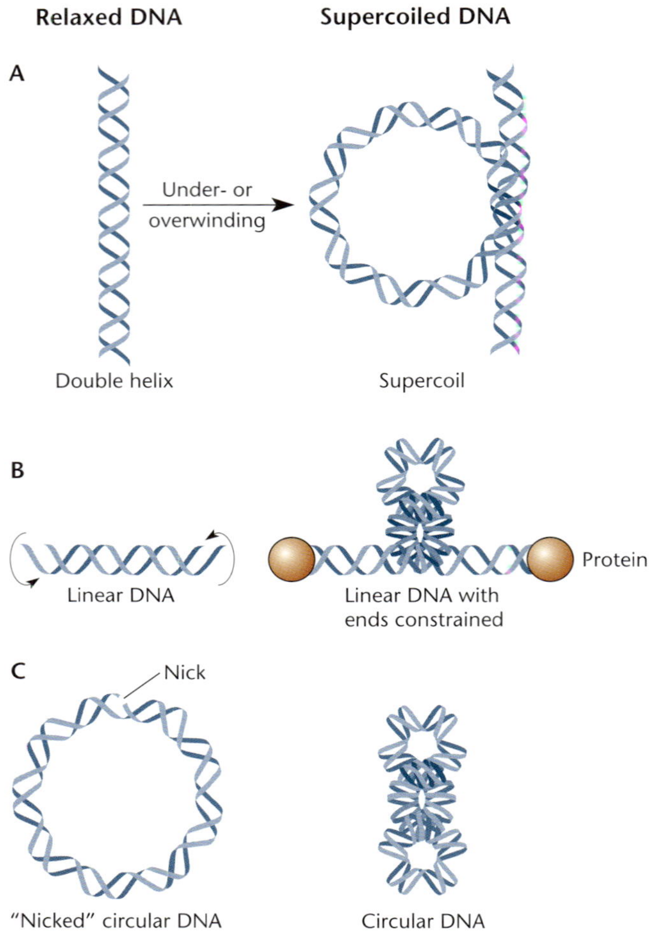

Figure 1.26 **(A)** Supercoiled DNA. **(B)** Twisting of the ends in opposite directions causes linear DNA to wrap up on itself. The supercoiling is lost if the ends of the DNA are not somehow constrained. **(C)** A break, or nick, in one of the two strands of a circular DNA relaxes the supercoils.

Postow et al., Suggested Reading). Although the exact mechanism or mechanisms that restrict topology are unresolved, this work indicates that the barriers to rotation are not fixed at certain places in the chromosome.

SUPERCOILING OF NATURAL DNAs

It is possible to estimate the extent of supercoiling of natural DNAs. According to the Watson-Crick structure, the two strands are wrapped around each other about once every 10.5 bp to form the double helix. Therefore, in a DNA of 2,100 bp, the two strands should be wrapped around each other about 2,100/10.5, or 200, times. In a supercoiled DNA of this size, however, the two strands are wrapped around each other either more or less than 200 times. If they are wrapped around each other more than once every 10.5 bp, the DNA is said to be **positively supercoiled**; if less than once every 10.5 bp, it is **negatively supercoiled**.

Most DNA in bacteria is negatively supercoiled, with an average of one negative supercoil for every 300 bp, although there are localized regions of higher or lower negative supercoiling. Also, in some regions, such as ahead of a transcribing RNA polymerase, the DNA may be positively supercoiled (see above).

SUPERCOILING STRESS

Some of the stress due to supercoiling of the DNA, which causes it to twist up on itself, can be relieved if the DNA is wrapped around something else, such as proteins. Sailors know about this effect: if you twist a rope in the right direction as you roll it up to store it, it does not try to unroll itself again when you are finished. Wrapping DNA around proteins in the cell is called constraining the supercoils. Unconstrained supercoils cause stress in the DNA, which can be relieved by twisting the DNA up on itself, as shown in Figure 1.26, and making the DNA more compact. The stress due to unconstrained supercoils can have other effects, as well, for example, helping to separate the strands of DNA during reactions such as replication, recombination, and initiation of RNA synthesis at promoters.

Topoisomerases

The supercoiling of DNA in the cell is modulated by topoisomerases (see Wang, Suggested Reading). Topoisomerases are discussed above, but not the molecular details of these enzymes. All organisms have these proteins, which manage to remove the supercoils from a circular DNA without permanently breaking either of the two strands. They perform this feat by binding to DNA, breaking one or both of the strands, and passing the DNA strands through the break before resealing it. As long as the enzyme holds the cut ends of the DNA so that they do not rotate, this process, known as **strand passage**, either introduces or removes supercoils in DNA.

The topoisomerases are classified into two groups, type I and type II (Figure 1.27). These two types differ in how many strands are cut and how many strands pass through the cut. The type I topoisomerases cut one strand and pass the other strand through the break before resealing the cut. The type II topoisomerases cut both strands and pass two other strands from somewhere else in the DNA, or even another DNA, through the break before resealing it. This basic difference changes how supercoiling is affected by these enzymes, as shown in Figure 1.27.

TYPE I TOPOISOMERASES

Bacteria have several type I topoisomerases. The major bacterial type I topoisomerase removes negative supercoils from DNA. In *E. coli* and *Salmonella enterica* serovar Typhimurium, the *topA* gene encodes this type I topoisomerase. As expected, DNA isolated from *E. coli* with a *topA* mutation is more negatively supercoiled than normal.

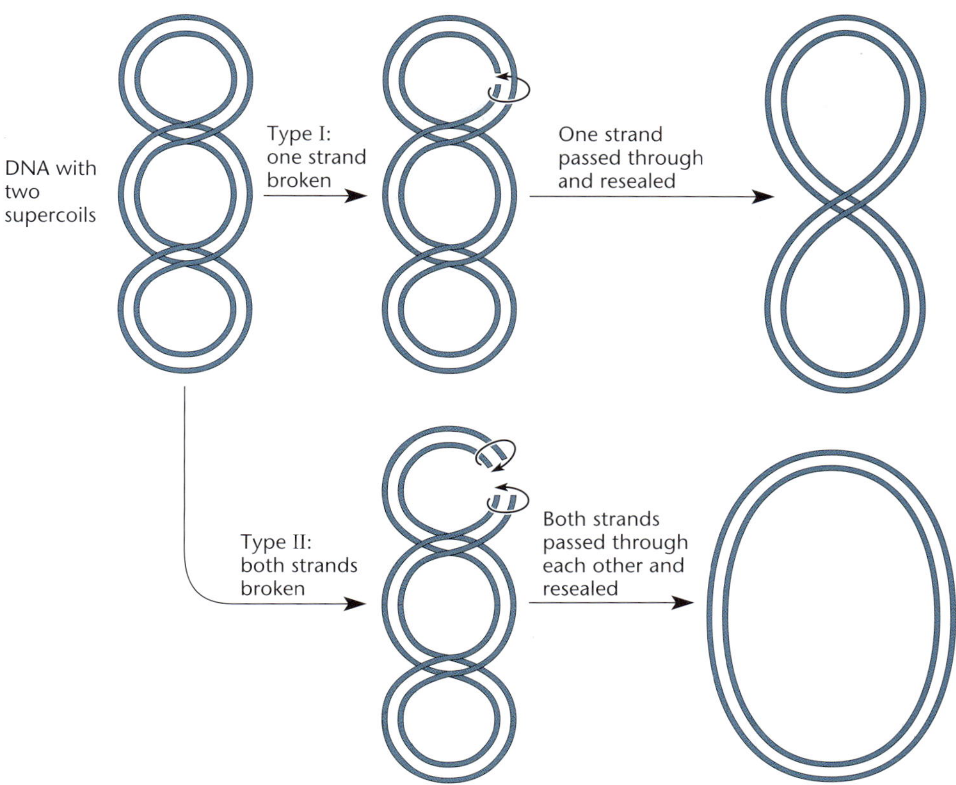

Figure 1.27 Action of the two types of topoisomerases. The type I topoisomerases break one strand of DNA and pass the other strand through the break, removing one supercoil at a time. The type II topoisomerases break both strands and pass another part of the same DNA through the breaks, introducing or removing two supercoils at a time.

TYPE II TOPOISOMERASES

Bacteria also have more than one type II topoisomerase. Because type II topoisomerases can break both strands and pass two other DNA strands through the break, they can either separate two linked circular DNA molecules or link them up. Linkage sometimes happens after replication or recombination. One major type II topoisomerase in *E. coli*, Topo IV (see above), decatenates daughter chromosomes after DNA replication, releasing the major source of cohesion between the chromosomes and allowing them to be segregated into the daughter cells. While most type II topoisomerases remove negative supercoils, a special type II topoisomerases in bacteria, called gyrase, plays an essential role in adding negative supercoils. Gyrase acts by first wrapping the DNA around itself and then cutting the two strands before passing another part of the DNA through the cuts, thereby introducing two negative supercoils. Adding negative supercoils increases the stress in the DNA and thus requires energy; hence, gyrase needs ATP for this reaction.

The gyrase of *E. coli* is made up of four polypeptides, two of which are encoded by the *gyrA* gene and two of which are encoded by *gyrB*. These genes were first identified by mutations that make the cell resistant to antibiotics that affect gyrase (Table 1.2 and Box 1.2). The GyrA subunits seem to be responsible for breaking the DNA and holding it as the strands pass through the cuts. The GyrB subunits have the ATP site that furnishes the energy for the supercoiling.

Table 1.2 Antibiotics that block replication

Antibiotic	Source	Target
Trimethoprim	Chemically synthesized	Dihydrofolate reductase
Hydroxyurea	Chemically synthesized	Ribonucleotide reductase
5-Fluorodeoxyuridine	Chemically synthesized	Thymidylate synthetase
Nalidixic acid	Chemically synthesized	*gyrA* subunit of gyrase
Novobiocin	*Streptomyces sphaeroides*	*gyrB* subunit of gyrase
Mitomycin C	*Streptomyces caespitosus*	Cross-links DNA

BOX 1.2

Antibiotics That Affect Replication and DNA Structure

Antibiotics are substances that block the growth of cells. Many antibiotics are naturally synthesized chemical compounds made by soil microorganisms, especially actinomycetes, that may help them compete with other soil microorganisms. There are also other ideas as to why bacteria make antibiotics, including for their use in intercellular communication, especially in highly organized structures like biofilms. Many antibiotics specifically block DNA replication or change the structure of DNA. Because some parts of the replication machinery have remained relatively unchanged throughout evolution, many of these antibiotics work against essentially all types of bacteria. Some even work against eukaryotic cells and so are used as antifungal agents and in antitumor chemotherapy.

Antibiotics That Block Precursor Synthesis

As discussed above, DNA is polymerized from the deoxynucleoside triphosphates. Any antibiotic that blocks the synthesis of these deoxynucleotide precursors will block DNA replication.

Inhibition of dihydrofolate reductase

Some of the most important precursor synthesis blockers are antibiotics that inhibit the enzyme dihydrofolate reductase. One such compound, trimethoprim, works very effectively in bacteria, and the antitumor drug methotrexate (amethopterin) inhibits the dihydrofolate reductase of eukaryotes. Methotrexate is used as an antitumor agent, among other uses such as in the treatment of inflammatory arthritis.

Antibiotics like trimethoprim that inhibit dihydrofolate reductase kill the cell by depleting it of tetrahydrofolate, which is needed for many biosynthetic reactions. This inhibition is overcome, however, if the cell lacks the enzyme thymidylate synthetase, which synthesizes dTMP; therefore, most mutants that are resistant to trimethoprim have mutations that inactivate the *thyA* thymidylate synthetase gene. The reason is apparent from the pathway for dTMP synthesis shown in Figure 1.5. Thymidylate synthetase is solely responsible for converting tetrahydrofolate to dihydrofolate when it transfers a methyl group from tetrahydrofolate to dUMP to make dTMP. The dihydrofolate reductase is the only enzyme in the cell that can restore the tetrahydrofolate needed for other biosynthetic reactions. However, if the cell lacks thymidylate synthetase, there is no need for a dihydrofolate reductase to restore tetrahydrofolate. Therefore, inhibition of dihydrofolate reductase by trimethoprim has no effect, thus making *thyA* mutant cells resistant to the antibiotic. Of course, if the cell lacks a thymidylate synthetase, it cannot make its own dTMP from dUMP and must be provided with thymidine in the medium so that it can replicate its DNA.

There is more than one mechanism by which cells can achieve trimethoprim resistance. They can have an altered dihydrofolate reductase to which trimethoprim cannot bind, or they can have more copies of the gene so that they make more enzyme than there is trimethoprim to inhibit it. Some plasmids and transposons carry genes for resistance to trimethoprim. These genes encode dihydrofolate reductases that are much less sensitive to trimethoprim and so can act even in the presence of high concentrations of the antibiotic.

Inhibition of ribonucleotide reductase

The antibiotic hydroxyurea inhibits the enzyme ribonucleotide reductase, which is required for the synthesis of all four precursors of DNA synthesis (Figure 1.5). The ribonucleotide reductase catalyzes the synthesis of the deoxynucleoside diphosphates dCDP, dGDP, dADP, and dUDP from the ribonucleoside diphosphates, an essential step in deoxynucleoside triphosphate synthesis. Mutants resistant to hydroxyurea have an altered ribonucleotide reductase.

Competition with deoxyuridine monophosphate

5-Fluorodeoxyuridine and the related 5-fluorouracil have monophosphate forms resembling dUMP, the substrate for thymidylate synthetase. By competing with the natural substrate for this enzyme, they inhibit the synthesis of deoxythymidine monophosphate. Mutants resistant to these compounds have an altered thymidylate synthetase. These are useful antibiotics for the treatment of fungal, as well as bacterial, infections.

Antibiotics That Block Polymerization of Deoxynucleotides

The polymerization of deoxynucleotide precursors into DNA would also seem to be a tempting target for antibiotics. However, there seem to be surprisingly few antibiotics that directly block this process. Most antibiotics that block polymerization do so indirectly, by binding to DNA or by mimicking the deoxynucleotides and causing chain termination, rather than by inhibiting the DNA polymerase itself.

Deoxynucleotide precursor mimics

Dideoxynucleotides are similar to the normal deoxynucleotide precursors, except that they lack a hydroxyl group on the 3′ carbon of the deoxynucleotide. Consequently, they can be incorporated into DNA, but then replication stops because they cannot link up with the next deoxynucleotide. These compounds are not useful antibacterial agents, probably because they are not phosphorylated well in bacterial cells. However, this property of prematurely terminating replication has made them the basis for one of the first methods for DNA sequencing (see chapter 13).

Cross-linking

Mitomycin C blocks DNA synthesis by cross-linking the guanine bases in DNA to each other. Sometimes the cross-linked bases are in opposing strands. If the two strands are attached to each other, they cannot be separated during replication. Even one cross-link in DNA that is not repaired prevents replication of the chromosome. This antibiotic is also a useful antitumor drug, probably for the same reason. DNA cross-linking also affects RNA transcription.

Antibiotics That Affect DNA Structure

Acridine dyes

The acridine dyes include proflavine, ethidium, and chloroquine. These compounds insert between the bases of DNA and thereby cause frameshift mutations and inhibit DNA synthesis. Their ability to insert themselves between the bases in DNA has made acridine dyes very useful in genetics and molecular biology. Some of these applications are discussed in later chapters. In general, acridine dyes are not useful as antibiotics because of their toxicity due to their ability to block DNA synthesis in the mitochondria of eukaryotic cells. Some members of this large family of antibiotics have long been used as antimalarial drugs because of their ability to block DNA synthesis in the mitochondria (kinetoplasts) of trypanosomes. This is the basis for the antimalarial activity of the tonic water in a gin and tonic.

Thymidine mimic

5-Bromodeoxyuridine (BUdR) is similar to thymidine and is efficiently phosphorylated and incorporated in its place into DNA. However, BUdR incorporated into DNA often mispairs and increases replication errors. DNA containing BUdR is also more sensitive to some wavelengths of ultraviolet (UV) light (which makes BUdR useful in enrichment schemes for isolating mutants [see chapter 3]). Moreover, DNA containing BUdR has a different density from DNA exclusively containing thymidine (another feature of BUdR that is useful in experiments).

Antibiotics That Affect Gyrase

Many antibiotics and antitumor drugs affect topoisomerases. The type II topoisomerase, gyrase, in bacteria is a target for many different antibiotics. Because this enzyme is similar among all bacteria, these antibiotics have a broad spectrum of activity and kill many types of bacteria.

GyrA inhibition

Nalidixic acid specifically binds to the GyrA subunit, which is involved in cutting the DNA and in strand passage. This activity makes nalidixic acid and its many derivatives, including oxolinic acid and chloromycetin, very useful antibiotics. Another antibiotic that binds to the GyrA subunit, ciprofloxacin, is used for treating gonorrhea, anthrax, and bacterial dysentery. However, because these antibiotics can induce prophages, they may actually make some diseases worse (see chapter 7).

The mechanism of killing by these antibiotics is not completely understood. They are known to cause degradation of DNA and can cause the DNA to become covalently linked to gyrase, presumably by trapping it in an intermediate state in the process of strand passage. Bacteria resistant to nalidixic acid have an altered *gyrA* gene.

GyrB inhibition

Novobiocin and its more potent relative coumermycin bind to the GyrB subunit, which is involved in ATP binding. These antibiotics do not resemble ATP, but by binding to the gyrase they somehow prevent ATP cleavage, perhaps by changing the conformation of the enzyme. Mutants resistant to novobiocin have an altered *gyrB* gene.

The Bacterial Genome

The discussion of the replication and structure of the bacterial genome ignores the complexity of the sequences and the functions they encode. Advances in DNA-sequencing technologies are drastically reducing both the time it takes to sequence DNA and the cost of DNA sequencing. We are now at a time when determining the genomic sequence of a newly discovered bacterium is often one of the earliest steps in trying to fully characterize the functions of a new species. In addition, there are also a large number of studies where DNA from a particular environment or a consortium of organisms is chosen for DNA sequencing instead of that of a single organism derived from pure culture. Bacterial and archaeal genomes are in some ways more amenable to genome analysis than eukaryotic genomes because they are relatively small, ranging from ca. 0.5 megabases (Mb) for some obligate parasites to around 10 Mb for some free-living bacteria. They contain few introns and much less repetitive DNA than eukaryotic genomes. Chapter 13 deals with ideas for how genomes are structured and the tools we use to study them.

1. DNA consists of two strands wrapped around each other in a double helix. Each strand consists of a chain of nucleotides held together by phosphates joining their deoxyribose sugars. Because the phosphate joins the third carbon of one sugar to the fifth carbon of the next sugar, the DNA strands have directionality, or polarity, and have distinct 5′ phosphate and 3′ hydroxyl ends. The two strands of DNA are antiparallel, so that the 5′ end of one is on the same end as the 3′ end of the other.

2. DNA is synthesized from the precursor deoxynucleoside triphosphates by DNA polymerase. The first phosphate of each nucleotide is attached to the 3′ hydroxyl of the next deoxynucleotide, giving off the terminal two phosphates to provide energy for the reaction.

3. DNA polymerases require both a primer and a template strand. The pairing of the bases between the incoming deoxynucleotide and the base on the template strand dictates which deoxynucleotide will be added at each step, with A always pairing with T and G always pairing with C. The DNA polymerase synthesizes DNA in the 5′-to-3′ direction, moving in the 3′-to-5′ direction on the template.

4. DNA polymerases cannot put down the first deoxynucleotide, so RNA is usually used to prime the synthesis of a new strand. Afterward, the RNA primer is removed and replaced by DNA, using upstream DNA as a primer. The use of RNA primers helps reduce errors by allowing editing.

5. DNA polymerase does not synthesize DNA by itself but needs other proteins to help it replicate DNA. These other proteins are helicases that separate the strands of the DNA, ligases to join two DNA pieces together, primases to synthesize RNA primers, and other accessory proteins to keep the DNA polymerase on the DNA and reduce errors.

6. Both strands of double-stranded DNA are replicated from the same end, so that the overall direction of DNA replication is from 5′ to 3′ on one strand and from 3′ to 5′ on the other strand. Because DNA polymerase can polymerize only in the 5′-to-3′ direction, it must replicate one strand in short pieces and ligate them afterward to form a continuous strand. The short pieces are called Okazaki fragments. The two DNA polymerases replicating the leading and lagging template strands remain bound to each other in a process called the trombone model of replication.

7. Lesions on DNA, proteins bound to DNA, and transcription can stall DNA replication. Primase can reinitiate replication to prevent the entire holoenzyme from stalling but leaving a gap that is repaired by other mechanisms. DNA helicases and other enzymes can be used to help DNA polymerase past bound proteins and to allow the polymerase to continue after it stalls from collisions with RNA polymerase.

8. The DNA in a bacterium that carries most of the genes is called the bacterial chromosome. The chromosome of most bacteria is a long, circular molecule that replicates in both directions from a unique origin of replication, *oriC*. Replication of the chromosomes initiates each time the cells reach a certain size. For fast-growing cells, new rounds of replication initiate before old ones are completed. This accounts for the fact that fast-growing cells have a higher DNA content than slower-growing cells.

9. Chromosome replication terminates, and the two daughter DNAs separate, when the replication forks meet. In some bacteria, multiple *ter* sites that act as "one-way gates" delay movement of the replication forks on the chromosome. Proteins that bind to the *ter* sites that are inhibitors of the DnaB helicase stop replication at these sites.

10. To separate the daughter DNAs after replication, dimerized chromosomes, created by recombination between the daughter DNAs, are resolved by XerC and XerD, a site-specific recombination system that promotes recombination between the *dif* sites on the daughter chromosomes. The FtsK protein is a DNA translocase that promotes XerC-XerD recombination at *dif* sites to prevent dimerized chromosomal DNA from being guillotined by the forming septum. Topo IV decatenates the intertwined daughter DNAs by passing the double-stranded DNAs through each other, and its activity is also regulated temporally and spatially via an interaction with FtsK.

11. The daughter chromosomes are segregated by condensing the DNAs through supercoiling by DNA gyrase and by condensins that hold the DNA in large supercoiled loops.

12. The FtsZ protein forms a ring at the midpoint of the cell, attracting other proteins, which form the division septum.

13. The Min proteins prevent the formation of FtsZ rings anywhere in the cell other than in the middle. Nucleoid occlusion proteins prevent the formation of FtsZ rings over nucleoids.

14. Initiation of a round of chromosome replication occurs once every time the cell divides. Initiation occurs when the ratio of active DnaA protein to origins of replication reaches a critical number. After replication initiates, its ATPase is upregulated by interaction with the β clamp and Hda, converting it to the inactive ADP-bound state to prevent reinitiation. In some bacteria, including *E. coli* and related enteric bacteria, new initiations are prevented by hemimethylation of the newly

Summary (continued)

replicated DNA at the origin and by sequestration until the replication fork has left the origin.

15. The chromosomal DNA of bacteria is usually one long, continuous circular molecule about 1,000 times as long as the cell itself. This long DNA is condensed in a small part of the cell called the nucleoid. In this structure, the DNA loops out of a central condensed core region. Some of these loops of DNA are negatively supercoiled. In *E. coli*, most DNAs have one supercoil about every 300 bases.

16. The enzymes that modulate DNA supercoiling in the cell are called topoisomerases. There are two types of topoisomerases in cells. Type I topoisomerases can remove supercoils one at a time by breaking only one strand and passing the other strand through the break. Type II topoisomerases remove or add supercoils two at a time by breaking both strands and passing another region of the DNA through the break. The enzyme responsible for adding the negative supercoils to DNA in bacteria is a type II topoisomerase called gyrase. Topo IV decatenates daughter DNAs after replication.

QUESTIONS FOR THOUGHT

1. Some viruses, such as adenovirus, avoid the problem of lagging-strand synthesis by replicating the individual strands of the DNA in the leading-strand direction simultaneously from both ends so that eventually the entire molecule is replicated. Why do bacterial chromosomes not replicate in this way?

2. Why are DNA molecules so long? Would it not be easier to have many shorter pieces of DNA? What are the advantages and disadvantages of a single long DNA molecule?

3. Why do cells have DNA as their hereditary material instead of RNA, like some viruses?

4. What effect would shifting a temperature-sensitive mutant with a mutation in the *dnaA* gene for initiator protein DnaA have on the rate of DNA synthesis? Would the rate drop linearly or exponentially? Would the slope of the curve be affected by the growth rate of the cells at the time of the shift? Explain.

5. The gyrase inhibitor novobiocin inhibits the growth of almost all types of bacteria. What would you predict about the gyrase of the bacterium *Streptomyces sphaeroides*, which makes this antibiotic? How would you test your hypothesis?

6. How do you think chromosome replication and cell division are coordinated in bacteria like *E. coli*? How would you go about testing your hypothesis?

7. Why is termination of chromosome replication so sloppy that the *ter* region is nonessential for growth and there has to be more than one *ter* site in each direction to completely stop the replication fork? What are the advantages of not having a definite site on the chromosome at which replication always terminates?

SUGGESTED READING

Aussel L, Barre F-X, Aroyo M, Stasiak A, Stasiak AZ, Sherratt D. 2002. FtsK is a DNA motor protein that activates chromosome dimer resolution by switching the catalytic state of the XerC and XerD recombinases. *Cell* **108:**195–205.

Bernhardt TG, de Boer PAJ. 2005. SlmA, a nucleoid-associated, FtsZ binding protein required for blocking septal ring assembly over chromosomes in *E. coli*. *Mol Cell* **18:**555–564.

Brézellec P, Hoebeke M, Hiet MS, Pasek S, Ferat JL. 2006. DomainSieve: a protein domain-based screen that led to the identification of *dam*-associated genes with potential link to DNA maintenance. *Bioinformatics* **22:**1935–1941.

Biller SJ, Burkholder WF. 2009. The *Bacillus subtilis* SftA (YtpS) and SpoIIIE DNA translocases play distinct roles in growing cells to ensure faithful chromosome partitioning. *Mol Microbiol* **74:**790–809.

Blakely G, May G, McCulloch R, Arciszewska LK, Burke M, Lovett ST, Sherratt DJ. 1993. Two related recombinases are required for site-specific recombination at *dif* and *cer* in *E. coli* K12. *Cell* **75:**351–361.

Britton RA, Lin DC, Grossman AD. 1998. Characterization of a prokaryotic SMC protein involved in chromosome partitioning. *Genes Dev* **12:**1254–1259.

Camara JE, Breier AM, Brendler T, Austin S, Cozzarelli NR, Crooke E. 2005. Hda inactivation of DnaA is the predominant mechanism preventing hyperinitiation of *Escherichia coli* DNA replication. *EMBO Rep* **6:**736–741.

Cortez D, Quevillon-Cheruel S, Gribaldo S, Desnoues N, Sezonov G, Forterre P, Serre M-CM. 2010. Evidence for a Xer/dif system for chromosome resolution in archaea. *PLoS Genet* **6:**e1001166.

Dervyn E, Suski C, Daniel R, Bruand C, Chapuis J, Errington J, Jannière L, Ehrlich SD. 2001. Two essential DNA polymerases at the bacterial replication fork. *Science* **294:**1716–1719.

Dohrmann PR, Correa R, Frisch RL, Rosenberg SM, McHenry CS. 2016. The DNA polymerase III holoenzyme contains γ and is not a trimeric polymerase. *Nucleic Acids Res* **44:**1285–1297.

Fournes F, Val M-E, Skovgaard O, Mazel D. 2018. Replicate once per cell cycle: replication control of secondary chromosomes. *Front Microbiol* **9**:1833.

Fricker AD, Peters JE. 2014. Vulnerabilities on the lagging-strand template: opportunities for mobile elements. *Annu Rev Genet* **48**:167–186.

Fujimitsu K, Senriuchi T, Katayama T. 2009. Specific genomic sequences of *E. coli* promote replicational initiation by directly reactivating ADP-DnaA. *Genes Dev* **23**:1221–1233.

Gabbai CB, Yeeles JTP, Marians KJ. 2014. Replisome-mediated translesion synthesis and leading strand template lesion skipping are competing bypass mechanisms. *J Biol Chem* **289**:32811–32823.

Galli E, Ferat J-L, Desfontaines J-M, Val M-E, Skovgaard O, Barre FX, Possoz C. 2019. Replication termination without a replication fork trap. *Sci Rep* **9**:8315. http://doi.org/10.1038/s41598-019-43795-2

Guy CP, Atkinson J, Gupta MK, Mahdi AA, Gwynn EJ, Rudolph CJ, Moon PB, van Knippenberg IC, Cadman CJ, Dillingham MS, Lloyd RG, McGlynn P. 2009. Rep provides a second motor at the replisome to promote duplication of protein-bound DNA. *Mol Cell* **36**:654–666.

Hayama R, Marians KJ. 2010. Physical and functional interaction between the condensin MukB and the decatenase topoisomerase IV in *Escherichia coli*. *Proc Natl Acad Sci USA* **107**:18826–18831.

Heller RC, Marians KJ. 2006. Replication fork reactivation downstream of a blocked nascent leading strand. *Nature* **439**:557–562.

Helmstetter CE, Cooper S. 1968. DNA synthesis during the division cycle of rapidly growing *Escherichia coli* B/r. *J Mol Biol* **31**:507–518.

Jean NK, Rutherford TJ, Löwe J. 2019. FtsK in motion reveals its mechanism for double-stranded DNA translocation. *bioRxiv* 1–24.

Joshi MC, Magnan D, Montminy TP, Lies M, Stepankiw N, Bates D. 2013. Regulation of sister chromosome cohesion by the replication fork tracking protein SeqA. *PLoS Genet* **9**:e1003673.

Kasho K, Katayama T. 2013. DnaA binding locus *datA* promotes DnaA-ATP hydrolysis to enable cell cycle-coordinated replication initiation. *Proc Natl Acad Sci USA* **110**:936–941.

Kohanski MA, Dwyer DJ, Hayete B, Lawrence CA, Collins JJ. 2007. A common mechanism of cellular death induced by bactericidal antibiotics. *Cell* **130**:797–810.

Kysela DT, Randich AM, Caccamo PD, Brun YV. 2016. Diversity takes shape: understanding the mechanistic and adaptive basis of bacterial morphology. *PLoS Biol* **14**:e1002565.

Le Bourgeois P, Bugarel M, Campo N, Daveran-Mingot M-L, Labonté J, Lanfranchi D, Lautier T, Pagès C, Ritzenthaler P. 2007. The unconventional Xer recombination machinery of streptococci/lactococci. *PLoS Genet* **3**:e117.

Liu N-J, Dutton RJ, Pogliano K. 2006. Evidence that the SpoIIIE DNA translocase participates in membrane fusion during cytokinesis and engulfment. *Mol Microbiol* **59**:1097–1113.

Mercier R, Petit MA, Schbath S, Robin S, El Karoui M, Boccard F, Espéli O. 2008. The MatP/*matS* site-specific system organizes the terminus region of the *E. coli* chromosome into a macrodomain. *Cell* **135**:475–485.

Moolman MC, Krishnan ST, Kerssemakers JWJ, van den Berg A, Tulinski P, Depken M, Reyes-Lamothe R, Sherratt DJ, Dekker NH. 2014. Slow unloading leads to DNA-bound β2-sliding clamp accumulation in live *Escherichia coli* cells. *Nat Commun* **5**:5820.

Neidhardt FC, Curtiss R III, Ingraham JL, Lin ECC, Low KB, Magasanik B, Reznikoff WS, Riley M, Schaechter M, Umbarger HE (ed). 1996. Escherichia coli *and* Salmonella*: Cellular and Molecular Biology*, 2nd ed. ASM Press, Washington, DC.

Olby R. 1974. *The Path to the Double Helix*. Macmillan Press, London, United Kingdom.

Pomerantz RT, O'Donnell M. 2008. The replisome uses mRNA as a primer after colliding with RNA polymerase. *Nature* **456**:762–766.

Postow L, Hardy CD, Arsuaga J, Cozzarelli NR. 2004. Topological domain structure of the *Escherichia coli* chromosome. *Genes Dev* **18**:1766–1779.

Reddy CA, Beveridge TJ, Breznak JA, Marzluf G, Schmidt TM, Snyder LR (ed). 2007. *Methods for General and Molecular Microbiology*, 3rd ed. ASM Press, Washington, DC.

Reyes-Lamothe R, Sherratt DJ, Leake MC. 2010. Stoichiometry and architecture of active DNA replication machinery in *Escherichia coli*. *Science* **328**:498–501.

Rudolph CJ, Upton AL, Stockum A, Nieduszynski CA, Lloyd RG. 2013. Avoiding chromosome pathology when replication forks collide. *Nature* **500**:608–611.

Sambrook J, Russell D. 2006. *The Condensed Protocols from Molecular Cloning: a Laboratory Manual*. Cold Spring Harbor Laboratory Press, Cold Spring Harbor, NY.

Sanders GM, Dallmann HG, McHenry CS. 2010. Reconstitution of the *B. subtilis* replisome with 13 proteins including two distinct replicases. *Mol Cell* **37**:273–281.

Shih Y-L, Rothfield L. 2006. The bacterial cytoskeleton. *Microbiol Mol Biol Rev* **70**:729–754.

Singleton MR, Dillingham MS, Wigley DB. 2007. Structure and mechanism of helicases and nucleic acid translocases. *Annu Rev Biochem* **76**:23–50.

Slater S, Wold S, Lu M, Boye E, Skarstad K, Kleckner N. 1995. *E. coli* SeqA protein binds *oriC* in two different methyl-modulated reactions appropriate to its roles in DNA replication initiation and origin sequestration. *Cell* **82**:927–936.

Thanbichler M. 2009. Closing the ring: a new twist to bacterial chromosome condensation. *Cell* **137**:598–600.

Thanbichler M, Shapiro L. 2006. MipZ, a spatial regulator coordinating chromosome segregation with cell division in *Caulobacter*. *Cell* **126**:147–162.

Valens M, Penaud S, Rossignol M, Cornet F, Boccard F. 2004. Macrodomain organization of the *Escherichia coli* chromosome. *EMBO J* **23**:4330–4341.

Valens M, Thiel A, Boccard F. 2016. The MaoP/*maoS* site-specific system organizes the Ori region of the *E. coli* chromosome into a macrodomain. *PLoS Genet* **12**:e1006309.

Vecchiarelli AG, Li M, Mizuuchi M, Hwang LC, Seol Y, Neuman KC, Mizuuchi K. 2016. Membrane-bound MinDE complex acts as a toggle switch that drives Min oscillation coupled to cytoplasmic depletion of MinD. *Proc Natl Acad Sci USA* **113**:E1479–E1488.

Viollier PH, Thanbichler M, McGrath PT, West L, Meewan M, McAdams HH, Shapiro L. 2004. Rapid and sequential movement of individual chromosomal loci to specific subcellular locations during bacterial DNA replication. *Proc Natl Acad Sci USA* **101**:9257–9262.

Wang JC. 1996. DNA topoisomerases. *Annu Rev Biochem* **65**:635–692.

Watson JD. 1968. *The Double Helix*. Atheneum, New York, NY.

Watson JD, Crick FHC. 1953. Molecular structure of nucleic acids; a structure for deoxyribose nucleic acid. *Nature* **171**:737–738.

Windgassen TA, Leroux M, Satyshur KA, Sandler SJ, Keck JL. 2018. Structure-specific DNA replication fork recogntion directs helicase and replication restart activities of the *Escherichia coli* PriA helicase. *Proc Natl Acad Sci USA* **115**:E9075–E9084.

Wu LJ, Ishikawa S, Kawai Y, Oshima T, Ogasawara N, Errington J. 2009. Noc protein binds to specific DNA sequences to coordinate cell division with chromosome segregation. *EMBO J* **28**:1940–1952.

Yamaichi Y, Niki H. 2004. *migS*, a *cis*-acting site that affects bipolar positioning of *oriC* on the *Escherichia coli* chromosome. *EMBO J* **23**:221–233.

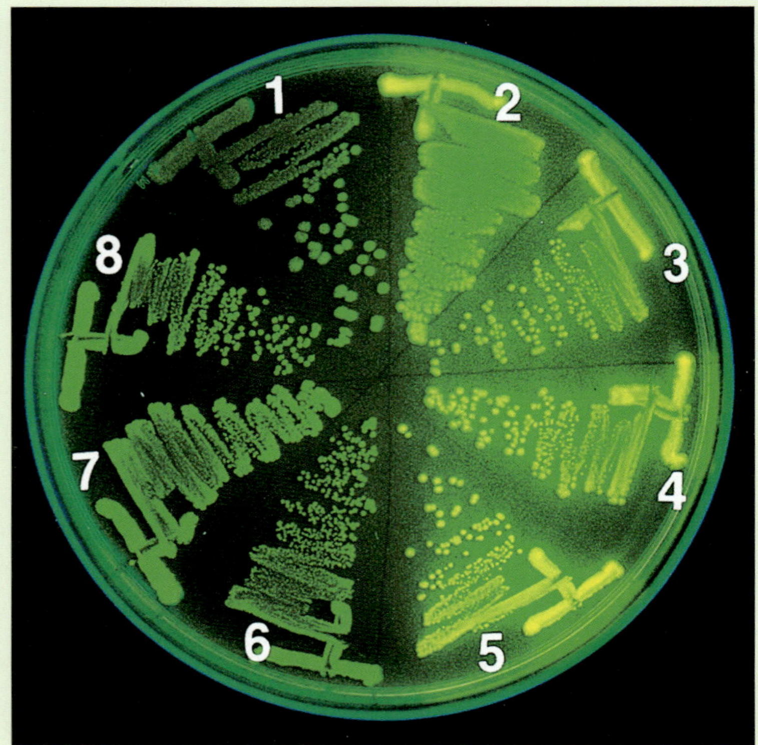

Fluorescence of transformants expressing MBP-GFP hybrid proteins. MC4100 transformed with the following: 1, pMGP2; 2, pMGC2; 8, MM52 [*secA*(Ts)] transformed with pMGP2; 3, pMGC2; 7, CK2163 (*secB*) transformed with pMGP2; 4, pMGC2; 6, IQ85 [*secY*(Ts)] transformed with pMGP2; and 5, pMGC2. From Feilmeier et al. 2000 (see Suggested Reading).

Bacterial Gene Expression: Transcription, Translation, Protein Folding, and Localization

2

UNCOVERING THE MECHANISM OF PROTEIN SYNTHESIS, and therefore of gene expression, was one of the most significant accomplishments in the history of science. The process of gene expression is called the **central dogma** of molecular biology, which states that information stored in DNA is copied into RNA and then translated into protein. We now know of many exceptions to the central dogma. For example, information can sometimes flow in the reverse direction, from RNA to DNA. The information in RNA also can be changed after it has been copied from the DNA. Moreover, the information in DNA may be expressed differently depending on where it is in the genome. Despite these exceptions, however, the basic principles of the central dogma remain sound.

This chapter outlines the process of gene expression and protein synthesis, with a brief discussion of how proteins can be differentially localized after synthesis. The discussion is meant to be only a broad overview, with special emphasis on topics essential to an understanding of the chapters that follow and on subjects unique to bacteria. For more detailed treatments, consult any modern biochemistry textbook.

Overview

DNA carries the information for the synthesis of RNA and proteins in regions called **genes**. The first step in expressing a gene is to **transcribe** an RNA copy from one strand in that region. The word "transcription" is descriptive, because the information in RNA is copied from DNA in the same language, which is written in a sequence of nucleotides. If the gene carries information for a protein, this RNA transcript is called **messenger RNA (mRNA)**. An mRNA is a messenger because it carries the information encoded in a gene to a **ribosome**, which is the main machinery for protein synthesis. Once on the ribosome, the information in the mRNA can be **translated** into the protein. Translation is another descriptive word, because one language—a sequence of nucleotides in DNA and RNA—is translated into a different language—a sequence of amino acids in a protein. The mRNA is translated as it moves through the ribosome, 3 nucleotides at a time. Each 3-nucleotide sequence, called a **codon**, carries information for a specific amino acid. The assignment of each of the possible codons to an amino acid is called the **genetic code**.

The actual translation from the language of nucleotide sequences to the language of amino acid sequences is performed by small RNAs called **tRNAs** and enzymes called **aminoacyl-tRNA synthetases** (aaRSs). The aaRS enzymes attach specific amino acids to their matching tRNAs. Each aminoacylated tRNA (aa-tRNA) specifically pairs with a codon in the mRNA as it moves through the ribosome, and the amino acid carried by the tRNA is added to the growing protein. The tRNA pairs with the codon in the mRNA through a 3-nucleotide sequence in the tRNA called the **anticodon** that is complementary to the codon in the mRNA. The base-pairing rules for codons and anticodons are basically the same as the base-pairing rules for DNA replication, and the pairing is antiparallel. The only major differences are that RNA has uracil (U) rather than thymine (T) and that the pairing between the last of the 3 bases in the codon and the first base in the anticodon is less stringent.

This basic outline of gene expression leaves many important questions unanswered. How does mRNA synthesis begin and end at the correct places and on the correct strand in the DNA? Similarly, how does translation start and stop at the correct places on the mRNA? What actually happens to the tRNA and ribosomes during translation? What happens to the mRNA and proteins after they are made? The answers to these questions and many others are important for the interpretation of genetic experiments, so we will discuss the structure of RNA and proteins and the processes by which they are synthesized in more detail.

The Structure and Function of RNA

In this section, we review the basic components of RNA and how it is synthesized. We also review how structure varies among different types of cellular RNAs and the role each type plays in cellular processes.

Types of RNA

There are several different classes of RNA in cells. Some of these, including mRNA, rRNA, and tRNA, are involved in protein synthesis. Each of these types of RNA has special properties, which are discussed below. Others are involved in regulation, replication, and protein secretion.

RNA Precursors

RNA is similar to DNA in that it is composed of a chain of nucleotides. However, RNA nucleotides contain the sugar ribose instead of deoxyribose. These five-carbon sugars differ in the second carbon, which is attached to a hydroxyl group in ribose rather than the hydrogen found in deoxyribose (see Figure 1.2). Figure 2.1A shows the structure of a **ribonucleoside triphosphate** (rNTP), which is the form that is used as a precursor during RNA synthesis.

RNA and DNA chains also vary in the bases that are present. Three of the bases—adenine, guanine, and cytosine—are the same, but RNA has uracil instead of the thymine found in DNA (Figure 2.1B). The RNA bases also can be modified after they are incorporated into an RNA chain, as discussed below.

Figure 2.1C shows the basic structure of an RNA polynucleotide chain. As in DNA, RNA nucleotides are held together by phosphates that join the 5′ carbon of one ribose sugar to the 3′ carbon of the next. This arrangement ensures that, as with DNA chains, the two ends of an RNA polynucleotide chain will be different from each other, with the 5′ end terminating in a phosphate group and the 3′ end terminating in a hydroxyl group. The 5′ end of a newly synthesized RNA chain has three phosphates attached to it because transcription initiates with an rNTP. As each new rNTP is added to the growing RNA chain, two phosphate groups are released so that the sugar phosphate backbone alternates between the ribose (to which the base is attached) and a single phosphate group.

According to convention, the sequence of bases in RNA is given from the 5′ end to the 3′ end, which is the direction in which the RNA is synthesized, by addition of the 5′ α-phosphate of an incoming nucleoside triphosphate to the 3′ hydroxyl end of the growing RNA chain. Also by convention, regions in RNA that are closer to the 5′ end in a given sequence are referred to as **upstream**, and regions that are closer to the 3′ end are referred to as **downstream**, because RNA is both synthesized and translated in the 5′-to-3′ direction.

RNA Structure

Except for the sequence of bases and minor differences in the pitch of the helix, little distinguishes one DNA molecule from another. However, RNA chains generally have more structural properties than DNA and often are folded into complex structures that have important biological roles. Extensive base modifications can further change the structure of the RNA molecule.

PRIMARY STRUCTURE

All RNA transcripts are synthesized in the same way, from a DNA template. Only the sequences of their nucleotides and their lengths are different. The sequence of nucleotides in RNA is the **primary structure** of the RNA. In some cases, the primary structure of an RNA is changed after it is transcribed from the DNA (see "RNA Processing and Modification" below).

SECONDARY STRUCTURE

Unlike DNA, RNA is usually single stranded. However, pairing between the bases in different regions of the molecule may cause it to fold up on itself to form double-stranded regions. Such double-stranded regions are called the **secondary structure** of the RNA. All RNAs,

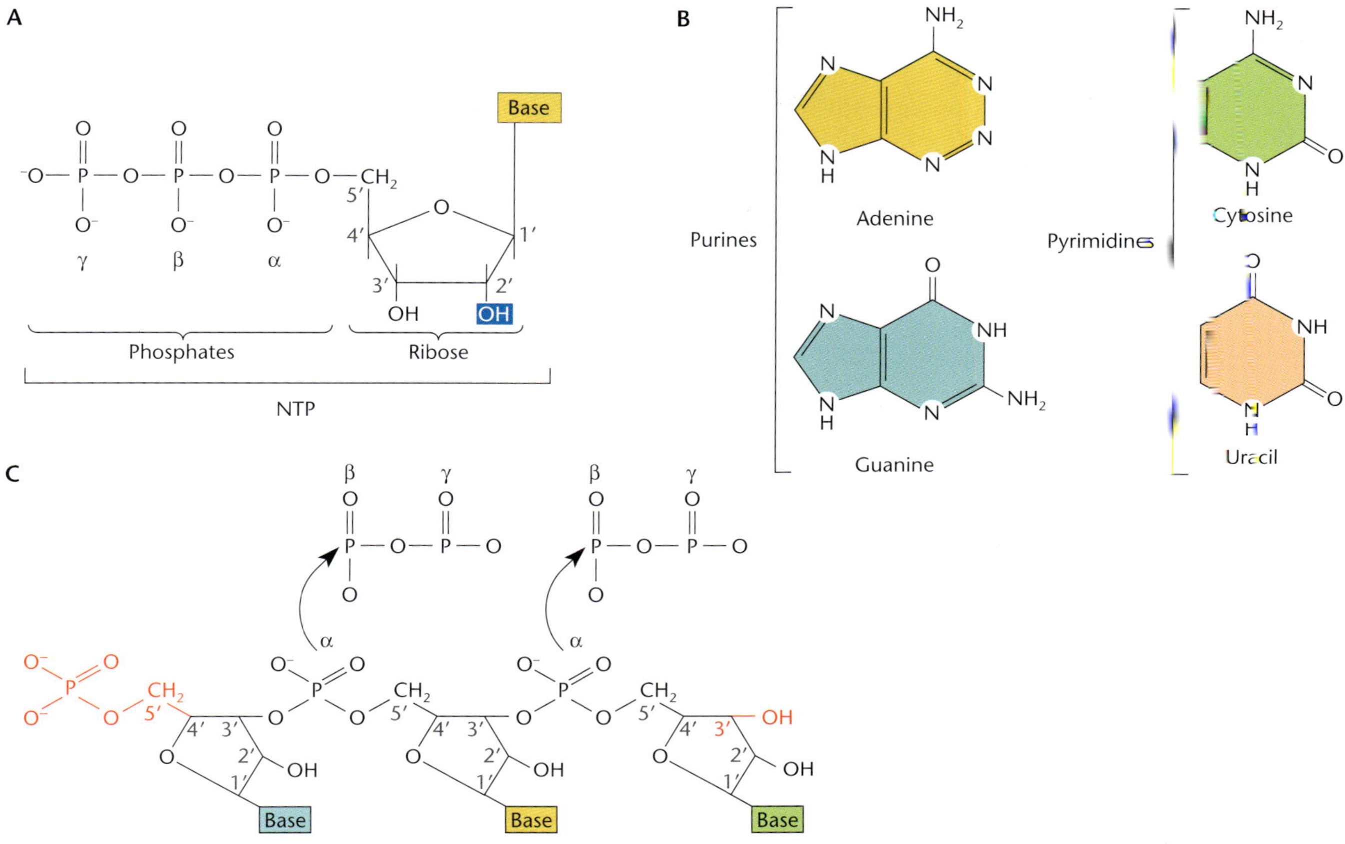

Figure 2.1 RNA precursors. **(A)** A ribonucleoside triphosphate (rNTP) (the form of NTP used as a precursor for RNA) contains a ribose sugar, a base, and three phosphates. **(B)** The four bases in RNA. **(C)** An RNA polynucleotide chain with the 5′ and 3′ ends shown in red.

including mRNAs, probably have extensive secondary structure.

Figure 2.2 shows an example of RNA secondary structure in which the sequence 5′-AUCGGCA-3′ has paired with the complementary sequence 5′-UGCUGAU-3′ somewhere else in the molecule. As in double-stranded DNA, the paired strands of RNA are antiparallel, i.e., pairing occurs only when the two sequences are complementary when read in opposite directions (5′ to 3′ and 3′ to 5′) and the double-stranded RNA forms a helix that is similar to a DNA-DNA helix, capped with a few unpaired residues called a loop; the helix plus the loop are sometimes called a hairpin. However, the pairing rules for double-stranded RNA are slightly different from the pairing rules for DNA. As in DNA, G pairs with C, and A pairs with U (which replaces T in RNA). In RNA, guanine can pair not only with cytosine, but also with uracil. Because these G-U "wobble" pairs do not share hydrogen bonds, as indicated in the figure, they contribute less substantially to the stability of the double-stranded RNA. Additional "non-Watson-Crick" pairings are also found in RNA; these often involve

Figure 2.2 Secondary structure in an RNA. **(A)** The RNA folds back on itself to form a helical element sometimes called a hairpin. The presence of a GU pair (in parentheses) does not disrupt the structure. **(B)** Different regions of the RNA can also pair with each other to form a pseudoknot (participating residues shown in red). In the example, the loop of the hairpin pairs with another region of the RNA.

A Helix

5′ A-C-U-C-A-U-A-U-C-C-G-G-C 3′

B Pseudoknot

3′ U-C-A-C-A

5′ A-C-U-C-A-U-A-U

surfaces of the nucleoside other than the normal hydrogen-bonding edge.

Each base pair that forms in the RNA makes the secondary structure more stable. Consequently, the RNA generally folds so that the greatest number of continuous base pairs can form. The stability of a structure can be predicted by adding up the energy of all of the hydrogen bonds that contribute to the structure. By eye, it is often difficult to predict which regions of a long RNA will pair to give the most stable structure. Computer software (e.g., mfold [http://mfold.rna.albany.edu/?q=mfold]) is available that, given the sequence of bases (primary structure) of the RNA, can predict the most stable secondary structure; however, the structure of complex RNAs is difficult to predict computationally because of interrupted base-pairing and non-Watson-Crick interactions.

TERTIARY STRUCTURE

Double-stranded regions of RNAs generated by base pairing are stiffer than single-stranded regions. As a result, an RNA that has secondary structure will have a more rigid shape than one without double-stranded regions. Also, the intermingled paired regions cause the RNA to fold back on itself extensively, facilitating additional tertiary interactions. One type of tertiary interaction occurs when an unpaired region (such as the loop of a hairpin like that shown in Figure 2.2) pairs with another region of the same RNA molecule. A structure like this is called a pseudoknot (rather than a real knot) because it is held together only by hydrogen bonds. Together, these interactions give many RNAs a well-defined three-dimensional shape, called their **tertiary structure**. Proteins or other cellular constituents often recognize RNAs by their tertiary structures.

RNA Processing and Modification

The folding of an RNA molecule as a result of secondary and tertiary structure represents a **noncovalent change**, because only hydrogen bonds or electrostatic interactions, not chemical (**covalent**) bonds, are formed or broken. However, once the RNA is synthesized, **RNA processing** and **RNA modification** can introduce covalent changes.

RNA processing involves forming or breaking phosphate bonds in the RNA after it is made. For example, the terminal phosphates at the 5′ end may be removed, or the RNA may be cut into smaller pieces and even relegated into new combinations, requiring the breaking and making of new phosphate bonds. In one of the most extreme cases of RNA processing, called RNA editing, nucleotides can be excised from or added to mRNA after it has been transcribed from DNA.

RNA modification, in contrast, involves altering the bases or sugars of RNA. Examples include methylation of the bases or sugars of **rRNA** (ribosomal RNA) and enzymatic alteration of the bases of **tRNA** (transfer ribonucleic acid). In eukaryotes, "caps" of methylated nucleotides are added to the 5′ ends of some types of mRNA. In bacteria, mRNAs are not capped, and only the stable rRNAs and tRNAs are extensively modified.

Transcription

Transcription is the synthesis of RNA by copying information from a DNA template. The overall processes of transcription are fairly similar in all organisms, but it is best understood in bacteria.

Structure of Bacterial RNA Polymerase

The transcription of DNA into RNA is carried out by **RNA polymerase**. In bacteria, the same RNA polymerase makes all the cellular RNAs, including rRNA, tRNA, and mRNA (but excluding the RNA primers used in DNA replication). There are approximately 2,000 molecules of RNA polymerase in each bacterial cell under slow growth conditions, with higher numbers during rapid growth. Only the primer RNAs of Okazaki fragments are made by a different RNA polymerase (primase) during DNA replication. In contrast, eukaryotes have three nuclear RNA polymerases, as well as a mitochondrial RNA polymerase, that make their cellular RNAs.

Figure 2.3 shows a schematic structure of *Escherichia coli* RNA polymerase, which has six subunits and a molecular mass of more than 400,000 Daltons (Da) (400 kDa), making it one of the largest enzymes in the *E. coli* cell. The core enzyme consists of five subunits: two identical α subunits, two very large subunits called β and β', and the ω subunit. The α, β, and β' subunits are essential parts of the RNA polymerase, and the ω subunit helps in its assembly. A sixth subunit, the σ factor, is required only for initiation and cycles off the enzyme after initiation of transcription. Without the σ factor, the RNA polymerase is called the **core enzyme**; with σ, it is called the **holoenzyme**.

Crystal structures of RNA polymerase from *Thermus aquaticus* and *E. coli* have revealed the structure shown in Figure 2.4. The overall shape resembles a crab claw. Regions of the β and β' subunits form the pincers of the crab claw (as shown in Figure 2.3). The two α subunits, αI and αII, are on the opposite end from the claw, making up the rear end when the enzyme is transcribing DNA. The carboxy-terminal regions of the α polypeptides hang out from the enzyme, where they can contact other proteins or DNA regions upstream of the promoter and stimulate transcription initiation (see below). The ω subunit is also on this end of the RNA polymerase, wrapped around the β' subunit (Figure 2.3).

The σ factor is crucial for directing RNA polymerase to recognize the **promoter** sequence, which is the site on

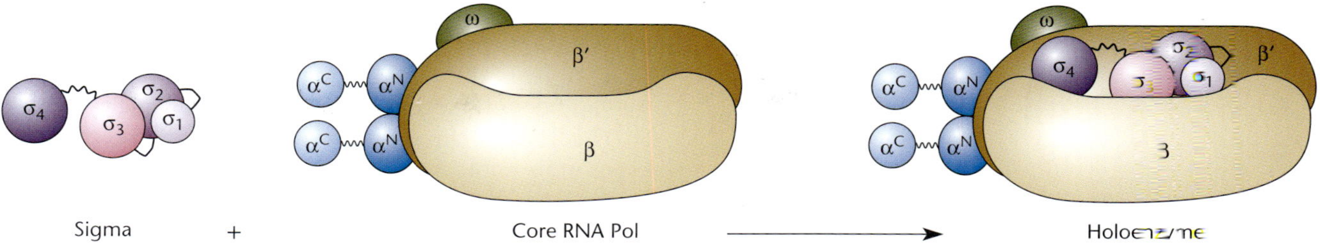

Figure 2.3 The structure of bacterial RNA polymerase. The core enzyme is composed of the β, β′, ω, and two copies of the α protein. Addition of the σ protein results in formation of the holoenzyme, which is the form required for initiation of transcription. The functions of some of the domains of σ, σ_1, and $\sigma_{3.2}$ are discussed in this chapter.

the DNA at which transcription initiates. When the σ factor is bound to the complex to form the holoenzyme, it wraps around the front end of the core enzyme in such a way that it can contact the DNA as it enters the open claw. One of the σ domains, domain σ_2, contacts the β′ pincer and is in position to bind to one part of the promoter sequence on the DNA (the −10 region), while two other domains, σ_3 and σ_4, contact the β subunit further upstream in the active-center channel in such a way that the σ_4 domain is in position to contact the −35 region of the promoter (see below). The RNA polymerases of all bacteria are probably very similar in sequence and composition. One interesting difference is that in some types of bacteria, the β and β′ subunits are attached to each other to form an even larger polypeptide. The eukaryotic and archaeal core RNA polymerases have more subunits and have very different sequences, but their basic overall structures are very similar to that of bacterial RNA polymerase.

Overview of Transcription

Much like DNA polymerase (see chapter 1), RNA polymerase makes a complementary copy of a DNA template, building a chain of RNA by attaching the 5′ phosphate of a ribonucleotide to the 3′ hydroxyl of the one preceding it in the growing chain (Figure 2.5A). However, in contrast to DNA polymerases, RNA polymerases do not need a preexisting primer to initiate the synthesis of a new chain of RNA. To begin transcription, the RNA polymerase holoenzyme binds to the promoter sequence and separates the strands of the DNA, exposing the bases. Unlike DNA polymerases, which require helicases to separate the strands, the RNA polymerase can complete this step by itself. Then, an rNTP complementary to the nucleotide at the **transcription start site** enters the complex through a channel in the RNA polymerase and pairs with the template; this first rNTP is usually ATP or GTP. The second rNTP comes in, and if it is complementary to the next base in the DNA template, it is retained, but if it is not

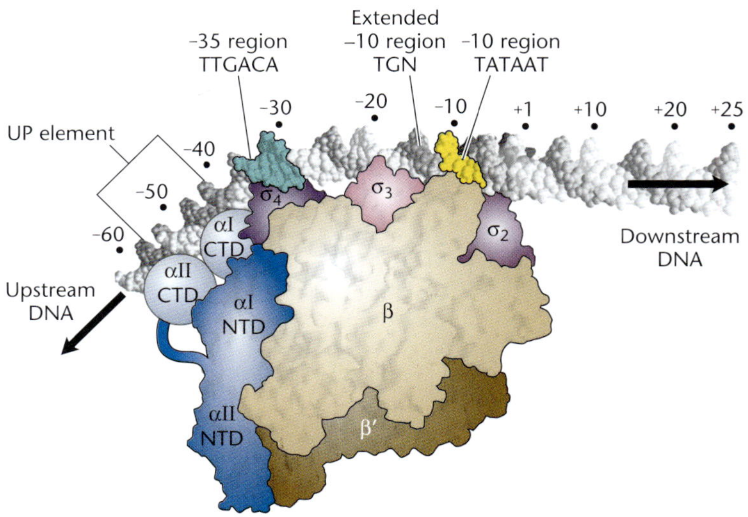

Figure 2.4 Crystal structure of bacterial RNA polymerase and σ interactions with promoter DNA. Subunits of RNA polymerase are shown. The α subunits are divided into the α subunit amino-terminal domains (αNTD), which associate with the β and β′ subunits, and the α subunit carboxy-terminal domains (αCTD), which can bind to the UP element in the DNA. The structures of the αCTDs are not resolved and are shown as balls. σ is divided into four domains that extend along the promoter DNA, interacting with different regions of the promoter. From Murakami KS, Masuda S, Campbell EA, et al. *Science* **296**:1285–1290, 2002. Modified with permission from AAAS.

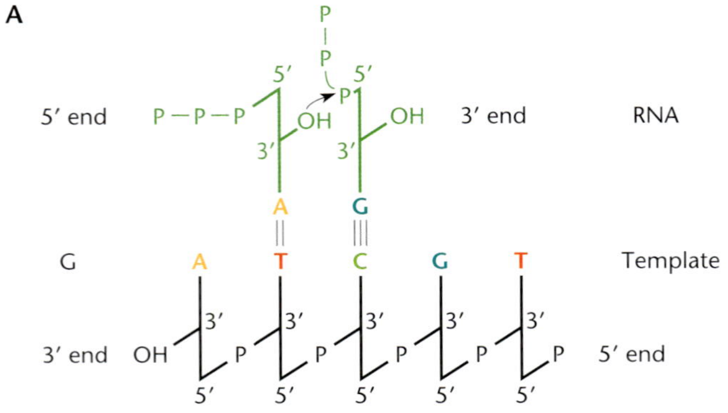

A

B

DNA { 5′ AGCACGACCTACGCAT 3′ Coding strand
3′ TCGTGCTGGATGCGTA 5′ Template strand }

RNA 5′ AGCACGACCUACGCAU 3′

Figure 2.5 RNA transcription. **(A)** The polymerization reaction, in which incoming nucleoside triphosphates (NTPs) pair with the template strand of DNA during transcription and are joined to generate the RNA chain. The β and γ phosphates of each incoming NTP (other than the initiator NTP) are removed as pyrophosphate (PP_i). **(B)** The coding strand (or nontemplate strand) of the DNA has the same sequence as the mRNA (with T residues in the DNA replaced with U residues in the RNA). The template strand is the DNA strand to which the mRNA is complementary.

complementary, it is rejected. RNA polymerase then catalyzes the reaction in which the α phosphate of the second nucleotide joins with the 3′ hydroxyl of the first nucleotide. Then, the third nucleotide comes in and pairs with the next nucleotide in the template, and so forth. In this way, RNA polymerase makes a complementary copy, i.e., **transcribes** the sequence of one strand of DNA into RNA. As shown in Figure 2.5B, the strands of DNA in a region that is transcribed are named to reflect the sequence of the RNA made from that region. The **template strand** of DNA that is copied is also called the **transcribed strand**. The other strand of the DNA, which has the same sequence as the RNA copy, is called the **nontemplate strand** or **coding strand**, since it has the same sequence as the mRNA that encodes the protein (even if the RNA that is made does not actually encode a protein). The sequence of a gene is usually written as the sequence of the nontemplate, or coding, strand.

PROMOTERS

RNA transcripts are copied only from selected regions of the DNA, rather than from the whole molecule; therefore, the RNA polymerase holoenzyme can start making an RNA chain from a double-stranded DNA only at certain sites. These DNA regions are called **promoters**, and the RNA polymerase recognizes a particular nucleotide (usually a T or C) in the promoter region of the template strand as a **transcription start site**, shown as +1 in Figure 2.6. Thus, the first base in the chain is usually an A or a G laid down opposite to a T or C, respectively.

The RNA polymerase holoenzyme recognizes different types of promoters on the basis of which type of σ factor it contains. The most common promoters are those recognized by the RNA polymerase with the σ called σ^{70} in *E. coli*. The σ factors are often named for their size, and this one has a molecular mass of 70,000 Da (70 kDa). Replacement of σ^{70} with a different σ factor results in an RNA polymerase holoenzyme that recognizes a different set of promoters; this will be discussed in later chapters on gene regulation.

Promoters recognized by holoenzymes containing the same σ are not identical to each other, but they do share certain sequences, known as **consensus sequences**, by which they can be distinguished. Figure 2.6 shows the consensus sequence of promoters recognized by holoenzymes containing σ^{70} in *E. coli*, which illustrates a common pattern for promoter structure. The promoter sequence has two important regions: a short AT-rich region centered about 10 bp upstream of the transcription start site, known as the **–10 sequence**, and a second region centered about 35 bp upstream of the start site, called the **–35 sequence**. The σ^{70} factor usually must bind to both sequences to start transcription (see below) but does not require that the DNA have a perfect match to these consensus sequences; binding to the promoter occurs only when σ^{70} is in the holoenzyme complex. Sequence-specific binding to the promoter determines not only the site at which transcription will initiate, but also the direction the RNA polymerase will move along the DNA (in other words, which strand of the DNA will be transcribed from a given region).

THE STEPS OF TRANSCRIPTION

Figure 2.7 shows an overview of the steps of transcription. The RNA polymerase holoenzyme recognizes a promoter and begins transcription with a **ribonucleoside triphosphate**. As the RNA chain begins to grow, the RNA polymerase holoenzyme releases its σ factor, and the five-subunit core enzyme continues to move along the template DNA strand in the 3′-to-5′ direction, synthesizing RNA in the 5′-to-3′ direction. Inside an opening in the DNA helix approximately 17 bases long, called the **transcription bubble**, the elongating RNA and the template DNA strand pair with each other to form a DNA-RNA hybrid of approximately 8 or 9 bp with a double-helix structure similar to that of a double-stranded DNA molecule. As RNA polymerase moves along the DNA, the upstream portion of the DNA-RNA helix separates as new ribonucleotides are incorporated into the 3′ end of the growing RNA chain, and the 8- to 9-bp hybrid is maintained. The resulting RNA product emerges from RNA polymerase through a channel, and

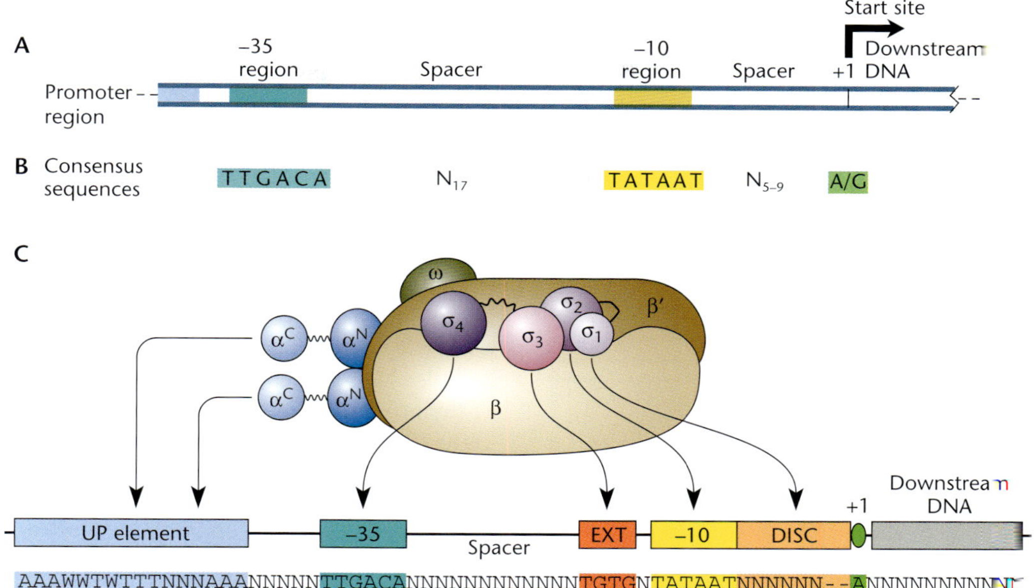

Figure 2.6 **(A)** Typical structure of a σ^{70} bacterial promoter. **(B)** The consensus sequences of a σ^{70} bacterial promoter. RNA synthesis typically starts with an A or a G, and no primer is required. N, any nucleotide. **(C)** Positions of interaction between RNA polymerase and promoter DNA.

the DNA strands behind the RNA polymerase rebind to each other. The RNA polymerase continues to move along the DNA template until it reaches a **terminator**, which signals the RNA polymerase to release both the DNA template and the RNA transcript.

Details of Transcription

It was once assumed that after initiation occurs, the RNA polymerase moves along the DNA at a uniform rate, polymerizing nucleotides into RNA. However, it is now known that the RNA polymerase often starts making RNA and then repeatedly aborts, synthesizing a number of short RNAs in a process called **abortive initiation** before finally leaving the promoter. Even after transcription is under way, RNA polymerase often pauses and sometimes even backs up before continuing. In this section, we discuss in more detail each of the steps in transcription (Figure 2.8), which have been established over many years by a large number of researchers. We discuss these steps one at a time because each of them is the basis for regulatory mechanisms that are discussed in later chapters.

PROMOTER RECOGNITION

In the first step (Figure 2.9), the RNA polymerase core enzyme binds to a σ factor to form the holoenzyme. The bound σ factor then directs the complex to the correct promoter in a process called **promoter recognition** or **binding**. The σ factor must be able to recognize the promoter even though the DNA in the promoter is still in a double-stranded state. Sigma factors consist of a number of domains held together by flexible linkers. Most σ factors are related to σ^{70}, and their domains play similar roles in recognizing their specific promoters. Figure 2.6C shows the conserved regions of the σ^{70} family of sigma factors and the roles played by some of the conserved domains in promoter recognition and initiation of transcription.

One domain of the bound σ, σ_4, recognizes the –35 sequence when it is still in the double-stranded state. Another σ domain, σ_2, binds to the AT-rich –10 sequence. Some σ^{70} promoters have additional sequence elements that interact with other domains of the RNA polymerase (Figures 2.6 and 2.10). The efficiency of binding of RNA polymerase to a promoter can be enhanced by sequences upstream of the promoter, called UP (for upstream) elements, to which the carboxy terminus of the α subunits, called αCTD (for α subunit carboxyl-terminal domain), can bind and help stabilize the binding of RNA polymerase to the DNA. A flexible domain that links αCTD and the amino-terminal domain of α (αNTD) allows αCTD to reach the UP element on the DNA. Also, some promoters have what is called an extended –10 sequence (TGN, located immediately upstream of the –10 sequence to give the sequence TGNTATAAT). This sequence is recognized by the σ_3 domain and is often found in promoters that lack a –35 sequence that is efficiently recognized by σ_4. The similarity of a promoter sequence to the consensus sequences for a particular σ, in combination with other elements that interact with other domains of RNA polymerase, dictates the efficiency with which a promoter is recognized by holoenzyme containing that σ.

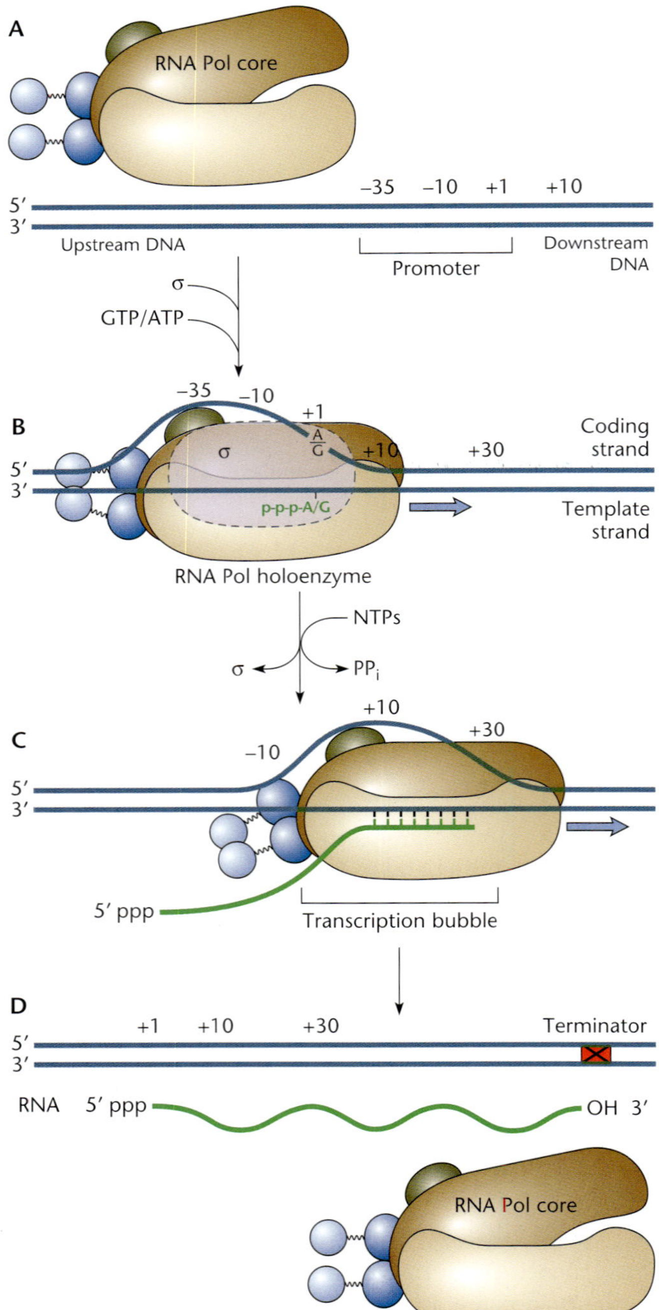

Figure 2.7 diagram labels:

A
RNA Pol core
5′
3′
−35 −10 +1 +10
Upstream DNA
Promoter
Downstream DNA
σ
GTP/ATP

B
−35 −10
+1
A
G
σ
+10
+30
Coding strand
Template strand
p-p-p-A/G
RNA Pol holoenzyme
NTPs
σ
PP_i

C
+10
+30
−10
5′ ppp
Transcription bubble

D
+1 +10 +30 Terminator
RNA 5′ ppp
OH 3′
RNA Pol core

Figure 2.7 Transcription begins at a promoter and ends at a transcription terminator. **(A)** The RNA polymerase core must bind σ factor to recognize a promoter. **(B)** Transcription begins when the strands of DNA are opened at the promoter, and the first ribonucleoside triphosphate (rNTP), usually ATP or GTP, enters the active site opposite the +1 nucleotide in the template strand. **(C)** As RNA polymerase moves along the DNA, polymerizing ribonucleotides into RNA (green), it forms a transcription bubble containing an RNA-DNA double-stranded hybrid, which helps to hold the RNA polymerase on the DNA. The sigma factor is released after RNA polymerase leaves the promoter, and transcription by the RNA polymerase core enzyme continues. **(D)** The RNA polymerase stops transcription, comes off the DNA, and releases the newly synthesized RNA at a transcription terminator.

ISOMERIZATION

When RNA polymerase holoenzyme first binds to the promoter, the DNA is double stranded. The resulting complex is called the RP_c, because the DNA strands are still "closed." In the next step, the β′ pincer of the crab claw closes around the DNA to form the active-site channel around the template strand of the DNA. This allows the σ_2 region to separate the strands of DNA at the −10 region and bind to the nontemplate strand in a process called **isomerization** (Figures 2.8 and 2.9). Recall that AT base pairs are less stable than GC pairs, so the AT-rich −10 sequence is relatively easy to melt. The complex is now called the open complex (RP_o), since the strands of DNA at the −10 region of the promoter are "open." The +1 nucleotide of the template strand is held in the active-site channel, where the polymerization reaction is about to occur.

INITIATION

In the initiation process, a single nucleoside triphosphate (usually an ATP or guanosine triphosphate [GTP]) enters through the secondary channel and pairs with nucleotide +1 (usually a T or C, respectively) in the template strand in the active site of the enzyme. Then, a second nucleoside triphosphate enters, and if that nucleoside triphosphate can base pair with the +2 nucleotide of the template strand, a phosphodiester bond forms between its α phosphate and the 3′ hydroxyl of the first nucleotide, releasing two phosphates in the form of pyrophosphate (Figure 2.5A). This is called the **initiation complex** or **initial transcription complex** and is the step at which the antibiotic rifampin can block transcription (see Box 2.1). Rifampin binds to RNA polymerase in the β-subunit face of the active-site channel in such a way that the growing RNA encounters it when it reaches a length of only 2 or 3 nucleotides, preventing further growth of the RNA chain and freezing the RNA polymerase on the promoter. This explains why rifampin blocks only the initiation of transcription.

Even in the absence of the antibiotic rifampin, the RNA polymerase is not yet free to continue transcription. When the RNA chain grows to a length of about 10 nucleotides,

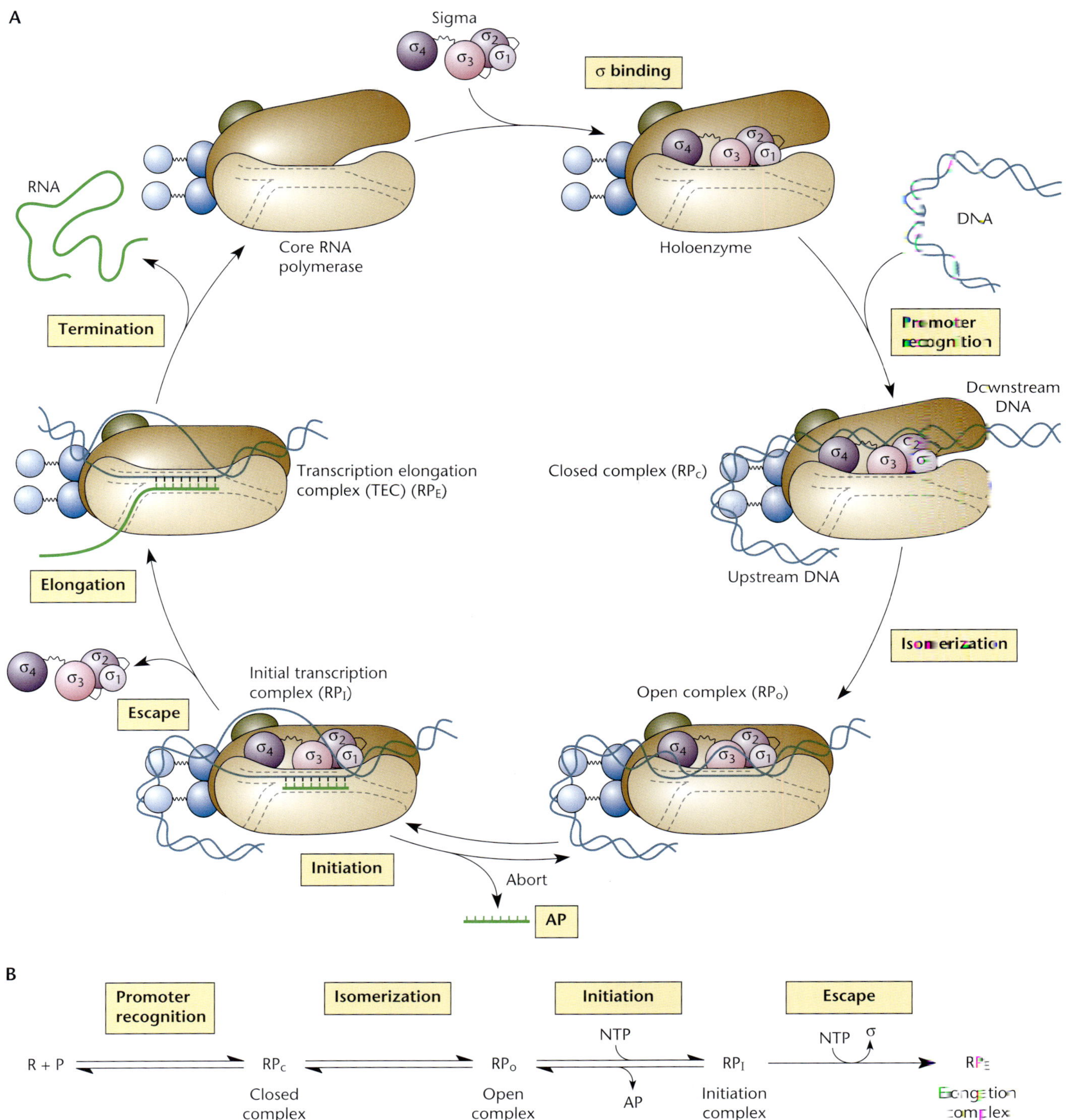

Figure 2.8 Overview of transcription. **(A)** The transcription cycle. Each step is discussed separately in the text. **(B)** Summary of the major steps in transcription initiation. R, RNA polymerase; P, promoter DNA; AP, abortive products; dsDNA, double-stranded DNA. Modified from Geszvain K, Landick R, *in* Higgins NP (ed), *The Bacterial Chromosome* (ASM Press, Washington, DC, 2005).

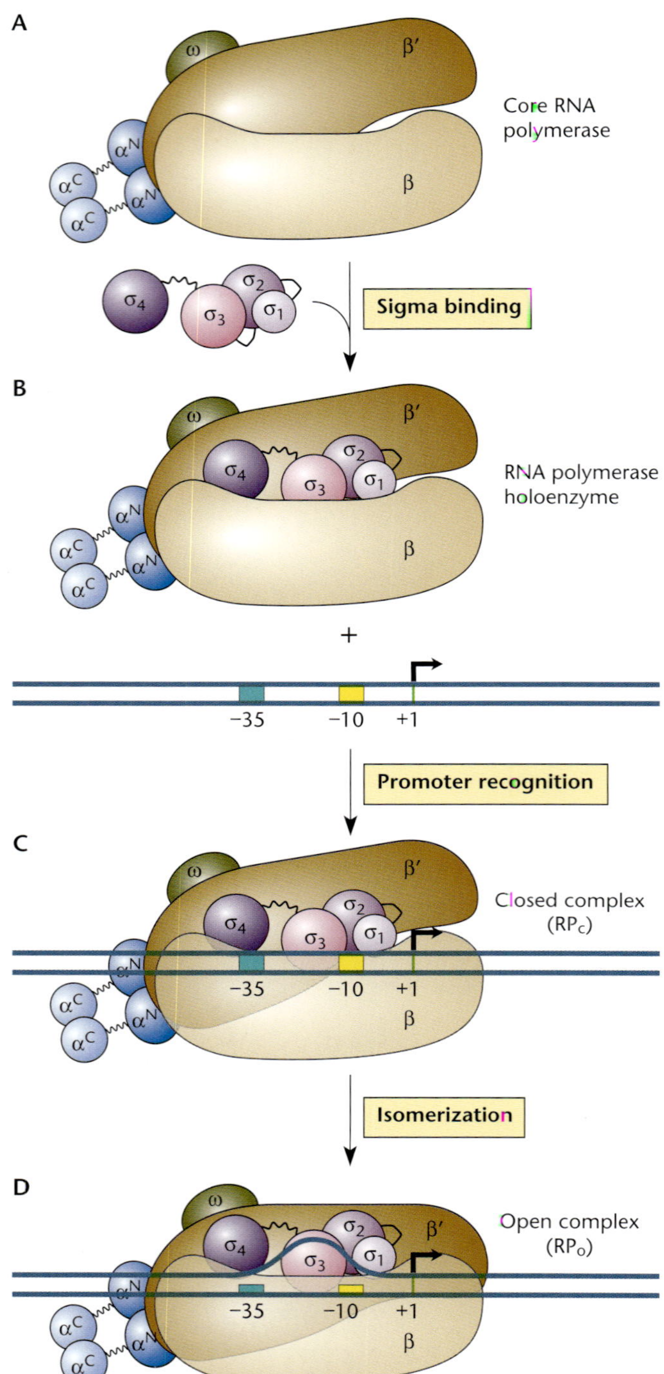

Figure 2.9 Transcription initiation. **(A)** Binding of σ to RNA polymerase core. **(B)** RNA polymerase holoenzyme binds to promoter DNA. **(C)** The initial RNA polymerase-promoter complex contains fully double-stranded DNA and is called the closed complex (RP_C). **(D)** The RNA polymerase-promoter complex isomerizes to form the open complex (RP_O).

it encounters the $\sigma_{3.2}$ loop, which blocks the site in RNA polymerase through which the newly synthesized RNA will emerge, a region called the exit channel (Figure 2.11). This causes transcription to stop, often releasing a short transcript about 10 nucleotides in length. This process is called **abortive initiation** and occurs to various degrees on many promoters for reasons that are not well understood. Eventually, a growing (or nascent) transcript pushes the $\sigma_{3.2}$ loop aside and enters the exit channel, usually causing the σ factor to be released from the core RNA polymerase. At this point, designated **promoter escape**, RNA polymerase has left the promoter site and has entered the elongation phase, during which the transcription bubble is enlarged to 17 bp, the complex is stabilized, and synthesis of RNA proceeds efficiently as the enzyme moves along the DNA template.

ELONGATION

Figure 2.12 shows the **transcription elongation complex** (TEC) in the process of elongating the RNA transcript. Most of the features are mentioned above, including the approximately 17-bp transcription bubble where the two strands of DNA are separated, the approximately 8- to 9-bp RNA-DNA hybrid that forms in the active site is maintained, and the newly synthesized RNA strand separates from the DNA template strand and emerges through the RNA exit channel. The RNA polymerase is capable of synthesizing RNA at a rate of 30 to 100 nucleotides per second. However, it sometimes pauses and even slides backward (**backtracks**). This phenomenon often occurs when a helical domain, or hairpin, forms in the RNA as it exits the RNA exit channel, when the newly synthesized RNA contains inverted-repeated sequences. It is not clear why hairpins cause pausing and backtracking, but they may pull the RNA polymerase backward or bind to it and change its conformation. Backtracking creates special problems for the TEC. When the RNA polymerase is forced backward, it pushes the 3′ end of the newly synthesized RNA forward, driving it into the secondary channel through which the nucleotides enter, as shown in Figure 2.13. It would remain this way, permanently blocked, except for the action of two proteins called GreA and GreB. These proteins insert their N termini into the secondary channel and cleave the 3′ end of the RNA in the channel until it is in its proper place in the active site so that transcription can continue.

RNA polymerase pausing and backtracking reduce the rate of transcription overall and create the necessity for the Gre proteins. Selective pausing may help the folding of the newly synthesized RNA or the protein being translated from the RNA (see below). Some genes whose RNA products must be made in large amounts, such as the rRNA genes, have special mechanisms to reduce pausing and backtracking. The rRNA genes have sequences called antitermination sites that recruit protein factors that bind to the RNA polymerase and preempt the pausing effect

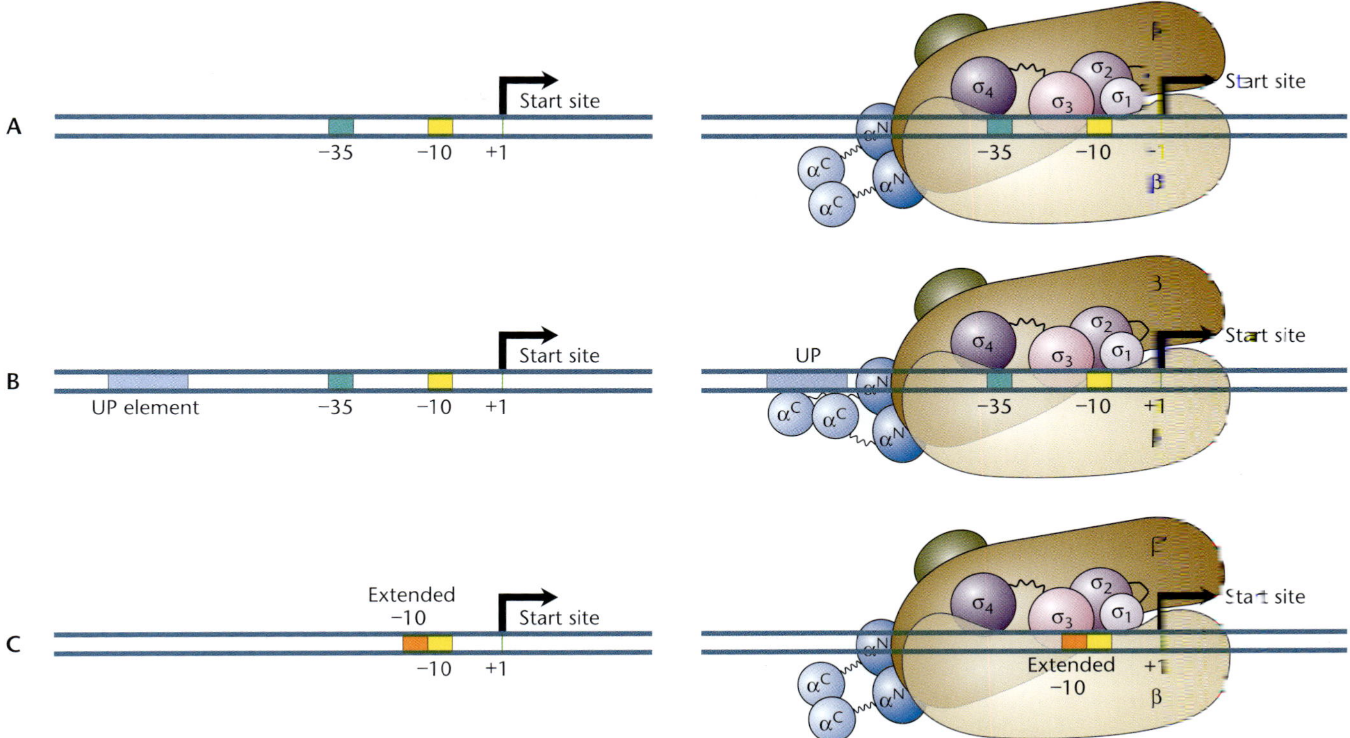

Figure 2.10 Interactions between RNA polymerase subunits and promoter elements. **(A)** Interaction of σ^{70} at −10 and −35 regions. **(B)** Flexible linkers between the α subunit carboxyl-terminal domain (αCTD) and the amino-terminal domain of α (αNTD) domains allow binding of αCTDs to UP elements in the DNA. **(C)** σ_3 binding to an extended −10 region. Modified from Dove SL, Hochschild A, *in* Higgins NP (ed), *The Bacterial Chromosome* (ASM Press, Washington, DC, 2005).

of RNA hairpins that form in the emerging rRNA and avoid premature termination of transcription (see below). Pausing also plays an important role in a variety of mechanisms for gene regulation (see chapter 11).

TERMINATION OF TRANSCRIPTION

Once the RNA polymerase has initiated transcription at a promoter, it continues along the DNA, polymerizing ribonucleotides into a growing RNA chain, until it encounters a **transcription termination** signal. These termination sites are not necessarily at the ends of each individual gene. In bacteria, more than one gene is often transcribed into a single RNA, so a transcription termination site does not occur until the end of the cluster of genes that are transcribed together. Even if only a single gene is being transcribed, the transcription termination site may occur far downstream of the protein-coding region of the gene.

Bacterial RNA polymerase responds to two basic types of transcription termination signals, designated **factor-independent** (or **intrinsic**) and **factor-dependent** terminators (see Washburn and Gottesman, Suggested Reading). As their names imply, these types are distinguished by whether they work with just RNA polymerase and DNA alone or need other factors before they can terminate transcription. Both types of termination signal require participation of the newly transcribed RNA to promote termination, which means that RNA polymerase must transcribe the terminator region before termination can occur.

Factor-Independent Termination

A typical factor-independent (or intrinsic) transcription terminator, shown in Figure 2.14, consists of two parts. The first is an inverted repeat. When an inverted-repeat DNA sequence is transcribed into RNA, the RNA can form a hairpin because the two parts of the repeat are complementary to each other. The inverted repeat is followed by a short string of A's in the template strand, which results in synthesis of a series of U's in the RNA. Transcription usually terminates somewhere in the string of A's in the DNA, leaving a string of U's at the 3′ end of the RNA terminated using this mechanism

Although the details of the process remain under study, the different elements of the terminator work together to promote termination. The transcription of the U-rich RNA from the A-rich template causes the RNA polymerase to pause, which allows time for the GC-rich hairpin to form in the emerging RNA transcript. The hairpin

BOX 2.1

Antibiotic Inhibitors of Transcription

Some of the components of the transcription apparatus are the targets of antibiotics used in the treatment of bacterial infections and in tumor therapy. Some of these antibiotics are made by soil bacteria and fungi, and some have been synthesized chemically.

Inhibitors of rNTP Synthesis

Some antibiotics that inhibit transcription do so by inhibiting the synthesis of the rNTPs. An example is azaserine, which inhibits purine biosynthesis. Azaserine and other antibiotics that block the synthesis of the ribonucleotides are usually not specific to transcription, since the ribonucleotides, including ATP and GTP, have many other uses in the cell. This lack of specificity limits the usefulness of these antibiotics for studying transcription, although some of them have other uses.

Inhibitors of Transcription Initiation

Rifamycin and its more commonly used derivative, rifampin, block transcription by binding to the β subunit of RNA polymerase and specifically blocking the initiation of RNA synthesis. The antibiotic binds in the active-site channel of RNA polymerase and limits growth of the RNA chain to a few nucleotides. The property of blocking only transcription initiation has made these antibiotics very useful in the study of transcription. For example, they have been used to analyze the steps in initiation of RNA synthesis and to study the stability of RNA in the cell. These antibiotics are useful therapeutic agents in the treatment of tuberculosis and other difficult-to-treat bacterial infections because they inhibit the RNA polymerases of essentially all types of bacteria, but not the RNA polymerases of eukaryotes, so they are not toxic to humans and animals. Accordingly, many derivatives have been made from them.

In rifampin-resistant mutants, one or more amino acids in the β subunit of RNA polymerase lining the active-site channel have been changed so that rifampin can no longer bind but the RNA polymerase still functions. Chromosomal mutations that confer resistance to rifampin are fairly common and have limited the usefulness of these antibiotics somewhat.

Inhibitors of RNA Elongation and Termination

Streptolydigin also binds to the β subunit of the RNA polymerase of bacteria but can block RNA synthesis after it is under way. It has a weaker affinity for RNA polymerase than does rifampin, so it blocks transcription only when added at higher concentrations, which limits its usefulness. Bicyclomycin targets the transcription terminator protein ρ and prevents ρ-dependent transcription termination (but not factor-independent transcription termination). Use of bicyclomycin as an antibiotic depends on the importance of ρ-dependent transcription termination in the target organism.

Inhibitors that Affect the DNA Template

Actinomycin D and bleomycin block transcription by binding to the DNA. After bleomycin binds, it also nicks the DNA. While such drugs have been useful for studying transcription in bacteria, they are not very useful in antibacterial therapy because they are not specific to bacteria and are very toxic to humans and animals. They are, however, used in antitumor therapy.

structure then causes the RNA polymerase to be released by an unknown mechanism, but it is likely to involve hairpin-induced conformational changes and destabilization of the DNA-RNA hybrid, which is facilitated by the fact that the AU base pairs that form in the DNA-RNA hybrid are less stable than a normal DNA-RNA that includes GC base pairs. RNA polymerase releases the RNA transcript and the DNA, terminating transcription, and the released transcript contains the hairpin structure at its 3′ end.

Factor-Dependent Termination

While factor-independent terminators are easily recognizable, the factor-dependent transcription terminators have very little sequence in common with each other and therefore are not readily apparent. The major termination factor in *E. coli* is called Rho (ρ). The ρ factor can be found in most types of bacteria, and this type of termination is likely to be widely conserved.

Any model for how the ρ factor terminates transcription at ρ-dependent termination sites has to incorporate the following facts. First, ρ usually causes the termination of RNA synthesis only if the RNA is not being translated. In bacteria, which lack a nuclear membrane, translation can begin on a nascent RNA before transcription is complete (see the introduction). Second, ρ is an RNA-dependent ATPase that cleaves ATP to get energy, but its ATPase activity is dependent on the presence of RNA. Finally, ρ is also an RNA-DNA helicase. It is similar to the DNA helicases that separate the strands of DNA during replication, but it unwinds only a double helix with RNA in one strand and DNA in the other.

Figure 2.15 illustrates a current model for how ρ terminates transcription. The ρ protein forms a hexameric (six-member) ring made up of six subunits encoded by the *rho* gene. This ring binds to a sequence in the RNA called

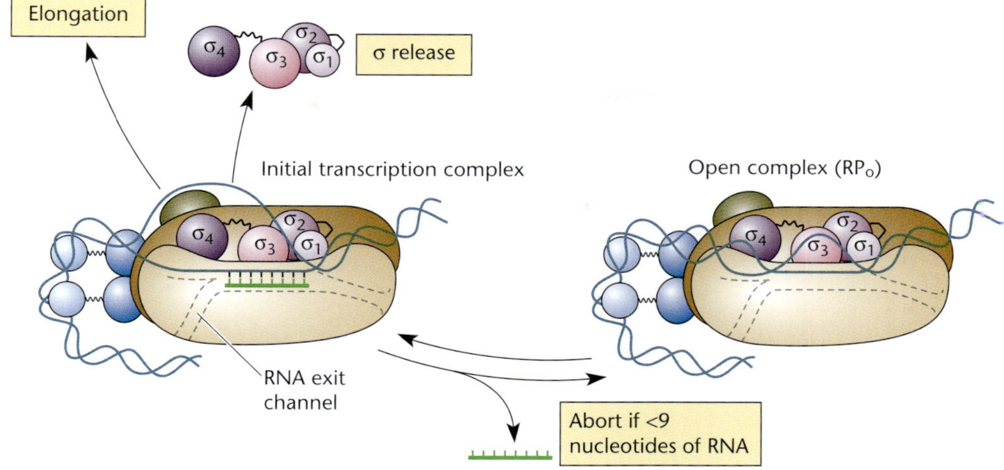

Figure 2.11 Abortive transcription and RNA polymerase escape from the promoter. RNA polymerase can escape from the promoter only if more than 10 or 11 nucleotides are polymerized. At 12 nucleotides, the RNA transcript displaces the $\sigma_{3.2}$ region, which blocks the active-site channel. Abortive transcription results when short (<9 nucleotides) newly synthesized transcripts are released, and the complex returns to the open complex (RP_o) state. With RNA polymerase escape, σ is released, and transcription elongation can continue.

the *rut* (for rho utilization) site. However, ρ can bind to a *rut* site in the RNA only if the RNA in this region is not occupied by ribosomes during translation of an mRNA; for example, ρ can bind if translation has terminated upstream at a termination codon (see below). The *rut* sites are not very distinctive but are about 40 nucleotides long and have many C's and have minimal secondary structure in the mRNA. Once ρ has bound to a *rut* site through the outside of its ring, the RNA then passes through the central hole, and the ring then moves along the RNA in the 5′-to-3′ direction, following the RNA polymerase. Energy for this movement is provided by the cleavage of ATP to adenosine diphosphate (ADP) by the ATPase activity of ρ. The ρ factor ring rotates down the RNA behind the RNA polymerase at a speed of about 60 nucleotides per second. However, RNA polymerase is capable of transcribing at 100 nucleotides per second, so ρ factor can catch up only if the RNA polymerase pauses at a ρ-dependent termination site. Then, the ρ factor can catch up to the RNA polymerase, and its RNA-DNA helicase activity disrupts the RNA-DNA hybrid in the transcription bubble, stopping transcription and releasing the RNA polymerase from the DNA template. While this model accounts for most of the known activities of ρ, it leaves unanswered the question of how ρ can access the RNA-DNA helix, which is inside the RNA polymerase. One possibility is that the RNA polymerase partially opens up when it is paused at a ρ-dependent termination site.

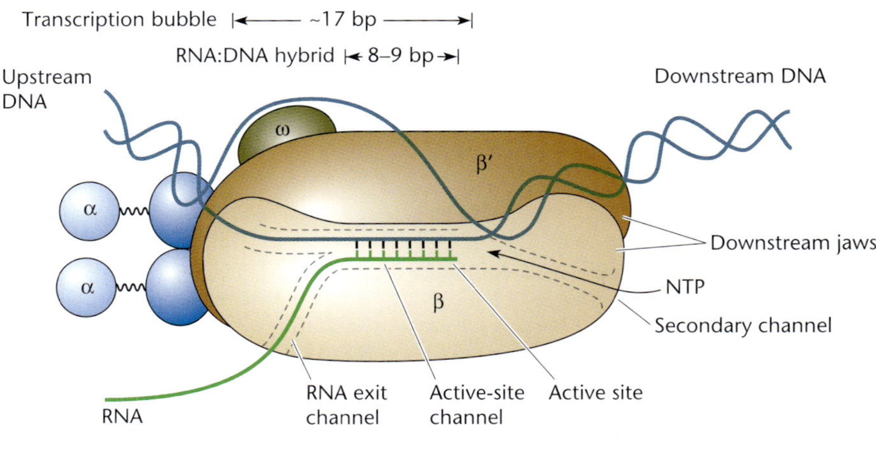

Figure 2.12 The transcription elongation complex (TEC). During elongation, nucleoside triphosphates (NTPs) enter through the secondary channel and are polymerized at the active site; the nascent RNA exits through the RNA exit channel. Modified from Geszvain K, Landick R, *in* Higgins NP (ed), *The Bacterial Chromosome* (ASM Press, Washington, DC, 2005).

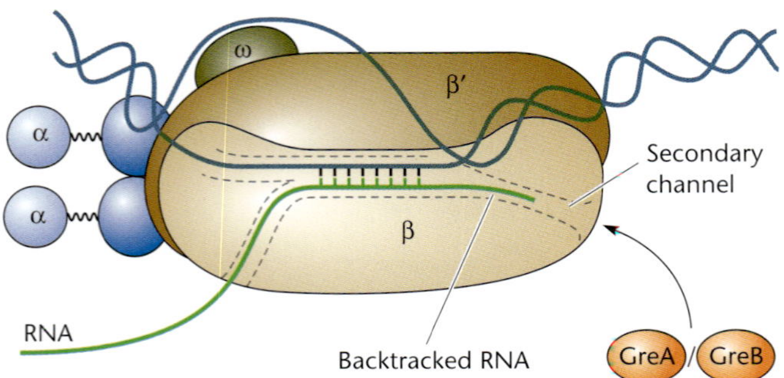

Figure 2.13 Backtracked transcription elongation complex (TEC). Backward movement of RNA polymerase results in placement of the 3′ end of the nascent RNA within the secondary channel, which prevents entry of nucleoside triphosphate (NTP) substrates. GreA and GreB can enter the secondary channel to cleave the nascent RNA, which repositions the 3′ end of the transcript in the active site, allowing transcription elongation to continue.

The coupling of transcription termination to translation blockage promotes termination of transcription when protein synthesis is interrupted. However, ρ-dependent termination is not very efficient, and transcription continues through a ρ-sensitive pause site as much as 50% of the time. ρ-dependent termination not only occurs at the ends of transcribed regions, but also accounts for ρ-dependent polarity (see "Polar Effects on Gene Expression" below).

rRNAs and tRNAs

Transcription of the genes for all RNAs in the cell follows the same basic process. However, rRNAs and tRNAs play special roles in protein synthesis, so their fates after transcription differ from that of mRNAs.

Figure 2.14 Transcription termination at a factor-independent termination site. **(A)** DNA sequence of a typical terminator site. **(B)** Sequence and structure of the RNA hairpin that forms in the nascent RNA as it emerges from RNA polymerase, which (in combination with the weak RNA-DNA hybrid) causes RNA polymerase to dissociate from the template DNA and release the RNA product.

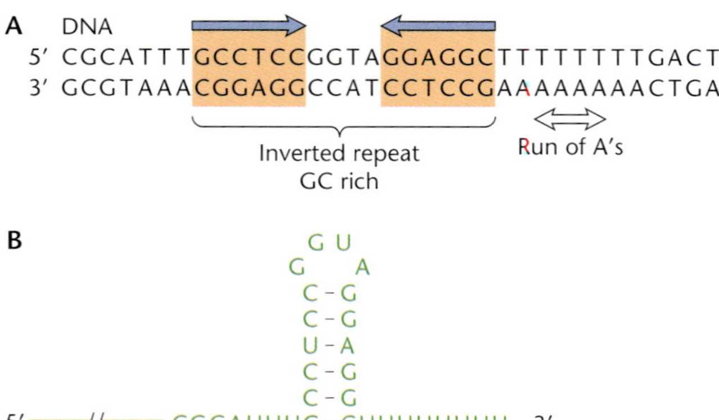

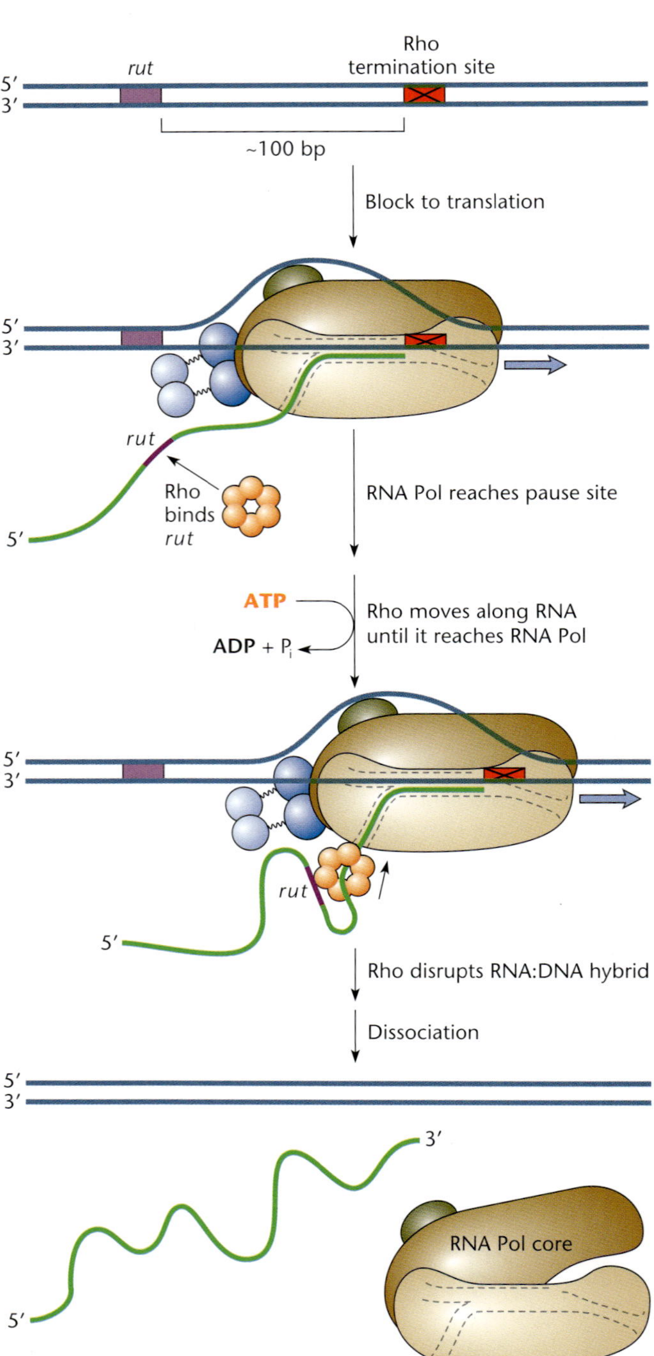

Figure 2.15 Model for factor-dependent transcription termination at a ρ-sensitive pause site. The ρ factor attaches to the RNA at a *rut* site if the RNA is not being translated (for example, if the ribosome has stopped at a termination codon) and forms a hexameric ring around the RNA. It then moves along the RNA with the cleavage of ATP until it catches up with RNA polymerase paused at a ρ-sensitive pause site. The helicase activity of the ρ factor then dissociates the RNA-DNA hybrid in the transcription bubble, causing the RNA polymerase and the RNA to be released.

The ribosomes are some of the largest structures in bacterial cells and are composed of both proteins and RNA. Bacterial ribosomes contain three types of rRNA: 16S, 23S, and 5S. The S value (from Svedberg, the name of the person who pioneered this way of measuring the sizes of molecules) is a measure of how fast a molecule sediments in an ultracentrifuge. In general, the higher the S value, the larger the RNA. The designation has persisted, even though this method of measuring molecular size is rarely used.

The rRNAs are among the most highly evolutionarily conserved of all the cellular constituents, as indeed are many of the components of the translational machinery. For this reason, they have formed the basis for molecular phylogeny (Box 2.2). Comparisons of the sequences of rRNAs and other constituents of the translation apparatus from different species permit estimates to be made of how long ago these constituents separated evolutionarily.

In addition to their structural role in the ribosome, the rRNAs play a direct role in translation. The 23S rRNA is the peptidyltransferase enzyme, which joins amino acids into protein on the ribosome. The 23S rRNA therefore acts as a **ribozyme**, an RNA enzyme (see below). The 16S RNA lacks enzymatic activity but plays crucial roles in initiation and termination of translation, as well as in decoding of the sequence of the mRNA.

The rRNAs and tRNAs make up the bulk of the RNA in cells because of their central role in protein synthesis. In a rapidly growing bacterial cell, much of the total RNA synthesis is devoted to making these RNAs. Also, the rRNAs and tRNAs are far more resistant to degradation than mRNA. With this combination of a high synthesis rate and high stability, the rRNAs and tRNAs together can amount to more than 95% of the total RNA in a rapidly growing bacterial cell.

Not only do the rRNAs physically associate in the ribosome, but they also are synthesized together as long precursor RNAs containing all three forms of rRNA separated by so-called spacer regions. The precursors often contain one or more tRNAs, as well (Figure 2.16), while other tRNAs are encoded in operons that do not include

BOX 2.2

Molecular Phylogeny

The translation apparatus is the most highly conserved of all the cellular components. The structures of ribosomes, translation factors, aaRS enzymes, tRNAs, and the genetic code itself have changed remarkably little in billions of years of evolution. This is why these components have been used extensively in molecular phylogeny. By comparing the sequences of the rRNAs and other components of the translation apparatus and determining how much they have diverged, it has been possible to establish phylogenetic trees that include all organisms on Earth. The high level of conservation probably also explains why so many different antibiotics target the translation apparatus compared to other cellular components. An antibiotic designed to inhibit translation in one type of bacteria will probably inhibit translation in many other types of bacteria.

The conservation of components of the translation apparatus is so high that "rooted" evolutionary trees can be made that include eukaryotes and archaea (see the introduction). Such trees are usually not too different from what has been obtained from physiological and other comparisons, but there are sometimes surprises. Also, the sequence of the translation elongation factors led to the suggestion that the archaea are more closely related to eukaryotes than they are to other bacteria, prompting the change of their name to archaea from the original designation "archaebacteria."

Many of the initiation and elongation factors in bacteria have counterparts in archaea and eukaryotes. Nevertheless, the major differences in the translation apparatus come in the translation initiation factors. While bacteria have only three initiation factors (some of them have more than one form), archaea and eukaryotes have many more. As is the case with other cellular functions, archaea share more of their initiation factors with eukaryotes than they do with bacteria. Also, some of the initiation factors, while conserved, seem to have somewhat different functions in the three kingdoms of life. These differences may reflect differences in the initiation sites for translation.

References

Ganoza MC, Kiel MC, Aoki H. 2002. Evolutionary conservation of reactions in translation. *Microbiol Mol Biol Rev* **66:**460–485.

Hug LA, Baker BJ, Anantharaman K, Brown CT, Probst AJ, Castelle CJ, Butterfield CN, Hernsdorf AW, Amano Y, Ise K, Suzuki Y, Dudek N, Relman DA, Finstad KM, Amundson R, Thomas BC, Banfield JF. 2016. A new view of the tree of life. *Nat Microbiol* **1:**16048.

Iwabe N, Kuma K, Hasegawa M, Osawa S, Miyata T. 1989. Evolutionary relationship of archaebacteria, eubacteria, and eukaryotes inferred from phylogenetic trees of duplicated genes. *Proc Natl Acad Sci USA* **86:**9355–9359.

Owen RJ. 2004. Bacterial taxonomics: finding the wood through the phylogenetic trees. *Methods Mol Biol* **266:**353–383.

Pace NR. 2009. Mapping the tree of life: progress and prospects. *Microbiol Mol Biol Rev* **73:**565–576.

Woese CR. 1987. Bacterial evolution. *Microbiol Rev* **51:**221–271.

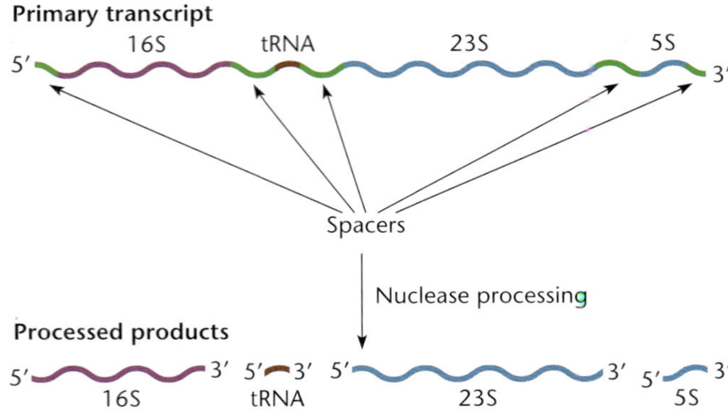

Figure 2.16 Precursor of rRNA. The precursor transcript (top) contains the 16S, 23S, and 5S rRNAs, as well as one or more tRNAs. RNases cut the individual rRNAs and tRNAs out of the precursor after it is synthesized.

rRNA sequences. The individual rRNAs and tRNAs are released from the precursor RNAs by **ribonucleases (RNases).** Some of these RNases participate in both rRNA and tRNA processing and RNA degradation, while others are dedicated to a single function; for example, RNase P generates the precise 5′ end of tRNAs. At some point during the processing, individual nucleotides within the rRNA and tRNA precursor molecules are also modified to make the mature rRNAs and tRNAs.

The faster a cell grows, the more protein it needs to make. Ribosomes are the site of protein synthesis; therefore, cells can increase their growth rate only if they increase the number of ribosomes. In most bacteria, the coding sequences for the rRNAs are repeated in several copies in the genome. Duplication of these genes leads to higher rates of rRNA synthesis in these bacteria. Although the precursor RNAs encoded by these different copies produce identical rRNAs, the rRNA gene clusters often contain different tRNAs and spacer regions.

MODIFICATION OF RNA

Some of the RNAs in cells are modified after they are made. For example, specific nucleotides in the rRNAs are methylated. The tRNAs are probably the most highly processed and modified RNAs in cells (see Björk and Hagervall, Suggested Reading). Figure 2.17 shows a "mature" tRNA that was originally cut out of a much longer molecule that may also have included the rRNAs as well as other tRNAs. Some of the bases within the tRNA are modified by specific enzymes that produce altered bases, such as pseudouracil and dihydrouracil. A CCA sequence is found on the 3′ end of all tRNAs. In some cases this is encoded in the gene, and therefore is part of the precursor RNA, while in others it is added posttranscriptionally by an enzyme called CCA nucleotidyltransferase.

Figure 2.17 Structure of mature tRNAs. **(A)** Standard clover leaf representation of tRNA, showing the base pairing that holds the molecule together and some of the standard modifications. D is the modified base dihydrouridine. tRNAs also contain thymine (T) and pseudouracil (Ψ) among other modifications. The amino acid is attached to the terminal A of the CCA at the 3′ end. **(B)** Folding of tRNA into its tertiary structure. The discriminator base (black dot), immediately upstream of the CCA, is important for tRNA recognition by the correct aminoacyl-tRNA synthetase (aaRS).

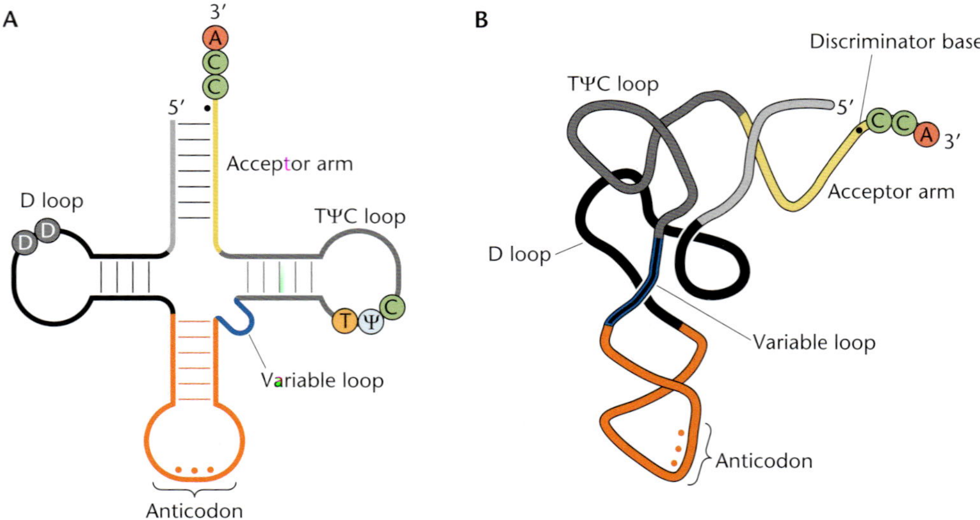

RNA Degradation

Different classes of RNAs have very different survival times in the cell. Highly structured RNAs, like rRNAs and tRNAs, are very stable and may persist through several rounds of cell division. Individual rRNAs are stabilized when they are assembled into ribosomal particles (see below), while tRNAs are stabilized because they are generally present in a complex with their cognate aaRS or with translation **elongation factor Tu (EF-Tu)** (see below). In contrast, most bacterial mRNAs are very unstable, with an average **half-life** in *E. coli* of 1 to 3 minutes; this term refers to the time required for the amount of an RNA to decrease to 50% of its initial level. The short half-life of mRNAs in bacteria contrasts with the situation in eukaryotes, where mRNAs are often very stable (with a half-life of hours). Efficient mRNA degradation is important for gene regulation and also releases nucleotides for use in new rounds of transcription. A variety of RNases participate in mRNA degradation, and the profiles of RNases vary somewhat in different groups of bacteria.

RNases

There are two major classes of RNases (Table 2.1 and Figure 2.18). **Endoribonucleases** cleave the sugar-phosphate backbone of the RNA within the RNA chain to generate two smaller RNA products, one with a 3′ hydroxyl and the other with a 5′ monophosphate. **Exoribonucleases** digest the RNA processively, removing 1 nucleotide at a time, starting at a free end. Most organisms have multiple endo- and exoribonucleases, and some subtypes of these enzymes are found in many different organisms, while others are present in only certain groups of organisms. In *E. coli*, all exoribonucleases that have been identified bind to the 3′ end of an RNA substrate and digest the RNA in a 3′-to-5′ direction. In contrast, *Bacillus subtilis* and a number of other organisms contain both 3′-5′ exoribonucleases and 5′-3′ exoribonucleases. At least one of the *B. subtilis* 5′-3′ RNases (RNase J2) has both exoribonucleolytic and endoribonucleolytic activities (Table 2.1).

MODULATION OF RNase ACTIVITY

The susceptibility of an RNA to different RNases can be affected by structural features of the RNA. RNA 3′ ends generated by termination of transcription at a factor-independent terminator contain an RNA hairpin, which inhibits binding of 3′-5′ exoribonucleases (Figure 2.18). Degradation of RNAs of this type is often initiated by endonucleolytic cleavage, which removes the 3′ end of the RNA and allows the 5′ region of the molecule to be degraded. Degradation of the 3′ fragment can be initiated by polyadenylation of the 3′ end of the RNA by polyadenylate [poly(A)] polymerase, encoded by the *pcnB* gene. Addition of the poly(A) tail provides a "landing zone" for 3′-5′ exoribonucleases, which can initiate degradation of the poly(A) sequence and then continue to move through the terminator hairpin. This may be facilitated by colocalization of poly(A) polymerase and polynucleotide phosphorylase (PNPase; Table 2.1), one of the major 3′-5′ exonucleases, with other RNases into a complex called the degradosome. Note that polyadenylation of an mRNA in eukaryotes generally results in stabilization of the mRNA, while polyadenylation of an RNA in bacteria results in rapid degradation. Degradation of the 3′ fragment generated by endonucleolytic cleavage can also be directed by 5′-3′ exoribonucleases in organisms like *B. subtilis* that have this activity (see Condon, Suggested Reading). It is interesting to note that the 5′ ends of transcripts newly synthesized by RNA polymerase contain a triphosphate (from the initiating nucleotide), whereas the 5′ ends of RNAs generated by endonuclease cleavage contain monophosphates. The presence of a triphosphate protects the RNA, and this triphosphate can be removed by a dedicated enzyme, designated RppH, which enhances susceptibility to degradation by 5′-3′ exonucleases (see Hui et al., Suggested Reading).

Susceptibility to degradation can be used as a mechanism to regulate gene expression, because rapid degradation of an mRNA results in reduced synthesis of its protein product. Modulation of RNA stability can occur through changes in the RNA structure that affect RNase

Table 2.1 Enzymes involved in mRNA processing and degradation

Enzyme	Substrate(s)	Description
RNase E	mRNA, rRNA, tRNA	Endonuclease, highly conserved in all *Proteobacteria* and some *Firmicutes* (not *B. subtilis*)
RNase III	rRNA, polycistronic mRNA	Endonuclease, cleaves double-stranded RNA in some stem-loops; found in most bacteria
RNase P	Polycistronic mRNA, tRNA precursors	Ribozyme, necessary to process 5′ ends of tRNAs
RNase G	5′ end of 16S rRNA, mRNA	Endonuclease, replaces RNase E in some bacteria
RNases J1 and J2	mRNA, rRNA	5′-3′ exonuclease, endonuclease; found in most *Firmicutes* and some *Proteobacteria* bacteria (not *E. coli*)
Poly(A) polymerase	mRNA	Found in most bacteria
PNPase	mRNA, poly(A) tails	3′-5′ exonuclease, found in all bacteria

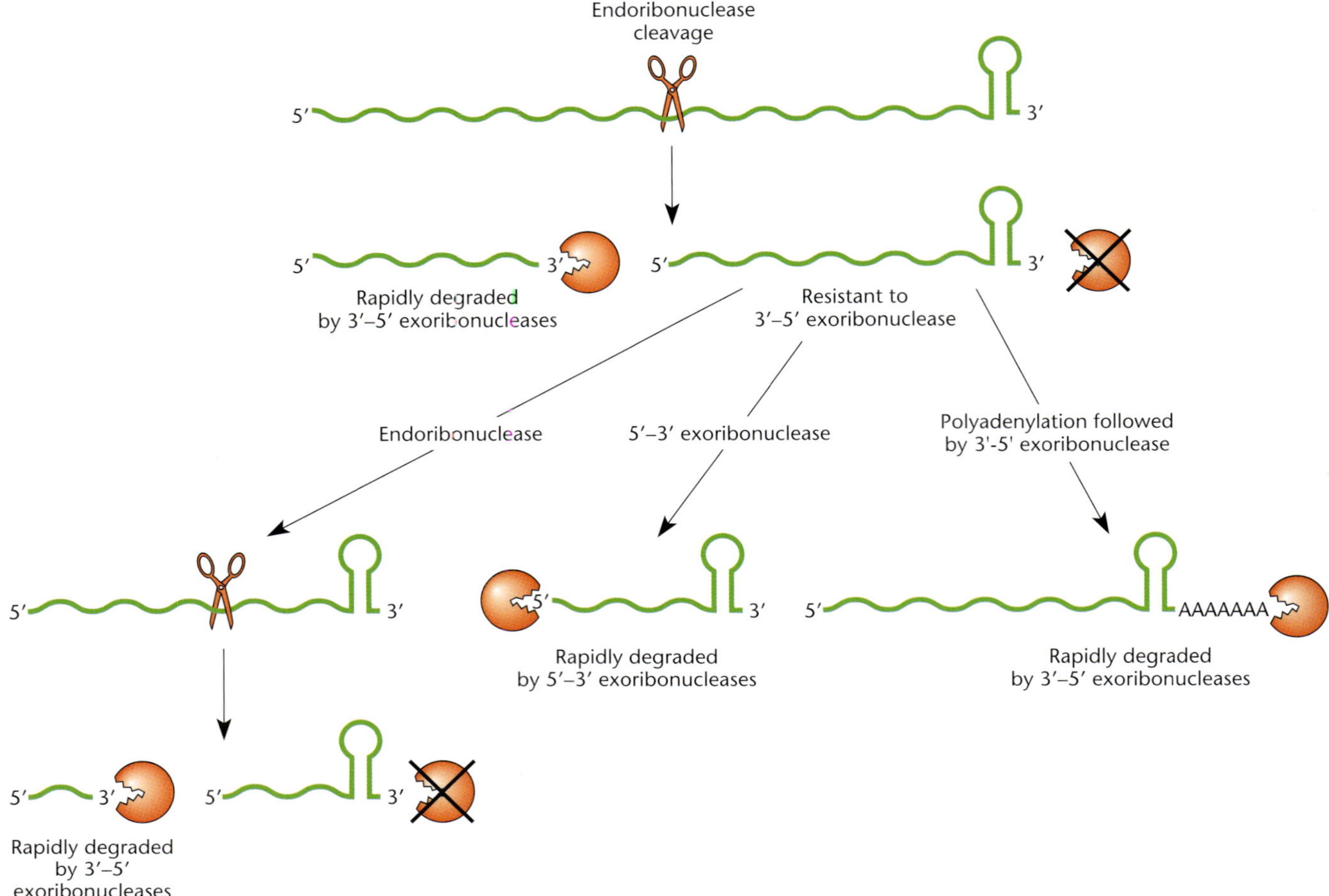

Figure 2.18 Pathways for RNA degradation. RNA transcripts that are generated by termination at a factor-independent terminator contain a hairpin at the 3′ end, which inhibits degradation by 3′-5′ exoribonucleases. Degradation is often initiated by cleavage by an endonuclease, followed by rapid exonucleolytic digestion from the new 3′ end. The stable 3′ fragment (which retains the terminator hairpin) can be cleaved again by an endoribonuclease or can be degraded by a 5′-3′ exoribonuclease in organisms that have this class of enzyme. Alternatively, poly(A) polymerase can add a poly(A) tail to the 3′ end of the RNA, which allows binding of a 3′-5′ exoribonuclease and degradation.

binding by binding of a regulatory protein to the RNA or by binding of a regulatory RNA. Mechanisms of this type are discussed in chapter 11.

The Structure and Function of Proteins

Proteins do most of the work of the cell. While there are a few RNA enzymes (ribozymes), most of the enzymes that make and degrade energy sources and make cell constituents are proteins. Also, proteins contribute to much of the structure of the cell. Because of these diverse roles, there are many more types of proteins than there are types of other cell constituents. Even in a relatively simple bacterium, there are thousands of different types of proteins, and most of the DNA sequences in bacteria are dedicated to genes that encode proteins.

Protein Structure

Unlike DNA and RNA, which consist of a chain of nucleotides held together by phosphodiester bonds between the sugars and phosphates, proteins consist of chains of 20 different amino acids held together by **peptide bonds** (see Figure 2.19). The peptide bond is formed by joining the amino group (NH_2) of one amino acid to the carboxyl group (COOH) of the previous amino acid. These amino acids in turn are attached to other amino acids by the same type of bond, making a chain. A short chain of

H—N—C—C—OH (R on the α carbon) + H—N—C—C—OH (R on the α carbon)

H_2O ← Peptide bond formation

H—N—C—C—N—C—C—OH

Figure 2.19 Two amino acids joined by a peptide bond. The bond connects the amino group on the second amino acid to the carboxyl group on the preceding amino acid. R is the side group of the amino acid that differs in each type of amino acid.

amino acids is called an **oligopeptide**, and a long chain is called a **polypeptide**.

Like RNA and DNA, polypeptide chains have directionality and a way to distinguish the ends of the chain from each other. In polypeptides, the direction is defined by their amino and carboxyl groups. One end of the chain, the **amino terminus**, or **N terminus**, has an unattached amino group. The amino acid at this end is called the **N-terminal amino acid**. On the other end of the polypeptide, the final carboxyl group is called the **carboxy terminus**, or **C terminus**, and the amino acid is called the **C-terminal amino acid**. As we shall see, proteins are synthesized from the N terminus to the C terminus.

Protein structure terminology is the same as that for RNA structures. Proteins have primary, secondary, and tertiary structures, as well as quaternary structures. All of these are shown in Figure 2.20.

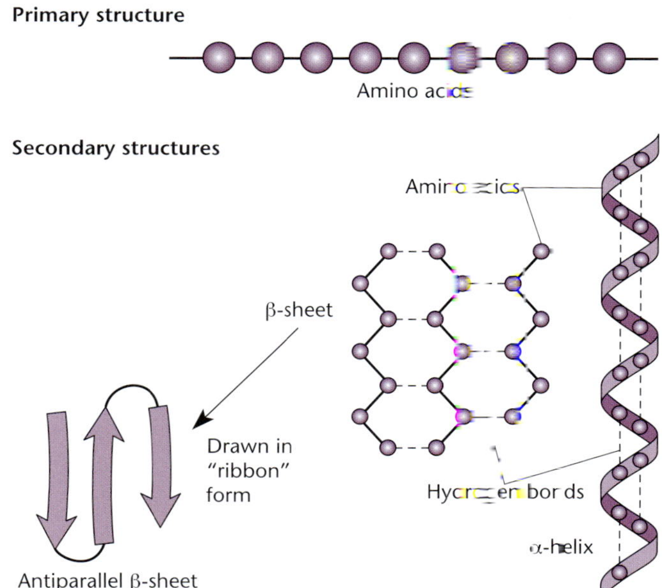

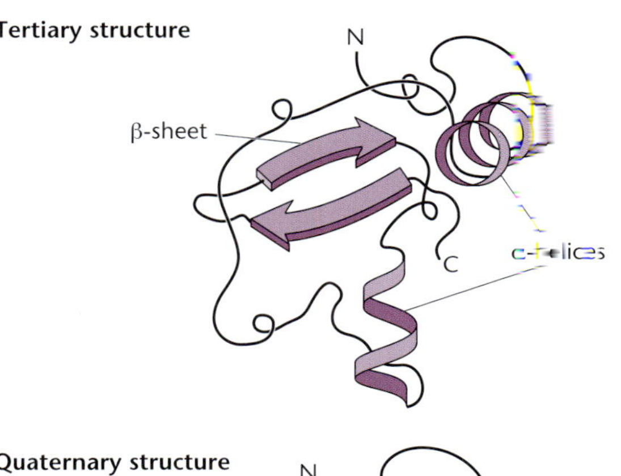

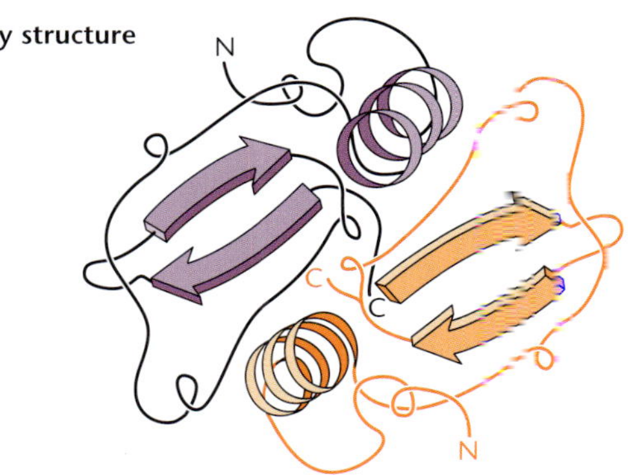

Figure 2.20 Primary, secondary, tertiary, and quaternary structures of proteins.

PRIMARY STRUCTURE

Primary structure refers to the sequence of amino acids and the length of a polypeptide. Because polypeptides are made up of 20 amino acids instead of just 4 nucleotides, as in RNA, many more primary structures are possible for polypeptides than for RNA chains. The sequence of amino acids in a polypeptide is dictated by the sequence of nucleotides in the mRNA used as the template for synthesis of that protein.

SECONDARY STRUCTURE

Also like RNA, polypeptides can have a secondary structure, in which parts of the chain are held together by hydrogen bonds. However, because many more types of interactions are possible between amino acids than between nucleotides, the secondary structure of a polypeptide is more difficult to predict. The two basic forms of secondary structures in polypeptides are α-helices, where a short region of the polypeptide chain forms a helix due to the interaction of each amino acid with the one before and the one after it, and β-sheets, in which stretches of amino acids interact with other stretches to form sheetlike structures (Figure 2.20). These types of structured regions are often joined together by more flexible regions known as linkers. Computer software is available to help predict which secondary structures of a polypeptide are possible on the basis of its primary structure. However, these programs are not entirely reliable, and techniques like X-ray crystallography and nuclear magnetic resonance spectroscopy provide much more detailed information about the secondary structure of a polypeptide.

TERTIARY STRUCTURE

Polypeptides usually also have a well-defined tertiary structure, in which they fold up on themselves with hydrophobic amino acids (such as leucine and isoleucine), which are not very soluble in water, on the inside and charged amino acids (such as glutamate and lysine), which are more water soluble, or hydrophilic, on the outside. We discuss the structure of proteins in more detail in "Protein Folding and Degradation" below.

QUATERNARY STRUCTURE

Proteins made up of more than one polypeptide chain also have **quaternary structure**. Such proteins are called **multimeric proteins**. When the polypeptides are the same, the protein is a **homomultimer**. When they are different, the protein is a **heteromultimer**. Other names reflect the number of polypeptides in the protein. For example, the term **homodimer** describes a protein made of two identical polypeptides, whereas **heterodimer** describes a protein made of two different polypeptides. The names trimer, tetramer, and so on refer to increasing numbers of polypeptides. Hence, the ρ transcription termination factor is a homohexamer (see above).

The polypeptide chains in a protein are usually held together by hydrogen bonds. The only covalent chemical bonds in most proteins are the peptide bonds that link adjacent amino acids to form the polypeptide chains. As a result, if a multimeric protein is heated, it falls apart into its individual polypeptide chains. However, some proteins are unusually stable; these include extracellular enzymes, which must be able to function in the harsh environment outside the cell. Such proteins are often also held together by **disulfide bonds** between cysteine amino acids in the protein.

Translation

The translation of the sequence of nucleotides in mRNA into the sequence of amino acids in a protein occurs on the ribosome. The overall process of translation is highly conserved in bacteria, archaea, and eukaryotes, and the machinery is also highly conserved. As mentioned in "rRNAs and tRNAs" above, the ribosome is one of the largest and most complicated structures in cells, consisting of three different RNAs and over 50 different proteins in bacteria. It is also one of the major constituents of the bacterial cell, and much of the biosynthetic capacity of the cell goes into making ribosomes. Each cell contains thousands of ribosomes, with the actual number depending on the growth conditions. It is also one of the most evolutionarily highly conserved structures in cells, having remained largely unchanged in shape and structure from bacteria to humans. The key role of protein synthesis in the cell has led to the development of important antibiotics that target the translational machinery (Box 2.3).

The ribosome is an enormous enzyme that performs the complicated role of polymerizing amino acids into polypeptide chains, using the information in mRNA as a guide. As such, a better name for it might have been protein polymerase, by analogy to DNA and RNA polymerases. The historical name "ribosome" was coined before its function was known, because it is large enough to have been visualized under the electron microscope and so it was called a "some" (for body) and "ribosome" because it contains ribonucleotides. The recent determination of the structure of the ribosome (see below) has led to important insights into how it performs its function of polymerizing amino acids.

Structure of the Bacterial Ribosome

Figure 2.21 shows the components of a bacterial ribosome. The complete ribosome, called the 70S ribosome in bacteria, consists of two subunits, the 30S subunit, which contains one molecule of 16S rRNA, and the 50S subunit, which contains one molecule each of 23S and 5S rRNA. Each subunit also contains **ribosomal proteins**; the 30S subunit contains approximately 21 different proteins (S1, S2, etc., where S indicates small subunit proteins), while the 50S subunit contains approximately 31 different proteins (L1, L2, etc., where L indicates large subunit proteins). Ribosomes from different bacterial species have very similar compositions, with small differences in the number of ribosomal proteins. Like the names for the different types of rRNA, the names of ribosomal subunits are derived from their sedimentation rates during ultracentrifugation. The 30S and 50S subunits normally exist separately in the cell and come together to form the complete 70S ribosome only when they are translating an mRNA. Note that sedimentation values are not additive; the complete ribosome is 70S in size, despite being composed of subunits that are 30S and 50S in size. This is because the two subunits fit together into a tight complex.

The two ribosomal subunits play very different roles in translation. To initiate translation, the 30S subunit first binds to the mRNA. Then, the 50S ribosome binds to the 30S subunit to make the 70S ribosome. From that point on, the 30S subunit mostly helps to select the correct **aminoacyl-tRNA (aa-tRNA)** for each codon while the 50S

BOX 2.3

Antibiotic Inhibitors of Translation

The translation apparatus of bacteria is a particularly tempting target for antibacterial drugs because it is somewhat different from the eukaryotic translation apparatus and is highly conserved among bacteria. Antibiotics that inhibit translation are among the most useful of all the antibiotics and are listed in the table, which also lists their targets and sources. Toxicity in some cases may result from the similarity of bacterial and mitochondrial ribosomes. Some of these antibiotics are also very useful in combating fungal diseases and in cancer chemotherapy.

Antibiotics that block translation

Antibiotic	Source	Target
Puromycin	*Streptomyces alboniger*	Ribosomal A site
Kanamycin	*Streptomyces kanamyceticus*	16S rRNA
Neomycin	*Streptomyces fradiae*	16S rRNA
Streptomycin	*Streptomyces griseus*	30S ribosome
Thiostrepton	*Streptomyces azureus*	23S rRNA
Gentamicin	*Micromonospora purpurea*	16S rRNA
Tetracycline	*Streptomyces rimosus*	Ribosomal A site
Chloramphenicol	*Streptomyces venezuelae*	Peptidyltransferase
Erythromycin	*Saccharopolyspora erythraea*	23S rRNA
Fusidic acid	*Fusidium coccineum*	EF-G
Kirromycin	*Streptomyces collinus*	EF-Tu

Inhibitors that Mimic tRNA

Puromycin mimics the 3′ end of tRNA with an amino acid attached (aa-tRNA). It enters the ribosome as does an aa-tRNA, and the peptidyltransferase attaches it to the growing polypeptide. However, it does not translocate properly from the A site to the P site, and the peptide with puromycin attached to its carboxyl terminus is released from the ribosome, terminating translation.

Studies with puromycin have contributed greatly to our understanding of translation. The model of the A and P sites in the ribosome and the concept that the 50S ribosome contains the enzyme for peptidyl bond formation came from studies with the antibiotic. Puromycin is not a very useful antibiotic for treating bacterial diseases, however, because it also inhibits translation in eukaryotes, making it toxic in humans and animals. It is, however, one of the few antibiotics that is useful in archaeal genetics, with the availability of resistance cassettes.

Inhibitors that Bind to the 23S rRNA

Chloramphenicol

Chloramphenicol inhibits translation by binding to ribosomes and preventing the binding of aa-tRNA to the A site. It might also inhibit the peptidyltransferase reaction, preventing the formation of peptide bonds. Structural studies have shown that chloramphenicol binds to specific nucleotides in the 23S rRNA, although ribosomal proteins are also part of the binding site.

Chloramphenicol is effective at low concentrations and therefore has been one of the most useful antibiotics for studying cellular functions. For example, it has been used to determine the time in the cell cycle when proteins required for cell division and for initiation of chromosomal replication are synthesized. It is also quite useful in treating bacterial diseases, since it is not very toxic for humans and animals because it is fairly specific for the translation apparatus of bacteria. It can cross the blood-brain barrier, making it useful for treating diseases of the central nervous system, such as bacterial meningitis. Chloramphenicol is bacteriostatic, which means that it stops the growth of bacteria without actually killing them. Such antibiotics should not be used in combination with antibiotics such as penicillin that depend on cell growth for their killing activity, since they neutralize the effect of these other antibiotics.

It takes multiple mutations in ribosomal proteins to make bacteria resistant to chloramphenicol, so resistant mutants are very rare. Some bacteria have enzymes that inactivate chloramphenicol. The genes for these enzymes are often carried on plasmids and transposons, interchangeable DNA elements that are discussed in chapters 4 and 8. The best-characterized chloramphenicol resistance gene is the *cat* gene of transposon Tn*9*, whose product is an enzyme that specifically acetylates (adds an acetyl group to) chloramphenicol, thereby inactivating it. The *cat* gene has been used extensively as a reporter gene to study gene expression in both bacteria and eukaryotes and has been introduced into many plasmid cloning vectors.

Macrolides

Erythromycin is a member of a large group of antibiotics called the macrolide antibiotics, which have large ring structures. These antibiotics may also inhibit translation by binding to the 23S rRNA and blocking the exit channel of the growing polypeptide. This causes the polypeptide to be released prematurely at either the peptidyltransferase reaction or the translocation step, causing the peptidyl-tRNA to dissociate from the ribosome.

Erythromycin and other macrolide antibiotics have been among the most useful antibiotics. They are effective mostly against *Firmicutes* but are also useful in treating some other bacterial diseases, including infections by *Legionella* and *Rickettsia*. Resistance to macrolide antibiotics occurs in a number of ways. One way is methylation of a specific adenine base in the 23S rRNA, which prevents binding of the antibiotic, by enzymes called the Erm methylases. Some mutational changes in the 23S rRNA can also confer resistance to these antibiotics. Resistance can also occur by acquisition of genes that encode

(continued)

BOX 2.3 (continued)

Antibiotic Inhibitors of Translation

efflux pumps that remove the antibiotic from the cell. New derivatives of the antibiotics must be made constantly to stay ahead of the advancing bacterial resistance.

Thiostrepton

Thiostrepton and other thiopeptide antibiotics block translation by binding to 23S rRNA in the region of the ribosome involved in the peptidyltransferase reaction and preventing the binding of EF-G. Thiostrepton is specific to *Firmicutes* because it does not cross the outer membrane of Gram-negative bacterial cells.

Most thiostrepton-resistant mutants are missing the L11 ribosomal protein from the 50S ribosomal subunit. This protein seems not to be required for protein synthesis but plays a role in guanosine tetraphosphate synthesis (see chapter 12). Other mutations confer resistance by changing nucleotides in the 23S rRNA close to where the antibiotic binds. Genes derived from plasmids and transposons can confer resistance by directing specific methylation of 23S rRNA. Eukaryotes may be insensitive to the antibiotic because the analogous ribose sugars of the eukaryotic 28S rRNAs are normally extensively methylated.

Inhibitors of Binding of aa-tRNA to the A Site

Tetracycline was one of the first antibiotics identified. It may inhibit translation by allowing the aa-tRNA–EF-Tu complex to bind to the A site of the ribosome and allowing the GTP on EF-Tu to be cleaved to GDP, but then inhibiting the next step, causing a futile cycle of binding and release of the aa-tRNA from the A site.

Tetracycline has been a very useful antibiotic for treating bacterial diseases, although it is somewhat toxic to humans because it also inhibits the eukaryotic translation apparatus. Unfortunately, overuse has led to the spread of resistance, and it is no longer useful against many infections. In some types of bacteria, ribosomal mutations confer low levels of resistance to tetracycline by changing protein S10 of the ribosome. However, most clinically important resistance to tetracycline and its derivatives is acquired on plasmids and transposons. One of these genes, *tetM*, carried by Tn*916* and its relatives (see chapter 5), encodes an enzyme that confers resistance by methylating certain bases in the 16S rRNA. Other tetracycline resistance genes, such as the *tet* genes carried by transposon Tn*10* and plasmid pSC101 of *E. coli*, confer resistance by pumping tetracycline out of the cell. One of the more interesting types of resistance to tetracyclines is due to the so-called ribosome protection proteins, represented by TetO and TetQ, which bind to the A site of the ribosome and release tetracycline from its binding site. This is the type of resistance exhibited by the soil bacteria that make tetracycline.

Inhibitors of Translocation

Aminoglycosides

Kanamycin and its close relatives neomycin and gentamicin are members of a larger group of antibiotics, the aminoglycoside antibiotics, which also includes streptomycin. They seem to affect translocation by binding to the A site of the ribosome. Aminoglycosides have a very broad spectrum of action, and some of them inhibit translation in plant and animal cells, as well as in bacteria. However, their toxicity, especially during sustained use, and high rates of resistance somewhat limit their usefulness as therapeutic agents.

Bacterial mutants resistant to aminoglycosides arise primarily due to genes exchanged on transposons and plasmids. The products of some of these genes inactivate the aminoglycosides by phosphorylating, acetylating, or adenylating (adding adenosine to) them. For example, the *neo* gene for kanamycin and neomycin resistance, from transposon Tn*5*, phosphorylates these antibiotics.

Fusidic Acid

Fusidic acid specifically inhibits EF-G (called EF-2 in eukaryotes), probably by preventing its dissociation from the ribosome after GTP cleavage. It has been very useful in studies of the function of ribosomes. In *E. coli*, mutations that confer resistance to fusidic acid are in the *fusA* gene, which encodes EF-G. Unexpectedly, some acetyltransferases that confer resistance to chloramphenicol also bind to and inactivate fusidic acid.

subunit does most of the work of forming the peptide bonds and translocating the tRNAs from one site on the ribosome to another (see below). The 70S ribosome moves along the mRNA, allowing tRNA anticodons to pair with the mRNA codons and translate the information from the nucleic acid chain into a polypeptide. After the polypeptide chain is completed, the ribosome separates again into the 30S and 50S subunits. The role of the subunits is discussed in more detail below.

A variety of physical techniques, combined with much indirect information accumulated over the years from genetics and biochemistry, have revealed many details of the overall structure of the ribosome. The crystal structures of the individual subunits and the entire

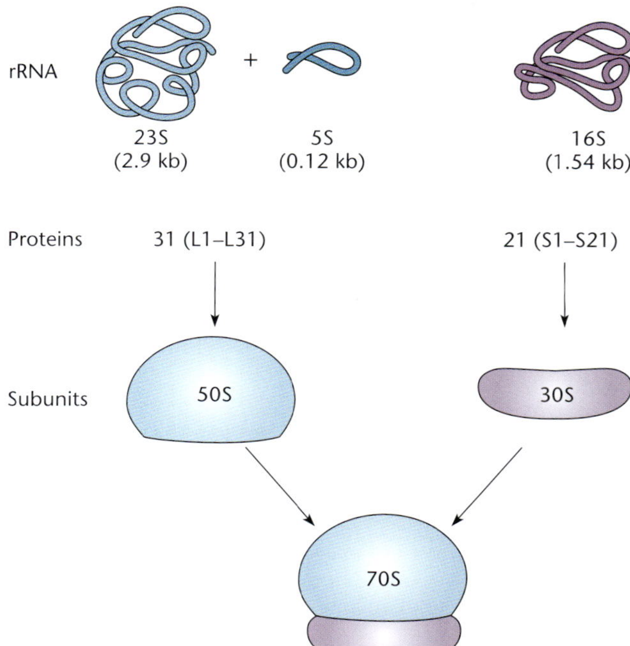

Figure 2.21 The composition of a bacterial ribosome containing one copy each of the 16S, 23S, and 5S rRNAs, as well as many proteins. The proteins of the large 50S subunit are designated L1 to L31. The proteins of the small 30S subunit are designated S1 to S21. The 30S and 50S subunits combine to form the 70S ribosome, which carries out protein synthesis.

70S ribosome have been determined and correlated with the earlier indirect information. Many laboratories participated in this project, and this awesome achievement will go down in history as one of the major milestones in molecular biology, recognized by the Nobel Prize. We can review only a few of the most salient features here.

The two subunits of the ribosome are frequently represented as ovals, with a flat side that binds to the other subunit, leaving a small gap between them. It is through this gap that the mRNA moves, and the aa-tRNAs enter, interact with the mRNA, and pass through the ribosome, contributing their amino acids to the growing polypeptide chain. The newly synthesized polypeptide chain exits through a channel running through the 50S subunit. This channel is long enough to hold a chain of about 70 amino acids, so a polypeptide of this length must be synthesized before the N-terminal end of a protein first emerges from the ribosome.

The rRNAs play many of the most important roles in the ribosome, and the ribosomal proteins seem to be present mostly to give rigidity to the structure, helping to hold the rRNAs in place. This has contributed to speculation that RNAs were the primordial enzymes and that proteins came along later in the earliest stages of life on Earth. The 23S rRNA, rather than a ribosomal protein, acts as the peptidyltransferase enzyme that performs the enzymatic function that forms the peptide bonds between the carboxyl end of the growing polypeptide and the amino group of the incoming amino acid. Thus, 23S rRNA is an RNA enzyme, or ribozyme. The 23S rRNA also forms most of the channel in the 50S subunit through which the growing polypeptide passes. The 16S RNA plays crucial roles in translation initiation and in matching each incoming aa-tRNA with the mRNA. A structure of the ribosome is shown in Figure 2.22.

Overview of Translation

The information in mRNA is interpreted by the pairing of consecutive triplets of nucleotides (codons) in the mRNA with the complementary anticodon sequences of the corresponding tRNAs. This pairing takes place on the ribosome, and accuracy of translation also depends on matching of a specific tRNA (with the appropriate anticodon) to the corresponding amino acid by a set of enzymes called **aminoacyl-tRNA synthetase (aaRSs)**. aa-tRNAs generated by the aaRS enzymes are delivered to the ribosome by a protein factor called elongation factor-Tu (EF-Tu). During translation, the ribosome moves along the mRNA, and tRNAs interact with the mRNA at three distinct sites, designated the A (aminoacyl) site, the P (peptidyl) site, and the E (exit) site (Figure 2.23). Each aa-tRNA is brought into the A site first, where its anticodon is tested for complementarity to the mRNA codon present at that site. If the anticodon is complementary to the codon, the tRNA is retained; if the anticodon does not match the codon, the tRNA is rejected, and a new aa-tRNA is brought into the A site. A match between the anticodon and the codon in the A site triggers the tRNA in the A site to interact with the tRNA already present in the P site. The P site tRNA carries the growing polypeptide chain, and the next step is transfer of the growing peptide chain from the P site tRNA to the A site tRNA. The amino acid carried on the A site tRNA is attached to the carboxyl end of the polypeptide, and the polypeptide (now 1 amino acid longer) is now linked to the A site tRNA. The now unattached P site tRNA moves to the E site of the ribosome and the A site tRNA (which now carries the polypeptide) moves to the P site of the ribosome, which results in an empty A site. Each tRNA retains contact with the mRNA through anticodon-codon pairing so that the mRNA moves through the ribosome in concert with the tRNAs. This results in placement of the next codon of the mRNA in the empty A site, which is now available for entry of another aa-tRNA. Movement of the unattached P site tRNA into the E site results in ejection of the previous unattached tRNA from the E site, which allows the next tRNA to enter the cycle. The details of this process are described below.

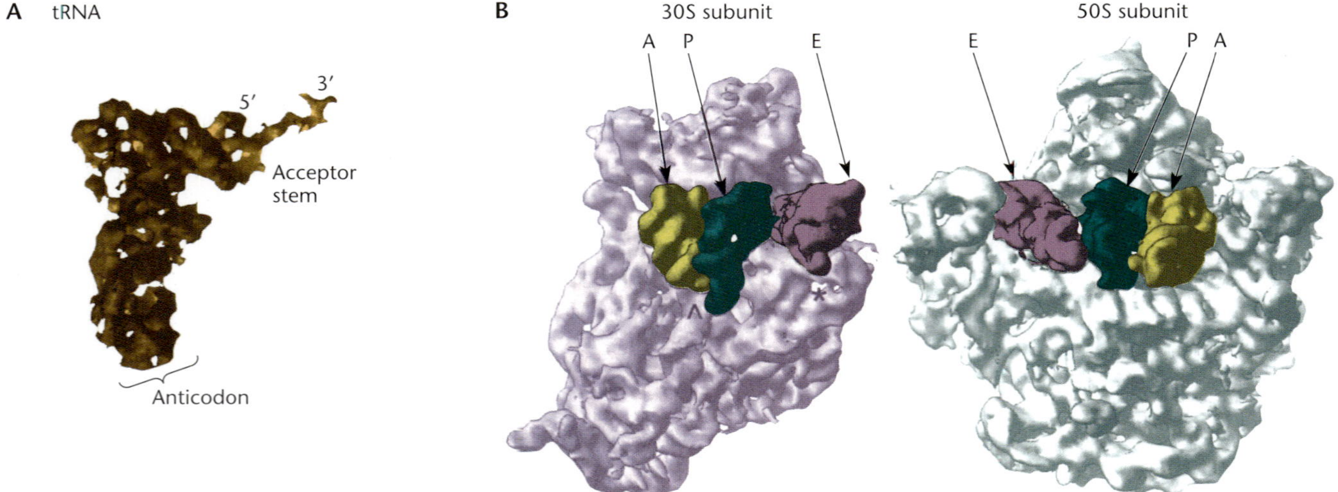

Figure 2.22 Crystal structures of a tRNA and the ribosome. **(A)** Structure of a tRNA. The anticodon loop is at the bottom, and the 3′ acceptor end, where the amino acid or growing polypeptide is attached, is at the top. (From Yusupov MM, Yusupova GZ, Baucom A, et al, *Science* **292**:883–896, 2001. Modified with permission from AAAS.) **(B)** The two subunits of the ribosome separated and rotated to show the channel between them through which the tRNAs move. The 30S subunit is on the left, and the 50S subunit is on the right. The tRNAs bound at the A, P, and E sites are indicated in yellow, green, and purple, respectively. (From Cate JH, Yusupov MM, Yusupova G, et al, *Science* **285**:2095–2014, 1999. Modified with permission from AAAS.)

Details of Protein Synthesis

In this section, we discuss the process of translation in more detail. First, we discuss reading frames, which determine which nucleotides in an mRNA are recognized as codons, and tRNA aminoacylation, which is responsible for correct matching of a tRNA with its cognate amino acid. Then, we discuss translation initiation, or how the 30S subunit finds the starting point of a coding sequence in the mRNA. Next, we discuss translation elongation, or what happens as the 70S ribosome moves along the mRNA, translating the information in the mRNA in the form of nucleotides into amino acids. Finally, we discuss how translation is terminated.

READING FRAMES

Each 3-nucleotide sequence, or codon, in the mRNA encodes a specific amino acid, and the assignment of the codons is known as the genetic code (Table 2.2). Because there are 3 nucleotides in each codon, an mRNA can be translated in three different frames in each region. Initiation of translation at a specific **initiation codon** establishes the **reading frame of translation**, so that in most cases only a single reading frame is utilized. Once translation has begun, the ribosome moves 3 nucleotides at a time through the coding part of the mRNA. If the translation is occurring in the proper frame for protein synthesis, we say the translation is in the **zero frame** for that protein. If translation is occurring in the wrong reading frame, it can be displaced either back by 1 nucleotide in each codon (the −1 frame) or forward by 1 nucleotide (the +1 frame). In a few instances, translational frameshifts that change the reading frame even after translation has initiated can occur. Frameshift mutations (see chapter 3) cause incorrect reading of an mRNA because of the insertion or deletion of nucleotides in the DNA sequence, which is then copied into mRNA. Translational frameshifting occurs when the ribosome shifts its position on the mRNA without a change in the mRNA sequence itself.

TRNA AMINOACYLATION

Before translation can begin, a specific amino acid is attached to each tRNA by its **cognate aaRS** (Figure 2.24). Each of these enzymes specifically recognizes only one amino acid and one class of tRNA—hence the name cognate. How each aaRS recognizes its own tRNA varies, but the anticodon (i.e., the three tRNA nucleotides that base pair with the complementary mRNA sequence [Figure 2.25]) is not the only determinant. Other variations in the tRNA structure, such as the identity of the discriminator base immediately upstream of the CCA at the 3′ end (Figure 2.17) and the size of the variable loop, also contribute to aaRS recognition specificity. In some cases, if the anticodon in a given tRNA is mutated, the cognate aaRS still attaches the amino acid for the original tRNA, and that amino acid is inserted at a different codon in the mRNA. This is the basis of nonsense suppression,

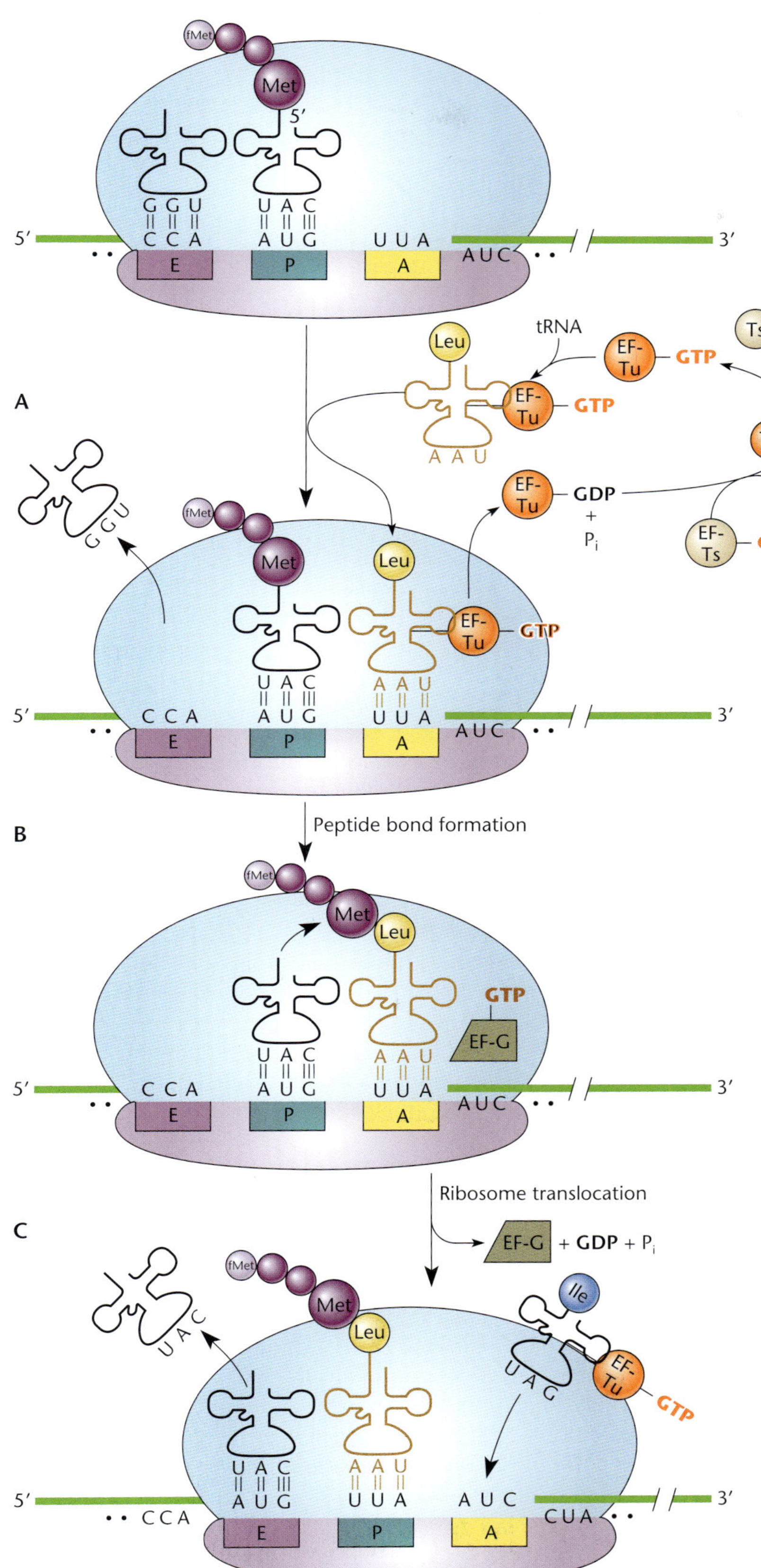

Figure 2.23 Overview of translation. **(A)** The ribosomal A (aminoacyl) site is empty, the growing peptide chain is attached to the P (peptidyl) site tRNA, and the tRNA that previously contained the peptide chain is in the E (exit) site. **(B)** The tRNA bound to its amino acid and complexed with elongation factor Tu (EF-Tu) and guanosine triphosphate (GTP) comes into the empty A site and remains there if its anticodon matches the mRNA codon at that site; the E site tRNA leaves the ribosome. EF-Tu–GTP is converted to EF-Tu–GDP and released from the ribosome, then is recycled to EF-Tu–GTP by EF-Ts. **(C)** Peptidyl transferase (23S rRNA in the 50S ribosome) catalyzes peptide bond formation between the carboxyl end of the growing polypeptide carried by the P site tRNA and the amino end of the amino acid carried by the A site tRNA. **(D)** Translation elongation factor G (EF-G) catalyzes translocation of the A site tRNA to the P site, making room at the A site for another aminoacyl-tRNA. The previous P site tRNA, now stripped of its polypeptide, moves to the E site before exiting the ribosome.

Table 2.2 The genetic code

First position (5′ end)	Second position U	C	A	G	Third position (3′ end)
U	Phe	Ser	Tyr	Cys	U
	Phe	Ser	Tyr	Cys	C
	Leu	Ser	Stop	Stop	A
	Leu	Ser	Stop	Trp	G
C	Leu	Pro	His	Arg	U
	Leu	Pro	His	Arg	C
	Leu	Pro	Gln	Arg	A
	Leu	Pro	Gln	Arg	G
A	Ile	Thr	Asn	Ser	U
	Ile	Thr	Asn	Ser	C
	Ile	Thr	Lys	Arg	A
	Met	Thr	Lys	Arg	G
G	Val	Ala	Asp	Gly	U
	Val	Ala	Asp	Gly	C
	Val	Ala	Glu	Gly	A
	Val	Ala	Glu	Gly	G

which is discussed in chapter 3. The amino acid is attached to the terminal A residue on the tRNA, and the energy for the reaction is provided by cleavage of ATP. Finally, the tRNA with its amino acid is bound to the protein EF-Tu, which assists in delivery of the aa-tRNA to the ribosome.

Figure 2.24 Aminoacylation of a tRNA by its cognate aminoacyl-tRNA synthetase (aaRS). ATP is used as a source of energy, and the amino acid is attached to the adenosine residue at the 3′ end of the tRNA. Each amino acid utilizes a dedicated aaRS.

$$\text{Amino acid} + \text{tRNA} + \text{ATP} \xrightarrow[\text{synthetase}]{\text{Aminoacyl-tRNA}} \text{Aminoacyl-tRNA} + \text{AMP} + \text{PP}_i$$

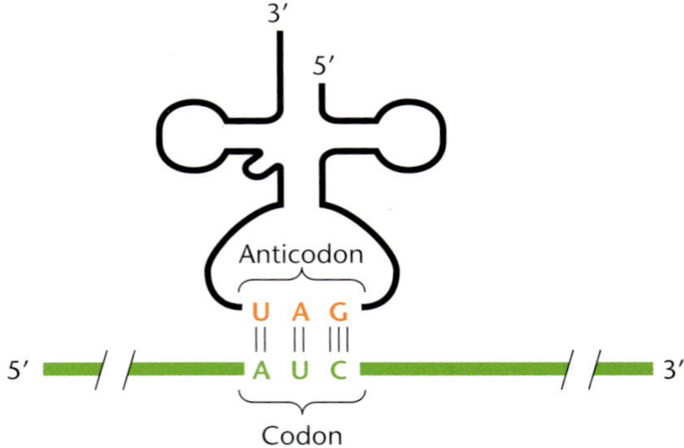

Figure 2.25 Complementary pairing between a tRNA anticodon and an mRNA codon. The codon (green) is shown 5′-3′, and the tRNA is shown 3′-5′ to allow antiparallel pairing between the codon and the tRNA anticodon (orange).

TRANSLATION INITIATION REGIONS

In the chain of thousands of nucleotides that make up an mRNA, the ribosome must bind and initiate translation at the correct site. If the ribosome starts working at the wrong position, the protein will have the wrong N-terminal amino acids and/or the mRNA will be translated out of frame and all of the amino acids will be wrong. Hence, mRNAs have sequences called **translational initiation regions (TIRs)** or **ribosome-binding sites (RBSs)** that flag the correct first codon for the ribosome. In spite of extensive research, it is still not possible to predict with 100% accuracy whether a sequence is a TIR. However, some general features of TIRs are known.

Initiation Codons

All bacterial and archaeal TIRs have an initiation codon, which is recognized by a dedicated **initiator tRNA**. This tRNA is always aminoacylated with methionine by methionyl-tRNA synthetase, and the methionine is further modified by addition of a formyl group (Figure 2.26). The three bases in initiator codons are usually AUG or GUG but in rare cases are UUG or CUG. Regardless of which amino acid these sequences call for in the genetic code (Table 2.2), they encode methionine (actually formylmethionine) as the N-terminal amino acid if they are serving as initiation codons. After translation, this methionine is often cut off by an aminopeptidase (see below). Notice that for the initiation codons, the first position of the codon can mispair with the tRNA anticodon, which always matches AUG; this differs from "wobble" during translation elongation, which involves mismatches at the third position of the codon (see below).

The initiation codon does not have to be the first sequence in the mRNA chain. In fact, the 5′ end of the mRNA may be some distance from the initiation codon of the first coding region in a transcript; this intervening region is called the **5′ untranslated region** (5′-UTR) or

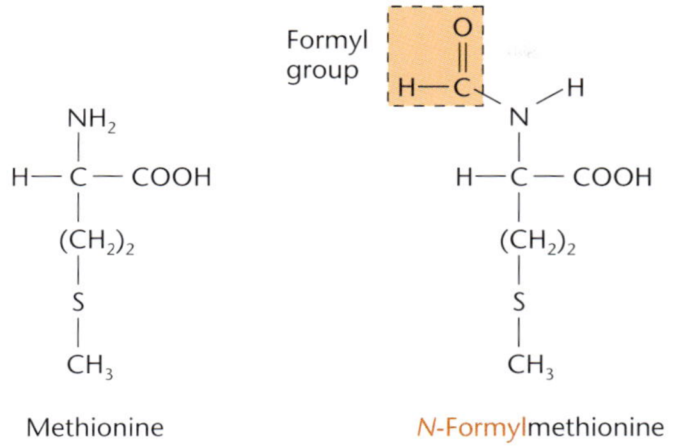

Figure 2.26 Conversion of methionine (Met) to *N*-formylmethionine (fMet) by transformylase.

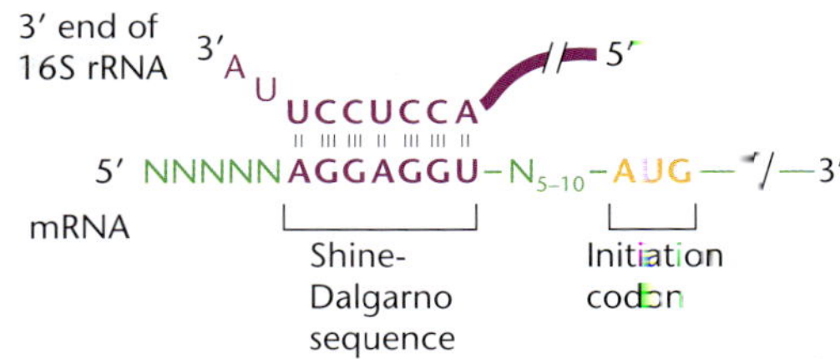

Figure 2.27 Structure of a typical bacterial translational initiation region (TIR) showing the pairing between the Shine-Dalgarno (S-D) sequence in the mRNA and a short sequence complementary to the S-D sequence that is located close to the 3′ end of the 16S rRNA. The initiation codon, typically AUG or GUG, is 5 to 10 bases downstream of the S-D sequence. N represents any base.

leader region. In bacteria, many transcripts contain multiple coding regions. The translation initiation complex binds internally to the mRNA to identify the TIR for each coding region. This is different from translation in eukaryotes, where the translation initiation complex usually binds to the 5′ end of the transcript and initiates translation at the first AUG codon it encounters that is accessible (see below).

Shine-Dalgarno Sequences

Not all methionine codons serve as initiation codons, because methionine can be found within polypeptide sequences and not only at the N-terminal end. Furthermore, the fact that some of the initiation codons (like GUG) code for amino acids other than methionine when internal to a coding region demonstrates that the presence of one codon is clearly not enough to define a TIR. These sequences also may occur out of frame or in an mRNA sequence that is not translated at all. Obviously, sequences in addition to these three bases must help to define them as a place to begin translation.

Most bacterial genes have a second component for recognizing a translational start site within the TIR. This sequence, named the **Shine-Dalgarno (S-D) sequence** after the two scientists who first noticed it, is located 5 to 10 nucleotides on the 5′ side (upstream) of the initiation codon and is complementary to a short sequence near the 3′ end of the 16S RNA. Figure 2.27 shows an example of a typical bacterial TIR with a characteristic S-D sequence. By pairing with their complementary sequences on the 16S rRNA, S-D sequences help define TIRs by properly aligning the mRNA on the ribosome. However, these sequences are not always easy to identify because they can be very short and exhibit considerable variability. Moreover, not all bacterial genes have S-D sequences. The initiation codon sometimes resides at the extreme 5′ end of the mRNA (in "leaderless mRNAs"), leaving no room for an S-D sequence. In such cases, translation initiation may occur by a somewhat different mechanism (see below).

Because of this lack of universality, often the only way to be certain that translation is initiated at a particular initiation codon is to sequence the N terminus of the polypeptide to see if the N-terminal amino acids correspond to the codons immediately following the putative initiation codon. Protein sequencing will also reveal whether the methionine encoded by the initiation codon remains on the mature protein or is removed.

INITIATOR tRNA

Translation initiation requires a dedicated aa-tRNA, the formylmethionine initiator tRNA (fMet-$tRNA^{fMet}$). This unique aa-tRNA has a formyl group attached to the amino group of the methionine (Figure 2.26), making it resemble a peptidyl-tRNA rather than a normal aminoacyl-tRNA because the amino group on the amino acid is blocked. This causes fMet-$tRNA^{fMet}$ to bind to the P site rather than the A site of the ribosome, which is an important step in initiation, as discussed below, and prevents its use as an elongator tRNA. The initiator fMet-$tRNA^{fMet}$ is synthesized somewhat differently from the other aminoacyl-tRNAs. This special tRNA uses the aminoacyl-tRNA synthetase of the normal $tRNA^{Met}$ to attach methionine to the $tRNA^{fMet}$. Then, an enzyme called **transformylase** adds a formyl group to the amino group of the methionine on the $tRNA^{fMet}$ to form fMet-$tRNA^{fMet}$.

STEPS IN INITIATION OF TRANSLATION

The currently accepted view of the steps in the initiation of translation at a TIR is outlined in Figure 2.28. Initiation requires three different initiation factors, IF1, IF2, and IF3, in addition to fMet-$tRNA^{fMet}$. These initiation factors interact mostly with the initiator tRNA and the P site of the 30S ribosomal subunit.

For initiation to occur, the 70S ribosome must first be separated (or dissociated) into its smaller 30S and 50S subunits. This dissociation occurs after the termination step of translation (see below). The IF3 initiation factor binds to the 30S subunit and helps to keep the subunits dissociated.

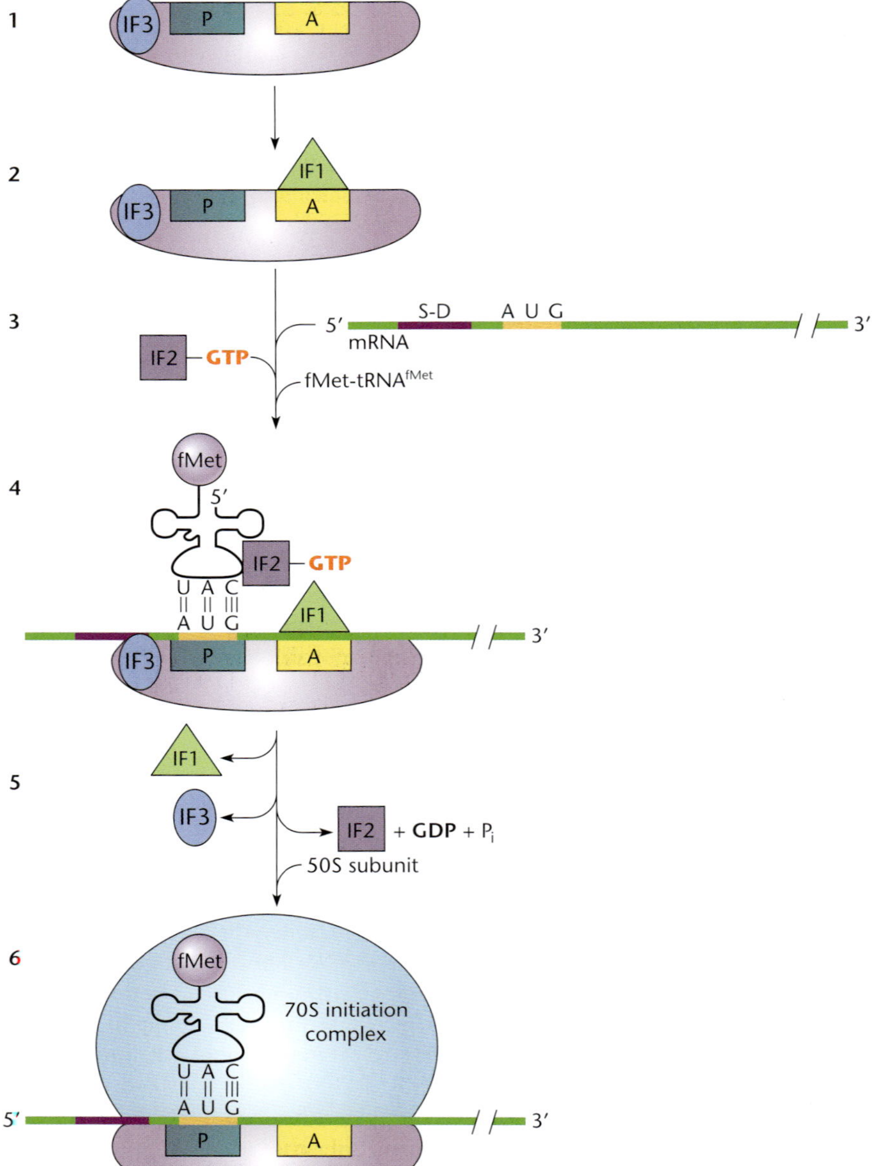

Figure 2.28 Initiation of translation. **(1)** The IF3 factor binds the 30S subunit to keep it dissociated from the 50S subunit during initiation. **(2)** IF1 binds to the A site to block the site and prevent tRNA binding. **(3)** A complex is formed between formylmethionine tRNA (fMet-tRNA^{fMet}), IF2, and guanosine triphosphate (GTP). **(4)** The fMet-tRNA^{fMet}-IF2-GTP complex binds to the P site of the 30S subunit and the mRNA translational initiation region (TIR) site. **(5)** IF1 and IF3 are released, the cleavage of GTP on IF2 correctly positions the fMet-tRNA^{fMet} on the P site initiation codon, and the 50S subunit binds. **(6)** The 70S initiation complex is ready to accept an aminoacyl-tRNA at the A site.

Therefore, ribosomes are continuously cycling between the 70S ribosome and the 30S and 50S subunits, depending on whether they are active in translation. This is called the **ribosome cycle**.

Once the subunits are dissociated, IF1 binds to the A site on the 30S subunit to prevent the fMet-tRNA^{fMet} from inadvertently binding to this site. IF2, in conjunction with GTP, binds to fMet-tRNA^{fMet} to form a ternary (three-member) complex, which binds to the mRNA and P site of the 30S ribosomal subunit. The initial binding of fMet-tRNA^{fMet} does not depend on an initiator codon in the P site. However, IF2, with the help of IF3, adjusts the fMet-tRNA^{fMet} and the mRNA initiator codon so that the binding becomes codon-specific. IF1 and IF3 are then released, and IF2 promotes the association of this initiation complex with the 50S subunit of the ribosome. IF2 is then released, with the cleavage of GTP to guanosine diphosphate (GDP). The newly formed 70S ribosome is now ready for translation, and another aa-tRNA can enter the A site.

Translation Initiation from Leaderless mRNAs

As mentioned above, a few mRNAs in bacteria do not have standard TIRs with leader regions containing S-D sequences. In these rare mRNAs, the initiator codon can be right at or very close to the 5′ end of the mRNA. How the ribosome recognizes such an initiator codon and initiates translation is not fully understood, but the mechanism seems to be very different from that of initiation at a normal TIR. There is some evidence that a complex first forms between fMet-tRNA^{fMet}, IF2, and the small subunit of the ribosome. This complex may then help recognize the initiation codon in the absence of upstream sequences to help distinguish the initiation codon. Other evidence suggests that the 70S ribosome itself recognizes the leaderless initiator codon.

TRANSLATION INITIATION IN ARCHAEA AND EUKARYOTES

Translation initiation in the archaea is similar to that in the bacteria. Like bacteria, archaea use well-defined TIRs with leader sequences and formylmethionine for initiation of translation. In contrast, eukaryotes do not seem to use special sequence elements and instead usually use the first AUG from the 5′ end of the mRNA as the initiation codon. Sequences around this initiator AUG may also be important for its recognition, and secondary structure in the mRNA may mask other AUG sequences that could potentially be used as initiator codons. Although eukaryotes have a special methionyl tRNA that responds to the first AUG codon, called Met-tRNA$_i$, the methionine attached to the eukaryotic initiator tRNA is never formylated. As in bacteria, however, the first methionine is usually removed by an aminopeptidase after the protein is synthesized. Eukaryotes and archaea also seem to use many more initiation factors and elongation factors than do bacteria. The archaea use formylated methionine and S-D sequences like bacteria, but their initiation factors are more similar to those in eukaryotes. Although the exact roles of most of these initiation factors are unknown in archaea, many are obviously related to the initiation and translation factors of bacteria. It therefore appears that the mechanism of translation initiation in the archaea is a sort of hybrid between that in bacteria and that in eukaryotes.

TRANSLATION ELONGATION

During translation elongation, the ribosome moves 3 nucleotides at a time along the mRNA in the 5′-to-3′ direction, allowing tRNAs carrying amino acids (aa-tRNAs) to pair with the mRNA. Each tRNA has a specific 3-nucleotide anticodon sequence in one of its loops, and these nucleotides must be complementary to the mRNA codon for the tRNA to be bound tightly to the ribosome (Figure 2.25). As in other nucleic acids, pairing is antiparallel, so that the two RNA sequences are complementary when read in opposite directions. In other words, the 3′-to-5′ sequence of the anticodon must be complementary to the 5′-to-3′ sequence of the codon.

Entry of aa-tRNAs bound by EF-Tu into the ribosome is random. If the anticodon is complementary to the mRNA codon at the A site (Figure 2.23), codon-anticodon pairing stimulates a structural transition of the ribosome that promotes the next step. A mispaired tRNA is not stabilized and is released from the ribosome. This pairing of only three bases is sufficient to direct the correct tRNA to the A site on the ribosome; in fact, sometimes the pairing of only two bases is sufficient to direct the anticodon-codon interaction (see "Wobble" below). Accurate codon-anticodon pairing is monitored by specific residues in 16S rRNA that form the **decoding site**. The tRNA is positioned across the 30S and 50S subunits of the ribosome so that the anticodon loop is in communication with the mRNA in the 30S subunit and the acceptor end of the aa-tRNA containing the bound amino acid is in communication with the 23S rRNA in the 50S subunit. The conformational change in the ribosome that is triggered by accurate codon-anticodon pairing results in hydrolysis of the GTP on EF-Tu to GDP and release of EF-Tu–GDP from the ribosome. EF-Tu–GDP is recycled into EF-Tu–GTP by the action of another protein factor, EF-Ts (Figure 2.23).

After the matching aa-tRNA is bound at the A site of the ribosome, the **peptidyltransferase** catalyzes the formation of a peptide bond between the incoming amino acid at the A site and the growing polypeptide at the adjacent P site (Figure 2.29). This reaction links the carboxy terminus of the growing peptide chain (attached to the P site tRNA) to the amino group of the amino acid attached to the A site tRNA. The peptidyltransferase activity is catalyzed by the 23S rRNA ribozyme.

After peptide bond formation, the P site tRNA no longer has anything attached to it, while the A site tRNA carries the polypeptide chain. Another enzyme, **translation elongation factor G (EF-G)**, in complex with GTP, then enters the ribosome and moves (or translocates) both the polypeptide-containing tRNA and the mRNA from the A site to the P site, moving the deacylated tRNA from the P site to the E site and making room for another aa-tRNA to enter the A site. Translocation requires hydrolysis of the GTP bound to EF-G, which also results in release of EF-G from the ribosome (Figure 2.23). The deacylated tRNA in the E site later exits the ribosome (possibly in conjunction with entry of the next aminoacyl-tRNA into the A site). The mRNA maintains contact with the tRNAs as they move through the ribosome and therefore also moves through the ribosome 3 nucleotides at a time.

The translation of even a single codon in an mRNA requires a lot of energy. First, ATP must be hydrolyzed for an aaRS to attach an amino acid to a tRNA (Figure 2.24). Second, EF-Tu requires that a GTP be hydrolyzed to GDP before it can be released from the ribosome after each tRNA is bound, and another GTP is used for recycling of

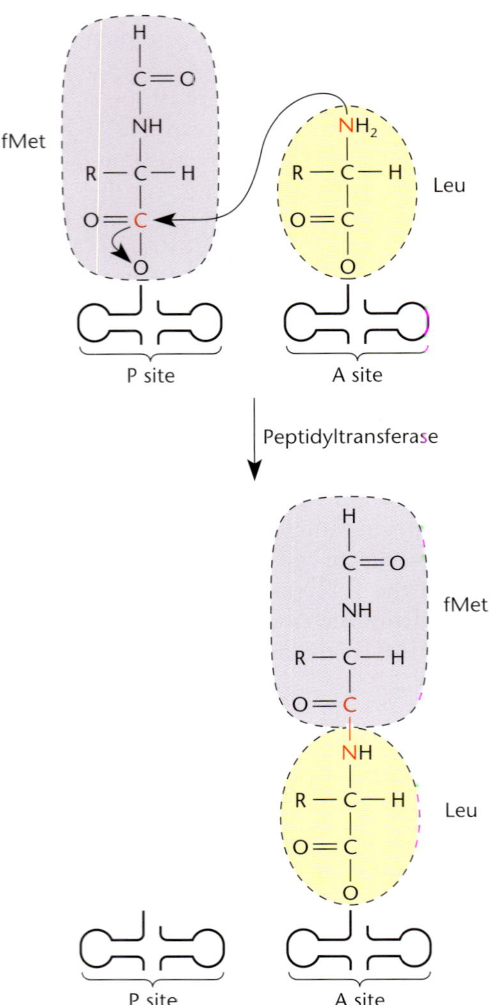

Figure 2.29 The peptidyltransferase reaction catalyzes dissociation of the carboxyl end of the formyl-methionine from the P site tRNA^{fMet} and peptide bond formation with the amino end of the amino acid on the A site tRNA (leucine in the diagram). This results in a P site tRNA with no amino acid and an A site tRNA containing two amino acids.

EF-Tu–GDP to EF-Tu–GTP by EF-Ts (Figure 2.23B). Yet another GTP must be hydrolyzed to GDP for EF-G to move the tRNA with the attached polypeptide to the P site (Figure 2.23D). In all, the energy of at least four nucleoside triphosphates is required for each step of translation.

TRANSLATION TERMINATION

During elongation, translation proceeds along the mRNA, one codon at a time, until the ribosome encounters one of three special codons, UAA, UAG, or UGA, that serve as translational stop signals. These codons, designated **termination codons**, **stop codons**, or **nonsense codons** (because they do not encode an amino acid), have no corresponding tRNA (Table 2.2). When a termination codon enters the A site of a translating ribosome, translation stops because no aa-tRNA can match the codon. Similar to the positioning of translation initiation codons, the termination codon that terminates translation may not be at the 3′ end of the mRNA molecule. The region between the termination codon and the 3′ end of the mRNA (or a downstream coding sequence for another polypeptide in a transcript that encodes multiple proteins [see below]) is called the **3′ untranslated region** (3′-UTR).

RELEASE FACTORS

In addition to a termination codon, termination of translation requires **release factors**. These proteins recognize the termination codons in the ribosomal A site and promote the release of the polypeptide from the tRNA and the ribosome from the mRNA. In *E. coli*, there are two translation release factors, called RF1 and RF2, that recognize specific termination codons; RF1 responds to UAA and UAG, and RF2 responds to UAA and UGA. Another factor, called RF3, helps to release these factors from the ribosome after termination. Eukaryotes have only one release factor, which responds to all three termination codons. Some types of bacteria and mitochondria also have only one release factor, but those that do generally use UGA to encode an amino acid and not as a termination codon (Box 2.4).

Figure 2.30 outlines the process of **translation termination**. After translation stops at the termination codon, the A site is left unoccupied because there is no tRNA to pair with the termination codon. The release factors bind to the A site of the ribosome instead. They then cooperate with EF-G and **ribosome release factor (RRF)** to cleave the polypeptide chain from the tRNA and release it and the mRNA from the ribosome. An attractive model to explain how this could happen is suggested by the observation that the release factors mimic aa-tRNA (Box 2.4). If the release factor is occupying the A site, then the peptidyltransferase might try to transfer the polypeptide chain to the release factor rather than to the amino acid on a tRNA normally occupying the A site. When EF-G then tries to translocate the release factor with the polypeptide attached to the P site, it may trigger a series of reactions that release the polypeptide. The role of ribosome release factor in this process is uncertain, but it might be involved in releasing the mRNA after termination.

REMOVAL OF THE FORMYL GROUP AND THE N-TERMINAL METHIONINE

Normally, polypeptides do not have a formyl group attached to their N termini. In fact, they usually do not even have methionine as their N-terminal amino acid. The formyl group is removed from the completed polypeptide by a special enzyme called **peptide deformylase** (Figure 2.31). The N-terminal methionine or entire

BOX 2.4

Mimicry in Translation

The ribosome is a very busy place during translation, with numerous factors and tRNAs cycling quickly through the A and P sites. Different factors have to enter the ribosome for each of the steps and then leave when they have finished their functions. One way the complexity of the system seems to be reduced is by having the various factors and tRNAs mimic each other's structure, which allows them to bind to the same sites on the ribosome. For example, the translation factor EF-G seems to be roughly the same shape as the translation factor EF-Tu bound to an aa-tRNA. This may allow EF-G to enter the A site, displace the tRNA (now attached to the growing polypeptide), and move it to the P site. Another example is the mimicry between the tRNAs and the release factors. The release factors resemble tRNAs in shape, but they seem to bind to specific terminator codons through interactions between amino acids in the release factors and nucleotide bases in the termination codon, rather than through base pairing between the codon and the anticodon on a tRNA. When the peptidyltransferase attempts to transfer the polypeptide to the release factor in the A site, it sets in motion the string of events that cause translation to be terminated and the polypeptide and mRNA to be released from the ribosome. It is an attractive idea that the release factors replaced what were once terminator tRNAs that responded to these terminator codons. Perhaps, in the earliest forms of life, everything in translation was done by RNA; now, RNA is used to make proteins, and the proteins, being more versatile, play many of the roles previously played by RNA.

References

Clark BFC, Thirup S, Kjeldgaard M, Nyborg J. 1999. Structural information for explaining the molecular mechanism of protein biosynthesis. *FEBS Lett* **452:**41–46.

Nakamura Y, Ito K. 2011. tRNA mimicry in translation termination and beyond. *Wiley Interdiscip Rev RNA* **2:**647–668.

Nyborg J, Nissen P, Kjeldgaard M, Thirup S, Polekhina G, Clark BFC. 1996. Structure of the ternary complex of EF-Tu: macromolecular mimicry in translation. *Trends Biochem Sci* **21:**81–82.

Figure 2.30 Termination of translation at a nonsense codon. In the absence of a tRNA with an anticodon capable of pairing with the nonsense codon at the A site, the ribosome stalls, and a specific release factor interacts with the A site, possibly through a specific interaction between amino acids in the release factor and the UAA nonsense codon. Translocation by translation elongation factor G (EF-G) (brown) causes dissociation of the ribosome and release factor from the mRNA, with the assistance of the ribosome release factor.

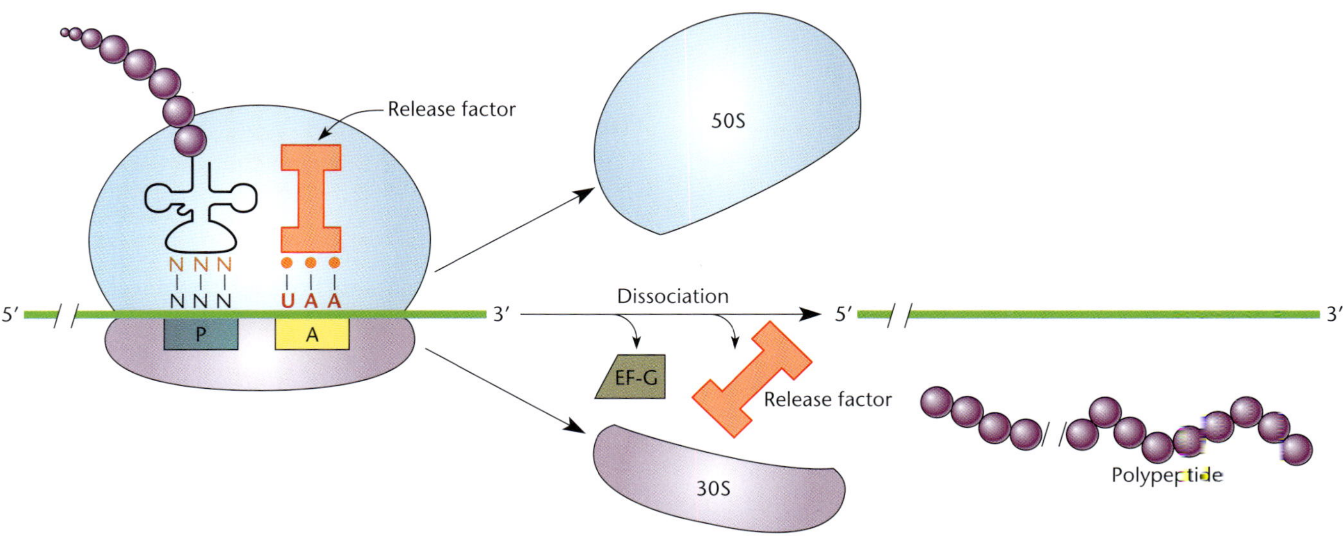

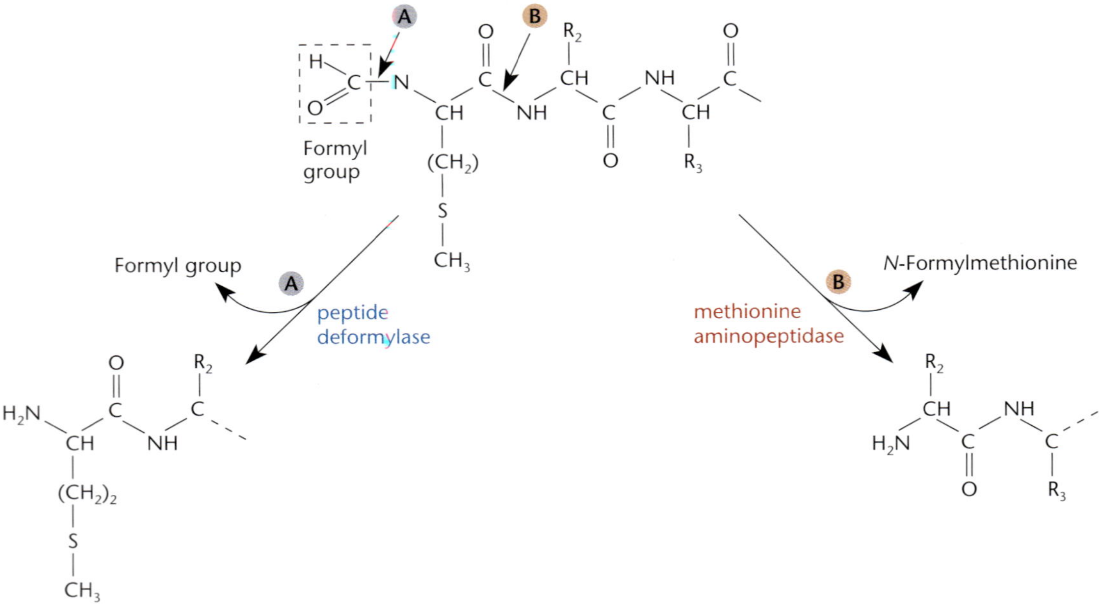

Figure 2.31 Removal of the N-terminal formyl group by peptide deformylase **(A)** and of the N-terminal methionine by methionine aminopeptidase **(B)**.

N-formylmethionine is also often removed by an enzyme called **methionine aminopeptidase**, so that methionine is usually not the N-terminal amino acid in a mature polypeptide.

trans-TRANSLATION (tmRNA)

A problem occurs when the ribosome reaches the 3′ end of an mRNA without encountering a termination codon. This might happen fairly often, because mRNA is constantly being cleaved or degraded (often from the 3′ end), and transcription may terminate prematurely, resulting in a truncated mRNA. The release factors can function only at a termination codon, so the ribosome will stall on a truncated mRNA. Not only would this cause a traffic jam and sequester ribosomes in a nonfunctional state, but also the protein that is made will be defective because it is shorter than normal, and accumulation of defective proteins may cause problems for the cell. This is where a small RNA called **transfer-messenger RNA (tmRNA)** comes to the rescue. As the name implies, tmRNA is both a tRNA and an mRNA, as shown in Figure 2.32. It can be aminoacylated with alanine by alanyl-tRNA synthetase like an alanyl-tRNA, and it also contains a short **open reading frame (ORF)** that terminates in a termination codon like an mRNA. If the ribosome reaches the end of an mRNA without encountering a termination codon, tmRNA (together with an accessory protein) enters the A site of the stalled ribosome, and alanine is inserted as the next amino acid of the polypeptide. Then, by a process that is not well understood, the ribosome shifts from translating the ORF on the mRNA to translating the ORF on the tmRNA, where it soon encounters the termination codon at the end of the ORF. The release factors then release the ribosome and the truncated polypeptide, which now contains a short additional "tag" sequence of about 10 amino acids encoded by the tmRNA. The tag sequence that has been attached to the carboxy end of the truncated polypeptide is recognized by the ClpXP protease (see below), which degrades the entire defective polypeptide. In some cases, tmRNA-mediated degradation may play a regulatory role, allowing the targeted degradation of proteins until they are needed (see Keiler and Feaga, Suggested Reading).

The Genetic Code

As mentioned above, the genetic code determines which amino acid will be inserted into a protein for each 3-nucleotide set, or codon, in the mRNA. More precisely, the genetic code is the assignment of each possible combination of 3 nucleotides to 1 of the 20 amino acids or to serve as a signal to stop translation. The code is universal, with a few minor exceptions (Box 2.5), meaning that it is the same in all organisms from bacteria to humans. The assignment of each codon to its amino acid appears in Table 2.2.

REDUNDANCY

In the genetic code, more than one codon often encodes the same amino acid. This feature of the code is called **redundancy**. There are 64 (4 × 4 × 4) possible codons that

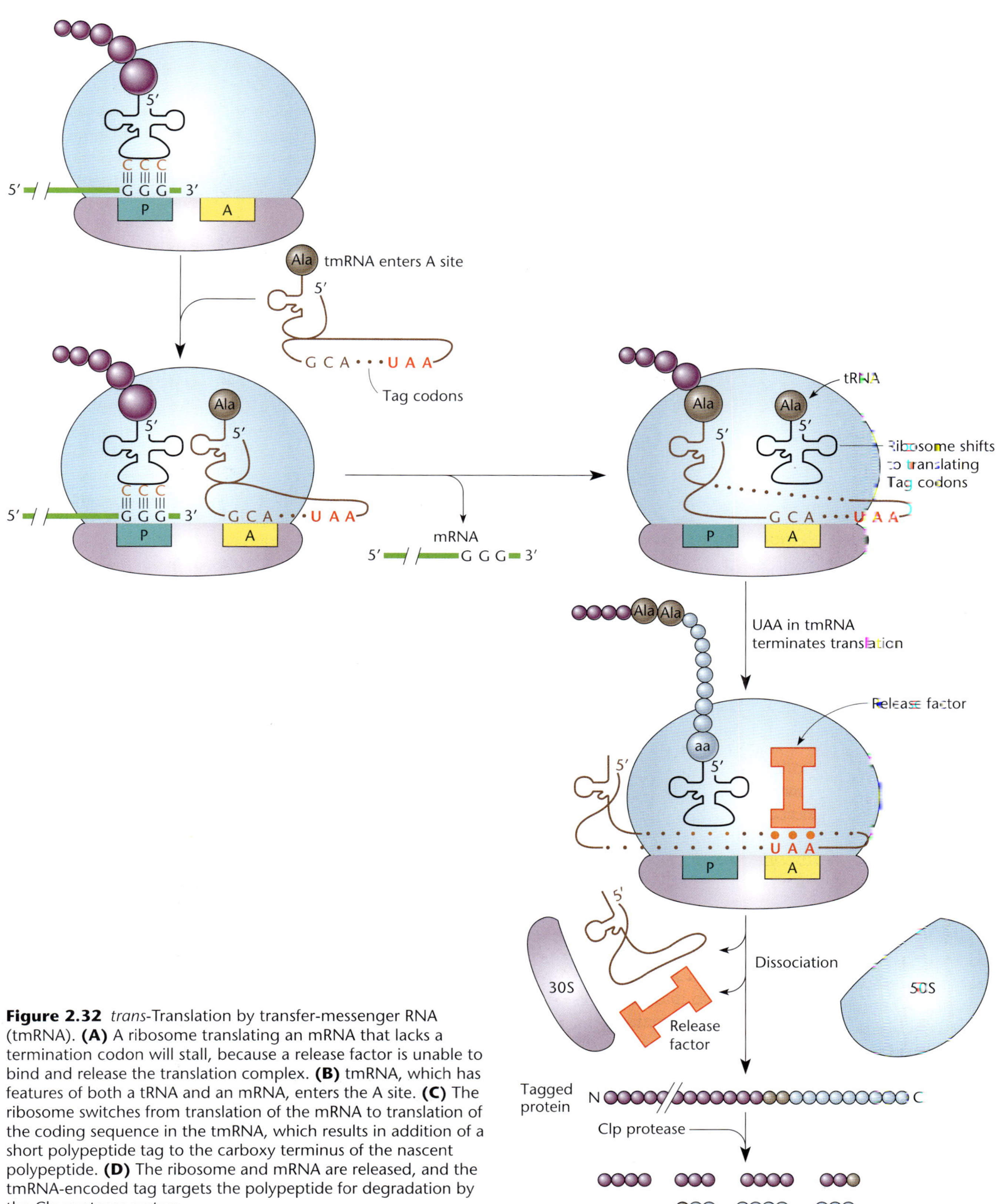

Figure 2.32 *trans*-Translation by transfer-messenger RNA (tmRNA). **(A)** A ribosome translating an mRNA that lacks a termination codon will stall, because a release factor is unable to bind and release the translation complex. **(B)** tmRNA, which has features of both a tRNA and an mRNA, enters the A site. **(C)** The ribosome switches from translation of the mRNA to translation of the coding sequence in the tmRNA, which results in addition of a short polypeptide tag to the carboxy terminus of the nascent polypeptide. **(D)** The ribosome and mRNA are released, and the tmRNA-encoded tag targets the polypeptide for degradation by the Clp protease system.

BOX 2.5

Exceptions to the Code

One of the greatest scientific discoveries of the 20th century was that of the universal genetic code. Whether human, bacterium, or plant, for the most part, all organisms on Earth use the same three bases in nucleic acids to designate each of the amino acids. However, although the code is almost universal, there are exceptions to this general rule. In some situations, a codon can mean something else. We gave the example of initiation codons that encode different amino acids when internal to a gene than they do at the beginning of a gene, where they invariably encode methionine (see the text). Also, some organelles and primitive microorganisms use different codons for some amino acids. For example, in mammalian mitochondria, UGA, which is normally a termination codon, instead designates tryptophan. Also, some protozoans use the termination codons UAA and UAG for glutamine. In these organisms, UGA is the only termination codon. Some yeasts of the genus *Candida*, the causative agent of thrush, ringworm, and vaginal yeast infections, recognize the codon CUG as serine instead of the standard leucine. In bacteria, the only known exceptions to the universal code involve the codon UGA, which encodes the amino acid glutamine in some bacteria of the genus *Mycoplasma*, which are responsible for some plant and animal diseases.

Some exceptions to the code occur only at specific sites in the mRNA. For example, UGA encodes the rare amino acid selenocysteine in some contexts. This amino acid exists at one or a very few positions in certain bacterial and eukaryotic proteins. It has its own unique aaRS, translation elongation factor (analogous to EF-Tu), and tRNA, to which the amino acid serine is added and then converted into selenocysteine. This tRNA then inserts the amino acid selenocysteine at certain UGA codons, but only at a very few unique positions in proteins and not every time a UGA appears in frame. How, then, does the tRNA distinguish between these sites and the numerous other UGAs, which usually signify the end of a polypeptide? The answer seems to be that the selenocysteine-specific EF-Tu has extra sequences that recognize the mRNA sequences around the selenocysteine-specific UGA codon, and only if the UGA codon is flanked by these particular sequences will this EF-Tu allow its tRNA to enter the ribosome. It is a mystery why the cell goes to so much trouble to insert selenocysteine in a specific site in only a very few proteins. In some instances where selenocysteine was replaced by cysteine, the mutated protein still functioned, albeit less efficiently. However, it may be required in the active center of some enzymes involved in anaerobic metabolism, and this amino acid has persisted throughout evolution, existing in organisms from bacteria to humans.

Another striking deviation from the code is found in the methanogenic archaea (archaea that produce methane). These bacteria insert the lysine analog pyrrolysine at the normal termination codon UAG. Unlike selenocysteine, which is chemically derived from serine already on its tRNA, pyrrolysine is synthesized and then loaded onto a dedicated tRNA by a dedicated aaRS. It therefore qualifies as the 22nd amino acid. Its aa-tRNA uses the normal EF-Tu and is inserted whenever the codon UAG appears within the mRNA.

Recently, efforts have been made to purposely reprogram the genetic code to allow targeted insertion of nonnatural amino acids into specific sites of individual target proteins with the goal of generating new classes of proteins with novel activities. This work relies on engineering systems to attach a new type of amino acid to a dedicated tRNA so that the tRNA will insert its amino acid only at a specific codon within the target protein. Approaches of this type will enable both specific labeling of proteins in the cell and development of new enzymatic activities.

Other exceptions violate the rule that the code is read three bases at a time until a termination codon is encountered. This happens with high-level frameshifting and readthrough of termination codons. In high-level frameshifting, the ribosome can back up one base or go forward one base before continuing translation. High-level frameshifting usually occurs where there are two cognate codons next to each other in the RNA, for example, in the sequence UUUUC, where both UUU and UUC are phenylalanine codons that are presumably recognized by the same tRNA through wobble. The ribosome with the tRNA bound can slip forward or backward by one nucleotide before it continues translating in the new reading frame, creating a frameshift. Sites at which high-level frameshifting occurs are designated "shifty sequences" and usually have common features. They often have a secondary structure, such as a pseudoknot, in the RNA (see Figure 2.2) just downstream of the frameshifted region, which causes the ribosome to pause. They also may have a sequence similar to an S-D sequence just upstream of the frameshifted site to which the ribosome then binds through its 16S rRNA, shifting the ribosome 1 nucleotide on the mRNA and causing the frameshift. Sometimes both the normal protein and the frameshifted protein, which has a different carboxyl end, can function in the cell. Examples are the *E. coli* DNA polymerase accessory proteins γ and τ, which are both products of the *dnaX* gene (see Table 1.1) but differ because of a frameshift that results in the formation of a truncated protein. Frameshifting can also allow the readthrough of termination codons to make "polyproteins," as occurs in many retroviruses, such as human immunodeficiency virus (the acquired immune deficiency syndrome [AIDS] virus). Moreover, high-level frameshifting can play a regulatory role, for example, in the regulation of the RF2 gene in *E. coli*. The RF2 protein causes release of the ribosome at the termination codons UGA and UAA (see "Translation Termination" in the text). The gene for RF2 in *E. coli* is arranged so that its function in translation termination can be used to regulate its own synthesis through frameshifting (see chapter 11). How long the ribosome pauses at a UGA codon depends on the amount of RF2 in the cell. If there is a lot of RF2 in the cell, the pause is brief and the polypeptide is quickly released by RF2. If there is less RF2, the ribosome will pause longer, allowing time for a –1 frameshift. The RF2 protein is translated in the –1 frame, so this is the correct frame for translation of RF2, and more RF2 will be made if there is not enough for rapid termination.

In the most dramatic cases of frameshifting, the ribosome can hop over large sequences in the mRNA and then continue translating. This is known to occur in gene *60* of bacteriophage T4 and the *trpR* gene of *E. coli.* Somehow, the ribosome stops translating the mRNA at a certain codon and "hops" to the same codon further along. Presumably, the secondary and tertiary structures of the mRNA between the two codons cause the ribosome to hop. In the case of gene *60* of T4, the hopping occurs almost 100% of the time, and the protein that results is the normal product of the gene. In the *E. coli trpR* gene, the hopping is less efficient, and the physiological significance of the hopped form is unknown.

High-level readthrough of termination codons can also give rise to more than one protein from the same ORF. Instead of stopping at a particular termination codon, the ribosome sometimes continues making a longer protein, in addition to the shorter one. Examples are the synthesis of the head proteins in the RNA phage Qβ and the synthesis of Gag and Pol proteins in some retroviruses. Many plant viruses also make readthrough proteins. Again, it seems to be the sequence around the termination codon that promotes high-level readthrough. However, it is important to emphasize that these are all exceptions, and normally, the codons on an mRNA are translated faithfully one after the other from the TIR until a termination codon is encountered.

References

Baranov PV, Fayet O, Hendrix RW, Atkins JF. 2006. Recoding in bacteriophages and bacterial IS elements. *Trends Genet* **22:**174–181.

Böck A, Forchhammer K, Heider J, Leinfelder W, Sawers G, Veprek B, Zinoni F. 1991. Selenocysteine: the 21st amino acid. *Mol Microbiol* **5:**515–520.

Gaston MA, Jiang R, Krzycki JA. 2011. Functional context, biosynthesis, and genetic encoding of pyrrolysine. *Curr Opin Microbiol* **14:**342–349.

Maldonado R, Herr AJ. 1998. Efficiency of T4 gene *60* translational bypassing. *J Bacteriol* **180:**1822–1830.

Young TS, Schultz PG. 2010. Beyond the canonical 20 amino acids: expanding the genetic lexicon. *J Biol Chem* **285:**11039–11044.

can be made of 4 different nucleotides taken 3 at a time. Thus, without redundancy, there would be far too many codons for only 20 amino acids. As shown in Table 2.2, some amino acids are encoded by a single codon (e.g., tryptophan), while others use as many as six different codons (e.g., arginine).

WOBBLE

Codons that encode the same amino acid often differ only by their third base, which is why they tend to be together in the same column when the code is presented as in Table 2.2. This pattern of redundancy in the code is due to less stringent pairing, or **wobble**, between the last (3′) base in the codon on the mRNA and the first (5′) base in the anticodon on the tRNA (remember that RNA sequences are always given 5′ to 3′ and that the pairing of strands of RNA, like that of DNA, is antiparallel [Figure 2.25]). As a consequence of wobble, the same tRNA can pair with more than one of the codons for a particular amino acid, so there can be fewer types of tRNA than there are codons. For example, even though there are two codons for lysine, AAA and AAG, *E. coli* has only one tRNA for lysine, which, because of wobble, can pair with both lysine codons.

While the term "wobble" may suggest that the process is random or that there is a lack of stringency inherent in the system, this is not really the whole story. As indicated earlier in the chapter, base pairing in RNA is different than in DNA, and modification of the anticodon itself contributes to alternative base-pairing rules, resulting in a process in which fewer tRNAs can be utilized to accurately recognize more codons (Figure 2.33). For example, wobble allows a G in the first position of the anticodon to pair with either a C or a U in the third position of the codon but not with an A or a G; this explains why UAU and UAC, but not UAA or UAG, are codons for tyrosine and can be recognized by a single tRNA with a GUA anticodon sequence. Similarly, a U in the first position of the anticodon can pair with either an A or a G in the third position of the codon (corresponding to the fact that both CAA and CAG are glutamine codons and can be recognized by a single tRNA with a UUG anticodon sequence). The rules for wobble are complicated by the fact that the bases in tRNA are sometimes modified, and a modified base in the first position of an anticodon can have altered pairing properties. Inosine, which is a purine base found only in tRNA, can pair with any residue, so a single tRNA with inosine at the first position of the

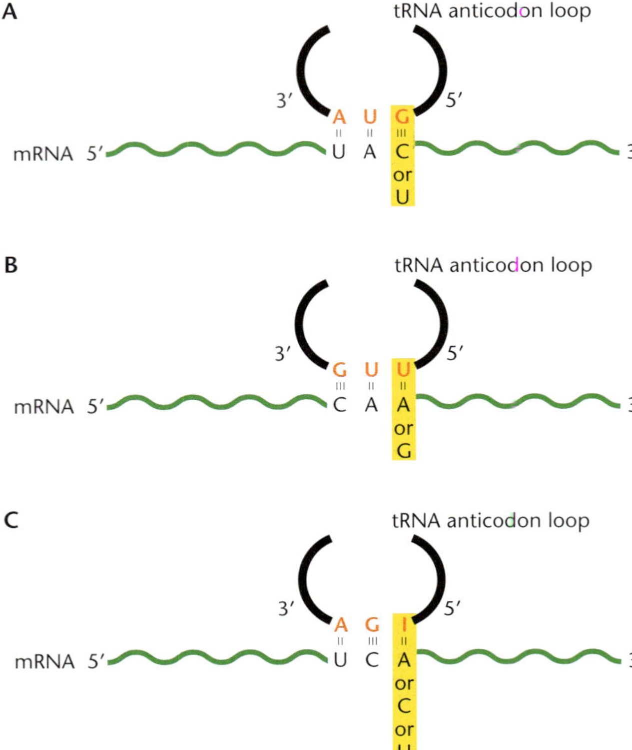

Figure 2.33 Wobble pairing between the anticodon on the tRNA and the codon in the mRNA. Non-Watson-Crick pairing interactions are possible in the third position of the codon. Alternative pairings for the anticodon base are shown: **(A)** guanine to cytosine or uracil; **(B)** uracil to adenine or guanine; and **(C)** inosine (a purine base found only in tRNAs) to adenine, cytosine, or uracil. This allows a single tRNA to recognize multiple codons.

anticodon can recognize multiple codons (UCU, UCA, and UCC in Figure 2.33C, all of which encode serine). In other cases, an organism may use multiple tRNAs to recognize different codons that specify the same amino acid.

TERMINATION CODONS

As noted above, not all codons stipulate an amino acid; of the 64 possible nucleotide combinations, only 61 actually encode an amino acid. The other three (UAA, UAG, and UGA) are termination (or nonsense) codons in most organisms. The termination codons are usually used to terminate translation at the end of genes (see "Translation Termination" above).

AMBIGUITY

In general, each codon specifies a single amino acid, but some can specify a different amino acid, depending on where they are in the mRNA. For example, the codons AUG and GUG encode formylmethionine if they are at the beginning of the coding region but encode methionine or valine, respectively, if they are internal to the coding region. The codons CUG, UUG, and even AUU also sometimes encode formylmethionine if they are at the beginning of a coding sequence.

The codon UGA is another exception. This codon is usually used for termination but encodes the amino acid selenocysteine in a few positions in genes (Box 2.5) and encodes tryptophan in some types of bacteria. Similarly, UAG is usually used for termination but can be used to encode the novel amino acid pyrrolysine in certain organisms.

CODON USAGE

Just because more than one codon can encode an amino acid does not mean that all the codons are used equally in all organisms. The same amino acid may be preferentially encoded by different codons in different organisms. This codon preference may reflect higher concentrations of certain tRNAs or may be related to the base composition of the DNA of the organism. While mammals have an average G+C content of about 50% (so that there are about as many AT base pairs in the DNA as there are GC base pairs), some bacteria and their viruses have very high or very low G+C contents. How the G+C content can influence codon preference is illustrated by some members of the genera *Pseudomonas* and *Streptomyces*. These organisms have G+C contents of almost 75%. To maintain such high G+C contents, the codon usage of these bacteria favors the codons that have the most G's and C's for each amino acid.

Polycistronic mRNA

In eukaryotes, each mRNA normally encodes only a single polypeptide. In contrast, in bacteria and archaea, one mRNA can encode either one polypeptide (**monocistronic mRNAs**) or more than one polypeptide (**polycistronic mRNAs**). Polycistronic mRNAs must have a separate TIR for each coding sequence to allow them to be translated.

The name "polycistronic" is derived from "cistron," which is the genetic definition of the coding region for each polypeptide, and "poly," which means many. Similarly, "monocistronic" is derived from "mono," which means one. Figure 2.34 shows a typical polycistronic mRNA in which the coding sequence for one polypeptide is followed by the coding sequence for another. The space between two coding regions can be very short, and the coding sequences may even overlap. For example, the coding region for one polypeptide may end with the termination codon UAA, but the last A may be the first nucleotide of the initiator codon AUG for the next coding region. Even if the two coding regions overlap, the two polypeptides on an mRNA can be translated independently by different ribosomes.

Polycistronic mRNAs do not exist in eukaryotes, in which, as described above, TIRs are much less well de-

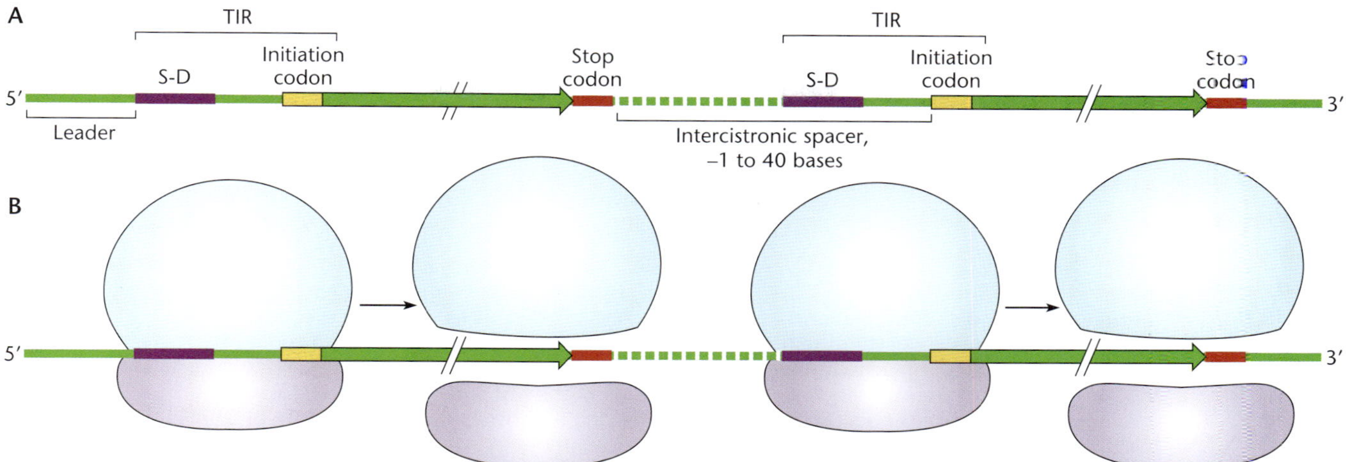

Figure 2.34 Structure of a polycistronic mRNA. **(A)** The coding sequence for each polypeptide is between the initiation codon and the stop codon. The region 5′ of the first initiation codon is called the leader sequence, and the untranslated region between a stop codon for one gene and the next initiation codon is known as the intercistronic spacer. **(B)** The association of the 30S and 50S ribosomes at a translational initiation region (TIR) and their dissociation at a stop codon. New 30S and 50S subunits associate at a downstream TIR.

fined and translation usually initiates at the AUG codon closest to the 5′ end of the mRNA. In eukaryotes, the synthesis of more than one polypeptide from the same mRNA usually results from differential splicing of the mRNA or from high-level frameshifting during the translation of one of the coding sequences (see "Reading Frames" above); there are also specialized events in which an RNA element called an internal ribosome entry sequence directs binding of a ribosome to a site within the RNA. Polycistronic RNA leads to phenomena unique to bacteria, i.e., **translational coupling** and **polarity**, which are described below.

TRANSLATIONAL COUPLING

Two or more polypeptides encoded by the same polycistronic mRNA can be translationally coupled. Two genes are **translationally coupled** if translation of the upstream gene affects the efficiency of the translation of the gene immediately downstream.

Figure 2.35 shows an example of how two genes could be translationally coupled. The TIR including the AUG initiation codon of the second gene is sequestered in a hairpin on the mRNA, so it cannot be recognized by an initiating ribosome. However, a ribosome arriving at the UAA stop codon for the first gene can open up this secondary structure, allowing another ribosome to bind to the downstream TIR and initiate translation of the second gene. Thus, translation of the second gene in the mRNA depends on the translation of the first gene. Mutations that disrupt translation of the upstream coding sequence (e.g., nonsense mutations or frameshift mutations that result in premature termination because the ribosome encounters nonsense codons in the new reading frame) therefore affect not only the gene in which they are located, but also the translationally coupled downstream gene.

POLAR EFFECTS ON GENE EXPRESSION

Some mutations that affect the expression of a gene in a polycistronic mRNA can have secondary effects on the transcription of downstream genes. Such mutations are said to exert a polar effect on gene expression. Several types of mutations can result in polar effects. One type of mutation that can cause a polar effect is an insertion mutation that carries a factor-independent transcriptional terminator. For example, if a transposon "hops" into a polycistronic transcription unit, the transcriptional terminators on the transposon may prevent the transcription of genes downstream of the insertion site in the same polycistronic transcription unit. Likewise, a "knockout" of a gene by insertion of an antibiotic resistance gene with a transcriptional terminator causes a polar effect on the genes downstream in the same transcription unit.

A second way a mutation in an upstream coding sequence can affect transcription of a downstream coding sequence is through effects on ρ-dependent termination of transcription. Recall that translation of mRNAs in bacteria normally occurs simultaneously with transcription and that the mRNA is translated in the same 5′-to-3′ direction as it is synthesized. Moreover, ribosomes often load onto a TIR as soon as it is vacated by the preceding ribosome, so that the mRNA is coated with translating ribosomes. If a nonsense mutation causes premature dissociation of ribosomes, the

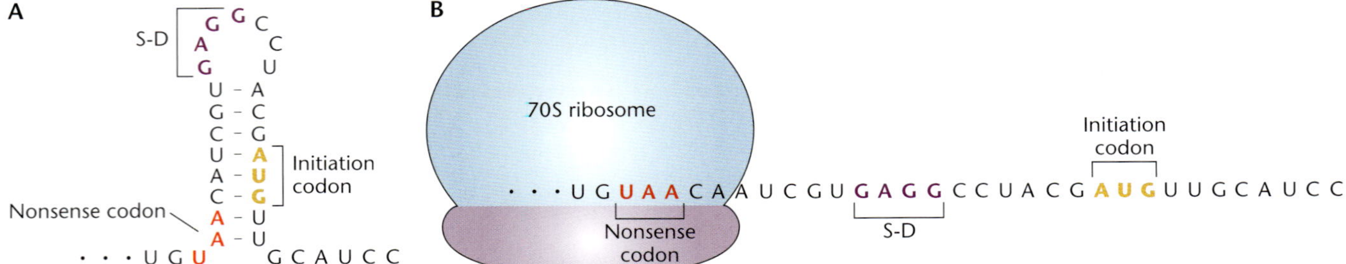

Figure 2.35 Model for translational coupling in a polycistronic mRNA. **(A)** The secondary structure of the RNA sequesters the translational initiation region (TIR) of the second coding sequence (Gene 2) and blocks translation initiation (note that the Shine-Dalgarno shown is not the complete consensus sequence). **(B)** Translation of the first coding sequence (Gene 1) results in disruption of the secondary structure, allowing a ribosome to access the TIR of Gene 2 to translate the second coding sequence.

abnormally naked mRNA downstream of the mutation may be targeted by the transcription termination factor ρ, which may find an exposed *rut* sequence in the mRNA and cause transcription termination, as shown in Figures 2.15 and 2.36. The nonsense mutation therefore prevents the expression of the downstream gene by preventing its transcription. Such ρ-dependent polarity effects occur only if a *rut* sequence recognizable by ρ and a ρ-dependent terminator lie between the point of the mutation and the next downstream TIR.

Superficially, translational coupling and polarity due to transcription termination have similar effects; in both cases, blocking the translation of one coding sequence affects the synthesis of another polypeptide encoded downstream on the same mRNA. However, the molecular bases of the two phenomena are completely different.

Protein Folding and Degradation

Translating the information in an mRNA into a polypeptide chain is only the first step in making an active protein. To be active, the polypeptide must fold into its final conformation. This is the most stable state of the protein and is determined by the primary structure of its polypeptides. Whereas some proteins fold efficiently into their active states, other proteins may need the assistance of other factors to increase the rate of folding into the active state and to prevent misfolding into an inactive state.

Protein Chaperones

Proteins called **chaperones** help other proteins fold into their final conformations. Some chaperones are dedicated to the folding of only one other protein, while others are general chaperones that help many different proteins to fold. We discuss only general chaperones here.

THE DnaK PROTEIN AND OTHER Hsp70 CHAPERONES

The **Hsp70** family of chaperones is the most prevalent and ubiquitous type of general chaperone, existing in all types of cells with the possible exception of some archaea (see Bukau and Horwich, Suggested Reading). These chaperones are highly conserved evolutionarily. Chaperones in this family are called the Hsp70 proteins because they are about 70 kDa in size and because more of them are made (along with many other proteins) if cells are subjected to a sudden increase in temperature, or "heat shock" (see chapter 12); other stresses that denature proteins (such as ethanol) can have the same effect. Synthesis of chaperones increases after such stresses to help refold proteins that have been denatured by the environmental stress, although they also help to fold proteins under normal conditions. The Hsp70 type of chaperone was first discovered in *E. coli*, where it was given the name **DnaK** because it is required to assemble the DNA replication apparatus of phage λ and so is required for λ DNA replication. This name for the Hsp70 chaperone in *E. coli* is still widely used in spite of being a misnomer, because the chaperone has nothing directly to do with DNA but functions more generally in protein folding. In its role as a heat shock protein, the DnaK protein of *E. coli* also functions as a cellular thermometer, regulating the synthesis of other proteins in response to heat shock (see chapter 12).

To understand how Hsp70 chaperones, including DnaK, help fold proteins, it is necessary to understand something about the structure of most proteins. Proteins are made up of chains of amino acids that are folded up into well-defined structures, which are often rounded or globular. The amino acids that make up proteins can be charged, polar, or hydrophobic (see the inside front cover for a list). Amino acids that are charged (either acidic or

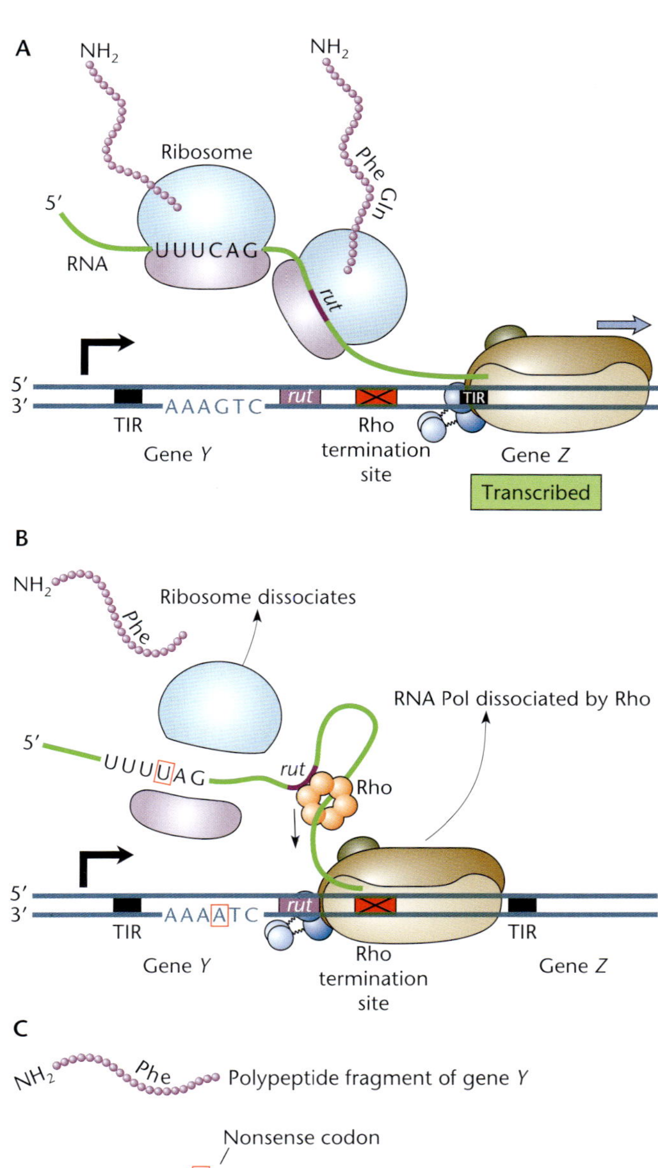

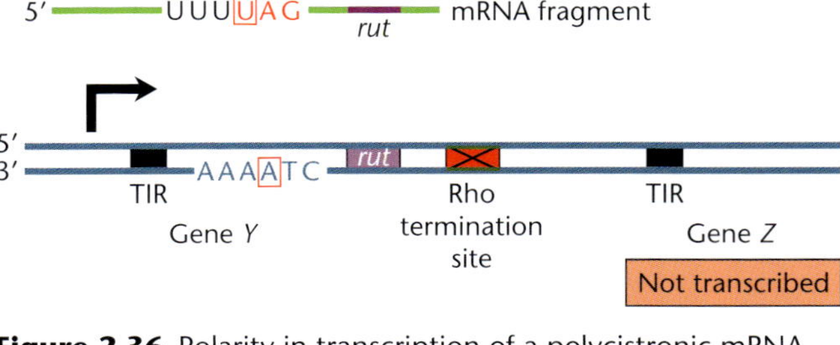

Figure 2.36 Polarity in transcription of a polycistronic mRNA transcribed from p_{YZ}. **(A)** The *rut* site in gene *Y* is normally masked by ribosomes translating the gene *Y* mRNA. **(B)** If translation is blocked in gene *Y* by a mutation that changes the codon CAG to UAG (boxed in red), the ρ factor can bind to the mRNA and cause transcription termination before the RNA polymerase reaches gene *Z*. **(C)** Only fragments of the gene *Y* protein and mRNA are produced, and gene *Z* is not even transcribed into mRNA.

basic) or polar tend to be more soluble in water and are called hydrophilic (water loving). Amino acids that are not charged or polar are hydrophobic (water fearing) and tend to be in the inside of the globular protein among other hydrophobic amino acids and away from the water on the surface. If the hydrophobic amino acids are exposed, they tend to associate with hydrophobic amino acids on other proteins and cause the proteins to precipitate. This is essentially what happens when you cook an egg. High temperatures cause the proteins in the egg to unfold, exposing their hydrophobic regions, which then associate with each other, causing the proteins to precipitate into a solid white mass.

The Hsp70-type chaperones help proteins fold by binding to the hydrophobic regions in denatured proteins and nascent proteins as they emerge from the ribosome and keeping these regions from binding to each other prematurely as the protein folds. The Hsp70 proteins have an ATPase activity that, by cleaving bound ATP to ADP, helps the chaperone to sequentially bind to, and dissociate from, the hydrophobic regions of the protein they are helping to fold. The Hsp70-type chaperones are directed in their protein-folding role by smaller proteins called **cochaperones**. The major cochaperones in *E. coli* were named DnaJ and GrpE, again for historical reasons. The DnaJ cochaperone helps DnaK to recognize some proteins and to cycle on and off of the proteins by regulating its ATPase activity. It can also sometimes function as a chaperone by itself. The GrpE protein is a nucleotide exchange protein that helps regenerate the ATP-bound form of DnaK from the ADP-bound form, allowing the cycle to continue.

TRIGGER FACTOR AND OTHER CHAPERONES

Given the prevalence and central role of DnaK in the cell, it came as a surprise that *E. coli* mutants that lack DnaK still multiply, albeit slowly. In fact, the only reason they are sick at all is because they are making too many copies of the other heat shock proteins, since DnaK also regulates the heat shock response (see chapter 12). One reason why cells lacking DnaK are not dead is that other chaperones can substitute for it. One of these is **trigger factor**. This type of chaperone has so far been found only in bacteria, and much less is known about it. It binds close to the exit pore of the ribosome and helps proteins fold as they emerge from the ribosome. It is also a **prolyl isomerase**. Of all the amino acids, only proline has an asymmetric carbon, which allows it to exist in two isomers. Trigger factor can convert the prolines in a protein from one isomer to the other. There are many other examples of chaperones that act as prolyl isomerases.

Another set of chaperones, including ClpA, ClpB, and ClpX, form cylinders and unfold misfolded proteins by sucking them through the cylinder. This takes energy, which is derived from cleavage of ATP. Some of them, including ClpA and ClpX, can also feed the unfolded proteins directly into an associated protease called ClpP, which degrades the unfolded protein. Association with ClpP switches the function of ClpA and ClpX from protein folding to protein degradation. ClpB, another cylindrical chaperone, does not associate with a protease but seems to cooperate with the small heat shock proteins IbpA and IbpB to help redissolve precipitated proteins so that they can be refolded by DnaK (see Mogk et al., Suggested Reading).

CHAPERONINS

In addition to the relatively simple protein chaperones, cells contain much larger structures that help proteins fold. These large structures are called **chaperonins**, and they exist in all forms of life, including the archaea and eukaryotes. They are composed of two large cylinders with hollow chambers held together back to back with openings at their ends (Figure 2.37). They help fold a misfolded protein by taking it up into one of the chambers. A cap, called a **cochaperonin**, is then put on the chamber, and the protein folds within the more hospitable environment of the chamber. A more detailed model for what happens in the chamber and how this helps a protein fold is suggested by the structure (see Wang and Boisvert, Suggested Reading). When the misfolded protein is first taken up, the lining of the chamber consists of mostly hydrophobic amino acids that bind to the exposed hydrophobic regions of the misfolded protein. When the cochaperonin cap is put onto this chamber (the *cis*-chamber) and the bound ATP is cleaved to ADP, the lining may switch to being mostly hydrophilic amino acids, driving the more hydrophobic regions of the misfolded protein to the interior of the protein, where they usually reside in the folded protein. Binding of ATP and an unfolded protein to the other chamber (the *trans*-chamber) causes the cap to come off the *cis*-chamber, releasing the folded protein and preparing the other chamber to take up the misfolded protein and take its turn being the *cis*-chamber. This process takes a lot of energy, and a number of ATP molecules (one for each subunit of one chamber of the chaperonin, which is seven in *E. coli* [see below]) are cleaved to ADP in each cycle. It is a mystery why chaperonins have two chambers and why the folding has to alternate between the two chambers. Chaperonins composed of only one chamber function more poorly, although they might retain some of their activity (see Sun et al., Suggested Reading). The communication to one chamber that ATP and unfolded protein are bound to the opposite chamber is one of the most remarkable examples of **allosterism** known, especially considering that the chambers, while touching back to back, are composed of separate polypeptides.

The first chaperonin was discovered in *E. coli*, where it was named GroEL because it helps assemble the E protein of phage λ into the phage head. The GroEL chaperonin consists of 14 identical polypeptides (7 making up each cylinder) of 60 kDa. Its cochaperonin cap is called GroES, which is also made up of 7 subunits, each 10 kDa in size. Unlike DnaK and the other chaperones, GroEL and GroES are required for *E. coli* growth, even at lower temperatures. The GroEL chaperonin is known to be required for folding of some essential proteins, which explains why GroEL is essential. The chaperonins come in two general types called the group I and group II chaperonins. The group I chaperonins, related to GroEL, are composed of 60-kDa subunits and are found in all bacteria and in the mitochondria and chloroplasts of eukaryotes; this makes sense, since these organelles are derived

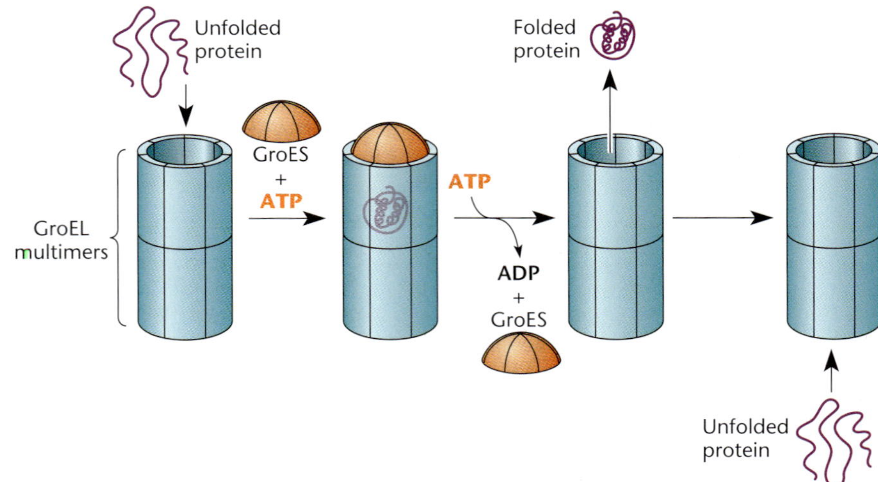

Figure 2.37 Chaperonins. The GroEL (Hsp60)-type chaperonin multimers form two connected cylinders. A denatured or unfolded protein enters the chamber in one of the cylinders, and the chamber is capped by the cochaperonin GroES (Hsp10). The denatured protein can then be helped to fold in the chamber and is released. Entry of a second unfolded protein into the second cylinder helps to trigger release of the folded protein.

from bacteria (see the introductory chapter). These chaperonins and their cochaperonins are induced by heat shock and other stresses, so they are called the Hsp60 proteins and Hsp10 proteins for heat shock 60-kDa and 10-kDa proteins (see Bukau and Horwich, Suggested Reading). The group II chaperonins are found in the archaea and in the cytoplasm of eukaryotes. They have very little amino acid sequence in common with the group I chaperonins and are not composed of identical subunits (i.e., they are mixed multimers) and often have eight or more polypeptide subunits per cylinder. Furthermore, if they have a cochaperonin cap, it might be attached to the opening of the chamber rather than being detachable, like GroES. They may also be more dedicated and fold only a small subset of proteins, including actin in the eukaryotic cytoplasm. Nevertheless, the two types form similar cylindrical structures and presumably use a similar two-stroke mechanism to help fold proteins. Note that this is yet another example where the archaea and eukaryotes are similar to each other and different from the bacteria (see the introductory chapter).

Protein Degradation

As noted above, protein folding can be coupled to protein degradation through the association of chaperones such as ClpA and ClpX with proteases such as ClpP. Complexes such as ClpAP and ClpXP are members of the family of ATP-dependent proteases, which use cleavage of ATP to provide the energy for protein unfolding and proteolysis (see Baker and Sauer, Suggested Reading). These enzymatic machineries are important not only for the destruction of misfolded and denatured proteins, but also for regulated proteolysis to target specific substrate proteins under specific conditions. Their activity can be directed by the presence of a specific degradation tag, a short protein sequence that increases affinity for a specific protease; these tags are similar to the sequence added by tmRNA to target incomplete proteins for degradation (see above). This tag may be present all the time within the target protein, but it becomes available for recognition only under certain conditions, which allows the target protein to be stable under some conditions and unstable under other conditions. Proteolysis can also be controlled by adaptor proteins that deliver specific target proteins to specific proteases. Regulated proteolysis is discussed as a mechanism for gene regulation in chapters 11 and 12.

The *Actinobacteria*, including important pathogens such as *Mycobacterium tuberculosis*, utilize a proteasome structure similar to that used in eukaryotes to mediate degradation of specific protein targets. This structure is barrel-shaped and carries out both ATP-dependent and ATP-independent proteolysis. Eukaryotes use a specific protein tag called ubiquitination to direct protein substrates to the proteasome; in *Actinobacteria*, this is replaced by a different protein modification called pupylation (see Becker and Darwin, Suggested Reading).

Protein Localization

About one-fifth of the proteins made in a bacterium do not remain in the cytoplasm and instead are transported or exported into or through the surrounding membranes. The terminology is often used loosely, but we refer to proteins that leave the cytoplasm as being **transported**. The process of transferring them through one or both membranes is **secretion**. If they are transferred through both membranes to the exterior of the cell, they are **exported**. Correspondingly, proteins that remain in either the inner or outer membrane are **inner membrane proteins** or **outer membrane proteins**, while those that remain in the periplasmic space are **periplasmic proteins**. Proteins that are passed all the way out of the cell into the surrounding environment are **exported proteins**.

By far the largest group of proteins that are transported from the cytoplasm are destined for the inner membrane. Inner membrane proteins often extend through the membrane a number of times and have some stretches that are in the periplasm and other stretches that are in the cytoplasm. The stretches that traverse the membrane have mostly uncharged, nonpolar (hydrophobic [see the inner cover]) amino acids, which make them more soluble in the membranes. A stretch of about 20 mostly hydrophobic amino acids is long enough to extend from one side of the bipolar lipid membrane to the other, and such stretches in proteins are called the **transmembrane domains**. The less hydrophobic stretches between them are called the **cytoplasmic domains** or **periplasmic domains**, depending on whether they extend into the cytoplasm on one side of the membrane or into the periplasm on the other side. Proteins with domains on both sides of the membrane are called **transmembrane proteins** and are very important, because they allow communication from outside the cell to the cytoplasm. Some transmemembrane proteins that play such a communicating role are discussed in chapter 12.

The Translocase System

Transported proteins usually contain many amino acids that are either polar or charged (basic or acidic), which makes it difficult for them to pass through the membranes. They must be helped in their translocation through the membrane by other specialized proteins. Some of these proteins form a channel in the membrane. Some transported proteins make their own dedicated channel, but most use the more general channel called the **translocase**, so named because its function is to translocate proteins.

A current picture of the structure of the translocase that helps proteins pass through the inner membrane, as well as how it works, is outlined in Figure 2.38. We can

A Export channel

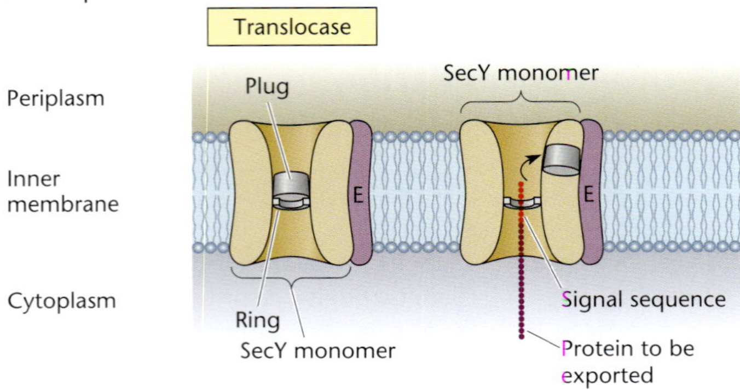

B Posttranslational export

Gram-negative outer membrane

Periplasm

Lep cleaves signal sequence

Protein is secreted or folded

Inner membrane

E

E

E

Cytoplasm

SecA

ATP

P_i

SecA

ADP

Lep

SecB

SecB

Completed protein for export

C Cotranslational transport

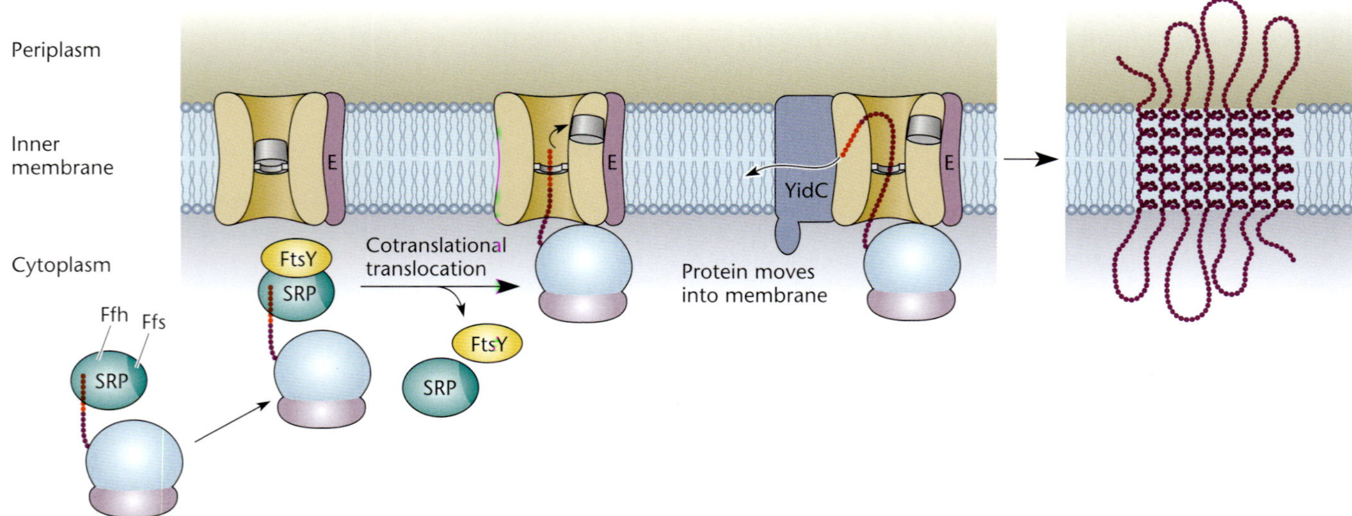

predict some of the features this channel must have. It must have a relatively hydrophilic inner channel through which charged and polar amino acids can pass. It also must normally be closed and should open only when a protein is passing through it; otherwise, other proteins and small molecules would leak in and out of the cell through the channel. Even the leakage of molecules as small as protons cannot be tolerated, because it would destroy the proton motive force. The channel is made up of one each of three proteins, SecY, SecE, and SecG, and is therefore called the **SecYEG channel** or **SecYEG translocase**. These three proteins form a heterotrimer made up of one each of the three different polypeptides. The SecY protein is by far the largest of the three proteins and forms the major part of the channel, while the other two proteins play more ancillary, albeit important, roles. One heterotrimer can form a large enough channel to let an unfolded protein through (see van den Berg et al., Suggested Reading), but it seems likely that more than one of these heterotrimers is involved.

Besides forming the major part of the channel, a region of the SecY protein forms a hydrophobic "plug," which opens only when a protein is passing through (Figure 2.38). The binding of a **signal sequence** (see below) in a protein to be transported causes the plug to move over toward SecE on the side of the channel, opening the channel. As a result, only proteins that have a *bona fide* signal sequence can be translocated.

The Signal Sequence

As mentioned above, the defining feature of proteins that are to be transported into the inner membrane or beyond by the SecYEG channel is the presence at their N termini of a signal sequence. The nature and fate of this signal sequence depend upon the ultimate destination of the transported protein. For proteins that are to be transported through the inner membrane into the periplasm and beyond, the signal sequence is approximately 20 amino acids long and consists of a basic region at the N terminus, followed by a mostly hydrophobic region and then a region with some polar amino acids. In contrast, most proteins whose final destination is the inner membrane merely use their first N-terminal transmembrane domain as a signal sequence.

If the protein is to be secreted through the membrane, the signal sequence is removed by a protease as the protein passes through the SecYEG channel (Figure 2.38B). The most prevalent of the proteases that clip off signal sequences in *E. coli* is the **Lep protease** (for leader peptide protease), but there is at least one other, more specialized protease called LspA, which removes the leader sequence from some lipoproteins destined for the outer membrane. Proteins that are destined to be transported beyond the inner membrane but have just been synthesized and so still retain their signal sequences are called presecretory proteins. When the short signal sequence is removed in the SecYEG channel, the presecretory protein becomes somewhat shorter before it reaches its final destination in the periplasm or the outer membrane or outside the cell.

The Targeting Factors

The targeting factors recognize proteins to be transported into or through the inner membrane and help target them to the membrane. Which type of signal sequence a protein has determines which of the targeting factors directs it to the SecYEG translocon. Enteric bacteria like *E. coli* have at least two separate systems that target proteins to and through the membranes. The **SecB** system is dedicated to proteins that are directed through the inner membrane into the periplasm or exported from the cell. The **signal recognition particle (SRP) system** seems to be dedicated to proteins that are mostly destined to reside in the inner membrane. Another protein, SecA, participates in both pathways, at least for some proteins; it is found in all bacteria, but not in archaea or eukaryotes, although in eukaryotes other proteins may play a similar role.

THE SecB PATHWAY

Proteins that have a removable signal sequence and are transported through the inner membrane into the periplasm or beyond are most often targeted by the SecB system in *E. coli* and the other bacteria that have it. The SecB

Figure 2.38 Protein transport systems. **(A)** Cutaway view of the secretion *sec* channel. SecY, SecE, and SecG (not shown) form the translocase. SecY forms the channel, ring, and plug. The signal sequence of the transported protein moves the plug toward SecE. **(B)** Posttranslational secretion by the SecB-SecA system. SecB keeps the protein unfolded until it binds to SecA, which interacts with SecY. The signal sequence is removed, in this case by Lep protease. The exported protein is folded in the periplasm or may be secreted across the outer membrane by one of the dedicated secretion systems. **(C)** Cotranslational transport by the signal recognition particle (SRP) system. SRP binds to the first transmembrane domain as it emerges from the ribosome and then binds to the FtsY docking protein, bringing the ribosome to interact with SecY. The protein is translated, driving it into the SecYEG channel. The transmembrane domains of the protein somehow escape through the side of the channel into the membrane, in some cases with the help of the YidC protein, as shown.

protein is a specialized chaperone that binds to presecretory proteins either cotranslationally (e.g., as soon as the N-terminal region of the polypeptide emerges from the ribosome) or after they are completely synthesized, thereby preventing them from folding prematurely and ensuring that the signal sequence is exposed. The SecB chaperone passes the unfolded protein to SecA, which facilitates the association of the protein with the SecYEG channel, perhaps by binding simultaneously to the signal sequence and to a SecYEG heterotrimer (Figure 2.38B). Some bacteria have paralogs of SecA that may be dedicated to transporting only one or very few proteins. After SecA binds to the channel, the cleavage of ATP to ADP on SecA provides the energy to drive the protein into the channel, aided by the proton motive force of the membrane. As the protein passes through the channel, it loses its signal sequence, as shown in the figure. SecB is not an essential protein, and the cell can use DnaK or other general chaperones as substitutes for SecB to help transport some proteins.

THE SRP PATHWAY

The SRP pathway in bacteria generally targets proteins that are to remain in the inner membrane. It consists of a particle (the SRP) made up of both a small 4.5S RNA, encoded by the *ffs* gene, and at least one protein, Ffh, as well as a specific receptor on the membrane, called FtsY in *E. coli*, to which the SRP binds. FtsY is sometimes referred to as the docking protein because it "docks" proteins targeted by the SRP pathway to the SecYEG channel in the membrane. The *ftsY* (filament temperature-sensitive Y) gene was originally identified through temperature-sensitive mutations that cause *E. coli* not to divide properly and to form long filaments of many cells linked end to end at higher temperatures, but its role in cell division is indirect.

Figure 2.38C illustrates how the SRP system works. The SRP binds to the first hydrophobic transmembrane sequence of an inner membrane protein as this region of the protein emerges from the ribosome. The complex binds to the membrane, and synthesis of the protein continues, feeding the protein directly into the SecYEG translocon as the protein emerges from the ribosome. The energy of translation due to cleavage of GTP to GDP drives the polypeptide out of the ribosome into the SecYEG translocon, replacing the role of SecA, although the SecA protein might still be required for transmembrane proteins with long periplasmic domains.

The process of translating a protein as it is inserted into the translocon is called **cotranslational translocation**. There is a good reason why proteins destined for the inner membrane are cotranslated with their insertion into the translocon in the membrane while proteins targeted by the SecB pathway can first be translated in their entirety and then inserted into the translocon. Inner membrane proteins are much more hydrophobic than exported proteins and would form an insoluble aggregate in the aqueous cytoplasm if they were translated in their entirety before being transported into the membrane (see Lee and Bernstein, Suggested Reading).

What happens after an inner membrane protein enters the SecYEG channel is less clear. The transmembrane domains of the protein must escape the SecYEG channel and enter the surrounding membrane, while the periplasmic and cytoplasmic domains must stay in the correct compartments. Presumably, the SecYEG channel has a lateral gate that opens and allows the transmembrane domains of the protein to escape into the membrane. Another inner membrane protein called YidC might help in this process (Figure 2.38C) (see Xie and Dalbey, Suggested Reading). YidC seems to be required for the lateral escape of some proteins but not others. Some inner membrane proteins bypass SecYEG altogether and require only YidC to enter the inner membrane.

Sec Systems of Archaea and Eukaryotes

Archaea and eukaryotes do not have SecB or SecA and use the SRP system to translocate all exported proteins. Although they lack SecA, they may have other systems that help direct already translated proteins to the translocon. The translocon itself was first discovered in eukaryotes and is composed of three proteins that form similar structures in all three kingdoms of life. The amino acid sequences of the SecY and SecE subunits are similar in all three kingdoms; only the sequence of the third subunit (SecG in bacteria) is very different in eukaryotes and archaea, where it may have different functions. While eukaryotes have other such channels, the translocase, which helps transported proteins to enter the endoplasmic reticulum of eukaryotic cells, is the one most similar to the SecYEG channel of bacteria.

The SRP system was also first described in eukaryotes, where it is much larger, consisting of a 300-nucleotide RNA and eight proteins, six in the SRP and two in the docking protein, called the SRP receptor. However, some of the proteins in eukaryotes are very similar to those in bacteria, such as the 54-kDa SRP protein in eukaryotes, which is similar to the Ffh protein in the SRP of bacteria. The SRP system of eukaryotes targets both membrane and presecretory proteins to the endoplasmic reticulum.

The Tat Secretion Pathway

Only proteins that have not yet folded and so are still long, flexible polypeptides can be transported by the narrow SecYEG channel. Once folded into their final three-dimensional structure, proteins are much too wide to fit through the channel. However, some proteins must fold in the cytoplasm before they can be transported. These include membrane proteins that contain redox factors, such as molybdopterin and FeS clusters, which are synthesized in the cytoplasm and can be inserted into the protein only after it has been folded. Other examples of proteins that are transported after they are folded are some heterodimers in which only one member has a signal sequence, so the other partner would be left behind if the signal sequence-containing polypeptide is transported before the two polypeptides have combined and folded. Folded proteins and complexes are transported by the Tat system.

STRUCTURE OF THE TAT SYSTEM

The Tat system of *E. coli* has three subunits, TatA, TatB, and TatC, while that of *B. subtilis* has only two, with TatA and TatB seemingly combined into one larger subunit. In *E. coli*, TatB and TatC bind the signal peptide on the protein to be transported and then recruit TatA, which forms the channel in the membranes. In this way, the channel may form only when there is a protein to be transported. Unlike the Sec translocon, which uses both the energy of ATP cleavage by SecA and the proton motor force to drive the protein through the channel, the Tat system may use only the latter.

The Tat Signal Sequence

The signal sequence recognized by the Tat system is structurally similar to the signal sequence recognized by the Sec system, with a positively charged region followed by a longer hydrophobic region and a polar region, and is cleaved from the protein as it passes through the channel. However, it is somewhat longer, especially in the positively charged region. Two of the positively charged amino acids at the junction of the charged and hydrophobic regions are usually arginines, which give the Tat system its name (twin-arginine transport). The arginines are found in the motif S-R-R, although the first of the twin arginines is sometimes a lysine (K). This sequence is followed by two hydrophobic amino acids, usually F and L, and then often by a K.

The presence of this particular signal sequence at the N terminus of a newly synthesized protein targets the protein for transport by the Tat system rather than by the SecYEG channel. This raises an interesting question. How does the system know the protein has already folded properly and contains all the needed cofactors, etc., so that it is time to transport it? The Tat system needs a "quality control" system to ensure that it transports only properly folded proteins and does not transport proteins that are unfolded or only partially folded. This quality control system should also be specific for each protein to be transported, since each type of folded protein has a unique structure (see above). *E. coli* solves this problem by encoding dedicated proteins that specifically bind to the Tat signal sequence of only one type of protein and come off only when that protein has folded properly (see Palmer et al., Suggested Reading).

Tat Systems in Other Organisms

Most bacteria and archaea, as well as the chloroplasts of plants (which descended from cyanobacteria [see the introduction]), have a Tat secretion system, although they might differ from that of *E. coli* in the number of subunits. Some bacteria, including *B. subtilis*, have two or more Tat systems. Some of them are dedicated to the transport of only one or very few proteins.

Disulfide Bonds

Another characteristic of proteins that are exported to the periplasm or secreted outside the cell is that many of them have disulfide bonds between cysteines (see the inside front cover). In other words, two of the cysteines in the protein are held together by covalent bonds between their sulfides. The sulfur atom of a cysteine in a disulfide bond is in its oxidized form because one of its electrons is shared by the two sulfurs, while the sulfur atom of an unbound cysteine is in its reduced form because it has an extra electron. These disulfide bonds can be between two cysteines in the same polypeptide chain or between cysteines in different polypeptide chains. Exported proteins need the covalent disulfide bonds to hold them together in the harsh environments of the periplasm and outside the cell. Failure to form the correct disulfide bonds or formation of disulfide bonds between the wrong cysteines can result in inactivity of the protein

The disulfide bonds in proteins are formed by enzymes called **disulfide oxidoreductases** (DsbA, DsbB, etc.) as the proteins pass through the oxidizing environment in the periplasmic space between the inner and outer membranes of bacteria that contain outer membranes (or at the outer cell surface of bacteria that lack the outer membrane and periplasm). Proteins that are found inside the cell in the cytoplasm lack disulfide bonds because of the "reducing atmosphere" inside the cytoplasm due to the presence of high concentrations of small reducing molecules, such as glutathione, thioredoxin, and bacillithiol. In fact, the appearance of disulfide bonds in some cytoplasmic regulatory proteins is taken as a signal by the cell that oxidizing chemicals are accumulating in the cell and that proteins should be made to combat the potentially lethal oxidative chemical stress.

Protein Secretion and Export

Some proteins are transported through the membrane(s) to the outside of the cell, where they can remain attached, enter the surrounding medium, or even directly enter another cell. The process differs markedly between cells that have or lack an outer membrane. Bacteria that have an outer membrane require translocation from the periplasm through the extremely hydrophobic outer membrane. Because of the additional challenge created by the outer membrane, these bacteria have developed elaborate specialized structures to export proteins. Many of these play important roles in bacterial pathogenesis, so they have attracted considerable attention.

Protein Secretion Systems in Bacteria with an Outer Membrane

At least six basic types of protein secretion systems, imaginatively named types I to VI, have been identified. All of these secretion systems rely on channels in the outer membrane (called β-barrels or secretins) formed from β-sheets organized in a ring (see Figure 2.20 for an explanation of protein secondary and tertiary structures). The β-barrels are assembled so that the side chains of charged and polar amino acids tend to be in the center of the barrel, where they are in contact with hydrophilic proteins that are passing through, while the side chains of hydrophobic amino acids are on the outside of the barrel in contact with the very hydrophobic surrounding membrane. Assembly of the β-barrels requires a complex of proteins called the Bam complex (BamA, -B, -C, etc.) and periplasmic chaperones, including Skp.

Having channels in the outer membrane presents some of the same problems associated with having channels in the cytoplasmic membrane, such as the SecYEG channel. For example, how do they select the proteins that are to go through without letting others through, and how do they keep smaller molecules from going in and out? This process is called **channel gating**; the gate is open only when the protein being exported passes through. A second issue is the source of the energy to export a protein through the outer membrane. There is no ATP or GTP in the periplasmic space to provide energy, and the outer membrane is not known to have a proton gradient across it to create an electric field. In this section, we describe mechanisms used by the various secretion systems for solving these problems and mention some examples of proteins exported by each of the systems.

TYPE I SECRETION SYSTEMS

Type I secretion systems (T1SS) secrete a protein directly from the cytoplasm to the outside of the cell (Figure 2.39). They are different from the other types of secretion systems and more closely related to a large family of ATP-

Figure 2.39 Schematic representation of the type I, II, III, and IV protein secretion systems. The examples shown are for type I (hemolysin A [HlyA] of *Escherichia coli*), type II (pullulanase of *Klebsiella oxytoca*), type III (Yop of *Yersinia*), and type IV (*vir* of *Agrobacterium tumefaciens*). EM, extracellular milieu; OM, outer membrane; Peri, periplasmic space; IM, inner membrane; Cyto, cytoplasm. The arrows indicate which pathways use the Sec and Tat pathways through the inner membrane. Modified from Henderson IR, Navarro Garcia F, et al, *Microbiol Mol Biol Rev* **68:**692–744, 2004.

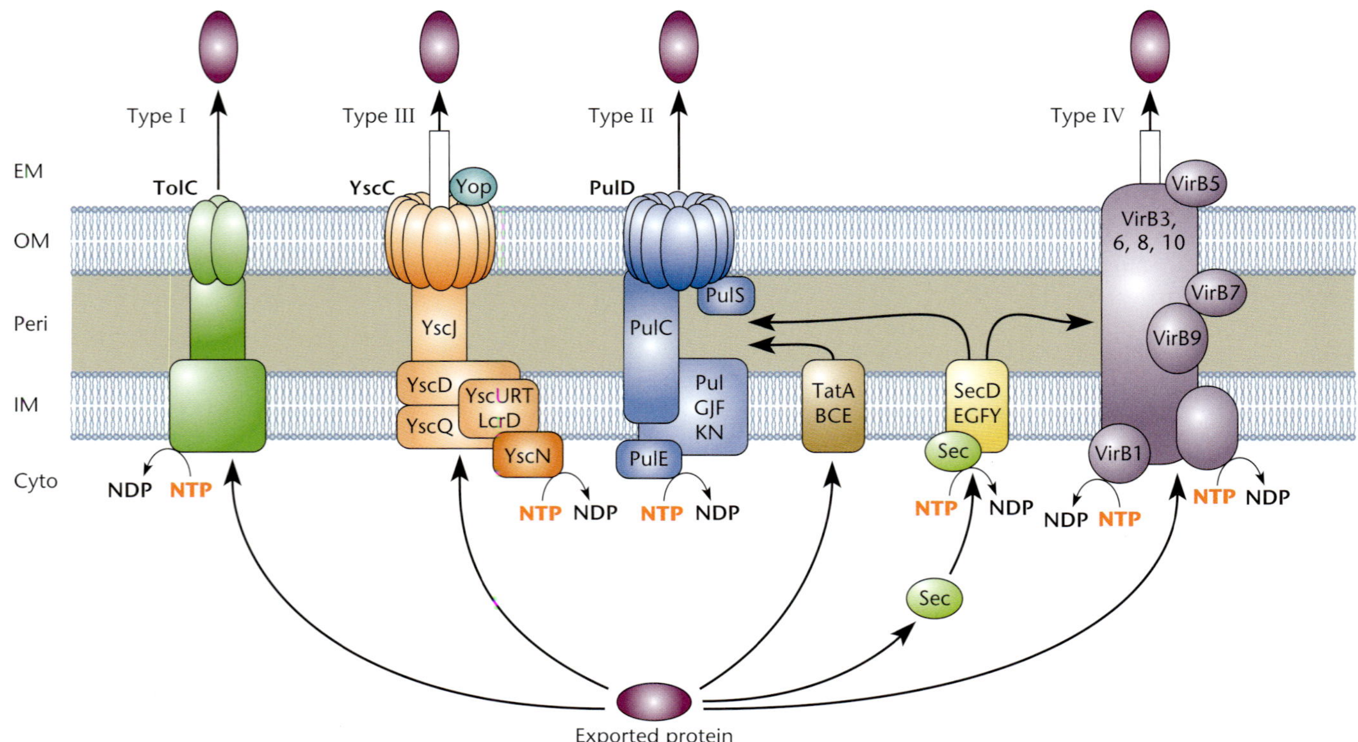

binding cassette (ABC) transporters that export small molecules, including antibiotics and toxins, from the cell. The ABC transporters tend to be more specialized, exporting only certain molecules from the cell. To get the protein through the inner membrane, T1SS use a dedicated system that consists of two proteins, an ABC-type protein in the inner membrane and an integral membrane protein that bridges the inner and outer membranes. To get through the outer membrane, T1SS use a multiuse protein, TolC, that forms the β-barrel channel in the outer membrane. Because the TolC channel has other uses and also exports other molecules, including toxic compounds, from the cell, it is recruited to the T1SS only when the specific protein is to be secreted. When the molecule to be secreted binds to the ABC protein, the integral membrane protein recruits TolC, which then forms the β-barrel in the outer membrane. The cleavage of ATP by the ABC protein presumably provides the energy to push the secreted protein all the way through the TolC channel to the outside of the cell.

The classical example of a protein secreted by a T1SS is the HylA hemolysin protein of pathogenic *E. coli*. This toxin inserts itself into the plasma membrane of eukaryotic cells, creating pores that allow the contents to leak out. It uses a dedicated T1SS composed of HylB (the ABC protein) and HylD (the integral membrane protein). Because HylA is not transported through the inner membrane by either the SecYEG channel or the Tat system, it does not contain a cleavable N-terminal signal sequence. Instead, like all proteins secreted by T1SS, it has a sequence at its carboxyl terminus that is recognized by the ABC transporter but, unlike a signal sequence, is not cleaved off as the protein is exported.

The TolC channel has been crystallized and its structure determined (see Koronakis et al., Suggested Reading). This structure has provided interesting insights into the structure of β-barrels in general and how they can be gated and opened to transport specific molecules. Briefly, three TolC polypeptides come together to form the channel through the outer membrane. Each of these monomers contributes four transmembrane domains to form a β-barrel that is always open on one side of the outer membrane, the side on the outside of the cell. In addition, each monomer has four longer α-helical domains that are long enough to extend all the way across the periplasm. These four α-helical domains contribute to the formation of a second channel that is aligned with the first channel and traverses the periplasm. Because of these two channels, the secreted protein can be transported all the way from the inner membrane to the outside of the cell. In addition, the channel in the periplasm can open and close and therefore "gate" the channel. When a protein is being transported and the TolC channel is recruited, the α-helical domains of the periplasmic channel may rotate, which untwists them and opens the gate on the periplasmic side. The molecule is then secreted all the way through both channels to the outside of the cell.

TYPE II SECRETION SYSTEMS

Type II secretion systems (T2SS) are very complex, consisting of as many as 15 different proteins (Figure 2.39). Most of these proteins are in the inner membrane and periplasm, and only 1 is in the outer membrane, where 12 of the secretin polypeptides come together to form a large β-barrel with a pore large enough to pass already folded proteins. The formation of this channel requires the participation of normal cellular lipoproteins that may become part of the structure. The secretin protein has a long N terminus that extends through the periplasm to make contact with other proteins of the T2SS in the inner membrane. This periplasmic portion of the secretin may also gate the channel, as with the TolC channel.

Even though many of the components of the T2SS are in the inner membrane, they use either the SecYEG channel or the Tat pathway to get their substrates through the inner membrane. Therefore, proteins transported by this system have either the Sec type or the Tat type of cleavable signal sequences at their N termini. Protein folding is usually completed in the periplasm before transport through the outer membrane. Some of the periplasmic and inner membrane proteins of the secretion system are related to components of pili and have been called pseudopilin proteins (see chapter 4). It has been proposed that the formation and retraction of these pseudopili work like a piston to push the protein through the secretin channel in the outer membrane to the outside of the cell. In this way, the energy for secretion could come from the inner membrane or the cytoplasm, as shown in the figure, since, as mentioned above, there is no source of energy in the periplasm. In support of this model, the pseudopili have been seen to produce pili outside an *E. coli* cell when the gene for the pilin-like protein was cloned and overproduced in *E. coli*.

Some examples of proteins secreted by T2SS are the pullulanase of *Klebsiella oxytoca* and the cholera toxin of *Vibrio cholerae*. The pullulanase degrades starch, and the cholera toxin is responsible for the watery diarrhea associated with the disease cholera (see chapter 12). The cholera toxin is composed of two subunits, A and B, and after transport by the SecYEG channel, one of the A and five of the B subunits assemble in the periplasm, followed by secretion through the secretin channel and into the intestine of the vertebrate host. The associated B subunit then assists the A subunit into mucosal cells, where it ADP-ribosylates (adds ADP) to a membrane protein that regulates the adenylate cyclase. This disrupts the signaling pathways and causes diarrhea. T2SS are also related to some DNA transfer systems used in transformation (see chapter 6) and are closely related to the systems that assemble type IV pili on the cell surface (see below).

TYPE III SECRETION SYSTEMS

Type III secretion systems (T3SS) are probably the most impressive of the secretion systems (see Galan and Waksman, Suggested Reading). They form a syringe-like structure composed of about 20 proteins, which takes up virulence proteins called effectors from the cytoplasm of the bacterium and injects them directly through both membranes into a eukaryotic cell (Figure 2.39). For this reason, they are sometimes called **injectisomes**. They exist in many Gram-negative animal pathogens, including *Salmonella* and *Yersinia*, but are also found in many plant pathogens, including *Erwinia* and *Xanthomonas*. One striking feature of T3SS is how similar they are in both animal and plant pathogens. Where they differ is in the protuberance, called the needle, that penetrates the eukaryotic cell to allow injection through the wall into the host cell cytoplasm. This difference is expected, because animal and plant cells are surrounded by very different cell surfaces.

T3SS are usually encoded on **pathogenicity islands** (see chapter 12), and their genes are induced only when the bacterium encounters its host or if they are cultivated under conditions that are designed to mimic the host. The effector proteins they inject are also encoded by the same DNA element, and their genes are turned on at the same time. The part of the injectisome that traverses the outer membrane is composed of a secretin protein related to those of the T2SS. It also forms a β-barrel composed of about a dozen secretin subunits. Like the secretins of the T2SS, these might require normal bacterial lipoproteins, as well as other components of the secretion machinery, to assemble the channel in the membrane.

Effector proteins to be secreted by at least some T3SS contain a short sequence located on the N terminus of the protein; unlike the cleavable signal sequences used by the Sec and Tat systems, this signal is not cleaved off when the protein is secreted. Many of the effector proteins injected into eukaryotic cells are involved in subverting the host defenses against infection by bacteria. This can be illustrated by *Yersinia pestis*, the bacterium that causes bubonic plague and in which T3SS were first discovered. In animals, one of the first lines of defense against infecting bacteria is the macrophages, phagocytic white blood cells that engulf invading bacteria and destroy them by emitting a burst of oxidizing compounds. However, when a macrophage binds to a *Yersinia* cell, the bacterium injects effectors called Yop proteins into the macrophage cell before it can be engulfed. Once in the eukaryotic cell, these effectors disarm the cell by interfering with its signaling systems and thus preventing the macrophage from engulfing the bacterium. For example, one of the Yop proteins is a tyrosine phosphatase, which removes phosphates from proteins in a signal transduction system in the macrophage, blocking the signal to take up the bacterium and preventing the burst of oxidizing compounds. Some T3SS even inject proteins that provide receptors on the cell surface to which the bacterium can adsorb in order to enter the eukaryotic cell.

TYPE IV SECRETION SYSTEMS

Type IV secretion systems (T4SS) utilize a secretin-like protein (VirB9) that forms a β-barrel channel in the outer membrane and extends into the periplasm, where it makes contact with proteins in the inner membrane. However, unlike true secretins, it seems to require another outer membrane protein, VirB7, to make a channel. The VirB9 protein is covalently attached to the VirB7 protein, which in turn is covalently attached to the lipid membrane, making the structure very stable. A coupling protein (VirD4) binds specific proteins and targets them to the channel. The energy for secretion comes from the cleavage of ATP or GTP in the cytoplasm by channel-associated proteins (Figure 2.39).

T4SS are discussed in chapters 5 and 6, because they are also involved in DNA transfer during conjugation and transformation. The T-DNA transfer system of *Agrobacterium tumefaciens* has served as the prototype T4SS and is the one about which the most is known and to which all others are compared. Accordingly, the genes and proteins of other T4SS are numbered after their counterparts in the T-DNA transfer system, named the *vir* genes because of their role in virulence in plants. Some of the genes in the T-DNA transferred into plant cells cause growth of the plant cell, leading to the formation of tumors called crown galls. Others trigger the plant cells to make unusual compounds, called opines, that can be used by the bacterium as a carbon, nitrogen, and energy source (see Box 5.1). Like other T4SS, the *Agrobacterium* system also directly injects proteins into the plant cell, which makes it a *bona fide* protein secretion system.

TYPE V SECRETION SYSTEMS: AUTOTRANSPORTERS

All of the secretion systems discussed above use some sort of structure formed of β-sheets assembled into a ring called a β-barrel to get them through the outer membrane. Some of these β-barrels are part of the secretion apparatus itself, while others, like TolC, are recruited from other functions in the cell. However, in type V systems, secreted proteins carry their own β-barrel with them in the form of a domain of the protein that can create a β-barrel when it gets to the outer membrane. These proteins are called **autotransporters** because they transport themselves.

The mechanism used by autotransporters is illustrated in Figure 2.40, which also shows their basic structure.

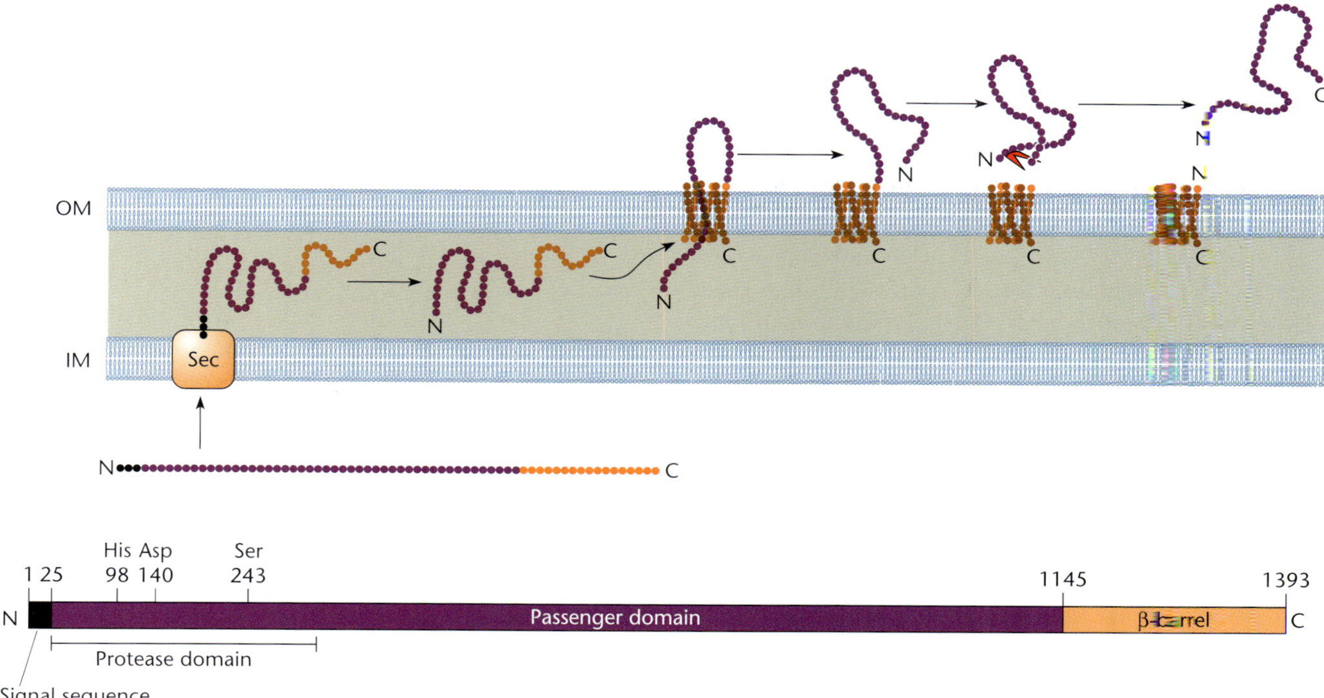

Figure 2.40 Structure and function of a typical autotransporter. A *Haemophilus influenzae* adhesin is shown; the length in amino acids of each domain, where known, is indicated by the number above the structure, as are some of the important amino acids in the protease domain. The transporter domain at the C terminus that forms a β-barrel in the outer membrane is shown in orange; the passenger domain and the protease domain that cleaves the passenger domain off the transporter domain outside the cell are shown in purple. The flexible linker domain is not indicated. The signal sequence that is cleaved off when the protein passes through the SecYEG (Sec) channel in the inner membrane is shown in black. OM, outer membrane; IM, inner membrane. Modified from Surana NK, Cotter SE, Yeo H-J, et al, *in* Waksman G, Caparon M, Hultgren S (ed), *Structural Biology of Bacterial Pathogenesis* (ASM Press, Washington, DC, 2005).

Most autotransporters consist of four domains, the translocator domain at the C terminus that forms a β-barrel in the outer membrane, an adjacent flexible linker domain (not shown) that may extend into the periplasm, a passenger domain that contains the functional part of the autotransported protein, and sometimes a protease domain that may cleave the passenger domain off the translocator domain after it passes through the channel formed by the translocator domain.

Autotransporters are typically transported through the inner membrane using the SecYEG channel, so they have a signal sequence that is cleaved off as they pass into the periplasm. Their translocator domain then enters the outer membrane, where it forms a 12-stranded β-barrel. This assembly does not occur by itself but requires the same accessory factors used for the assembly of many secretins, including the periplasmic chaperone, Skp, and the Bam complex (see above). The flexible linker domain guides the passenger domain into and through the channel to the outside of the cell. The passenger domain can then be cleaved off by its own protease domain or remain attached to the translocator domain and protrude outside the cell, depending on the function of the passenger domain. The source of the energy for autotransportation is unclear, since, as mentioned, there is no ATP or GTP in the periplasm and the outer membrane does not have a membrane potential. One possibility is that the autotransporter arrives at the periplasm in a “cocked” or high-energy state that drives its own transport.

The prototypical autotransporter is the immunoglobulin A protease of *Neisseria gonorrhoeae*. It is involved in evading the host immune system by cleaving IgA antibodies on mucosal surfaces. Most known autotransporters are large virulence proteins that perform various roles in bacterial pathogenesis or in helping to evade the host immune system. The IcsA protein of *Shigella flexneri*, a cause of bacterial dysentery, is localized to the outer membrane, where it recruits a host actin-regulating protein, which in turn recruits another host complex that polymerizes host actin into filaments, pushing the bacterium through the eukaryotic cell cytoplasm as part of the infection mechanism, a process called actin-based motility.

Chaperone-Usher Secretion

Chaperone-usher secretion is related to type V secretion. This type of secretion is often used to assemble some types of pilins on the cell surface, such as the P pilus of uropathogenic *E. coli*. The secretion system consists of three proteins, a β-barrel-forming protein in the outer membrane called the usher, a periplasmic protein called the chaperone, and the pilin subunit to be assembled on the cell surface. The pilin protein is transported through the inner membrane by the SecYEG channel and therefore has a cleavable signal sequence. Once in the periplasm, the pilin protein is bound by the dedicated periplasmic chaperone. However, rather than merely helping it fold like other chaperones, this chaperone actually contributes a strand to the pilus protein that completes one of the folds of the pilin protein and makes it much more stable (see Waksman and Hultgren, Suggested Reading). The complex of pilin protein and chaperone is then targeted to the usher channel in the outer membrane, where it is assembled into the growing pilus. Again, there is the problem of where the energy for pilus assembly comes from, since the assembly of the pilus occurs at the inner face of the outer membrane after the pilin protein has been transported through the inner membrane and cytoplasm. One idea is that the periplasmic chaperone holds the pilin protein in a high-energy state, and its eventual folding at the usher drives the assembly process.

TYPE VI SECRETION SYSTEMS

Type VI secretion systems (T6SS) are very large, with up to 21 proteins encoded within the gene cluster, 12 of which are highly conserved and are thought to play structural roles in the secretion apparatus. At least some of the others unique to each system may encode effectors specifically transported by that system. An intriguing observation has been made that some of the highly conserved proteins structurally resemble components of phage tails (see chapter 7). The highly conserved protein Hcp resembles gp19, the major tail protein of phage T4, and another, VgrG, resembles the syringe of the baseplate of phage T4, consisting of gp27 and gp5. This has led to speculation that T6SS are derived from phage genes and act as inverted phage tails that now eject proteins from the bacterial cell instead of injecting DNA (and sometimes proteins) into the cell.

These systems are required for the secretion of proteins involved in pathogenesis and symbiosis, but they are also found in marine and soil bacteria, where they may have roles in biofilm formation and cell-to-cell communication. There can be little doubt of the importance of these systems, since mutations in their genes can have dramatic phenotypes, but many questions remain about their structure and function.

Protein Secretion in Bacteria That Lack an Outer Membrane

So far, we have limited our discussion of protein transport to the mechanisms used by bacteria that contain an outer lipid bilayer membrane (commonly referred to as Gram-negative bacteria). However, a large group of bacteria, notably the *Firmicutes* and *Actinobacteria*, lack an outer membrane (and are often referred to as the Gram-positive bacteria). Organisms in this group have evolved different types of secretion systems.

INJECTOSOMES

Pathogenic bacteria that lack the secretion systems described above require alternative mechanisms to translocate virulence effectors into eukaryotic cells. Some AB-type toxins, such as diphtheria toxin, and clostridial neurotoxins, such as botulinum toxin, are self-translocating, similar to the autotransporter systems described above. However, some bacteria, such as *Streptococcus pyogenes*, inject a virulence effector into the mammalian target cell by a mechanism functionally analogous to the T3SS (see Madden et al., Suggested Reading). This has been named an injectosome to distinguish it from the standard type III injectisome (note the different spellings). As shown in Figure 2.41, the function of the injectosome requires cotranslational translocation of the effector by the Sec-dependent secretion system across the bacterial membrane before it translocates the effector through the cell wall and directly across the membrane of the eukaryotic target cell.

Sortases

In bacteria that lack a bilipid outer membrane, the cell wall is available to the external surface of the cell. These bacteria take advantage of this by attaching some proteins directly to their cell wall so that the proteins will be exposed on the cell surface. Proteins destined for covalent attachment to the outer surface of the cell wall are the targets of a type of cell wall-sorting enzyme called a **sortase**. The general sortase, sortase A (StrA), sometimes called the housekeeping sortase, may attach as many as 40 different proteins to the cell wall. Others are more specific for certain proteins.

A sortase is able to create covalent attachments between peptides by catalyzing a transpeptidation reaction. Figure 2.42 illustrates a typical StrA pathway. Surface proteins that are StrA targets contain an N-terminal signal sequence and a 30- to 40-residue C-terminal sorting signal, which is composed of a pentapeptide cleavage site, LPXTG, and a hydrophobic domain that together constitute the cell wall-sorting signal (Cws) (Figure 2.42A). The N-terminal signal sequence of the sortase target protein directs the protein to the membrane translocase, where the signal sequence is removed (Figure 2.42B). Af-

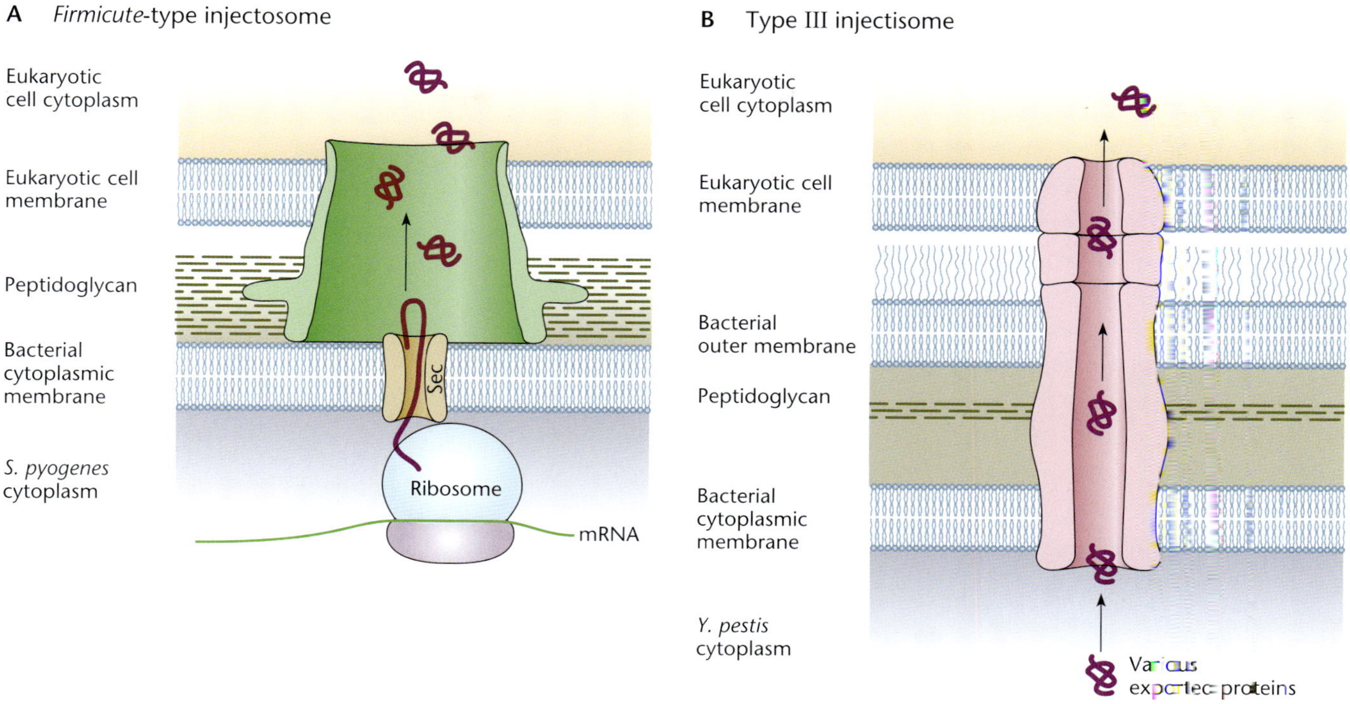

Figure 2.41 Comparison of the *Firmicute*-type injectosome **(A)** and the type III injectisome **(B)**. Modified from Tweten RK, Caparon M, *in* Waksman G, Caparon M, Hultgren S (ed), *Structural Biology of Bacterial Pathogenesis* (ASM Press, Washington, DC, 2005).

ter the protein has been translocated across the cytoplasmic membrane, the sortase cuts between the threonine (T) and glycine (G) in the pentapeptide sorting signal and then covalently links the carboxyl group of the threonine to a specific cysteine in the sortase C terminus. The sortase then attaches it to the terminal amino acid in the lipid II interlinking peptide that links two peptide cross-links in the cell wall. When the *N*-acetylmuramic acid (MurNac) in this lipid II is incorporated into the cell wall, the protein becomes covalently attached to the cell wall.

Sortase subfamilies are defined based on their taxonomic distributions and differences in the sorting signal and the peptide to which it can be attached (see Hendrickx et al., Suggested Reading). A particularly interesting example is the assembly of pili from their subunits on the cell surfaces. Pili are composed of one major subunit, often called pilin, and one or more minor subunits. In the simplest example, pili composed of pilin and only one other subunit, the minor subunit at the tip is attached to the first pilin subunit by a specialized sortase, and this first subunit is then attached to a number of other pilin subunits by the same specialized sortase. Finally, the last subunit is attached to the cell wall by the housekeeping sortase A, which attaches most other proteins to the cell wall. This makes for a very stable pilus and secures its attachment to the cell wall.

Regulation of Gene Expression

The previous sections have reviewed how a gene is expressed in the cell, from the time mRNA is transcribed from the gene until the protein product of the gene reaches its final destination in or outside the cell and has its effect. In general, genes are expressed in the cell only when the product is needed by the cell and then only as much as is required to make the amount of product needed by the cell. This saves energy and prevents the products of different genes from interfering with each other. The process by which the output of genes is changed depending on the state of the cell is called the **regulation of gene expression** and can occur at any stage in the expression of the gene (see chapter 11).

Genes whose products regulate the expression of other genes are called **regulatory genes**. The product of a regulatory gene can either inhibit or stimulate the expression of a gene. If it inhibits expression, the regulation is negative; if it stimulates expression, the regulation is positive. Some regulatory gene products are both positive and negative regulators depending on the situation. The product of a regulatory gene can regulate the expression of only one other gene, or it can regulate the expression of many genes. The set of genes regulated by the same regulatory gene product is called a **regulon**. If a gene product regulates its own expression, it is said to be **autoregulated**.

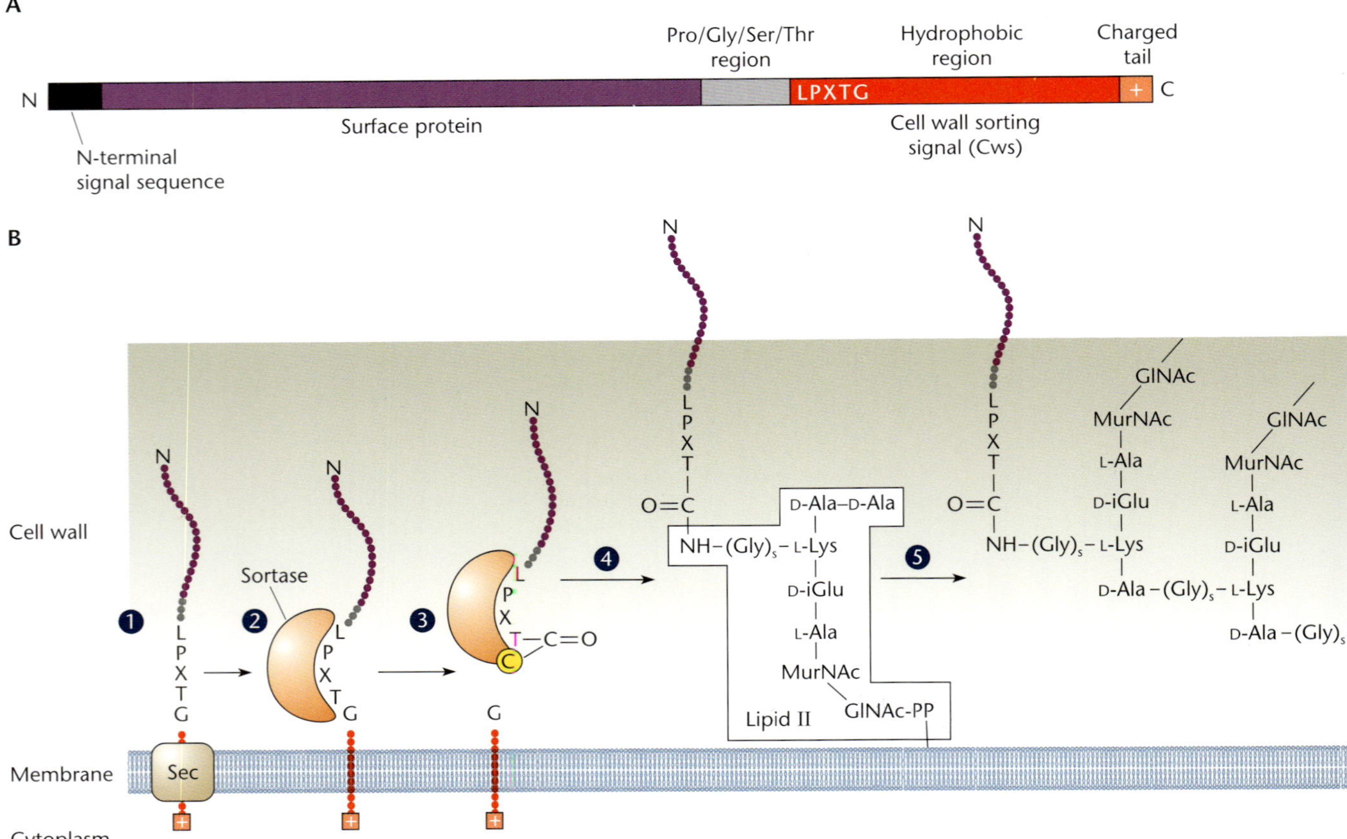

Figure 2.42 The sortase A pathway. **(A)** Typical sortase substrate. The protein is composed of an N-terminal signal peptide and a C-terminal cell wall-sorting signal (Cws). The Cws contains a conserved LPXTG motif followed by a hydrophobic stretch of amino acids and positively charged residues at the C terminus. **(B)** Model for the cell wall sortase A pathway in *Staphylococcus aureus*. **(1)** The full-length surface protein precursor is secreted through the cytoplasmic membrane via an N-terminal signal sequence. **(2)** A charged tail (+) at the C terminus of the protein may serve as a stop transfer signal. Following cleavage of this secretion signal, a sortase enzyme cleaves the protein between the threonine and glycine residues of the LPXTG motif, forming a thioacyl-enzyme intermediate to a specific cysteine in the sortase **(3)**. It is then attached to the free amine of the five-glycine cross-bridge of lipid II **(4)** before transfer into the cell wall **(5)**. The Pro-Gly-Ser-Thr region may help it through the thick cell wall so that it is expressed on the cell surface. PP is the site of MurNAc pentapeptide attachment to bactoprenyl in the membrane in lipid II. Modified from Connolly KM, Clubb RT, *in* Waksman G, Caparon M, Hultgren S (ed), *Structural Biology of Bacterial Pathogenesis* (ASM Press, Washington, DC, 2005).

We discuss the molecular mechanisms of regulation of gene expression in much more detail in chapter 11, but in this chapter, we briefly review some basic concepts needed to understand the following chapters.

Transcriptional Regulation

Expression of a gene is often regulated by controlling the amount of mRNA that is made from the gene. This is called **transcriptional regulation**. It makes sense to regulate gene expression at this level, as it is wasteful to make mRNAs if the expression of the gene is going to be inhibited at a later stage. Also, bacterial genes are often arranged in a polycistronic unit, or **operon**; if the genes in this unit are involved in a related function, they can all be regulated simultaneously by regulating the synthesis of the polycistronic mRNA of that operon.

Regulation of transcription of an operon usually occurs at the initial stages of transcription, at the promoter. Whether or not a gene is expressed depends on whether the promoter for the gene is used to make mRNA. Transcriptional regulation at the promoter for a gene can be determined by specific recognition of the promoter by

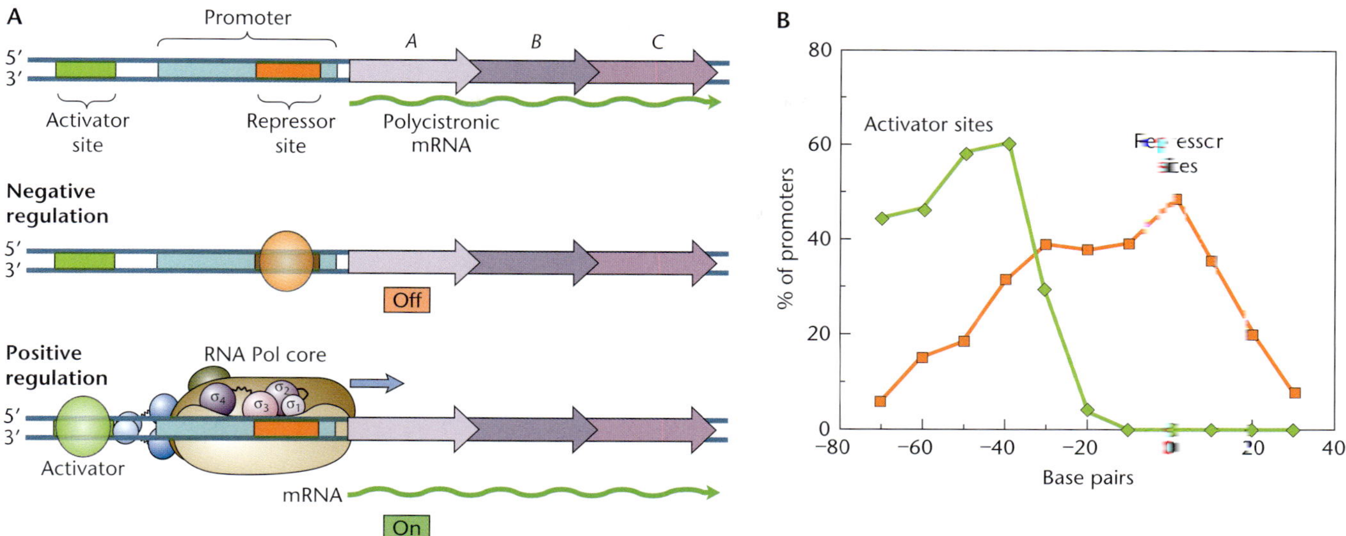

Figure 2.43 **(A)** The two general types of transcriptional regulation. In negative regulation, a repressor binds to a repressor-binding site (or operator) and turns expression of the operon off. In positive regulation, an activator protein binds upstream of the promoter and turns expression of the operon on. **(B)** Graph showing the most common locations of activator sites relative to repressor sites. Activator sites are usually farther upstream. Each data point indicates the middle of the known region on the DNA where a regulatory protein binds. Zero on the *x* axis marks the start point of transcription. Modified from Collado-Vides J, Magasanik B, Gralla JD, *Microbiol Rev* **55**:371–394, 1991.

RNA polymerase holoenzyme containing an alternative sigma factor. Regulation can also use regulatory proteins that can act either negatively or positively, depending on whether the regulatory gene product is a transcriptional **repressor** or a transcriptional **activator**, respectively. The difference between regulation of transcription by repressors and activators is illustrated in Figure 2.43. A repressor binds to the DNA at an **operator** sequence close to, or even overlapping, the promoter and prevents RNA polymerase from using the promoter, often by physically obstructing access to the promoter by the RNA polymerase. An activator, in contrast, usually binds upstream of the promoter at an **activator site**, where it can help the RNA polymerase bind to the promoter or help open the promoter after the RNA polymerase binds. Sometimes, a transcriptional regulator can be a repressor on some promoters and an activator on other promoters, depending on where it binds relative to the start site of transcription.

The activity of a regulatory protein can itself be modified by the binding of small molecules called **effectors** which affect its activity (note that the term effector is also used for proteins that are translocated by pathogens into eukaryotic cells). Effectors used in gene regulation are often molecules that can be used by the cell if the regulated operon is expressed or essential metabolites that do not have to be made by the cell if their concentration is already high. If the small-molecule effector causes transcription of the operon to be turned on (for example, by binding to a repressor and changing its structure so that the repressor can no longer bind to the DNA), the small molecule is called an **inducer**. If binding of the effector to a repressor causes the operon to be turned off, the small molecule is called a **corepressor**. The activity of regulatory proteins can also be modulated by posttranslational modification (e.g., phosphorylation [see Box 12.3]) or by interaction with other proteins or RNAs.

Not all transcriptional regulation occurs at the promoter, however. Sometimes transcription starts and then stops prematurely, resulting in synthesis of a truncated mRNA that does not include the protein-coding sequence. Such regulation is called attenuation of transcription. These and other mechanisms of transcriptional regulation are discussed in subsequent chapters.

Posttranscriptional Regulation

Expression of a gene can be regulated at later stages in gene expression (see chapter 11). For example, the mRNA may be degraded by RNases as soon as it is made, before it can be translated. In **translational regulation**, translation of the mRNA can be regulated to determine how much of the protein product is made. **Posttranslational regulation** results in regulation of the activity of a protein product of a gene by degradation by proteases or modifications, such as phosphorylation or methylation,

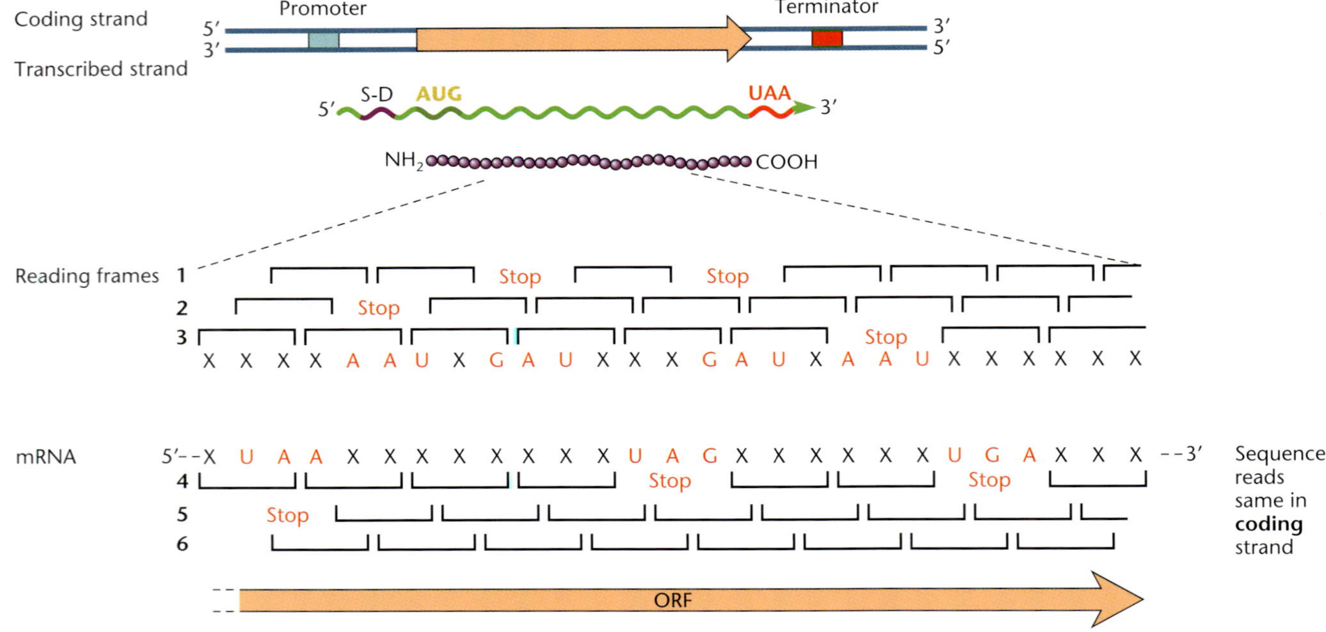

Figure 2.44 Relationship between gene structure in DNA and the coding sequence in mRNA. Each of the different reading frames in the two strands of DNA may contain open reading frames (ORFs), but generally, only one ORF in each region (reading frame 6 in the diagram) is translated to yield a polypeptide product.

depending on the conditions in which the cell finds itself. In addition, the product of a pathway may inhibit the activity of an enzyme in the pathway by a process called **feedback inhibition**. In general, a type of regulation of gene expression that operates after the mRNA for a gene has been made is called **posttranscriptional regulation**. Specific examples of posttranscriptional regulation are also discussed in subsequent chapters.

What You Need To Know

We have introduced a lot of detail in this chapter, so it is worth reviewing some of the most important concepts and terms. As with any field, molecular genetics has its own jargon, and in order to follow a paper or seminar that includes some molecular genetics, familiarity with this jargon is very helpful.

Figure 2.44 shows a typical gene with a promoter and transcription terminator. The mRNA is transcribed beginning at the promoter and ending at the transcription terminator. The direction on the DNA or RNA is indicated by the direction of the phosphate bonds between the carbons on the ribose or deoxyribose sugars in the backbone of the polynucleotide. These carbons are labeled with a prime to distinguish them from the carbons in the bases of the nucleotides. On one end of the RNA, the 5′ carbon of the terminal nucleotide is not joined to another nucleotide by a phosphate bond. Therefore, this is called the **5′ end**. Similarly, the other end is called the **3′ end**, because the 3′ carbon of the last nucleotide on this end is not joined to another nucleotide by a phosphate bond. The direction on DNA or RNA from the 5′ end to the 3′ end is called the **5′-to-3′ direction**. An RNA polymerase molecule synthesizes mRNA in the 5′-to-3′ direction, moving 3′ to 5′ on the transcribed strand (or template strand) of DNA. The opposite strand of DNA from the transcribed strand has the same sequence and 5′-to-3′ polarity as the RNA, so it is called the coding strand (or nontemplate strand). Sequences of DNA in the region of a gene are usually shown as the sequence of the coding strand. A sequence that is located in the 5′ direction of another sequence on the coding strand is upstream of that sequence, while a sequence in the 3′ direction is downstream. Therefore, the promoter for a gene and the S-D sequences are both upstream of the initiation codon, while the termination codon and the transcription termination sites are both downstream.

The positions of nucleotides in a promoter region are numbered as shown in Figure 2.6. The position of the first nucleotide in the RNA is called the start point and is given the number +1; the distance in nucleotides from this point to another point is numbered negatively or positively, depending on whether the second site is upstream or downstream of the start point, respectively. Note that these definitions can be used to describe only a region of DNA that is known to encode an RNA or protein, where we know which is the coding strand and which is the transcribed strand. Otherwise, what is upstream on one strand of DNA is downstream on the other strand.

Because mRNAs are both made and translated in the 5′-to-3′ direction, an mRNA can (and usually will) be translated while it is still being made, at least in bacteria and archaea, in which there is no nuclear membrane separating the DNA from the cytoplasm, where the ribosomes reside. We have discussed how this can lead to phenomena unique to bacteria, such as ρ-dependent polarity, and it is used to regulate expression of some genes in bacteria (see chapter 11).

It is important to distinguish promoters from TIRs and to distinguish transcription termination sites from translation termination sites. Figure 2.44 illustrates this difference. Transcription begins at the promoter and defines the 5′ end of the mRNA, but the place where translation begins, the TIR, can be some distance from the 5′ end. The untranslated region on the 5′ end of an mRNA upstream of the TIR is called the 5′ untranslated region or leader region and can be quite long. Similarly, a nonsense codon in the reading frame for the protein is a translation terminator, not a transcription terminator. The transcription terminator, and therefore the 3′ end of the mRNA, may be some distance downstream from the nonsense codon that terminates translation of the mRNA. The distance from the last termination codon to the 3′ end of the mRNA is the 3′ untranslated region. Polycistronic mRNAs encode more than one polypeptide. These mRNAs have a separate TIR and termination codon for each gene and can have noncoding or untranslated sequences upstream of, downstream of, and between the genes. Eukaryotes generally do not have polycistronic mRNAs, which is related to the dependence on ribosome binding to the 5′ end of the mRNA for translation initiation.

Open Reading Frames

The concept of an open reading frame, or ORF, is very important, particularly in this age of genomics. As discussed above, a reading frame in DNA is a succession of nucleotides in the DNA taken three at a time, the same way the genetic code is translated. Each DNA sequence has six reading frames, three on each strand, as illustrated in Figure 2.44. An ORF is a string of potential codons for amino acids in DNA unbroken by termination codons in one of the reading frames. Computer software can show where all the ORFs in a sequence are located, and most DNA sequences have many ORFs on both strands, although most of them are short. The region shown in Figure 2.44 contains many ORFs, but only the longest, in frame 6, is likely to encode a polypeptide. However, the presence of even a long ORF in a DNA sequence does not necessarily indicate that the sequence encodes a protein, and fairly long ORFs often occur by chance. Furthermore, it has become evident recently that even very short ORFs can encode short peptides with important biological functions.

If an ORF does encode a polypeptide, it will begin with a TIR, but as discussed above, TIRs are sometimes difficult to identify. Clues to whether an ORF is likely to encode a protein may come from the choice of the third base in the codon for each amino acid in the ORF. Because of the redundancy of the code, an organism has many choices of codons for each amino acid, but each organism prefers to use some codons over others (see "Codon Usage" above) (Table 2.2).

A more direct way to determine if an ORF actually encodes a protein is to ask which polypeptides are made from the DNA in an *in vitro* transcription-translation system. These systems use extracts of cells, typically of *E. coli*, from which the DNA has been removed but the RNA polymerase, ribosomes, and other components of the translation apparatus remain. When DNA with the ORFs under investigation is added to these extracts, polypeptides can be synthesized from the added DNA. If the size of one of these polypeptides corresponds to the size of an ORF on the DNA, the ORF probably encodes a protein. Another way to determine if an ORF encodes a protein is to make a translation fusion of a reporter gene to the ORF and to determine whether the reporter gene is expressed (see below).

Transcriptional and Translational Fusions

Probably the most convenient way to determine which of the possible ORFs on the two strands of DNA in a given region are translated into proteins is to make **transcriptional** and **translational fusions** to the ORFs. These methods make use of **reporter genes**, such as *lacZ* (β-galactosidase), *gfp* (green fluorescent protein), *lux* (luciferase), or other genes whose products are easy to detect. Figure 2.45 illustrates the concepts of transcriptional and translational fusions.

An ORF can be translated only if it is transcribed into RNA. Transcriptional fusions can be used to determine whether this has occurred. To make a transcriptional fusion, a reporter gene containing its TIR sequence but without its own promoter is fused immediately downstream of the promoter of the gene to be tested. If the promoter is active, and its gene is transcribed into mRNA, the reporter gene will also be transcribed, and the reporter gene product will be detectable in the cell. Transcriptional fusions also offer a convenient way of determining how much mRNA is made on a coding sequence. In general, the more reporter gene product that is made in the transcriptional fusion, the more mRNA was made that was directed by the upstream sequence. Translation of the mRNA depends

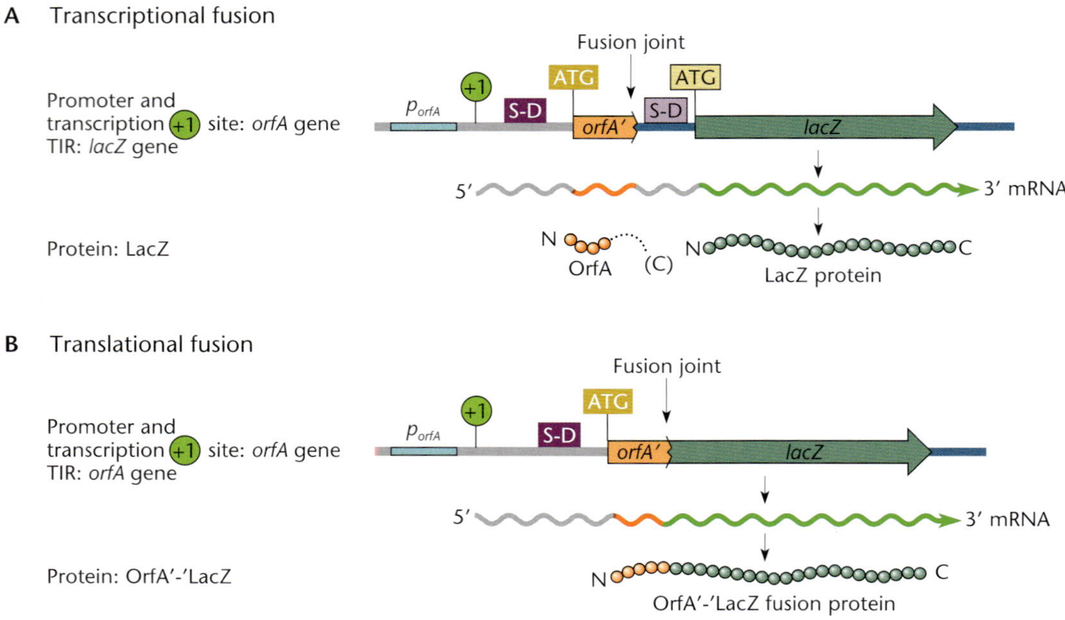

Figure 2.45 Transcriptional and translational fusions to express a *lacZ* reporter (which encodes β-galactosidase). In both types of fusion, transcription begins at the +1 site at the p_{orfA} promoter upstream of the OrfA coding sequence, and the levels of mRNA generated are dependent on the activity of this promoter. **(A)** In a transcriptional fusion, both the upstream OrfA coding region and the downstream *lacZ* reporter gene are included in the mRNA, and translation can initiate from both TIRs. Only the TIR for *lacZ* is used to generate β-galactosidase protein. The translation of the upstream OrfA continues until it encounters a termination codon in frame, as indicated by the dotted line, and therefore the activity of the *orfA* TIR does not contribute to β-galactosidase synthesis. Levels of β-galactosidase indicate the activity of the p_{orfA} promoter. **(B)** In a translational fusion, the mRNA includes the *orfA* TIR but the *lacZ* portion lacks its own TIR. Translation of the mRNA initiates at the TIR upstream of OrfA to make a fusion protein containing the remaining portion of the upstream OrfA coding sequence fused (in frame) to the LacZ reporter protein. The prime symbols indicate that part of each protein may be deleted. Levels of β-galactosidase indicate the activity of both the p_{orfA} promoter and the *orfA* TIR.

on the activity of the TIR from the reporter gene and is usually consistent regardless of the identity of the upstream sequence. Examples of the use of transcriptional fusions in studying the regulation of operons are given in subsequent chapters.

In a translational fusion, the reporter gene lacks both a promoter and a TIR and its coding sequence is fused immediately downstream of the TIR of the gene under investigation; it is crucial that the two coding sequences are fused in such a way that they are translated in the same reading frame and there are no termination codons between them. Translation beginning at the TIR of the upstream coding sequence will proceed through the reporter gene coding sequence, resulting in a **fusion protein** that contains both polypeptide sequences. The reporter gene product can then be assayed as before to determine how much of the fusion protein has been made. The reporter gene product must retain its activity even when fused to the potential polypeptide encoded by the upstream ORF; otherwise, it will not be detectable. Many reporter genes have been chosen because their products remain active when fused to other polypeptides. Translational fusions are also often used to attach affinity tags to proteins to use in their purification.

Summary

1. RNA is a polymer made up of a chain of ribonucleotides. The bases of the nucleotides—adenine, cytosine, uracil, and guanine—are attached to the five-carbon sugar ribose. Phosphate bonds connect the sugars to make the RNA chain, attaching the third (3′) carbon of one sugar to the fifth (5′) carbon of the next sugar. The 5′ end of the RNA is the nucleotide that has a free phosphate attached to the 5′ carbon of its sugar. The 3′ end has a free hydroxyl group at the 3′ carbon, with no phosphate attached. RNA is both made and translated from the 5′ end to the 3′ end.

2. After they are synthesized, RNAs can undergo extensive processing and modification. Processing occurs when phosphate bonds are broken or new phosphate bonds are formed. Modification occurs when the bases or the sugars of the RNA are chemically altered, for example, by methylation. In bacteria, rRNAs and tRNAs, but not mRNAs, are extensively modified.

3. The primary structure of an RNA is its sequence of nucleotides. The secondary structure is formed by hydrogen bonding between bases in the same RNA to give localized double-stranded regions. The tertiary structure is the three-dimensional shape of the RNA due to the stiffness of the double-stranded regions of the secondary structure. All RNAs, including mRNA, rRNA, and tRNA, probably have secondary and tertiary structures.

4. The enzyme responsible for making RNA is called RNA polymerase. One of the largest enzymes in the cell, the bacterial RNA polymerase core enzyme has five subunits plus another detachable subunit, the σ factor, which comes off after the initiation of transcription. The ω subunit helps in its assembly. The core enzyme is active in transcription elongation but requires addition of σ to form the holoenzyme to initiate transcription.

5. Transcription begins at well-defined sites on DNA called promoters. The type of promoter used depends on the type of σ factor bound to the RNA polymerase.

6. Transcription stops at sequences in the DNA called transcription terminators, which can be either factor dependent or factor independent. The factor-independent terminators have a string of A's (transcribed as U's in the RNA) that follows a sequence that forms an inverted repeat. The RNA transcribed from that region folds back on itself to form a stem-loop, or hairpin, which causes the RNA molecule to fall off the DNA template. The factor-dependent terminators do not have a well-defined sequence. The ρ protein is the best-characterized termination factor in *E. coli*. It forms a ring that binds to and encircles the RNA, moving toward the RNA polymerase. If the RNA polymerase pauses at a ρ termination site, the ρ factor catches up to it and causes it to dissociate from the DNA and release the RNA product.

7. Most of the RNA in the cell falls into three classes: messenger (mRNA), ribosomal (rRNA), and transfer (tRNA). mRNA is very unstable, existing for only a few minutes before being degraded. rRNAs in bacteria are further divided into three types: 16S, 23S, and 5S. Both rRNA and tRNA are very stable and account for about 95% of the total RNA. Other RNAs include the primers for DNA replication and small RNAs involved in regulation or RNA processing.

8. Ribosomes, the site of protein synthesis, are made up of two subunits, the 30S subunit and the 50S subunit, which contain the rRNAs as well as approximately 50 proteins. The 16S rRNA is in the 30S subunit, while the 23S and 5S rRNAs are in the 50S subunit.

9. Polypeptides are chains of the 20 amino acids that are held together by peptide bonds between the amino group of one amino acid and the carboxyl group of another. The amino terminus (N terminus) of the polypeptide has the amino acid with an unattached amino group. The carboxy terminus (C terminus) of a polypeptide has the amino acid with a free carboxyl group.

10. Translation is the synthesis of polypeptides, using the information in mRNA to direct the sequence of amino acids. During translation, the mRNA moves in the 5′-to-3′ direction along the ribosome 3 nucleotides at a time. Three reading frames are possible, depending on how the ribosome is positioned at each triplet.

11. The genetic code is the assignment of each possible 3-nucleotide codon sequence in mRNA to 1 of 20 amino acids. The code is redundant, with more than one codon sometimes encoding the same amino acid. Because of wobble, the first position of the tRNA anticodon (written 5′ to 3′) does not have to behave by the standard base-pairing complementarity to the third position of the antiparallel codon sequence, and other pairings are possible.

12. Initiation of translation occurs at TIRs on the mRNA that consist of an initiation codon, usually AUG or GUG, and often an S-D sequence, a short sequence that is complementary to part of the 16S rRNA and precedes the initiation codon.

13. In bacteria, the first tRNA to enter the ribosome is a special methionyl-tRNA called fMet-tRNA^{fMet}, which carries the amino acid formylmethionine. After the polypeptide has been

Summary (continued)

synthesized, the formyl group and often the first methionine are removed.

14. Translation termination occurs when one of the termination or nonsense codons, UAA, UAG, or UGA, is encountered as the ribosome moves down the mRNA. Proteins called release factors are also required for release of the polypeptide.

15. The primary structure of a polypeptide is the sequence of amino acids in the polypeptide. Proteins can be made up of more than one polypeptide chain, which can be the same as or different from each other. The secondary structure results from hydrogen bonding of the amino acids to form α-helical regions and β-sheets. Tertiary structure refers to how the chains fold up on themselves, and quaternary structure refers to one or more different polypeptide chains folding up on each other.

16. Proteins that help other proteins fold are called chaperones. The most ubiquitous chaperones are the Hsp70 chaperones, called DnaK in *E. coli*, which are very similar in all types of cells from bacteria to humans. These chaperones bind to the hydrophobic regions of proteins and prevent them from associating prematurely. They are aided by their smaller cochaperones, DnaJ and GrpE, which help in binding to proteins and cycling ADP off the chaperone, respectively. Other proteins, called Hsp60 chaperonins, also help proteins fold, but by a very different mechanism. One chaperonin, called GroEL in *E. coli*, forms large cylindrical structures with internal chambers that take up unfolded proteins and help them refold properly. A cochaperonin called GroES forms a cap on the cylinder after the unfolded protein is taken up. Chaperonins like GroEL are found in bacteria and in the organelles of eukaryotes and are called group I chaperonins. Another type, group II chaperonins, is found in the cytoplasm of eukaryotes and in archaea. They have a similar structure but a very different amino acid sequence.

17. The process of passing proteins through membranes is called transport. Proteins that pass through the inner membrane into the periplasm and beyond are said to be exported. Proteins that pass out of the cell are secreted.

18. Proteins can also be held together by disulfide linkages between cysteines in the protein. Generally, only proteins that are exported into the periplasm or out of the cell have disulfide bonds. These disulfide bonds are made by oxidoreductases in the periplasm of bacteria with an outer membrane.

19. Bacterial cells utilize a variety of mechanisms to target proteins to different locations, such as the cell membrane, the periplasm, or outside the cell. Some of these systems are used for a wide range of proteins, while others are specific to individual sets of proteins.

20. The expression of genes is regulated, depending on the conditions in which the cell is found. This regulation can be either transcriptional or posttranscriptional. Transcriptional regulation can be either negative or positive, depending on whether the regulatory protein is a repressor or an activator, respectively. A repressor binds to an operator, which is usually close to the promoter, and prevents transcription from the promoter. An activator binds to an activator sequence that is usually upstream of the promoter and increases transcription from the promoter. Transcriptional regulation can also occur after the RNA polymerase leaves the promoter, as in attenuation or antitermination of transcription. Posttranscriptional regulation can occur at the level of stability of the mRNA, translation of the mRNA, or processing, modification, or degradation of the gene product.

21. The strand of DNA from which the mRNA is made is the transcribed, or template, strand. The opposite strand, which has the same sequence as the mRNA, is the coding, or nontemplate, strand.

22. A sequence 5′ on the coding strand of DNA relative to a particular element is said to be upstream, whereas a sequence 3′ to that element is downstream.

23. The TIR sequence of a gene does not necessarily occur at the beginning of the mRNA. The 5′ end of the mRNA is called the 5′ untranslated region or leader region. Similarly, the sequence downstream of the termination codon is the 3′ untranslated region.

24. Because mRNA is both transcribed and translated in the 5′-to-3′ direction, translation can begin before synthesis of the mRNA is complete in bacteria, which have no nuclear membrane.

25. Bacteria and archaea often make polycistronic mRNAs with more than one polypeptide coding sequence on an mRNA. This can result in polarity of transcription and translational coupling, phenomena unique to these domains, where mutations in the 5′ coding region of an mRNA can affect the expression of genes in the 3′ region.

26. An ORF is a string of amino acid codons in DNA unbroken by a termination codon. *In vitro* transcription-translation systems or transcriptional and translation fusions are often required to prove that an ORF in DNA actually encodes a protein.

27. Gene fusions have many uses in modern molecular genetics. They can be either transcriptional or translational fusions. In a transcriptional fusion, the downstream reporter gene is transcribed onto the same mRNA as the upstream gene, but the reporter gene coding region is translated from its own TIR, so expression of the downstream reporter gene is dependent on the activity of the promoter of the upstream gene but not the translational signals of the upstream gene. In a translational fusion, the two coding regions are fused to each other, so expression of the downstream reporter gene is dependent

Summary (continued)

on the activities of both the promoter and TIR of the upstream gene.

28. Many naturally occurring antibiotics target components of the transcription and translation machinery. Some of the most commonly used are rifampin, streptomycin, tetracycline, thiostrepton, chloramphenicol, and kanamycin. In addition to their uses in treating bacterial infections, tumor chemotherapy, and biotechnology, antibiotics have also helped us understand the mechanisms of transcription and translation. In addition, the genes that confer resistance to these antibiotics have served as selectable genetic markers and reporter genes in molecular genetic studies of organisms in all domains of life.

QUESTIONS FOR THOUGHT

1. Which do you think came first in the very earliest life on Earth, DNA, RNA, or protein? Why?

2. Why is the genetic code universal?

3. Why do you suppose prokaryotes have polycistronic mRNAs but eukaryotes do not?

4. Why do you suppose mitochondrial genes show differences in their genetic code from chromosomal genes in eukaryotes?

5. Why is selenocysteine inserted into proteins of almost all organisms but into only a few sites in a few proteins in these organisms?

6. Why do so many antibiotics inhibit the translation process as opposed to, say, amino acid biosynthesis?

7. Why do you think chaperonins have two linked chambers and alternate the folding of proteins between the two chambers?

8. List all the reasons you can think of why bacteria would regulate the expression of their genes.

9. Why do bacteria have so many different mechanisms for localization of proteins to different sites?

SUGGESTED READING

Agashe VR, Guha S, Chang H-C, Genevaux P, Hayer-Hartl M, Stemp M, Georgopoulos C, Hartl FU, Barral JM. 2004. Function of trigger factor and DnaK in multidomain protein folding: increase in yield at the expense of folding speed. *Cell* **117:**199–209.

Bae B, Feklistov A, Lass-Napiorkowska A, Landick R, Darst SA. 2015. Structure of a bacterial RNA polymerase holoenzyme open promoter complex. *eLife* **4:**e08504.

Baker TA, Sauer RT. 2006. ATP-dependent proteases of bacteria: recognition logic and operating principles. *Trends Biochem Sci* **31:**647–653.

Ban N, Nissen P, Hansen J, Capel M, Moore PB, Steitz TA. 1999. Placement of protein and RNA structures into a 5 A-resolution map of the 50S ribosomal subunit. *Nature* **400:**841–847.

Becker SH, Darwin KH. 2017. Bacterial proteasomes: mechanistic and functional insights. *Microbiol Mol Biol Rev* **81:**e00036–16.

Björk GR, Hagervall TG. 25 July 2005, posting date. Transfer RNA modification. *EcoSal Plus* 2005 doi:10.1128/ecosalplus.4.6.2.

Browning DF, Busby SJW. 2004. The regulation of bacterial transcription initiation. *Nat Rev Microbiol* **2:**57–65.

Bukau B, Horwich AL. 1998. The Hsp70 and Hsp60 chaperone machines. *Cell* **92:**351–366.

Condon C. 2007. Maturation and degradation of RNA in bacteria. *Curr Opin Microbiol* **10:**271–278.

Feilmeier BJ, Iseminger G, Schroeder D, Webber H, Phillips GJ. 2000. Green fluorescent protein functions as a reporter for protein localization in *Escherichia coli. J Bacteriol* **182:**4068–4076.

Freudl R. 2013. Leaving home ain't easy: protein export systems in Gram-positive bacteria. *Res Microbiol* **164:**664–674.

Galan JE, Waksman G. 2018. Protein-injection machines in bacteria. *Cell* **172:**1306–1318.

Gualerzi CO, Pon CL. 2015. Initiation of mRNA translation in bacteria: structural and dynamic aspects. *Cell Mol Life Sci* **72:**4341–4367.

Hendrickx APA, Budzik JM, Oh SY, Schneewind O. 2011. Architects at the bacterial surface: sortases and the assembly of pili with isopeptide bonds. *Nat Rev Microbiol* **9:**166–176.

Hui MP, Foley PL, Belasco JG. 2014. Messenger RNA degradation in bacterial cells. *Annu Rev Genet* **48:**537–559.

Keiler KC, Feaga HA. 2014. Resolving nonstop translation complexes is a matter of life or death. *J Bacteriol* **196:**2123–2130.

Koronakis V, Eswaran J, Hughes C. 2004. Structure and function of TolC: the bacterial exit duct for proteins and drugs. *Annu Rev Biochem* **73:**467–489.

Korostelev A, Noller HF. 2007. The ribosome in focus: new structures bring new insights. *Trends Biochem Sci* **32:**434–441.

Lee HC, Bernstein HD. 2001. The targeting pathway of *Escherichia coli* presecretory and integral membrane proteins is specified by the hydrophobicity of the targeting signal. *Proc Natl Acad Sci USA* **98:**3471–3476.

Li L, Park E, Ling J, Ingram J, Ploegh H, Rapoport TA. 2016. Crystal structure of a substrate-engaged SecY protein-translocation channel. *Nature* **531:**395–399.

Madden JC, Ruiz N, Caparon M. 2001. Cytolysin-mediated translocation (CMT): a functional equivalent of type III secretion in Gram-positive bacteria. *Cell* **104:**143–152.

Marbaniang CN, Vogel J. 2016. Emerging roles of RNA modifications in bacteria. *Curr Opin Microbiol* **30:**50–57.

McGary K, Nudler E. 2013. RNA polymerase and the ribosome: the close relationship. *Curr Opin Microbiol* **16:**112–117.

Meinnel T, Sacerdot C, Graffe M, Blanquet S, Springer M. 1999. Discrimination by *Escherichia coli* initiation factor IF3 against initiation on non-canonical codons relies on complementarity rules. *J Mol Biol* **290:**825–837.

Mikula KM, Leo JC, Łyskowski A, Kedracka-Krok S, Pirog A, Goldman A. 2012. The translocation domain in trimeric autotransporter adhesins is necessary and sufficient for trimerization and autotransportation. *J Bacteriol* **194:**827–838.

Mogk A, Deuerling E, Vorderwülbecke S, Vierling E, Bukau B. 2003. Small heat shock proteins, ClpB and the DnaK system form a functional triade in reversing protein aggregation. *Mol Microbiol* **50:**585–595.

Noeske J, Wasserman MR, Terry DS, Altman RB, Blanchard SC, Cate JH. 2015. High-resolution structure of the *Escherichia coli* ribosome. *Nat Struct Mol Biol* **22:**336–341.

Olivares AO, Baker TA, Sauer RT. 2016. Mechanistic insights into bacterial AAA+ proteases and protein-remodelling machines. *Nat Rev Microbiol* **14:**33–44.

Palmer T, Sargent F, Berks BC. 2005. Export of complex cofactor-containing proteins by the bacterial Tat pathway. *Trends Microbiol* **13:**175–180.

Ramakrishnan V. 2014. The ribosome emerges from a black box. *Cell* **159:**979–984.

Ruff EF, Record MT Jr, Artsimovitch I. 2015. Initial events in bacterial transcription initiation. *Biomolecules* **5:**1035–1062.

Sun Z, Scott DJ, Lund PA. 2003. Isolation and characterisation of mutants of GroEL that are fully functional as single rings. *J Mol Biol* **332:**715–728.

van den Berg B, Clemons WM, Jr, Collinson I, Modis Y, Hartmann E, Harrison SC, Rapoport TA. 2004. Xray structure of a protein-conducting channel. *Nature* **427:**36–44.

Waksman G, Hultgren SJ. 2009. Structural biology of the chaperone-usher pathway of pilus biogenesis. *Nat Rev Microbiol* **7:**765–774.

Wang J, Boisvert DC. 2003. Structural basis for GroEL-assisted protein folding from the crystal structure of (GroEL-KMgATP)14 at 2.0A resolution. *J Mol Biol* **327:**843–855.

Washburn RS, Gottesman ME. 2015. Regulation of transcription elongation and termination. *Biomolecules* **5:**1063–1078.

Xie K, Dalbey RE. 2008. Inserting proteins into the bacterial cytoplasmic membrane using the Sec and YidC translocases. *Nat Rev Microbiol* **6:**234–244.

Zuker M. 2003. Mfold Web server for nucleic acid folding and hybridization prediction. *Nucleic Acids Res* **31:**3406–3415.

Lederberg mades strides in the lab throughout the 1950s. Courtesy of the Esther M. Zimmer Lederberg Memorial Website estherlederberg.com.

Bacterial Genetic Analysis: Fundamentals and Current Approaches

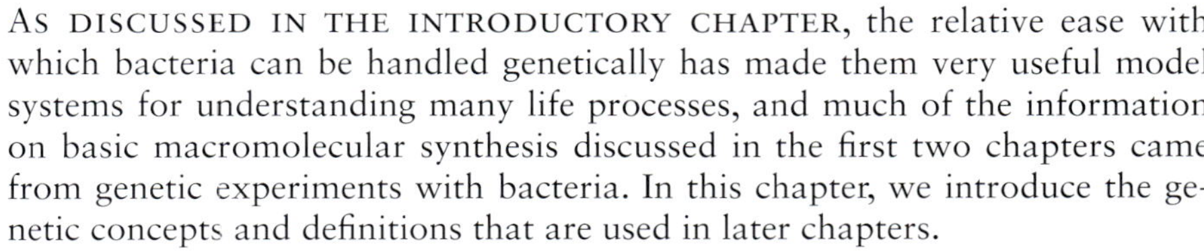

AS DISCUSSED IN THE INTRODUCTORY CHAPTER, the relative ease with which bacteria can be handled genetically has made them very useful model systems for understanding many life processes, and much of the information on basic macromolecular synthesis discussed in the first two chapters came from genetic experiments with bacteria. In this chapter, we introduce the genetic concepts and definitions that are used in later chapters.

Definitions

In genetics, as in any field of knowledge, we need definitions. However, words do not mean much when taken out of context, so here we define only the most basic terms. We will define other important terms as we go along.

Terms Used in Genetics

The words in the next few headings are common to all types of genetic experiments, whether with bacterial, archaeal, or eukaryotic systems, with some small variations.

MUTANT

The word **mutant** refers to an organism that is the direct offspring of a normal member of the species (the **wild type**) but is genetically different, so the difference is inherited by its offspring. Organisms of the same species isolated from nature that have different properties are usually not called mutants but, rather, **variants** or **strains**, because even if one of the strains has recently arisen from the other in nature, we have no way of knowing which one is the mutant and which is the wild type. Organisms that differ from the wild type because of a reversible programmed event, such as an inversion promoted by a DNA invertase, are usually not called mutants but, rather, **phases**, since the event is reversible and the population often contains representatives of both types.

PHENOTYPE

The phenotypes of an organism are all the observable properties of that organism. Usually, in genetics, the term **phenotype** means **mutant phenotype**, or the characteristics of the mutant organism that differ from those of the wild type. The corresponding normal property is sometimes referred to as the **wild-type phenotype**.

GENOTYPE

The **genotype** of an organism is the actual sequence of its DNA. If two organisms have the same genotype, they are genetically identical. Identical twins have almost the same genotype. If two organisms differ by only one mutation or a small region of their DNA, they are said to be **isogenic** except for that mutation or region.

MUTATION

A **mutation** is any heritable change in the DNA sequence. By heritable, we mean that the DNA with the changed sequence can be faithfully replicated. Practically every imaginable type of change is possible, and all changes are called mutations. However, the word "heritable" must be emphasized. Damage or changes to DNA that are repaired or that are restored to the original sequence are not inherited and therefore are not mutations. Only a permanent change in the sequence of deoxynucleotides in DNA constitutes a mutation. For the same reason, reversible sequence changes due to programmed events, such as inversions due to DNA invertases (see chapter 8), are usually not referred to as mutations.

ALLELE

Different forms of the same gene are called **alleles**. For example, if one form of a gene has a mutation and the other has the wild-type sequence, the two forms of the gene are different alleles of the same gene. In this case, one gene is a mutant allele and the other gene is the wild-type allele. Different sequence changes in the same gene are also referred to as different alleles of that gene. Diploid organisms can have two different alleles of the same gene, one on each homologous chromosome. The term "allele" can also refer to genes with the same or similar sequences that appear at the same chromosomal location in closely related species. However, similar gene sequences occurring in different chromosomal locations are not alleles; rather, they are **copies** or **duplications** of the gene.

USE OF GENETIC DEFINITIONS

The following example illustrates the use of the definitions in the previous section. For an explanation of the methods, see the introductory chapter.

A culture of *Pseudomonas fluorescens* normally grows as bright-green colonies on agar plates. However, suppose that one of the colonies is colorless. It probably arose through multiplication of a mutant organism. The mutant phenotype is "colorless colony," and the corresponding wild-type phenotype is "green colony." The mutant bacterium that formed the colorless colony probably had a mutation in a gene for an enzyme required to make the green pigment. For example, the mutation may be a base pair change in the gene that causes the insertion of a wrong amino acid in the polypeptide, which results in loss of function of the resulting enzyme. Thus, the mutant and wild-type bacteria have different alleles of this gene, and we can refer to the gene in the colorless-colony-forming bacteria as the **mutant allele** and the gene in the green-colony-forming bacteria as the **wild-type allele**.

In the example above, we know that a mutation has occurred only because of the lack of color. However, recall that any heritable change in the DNA sequence is a mutation, so mutations can occur without changing the organism's phenotype. Some of them do not even change the amino acid sequence of a protein in whose gene they lie because of the redundancy of the genetic code. Changes of this type, called **silent mutations**, are usually found by DNA sequencing.

Genetic Names

There are some commonly accepted rules for naming mutants, phenotypes, and mutations in bacteria, although different publications sometimes use different notations. We use the terms recommended by the American Society for Microbiology.

NAMING MUTANT ORGANISMS

The mutant organism can be given any name, as long as the designation does not refer specifically to the phenotype or the gene thought to have been mutated. This rule helps to avoid confusion if the gene with the mutation is introduced into another strain or if other mutations occur or are transferred into the original organism. Quite often, someone who has isolated a mutant names the strain after himself or herself, giving it his or her initials and a number (e.g., *Escherichia coli* AB2497). This method of naming informs others where they can obtain the mutant strain and get advice about its properties. If another mutation alters the mutant strain, the new strain is usually given a related name, such as *E. coli* AB2498. However, it is usually best not to be too descriptive in naming strains, since it will be confusing later if we were wrong about their genetic properties.

NAMING GENES

Bacterial genes are designated by three lowercase italic letters that usually refer to the function of the gene's product, when it is known. For example, the name *his* refers to genes whose product is an enzyme required to synthesize the amino acid histidine or other products involved in histidine acquisition or incorporation into protein (e.g., histidine transporters, histidyl-tRNA synthetase [which attaches histidine to $tRNA^{His}$ for use in protein synthesis], and $tRNA^{His}$). The name *hut* refers to genes required to utilize the amino acid histidine as a carbon and nitrogen source to differentiate between genes involved in histidine incorporation and its breakdown. Individual genes that encode different products involved in the same function or enzymatic pathway are designated by a capital letter that follows the three lowercase letters. For example, the *hisA* and *hisB* genes encode different polypeptides, both of which are required to synthesize histidine. A mutation that inactivates either gene will make the cell unable to synthesize histidine. When possible, it is helpful to name newly identified genes by referring to related genes in a model system such as *E. coli*, so that *hisA* genes in different organisms all encode the same enzymatic function and also are related in sequence.

NAMING MUTATIONS

Hundreds of different types of mutations can occur in a single gene, so all alleles of a particular gene are given a specific allele number. For example, *hisA4* refers to the *hisA* gene with mutation number 4, and the *hisA* gene with mutation number 4 is referred to as the *hisA4* allele. Nowadays, the actual amino acid change in a protein due to a mutation in its gene is often obtained by DNA sequencing. If this is the case, it is common to designate the mutation by the actual amino acid change. For example, H14Q means that the mutation has changed amino acid number 14 in the polypeptide product of the gene counting from the formylmethionine at the N terminus (see chapter 2) from the histidine found in the wild type to a glutamine. For some model organisms, like *E. coli*, the genus *Salmonella*, and the genus *Bacillus*, there are official registries where there are unique allele numbers for each mutant, which are maintained even if the gene name changes. More information on these numbering systems can be found at the major genetic stock centers.

If a mutation is known to completely inactivate the product of a gene, it can be called a **null mutation**. However, this designation must be used with care, because the product of a gene can sometimes retain some of its activity if the mutation is leaky or if only a part of the gene remains (see below). In some organisms, it is now possible to precisely delete all or almost all of the coding region of a gene so that no chance remains that the product has any activity. In such organisms, this has become the new standard for claiming a null mutation. Different nomenclatural rules apply if a mutation is a deletion or an insertion. We defer a discussion of these rules until we discuss these types of mutations (see below).

NAMING PHENOTYPES

Phenotypes are also denoted by three-letter names, but the letters are not italicized and the first letter is capitalized. In addition, superscripts are often used to distinguish mutant from wild-type phenotypes. For example, His$^-$ describes the phenotype of an organism with a mutated *his* gene that cannot grow without histidine in its environment. The corresponding wild-type organism grows without histidine, so it is phenotypically His$^+$. Another example, Rif^r, describes resistance to the antibiotic rifampin, which blocks RNA synthesis (see chapter 2). Some specific mutations in the *rpoB* gene, which encodes a subunit of the RNA polymerase, make the cell resistant to the antibiotic. The corresponding wild-type phenotype is rifampin sensitivity, or Rif^s.

Useful Phenotypes in Bacterial Genetics

What phenotypes are useful for genetic experiments depends on the organism being studied. For bacterial genetics, the properties of the colonies formed on agar plates often are the most useful phenotypes, since individual cells with distinguishing characteristics are difficult to isolate, even if they can be seen under a microscope (see the introductory chapter).

The visual appearance of colonies sometimes provides useful mutant phenotypes, such as the colorless colony discussed above. Colonies formed by mutant bacteria might also be smaller than normal or smooth instead of wrinkled. The mutant bacterium may not multiply to form a colony at all under some conditions, or conversely, it may multiply when the wild type cannot.

Many mutant phenotypes have been used to study cellular processes, such as DNA recombination and repair, mutagenesis, and development. The following sections describe a few of the more commonly used phenotypes. In later chapters, we discuss many more types of mutants and demonstrate how mutations can be used to study key processes.

Auxotrophic and Catabolic Mutants

Historically, some of the most useful bacterial mutants are **auxotrophic mutants**, or **auxotrophs**. These mutants have mutations in genes whose products are required to make an essential growth substance; they therefore cannot multiply unless that substance is provided as a growth supplement. The original strain from which the mutant was derived can make the substance and is sometimes called the **prototroph**. For example, a His$^-$ auxotrophic mutant cannot grow unless the medium is supplemented with the amino acid histidine, while the wild type can grow without added histidine and so is a histidine prototroph. Similarly, a Bio$^-$ auxotrophic mutant cannot grow without the addition of the vitamin biotin, which is not needed by the corresponding prototroph.

Another common type of mutant is one that cannot use a particular substance for growth that can be used by the wild type. For example, the original bacterium may be able to use the sugar maltose as a sole carbon and energy source, but a Mal$^-$ mutant derived from it must be provided with a different carbon and energy source, such as glucose, in order to grow. Other examples are mutants that can no longer use a particular amino acid as a nitrogen source or a particular phosphate-containing compound as a source of phosphate. Such mutants are sometimes called catabolic mutants because they cannot make essential catabolites or derive energy from a particular substance, although no common name is used to include all such mutants. Even though auxotrophic and catabolic mutants seem to be opposites, the molecular basis is similar. In both types, a mutation has altered a gene encoding an enzyme of a metabolic pathway, thereby inactivating the enzyme. The only difference is that in the first case, the inactivated enzyme was in a **biosynthetic** pathway, which is required to synthesize a substance, while in the second case, the inactivated enzyme was in a **catabolic** pathway, which is required to degrade a substance to use it as a carbon and energy source, a nitrogen source, or a phosphate source.

ISOLATING AUXOTROPHIC MUTANTS

Figure 3.1 shows a simple method for isolating mutants auxotrophic for histidine and biotin. In this experiment, bacterial colonies were plated and allowed to form colonies on plate 1, which contains all the nutrients the bacteria need, including histidine and biotin. Bacteria in eight of these colonies were picked up with a loop and transferred onto two other plates. These plates are the same as plate 1, except that plate 2 lacks biotin but has histidine and plate 3 lacks histidine but has biotin. The bacteria from most of the colonies can multiply on all three types of plates. However, the bacteria in colony 2 grow on plate 2 but not on plate 3; this indicates a mutant that requires added histidine, that is, they are His$^-$. These mutant cells do not require biotin (as indicated by their ability to grow in its absence), so they are Bio$^+$. Similarly, the bacteria in colony 6 are Bio$^-$ but His$^+$, since they can grow on plate 3 but not on plate 2. Under real conditions, mutants that require histidine or biotin would not be this frequent, and hundreds of thousands of colonies would have to be tested to find one mutant that required histidine, biotin, or indeed any growth supplement not required by the wild type.

In principle, it should be possible to find auxotrophic mutants unable to synthesize any compound required for growth or unable to use any carbon and energy source. However, auxotrophic mutants must be supplied with the compound they cannot synthesize, and these compounds must enter the cell. Many bacteria cannot take in some compounds that have a high electrical charge, such as nucleotides, so some types of auxotrophs are very difficult to isolate.

Conditional-Lethal Mutants

As mentioned above, auxotrophic and catabolic mutants can be isolated because they have mutations in genes whose products are required only under certain conditions. The cells can be grown under conditions where the product of the mutated gene is not required and tested under conditions where it is required. However, many gene products of the cell are essential for growth no matter what conditions the bacteria find themselves in. The genes that encode such functions are called **essential genes**. Examples of essential genes are those for DNA and RNA polymerase, ribosomal proteins, DNA ligase, and some helicases. Cells with mutations that inactivate essential

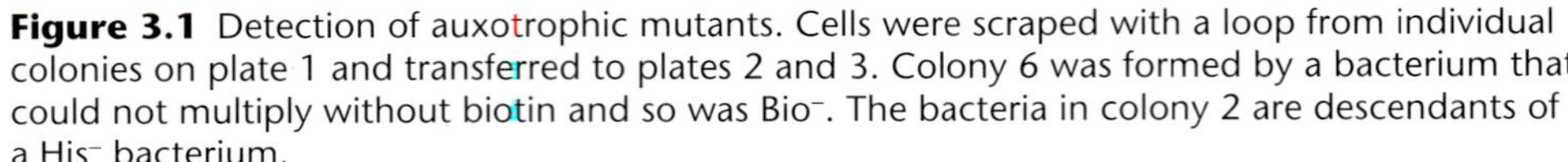

Figure 3.1 Detection of auxotrophic mutants. Cells were scraped with a loop from individual colonies on plate 1 and transferred to plates 2 and 3. Colony 6 was formed by a bacterium that could not multiply without biotin and so was Bio$^-$. The bacteria in colony 2 are descendants of a His$^-$ bacterium.

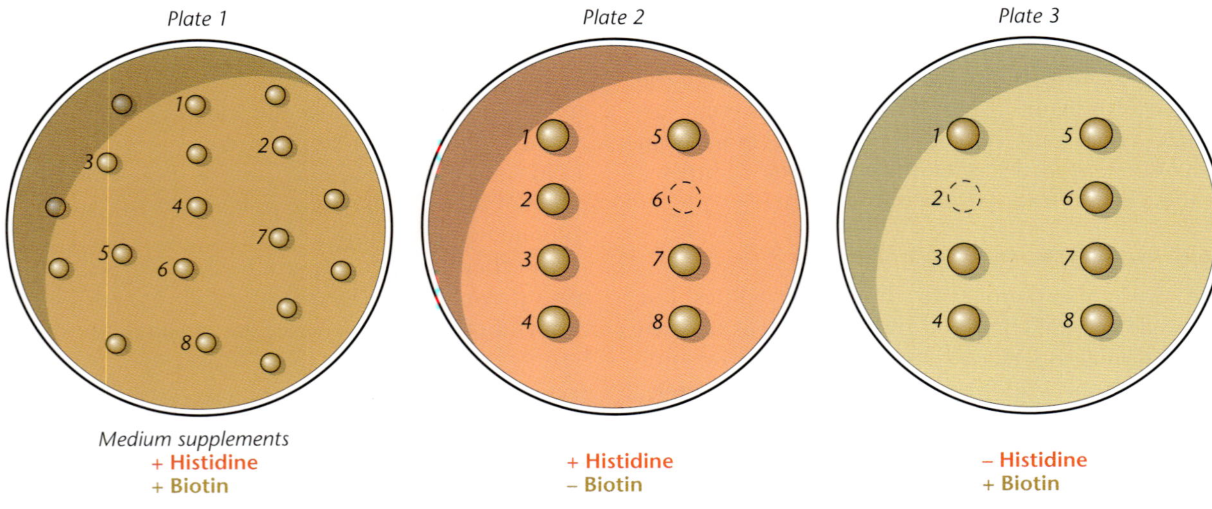

genes cannot be isolated unless the mutations inactivate the gene under only some conditions. Hence, any mutants that are isolated must have **conditional-lethal mutations**, because these DNA changes inactivate the gene product only under some conditions, whereas the gene product is functional under other conditions that allow growth of the mutant.

TEMPERATURE-SENSITIVE MUTANTS

The most generally useful conditional-lethal mutations in bacteria are mutations that make the mutant **temperature sensitive** for growth. Usually, such mutations change an amino acid of a protein so that the protein no longer functions at higher temperatures but still functions at lower temperatures. The higher temperatures are called the **nonpermissive temperatures** for the mutant, whereas the temperatures at which the protein still functions are the **permissive temperatures** for the mutant.

Mutations can affect the temperature stability of proteins in various ways. Often, an amino acid required for the protein's stability at the nonpermissive temperature is changed, causing the protein to unfold, or denature, partially or completely. The protein could then remain in the inactive state or be destroyed by cellular proteases that remove abnormal proteins (see chapter 2). If the protein is not degraded, it can sometimes spontaneously renature (refold) when the temperature is lowered; growth can then resume immediately. The temperature sensitivity due to the mutation is reversible, which can be very useful for enrichment (see below) or for studying the function of the gene product. With other mutations, the protein is irreversibly denatured and must be resynthesized before growth can resume. Such mutations can be lethal if they occur in a protein required for gene expression, such as RNA polymerase or a translation factor, since then the functional protein is required to make itself, and it therefore cannot be remade.

The temperature ranges used to isolate temperature-sensitive mutations depend on the organism. Cellular proteins generally have evolved to function over a wide range of temperatures. However, different species of bacteria differ greatly in their preferred temperature ranges. For example, a "mesophilic" bacterium, such as *E. coli*, may grow well at a range of temperatures from 20 to 42°C. In contrast, a "thermophilic" bacterium, such as *Geobacillus stearothermophilus*, may grow well only between 42 and 60°C. For *E. coli*, a temperature-sensitive mutation may leave a protein functional at 30°C, the permissive temperature, but not at 42°C, the nonpermissive temperature. For *G. stearothermophilus*, the temperature-sensitive mutation may leave a protein active at the permissive temperature of 45°C but render it nonfunctional at the nonpermissive temperature of 55°C. Temperature-sensitive mutations are extremely useful for studying the function of the product of a gene because the product of a gene can be rapidly inactivated, even if no inhibitor of the function is known.

Isolating Temperature-Sensitive Mutants

In principle, temperature-sensitive mutants are as easy to isolate as auxotrophic mutants. If a mutation that makes the cell temperature sensitive occurs in a gene whose protein product is required for growth, the cells stop multiplying at the nonpermissive temperature. To isolate such mutants, the bacteria are incubated on a plate at the permissive temperature until colonies appear, and then the colonies are transferred to another plate and incubated at the nonpermissive temperature. Bacteria that can form colonies at the permissive temperature but not at the nonpermissive temperature are temperature-sensitive mutants. However, temperature-sensitive mutants in a particular gene are usually much rarer than auxotrophic mutants. Many changes in a protein will inactivate it, but very few will make a protein functional at one temperature and nonfunctional at another. A more serious problem is that all temperature-sensitive mutants have the same phenotype (failure to grow at the nonpermissive temperature), independent of the essential gene they are in, and finding one that has a mutation in a particular gene can mean screening thousands of temperature-sensitive mutants with mutations in a wide variety of essential genes. Doing some form of regional mutagenesis to limit mutations to a single gene (see below) can help, and some amino acid changes are more apt to make a protein temperature sensitive than others. Nevertheless, it has proven difficult to predict what particular amino acid changes will make a protein temperature sensitive. A better way is to randomly mutagenize the gene by a type of regional mutagenesis and then screen for mutations that make the protein temperature sensitive. The frequency of occurrence of different types of mutations and methods for doing site-specific mutagenesis are discussed later in this chapter and in subsequent chapters.

COLD-SENSITIVE MUTANTS

Cells with proteins that fail to function at low temperatures are called **cold-sensitive mutants**. Mutations that make a bacterium cold sensitive for growth are rarer than temperature-sensitive (heat-sensitive) mutations and are often in genes whose products must enter a larger complex, such as the ribosome. The increased molecular movement at the higher temperature may allow the mutated protein to wriggle into the complex despite its altered shape, but it is unable to do so at lower temperatures. Such mutations often show a phenotype only after a long delay, so they are generally less useful than mutations that make a protein unstable at higher temperatures. Both mutations are in a sense temperature sensitive, but the term is usually reserved for heat-sensitive mutations.

NONSENSE MUTATIONS

Mutations that change a codon in a gene to one of the three nonsense codons—UAA, UAG, or UGA (in most bacteria)—can also be conditional-lethal mutations. A **nonsense mutation** can be conditionally lethal because it causes translation to stop within the gene unless the cell has a "nonsense suppressor" tRNA that allows insertion of an amino acid at the position of the nonsense codon, as explained later in this chapter. Nonsense mutations are in contrast to **missense mutations**, where the codon changes to represent a different amino acid which does not terminate the polypeptide

CONDITIONAL EXPRESSION OF THE WILD-TYPE ALLELE

Another way in which mutations in essential genes can be isolated is when a second wild-type copy of the gene is provided in *trans* (from another site in the cell), which allows synthesis of the gene product. This is most useful when expression of the wild-type copy can be controlled, for example, by replacing its own promoter with a promoter whose activity can be controlled (see chapters 2 and 11). In this situation, growth can occur when the promoter is active (e.g., when the appropriate inducer molecule is present), and loss of promoter activation results in loss of synthesis of the functional product, which is depleted as the cells grow and divide until the cells are no longer able to grow. Alternatively, the presence of the second copy of the gene can be controlled by placing this gene on a plasmid with temperature-sensitive replication so that the plasmid (and therefore the wild-type allele) is present during growth at the permissive temperature and is lost from the cells during growth at the nonpermissive temperature. The functional gene product is depleted during growth, allowing observation of the resulting phenotype.

Another mechanism for conditional expression of a wild-type version of a gene involves controlling the degradation of the gene product. This has become common with specific tags in eukaryotic proteins that can be used to degrade a protein under certain conditions but has also been used in clever ways in bacteria, for example, with the ribosome rescue transfer-messenger RNA (tmRNA) system that targets truncated proteins to the ClpXP proteasome (see chapter 2). Adding a modified tag to the C-terminal end of a protein can allow it to be controlled for degradation using inducible expression of the cognate adaptor protein and endogenous tmRNA system, through the use of a tag and adaptor pair from other bacteria, or by use of another tag/adaptor system that targets the tagged protein to a specific proteolysis pathway (see McGinness et al. and Griffith and Grossman, Suggested Reading).

Resistant Mutants

Among the most useful and easily isolated types of bacterial mutants are resistant mutants. If a substance kills or inhibits the growth of a bacterium, mutants resistant to the substance can often be isolated merely by plating the bacteria in the presence of the substance. Only mutants in the population that are resistant to the substance will multiply to form a colony.

The numerous mechanisms of resistance depend on the basis for toxicity and on the options available to prevent the toxicity (examples are given in Table 3.1). For example, the mutation may destroy a cell surface receptor to which the toxic substance must bind to enter the cell. If the substance cannot enter the mutant cell, it cannot kill the cell. A bacteriophage that requires a specific cell surface receptor for binding or DNA delivery provides a good example of this. Alternatively, a mutation might change the "target" affected by the substance inside the cell. For example, an antibiotic might normally bind to a ribosomal protein and affect protein translation. However, if the antibiotic cannot bind to a mutant (but still functional) protein, it cannot kill the cell. An example of such a resistance mutation is a mutation to streptomycin resistance in *E. coli* (Table 3.1). The antibiotic streptomycin binds to the 16S rRNA in the 30S subunit of the ribosome and blocks translation. However, some mutations in the gene for the S12 protein, *rpsL* (for ribosomal protein small-subunit number 12), prevent streptomycin from binding to the ribosome but do not inactivate the S12 protein. These mutations therefore confer streptomycin resistance (Str^r) on *E. coli.* In some cases, the substance added to the cells is not toxic until one of the cell's own enzymes changes it. A mutation inactivating the enzyme that converts the nontoxic substance into the toxic one could make the cell resistant to that substance. Finally, a compound that disrupts a normal physiological process, such as through feedback inhibition of a key enzyme, can block that process, and mutations in the enzyme that prevent feedback inhibition can allow resistance.

Inheritance in Bacteria

Salvador Luria and Max Delbrück were among the first people to attempt to study bacterial inheritance quantitatively, and they published this work in a now-classic paper in 1943 (see Luria and Delbrück, Suggested Reading). As discussed in the introductory chapter, the experiments and reasoning of Luria and Delbrück helped debunk what was then a popular misconception among bacteriologists. At

Table 3.1 Some resistance mutations

Agent	Toxicity	Resistance mutation
Bacteriophage T1	Infects and kills	Inactivates TonA or TonB outer membrane proteins; phage cannot bind to cell
Rifampin	Binds to RNA polymerase; inhibits transcription	Changes β subunit of RNA polymerase so that rifampin no longer binds
Streptomycin	Binds to ribosomes; inhibits translation	Changes ribosomal protein S12 so that streptomycin no longer binds to the ribosome
Chlorate	Converted to chlorite, which is toxic	Inactivates nitrate reductase, which converts chlorate to chlorite
High concentrations of valine in absence of isoleucine	Feedback inhibits acetolactate synthase; cell starves for isoleucine	Activates a valine-insensitive acetolactate synthase

the time, it was generally thought that bacteria are different from other organisms in their inheritance. It was generally accepted that heredity in higher organisms followed "Darwinian" principles. According to Charles Darwin, random mutations occur, and if one happens to confer a desirable phenotype, organisms with the mutation are selected by the environment and become the predominant members of the population. Undesirable, as well as desirable, mutations continuously occur, but only the desirable mutations are passed on to future generations.

However, many bacteriologists thought that heredity in bacteria followed different principles. They thought that bacteria, rather than changing as the result of random mutations, somehow "adapt" to the environment by a process of directed change, after which the adapted organism would pass the adaptation on to its offspring. This process is called Lamarckian inheritance, named after the geneticist who proposed it, and acceptance of it was encouraged by the observation that all the bacteria in a culture exposed to a toxic substance seem to become resistant to that substance. The key question was whether exposure to the substance caused resistance to emerge (by Lamarckian inheritance) or simply selected for spontaneous resistant mutants already present in the population (by Darwinian inheritance).

The Luria and Delbrück Experiment

The Luria and Delbrück experiment was designed to differentiate between two hypotheses for how mutants arise in bacterial cultures: the **random-mutation hypothesis** and the **directed-change hypothesis**. The random-mutation hypothesis predicts that mutants appear randomly prior to the addition of the selective agent (as proposed by Darwin), whereas the directed-change hypothesis predicts that mutants appear only in response to the selective agent (as proposed by Lamarck).

One distinction between the two hypotheses is reflected in the distribution of the numbers of mutants in a series of cultures. If the random-mutation hypothesis is correct, mutations that occur early in the growth of a culture will have a disproportionate effect on the number of mutants in the culture, and the fraction of mutants should vary widely from culture to culture. Figure 3.2 illustrates this principle. In culture 1, only one mutation occurred, but this mutation gave rise to 15 total resistant mutants (the original mutant plus 14 offspring) because it occurred early in the growth of the culture. In culture 2, two mutations arose, but they gave rise to only 10 total resistant mutants (the original two mutants plus 8 offspring) because they occurred later in the growth of the culture. In their experiments, Luria and Delbrück used *E. coli* as the bacterium and bacteriophage T1 as the selective agent. As shown in Table 3.1, phage T1 kills wild-type *E. coli*, but a mutation in either one of the genes for two outer membrane proteins called TonA and TonB can make these cells resistant to killing by the phage. If bacteria are spread on an agar plate in the presence of the phage, only those cells that are resistant to the phage multiply to form a colony. All the others are killed. The number of colonies on the plate is therefore a measure of the number of bacteria resistant to the bacteriophage in the culture.

Figure 3.3 shows the two experiments that Luria and Delbrück performed, and Table 3.2 gives some representative results. It can be seen that the two experiments give very different results, even though they seem superficially similar. In experiment 1, the authors started one culture of bacteria. After incubating it, they took out a number of small aliquots and plated them with and without phage T1 to determine the number of resistant mutants, as well as the total number of bacteria, in the culture. They then calculated the fraction of resistant mutants. In experiment 2, they started a large number of smaller cultures. After incubating these cultures, they determined the number of resistant mutants and the total number of bacteria in each culture. It can easily be seen that in experiment 1, the number of resistant mutants in each aliquot is almost the same, subject only to sampling errors and statistical fluctuations. However, in experiment 2, a very large variation in the number of resistant bacteria

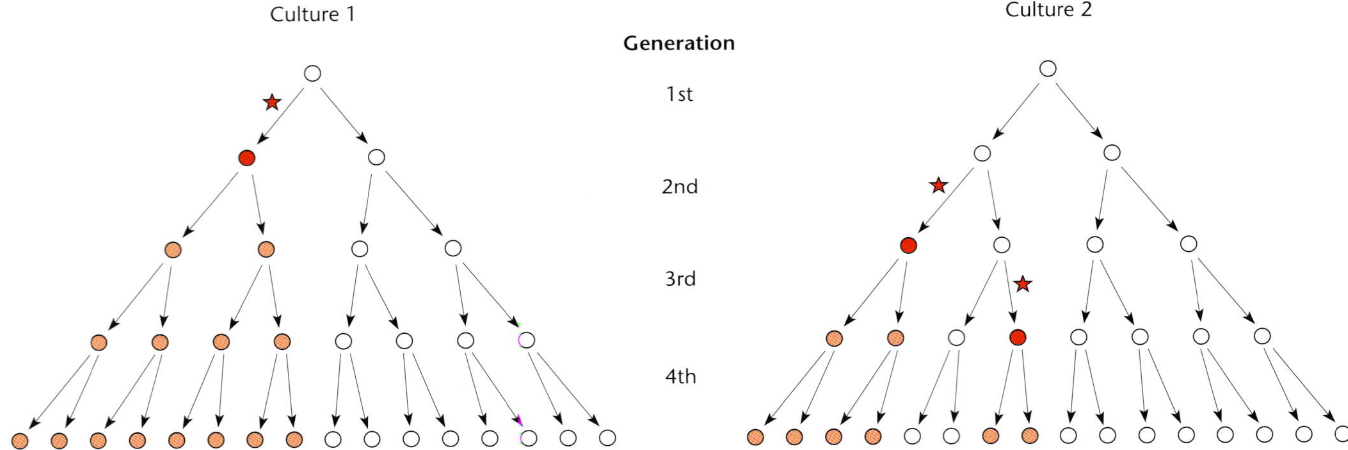

Figure 3.2 The number of mutants in a culture is not proportional to the number of mutations that have occurred in the culture, because earlier mutations give rise to more mutant progeny. Only one mutation (indicated by star) occurred in culture 1, but the original mutant cell (red) gave rise to 14 mutant progeny (orange) because the mutation occurred in the first generation. In culture 2, two mutations occurred, one in the second generation and one in the third. However, because these mutations occurred later, the mutated cells gave rise to only 8 mutant progeny.

per culture was found. Some cultures had no resistant mutants, while some had many. One culture even had 107 resistant mutants. Luria and Delbrück referred to this and the other mutant-rich cultures as "jackpot" cultures. Apparently, these are cultures in which a mutation to resistance occurred very early in the growth of the culture. Hence, these results fulfill the predictions of the random-mutation hypothesis. In contrast, the directed-change hypothesis predicts that the results of the two experiments should be the same, and no jackpot cultures should appear in the second experiment.

Mutants Are Clonal

The Luria and Delbrück experiments demonstrated that, even in bacteria, mutations occur randomly and are then selected upon by the environment rather than arising in response to the selective pressure. However, their analysis was fairly sophisticated mathematically and so was not generally understood. Other, simpler demonstrations showed that bacterial mutants were **clonal**, another prediction of the random-mutation hypothesis. These experiments showed that once a bacterial mutant appeared, all of the descendants of that mutant were also mutant in the same way, even in the absence of the selective pressure. One way this was demonstrated was to show that *E. coli* mutants resistant to the phage T1 grow together as a colony on a plate, even in the absence of the phage (see Newcombe, Suggested Reading). When millions of nonresistant *E. coli* bacteria were spread on a plate and were allowed to grow for a short time prior to spraying the plate with the phage and incubating, a few colonies appeared, due to the multiplication of a few mutants resistant to the phage that had arisen prior to applying the phage (Figure 3.4). However, if the bacterial cells were allowed to grow briefly, as above, but then were spread around on the plate prior to spraying the plate with the phage, many more resistant colonies appeared. This demonstrated that the resistant mutants were all concentrated in a few colonies as descendants of the original mutants until they were dispersed by the spreading, i.e., that they were clonal.

Esther and Joshua Lederberg's Experiment

The experiments that really buried the directed-change hypothesis, at least as the sole explanation for some types of resistant bacterial mutants, were the replica-plating experiments (see below) pioneered by Esther and Joshua Lederberg (see Lederberg and Lederberg, Suggested Reading). These experiments showed in a simple way that bacterial mutants resistant to an antibiotic arose in the absence of the antibiotic and that the resistance was stably inherited by the descendants of the original mutant. They spread millions of bacteria on a plate in the absence of an antibiotic and allowed the bacteria to form a lawn during overnight incubation. Bacteria from this plate were then replicated onto another plate containing the antibiotic. After incubating the antibiotic-containing plate, Esther and Joshua Lederberg could determine where antibiotic-resistant mutants had arisen on the original plate by aligning the two plates and marking the regions on the first plate

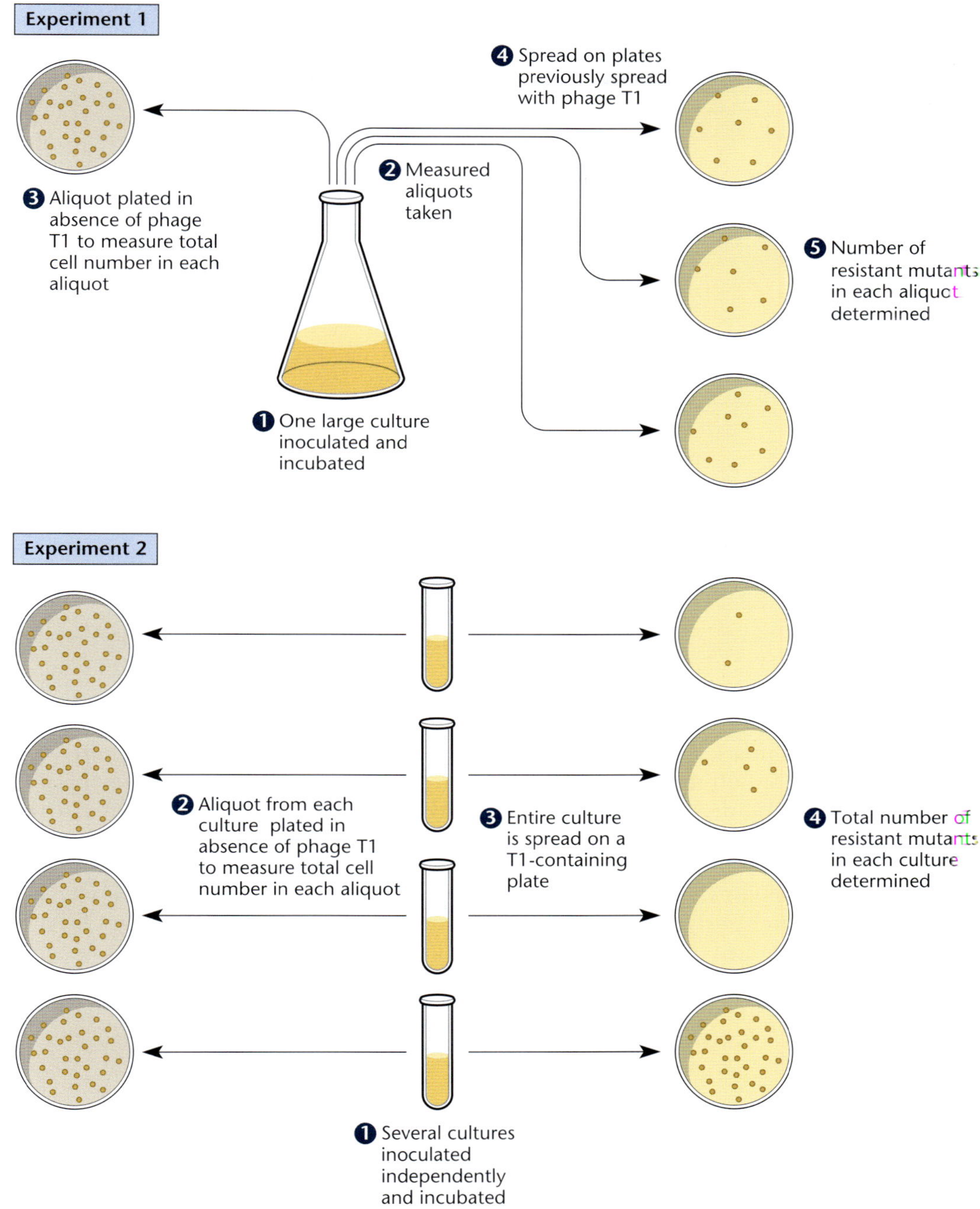

Figure 3.3 The Luria and Delbrück experiment. In experiment 1, a single flask containing standard medium was inoculated with bacteria and incubated overnight. The total number of cells per ml was determined by plating an aliquot in the absence of phage T1, and the number of T1-resistant cells was determined by plating in the presence of the phage. For experiment 1, all of the aliquots were taken from this single culture. In experiment 2, both the total number of cells and the number of resistant mutants were determined in each of a number of smaller cultures that had been started with a few nonmutant bacteria and incubated overnight. See the text for details.

Table 3.2 The Luria and Delbrück experiment

Experiment 1		Experiment 2	
Aliquot no.	**No. of resistant bacteria**	**Culture no.**	**No. of resistant bacteria**
1	14	1	1
2	15	2	0
3	13	3	3
4	21	4	0
5	15	5	0
6	14	6	5
7	26	7	0
8	16	8	5
9	20	9	0
10	13	10	6
		11	107
		12	0
		13	0
		14	0
		15	1
		16	0
		17	0
		18	64
		19	0
		20	35

where antibiotic-resistant mutants had grown on the second. They scraped these regions off the original plate, suspended and diluted the bacteria they had scraped off this region, and plated them without antibiotic to repeat the experiment. This time, when the bacteria were replica plated onto a plate with antibiotic, there were many more resistant mutants than previously. Eventually, by repeating this process, they obtained a pure culture of bacteria, all of which were resistant to the antibiotic even though they had never been exposed to it. The only possible explanation was that the bacteria must have acquired the resistance prior to exposure to the antibiotic and passed the resistance on to their offspring. Thus, the experiments dispelled the idea that *E. coli* mutations are directed, at least as far as resistance to this antibiotic is concerned.

Mutation Rates

As defined above, a mutation is any heritable change in the DNA sequence of an organism, and we usually know that a mutation has occurred because of a phenotypic change in the organism. The **mutation rate** can be loosely defined as the chance of mutation to a particular phenotype in a certain time interval. Mutation rates can differ because mutations to some phenotypes occur much more often than mutations to other phenotypes. The overall mutation rate is the sum of all the mutation rates for each mutation that can cause the phenotype. As a consequence, the mutation rate is relatively high when many different possible mutations in the DNA can give rise to a particular phenotype. However, if only a very few types of mutations can cause a particular phenotype, the mutation rate for that phenotype is relatively low. This can be illustrated by comparing the mutation rates for the phenotypes His$^-$ and streptomycin resistance (Str^r). The products of 11 genes are required for histidine biosynthesis, and each enzyme has hundreds of amino acids, many of which are required for activity. Changing any of these required amino acids inactivates the enzyme, as do deletions or other disruptions of any one of the 11 genes. In contrast, streptomycin resistance results from a specific change in one of very few amino acids in a single ribosomal protein, S12, that changes the protein so that it no longer affects binding of streptomycin to the ribosome but leaves its function intact. This makes the mutation rate for streptomycin resistance very low. Hence, mutation to Str^r occurs spontaneously in about 1 in 10^9 to 10^{10} cells,

Figure 3.4 Mutants are clonal. In experiment 1, cells were spread onto a plate, grown briefly, then sprayed with phage T1 to select for T1-resistant colonies. In experiment 2, the cells were treated the same way, but immediately before spraying with the phage, they were spread around the plate. Many more colonies appeared in experiment 2, indicating that the extra mutants were derived from clonal populations present on the plate before the second spreading.

whereas a mutation to His⁻ occurs in about 1 in 10^6 to 10^7 cells, or approximately 1,000 times as often. Therefore, the mutation rate for a particular phenotype gives an early clue to what kinds of mutations can cause the phenotype and whether multiple genes may be involved. An extremely high mutation rate for a particular phenotype (1 in 10^3 to 10^4) may indicate not a mutation but, rather, the loss of a plasmid or prophage or the occurrence of some programmed recombination event, such as inversion of an invertible sequence; on the other hand, an extremely low mutation rate (1 in 10^{12} to 10^{13}) might suggest that two or more independent mutations are required.

Calculating Mutation Rates

The mutation rate is usually defined as the chance of a mutation occurring each time a cell grows and divides. This is a reasonable definition because, as discussed in chapter 1, DNA replicates once each time the cell divides, and most mutations occur as mistakes that are made during this process. The number of times a cell grows and divides in a culture is called the number of **cell divisions**. This is not to be confused with the division time, or **generation time**, which is the time it takes for a cell to grow and divide. Not all bacterial cells divide as they multiply, and some form long filaments, or hyphae. Nevertheless, we will refer to cell divisions as the number of times in a culture a cell has grown to maturity and doubled its mass enough to form two cells, since the DNA must have duplicated during this time. The mutation rate is then the number of mutations to a particular phenotype that have occurred in a growing culture divided by the total number of cell divisions that have occurred in the culture during the same time.

DETERMINING THE NUMBER OF CELL DIVISIONS

The total number of cell divisions that have occurred in an exponentially growing culture is easy to calculate and is simply the total number of cells in the culture minus the number of cells in the starting inoculum. To understand this, see Figure 3.5. In this illustration, a culture that was started from one cell multiplies to form eight cells in seven cell divisions. This number equals the final number of cells (8) minus the number of cells at the beginning (1). In general, the number of cell divisions that have occurred in the culture equals $N_2 - N_1$ if N_2 equals the number of cells at time 2 and N_1 equals the number of cells at time 1.

Therefore, from the definition above, the mutation rate (a) is given by the following equation:

$$a = m_2 - m_1/N_2 - N_1$$

where m_2 and m_1 are the number of mutations in the culture at time 2 and time 1, respectively.

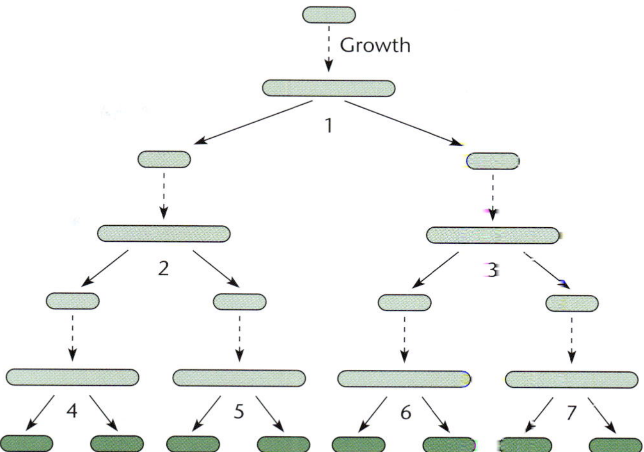

Figure 3.5 The number of cell divisions (7) equals the total number of cells in an exponentially growing culture (8) minus the number at the beginning (1). Dashed arrows represent growth; solid arrows represent cell division.

Usually, a culture is started with only a few cells with no mutations and ends with many cells, so we can often ignore the mutations and the number of cells at the beginning and just call the number of cell divisions N, where N is the total number of cells in the culture. Then, we can just measure the number of cells and the number of mutants at the end of the growth period, and the mutation rate equation can be simplified to

$$a = m/N$$

where m is the number of mutations that have occurred in the culture, and N is the number of bacteria.

DETERMINING THE NUMBER OF MUTATIONS THAT HAVE OCCURRED IN A CULTURE

From the equations above, it looks as though it might be easy to calculate the mutation rate. The total number of mutations is simply divided by the total number of cells. The problem comes in determining the number of mutational events that have occurred in a culture, because mutant cells, not mutational events, are usually what are detected. Recall from Figure 3.2 that one mutant cell resulting from a single mutational event can give rise to many mutant cells, depending on when the mutation occurred during the growth of the culture. Therefore, one cannot determine the number of mutations in a culture merely by counting the mutant cells. However, in some cases, the number of mutant cells can form the basis of a calculation of the number of mutational events and, by extension, the mutation rate. Some examples of such situations are given below.

Calculating the Mutation Rate from the Data of Luria and Delbrück

Luria and Delbrück used data like those shown in Table 3.2 to calculate the mutation rate for T1 phage resistance. They assumed that even though the number of mutants per culture varies a great deal and does not follow a normal distribution, the number of mutations per culture should do so, because each cell has the same chance of acquiring a mutation to T1 phage resistance each time it grows and divides. For convenience, the **Poisson distribution** can be used to approximate the normal distribution in a case like this, where there are a lot of bacteria in the population and the probability of a mutation in each bacterium is extremely low. According to the Poisson distribution, if P_i is the probability of having i mutations in a culture, then

$$P_i = m^i e^{-m}/i!$$

where m is the average number of mutations per culture, the number they wanted to know. Therefore, if they knew how many cultures had a certain number of mutations, they could calculate the average number of mutations per culture. However, this is not as obvious as it seems. The data give the number of T1 phage-resistant mutants per culture but do not indicate how many of the cultures had one, two, three, or more mutations. For example, cultures with one mutant probably had one mutation, but others, even the one with 107 mutants, might also have had only one mutation that occurred early during the growth of the culture. Only the number of cultures with zero mutations seems clear—those with zero mutants, or 11 of 20. Therefore, the probability of having zero mutations equals 11/20. Applying the formula for the Poisson distribution, the probability of having zero mutations is given by the equation

$$11/20 = m^0 e^{-m}/0! = 1 \times e^{-m}/1 = e^{-m}$$

and $m = -\ln 11/20 = 0.60$. Therefore, in this experiment, an average of 0.60 mutations occurred in each culture. From the equation for the mutation rate,

$$a = m/N = 0.60/(5.6 \times 10^8) = 1.07 \times 10^{-9}$$

Therefore, there are 1.07×10^{-9} mutations per cell division if there are a total of 5.6×10^8 total bacteria per culture. In other words, a mutation for phage T1 resistance occurs about once every 10^9 (or every billion) times a cell divides.

There are a number of problems with measuring mutation rates this way, as indeed there are with any way of measuring mutation rates. One issue is **phenotypic lag**, which is a problem with T1 phage resistance in *E. coli* and with most phenotypes caused by loss of a gene product. When a mutation first inactivates either the *tonA* or *tonB* gene in the chromosome of a growing bacterium, the TonA and TonB proteins are still both in the outer membrane and the cell is still sensitive to the phage. Although the product of the mutated gene will no longer be made in the active form as a result of the mutation, it could take a few generations to dilute the protein that was present before the mutation occurred out of the membrane, depending on how many copies of each protein originally exist in the membrane. This is compounded by the fact that, in an exponentially growing culture, half of all the mutations are likely to have occurred in the last generation time, when the cell number is highest, so a significant proportion of all mutations could be missed. Accordingly, some of the cultures with no resistant mutants presumably had bacteria with a *tonA* or *tonB* mutation, but the presence of preexisting wild-type protein in the membrane allows the cells to be killed by the phage, and therefore they are not counted.

Another problem with this method for calculating the mutation rate is that it is wasteful in that it ignores most of the data and considers only the cultures with no mutants. In their classic paper, Luria and Delbrück also derived an equation to estimate the mutation rate by using the number of mutants in all of the cultures. This equation is widely used and is expressed as

$$r = aN_t \ln(N_t Ca)$$

where r is the average number of mutants per culture, C is the number of cultures, N_t is the average number of bacteria per culture, and a is the mutation rate. Note that a, the mutation rate, appears on both sides of the equation, so this equation must be solved numerically, for example, by successive approximations. In their paper, they include a table showing the solutions for different numbers of cultures that can be used to solve for a, the mutation rate. Others have subsequently proposed other methods to measure mutation rates from such data (see Lea and Coulson and Jones et al., Suggested Reading).

Calculating the Mutation Rate from the Number of Clones of Mutants

In some cases, it might be possible to use the fact that the multiplication of mutants is clonal to calculate the mutation rate. As discussed above, since mutants of some types of bacteria multiplying on an undisturbed plate stay together to form a colony, the number of mutant colonies on the undisturbed plate is a measure of the number of mutations that have occurred. The bacteria on a parallel plate can then be washed off and plated to determine the total number of bacteria on each plate. Again, this way of determining the number of mutations is strongly influenced by phenotypic lag for some types of mutants because the mutant phenotype is not observed for several generations after the mutation occurred.

Calculating the Mutation Rate from the Rate of Increase in the Proportion of Mutants

Another prediction of the random-mutation hypothesis is that the number of mutants should increase faster than the total population. In other words, the fraction of mutants in the population should increase as the population grows. At first, it seems surprising that the total number of mutants increases faster than the total population until one thinks about where mutants come from. If the multiplication of old mutants were the only source of mutants, the fraction of mutants would remain constant or even drop if the mutants did not multiply as rapidly as the normal type (which is often the case). However, new mutations occur constantly, and the resultant mutants are also multiplying. Therefore, new mutations are continuously adding to the total number of mutants.

This fact can also be used to measure mutation rates. The higher the mutation rate, the faster the proportion of mutants will increase as the culture multiplies. In fact, if we plot the fraction of mutants (M/N) against time (in doubling times of the culture), as in Figure 3.6, the slope of this curve is the forward mutation rate. The fact that other mutations are causing some mutants to become wild type again (in a process known as reversion [see below]) also affects the results shown in Figure 3.6. However, reversion of mutants becomes significant only when the number of mutants is very large and the number of mutants multiplied by the mutation rate back to the wild type (called the reversion rate) begins to approximate the rate at which new mutants are being formed, which, as we showed above, is the forward mutation rate times the number of nonmutant bacteria. At earlier stages of culture growth, the vast majority of bacteria are nonmutant, so forward mutations are occurring much more often than reversion mutations, and the mutation rate is much higher than the reversion rate. Also, for reasons discussed later in this chapter, reversion rates are often much lower than forward mutation rates. Therefore, the contribution to the mutation rate of the reversion of mutants to the wild type can generally be ignored, at least in the early stages of growing a culture.

Figure 3.6 The fraction of mutants increases as a culture multiplies, and the slope is the mutation rate (*a*). *M* is the number of mutants, *N* is the total number of cells, and Time is the time in generation times, which is the total time elapsed divided by the time it takes the culture to double in mass (i.e., the doubling time [*g*]).

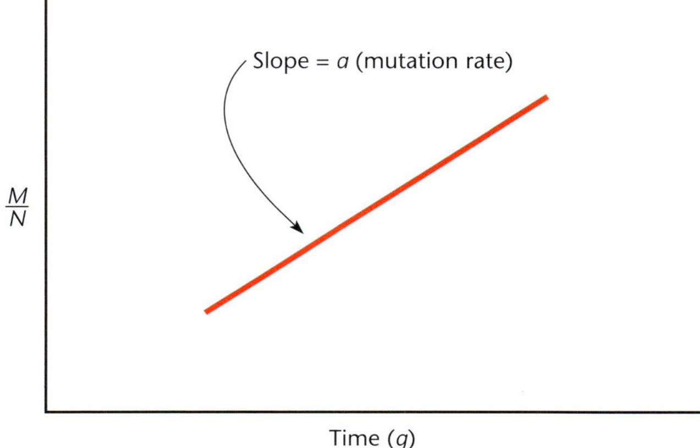

In theory, plots such as that shown in Figure 3.6 could be used in calculating mutation rates. The method does have many advantages. It is not affected by phenotypic lag, although it is affected by the relative growth rates of the mutants. In practice, however, the method is not very practical in most laboratory situations. Mutation rates are usually low, and the numbers of bacteria we can conveniently work with are relatively small, so each new mutation makes too large a contribution to the number of mutants, and we do not get a straight line. To make this method practicable, we would have to work with trillions of bacteria growing in a large chemostat, in which the medium is continuously replaced and the bacteria are removed as fast as they multiply.

PRACTICAL IMPLICATIONS OF POPULATION GENETICS

The fact that the proportion of mutants of all types increases as the culture grows presents both opportunities and problems in genetics. This fact can be advantageous in the isolation of a rare mutant, such as one resistant to streptomycin. If we grow a culture from a few bacteria and plate 10^9 bacteria on agar containing streptomycin, we might not find any resistant mutants, since mutations to streptomycin resistance occur at a frequency of only about 1 in 10^{10} cell divisions. However, if we add a large number of bacteria to fresh broth, grow the broth culture to saturation, dilute a sample of the culture in fresh medium, and then repeat this process a few times, the fraction of streptomycin-resistant mutants will increase. Then, when we plate 10^9 bacteria, we may find many streptomycin-resistant mutants.

Because the fraction of all types of mutants increases as the culture multiplies, if we allow a culture to go through enough generations, it will become a veritable "zoo" of different kinds of mutants—virus resistant, antibiotic resistant, auxotrophic, and so on. To deal with this problem, most researchers store cultures under nongrowth conditions (e.g., as spores or lyophilized cells or in a freezer) that still maintain cell viability. An alternative is to periodically colony-purify bacteria in the culture to continuously isolate the progeny of a single cell (see the introductory chapter). The progeny of a single cell are not likely to be mutated in a way that could confound our experiments later.

Types of Mutations

As defined above, any heritable change in the sequence of nucleotides in DNA is a mutation. A single base pair may be changed, deleted, or inserted; a large number of base pairs may be deleted or inserted; or a large region of the DNA may be duplicated or inverted. Regardless of how many base pairs are affected, a mutation is considered to be a **single mutation** if only one error in replication, recombination, or repair has altered the DNA sequence.

As discussed above, to be considered a mutation, the change in the DNA sequence must be heritable. In other words, the DNA must be able to replicate normally but with the changed sequence. Damage to DNA by itself is not a mutation, but a mutation can occur when the cell attempts to repair damage or replicate over it and a strand of DNA is synthesized that is not completely complementary to the original sequence. The wrong sequence is then replicated, and the resulting DNA carries a mutation. Lethal changes in the DNA sequence (as also mentioned above) do occur but ordinarily cannot be scored as mutations, because these cells by definition do not survive, and other techniques are needed for the study of this kind of mutation (as described above). Sometimes, a situation can be established to search for mutations that are only lethal in combination with another mutation that results, for example, in the absence of another gene product. This is called **synthetic lethality** and has been used to search for genes whose products can substitute for each other. We talk about selections in subsequent sections of this chapter.

Properties of Mutations

The properties and causes of the different types of mutations are probably not very different in all organisms, but they are more easily studied with bacteria. A geneticist can often make an educated guess about what type of mutation is causing a mutant phenotype merely by observing some of its properties.

One property that distinguishes mutations is whether they are **leaky**. The term "leaky" means something very specific in genetics. It means that in spite of the mutation, the gene product still retains some activity and therefore may result in a partial phenotype.

Another distinguishing property of mutations is whether they **revert**. If the sequence has been changed to a different sequence, it can often be changed back to the original sequence by a subsequent mutation. The organism in which a mutation has reverted is called a **revertant**, and the **reversion rate** is the rate at which the mutated sequence in DNA returns to the original wild-type sequence.

Usually, the reversion rate is much lower than the mutation rate that gave rise to the mutant phenotype. As an illustration, consider the previously discussed example, histidine auxotrophy (His$^-$). Any mutation that inactivates any of the approximately 11 genes whose products are required to make histidine will cause a His$^-$ phenotype. Since thousands of changes can result in this phenotype, the mutation rate for His$^-$ is relatively high. However, once a *his* mutation has occurred, the mutation can in general revert only through a change in the mutated sequence that restores the original sequence, to generate a **true revertant**, or one that inserts a different amino acid at this site that allows function of the gene product. Without sequencing the DNA, it is difficult to know whether the original sequence has been restored or another sequence that is acceptable at the site has been inserted, and it may be best to refer to such His$^+$ bacteria as **apparent revertants**, since we do not know yet whether they are true revertants. Because of the specific event required, the reversion rate of a particular His$^-$ mutant to His$^+$ revertants would be expected to be thousands of times lower than the forward mutation rate to His$^-$. If it is not, we suspect that the mutation is being suppressed rather than reverted (see below). That is why it is useful to refer to them as apparent revertants until we know their molecular basis.

Some types of apparent revertants are very easy to detect. For example, His$^+$ apparent revertants can be obtained by plating large numbers of His$^-$ mutants on a plate with all the growth requirements except histidine. Most of the bacteria cannot multiply to form a colony. However, any His$^+$ apparent revertants in the population will multiply to form a colony. The appearance of His$^+$ colonies when large numbers of a His$^-$ mutant are plated would be evidence that the *his* mutation can revert, provided the possibility of suppression has been eliminated (see below).

Base Pair Changes

A **base pair change** is when one base pair in DNA, for example, a GC pair, is changed into another base pair, for example an AT pair. Base pair changes can be further classified as **transitions** or **transversions** (Figure 3.7). In a transition, the purine (A or G) in a base pair is replaced by the other purine and the pyrimidine (C or T) is replaced by the other pyrimidine. Thus, a TA pair would become a CG pair or a CG pair would become a TA pair. In a

Figure 3.7 Transitions versus transversions. The mutations are shown in gold shading.

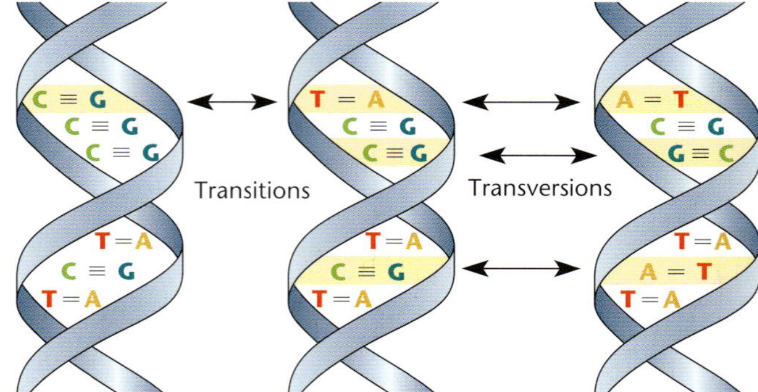

transversion, in contrast, the purines are replaced by pyrimidines, and vice versa. For example, a TA could become an AT or a GC, or a CG could become an AT or a GC.

CAUSES OF BASE PAIR CHANGES

Base pair changes in DNA can occur spontaneously or be induced by external factors, such as irradiation or chemical damage. The wrong base can be inserted during replication, or a damaged base can result in mispairing during replication or repair. If mutations occur without known external factors, such as deliberate application of ionizing radiation or mutagenic chemicals, they are referred to as **spontaneous mutations**, even if they are caused by internal factors, such as reactive oxygen species generated during oxidative metabolism (see below). In this section, we discuss only spontaneous mutations; we reserve discussion of mutations due to external factors, such as irradiation and damage due to added chemicals, called **induced mutations**, for chapter 10.

Base Pair Changes Due to Mispairing during Replication

Base pair changes can be the result of base-pairing mistakes in replication, recombination, or repair, whenever DNA polymerase attempts to synthesize a complementary copy of DNA. Figure 3.8A shows an example of mispairing during replication. In this example, a T instead of the usual C is mistakenly placed opposite a G as the DNA replicates. In the next round of replication, this T usually pairs correctly with an A, causing a GC-to-AT transition in one of the two daughter DNAs. Mistakes in pairing may occur because the bases are sometimes in a different form, called the **enol** form, which causes them to pair differently (Figure 3.8B).

What type of base pair change occurs depends on which types of bases mispaired. Mispairing between a purine and a pyrimidine causes a transition, whereas mispairing between two purines or two pyrimidines causes a transversion. Because a pyrimidine in the enol form still pairs

Figure 3.8 **(A)** A mispairing during replication can lead to a base pair change in the DNA. The original DNA is shown as a dark blue line, the first generation of newly replicated DNA is shown as a medium blue line, and the second generation of newly replicated DNA is shown as a light blue line. The mispaired base is shaded in blue, and the resulting A-T base pair is shaded in yellow. **(B)** Pairing between guanine and an alternate enol form of thymine can cause G-T mismatches.

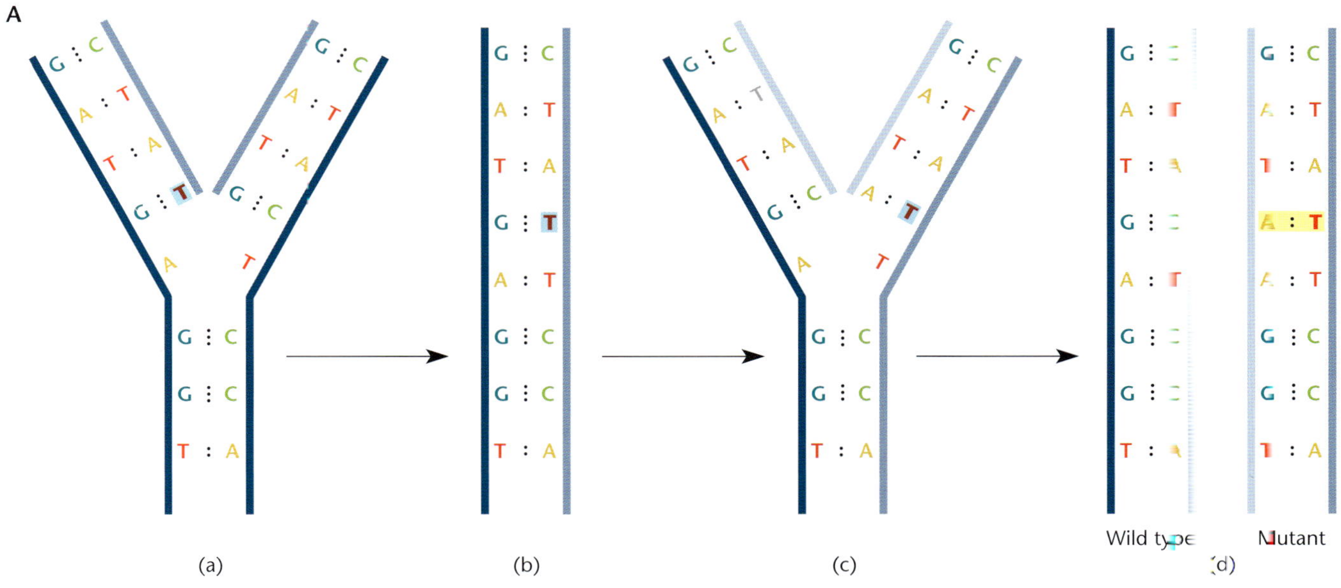

with a purine and a purine in the enol form still pairs with a pyrimidine, mispairing during replication usually leads to transition mutations. Furthermore, all four bases can undergo the shift to the enol form, and either the base in the DNA template or the incoming base can be in the enol form and cause mispairing. Thus, the thymine in the enol form pictured in Figure 3.8B might be in the template, in which case the transition would be AT to GC, or it could be the incoming base, resulting in a GC-to-AT transition (as shown).

Mistakes during replication leading to mutations are not completely random, and some sites are much more prone to base pair changes than are others. Mutation-prone sites are called **hot spots** and can have many causes, including the deamination of methylated bases (see below). Mispairing occurs fairly often during replication, and it is an obvious advantage for the cell to reduce the number of base pair change mutations that occur during replication. In chapter 1, we discussed some of the mechanisms cells use to reduce these base pair changes.

Base Pair Changes Due to Spontaneous Deamination

The **deamination** of the bases in DNA, or the removal of an amino group from one of the bases in the DNA, can also cause base pair changes. This deamination can occur spontaneously, especially at higher temperatures. Cytosine is particularly susceptible to deamination, and the damage is apt to be mutagenic, since deamination of cytosine results in the formation of uracil (Figure 3.9A), which pairs with adenine instead of guanine (see chapters 1 and 2 for structures). Therefore, unless the uracil derived from deamination of cytosine is removed from the DNA, it will cause a GC-to-AT transition the next time the DNA replicates.

Because of the problems caused by deamination of cytosine, cells have evolved a special mechanism for removing uracil from DNA whenever it appears (Figure 3.9B). An enzyme called uracil-*N*-glycosylase, the product of the *ung* gene in *E. coli*, recognizes the uracil as unusual in DNA and removes the uracil base. The DNA strand in the region where the uracil was removed is then degraded and resynthesized, and usually, the correct cytosine is inserted opposite the guanine. As expected, *ung* mutants of *E. coli* show high rates of spontaneous mutagenesis, and most of the mutations are GC-to-AT transitions. If the cytosine has a methyl group at the 5 position of the pyrimidine ring, a common modification in many organisms, the cytosine will become thymine when deaminated and will not be recognized as unusual and removed by the Ung enzyme, causing a hot spot for spontaneous mutagenesis.

Figure 3.9 Removal of deaminated cytosine (uracil) from DNA. **(A)** Comparison of cytosine and uracil; the amino group that is removed from cytosine by deamination is shown in red. **(B)** Uracil-*N*-glycosylase removes the uracil, and the DNA strand is partially degraded and resynthesized with cytosine at that position.

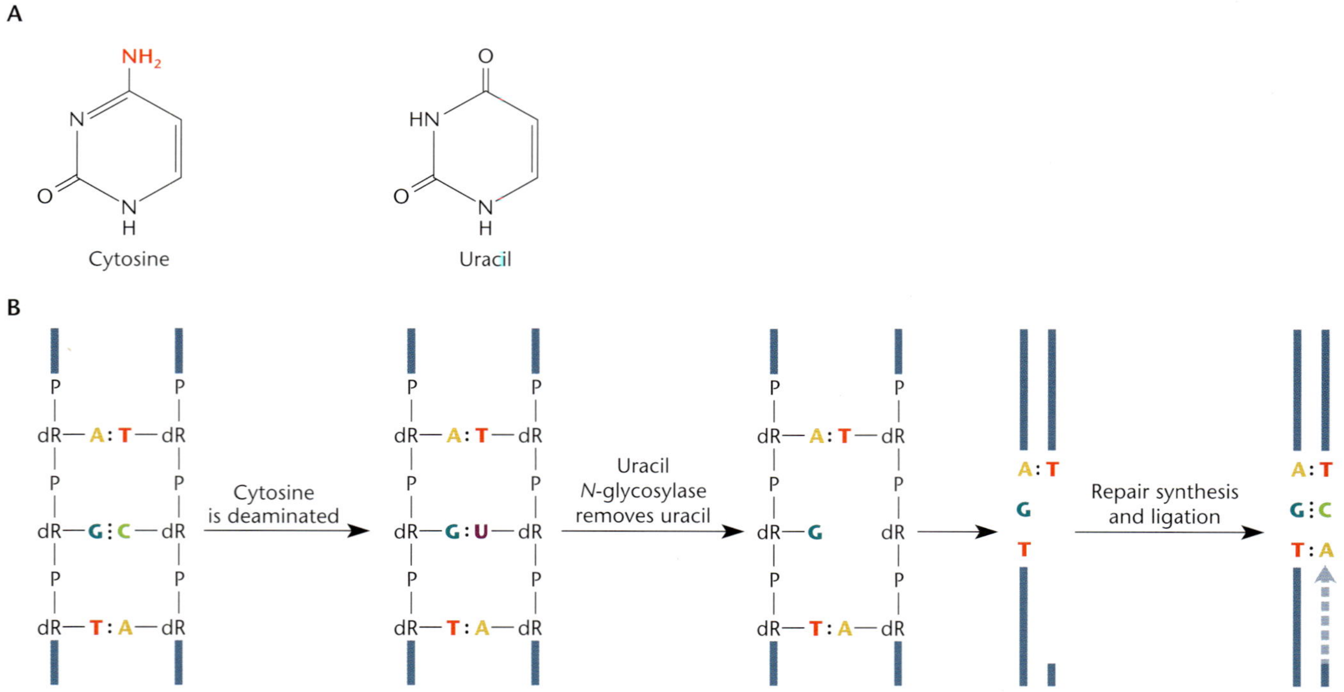

All organisms have the problem of cytosine deamination in their DNA, possibly partially explaining why the testicles of warm-blooded animals, including mammals, are external, where the average temperature is lower and deaminations are less frequent, since they are triggered by heat.

Base Pair Changes Due to Oxidation of Bases

Reactive forms of oxygen, such as peroxides and free radicals, are given off as by-products of oxidative metabolism by aerobic organisms, and these forms can react with and alter the bases in DNA. A common example is the altered guanine base 8-oxoG, which sometimes mistakenly pairs with adenine instead of cytosine, causing GC-to-TA or AT-to-CG transversion mutations, depending on whether the guanine has been altered in the DNA or in the incoming guanosine deoxynucleotide. We, as aerobic organisms, are also subject to this type of DNA damage, which is why consumption of foods rich in antioxidants, which are oxidized in place of our DNA, may reduce the incidence of cancer and other diseases (see Box 10.1). Repair systems specific to damage such as deamination and oxidation are discussed in more detail in chapter 10.

CONSEQUENCES OF BASE PAIR CHANGES

Whether a base pair change causes a detectable phenotype depends, of course, on where the mutation occurs and what the actual change is. Even a change in an open reading frame (ORF) that encodes a polypeptide may not result in an altered protein. If the mutated base is the third position in a codon, the amino acid inserted into the protein may not be different because of the degeneracy of the code (see chapter 2). Mutations in the coding region of a gene that do not change the amino acid sequence of the polypeptide product are called silent mutations. The change also may occur in a region that encodes an RNA rather than a polypeptide or in a regulatory sequence, such as an operator or promoter (see chapters 2 and 11). Alternatively, the mutation may occur in a region that has no detectable function. We first discuss mutations that change the coding region of a polypeptide.

Missense Mutations

Because most of the DNA of bacteria is devoted to encoding proteins, most base pair changes in bacterial DNA cause one amino acid in a polypeptide to be replaced by another. These mutations are called **missense mutations** (Figure 3.10). However, a missense mutation that changes an amino acid in a protein does not always inactivate the protein. If the original and new amino acids have similar properties, the change may have little or no effect on the activity of a protein. For example, a missense mutation that changes an acidic amino acid, such as glutamate, into another acidic amino acid, such as aspartate, may have less

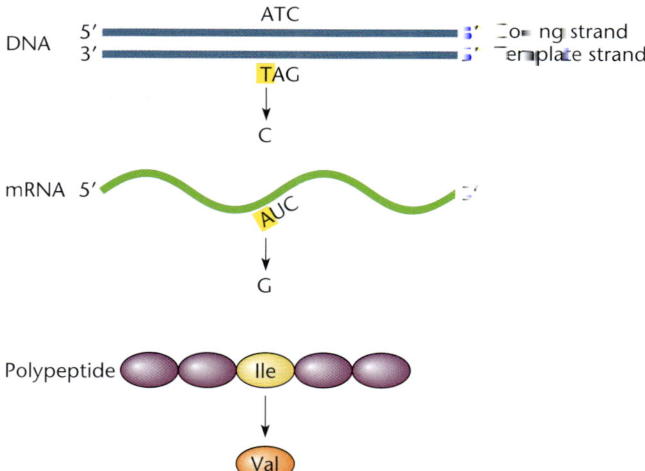

Figure 3.10 Missense mutation. A mutation that changes T to C in the DNA template strand results in an A-to-G change in the mRNA. The mutant codon GUC is translated as valine instead of isoleucine, resulting in the insertion of valine at that position of the polypeptide.

of an effect on the functioning of the protein than does a mutation that substitutes a basic amino acid, such as arginine, for an acidic one. The consequences also depend on which amino acid is changed. Certain amino acids in any given protein sequence are more essential to activity than others, and a change at a crucial position can have a much greater effect than a change elsewhere. Investigators often use this fact to determine which amino acids are essential for activity in different proteins.

Nonsense Mutations

As noted above, base pair changes that produce one of the nonsense codons (UAA, UAG, or UGA) in the ORF for the protein are called **nonsense mutations**. Because the nonsense codons are normally used to signify the end of a gene, they are normally recognized by release factors (see chapter 2), which cause release of the translating ribosome and the polypeptide chain. Therefore, if a mutation to one of the nonsense codons occurs in an ORF for a protein, the protein translation terminates prematurely at the site of the nonsense codon, and the shortened or truncated polypeptide is released from the ribosome (Figure 3.11). For this reason, nonsense mutations are sometimes called "chain-terminating mutations." These mutations almost always inactivate the protein product of the gene in which they occur, especially if they occur early in the coding region for the protein. If, however, they occur in a noncoding region of the DNA or in a region that encodes an RNA rather than a protein, such as a gene for a tRNA, they are indistinguishable from other base pair changes.

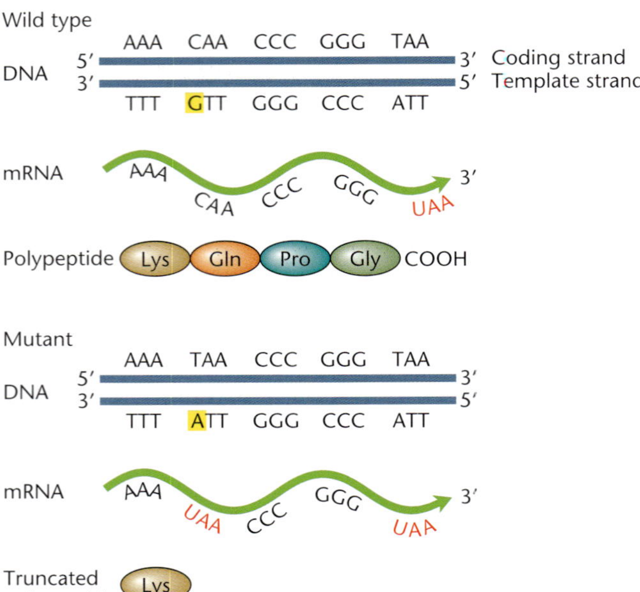

Figure 3.11 Nonsense mutation. Changing the CAA codon, encoding glutamine (Gln), to UAA, a nonsense codon, causes truncation of the polypeptide gene product.

The three nonsense codons—and their corresponding mutations—are sometimes referred to by color designations: **amber** for UAG, **ochre** for UAA, and **opal** for UGA. These names have nothing to do with the effects of the mutation. Rather, when nonsense mutations were first discovered, their molecular basis was unknown. The investigators thought that descriptive names might be confusing later on if their interpretations were wrong, so they followed the lead of physicists with their "quarks" and "barns." The first nonsense mutations to be discovered, to UAG, were called amber mutations. The legend is that they were named after a graduate student at the California Institute of Technology, Harry Bernstein (or his grandmother), who was involved in the first clear experiments describing these mutations (Bernstein means amber in the German language). Following suit, UAA and UGA mutations were also named after colors—ochre and opal, respectively.

PROPERTIES OF BASE PAIR CHANGE MUTATIONS

Base pair changes are often leaky, which means that the mutation results in partial activity. A substituted amino acid may not work as well as the original at that position in the chain, but the protein often retains some activity. Even nonsense mutations are usually somewhat leaky, because even without a nonsense suppressor, sometimes an amino acid is inserted for a nonsense codon, albeit at a low frequency. In wild-type *E. coli*, UGA tends to be the most leaky, followed by UAG; UAA tends to be the least leaky.

Base pair mutations also revert. If the base pair has been changed to a different base pair, it can also be changed back to the original base pair by a subsequent mutation (a true reversion). Being base pair changes, nonsense mutations can revert. However, apparent revertants of nonsense mutations often have suppressor mutations in tRNA genes, and this property of being suppressed by nonsense suppressor mutations in tRNA genes can be used to identify nonsense mutations (see below). Moreover, base pair changes are a type of **point mutation** because they map to a particular "point" on the DNA, as discussed below.

Frameshift Mutations

A high percentage of all spontaneous mutations are **frameshift mutations** (Figure 3.12). This type of mutation occurs when one or a few base pairs are removed from or added to the DNA; if these mutations occur in an ORF that encodes a polypeptide, this can cause a shift in the reading frame. Because the code is three lettered, any addition or subtraction that is not a multiple of 3 causes a frameshift in the translation of the remainder of the gene. For example, adding or subtracting 1, 2, or 4 base pairs (bp) causes a frameshift, but adding or subtracting 3 or 6 bp does not. Mutations that remove or add base pairs are often called frameshift mutations even if they do not occur in an ORF and do not actually cause a frameshift in the translation of a polypeptide. A more general name would be single (or double, etc.) base additions or deletions, since it does not imply that the mutation is in a reading frame for a protein.

CAUSES OF FRAMESHIFT MUTATIONS

Spontaneous frameshift mutations often occur where there is a short repeated sequence that can slip during DNA replication. The slippage can occur either in the template DNA or in the nucleotide being added. As an example, Figure 3.13 shows a string of AT base pairs in

Figure 3.12 Frameshift mutation. The wild-type mRNA is translated as glutamine (Gln)-serine (Ser)-arginine (Arg), etc. Insertion of an A (shaded in yellow) would shift the reading frame so that the codons would be translated glutamine (Gln)-isoleucine (Ile)-proline (Pro), etc., with all downstream amino acids being changed.

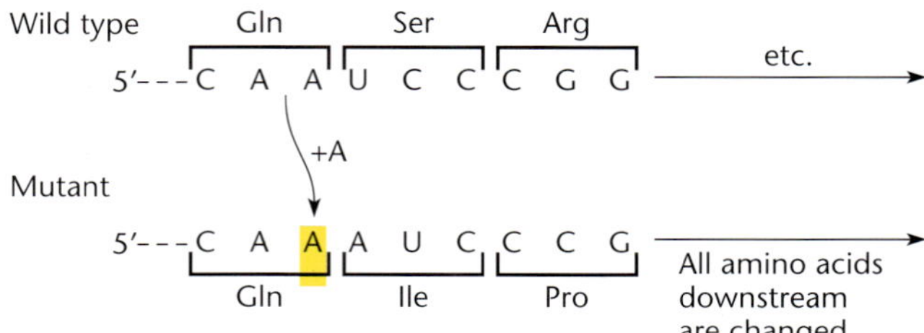

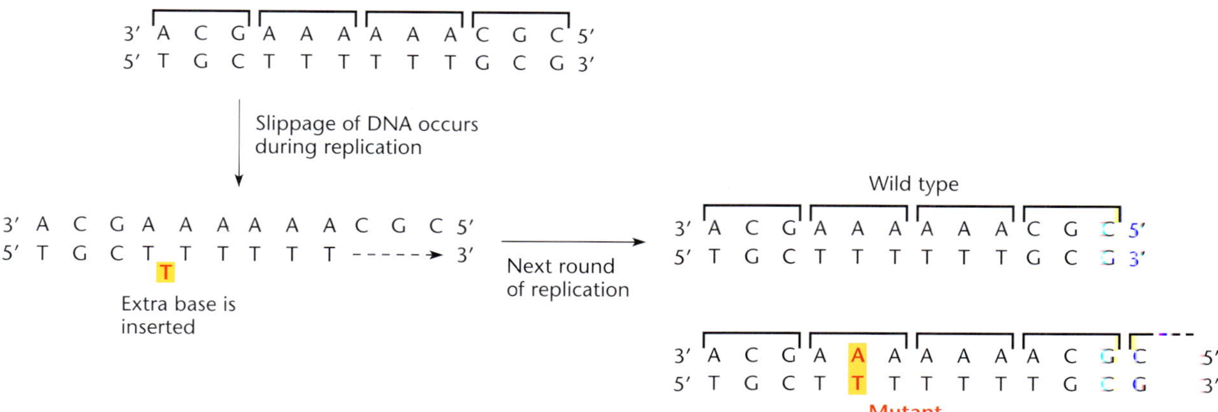

Figure 3.13 Slippage of DNA at a repeated sequence (for example, a series of A-T base pairs) can cause a frameshift mutation because of insertion of an extra residue during replication by DNA polymerase.

the DNA. Since any one of the A's in one strand can pair with any T in the other strand, the two strands could slip with respect to each other, as in the illustration. Slippage during replication could occur in the template before the base was added, and an extra AT base pair could appear in one of the resulting DNA products, as shown. Alternatively, the slippage could leave one T unpaired, and an AT base pair would be left out of one of the DNA products.

PROPERTIES OF FRAMESHIFT MUTATIONS

Frameshift mutations in a coding sequence are usually not leaky and almost always inactivate the protein, because every amino acid in the protein past the point of the mutation is wrong. The protein is usually also truncated, because nonsense codons are likely to be encountered while the gene is being translated in the wrong frame. In general, because 3 of the 64 codons are nonsense codons, one of them should be encountered by chance about every 20 codons when the region is being translated in the wrong frame.

Another property of frameshift mutations is that they can revert. If a base pair has been subtracted, one can be added to restore the correct reading frame, and vice versa. Frameshift mutations can also be suppressed (see below) by the addition or subtraction of a base pair close to the site of the original mutation that restores the original reading frame. This means of restoring the function of a protein inactivated by a frameshift mutation is discussed below and was used to determine the characteristics of the genetic code. Finally, frameshift mutations are a type of point mutation in that they map to a single point on the DNA.

Some types of pathogenic bacteria apparently take advantage of the frequency and high reversion rate of frameshift mutations to avoid host immune systems. In such bacteria, genes required for the synthesis of cell surface components that are recognized by host immune systems often have repeated sequences that favor frameshift mutations. Consequently, these genes can be frequently turned off and on by frameshift mutations and subsequent reversion, allowing some members of the population to escape immune detection. For example, frameshift mutations may aid in the synthesis of virulence gene products by *Bordetella pertussis*, the causative agent of whooping cough. Frameshift mutations are also used to reversibly inactivate genes of *Haemophilus influenzae* and *Neisseria gonorrhoeae*, which cause spinal meningitis and gonorrhea, respectively.

Deletion Mutations

Another common type of spontaneous mutation is the **deletion mutation**, in which entire stretches of the DNA have been removed. As much as 5% of all spontaneous mutations can be deletions in *E. coli*, and they can often be very long, removing thousands of base pairs and possibly many genes, provided that none of the deleted genes encode proteins that are essential for growth.

CAUSES OF DELETIONS

Deletions are usually caused by **recombination** between directly repeated sequences in different locations of the DNA. In recombination, the strands of two DNA molecules are broken and rejoined in new combinations—hence the name recombination (see chapters 8 and 9). Homologous recombination occurs at sites of identical sequences in the two DNAs because the strands of the two DNAs must be complementary to each other to initiate recombination. This ensures that the recombination usually occurs at the same place in the two DNAs, where their sequences are the same. However, sometimes similar or identical sequences exist in more than one location in the DNA, and recombination can sometimes mistakenly occur between these two different locations. Such recombination is called **ectopic recombination** (Greek for out of place) because it is occurring outside the correct place. Deletions result from ectopic recombination if the two similar sequences involved in ectopic recombination are **direct repeats**, that is, the two sequences are similar or identical when read in the 5′-to-3′ direction on the same strand of DNA.

As shown in Figure 3.14, ectopic recombination between directly repeated sequences in the DNA can give rise to deletions in two ways, depending on whether the directly repeated sequences are on different DNA molecules or the same DNA molecule. Part I shows how a deletion can occur when the different copies of a directly repeated sequence occur in different DNA molecules, for example, in the daughter DNAs created during replication. Different copies of the directly repeated sequences mistakenly pair with each other and recombine, deleting the sequence between them on one of the two daughter DNA molecules. Notice that after recombination, the sequence between the two repeats is not deleted in both molecules but instead is repeated on the other daughter DNA molecule, creating a tandemly duplicated sequence in this DNA. We discuss tandem duplications later in this section. Recombination between different regions of two DNA molecules is sometimes called "unequal crossing over," because the two DNAs are not equally aligned during the recombination.

The outcome is different if the ectopic recombination that creates a deletion occurs between direct repeats on the same DNA, as shown in Figure 3.14, part II. Now, the intervening sequence "loops out," and recombination removes the looped-out sequences between the two repeats, as shown. For purposes of illustration, the directly repeated sequences shown in the figure are much shorter than would normally be required for recombination. Usually, direct repeats that promote ectopic recombination are much larger in *E. coli* and other types of bacteria. The actual molecular mechanism of homologous recombination is addressed in chapter 9. Bacterial DNA contains several types of repeats, the longest of which include mobile elements like insertion sequence (IS) elements (see chapter 8) and the rRNA genes, which are thousands of base pairs long and are often repeated in many places in the DNA.

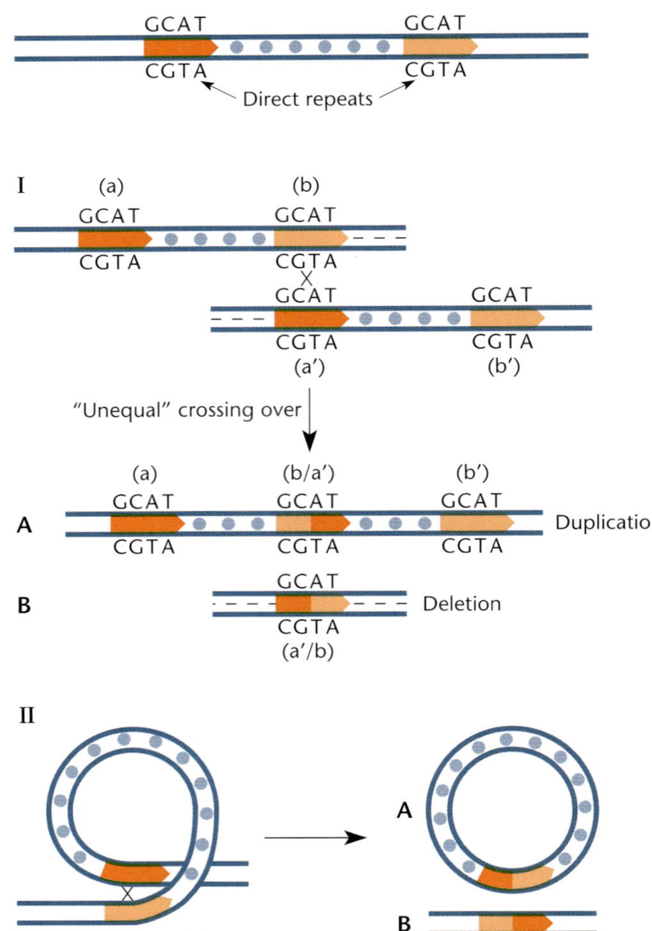

Figure 3.14 Ectopic recombination between directly repeated sequences can cause deletion and tandem-duplication mutations. **(I)** The recombination can occur between repeated sequences in different DNAs, resulting in a duplication **(A)** or a deletion **(B)**. **(II)** Alternatively, it can occur between repeated sequences in the same DNA, resulting in a deletion **(B)** and the looped-out deleted segment **(A)**. The direct repeats involved in recombination are often hundreds of base pairs long, but repeats of only 4 bp are shown for convenience.

PROPERTIES OF DELETION MUTATIONS

Deletions have very distinctive properties. They are usually not leaky; deleting part or all of a gene usually totally inactivates the gene product. Mutations that inactivate two or more genes simultaneously are most often deletions. Moreover, deletion mutations sometimes fuse one gene to another or put one gene under the control of different regulatory sequences. However, the most distinctive property of long deletion mutations is that they never revert. Every other type of mutation reverts at some frequency, but for a deletion to revert, the missing sequence would somehow have to be found and reinserted, which is essentially an impossible event in a culture of bacteria composed of genetically identical bacteria all harboring the deletion. Deletions also behave differently from point mutations in genetic crosses, not mapping to a single point.

Isolating Deletion Mutants

Deletion mutations have been very important historically in genetic experiments because of the ease with which they can be used to map multiple point mutations in a gene. We give a few examples later in this book, including later in this chapter, of how they were used to map point mutations in the *thyA* gene of *E. coli*. The properties of deletion mutations make them relatively easy to identify among other types of mutations. One useful property that has been used historically is the ability of deletions to inactivate more than one gene simultaneously. Even though deletions are fairly rare, on the order of 5% of spontaneous mutations in *E. coli*, a single deletion that inactivates two nearby genes occurs much more frequently than two independent point mutations, such as base pair changes, that inactivate the same two genes. Two independent point mutations occur only as frequently as the product of the frequencies of each of the point mutations taken separately. Thus, if mutants are selected for having inactivated one gene, those that also inactivated a nearby gene are probably deletions extending into the nearby gene. This principle was used to map *E. coli lacI* mutations before DNA sequencing was available. The *lac* operon was moved close to the *tonB* gene, and mutations for T1 phage resistance (due to inactivation of *tonB*, which encodes the receptor for T1 attachment; see chapter 7) that also made the cells constitutive for the *lac* operon were mostly deletions extending from *tonB* into *lacI* and could be used to map *lacI* mutations isolated by various other types of selections.

Another property of deletions that has been used historically is that they can fuse one gene to another, sometimes putting the downstream gene under the control of the regulatory region of the upstream gene. The downstream gene will then be expressed when the upstream gene is expressed and not when the downstream gene is normally expressed. Techniques for generating precise targeted deletions are discussed in chapter 13.

NAMING DELETION MUTATIONS

Deletion mutations are named differently from other mutations. The Greek letter Δ (delta), for *del*etion, is written in front of the gene designation and allele number, e.g., Δ*his8*. Often, deletions remove more than one gene, and so, if known, the deleted regions are shown, followed by a number to indicate the particular deletion. For example, Δ(*lac-proAB*)195 is deletion number 195 extending through the *lac* and adjacent *proAB* genes on the *E. coli* chromosome. Often, a deletion removes one or more known genes but extends into a region of unknown genes, so that the endpoints of the deletion are not known. In this case, the deletion is often named after the known gene. For example, the Δ*his8* deletion may delete the entire *his* operon but may also extend an unknown distance into neighboring genes.

With DNA sequencing, it is often possible to know the precise endpoints of a deletion. If a deletion is completely included in the coding region for one protein, the endpoints can be given in the name. For example, Δ*trpA*(H4-Q16) deletes from the 4th amino acid in the TrpA protein, a histidine, to the 16th, a glutamine. If it is an in-frame deletion, i.e., has deleted a multiple of 3 bases (see above), the amino acids encoded after the deletion will be unchanged, and the only change in the protein will be the precise removal of the amino acids encoded in the deleted region.

Tandem-Duplication Mutations

In a duplication mutation, a sequence in one region of the DNA is copied so that it then exists in two copies in the DNA. The most common type of duplication, a **tandem duplication**, consists of a sequence immediately followed by its duplicate. Tandem duplications occur fairly frequently and can be very long. Other types of duplications can be caused by the insertion into the chromosome of DNA elements containing parts of the genome, for example, transducing phages (see chapter 7). In this case, the second copy is located at a different site in the chromosome. These are special cases and will be discussed in subsequent chapters.

CAUSES OF TANDEM DUPLICATIONS

Like deletions, tandem duplications can result from ectopic recombination between directly repeated sequences in different DNA molecules. In fact, as shown in Figure 3.14, they are probably often created at the same time as a deletion. Pairing between two directly repeated sequences in different DNAs, followed by recombination, can give rise to a tandem duplication and a deletion as the two products. These products segregate into different progeny cells after replication and segregation.

PROPERTIES OF TANDEM-DUPLICATION MUTATIONS

Although tandem duplications often arise during the same ectopic recombination event as deletions, their properties are very different. Tandem-duplication mutations that occur within a single gene usually inactivate the gene and are not leaky. However, if the duplicated region is long enough to include one or more genes, no genes are inactivated, including the genes in which the recombination occurred, which contain the **duplication junctions**. This conclusion may seem surprising, but consider the example shown in Figure 3.15. Direct repeats in genes A and C on different DNAs pair with each other. The repeats in the two DNAs then recombine with each other, creating a duplication in one DNA (and a deletion in the other). Only part of gene A exists in the downstream duplicate copy, but an entire gene A exists in the upstream one. Conversely, only part of gene C exists in the upstream duplicate, but the entire C gene exists in the downstream duplicate. There are now two copies of gene B, both of

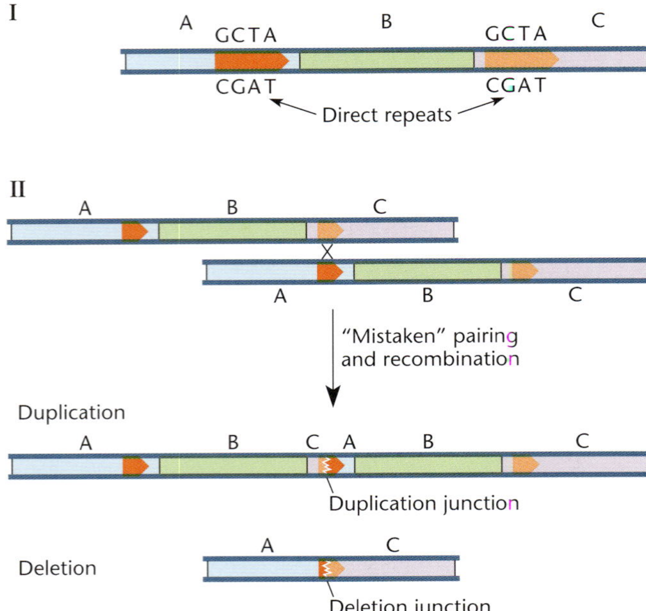

Figure 3.15 Formation of a long tandem-duplication mutation does not inactivate any genes. **(I)** Direct repeats in genes A and C. **(II)** Ectopic recombination between the direct repeats in genes A and C leads to a tandem duplication (and a deletion in the second molecule). A complete gene A exists in the upstream copy and a complete gene C in the downstream copy. Gene B now exists in two copies, and one copy of gene B could be under the control of the promoter for gene C. The direct repeats shown are a representation of much longer sequences.

which are unaltered. Therefore, intact genes A, B, and C still exist after the duplication, and there may be little phenotypic indication that a mutation had even occurred. Note that because duplications of this type do not inactivate genes, bacterial cells can survive very long duplications, even those that include many essential genes.

The most characteristic property of tandem duplications is that they are very unstable and revert at a high frequency. Even though the ectopic recombination events that lead to a duplication are usually rare, recombination anywhere within the duplicated segments can delete them, restoring the original sequence. The instability of tandem duplications is often the salient feature that allows their identification. If the recombination systems of the bacterium are inactivated after a tandem duplication has occurred, the tandem duplications will be stabilized, which makes it easier to study them.

Identifying Mutants with Tandem-Duplication Mutations

As mentioned above, because no genes are inactivated in longer tandem duplications, usually there is little phenotypic indication that a mutation has even occurred. However, in some cases, there may be a phenotype associated with having two copies of one or more of the genes included in the duplicated region. An excess of some gene products can affect the appearance of colonies on plates. Also, like deletions, duplications sometimes fuse two genes, putting expression of one gene under the control of the regulatory region of a different gene. In the example in Figure 3.15, part of gene A and all of gene B have been fused to the beginning of gene C, which might put genes A and B under the control of the promoter for gene C, resulting in an altered pattern of gene expression.

ROLE OF TANDEM-DUPLICATION MUTATIONS IN EVOLUTION

Tandem-duplication mutations play an important role in evolution. Ordinarily, a gene cannot change without loss of its original function, and if the lost function was a necessary one, the organism will not survive. However, when a duplication has occurred, there are two copies of the genes in the duplicated region, and now one of these is free to evolve to a different function. This mechanism allows organisms to acquire more genes and become more complex.

Inversion Mutations

Sometimes a DNA sequence is not removed or duplicated but, rather, is flipped, or **inverted**, relative to the surrounding sequences. After such an **inversion**, all the genes in the inverted region face in the opposite orientation relative to the upstream and downstream regions. Programmed inversions can play roles in regulation in bacteria, but because they are not true mutations but, rather, a type of site-specific recombination that is often reversible, they are considered in chapter 8.

CAUSES OF INVERSIONS

Like deletions and tandem duplications, inversions are caused by ectopic recombination between repeated sequences. However, now the repeats at which the ectopic recombination occurred are inverted rather than direct (Figure 3.16). As discussed in chapter 1, inverted-repeat sequences read the same way in the 5′-to-3′ direction on opposite strands. Also, the ectopic recombination that produces inversions must occur between two inverted-repeat regions on the same DNA molecule. Alignment of the two repeated sequences results in looping out of the DNA between the repeats (ABCD), and recombination between the repeats causes the looped out region to be flipped relative to the repeats and the surrounding sequence (X and Y). This changes the order from the original XABCDY to XDCBAY.

PROPERTIES OF INVERSION MUTATIONS

Unlike deletions, inversion mutations can generally revert. Recombination between the inverted repeats that

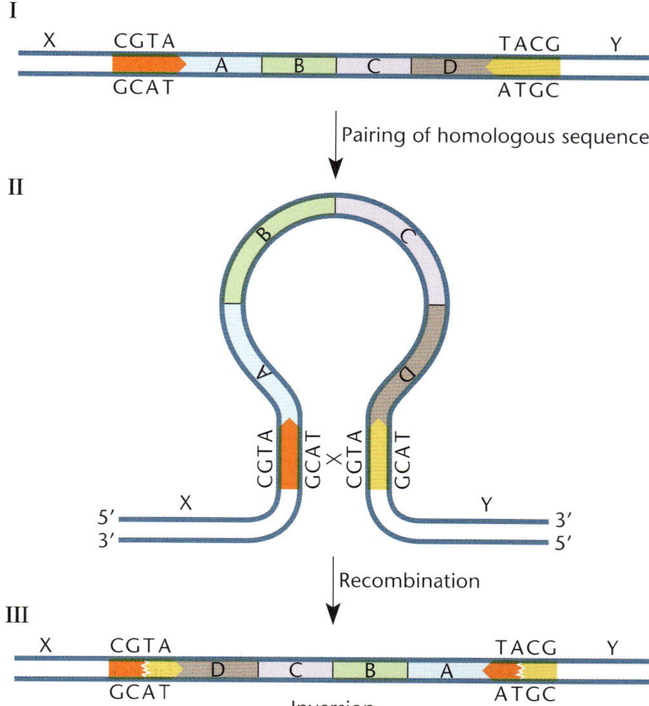

Figure 3.16 Recombination between inverted repeats can cause inversion mutations. The order of genes within the inversion is reversed after the recombination. The inverted repeats are usually ≥50 bp long, but inverted repeats of only 4 bp are shown for convenience.

caused the mutation will "reinvert" the affected sequence, re-creating the original arrangement (XABCDY). However, if the ectopic recombination that originally caused the inversion occurred between very short inverted repeats or repeats that are not exactly the same, then the recombination event that caused the inversion would have to occur between the exact bases involved in the first recombination to restore the correct sequence. Such a recombination could be a very rare event, and reversions of such an inversion mutation would be very rare.

Like tandem duplications, inversion mutations often cause no phenotype. If the inversion involves a longer sequence, including many genes, generally the only affected regions are those that include the **inversion junctions**, where the recombination occurred. In many cases the inverted repeats where the recombination occurred are due to repeated sequences between genes, so no genes are inactivated by the inversion. The genes in the inverted region are still intact, although they are in the opposite orientation relative to the other genes in the chromosome. While even very long inversion mutations often cause no obvious phenotypes, they may cause other longer-term problems for the cell by inverting certain polar sequences, such as the *ter* sites involved in replication termination and the KOPS sequences discussed in chapter 1. Perhaps because of this, long inversions seem to have been selected against in nature, since the arrangements of genes in even quite distantly related bacteria are often remarkably similar (Box 3.1). Of note, inversions that maintain the polar nature of chromosomal features, that is, inversions around the terminus or origin region, are somewhat common in bacteria.

Identification of Inversion Mutations

Like deletions, inversion mutations sometimes fuse one gene to another gene. This property often provides the mechanism for detecting them. This is often why cells can use programmed inversions to regulate genes, since promoters in the inversion can transcribe a gene in one orientation but not the other or transcribe different genes depending on their orientation (see chapter 8).

NAMING INVERSIONS

A mutation known to be an inversion is given the letters IN followed by the genes in which the inversion junctions occur, provided that they are known, followed by the number of the mutation. For example, IN(*purE-trpA*)*3* is inversion number 3, in which the inverted region extends from somewhere near the gene *purB* to somewhere near the gene *trpA*, so both genes and the genes between them are included in the inversion.

Insertion Mutations

An **insertion mutation** occurs when a piece of DNA inserts into the chromosome. Smaller insertions of just a few base pairs are usually referred to as frameshift mutations, but sometimes insertions occur that are much too large to be caused by replication slippage. These mutations are usually caused by insertion sequence (IS) elements that spontaneously move, or "hop," from another place in the DNA. IS elements are small transposable elements—usually only about 1,000 bp long—that exist in multiple copies in many bacterial genomes and carry genes for enzymes to promote their own movement. Because they exist in multiple copies, IS elements are also often the sites of ectopic recombination that creates deletion, duplication, and inversion mutations. We discuss transposable elements in chapter 8.

PROPERTIES OF INSERTION MUTATIONS

Insertion of DNA into a gene almost always inactivates the gene; therefore, insertion mutations are usually not leaky. Transposable elements, such as IS elements, also contain many transcription termination sites, so their insertion results in polarity (see chapter 2), which prevents the transcription of genes downstream of the gene into which the element has inserted if those genes are normally cotranscribed with the disrupted gene. They also map, like

BOX 3.1

Inversions and the Genetic Map

Even a single large inversion mutation causes a dramatic change in the genetic map, or the order of genes in the DNA, of an organism. The order of all the genes is reversed between the sites of the recombination that led to the inversion. We would also expect inversions to be fairly frequent, because repeated sequences exist in many places in bacterial genomes, and many of them are inverted with respect to each other. In spite of this, inversions seem to have occurred very infrequently in evolution. As evidence, consider the genetic maps of *Salmonella enterica* serovar Typhimurium and *E. coli*. These bacteria presumably diverged millions of years and billions of generations ago and are very different in many ways, including in the diseases they cause. Nevertheless, the maps are very similar, except for one short inverted sequence between about 25 and 27 min on the *E. coli* map. The apparent reason why large inversions have not occurred is that some sequences in the chromosome cannot be inverted without seriously disadvantaging the organism. These are sequences that are polar and do not read the same on both strands in the 5′-to-3′ direction, so if they are inverted with respect to the direction of replication, they will be nonfunctional. Some examples we know of in *E. coli* are the KOPS sequences that direct the FtsK DNA translocase toward the *dif* sequences at the terminus of replication to remove chromosome dimers after replication (see chapter 1) and χ sites that promote recombination to restart collapsed replication forks (see chapter 9).

References

Mahan MJ, Roth JR. 1991. Ability of a bacterial chromosome segment to invert is dictated by included material rather than flanking sequence. *Genetics* **129**:1021–1032.

Maisel L. 2010. Barriers to the formation of inversion rearrangements in *Salmonella*, p 233–244. *In* Maloy S, Hughes KT, Casadesus J (ed), *The Lure of Bacterial Genetics: A Tribute to John Roth*. ASM Press, Washington, DC.

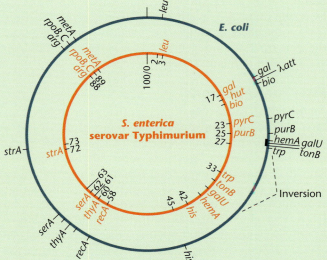

Genetic maps of *S. enterica* serovar Typhimurium and *E. coli*, showing a high degree of conservation. The region from *hemA* to the 40-min position is inverted in *E. coli* relative to that in serovar Typhimurium.

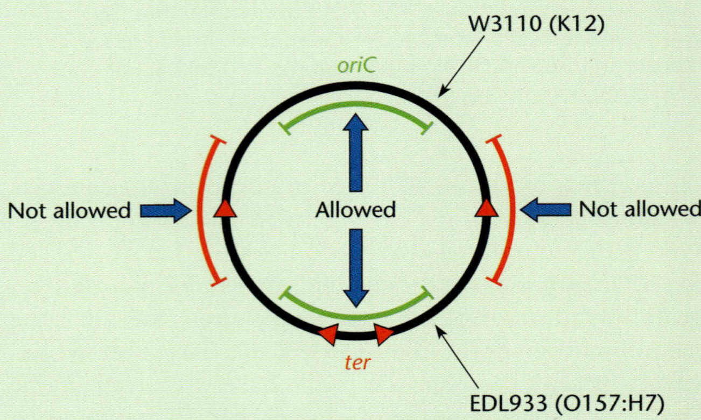

Limitations to chromosomal DNA inversions, which occur primarily near the origin and terminus regions. Comparison of *E. coli* K-12 strains MG1655 and W3110 revealed inversions across ribosomal operons near *oriC*. Comparison of *E. coli* K-12 strains with the first pathogenic O157:H7 strain that was sequenced revealed an inversion around the terminus.

a point mutation, to a certain place in the DNA. Finally, insertion mutations do revert, but only rarely. To revert, the inserted DNA must be removed precisely, with no DNA sequences remaining. Many types of transposable elements copy a short sequence when they integrate into the DNA (see chapter 8), and recombination between these short repeated sequences can precisely excise the element. However, these repeated sequences are too short to promote recombination very frequently. The unusual properties of insertion mutations—totally inactivating the gene, mapping like a point mutation, and seldom reverting—led to their discovery (see chapter 8).

IDENTIFICATION OF INSERTION MUTATIONS

A significant percentage of all spontaneous mutations are insertion mutations, but their phenotypes are difficult to distinguish from those of other types of mutations. Often, a mutation is not known to be an insertion mutation until the DNA is isolated and/or sequenced. Insertion of transposable elements significantly changes the size of DNAs into which they insert, making them relatively easy to map physically. Some transposable elements, called transposons, carry a selectable gene, such as one for antibiotic resistance; these transposons can serve as useful tools in genetics experiments, as the insertion of such a transposon into a cell's DNA makes the mutant cell antibiotic resistant and makes the insertion easy to map. However, if the transposon has just moved from one DNA in the cell to another DNA in the same cell, the phenotype of antibiotic resistance will remain unchanged, and it is not obvious that a mutation has occurred. To isolate transposon insertion mutations, one must introduce the transposon into the cell from outside. A common way to do this is to introduce the transposon into the cell in a DNA that cannot replicate in that cell, i.e., a **suicide vector**. Then, the cell can become resistant to the antibiotic only if the transposon hops from the original DNA into the chromosome or some other DNA that can replicate in the cell. The methods of transposon mutagenesis are central to bacterial molecular genetics and biotechnology and are discussed in some detail in chapter 8.

NAMING INSERTION MUTATIONS

An insertion mutation in a particular gene is represented by the gene name, two colons, and the name of the insertion. For example, *galK*::Tn*5* denotes the insertion of the transposon Tn*5* into the *galK* gene. Because an insertion can occur at any one of many positions within the gene (and these could result in different phenotypes), each insertion is also indicated with an allele number to distinguish them from each other (e.g., *galK35*::Tn*5*). Because of their usefulness, insertions of just a selectable gene, for example, for antibiotic resistance, are often made using recombinant DNA techniques. When insertion mutations are so constructed, they are denoted with the capital Greek letter Ω (omega) followed by the name of the insertion. For example, *galK*Ω::*kan* has a kanamycin resistance gene inserted into the *galK* gene.

Reversion versus Suppression

Reversion mutations are often easily detected through the restoration of a mutated function. As discussed above, a true reversion actually restores the original sequence of a gene. However, sometimes the function that was lost because of the original mutation can be restored by a second mutation elsewhere in the DNA. Whenever a second mutation somewhere else in the DNA relieves the effect of a mutation, the mutation has been **suppressed** rather than **reverted**, and the second mutation is called a **suppressor mutation**. The following sections present some of the mechanisms of suppression.

Intragenic Suppressors

Suppressor mutations in the same gene as the original mutation are called **intragenic suppressors**, from the Latin prefix "intra," meaning "within." These mutations can restore the activity of a mutant protein by a variety of mechanisms. For example, the original mutation may have introduced an unacceptable amino acid change that inactivated the protein, and changing another amino acid somewhere else in the polypeptide could restore the activity of the protein, thereby compensating for the effect of the first mutation. This form of suppression can sometimes indicate an interaction between the two amino acids in the protein, but other explanations, such as protein folding effects, are also possible. Many intragenic suppressors are **allele-specific suppressors** because only certain original mutations can be suppressed by a particular secondary mutation.

The suppression of one frameshift mutation by another frameshift mutation in the same gene is another example of intragenic suppression. If the original frameshift resulted from the removal of a base pair, the addition of another base pair close by could return translation to the correct frame. The second frameshift can restore the activity of the protein product if the translating ribosomes do not encounter any nonsense codons or insert any amino acids that significantly alter the activity of the protein during translation in the new reading frame that is used between the two frameshift mutations.

Intergenic Suppressors

Intergenic (or extragenic) suppressors do not occur in the same gene as the original mutation. The prefix "inter" comes from the Latin for "between" and therefore indicates the interaction between two different genes, and the prefix "extra" means "outside" and therefore indicates a suppressor that is in a different gene. Both of these terms are used in the literature, but we will use "intergenic" for consistency. There are many ways in which intergenic suppression can occur. The suppressing mutation may restore the activity of the mutated gene product or provide another gene product to take its place. Alternatively, it may alter another gene product with which the original gene product must interact in a complementary way so that the two mutated gene products can again interact properly.

One common way an intergenic suppressing mutation may restore the viability of the mutant cell is by preventing the accumulation of a toxic intermediate. If the gene

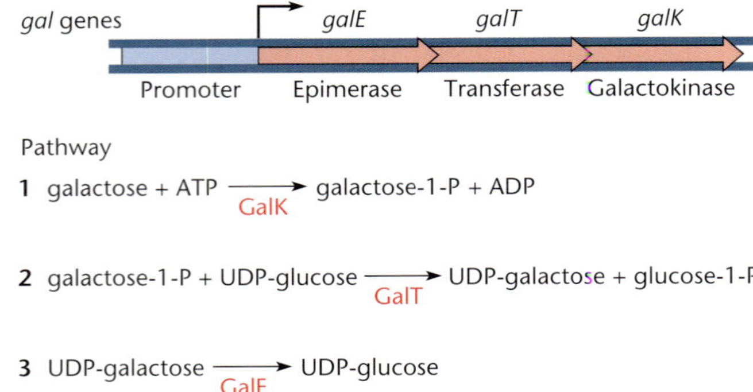

Figure 3.17 The pathway to galactose utilization in *E. coli* and most other organisms. *galK* mutations suppress the toxicity of galactose in mutants with *galE* mutations because they prevent the accumulation of the toxic intermediates galactose-1-phosphate and UDP-galactose when cells are growing in the presence of galactose.

for a step in a biochemical pathway is mutated, a toxic intermediate in that pathway can accumulate, causing cell death. However, a suppressing mutation in another gene of the pathway may prevent the accumulation, allowing the cell to survive even though it still has the original mutation.

The suppression of *galE* mutations by *galK* mutations provides an illustration of such an intergenic suppressor. Many types of cells, including human cells, use the pathway shown in Figure 3.17 to convert galactose into glucose in order to use the glucose as a carbon and energy source. The first step is to convert the galactose into galactose-1-phosphate by a kinase, the product of the *galK* gene. Subsequent steps catalyzed by the products of the *galT* and *galE* genes then form uridine diphosphate (UDP)-galactose and UDP-glucose, respectively, with the release of glucose. Cells with *galE* mutations are galactose sensitive (galactosemic), i.e., their growth is inhibited by galactose in the medium because galactose-1-P and UDP-galactose accumulate and are toxic. Cells with mutations in *galE* therefore not only are unable to grow with galactose as the sole carbon and energy source, but they also are unable to grow in the presence of both galactose and glucose, because of the accumulation of the toxic intermediates. In humans, the hereditary human condition of galactosemia is due to a deficiency in the human equivalent of *E. coli* GalE.

The galactose sensitivity of *galE* mutants makes it easy to isolate *galK* mutants. Selection for revertants of *galE* mutants that are able to grow in the presence of galactose results primarily in cells that have not actually reverted the *galE* mutation. In most of them, the *galE* mutation is still present, but it is being suppressed by a *galK* mutation, since the *galK* mutations occur much more frequently. This is because any mutation in the *galK* gene that inactivates the galactose kinase will prevent the accumulation of the toxic intermediates, but only very few base pair changes can revert the *galE* mutation to restore GalE function. Furthermore, galactose-resistant mutants with *galK* suppressors can be distinguished from true *galE* revertants because the true *galE* revertants are able to grow with galactose as the sole carbon and energy source (i.e., they are Gal^+), whereas the *galE galK* double mutants cannot grow with galactose as the sole carbon and energy source (i.e., they are still Gal^-), and another carbon source, such as glucose, must also be provided for them to grow and form a colony.

NONSENSE SUPPRESSORS

Nonsense suppressors are another type of intergenic suppressor. A **nonsense suppressor** is usually a mutation in a tRNA gene that changes the anticodon of the tRNA product of the gene so that it recognizes a nonsense codon. In the example shown in Figure 3.18, a mutation occurs in the gene for a $tRNA^{Gln}$ with the anticodon 5′-CUG-3′ (3′-GUC-5′, so that it normally recognizes the glutamine codon 5′-CAG-3′), causing the anticodon to become 5′-CUA-3′ (3′-AUC-5′). This altered anticodon can pair with the nonsense codon UAG instead of CAG. However, the anticodon mutation does not significantly change the tertiary shape of the tRNA or other recognition determinants for the cognate aminoacyl-tRNA synthetase, GlnRS, which still aminoacylates it with glutamine. Therefore, this mutated tRNA will insert glutamine into a growing peptide chain at the position of a UAG nonsense codon. This can lead to synthesis of an active polypeptide and suppression of the nonsense mutation. Nonsense suppressors are considered a form of **informational suppressors**

Figure 3.18 Formation of a nonsense suppressor tRNA. **(A)** Gene *X* (turquoise) and gene *Y* (purple) contain CAG codons, encoding glutamine. Other codons are not shown. The bacterium also has two different tRNA genes that normally insert glutamine at a CAG codon. Only the anticodons of the tRNAs are shown. **(B)** A mutation occurs in gene *X*, changing a CAG codon to UAG and causing the synthesis of a truncated polypeptide. **(C)** A suppressor mutation in the gene for one of the two tRNAs changes its anticodon so that it now pairs with the nonsense codon UAG. The translational machinery now sometimes inserts glutamine for the UAG nonsense codon in gene *X*, allowing synthesis of the complete polypeptide. The anticodon of the other tRNA still pairs with the CAG codon, allowing synthesis of the gene *Y* protein and the products of other genes that carry the CAG codon.

A Wild type

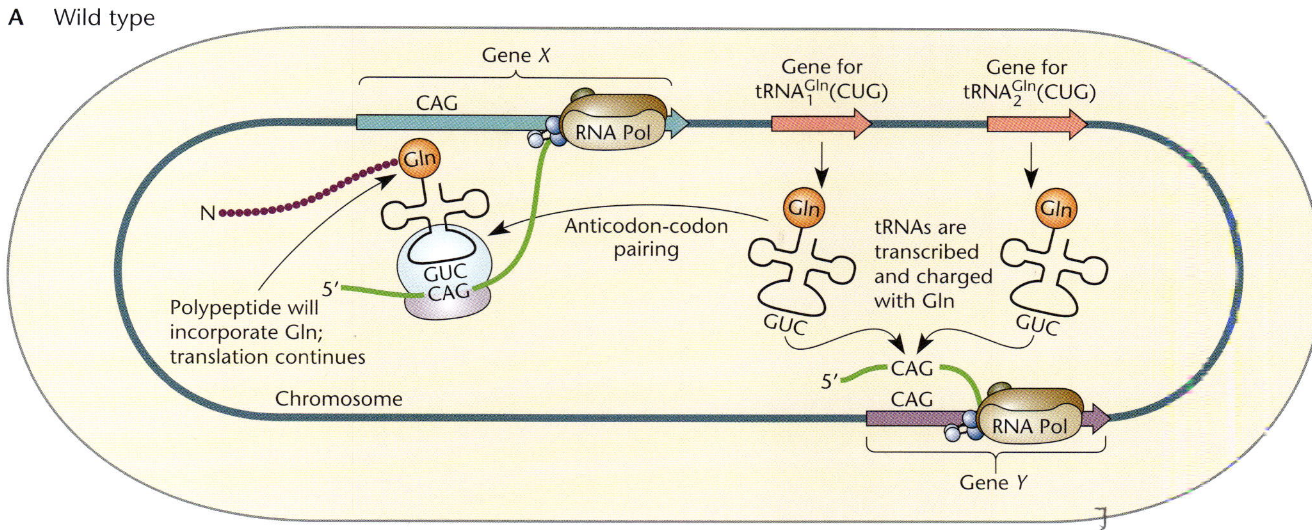

B Mutant A with nonsense mutation in gene X

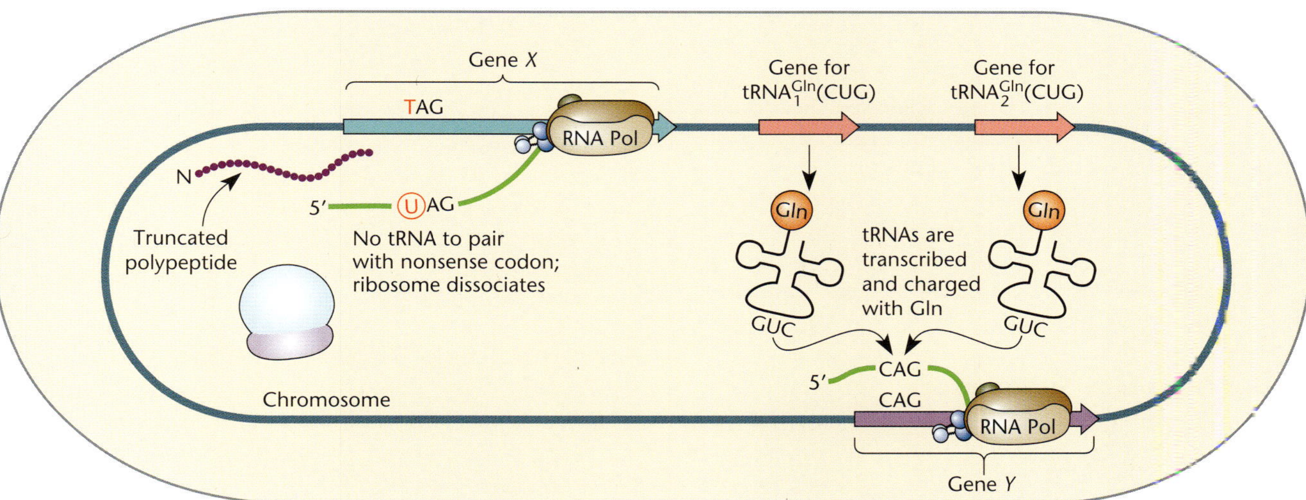

C Mutant B with nonsense mutation in gene X and nonsense suppressor mutation in gene for tRNA$_1^{Gln}$

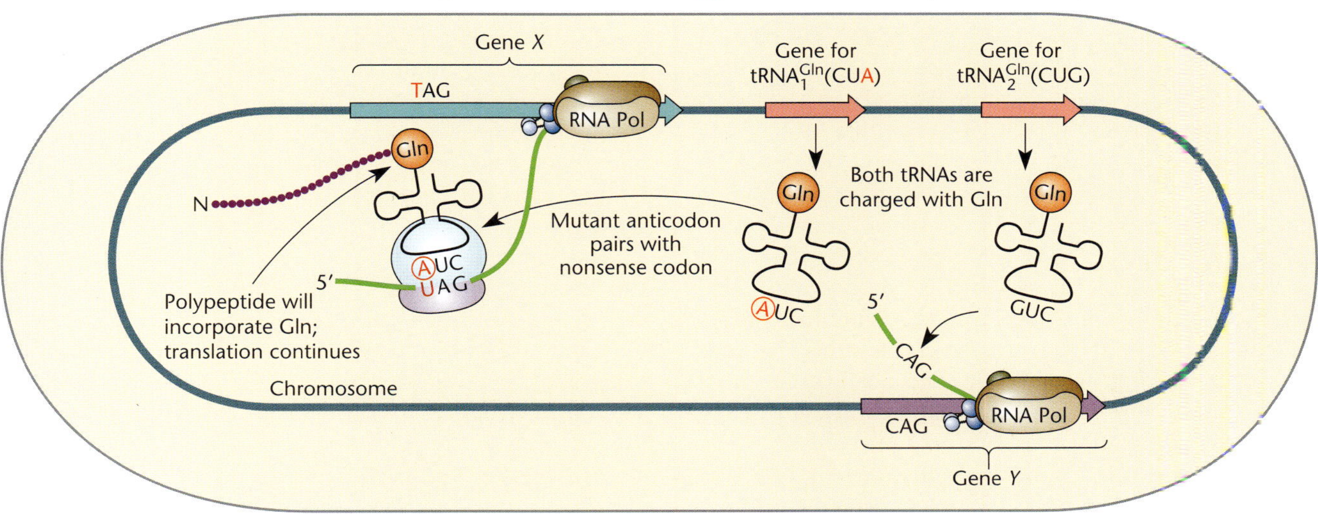

Table 3.3 All *E. coli* UAG nonsense suppressor tRNAs

Suppressor name	Gene name	tRNA	Change in anticodon[a]	Suppressor type
supD	*serU*	tRNA^{Ser}	CGA→CUA	Amber (UAG)
supE	*glnV*	tRNA^{Gln}	CUG→CUA	Amber (UAG)
supF	*tyrT*	tRNA^{Tyr}	GUA→CUA	Amber (UAG)
supB	*glnU*	tRNA^{Gln}	UUG→UUA	Ochre/amber (UAA/UAG)
supL	*lysT*	tRNA^{Lys}	UUU→UUA	Ochre/amber (UAA/UAG)

[a]Anticodon shown 5′ to 3′.

because they function by changing the gene expression machinery of the cell.

The mutated tRNA is called a **nonsense suppressor tRNA**, and nonsense suppressors themselves are often referred to as amber suppressors, ochre suppressors, or opal suppressors depending on whether they suppress UAG, UAA, or UGA mutations, respectively. Table 3.3 lists several *E. coli* nonsense suppressor tRNAs. Nonsense suppressors can be considered partially allele specific, because they suppress only one type of allele of a gene, that is, one with a particular type of nonsense mutation. However, a given nonsense suppressor can suppress many different nonsense mutations that have the matching nonsense mutation. In contrast, for example, the *galK*-suppressing mutations discussed above suppress any *galE* mutation and so are not allele specific.

Types of Nonsense Suppressors

A tRNA gene cannot always be mutated to form a nonsense suppressor. Generally, if there is only one type of tRNA, encoded by a single gene, to respond to a particular codon, the gene for that tRNA cannot be mutated to make a suppressor tRNA. The original codon to which the tRNA responded would be "orphaned," and no tRNA would respond to it wherever it appeared in an mRNA; protein synthesis would therefore stop, and the cell would die. When a tRNA can be mutated to a nonsense suppressor tRNA, it is because there is another tRNA that can respond to the same original codons. We have already mentioned how a tRNA gene included in a tandem duplication can be mutated to a nonsense suppressor because another copy of the tRNA gene exists in the duplication. However, such suppressors are very unstable and often revert to nonsuppressors due to recombination between the duplicated sequences. If stable nonsense suppressors can occur, it is usually because cells quite often have more than one tRNA, encoded by different genes, that can respond to the same codons. In the example in Figure 3.18, two different tRNAs encoded by different genes recognize the codon CAG, one of which continues to recognize CAG after the other has been mutated to recognize UAG.

Wobble (see Figure 2.33) offers an exception to the rule that a tRNA can be mutated to a stable nonsense suppressor only if there is another tRNA with an anticodon that matches the original codon. Because of wobble, the same tRNA can sometimes respond both to its original codon and to one of the nonsense codons. For example, in a particular organism, there may be only one tRNA with the anticodon 5′-CCA-3′ (3′-ACC-5′) that recognizes the codon for tryptophan, 5′-UGG-3′. However, from the wobble rules in Figure 2.33, if the anticodon is mutated to 5′-UCA-3′ (3′-ACU-5′), the U in the 5′ position of the anticodon will pair with either G or A in the 3′ position of the codon, and the tRNA might be able to recognize both the tryptophan codon UGG and the nonsense codon UGA, so that the suppressor strain can still recognize UGG and be viable. Wobble also allows the same suppressor tRNA to recognize more than one nonsense codon. In *E. coli*, all naturally occurring ochre suppressors also suppress amber mutations (Table 3.3). From the wobble rules, we know that a suppressor tRNA with the anticodon 5′-UUA-3′ (3′-AUU-5′) should recognize both the UAG and UAA nonsense codons in mRNA because the U in the 5′ position of the anticodon can pair with either G or A in the 3′ position of the codon. Note that in Table 3.3, anticodons are written 5′-3′ according to convention, even though they pair with the codon 3′-5′, because pairing of the codon and anticodon are antiparallel.

Consequences of Nonsense Suppression

Strains containing a nonsense-suppressor tRNA gene are seldom completely normal for a number of reasons. Normal translation may be less efficient because of reduction in the amount of normal tRNA to recognize its matching codon. Furthermore, if the nonsense suppressor is suppressing a nonsense codon in a gene, the polypeptide synthesized as the result of a nonsense suppressor may not be fully active. Usually, the amino acid inserted at the site of a nonsense mutation is not the same amino acid that was encoded by the original gene. This changed amino acid may cause the polypeptide to be less active or temperature sensitive, which can make the cell sick or temperature sensitive if the polypeptide is essential for growth.

Another reason cells can be sick if the suppressor is suppressing a nonsense mutation in an essential gene is that nonsense suppression is never complete. The nonsense codons are also recognized by release factors, which free the polypeptide from the ribosome (see chapter 2). Therefore, whether a nonsense codon is suppressed by the suppressor tRNA and translation continues or termination occurs and a shortened polypeptide is released depends on the outcome of a race between the release factors and the suppressor tRNA. If the tRNA can base pair with the nonsense codon before the release factors terminate translation at that point, translation will continue. Sequences around the nonsense codon influence the outcome of this race and determine the efficiency of suppression of nonsense mutations at particular sites.

Nonsense suppressor strains can be sick even if they are not suppressing a nonsense mutation anywhere in the genome. The major reason is that the coding regions (ORFs) for polypeptides normally end in one or more nonsense codons (acting appropriately as a translational stop signal in this situation), and the nonsense suppressor allows translation past the ends of normal ORFs. This results in longer than normal polypeptides, and the extra material at the carboxyl-terminal end of a polypeptide might interfere with its function. However, because nonsense suppressors are never 100% efficient, some of the correct polypeptides are always synthesized. Moreover, since the efficiency of suppression depends on the sequence of nucleotides around the nonsense codon, the sequences at the ends of genes presumably have a "context" that favors termination rather than suppression. The problem is further minimized because coding regions also often naturally have multiple different nonsense codons at their end, likely as an evolved adaptation to make translation termination more efficient. Multiple stop codons may limit the accidental readthrough of the nonsense codon or could avoid suppression by any particular tRNA suppressor. Nevertheless, tRNA suppressors can be isolated only in some single-celled organisms, including bacteria and fungi. In mammals and other eukaryotes, a system called nonsense-mediated decay destroys transcripts containing in-frame nonsense codons and prevents the accumulation of truncated proteins that could be toxic to the organism.

Genetic Analysis in Bacteria

One of the cornerstones of modern biological research is genetic analysis. Gregor Mendel probably performed the first definitive genetic analysis more than 150 years ago when he crossed wrinkled peas with smooth peas and counted the progeny of each type. The methods of genetic analysis have become considerably more sophisticated since then and are still central to research in cell and developmental biology. The first information about many basic cellular and developmental processes often comes from a genetic analysis of the process. The advantages of the genetic approach are that it requires few assumptions about the molecular basis for a biological process and can be applied to any type of organism, even ones about which little or nothing is known.

More modern techniques have revolutionized how genetic analysis is done. Now that DNA sequencing of an entire bacterial genome takes only a few days and costs less than $1,000 and more than 100,000 bacterial genomes have been sequenced, it is often possible to map mutations to a precise location in the genome of a bacterium about which little is known. Also, because of the large number of genes that have been sequenced and annotated, we can often tentatively identify the function of a gene product on the basis of similarities in sequence and structure to those that have already been characterized, a process called annotation (see chapter 1[illegible]). Mutations such as transposon insertions can be easily located on the genome of the bacterium using PCR and DNA sequencing (see chapters 8 and 13), and the gene in which they occur can be identified. Also, marker rescue or complementation cloning (see below) often makes it possible to locate a mutation on a sequenced genome or to prove that a mutation identified by sequencing is responsible for a phenotype. Probably, no one ever again will undergo the laborious task of constructing the genetic map of a bacterium or phage, or indeed that of any other organism, from scratch. Nevertheless, genetic analysis, including techniques of reverse genetics, is often still the only way to answer some of the most important questions in biology. For example, it is often the only way to determine how many gene products are involved in a function and to obtain a preliminary idea of the role of each gene product in the function. Suppressor analysis also offers one of the best ways to ascertain which gene products interact with each other in performing the function. Genetic analysis is covered in general genetics textbooks, and we review the basic principles here only as they apply particularly to bacteria.

Isolating Mutants

As discussed in the introductory chapter, a classical genetic analysis begins with finding mutants in which the function is altered. This process is called the **isolation of mutants** because the mutant organisms are somehow found and separated, or "isolated," from the myriad normal or nonmutant organisms with which they are associated. As discussed in the introductory chapter, a major reason why bacteria are such excellent genetic subjects is the relative ease with which mutants can be isolated. Bacteria are generally haploid, meaning that they have only one allele of each gene. This makes the effects of even recessive mutations immediately apparent. Bacteria also multiply asexually, not requiring crosses to make progeny. To make large populations of genetically identical bacteria, we do not need to clone them; they clone themselves when they multiply. Bacteria are also small, and for some types of bacteria, numbers equivalent to the entire human population on Earth can be placed on a single petri plate, making possible the isolation of even very rare mutants.

MUTAGENESIS

The first step in obtaining a collection of mutants for a genetic analysis is to decide whether to allow the mutations to occur spontaneously or to mutagenize the organism. Spontaneous mutations occur normally as mistakes in DNA replication, but the frequency of mutations can be greatly increased by treating the cells with certain chemicals or types of irradiation. Treatments that cause mutations are said to be **mutagenic**, and agents that cause mutations are **mutagens**. In general, treatments that damage DNA are mutagenic, although they differ greatly in their mutagenicity and what types of mutations they cause. The molecular basis for many types of mutagenesis is discussed in more detail in chapter 10.

Both spontaneous and induced mutations have advantages in a genetic analysis. To decide whether to mutagenize the cells and, if so, which mutagen to use, we must first ask how frequent the mutations are likely to be. Spontaneous mutations are usually rarer than induced mutations, so mutants with spontaneous mutations are more difficult to isolate. Therefore, to isolate very rare types of mutants or ones for which there is no good selection method (see below), we might have to use a mutagen. On the other hand, use of a mutagen increases the probability of multiple mutations occurring in the same cell, and the presence of multiple mutations can confuse the analysis later.

One major advantage of inducing mutations by using mutagens is that a particular mutagen often induces only a particular type of mutation. Spontaneous mutations can be base pair changes, frameshifts, duplications, insertions, or deletions. However, the acridine dye mutagens, such as acriflavine, cause only frameshift mutations, and base analogs, such as 2-aminopurine, cause only base pair changes. Transposon insertion mutations (see chapter 8) have particular advantages because they are easier to map and transfer from strain to strain. However, they are usually inactivating, causing null mutations, which cannot be isolated in a gene that is essential. Therefore, the decision of whether to isolate spontaneous mutations or use a particular mutagen depends on how frequent the mutants are likely to be and what types of mutations are desired.

INDEPENDENT MUTATIONS

For an effective genetic analysis, mutants defective in a function should have mutations that are as representative as possible of all the mutations that can cause the phenotype. If the strains in a collection of mutants each carry a different mutation, we can get a better idea of how many genes can be mutated to give the phenotype and how many types of mutations can cause the phenotype. A general rule is that if some genes are represented by only a single mutation, then, by the Poisson distribution discussed above, there are likely to be many other genes in which mutations could give the same phenotype but which have been missed because not enough mutants have been analyzed. The attempt to identify all the genes whose products are involved in a particular phenotype (or in some cases, all the possible sequence changes in a gene that cause a particular phenotype) is called **saturation genetics**.

There are two ways to ensure that a collection of mutations that cause a particular phenotype is as representative as possible. One way is to avoid picking **siblings**, which are organisms that are descendants of the same original mutant. If two mutants are siblings, they always have the same mutation. The best way to avoid picking siblings is to isolate only one mutant from each of a number of different cultures, all of which are started from nonmutant bacteria. If two mutants arose in different cultures, their mutations must have arisen independently, and they could not be siblings.

Another way to avoid isolating mutants with the same mutation is to isolate mutants from cultures mutagenized with different mutagens. All mutagens have preferred hot spots and tend to mutagenize some sites more than others (see chapter 10). If all the mutants are spontaneous or are obtained with the same mutagen, many of them have mutations in the same hot spot, but mutants obtained with different mutagens tend to have different mutations.

Regional and Site-Specific Mutagenesis

If the region of the chromosome to be mutagenized is known, more modern techniques can be used to introduce mutations into only that region of the DNA. **Regional mutagenesis** is designed to limit mutagenesis to a selected region of the chromosome, and **site-specific mutagenesis** introduces a precise genetic change at one site. Many of these techniques involve introducing oligonucleotides or DNA clones that have been synthesized or altered in the test tube into the cell. For site-specific mutagenesis, the oligonucleotide is synthesized to be completely complementary to the region of the chromosome with one desired change, whereas for regional mutagenesis, a mixture of nucleotides is used during the synthesis so that the final oligonucleotide has an average of one difference per molecule. Alternatively, double-stranded DNA fragments can be made by PCR amplification of the desired sequence, under conditions where the thermostable DNA polymerase is particularly mistake-prone, to introduce random mutations into the DNA fragment. If the oligonucleotide or DNA fragment introduced into the cell is mostly complementary to a region of the chromosome, it can replace that region of the chromosome and introduce the desired change in the sequence. This process is most efficient in those types of bacteria for which **recombineering** has been developed. Recombineering systems use the recombination systems taken from bacteriophages to recombine single-stranded oligonucleotides or double-stranded DNA fragments with the chromosome at high frequency (see chapters 9 and 13). However, the normal recombination systems of the cell can often be used. Such techniques can also be used to introduce a selectable marker, for example, for resistance to an antibiotic, into a gene or to delete a gene in its entirety, to be certain that the gene product has no residual activity. Such methods go under the general name of **gene replacements**. We return to such techniques below.

IDENTIFICATION OF MUTANTS

Even after mutagenesis or region-specific mutagenesis with even the most efficient recombineering system, mutants are rare and still must be found among the myriad individuals that remain normal for the function. The process of finding mutants is called **screening**. Screening for mutants is usually the most creative part of a genetic analysis. One must anticipate the phenotypes that might be caused by mutations in the genes for a particular function in order to screen for them. This is where geneticists earn their "pay," because predicting what types of mutations are possible and how to identify them often requires intuition, as well as rational thinking, but it is one of the more enjoyable aspects of genetics. Specific examples of screening for various types of mutants are discussed in later chapters.

Screening for mutant bacteria usually involves finding **selective conditions** that can be used to distinguish the mutants from the original type. These are usually conditions under which either the mutant or the wild type cannot multiply to form a colony. Agar plates and media with selective conditions are called **selective plates** and **selective media**, respectively.

Selection of Mutants

Some types of mutants can be selected from among the normal or wild type. In what is sometimes called a **positive selection**, selective conditions are chosen under which the mutant but not the original wild type can multiply. This is a powerful approach for the study of cellular functions and works particularly well with many types of bacteria and most phages because so many can be put on a single selective plate, so that even extremely rare mutants can be found. We have already mentioned some types of selection. Mutants with *tonA* or *tonB* mutations can be selected merely by plating *E. coli* on plates on which T1 phages have been spread. Only those cells that have a *tonA* or *tonB* mutation can multiply to form a colony, because those cells lack the receptor for phage binding and therefore cannot be infected and killed. Figure 3.19 shows an example of a selection for His$^+$ revertants of a *his* mutation. Cells containing a mutation in

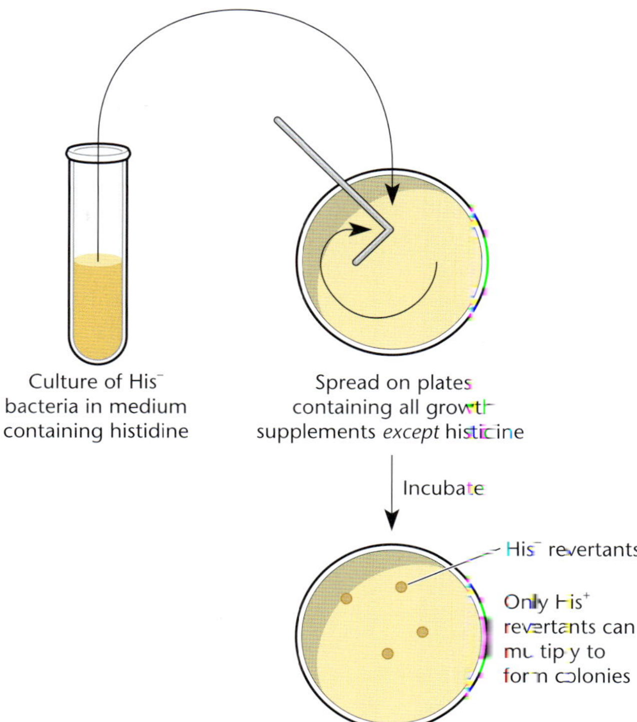

Culture of His$^-$ bacteria in medium containing histidine
Spread on plates containing all growth supplements *except* histidine
Incubate
His$^-$ revertants
Only His$^+$ revertants can multiply to form colonies

Figure 3.19 Selection of a His$^+$ revertant. A sample of a His$^-$ mutant culture is plated on minimal medium with all the growth requirements except histidine. Any colonies that form after the plate is incubated are due to apparent His$^+$ revertants that can multiply without histidine in the medium.

one of the histidine biosynthesis genes are grown in the presence of histidine and then plated on medium that lacks histidine, and those that can multiply to form colonies are His$^+$ revertants. The His$^-$ cells may survive in the absence of histidine but are unable to grow to form a colony. The His$^+$ cells could be tested to see if the reversion occurred at the site of the original mutation or if the *his* mutation is being suppressed by a mutation elsewhere in the genome (see below).

Cells with a *galK* mutation are normally not selectable in a wild-type strain because *galK* mutations make cells unable to multiply with galactose as the sole carbon and energy source. However, the selection of *galK* mutants can be achieved by plating a *galE* mutant on medium containing galactose plus another carbon source, such as glucose (see above). Under these conditions, most of the cells that grow to form a colony are *galK* mutants, because inactivation of *galK* prevents the accumulation of toxic intermediates (Figure 3.17).

The possible applications of using selection to isolate mutants are limitless. For example, in spite of our extensive understanding of protein structure and function, it is still often impossible to predict what amino acid changes in a protein will have a particular effect on the protein, for example, make it temperature sensitive. Mutational analysis allows us to ask the cell what amino acid changes will have that effect if we can establish conditions under which the activity of the protein is toxic. Then, we can randomly mutagenize the gene for the protein, using region-specific mutagenesis, so that the protein, with almost every amino acid change possible, will be represented in some of the cells. We then plate the cells at the nonpermissive high temperature. Those cells that grow to form a colony could have a temperature-sensitive mutation in the toxic protein, provided the cells do not grow at the lower permissive temperature. While selection is often not an option for many types of mutants (see below), identifying mutants is much easier with a selection, and geneticists expend much effort trying to design selections.

Screening for Mutants without a Selection

Most types of mutants cannot be selected. In that case, conditions under which the wild type but not the mutant can grow must be used. This type of mutant is the most common, because most of an organism's gene products help it to multiply, so mutations that inactivate a gene product are more likely to make the organism unable to multiply under a given set of conditions rather than able to multiply when the wild type cannot. These have been called negative selections, but they are not really selections at all; they are screens (see below), because the selective conditions are being used to screen for the mutants rather than to eliminate all other organisms that are not mutated to give the desired phenotype.

To screen for mutants without a selection, the bacteria are first plated on a nonselective medium, conditions under which both the mutant and the wild type can multiply. When the colonies have developed, some of the bacteria in each colony are transferred to a selective plate to determine which colonies contain mutant bacteria that cannot multiply to form a colony under those conditions. Once such a colony has been identified, the mutant bacterial strain can be retrieved from the corresponding colony on the original nonselective plate. Figure 3.1 shows the detection of two types of auxotrophic mutants, His$^-$ (unable to make histidine) and Bio$^-$ (unable to make the vitamin biotin), by such a screen.

Replica plating. Because of the general rarity of mutants, many colonies usually have to be screened to find a mutant. **Replica plating** can be used to streamline this process (Figure 3.20). Several hundred bacteria are spread on a nonselective plate, and the plate is incubated for a short time to allow microcolonies to form. A replica (i.e., a copy of the distribution of microcolonies on the initial plate) is then made of the plate by inverting the plate and pressing it down over a piece of sterile fuzzy cloth, such as velveteen. This transfers cells from each microcolony onto the cloth. Then, a selective plate is inverted and pressed down over the same cloth so that the cells from the microcolonies are transferred from the cloth to the selective plate at the same position on the plate that they were in on the original plate. The process is repeated to transfer cells from the cloth to a fresh nonselective plate. After the selective and nonselective plates have been incubated, they are compared to identify colonies that appear on the nonselective plate but not on the selective plate. The colonies missing from the selective plate presumably contain descendants of a mutant bacterium that are unable to multiply on the selective plate. The mutant bacteria can then be taken from the colony on the nonselective plate. We mentioned above how replica plating was developed and used by Esther and Joshua Lederberg to demonstrate that bacteria behave by the principles of Darwinian inheritance.

If a type of bacterial mutant being sought is rare, finding it by screening can be very laborious, even with replica plating. Even with the best-behaved types of bacteria, no more than about 500 bacteria can be spread on a plate and still give discrete colonies. Thus, for example, if the mutant occurs at a frequency of 1 in 10^6, more than 2,000 plates might have to be replicated to find a mutant. Anything that increases the frequency of mutants in the population will make the task easier. This can include mutagenesis, as discussed above and in chapter 10, and various tricks that increase the frequency of mutants in the population before screening.

Enrichment. Many fewer colonies need to be screened to find a mutant if the frequency of mutants is first increased through mutant **enrichment**. This method depends on the use of antibiotics, such as ampicillin, that kill growing but not nongrowing cells. Ampicillin inhibits cell wall synthesis, and when a growing bacterial cell tries to divide, it lyses (see chapter 2). A mutant cell that is not growing while the ampicillin is present does not lyse and is not killed. To be enriched, the mutant must be able to resume multiplying after it is removed from the selective conditions.

To enrich for mutants that cannot grow under a particular set of selective conditions, the population of mutagenized cells is placed under the selective conditions in which the desired mutants stop growing. Meanwhile, the nonmutant, wild-type cells continue to multiply. The antibiotic is then added to kill any multiplying cells. The cells are filtered or centrifuged to remove the antibiotic and transferred to nonselective conditions. The mutant cells will survive preferentially because they were not growing in the presence of the antibiotic; therefore, they will become a higher percentage of the population. Some nonmutant cells will always survive the enrichment if

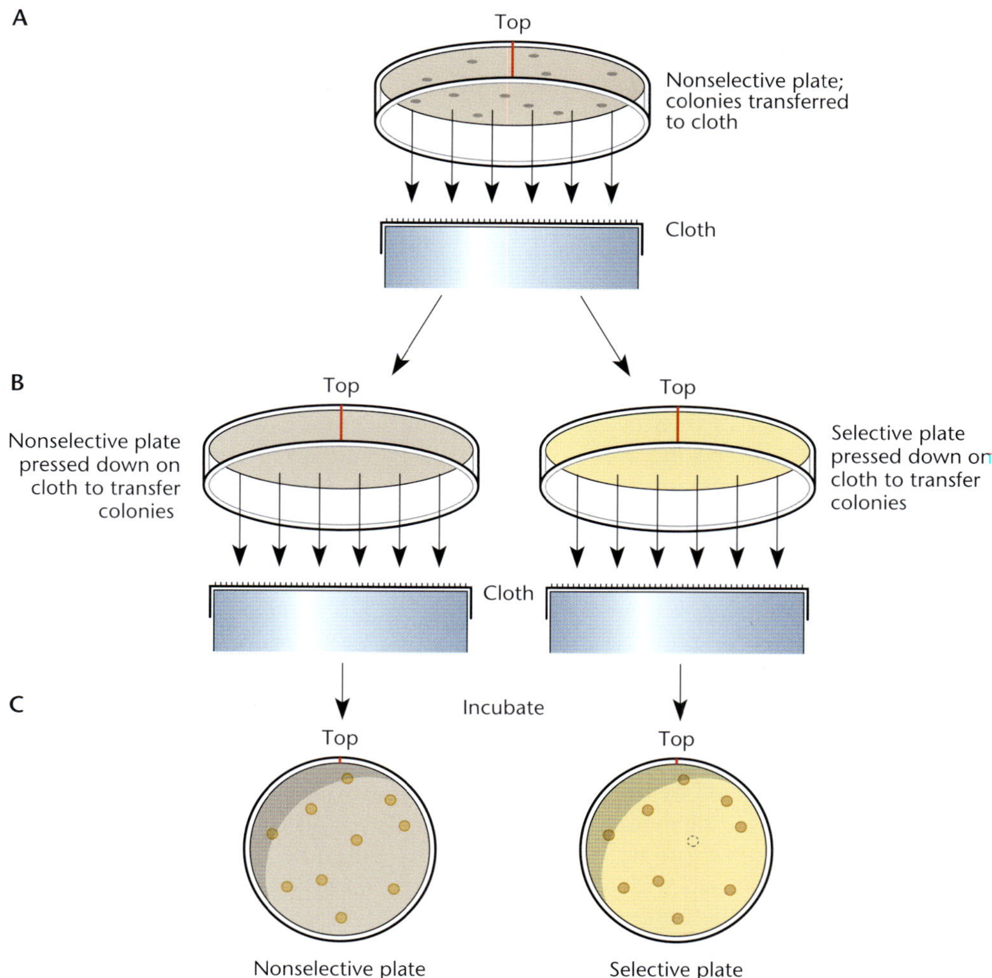

Figure 3.20 Replica plating. **(A)** A few hundred bacteria are spread on a nonselective plate, and the plate is incubated to allow colonies to form. The plate is then inverted over velveteen cloth to transfer the colonies to the cloth. **(B)** Cells from the cloth are then transferred to both a selective and a nonselective plate, both of which are marked (red line) to identify the orientation relative to each other and then incubated. **(C)** Both plates after incubation. The dotted circle indicates the position of a colony missing from the selective plate. See the text for details.

they were not growing when the antibiotic was present. However, even if the enrichment makes the mutant only 100 times more frequent, only 1/100 as many colonies and therefore 1/100 as many plates must be replicated to find a mutant after an enrichment. In the example given above, after enrichment, we would need to replicate only 20 plates instead of 2,000 to find a mutant.

Genetic Characterization of Mutants

Once we have our collection of mutants with the desired phenotypes, we wish to further characterize the mutations responsible for those phenotypes. First, we want to know where they are in the genome and what genes they are in. Locating them in the entire genome is like finding a needle in a haystack, in this case, a single base pair change or other alteration in the millions of base pairs in the entire genome of the bacterium. Until fairly recently, the only way to do this was by genetic mapping. However, genetic mapping requires that the genetic map of the organism be available. Constructing genetic maps is very laborious, and extensive genetic maps have been constructed for only a few different types of bacteria, including *E. coli*, *Bacillus subtilis*, and *Streptomyces coelicolor*. We discuss the methods of genetic mapping below. Nowadays, it is easier to have the genome of the bacterium sequenced. If it has been sequenced already, we can compare the genome sequence of the mutant to that of the wild-type strain and identify the genetic loci that contain sequence changes. If the genome of this bacterium has not been sequenced already, we will need to first identify as many genes as possible in the sequence by annotation (see chapter 13). One way to locate the mutations in the genome sequence is to clone the region of the mutations by a method such as marker rescue or complementation (see below). The clones containing the region of the mutations can then be sequenced and located in the annotated sequence of the genome, or if the mutations have been obtained by transposon mutagenesis, it may be possible to sequence out from the inserted transposon into the surrounding bacterial DNA using the known sequence of the transposon to design primers. The surrounding genomic-DNA sequences can then be located on the genome sequence using available software. However, as mentioned, transposon insertions cause only null or loss-of-function mutations. If the phenotype requires an amino acid change or some other more subtle change in the sequence, mutations to that phenotype cannot be obtained by transposon mutagenesis.

Even if the entire DNA sequence of the genome of the bacterium being studied is known and it is possible to locate mutations on the genome by DNA sequencing alone, we still must do genetic analysis of the mutations. For example, it is desirable to have independent verification that the mutations identified by sequencing alone are causing the phenotype. Also, the mutant strain may contain multiple sequence changes, and we need to know which of these changes is associated with the phenotype. We also want to know that the polarity of the mutations on another gene is not responsible for the phenotypes, depending on what kind of mutations they are, or we may want to continue the analysis by seeing if there are any suppressors of the mutations to identify other proteins or functions with which the product of the mutated gene might interact, etc. All of these require genetic manipulations, so in this section we spend some time on genetic characterizations and how they are interpreted.

The two major tools of genetic characterization of mutations are recombination and **complementation**. While recombination and complementation can have seemingly similar outcomes, their molecular bases are completely different and they even occur in different parts of the cell. Recombination is defined as the breakage and rejoining of two DNA molecules in new combinations, while complementation is the ability of the products of different mutant genes or alleles of genes to interact with or substitute for each other. While recombination can be used to locate mutations in DNA, complementation can be used to assign mutations to genes or to the parts of genes encoding individual domains of the same gene product and to give preliminary indications of the functions of gene products.

LOCATING MUTATIONS BY RECOMBINATION

We have already mentioned recombination in connection with how deletions, duplications, and inversions are formed and as a way of doing site-specific mutagenesis. Recombination can be used for mapping, region-specific mutagenesis, and gene replacements because cells have mechanisms to recombine their DNA in new combinations. The site of breakage and rejoining is called a **crossover**. Recombination can be either **site-specific recombination** or **general recombination**. Site-specific recombination is discussed in detail in chapter 8. It uses specialized enzymes that cut and religate DNA, but only at unique sequences, so it is not generally useful for genetic characterizations and is not discussed here. Most genetic manipulations require generalized recombination, also called **homologous recombination**, because it can occur anywhere but usually occurs only between two DNA regions that have the same or homologous sequences (*homo-logos* means "same-word" in Greek). Generalized or homologous recombination, referred to below as generalized recombination, or simply recombination, probably occurs naturally in all organisms and serves the dual purpose of increasing genetic diversity within a species and restarting replication forks that have stalled at damage in the DNA (chapter 9).

Generalized recombination is quite complex and uses a number of different enzymes and pathways, depending on the situation. These details are discussed further in chapter 9. However, for now, the simplified model of recombination used in general genetics textbooks is sufficient (Figure 3.21). According to the simplified model, a strand of one or both of the two DNAs that will recombine invades the other DNA, and their opposing strands pair in a region where their nucleotide sequences are homologous (GCATA/CGTAT in Figure 3.21, although the homology will be much longer). The requirement for base pairing ensures that recombination occurs in the same place in the two DNAs, since generally it is only in homologous regions that the opposing strands of the two DNAs are complementary and able to base pair. This invasion by one or both DNAs allows the paired strands of the two DNAs to cross over, bridging the two molecules, as shown. The bridged DNAs are then processed by DNA breaks and ligation events to give rise to recombinant DNAs. The details of this process vary depending on the situation and are covered in chapter 9. Note in the figure that after the crossover event, the distal markers are switched, although this is not a necessary outcome of such a recombination event, as we show below.

Consequences of Recombination in Bacteria

Most models of recombination shown in general genetics textbooks, such as the one in Figure 3.21, are between two complete linear DNA molecules like those in the

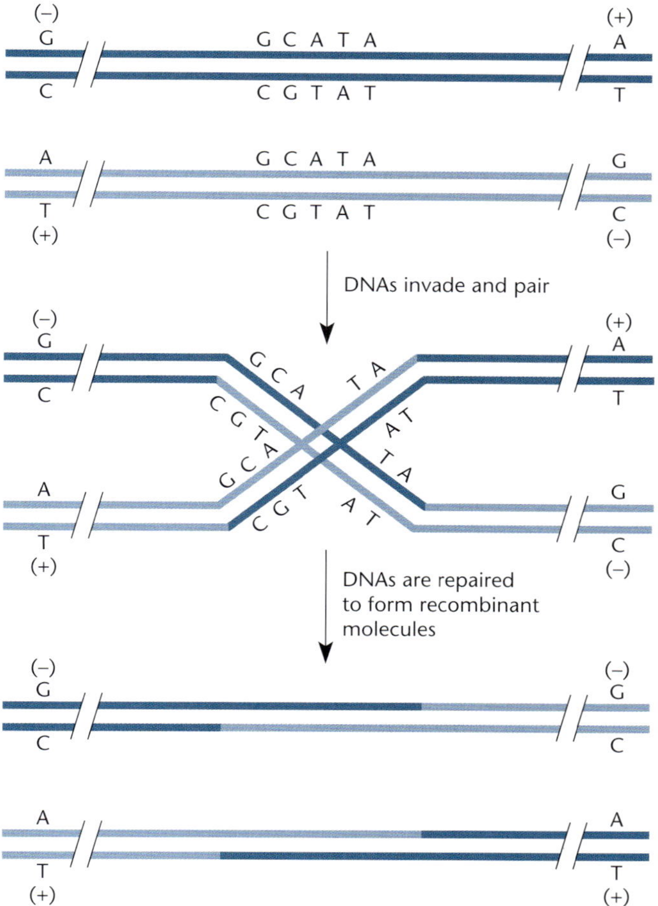

Figure 3.21 A simplified diagram of recombination between two genetic markers. One strand of each or of both DNAs invades the other DNA, pairing with its complementary sequence. This joins the two DNAs. The structure is then resolved so that sequences from one DNA can become associated with the other DNA and lead to a rearrangement of genetic markers between the two DNAs. The positions of bases are shown only where the mutations have occurred and where the crossing over occurs. (–), mutation; (+), wild type.

chromosomes of eukaryotic organisms or in phages or other viruses with linear DNAs. However, in bacteria, recombination between the DNAs of different organisms usually occurs between a piece of DNA from one strain of a bacterium, called the **donor strain**, and the entire chromosome of another strain, called the **recipient strain**. If the chromosomal DNA sequences from the two strains are almost identical, which usually means they are members of the same species of bacterium, the piece of DNA from the donor is homologous to some region of the chromosome of the recipient. The two DNAs can then pair in this region, and recombination can occur in the recipient strain. The piece of DNA that comes from the donor strain can enter by any one of a number of mechanisms, including conjugation, transformation, and transduction (the molecular mechanisms of which are discussed in chapters 5, 6, and 7). The piece of chromosomal DNA from the donor strain that participates in the recombination can also be either linear, as occurs during transformation or transduction, or circular if the fragment of the donor chromosome has been cloned into a circular plasmid or some other circular cloning vector. The consequences of the recombination are very different, depending on whether the piece of donor DNA is linear or in a circular molecule.

Figure 3.22 shows the different consequences when the piece of chromosomal DNA from the donor strain is linear and when it is included in a circular DNA. If the DNA is linear, as in Fig. 3.22A, two invasion events will occur, one at each end of the linear DNA, initiating DNA replication events on the circular chromosome as explained in more detail in chapter 9. Replication will establish these crossovers, replacing this region of the DNA of the recipient with the DNA of the donor. In other words, it will leave a patch of DNA from the donor to replace the corresponding DNA from the recipient. For simplicity, the sister DNA molecule that remains unmodified is not shown.

A very different situation prevails if the piece of donor chromosomal DNA is included in a circular DNA, such as a plasmid cloning vector, as shown in Fig 3.22B. Then, a single crossover will integrate the circular DNA into the chromosome of the recipient, and the region of chromosomal DNA from the donor that was in the circular vector will be duplicated, with one copy on either side of (i.e., bracketing) the circular cloning vector. If the donor sequence differed from the recipient in one or more places, the donor sequences can be in one copy or the other, depending on where the crossover occurred. This creates a partial diploid that can be used for complementation tests (see below). However, it is an unstable situation, because the chromosomal sequence where the recombination occurred exists as a direct repeat on either side of the integrated plasmid, and a second crossover between the two direct repeats can loop out (delete) the plasmid, leaving behind a patch of DNA sequences from the donor DNA, depending on where the first and second crossovers occur. Therefore, recombination with either linear or circular DNA can be used to do gene replacements.

The example in Figure 3.23 shows a common application using recombination with a circular plasmid to do a gene replacement. In the example, a kanamycin resistance **gene cassette** that contains a gene for resistance to the antibiotic kanamycin (Kan^r) has been introduced into a chromosomal gene cloned in a plasmid cloning vector that carries a gene for ampicillin resistance (Amp^r). This plasmid is then introduced into the cell, and the homologous chromosomal sequences on the plasmid and in the chromosome can recombine to integrate the plasmid into the

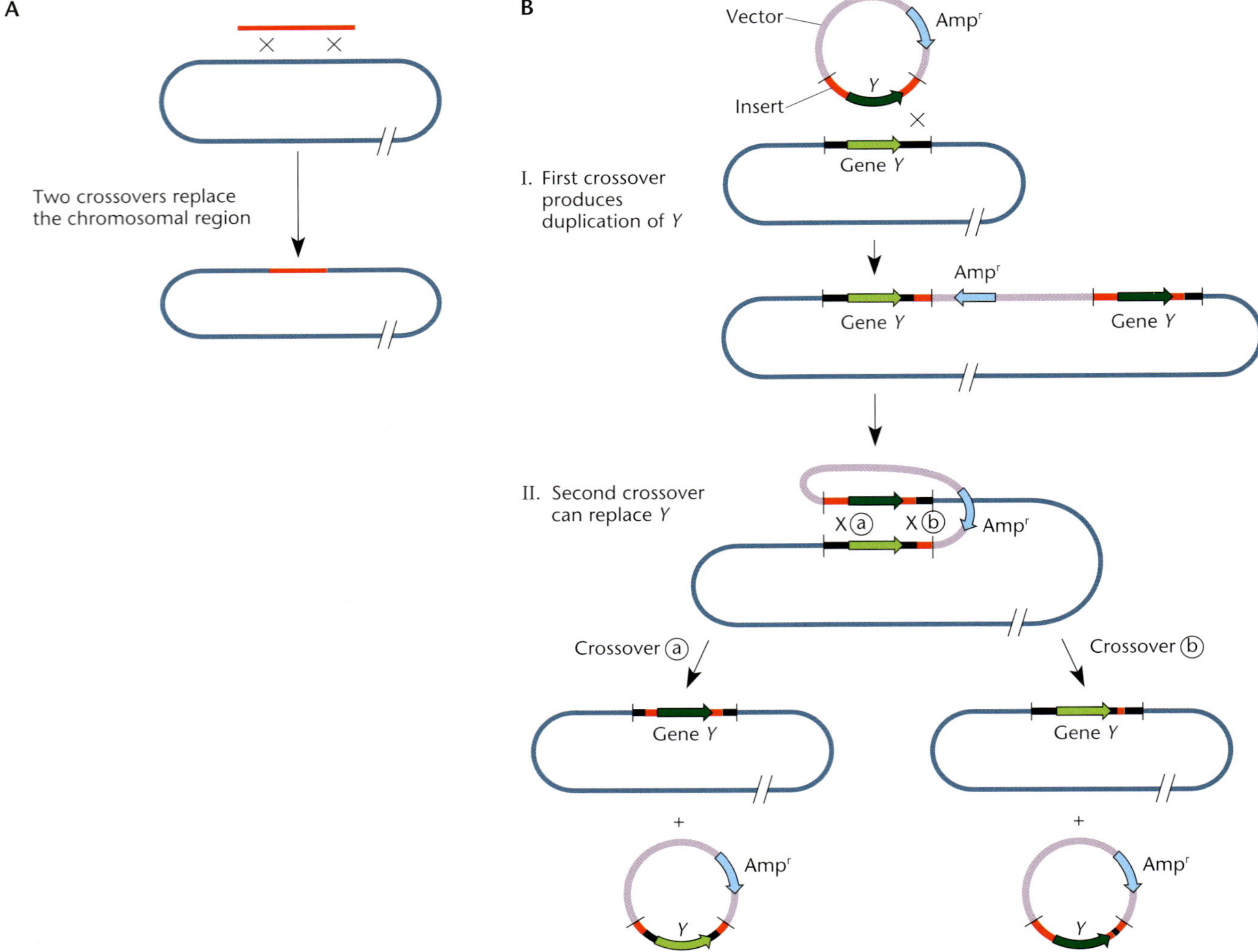

Figure 3.22 Different consequences of recombination between linear and circular DNAs in bacteria and its use for region-specific mutagenesis. **(A)** The introduced DNA from the donor is linear. Two crossovers can replace a sequence in the chromosome (shown in blue) with the altered sequence of the donor (shown in red). **(B)** The introduced donor DNA carrying an altered gene *Y* (dark green) is cloned in a circular plasmid carrying a gene for ampicillin resistance (light blue). **(I)** A single crossover between the cloned DNA and the corresponding homologous region in the chromosome with a normal gene *Y* (light green) inserts the plasmid, which is now bracketed by the chromosomal region. **(II)** A second crossover can loop out and delete the plasmid, leaving only one copy of gene *Y*, either the mutated copy from the donor or the normal gene from the recipient.

chromosome. The cell is now both Kan^r and Amp^r. A second recombination or crossover loops out the plasmid. Depending on where the second crossover occurs, it can leave the region of the gene with the Kan^r cassette, resulting in a cell that is only Kan^r and not Amp^r. In practice, the plasmid cloning vector should not be able to replicate in the cell, i.e., it should be a suicide vector; otherwise, all of the cells become Kan^r and Amp^r even if no recombination has occurred. It is also desirable to have some way to select for cells that have lost the plasmid, in other words, to select for the second crossover; otherwise, we are faced with the laborious task of screening many cells for one in which the second crossover has occurred.

Genetic Markers

Events such as those described above can happen whenever two bacteria exchange DNA, provided the two DNAs have almost identical sequences in the region where the crossovers occur. We would have no way of detecting them if the sequences of the DNAs from the donor and

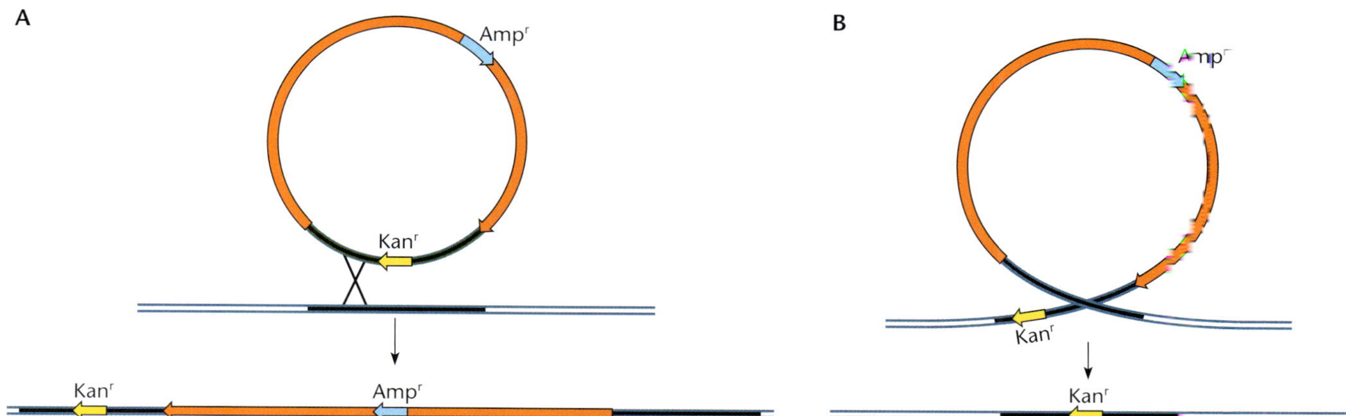

Figure 3.23 Using recombination to introduce an antibiotic resistance cassette into a chromosomal gene. **(A)** With a single crossover, the cloning vector integrates into the chromosome and the cells become ampicillin resistant (Amp^r) and kanamycin resistant (Kan^r). **(B)** With a second crossover, the cells lose the cloning vector and become Amp^s. Depending on where this second crossover occurs, the cells may be left with only the cloned sequence with the kanamycin resistance cassette and be resistant to kanamycin alone. See the text for details. Plasmid cloning vector sequences are in orange. The cloned region of the chromosome is in dark blue.

recipient were identical to each other. However, if they have some differences between them in the form of mutations, whether point mutations, deletions, or insertions, we can often use these sequence differences to detect recombination and locate the sites of the mutations on the DNA. If the mutations that create the sequence differences in the DNAs cause observable phenotypes in the organism, we can use these phenotypes to determine that recombination has occurred. In so doing, we are using the sequence differences as **genetic markers**. For example, the Kan^r cassette in the donor DNA in the example shown in Figure 3.23 is a genetic marker. The sequence of the donor DNA is different as a result of the insertion of the cassette, and we use this difference as a genetic marker to determine if recombination has occurred. When the DNA that has formed as a result of a crossover **segregates** into progeny, these progeny are genetically different and, correspondingly, phenotypically different from the recipient. In general, progeny that are genetically different from their parents as a result of a recombination are called **recombinant types**. Progeny that are genetically identical to one or the other of the two parents are **parental types**.

Because in bacterial crosses, only a part of the chromosome of the donor enters the recipient cell, recombinant types can appear only in the recipient cell. If the sequence of the recipient cell at the position of the genetic marker is replaced by the sequence of the donor cell, we know that recombination has occurred and the recipient bacterium has become recombinant for that marker. The nomenclature is much like that for the phenotypes of mutations. For example, if only the recipient has a *his* mutation that makes it His$^-$, the recombinant that gets this region from the donor will become His$^+$ and is referred to as a His$^+$ recombinant or simply a *his* recombinant. If only the recipient has a mutation in RNA polymerase that makes it rifampin resistant (Rif^r [chapter 2]) and the donor DNA comes from a strain that is rifampin sensitive (Rif^s), the recombinant for this marker will be Rif like the donor, and the recombinants for the marker will be Rif^s recombinants or simply *rif* recombinants. In general, when the recipient cell gets the allele of the donor it has become recombinant for that marker. Whether recombinant types appear in the recipient cells for each of the markers that are different in the donor and recipient strains tells us something about where the mutations that caused the genetic markers are on the DNA of that species. Later in the chapter, we discuss how such genetic evidence is interpreted for the different types of crosses in bacteria.

Marker rescue

One of the most important historical uses of recombination is in **marker rescue**, in which a genetic marker in the chromosome is "rescued," or becomes recombinant, by recombination with a piece of chromosomal DNA that has entered the cell that does not have that marker. How the piece of DNA enters the cell depends on the type of bacterium being studied. Marker rescue is useful for mapping and cloning, because if some of the cells become recombinant for a genetic marker, it tells us that the piece of chromosomal DNA from the donor must have come from the same region of the chromosome that has the mutation. To do marker rescue, we must have a way of selecting recombinants for the marker because they are usually rare. For example, in the case of auxotrophic

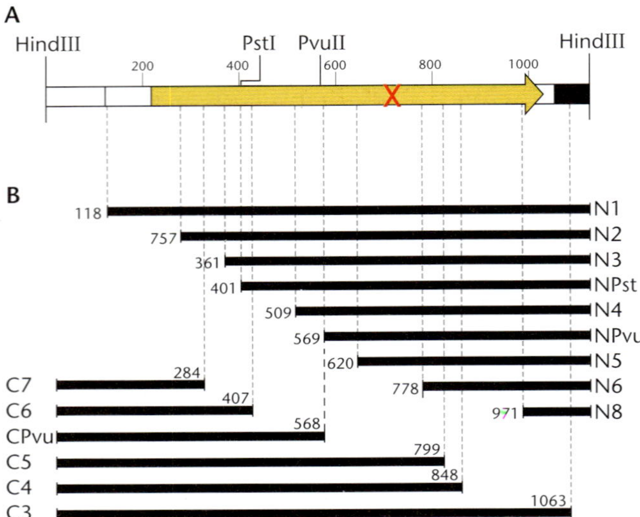

Figure 3.24 Using marker rescue to locate a mutation in the physical map of the *thyA* gene of *E. coli.* **(A)** Physical map of the gene showing the positions of some restriction sites, the numbers of base pairs from the 5′ end of the DNA fragment, and the position of a mutation (red X). **(B)** The mutation in the *thyA* gene in the chromosome was mapped using nested deletions extending various distances into the cloned *thyA* gene of *E. coli* from the N-terminal and C-terminal coding sequences. The clones were introduced into the mutant cells, and the position of the point mutation was determined by whether Thy$^+$ recombinants could be selected. The solid bars show the regions retained in each of the constructs. N constructs contain deletions that remove the 5′ end of the gene fragment; C constructs contain deletions that remove the 3′ end of the gene fragment. Modified from Belfort M, Pedersen-Lane J, *J Bacteriol* **160:**371–378, 1984.

mutations, this would mean growth media lacking the required supplement. Figure 3.24 shows how **nested deletions** of the *thyA* gene of *E. coli* were used to map mutations within the *thyA* gene by marker rescue. This method is particularly useful to map mutations in a gene that affect different functions of the protein product and therefore define different domains of the protein. It also can be used to identify clones that come from a particular region of the chromosome, e.g., to clone a gene or region. Marker rescue has advantages and disadvantages over complementation for cloning chromosomal genes in cloning vectors. We contrast cloning by complementation and by marker rescue below.

Complementation Tests

The other general method for characterizing mutants is complementation. Rather than depending on breaking and joining DNA in new combinations, complementation depends on the functional interaction of gene products made from different DNAs. Complementation allows us to determine how many gene products are represented by a collection of mutations and to obtain preliminary information about the functions affected by the mutations. Basically, we are asking whether a copy of a region of the chromosome can provide the function inactivated by a mutation, i.e., complement the mutation. If two mutations in different DNAs inactivate different functions, each can provide the function the other cannot, and the two mutations complement each other.

To perform a complementation test, we must have two copies of the regions of DNA containing one or more mutations in the same cell and see what effect this has on the phenotypes of the mutations. With a **diploid** organism, like us and most other multicellular organisms that contain two homologous chromosomes of each type, this is no problem, since diploids normally have two copies of each gene. With phages and other viruses, it is also no problem, because we can infect cells with two different mutant viruses simultaneously. However, with bacteria, which are naturally **haploid**, with only one copy of each gene, complementation tests are more difficult. Rather than being made complete diploids, bacteria can be made **partial diploids** (or **merodiploids**) by introducing a small region of the chromosome of one strain into another strain, using plasmids or prophages that can stably coexist with the chromosome, or by integrating a second copy of the DNA region of interest into a new site within the chromosome. A common way of doing this is to introduce a plasmid cloning vector containing a clone of the region of interest into the cell. If the plasmid can replicate in the cell and if the gene or genes on the plasmid clone are expressed, they can be tested for the ability to complement mutations in the chromosome.

Plasmid clones can sometimes be used for complementation tests even if the plasmid vector cannot replicate in the cell. As we showed in Figure 3.22B, if the plasmid integrates into the chromosome by a single crossover between the cloned sequence and the corresponding chromosomal sequence, the region of the chromosome will be duplicated. This will create a partial diploid of the cloned region and might give apparent complementation, but care must be taken that the mutations have not been lost by the recombination that integrated the plasmid; also, this integration event is not stable, because a second recombination event can result in loss of the second copy. A safer way is to clone the chromosomal region of interest into a lysogenic phage or a transposon that then can be integrated into a different place in the chromosome. We can also sometimes see complementation between mutations on tandem duplications, provided that a way can be found to introduce different mutations into the copies of the duplication. Also, in some cases, it might be possible to observe complementation by an introduced DNA that exists only temporarily in the cell after some types of matings; the resulting strains are called **transient diploids**. Transient diploids are, by definition, not stable, but they might last long enough for some types of complementation tests.

RECESSIVE OR DOMINANT

Complementation can be used to tell if a mutation is **recessive** or **dominant** to the wild-type allele. A recessive mutation is subordinate to the wild type in the sense that an allele with the mutation does not exert its phenotype if the wild-type allele of the gene is present in the same cell. In contrast, a dominant mutation exerts its phenotype even if the wild-type allele is present. Recessive mutations generally inactivate the gene product, i.e., they are loss-of-function mutations, while dominant mutations are often **gain-of-function** mutations if they subtly change the gene product so that it can perform a function that the wild type cannot perform. If the mutant allele can function in a situation where the wild type cannot function, it is referred to as a gain-of-function allele. In eukaryotic genetics, alleles with altered function are typically called hypomorphic or hypermorphic alleles if they decrease or increase the activity of the allele, respectively, but this nomenclature has not caught on in bacterial genetics. Recessive mutations are much more common than dominant mutations because many more types of changes in the DNA inactivate the gene product than change it in some subtle way. Whether a particular type of mutation in a gene that gives a particular phenotype is dominant or recessive can tell us something about the normal functioning of the gene product.

To determine whether a mutation is recessive or dominant, we need to make a partially diploid cell that has both the wild-type allele and the mutant allele and determine whether the wild-type phenotype or the mutant phenotype prevails. To illustrate the difference between recessive and dominant mutations, we can return to the example of the *his* pathway. Most mutations that make the cell His$^-$ have inactivated one of the enzymes required to make histidine. These mutations are all recessive to the wild type because, in the presence of the wild-type allele for each of the genes, all the enzymes required to make histidine are made, and the cell will be His$^+$, the phenotype of the wild-type alleles. However, assume that there is an inhibitor of the pathway, such as an analog of histidine that binds to the first enzyme of the pathway and inactivates it. In the presence of this inhibitor, the cell is also unable to make histidine and will be His$^-$, so the phenotype caused by the wild-type allele in the presence of the inhibitor is His$^-$. However, a mutation in the gene for the enzyme can make the enzyme insensitive to the inhibitor, so that the mutant cell can make histidine even in the presence of the inhibitor. The phenotype of the cell containing this mutation is His$^+$ even in the presence of the inhibitor. Furthermore, the mutant enzyme might continue to function and make histidine in the presence of the inhibitor even if the sensitive wild-type enzyme is present. If so, in a diploid containing both the mutant and wild-type alleles, the phenotype would be His$^+$ in the presence of the inhibitor, the phenotype of the mutant allele, and the resistance mutation is dominant.

cis/*trans* TESTS

One of the most important uses of complementation is to determine whether a mutation is ***trans* acting** or ***cis* acting**. These prefixes come from Latin and mean "on the other side" and "on this side," respectively. A *trans*-acting mutation usually affects a diffusible gene product, either a protein or an RNA. If the mutation affects a protein or RNA product, it can be complemented, and it does not matter which DNA has the mutation in a complementation test because the gene product is free to diffuse around the cell (i.e., the mutation acts in *trans*). A *cis*-acting mutation usually changes a site on the DNA, such as a promoter or an origin of replication. If the mutation affects a site on the DNA, it affects only the DNA region containing that mutation and cannot be complemented (i.e., it acts in *cis*). In our example of the histidine synthesis genes, mutations (either recessive or dominant) that affect the enzymes that make histidine are *trans* acting and can be complemented, while a promoter mutation that prevents transcription of the genes for histidine synthesis is *cis* acting and cannot be complemented. In subsequent chapters, we discuss how *cis-trans* tests have been used to analyze the regulation of gene expression and other cellular functions.

ALLELISM TESTS

A common use of complementation is to determine how many different genes (or regions encoding a particular gene product) can be mutated to give a particular phenotype. Another name for this is an **allelism test**, because we are asking whether any two mutations are allelic, i.e., whether they affect the same gene (see above). Allelism tests can be performed only with recessive mutations and not with dominant mutations or *cis*-acting mutations, because the last two types of mutations cannot be complemented.

Returning to our example of histidine biosynthesis, assume that we have isolated a collection of mutants, all of which exhibit the His$^-$ phenotype and want to know how many genes they represent. This should tell us how many separate polypeptides are required to make the enzymes that make the amino acid histidine in the cell and therefore allow the cell to multiply in the absence of histidine in the medium. Some enzymes are composed of more than one different polypeptide, each encoded by a different gene. A different gene or complementation group, composed of all the mutations that do not complement each other, encodes each polypeptide, and if our collection of mutations is large and varied enough, each of these genes should be inactivated by at least one of our mutations. The allelism test is performed on the mutations two at a time, as illustrated in Figure 3.25A and B. If the two mutations are in different genes, as shown in Figure 3.25A, each DNA can furnish the polypeptide that cannot be furnished by the other, so that all the polypeptides are present and the diploid cell is phenotypically His$^+$. If, however, the two

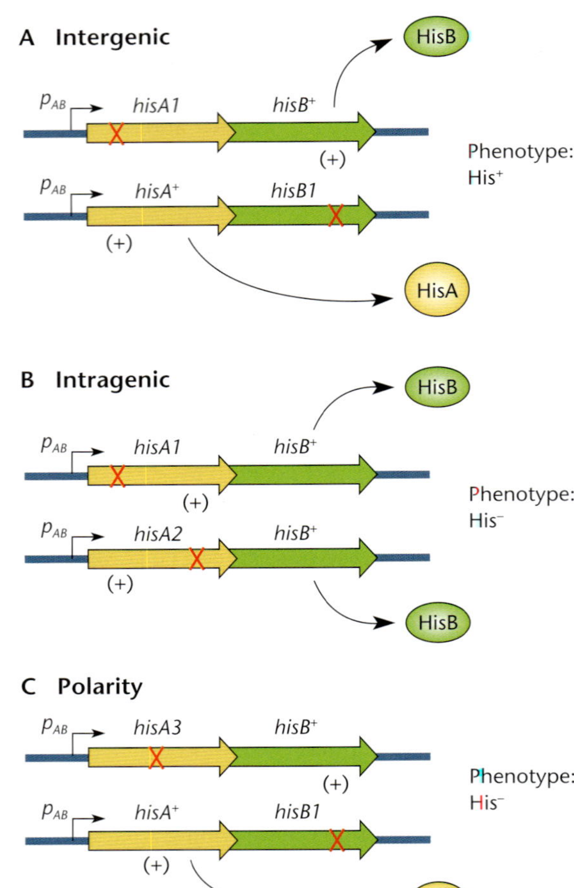

Figure 3.25 Complementation tests for allelism. Four mutations, *hisA1*, *hisA2*, *hisA3*, and *hisB1*, that make the cell require histidine are being tested to determine whether they are allelic or are in different genes. The mutations are in different DNAs in the same cell in different combinations. **(A)** The *hisA1* and *hisB1* mutations are in different genes whose products are required for synthesis of histidine. The DNA with the *hisA1* mutation can make HisB, and the DNA with the *hisB1* mutation can make HisA; hence, the two mutations complement each other, and the cell is His$^+$ because of intergenic complementation. **(B)** The *hisA1* and *hisA2* mutations are in the same gene, i.e., are allelic. Neither DNA can make HisA; hence, the cell is His$^-$. **(C)** Polarity can complicate the allelism test. If mutation *hisA3* is a nonsense mutation in gene *A* that is polar on gene *B*, neither DNA will make HisB, and the two mutations will not complement each other. Nevertheless, the two mutations are not allelic.

mutations are allelic in a region coding for the same gene product, as shown in Figure 3.25B, neither DNA can make that gene product, the two mutations cannot complement each other, and the cells remain phenotypically His$^-$. We can then extend this analysis to include the other mutations in our collection, two at a time, to place them in complementation groups and determine how many total genes or complementation groups are represented in the collection of mutations.

Table 3.4 Interpretation of complementation tests

Test result	Possible explanations
x and *y* complement	Mutations are in different genes
	Intragenic complementation has occurred[a]
x and *y* do not complement	Mutations are in the same gene
	One of the mutations is dominant
	One of the mutations affects a regulatory site or is polar

[a]See the text for an explanation of intragenic complementation. This is a less likely explanation than the mutations being in different genes.

Usually, the interpretation of complementation tests is straightforward and the number of complementation groups equals the number of gene products required for a biological function. However, sometimes, the interpretation of complementation tests for allelism is not so simple, and mutations that complement each other can be in the same gene, and those that do not complement each other might be in different genes. Table 3.4 outlines the interpretation of complementation experiments and their possible complications, which are discussed below.

Intragenic Complementation

When complementation occurs between two mutations that are in the same gene, it is called **intragenic complementation**. This usually occurs only with some types of mutations and if the protein product of the gene is a homodimer or higher homomultimer that contains more than one polypeptide product of the same gene (see chapter 2 for definitions of homodimer and homomultimer). Also, the polypeptide must have more than one functional domain, with different activities of the polypeptide residing in the different domains.

One classic example of intragenic complementation that is used in a number of common cloning vectors is *E. coli lacZ* α complementation (Figure 3.26). Cells that produce the intact LacZ polypeptide in addition to the LacY product are Lac$^+$ and form blue colonies during growth on media containing the indicator dye X-Gal, whereas Lac$^-$ colonies form white colonies. If cells contain a deletion in the *lacZ* that results in production of an inactive LacZ protein, the colonies are phenotypically Lac$^-$. Introduction of a plasmid containing the intact *lacZ* gene into cells containing the *lacZ* deletion results in cells that are Lac$^+$ because the plasmid-borne *lacZ* gene complements the mutant chromosomal *lacZ* gene. Figure 3.26B shows a more complicated situation in which two different *lacZ* deletions are used. The *lacZ*ΔM15 mutation results in production of the ω fragment of LacZ, which is missing amino acids 11 to 41 and is inactive. The *lacZ*Δα mutation results in production of the α fragment of LacZ, which contains only amino acids 3 to 92 and is also inactive. However, production of both the α and ω fragments of LacZ in the

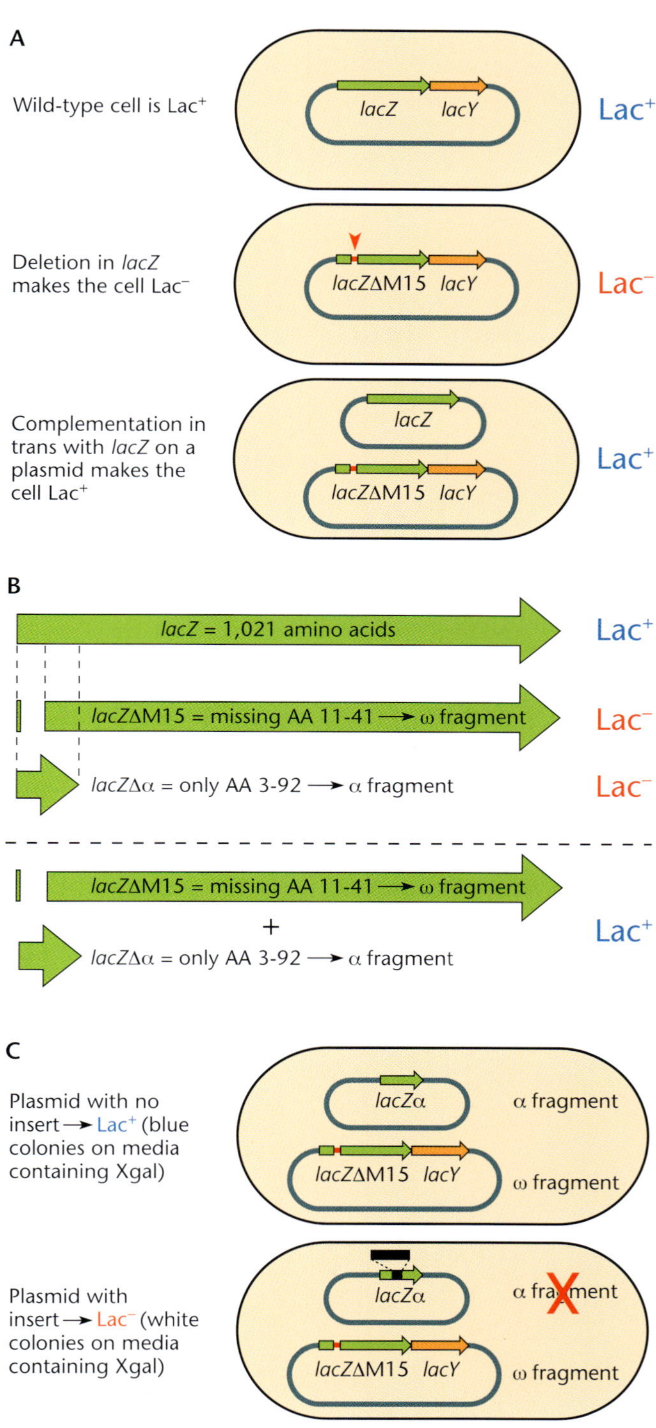

same cell results in formation of a complex that has β-galactosidase activity and results in formation of blue colonies on media containing X-Gal. This is the result of intragenic complementation between the two mutant copies of *lacZ* (see Shuman and Silhavy, Suggested Reading).

The *lacZ* α complementation has been used as an indicator for insertion of cloned fragments into plasmid and phage vectors. As illustrated in Figure 3.26C, the chromosome of the host strain contains the *lacZ*ΔM15 mutation, which results in production of the ω fragment of LacZ. Introduction of the vector, which contains an intact *lacZ*Δα gene, results in production of the α fragment of LacZ and formation of blue colonies. Insertion of a cloned DNA fragment at restriction endonuclease cleavage sites within the *lacZ*Δα segment disrupts *lacZ*Δα and prevents production of the α fragment, resulting in formation of white colonies on media containing X-Gal. This allows rapid identification of colonies containing plasmids that have a new piece of DNA inserted into the vector.

Polarity

Another potential complication of complementation tests is **polarity** (see chapter 2 for a description of polarity). Polarity can have an effect opposite that of intragenic complementation in that it can prevent complementation between two mutations, even though they are in different genes and thus are not allelic. As discussed in chapter 2, a mutation in one gene that terminates translation or transcription can prevent the expression of (be polar on) another gene if the two genes are in the same operon so that they are transcribed onto the same mRNA.

We can use our hypothetical *his* operon to illustrate how polarity affects complementation tests (Figure 3.25C). Assume that mutation *hisA3* in gene A is a nonsense mutation that is polar on gene B, because the two genes are

Figure 3.26 *E. coli lacZ* α intragenic complementation. **(A)** The *lacZ*ΔM15 deletion results in a nonfunctional LacZ protein, and the cells are phenotypically Lac−. Introduction of a plasmid containing the wild-type *lacZ* allows synthesis of active LacZ product, and the cells are phenotypically Lac+. **(B)** The *lacZ*ΔM15 mutation results in loss of amino acids 11 to 41 of LacZ, and the resulting product is the ω fragment. The *lacZ*Δα mutation results in production of the α protein fragment containing amino acids 3 to 92. Both the α and ω fragments are nonfunctional, but expressing both fragments in the same cell results in a functional β-galactosidase enzyme, and the cells are phenotypically Lac+. **(C)** Use of a host strain with the *lacZ*ΔM15 mutation and a cloning vector with the *lacZ*Δα mutation allows differentiation of empty vectors from vectors containing a new DNA insert. Plasmids with no insert allow synthesis of the α fragment and result in formation of Lac+ colonies, whereas plasmids with a cloned DNA fragment inserted into the vector do not produce the α fragment and result in formation of Lac− colonies. Lac+ cells form blue colonies on media containing the β-galactosidase indicator X-Gal, and Lac− cells produce white colonies, allowing easy identification of colonies containing plasmids with an inserted DNA fragment. AA, amino acids.

cotranscribed into the same mRNA starting at a promoter upstream of gene A. If the *hisA3* mutation is a nonsense mutation that is polar on the *hisB* gene, it will not complement the *hisB1* mutation on a different DNA because neither DNA can make the HisB polypeptide. Whether the polarity is due to transcriptional termination in gene A downstream of the *hisA1* mutation or to translational coupling between *hisA* and *hisB*, the effect will be the same, i.e., no HisB polypeptide is made. This complicates our interpretation of the experiment in a number of ways. For one thing, from this experiment alone, we cannot even conclude that the HisA polypeptide is required for histidine biosynthesis. The *hisA3* mutation makes the cell His^-, but because of polarity, it could be the polypeptide encoded by the *hisB* gene that is required, and the *hisA3* mutation is only preventing its synthesis by being polar on it. Alternatively, both HisA and HisB could be required to synthesize histidine, since neither is being made. Further complementation tests are required to answer these questions. If we introduce a plasmid containing a clone that expresses only the HisA polypeptide and the cell is still His^-, the HisB polypeptide must be required, because the *hisA3* mutation is preventing the synthesis of both the HisA and the HisB polypeptides from the chromosome, but the clone provides the HisA polypeptide by complementation. On the other hand, if the complementing clone expresses only HisB and the cell is still His^-, the HisA polypeptide must also be required. Such complementation tests are the standard way of establishing that a mutation in a gene is responsible for a particular phenotype and not the absence of the product of another gene downstream of it in the operon due to polarity. To avoid the complications of polarity, genes are often inactivated by in-frame deletions or insertions or other mutations that cannot be polar on other genes. Gene modification cassettes have been engineered so that transposon insertion mutations can be converted into short in-frame insertions (scars) lacking transcription termination sites by using site-specific recombination systems to delete most of the transposon sequences after the transposon has inserted (see chapter 13).

Note that the effect of polarity on the expression of downstream genes is *cis* acting. In our example, the *hisA3* mutation can prevent the synthesis of the HisB polypeptide from the same DNA, the chromosome, but not from the complementing clone expressing HisB from the plasmid.

CLONING BY MARKER RESCUE AND COMPLEMENTATION

There are a number of reasons why we may want to clone a gene in which mutations give a certain phenotype. We have mentioned how the clone could direct us to the region in the genome in which the mutations lie, i.e., map the mutations. We could have the entire mutant genome sequenced to locate the mutation. However, cloning the mutated gene would give independent confirmation that we have located the responsible mutation. Once we have located the gene, comparisons to other gene sequences might give us clues to which biochemical activities it has. However, many genes in a bacterial chromosome have no known homologs, and even if homologs can be found by searches such as BLAST, such annotation can carry us only so far. To determine directly the biochemical activity and structure of a protein, we usually have to clone the gene to express and purify its product. The easiest way to go about the cloning depends on the situation. If we already know the sequence of the gene, we can use PCR to amplify the gene directly from the chromosome and clone the amplified fragment using methods outlined in chapter 1. However, in many cases, we need to identify clones from a library of the entire chromosome.

Either marker rescue or complementation can be used to identify clones of a gene for which we have mutations affecting a particular phenotype. Both approaches have their advantages and disadvantages, depending on the situation. Figures 3.27 and 3.28 contrast the use of complementation and marker rescue to identify clones of the gene for thymidylate synthase (*thyA*) in a bacterium for which we have a *thyA* mutant. This enzyme is needed to

Figure 3.27 Identification of clones of the *thyA* gene of *E. coli* by complementation. **(A)** A *thyA* mutant of *E. coli* is transformed by a library of DNA from wild-type Thy^+ *E. coli*, and the transformants are selected directly on plates lacking thymidine but containing the antibiotic to which the cloning vector confers resistance. **(B)** Thy^+ transformants contain a clone of the *thyA* gene that directs synthesis of thymidylate synthase from the plasmid, thus complementing the *thyA* mutation in the chromosome.

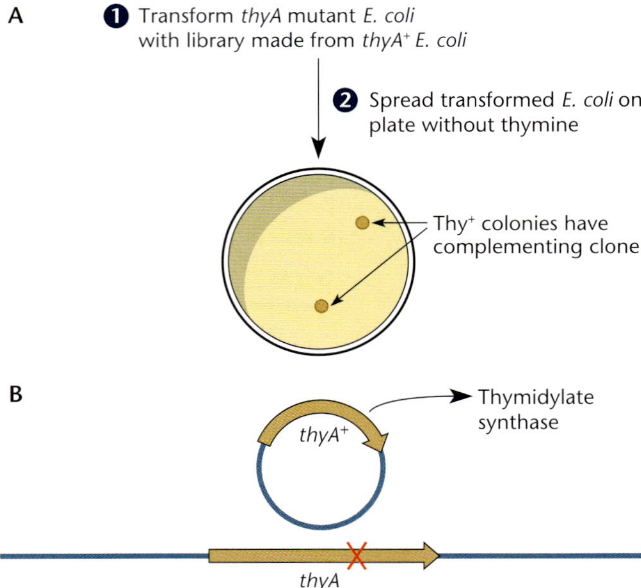

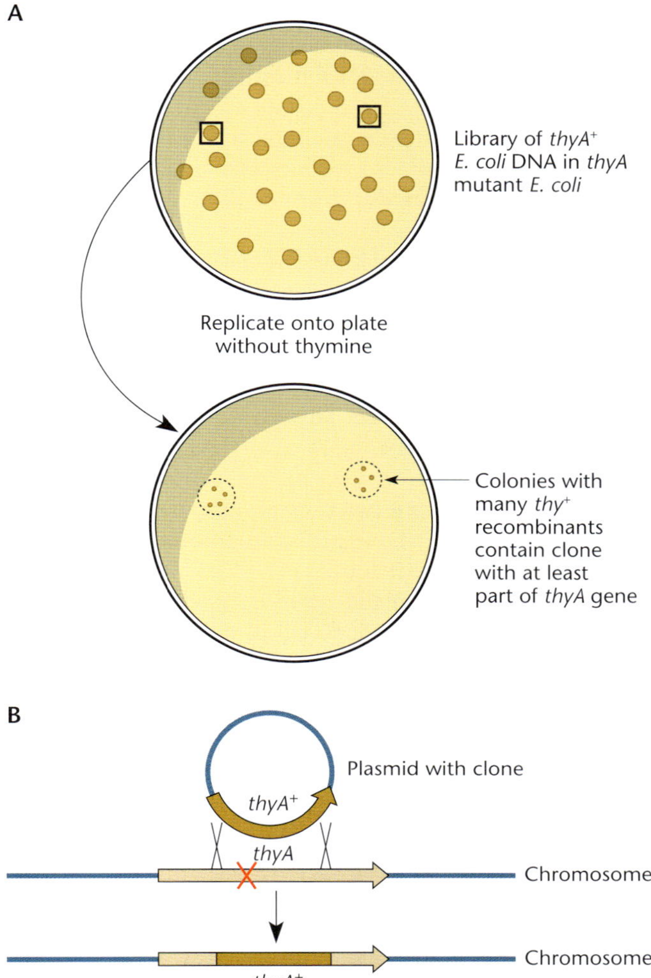
A
Library of thyA+ E. coli DNA in thyA mutant E. coli
Replicate onto plate without thymine
Colonies with many thy+ recombinants contain clone with at least part of thyA gene
B
Plasmid with clone
thyA+
thyA
Chromosome
Chromosome
thyA+

Figure 3.28 Use of marker rescue to identify a clone containing at least part of the *thyA* gene of *E. coli.* **(A)** A plasmid library of clones from wild-type Thy+ *E. coli* is transformed into a *thyA* mutant of *E. coli,* and the transformants are plated on nonselective plates with medium containing thymidine. After the colonies develop, the plate is replicated onto a selective plate without thymidine. Colonies that show some Thy+ recombinants on the selective plate can be picked from the original nonselective plate (boxed). **(B)** The part of the *thyA* gene on the cloning vector has recombined with the *thyA* mutant gene on the chromosome to produce Thy+ recombinants.

synthesize dTMP from dUMP, so a *thyA* mutant will not be able to replicate its DNA and multiply to form a colony unless thymine is provided in the medium (see chapter 1). As shown in Figure 3.27, we obtained or made a library of the chromosome of the wild-type (*thyA*+) bacterium and wanted to use complementation to identify which of the clones contains the *thyA* gene. We introduced this library into a mutant strain and plated the bacteria on medium lacking thymine. Any bacteria that multiplied to form a colony might contain a plasmid clone that was expressing the *thyA* gene, which complemented the *thyA* mutation in the chromosome.

Identifying clones by marker rescue recombination usually requires an additional step. If we plate the bacteria containing the library directly on plates lacking thymine, the clone will have been rearranged by the recombination that gave rise to the Thy+ recombinants. One way to avoid this problem is to first plate the bacteria on medium with thymine so the bacteria will grow to form colonies whether or not the mutation is being marker rescued or complemented. The cells on these plates could then be replicated or patched onto selective plates containing all the necessary growth supplements but without thymine. Any replicated patches that contain many more Thy− colonies than average may be due to bacteria that received a clone containing at least part of the *thyA* gene from the donor, as shown in Figure 3.28. In the case of marker rescue, only some of the cells that took up the clone became Thy+ by recombination with the clone, but if the clone complemented the *thyA* mutation in the chromosome, all of the cells that took up the clone would have become Thy+. In the example, only a few of the cells in the colony were able to grow on the selective plate, so marker rescue was suspected.

Both approaches to cloning have their advantages and disadvantages. One advantage of marker rescue is that the piece of DNA need not be in a cloning vector that replicates in that cell and the gene need not be expressed in the cell. The DNA fragment in the clone also need not include the entire gene, only the part of it containing the region of the mutation. The disadvantage is that the fragment of the chromosome that participated in the marker rescue no longer exists in the cell, having been irreversibly altered by the recombination that rescued the mutation. If the piece of DNA was cloned in a multicopy cloning vector that can replicate in that cell, we can get the clone from a different plasmid in that cell that did not participate in the recombination. Otherwise, we need to return to the original nonselective plate to get the clone that demonstrated marker rescue from the original transformant before it was altered by the recombination. Also, if the clone does not contain the entire *thyA* gene, we may have to construct it from neighboring genome sequences to express the gene product.

The major advantage of cloning by complementation is that we can introduce our library directly into the cells by transformation and select cells in which the mutation is complemented. However, the major advantage of cloning by complementation is also its major disadvantage. To identify clones by complementation, the DNA fragments must be cloned in a vector that can replicate in those cells. Most plasmid cloning vectors have a fairly narrow host range, but some broad-host-range vectors that will replicate in many types of bacteria have been developed. We talk about the host range of plasmids in chapter 4. Also, the entire gene must be included in the clone, and the gene must be expressed in those cells. This means that the gene must be transcribed from a functional promoter and translated from a functional translation initiation site in the cells. However, the characteristics of a functional promoter and translation initiation site can differ between species of bacteria, and genes from distantly related organisms often cannot be expressed in a particular host, so this method is generally useful only for identifying clones in the "host of origin," for example, a bacterial gene in the bacterium it came from or a close relative.

Another potential complication of complementation cloning is **copy number effects**. If we express the gene from a multicopy plasmid or other vector, we might make many more copies of the gene product than normally exist in the cell. This can affect the ability of the cloned gene to complement and may even change the phenotypes in unpredictable ways. Sometimes, a gene product other than the one that was mutated can give apparent complementation when it is expressed from a cloning vector. In our example, there might be another gene for thymidylate synthase in the chromosome that is normally not expressed, perhaps because it lacks a functional promoter. A promoter in the cloning vector might direct transcription of the gene, leading to apparent complementation of the *thyA* mutation even though the clone is of a different gene. Sometimes the overproduction of a different gene product, such as occurs from a multicopy cloning vector, can give apparent complementation even if its product seems unrelated. If a mutation is complemented by a clone of a different gene whose product is being overproduced, this is called **multicopy suppression**. Identification of other proteins that can substitute for a mutated protein when expressed in greater than normal amounts is interesting and can give clues to the function of the mutated protein. However, this does not result in cloning the gene of interest, our original goal. For reasons such as these, it is often preferable to express cloned genes for complementation from a cloning vector that replicates with a low copy number or that integrates in a single copy somewhere else in the chromosome, for example, a lysogenic phage or specific transposon integration site. In any case, care must be taken that any apparent clone isolated by complementation actually contains a wild-type copy of the mutated gene.

Genetic Crosses in Bacteria

We have discussed the molecular basis for genetic mapping and complementation tests without going into any detail about how DNA can be introduced into bacterial cells and how data can be analyzed using the different means of genetic exchange. As discussed in the introductory chapter, the three means of genetic exchange in bacteria are **conjugation**, **transformation**, and **transduction**. We could add **electroporation** to this list, but for genetic purposes, it is just another means of doing transformation in bacteria that are not naturally transformable. We talk about the molecular mechanisms of each of these forms of genetic transfer in chapters 5, 6, and 7. Here, we discuss their uses in genetic experiments and how data obtained by using them are analyzed.

SELECTED AND UNSELECTED MARKERS

What all the means of gene exchange in bacteria have in common is that they are not 100% efficient. Unlike us and most other known multicellular eukaryotes, bacteria do not have an obligate sexual cycle and do not need to mate to reproduce. They just divide. If one of the means of genetic exchange is available, a few of them might exchange DNA with each other and become recombinants. As in searches for mutants, we have to find the few bacteria that have recombined among the many that have not.

Fortunately, bacteria that have participated in a genetic transfer event can often be detected, even if they are very rare. The process is much like selecting mutants. Conditions are established under which only bacteria recombinant for one of the markers can multiply, independent of whether they are recombinant for any of the other markers. The recombinants for this marker, called the **selected marker**, are then tested to see if they are also recombinant for the other markers, called the **unselected markers**. Which marker is chosen to be the selected marker is a matter of convenience. The example in Figure 3.29 provides an illustration. In this example, a piece of DNA from one strain that has one of the mutations has been transferred into another strain that has another mutation. Both mutations inactivate a gene whose product is required for growth under some conditions. Therefore, the recipient strain is not able to grow under the selective conditions, because it has the m_y mutation. However, if crossovers occur between the incoming DNA and the homologous region in the recipient DNA, recombinant-type progeny can arise in which the sequence of the donor DNA, which does not have the m_y mutation (indicated by "+" in Figure 3.29), has replaced the sequence of the recipient strain, which did have the mutation, allowing the recipient strain to grow under the selective conditions. We decided to use the m_y marker as the selected marker to select recipient bacteria in which crossovers have occurred that make the strain recombinant for the selected marker, because they can be selected. Bacteria selected for being recombinant for this marker must have participated in a mating event, no matter how rare these mating events may be.

Once we have selected for recipient bacteria that have participated in a mating event because they have become recombinant for the m_y marker, we can test them to see if they are also recombinant for the m_z marker, the unse-

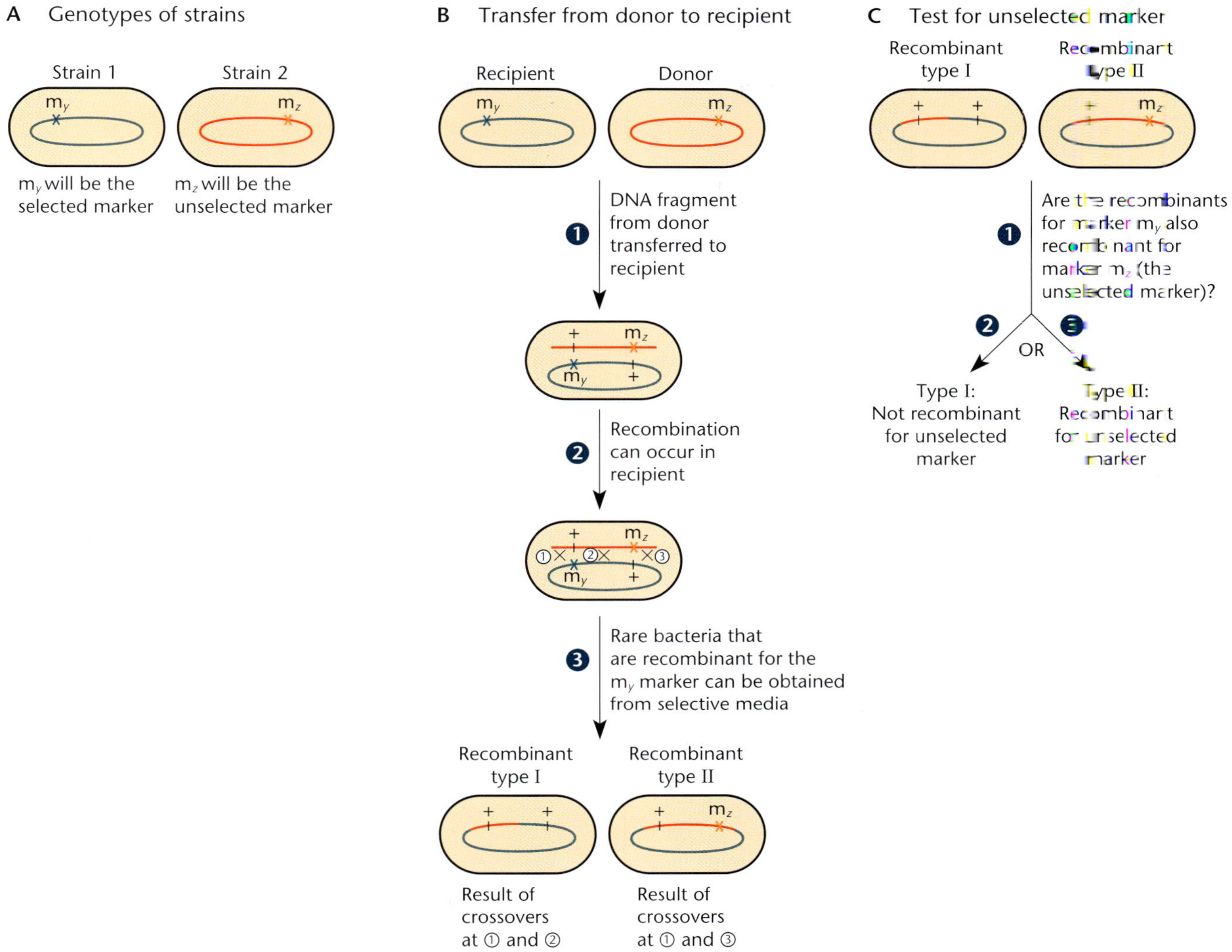

Figure 3.29 Selected versus unselected markers in a bacterial cross. Replacement by recombination of the sequence of the recipient DNA (blue) by the sequence of the donor DNA (red) creates a recombinant type. Recipient bacteria, recombinant for the selected marker, are tested to see how many are also recombinant for a second unselected marker. In the example, m_y is the selected marker and m_z is the unselected marker, chosen because it is easier to select recombinants for the m_y marker. **(A)** Genotypes of the strains. **(B)** A fragment of the donor DNA (wild type [+] for m_y) is transferred into the recipient cell. Recombination occurs between the incoming donor DNA and the recipient DNA, replacing regions of the recipient DNA with donor DNA. The recipients in which the wild-type m_y region of the donor has replaced that of the recipient are selected and purified on selective plates. **(C)** The recipients that have been selected for being recombinant for the m_y marker are tested to see if they are also recombinant for the m_z marker (recombinant type II). For details, see the text.

lected marker. The recipient will have become recombinant for the other marker if it now has the m_z mutation and is unable to grow without the growth supplement that m_z mutants require, since that is the sequence of the donor DNA, which had the m_z mutation. In general, a recipient has become recombinant for a marker if it has the marker sequence of the donor. If any of the recipients that were selected for being recombinant for the selected marker have also become recombinant for the unselected marker, the donor DNA from both regions must have entered the same recipient cell and both regions must have recombined with the recipient chromosome. If the DNA transfer events are rare enough, it is highly unlikely that two DNAs transferred in separately, and the two regions must have come in on the same DNA molecule. Therefore, the frequency with which bacteria selected for one marker have become recombinant for other markers gives us information about where the mutations that gave rise to the phenotypic differences were on the chromosome relative to each other. Note that once we have selected for one marker, we do not need to screen nearly as many bacteria to get meaningful data about the frequency of recombinants for the unselected markers, and we can test them directly for the other markers without a selection. Specific examples of this process, using the various methods of gene exchange, which differ in how data are interpreted, are given below.

Mapping of Bacterial Markers by Transduction and Transformation

The transfer of DNA from one bacterial cell to another by transformation or transduction can be used to map genetic markers. Recipient bacteria that have received DNA from the donor by transformation or transduction are called **transformants** or **transductants**, respectively. Even though transformation and transduction occur by very different mechanisms, the processes by which genetic data are processed with them are quite similar, so they are treated together here. The major difference is that single-stranded DNA enters the cell during natural transformation, whereas double-stranded DNA enters during transduction. Even this difference may not matter, because the double-stranded DNA introduced by transduction is probably quickly converted into single-stranded DNA by an enzyme called the RecBCD nuclease in *E. coli*, as discussed in chapter 9. More details of how transformation and transduction occur are given in chapters 6 and 7, respectively.

TRANSFORMATION

In bacterial transformation, only the DNA from one bacterium enters another bacterium. This happens normally in some types of bacteria, which therefore are said to be **naturally competent** or **naturally transformable**. This means that they have the natural ability to pick up DNA under some conditions. Some of them pick up only DNA of the same species, while others can take up any DNA. While the purpose of natural transformation is the subject of intense debate, it may have a number of roles, including DNA repair, increasing genetic or antigenic variability, or even serving as a food source during development. In any case, it is very useful in the genetics of the bacteria that have this capability, including *B. subtilis*, one of the major model systems.

Even if a type of bacterium is not naturally competent, it can sometimes be made competent by chemical treatments or electroporation. Developing a transformation system is a high priority for the study of any type of bacterium at the molecular level. The advantages of transformation are that DNA that has been manipulated in the test tube, or even synthesized chemically, can be introduced into cells, which makes it essential for many types of molecular genetic applications.

TRANSDUCTION

In transduction, DNA from the donor bacterium is packaged, or encapsulated, in a bacteriophage head, so that when the bacteriophage infects the recipient bacterium, the donor bacterial DNA is injected. There are two types of transduction, specialized and generalized. In specialized transduction, part of the phage DNA is replaced with chromosomal DNA, so the transducing particles that transduce chromosomal genes also carry phage DNA, and only certain regions of the chromosome are carried. In generalized transduction, the transducing particles have only chromosomal DNA, which can come from any region of the chromosome. Specialized transduction is generally less useful for mapping most markers, since only some regions of the chromosome are packaged in a particular phage. We discuss only mapping by generalized transduction here and refer to it simply as transduction.

An illustration of using transduction to generate a recombinant is shown in Figure 3.30. Very few of the phages pick up chromosomal DNA, and even fewer have picked up the region of the genetic marker (a region of the *trp* gene in this example). Once the DNA is injected into the recipient cell, it still has to recombine with the corresponding region of the chromosome to make a recombinant for that marker, all of which makes recombinants for a particular marker very rare. However, even though they are very rare, recombinants can be identified with a powerful selection. Not all types of phages work well for generalized transduction, although sometimes, a nontransducer can be converted to a transducing phage by genetic manipulation. The properties of a good general transducing phage are discussed in chapter 7. Because of their continued usefulness, much effort is expended to find transducing phages for types of bacteria for which they are not available.

ANALYZING DATA OBTAINED BY TRANSFORMATION AND TRANSDUCTION

Mapping by transformation or transduction is based on whether the regions of markers can be carried in on the same piece of DNA, i.e., can be **cotransformed** or **cotransduced**. If a strain that has become recombinant for the selected marker sometimes also becomes recombinant for another unselected marker, the regions of the two markers are **cotransformable** or **cotransducible**, respectively. Both methods transfer only a relatively small piece of DNA and are inefficient enough that the donor regions of markers that are cotransformable or cotransducible must have come in together on the same piece of DNA and therefore are very close to each other on the DNA relative to the length of the bacterial genome.

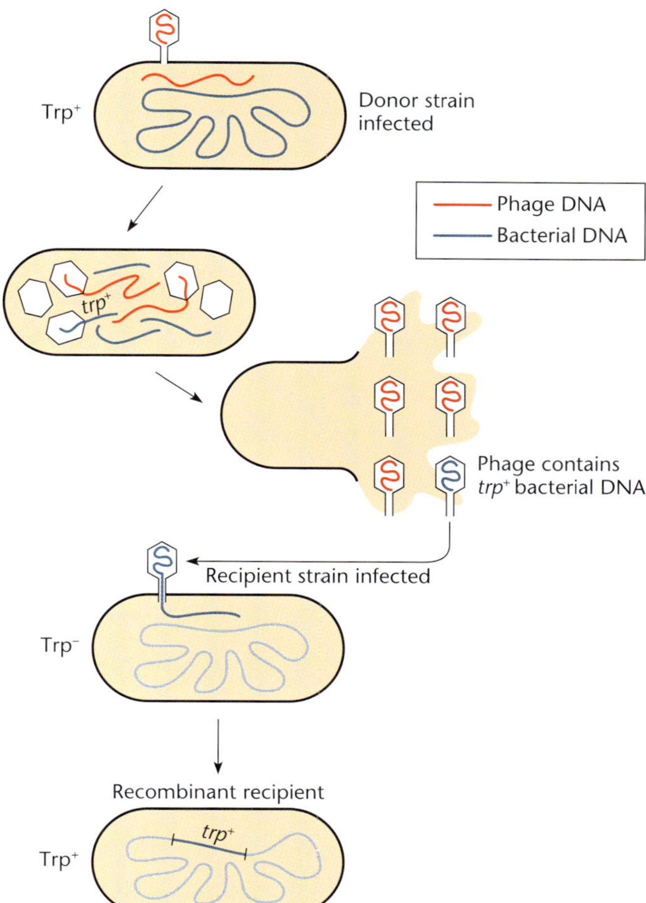

Figure 3.30 Example of generalized transduction. A phage infects a Trp$^+$ bacterium, and in the course of packaging DNA into heads, the phage mistakenly packages some bacterial DNA containing the *trp* region (blue) instead of its own DNA (red) into a head. In the next infection, this transducing phage injects the Trp$^+$ bacterial DNA instead of phage DNA into the Trp$^-$ bacterium. If the incoming DNA recombines with the chromosome of the newly infected cell, a Trp$^+$ recombinant transductant may arise.

Not only does the appearance of cotransformants or cotransductants that are recombinant for two markers signify that the markers are close to each other in the DNA, but also, the higher the percentage of recombinants for the selected marker that are also recombinant for the unselected marker, the closer together the two markers are likely to be. The percentage of the total transformants or transductants selected for one marker that are also recombinant for the other marker is called the **cotransformation frequency** or **cotransduction frequency** of the two markers, respectively. In principle, this frequency between two markers should be a constant for any two markers and should be independent of which of the two markers is the selected marker and which is the unselected marker. A cross with the selected and unselected markers reversed is called a **reciprocal cross**. In the next section, we illustrate how mapping data from transductional crosses are interpreted by using an actual example. Similar reasoning would apply to transformational crosses.

MAPPING BY COTRANSDUCTION FREQUENCIES

In our example, phage P1 is used in *E. coli* for transduction to map a mutation to rifampin resistance in the gene for the β subunit of RNA polymerase (chapter 2) relative to a mutation in the *argH* gene, whose product is required to make the amino acid arginine, and a mutation in the *metA* gene required to make the amino acid methionine. The first step is to select for one of the markers and to determine if any of the other markers are cotransducible with the selected marker. For practical reasons, it is easier to use either the *argH* or the *metA* marker and not the *rif* marker as the selected marker. Rifampin resistance is difficult to select, because it takes many generations to be expressed. Any transductants to rifampin resistance will remain sensitive for many generations, because the rifampin-sensitive RNA polymerase molecules bind to promoters and block the resistant ones until the sensitive ones are diluted out by many cell divisions. For our example, we shall use a donor that is Arg$^+$ and Met$^-$ and a recipient that is Rif^r and select the *argH* marker by plating on medium without arginine or rifampin but with methionine, so that recombinants can grow, whether or not they are Met$^-$ or Met$^+$ or rifampin sensitive or rifampin resistant. We then test the Arg$^+$ transductants to see if any are also recombinant for the other markers.

Figure 3.31A illustrates the cotransduction of the *argH* and *rif* markers, and Table 3.5 gives some representative data. In this example, 33 (22+11) of 96, or 34%, of the Arg$^+$ transductants are also Rif^s and thus have the allele of the donor. Thus, the two markers are cotransducible, and the cotransduction frequency is about 34%. We can estimate how close together on the *E. coli* DNA the *argH* and *rif* markers would have to be in order to be cotransducible by the P1 phage, a phage that is commonly used with *E. coli*. The chromosome is 100 minutes (min) long, and the P1 phage head accounts for only about 2% of that length; therefore, the two markers must be less than 2 min apart. Translating this distance into base pairs of DNA, the *E. coli* chromosome is about 4.5×10^6 bp long, and the P1 phage head holds only $0.02 \times 4.5 \times 10^6$, or 90,000, bp. Thus, to be cotransducible, two markers in the DNA must be less than 90,000 bp, or 2 min, apart on the 100-min map (see below).

From the data in Table 3.5, we can use the relative cotransduction frequencies to begin to order the *argH* and *rif* markers with respect to the *metA* marker. As mentioned above, the closer two markers are to each other, the more apt they are to be carried in the same phage head and the less likely there is to be a crossover between

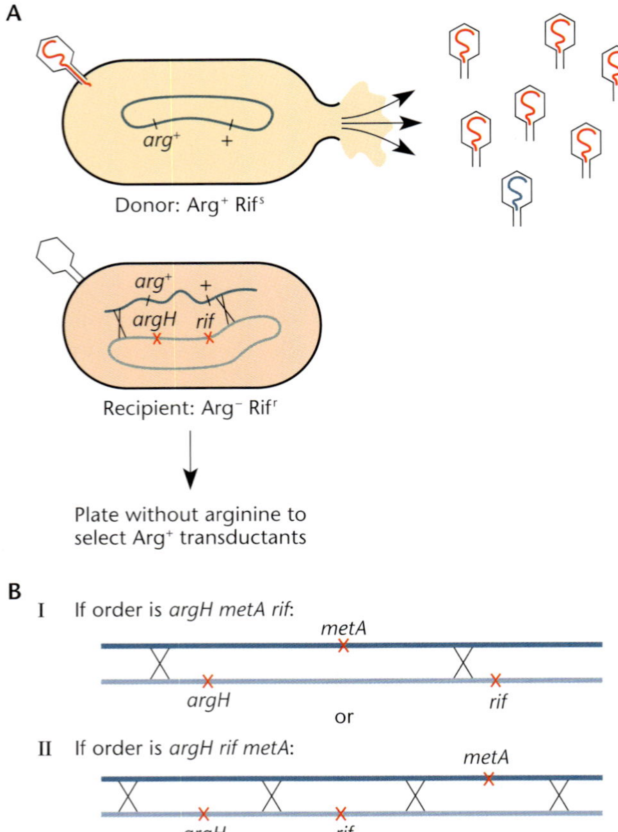

Figure 3.31 Cotransduction of bacterial genetic markers. **(A)** Two-factor cross. The regions of the *argH* and *rif* mutations are in close enough proximity that both regions can be carried on a piece of DNA that fits into a phage head (not shown to scale). The donor cell is wild type for the *argH* gene and is rifampin sensitive, while the recipient carries an *argH* mutation and is rifampin resistant. After transduction, the cells are selected for the ability to grow in the absence of arginine, and some of the Arg^+ transductants are also rifampin sensitive. The red lines indicate phage genomes in progeny phage; the dark blue lines indicate donor bacterial DNA in transducing particles; the light blue circle indicates chromosomal DNA in the recipient cell. **(B)** Three-factor cross to test if the order is I, *argH-metA-rif*, or II, *argH-rif-metA*. The number of crossovers required to make the rarest recombinant type (Arg^+ Rif^r Met^-) is predicted to be two for order I and four for order II. The rarest class is predicted to require four crossovers, suggesting that order II is correct. See the text for details.

Table 3.5 Typical transductional data from a three-factor cross[a]

Recombinant phenotype	No. of recombinants
Arg^+-Met^+-Rif^r	61
Arg^+-Met^+-Rif^s	22
Arg^+-Met^--Rif^r	2
Arg^+-Met^--Rif^s	11
Total	96

[a]Donor: Arg^+-Met^--Rif^s; recipient: Arg^--Met^+-Rif^r.

them. Therefore, the closer to each other they are, the higher the cotransduction frequency. In the example, 13 (2 plus 11) of the Arg^+ transductants are Met^-, so the cotransduction frequency of the *argH* and *metA* markers is ~14%, which is less than the 34% cotransduction frequency of the *arg* and *rif* markers. Therefore, the *arg* marker is closer to the *rif* marker than it is to the *met* marker, and the order seems to be *argH-rif-metA*. It also seems possible that the order is *rif-argH-metA*, with *argH* in the middle. However, both the *rif* marker and the *metA* marker seem to be on the same side of the *argH* marker, because most of the Met^- transductants are also Rif^s (11 of 13), as though the ones that received the *metA* region were more apt to also receive the *rif* region, rather than less apt, as they would be if they were on opposite sides.

Ordering Mutations by Three-Factor Crosses

A careful determination of cotransduction frequencies can reveal the order of markers in the DNA. However, three-factor crosses offer a less ambiguous way to determine marker order. This analysis is based on how many crossovers are required to make a certain recombinant type with a given order of markers. We have already pointed out that a single crossover between a short linear piece of the chromosome and the entire chromosome will break the chromosome and be lethal. Therefore, in a transductional cross with a linear DNA from the donor, a minimum of two crossovers are required to replace the chromosome sequence with the sequence on the incoming donor DNA and form a recombinant type. In general, in such a cross, any viable recombinant types must have originated from an even number of crossovers.

We can also use the data in Table 3.5 to illustrate the ordering of bacterial markers by a three-factor-cross analysis. There are four recombinant types possible in a cross of this type, listed in Table 3.5. With any particular order of the three markers, three of the four recombinant types listed require only two crossovers, and the fourth requires four crossovers. The recombinant type that requires four crossovers should be rarer than the others. In Table 3.5, the rarest recombinant type is clearly Arg^+ Met^- Rif^r, since only 2 of the 96 Arg^+ transductants were of this type. Figure 3.31B shows how many crossovers should be required to make this recombinant type if the order is *argH-metA-rif* or if the order is, as we suspect, *argH-rif-metA*. Only two crossovers are required to make this recombinant type if the order is *argH-metA-rif* (with the *metA* marker in the middle), while four crossovers are required if the order is *argH-rif-metA*. Thus, it seems clear that the order is *argH-rif-metA*, which is also consistent with the order we obtained based on cotransduction frequencies alone. Since resistance to rifampin is due to a mutation in the gene for the β subunit of RNA polymerase (see chapter 2), these data indicate that the gene for the β subunit of RNA polymerase lies between the *argH* and *metA* genes in the chromosome of *E. coli*.

Other Uses of Transformation and Transduction

There are many uses for transformation and transduction in bacterial genetics besides genetic mapping. They are all indispensable in molecular genetic studies in bacteria and explain the desirability of developing a transformation system and/or a transducing phage for the type of bacterium being studied.

STRAIN CONSTRUCTION

One of the major uses of transformation and transduction is in constructing **isogenic** bacterial strains, which differ only in a small region of the chromosome. Even different strains of the same species can differ at a number of genetic loci. To compare two strains in an experiment, it is necessary to compare isogenic strains; otherwise, there is no way to be certain that a phenotypic difference between two strains is due to a particular genetic difference as opposed to other unknown differences between the strains. Therefore, meaningful experiments often require constructing isogenic strains. Transformation and transduction introduce only a small region of the chromosome, so any differences between the original recipient strain and a transductant must have been carried in on the same piece of DNA. If other genetic differences are contributing to phenotypic differences between the two strains, they must be very closely linked to the mutation being introduced.

Sometimes, we can use transformation or transduction to move mutations into a strain, even if the mutation has no easily selectable phenotype. For example, we might use a closely linked transposon carrying an antibiotic resistance gene to move such a mutation. An invaluable early tool for this purpose was a collection of *E. coli* strains with marked transposon insertions around the genome. Therefore, the site of any mutation is cotransducible with at least one of the transposon insertions. A newer collection of gene knockouts in every nonessential gene in *E. coli* proves even more valuable (see Baba et al., Suggested Reading). Using these types of collections to move a mutation into new genetic backgrounds is a three-step process. First, we use the collection as a donor for transduction, selecting the antibiotic resistance gene on the transposon and testing a number of the transductants for the mutation. A recipient strain that has lost the mutation due to cotransduction will probably have the transposon integrated close to the former site of the mutation. We can then use that strain as a donor to transduce the mutant strain carrying the mutation again. This time, if the transposon insertion is cotransducible with the mutation, some of the antibiotic-resistant transductants will have lost the mutation and some will still have it. We choose one of those that still have the mutation as a donor to easily transduce the mutation into as many other strains as we wish, even if the mutation has no easily selectable phenotype, merely by selecting for the antibiotic resistance on the transposon.

REVERSION VERSUS SUPPRESSION

Another important use of transformation or transduction is to distinguish reversion mutations from suppressor mutations. As mentioned above, suppressor mutations can be very revealing of the role of a gene product and can help identify other proteins and functions with which it might interact, but reversion and suppression can be difficult to distinguish based on their phenotypes alone. Both reversion mutations and suppressor mutations can allow a mutant strain to multiply under conditions where it could not multiply previously, but they can be distinguished by a

Figure 3.32 Generic test for reversion versus suppression. **(A)** The mutation had reverted, giving the His$^+$ phenotype. The light blue (+) shows the site of the reversion mutation. When the revertant strain is crossed with the wild type (in dark blue), no His$^-$ recombinants appear in the progeny. **(B)** A suppressor mutation, *supX*, has suppressed the mutation, giving the His$^+$ phenotype. When the suppressed strain is crossed with the wild type (in dark blue), some His$^-$ recombinants arise. The sites of the *his* mutation and suppressor mutation in the wild type are shown as pluses, and the site of the suppressor mutation is shown by the black box.

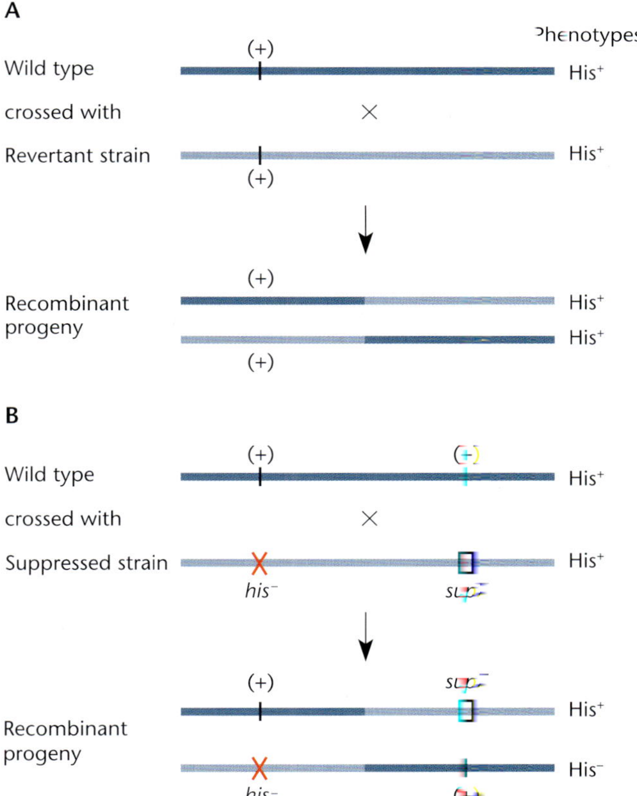

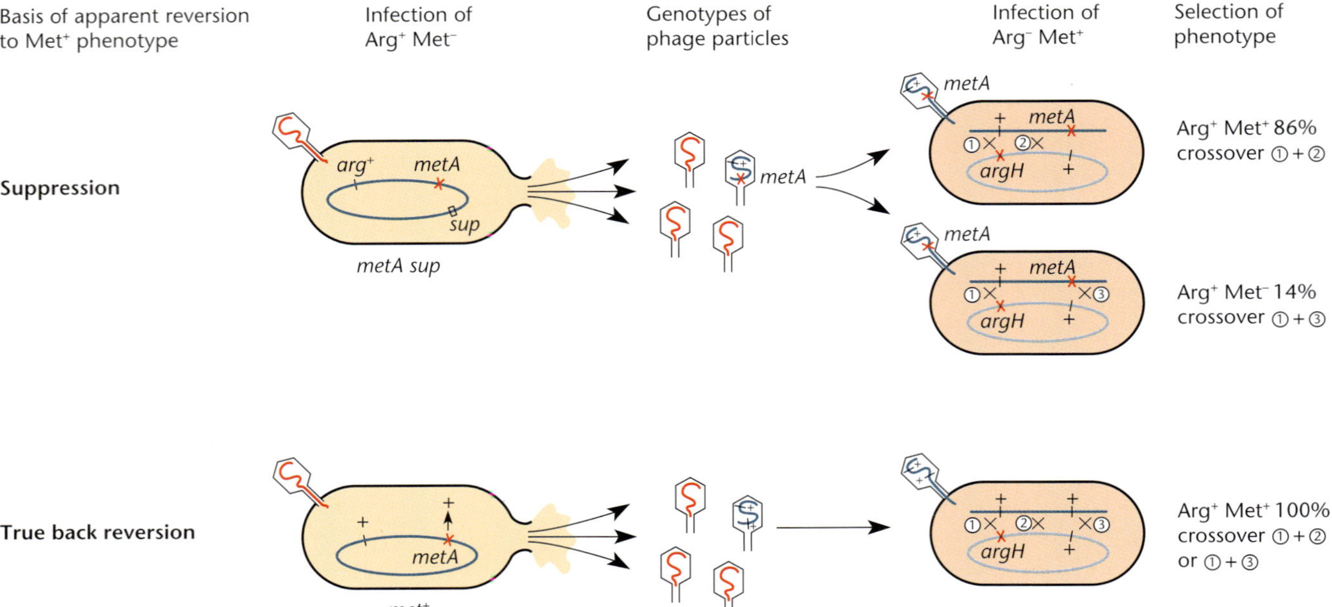

Figure 3.33 Using transduction to distinguish reversion from suppression. If the *metA* mutation has been suppressed, about 14% of the Arg$^+$ transductants will be Met$^-$. If the mutation has reverted, the *metA* mutation no longer exists, and none of the Arg$^+$ transductants will be Met$^-$.

simple genetic test. Until their molecular basis is known, it is best to refer to them as apparent revertants, since on the surface they appear to be revertants.

Figure 3.32 shows the generic recombination test for suppression, applied to apparent revertants of a *his* mutation obtained as shown in Figure 3.19. The apparent revertant is crossed with another strain that has never had the *his* mutation and presumably does not have the possible suppressor mutation. Sometimes recombination occurs between the site of the original mutation and the site of the possible suppressor mutation. If the His$^+$ apparent revertants are true revertants, the original *his* mutation no longer exists, and all the recombinants are still His$^+$. However, if the original mutation has been suppressed, the suppressor mutation will sometimes be separated from the original mutation, which still exists, and some of the recombinants will be phenotypically His$^-$.

Any of the means of genetic exchange in bacteria can be used to distinguish reversion from suppression. However, transduction and transformation are particularly useful in this regard. Basically, we use the apparent revertant as a donor, selecting for a nearby cotransducible marker. If any of the transductants have the phenotype of the original mutation, the original mutation has been suppressed rather than having reverted. In the example shown in Figure 3.33, the *metA* mutation in the donor strain has apparently been reverted by being plated on medium without methionine, and the *argH* marker is being selected to test for suppression. If any of the Arg$^+$ transductants are Met$^-$ and cannot grow without methionine, the *metA* mutation has been suppressed rather than having reverted. For example, the *metA* mutation might have been a nonsense mutation, and a tRNA mutation elsewhere in the chromosome is making the cell Met$^+$ by creating a nonsense suppressor.

Genetic Mapping by Hfr Crosses

The other way bacteria can exchange chromosomal DNA between strains is by conjugation. We discuss the molecular mechanism of conjugation in chapter 5, but here we review how it has been used for genetic mapping. Conjugation is the transfer of DNA from one bacterium to another by a **plasmid**. Plasmids that can transfer themselves from one cell to another are called **self-transmissible plasmids**. Bacteria that have obtained DNA by conjugation are called **transconjugants**. Usually, conjugation transfers only the self-transmissible plasmid from one cell to another, but transmission of chromosomal DNA can occur if the plasmid has integrated into the chromosome or if chromosomal DNA has somehow been incorporated into the plasmid. A bacterial strain with a self-transmissible plasmid integrated into its chromosome is called an **Hfr strain**, for high-frequency recombination. As mentioned in the introductory chapter, this phenomenon was first detected in 1946 by Joshua Lederberg and Edward Tatum when they observed recombinant types after mixing some strains of *E. coli* with other strains (see Lederberg and Tatum, Suggested Reading). We now know that one of the strains contained a self-transmissible plasmid called the fertility factor, or F plasmid, a genetic element discovered by Esther Lederberg. In a few bacteria in the population, the F plasmid had integrated into the chromosome, and these bacteria were transferring chromosomal DNA into the other strain, leading to a high frequency of recombinant types. In retrospect, it was doubly fortuitous that some of the strains used by Lederberg and Tatum contained the F plasmid. Their experiment would not have succeeded if none of the strains they used had contained a self-transmissible plasmid. Also, any plasmid other than the F plasmid would not have worked as well, since, as discussed in chapter 5, the F plasmid they used is a mutant that is always ready to transfer.

Figure 3.34 illustrates the process by which chromosomal DNA is transferred in a mating between a donor Hfr strain containing an integrated F plasmid and a recipient strain that does not contain the F plasmid (F⁻). On contact with a recipient cell, the DNA in the donor is nicked at a site in the integrated plasmid, and one strand is displaced into the recipient cell. Normally, only plasmid DNA would be transferred, but because the plasmid is integrated into the chromosome, the chromosomal DNA is also transferred, beginning with chromosomal sequences at one side of the integrated plasmid and continuing around the chromosome in one direction, with the direction depending on the orientation of the integrated plasmid. In theory, if the transfer continued long enough (in *E. coli*, approximately 100 min at 37°C), it would eventually come full circle and transfer the entire 1-mm-long chromosome, finally reaching the chromosomal sequences on the other side of the plasmid and continuing into the plasmid sequences, ending at the *oriT* site in the plasmid where it started. However, transfer of the entire chromosome is nearly impossible, probably because the union between the cells is frequently broken and/or because the DNA is often broken during conjugation at nicks in the transferred strand. As a consequence, markers further from the origin of transfer are transferred less frequently. The ordered transfer of DNA markers and the decay in their frequency of transfer based on where they are in the genome allow us to use conjugation to map markers, as we discuss below.

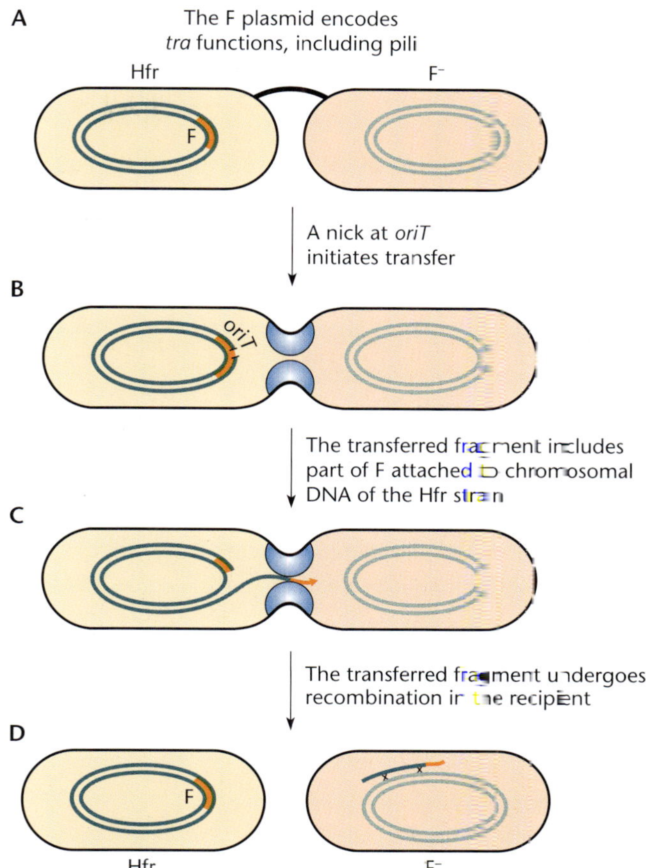
A
The F plasmid encodes
tra functions, including pili
Hfr
F⁻
F
A nick at *oriT*
initiates transfer
B
oriT
The transferred fragment includes
part of F attached to chromosomal
DNA of the Hfr strain
C
The transferred fragment undergoes
recombination in the recipient
D
F
Hfr
F⁻

Figure 3.34 Transfer of chromosomal DNA by an integrated plasmid. Formation of mating pairs **(A)**, nicking of the F *oriT* sequence **(B)**, and transfer of the 5′ end of a single strand of DNA **(C)** proceed as in transfer of the plasmid. Transfer of the covalently linked chromosomal DNA also occurs as long as the mating pair is stable. Because complete chromosome transfer rarely occurs, most recipient cells remain F⁻, even after long matings. Replication in the donor usually accompanies DNA transfer. Some replication of the transferred single strand may also occur. Once in the recipient cell, the transferred DNA may recombine with homologous sequences in the recipient chromosome **(D)**.

Mapping by Hfr crosses requires a genetic map of the bacterium being mapped, and genetic maps have been constructed for only a few model bacteria, including *E. coli*, *Salmonella*, and *S. coelicolor*. Figure 3.35 shows a partial genetic map of *E. coli* and the location and direction of transfer of some of the integrated F plasmids in the Hfr strains that were used to construct the map. By convention, the direction of transfer for each Hfr strain is indicated by the point of the arrow, and the donor chromosome enters the recipient cell like an arrow starting at this point. Hfr crosses are seldom used anymore to map mutations on the chromosome, even if a genetic map is available for the bacterium being studied, since it is easier to use the DNA sequence of the entire bacterial genome and map mutations on the genome sequence by techniques such as marker rescue with clones or with PCR-generated fragments. Nevertheless, it is worth summarizing the procedures that have been used to construct genetic maps using Hfr crosses.

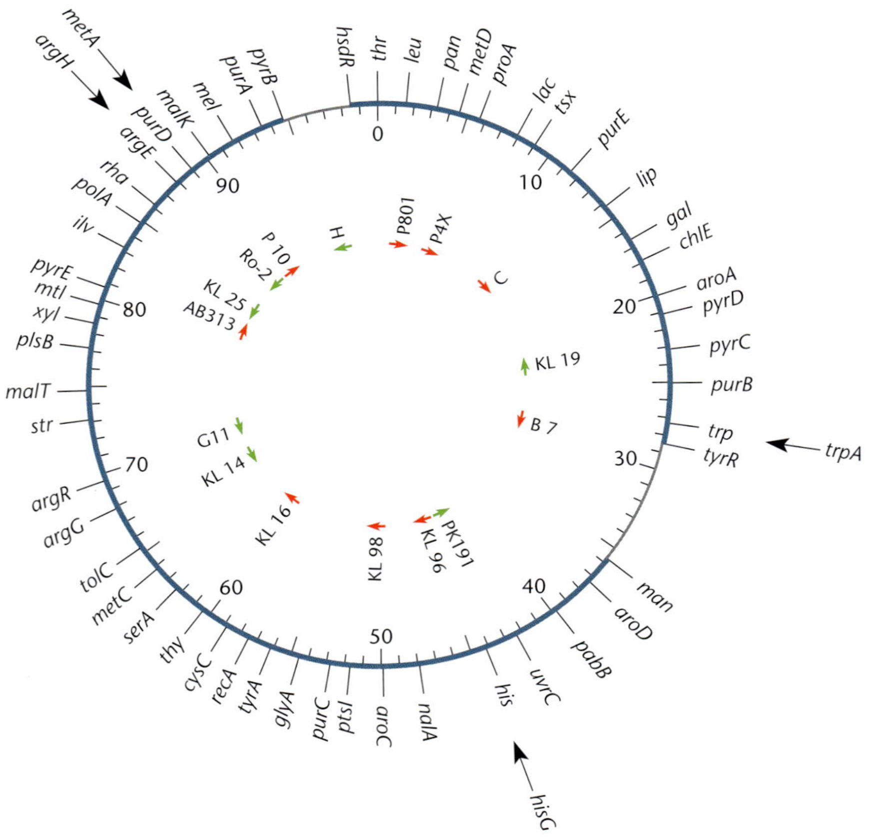

Figure 3.35 Partial genetic linkage map of *E. coli* showing the positions (black arrows) of the markers used for the Hfr gradient of transfer in Figure 3.37. The red (clockwise) and green (counterclockwise) arrows indicate the positions of integration and directions of transfer of the F plasmid in some Hfr strains, including PK191 (located near the position of *hisG* at 44 min and transferring counterclockwise). By convention, the chromosomal DNA is shown transferred like an arrow being shot into the recipient cell, beginning from the tip of the arrow. Modified from Bachmann BJ, Low B, Taylor AL, *Bacteriol Rev* **40:**116–167, 1976.

MAPPING BY GRADIENT OF TRANSFER

Mapping by Hfr crosses is like mapping by transformation and transduction in that only part of the donor strain is transferred into the recipient, and recombination can occur only in the recipient strain. Also, one marker is selected, and recombinants for that marker are tested to see if they are also recombinant for the other markers. However, we must select a marker that is transferred before the other markers, and the data are interpreted differently. Rather than depending on the fact that only a small piece of DNA is transferred, genetic mapping by Hfr crosses depends on the fact that the DNA is transferred from the *oriT* on the integrated plasmid and the transfer is periodically disrupted. Therefore, the further a marker is from the integrated plasmid in the direction of transfer, the less frequently it is transferred, creating a **gradient of transfer**. Furthermore, there is an equal probability that each surviving mating pair will be disrupted in the next time interval. This creates an exponential "decay" in the transfer of markers based on how far they are from the origin of transfer, so if we plot the frequencies of recombination of known markers versus their distances from the site of a selected marker in the direction of transfer on semilog paper, we get a more or less straight standard curve or line. If we then put the frequencies of recombinant types for an unmapped marker on this standard curve and read down to the distance from the selected marker, we will have determined the distance of the unknown marker from the selected point on the genome.

To show how this works in practice, Table 3.6 lists some data for a typical Hfr cross, and Figures 3.36 and 3.37 illustrate how these data are analyzed. In this example, we used the PK191 Hfr strain shown in the genetic map in Figure 3.35, which transfers counterclockwise from about 42 minutes to map the *rif* genetic marker, which, as mentioned above, confers resistance to the antibiotic rifampin (Rif^r). As our known markers, we use the *hisG*, *argH*, and *trpA* mutations, all of which are in the

Table 3.6 Typical results of an Hfr cross

Selected marker	% Recombinant for unselected marker:			
	hisG	***trpA***	***argH***	***rif***
hisG	100	1	7	6
trpA	33	100	29	31
argH	28	12	100	89

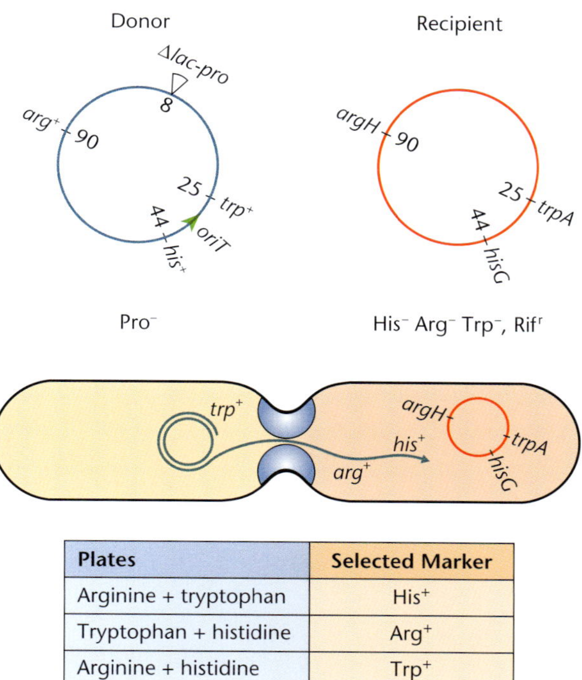

Plates	Selected Marker
Arginine + tryptophan	His⁺
Tryptophan + histidine	Arg⁺
Arginine + histidine	Trp⁺

Figure 3.36 Mapping by Hfr crosses. The phenotypes and positions of the markers used for mapping in the genetic maps of the donor and recipient bacteria are shown. The chromosome is transferred from the donor to the recipient, starting at the position of the integrated self-transmissible plasmid in PK191 (arrowhead). The supplements to the plating media used to select and test for recombinants of each of the markers are also shown.

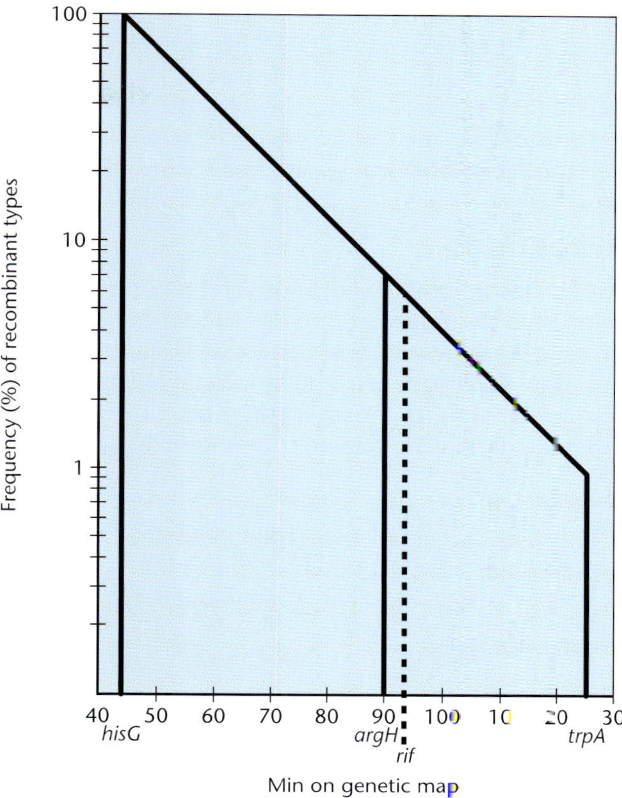

Figure 3.37 Mapping by gradient of transfer during an Hfr cross. The ordinate shows the frequency of each unselected marker, with *his* as the selected marker. The abscissa is the distance in minutes of each unselected marker from the *his* marker from the map in Figure 3.35. The dashed line shows an estimate of the position of the *rif* marker based on plotting the percentage of Rifˢ recombinants from the data in Table 3.6 and reading down to the abscissa.

recipient, as is the unknown *rif* mutation. In an Hfr cross, the Hfr donor strain must be **counterselected** by plating the cross on selective medium on which the donor cannot grow; otherwise, the donor would grow on the plates and obscure any recombinants. We use the fact that the donor is Pro⁻ and requires proline for growth, due to the *lac-pro* deletion shown in Figure 3.36, to **counterselect** it. The partial *E. coli* genetic map in Figure 3.35 shows the positions on the chromosome of all of the markers we are using except the *rif* marker, which we are trying to map. We first select the *his* marker, since that is transferred first by this Hfr strain, followed by the *argH* marker and then the *trp* marker, as shown in Figure 3.36. We then test recombinants for the *hisG* marker (His⁺) to see what percentage are also recombinant for the *argH* marker (Arg⁺) and what percentage are also recombinant for the *trpA* marker (Trp⁺). If we take the frequency of recombinants for the *his* marker to be 100% and plot the frequency with which these recombinants have also become recombinant for each of the other markers versus their distances on the genetic map from the *his* marker, we get a more or less straight line, as illustrated in Figure 3.37. The frequency of recombinants for the *rif* marker (i.e., the frequency of His⁺ recombinants that are also Rifˢ) is then plotted on the standard curve, reading down to its position on the genetic map. As expected, it maps close to the *argH* marker, since we had determined earlier that it is cotransducible with *argH*.

We can also see from the data in Table 3.6 what would have happened if we selected one of the other markers, for example, the *trpA* marker that came in after the others in this particular Hfr cross. Each bacterium that became recombinant for the *trpA* marker must have also received the regions of the other markers, since they were transferred before the *trpA* region (Figure 3.36). Now, approximately 30% of the other markers are recombinant, independent of where they are on the genome, which gives us little specific information about their map positions. Apparently, 30% is the frequency with which a piece of DNA will recombine with the chromosome once it enters the cell. If we selected the *argH* marker, a very high percentage (89%) are recombinant for the *rif* marker, because they are close to each other or closely genetically **linked**. The distances on the DNA between most of these markers is so great that many crossovers occur between them. Only if they are very close, as is the case with the *argH* and *rif* markers, are crossovers between them infrequent enough that we can see any genetic linkage.

Perspective

Genetic analysis provided the powerful tool that allowed for most of the information that you will read in this textbook. While many of the landmark discoveries in all of biology were made with these tools, many are still in use today in laboratories across the world. Examples are given in each chapter of how relatively simple experiments paired with clear logic can pick apart even extremely complicated systems. Considering these examples will hopefully provide inspiration for addressing your own questions in the laboratory. New sets of tools are also being developed and are the focus of chapter 13, which is primarily dedicated to genomics, a term generally used to describe techniques that are only made possible by knowing the entire DNA sequence of an organism's genome. As described throughout this chapter and the textbook, genetic tools identified in model bacterial systems have also been adapted to provide the foundation for molecular approaches in all other organisms, including higher eukaryotes. Understanding how these tools were developed, and how they have been applied, makes it much easier to appreciate current and future applications.

1. A mutation is any heritable change in the sequence of the DNA of an organism. A leaky mutation in a coding part of the DNA leaves some residual activity of the gene product; a null mutation totally inactivates the product of the gene. A gain-of-function mutation gives the gene product an activity it did not have previously or leads to the expression of a silent gene. The organism with a mutation is called a mutant, and that organism's mutant phenotype includes all of the characteristics of the mutant organism that are different from the wild-type, or normal, organism.

2. The mutation rate is the chance of occurrence of a mutation to a particular phenotype each time a cell divides. The mutation rate offers clues to the molecular basis of the phenotype, e.g., whether it can be caused by a null mutation or only by a gain-of-function mutation. Quantitative determination of mutation rates is often difficult, because it is necessary to determine how many mutations have occurred in a growing culture, which is usually not equal to the number of mutants in the culture. One mutation can give rise to a large number of mutants, depending on when it occurred in the growth of the culture. Also, most ways of determining mutation rates are influenced by phenotypic lag and/or the relative growth rates of mutants.

3. One important conclusion from population genetics is that the fraction of mutants of all types increases as the population grows. This causes practical problems in genetics, which can be partially overcome by storing organisms without growth or by periodically colony purifying one or very few organisms before mutations have had time to accumulate.

4. The type of mutation causing a phenotype can often be ascertained from the properties of the mutation. Base pair changes revert and are often leaky. Frameshift mutations, due to the addition or deletion of one or very few base pairs, also revert but are seldom leaky. Longer deletion mutations do not revert and are seldom leaky. They also often inactivate more than one gene simultaneously and can fuse one gene to another. Tandem-duplication mutations revert at a high frequency and often have no observable phenotypes, except that they can fuse one gene to another or cause phenotypes due to the overproduction of the product of a gene included in the duplication. Inversion mutations also often have no observable phenotype and often revert. Large insertion mutations seldom revert, are usually not leaky, and are usually due to the transposition of an insertion element or other type of transposon.

5. Deletions, inversions, and tandem-duplication mutations can be caused by recombination between repeated sequences in the DNA called ectopic recombination. Deletion and tandem-duplication mutations are caused by recombination between directly repeated sequences in the DNA, and inversions are caused by recombination between inverted repeats. Deletions and tandem duplications arise as reciprocal recombination products between directly repeated sequences in different DNA molecules. Sometimes, deletions and inversions are caused by site-specific recombinases. If so, they are referred to not as mutations, but as programmed deletion or inversion events.

6. If a secondary mutation restores the original phenotype by returning the DNA to its original sequence, the original muta-

Summary (continued)

tion has reverted. If a secondary mutation somewhere else in the DNA restores function, the original mutation has been suppressed. Suppressors can be either intragenic or intergenic, depending on whether they occur in the same gene or in a different gene from the original mutation, respectively. An example of an intragenic suppressor is a frameshift mutation within the same gene that restores the reading frame shifted by another frameshift mutation. An example of an intergenic suppressor is a mutation in a tRNA gene that changes the tRNA so that it recognizes one or more of the nonsense codons, allowing translation of nonsense mutations in other genes. Suppressors can also be allele specific if they suppress only a particular type of mutation, for example, an amber suppressor, which suppresses only UAG mutations, independent of where they occur.

7. To isolate a mutant means to separate the mutant strain from the many other members of the population that are normal for the phenotype. Bacteria have advantages in genetic analysis, because of the ease of isolating mutants. They multiply asexually, they are usually haploid, and large numbers can multiply on a single petri plate.

8. Mutations can be either spontaneous or induced. Spontaneous mutations often occur as mistakes while the DNA is replicating, while induced mutations are deliberately caused by using mutagenic chemicals, irradiation, or transposons. Induction of mutations with mutagens has the advantage that mutations are more frequent and that a specific mutagen often causes only a specific type of mutation. Region-specific mutagenesis can be used to mutate a particular region of the chromosome if the sequence is known. Recombination can be used to replace chromosomal sequences with sequences manipulated in the test tube. In the types of bacteria for which they have been developed, recombineering systems, which use the recombination enzymes of phages to increase recombination frequencies, can be used for region-specific mutagenesis.

9. Screening for a mutant means devising a way to distinguish the mutant from the normal, or wild, type. Selecting a mutant means devising conditions under which either only the mutant or the wild type can multiply. Most types of mutants cannot be selected. However, enrichment can sometimes be used to increase the frequency of the mutant in a population by killing the wild-type multiplying cells under the nonpermissive conditions, provided the arrest of growth of the mutant is reversible.

10. The three means of genetic exchange in bacteria are transformation (electroporation), transduction, and conjugation. In transformation and electroporation, only DNA is taken up by cells. In transduction, bacterial DNA is encapsulated in a phage head during multiplication of the phage and then injected into another bacterium upon reinfection. In conjugation, a self-transmissible plasmid transfers DNA from one bacterium to another. Sometimes, bacterial DNA can also be transferred by conjugation if the plasmid has taken up the bacterial DNA or if the plasmid has integrated into the chromosome (an Hfr strain). Gene exchange makes it possible to do genetic mapping and other types of genetic analysis with bacteria. Because only a small piece of DNA is transferred by transformation and transduction, these methods are useful for constructing isogenic strains that differ at only one locus and distinguishing suppressors from revertants. Hfr mapping has been used to locate a mutation on the entire chromosome where a genetic map is available but has largely been replaced by other methods.

11. In generalized, or homologous, recombination, two DNAs with the same sequence will recombine following spontaneous breaks in one of the DNAs, which then invades the homolog to rejoin in new combinations. Progeny organisms that are different genetically from their parents as a result of recombination are called recombinant types, while progeny that are the same as one of the parents are parental types. Marker rescue recombination can be used to tell if a mutation lies within a particular region of the chromosome or whether a clone includes the region of a mutation. Recombination frequencies can also be used to map genetic markers on the chromosome using the various means of genetic exchange. Recombinants for one of the markers (the selected marker) are tested for recombination of the other markers (the unselected markers). After mating, a marker in the recipient that is replaced by the corresponding sequence of the donor is said to be recombinant for that marker. The frequency of recombination for the unselected markers can give us information about where the markers are on the chromosome.

12. In complementation, a gene product encoded by one copy of a gene in the cell can substitute for the product of another copy of the same gene that has been inactivated by a mutation. Complementation tests reveal whether different mutations inactivate different gene products and how many separate gene products contribute to a phenotype. It can also be used to clone genes, to identify regions of genes encoding separate domains of a protein, and to determine if a mutation is dominant or recessive or is *cis* acting or *trans* acting.

QUESTIONS FOR THOUGHT

1. A single-inversion mutation greatly alters a genetic map or the order of genes in the DNA, yet the genetic maps of *Salmonella enterica* serovar Typhimurium and *E. coli* are similar. Can you think of reasons why genetic maps might be preserved in evolution?

2. Luria and Delbrück showed that mutations to T1 phage resistance in *E. coli* were not directed and occurred randomly while the bacteria were multiplying. However, this does not mean that all types of mutations are random. Can you propose a mechanism by which "directed" mutations might occur?

3. Nonsense suppressors are possible in single-cell organisms, but not in multicellular organisms. One possible explanation is that readthrough of nonsense codons at the end of coding sequences might be too much of a burden for higher organisms. Can you think of any other possible explanations?

SUGGESTED READING

Baba T, Ara T, Hasegawa M, Takai Y, Okumura Y, Baba M, Datsenko KA, Tomita M, Wanner BL, Mori H. 2006. Construction of *Escherichia coli* K-12 in-frame, single-gene knockout mutants: the Keio collection. *Mol Syst Biol* **2:**0008.

Griffith KL, Grossman AD. 2008. Inducible protein degradation in *Bacillus subtilis* using heterologous peptide tags and adaptor proteins to target substrates to the protease ClpXP. *Mol Microbiol* **70:**1012–1025.

Jones ME, Thomas SM, Clarke K. 1999. The application of a linear algebra to the analysis of mutation rates. *J Theor Biol* **199:**11–23.

Lea DE, Coulson CA. 1949. The distribution of the numbers of mutants in bacterial populations. *J Genet* **49:**264–285.

Lederberg J, Lederberg EM. 1952. Replica plating and indirect selection of bacterial mutants. *J Bacteriol* **63:**399–406.

Lederberg J, Tatum EL. 1946. Gene recombination in *Escherichia coli*. *Nature* **158:**558.

Luria SE, Delbrück M. 1943. Mutations of bacteria from virus sensitivity to virus resistance. *Genetics* **28:**491–511.

McGinness KE, Baker TA, Sauer RT. 2006. Engineering controllable protein degradation. *Mol Cell* **22:**701–707.

Newcombe HB. 1949. Origin of bacterial variants. *Nature* **164:**150–151.

Shuman HA, Silhavy TJ. 2003. The art and design of genetic screens: *Escherichia coli*. *Nat Rev Genet* **4:**419–431.

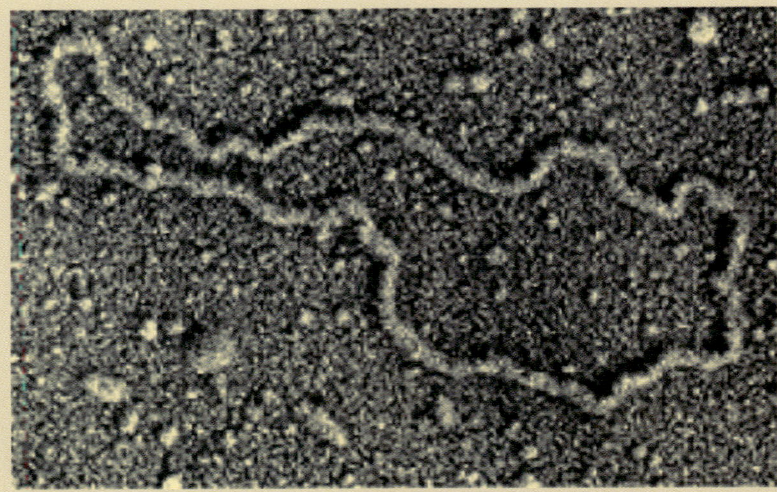

Electron photomicrograph of the first plasmid used for DNA cloning, pSC101 (see Cohen SN, Suggested Reading).

4

Plasmids

What Is a Plasmid?

Bacterial cells often contain **plasmids**, in addition to the chromosome. Plasmid DNA molecules are found in essentially all types of bacteria and, as discussed below, play a significant role in bacterial adaptation and evolution. They also serve as important tools in studies of molecular biology. We address such uses later in this chapter.

Plasmids, which vary widely in size from a few thousand to hundreds of thousands of base pairs (bp) (a size comparable to that of the bacterial chromosome), are most often circular molecules of double-stranded DNA but can also be linear. The number of copies per cell also varies among plasmids, and bacterial cells can harbor more than one type of plasmid at a time. Thus, a cell can harbor two or more different types of plasmids, with hundreds of copies of some plasmid types and only one or a few copies of other types.

Like chromosomes, plasmids encode proteins and RNA molecules and replicate as the cell grows, and the replicated copies are usually distributed into each daughter cell when the cell divides. Plasmids even share some of the same types of functions for accurate partitioning (Par functions) and site-specific recombinases with the host chromosome (see below). By one definition, any independently replicating element in the cell that does not contain genes essential for bacterial growth (the so-called housekeeping genes) is called a plasmid. Plasmids probably persist because they very often provide gene products that can benefit the bacterium under certain circumstances. Consequently, isolates of bacteria taken from the environment often will lose some or all plasmids over time when cultured in the laboratory. There are a number of examples where a plasmid has taken on many of the attributes of a chromosome, such as larger size and encoding multiple housekeeping genes. For example, the pSymB plasmid of some *Sinorhizobium* species is about half as big as the chromosome and carries essential genes, including a gene for an arginine tRNA and the *minCDE* genes involved in division site selection. Also, *Vibrio cholerae* has two large DNA molecules, both of which carry essential genes. *Agrobacterium tumefaciens* has two large DNA molecules, one circular and the other linear, both of which carry essential genes. In cases where a plasmid is almost as big as a chromosome and carries essential genes, which one is the chromosome and which is a plasmid? In all known cases, one of the large DNA molecules has a typical bacterial origin of replication with an *oriC* site and closely linked *dnaA*, *dnaN*, and *gyrA* genes (among others), while the other DNA molecule has a typical plasmid origin with *repABC*-like genes more characteristic of plasmids.

Naming Plasmids

Before methods for physical detection of plasmids became available, plasmids made their presence known by conferring phenotypes on the cells that harbor them. Consequently, many plasmids were named after the genes they carry. For example, R-factor plasmids contain genes for resistance to several antibiotics (hence the name R, for resistance). These were the first plasmids discovered, when *Shigella* and *Escherichia coli* strains resistant to a number of antibiotics were isolated from the fecal microbiota of patients in Japan in the late 1950s. The ColE1 plasmid, from which many of the commonly used cloning vectors were derived, carries a gene for the protein colicin E1, a bacteriocin that kills bacteria that do not carry the plasmid. The Tol plasmid contains genes for the degradation of toluene, and the Ti plasmid of *A. tumefaciens* carries genes for tumor induction in plants (see Box 5.3). This system of nomenclature has led to some confusion, because plasmids carry various genes besides the ones for which they were originally named and because similar plasmids can contain very different sets of genes. Many plasmids have also been altered beyond recognition in the laboratory to make plasmid cloning vectors (see below) and for other purposes.

To avoid further confusion, the naming of plasmids is now standardized. Plasmids are given number-and-letter names much like bacterial strains. A small "p," for plasmid, precedes capital letters that describe the plasmid or sometimes give the initials of the person or persons who isolated or constructed it. These letters are often followed by numbers to identify the particular construct. When the plasmid is further altered, a different number is assigned to indicate the change. For example, plasmid pBR322 was constructed by Bolivar and Rodriguez from the ColE1 plasmid and is derivative number 322 of the plasmids they constructed. pBR325 is pBR322 with a chloramphenicol resistance gene inserted. The new number, 325, distinguishes this plasmid from pBR322.

Functions Encoded by Plasmids

Plasmids can encode a few or hundreds of different proteins, resulting in vast differences in their sizes. However, as mentioned above, plasmids rarely encode gene products that are always essential for growth, such as RNA polymerase, ribosomal subunits, or enzymes of the tricarboxylic acid cycle. Instead, plasmid genes usually give bacteria a selective advantage under only some conditions.

Table 4.1 lists a few naturally occurring plasmids and some traits they encode, as well as the host in which they were originally found. As mentioned above, gene products encoded by plasmids include enzymes for the utilization of unusual carbon sources, such as toluene, resistance to substances such as heavy metals and antibiotics, synthesis of antibiotics, and synthesis of toxins and proteins that allow the successful infection of higher organisms. Plasmids, combined with their hosts, have been an invaluable tool for investigation of other organisms. Table 4.2 lists the major classes of plasmids used in *E. coli* and some of their relevant features for molecular genetics.

Table 4.1 Some naturally occurring plasmids and the traits they carry

Plasmid	Trait	Original source
ColE1	Bacteriocin that kills *E. coli*	*E. coli*
Tol	Degradation of toluene and benzoic acid	*Pseudomonas putida*
Ti	Tumor initiation in plants	*A. tumefaciens*
pJP4	2,4-D (2,4-dichlorophenoxyacetic acid) degradation	*Alcaligenes eutrophus*
pSym	Nodulation on roots of legume plants	*Sinorhizobium meliloti*
SCP1	Antibiotic methylenomycin biosynthesis	*Streptomyces coelicolor*
RK2	Resistance to ampicillin, tetracycline, and kanamycin	*Klebsiella aerogenes*

It is interesting to speculate about why certain types of genes are found on plasmids and others are found on the chromosome. It is easy to understand why genes that directly favor the plasmid would be encoded on the element, and this chapter goes into the details of some of these systems (see Thomas, Suggested Reading). However, there are many genes that favor the host and plasmid equally, and it is curious that certain broad classes of genes are normally encoded on the chromosome while others are on plasmids. One idea holds that plasmids tend to harbor genes that are locally advantageous (see Eberhard, Suggested Reading). For example, genes that encode such things as heavy metal resistance or antibiotic resistance may be advantageous only in certain transient situations in certain specific places. This selection might be sufficient to allow them to be maintained in some hosts in the environment even though selection is not pervasive enough to allow them to become associated with a lineage of bacteria. It would be impossible for every bacterium to maintain genes that could be advantageous for every environment. However, there is a strong selective advantage for plasmids to accumulate in hosts in these environments. Across nature, plasmids allow bacteria to occupy a larger variety of ecological niches. This also explains why a large number of plasmids exist with the capacity to direct their own transfer between bacteria in a process called conjugation (see chapter 5) and

Table 4.2 General classes of plasmids commonly used in *E. coli*

Founding replicon	Common examples	Host range	Comments
pMB1/ColEI	pBR322, pUC vectors, pGEM vectors, pBluescript vectors	Narrow	pBR322 is a low-copy-number vector (~20 copies/cell) with derivatives that are very-high-copy-number vectors (>300 copies/cell).
p15A	pACYC177, pACYC184	Narrow	The pACYC vectors are low-copy-number vectors (~15/cell). p15A is similar to but compatible with the pMB1/ColEI replicon.
pSC101	pSC101	Narrow	Low-copy-number vector (~5 copies/cell); good for toxic genes; temperature-sensitive derivatives exist.
F plasmid	pBeloBAC11	Narrow	The original fertility (F) plasmid; the replication origin is utilized in BACs.
RK2 (RP4)	pSP329, pCM62, pCM66	Broad	IncP group
RSF1010	pJRD215, pSUP104, pSUP204	Broad	IncQ group
pSa	pUCD2	Broad	IncW group
R6K	R6K	Broad	IncX group
pBBR1MCS	pBBR1MCS-2, pBBR1MCS-3, etc.	Broad	Undefined Inc group

why many plasmids are capable of replicating in different types of bacteria. As we will see below, some plasmids are also likely to be maintained not because of the benefits they provide the host bacterium, but because the plasmid has a system to actively harm hosts that lose the plasmid (see "Toxin-Antitoxin Systems and Plasmid Maintenance" below).

Plasmid Structure

Most plasmids are circular, although linear plasmids also exist. In a circular plasmid, all of the nucleotides in each strand are joined to another nucleotide on each side by covalent bonds to form continuous strands that are wrapped around each other. Such DNAs are said to be **covalently closed circular**, and because there are no ends to rotate, the plasmid can maintain supercoiled stress. As discussed in chapter 1, nonsupercoiled DNA strands exhibit a helical periodicity of about 10.5 bp, as predicted from the Watson-Crick double-helical structure of DNA. In contrast, in a DNA that is supercoiled, the two strands are wrapped around each other either more or less often than once in about 10.5 bp. If they are wrapped around each other more often than once every 10.5 bp, the DNA is positively supercoiled; if they are wrapped around each other less often, the DNA is negatively supercoiled. Like the chromosome, covalently closed circular plasmid DNAs are usually negatively supercoiled (see chapter 1). Because DNA is stiff, the negative supercoiling introduces stress, and this stress is partially relieved by the plasmid wrapping up on itself, as illustrated in Figure 4.1A. This makes the plasmid more compact, so that it migrates more quickly in an agarose gel (Figure 4.1B). In the cell, the DNA wraps around proteins, which relieves some of the stress. The remaining stress facilitates some reactions involving the plasmid,

Figure 4.1 Supercoiling of a covalently closed circular plasmid. **(A)** A break in one strand relaxes the DNA, eliminating the supercoiling and making the DNA less compact. **(B)** Schematic diagram of an agarose gel showing that the covalently closed supercoiled circles run faster on a gel than the nicked relaxed circles. Depending on the conditions, linear DNA and covalently closed circular (CCC) DNA run in approximately the same position as nicked relaxed circles of the same length. The arrow shows the direction of migration.

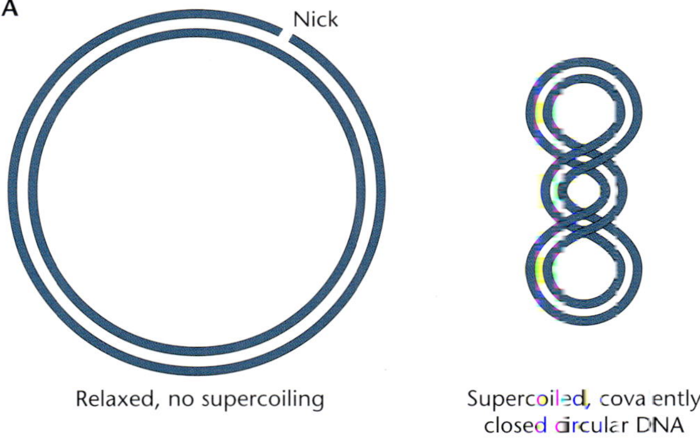

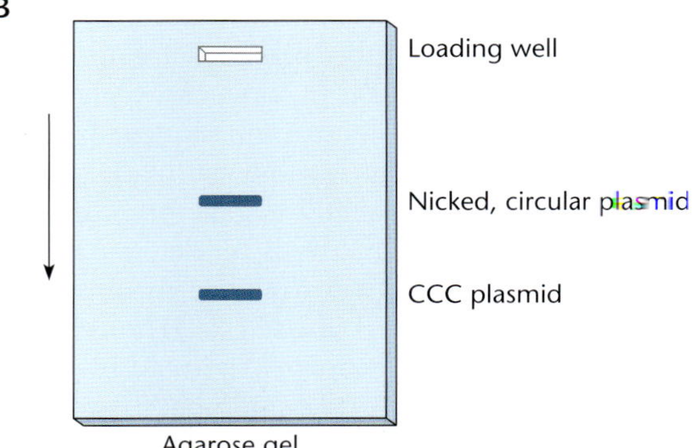

such as separation of the two DNA strands for replication or transcription.

PURIFYING PLASMIDS

The structure of plasmids can be used to purify them away from chromosomal DNA in the cell. Cloning manuals give detailed protocols for these methods (see chapter 1, Suggested Reading), but we review them briefly here. Many purification procedures are based on the relatively small size of most plasmids. Bacterial cells are usually lysed with a combination of a strong base (sodium hydroxide), the detergent sodium dodecyl sulfate (SDS) to solubilize the membranes, and a component to denature proteins. Following this treatment, a high-salt (potassium acetate) solution is added. Under these conditions, the chromosome precipitates with the cell debris and SDS upon centrifugation, while the small supercoiled plasmids remain in solution. Following these steps, a variety of procedures are used to further purify the plasmid DNA from the proteins and to concentrate the plasmid preparation. Historically, a solution of phenol and chloroform was added to the solution to separate the DNA from the proteins. After mixing of the aqueous lysate with phenol and chloroform, the components quickly separate into phases, similar in concept to oil and water. Plasmid DNA is separated from the proteins because the polar DNA molecules remain in the aqueous phase while the proteins remain at the interface of the solutions. Following extraction, plasmid DNA is concentrated using precipitation with a salt and alcohol solution and finally resuspended in water or a buffered solution. For convenience today, commercially available spin columns are usually used instead of separation using the phenol-chloroform solution. The plasmid solution is added to a resin matrix in a column, which binds DNA. An alcohol solution is used to wash away the residual proteins and salts from the column, and the plasmid DNA is finally eluted from the column using water or a buffered solution. The isolated plasmid DNA can then be visualized on a gel, digested with a restriction endonuclease, or sequenced, among other applications (see chapter 13).

The methods discussed above work well with plasmids that have many copies per cell and are not too large. However, large, low-copy-number plasmids are more difficult to isolate or even to detect. In some cases, plasmids have been discovered by whole-genome sequencing of a bacterium when a series of DNA sequence reads assemble into a unique contiguous DNA structure separate from the host genome. Visual methods for detecting large plasmids involve separating them from the chromosome directly by electrophoresis on agarose gels. The cells are often lysed directly on the agarose gel to avoid breaking the large plasmid DNA. The plasmid, because of its unique size, makes a sharp band on the gel, distinct from that due to chromosomal DNA, which is usually fragmented randomly and so appears as a diffuse smear of molecules of varying sizes. Also, methods such as pulsed-field gel electrophoresis have been devised to allow the separation of long pieces of DNA based on size. These methods depend on periodic changes in the direction of the electric field. The molecules attempt to reorient themselves each time the field shifts, and the longer molecules take longer to reorient than the shorter ones and thus move more slowly through the gel. Such methods have allowed the separation of DNA molecules hundreds of thousands of base pairs long and the detection of very large plasmids.

Properties of Plasmids

Replication

To exist free of the chromosome, plasmids must have the ability to replicate independently. DNA molecules that can replicate autonomously in the cell are called **replicons**. Plasmids, phage DNA, and the chromosomes are all replicons, at least in some types of cells.

To be a replicon in a particular type of cell, a DNA molecule must have at least one origin of replication, or *ori* site, where replication begins (see chapter 1). In addition, the cell must contain the proteins that enable replication to initiate at this site. Plasmids encode only a few of the proteins required for their own replication. In fact, many encode only one of the proteins needed for initiation at the *ori* site. All of the other required proteins—DNA polymerases, ligases, primases, helicases, and so on—are borrowed from the host.

Each type of plasmid replicates by one of two general mechanisms, which is determined along with other properties by its *ori* region (see "Functions of the *ori* Region" below). The plasmid replication origin is often named *oriV* for *ori* vegetative, to distinguish it from *oriT*, which is the site at which DNA transfer initiates in plasmid conjugation (see chapter 5). Most of the evidence for the mechanisms described below came from electron microscope observations of replicating plasmid DNA.

THETA REPLICATION

Some plasmids begin replication by opening the two strands of DNA at the *ori* region, creating a structure that looks like the Greek letter θ—hence the name **theta replication** (Figure 4.2A and B). In this process, an RNA primer begins replication, which can proceed in one or both directions around the plasmid. In the first case (unidirectional replication), a single replication fork moves around the molecule until it returns to the origin, and then the two daughter DNAs separate. In the other case (bidirectional replication), two replication forks move out from the *ori* region, one in each direction, and replication is complete (and the two daughter DNAs separate) when the two forks meet somewhere on the other side of the molecule. This is the type of replication used by circular bacterial chromosomes and also plasmids commonly used in *E. coli* and other *Proteobacteria*, including ColE1, RK2, and F, as well as the bacteriophage P1.

ROLLING-CIRCLE REPLICATION

Other types of plasmids replicate by very different mechanisms. One of these mechanisms is called **rolling-circle (RC) replication** because it was first discovered in a type of phage where the template circle seems to "roll" while a copy of the plasmid is made and processed to unit length. Plasmids that replicate by this mechanism are sometimes called **RC plasmids**. This type of plasmid is widespread and is found in the largest groups of bacteria, as well as archaea.

In an RC plasmid, replication occurs in two stages. In the first stage, the double-stranded circular plasmid DNA replicates to form another double-stranded circular DNA and a single-stranded circular DNA. This stage is analogous to the replication of the DNA of some single-stranded DNA phages (see chapter 7) and to DNA transfer during plasmid conjugation (see chapter 5). In the second stage, the complementary strand is synthesized using the single-stranded DNA as a template to make another double-stranded DNA.

The details of the RC mechanism of plasmid replication are shown in Figure 4.2C. First, the Rep protein recognizes and binds to a palindromic sequence that contains the double-strand origin (DSO) on the DNA. Binding of the Rep protein to this sequence might allow the formation of a cruciform structure by base pairing between the inverted-repeat sequences in the cruciform. Once the cruciform forms, the Rep protein can make a nick in the sequence. It is important for the models that the Rep protein is also known to function as a dimer, at least in some plasmids. After the Rep protein has made a break in the DSO sequence, it remains covalently attached to the phosphate at the 5′ end of the DNA at the nick through a tyrosine in one copy of the Rep protein in the dimer, as shown. DNA polymerase III (the replicative polymerase [see chapter 1]) uses the free 3′ hydroxyl end at the break as a primer to replicate around the circle, displacing one of the strands. It may use a host helicase to help separate the strands, or the Rep protein itself may have the helicase activity, depending on the plasmid. Once the circle is complete, the 5′ phosphate is transferred from the tyrosine on the Rep protein to the 3′ hydroxyl on the other end of the displaced strand, producing a single-stranded circular DNA. This process is called a phosphotransferase reaction and requires little energy. The same reaction is used to re-form a circular plasmid after conjugational transfer (see chapter 5).

It is less certain what happens to the newly formed double-stranded DNA when the DNA polymerase III has made its way all around the circle and gets back to the site of the DSO. Why does it not just keep going, making a longer molecule with individual genomes linked head to tail in a structure called a concatemer? Such structures are created when some phage and plasmid DNAs replicate by an RC mechanism (see chapter 7). One idea is that the DNA polymerase III does keep going past the DSO for a short distance, creating another double-stranded DSO. The other copy of the Rep protein in the dimer may then nick the newly created DSO, transferring the 5′ end to itself as described above. This might inactivate the Rep protein, releasing it with a short oligonucleotide attached. Other reactions, probably involving host DNA ligase, then cause the nick to be resealed, resulting in a circular double-stranded DNA molecule.

The displaced circular single-stranded DNA now replicates by a completely different mechanism using only host-encoded proteins. The RNA polymerase first makes a primer at a different origin, the single-strand origin (SSO), and this RNA then primes replication around the circle by DNA polymerase III. However, the RNA polymerase does not make this primer until the single-stranded DNA is completely displaced during the first stage of replication. This delay is accomplished by locating the SSO immediately counterclockwise from the DSO (Figure 4.2C), so that the SSO does not appear in the displaced DNA until the displacement of the single-stranded DNA is almost complete. After the entire complementary strand has been synthesized, the DNA polymerase I removes the RNA primer with its flap exonuclease activity while simultaneously replacing it with DNA, and host DNA ligase joins the ends to make another double-stranded plasmid. The net result is two new double-stranded plasmids synthesized from the original double-stranded plasmid.

In order for the complementary strand of the displaced single-stranded DNA to be synthesized, the RNA polymerase of the host cell must recognize the SSO on the DNA. In some hosts, the SSO is not well recognized, and single-stranded DNA accumulates. For this reason, some RC plasmids were originally called single-stranded DNA plasmids, although we now know that this is not their normal state. RC plasmids capable of replication in a wide variety of bacteria presumably have an SSO that is recognized by the RNA polymerases of these hosts, which allows them to make the complementary strand of the displaced single-stranded DNA in a variety of hosts.

The Rep protein is used only once for every round of plasmid DNA replication and is destroyed after the round is completed. This allows the replication of the plasmid to be controlled by the amount of Rep protein in the cell and keeps the total number of plasmid molecules in the

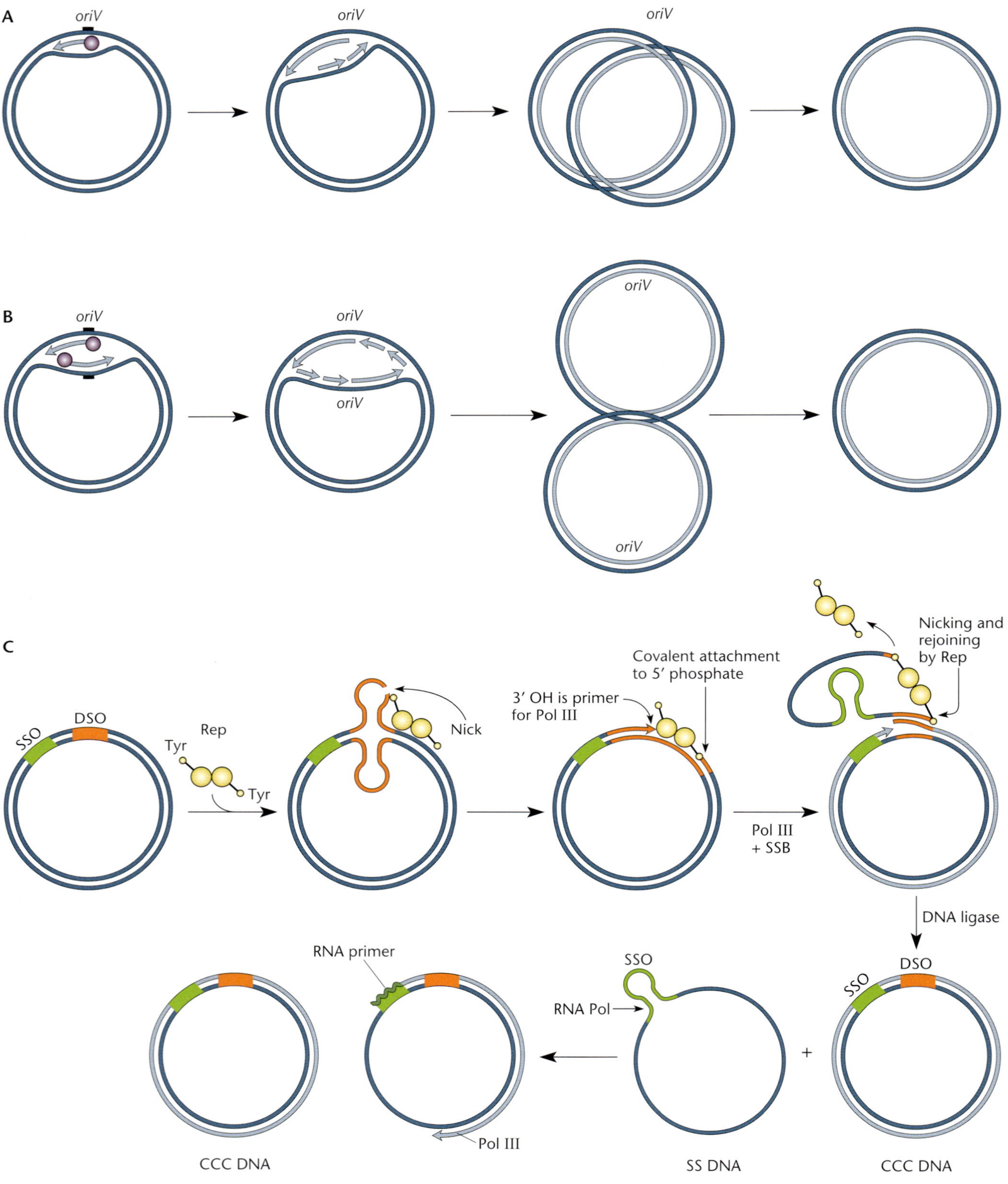
A
oriV
oriV
oriV
B
oriV
oriV
oriV
oriV
oriV
C
SSO
DSO
Tyr
Rep
Tyr
Nick
3′ OH is primer
for Pol III
Covalent attachment
to 5′ phosphate
Nicking and
rejoining
by Rep
Pol III
+ SSB
DNA ligase
RNA primer
SSO
RNA Pol
DSO
SSO
Pol III
CCC DNA
SS DNA
CCC DNA

cell within narrow limits. A little later in this chapter, we discuss how the copy numbers of other types of plasmids are controlled.

REPLICATION OF LINEAR PLASMIDS

As mentioned above, some plasmids and bacterial chromosomes are linear rather than circular (Box 4.1). In general, linear DNAs face two problems in all organisms. One issue with linear DNAs is that the cell must have a way to distinguish the "normal" DNA ends at the ends of the linear fragments from ends formed when DNA double-strand breaks occur, which would otherwise be lethal to the cell and must be repaired quickly. A second problem with linear plasmids and chromosomes has to do with replicating the lagging-strand template, the strand that ends with a 5′ phosphate, all the way to the end of the DNA. This has been called the "primer problem" because DNA polymerases cannot initiate the synthesis of a new strand of DNA. They can only add nucleotides to a preexisting primer, and in a linear DNA, there is no upstream primer on this strand from which to grow. Different linear DNAs solve the primer problem in different ways. Some linear plasmids have hairpin ends, with the 5′ and 3′ ends joined to each other. These plasmids replicate from an internal origin of replication to form dimeric circles composed of two plasmids joined head to tail to form a circle, as shown in the figure in Box 4.1. These dimeric circles are then resolved into individual linear plasmid DNAs with closed hairpins at the ends. The hairpin ends are presumably not recognized as DNA double-strand breaks, because they are not targets for exonucleases in the cell. A completely different mechanism is used to maintain linear plasmids in some systems. With this mechanism, a special enzyme called a terminal protein attaches to the 5′ ends of the plasmid DNA (Box 4.1).

Functions of the *ori* Region

In most plasmids, the genes for proteins required for replication are located very close to the *ori* sequences at which their products act. Thus, only a very small region surrounding the plasmid *ori* site is required for replication. As a consequence, the plasmid still replicates if most of its DNA is removed, provided that the *ori* region remains and the plasmid DNA is still circular. Smaller plasmids are easier to use as cloning vectors, as discussed below, so often the only part of the original plasmid that remains in a cloning vector is the *ori* region.

In addition, the genes in the *ori* region often determine many other properties of the plasmid. Therefore, any DNA molecule with the *ori* region of a particular plasmid will have most of the characteristics of that plasmid. The following sections describe the major plasmid properties determined by the *ori* region.

HOST RANGE

The **host range** of a plasmid includes the types of bacteria in which the plasmid can replicate; it is usually determined by the *ori* region. Some plasmids, such as those with *ori* regions of the ColE1 plasmid type, including pBR322, pET, and pUC, have a **narrow host range** (Table 4.2). These plasmids replicate only in *E. coli* and some other closely related bacteria, such as *Salmonella* and *Klebsiella* species. Work with plasmids has historically been biased by work in *E. coli*, but presumably there are also types of plasmids that will replicate only in other closely related groups of bacteria that would also technically qualify as possessing a narrow host range. In contrast, plasmids with a **broad host range** include the RK2, RSF1010, and pBBR1MCS plasmids, as well as the RC plasmids (Table 4.2). The host ranges of these plasmids are truly remarkable. Plasmids with the *ori* region of RK2 can replicate in most types of *Proteobacteria*, and RSF1010-derived plasmids even replicate in the highly diverged *Firmicutes*. Some plasmids used for expressing genes in various examples from the *Firmicutes* are shown in Table 4.3.

It is perhaps surprising that the same plasmid can replicate in bacteria that are so distantly related to each other. Broad-host-range plasmids must encode all of their own proteins required for initiation of replication, and therefore, they do not have to depend on the host cell for any of

Figure 4.2 Some common schemes of plasmid replication. **(A)** Unidirectional replication. The origin region is designated *oriV*. Replication terminates when the replication fork gets back to the origin. **(B)** Bidirectional replication. Replication terminates when the replication forks meet somewhere on the DNA molecule opposite the origin. **(C)** Rolling-circle (RC) replication. A nick is made at the double-strand origin (DSO) by the plasmid-encoded Rep protein, which remains bound to the 5′ phosphate end at the nick. The free 3′ OH end then serves as a primer for the DNA polymerase III (Pol III) that replicates around the circle, displacing one of the old strands as a single-stranded DNA. The Rep protein then makes another nick, releasing the single-stranded circle, and also joins the ends to form a circle by a phosphotransferase reaction (see the text). The DNA ligase then joins the ends of the new DNA to form a double-stranded circle. The host RNA polymerase makes a primer on the single-stranded DNA origin (SSO), and Pol III replicates the single-stranded (SS) DNA to make another double-stranded circle. DNA Pol I removes the primer, replacing it with DNA, and ligase joins the ends to make another double-stranded circular DNA. CCC, covalently closed circular; SSB, single-strand-DNA-binding protein. Modified from Khan SA, *in* Funnell BE, Phillips GJ (ed), *Plasmid Biology* (ASM Press, Washington, DC, 2004).

BOX 4.1

Linear Chromosomes and Plasmids in Bacteria

Not all bacteria have circular chromosomes and plasmids. Some, including *Borrelia burgdorferi* (the causative agent of Lyme disease), *Streptomyces* spp., *Agrobacterium tumefaciens*, and *Rhodococcus fascians*, have linear chromosomes and often multiple linear plasmids. As mentioned in the text, the replication of the ends of linear DNAs presents special problems because DNA polymerases cannot prime their own replication. This means that they cannot replicate all the way to a 3′ end in a linear DNA. If RNA at the end of a linear DNA primes the last Okazaki fragment and the RNA primer is then removed, there is no upstream primer DNA to prime its replacement with DNA as there is in a circular DNA. Eukaryotic chromosomes, which are linear, solve this problem by having special DNA regions called telomeres at their ends. Most telomeres do not need complementary sequences to be synthesized from the template as during normal DNA replication. Most use an enzyme called telomerase. Telomerase contains an RNA that is complementary to the repeated sequences at the ends of the DNA. This enzyme makes reiterated copies of the repeated telomeric sequences at the ends. When the linear chromosome replicates, some of these repeated sequences at the 3′ end are lost, but this is not a problem, because they will be resynthesized by the telomerase before the DNA replicates again.

Telomeres solve the problem of replicating linear DNAs without losing sequence information. However, bacteria with linear chromosomes are not known to have telomeres made by telomerases and must use different strategies. Two different strategies for dealing with linear chromosomes in bacteria are exemplified in *Streptomyces* and *Borrelia*. In both examples, the chromosome replicates from an *ori* sequence located toward the middle of the chromosome, from where replication occurs bidirectionally using the DnaA initiator protein in a system that is likely similar to those of other bacteria. However, the ways in which the linear ends are maintained in these organisms are very different.

The very large linear chromosome of *Streptomyces* has inverted-repeat sequences at its ends and a protein, terminal protein, attached to the 5′ ends. Replication to the 3′ ends of the chromosome is thought to involve both these inverted repeats and the terminal protein in a process called "patching." After the linear DNA replicates, the 3′ end of each DNA remains single stranded, which allows a complex to form between the terminal protein and the inverted-repeat sequences. This complex is capable of producing a DNA primer that can be extended by other DNA polymerase in the cell.

The mechanism used by *Borrelia* to replicate its linear chromosome is very different from that found in *Streptomyces* spp. and is illustrated in the figure. In *Borrelia*, the 5′ phosphate and 3′ OH at each end of its linear chromosome are joined to each other to form hairpins, as shown. When replication initiated in the center of the chromosome gets to the ends, the linkage between the 5′ phosphate and 3′ OH hairpins forms a dimerized chromosome, with two copies of the chromosome forming a circle containing two copies of the chromosome linked end to end, as shown. An enzyme called a hairpin telomere resolvase protein (ResT) (in analogy to the telomerase of eukaryotes, even though it does not work in the same way) then re-creates the original hairpin ends from these double circles by making a staggered break in the two strands where the original ends were and then rejoining the 3′ end of one strand to the 5′ end of the other strand to form a hairpin. The ResT enzyme works

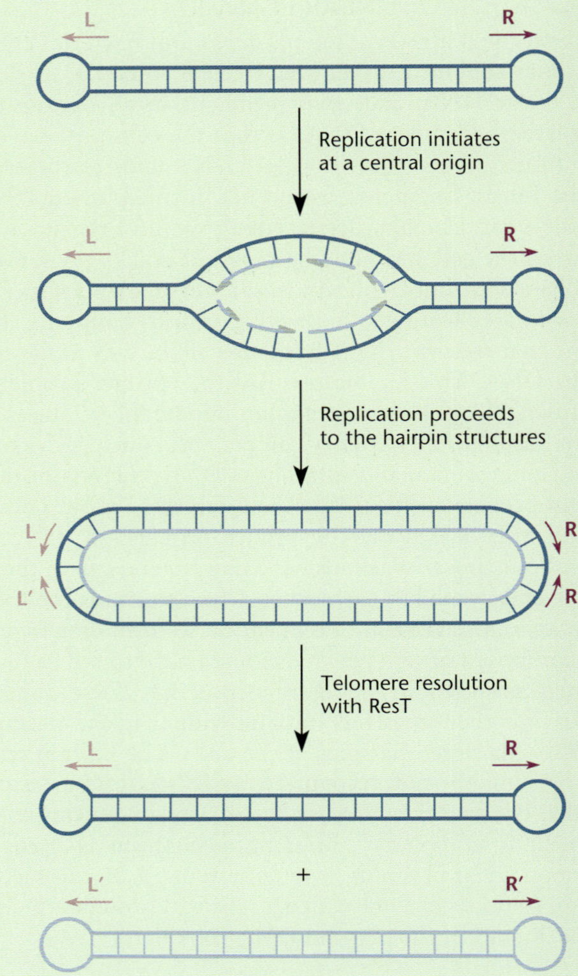

somewhat like some topoisomerases and tyrosine recombinases (see chapter 8) in that the breaking and rejoining process goes through a 3′ phosphoryltyrosine intermediate, where the 3′ phosphate end is covalently joined to a tyrosine (Y [see inside front cover]) before it is joined to the 5′ hydroxyl end.

Bacteria that have linear chromosomes may contain linear or circular plasmids, and the same is true for bacteria with circular chromosomes. This suggests that minimal cellular adaptation may be needed to maintain a chromosome in either the linear or circular form. Experiments in *E. coli* have provided a clever and clear way to help address this question. A lysogenic *Siphoviridae* (lambda-like) bacteriophage called N15 that was found in an isolate of *E. coli* uses the same mechanism for maintaining its linear genome as that found in *Borrelia*, with a *cis*-acting telomere region called *tos* and a *trans*-acting resolvase protein, TelN. Amazingly, it was found that the normally circular *E. coli* chromosome could be converted into a linear chromosome simply by moving the *tos* region into the terminus region of *E. coli* and expressing the TelN protein; no other adaptations were needed. The *E. coli* strain replicated normally and showed essentially no changes in gene expression. As would be predicted, the *E. coli* strain with a linear genome no longer required the Xer dimer resolution system that is normally needed to resolve circular dimer chromosomes. Supporting the finding that supercoiling is constrained into multiple domains in the chromosome, the *E. coli* strain with the linear chromosome still required the topoisomerases Topo IV and gyrase for its replication.

Linear chromosome ends also need to be differentiated from lethal double-strand break damage that must be recognized and repaired in the chromosome. Having a hairpin structure and/or a terminal protein at the end of the chromosome likely allows the cellular machinery to distinguish normal linear chromosome ends from random double-stranded breaks in *Borrelia* and *Streptomyces*.

References

Cui T, Moro-oka N, Ohsumi K, Kodama K, Ohshima T, Ogasawara N, Mori H, Wanner B, Niki H, Horiuchi T. 2007. *Escherichia coli* with a linear genome. *EMBO Rep* **8:**181–187.

Mir T, Huang SH, Kobryn K. 2013. The telomere resolvase of the Lyme disease spirochete, *Borrelia burgdorferi*, promotes DNA single-strand annealing and strand exchange. *Nucleic Acids Res* **41:**10438–10448.

Yang C-C, Tseng S-M, Chen CW. 2015. Telomere-associated proteins add deoxynucleotides to terminal proteins during replication of the telomeres of linear chromosomes and plasmids in *Streptomyces*. *Nucleic Acids Res* **43:**6373–6383.

these functions. They also must be able to express these genes in many types of bacteria. Apparently, the promoters and ribosome initiation sites for the replication genes of broad-host-range plasmids have evolved so that they can be recognized in a wide variety of bacteria.

Determining the Host Range

The actual host ranges of most plasmids are unknown because it is sometimes difficult to determine if a plasmid can replicate in other hosts. First, we must have a way of introducing the plasmid into other bacteria. Transformation systems (see chapter 6) have been developed for some, but not all, types of bacteria, and if one is available, it can be used to introduce plasmids into the bacterium. Electroporation can often be used to introduce DNA into cells. Plasmids that are self-transmissible or mobilizable (see chapter 5) can sometimes be introduced into other types of bacteria by conjugation, a process in which DNA is transferred from one cell to another using plasmid-encoded transfer functions.

Even if we can introduce the plasmid into other types of bacteria, we still must be able to select cells that have received the plasmid. Most plasmids, as isolated from nature, are not known to carry a convenient selectable gene, such as one for resistance to an antibiotic, and even if they do, the selectable gene may not be expressed in the other bacterium, since most genes are not expressed in bacteria distantly related to those in which they were originally found. Sometimes we can introduce a selectable gene, chosen because it is expressed in many hosts, into the plasmid. For example, the kanamycin resistance gene, first found in the Tn*5* transposon, is expressed in most *Proteobacteria*, making them resistant to the antibiotic kanamycin. We can either clone a marker gene into the plasmid using molecular biology techniques or introduce a transposon carrying a selectable

Table 4.3 Plasmids used in *B. subtilis*

Plasmid	Use	*B. subtilis* ori	*E. coli* ori	*B. subtilis* drug resistance	*E. coli* drug resistance
pUB110	Cloning vector	pUB110		Neo^r	
pMK3	Cloning vector	pMK3		Cam^r	
pDG148	Shuttle vector Inducible expression	pUB110	pBR322	Neo^r	Amp^r

marker into the plasmid by methods discussed in chapter 8.

If all goes well and we have a way to introduce the plasmid into other bacteria, and the plasmid carries a marker that is likely to be expressed in other bacteria, we can see if the plasmid can replicate in bacteria other than its original host. Care must also be taken to ensure that the plasmid has not recombined into the host chromosome. Since the mechanisms for introducing DNA into different types of bacteria differ and because there are many barriers to plasmid transfer between species, determining the host range of a new plasmid is a very laborious process. Therefore, the host ranges of plasmids are often extrapolated from only a few examples.

REGULATION OF COPY NUMBER

Another characteristic of plasmids that is determined mostly by their *ori* region is the **copy number**, or the average number of molecules of that particular plasmid per cell. More precisely, we define the copy number as the number of copies of the plasmid in a newborn cell immediately after cell division. Copy number control must have been an important early step in the evolution of plasmids. All plasmids must regulate their replication; if the replication level is too high, they would become too great a burden for the host, whereas if their replication is too low, it would not keep up with the cell replication, and they would be progressively lost during cell division. Some plasmids, such as the F plasmid of *E. coli*, replicate only about once during the cell cycle. Naturally, all plasmids have a somewhat low copy number, but plasmids have also been engineered to allow a much higher copy number per cell to facilitate biotechnology. The copy number information for some plasmids is shown in Table 4.2.

The regulation mechanisms used by plasmids with higher copy numbers often differ greatly from those used by plasmids with lower copy numbers. Plasmids that have high copy numbers, such as the modified derivatives of the ColE1 plasmid origin, need only have a mechanism that inhibits the initiation of plasmid replication when the number of plasmids in the cell reaches a certain level. Consequently, these molecules are called **relaxed plasmids**. In contrast, low-copy-number plasmids, such as F, must replicate only once or very few times during each cell cycle and so must have a tighter mechanism for regulating their replication. Hence, they are called **stringent plasmids**. Much more is understood about the regulation of replication of relaxed plasmids than about the regulation of replication of stringent plasmids.

The regulation of relaxed plasmids falls into three general categories. Some plasmids are regulated by an antisense RNA, sometimes called a countertranscribed RNA (ctRNA) because it is transcribed from the same region of the plasmid DNA as an RNA essential for plasmid replication but from the opposite strand. The ctRNA is therefore complementary to the essential RNA and is able to hybridize to the essential RNA and inhibit its function. In many cases, the ctRNA inhibits the translation of a protein essential for replication. The ctRNA of these plasmids is often assisted in its inhibitory role by a protein. Other plasmids are regulated by a ctRNA alone. Still others are regulated by a protein alone, which binds to repeated sequences in the plasmid DNA called iterons, thereby inhibiting plasmid replication. Examples of these three types of regulation are discussed below.

INCOMPATIBILITY

Another function of plasmids that is controlled by the *ori* region is incompatibility. Incompatibility refers to the inability of two plasmids to coexist stably in the same cell. Many bacteria, as they are isolated from nature, contain more than one type of plasmid. These plasmid types coexist stably in the bacterial cell and remain there even after many cell generations. In fact, bacterial cells containing multiple types of plasmids are not cured of each type of plasmid any more frequently than if the other type of plasmid was not there.

However, sometimes two plasmids of different types cannot coexist stably in the same cell. In this case, one or the other plasmid is lost as the cells multiply; this loss is more frequent than would occur if the two plasmid types were not occupying the cells together. If two plasmids cannot coexist stably, they are said to be members of the same **incompatibility (Inc) group** (see Box 4.2). If two plasmids can coexist stably, they belong to different Inc groups. There are a number of ways in which plasmids can be incompatible. One way is if they can each regulate the other's replication. There may be hundreds of different Inc groups, and plasmids are usually classified by the Inc group to which they belong. For example, RP4 (also called RK2) is an IncP (incompatibility group P) plasmid. In contrast, RSF1010 is an IncQ plasmid; it can therefore be stably maintained with RP4 because it belongs to a different Inc group but cannot be stably maintained with another IncQ plasmid.

Incompatibility Due to Shared Replication Control

One way in which two plasmids can be incompatible is if they share the same mechanism of replication control. The replication control system does not recognize the two as different, so either plasmid may be randomly selected for replication. At the time of cell division, the total copy number of the two plasmids will be the same, but one may be represented much less than the other. Figure 4.3 illustrates this by contrasting the distributions at cell division of plasmids of the same Inc group with plasmids of different Inc groups. Figure 4.3A shows a cell containing

BOX 4.2

Determining the Inc Group

To classify a plasmid by its Inc group, we must determine if it can coexist with other plasmids of known Inc groups. In other words, we must measure how frequently cells are cured of the plasmid when it is introduced into cells carrying another plasmid of a known Inc group. However, we can know that cells have been cured of a plasmid only when it encodes an easily testable trait, such as resistance to an antibiotic. Then, the cells become sensitive to the antibiotic if the plasmid is lost.

The experiment shown is designed to measure the curing rate of a plasmid that contains the Cam^r gene, which makes cells resistant to the antibiotic chloramphenicol. To measure the frequency of plasmid curing, we grow the plasmid-containing cells in medium with all the growth supplements and no chloramphenicol. At different times, we take a sample of the cells, dilute it, and plate the dilutions on agar containing the same growth supplements but, again, no chloramphenicol. After incubation of the plates, we replicate the plate onto another plate containing chloramphenicol (see chapter 3). If we do not observe any growth of a colony "copied" from the master plate, the bacteria in that colony must all have been sensitive to the antibiotic, and hence, the original bacterium that had multiplied to form the colony must have been cured of the plasmid. The percentage of colonies that contain no resistant bacteria is the percentage of bacteria that were cured of the plasmid at the time of plating.To apply this test to determine if two plasmids are members of the same Inc group, the two plasmids must contain different selectable genes, for example, genes encoding resistance to different antibiotics. Then, one plasmid is introduced into cells containing the other plasmid, and resistance to both antibiotics is selected. Then, cells containing both plasmids are incubated without either antibiotic and finally tested on antibiotic-containing plates, as described above in the example with chloramphenicol. The only difference is that the colonies are transferred onto two plates, each containing one or the other antibiotic. If the percentage of cells cured of one or the other plasmid is no higher than the percentage cured of either plasmid when it was alone, the plasmids are members of different Inc groups. We continue to apply this test until we find a known plasmid, if any, that is a member of the same Inc group as our unknown plasmid.

Technically, even two plasmids in the same Inc group could be maintained if they have a high copy number (which reduces the probability of plasmid loss) and had distinct antibiotic resistances that could be selected at the same time. However, this is not advisable, because it can select for plasmid fusion events via recombination.

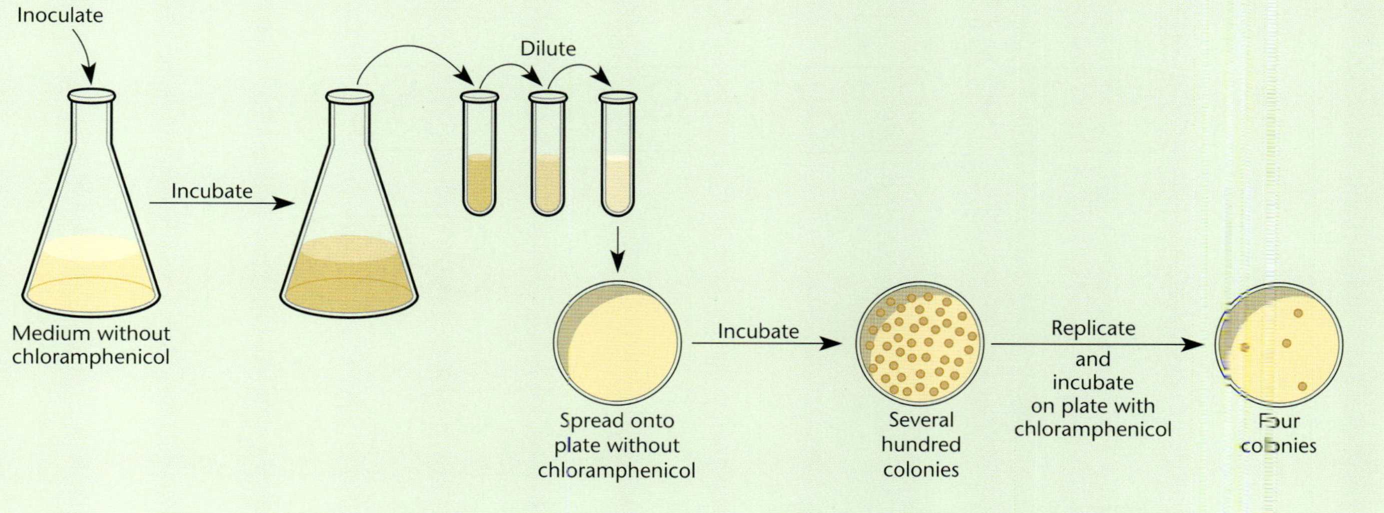

two types of plasmids that belong to different Inc groups and use different replication control systems. In the illustration, the two plasmids exist in equal numbers before cell division, but after division, the two daughter cells are not likely to get the same number of each plasmid. However, in the new cells, each plasmid replicates to reach its standard copy number, so that at the time of the next division, both cells again have the same numbers of both of the plasmids. This process is repeated each generation, so very few cells will be cured of either plasmid.

Now, consider the situation illustrated in Figure 4.3B, in which the cell has two plasmids that belong to the same

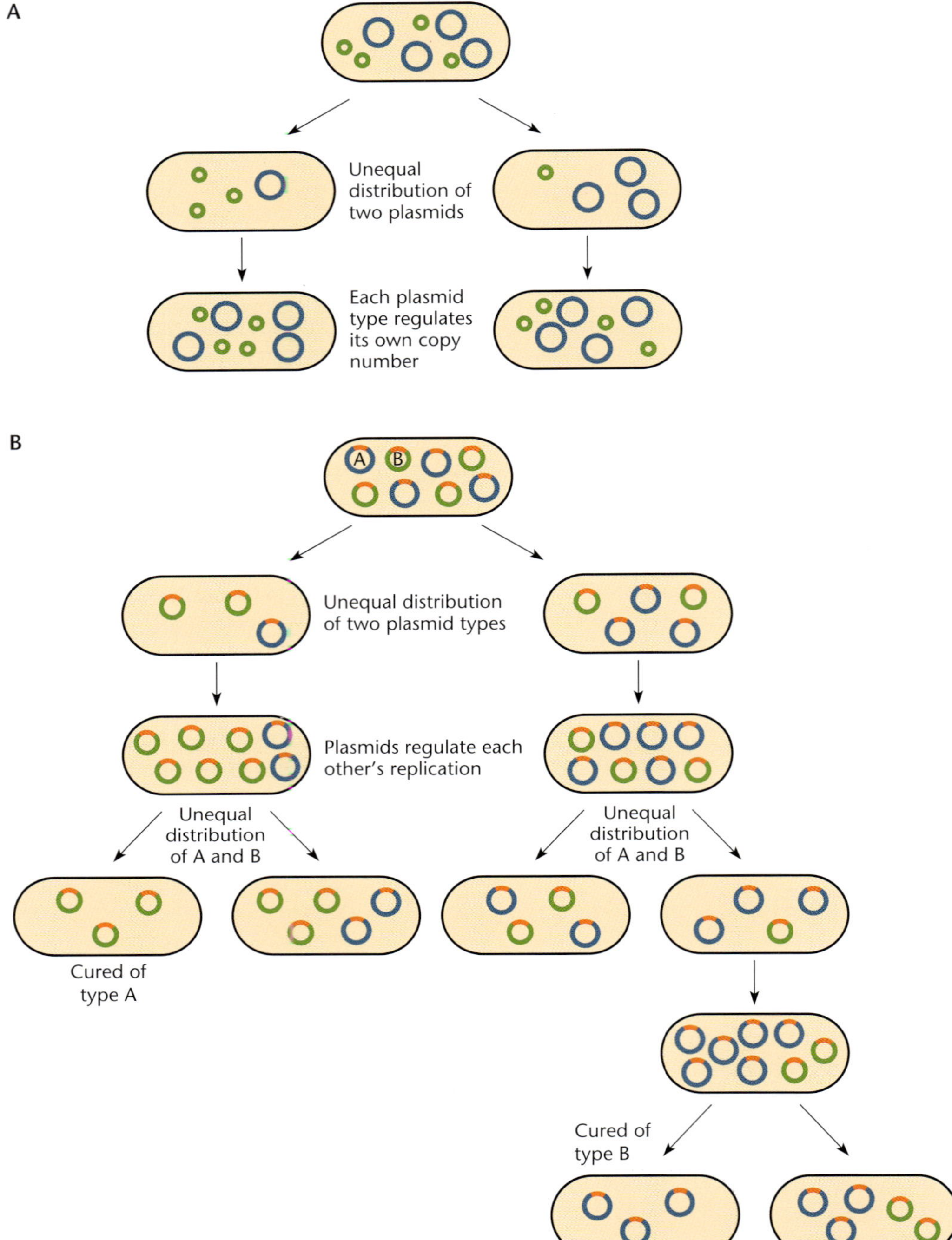

Figure 4.3 Coexistence of two plasmids from different Inc groups. **(A)** After division, both plasmids replicate to reach their copy numbers. **(B)** Curing of cells of one of two plasmids when they are members of the same Inc group. The plasmids are different, indicated in blue and green, but share the same incompatibility functions, indicated in orange. The sum of the two plasmids is equal to the copy number, but one may be underrepresented and lost in subsequent divisions. Eventually, most of the cells contain only one or the other plasmid.

Inc group and therefore share the same replication control system (indicated in orange). As in the first example, both plasmids originally exist in equal numbers, but when the cell divides, it is unlikely that the two daughter cells will receive the same number of the two plasmids. Note that in the original cell, the copy number of each plasmid is only half its normal number; both plasmids contribute to the total copy number, since they both have the same *ori* region and inhibit each other's replication. After cell division, the two plasmids replicate until the total number of plasmids in each cell equals the copy number. The underrepresented plasmid (recall that the daughters may not receive the same number of plasmids if the plasmid is high copy number) does not necessarily replicate more than the other plasmid, so that the imbalance of plasmid numbers might remain or become even worse. At the next cell division, the underrepresented plasmid has less chance of being distributed to both daughter cells, since there are fewer copies of it. Consequently, in subsequent cell divisions, the daughter cells are much more likely to be cured of one or the other of the two plasmid types by chance alone.

Incompatibility due to copy number control is probably more detrimental to low-copy-number plasmids than to high-copy-number plasmids. If the copy number is only 1, then only one of the two plasmids can replicate; each time the cell divides, a daughter is cured of one of the two types of plasmids.

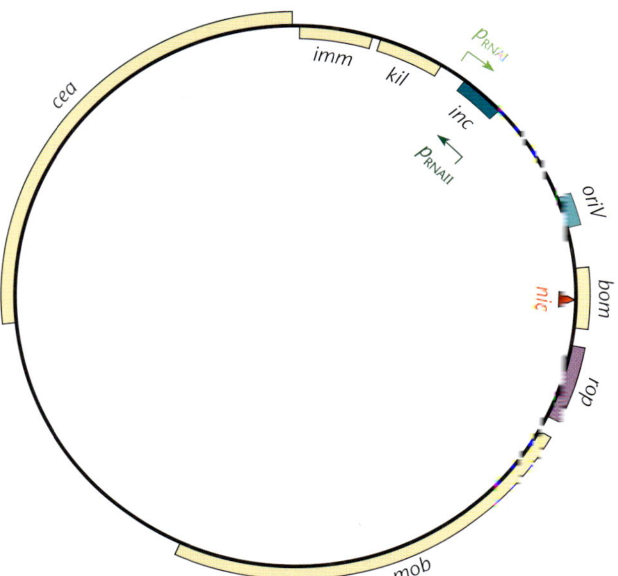

Figure 4.4 Genetic map of plasmid ColE1. The plasmid is 6,646 bp long. On the map, *oriV* is the origin of replication; p_{RNAII} is the promoter for the primer RNA II, *inc* encodes RNA I, *rop* encodes a protein that helps regulate the copy number, *bom* is a site that is nicked at *nic*, *cea* encodes colicin ColE1, and *mob* encodes functions required for mobilization (discussed in chapter 5). Modified from Luria SE, Suit JL, *in* Neidhardt FC, et al, Escherichia coli *and* Salmonella typhimurium: *Cellular and Molecular Biology* (American Society for Microbiology, Washington, DC, 1987).

Incompatibility Due to Partitioning

Two plasmids can also be incompatible if they share the same Par (partitioning) system. Par systems help segregate plasmids or chromosomes into daughter cells upon cell division (see below). Normally this helps ensure that both daughter cells get at least one copy of the plasmid and neither daughter cell is cured of the plasmid. If coexisting plasmids share the same Par system, one or the other is always distributed into the daughter cells during division. However, sometimes one daughter cell receives one plasmid type and the other cell gets the other plasmid type, producing cells cured of one or the other plasmid. We discuss what is understood about the mechanisms of partitioning below.

Plasmid Replication Control Mechanisms

The mechanisms used by some plasmids to regulate their copy number have been studied in detail. Some of the better-understood mechanisms are reviewed in this section.

ColE1-DERIVED PLASMIDS: REGULATION OF PROCESSING OF PRIMER BY COMPLEMENTARY RNA

The mechanism of copy number regulation of the plasmid ColE1 was one of the first to be studied. Figure 4.4 shows a partial genetic map of the original ColE1 plasmid. This plasmid has been put to use in numerous molecular biology studies, and many cloning vectors have been derived from it or its close relative, pMB1 (Table 4.2). These vectors include the commonly used pBR322 plasmid and plasmids with modified forms of the pMB1/ColE1 origin, such as the pUC plasmids, the pBAD plasmids, and the pET series of plasmids discussed below and in Box 7.2. Expression control using the p_{BAD} promoter that is found in the pBAD plasmids is described in chapter 11. Although the genetic maps of these cloning vectors have been changed beyond recognition for the pMB1/ColE1 vectors, they all retain the basic properties of the original ColE1 *ori* region, and hence, they share many of its properties, including the mechanism of replication regulation. However, these derivatives often have modifications to vastly increase the copy number of the vector to a level greater amounts of plasmid DNA to be easily isolated (Table 4.2).

The mechanism of regulation of ColE1-derived plasmids is shown in Figure 4.5. Replication is regulated mostly through the effects of a small plasmid-encoded RNA called RNA I. This small RNA inhibits plasmid replication by interfering with the processing of another RNA called RNA II, which forms the primer for plasmid DNA replication. In the absence of RNA I, RNA II forms an RNA-DNA hybrid at the replication origin. RNA II is

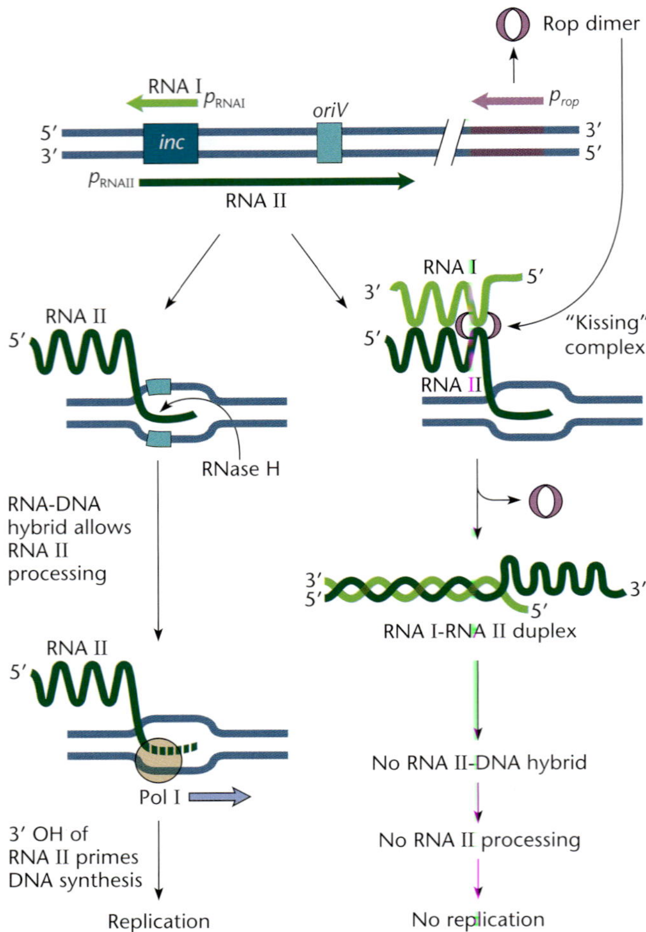

Figure 4.5 Regulation of the replication of ColE1-derived plasmids. RNA II must be processed by RNase H before it can prime replication. "*oriV*" indicates the transition point between the RNA primer and DNA. RNA I binds to RNA II and inhibits the processing, thereby regulating the copy number. p_{RNAI} and p_{RNAII} are the promoters for RNA I and RNA II transcription, respectively. The Rop protein dimer enhances the initial pairing of RNA I and RNA II.

then cleaved by the RNA endonuclease RNase H, releasing a 3′ hydroxyl group that serves as the primer for replication first catalyzed by DNA polymerase I. Unless RNA II is processed properly, it does not function as a primer, and replication does not ensue.

RNA I inhibits DNA replication through interference with RNA II primer formation by forming a double-stranded RNA with it, as illustrated in Figure 4.6. It can do this because the two RNAs are transcribed from opposite strands in the same region of DNA. Figure 4.6 illustrates how any two RNAs transcribed from the same region of DNA but from opposite strands are complementary. The regulatory capacity of small complementary RNAs was first shown in this system, but small RNAs are now known to be very important as a mechanism for controlling gene expression in many groups of bacteria (see chapter 11). Initially, the pairing between RNA I and RNA II occurs through short exposed regions on the two RNAs that are not occluded by being part of secondary structures. This initial pairing is very weak and therefore has been called a "kissing" complex. The plasmid-encoded protein named Rop (Figure 4.5) helps stabilize the kissing complex, although it is not essential. The kissing complex can then extend into a "hug," with the formation of the double-stranded RNA, as shown. Formation of the double-stranded RNA prevents RNA II from forming the secondary structure required for it to hybridize to the DNA before being processed by RNase H to form the mature primer.

Even though Rop is known to help RNA I to pair with RNA II and therefore help inhibit plasmid replication, it is not clear how Rop works, nor whether the protein is essential to maintain copy number control. Mutations that inactivate Rop cause only a moderate increase in the plasmid copy number.

The mechanism involving interactions between RNA I and RNA II provides an explanation for how the copy

Figure 4.6 Pairing between an RNA and its antisense RNA. **(A)** An antisense RNA is made from the opposite strand of DNA in the same region. **(B)** The two RNAs are complementary and can base pair with each other to make a double-stranded RNA.

number of ColE1 plasmids is maintained. Since RNA I is synthesized from the plasmid, more RNA I is made when the concentration of the plasmid is high. A high concentration of RNA I interferes with the processing of most of the RNA II, and replication is inhibited. The inhibition of replication is almost complete when the concentration of the plasmid reaches about 16 copies per cell, the copy number of the original ColE1 plasmid.

We can predict from the model what the effect of mutations in RNA I should be. Formation of the kissing complex involves pairing between very small regions of RNA I and RNA II. However, these regions must be completely complementary for this pairing to occur and for plasmid replication to be inhibited. Changing even a single base pair in this short sequence makes the mutated RNA I no longer complementary to the RNA II of the original nonmutant ColE1 plasmid, so it is no longer able to "kiss" it and regulate its replication. However, a mutation in the region of the plasmid DNA encoding RNA I also changes the sequence of RNA II made by the same plasmid in a complementary way, since they are encoded in the same region of the DNA, but from the opposite strands. Therefore, the mutated RNA I should still form a complex with the mutated RNA II made from the same mutated plasmid and prevent its processing; it just cannot interfere with the processing of RNA II from the original nonmutant plasmid. Therefore, a single-base-pair mutation in the RNA I coding region of the plasmid should effectively change the Inc group of the plasmid to form a new Inc group, of which the mutated plasmid is conceivably the sole member! In fact, the naturally occurring plasmids ColE1 and its close relative p15A, from which other cloning vectors, such as pACYC177 and pACYC184, have been derived, are members of different Inc groups, even though they differ by only 1 base in the kissing regions of their RNA I and RNA II.

R1 AND ColIB-P9 PLASMIDS: REGULATION OF TRANSLATION OF Rep PROTEIN BY COMPLEMENTARY RNA

The ColE1-derived plasmids are unusual in that they do not require a plasmid-encoded protein to initiate DNA replication at the *oriV* region, and they use only an RNA primer synthesized from the plasmid. Most plasmids require a plasmid-encoded protein, often called Rep, to initiate replication. The Rep protein is required to separate the strands of DNA at the *oriV* region, often with the help of host proteins, including DnaA (see chapter 1). Opening the strands is a necessary first step that allows the replication apparatus to assemble at the origin. The Rep proteins are very specific in that they bind only to the *oriV* of the same type of plasmid because they bind to certain specific DNA sequences within *oriV*. The amount of Rep protein is usually limiting for replication, meaning that there is never more than is needed to initiate replication. Therefore, the copy number of the plasmid can be controlled, at least partially, by controlling the synthesis of the Rep protein.

The R1 Plasmid

One type of plasmid that regulates its copy number by regulating the amount of a Rep protein is the R1 plasmid, a member of the IncFII family of plasmids. Like ColE1 plasmids, this plasmid uses a small complementary RNA to regulate its copy number, and this small RNA forms a kissing complex with its target RNA (see Kolb et al., Suggested Reading). Also like ColE1 plasmids, the more copies of the plasmid in the cell, the more of this antisense RNA is made and the more plasmid replication is inhibited. However, rather than inhibiting primer processing, the R1 plasmid uses its complementary RNA to inhibit the translation of the mRNA that encodes the Rep protein and thereby inhibits the replication of the plasmid DNA.

Figure 4.7 illustrates the regulation of R1 plasmid replication in more detail. The plasmid-encoded RepA protein is the only plasmid-encoded protein that is required for the initiation of replication. The *repA* gene can be transcribed from two promoters. One of these promoters, called p_{copB}, transcribes both the *repA* and *copB* genes, making an mRNA that can be translated into the proteins RepA and CopB. The second promoter, p_{repA}, is within the *copB* gene and so makes an RNA that can encode only the RepA protein. The p_{repA} promoter is repressed by the CopB protein and is therefore turned on only immediately after the plasmid enters a new cell and before any CopB protein is made. This results in a short burst of synthesis of RepA from p_{repA} after the plasmid enters a cell, which causes the plasmid to replicate until it reaches its copy number. Then, the p_{repA} promoter is repressed by newly synthesized CopB protein, and the *repA* gene can be transcribed only from the p_{copB} promoter, which directs synthesis of both RepA and the CopB repressor.

Once the plasmid has attained its copy number, the synthesis of RepA, and therefore the replication of the plasmid, is regulated by the small CopA RNA. The *copA* gene is transcribed from its own promoter, and the RNA product affects the stability of the mRNA made from the p_{copB} promoter. Because the CopA RNA is made from the same region encoding the translation initiation region (TIR) for the *repA* gene, but from the other strand of the DNA, the two RNAs are complementary and can pair to make double-stranded RNA. Then, an RNase called RNase III, a chromosomally encoded enzyme that cleaves some double-stranded RNAs (see chapter 2), cleaves the CopA-RepA duplex RNA.

A Plasmid genetic organization

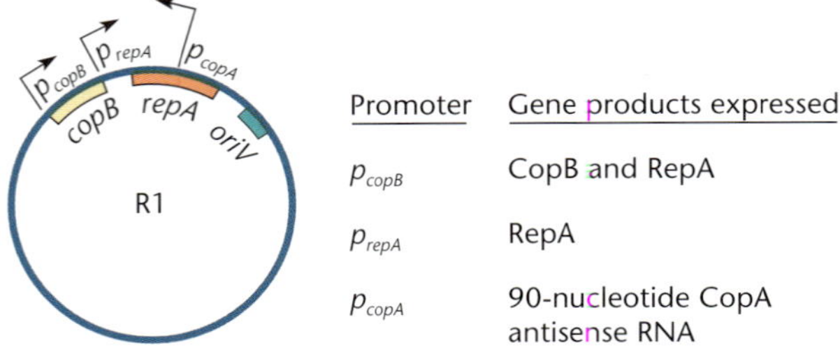

B Replication occurs after plasmid enters cells

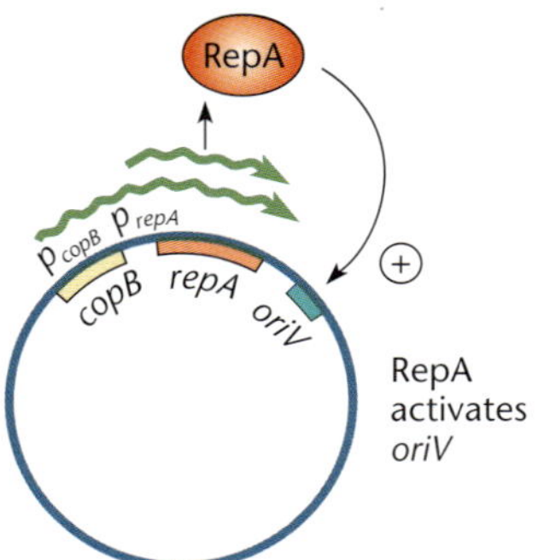

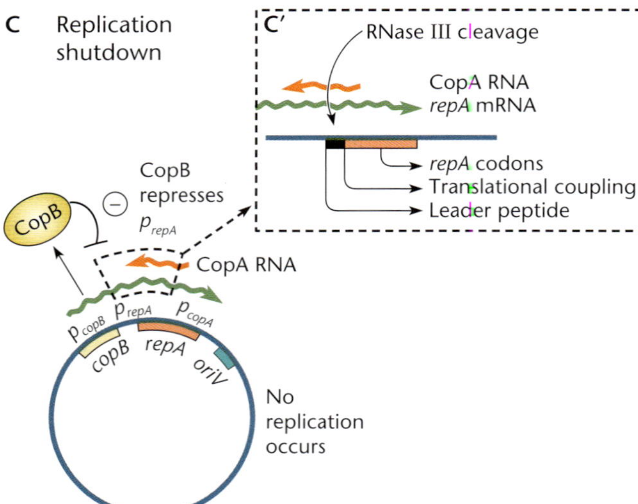

Figure 4.7 Regulation of replication of the IncFII plasmid R1. **(A)** Locations of promoters, genes, and gene products involved in the regulation. **(B)** Immediately after the plasmid enters the cell, most of the *repA* mRNA is made from promoter p_{repA} until the plasmid reaches its copy number. **(C)** Once the plasmid reaches its copy number, the CopB protein represses transcription from p_{repA}. Now, *repA* is transcribed only from p_{copB}. **(C′)** The antisense RNA CopA hybridizes to the leader peptide coding sequence in the *repA* mRNA, and the double-stranded RNA is cleaved by RNase III. This prevents translation of RepA, which is translationally coupled to the translation of the leader peptide.

The reasons why cleavage of this RNA prevents the synthesis of RepA are a little complicated. The 5′ leader region of the mRNA, upstream of where the RepA protein is encoded, encodes a short leader polypeptide that has no function of its own but simply exists to be translated. The translation of RepA is coupled to the translation of this leader polypeptide (see chapter 2 for an explanation of translational coupling). Cleavage of the mRNA by RNase III in the leader region interferes with the translation of this leader polypeptide and, by blocking its translation, also blocks translation of the downstream RepA. Therefore, by having the CopA RNA activate cleavage of the mRNA for the RepA protein upstream of the RepA coding sequence, the plasmid copy number is controlled by the amount of CopA RNA in the cell, which in turn depends on the concentration of the plasmid. The higher the concentration of the plasmid, the more CopA RNA is made and the less RepA protein is synthesized, maintaining the concentration of the plasmid around the plasmid copy number.

The ColIb-P9 Plasmid

Yet another level of complexity of the regulation of copy number by a complementary RNA is provided by the ColIb-P9 plasmid (Figure 4.8) (see Asano and Mizobuchi, Suggested Reading). As in the R1 plasmid, the Rep protein-encoding gene (called *repZ* in this case) is translated downstream of a leader peptide open reading frame, called *repY*, and the two are also translationally coupled. The translation of *repY* opens an RNA secondary structure that normally occludes the Shine-Dalgarno (SD) sequence of the TIR of *repZ*. A sequence in the secondary structure then can pair with the loop of a hairpin upstream of *repY*, forming a pseudoknot (see Figure 2.2 for an example of a pseudoknot), thus permanently disrupting the secondary structure and leaving the SD sequence for *repZ* exposed. A ribosome can then bind to the TIR for *repZ* and translate the initiator protein. The small complementary Inc RNA pairs with the loop of the upstream hairpin and prevents hairpin formation, leaving the SD sequence of the *repZ* coding sequence blocked and preventing translation of *repY*.

THE pT181 PLASMID: REGULATION OF TRANSCRIPTION OF THE *rep* GENE BY A SMALL COMPLEMENTARY RNA

Not all plasmids that have an antisense RNA to regulate their copy numbers use it to inhibit translation or primer processing. Some plasmids, including the *Staphylococcus* plasmid pT181, use antisense RNAs to regulate transcription of the *rep* gene, in this case called *repC*, through a process called attenuation (Figure 4.9) (see Novick et al., Suggested Reading). The pT181 plasmid replicates by an RC mechanism, and the RepC protein is required to initiate replication of the leading strand at *oriV*. Also, the RepC

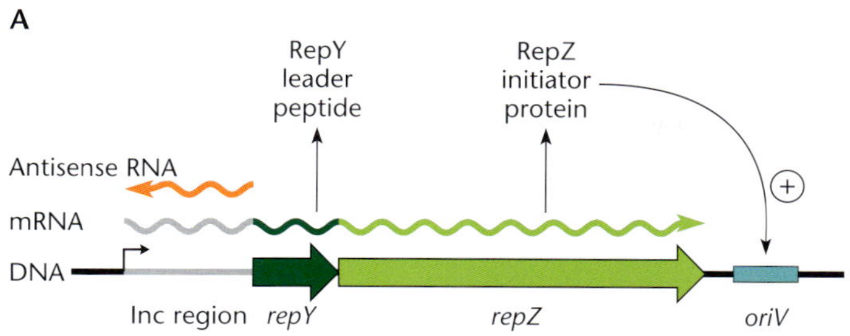

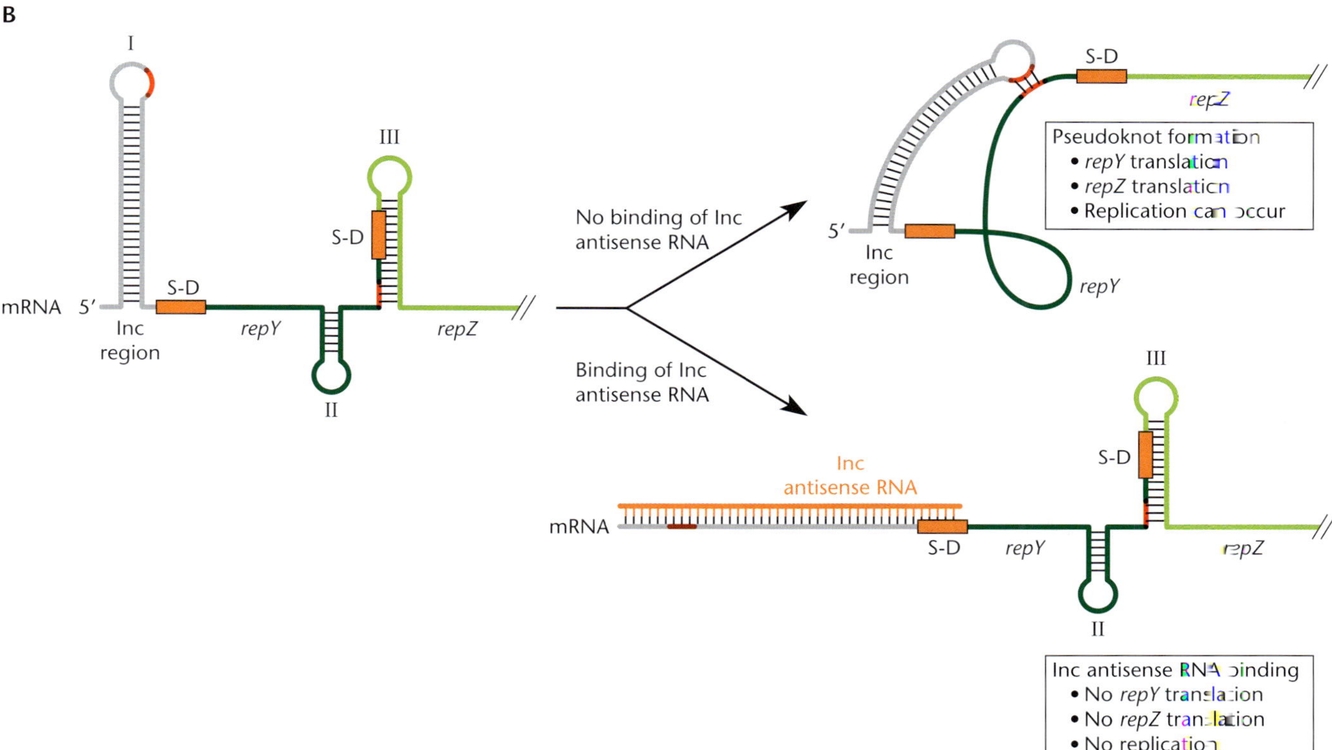

Figure 4.8 Regulation of plasmid CoIIb-P9 copy number by antisense RNA inhibition of pseudoknot formation. **(A)** The minimal replicon with the *repY* (leader peptide) (dark green) and *repZ* (light green) genes is shown. The Inc region encodes both the 5′ end of the *repYZ* mRNA and the antisense RNA. **(B)** The *repY* and *repZ* genes are translationally coupled. On the mRNA, the *repY* S-D sequence is exposed, whereas structure III sequesters the *repZ* S-D sequence and thereby prevents *repZ* translation. Shown in red are regions in structures I and III that are complementary and so can pair, resulting in pseudoknot formation. Unfolding of structure II by the ribosome stalling at the end of the *repY* gene results in the formation of a pseudoknot by base pairing between the complementary sequences and allows the ribosome to access the *repZ* S-D sequence. Binding of Inc antisense RNA to the loop of structure I directly inhibits formation of the pseudoknot, and the subsequent Inc RNA-mRNA duplex inhibits RepY translation, and consequently RepZ translation, since the two are translationally coupled. Modified from Brantl S, *in* Funnell BE, Phillips GJ (ed), *Plasmid Biology* (ASM Press, Washington, DC, 2004).

protein is inactivated each time the DNA replicates (see above). This makes the RepC protein rate limiting for replication, i.e., the more RepC protein there is, the more plasmids are made. The antisense RNA binds to the mRNA for the RepC protein as the mRNA is being made and prevents formation of a secondary structure. This secondary structure would normally prevent the formation of a hairpin that is part of a factor-independent transcriptional terminator (see chapter 2). Therefore, if the secondary structure does not form, the hairpin forms and transcription terminates (i.e., is attenuated). Transcriptional regulation by attenuation is discussed in more detail in chapter 11.

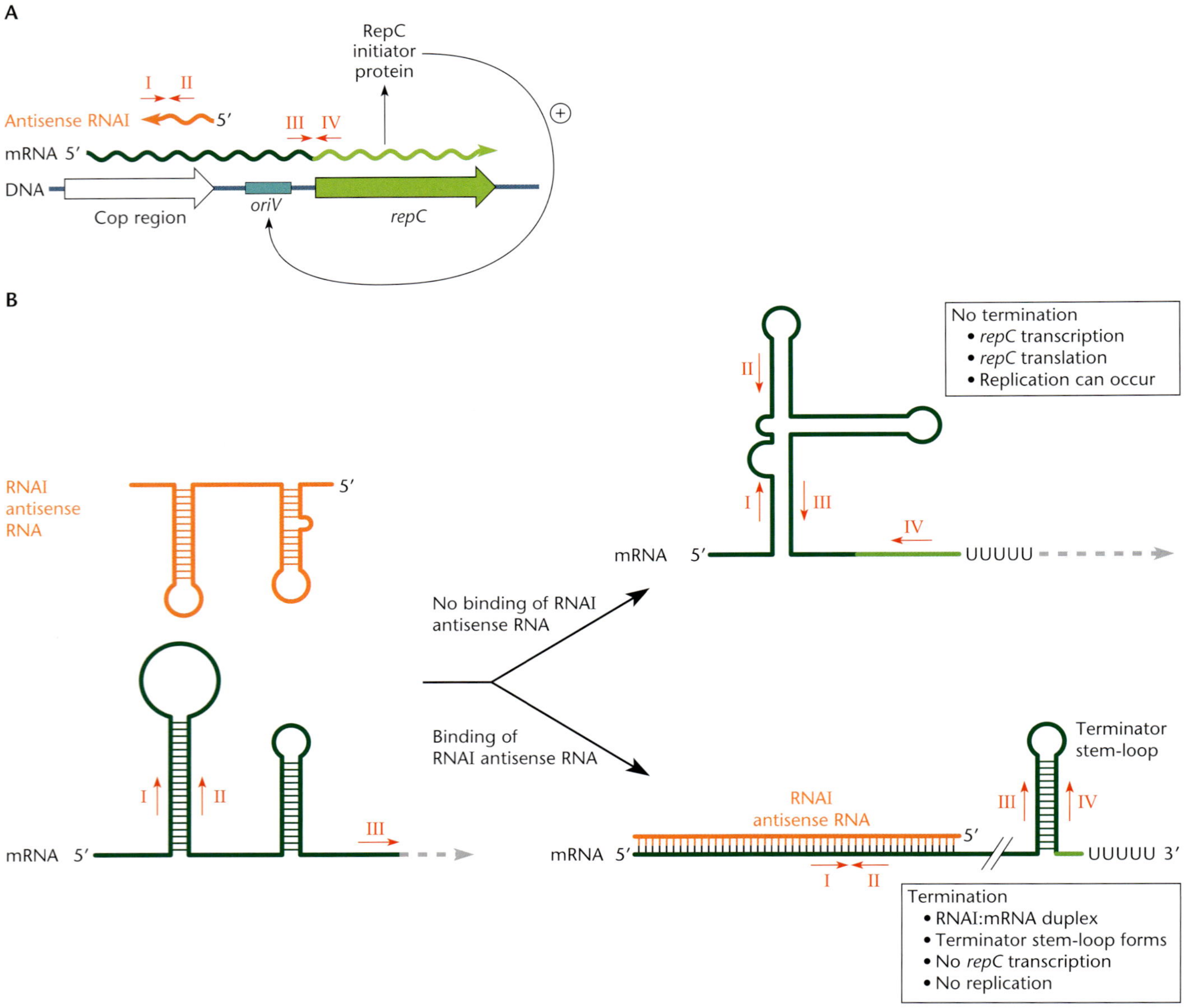

Figure 4.9 Regulation of plasmid pT181 copy number by antisense RNA regulation of transcription of the *repC* gene. **(A)** The genetic structure of the minimal replicon of pT181. Shown are the mRNA that encodes the RepC protein that initiates leading-strand replication at *oriV*; the Cop region that encodes the antisense RNA (RNA I) that regulates the copy number; and regions in the mRNA and antisense RNA, indicated by arrows labeled I, II, III, and IV, that can pair to form alternative secondary structures. **(B)** Formation of an antisense RNA-mRNA duplex regulates RepC expression by a transcriptional attenuation mechanism. The antisense RNA I can form a duplex with the 5′ end of the mRNA that encodes RepC and disrupt a secondary structure, allowing instead the formation of a terminator loop that causes transcription termination.

This regulation works well only because the antisense RNA is so unstable that its concentration drops quickly if the copy number of the plasmid decreases, allowing fine-tuning of the replication of the plasmid with the copy number. In other plasmids, the antisense RNA is much more stable. These plasmids also use a transcriptional repressor to regulate transcription of the *rep* gene.

THE ITERON PLASMIDS: REGULATION BY PLASMID COUPLING

Many commonly studied plasmids use a very different mechanism to regulate their replication. These plasmids are called iteron plasmids because their *oriV* regions contain several repeats of a certain set of DNA bases called an **iteron sequence**. The iteron plasmids include pSC101, F,

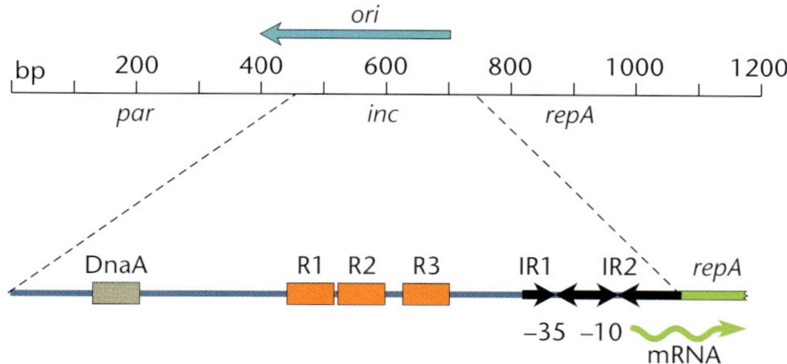

Figure 4.10 The *ori* region of pSC101. R1, R2, and R3 are the three iteron sequences (CAAAGGTCTAGCAGCAGAATTTACAGA for R3) to which RepA binds to handcuff two plasmids. RepA autoregulates its own synthesis by binding to the inverted repeats IR1 and IR2. The location of the partitioning site, *par* (see "Partitioning"), and the binding sites for the host protein DnaA are also shown.

R6K, P1, and the RK2-related plasmids (Table 4.2). The iteron sequences of these plasmids are typically 17 to 22 bp long and exist in about three to seven copies in the *ori* region. In addition, there are usually additional copies of these repeated sequences a short distance away that contribute to lowering the plasmid copy number.

One of the simplest of the iteron plasmids is pSC101. For our purposes, the essential features of the *ori* region of this plasmid (Figure 4.10) are the gene *repA*, which encodes the RepA protein required for initiation of replication, and three repeated iteron sequences, R1, R2, and R3, where RepA binds to regulate the copy number. The RepA protein is the only plasmid-encoded protein required for the replication of the pSC101 plasmid and many other iteron plasmids. It serves as a positive activator of replication, much like the RepA protein of the R1 plasmid. The host chromosome encodes the other proteins that either bind to this region or otherwise act to allow initiation of replication; they include DnaA, DnaB, DnaC, and DnaG (see chapter 1).

Iteron plasmid replication is regulated by two superimposed mechanisms. The first is control of RepA synthesis. Most commonly, the RepA protein represses its own synthesis by binding to its own promoter region and blocking transcription of its own gene. Therefore, the higher the concentration of plasmid, the more RepA protein is made and the more it represses its own synthesis. Thus, the concentration of RepA protein is maintained within narrow limits, and the initiation of replication is strictly regulated. This type of regulation, known as **transcriptional autoregulation**, is discussed in chapter 11.

However, this mechanism of regulation by itself is not sufficient to regulate the copy number of the plasmid, especially that of low-copy-number stringent plasmids, such as F and P1. Iteron plasmids must have another mechanism to regulate their copy numbers within narrow limits.

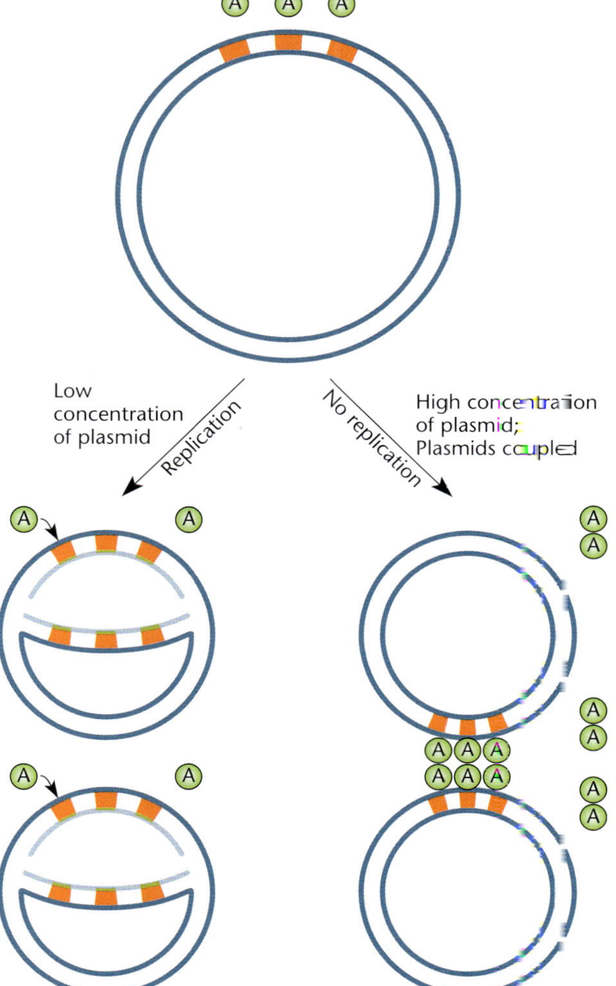

Figure 4.11 The "handcuffing" or "coupling" model for regulation of iteron plasmids. At low concentrations of plasmids, the RepA protein binds to only one plasmid at a time, initiating replication. At high plasmid and RepA concentrations, the RepA protein may dimerize and bind to two plasmids simultaneously, handcuffing them and inhibiting replication.

This other form of regulation has been hypothesized to be the coupling of plasmid molecules through the Rep protein and their iteron sequences (see McEachern et al., Suggested Reading). The **coupling hypothesis** for regulation of plasmid replication is illustrated in Figure 4.11. When the concentration of plasmids is high enough, they interact with each other via bound RepA proteins. This inhibits the replication of both coupled plasmids. The coupling mechanism allows plasmid replication to be controlled, not only by how much RepA protein is present in the cell, but also by the concentration of the plasmid itself or, more precisely, the concentration of the iteron sequences on the plasmid. Direct support for the coupling model in the replication control of iteron plasmids has come from electron micrographs of purified iteron plasmids mixed with the purified RepA protein for that plasmid. In these pictures, two plasmid molecules can often be seen coupled by RepA protein. *In vitro* and *in vivo* work also supports coupling as an important mechanism to prevent plasmid overreplication (see Das et al., Suggested Reading).

HOST FUNCTIONS INVOLVED IN REGULATING PLASMID REPLICATION

As mentioned above, in addition to Rep, many plasmids require host proteins to initiate replication. For example, some plasmids require the DnaA protein, which is normally involved in initiating replication of the chromosome, and have *dnaA* boxes in their *oriV* regions to which DnaA binds (Figure 4.10). The DnaA protein may also directly interact with the Rep proteins of some plasmids. This may explain why some broad-host-range plasmids, such as RP4, make two types of Rep proteins. The different forms of the Rep protein might better interact with the DnaA proteins of different species of bacteria (see Caspi et al., Suggested Reading). The DnaA protein is involved in coordinating replication of the chromosome with cell division (see chapter 1); making their replication dependent on DnaA may allow plasmids to better coordinate their own replication with cell division. Like the chromosome origin (*oriC*), some *E. coli* plasmids also have Dam methylation sites close to their *oriV*. These methylation sites presumably help to further coordinate their replication with cell division. As with the chromosomal origin of replication, both strands of DNA at these sites must be fully methylated for initiation to occur. Immediately after initiation, only one strand of these sites is methylated (hemimethylation), delaying new initiations at the sites (see the discussion of sequestration of chromosome origins in chapter 1). Despite substantial progress, however, the method by which the replication of very stringent plasmids, such as P1 and F (with a copy number of only 1), is controlled to within such narrow limits is still something of a mystery and the object of current research.

Mechanisms To Prevent Curing of Plasmids

Cells that have lost a plasmid during cell division are said to be **cured** of the plasmid. Several mechanisms prevent curing, including toxin-antitoxin systems (Box 4.3), site-specific recombinases that resolve multimers, and partitioning systems. The last two are reviewed below.

RESOLUTION OF MULTIMERIC PLASMIDS

The possibility that a cell will lose a plasmid during cell division is increased if the plasmids form dimers or higher multimers during replication. A dimer consists of two individual copies of the plasmid molecules linked head to tail to form a larger circle, and a multimer links more than two such monomers. Such dimers and multimers probably occur as a result of recombination between monomers. Recombination between two monomers forms a dimer, and subsequent recombination can form higher and higher multimers. Also, replication of RC plasmids can form multimers if termination after each round of replication is not efficient. Multimers may replicate more efficiently than monomers, perhaps because they have more than one origin of replication, so they tend to accumulate if the plasmid has no effective way to remove them. The formation of multimers creates a particular problem when the plasmid attempts to segregate into the daughter cells on cell division. One reason is that multimers lower the effective copy number. Each multimer segregates into the daughter cells as a single plasmid, and if all of the plasmid is taken up in one large multimer, it can segregate into only one daughter cell. Also, the presence of more than one *par* site on the multimer may cause it to be pulled to both ends of the cell at once (see below), much like a dicentric chromosome can lead to nondisjunction in higher organisms. Therefore, multimers greatly increase the chance of a plasmid being lost during cell division.

To avoid this problem, many plasmids have site-specific recombination systems (see chapter 8) that resolve multimers as soon as they form. These systems can be either chromosomally encoded or encoded by the plasmid itself. A site-specific recombination system promotes recombination between specific sites on the plasmid if the same site occurs more than once in the molecule, as it would in a dimer or multimer. This recombination has the effect of resolving multimers into separate monomeric plasmid molecules.

A well-studied example of a plasmid-encoded site-specific recombination system is the Cre-*loxP* system encoded by phage P1. This phage is capable of lysogeny, and its prophage form is a plasmid, subject to all the problems faced by other plasmids, including multimerization due to recombination. The Cre protein, a tyrosine (Y) recombinase, promotes recombination between two *loxP* sites on a dimerized plasmid, resolving the dimer into two monomers. This system is very efficient and relatively simple and has been useful in a number of studies, including demonstrations of the interspecies transfer of proteins (see chapter 5). It has also been used as a model for Y recombinases, since the recombinase has been crystallized with its *loxP* DNA substrate. Y recombinases and their mechanism of action are discussed in chapter 8.

The best-known examples of host-encoded site-specific recombination systems used to resolve plasmid dimers are the *cer*-XerC/D and the *psi*-XerC/D site-specific recombination systems used by the ColE1 plasmid and the pSC101 plasmid, respectively. The XerC/D system is mentioned in chapter 1 in connection with segregation of

BOX 4.3

Toxin-Antitoxin Systems and Plasmid Maintenance

Extrachromosomal elements like plasmids can easily be lost during division. As described in the text, numerous adaptations beyond encoding beneficial functions have evolved in plasmids to guard against loss, including copy number control, dimer resolution systems, and partitioning systems. However, plasmids have an additional trick for being maintained on plasmids, called toxin-antitoxin systems. In what seems like revenge, some plasmids encode proteins that kill a cell if it is cured of that plasmid. Such functions are relatively common on mobile plasmids, including the F plasmid, the R1 plasmid, and the P1 prophage, which replicates as a plasmid, and even on integrating conjugating elements (see chapter 5). These systems have been called plasmid addiction systems because they cause the cell to undergo severe withdrawal symptoms and die if it is cured of the plasmid to which it is addicted.

Plasmid addiction systems all use basically the same strategy. They consist of two components, which can be either proteins or RNA. One component functions as a toxin, and the other functions as an antitoxin or antidote. While the cell contains the plasmid, both the toxin and the antitoxin are made, and the antitoxin blocks the effect of the toxin, either by binding to the toxin and inactivating it directly or by somehow indirectly alleviating its effect. Once the cell is cured of the plasmid, neither the toxin nor the antitoxin continues to be made. The toxin, however, is more stable than the antitoxin, so eventually, the antitoxin is degraded and the cured cells contain only the toxin. Without the antitoxin to counteract it, the toxin interrupts one of the essential processes in the cell. Five types of toxin-antitoxin systems have been described (see the figure).

Type I: An antisense RNA prevents the expression of a protein toxin. If the toxin-antitoxin coding genes are lost, the mRNA for the toxin is no longer neutralized by the more labile antisense RNA, allowing the toxin to be translated from the mRNA (see Fozo et al., References).

Type II: A toxic protein is neutralized by the direct binding of an antitoxin protein. If the toxin-antitoxin coding genes are lost, the more labile antitoxin is more quickly degraded than the toxin, allowing the toxic protein to function (see Leplae et al., References).

Type III: An antitoxin RNA directly binds the toxin protein, thereby neutralizing it. If the toxin-antitoxin coding genes are lost, the more labile antitoxin RNA is more quickly degraded than the toxin protein, allowing it to function (see Blower et al., References).

Type IV: An antitoxin protein prevents the toxin from binding and incapacitating an essential protein in the cell. If the toxin-antitoxin coding genes are lost, the more labile antitoxin protein is more quickly degraded than the toxin protein, allowing it to incapacitate the essential component (see Masuda et al., References).

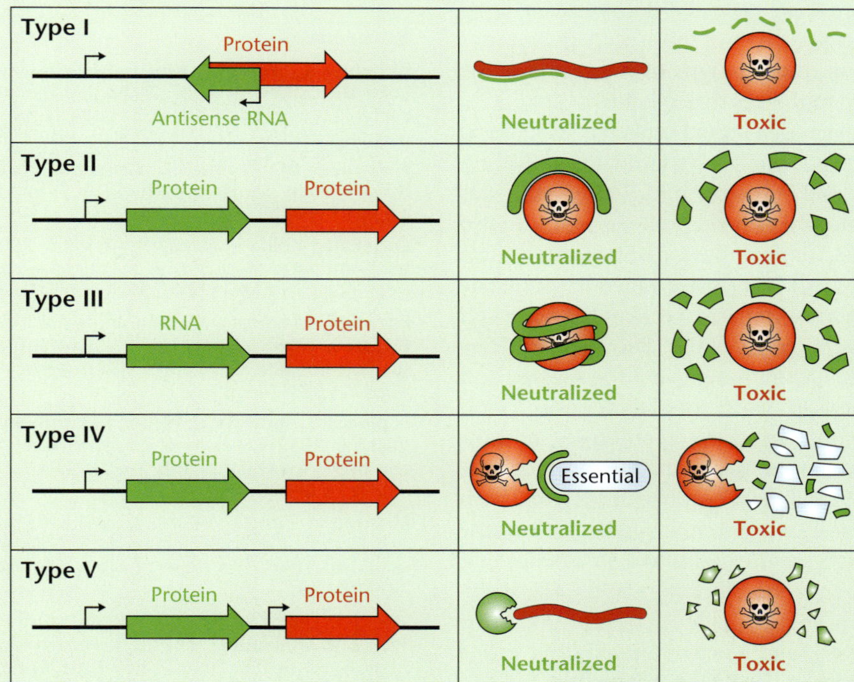

(continued)

Toxin-Antitoxin Systems and Plasmid Maintenance

Type V: An antitoxin protein degrades the mRNA for the toxin. If the toxin-antitoxin coding genes are lost, the labile antitoxin protein is quickly degraded, stabilizing the mRNA for the toxin protein, allowing for its toxic effect (see Wang et al., References).

The toxin proteins can affect any one of many essential processes in bacteria, depending on the system. For example, the toxin protein of the F plasmid, Ccd, alters DNA gyrase so that it causes double-strand breaks in the DNA. The Hok protein of plasmid R1 destroys the cellular membrane potential, causing loss of cellular energy. The Doc toxin of phage P1 inhibits translation elongation through an interaction with the 30S ribosomal subunit. Even some restriction endonucleases can act as a toxin-antitoxin system when the modification component that methylates DNA recognition sequences is more labile than the endonuclease component.

Revenge aside, it is hard to understand the evolutionary advantage of killing cells after plasmid loss. There may be two interrelated answers to why mobile plasmids may contain toxin-antitoxin systems. One is that plasmid loss would likely lead to better growth of the plasmidless cell, as it is relieved of the burden of replicating and expressing the plasmid DNA. Therefore, the newly cured cells could otherwise overgrow the plasmid-containing cells in the same population, and the toxin-antitoxin systems take away this advantage. A second benefit to having the toxin-antitoxin system on a mobile element is that the cells affected by the toxin could be saved from the toxic effect of plasmid loss if they again obtained the mobile plasmid from an adjacent cell. Therefore, part of the advantage may not be killing by the toxin, but instead, a type of paralysis until the mobile plasmid can return to the cell that lost the plasmid.

There is a good rationale for plasmid addiction systems in that they prevent cells cured of the plasmid from accumulating and thus help ensure survival of the plasmid. Therefore, it was a surprise to discover that similar toxin-antitoxin systems also occur in the chromosome. Some of these are on exchangeable DNA elements, such as genetic islands and superintegrons (see chapter 8), and may play a role similar to that of the addiction systems of plasmids, preventing loss of the DNA element. They could be considered selfish genes that prevent themselves, and therefore the DNA element in which they reside, from being lost from the cell. However, other toxin-antitoxin modules seem to be encoded by normal genes in the chromosome. Two examples of these are the MazEF and RelBE systems, both found in *E. coli* K-12. These two systems work in remarkably similar ways. MazF and RelE are the toxins and are RNases that cleave mRNA in the ribosome and block translation, killing the cell. MazE and RelB are the antitoxins, which bind to the toxins MazF and RelE, respectively, and inactivate them. The toxins and antitoxins are not made if translation is inhibited, but the toxin is more stable and longer lived than the antitoxin. These systems could therefore be considered suicide modules in that they cause the cell to kill itself if translation is inhibited (see Engelberg-Kulka et al., References). A number of hypotheses have been proposed to explain the existence of these suicide modules. One is that they help prevent the spread of phages that inhibit host translation. Another is that they help shut down cellular metabolism in response to starvation and help ensure the long-term survival of some of the cells. Cells that have shut down cellular metabolism are sometimes referred to as **persister** cells, which can also tolerate antibiotic treatment, allowing them to survive even though they technically do not encode a mechanism for resistance to the antimicrobial.

Another apparent suicide system in *B. subtilis* lends credence to the idea that the purpose of suicide systems might be to kill some bacteria so that others may live (see Ellermeier et al., References). This system is much more complex than the others we have mentioned and consists of many genes in two operons, *skf* and *sdp*. To summarize, when a population of *B. subtilis* cells is starved for nutrients, some of them begin to sporulate (see chapter 12). These cells produce a toxin that kills other cells that were slow to start sporulating. The nutrients released by the killed cells are then consumed by the cells producing the toxin, which can then reverse their sporulation process. This buys the sporulating cells more time, in case the situation changes and they do not really need to sporulate, allowing them to avoid the need to employ a drastic measure to ensure survival.

References

Blower TR, Short FL, Rao F, Mizuguchi K, Pei XY, Fineran PC, Luisi BF, Salmond GPC. 2012. Identification and classification of bacterial type III toxin-antitoxin systems encoded in chromosomal and plasmid genomes. *Nucleic Acids Res* **40:**6158–6173.

Ellermeier CD, Hobbs EC, Gonzalez-Pastor JE, Losick R. 2006. A three-protein signaling pathway governing immunity to a bacterial cannibalism toxin. *Cell* **124:**549–559.

Engelberg-Kulka H, Hazan R, Amitai S. 2005. *mazEF*: a chromosomal toxin-antitoxin module that triggers programmed cell death in bacteria. *J Cell Sci* **118:**4327–4332.

Fozo EM, Makarova KS, Shabalina SA, Yutin N, Koonin EVE, Storz G. 2010. Abundance of type I toxin-antitoxin systems in bacteria: searches for new candidates and discovery of novel families. *Nucleic Acids Res* **38:**3743–3759.

Leplae R, Geeraerts D, Hallez R, Guglielmini J, Drèze P, Van Melderen L. 2011. Diversity of bacterial type II toxin-antitoxin systems: a comprehensive search and functional analysis of novel families. *Nucleic Acids Res* **39:**5513–5525.

Masuda H, Tan Q, Awano N, Wu K-P, Inouye M. 2012. YeeU enhances the bundling of cytoskeletal polymers of MreB and FtsZ, antagonizing the CbtA (YeeV) toxicity in *Escherichia coli. Mol Microbiol* **84:**979–989.

Wang X, Lord DM, Cheng H-Y, Osbourne DO, Hong SH, Sanchez-Torres V, Quiroga C, Zheng K, Herrmann T, Peti W, Benedik MJ, Page R, Wood TK. 2012. A new type V toxin-antitoxin system where mRNA for toxin GhoT is cleaved by antitoxin GhoS. *Nat Chem Biol* **8:**855–861.

the chromosome of *E. coli*. The XerC and XerD proteins are part of a site-specific recombinase that acts on a site, *dif*, close to the terminus of replication of the chromosome to resolve chromosome dimers generated during chromosome replication. The plasmids have commandeered this site-specific system of the host to resolve their own dimers by having sites at which the recombinase can act. The site on the ColE1 plasmid is called *cer*, and the site on pSC101 is called *psi*. As described in chapter 1, resolution of dimer chromosomes at *dif* sites requires an interaction between XerC/D and the FtsK protein. In contrast, sites on plasmids do not share this requirement because of differences in the sequence where recombination occurs; however, these plasmid sites have a different requirement. The *cer* site on the ColE1 plasmid is not recognized as such but is recognized only if two other host proteins, called PepA and ArgR, are bound close by in the DNA, as shown in Figure 4.12. A similar situation occurs in the pSC101 plasmid, but there the auxiliary host proteins are PepA and ArcA~P (phosphorylated ArcA), which binds close to where ArgR binds in *cer*. Apparently, these other proteins bind to XerC/D recombinase at the plasmid *cer* or *psi* sites and help orient it for the recombination process. However, it is not clear how these particular host proteins came to play this role. The only thing these accessory proteins have in common is that they all normally bind to DNA because they are involved in regulating transcription in the host. This is yet another example of a case where plasmids commandeer host functions for their own purposes, in this case, a site-specific recombination system normally used for resolving chromosome dimers, as well as transcriptional regulatory proteins that are used for host cell gene regulation. XerC/D is also a Y recombinase, and its mechanism of action is discussed in more detail in the section on tyrosine recombinases in chapter 8.

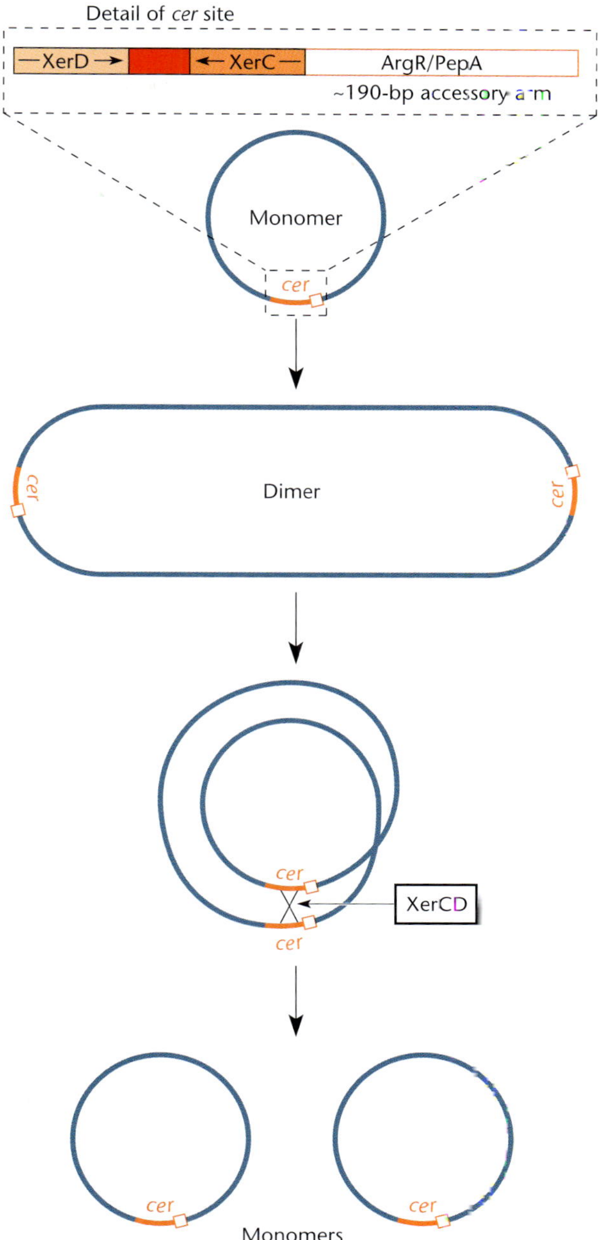

Figure 4.12 The Xer functions of *E. coli* catalyze site-specific recombination at the ColE1 plasmid *cer* site to resolve plasmid dimers. The region of binding for the host proteins ArgR and PepA is shown.

PARTITIONING

A very effective mechanism that plasmids use to avoid being lost from dividing cells is **partitioning** systems. These systems ensure that at least one copy of the plasmid is present in each daughter cell after cell division. The functions involved in these systems are called **Par functions**, and in many ways they are analogous to the Par functions involved in chromosome segregation. In fact, the discovery of Par systems in plasmids preceded their discovery in chromosome segregation.

The Par Systems of Plasmids

The Par systems have been studied extensively in plasmids. The Par systems of low-copy-number plasmids fall into at least two groups whose members are related to each other by sequence and function. At least one of these groups of plasmid Par functions is also related to the putative chromosomal Par systems of some bacteria, as described in chapter 1. One group is represented by the Par system of the R1 plasmid of *E. coli*, and the other, much larger group is represented by the Par systems of the F, P1, and broad-host-range RK2 plasmids, among others. The latter is also the group to which the chromosomal partitioning systems from the bacteria *Bacillus subtilis* and *Caulobacter crescentus* belong. The two groups of partitioning systems differ in the details of how they achieve the feat of plasmid partitioning. The R1 plasmid partitioning system is addressed first, since it is the better understood of the two systems. It is also the best example to date of a dynamic filament-forming structure in bacteria that can move cellular constituents around, analogously to the actin filaments of eukaryotes.

THE R1 PLASMID PAR SYSTEM

The mechanism of partitioning by the R1 plasmid is illustrated in Figure 4.13. The partitioning system consists of two protein-coding genes, *parM* and *parR*, as well as a centromere-like *cis*-acting site, *parC*. The actin-like ParM protein forms a filament that pushes the plasmids to the cell poles (see Campbell and Mullins, Suggested Reading). The polymerization process occurs quickly, rapidly pushing the plasmids to the cell poles (Figure 4.13). This is followed by depolymerization, leaving the plasmids at the poles. The polymerization and depolymerization process is regulated by the ATP-bound state of ParM; in the test tube, ParM proteins form a stable filament in the presence of nonhydrolyzable ATP, indicating that the shift from the ATP-bound state to the ADP-bound state is important in the depolymerization process and the dynamic nature of the filaments. The prevailing model that supports the available data holds that polymerization of ParM-ATP occurs under the cap-like ParR-*parC* complex, propelling the complex toward the cell pole. ParM-bound ATP added at the ParR-*parC* complex is hydrolyzed to ADP within the filament. The filament of ParM-ADP is thought to be stable only when a ParM-bound ATP remains at the ParR-*parC* cap. Once the ParM-ATP at the ParR-*parC* complex hydrolyzes ATP to ADP, the filament is no longer protected and completely collapses, leaving the plasmids at the poles of the cell.

THE P1 AND F PLASMID Par SYSTEMS

The larger group of plasmid-partitioning systems is related to the systems of the P1 prophage plasmid and the

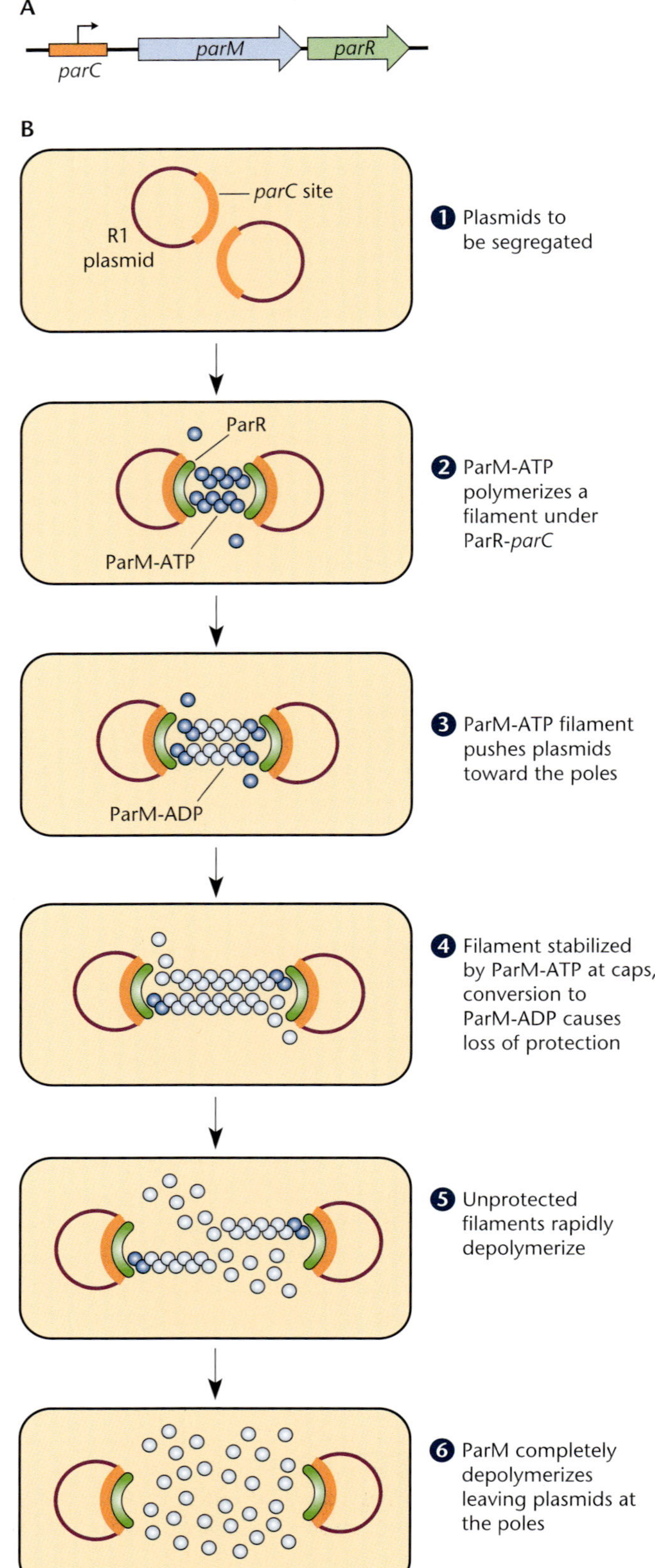
A
parC
parM
parR
B
parC site
R1
plasmid
1 Plasmids to be segregated
ParR
ParM-ATP
2 ParM-ATP polymerizes a filament under ParR-parC
ParM-ADP
3 ParM-ATP filament pushes plasmids toward the poles
4 Filament stabilized by ParM-ATP at caps, conversion to ParM-ADP causes loss of protection
5 Unprotected filaments rapidly depolymerize
6 ParM completely depolymerizes leaving plasmids at the poles

Figure 4.13 Model for partitioning of the R1 plasmid. **(A)** Structure of the *par* locus of R1, showing the positions of the *parM* and *parR* genes, as well as the *cis*-acting *parC* site. The transcription start site is in the *parC* site. **(B)** ParR binds to *parC*, making a site of ParM-ATP nucleation. Filaments grow by adding successive ParM-ATP subunits under the ParR-*parC* "cap," where growth of the filament pushes the plasmid containing *parC* to the cell poles. ParM hydrolysis gradually converts it to ParM-ADP, but the filament remains stable as long as ParM-ATP molecules reside at the ParR-*parC* cap. Loss of the terminal ParM-ATP under the cap destabilizes the filament, causing it to be disassembled and leaving the plasmids at the cell poles. ParM can then be recharged with ATP before the plasmids are partitioned again prior to the next cell division.

F plasmid. Also, some bacteria, including *B. subtilis* and *C. crescentus*, use a similar Par system to aid in partitioning and organizing their chromosomes (see chapter 1). The composition of these Par systems is largely similar to that of the Par system of the R1 plasmid in that they usually consist of two proteins and an adjacent *cis*-acting site on the DNA to which one of the proteins, often called ParB, binds. Also, the other protein, often called ParA, has ATP-binding motifs. However, the ATP-binding protein has no homology to actin or any other cytoskeletal protein and belongs to a special family of proteins called the P-loop ATPases. It is distantly related to MinD, a protein that helps select the division site in bacteria and, in *E. coli* at least, shows dynamic localization during the cell cycle (see chapter 1). In the F plasmid, the corresponding proteins are called SopA and SopB, and the site is called *sopC* (*sop*, for stability of plasmid). For simplicity, we refer to them as ParA, ParB, and *parS*. In some plasmids, the Par systems are encoded in the same region of the plasmid as the replication proteins; in rare instances, the system may lack an autonomous ParB protein, and perhaps a larger ParA-related protein may play both roles.

The mechanism of action of these Par systems appears to be fundamentally different from that of the system described above. Microscopic analyses of plasmid partitioning with these systems indicate that it involves directed movement across the nucleoid. A model that accounts for the available observations holds that instead of being pushed, as in the R1 system, the ParB-*parS* complex is actually pulled by an attraction to ParA that polymerizes at random places on the nucleoid. In one of the models, this would involve the ParB-*parS* complex being pulled as it interacts with patches of ParA-ATP, coaxing it to hydrolyze ATP to the ParA-ADP form that no longer binds DNA (see Vecchiarelli et al., Suggested Reading) (Figure 4.14). Repeated cycles of ParA assembly and disassembly would tend to maximize the distribution of plasmid copies to opposite cell halves, so that upon cell division, the plasmids are appropriately segregated. In this model, the partition system can be separated into two interacting subsystems. In the first subsystem, ATP-bound ParA dimers bind chromosomal DNA at random positions nonspecifically but cooperatively, leading to ParA-bound regions. In the second part of the system, dimers of ParB bind specifically to the *parS* site on the plasmid and

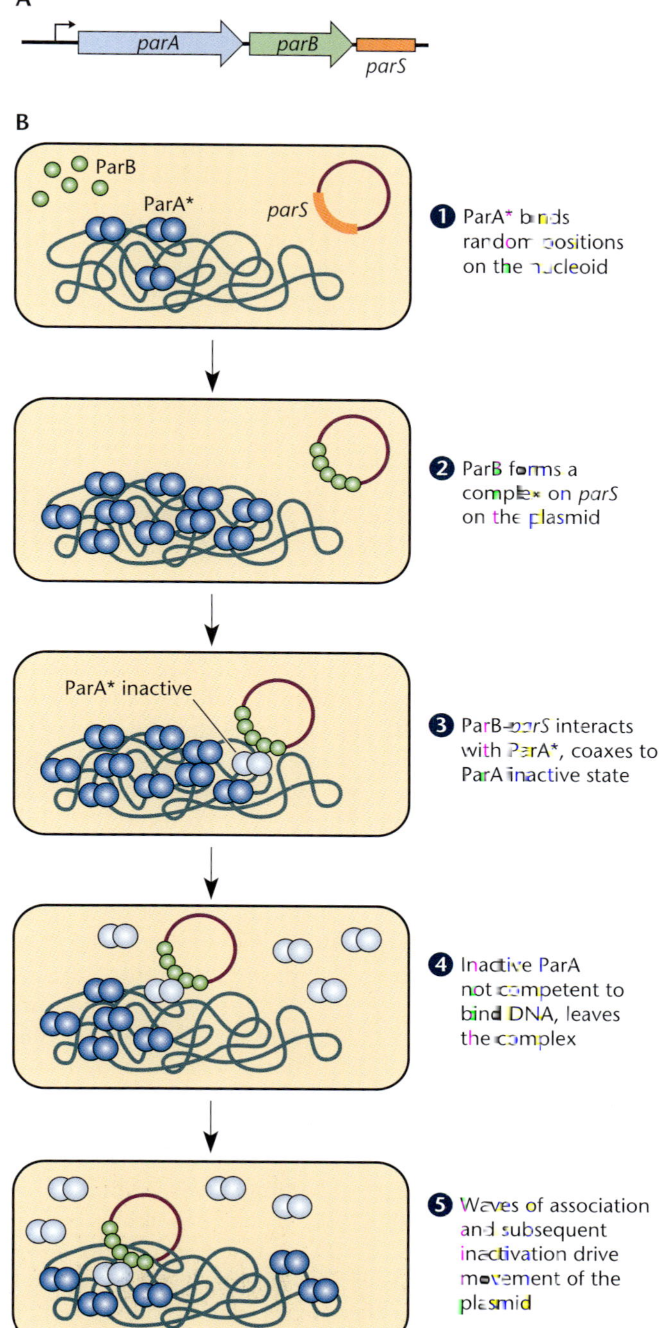
A
parA
parB
parS
B
ParB
ParA*
parS
1 ParA* binds random positions on the nucleoid
2 ParB forms a complex on parS on the plasmid
ParA* inactive
3 ParB-parS interacts with ParA*, coaxes to ParA inactive state
4 Inactive ParA not competent to bind DNA, leaves the complex
5 Waves of association and subsequent inactivation drive movement of the plasmid

Figure 4.14 Model for partitioning by *par* systems on P1, F, and RK2. **(A)** Structure of the *par* locus on P1, F, and RK2 showing the positions of the *parA* and *parB* genes and the *cis*-acting *parS* site. Transcription for the promoter (P*parAB*) is controlled by the ParA protein. **(B)** Plasmids containing the *parS* site bound by ParB are pulled across the chromosome as they mediate the depolymerization of ParA across the nucleoid. ParA-ATP (ParA*) binds randomly as a dimer to the nucleoid and polymerizes in three dimensions, while ParB specifically associates with *parS* sites on the plasmid. Interaction between the ParB-*parS* complex and ParA stimulates ParA to hydrolyze ATP, converting it to ParA-ADP, which dissociates from the DNA. The ParB-*parS* complex is then free to associate with the next dimer of ParA-ATP, causing net movement of the ParB-*parS* complex-containing plasmid. ParA-ADP can then convert to its ParA-ATP form and slowly take on a form capable of associating at distal locations across the nucleoid.

nucleate the binding of new ParB dimers around this site. The ParB-*parS* complex on the plasmid then becomes competent for associating with ParA on the chromosome. The interaction with ParB signals ParA-ATP dimers to hydrolyze ATP to ADP, inactivating the protein, and ParA-ADP subsequently releases from the DNA. As the first dimer of ParA is released, the ParB-*parS* complex then binds to the next ParA dimer. This behavior would allow the ParB-*parS* complex to be pulled along the nucleoid as it goes through cycles of ParA binding on the receding region of ParA-bound DNA. While ParA will again bind ATP, it must first go through a slow conformational change before it again becomes competent to bind with its nonspecific DNA-binding activity. This delay ensures that an existing patch of ParA-ATP can be chased by the ParB-*parS* complex to mobilize the plasmid across the chromosome. In this way, the Par proteins would go through a type of dynamic behavior reminiscent of the activity of the MinE and MinD proteins but on a local level across the nucleoid instead of a behavior that spans the entire length of the cell. Such a system will naturally distribute *parS*-containing plasmids away from one another in the cell.

INCOMPATIBILITY DUE TO PLASMID PARTITIONING

If two plasmids share the same partitioning system, they will be incompatible, even if their replication control systems are different. Incompatibility due to shared partitioning systems makes sense, considering the models presented above. If two plasmids that are otherwise different share the same Par system, one plasmid of each type can be directed to opposite ends of the cell before the cell divides. In this way, one daughter cell can get a plasmid of one type while the other daughter cell gets the plasmid of the other type, and cells are cured of one or the other plasmid. However, even though shared partitioning systems can cause incompatibility, this is usually not the sole cause of their incompatibility. Usually, cells with the same partitioning system also share the same replication control system, since the two are often closely associated on the plasmid; therefore, the incompatibility is due to both systems. In fact, in some cases, the replication control genes and the partitioning genes are intermingled around the origin of plasmid replication.

Plasmid Cloning Vectors

As discussed in chapter 1, a cloning vector is an autonomously replicating DNA (replicon) into which other DNAs can be inserted. Any DNA inserted into the cloning vector replicates passively with the vector, so that many copies (clones) of the original piece of DNA can be obtained. Plasmids offer many advantages as cloning vectors, and many types of plasmids have been engineered to serve as plasmid cloning vectors.

As described above, the physical properties of plasmids allow them to be purified easily for manipulation and reintroduction into bacteria. Because very few functions need to reside on the plasmid itself, they can be small, which, in addition to allowing them to be manipulated more easily, also reduces the burden on the cell. Given the universality of the genetic code, they offer a way of expressing proteins from other bacteria or from other types of organisms. In fact, in one of the first cloning experiments, a frog gene was cloned into plasmid pSC101 (see Cohen et al., Suggested Reading). Cloning vectors have a wide variety of uses that make them important for the study of bacteria or as tools for a better understanding of a broad range of other organisms. For example, sometimes they are used as expression constructs to express a protein for study directly in the host, while at other times they may be used to overexpress a protein to facilitate isolating large amounts of the protein for work outside the cell, such as for structural studies, biochemical studies, or commercial applications. Plasmid constructs are very commonly constructed using *E. coli* as a host; these constructs may later be moved into less tractable organisms, where they can also replicate, in what are called shuttle vectors (see below).

ANATOMY OF A PLASMID CLONING VECTOR

Most plasmids, as they are isolated from nature, are too large to be convenient as cloning vectors and/or often do not contain easily selectable genes that can be used to move them from one host to another. Commonly used vectors have a number of attributes that make them more convenient to work with. Some of these features are explained below.

ORIGINS OF REPLICATION

The most basic feature of a plasmid cloning vector is an origin of replication that allows it to replicate independently of the chromosome. In early genetic experiments, the minimal features required for autonomous replication could be isolated with the help of a selectable gene product. For example, the origin of DNA replication region that allows autonomous replication can be cloned when it resides on the same DNA fragment as a selectable marker, such as a gene allowing antibiotic resistance. As shown in Figure 4.15, the plasmid is cut into several pieces with a restriction endonuclease (indicated by the arrows), and the pieces are ligated (joined) to another piece of DNA that has a selectable marker, such as resistance to ampicillin (Amp^r). For the experiment to work, the second piece of DNA cannot have a functional origin of replication. The ligated mixture is then introduced into bacterial cells by transformation, and the antibiotic-

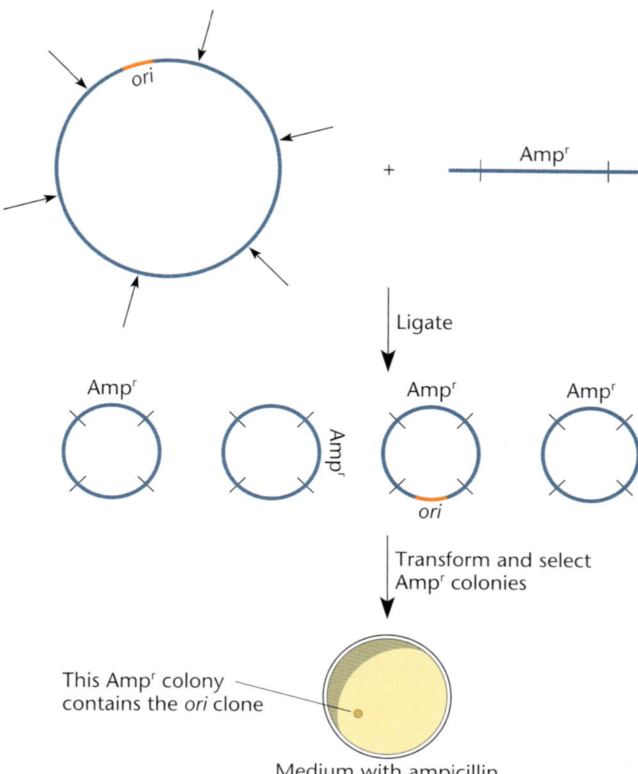

Figure 4.15 Finding the origin of replication (*ori*) in a plasmid. Random pieces of the plasmid are ligated to a piece of DNA containing a selectable gene but no origin of replication and introduced into cells. Cells that can form a colony on the selective plates contain the selectable gene ligated to the piece of DNA containing the origin.

resistant transformants are selected by plating the mixture on agar plates containing growth medium and the antibiotic. The only DNA molecules able to replicate and also to confer antibiotic resistance on the cells are hybrids with both the *ori* region of the plasmid and the piece of DNA with the antibiotic resistance gene. Therefore, only cells harboring these hybrid molecules can grow on the antibiotic-containing medium. Using similar techniques, segregation systems could also be isolated as DNA regions that when cloned into these plasmid constructs would stabilize their maintenance after many generations of growth. Given the broad availability of inexpensive DNA sequencing, DNA amplification using PCR has streamlined the manipulation of DNA, but these techniques are likely to continue to be important as more tools are needed for newly identified types of bacteria that lack a system that has been previously characterized.

SELECTABLE GENES

All plasmids exert some kind of load on the host and require a selectable gene to allow maintenance of the vector. In addition, a selectable marker is needed to select for cells that have received the plasmid, because of the low efficiency of the procedures used to introduce DNA. In some cases, it is useful to have plasmids with different selectable genetic markers, because more than one plasmid may need to be maintained at one time. Antibiotic resistance genes have historically been taken from transposons and other plasmids. Some antibiotic resistance genes that have been introduced into cloning vectors are the chloramphenicol resistance (Cam^r) gene of transposon Tn*9*, the tetracycline resistance (Tet^r) gene of plasmid pSC101, the Amp^r gene of transposon Tn*3*, and the kanamycin resistance (Kan^r) gene of transposon Tn*5*. The antibiotic resistance gene that is chosen depends on the uses to which the cloning vector will be put. Some antibiotic resistance genes, such as the Tet^r gene from pSC101, are expressed only in some types of bacteria closely related to *E. coli*, while others, such as the Kan^r gene from Tn*5*, are expressed in most *Proteobacteria*; similar resistance genes have been identified for use in *Firmicutes* and other systems.

UNIQUE RESTRICTION SITES

Since many applications of plasmid cloning vectors require that clones be introduced into restriction endonuclease cleavage sites, it is necessary that a cloning vector have some restriction sites that are unique. If a site is unique, the cognate restriction endonuclease cuts the vector at only that one site when it is used to cut the plasmid. One can then clone pieces of foreign DNA into the unique site, and the cloning vector will remain intact. In some plasmid cloning vectors, the unique sites are located in a selectable gene, so that insertion of a foreign piece of DNA in the site inactivates the selectable gene. This is called **insertional inactivation** and is discussed below. Normally, during a cloning operation, only a small percentage of the cloning vector molecules pick up a foreign DNA insert. If those that have picked up an insert no longer confer the selectable trait, for example, resistance to an antibiotic, they can be more easily identified. Many cloning vectors also have unique restriction sites for many different restriction endonucleases all grouped into one small region on the plasmid called a polyclonal or multiple restriction site. This offers the convenience of choosing among a variety of restriction endonucleases for cloning, and the cloned DNA is always inserted at the same general place in the vector, independent of the restriction endonuclease used. Polyclonal sites can also be used for **directional cloning**. If the cloning vector is cut by two different restriction endonucleases with unique sites within the polyclonal site, the resulting overhangs cannot pair to recyclize the plasmid. The plasmid can recyclize only if it picks up a piece of foreign DNA. If the piece of DNA to be cloned has overhangs for the two different sites at its ends, it is usually cloned in only one orientation into the polyclonal site.

The unique restriction sites can also be placed so that genes cloned into them will be expressed from promoters and translation initiation signals on the plasmid. Plasmids with these features are called expression vectors and can be used to express foreign genes in *E. coli* and other convenient hosts. Such vectors can also be used to attach affinity tags to proteins to aid in their purification. Expression vectors and affinity tags are discussed in connection with translational and transcriptional fusions in chapter 2. Expression vectors have also been adapted from bacteriophage regulation systems, as discussed in chapter 7.

Examples of Plasmid Cloning Vectors

A number of plasmid cloning vectors have been engineered for special purposes. Almost all of these plasmids have at least some of the features mentioned above for a desirable cloning vector.

1. They are small, so that the plasmid can be easily isolated and introduced into various bacteria.
2. They have relatively high copy numbers, so that the plasmid DNA can be easily purified in sufficient quantities.
3. They carry easily selectable traits, such as a gene conferring resistance to an antibiotic, which can be used to select cells that contain the plasmid.
4. They have one or a few sites for specific restriction endonucleases, which cut DNA and allow the insertion of foreign DNA segments. Also, these sites usually occur in genes that can be easily screened for to facilitate the detection of plasmids with foreign DNA inserts by insertional inactivation.

Many plasmid cloning vectors have other special properties that aid in particular experiments. For example, some contain the sequences recognized by phage-packaging systems (*pac* or *cos* sites), so that they can be packaged into phage heads (see chapter 7). Expression vectors can be used to produce foreign proteins in bacteria. Mobilizable plasmids have mobilization (*mob*) sites, so they can be transferred by conjugation to other cells (see chapter 5). Some broad-host-range vectors have *ori* regions that allow them to replicate in many types of bacteria or even in organisms from different kingdoms. Shuttle vectors contain more than one type of replication origin, so they can replicate in multiple unrelated organisms (see below). These and some other types of specialty plasmid cloning vectors are discussed in more detail below and in later chapters.

pUC PLASMIDS

Some of the most commonly used plasmid cloning vectors are the pUC vectors and vectors derived from them. One pUC vector, pUC18, is shown in Figure 4.16. The pUC plasmids are very small (only ~2,700 bp of DNA) and, as explained above, have been modified to have copy numbers in the hundreds, making them relatively easy to purify. They also have the easily selectable Amp^r gene. One of the most useful features of these plasmids is the ease with which they can be used for insertional inactivation. They encode the N-terminal region of the *lacZ* gene product, called the α-peptide, which is not active in the cell by itself but complements the C-terminal portion of the protein, called the *lacZ* β-polypeptide, to make active β-galactosidase by a process termed α-complementation (see chapter 3). This enzyme cleaves the indicator dye 5-bromo-4-chloro-3-indolyl-β-D-galactopyranoside (X-Gal) to produce a blue color. Some host strains, such as *E. coli* JM109, have been engineered to produce the β-polypeptide of *lacZ* but not the α-peptide. As a consequence, *E. coli* JM109 lacking a plasmid forms white colonies on plates containing X-Gal, but cells containing a pUC plasmid form blue colonies on X-Gal plates because the two segments, the α-peptide encoded in the plasmid and the β-peptide encoded in the chromosome, are capable of interacting to form a functional product capable of cleaving the X-Gal substrate (see Figure 3.26). The pUC plasmids have a multicloning site containing the recognition sequences for many different restriction endonucleases within the coding region for the α-peptide (Figure 4.16). If a foreign DNA fragment is cloned into any one of these sites, the bacterium does not make the α-peptide, and the colonies are white on X-Gal plates; bacteria containing plasmids with inserts are therefore easy to identify. These plasmids are also transcription vectors, because the promoter, p_{lac} (Figure 4.16), that drives expression of the α-peptide can also be used to express what is encoded in a piece of DNA directionally cloned into the multicloning site on the plasmid. The *lac* promoter is also inducible and only expressed if an inducer, such as isopropyl-β-D-thiogalactopyranoside (IPTG) or lactose, is added (see chapter 11). Thus, the cells can be propagated before the synthesis of the gene product is induced, a feature that is particularly desirable if the gene product is toxic to the cell. Genes cloned into one of the multicloning sites in the *lacZ* gene can also be translated from the *lacZ* translation initiation signal on the plasmid, provided that there are no intervening nonsense codons and the gene is cloned in the same reading frame as the upstream *lacZ* sequences (see chapter 2).

CONDITIONAL VECTORS

In some cases, it is useful to have a vector that can replicate only under certain conditions. Low-copy-number vectors with temperature-sensitive replication are often useful because they facilitate plasmid curing. The vector with temperature-sensitive replication is introduced at

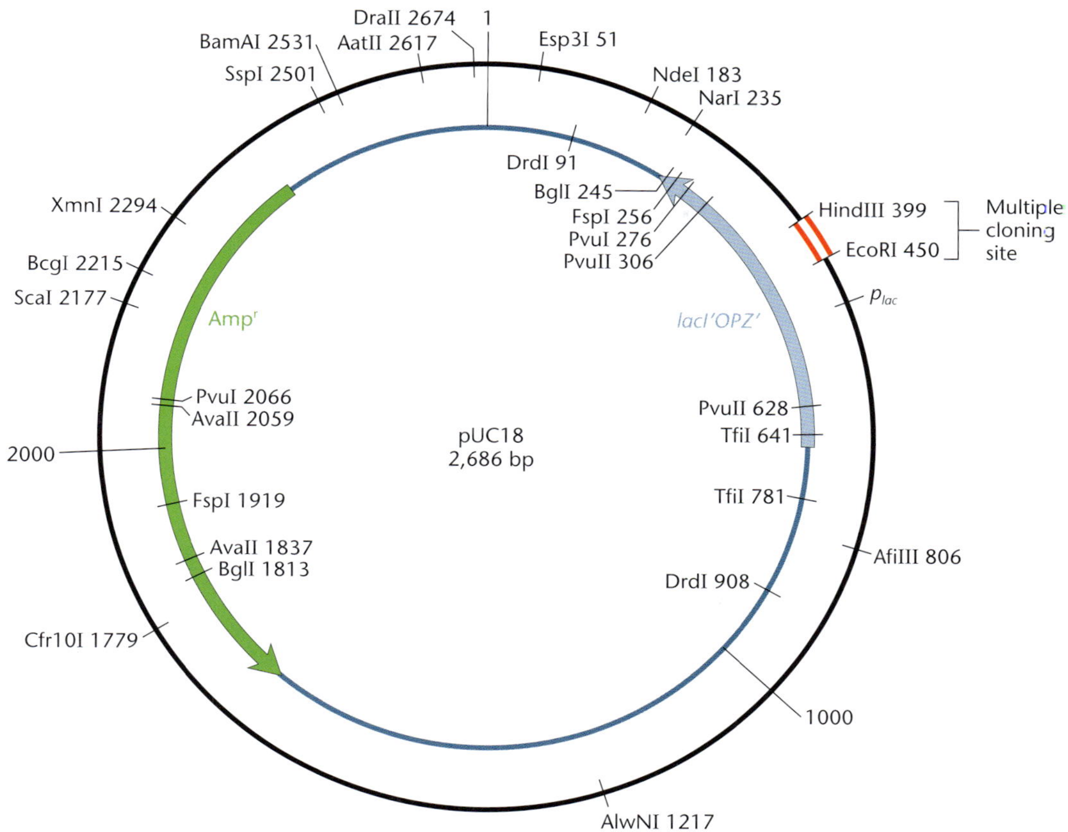

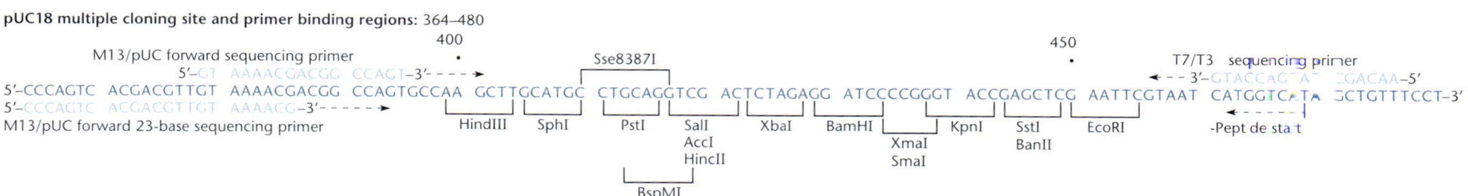

Figure 4.16 pUC expression vector. A gene cloned into one of the restriction sites in the multiple-cloning site almost invariably disrupts the coding sequence for the *lacZ* α-peptide. If it is inserted in the correct orientation, the gene is transcribed from the *lac* promoter (p_{lac}). If the open reading frame for the gene is in the same reading frame as that for the *lacZ* coding sequence, the gene is also translated from the *lacZ* translation initiation region (TIR), and the N-terminal amino acids of *lacZ* become fused to the polypeptide product of the gene.

the permissive temperature of 30°C in *E. coli*. When the cell population is shifted to the nonpermissive temperature of 40°C, replication stops and the plasmid is lost by dilution as the cells divide. Temperature-sensitive vectors can be useful as tools for integrating DNA sequences into the chromosome for gene fusions, as described in chapter 3. In some cases, it is useful to have a plasmid system in which replication occurs only in a certain host background. For example, as described in chapter 8, an *in vitro* transposition system in which a mobile DNA transposon containing an antibiotic resistance gene moves from a donor DNA plasmid into a different target plasmid as a way to subject it to insertional mutagenesis has been developed. In such a procedure, it is convenient to transform all of the DNA products under conditions where only the donor plasmid can replicate. One popular system utilizes the plasmid R6K origin of replication, called γ (see Metcalf et al., Suggested Reading). The γ origin requires the *trans*-acting protein π for replication. Normally, the π protein is encoded by a gene called *pir* on the plasmid. However, in this system, only the *cis*-acting origin remains on the plasmid. Because the π protein is not normally found in *E. coli*, these vectors replicate only in a conditional host where the *pir* product is expressed from the host chromosome. The *pir* gene is either inserted into a neutral site or is introduced through the use of phage λ. Under this system, the plasmid is maintained at a copy number of about 15 with the wild-type *pir* gene. However, a strain can also be used with a mutant *pir* gene, *pir-116*, which maintains the plasmid vector with a copy number of about 250, making it more amenable to plasmid purification.

BACTERIAL ARTIFICIAL CHROMOSOME VECTORS

One problem with using high-copy-number cloning vectors such as pUC vectors is that the clones are very unstable, particularly if they contain large DNA inserts. If the clone exists in many copies, ectopic recombination between repeated sequences in the copies can rearrange the sequences in the clone (see chapter 3). This is a particular problem in some applications, for example, cloning entire operons from bacteria, where it is necessary to obtain plasmid libraries containing very large inserts. For this reason, **bacterial artificial chromosome (BAC) cloning vectors** have been designed (Figure 4.17). These plasmid vectors are based on the F plasmid origin of replication and have a copy number of only 1 in *E. coli*. They can accommodate very large inserts, on the order of 300,000 bp of DNA. This was expected, since it was known that F′ factors, naturally occurring plasmids in which the F plasmid has incorporated a large region of the *E. coli* chromosome (see chapter 5), can be very large and are quite stable, especially in a *recA* host.

The original pBAC vector, shown in Figure 4.17 (see Shizuya et al., Suggested Reading), contains the F plasmid origin of replication and partitioning functions and a selectable chloramphenicol resistance gene. It also contains unique HindIII and BamHI restriction endonuclease cleavage sites into which large DNA fragments can be introduced, as well as a number of other features that are helpful in the cloning and sequencing of large fragments. This allows the DNA inserts to be excised from the cloning vector without (usually) cutting the DNA insert, as well. More contemporary vectors use sites recognized by an unusual class of enzymes called homing endonucleases, which have recognition sequences of around 30 bp. Even though these sites are not as stringently recognized as those recognized by restriction endonucleases, they still virtually eliminate the possibility that the site will be found in the DNA that is intended to be cloned.

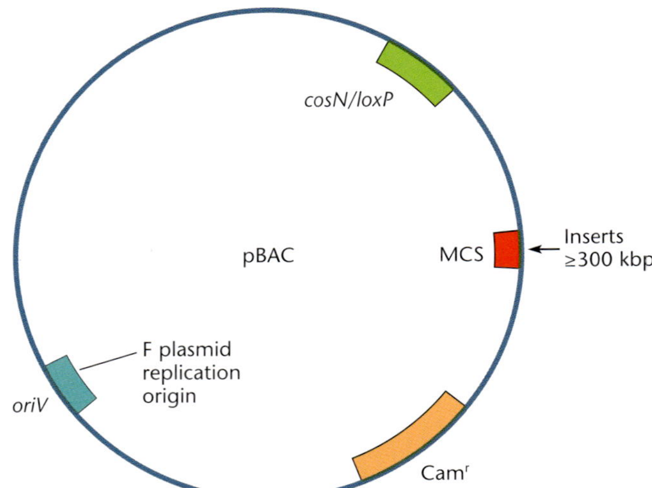

Figure 4.17 pBAC cloning vector for cloning large pieces of DNA. The multiple cloning site (MCS) where clones are inserted is shown. Also shown are the *loxP* and *cosN* sites, where the plasmid can be cut by the Cre recombinase or λ terminase, respectively, for restriction mapping of the insert. These recognition sites are long enough that they almost never occur by chance in the insert.

Broad-Host-Range Cloning Vectors

As explained above, many of the common *E. coli* cloning vectors, such as pBR322, the pUC plasmids, and the pET plasmids, have modified *ori* regions derived from ColE1. These modified ColE1 derivatives maintain the very narrow host range found with the original plasmid (Table 4.2) and replicate only in *E. coli* and a few of its close relatives. However, some cloning applications require a plasmid cloning vector that replicates in other types of bacteria, so cloning vectors have been derived from the broad-host-range plasmids RSF1010 and RK2, which replicate in most Gram-negative *Proteobacteria*. In addition to the broad-host-range *ori* region, these cloning vectors sometimes contain a *mob* site, which can allow them to be mobilized into other bacteria (see chapter 5). This trait is very useful, because ways of introducing DNA other than conjugation have not been developed for many types of bacteria, although electroporation works for many (see chapter 6).

SHUTTLE VECTORS

Sometimes, an experiment requires that a plasmid cloning vector be transferred from one organism into another. If the two organisms are distantly related, the same plasmid *ori* region is not likely to function in both organisms. Such applications require the use of **shuttle vectors**, so named because they can be used to "shuttle" genes between the two organisms. A shuttle vector has two origins of replication, one that functions in each organism. Shuttle vectors also must contain selectable genes that can be expressed in each organism.

In most cases, one of the organisms in which the shuttle vector can replicate is *E. coli*. The genetic tests can be performed with the other organism, but the plasmid can be purified and otherwise manipulated by the refined methods developed for *E. coli*.

Most plasmid replication functions and antibiotic resistance genes derived from *E. coli* are nonfunctional in distantly related bacteria, such as *B. subtilis*, which has led to the development of a series of shuttle vectors (Table 4.3), whereas still others can be used even in eukaryotes. For example, plasmid YEp13 (Figure 4.18) has the replication origin of the 2-μm circle, a plasmid found in the yeast *Saccharomyces cerevisiae*, which allows it to replicate in *S. cerevisiae*. It also has the pBR322 *ori* region and thus can replicate in *E. coli*. In addition, the plasmid contains the yeast gene *leu2*, which can be selected in yeast, as well as an Amp^r gene, which confers ampicillin resistance in *E. coli*. Similar shuttle vectors that can replicate in mammalian or insect cells and *E. coli* have been constructed. Some of these plasmids have the replication origin of the animal virus simian virus 40 (SV40) and the ColE1 origin of replication.

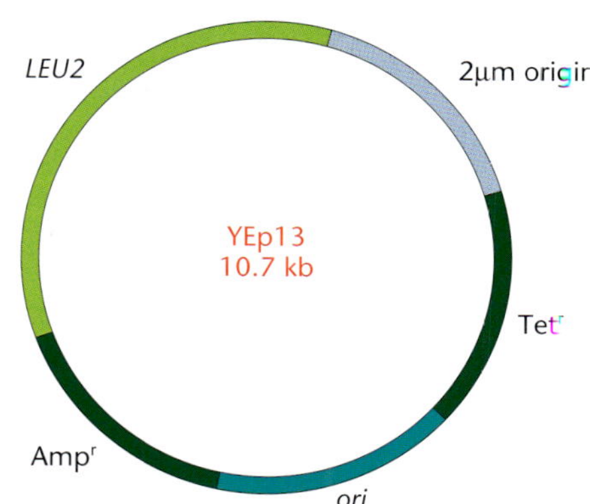

Figure 4.18 Shuttle plasmid YEp13. The plasmid contains origins of replication that function in the yeast *Saccharomyces cerevisiae* and the bacterium *Escherichia coli*. It also contains genes that can be selected in *S. cerevisiae* and *E. coli*.

1. Plasmids are DNA molecules that exist free of the chromosome in the cell. Most plasmids are circular, but some are linear. The sizes of plasmids range from a few thousand base pairs to almost the length of the chromosome itself. Probably the best distinguishing characteristic of a plasmid is that it has a typical plasmid origin of replication with an adjacent gene for a Rep protein rather than a chromosome origin with an *oriC* gene, along with a *dnaA* gene and other genes typical of the chromosomal origin of replication.

2. Plasmids usually carry genes for proteins that are necessary or beneficial to the host under some situations but are not essential under all conditions. Evolution probably selected for plasmids carrying nonessential or locally beneficial genes because it allows the chromosome to remain smaller but still allows bacterial populations to respond quickly to changes in the environment.

3. Plasmids replicate from a unique origin of replication, called *oriV*. Many of the characteristics of a given plasmid derive from the region surrounding *oriV*. They include the mechanism of replication, copy number control, partitioning, and incompatibility. If other genes are added to or deleted from the plasmid, it will retain most of its original characteristics, provided that the *ori* region remains.

4. Many plasmids replicate by a theta mechanism, with replication forks moving from a unique origin with leading and lagging template strands, much like circular bacterial chromosomes.

5. Some plasmids use a rolling circle (RC) mechanism, similar to that used to replicate some phage DNAs and during bacterial conjugation. In RC replication, the plasmid is nicked at a unique site, and the Rep protein remains attached to the 5′ end at the cut through one of its tyrosines. The free 3′ end is used as a primer to replicate around the circle, displacing one of the strands. When the circle is complete, the 5′ phosphate is transferred from the Rep protein to the 3′ hydroxyl to form a single-stranded circle. The host ligase rejoins the ends to form a double-stranded circular DNA. A strand complementary to the single-stranded circle is made, using a different origin, to form two double-stranded circular DNAs.

6. Replication of linear plasmids uses different mechanisms. Some have hairpin ends and replicate from an internal origin around the ends to form dimeric circles that are then processed to form two linear plasmids. Others have a terminal protein bound at both 5′ ends and extensive inverted-repeat sequences at their ends.

7. The copy number of a plasmid is the number of copies of the plasmid per cell immediately after cell division.

8. Different types of plasmids use different mechanisms to regulate their initiation of replication and therefore their copy numbers. Some plasmids use small complementary RNAs

Summary (continued)

transcribed from the other strand in the same region (countertranscribed, or ctRNAs) to regulate their copy numbers. In ColE1-derived plasmids, the ctRNA, called RNA I, interferes with the processing of the primer for leading-strand replication, called RNA II. In other cases, including the R1, ColIB-P9, and pT181 plasmids, the ctRNA interferes with the expression of the Rep protein required to initiate plasmid DNA replication.

9. Iteron plasmids regulate their copy numbers by two interacting mechanisms. They control the amount of the Rep protein required to initiate plasmid replication, and the Rep protein also couples plasmids through their iteron sequences.

10. Some plasmids have a special partitioning mechanism to ensure that each daughter cell gets one copy of the plasmid as the cells divide. These partitioning systems usually consist of two genes for proteins and a *cis*-acting centromere-like site.

11. If two plasmids cannot stably coexist in the cells of a culture, they are said to be incompatible or to be members of the same Inc group. They can be incompatible if they have the same copy number control system or the same partitioning functions.

12. The host range for replication of a plasmid is defined as all the different organisms in which the plasmid can replicate. Some plasmids have very broad host ranges and can replicate in a wide variety of bacteria. Others have very narrow host ranges and can replicate only in very closely related bacteria.

13. Plasmids have been engineered for use as cloning vectors. They make particularly desirable cloning vectors for some applications because they do not kill the host, can be small, and are easy to isolate. Some plasmids can carry large amounts of DNA and are used to make bacterial artificial chromosomes for cloning large fragments of DNA.

QUESTIONS FOR THOUGHT

1. Why are genes whose products are required for normal growth not carried on plasmids? List some genes that you would not expect to find on a plasmid and some genes you might expect to find on a plasmid.

2. Why do you suppose some plasmids have broad host ranges for replication? Why is it that not all plasmids have broad host ranges?

3. Why are the F plasmid and P1 compatible with one another even though they use the same system for partitioning?

4. How would you find the genes required for replication of the plasmid if they are not all closely linked to the *ori* site?

5. How would you determine which of the replication genes of the host *E. coli* (e.g., *dnaA* and *dnaC*) are required for replication of a plasmid you have discovered?

6. The R1 plasmid has a leader polypeptide translated upstream of the gene for RepA, and cleavage of the mRNA by RNase III occurs in the coding sequence for this leader polypeptide. This blocks the translation of the leader polypeptide and also the translation of the downstream *repA* gene to which it is translationally coupled. Do you think it would have been easier just to have the cleavage occur in the coding sequence for the RepA protein itself? Why or why not?

7. Try to design a mechanism that uses inverted-repeat sequences at the ends to replicate to the ends of a linear plasmid without the DNA getting shorter each time it replicates.

SUGGESTED READING

Aakre CD, Phung TN, Huang D, Laub MT. 2013. A bacterial toxin inhibits DNA replication elongation through a direct interaction with the β sliding clamp. *Mol Cell* **52:**617–628.

Asano K, Mizobuchi K. 2000. Structural analysis of late intermediate complex formed between plasmid ColIb-P9 Inc RNA and its target RNA. How does a single antisense RNA repress translation of two genes at different rates? *J Biol Chem* **275:**1269–1274.

Bagdasarian M, Lurz R, Rückert B, Franklin FCH, Bagdasarian MM, Frey J, Timmis KN. 1981. Specific-purpose plasmid cloning vectors. II. Broad host range, high copy number, RSF1010-derived vectors, and a host-vector system for gene cloning in *Pseudomonas*. *Gene* **16:**237–247.

Bao K, Cohen SN. 2001. Terminal proteins essential for the replication of linear plasmids and chromosomes in *Streptomyces*. *Genes Dev* **15:**1518–1527.

Baxter JC, Funnell BE. 2014. Plasmid partition mechanisms. *Microbiol Spectr* **2:**PLAS-0023-2014.

Campbell CS, Mullins RD. 2007. *In vivo* visualization of type II plasmid segregation: bacterial actin filaments pushing plasmids. *J Cell Biol* **179:**1059–1066.

Caspi R, Pacek M, Consiglieri G, Helinski DR, Toukdarian A, Konieczny I. 2001. A broad host range replicon with different requirements for replication initiation in three bacterial species. *EMBO J* **20:**3262–3271.

Cohen SN, Chang ACY, Boyer HW, Helling RB. 1973. Construction of biologically functional bacterial plasmids *in vitro*. *Proc Natl Acad Sci USA* **70:**3240–3244.

Das N, Valjavec-Gratian M, Basuray AN, Fekete RA, Papp PP, Paulsson J, Chattoraj DK. 2005. Multiple homeostatic mechanisms in the control of P1 plasmid replication. *Proc Natl Acad Sci USA* **102:**2856–2861.

Eberhard WG. 1989. Why do bacterial plasmids carry some genes and not others? *Plasmid* **21:**167–174.

Funnell BE, Phillips GJ (ed). 2004. *Plasmid Biology*. ASM Press, Washington, DC.

Garner EC, Campbell CS, Weibel DB, Mullins RD. 2007. Reconstitution of DNA segregation driven by assembly of a prokaryotic actin homolog. *Science* **315:**1270–1274.

Hamilton CM, Aldea M, Washburn BK, Babitzke P, Kushner SR. 1989. New method for generating deletions and gene replacements in *Escherichia coli*. *J Bacteriol* **171:**4617–4622.

Khan SA. 2000. Plasmid rolling-circle replication: recent developments. *Mol Microbiol* **37:**477–484.

Kobayashi K, et al. 2003. Essential *Bacillus subtilis* genes. *Proc Natl Acad Sci USA* **100:**4678–4683.

Kobryn K, Chaconas G. 2014. Hairpin telomere resolvases. *Microbiol Spectr* **2:**1–15.

Kolb FA, Malmgren C, Westhof E, Ehresmann C, Ehresmann B, Wagner EGH, Romby P. 2000. An unusual structure formed by antisense-target RNA binding involves an extended kissing complex with a four-way junction and a side-by-side helical alignment. *RNA* **6:**311–324.

Lewis K. 2010. Persister cells. *Annu Rev Microbiol* **64:**357–372.

Lim GE, Derman AI, Pogliano J. 2005. Bacterial DNA segregation by dynamic SopA polymers. *Proc Natl Acad Sci USA* **102:**17658–17663.

McEachern MJ, Bott MA, Tooker PA, Helinski DR. 1989. Negative control of plasmid R6K replication: possible role of intermolecular coupling of replication origins. *Proc Natl Acad Sci USA* **86:**7942–7946.

Meacock PA, Cohen SN. 1980. Partitioning of bacterial plasmids during cell division: a *cis*-acting locus that accomplishes stable plasmid inheritance. *Cell* **20:**529–542.

Metcalf WW, Jiang W, Daniels LL, Kim SK, Haldimann A, Wanner BL. 1996. Conditionally replicative and conjugative plasmids carrying lacZ alpha for cloning, mutagenesis, and allele replacement in bacteria. *Plasmid* **35:**1–13.

Mruk I, Kobayashi I. 2014. To be or not to be: regulation of restriction-modification systems and other toxin-antitoxin systems. *Nucleic Acids Res* **42:**70–86.

Novick RP, Hoppensteadt FC. 1978. On plasmid incompatibility. *Plasmid* **1:**421–434.

Novick RP, Iordanescu S, Projan SJ, Kornblum J, Edelman I. 1989. pT181 plasmid replication is regulated by a countertranscript-driven transcriptional attenuator. *Cell* **59:**395–404.

Peters JE. 2007. Gene transfer in Gram-negative bacteria, p 735–755. *In* Reddy CA, Beveridge TJ, Breznak JA, Marzluf GA, Schmidt TM, Snyder LR (ed), *Methods for General and Molecular Microbiology*, 3rd ed. ASM Press, Washington, DC.

Radloff R, Bauer W, Vinograd J. 1967. A dye-buoyant-density method for the detection and isolation of closed circular duplex DNA: the closed circular DNA in HeLa cells. *Proc Natl Acad Sci USA* **57:**1514–1521.

Ringgaard S, van Zon J, Howard M, Gerdes K. 2009. Movement and equipositioning of plasmids by ParA filament disassembly. *Proc Natl Acad Sci USA* **106:**19369–19374.

Shizuya H, Birren B, Kim U-J, Mancino V, Slepak T, Tachiiri Y, Simon M. 1992. Cloning and stable maintenance of 300-kilobase-pair fragments of human DNA in *Escherichia coli* using an F-factor-based vector. *Proc Natl Acad Sci USA* **89:**8794–8797.

Thomaides HB, Davison EJ, Burston L, Johnson H, Brown DR, Hunt AC, Errington J, Czaplewski L. 2007. Essential bacterial functions encoded by gene pairs. *J Bacteriol* **189:**591–602.

Thomas CM. 2000. Paradigms of plasmid organization. *Mol Microbiol* **37:**485–491.

Vagner V, Dervyn E, Ehrlich SD. 1998. A vector for systematic gene inactivation in *Bacillus subtilis*. *Microbiology* **144:**3097–3104.

Vecchiarelli AG, Han YW, Tan X, Mizuuchi M, Ghirlando R, Biertümpfel C, Funnell BE, Mizuuchi K. 2010. ATP control of dynamic P1 ParA-DNA interactions: a key role for the nucleoid in plasmid partition. *Mol Microbiol* **78:**78–91.

Yao S, Helinski DR, Toukdarian A. 2007. Localization of the naturally occurring plasmid ColE1 at the cell pole. *J Bacteriol* **189:**1946–1953.

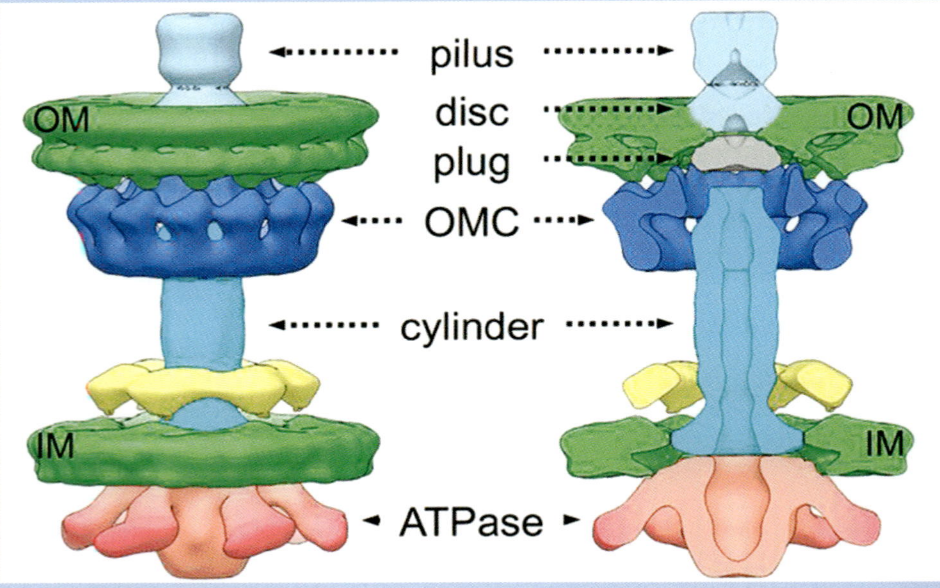

Three-dimensional surface rendering of the F plasmid-encoded plus complex and a cutaway view to display the architectural features. The position of the outer membrane (OM) and the associated outer membrane complex (OMC), as well as the position of the inner membrane (IM), are indicated. From Hu et al. 2019 (see Suggested Reading); for additional information, see Figure 3 in that reference.

5

Conjugation

A REMARKABLE FEATURE OF MANY PLASMIDS is the ability to transfer themselves and other DNA elements from one cell to another in a process called **conjugation**. Joshua Lederberg and Edward Tatum first observed this process in 1947, when they found that mixing certain strains of *Escherichia coli* with others resulted in strains that were genetically unlike either of the originals. As noted in the introduction, Lederberg and Tatum suspected that bacteria of the two strains exchanged DNA—that is, two parental strains mated to produce progeny unlike themselves but with characteristics of both parents. At that time, however, plasmids were unknown, and it was not until later that the basis for the mating was understood.

The plasmid that Lederberg and Tatum were unknowingly working with—called the fertility plasmid, or **F plasmid**—has been the focus of much of the research about the process of bacterial conjugation. The central role conjugation systems played in the development of bacterial genetics contrasts with the more notorious role these plasmids play as vectors for the spread of antibiotic resistance between bacteria. While historically studied with plasmids, conjugation is now also known to be used very frequently in the transfer of elements that are not maintained separately from the chromosome but instead are integrated into the chromosome and are therefore called **integrating conjugative elements** (ICEs). Their integration into the host chromosome typically uses the same type of recombination that is used for the integration of many bacteriophages. The same families of conjugation systems are found in ICEs and conjugal plasmids, indicating that over evolutionary time the same systems are alternatively utilized for intra- and extrachromosomal lifestyles (see Cury et al., Suggested Reading).

Overview

During plasmid conjugation, the two strands of an element separate in a process resembling rolling-circle replication (see "Mechanism of DNA Transfer during Conjugation in *Proteobacteria*" below), and one strand moves from the bacterium that originally contained the plasmid—the **donor**—into a **recipient** bacterium. The two single strands serve as templates for DNA replication concurrent with the process of DNA transfer to yield double-stranded DNA molecules in both the donor cell and the recipient cell. A recipient cell that has received DNA as a result of conjugation is called a **transconjugant**. A simplified view of conjugation is shown in Figure 5.1.

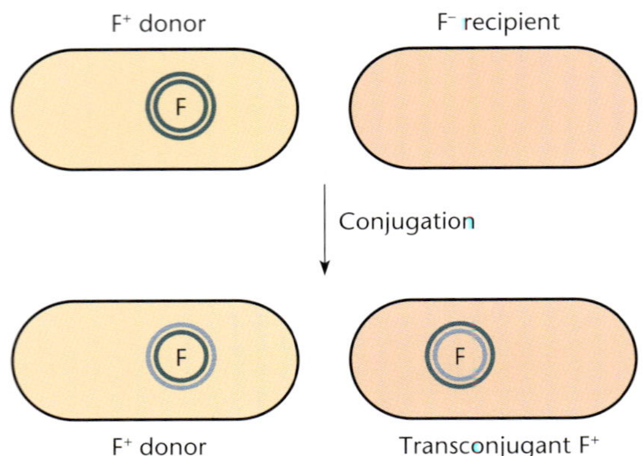

Figure 5.1 A simplified view of conjugation by a self-transmissible plasmid, the F plasmid. One strand of the plasmid is transferred from the donor to the recipient cell, the second strand is synthesized in the recipient cell, and the transferred strand is replaced in the donor cell by DNA replication, so that both the donor and recipient cells have the intact plasmid. A cell that has received the plasmid by conjugation is a transconjugant.

Many naturally occurring plasmids can transfer themselves. If so, they are said to be **self-transmissible**. The prevalence of conjugation systems suggests that plasmid conjugation is advantageous for plasmids without placing an undue burden on the bacterial host. Self-transmissible plasmids encode all the functions they need to move among cells, and often they also aid in the transfer of **mobilizable plasmids**. Mobilizable plasmids encode some, but not all, of the proteins required for transfer and consequently need the help of self-transmissible plasmids in the same donor cell to move to a new cell.

Any bacterium harboring a self-transmissible plasmid is a potential donor, because it can transfer DNA to other bacteria. In *Proteobacteria*, such cells produce a structure, called a **pilus**, that facilitates conjugation (discussed below). Bacteria that lack the self-transmissible plasmid are potential recipients, and conjugating bacteria are known as **parents**. Potential donor strains were historically referred to as **male strains**.

Self-transmissible plasmids appear to exist in essentially all types of bacteria (Guglielmini et al., 2011, Suggested Reading), but those that have been studied most exten-

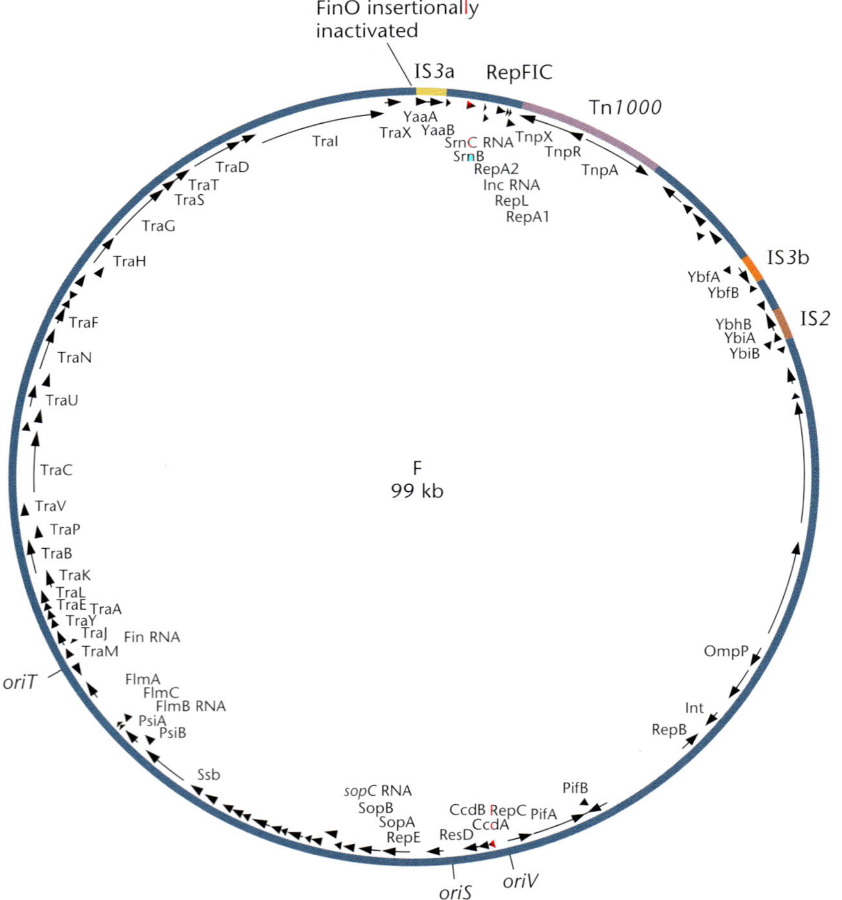

Figure 5.2 Partial genetic map of the ~100-kilobase pair (kbp) self-transmissible F plasmid. The regions of the insertion elements IS*3* (IS*3a* and IS*3b*) and IS*2* and transposon Tn*1000* are shown. *oriV* is the origin of replication; *oriT* is the origin of conjugal transfer. Multiple genes whose products' functions are known are shown (arrows) and are categorized in Table 5.1. Details are given in the text.

sively are from the *Proteobacteria* genera *Escherichia* and *Pseudomonas*; the *Firmicutes* genera *Enterococcus*, *Streptococcus*, *Bacillus*, and *Staphylococcus*; and the *Actinobacteria* genus *Streptomyces*. The best-known transfer systems are those of plasmids isolated from *Proteobacteria*. Therefore, we focus our attention first on these systems and the original conjugal plasmid, the F plasmid, and the important role it played in bacterial genetics. Conjugal plasmids found in other types of bacteria will be covered later in this chapter. Finally, we discuss multiple types of ICEs.

Classification of Self-Transmissible Plasmids and Integrating Elements

Bacteria have multiple types of transfer systems, which are encoded by the *tra* genes [see "Transfer (*tra*) Genes" below]. These groupings were originally aligned with the aforementioned plasmid incompatibility groups discussed in chapter 4, which are based on replication and segregation systems. With the large-scale analysis of genomic information, a broader picture has emerged in which almost all of the types of conjugation systems appear to share an ancient ancestor. Inferred ancestry is now used to group the related types of conjugation systems that often are named for the best-studied system within the group, regardless of whether the system was originally identified on a plasmid or an ICE.

The Fertility Plasmid

Our understanding of conjugation is largely based on work with the F plasmid, which was discovered first, and systems related to F. As mentioned above, the F plasmid has also played an essential role in the establishment of bacterial genetics as a field of study. The entire sequence of the ~100-kilobase (kb) F plasmid is known, and examining this sequence provides a window into the complexity of the process of conjugation (Figure 5.2). The genes required for transfer are called the ***tra*** genes, which make up a significant portion of the self-transmissible F plasmid genome. In addition to the *tra* gene products, there are many gene products that play supporting roles in the conjugation process or are otherwise advantageous to the host (Table 5.1). Interestingly, even in this highly studied plasmid, the functions of many of the gene products encoded by genes on the F plasmid remain unknown.

The hodgepodge organization of the F plasmid underscores the evolutionary path that is typical of large horizontally transferred elements. Examination of the F plasmid indicates that the sequences originated from many different sources, but beneath the patchwork is a complex and highly evolved system for transfer and maintenance of the plasmid (Figure 5.2). There is evidence for three replication origins on the F plasmid, RepFIA, RepFIB, and RepFIC. The origin responsible for replication

Table 5.1 Some F plasmid genes and sites

Function	Protein, site, or antisense RNA
Vegetative replication	Origin RepFIA/*oriV* site (RepC, RepE)
	Origin RepFIB/*oriS* site (RepB)
	Origin RepFIC/inactivated origin (RepA2, RepL, Inc [RNA], RepA1)
Regulation of conjugation	FinO, FinP (RNA), TraJ
Mating pair formation (Mpf component)	TraA, TraB, TraC, TraD, TraE, TraF, TraG, TraH, TraK, TraL, TraN, TraP, TraQ, TraV, TraW, TraX
DNA transfer and replication (Dtr component)	*oriT* site, TraI, TraM, TraU, TraY
Plasmid SOS inhibition	PsiA, PsiB
Surface exclusion	TraS, TraT
Plasmid partitioning	SopA, SopB, and SopC sites
Postsegregational killing	SrnC (RNA) and SrnB
	CcdA and CcdB
	FlmA, FlmC, FlmB (RNA)
Exclusion of T7	PifA, PifB
Transposon functions	IS*3a* (YaaA, YaaB)
	Tn*1000* (TnpX, TnpR, TnpA)
	IS*3b* (YbfA, YbfB)
	IS*2* (YbhB, YbiA, TbiB)
Known function, but unknown relation to conjugation	OmpP, Int, ResD, SsB

of the plasmid by theta replication is the RepFIA origin, which is also known as *oriV*. The RepFIB/*oriS* origin is not sufficient for reliable maintenance of the plasmid, and the RepFIC origin is inactivated by a Tn*1000* insertion element (Figure 5.2).

There are multiple systems encoded by genes on the plasmid that help stabilize the F plasmid. A partitioning system similar to those found on some bacterial chromosomes helps ensure that daughter cells each receive a copy of the plasmid, while multiple "postsegregational killing systems" sicken cells that lose the plasmid, which often die unless they receive a copy of the same plasmid (see chapter 4). Other systems that likely encourage maintenance of the plasmid in the host are a system that blocks the development of bacteriophage T7 and a system that helps transfer by preventing induction of the host SOS DNA damage response during conjugation.

Mechanism of DNA Transfer during Conjugation in *Proteobacteria*

Much has been learned over the years about the mechanistic details of plasmid conjugation, but many fundamental questions still remain.

Transfer (*tra*) Genes

About a third of the F plasmid is composed of *tra* genes (Figure 5.2). With so many genes required for conjugation and other transfer-related processes, it is understandable that self-transmissible plasmids are very large and that selection drives the evolution of mobilizable elements (see below).

The *tra* genes of a self-transmissible plasmid required for plasmid transfer can be divided into two components (Table 5.1). Some of the *tra* genes encode proteins involved in the processing of the plasmid DNA to prepare it for transfer. This is called the **Dtr component** (for DNA transfer and conjugal replication). The bulk of the *tra* genes encode proteins of the **Mpf component** (for mating pair formation). This large membrane-associated structure includes the pilus that is responsible for holding the mating cells together and the channel that forms between the mating cells through which the plasmid is transferred. Below, we outline what is known about these two components.

THE Mpf COMPONENT

The function of the Mpf component is to hold a donor cell and a recipient cell together during the mating process and to form a channel through which proteins and DNA are transferred during mating. It also includes the protein that communicates "news" of mating pair formation to the Dtr system, which initiates the transfer of plasmid DNA. A representation of the entire Mpf system of the F plasmid is shown in Figure 5.3.

The Pilus

The most dramatic feature of the Mpf structure is the pilus, a tube-like structure that sticks out of the cell surface (Figure 5.3). These pili are 10 nm or more in diameter with a central channel (see Hu et al., Suggested Reading). An ATP-driven process allows for extension and retraction of the large pilus structure. Each pilus is constructed of many copies of a single protein called the **pilin** protein. The pilin protein is synthesized with a long signal sequence that is removed as it passes through the membrane to assemble on the cell surface. In many systems, the pilin protein is also cyclized, with its head attached to its tail, which is unusual among proteins.

The structure of the pilus itself differs markedly among plasmid transfer systems. Differences in the length, rigidity, and diameter of the pilus are likely to be evolutionary adaptations to the specific environments where conjugation naturally takes place, but in the laboratory, these characteristics determine the conditions needed to allow mating pair formation and plasmid transfer to occur (i.e., in liquid cultures or on solid media). Plasmids like ColIb-P9 may be adapted to multiple environments; they have the capacity to make two types of pili, a long thin one and a short rigid one, with the former increasing the frequency of mating in liquid environments and the latter increasing the frequency of mating on solid surfaces. The pilus is also utilized as a receptor for some types of bacteriophages (see Box 5.1).

In the F plasmid, and probably other conjugal elements, the pilus is important for drawing cells together for mating pair formation. Once retracted, this will become the portal used for DNA transfer. Remarkably, it is still under debate whether DNA naturally travels through the extended pilus. The idea that transfer, at least under some circumstances, can occur without direct contact between cells has been supported by work with a novel system that allows transfer to be indirectly visualized under the microscope in real time (see Babic et al., 2008, Suggested Reading).

Coupling Proteins

The Mpf component of a donor cell is the first to make contact with a recipient cell (Figure 5.4). Then, the information that it has contacted another cell is communicated to the Dtr component before DNA transfer occurs. The communication between the Dtr and Mpf systems is provided by a **coupling protein**, which is part of the Mpf system. Coupling proteins provide the specificity for the transport process, so that only certain plasmids are transferred. Coupling proteins (TraD in the case of the F plasmid) are ATP-hydrolyzing enzymes that bind the membrane channel. Once an appropriate DNA substrate docks with the coupling protein, the coupling protein hydrolyzes ATP and, in turn, undergoes a conformational change

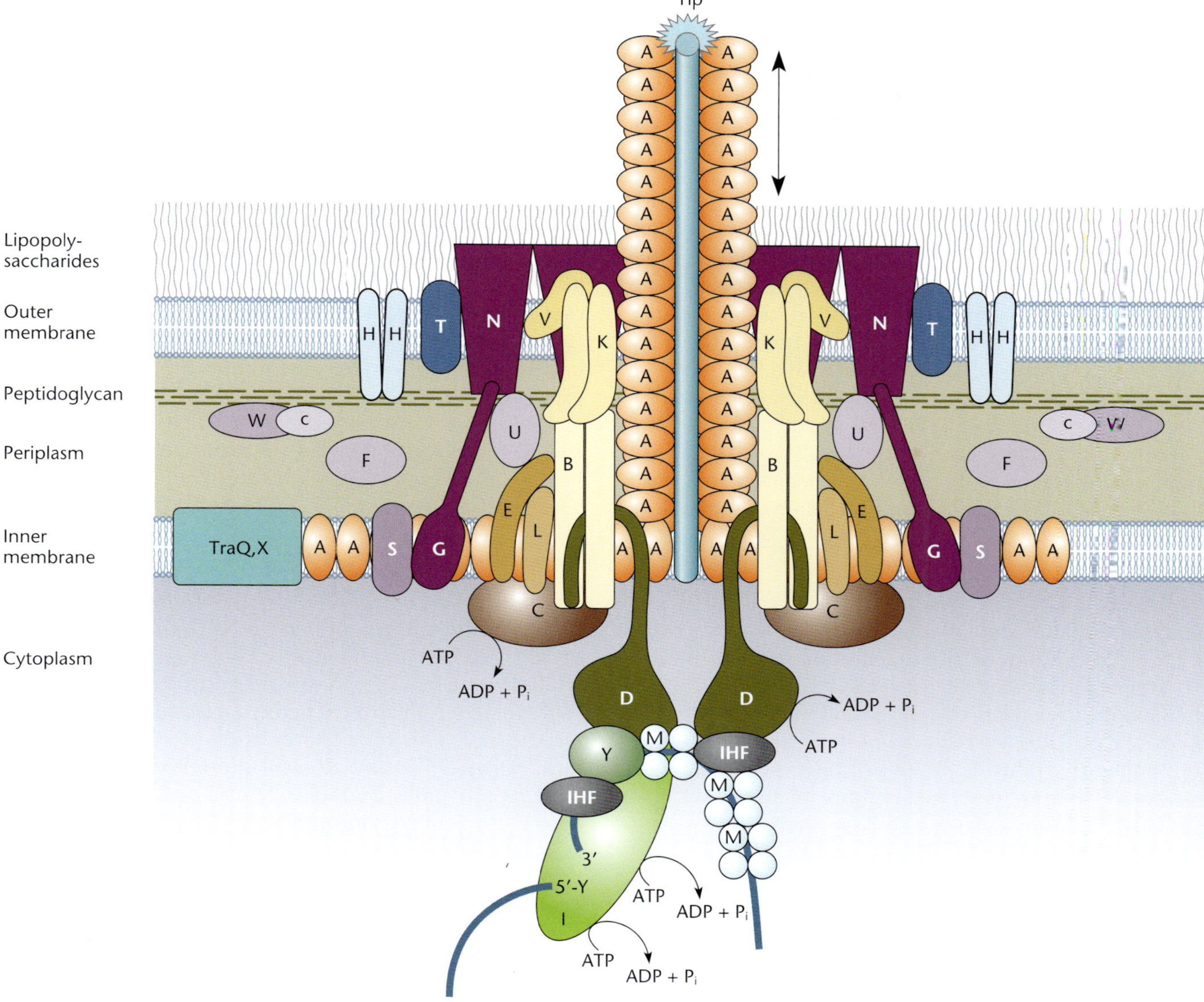

Figure 5.3 Representation of the F transfer apparatus. The pilus is assembled with five TraA (pilin) subunits per turn that are inserted into the inner membrane via TraQ and acetylated by TraX. The pilus is shown extending through a pore constructed of TraB and TraK, a secretin-like protein anchored to the outer membrane by the lipoprotein TraV. TraB is an inner membrane protein that extends into the periplasm and contacts TraK. TraL seeds the site of pilus assembly and attracts TraC to the pilus base, where it acts to drive assembly in an energy-dependent manner. A channel formed by TraA units is indicated, as is a specialized structure at the pilus tip that remains uncharacterized. The two-headed arrow indicates the opposing processes of pilus assembly and retraction. The Mpf proteins include TraG and TraN, which aid in mating pair stabilization (Mps), and TraS and TraT, which disrupt mating pair formation through entry and surface exclusion, respectively. TraF, TraH, TraU, TraW, and TrbC, which together with TraN are specific to F-like systems, appear to play a role in pilus retraction, pore formation, and mating pair stabilization. The relaxosome, consisting of TraY, TraM, TraI, and host-encoded IHF bound to the nicked DNA in *oriT*, is shown interacting with the coupling protein, TraD, which in turn interacts with TraB. The 5′ end of the nicked strand is shown bound to a tyrosine in TraI, and the 3′ end is shown as being associated with TraI in an unspecified way. The retained, unnicked strand is not shown. TraC, TraD, and TraI have ATP utilization motifs, represented by curved arrows; ATP is split into ADP and inorganic phosphate (P_i). Two curved arrows are shown with TraI, which utilizes ATP separately in two different regions of the protein for relaxase and helicase activities. Modified from Lawley T, Wilkins BM, Frost LS, *in* Funnell BE, Phillips GJ (ed), *Plasmid Biology* (ASM Press, Washington, DC, 2004).

BOX 5.1

Pilus-Specific Phages

Some types of phages only infect cells that express a certain type of conjugation-associated pilus on their surfaces. All phages adsorb to specific sites on the cell surface to initiate infection (see chapter 7), and some phages use the pilus of a self-transmissible plasmid as their adsorption site. Phages that adsorb to the conjugation pilus of a self-transmissible plasmid are called **pilus-specific phages** because they infect only donor cells. Pilus-specific phages are also known as male-specific phages because only "male" (e.g., F$^+$) cells produce the pilus. Examples of pilus-specific phages are M13 and R17, which infect only cells carrying the F plasmid, and Pf3 and PRR1, which infect cells containing plasmid RP4 and related plasmids.

The susceptibility of pilus-expressing cells to certain phages may in part explain why the *tra* genes of plasmids are usually tightly regulated. Most self-transmissible plasmids express a pilus only immediately after entering a cell and only intermittently thereafter (see "Example: Regulation of *tra* Genes in F Plasmids"). If cells containing the plasmid always expressed the pilus, a pilus-specific phage could spread quickly through the population, destroying many of the cells and with them the plasmids they contain. By only intermittently expressing a pilus, cells containing a self-transmissible plasmid limit their susceptibility to phages that use their pilus as an adsorption site. For bacteria that reside in animal hosts, pili might also be expected to be highly immunogenic; therefore, regulating the expression of proteins on the cell surface should also limit detection and clearance of the plasmid's bacterial host in the host animal. Expression of the *tra* genes could also be a significant burden on cellular resources and reduce host competitiveness if not regulated.

to activate the next steps of transfer. One or two other ATPases coordinate with the coupling protein to build the translocation channel and power substrates through the channel. Independently of the coupling protein, these other ATPases also are required to build and disassemble the conjugative pilus. Functions of the type IV secretion system ATPases are described in more detail below.

THE Dtr COMPONENT

The Dtr component of a self-transmissible plasmid is involved in preparing the plasmid DNA for transfer. A number of proteins make up this component, and the functions of many of them are known.

Relaxase

A central part of the Dtr component of plasmids is the **relaxase**, which is TraI in the F plasmid. Relaxase is a site-specific DNA endonuclease that makes a single-strand break, or "nick," at a specific *nic* site within the *oriT* sequence. Following a signal from the Mpf component, the relaxase initiates the transfer process. The way in which the relaxase works is illustrated in Figure 5.5. The relaxase breaks a phosphodiester bond at the *nic* site by transferring the bond from a deoxynucleotide to one of its own tyrosines. While covalently bound to the 5′ end of the DNA, the relaxase protein can pilot the strand into the recipient cell. The nicking reaction is called a transesterification reaction and requires very little energy because there is no net breakage or formation of new chemical bonds. In the recipient cell, the TraI relaxase protein recircularizes the plasmid by doing essentially the reverse of what it did in the donor cell. It binds to the two halves of the cleaved *oriT* sequence, holding them together while it transfers the phosphate bond from its tyrosine back to the 3′ hydroxyl deoxynucleotide in the DNA (Figure 5.5). This transesterification reaction reseals the nick in the DNA and releases the relaxase, which has done its job and is degraded. Transfer of a relaxase protein to recipient cells appears to be maintained in all conjugation systems studied to date (see below).

The Relaxosome

The relaxase protein in the donor cell is part of a larger structure called the **relaxosome**, which is made up of a number of proteins that are normally bound to the *oriT* sequence of the plasmid. The relaxase of the F plasmid uses accessory proteins, such as TraU, TraM, and TraY, for recognizing and coordinating activity with the Mpf proteins, although the exact mechanism of this signaling is still unclear. TraY and the host protein IHF are essential accessory proteins for nicking at *oriT* in the test tube, while TraM and TraD likely help coordinate activity with the transfer apparatus in living cells. In the F plasmid, the TraI protein that nicks at *oriT* also has a helicase activity that is required for conjugation. This helicase activity appears to play multiple roles during conjugation. In the donor cell, the helicase activity separates the strands for presentation to the Mpf component but probably also helps present a structure that can be recognized by DNA polymerase III, which is responsible for replicating the complementary strand. The helicase may also play additional roles in the recipient cell.

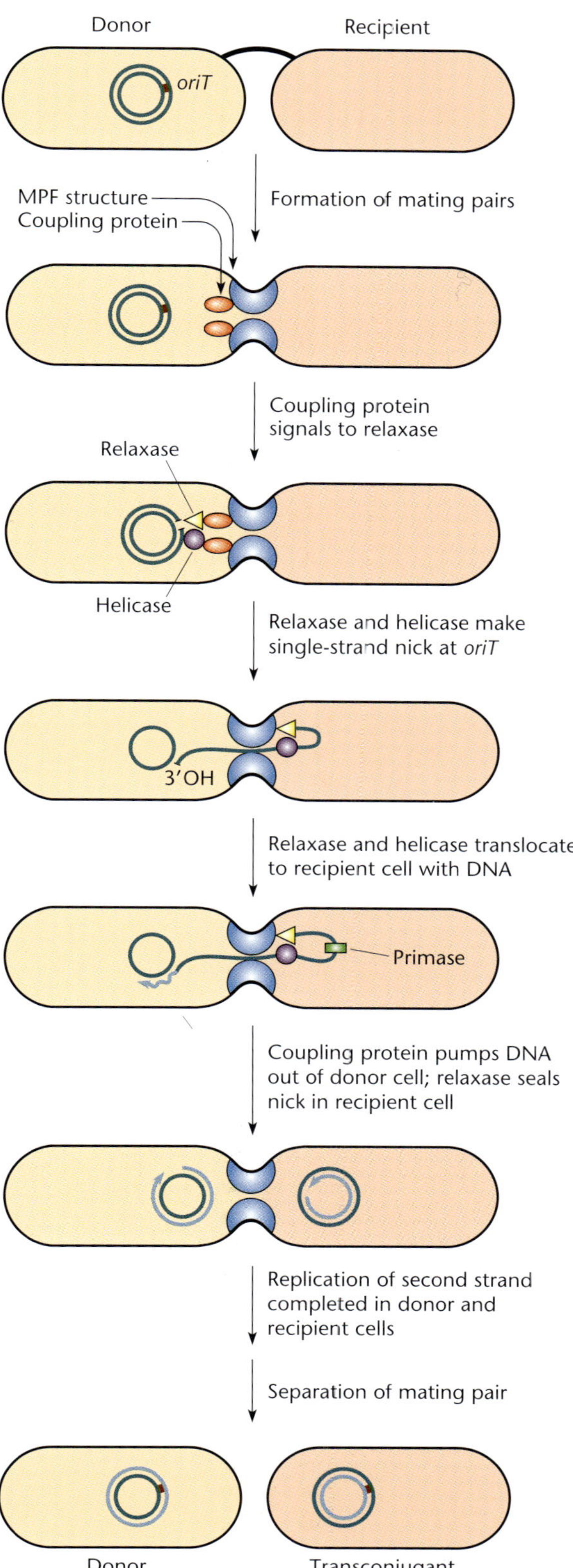

The *oriT* Sequence

The *oriT* site is not only the site at which plasmid transfer initiates, but is also the site at which the DNA ends rejoin to recircularize the plasmid. Plasmid transfer initiates specifically at the *oriT* site, because the specific relaxase encoded by one of the *tra* genes nicks DNA only at this sequence. Also, presumably, the plasmid-encoded helicase enters DNA only at this sequence to separate the strands. Regulation of *oriT* activity is coordinated through binding sites for IHF, TraY, and TraM. Cloning the minimal *oriT* sequence into a plasmid allows mobilization of the plasmid if the self-transmissible F plasmid also resides in the cell, as discussed below.

PRIMASE

In many conjugation systems, a component of the Dtr system made in the donor is the **primase**. Primases are needed for chromosomal DNA replication to make an RNA primer to initiate the synthesis on the lagging strand template during DNA replication (see chapter 1) and to prime plasmid replication (see chapter 4). However, at first, the role that a primase would play in the donor was not clear. A primase may not be necessary to prime replication in the donor cell, since the free 3′ hydroxyl end of DNA created at the nick in *oriT* could provide the primer for replication during transfer, similar to the priming of the first stage of replication of rolling-circle plasmids (see chapter 4). The second strand of the F plasmid is made in the recipient as it is spooled into the recipient cell in a process that is similar to lagging-strand DNA synthesis on a lagging strand template. Interestingly,

Figure 5.4 Mechanism of DNA transfer during conjugation, showing the Mpf functions in blue. The donor cell produces a pilus, which forms on the cell surface and which may contact a potential recipient cell and bring it into close contact or may help to hold the cells in close proximity after contact has been made, depending on the type of pilus. A pore then forms in the adjoining cell membranes. On receiving a signal from the coupling protein that contact with a recipient has been made, the relaxase protein initiates transfer by nicking the DNA at the *oriT* site in the plasmid. A plasmid-encoded helicase then separates the strands of the plasmid DNA. The relaxase protein, which has remained attached to the 5′ end of the single-stranded DNA, is then transported out of the donor cell through the channel directly into the recipient cell, dragging the attached single-stranded DNA along with it. The helicase activity pumps DNA out of the donor cell. Once in the recipient, the relaxase protein helps recyclize the single-stranded DNA. A primase, encoded either by the host or by the plasmid and injected with the DNA, then primes replication of the complementary strand to make the double-stranded circular plasmid DNA in the recipient. The 3′ end at the nick made by the relaxase in the donor can also serve as a primer, making a complementary copy of the single-stranded plasmid DNA remaining in the donor. Therefore, after transfer, both the donor and the recipient bacteria have a double-stranded circular copy of the plasmid. Details are given in the text.

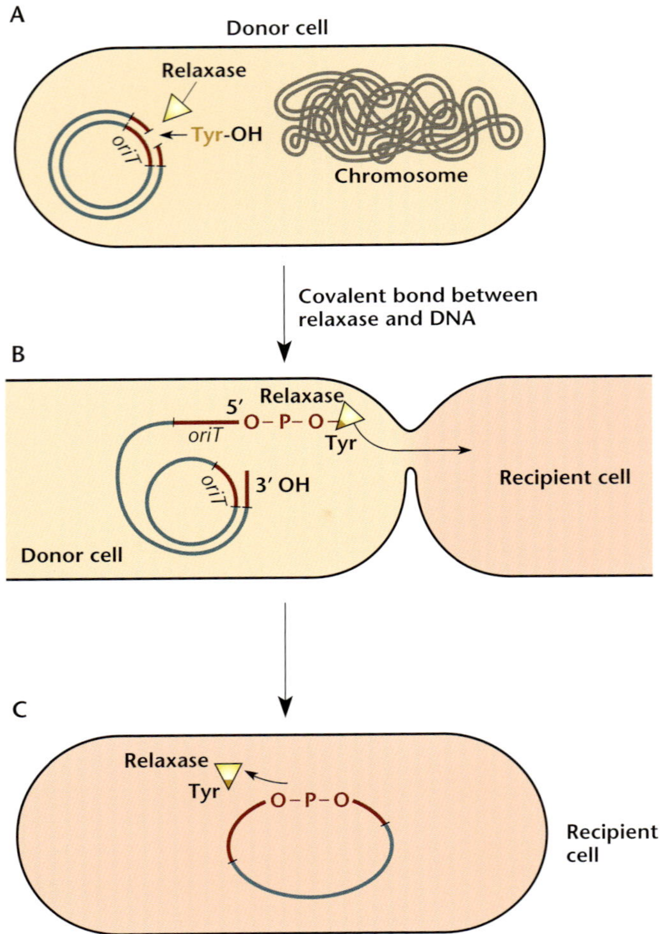

Figure 5.5 Reactions performed by the relaxase. **(A)** The relaxase nicks the DNA at a specific site in *oriT*, and the 5′ phosphate is transferred to one of its tyrosines in a transesterification reaction. **(B)** The relaxase is transferred to the recipient cell, dragging the DNA along with it. **(C)** In a reversal of the original transesterification reaction, the phosphate is transferred back to the 3′ hydroxyl of the other end of the DNA, recyclizing the DNA and releasing the relaxase.

priming of the F plasmid in recipient cells can occur without a primase protein expressed within the donor. Apparently, the primase can be expressed using the single-stranded DNA as it is transferred to the recipient. While this at first appears to violate what we learned about gene expression in chapter 2, the process is allowed by one or more unusual promoters that are capable of functioning when DNA is in the single-strand form. In other plasmids, such as RP4 and ColIb-P9, a protein primase produced in the donor cell is transported into the recipient cell, where it presumably primes lagging-strand DNA replication of the plasmid genome.

Why would a plasmid bother to make its own primase if it can use the host cell primase instead to initiate DNA replication? The answer may be that it does this to make it more promiscuous and able to transfer into a wider variety of bacterial species. Sometimes a promiscuous plasmid may find that it has transferred itself into a type of bacterium that is so distantly related to its original host that the primase in the bacterium does not recognize the sequences on the plasmid DNA necessary to prime the replication of the complementary strand. In that case, utilization of the plasmid-encoded primase allows replication to occur.

Efficiency of Transfer

One of the striking features of many transfer systems is their efficiency. Under optimal conditions, some plasmids can transfer themselves into other cells in almost 100% of cell contact events. This high efficiency has been exploited in the development of methods for transferring cloned genes between bacteria and in transposon mutagenesis, a process that requires highly efficient transfer of DNA. Such a method is described in Box 5.2.

REGULATION OF THE *tra* GENES

Many naturally occurring plasmids transfer with high efficiency for only a short time after they are introduced into cells and transfer only sporadically thereafter. Most of the time, the *tra* genes are repressed, and without the synthesis of pilin and other *tra* gene functions, the pilus is lost. For unknown reasons, the repression is relieved occasionally in some of the cells, allowing this small percentage of cells to transfer their plasmids at a given time.

This property of only periodically expressing their *tra* genes probably does not prevent the plasmids from spreading quickly through a population of bacteria that does not contain them. When a plasmid-containing population of cells encounters a population that does not contain the plasmid, the plasmid *tra* genes in one of the plasmid-containing cells are eventually expressed, and the plasmid transfers to another cell. Then, when the plasmid first enters a new cell, efficient expression of the *tra* genes leads to a cascade of plasmid transfer from one cell to another (a process sometimes referred to as zygotic induction). As a result, the plasmid soon occupies most of the cells in the population. This is the rationale behind **triparental mating**, a process where three parental stains are used in the mating mixture. The first strain contains a self-transmissible plasmid, the second contains the plasmid to be mobilized, and the third is the final recipient. Therefore, in this type of procedure, the *trans*-acting functions that mobilize the delivery vector are on a separate plasmid vector (compare this idea with the donor in Box 5.2, where the *trans*-acting conjugation functions are expressed from the chromosome). The efficiency of triparental mating comes from the burst of expression that is found when a conjugal plasmid enters a new host, maximizing transfer of the mobilizable plasmid into the final recipient.

BOX 5.2

Delivery of Conditional Plasmids by Conjugation

The efficiency of conjugation has been harnessed as a tool for delivering plasmid DNA from a host from which the plasmid can transfer to a strain where it cannot replicate. This is a useful technique for delivering mobile DNA elements like transposons that can move between positions in the genome. Transposons that insert randomly offer an important tool for knocking out genes by insertional inactivation, creating gene fusions in random locations, and other procedures discussed elsewhere in this text. Often, a researcher will want to select for an antibiotic resistance determinant that is encoded on the transposon as a means of identifying the bacteria in which the element has moved from a delivery plasmid into the chromosome. This procedure requires that the plasmid used to deliver the transposon cannot replicate in the target recipient bacterium. While transposon delivery is described here, the same procedures can be used to deliver a plasmid encoding a mutant allele of a gene into a host for recombination into the recipient chromosome for replacement by gene exchange.

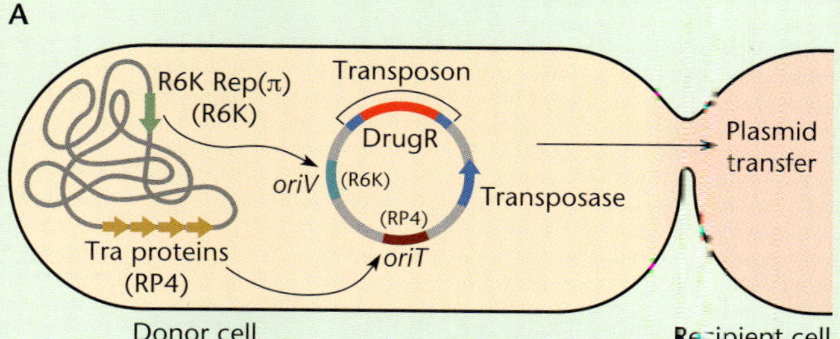

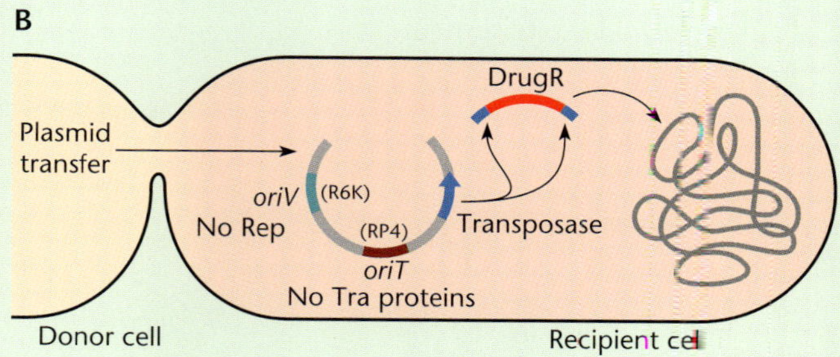

Processing Events in the Donor Cell (Panel A)

E. coli strains are available that are engineered to maintain and deliver plasmids efficiently by conjugation. A conditional plasmid (see chapter 4) can reside in the delivery strain because the host also expresses a *trans*-acting replication protein (Rep in the figure), like the π protein from plasmid R6K. Plasmids with the origin of replication (*oriV* in the figure) derived from plasmid R6K, but lacking the gene encoding the cognate replication protein, will be maintained only in a strain that expresses the replication protein of R6K. In one popular system, the transfer proteins (Tra proteins in the figure) from plasmid RP4 are expressed from the chromosome of the donor strain, and the *cis*-acting origin of transfer (*oriT* in the figure) from the RP4 plasmid resides in the delivery plasmid. This plasmid is therefore a hybrid with the *cis*-acting aspects of the R6K replication systems and RP4 transfer systems.

Processing Events in the Recipient Cell (Panel B)

After the delivery plasmid has moved into the recipient cell by conjugation, it is no longer able to be replicated and maintained because the recipient strain does not express the *trans*-acting replication protein. However, expression of the transposase gene that is encoded on the plasmid will allow the transposon to excise from the plasmid and move to other DNAs in the cell, like the chromosome. Because the plasmid is rapidly lost, expression of the transposon-encoded selectable marker, such as an antibiotic resistance gene (DrugR in the figure), will select for those bacteria in which the transposon has inserted into a DNA that is capable of stable replication (e.g., the chromosome or another plasmid). Additional transposition events are not possible, because the transposase gene is encoded on the plasmid that was lost from the recipient cell.

References

Metcalf WW, Jiang W, Wanner BL. 1994. Use of the *rep* technique for allele replacement to construct new *Escherichia coli* hosts for maintenance of R6K gamma origin plasmids at different copy numbers. *Gene* **138:**1–7.

Metcalf WW, Jiang W, Daniels LL, Kim S-K, Haldimann A, Wanner BL. 1996. Conditionally replicative and conjugative plasmids carrying lacZα for cloning, mutagenesis, and allele replacement in bacteria. *Plasmid* **35:**1–13.

Simon R, Priefer U, Puhler A. 1983. A broad host range mobilization system for *in vivo* genetic engineering: transposon mutagenesis in Gram negative bacteria. *Biotechnology* **1:**784–791.

Once a plasmid has established itself in a recipient cell, it inhibits the entry of additional copies of the same plasmid into the cell in a process called **surface exclusion**. In the F plasmid two proteins, TraS and TraT, are responsible for surface exclusion, but different molecular systems are responsible for this process in other conjugal elements. In some plasmids there are pheromone systems that are instead responsible for encouraging a potential host to serve as a recipient (See Dunny, Suggested Reading).

EXAMPLE: REGULATION OF *tra* GENES IN F PLASMIDS

Regulation of the *tra* genes in F plasmids has been studied more extensively than that of other types. This

A Genetic organization of *tra* region

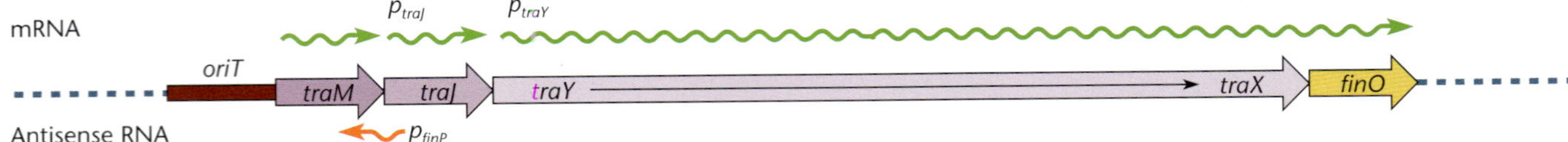

B Immediately after entry into cell

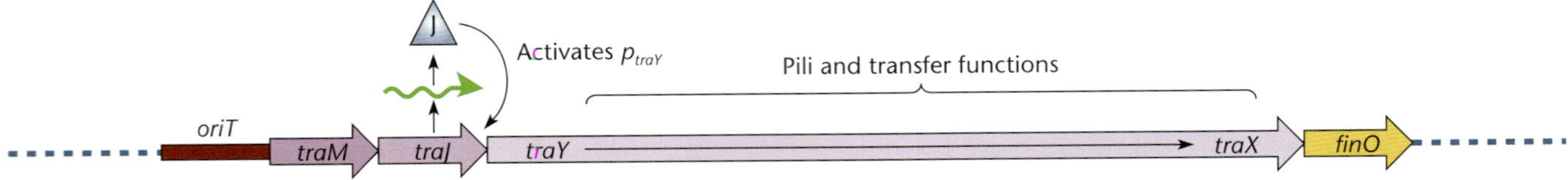

C After plasmid establishment

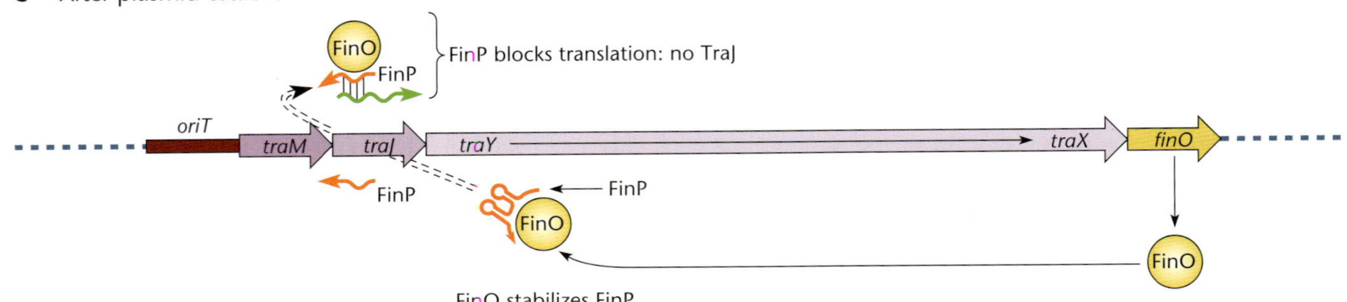

Figure 5.6 Fertility inhibition of the F plasmid. Only the relevant *tra* genes discussed in the text are shown. **(A)** Genetic organization of the *tra* region. **(B)** The *traJ* gene product is a transcriptional activator that is required for transcription of the other *tra* genes, *traY* to *traX*, and *finO* from the promoter p_{traY}. **(C)** Translation of the *traJ* mRNA is blocked by hybridization of an antisense RNA, FinP, which is transcribed from the complementary strand in the same region. The FinO protein stabilizes the FinP RNA. Details are given in the text.

regulation is illustrated in Figure 5.6. Transfer of these plasmids depends on TraJ, a transcriptional activator. A **transcriptional activator** is a protein required for initiation of RNA synthesis at a particular promoter (see chapters 2 and 11). If TraJ were always made, the other *tra* gene products would always be made, and the cell would always have a pilus. However, the translation of TraJ is normally blocked by the concerted action of the products of two plasmid genes, *finP* and *finO*, which encode an RNA and a protein, respectively. The FinP RNA is an antisense RNA that is transcribed constitutively from a promoter within and in opposite orientation to the *traJ* gene. Complementary pairing of the FinP RNA and the *traJ* transcript negatively regulates the operon by two mechanisms: the complementary pairing prevents translation of TraJ by blocking the ribosome-binding site, and the RNA also makes the transcript a target for degradation by RNase E (see chapter 2). The FinO protein stabilizes the FinP antisense RNA. When the plasmid first enters a cell, neither FinP RNA nor FinO protein is present, so TraJ and the other *tra* gene products are made. Consequently, a pilus is produced on the cell, and the plasmid can be transferred. Initially, the transferred plasmid is in a single-stranded state. However, priming in the recipient cell results in synthesis of the complementary strand to make the double-stranded form. After the plasmid has become established in the double-stranded state, the FinP RNA and FinO protein can be synthesized, the *tra* genes are repressed, and the plasmid can no longer transfer. Later, the *tra* genes are expressed only intermittently. The F plasmid is highly adapted to its host in that many host factors appear to help regulate conjugation. Regulation has been linked to supercoiling of the plasmid and many other systems. Much of the F plasmid is also sensitive to silencing by the HN-S DNA-binding protein, and HN-S may be an antagonist of TraJ.

The original discovery of the F plasmid may have resulted from a happy coincidence involving its *finO* gene. Because of an insertion mutation in the gene (IS3a in Figure 5.2), the F plasmid is itself a mutant that always expresses the *tra* genes. Consequently, a pilus almost always extends from the surfaces of cells that harbor this plasmid, and the F plasmid can always transfer, provided that recipient cells are available; the high efficiency of transfer facilitated the discovery of conjugation. Mutations that increase the efficiency of plasmid transfer, thereby increasing their usefulness in gene cloning and other applications, have been isolated in other commonly used transfer systems.

Interspecies Transfer of Plasmids

Many plasmids have transfer systems that enable them to transfer DNA between unrelated species. These are often referred to as broad-host-range plasmids and include R388, RP4, and pKM101. The RP4 plasmid can transfer itself or mobilize other plasmids from *E. coli* into essentially any of the *Proteobacteria*. Successful transfer and establishment of a plasmid in a new bacterial host requires that the conjugation system allow transfer into the host, but once there, the plasmid must also be capable of regulating DNA replication, copy number, and proper plasmid segregation as the new host divides (or, as we see below for ICEs, it must be capable of integrating into the host chromosome). The F plasmid can transfer to a variety of bacteria but is generally referred to as a narrow-host-range vector because its maintenance is largely limited to species highly similar to *E. coli*. Amazingly, some plasmids transfer at a low frequency into very different bacterial hosts like cyanobacteria and even actinobacteria, such as *Streptomyces* species. Plasmids have even been found to transfer across domains into plant cells. The F plasmid can transfer itself from *E. coli* into eukaryotic yeast cells. Transfer of bacterial DNA across domains is only relevant genetically if a mechanism exists for integration into the recipient cell genome and further adaptations allow expression of its genetic information.

The ability to transfer DNA into diverse hosts probably plays an important role in evolution in bacteria. Such transfer could help explain why genes with related functions are often very similar to each other regardless of the organisms that harbor them. These genes could have been transferred by promiscuous plasmids fairly recently in evolution.

The interspecies transfer of plasmids also has important consequences for the use of antibiotics in treating human and animal diseases. Many broad-host-range plasmids, including those of the IncP replication incompatibility group, such as RP4, and the IncW replication incompatibility group plasmid R388, were isolated in hospital settings. These large plasmids (commonly called R plasmids because they carry genes for antibiotic resistance) presumably have become prevalent in response to the indiscriminate use of antibiotics in medicine and animal agriculture. Antibiotic resistance genes are remarkably common in the environment, and a plausible source of resistance genes could be the bacteria that make antibiotics or other bacteria that live in close proximity to these organisms (see D'Costa et al., Suggested Reading). In chapter 8, we discuss transposons and other DNA elements that are usually responsible for mobilizing antibiotic resistance genes between chromosomes and conjugative elements.

Whatever their source, the emergence of R plasmids illustrates why antibiotics should be used only when they are absolutely necessary. In humans or animals treated indiscriminately with antibiotics, bacteria that carry R plasmids are selected from the normal microbiota, resulting in an increased proportion of resistant bacteria in the population. R plasmids can be quickly transferred into an invading pathogenic bacterium, making it antibiotic resistant. Consequently, the infection will be difficult to treat. These resistant organisms are also shed into the environment, further increasing the spread of resistance genes.

Conjugation and Type IV Secretion Systems Capable of Translocating Proteins

Interestingly, the processes of conjugation and certain type IV secretion systems are intimately related. While conjugation was understood first at the molecular level, it is reasonable to suggest that the DNA translocation system first evolved for use in protein translocation. After all, even DNA transfer requires the translocation of one or more proteins (see below). In the case of conjugation, it appears that the coupling protein acts to adapt a protein translocation machine for the transfer of DNA. One early method of determining if proteins expressed in donor cells can be transferred into recipient cells involved labeling the donor proteins with radioactivity prior to mating. Recipient cells could be specifically examined for labeled donor proteins using two methods: by killing the donor cells with a pilus-specific bacteriophage (see Box 5.1) or by using "minicells," which are devoid of nuclear material, as recipients (see Rees and Wilkins, Suggested Reading). Minicells are derived from a bacterial strain with a chromosome partitioning defect due to a mutation in the *min* genes, which results in generation of cells without chromosomes at high frequency (see chapter 1). In the case of the RP4 conjugal plasmid, the large Pri protein can be clearly detected in the cytoplasm of recipient cells. In other systems, more sensitive genetic techniques were needed to show that the relaxase is targeted for translocation to recipient cells through the pilus apparatus via specific recognition sequences found in the protein (see Parker and Meyer, Suggested Reading).

Conjugation is one adaptation of a type IV secretion system. In the case of bacterial pathogens, type IV secretion systems are often responsible for translocating proteins or other effector molecules to eukaryotic cells. In some cases, the other cell can be a eukaryotic plant cell, as in the case of the Ti plasmid, the causative agent of

BOX 5.3

Gene Exchange between Domains

Not only can some plasmids transfer themselves into other types of bacteria, but they can also sometimes transfer themselves into eukaryotes, that is, into organisms of a different domain of life.

Agrobacterium tumefaciens and Crown Gall Tumors in Plants

The first discovery of transfer of bacterial plasmids into eukaryotes occurred in the plant disease crown gall. Crown gall disease is caused by *A. tumefaciens*; it is identified by a tumor that appears on the plant, usually where the roots join the stem (the crown). Virulent strains of *A. tumefaciens* contain a plasmid called the tumor-inducing (or Ti) plasmid. The Ti plasmids of *A. tumefaciens* are in most respects normal bacterial self-transmissible plasmids. A typical Ti plasmid is shown in the figure. Like other self-transmissible plasmids, Ti plasmids encode conjugative-transfer functions (*tra* and *trb* genes) that enable them to transfer themselves between bacteria. What makes these plasmids remarkable is that they can also transfer part of themselves, called the T-DNA region, into plants.

The functions required for transfer of the T-DNA into plants are encoded by a region called the *vir* region (panel A). This

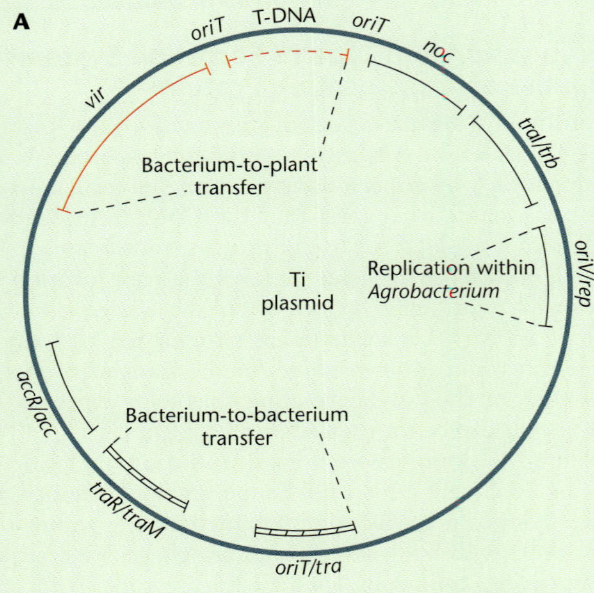

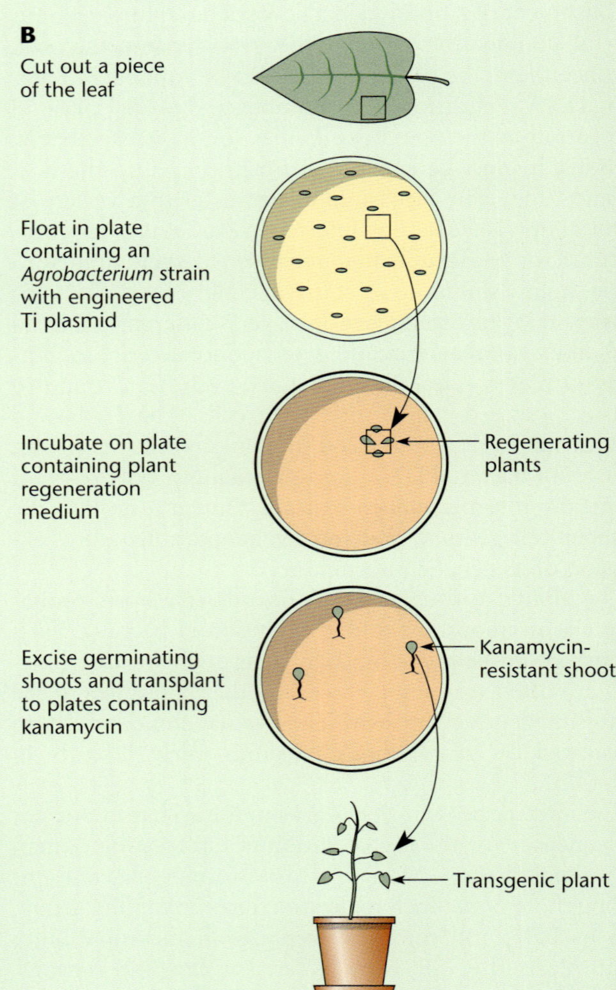

(A) The structure of a Ti plasmid showing the following regions (clockwise from the top): the T-DNA, bracketed by the border sequences, which resemble *oriT* sequences of conjugation systems (borders are not transferred in their entirety); the T-DNA contains the genes that are expressed in the plant to make opines and plant hormones (not shown); the *noc* genes, encoding enzymes for the catabolism of the opine nopaline in the bacterium; some *tra* and *trb* genes, for transfer into other bacteria; the *oriV* region, for replication of the plasmid in the bacterium; *oriT* and *tra* function genes, for transfer into other bacteria; *acc* genes, for catabolism of another opine; and *vir* genes, for transfer into plants. (B) A procedure for making a transgenic plant (for details, see the text).

region is distinct from the *tra* and *trb* regions, which are required for the transfer of the plasmid into other bacteria, but its functions are remarkably similar to those of both other Tra functions and other type IV secretion systems. DNA sequence analysis clearly demonstrates close similarity between many Vir proteins and Tra proteins. Like the *tra* region, the *vir* region encodes both an Mpf system, which elaborates a pilus, and a Dtr system, which processes the DNA for transfer. The pilus is composed of the pilin protein, which is the product of the *virB2* gene and is like the pilins of the pili of other type IV secretion systems. The Mpf system also includes a coupling protein, the product of the *virD4* gene, which communicates with the relaxosome, which includes the specific relaxase. The relaxase, the product of the *virD2* gene, cleaves the sequences bordering the T-DNA in the plasmid and remains covalently attached to the 5′ ends of the single-stranded T-DNA during transfer into the plant. The sequences at which the relaxase cuts these border sequences are similar to the *oriT* sequences of IncP plasmids, and the relaxase makes a cut in exactly the same place in the sequences where the relaxases of IncP plasmids cut their *oriT*.

The next stage is where the T-DNA transfer system begins to differ from ordinary conjugative-transfer systems. In addition to its role as a relaxase, the VirD2 protein contains amino acid sequences that target it to the plant cell nucleus once it is in the plant cell. These sequences, called nuclear localization signals, are essentially "passwords" that tell the plant cell that a particular plant nuclear protein should be transported into the nucleus after it has been translated in the cytoplasm. By imitating the password, the VirD2 protein tricks the plant into transporting it into the nucleus, dragging the attached T-DNA with it. Another plasmid-encoded protein, VirE2, is also transported into the plant cytoplasm (separately from T-DNA), where it coats the T-DNA and assists in delivery to the nucleus. Once in the nucleus, the T-DNA can insert into the plant genome by recombination. Once integrated into the plant DNA, the T-DNA of the plasmid encodes the synthesis of plant hormones that induce the plant cells to multiply and form tumors (galls) on the plant—hence the name "crown gall tumors." The T-DNA also encodes enzymes that synthesize unusual small molecules composed of an amino acid, such as arginine, joined to a carbohydrate, such as pyruvate. These compounds, called opines, are excreted from the tumor. The plant is able to express the genes on the T-DNA and make these compounds because the genes on the T-DNA have promoters and translational initiation regions that are recognized by the plant gene expression machinery, so they can be expressed in the plant. Meanwhile, back in the bacterium, the Ti plasmid carries genes that allow it to use the particular opine made by the T-DNA-modified plant as a carbon and nitrogen source. Very few types of bacteria can degrade opines, which gives the *Agrobacterium* species containing this particular Ti plasmid an advantage. In this way, the bacterium creates its own special "ecological niche" at the expense of the plant.

The discovery of T-DNA, made in the 1970s, allowed the construction of transgenic plants because any foreign genes cloned into one of the T-DNA regions of the Ti plasmid will be transferred into the plant along with the T-DNA and integrated into the plant DNA. The integrated foreign genes can alter the plant, provided that they are transcribed and translated in the plant. Panel B of the figure shows the general procedure for making a transgenic plant. A "disarmed" strain of *A. tumefaciens* in which the native T-DNA has been deleted is used. A small plasmid containing the gene to be transferred, a gene for kanamycin resistance, and a DNA sequence similar to the *oriT* site of conjugal plasmids is constructed. Single plant cells are inoculated, and after allowing time for T-DNA transfer, all bacteria and all nontransformed plant cells are killed using antibiotics. Transgenic plants with the T-DNA inserted in their chromosomes can be regenerated from the kanamycin-resistant cells. An entire industry has developed around this technology, and *Agrobacterium* species have been used to genetically engineer plants to make their own insecticides, to be more nutritious, and to survive more severe growth conditions.

References

Bates S, Cashmore AM, Wilkins BM. 1998. IncP plasmids are unusually effective in mediating conjugation of *Escherichia coli* and *Saccharomyces cerevisiae*: involvement of the Tra2 mating system. *J Bacteriol* **180**:6538–6543.

Brencic A, Winans SC. 2005. Detection of and response to signals involved in host-microbe interactions by plant-associated bacteria. *Microbiol Mol Biol Rev* **69**:155–194.

Buchanan-Wollaston U, Passiatore JE, Cannon F. 1987. The *mob* and *oriT* mobilization functions of a bacterial plasmid promote its transfer to plants. *Nature* **328**:172–175.

crown gall disease (Box 5.3). When type IV secretion is used in virulent bacteria, the pilus or another adhesin is used to hold the cells together during the transfer, which also involves special membrane structures through which the macromolecules must pass. As with DNA conjugation, the processes are very specific, and only some types of proteins or plasmid DNAs can be transferred. Nevertheless, it came as a surprise how closely related these two types of systems can be. The relatedness of type IV secretion to conjugation is dramatically illustrated by comparison of the T-DNA transfer system of *Agrobacterium tumefaciens*, called VirB, to some type IV secretion

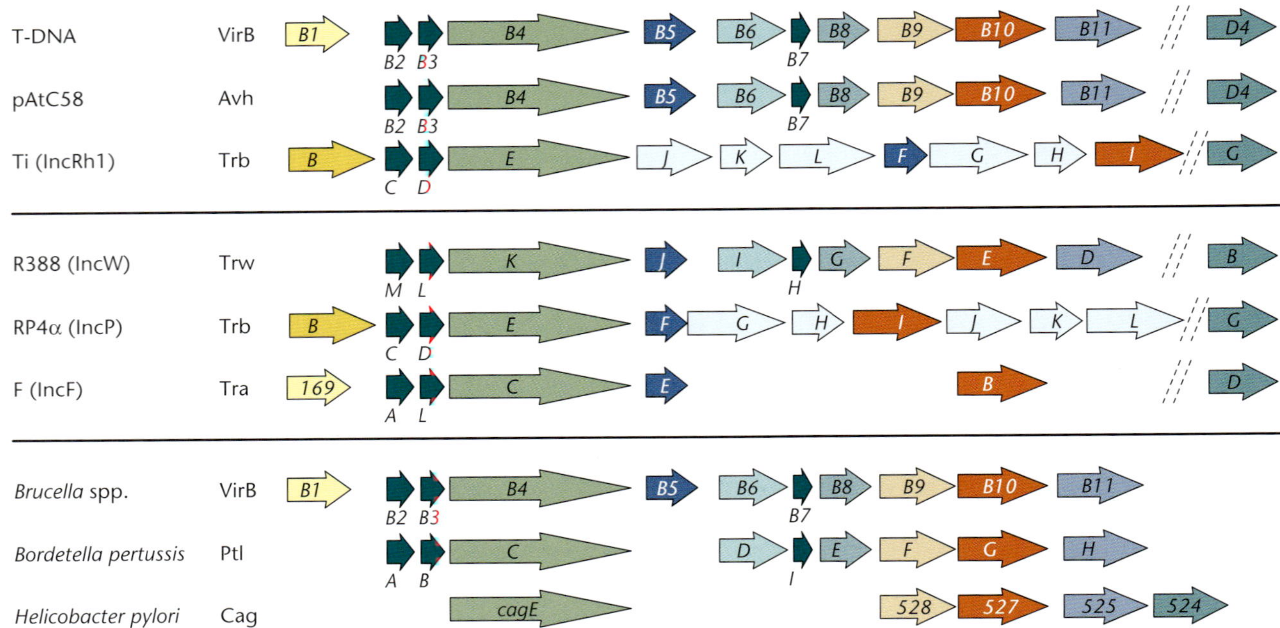

Figure 5.7 Gene arrangements of type IV secretion loci. Genes with homologs in the VirB system are shown in the same color. Genes set in light blue do not have homologs in the VirB system. **(Top)** *Agrobacterium tumefaciens* is the first species shown to carry three distinct type IV secretion systems whose substrates are defined. The VirB system transfers T-DNA and protein effectors to plants, while the Avh system transfers pATC58 and the Trb system transfers the Ti plasmid, respectively, to other bacteria. **(Middle)** Representative type IV secretion systems of other species that direct conjugal DNA transfer. **(Bottom)** Representative type IV secretion systems that direct protein transfer during the course of infection. Modified from Christie PJ, *in* Funnell BE, Phillips GJ (ed), *Plasmid Biology* (ASM Press, Washington, DC, 2004).

systems (Figure 5.7). As discussed in Box 5.3, the T-DNA transfer system transfers the T-DNA part of a plasmid from bacterial to plant cells. The T-DNA transfer channel (or VirB channel) also translocates several protein substrates, including a single-stranded DNA-binding protein, to plant cells whose jobs are to protect and guide the translocated T-DNA to the plant nucleus for integration into the plant nuclear genome. The VirB transfer system shares features with other plasmid conjugation systems in that it encodes a pilus, a relaxase, coupling proteins, and many other proteins involved in the transfer process. In fact, most of the proteins of the Tra systems of the F and R388 plasmids can be assigned homologs in the T-DNA transfer system (Figure 5.7). Moreover, some pathogenic bacteria that transfer proteins into eukaryotic cells as part of the disease-causing process also have analogous functions. One of the most striking similarities is to the CagA toxin-secreting system of *Helicobacter pylori*, implicated in some types of gastric ulcers and stomach cancer. This type IV toxin-secreting system has at least five protein homologs to the T-DNA system of the Ti plasmid of *Agrobacterium*. The Cag system delivers a toxin directly through the bacterial membranes and into the eukaryotic cell, where the toxin is phosphorylated on one of its tyrosines. In the phosphorylated state, the toxin causes many changes in the cell, including alterations in its actin cytoskeleton. Another pathogenic bacterium, *Bordetella pertussis*, the causative agent of whooping cough, also has a type IV secretion system that has nine proteins homologous to proteins in the T-DNA system. This system secretes the pertussis toxin through the outer membrane of the bacterial cell. Once outside the cell, the pertussis toxin assembles into a form that can enter the eukaryotic cell, where it can ADP-ribosylate key proteins, thereby interfering with signaling pathways and causing disease symptoms. A group of pathogens in the genus *Brucella* are the causative agents of brucellosis and require a set of genes homologous to the *virB* genes of *A. tumefaciens* for pathogenesis.

However, the most striking evidence that conjugation is related to type IV secretion has come from the virulence system of *Legionella pneumophila*, which causes Legionnaires' disease. Like many pathogenic bacteria, this bacterium can multiply in macrophages, specialized white blood cells that are designed to kill them (see Vogel et al., Suggested Reading). The bacteria are taken up by the macrophage but then secrete proteins that disarm the phagosomes that have engulfed them. Remarkably, the components of this type IV secretion system are analogous to the Tra functions of some self-transmissible plasmids, and this type IV system can even mobilize the plasmid RSF1010 at a low frequency (Figure 5.7).

Mobilizable Plasmids

Some plasmids are not self-transmissible but can be transferred by another, self-transmissible plasmid residing in the same cell. Plasmids that cannot transfer themselves but can be transferred by other plasmids are said to be mobilizable, and the process by which they are transferred is called **mobilization**. The simplest mobilizable plasmids merely contain the *oriT* sequence of a self-transmissible plasmid, since any plasmid that contains the *oriT* sequence of a self-transmissible plasmid can be mobilized by that plasmid. Expressed in genetic terminology, the Mpf and Dtr systems of the self-transmissible plasmid can act in *trans* on the *cis*-acting *oriT* site of the plasmid and mobilize it. As indicated in Box 5.2, plasmids encoding the *oriT* site of a self-transmissible plasmid have many applications in molecular genetics because they can be mobilized into other cells.

While we can construct such a plasmid, and it is mobilizable, minimally mobilizable plasmids containing only the *oriT* site of a self-transmissible plasmid do not seem to occur naturally. All mobilizable plasmids isolated so far encode their own Dtr systems, including their own relaxase and helicase. For historical reasons, the genes of the Dtr system of mobilizable plasmids are called the ***mob* genes**, and the region required for mobilization is called the ***mob* region**. The function of the *mob* gene products of mobilizable plasmids seems to be to expand the range of self-transmissible plasmids by which they can be mobilized (see below). A plasmid containing only the *oriT* sequence of a self-transmissible plasmid can be mobilized only by the *tra* system of that self-transmissible plasmid and not by those of other self-transmissible plasmids that do not share the same *oriT* site, while naturally occurring mobilizable plasmids can often be mobilized by a number of *tra* systems.

The process of mobilization of a plasmid by a self-transmissible plasmid is illustrated in Figure 5.8. The process is identical to the transfer of a self-transmissible plasmid, except that the Mpf system of the self-transmissible plasmid acts not only on its own Dtr system, but also on the Dtr system of the mobilizable plasmid. The self-transmissible plasmid forms a mating bridge with a recipient cell and communicates this information via its coupling protein, not only to its own relaxase, but also to the relaxase of a mobilizable plasmid that happens to be in the same cell. The relaxase of the mobilizable plasmid is responsible for processing at the *oriT* site and presentation to the Mfp system for transport to the recipient cell, as described above. The self-transmissible plasmid can be transferred at the same time that it mobilizes other plasmids. However, generally, either one plasmid or the other is transferred into a particular recipient cell, due to competition between the two plasmids for coupling protein.

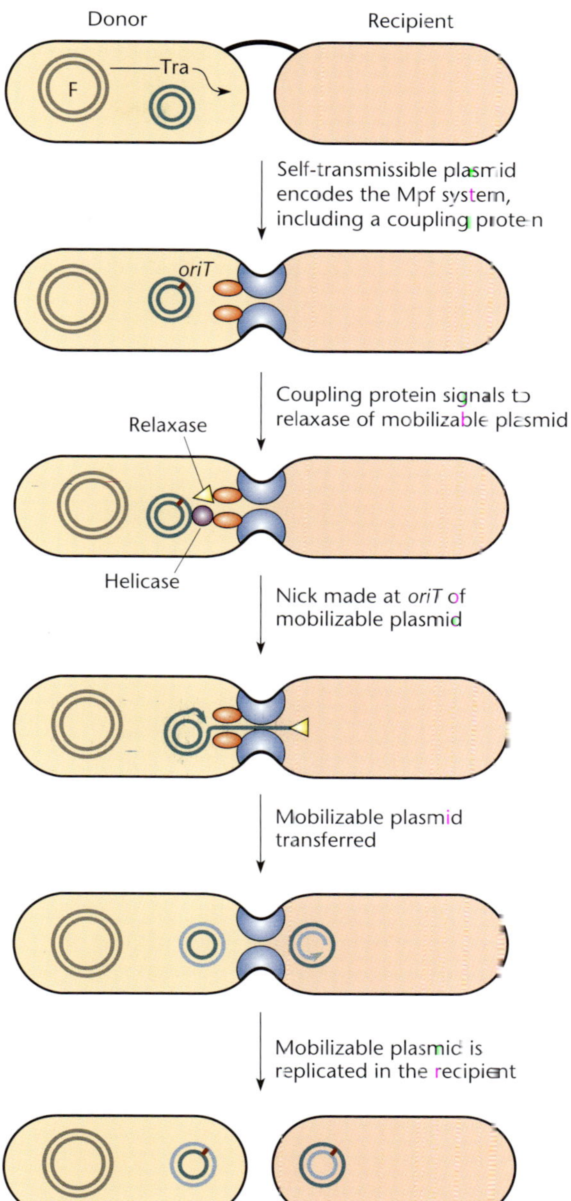

Figure 5.8 Mechanism of plasmid mobilization. The donor cell carries two plasmids, a self-transmissible plasmid, F, that encodes the *tra* functions that promote cell contact and plasmid transfer, and a mobilizable plasmid (blue). The *mob* functions encoded by the mobilizable plasmid make a single-stranded nick at *oriT* in the *mob* region. Transfer and replication of the mobilizable plasmid then occur. The self-transmissible plasmid may also transfer. Details are given in the text.

The key feature for being mobilized by another plasmid is recognition by the coupling protein of the other plasmid. Any plasmid encoding its own Dtr system can be mobilized by a coresident self-transmissible plasmid, provided that its Dtr system can communicate with the coupling protein of the Mpf system of the coresident plasmid. For plasmids that show the broadest ability to utilize other transfer systems, accessory proteins in addition to the relaxase are often important. The RSF1010 plasmid is a prime example, where its MobB accessory protein is critical to allow the plasmid-encoded MobA relaxase to recognize highly diverse coupling proteins.

Chromosome Transfer by Plasmids

Conjugation usually results in transfer of only a plasmid to another cell. However, plasmids sometimes transfer the chromosomal DNA of their bacterial host, a fact that has been put to good use in bacterial genetics. Without the transfer of genes, bacterial genetics is not possible, and conjugation is one of only three ways in which chromosomal and plasmid genes can be exchanged among bacteria (transformation [chapter 6] and transduction [chapter 7] are the others). In chapter 3, we discussed how these ways of exchanging genes between bacterial strains can be used to map genetic markers. In this chapter, we go into more detail about how plasmids transfer chromosomal DNA.

Formation of Hfr Strains of *E. coli*

Dedicated conjugal plasmids like the F plasmid can sometimes integrate into chromosomes, and when such plasmids attempt to transfer, they take the chromosome with them. These *E. coli* strains that can transfer their chromosomes because of an integrated plasmid are called **Hfr strains**, where Hfr stands for high-frequency recombination. The name derives from the fact that many recombinants can appear when such a strain is mixed with another strain of the same bacterium, as we will discuss.

The integration of F plasmids into the *E. coli* chromosome can occur by several different mechanisms, including recombination between sequences on the plasmid and sequences on the chromosome. For normal recombination to occur, the two DNAs must share a sequence (see chapter 9). Most F plasmid sequences are unique to the plasmid, but sometimes the plasmid and the *E. coli* chromosome share an **insertion element**, which is a common source of insertion mutations (see chapter 3). Figure 5.9 shows how recombination between the IS2 element in the F plasmid and an IS2 element in the chromosome of *E. coli* can lead to integration of the F plasmid by homologous recombination or site-specific recombination. Once integrated, the F plasmid is bracketed by two copies of the IS2 element. The bacterium is now an Hfr strain. The standard laboratory strain of *E. coli* contains 20 sites in

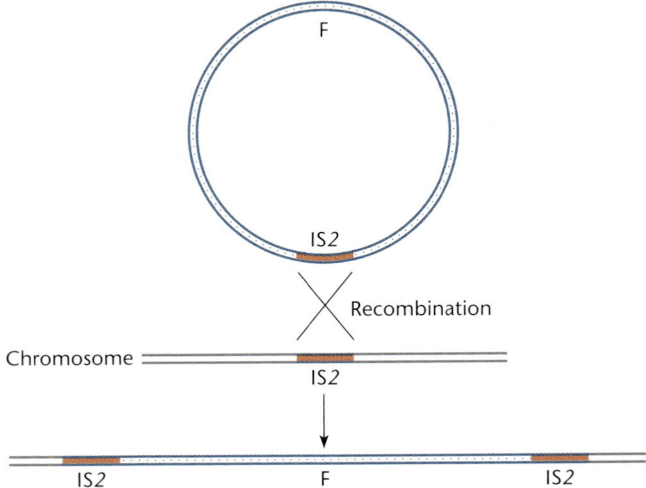

Figure 5.9 Integration of the F plasmid by recombination between IS*2* elements in the plasmid and in the chromosome, forming an Hfr cell. Integration can also occur through recombination with one of the two IS*3* elements or Tn*1000* sequences on the F plasmid and in the chromosome (see Figure 5.2).

the chromosome that are available for IS-mediated Hfr formation by the F plasmid, marking where IS elements are shared by the plasmid and the chromosome.

Transfer of Chromosomal DNA by Integrated Plasmids

We mentioned at the beginning of the chapter that self-transmissible plasmids were first detected in 1947 by Lederberg and Tatum, even though these investigators did not know what they were at the time. They observed recombinant types after mixing different strains of *E. coli*. Recombinant types differ from the two original, or parental, strains (see chapter 3) and in this case resulted from the transfer of chromosomal DNA from one strain to another by an F plasmid that had intermittently integrated into the chromosome and was transferring part of the chromosome to the other strain. In retrospect and as indicated earlier, it was fortuitous that the strains used by Lederberg and Tatum included an F plasmid that is incapable of repressing transfer.

The process of initiating the transfer of chromosomal alleles is the same as the process by which the transfer of the F plasmid itself is initiated (Figure 5.4). The plasmid expresses its *tra* genes, even though it is integrated into the chromosome, and a pilus is synthesized. On contact with a recipient cell, the coupling protein communicates with the relaxase, and the integrated plasmid DNA is transferred starting at the *oriT* site. One strand is displaced into the recipient cell, while the other strand is replicated. Now, however, after transfer of the portion of the *oriT* sequence and plasmid on one side of the nick, chromosomal DNA is also transferred into the new cell. If the transfer continued long enough (in *E. coli*, approximately 100 min at 37°C), the entire bacterial chromosome would eventually be transferred, ending with the remaining plasmid *oriT* sequences. However, transfer of the entire chromosome (and thus the whole integrated plasmid) is rare, because the union between the cells is fragile, and the transferred strand likely contains nicks that would halt the process. This fact is exploited for genetic mapping by Hfr crosses in a method called gradient of transfer (see chapter 3). Also, because the remainder of the plasmid is seldom transferred, the recipient cell will not itself become an F^+ cell. Figure 3.34 gives an overview of the process by which chromosomal DNA is transferred in an Hfr strain.

Chromosome Mobilization

Hfr strains where the F plasmid integrated at the position of an IS element in the *E. coli* chromosome played an essential part in early bacterial genetics. The F plasmid was also used to make Hfr strains in *Salmonella enterica* for conjugal mapping experiments. Presumably, conjugal plasmids could carry out a similar process in other bacterial species in nature. IS elements are common in bacteria and often exist in several copies in the chromosome and in plasmids; recombination between these common sequences could result in integration of a conjugal plasmid into the chromosome in other types of bacteria. While this process would allow mobilization of chromosomal information between bacteria, a process to recombine these sequences into the recipient cell genome would still be needed. Chromosome mobilization using the F plasmid has also found utility in modern synthetic biology as a tool to exchange large chromosomal DNA fragments in *E. coli* (Box 5.4).

Note that the Hfr process involving dedicated conjugal plasmids differs from the integration of conjugative elements discussed below. Typically, highly evolved ICEs have mechanisms that specifically prevent expression of conjugation functions until they have excised from the host chromosome, preventing the transfer of chromosomal alleles.

Prime Factors

Through the integration and subsequent excision of conjugal plasmids, chromosomal genes can be recombined onto the plasmid. These conjugal plasmids that now contain a portion of the host chromosome are called **prime factors**. When a prime factor moves to a new host, it will carry along those chromosomal genes that now reside on the plasmid. Prime factors are usually designated by the name of the plasmid followed by a prime symbol, for example, F′ factor. An R plasmid, such as RK2, that carries bacterial chromosomal DNA is an R′ factor.

GENERATION OF PRIME FACTORS

Prime factors are generated through the same recombination processes that generate Hfr strains. After an Hfr is formed, the plasmid can be restored as an independent element if the same recombination event that formed the Hfr occurs in reverse. Occasionally, however, instead of the same insertion sequence being used as the site of recombination, a different insertion sequence that resides on both the plasmid and the chromosome is used. As shown in Figure 5.10, this recombination event results in placement of a portion of the chromosome adjacent to the Hfr onto the plasmid, forming an F′ factor. The F′ factor carries the chromosomal DNA that lies between the recombining DNA sequences. A prime factor can form from recombination between any repeated sequences, including identical IS elements or genes for rRNA, which often exist in more than one copy in bacteria.

Note that a deletion forms in the chromosome when the prime factor loops out. Some of the genes deleted from the chromosome may have been essential for the growth of the bacterium. Nevertheless, the cells do not die, because the prime factor still contains the essential genes, which should be passed on to daughter cells when the plasmid replicates. However, cells that lose the prime factor will die if the deleted genes are essential for survival.

Prime factors can be very large, even larger than the chromosome of some species of bacteria. In general, the larger a prime factor is, the less stable it is. Maintaining large prime factors in the laboratory requires selection procedures designed so that cells die if they lose some or all of the prime factor. The real utility of prime factors in early genetics was realized by the ability to move these vectors into different host bacteria. Most prime factors are small and can be readily transferred in their entirety into new hosts. Because prime factors contain an entire self-transmissible plasmid, a cell that receives a complete prime factor becomes a donor and can transfer this DNA into other bacteria. Moreover, because prime factors are replicons with their own plasmid origin of replication, they can replicate in any new bacterium that falls within the plasmid host range (see chapter 4). The formation of F′ factors is a rare event; however, the products of this event can be selected. For example, gene products encoded adjacent to an Hfr can be selected in recipient cells that are incapable of homologous recombination.

COMPLEMENTATION TESTS USING PRIME FACTORS

Complementation tests can reveal whether two mutations are in the same or different genes and how many genes are represented by a collection of mutations that give the same phenotype (see chapter 3). Complementation tests can also provide needed information about the

BOX 5.4

Conjugation and Synthetic Genomics

Hfr strains provided the foundation for bacterial genetics in *Escherichia coli* over 70 years ago. More recently, an adaptation of this procedure has allowed an important tool in synthetic genomics for the construction of hybrid *E. coli* chromosomes. This procedure is termed conjugal assembly genome engineering.

As indicated in Box 5.2, the *cis*-acting *oriT* function and *trans*-acting Tra components of conjugation can be readily separated for mobilizing conditional plasmids of interest for delivering transposons or other tasks. Using recombination techniques described in chapter 9, the *cis*-acting *oriT* site can be precisely integrated at any one of many nonessential positions in the *E. coli* chromosome. This allows for the construction of synthetic modular Hfr strains where the conjugation functions are expressed in *trans* from a plasmid. As in standard Hfr experiments (see chapter 3), transfer will only initiate in one orientation from the *oriT* site, and the *cis*-acting *oriT* site itself is unable to recombine into the recipient because it is not flanked by homology in the recipient cell (see Figure 3.34). By using donor strains with the origin of transfer (*oriT*) placed at different positions in the *E. coli* chromosome, large fragments of the donor strain chromosome can be crossed into a recipient strain for replacing large regions of the recipient chromosome. The trick of the procedure is having the right selectable (and/or counterselectable markers) to increase the chances that only the specific region of the chromosome you are interested in is replaced.

This procedure allowed the sequential modification of the *E. coli* chromosome to reassign all 321of the UAG stop codons normally found in the lab strain to UAA stop codons. As explained in chapter 2, there are three stop codons, UAG, UAA, and UGA, in *E. coli* (and almost all organisms). In *E. coli*, the translation release factor RF1 is required to terminate translation at UAG stop codons. Successful replacement of all UAG stop codons was confirmed when it was shown that the gene encoding the RF1 protein, which is normally essential for cell survival, could be deleted in the strain background now lacking all UAG stop codons. In subsequent work, this strain was used to reassign UAG as a "sense" codon to encode nonstandard amino acids. Future work with this exciting new tool should allow for a variety of research and industrial applications, including synthesis of proteins containing novel types of amino acids that are not normally incorporated during protein synthesis.

As it turns out, it was easiest to first replace UAG codons in blocks, giving 32 strains with subsegments of the chromosome that were recoded; these sections were then crossed together into one background. Controlling the position where conjugal transfer started (i.e., the location of the *oriT* site) was as important as techniques to ensure that the replacement did not extend into regions of the chromosome that had already been replaced. The figure shows how a portion of a donor strain chromosome (shown in green) could be recombined into a recipient cell chromosome to replace a region of the chromosome shown in blue, without replacing a region shown in yellow (panel A). The position of the origin of transfer (*oriT*) is shown, as well as the position of two selectable markers placed in the donor and recipient chromosomes (P1 and P2). Also shown is the position of a marker that could be positively selected or negatively selected (P/N), allowing for selection for strains that have acquired or lost the marker by recombination.

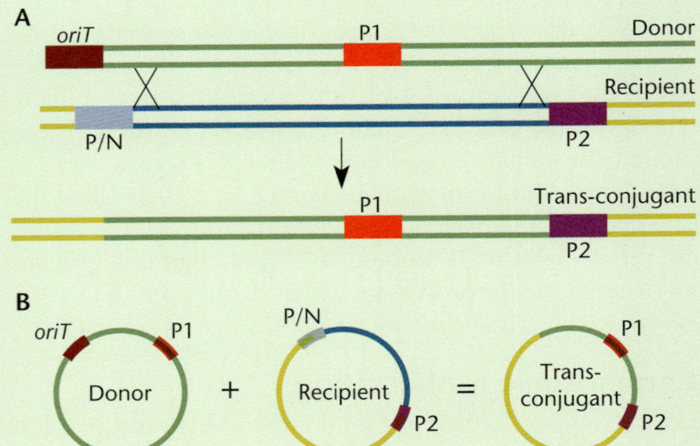

Panel A shows the details of where the markers sit and where a successful recombination event would occur ("X") for the experiment where P1 and P2 were selected and P/N was selected against. Panel B shows a schematic representation of the entire chromosomes. Note that the *oriT* will not be capable of recombining into the recipient chromosome because it lacks homology on one side, but the region immediately to the right of the *oriT* site in the diagram will recombine into the host based on negative selection with the P/N marker. Marker P1 will select for the green fragment of the chromosome in the recipient, while the P2 marker in the recipient strain will select against replacement of the chromosome region in yellow.

References

Isaacs FJ, Carr PA, Wang HH, Lajoie MJ, Sterling B, Kraal L, Tolonen AC, Gianoulis TA, Goodman DB, Reppas NB, Emig CJ, Bang D, Hwang SJ, Jewett MC, Jacobson JM, Church GM. 2011. Precise manipulation of chromosomes *in vivo* enables genome-wide codon replacement. *Science* **333**:348–353.

Lajoie MJ, Rovner AJ, Goodman DB, Aerni H, Haimovich AD, Kuznetsov G, Mercer JA, Wang HH, Carr PA, Mosberg JA, Rohland N, Schultz PG, Jacobson JM, Rinehart J, Church GM, Isaacs FJ. 2013. Genomically recoded organisms expand biological functions. *Science* **342**:357–360.

Ma N, Moonan D, Isaacs FJ. 2014. Precise manipulation of bacterial chromosomes by conjugative assembly genome engineering. *Nat Protoc* **9**:2285–2300.

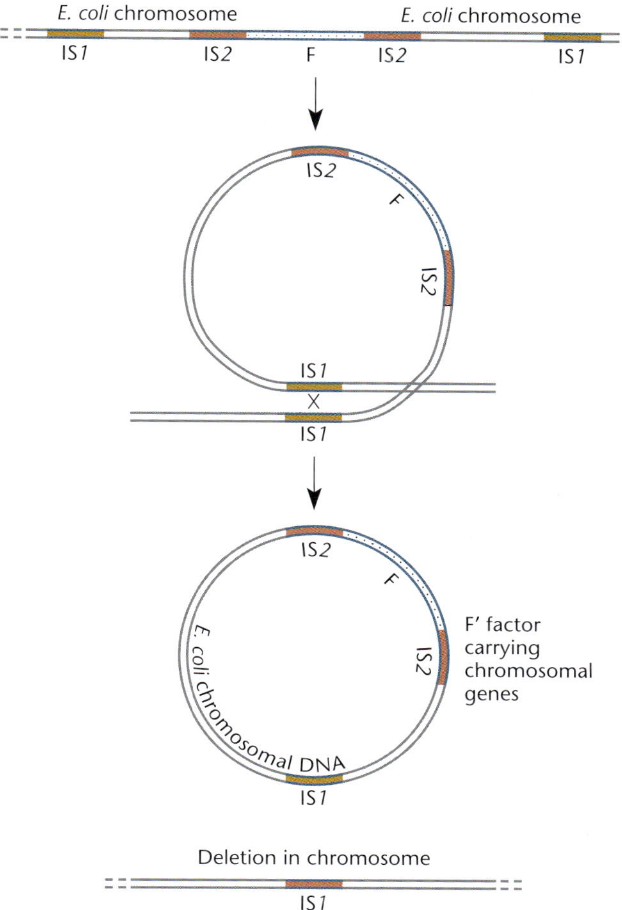

Figure 5.10 Generation of a prime factor by recombination. Recombination may occur between homologous sequences, such as IS sequences, in the chromosomal DNA outside the F factor. The resulting F′ factor contains chromosomal sequences, and the chromosome carries a deletion.

type of mutation being studied, whether the mutations are dominant or recessive and whether they affect a *trans*-acting function or a *cis*-acting site on the DNA. However, complementation tests require that two different alleles of the same gene be introduced into the same cell, and bacteria are normally haploid, with only one allele for each gene in the cell.

Because prime factors contain a region of the chromosome, they can be used to generate cells that are stable diploids for the region they carry. However, they contain only a short region of the chromosome and therefore are diploid for only part of the chromosome. Organisms that are diploid for only a region of their chromosome are called **partial diploids** or **merodiploids**.

ROLE OF PRIME FACTORS IN EVOLUTION

Prime factors formed with promiscuous, broad-host-range plasmids probably play an important role in bacterial evolution. Once chromosomal genes are on a broad-host-range promiscuous plasmid, they can be transferred into distantly related bacteria, where they are maintained as part of the broad-host-range plasmid replicon. They may then become integrated into the chromosome through various types of recombination.

Diversity in Transfer Systems

While we have focused on *Proteobacteria* up to this point, self-transmissible plasmids are found in a wide variety of bacteria, including *Cyanobacteria*, *Bacteroides*, *Firmicutes*, and *Actinobacteria* and even in *Archaea*. Conjugal transfer in *Firmicutes* appears to generally utilize the same type of system described above, but the original process appears to have evolved first in *Proteobacteria* (see Guglielmini et al. 2013, Suggested Reading). Examples in *Firmicutes* that have been investigated in some detail include pIP501, a promiscuous vector originally isolated from *Streptococcus agalactiae*, and the pheromone-responsive plasmids pCF10 and pAD1 from *Enterococcus faecalis* (Table 5.2). Differences from the systems found in *Proteobacteria* include additional hydrolase activities for traversing the thick layer of peptidoglycan found in *Firmicutes*. In addition, while elaborate conjugative pili are used in the case of conjugal plasmids found in *Proteobacteria*, the Mpf systems of conjugal plasmids found in *Firmicutes* appear to utilize adhesins to mediate contacts between the donor and recipient cells. Regardless of the genera where the self-transmissible element or mobilizable element is found, the plasmid has its own Dtr component. Table 5.2 shows some of the conserved proteins found in a variety of bacteria

Table 5.2 Selected conserved proteins in conjugal plasmids and ICEs from *Proteobacteria* and *Firmicutes*

Conjugal element	Host	Relaxase	ATPase	Coupling protein
F plasmid	*Escherichia coli*	TraI	TraC	TraD
Ti plasmid	*Agrobacterium tumefaciens*	VirD2	VirB4	VirD4
pIP501	*Streptococcus agalactiae*	TraA	TraE	TraJ
pCF10	*Enterococcus faecalis*	PcfG	PrgJ	PcfC
ICE*Bs*1	*Bacillus subtilis*	NicK	ConE	ConQ

(although some *Firmicutes* appear to utilize a distinct family of proteins, designated TcpA, instead of VirD4/TraD). As with the plasmids found in *Proteobacteria*, conjugal plasmids found in *Firmicutes* typically encode functions such as virulence determinants and genes encoding antibiotic resistance or novel degradation pathways that can aid their bacterial host.

Markedly different from the conjugal elements described thus far in this chapter are a set of elements that are found only in *Actinobacteria*, such as *Streptomyces* spp. The streptomycetes are important for their production of many of our most useful antibiotics. They are also capable of conjugation, which was discovered about the same time as conjugation in *E. coli*. One major difference from the other systems discussed in this chapter is that they transfer double-stranded DNA instead of single-stranded DNA. This system also requires only a single plasmid-encoded protein called TraB to transfer DNA from a donor to a recipient cell. This may stem from the morphology and life cycle of the organism, which facilitates cell-cell contact. Not unlike fungi, in one stage of their life cycle the streptomycetes grow in a branching filamentous arrangement of cells called a mycelium. Movement throughout the mycelium in a process called spreading requires a number of additional components, the Spd proteins. While poorly understood, the *cis*-acting feature in the plasmid that is acted on by TraB, called *clt*, is unrelated to the *oriT* sites discussed thus far. The *clt* element is a series of 8-bp repeats that are recognized by the TraB protein. TraB is distantly related to the FtsK DNA translocation protein that assists in the transfer of DNA between daughter cells (or into spores as in the case of SpoIIIE), as discussed in chapter 1. It is possible that these 8-bp repeats are recognized the same way that direction cues are recognized by the FtsK/SpoIIIE proteins in the chromosome. The 8-bp repeats are also found in the chromosome of the streptomycetes and probably also contribute to the ability of the TraB protein to mobilize transfer of chromosomal markers without the plasmid integrating into the chromosome.

Integrating Conjugative Elements

While conjugation was originally discovered and investigated with circular plasmids that are maintained separately from the host chromosome, it is now appreciated that conjugation is just as common in elements that integrate into the bacterial host chromosome. These elements, called integrating conjugative elements (ICEs), conjugate between bacteria as independent circular DNAs after they excise from the host chromosome. ICEs may have systems that allow limited DNA replication after excision from the chromosome to facilitate their maintenance and transfer (see below), but the stability of these elements stems from their ability to integrate into the host chromosome. Typically, these elements integrate into the host chromosome utilizing a tyrosine site-specific recombinase similar to the enzyme systems used for bacteriophage integration (chapter 8). However, other enzyme systems have also been found that allow ICEs to integrate into the chromosome (Table 5.3). In the majority of the cases, integration occurs at a single highly conserved site in the bacterial chromosome, such as a tRNA gene (as found with many bacteriophages), although other sites can be used. There are also multiple other notable examples that use different sites that share a weak consensus sequence or are simply rich in As and Ts (Table 5.3). As mentioned earlier, phylogenetic analysis of highly conserved components of the conjugation systems indicates that over evolutionary time, the same families of conjugation systems have often been utilized for conjugal plasmids and ICEs. As is true for conjugal plasmids, ICEs are important vectors for the transfer of antibiotic resistance and other features that benefit the host (Table 5.3).

ICEs are common vectors for the formation of islands of foreign DNA in bacterial chromosomes. In some cases where these islands were first discovered, they were believed to be large transposons and were subsequently named with a "Tn" designation, for transposon. However, in recognition of their distinct biology they are now categorized with an "ICE" designation (or, less typically, "CTn") and a designation for the host where they were first identified. Investigations into the molecular biology of individual types of ICEs are still somewhat new, but exciting work with a small number of elements is providing a picture of the mechanisms used to maintain and exchange these elements. While some themes are emerging in how all ICEs function, this work also reveals surprisingly diverse molecular adaptations that have evolved.

SXT/R391 ICE

The SXT and R391 elements were discovered in clinical settings from *Vibrio cholerae* O139 and *Providencia rettgeri*, respectively, owing to the drug resistances they encode. Initially R391 was thought to be a plasmid and was assigned an incompatibility group, but it was later shown that the SXT and R391 elements are related and insert into the *prfC* gene, which encodes the peptide release factor 3 (see chapter 2). In addition to being important vectors for the distribution of antibiotic and metal resistance, they have been shown to have regulatory effects on the host through functional diguanylate cyclases, which participate in cell signaling (see chapter 12). The backbones of these elements are conserved and encode functions such as the *xis* and *int* functions that catalyze excision and integration of the element, along with the conjugation functions. Other conserved functions include an error-prone DNA polymerase similar to that found in some conjugal plasmids and a λ *red*-like recombination system (*exo*/*bet*) as found on

Table 5.3 Examples of integrating conjugative elements (ICEs)

ICE	Original host	Size (Kb)	Attachment site	Other information
SXT	*Vibrio cholerae*	100	*prfC*	Encodes resistance to chloramphenicol, sulfamethoxazole, trimethoprim, and streptomycin
R391	*Providencia rettgeri*	89	*prfC*	Encodes resistance to mercury and kanamycin
ICE*Bs*1	*Bacillus subtilis*	20	tRNA^{LEU} gene	Provides evidence for replication in some ICEs, sensitive to multiple stimuli for inducing transfer
ICE*St*1	*Streptococcus thermophilus*	35	*fda*	Encodes a type II restriction-modification system and putative abortive infection system; provides evidence for *cis*-mobilizable elements
ICE*Ml*Sym	*Mesorhizobium loti*	502	tRNA^{PHE} gene	Allows plant symbiosis; encodes nodulation and nitrogen fixation functions
ICEpSAM2	*Streptomyces ambofaciens*	11	tRNA^{PRO} gene	Uses serine site-specific recombinases for integration/excision; transfers as double-stranded DNA
Tn*916*	*Enterococcus faecalis*	18	AT-rich regions	Encodes resistance to tetracycline; related elements transferred broadly in *Firmicutes*
CTnDOT	*Bacteroides* spp.	65	Limited number of nonidentical sites	Encodes resistance to tetracycline and erythromycin
Tn*GBS2*	*Streptococcus agalactiae* (group B *Streptococcus*)	34	Intergenic regions upstream of A promoters	Uses a DDE transposase for integration/excision

some bacteriophages (see chapter 9). Regulation of the SXT/R391 excision and transfer functions is controlled by at least three proteins, SetR, SetC, and SetD, and is reminiscent of λ phage regulation, including induction by SOS (see chapter 10). Activated RecA leads to the degradation of SetR via its coprotease activity, as found with LexA and λ repressor, *c*I (see chapter 7). Once repression from SetR is released, SetC and SetD act as positive regulators for the excision and transfer genes (Figure 5.11).

EXAMPLE: SURFACE EXCLUSION IN ICE WITH SXT/R391 ELEMENTS

The SXT and R391 elements are closely related at the gene level and utilize the same integration site in *prfC*;

Figure 5.11 Self-transmissible and mobilizable elements are found with integrating conjugative elements. Mobilizable genomic islands (MGI) appear to hijack the *setCDR* regulatory network and conjugation machinery of an integrating conjugative element (ICE) in the same cell. The self-transmissible ICE is located in its attachment site upstream of the *prfC* gene. The *setR* gene product represses expression of *setCD*. DNA damage relieves SetR repression when activated RecA acts as a coprotease causing autocleavage of SetR, allowing SetCD to be expressed and to positively activate the self-transmissible ICE. An MGI located in the attachment site downstream of *yicC* in the same cell hijacks the regulatory activation to allow expression of its own excision machinery (Int_{MGI} and RdfM) and utilizes the ICE conjugation machinery to spread to new hosts. Low-level expression of Int_{MGI} allows integration of an MGI in the new host even in the absence of the self-transmissible ICE element.

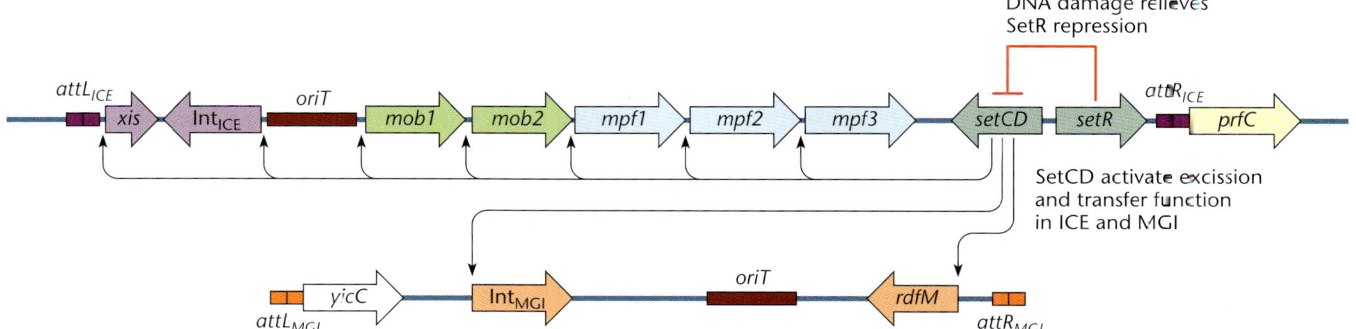

however, they also show evidence of diversification in their exclusion systems. Like many conjugal plasmids, ICEs utilize exclusion mechanisms that strongly inhibit the entry of a second element. In the case of the SXT/R391 family, this capacity resides in the TraG and Eex proteins, which act in the donor and recipient cell, respectively (see Marrero and Waldor, Suggested Reading). SXT elements utilize protein variants $TraG_S$ and Eex_S to exclude other SXT elements, and R391 elements utilize $TraG_R$ and Eex_R to exclude other R391 elements. Slight variations in these two proteins are responsible for allowing the exclusion of only the more similar class of elements. SXT and R391 elements (i.e., elements from different exclusion groups) can be found tandemly inserted into the *prfC* site in laboratory strains, and transfer from strains with these two elements reveals a low frequency of hybrid elements. Recombination to allow formation of hybrid elements can occur via the host RecA-mediated recombination or the ICE-encoded Bet/Exo recombination system.

EXAMPLE: SELF-TRANSMISSIBLE AND MOBILIZABLE ICE WITH SXT/R391 ELEMENTS

Also similar to conjugal plasmids, ICEs of the SXT/R391 family include both self-transmissible and mobilizable elements. Small (18 to 33 kb) elements called MGIs that contain the same *oriT* as the SXT/R391 elements can hijack the conjugation machinery of full SXT/R391 elements. The MGI elements utilize an attachment site in the *yicC* gene (instead of *prfA*), where integration and excision are catalyzed by an MGI-encoded integrase (Int_{MGI}) with a partner protein RdfM (Figure 5.11). Expression of these proteins is controlled by the SetCDR regulatory network from a coresident STX/R391 element; however, low expression of the Int_{MGI} integrase allows integration of the MGI element even without an STX/R391 element. MGI elements are naturally found in a variety of hosts that also can contain the SXT/R391 ICE. SXT/R391 ICE can also mobilize plasmids, as in the case of the plasmids CloDF13 and RSF1010. The SXT/R391 elements (and the MGI elements in the presence of an SXT/R391 element) have been shown to be capable of acting like an Hfr as described above and likely also play an important role in horizontal transfer between bacteria.

ICE*Bs*1

Probably the most well-understood ICE is the ICE*Bs*1 element. The ICE*Bs*1 element remained undiscovered for decades in the genetic workhorse for the *Firmicutes*, *Bacillus subtilis*. While the ICE*Bs*1 element appears to be found in all strains of *B. subtilis*, any benefit that it gives to the host remains undiscovered. Like phage λ (see chapter 7), ICE*Bs*1 requires two proteins, Int and Xis, to be excised. Excision of the element can be detected at a very low frequency under normal growth conditions, but robust excision via the *int* and *xis* gene products and transfer is controlled through systems that respond independently to DNA damage through the SOS response (see chapter 10) or quorum sensing (see chapter 12) through the RapI protein, which is encoded by the element. RapI belongs to a family of proteins found throughout *Bacillus* species that are regulated by small secreted peptides. RapI responds to a peptide that is processed from the *phrI* gene product, allowing the element to specifically gauge the density of ICE*Bs*I-containing hosts nearby.

EXAMPLE: AN ICE REPLICATION SYSTEM

In the initial work with ICEs, it was believed that these elements were replicated passively, through the replication of the host genome or during the process of conjugation. However, work with ICE*Bs*1 indicates that a regulated burst of rolling-circle DNA replication is induced after excision of the element. Following nicking at the ICE *oriT* by the NicK protein, the element-encoded helicase HelP works with host replication factors to allow replication of the element. As is typical for rolling-circle replication, second-strand synthesis is also needed, and in the case of ICE*Bs*1, it is initiated at two functionally redundant *cis*-acting *sso* sites in the element. While DNA replication is not required for conjugation to occur, the burst of replication may be important for maintenance of the element. Given that reintegration of the ICE*Bs*1 element may be slow, replication would allow sufficient copies of the element if the host cell divided while the element was excised.

EXAMPLE: TARGETING CONJUGATION IN ICE WITH ICE*Bs*1

B. subtilis tends to form chains of cells as it grows, and it has been shown that transfer of ICE*Bs*1 occurs most robustly at the poles, allowing the element to move efficiently though these chains of growing bacteria. This bias can be explained by the localization of one of the essential ATPases found in conjugation systems, ConE, which is the VirB4 homolog in ICE*Bs*1. The ConE protein is concentrated at the cell poles, but there is additional ConE protein dispersed around the entire cell (Figure 5.12). This distribution of ConE (and presumably the other unknown constituents of the pore complex) would enable the cells to transfer ICE*Bs*1 side to side, but the high concentration at the poles allows mating to occur very efficiently in cell chains from pole to pole (see Babic et al., 2011, Suggested Reading).

The ICE*Bs*1 element is capable of mobilizing certain plasmids in an unusual and incompletely understood way. Plasmids found to be mobilized by ICE*Bs*1 lack dedicated mobilization functions (i.e., a *cis*-acting *oriT* that is homologous to the *oriT* site found in ICE*Bs*1). Instead of transfer of these plasmids initiating at a canonical *oriT* site with

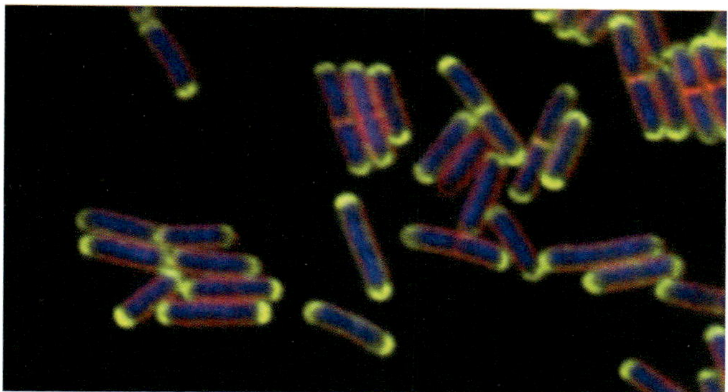

Figure 5.12 Fluorescence micrograph of *Bacillus subtilis* cells showing the localization of the cell membrane (red), DNA (blue), and ConE mating protein fused to green fluorescent protein (yellow-green). ConE is concentrated at the cell poles but is also found at lower levels around the entire cell periphery, presumably along with the other transfer components. This distribution is predicted to favor conjugation within the chain of bacterial cells while also allowing transfer to hosts not found in the chain. Modified from Berkman MB, et al, *J Bacteriol* **192:**38–45, 2020.

the ICE relaxase, transfer may initiate with the plasmid's Rep protein that is used for rolling-circle replication. Such an observation broadens our view of how ICEs can contribute to the transfer of information between bacteria.

cis-MOBILIZABLE ELEMENTS: EVOLUTION BY ACCRETION-MOBILIZATION

The ICE*St*1 and ICE*St*3 elements found in the 3′ end of the *fda* gene of the lactic acid bacterium *Streptococcus thermophiles* are distantly related to ICE*Bs*1. However, regulation of the ICE*St* elements differs from that of ICE*Bs*1 and remains unresolved. The conjugation and recombination genes in the ICE*St*1 and ICE*St*3 elements are nearly identical, but there are numerous other differences between the elements. Analysis of the ICE*St*1/ICE*St*3 elements revealed a type of mobilization different from that previously described in this chapter; this mechanism is significant for the formation and mobilization of **genomic islands** with ICE and other types of elements with a preferred attachment site. This process involves defective elements called *cis*-mobilizable elements (CIMEs) that generally range in size from 5 to 35 kb. CIMEs are widespread in *Firmicutes* localized in the *fda* attachment site but lack both the conjugation functions and the *cis*-acting *oriT* site.

As described throughout this chapter, elements that share the same *oriT* site can mobilize elements in *trans*. That is, the diffusible relaxase protein can act on an *oriT* site found on a plasmid or excised ICE in the cell, allowing it to be transferred via a cognate type IV secretion system. Analysis of ICE*St*1/ICE*St*3 elements revealed a different type of mobilization that functions in *cis* when two elements reside in the same *fda* attachment site; after they are excised together, both elements can be mobilized to a new host using one *oriT* site (Figure 5.13). In this process, an autonomous ICE from the same family that utilizes the same attachment site can integrate into the $attL_{CIME}$ or $attR_{CIME}$ of this element to form a composite element in the attachment site. Excision of this element can reestablish the autonomous ICE using the $attL_{ICE}$ and *attI′* site (not shown), or the composite element can be excised using the $attL_{ICE}$ and $attR_{CIME}$ sites (Figure 5.13). The excised composite element can now be mobilized by conjugation because the elements are on a contiguous DNA segment and can share the origin of transfer via *cis*-mobilization. A broader paradigm that is sometimes called **accretion-mobilization** is emerging where elements such as these could gather (or accrete) genetic information in the attachment site as elements integrate and eventually become defective, but subsequent insertion of active elements can mobilize the composite elements (see Bellanger et al., Suggested Reading).

Tn*916*

Tn*916* is a well-studied ICE encoding resistance to tetracycline that was originally discovered in *E. faecalis*. Tn*916* was the first element that was recognized to be capable of excising from the chromosome and transferring between host bacteria by conjugation. Under the current nomenclature, this element would be named using the convention used for ICE. Tn*916* and its relatives are known to be promiscuous, and they transfer into many representatives from the *Firmicutes* and even into some bacteria outside this family. The antibiotic resistance gene they carry, *tetM*, has also been found in many types of bacteria, and it is tempting to speculate that elements like Tn*916* are responsible for the widespread dissemination of the *tetM* gene. Other ICEs, such as CTnDOT, which also carries tetracycline resistance, have been found in *Bacteroides* species. These *Bacteroides* ICEs also mobilize other smaller chromosomal elements. Relatives are known to carry other antibiotic resistances as well. As indicated below, Tn*916* is unlike the ICEs discussed up to this point in that it is capable of integrating into a wide variety of different sites in the recipient chromosome.

EXAMPLE: SITE-SPECIFIC RECOMBINATION WITHOUT USING A SPECIFIC SITE

The excision process has been studied extensively for Tn*916*, and like most of the other elements requires Int and Xis (Figure 5.14) (see Marra et al., Suggested Reading). The integrase first makes a staggered break in the donor host DNA near the ends of the transposon to leave single-stranded ends 6 nucleotides long. The flanking sequences shown in Figure 5.14 are arbitrary because they differ depending on where the transposon was inserted. These single-stranded ends are not fully complementary to each

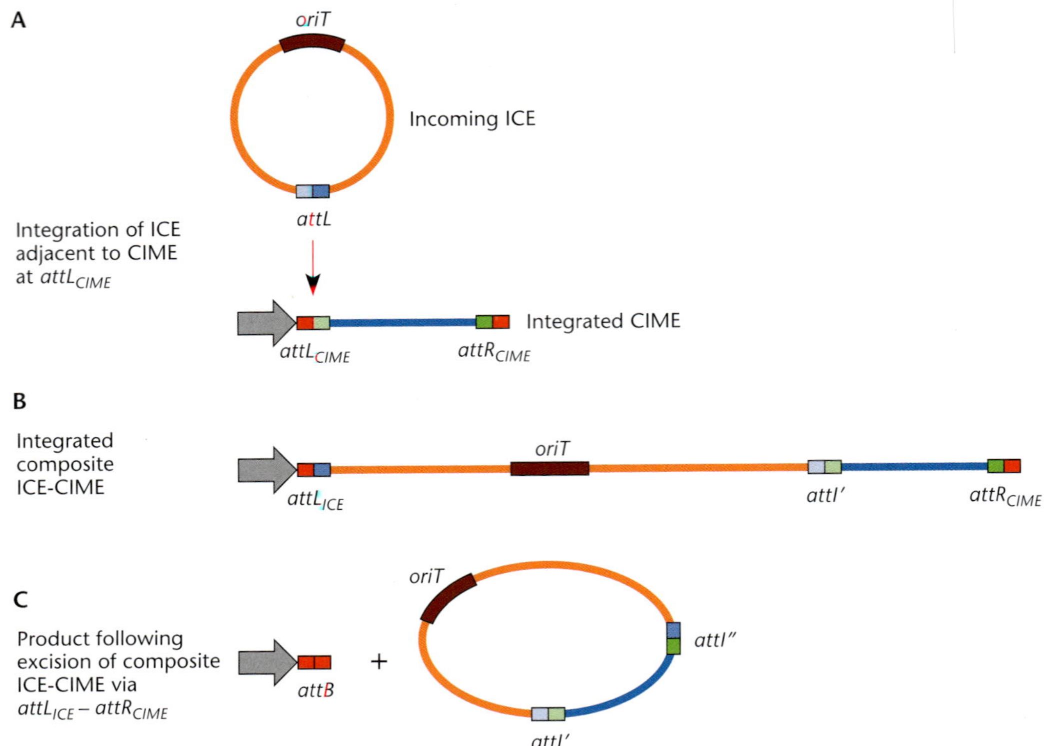

Figure 5.13 Self-transmissible integrating conjugative elements (ICE) can carry out a form of *cis*-mobilization of defective elements that utilize the same attachment site. **(A)** A *cis*-mobilizable element (CIME) in the chromosomal attachment site; this element has the *cis*-acting end sequences needed for excision and integration but none of the *trans*-acting proteins normally found in ICE. When an ICE that recognizes the same attachment site enters the cell, it can integrate adjacent to the CIME. **(B)** The ICE and CIME constitute a potential composite element with the active *attL* and *attR* sites flanking both units. Presumably, additional ICEs could continue to integrate into the same attachment site, allowing genetic information to accumulate at this locus, although ICE-encoded exclusion systems would limit this with intact ICE elements. **(C)** The machinery of a self-transmissible ICE can excise the entire composite element as shown, allowing the composite element to be conjugated into a new host. In the new host, the composite element, the ICE, or the CIME could integrate at the new attachment site (not shown).

other. Nevertheless, the ends seem to pair to form a circular intermediate of the element, including the unpaired region formed by the ends of the element, which are thought to form a heteroduplex sequence called the coupling sequence. It has proven difficult to confirm various aspects of this model because of the possibility of directed mismatch repair (see chapter 10) in the coupling sequence before or after transfer.

The *tra* genes are *orf13* to *orf23*, and the *oriT* sequence lies between *orf20* and *orf21*. Orf20 functions as the Tn*916* relaxase, and conjugation likely proceeds as described for other elements (see Rocco and Churchward, Suggested Reading). Through the process of conjugation, the 6-bp heteroduplex region becomes a homoduplex when the complementary strands are made in the donor and recipient cells. The Int protein of the element then integrates the element into the DNA of the recipient cell; this is similar to the way the Int protein, but not Xis, is required to integrate the λ phage genome into the chromosome (see chapter 7).

Tn*916* uses a clever mechanism to ensure that transfer occurs only after it is excised. The *tra* genes are arranged so that they can be transcribed only after the element has been excised. It accomplishes this by positioning the promoter, called p_{orf7} (Figure 5.14), so that the *tra* genes (*orf13* to *orf23*) are transcribed only when the element has been excised and has formed a circle. The Int protein may also bind to the *oriT* sequence, while the ICE is inserted in the chromosome, blocking access to the *oriT* sequence by the Tra functions and precluding transfer. While some elements are capable of transferring chromosomal markers, one can imagine distinct advantages to preventing conjugation until the element is excised. If Tn*916* transferred while still integrated in the chromosome, only

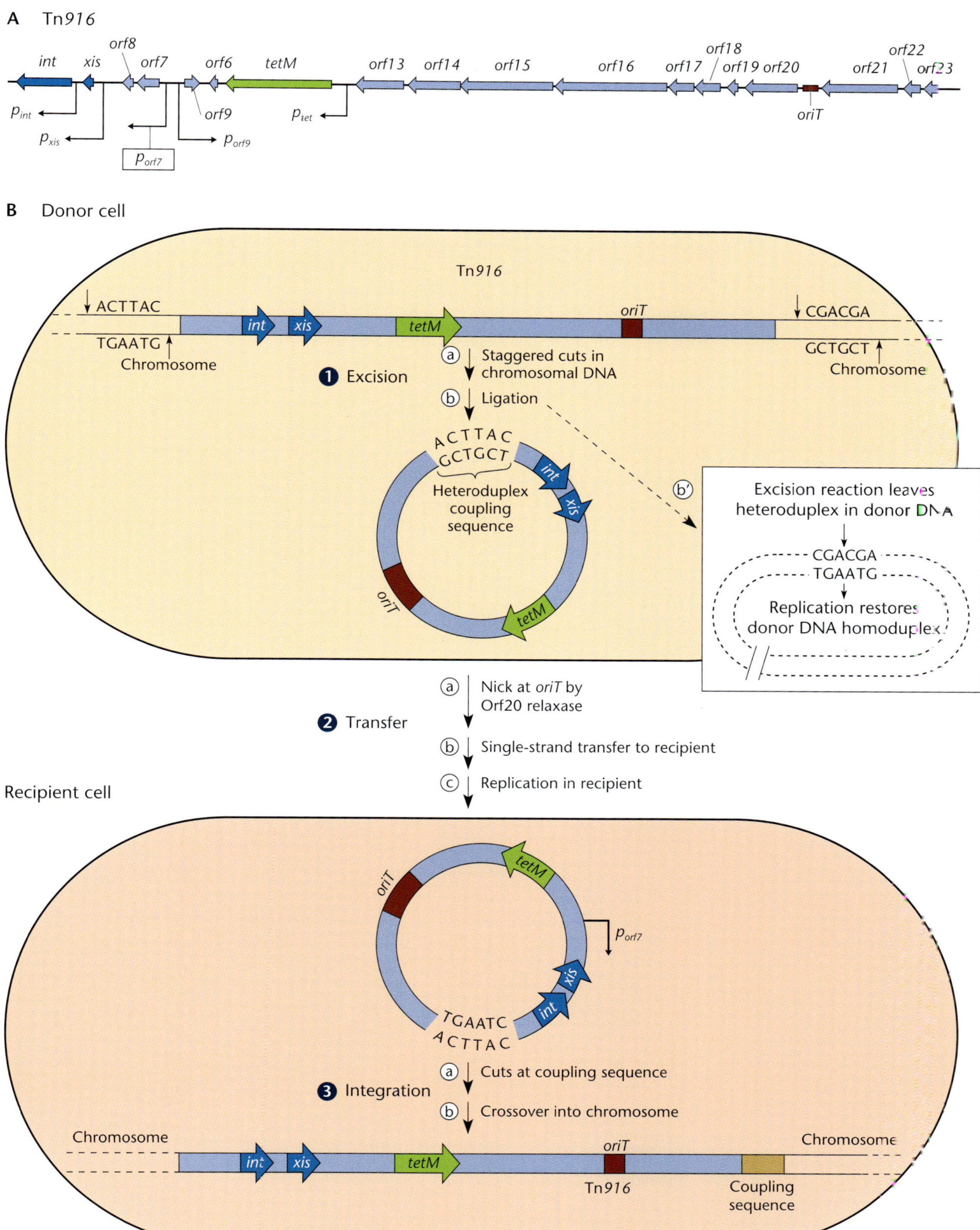

Figure 5.14 Genetic map and diagram of the integration and excision process of the integrating conjugative element (ICE) Tn*916*. **(A)** Genetic map of Tn*916*. **(B)** **(Top)** Mechanism used for excision of Tn*916* from the host chromosome; **(bottom)** mechanism used by Tn*916* to insert into the host chromosome following conjugation into a new host.

the part of the element on one side of the *oriT* site would enter the recipient. That is, transfer of the chromosomal information would be at the expense of transfer of the Tn*916* element because only a portion of the element would be mobilized to the new host.

The process of transfer of the *Bacteroides* conjugative elements, such as CTnDOT, shows similarities with the Tn*916* system (see Cheng et al., Suggested Reading). The transfer of both CTnDOT and Tn*916* is induced by subinhibitory concentrations of tetracycline in the medium in much the same way that opines induce transfer of the Ti plasmid. However, the CTnDOT element and other ICEs are more restricted in target site selection than Tn*916*, preferring sites with a 10-bp sequence that is similar to a sequence on the element.

Tn*GBS*1 and Tn*GBS*2

Tn*GBS* elements were originally identified in *S. agalactiae* (group B *Streptococcus*), and these elements broaden our definition of the molecular nature of ICEs. Unlike nearly all ICEs that use a tyrosine site-specific recombinase, Tn*GBS* elements utilize an enzyme typically associated with DNA transposons, the DDE recombinase (see chapter 8). DDE recombinases differ from most transposases in that the enzyme is capable of directing insertion into a specific position 15 or 16 bp upstream of the −35 region of promoters recognized by sigma A (see chapter 2). Tn*GBS* elements appear to have a limited host range and are typically found only in the genus *Streptococcus*. These elements also show replication following excision from the chromosome, and their host range may be limited not by the ability to enter other types of bacteria, but instead by poor stability in other bacteria because they cannot replicate outside *Streptococcus*. The mode of replication found with the Tn*GBS*1 and Tn*GBS*2 elements differs from the rolling-circle mechanism identified in ICE*Bs*1, which is also likely to occur in *Streptococcus* ICEs ICE*St*1/ICE*St*3. Instead, Tn*GBS*1 and Tn*GBS*2 appear to encode theta replication systems similar to those typically found in plasmids in the *Firmicutes*. Using synthetic constructs, it could be shown that replication of the element allows for much more efficient transfer of the element. Genes implicated in a theta mechanism of DNA replication are found in other ICEs that use tyrosine site-specific recombinases, suggesting that DNA replication may be important in stabilizing multiple types of ICE. Interestingly, the DDE recombinase utilized by Tn*GBS* elements to insert into the chromosome is not normally associated with targeting a specific position, and it will be interesting to know how the enzyme has adapted for this process.

1. Self-transmissible plasmids can transfer themselves to other bacterial cells, a process called conjugation. Some plasmids can transfer themselves into a wide variety of bacteria from different genera. Such plasmids are said to be promiscuous.

2. The plasmid genes whose products are involved in transfer are called the *tra* genes. The site on the plasmid DNA at which transfer initiates is called the origin of transfer (*oriT*). The *tra* genes can be divided into two groups, those whose products are involved in mating pair formation (Mpf) and those whose products are involved in processing the plasmid DNA for transfer (Dtr).

3. The Mpf component includes a sex pilus that extrudes from the cell and holds mating cells together. The pilus is the site to which male-specific phages adsorb. The Mpf system also includes the channel in the membrane through which DNA and proteins pass, as well as a coupling protein that lies on the channel, docks with the relaxase of the Dtr component, and translocates DNA through the channel.

4. The Dtr component includes the relaxase, which makes a nick within the *oriT* sequence and rejoins the ends of the plasmid in the recipient cell. The relaxase often contains a helicase activity, which separates the strands of DNA during transfer. The Dtr component also includes proteins that bind to the *oriT* sequence to form the multiprotein complex called the relaxosome and a primase that primes replication in the recipient cell and is sometimes transferred along with the DNA.

5. Most plasmids transiently express their Mpf *tra* genes immediately after transfer to a recipient cell and only intermittently thereafter.

6. Mobilizable plasmids cannot transfer themselves but can be transferred by other plasmids. Mobilizable plasmids encode only a Dtr component; they lack genes to encode an Mpf component. In the context of a mobilizable plasmid, the genes that encode the Dtr component are called the *mob* genes. A mobilizable plasmid can be mobilized by a self-transmissible plasmid only if the coupling protein of the self-transmissible plasmid can dock with the relaxase of the mobilizable plasmid. Because they lack an Mpf component, mobilizable plasmids can be much smaller than

Summary (continued)

self-transmissible plasmids, which makes them very useful in molecular genetics and biotechnology.

7. Hfr strains of bacteria have a self-transmissible plasmid integrated into their chromosomes. Hfr strains have historically been useful for genetic mapping in bacteria because they transfer chromosomal DNA in a gradient, beginning at the site of integration of the plasmid. Hfr crosses were an important early tool for ordering genetic markers on the entire genome.

8. Prime factors are self-transmissible plasmids that have picked up part of the bacterial chromosome. They can be used to make partial diploids for complementation tests. If a prime factor is transferred into a cell, the cell will be a partial diploid (merodiploid) for the region of the chromosome carried on the prime factor, making it useful for complementation experiments.

9. Self-transmissible plasmids are found in a wide variety of bacteria and also in archaea. Almost all of the systems use certain type IV secretion systems and Mpf and Dtr complexes and appear to be related. *Firmicutes*, which lack an outer membrane, use adhesins instead of a pilus to hold mating pairs together.

10. Integrating conjugative elements (ICEs) that transfer between host bacteria by conjugation appear to be as common as conjugal plasmids.

11. ICEs can display features like conjugal plasmids, including showing mechanisms of surface exclusion. ICEs also exist as both self-transmissible elements capable of their own transfer and versions that can be mobilized by hijacking the conjugation and/or integration systems of other elements.

12. ICE*Bs*1 is a well-studied element that regulates transfer with SOS induction or quorum sensing. Among other discoveries with this element is that some ICEs use a burst of replication after excision to allow transfer to be more efficient and ensure that a copy of the element exists to insert back into the host genome even if the host chromosome has divided.

13. ICE*St*1/ICE*St*3 elements utilize a form of *cis*-mobilization that allows the excision and transfer of composite elements that are likely important in forming genomic islands.

14. Tn*916* and CTnDOT are ICEs that utilize site-specific recombination to insert into the chromosome but are capable of recognizing many different insertion sites.

15. Tn*GBS*1 and Tn*GBS*2 elements further broaden our view of ICEs by using a DDE recombinase normally associated with DNA transposons that insert at many different sites, but the Tn*GBS* recombinase recognizes specific sites. Like some other ICEs, it replicates after excision but uses a theta mechanism, not a rolling-circle mechanism.

QUESTIONS FOR THOUGHT

1. Why are the *tra* genes whose products are directly involved in DNA transfer usually adjacent to the *oriT* site?

2. Why do you suppose plasmids with a certain *mob* site are mobilized by only certain types of self-transmissible plasmids?

3. Why do self-transmissible plasmids usually encode their own primase function?

4. What do you think is different about the cell surfaces of *Firmicutes* and *Proteobacteria* that causes only the self-transmissible plasmids of *Proteobacteria* to encode a pilus?

5. Why do many types of phages use the sex pilus of plasmids as their adsorption site?

6. Why are so many plasmids either self-transmissible or mobilizable? Why are so many promiscuous?

7. F primes can be selected by mating an Hfr into a strain that is incapable of doing homologous recombination. Explain how selecting a late marker early using an Hfr donor and recipient that are proficient at homologous recombination could accomplish the same thing.

8. What are some advantages to a conjugal system residing in the chromosome as an ICE versus being maintained as a conjugal plasmid?

9. Some ICEs undergo limited DNA replication following excision. Why might some replication be beneficial for an element even if it is maintained in the bacterial chromosome?

10. What are some advantages and disadvantages of an ICE being capable of utilizing a number of different sites instead of a single attachment site?

SUGGESTED READING

Babic A, Lindner AB, Vulic M, Stewart EJ, Radman M. 2008. Direct visualization of horizontal gene transfer. *Science* **319:**1533–1536.

Babic A, Berkmen MB, Lee CA, Grossman AD. 2011. Efficient gene transfer in bacterial cell chains. *MBio* **2:**e00027-11.

Bellanger X, Morel C, Gonot F, Puymege A, Decaris B, Guédon G. 2011. Site-specific accretion of an integrative conjugative element together with a related genomic island leads to *cis* mobilization and gene capture. *Mol Microbiol* **81:**912–925.

Berkmen MB, Lee CA, Loveday EK, Grossman AD. 2010. Polar positioning of a conjugation protein from the integrative and conjugative element ICE*Bs*1 of *Bacillus subtilis*. *J Bacteriol* **192:**38–45.

Carraro N, Burrus V. 2014. Biology of three ICE families: SXT/R391, ICEBs1, and ICESt1/ICESt3. *Microbiol Spectr* **2:**1–20.

Cheng Q, Paszkiet BJ, Shoemaker NB, Gardner JF, Salyers AA. 2000. Integration and excisions of a *Bacteroides* conjugative transposon, CTnDOT. *J Bacteriol* **182:**4035–4043.

Christie PJ, Atmakuri K, Krishnamoorthy V, Jakubowski S, Cascales E. 2005. Biogenesis, architecture, and function of bacterial type IV secretion systems. *Annu Rev Microbiol* **59:**451–485.

Clewell DB, Francia MV. 2004. Conjugation in Gram-positive bacteria, p 227–256. *In* Funnell BE, Phillips GJ (ed), *Plasmid Biology*. ASM Press, Washington, DC.

Covacci A, Telford JL, Del Giudice G, Parsonnet J, Rappuoli R. 1999. *Helicobacter pylori* virulence and genetic geography. *Science* **284:**1328–1333.

Cury J, Oliveira PH, de la Cruz F, Rocha EPC. 2018. Host range and genetic plasticity explain the co-existence of integrative and extrachromosomal mobile genetic elements. *Mol Biol Evol* **35:**2230–2239.

D'Costa VM, McGrann KM, Hughes DW, Wright GD. 2006. Sampling the antibiotic resistome. *Science* **311:**374–377.

Derbyshire KM, Hatfull G, Willetts N. 1987. Mobilization of the nonconjugative plasmid RSF1010: a genetic and DNA sequence analysis of the mobilization region. *Mol Gen Genet* **206:**161–168.

Dunny GM. 2007. The peptide pheromone-inducible conjugation system of *Enterococcus faecalis* plasmid pCF10: cell-cell signalling, gene transfer, complexity and evolution. *Philos Trans R Soc Lond B Biol Sci* **362:**1185–1193.

Goessweiner-Mohr N, Arends K, Keller W, Grohmann E. 2014. Conjugation in Gram-positive bacteria. *Microbiol Spectrum* 2(4): PLAS-0004-2013.

Grahn AM, Haase J, Bamford DH, Lanka E. 2000. Components of the RP4 conjugative transfer apparatus form an envelope structure bridging inner and outer membranes of donor cells: implications for related macromolecule transport systems. *J Bacteriol* **182:**1564–1574.

Guglielmini J, Quintais L, Garcillan-Barcia MP, de la Cruz F, Rocha EPC. 2011. The repertoire of ICE in prokaryotes underscores the unity, diversity, and ubiquity of conjugation. *PLoS Genet* **7:**e1002222.

Guglielmini J, de la Cruz F, Rocha EPC. 2013. Evolution of conjugation and type IV secretion systems. *Mol Biol Evol* **30:**315–331.

Hinerfeld D, Churchward G. 2001. Specific binding of integrase to the origin of transfer (*oriT*) of the conjugative transposon Tn*916*. *J Bacteriol* **183:**2947–2951.

Hu B, Khara P, Christie PJ. 2019. Structural bases for F plasmid conjugation and F pilus biogenesis in *Escherichia coli*. *Proc Natl Acad Sci USA* **116:**14222–14227.

Kurenbach B, Grothe D, Farias ME, Szewzyk U, Grohmann E. 2002. The *tra* region of the conjugative plasmid pIP501 is organized in an operon with the first gene encoding the relaxase. *J Bacteriol* **184:**1801–1805.

Lawley T, Wilkins BM, Frost LS. 2004. Bacterial conjugation in Gram-negative bacteria, p 203–226. *In* Funnell BE, Phillips GJ (ed), *Plasmid Biology*. ASM Press, Washington, DC.

Lederberg J, Tatum EL. 1946. Gene recombination in *Escherichia coli*. *Nature* **158:**558.

Manchak J, Anthony KG, Frost LS. 2002. Mutational analysis of F-pilin reveals domains for pilus assembly, phage infection, and DNA transfer. *Mol Microbiol* **43:**195–205.

Marra D, Pethel B, Churchward GG, Scott JR. 1999. The frequency of conjugative transposition of Tn*916* is not determined by the frequency of excision. *J Bacteriol* **181:**5414–5418.

Marra D, Scott JR. 1999. Regulation of excision of the conjugative transposon Tn*916*. *Mol Microbiol* **31:**609–621.

Marrero J, Waldor MK. 2005. Interactions between inner membrane proteins in donor and recipient cells limit conjugal DNA transfer. *Dev Cell* **8:**963–970.

Matson SW, Sampson JK, Byrd DRN. 2001. F plasmid conjugative DNA transfer: the TraI helicase activity is essential for DNA strand transfer. *J Biol Chem* **276:**2372–2379.

Parker C, Meyer RJ. 2007. The R1162 relaxase/primase contains two type IV transport signals that require the small plasmid protein MobB. *Mol Microbiol* **66:**252–261.

Rees CED, Wilkins BM. 1990. Protein transfer into the recipient cell during bacterial conjugation: studies with F and RP4. *Mol Microbiol* **4:**1199–1205.

Rocco JM, Churchward G. 2006. The integrase of the conjugative transposon Tn*916* directs strand- and sequence-specific cleavage of the origin of conjugal transfer, *oriT*, by the endonuclease Orf20. *J Bacteriol* **188:**2207–2213.

Schmidt JW, Rajeev L, Salyers AA, Gardner JF. 2006. NBU1 integrase: evidence for an altered recombination mechanism. *Mol Microbiol* **60:**152–164.

Schröder G, Lanka E. 2005. The mating pair formation system of conjugative plasmids: a versatile secretion machinery for transfer of proteins and DNA. *Plasmid* **54:**1–25.

Senghas E, Jones JM, Yamamoto M, Gawron-Burke C, Clewell DB. 1988. Genetic organization of the bacterial conjugative transposon, Tn*916*. *J Bacteriol* **170:**245–249.

Vogel JP, Andrews HL, Wong SK, Isberg RR. 1998. Conjugative transfer by the virulence system of *Legionella pneumophila*. *Science* **279:**873–876.

Winans SC, Walker GC. 1985. Conjugal transfer system of the IncN plasmid pKM101. *J Bacteriol* **161:**402–410.

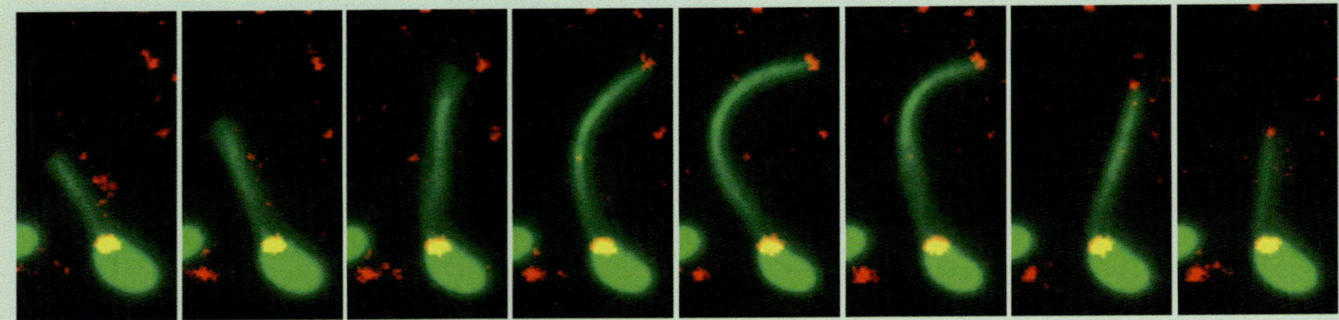

This series of images shows extension of a type IV pilus from a *Vibrio cholerae* cell (green) and contact being made between the tip of the pilus and a fragment of DNA (red). Retraction of the pilus brings the DNA to the cell surface and allows the DNA to enter the cell (see Ellison et al., Suggested Reading). Courtesy of the Dalia Lab, Indiana University.

6

Transformation

DNA CAN BE EXCHANGED AMONG BACTERIA IN THREE WAYS: conjugation, transduction, and transformation. Chapter 5 covers the mechanisms of conjugation, in which a plasmid or other self-transmissible DNA element transfers itself, and sometimes other DNA, into another bacterial cell. Transduction, which is discussed in chapter 7, utilizes phages to carry DNA from one bacterium to another. In this chapter, we discuss transformation, a process in which cells take up free DNA directly from their environment. Transformation is unique among genetic transfer mechanisms in that all required functions are encoded by the recipient cell; in most cases, the donor cell is dead, and its only role is to contribute its DNA after cell lysis. Because transformation involves uptake of naked DNA, it can be distinguished from other forms of genetic exchange because of its sensitivity to addition of extracellular DNase, which degrades free DNA. Addition of DNase cannot prevent conjugation (in which the transferred DNA is protected by direct cell-cell contact) or transduction (in which the transferred DNA is packaged into viral particles).

Transformation is one of the cornerstones of molecular genetics because it is often the best way to introduce experimentally altered DNA into cells. Transformation was first discovered in bacteria, but methods have been devised to transform many types of animal and plant cells, as well.

The terminology of genetic analysis by transformation is similar to that of conjugation and transduction (see chapter 3). DNA is derived from a **donor bacterium** and taken up by a **recipient bacterium**. Incoming plasmid or phage DNA may be able to establish replication on its own, and incoming chromosomal DNA must recombine with resident DNA in the cell, such as the chromosome, to form recombinant types. The cell that has taken up the incoming DNA is referred to as a **transformant**. The frequency of recombinant types for various genetic markers can be used for genetic analysis. Such genetic data obtained by transformation are analyzed similarly to those obtained by transduction. If the regions of two markers can be carried on the same piece of transforming DNA, the two markers are said to be **cotransformable**. The higher the **cotransformation frequency**, the more closely linked are the two markers on the DNA. The principles of using transduction and transformation for genetic mapping are outlined in chapter 3. In this chapter, we concentrate on the mechanism of transformation in various bacteria and its relationship to other biological phenomena.

Natural Transformation

Some types of bacteria are **naturally transformable**, which means that they can take up DNA from their environment without requiring special chemical or electrical treatments to make them more permeable. Even naturally transformable bacteria are not always capable of taking up DNA and in most cases do so only at certain stages in their life cycle or under certain growth conditions. Bacteria that can take up DNA are said to be **competent**, and bacteria that are naturally capable of reaching this state are said to be **naturally competent**.

Naturally competent transformable bacteria were originally found in several genera, including members of the *Firmicutes*, such as *Bacillus subtilis*, a soil bacterium, and *Streptococcus pneumoniae*, which causes throat infections, and *Proteobacteria*, such as *Haemophilus influenzae*, a causative agent of pneumonia and spinal meningitis, and *Neisseria gonorrhoeae*, which causes gonorrhea. More recently, natural transformation has been uncovered not only in close relatives of the initial model systems, but also in *Helicobacter pylori*, a stomach pathogen; *Acinetobacter baylyi*, another soil bacterium; marine cyanobacteria including *Synechococcus*; *Vibrio cholerae*, which causes cholera; *Thermus thermophilus*, an extreme thermophile; and *Deinococcus radiodurans*, an organism resistant to high levels of radiation. Genome sequencing has revealed that many organisms that have not been demonstrated to be naturally transformable contain some of the genes known to be involved in competence in other species, suggesting that some of these organisms may be transformable under certain conditions or that they have lost this property; the number of transformable organisms is increasing rapidly as new approaches for inducing competence are uncovered (see Johnston et al., Suggested Reading). The recognition that transformation is more widespread than previously thought suggests a broader impact on heterologous gene transfer and bacterial evolution.

Discovery of Transformation

Transformation was the first mechanism of bacterial gene exchange to be discovered. In 1928, Fred Griffith found that one form of the pathogenic pneumococci (now called *S. pneumoniae*) could be mysteriously "transformed" into another form. Griffith's experiments were based on the fact that *S. pneumoniae* makes two types of colonies with different appearances, one type made by pathogenic bacteria and the other type made by nonpathogenic bacteria that are incapable of causing infections. The colonies made by the pathogenic strains appear smooth on agar plates, because the bacteria excrete a polysaccharide capsule. The capsule apparently protects the bacteria and allows them to survive in vertebrate hosts, including mice, which they can infect and kill. However, rough-colony-forming mutants that cannot make the capsule sometimes arise from the smooth-colony formers, and these mutants are nonpathogenic in mice.

In his experiment, Griffith mixed dead *S. pneumoniae* cells that made smooth colonies with live nonpathogenic cells that made only rough colonies and injected the mixture into mice (Figure 6.1). Mice given injections of only the rough-colony-forming bacteria survived, but mice that received a mixture of dead smooth-colony formers and live rough-colony formers died. Furthermore, Griffith isolated live smooth-colony-forming bacteria from the blood of the dead mice. Concluding that the dead pathogenic bacteria gave off a "transforming principle" that changed the live nonpathogenic rough-colony-forming bacteria into the pathogenic smooth-colony form, he speculated that this transforming principle was the polysaccharide itself. Later, other researchers did an experiment in which they transformed rough-colony formers into smooth-colony formers by mixing the rough forms with extracts of the smooth-colony formers in a test tube and found that the surviving cells included smooth-colony formers, which indicated that the active fraction was present in the extracts and incubation in the mice was not required for transformation to occur. Then, about 16 years after Griffith did his experiments with mice, Oswald Avery and his collaborators purified the "transforming principle" from the extracts of the smooth-colony formers and showed

Figure 6.1 The Griffith experiment. **(A)** Type R (rough) nonencapsulated bacteria (black circles) are nonpathogenic and do not survive in the host. **(B)** Type S (smooth) encapsulated bacteria (black circles with green halos) are pathogenic and are recovered from the host. **(C)** Heat-killed type S bacteria (black circles with red halos) fail to kill the host and cannot be recovered. **(D)** Mixing heat-killed type S bacteria (black circles with red halos) with live type R bacteria (black circles) can convert the type R bacteria to the pathogenic capsulated form (black circles with green halos).

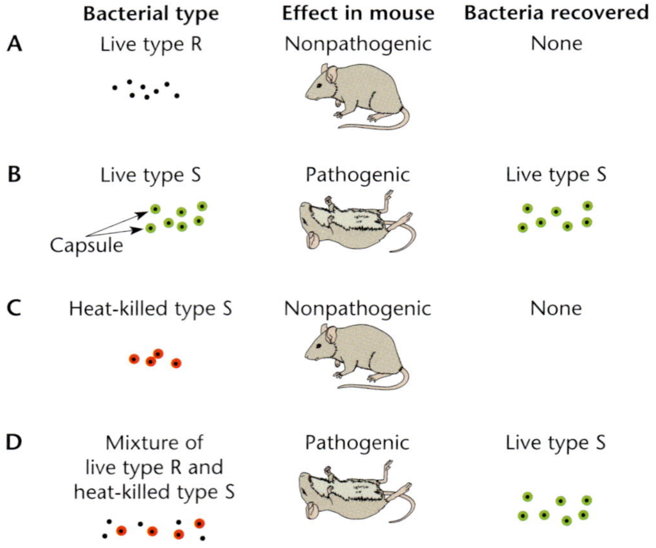

that the active fraction of the cell extracts is DNA (see Avery et al., Suggested Reading). Thus, Avery and colleagues were the first to demonstrate that DNA, and not protein, polysaccharide, or other factors in the cell, is the hereditary material (see the introductory chapter).

Overview of Natural Transformation

The general steps that occur in natural transformation differ somewhat in different systems. The best-characterized model systems are *B. subtilis* and *S. pneumoniae* (which lack an outer membrane and are sometimes described as Gram-positive based on diagnostic staining by the Gram stain procedure) and *H. influenzae* and *N. gonorrhoeae* (which contain an outer membrane and periplasmic space between the inner and outer membranes and are sometimes described as Gram-negative based on lack of staining by the Gram stain procedure). The mechanics of DNA uptake depends in part on whether the bacteria contain an outer membrane, although many of the protein components used are conserved. *H. pylori* is exceptional in that it uses a DNA uptake machinery different from that found in all other groups of bacteria that have been characterized. Regardless of the details, in all cases, the transforming DNA binds to the cell surface as double-stranded DNA, but only one strand traverses the cell membrane and enters into the cytoplasm. Once in the cell, the single-stranded transforming DNA might stably integrate into the genome by homologous recombination of the translocated single strand into the chromosome or other recipient DNA, reestablish itself as a plasmid after synthesis of the complementary strand, and recyclize using recombination, or be degraded.

DNA Uptake Mechanisms

The uptake of DNA is discussed in more detail below. Nearly all transformation systems characterized to date (with the notable exception of *H. pylori*) utilize a transformation-specific **pseudopilus** that resembles type IV pili that are characteristic of type II secretion systems and are similar to those used in some plasmid conjugation systems (see chapter 5); in contrast, *H. pylori* utilizes a mechanism related to type IV secretion systems related to those used for translocation of specific molecules to other cells (e.g., to host cells during pathogenesis). In all cases, DNA uptake is an active process that requires ATP hydrolysis to drive the DNA translocation machinery. Experimental approaches that have been used to measure DNA uptake are described in Box 6.1.

DNA UPTAKE IN *FIRMICUTES*

Two species of *Firmicutes* in which transformation has been particularly well studied are *B. subtilis* and *S. pneumoniae*. These organisms will bind and take up DNA from any organism, although the fate of the DNA within the cell depends on whether the incoming DNA can participate in recombination with DNA already present in the cell or can establish independent replication (e.g., plasmid DNA). The proteins involved in transformation in these bacteria were discovered on the basis of isolation of mutants that lack the ability to take up DNA. The genes affected in the mutants were named *com* (for *com*petence defective).

In *B. subtilis*, the *com* genes are organized into several operons. The products of several of these, including the *comA* and *comK* operons, are involved in regulation of competence (see "Regulation of Natural Competence" below). Others, including the products of genes in the *comE*, *comF*, and *comG* operons, form part of the machinery in the membrane that transports DNA into the bacterium. For historical reasons, the genes in these operons use a nonstandard type of nomenclature in which they are given two letters, the first for the operon and the second for the position of the gene in the operon. For example, *comFA* is the first gene of the *comF* operon, while *comEC* is the third gene of the *comE* operon. The corresponding protein products of the genes have the same name with the first letter capitalized, e.g., ComFA and ComEC, respectively.

The role played by some of the Com proteins in the competence machinery of *B. subtilis* is diagrammed in Figure 6.2A (see Chen and Dubnau and Johnston et al., Suggested Reading). The proteins encoded in the *comG* operon form the transformation pseudopilus. ComGC constitutes the major pilin protein; pilin processing is dependent on the ComC endopeptidase, and pseudopilus assembly requires the ComGB and ComGA proteins, which are embedded in the cell membrane. In *Streptococcus*, double-stranded DNA interacts directly with the pseudopilus outside the cell, probably by nonspecific electrostatic interactions between the negatively charged DNA backbone and the positively charged amino acids in the pilin proteins. In *B. subtilis*, initial interactions appear to be mediated by competence-specific cell wall teichoic acids (see Mirouze et al., Suggested Reading). Double-stranded DNA bound at the cell surface is cleaved by the NucA endonuclease into segments that are approximately 6 kb in length in *S. pneumoniae* and 15 kb in *B. subtilis*, and the segments are then believed to be brought through the cell wall to the cell membrane by retraction of the pseudopilus, which probably occurs via disassembly of the pilus subunits, by analogy with retraction of some pili used in conjugation (see chapter 5).

The first gene of the *comE* operon, *comEA*, encodes the protein that directly binds extracellular double-stranded DNA at the outer surface of the membrane. One strand of the DNA is then degraded, and the other strand is transported through the membrane and into the cytoplasm through the ComEC channel, using ComFA as an ATP-dependent DNA translocase, possibly as part of an ATP-binding cassette (ABC) transporter, in conjunction

BOX 6.1

Experimental Measurements of DNA Uptake

Biochemical approaches have been useful for measurements of the efficiency of DNA uptake.

The figure shows an experiment based on the fact that transport of free DNA into the cell makes the DNA insensitive to deoxyribonucleases (DNases), which cannot enter the cell because naturally competent cells take up only DNA and not proteins. Donor DNA is radioactively labeled by growing the cells in medium in which the phosphorus has been replaced with phosphorus-32, the radioisotope of phosphorus. The radioactive DNA is mixed with competent cells, and the mixture is treated with DNase at various times. Any DNA that is not degraded and survives intact must have been taken up by the cells, where it is protected from the DNase. The medium containing the cells is precipitated with acid and collected on a filter, and the radioactivity on the filter (which is due to undegraded DNA that must have been taken up by the cells) is counted and compared with the total radioactivity of the DNA that was added to the cells. Calculation of the percentage of DNA that is taken up gives the efficiency of DNA uptake. Experiments such as these have shown that some competent bacteria take up DNA very efficiently. Unlike genetic tests for competence, this procedure separates the actual uptake of DNA from the processes of recombination and degradation within individual cells.

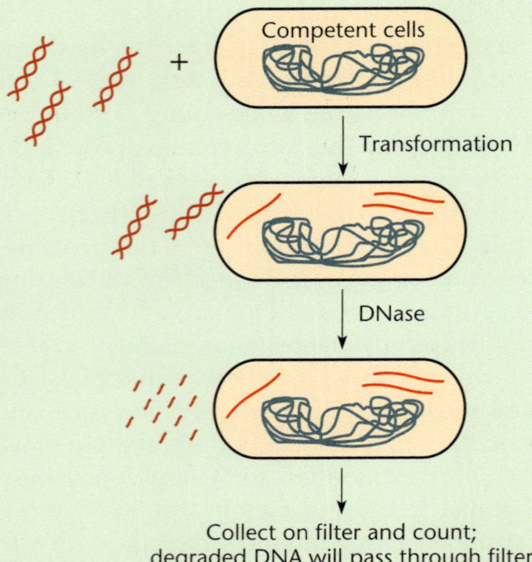

Determining the efficiency of DNA uptake during transformation. DNA inside the cell is insensitive to DNase. DNA that did not enter the cell is degraded and passes through a filter. The red segments indicate radioactively labeled DNA.

The same DNase resistance assay has been used to show that some types of bacteria (including *Neisseria* and *Haemophilus* spp.) take up DNA from only their own species, whereas others can take up DNA from any source. This is evident experimentally by testing whether DNA is protected from DNase after addition to competent cells. Those that take up any DNA will protect DNA from any source; in contrast, competent organisms that are selective protect DNA only from the same (or closely related) species, or heterologous DNA into which the appropriate uptake sequence has been artificially inserted using molecular techniques. Competition assays also can be used to verify DNA uptake specificity. Addition of high amounts of unlabeled DNA from the same species (and therefore containing uptake sequences) will inhibit uptake of radiolabeled DNA from that species if the organism utilizes uptake sequences. In contrast, addition of unlabeled DNA from a different source (and therefore lacking uptake sequences) will not inhibit uptake of the specific DNA if uptake is sequence-specific.

Figure 6.2 Structure of DNA uptake competence systems. **(A)** *Firmicutes.* **(B)** *Proteobacteria.* **(C)** *Helicobacter pylori.* Shown are some of the proteins involved and the channels they form. The nomenclature in panel A is based on *Bacillus subtilis*; in panel B, on *Neisseria gonorrhoeae*; and in panel C, on *H. pylori.* ComEA is a double-stranded DNA receptor protein in *B. subtilis.* The *B. subtilis* ComGC protein is analogous to the *Neisseria* PilE protein. The endonuclease that generates single-stranded DNA is EndA in *S. pneumoniae*; the corresponding protein in *B. subtilis* and *H. pylori* has not been identified. The *B. subtilis* ComEC protein is analogous to the *Neisseria* ComA protein and *H. pylori* ComEC. ComEC forms a transmembrane channel through which the single-stranded DNA passes. Energy is provided by an ATPase (ComFA in *B. subtilis*, PilT in *Neisseria*, unknown in *H. pylori*). ss, single stranded; SSB, competence-specific single-stranded DNA-binding protein. The DNA is shown running through the cell wall alongside the pseudopilus (ComG proteins in *B. subtilis*, PilE in *Neisseria*) but may be brought into the cell by retraction (disassembly) of the pseudopilus. Competence systems in both *B. subtilis* and *N. gonorrhoeae* are related to type II protein secretion systems; the competence system in *H. pylori* is related to type IV secretion systems.

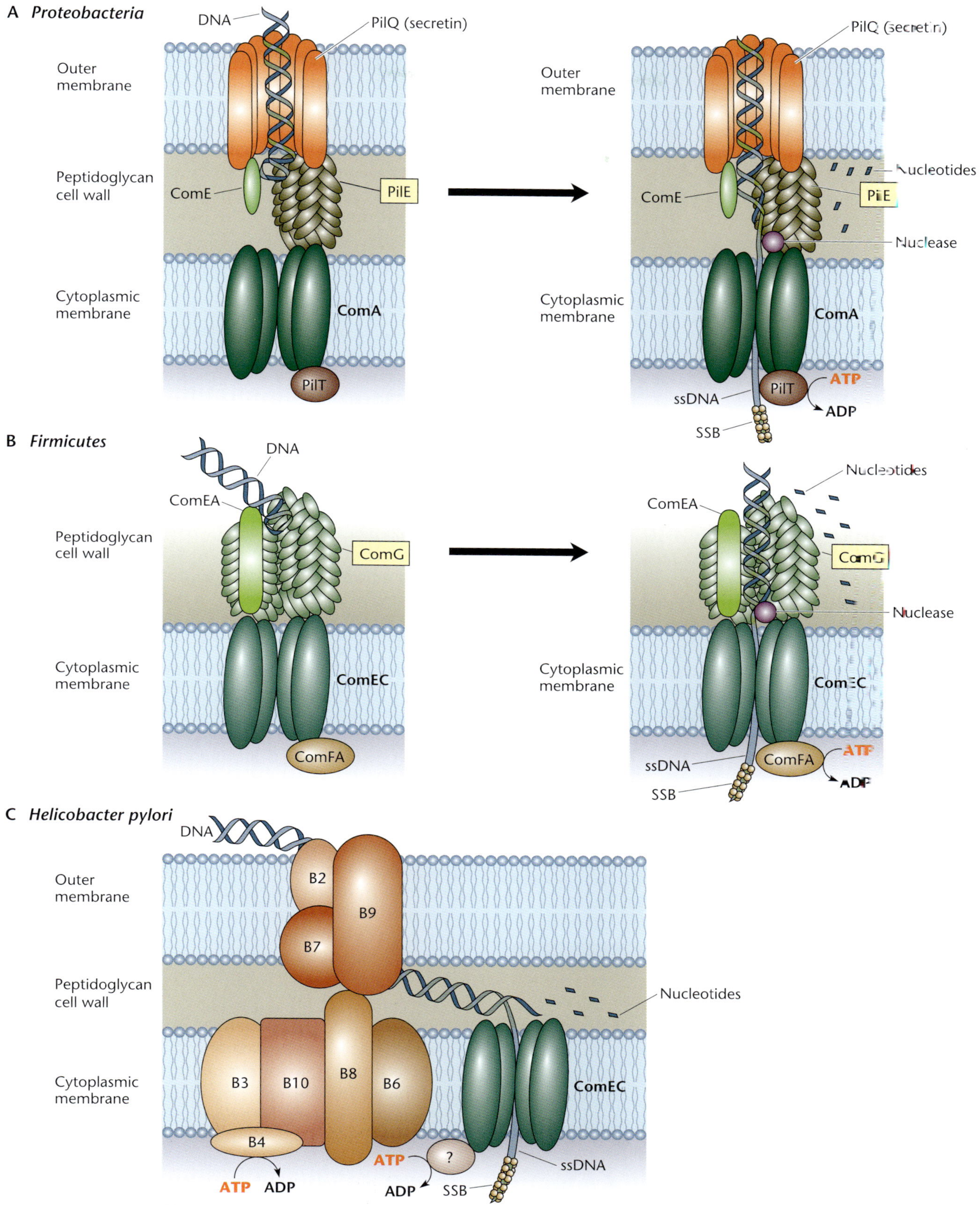
A Proteobacteria
DNA
PilQ (secretin)
Outer membrane
Peptidoglycan cell wall
ComE
PilE
Cytoplasmic membrane
ComA
PilT
PilQ (secretin)
Outer membrane
Nucleotides
ComE
PilE
Nuclease
Cytoplasmic membrane
ComA
ATP
PilT
ssDNA
ADP
SSB
B Firmicutes
DNA
ComEA
Peptidoglycan cell wall
ComG
Cytoplasmic membrane
ComEC
ComFA
Nucleotides
ComEA
ComG
Nuclease
Cytoplasmic membrane
ComEC
ssDNA
ComFA
ATP
SSB
ADP
C Helicobacter pylori
DNA
Outer membrane
B2
B9
B7
Peptidoglycan cell wall
Nucleotides
B8
Cytoplasmic membrane
B3
B10
B6
ComEC
B4
ATP
?
ssDNA
ATP
ADP
ADP
SSB

with ComEA and ComEC. Recent studies using transformation with fluorescently labeled DNA have yielded beautiful images of interaction of DNA with the competent cells and colocalization with ComFA (see Boonstra et al., Suggested Reading), as shown in Figure 6.3. DNA binding does not occur in noncompetent cells and is prevented by addition of DNase. The EndA nuclease is responsible for generation of single-stranded DNA in *S. pneumoniae*, but the corresponding activity in *B. subtilis* has not been identified. Single-stranded DNA is transported into the cell at the rate of 80 to 100 nucleotides per second, and in *S. pneumoniae*, the DNA has been shown to enter with 3′-to-5′ polarity; at that rate, a 10-kb segment of DNA would be transported in about 2 minutes. DNA uptake in *S. pneumoniae* utilizes proteins and mechanisms similar to those of transformation in *B. subtilis*, although the names of the *com* genes are often different (see Claverys et al., Suggested Reading).

DNA UPTAKE IN *PROTEOBACTERIA*

As mentioned above, a growing number of *Proteobacteria* are also being shown to be capable of natural competence. Well-characterized examples include the pathogens *Neisseria* spp., *Haemophilus* spp., and *Vibrio cholerae*. In *Neisseria* and *Haemophilus*, specific uptake sequences are required for the binding of DNA, so these species usually take up DNA only from the same or closely related species (see "Specificity of DNA Uptake" below). This differs from most other bacteria, which do not have specific uptake sequences.

Most *Proteobacteria* use a system similar to that described for the *Firmicutes*; as noted above, these systems are related to type II secretion systems (see chapter 2), which assemble type IV pili on the cell surface (Figure 6.2B).

Figure 6.3 Visualization of DNA uptake using fluorescent labels. Competent *B. subtilis* cells (expressing a green fluorescent protein tag) are mixed with DNA containing a red fluorescent label; bright red dots can be seen in association with green cells, but not with cells in the population that are not competent (dark brown). The association disappears in a noncompetent mutant (defective in the ComK regulator) or after addition of exogenous DNase. Adapted from Boonstra et al., 2018 (see Suggested Reading).

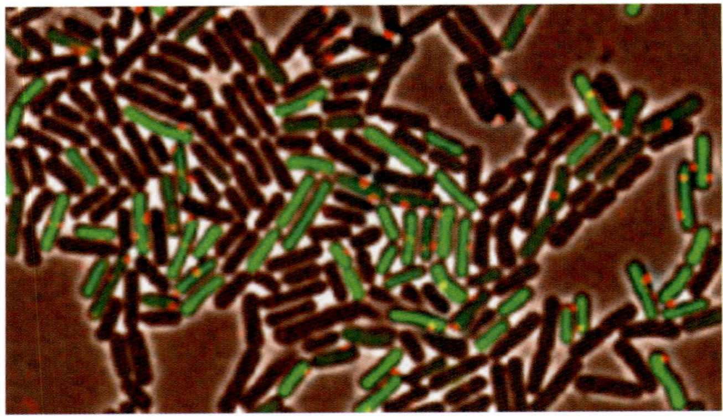

The major difference between the proteobacterial competence systems and the firmicutes systems is necessitated by the presence of an outer membrane in the *Proteobacteria*. The hydrophilic DNA must first pass through this hydrophobic outer membrane before it can reach the cytoplasmic membrane and enter the cytoplasm of the cell (see Obergfell and Seifert, Suggested Reading). Initial DNA binding occurs through interaction of double-stranded DNA with a pseudopilus (composed primarily of the PilE protein and assembled by PilG and PilF in *Neisseria*, which are analogous to ComGC, ComGB, and ComGF in *B. subtilis*). To facilitate DNA transfer through the outer membrane, the competence systems of *Proteobacteria* have an additional component, a pore through the outer membrane, made up of 12 to 14 copies of a secretin protein (called PilQ in *Neisseria*). This pore has a hydrophilic aqueous channel through which the double-stranded DNA can pass. The double-stranded DNA is delivered in the periplasm by ComE (analogous to *B. subtilis* ComEA) to the ComA channel in the inner membrane, which has sequences in common with the ComEC protein that forms a similar channel in *B. subtilis*. As in *B. subtilis*, one strand is degraded and one strand enters the cell through the ComA channel, possibly via an ATP-dependent hydrolase equivalent to *B. subtilis* ComFA, although the transport mechanism has not been identified.

COMPETENCE SYSTEMS BASED ON TYPE IV SECRETION SYSTEMS

As described above, most competence systems are based on type II secretion systems. In contrast, *H. pylori* utilizes a system based on type IV secretion-conjugation systems, like that of *Agrobacterium tumefaciens* discussed in chapter 5. *H. pylori* is an opportunistic pathogen involved in gastrointestinal diseases. The similarity between the competence system of *H. pylori* and type IV secretion-conjugation systems was discovered because of the similarity between the proteins in this system and the VirB conjugation proteins in *A. tumefaciens*, a plant pathogen, that transfer T-DNA from the Ti plasmid into plants (see Box 5.3). The Com proteins of *H. pylori* were therefore given letters and numbers corresponding to their orthologs in the T-DNA transfer system of the Ti plasmid in *A. tumefaciens*. Table 6.1 lists these Com proteins and their orthologs in the *Agrobacterium* Ti plasmid (see Karnholz et al., Suggested Reading). Apparently, type IV secretion pathway systems can move DNA either into the cell (for transformation) or out of the cell (for conjugation). Interestingly, in addition to its type IV transformation system, *H. pylori* has a *bona fide* type IV secretion system that secretes proteins directly into eukaryotic cells. However, even though the two systems are structurally related, they function independently of each other and have no proteins in common.

Table 6.1 Orthologous Com proteins of *H. pylori* and Vir proteins in *A. tumefaciens*

Helicobacter protein	Function	*Agrobacterium* ortholog
ComB2	Pseudopilus?	VirB2
ComB3	Unknown	VirB3
ComB4	ATPase	VirB4
ComB6	DNA binding?	VirB6
ComB7	Channel	VirB7
ComB8	Channel	VirB8
ComB9	Channel	VirB9
ComB10	Channel	VirB10

The nomenclature of competence and secretion systems is admittedly very confusing. To reiterate, type IV pili and type IV secretion systems are not related to each other. Type IV secretion systems utilize type II pili, and type II secretion systems are used to assemble type IV pili. There is also evidence that a type IV secretion system in *N. gonorrhoeae* is involved in releasing DNA into the environment, where it can be taken up by the type II secretion system DNA uptake machinery. When the secretion systems, transformation systems, and pili were being named, no one could have predicted their relationships to each other; this unfortunate confusion is the result.

Specificity of DNA Uptake

Most bacteria that are capable of natural transformation will take up DNA from any source, while others are highly selective (see Box 6.1 for a description of an experimental demonstration of selectivity). *Neisseria* and *Haemophilus* exhibit a very strong preference for DNA from the same species (see Mell and Redfield, Suggested Reading). This specificity is dependent on the presence of **uptake sequences**, which are short sequences that are present at high frequency throughout the genome (Figure 6.4). Uptake sequences are long enough that they almost never occur by chance in other DNAs. A minor pilin protein, ComP, has been shown to be responsible for recognition of the uptake sequence in *Neisseria*, and it is likely that an analogous protein provides this function in *Haemophilus*. *H. pylori*, which uses a different type of DNA uptake machinery, also exhibits high preference for uptake of its own DNA; however, no specific uptake sequences have been identified.

Possible reasons why some bacteria preferentially take up DNA from their own species while others take up any DNA are subjects of speculation and are discussed below (see "Role of Natural Transformation" below). Cells that take up DNA from different species can incorporate that DNA into their genomes only if the incoming DNA has high enough sequence similarity to resident DNA in the cell to promote homologous recombination or if the incoming DNA has the ability to promote independent replication (e.g., plasmid or bacteriophage DNA) (see "Plasmid Transformation and Phage Transfection of Naturally Competent Bacteria" below). The ability of DNA from a different species to be incorporated by recombination can be used as a measure of the close relationship between the species, at least in the genetic region that is tested for transformation activity.

As noted above, all known natural transformation systems require double-stranded DNA for the initial binding step and transport single-stranded DNA into the cytoplasm. Genetic experiments that take advantage of the molecular requirements for transformation can be used to study the molecular basis for transformation; in other words, transformation can be used to study itself. Evidence that DNA has transformed cells is usually based on the appearance of recombinant types after transformation. A recombinant type can form only if the donor and recipient bacteria differ in their genotypes and if the incoming DNA from the donor bacterium changes the genetic composition of the recipient bacterium. The chromosome of a recombinant type has the DNA sequence of the donor bacterium in the region of the transforming DNA.

Only double-stranded DNA can bind to the specific receptors on the cell surface, so double-stranded, but not single-stranded, DNA can transform cells and yield recombinant types. The fact that the DNA is converted into a single-stranded form in the course of the transformation process is also demonstrated by the observation that transforming DNA enters a phase during which it cannot be reisolated in a form that is active for transformation; this is termed the **eclipse phase**. An example of a genetic experiment that demonstrates this is shown in Box 6.2.

Figure 6.4 Sequence logos showing conservation of uptake sequences for natural transformation in **(A)** *Neisseria* and **(B)** *Haemophilus*. The size of the letters indicates the frequency with which this residue is found at a given position. Adapted from Mell and Redfield, 2014 (see Suggested Reading).

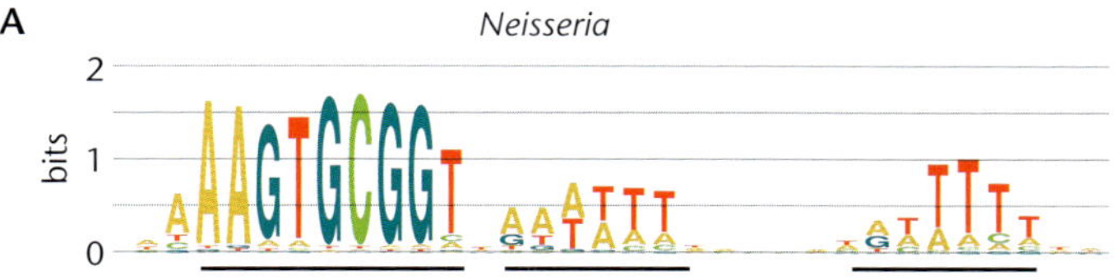

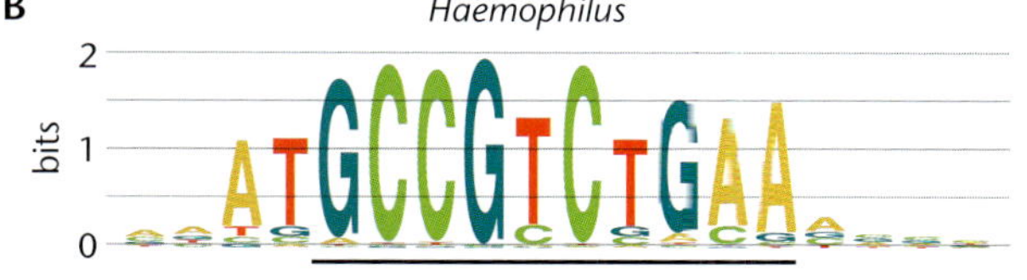

BOX 6.2

Genetic Evidence for Single-Stranded DNA Uptake

Natural transformation can be used to demonstrate that only double-stranded DNA can be bound in the initial step and that the double-stranded DNA is converted to single-stranded DNA immediately upon transport into the cytoplasm. The single-stranded DNA is then converted back into double-stranded DNA by recombination and replication.

An Arg⁻ mutant that requires arginine for growth is used as the recipient strain, and the corresponding Arg⁺ prototroph is the source of donor DNA. As shown in the figure, at various times after the donor DNA has been mixed with the recipient cells, the recipients are treated with DNase, which cannot enter the cells but destroys any DNA remaining in the medium. The DNase is removed and the surviving DNA in the recipient cells is then extracted and used for retransformation of new auxotrophic recipients, and Arg⁺ transformants are selected on agar plates without the growth supplement arginine. Any Arg⁺ transformants that arise in the second transformation experiment must have been due to double-stranded Arg⁺ donor DNA in the recipient cells.

Whether transformants are observed depends on the time point at which the DNA was extracted from the cells. When the DNA is extracted at time 1, while it is still outside the cells and accessible to the DNase, no Arg⁺ transformants are observed because the Arg⁺ donor DNA is all destroyed by the DNase. At time 2, some of the DNA is now inside the cells, where it cannot be degraded by the DNase, but this DNA is single stranded. It has not yet recombined with the chromosome, so Arg⁺ transformants are still not observed in the retransformation experiment. Only at time 3, when some of the donor DNA has recombined with the chromosomal DNA and so is again double stranded, do Arg⁺ transformants appear in the retransformation experiment. Thus, the transforming DNA enters the eclipse period for a short time after it is added to competent cells, as expected if it enters the cell in a single-stranded state.

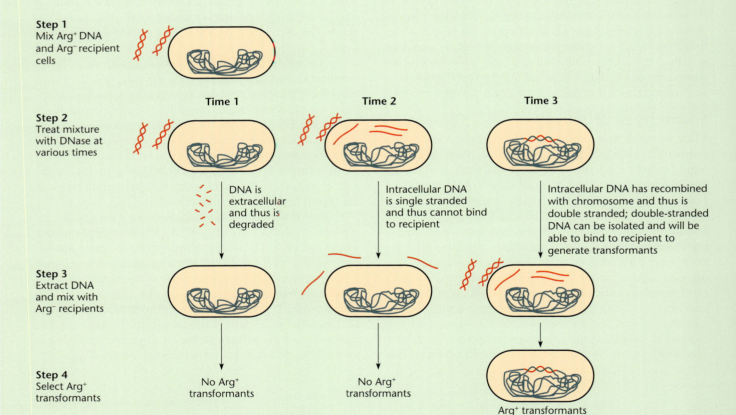

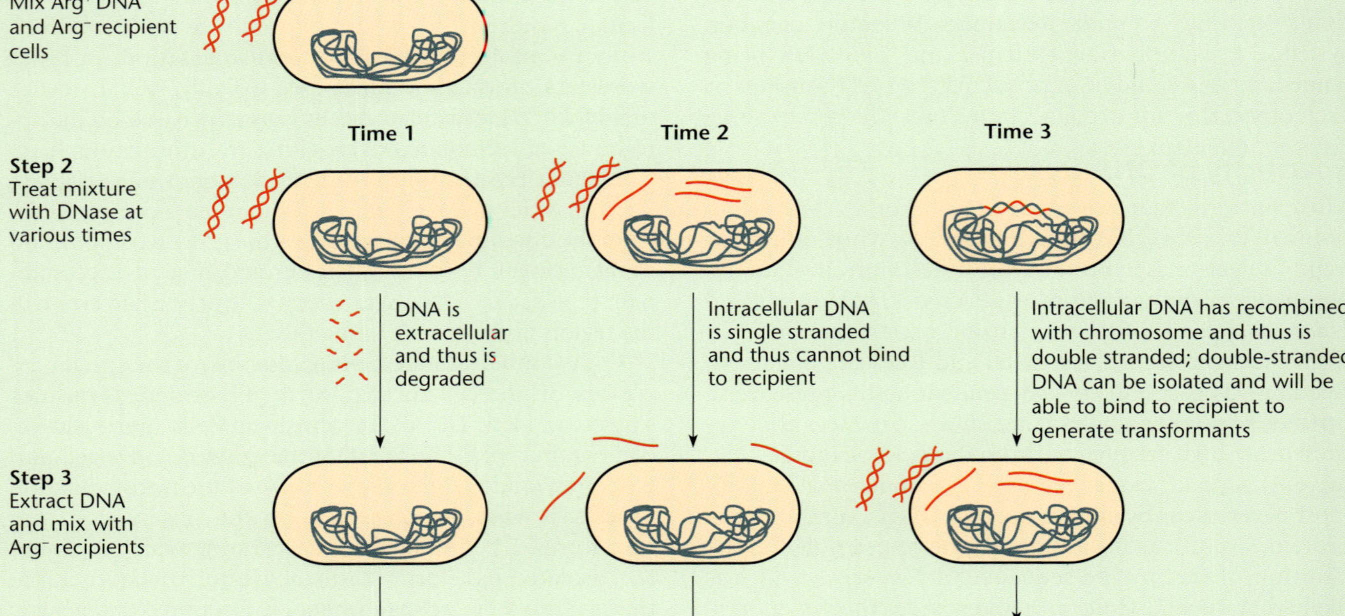
Step 1
Mix Arg⁺ DNA and Arg⁻ recipient cells
Time 1
Time 2
Time 3
Step 2
Treat mixture with DNase at various times
DNA is extracellular and thus is degraded
Intracellular DNA is single stranded and thus cannot bind to recipient
Intracellular DNA has recombined with chromosome and thus is double stranded; double-stranded DNA can be isolated and will be able to bind to recipient to generate transformants
Step 3
Extract DNA and mix with Arg⁻ recipients
Step 4
Select Arg⁺ transformants
No Arg⁺ transformants
No Arg⁺ transformants
Arg⁺ transformants

Genetic assay for the state of DNA during transformation. Only double-stranded donor DNA binds to the recipient cell to initiate transformation. After transformation is initiated, the donor DNA (red) is converted to the single-stranded form, and reisolation of the DNA from the transformants does not allow transformation of new recipient cells until after the donor DNA has been incorporated into the chromosome by recombination. The appearance of transformants in step 4 indicates that the transforming DNA was double stranded when it was isolated from the first recipient cells and therefore can be used to transform the second set of competent cells to yield Arg⁺ transformants.

DNA Processing after Uptake

Single-stranded DNA is highly susceptible to degradation. For protection, it is rapidly covered with single-stranded-DNA-binding protein (SSB). Some organisms use the normal SSB that functions during DNA replication, and others use a competence-induced additional form of SSB. A competence-specific DNA processing protein designated DprA is present in all organisms in which competence has been reported (see Johnston et al., Suggested Reading). DprA recruits the RecA recombination protein, which functions in the recombination of transforming DNA with other DNA molecules in the cell, including the chromosome, as well as generally in homologous recombination.

If the transformed DNA is highly similar in sequence to DNA that is resident in the cell, the RecA protein will mediate homologous recombination. The transforming DNA integrates into the cellular DNA in a homologous region using a process discussed in detail in chapter 9. If the donor DNA and recipient DNA sequences differ slightly in this region, recombinant types can appear. The maximum lengths of single-stranded DNA incorporated into the recipient chromosome are about 8 to 12 kb, as shown by cotransformation of genetic markers; this is limited by the size of the DNA fragment taken up into the cytoplasm, and the incorporation takes only a few minutes to be completed.

Some organisms that have the ability to become competent may exhibit very low frequency of recovery of transformants for chromosomal alleles. Low transformation efficiency can be affected by how efficiently DNA is transported into the cell, low activity of homologous recombination systems, or high intracellular exonuclease activity.

Natural Transformation as a Tool

As mentioned previously, natural transformation is a useful way to introduce DNA into the cell. This includes DNA that has been manipulated by molecular biology approaches or for transfer of DNA from one cell type to another.

PLASMID TRANSFORMATION AND PHAGE TRANSFECTION OF NATURALLY COMPETENT BACTERIA

Chromosomal DNA can efficiently transform any naturally competent bacterial cells that have high enough chromosomal similarity to allow homologous recombination. However, neither plasmids nor phage DNAs can be efficiently introduced into naturally competent cells for two reasons. First, most plasmids and phage DNAs must be double stranded to replicate. Natural transformation requires breakage of the double-stranded DNA and degradation of one of the two strands so that a linear single strand can enter the cell. Second, most plasmids (and many phages) must form complete circles in the new host cell prior to replication.

Introduction of multiple copies of the same plasmid DNA into a single cell can regenerate an intact plasmid molecule if each double-stranded plasmid molecule is cleaved randomly during the initial binding step and plasmid strands are selected randomly for entry into the cell. Overlapping single-stranded plasmid molecules can hybridize to regenerate a partially double-stranded molecule (Figure 6.5), and these molecules are substrates for cellular repair systems that fill in the missing DNA and

Figure 6.5 Transformation by plasmid DNA. DNA is linearized outside the cell (at site of blue and red arrows) to generate double-stranded molecules with different endpoints. DNA enters the cell as linear single strands. If multiple plasmid molecules enter the same cell, linear single-stranded pieces of plasmid DNA anneal, the remaining gaps are filled in by cellular DNA repair systems, and nicks are sealed by DNA ligase.

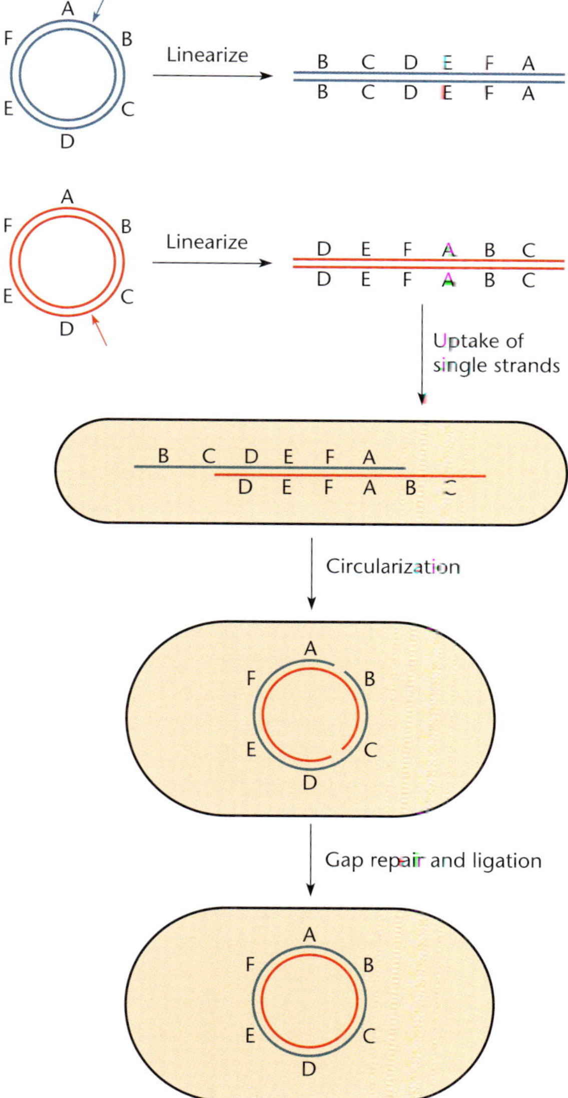

ligate the ends to regenerate a double-stranded circular plasmid. A similar process can generate intact phage genomes for replication.

IMPORTANCE OF NATURAL TRANSFORMATION FOR FORWARD AND REVERSE GENETICS

Natural transformation has many uses in molecular genetics. Transformation has been used in different types of bacteria to map genetic markers in chromosomes and to reintroduce DNA into cells after the DNA has been manipulated in the test tube. This has made naturally competent organisms, like *B. subtilis*, ideal model systems for molecular genetic studies. As noted above, the interpretation of genetic data obtained by transformation is similar to the interpretation of data obtained by transduction (see chapter 3). In the bacterium *A. baylyi*, transformation is so efficient that it can occur after simply spotting DNA restriction fragments or PCR-amplified DNA fragments onto streaks of recipient bacteria on plates. This offers opportunities to construct many different types of mutations in genes, including **loss-**, **gain-**, or **change-of-function mutations** (see Young et al., Suggested Reading). Such manipulations are more difficult in bacteria that do not have efficient natural competence systems.

CONGRESSION

The presence of multiple DNA-binding sites on a single competent cell allows the possibility of import of multiple DNA segments into a cell. This has important consequences for genetic mapping using transformation, as well as for the efficiency of transformation with plasmid DNA. In the case of genetic mapping, the conclusion that cotransformation of two markers is an indication of close physical linkage depends on the assumption that the two markers entered the cell on the same piece of DNA. Entry of markers on separate DNA segments is referred to as **congression**, and this occurs when cells are exposed to high concentrations of DNA so that multiple binding sites on a cell are simultaneously engaged in DNA uptake.

For genetic mapping experiments, congression can be avoided by using subsaturating concentrations of DNA (Figure 6.6). As increasing amounts of DNA are added to a transformation mixture, the number of transformants for a particular chromosomal marker (Arg$^+$ in the example) will increase. Simultaneously, the number of transformants that have also obtained an unlinked genetic marker (Met$^+$) will begin to increase, but only after the DNA concentration has reached a level at which each cell is likely to have taken up both the *arg* and *met* DNA fragments. Note that the number of Met$^+$ transformants does not exceed approximately 5% of the Arg$^+$ transformants, because once all of the DNA-binding sites are saturated, additional DNA has no effect.

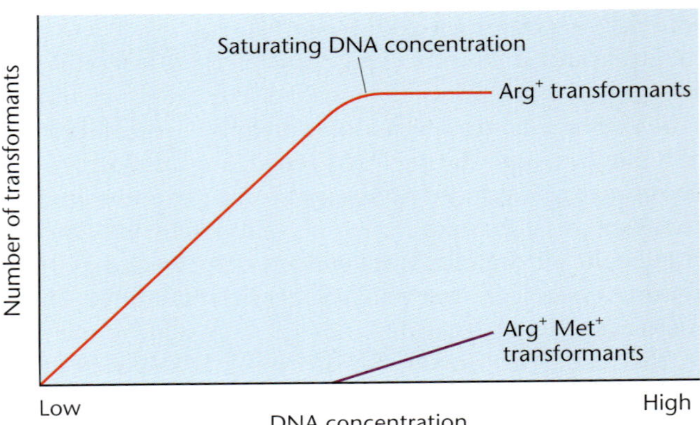

Figure 6.6 Import of multiple DNA fragments into a single cell by congression. The donor strain is Arg$^+$ Met$^+$, the recipient strain is Arg$^-$ Met$^-$, and the transformation mixture was plated on medium lacking arginine but containing methionine; Arg$^+$ colonies were then tested for the ability to grow in the absence of methionine. Addition of increasing concentrations of DNA results in increasing numbers of Arg$^+$ transformants (red line), until the DNA concentration reaches a level where the DNA-binding sites on the competent cells are saturated, at which point additional DNA has no effect on the number of transformants obtained. At low DNA concentrations, no transformants for the unlinked and nonselected Met$^+$ marker are observed. At high DNA concentrations, cells take up multiple DNA molecules, and incorporation of the unlinked Met$^+$ marker is observed, yielding Arg$^+$ Met$^+$ transformants (purple line).

Congression is useful for introduction of a nonselectable marker into a cell in the absence of a linked selectable marker. For example, if the donor cells are Arg$^+$ and Trp$^-$ and the recipient cells are Arg$^-$ and Trp$^+$, use of saturating concentrations of DNA and selection for Arg$^+$ will result in 5% Arg$^+$ Trp$^-$ transformants; in this case, recovery of the Trp$^-$ transformants requires that the transformation mixture be plated on medium lacking arginine (to select for Arg$^+$) and containing tryptophan (so that the Trp$^-$ transformants that result from congression can grow).

As noted above, transformation of plasmid DNA into naturally competent cells results in import of single-stranded DNA. Binding of multiple double-stranded plasmid molecules, and simultaneous uptake of the opposite strand from a different molecule through congression, increases the probability that the single strands will hybridize in the cell and regenerate an intact plasmid after repair of any gaps and nicks in the DNA.

Regulation of Natural Competence

Most naturally transformable bacteria express their competence genes only under certain growth conditions or in specific stages in the growth cycle. *S. pneumoniae* is maximally competent in the early exponential growth phase, and *B. subtilis* and *H. influenzae* become competent when nutrients are limited. Competence in *V. cholerae* is induced in the presence of chitin, which suggests that it occurs during colonization of shellfish in the natural environment (see Matthey and Blokesch, Suggested Reading). In contrast, *N. gonorrhoeae* appears to express its competence genes under all conditions (see Obergfell and Seifert, Suggested Reading). Each of the regulated systems uses different mechanisms to monitor its nutritional state and regulate competence gene expression.

COMPETENCE REGULATION IN *B. SUBTILIS*

The *B. subtilis comE*, *comF*, and *comG* operons are all under the transcriptional control of ComK, a transcription factor that is itself regulated by the ComP-ComA **two-component regulatory system**, analogous to those used to regulate many other systems in bacteria (see chapter 12). In systems of this type, one protein senses a signal and transmits that signal to a second protein via phosphorylation events. Information that the cell is running out of nutrients and the population is reaching a high density is registered by ComP, a **sensor kinase protein** in the membrane (Figure 6.7A). The signal for high cell density causes the sensor kinase protein to phosphorylate itself, i.e., to transfer a phosphate from ATP to a specific histidine residue within the ComP protein. ComP then transfers its phosphate to a specific aspartate residue in ComA, a **response regulator protein** in the cytoplasm. In the phosphorylated state, the ComA protein is a transcriptional activator (see chapters 11 and 12) for several genes, some of which are required for competence. One of the regulatory targets of phosphorylated ComA is another transcriptional activator, ComK, which activates transcription of over 100 genes, including the *com* genes that form the transformation machinery illustrated

Figure 6.7 Regulation of competence development by quorum sensing. **(A)** In *Bacillus subtilis*, the ComP protein in the membrane senses a high concentration of the ComX peptide and phosphorylates itself to generate ComP~P by transferring a phosphate from ATP. The phosphate is then transferred by ComP~P to ComA to form ComA~P, which allows the transcription of many genes, including those encoding ComK, the activator of the late *com* genes, and ComS, which prevents proteolytic degradation of ComK. In a separate pathway, the CSF peptide (blue), processed from the signal sequence of another protein (PhrC), is imported into the cell by the Spo0K oligopeptide permease and increases the amount of phosphorylated ComA (ComA~P) by inactivating the RapC phosphatase, which otherwise will dephosphorylate ComA~P. **(B)** In *S. pneumoniae*, a peptide signal (competence-stimulating peptide [CSP], purple) activates the ComD-ComE two-component system (analogous to ComP-ComA in *B. subtilis*), which directs expression of an initial early set of *com* genes that includes the *comX* gene. ComX is an alternate sigma factor that binds to core RNA polymerase and directs transcription of the late *com* genes.

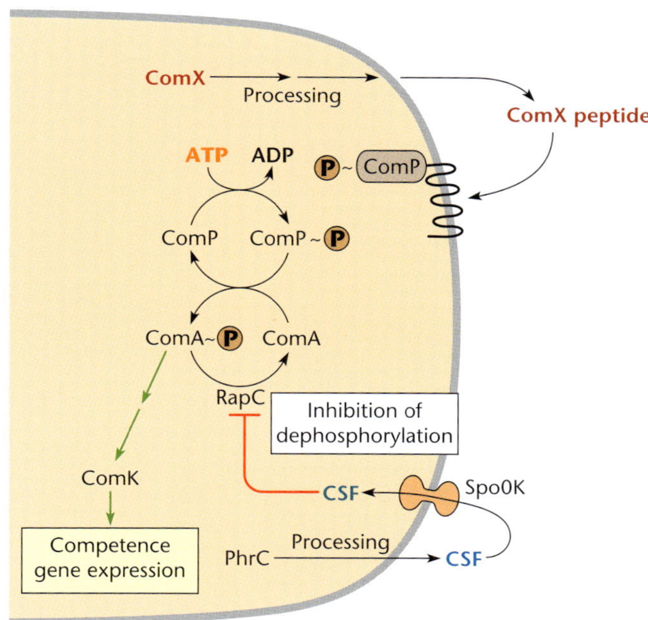

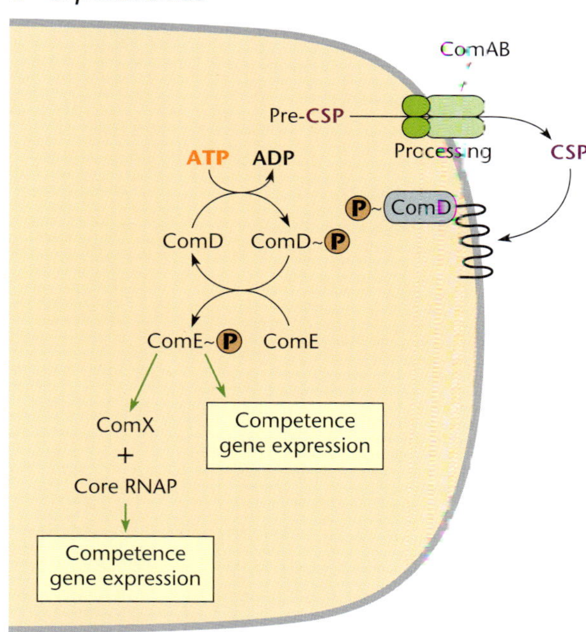

in Figure 6.2A. ComK also activates its own expression, a form of **autoregulation** that allows an initial signal to be amplified in the cell to generate a higher level of response. Another target of ComK activation is ComS, which regulates ComK posttranslationally. ComK is normally targeted for degradation by the MecA adaptor protein (see chapter 2 for discussion of protein turnover systems). ComS protects ComK from proteolytic degradation by interfering with the activity of MecA, which allows ComK to accumulate to high enough levels to activate the transcription of other *com* genes. This is an example of regulated proteolysis (see chapter 11), and the interplay between ComK and ComS results in a regulatory feedback loop as each protein affects the levels of the other.

Competence Pheromones

How does the cell know that other *B. subtilis* cells (and therefore released DNA molecules) are nearby and that it should induce competence? High cell density is signaled through small peptides called **competence pheromones** that are excreted by the bacteria as they multiply (see Bongiorni et al. and Potahill and Lazazzera, Suggested Reading). Cells become competent only in the presence of high concentrations of these peptides, and the concentration of the peptides in the medium is high only when the concentration of cells giving them off is high. The requirement for competence pheromones ensures that cells are able to take up DNA only when other *B. subtilis* cells (and therefore also dead cells that have released DNA) are nearby. This is one example of a phenomenon called **quorum sensing**, by which small molecules given off by cells send signals to other cells in the population that the cell density is high. Many such small molecules are known, including homoserine lactones that signal cell density in some groups of bacteria. Other examples of quorum sensing are discussed in chapter 12.

A second question is how the cell knows that the competence pheromone peptide came from other cells and was not produced internally. It does this by cutting the peptide out of a larger protein as the larger protein passes through the membrane of the cell in which it is synthesized. Once outside the cell, the peptide is diluted by the surrounding medium, and it can achieve high enough concentrations to induce competence only if the cell density is high because many surrounding cells are also giving off the peptide. In *B. subtilis*, the major competence pheromone peptide is called ComX, and the longer polypeptide it is cut out of is the product of the *comX* gene. Another gene, *comQ*, which is immediately upstream of *comX*, is also required for synthesis of the competence pheromone because its product is the protease enzyme that cuts the competence pheromone out of the longer polypeptide during export. Once the peptide has been cut out of the longer molecule and is released, it can bind to the ComP protein in the membranes of nearby cells and trigger ComP autophosphorylation.

At best, only about 10 to 15% of *B. subtilis* cells ever become competent, regardless of the growth conditions or concentration of peptide. Competent cells are not actively growing, so dedicating only a portion of the population to this state allows the population as a whole to explore different solutions to nutrient limitation. Division of the population into two subgroups in different physiological states is called **bistability** and in this case seems to be determined by autoregulation of the ComK activator protein. Bistable states are common biological phenomena, and competence in *B. subtilis* has been used as an experimental model for such phenomena (see Maamar and Dubnau, Suggested Reading).

Relationship between Competence, Sporulation, and Other Cellular States

As *B. subtilis* transitions to stationary phase, nutritional limitation triggers a variety of physiological responses. Depending on the conditions, some cells may acquire competence, while other cells may sporulate (see chapter 12). Sporulation, a developmental process found in some groups of bacteria, allows a bacterium to enter a dormant state and survive adverse conditions, such as starvation, irradiation, and heat. During sporulation, the bacterial chromosome is packaged into a resistant spore, where it remains viable until conditions improve and the spore can germinate into an actively growing bacterium. To coordinate sporulation and competence, *B. subtilis* cells produce other regulatory peptides (see Bongiorni et al., Suggested Reading). There are at least two such peptides (called Phr) that regulate ComA indirectly by inhibiting proteins called Rap proteins that dephosphorylate phosphorylated ComA and prevent it from binding to DNA and activating transcription. Like ComX, the Phr peptides are processed from the signal sequences of longer polypeptides, the products of the *phr* genes; unlike ComX, which interacts with ComP at the cell surface, the Phr peptides are transported into the cell by the oligopeptide permease Spo0K, which is utilized for both competence and sporulation (Figure 6.7A).

REGULATION OF COMPETENCE IN *S. PNEUMONIAE*

Unlike in *B. subtilis*, in which competence is induced in stationary phase, competence is induced during early exponential phase in *S. pneumoniae*, when nutrients are abundant. Both organisms utilize cell-cell signaling by small peptides, which in *S. pneumoniae* is called competence-stimulating peptide. Like *B. subtilis*, *S. pneumoniae* also uses a two-component system, designated ComDE. Binding of competence-stimulating peptide to the membrane-bound

ComD sensor kinase results in phosphorylation of its partner response regulator, ComE (Figure 6.7B); these are functionally analogous to ComP-ComA in *B. subtilis* (see Fontaine et al., Suggested Reading). Phosphorylated ComE activates the transcription of at least 20 competence genes, one of which encodes a new RNA polymerase sigma factor, σ^X (see chapters 2 and 11). Interaction of σ^X with core RNA polymerase directs the RNA polymerase to recognize a new set of promoter sequences that are found upstream of approximately 60 additional competence genes. These "late" competence genes encode most of the DNA-binding and transport machinery shown in Figure 6.2B. σ^X also directs transcription of its own gene, leading to amplification of the initial signal; this is analogous to ComK autoregulation in *B. subtilis*. Recent transcriptional analyses of *S. pneumoniae* competence have revealed previously unknown genes regulated by the competence system, providing new insights into how this system is integrated into other cellular processes (see Slager et al., Suggested Reading).

Both *B. subtilis* and *S. pneumoniae* use peptide signaling to monitor the presence of nearby related organisms, and both use two-component regulatory systems to transmit information about the extracellular peptide concentration to the gene expression machinery. They differ, however, in how they then control the expression of the late competence genes; these similarities and differences are illustrated in Figures 6.7 and 6.8.

COMPETENCE REGULATION IN *H. INFLUENZAE*

H. influenzae, like many other organisms, uses nutritional signals to regulate competence gene expression. The key regulator is a transcriptional activator called Sxy (Figure 6.8C). Sxy works in concert with the catabolite activator protein CAP (which is activated by cyclic AMP, levels of which increase when cells are limited for easily metabolized carbon sources; see chapter 12). Sxy and CAP bind together to sites upstream of promoters for competence genes to activate transcription. A regulatory protein similar to Sxy, designated TfoX, is involved in chitin-dependent regulation of competence gene expression in *V. cholerae* and is likely to also work in concert with CAP. A small RNA called TfoR, which is expressed in the presence of chitin, increases TfoX expression, thereby increasing competence gene expression (see chapter 11 for a discussion of small RNA-mediated regulation).

Figure 6.8 Comparison of competence regulatory mechanisms. Green arrows indicate transcriptional activation, and lightning bolts indicate different environmental signals. **(A)** In *Bacillus subtilis*, the environmental signal (ComX peptide, red lightning bolt) is sensed by the ComP sensor kinase (SK), which phosphorylates itself and then transfers a phosphate (P) to the ComA response regulator (RR). Phosphorylated ComA (ComA~P) turns on synthesis of a transcriptional activator (TA), ComK, which then turns on transcription of competence (*com*) genes. **(B)** *Streptococcus pneumoniae* utilizes a similar mechanism of peptide induction (purple lightning bolt) of a two-component system (SK, RR), and the phosphorylated response regulator turns on synthesis of a dedicated competence sigma factor (σ) that binds to core RNA polymerase to direct transcription of *com* genes. **(C)** *Haemophilus influenzae* utilizes a transcriptional activator (TA#1), designated Sxy, that is activated by an environmental signal (yellow lightning bolt); TA#1 acts in concert with a second transcriptional activator (TA#2), the carbon catabolite regulator CAP, which in the presence of cAMP (green lightning bolt) binds to DNA to activate transcription of *com* genes. Activation of *com* gene expression requires binding of both activator proteins and therefore both signals.

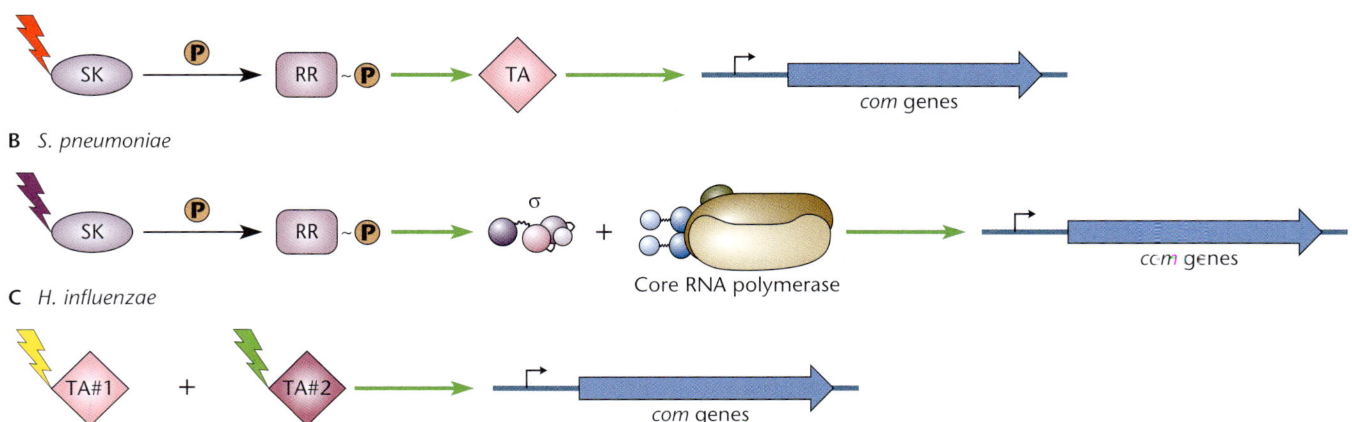

Identification of Competence in Other Organisms

As noted above, genome sequencing has revealed genes related to known competence genes in many organisms in which natural competence has not been demonstrated (see Johnston et al., Suggested Reading). Whereas in some cases these genes may have alternative functions (see Palchevskiy and Finkel, Suggested Reading), in other cases it is likely that competence is tightly regulated and the conditions for competence have not been identified, as these are highly variable for different organisms (see Attaiech and Charpentier, Suggested Reading). In addition, it is common to find that only certain strains of a particular species can be induced for competence.

One approach to identify conditions responsible for competence in new systems is to isolate mutants with higher competence, based on efficient incorporation of selectable antibiotic resistance markers. Studies of this type often yield mutations in genes that normally repress competence gene expression (see Sexton and Vogel, Suggested Reading). For example, laboratory strains of *B. subtilis* that are competent express ComS, which allows stabilization of ComK, while other strains that do not express ComS are not competent. A higher level of complexity is observed in some "wild" strains that contain a plasmid that encodes a small protein that inhibits the DNA uptake machinery; this may increase plasmid stability by preventing acquisition of other plasmids (see Konkol et al., Suggested Reading). Poor competence in organisms that appear to contain genes encoding the transformation machinery can sometimes be overcome by overexpression of the appropriate regulatory protein (e.g., ComK for *Bacillus* spp., ComX for *Streptococcus* spp., etc.). The ability to induce competence can be especially useful for biotechnological applications, such as industrial production of products, because of the ability to easily introduce new DNA constructs (see Jakobs and Meinhardt, Suggested Reading).

Role of Natural Transformation

The fact that so many gene products play a direct role in competence, requiring a large investment of energy to produce them, and that the presence of these genes is widespread, suggests that the ability to take up DNA from the environment is advantageous. Below, we discuss three possible advantages and the arguments for and against them.

NUTRITION

Organisms may take up DNA for use as a carbon and nitrogen source (see Mell and Redfield, Suggested Reading). One argument against this hypothesis is that taking up whole DNA strands for degradation inside the cell may be more difficult than degrading the DNA outside the cell and then taking up the nucleotides. In fact, noncompetent *B. subtilis* and *V. cholerae* cells excrete a DNase that degrades extracellular DNA so that the nucleotides can be taken up more easily. An argument against this hypothesis as a general explanation for transformation in all bacteria is that some bacteria take up only DNA of their own species; if DNA is used only for nutrition, there would be no reason to selectively take up only certain DNA, since DNA from other organisms should offer the same nutritional benefits. Moreover, the fact that competence develops only in a minority of the population, at least in *B. subtilis*, argues against the nutrition hypothesis, since all the bacteria in the population would presumably need the nutrients. Finally, not all competent organisms induce competence in response to nutritional limitation.

These arguments are attractive but do not rule out a role for nutrition in some cases. Bacteria may consume DNA of only their own species because of the danger inherent in taking up foreign DNAs, which might contain prophages, transposons, or other elements that could become parasites of the organism. Furthermore, consumption of DNA from the same species may be a normal part of colony development; cell death and cannibalism are thought to be part of some prokaryotic developmental processes (see chapter 12). These processes would require that only some of the cells in the population become DNA consumers, while the others become the "sacrificial lambs." The existence of specific cell-killing mechanisms that kill some cells in the population as *B. subtilis* enters the stationary phase lends support to such interpretations. Similarly, competent *S. pneumoniae* cells secrete a cell wall hydrolase that triggers lysis of noncompetent cells, resulting in release of DNA that can be taken up by the competent cells. It is interesting that although natural transformation has never been observed in *Escherichia coli*, homologs of competence genes are important for survival of the cells during stationary phase and for use of DNA as a nutritional source (see Palchevskiy and Finkel, Suggested Reading).

RECOMBINATIONAL REPAIR

Cells may take up DNA from other cells to repair damage to their own DNA. Figure 6.9 illustrates this hypothesis. In this example, a population of cells is exposed to UV irradiation. The radiation damages the DNA, causing pyrimidine dimers and other lesions to form (see chapter 10). DNA leaks out of some of the dead cells and is taken up by other competent bacteria. Because the damage to the DNA has not occurred at exactly the same places, undamaged incoming DNA segments can replace the damaged regions in the recipient, allowing at least some of the bacteria to survive. This scenario may explain why some bacteria take up DNA of only the same species, since in general, this is the only DNA that can recombine and thereby participate in the repair. Other forms of DNA

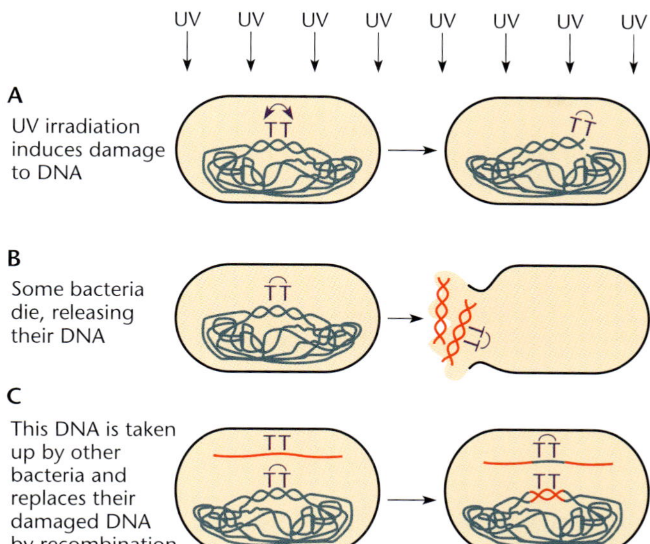

Figure 6.9 Repair of DNA damage by transforming DNA. Thymine dimers (T residues linked by curved line) are induced by UV irradiation. A region containing a thymine dimer is replaced by the same, but undamaged, sequence (TT) from the DNA released by another cell.

damage, such as double strand breaks, could also be repaired by the same model.

If natural transformation helps in DNA repair, we might expect that repair genes would be induced in response to developing competence and that competence would develop in response to UV irradiation or other types of DNA damage. In fact, in some bacteria, including *B. subtilis* and *S. pneumoniae*, the *recA* gene required for recombination repair is induced in response to the development of competence, along with DprA as described above (see Haijema et al. and Raymond-Denise and Guillen, Suggested Reading). However, in other bacteria, such as *H. influenzae*, the *recA* gene is not induced in response to competence. There is no evidence in most systems that competence genes are induced in response to DNA damage, suggesting that competence may provide an alternative solution to other repair mechanisms; an exception is *Legionella pneumophila*, which lacks an SOS-type DNA damage response (see chapter 10) but induces competence in response to DNA damage (see Charpentier et al., Suggested Reading). The need for DNA repair is an attractive explanation for why at least some types of bacteria develop competence.

GENETIC REASSORTMENT

The possibility that transformation allows exchange of genetic information between individual members of a species is also an attractive hypothesis but is difficult to prove. According to this hypothesis, transformation serves the same function that sex serves in higher organisms: it allows the assembly of new combinations of genes and thereby increases diversity and speeds up evolution. Bacteria do not have an obligatory sexual cycle; therefore, without some means of genetic exchange, any genetic changes that a bacterium accumulates during its lifetime are not necessarily exchanged with other members of the species to allow reassortment of genetic traits.

The gene exchange function of transformation is supported by the fact that cells of some naturally transformable bacteria release DNA as they grow. Some *Neisseria* species have a type IV secretion system that appears to be dedicated to active export of DNA from the cell into the medium. It is hard to imagine what function this DNA export could perform unless the exported DNA is taken up by other bacteria to promote gene transfer to neighboring cells.

Natural transformation has been shown to play an important role in pathogenesis in several systems (Box 6.3). In several *Neisseria* species, including *N. gonorrhoeae*, transformation may enhance antigenic variability, allowing the organism to avoid the host immune system (see Obergfell and Seifert, Suggested Reading). It is likely that some of this antigenic diversity results from recombination between DNAs brought together by transformation, while some arises from recombination between sequences within the chromosomal DNA of the bacterium itself.

Natural transformation also affects the formation (and persistence) of the clustered regularly interspaced short palindromic repeat (CRISPR) immunity system in bacteria. As discussed in chapter 13, CRISPR systems represent collections of DNA segments derived from past encounters with invading DNA (e.g., plasmids or phages) and provide the cell with a mechanism to destroy DNA with the same sequence if that DNA enters the cell again. Closely related strains of *Aggregatibacter actinomycetemcomitans* that retain or have lost competence function were shown to have a parallel presence or loss of CRISPR loci, suggesting that competence is tied to the ability to protect against invading DNA (see Jorth and Whiteley, Suggested Reading).

We still do not know why some types of bacteria are naturally transformable and others are not. It seems possible that most types of bacteria are naturally transformable at low levels and that we have not identified the appropriate laboratory conditions to induce competence for some of these organisms. As noted above, many organisms for which natural transformation has not been observed contain competence genes in their genomes, which suggests that they have the ability to become competent under some conditions or that the competence genes serve another function (as suggested for *E. coli*). Transformation may serve different purposes in different organisms. Perhaps

BOX 6.3

Role of Natural Transformation in Pathogens

Competence has been shown to affect the ability of a number of pathogens to cause disease. In some cases, this is likely to be due to avoidance of the host immune system by changing the antigens on their cell surfaces, whereas in others the basis for the effect on disease is less obvious.

The pili of *N. gonorrhoeae* are involved in attaching the bacteria to the host epithelial cells. These pili are highly antigenic and can undergo spontaneous alterations that can change the specificity of binding and confound the host immune system. *N. gonorrhoeae* appears to be capable of making millions of different pili.

The mechanism of pilin variation in *N. gonorrhoeae* is understood in some detail. The major protein subunit of the pilus is encoded by the *pilE* gene. In addition, silent copies of *pilE*, called *pilS*, lack promoters or have various parts deleted. These silent copies share some conserved sequences with each other and with *pilE* but differ in the so-called variable regions. Pilin protein is usually not expressed from these silent copies. However, recombination between a *pilS* gene and *pilE* can change the *pilE* gene and result in a somewhat different pilin protein. This recombination is a type of gene conversion, because reciprocal recombinants are not formed (see chapter 9). Interestingly, the availability of iron affects the frequency of antigenic variation. Many bacteria use the availability of iron as an indicator that they are in a eukaryotic host or in a particular tissue of that host, suggesting that the variation becomes activated in certain tissues.

Because *N. gonorrhoeae* is naturally transformable, not only can recombination occur between a *pilS* gene and the *pilE* gene in the same organism, but transformation also can allow even more variation through the exchange of *pilS* genes with other strains. Experiments indicate that pilin variation is affected by the presence of DNase, suggesting that transformation between individuals contributes to the variation. Also, experiments with marked *pil* genes indicate that transformation can result in the exchange of *pil* genes between bacteria. These experiments suggest, but do not prove, that transformation plays an important role in pilin variation during infection.

Natural competence also appears to be important for long-term colonization by *H. pylori*, a pathogen that causes gastric disease in humans. This organism usually colonizes the host for many years before causing disease symptoms, and the ability to persist in the host appears to be reduced in mutants with defects in key competence genes. Induction of competence gene expression by chitin in *V. cholerae* is also likely to promote increased genetic diversity during growth in the natural environment that enhances survival after infection of the host. It is likely that other examples like this will emerge as the number of organisms in which natural transformation is demonstrated expands.

References

Antonova ES, Hammer BK. 2014. Genetics of natural competence in *Vibrio cholerae* and other vibrios. *Microbiol Spectrum* **3**:VE-0010-2014.

Dorer MS, Cohen IE, Sessler TH, Fero J, Salama NR. 2013. Natural competence promotes *Helicobacter pylori* chronic infection. *Infect Immun* **81**:209–215.

Sechman EV, Rohrer MS, Seifert HS. 2005. A genetic screen identifies genes and sites involved in pilin antigenic variation in *Neisseria gonorrhoeae*. *Mol Microbiol* **57**:468–483.

Seifert HS, Ajioka RS, Marchal C, Sparling PF, So M. 1988. DNA transformation leads to pilin antigenic variation in *Neisseria gonorrhoeae*. *Nature* **336**:392–395.

transformation is used primarily for DNA repair in soil bacteria, such as *B. subtilis*, but is used primarily for nutrition in others and to increase genetic variability in obligate pathogens, such as *N. gonorrhoeae*.

Artificially Induced Competence

Many types of bacteria are not naturally transformable, at least not at easily detectable levels. Left to their own devices, these bacteria do not take up DNA from the environment. However, even these bacteria can sometimes be made competent by certain chemical treatments without the use of dedicated genetic systems, or DNA can be forced into them by a strong electric field in a process called electroporation.

Chemical Induction

Treatment with calcium ions (see Cohen et al., Suggested Reading) or related chemicals, such as rubidium ions, can make some bacteria competent; examples include *E. coli* and *Salmonella* spp., as well as some *Pseudomonas* spp. The reason for this is not fully understood, but it is likely that the calcium ions perturb the cell surface, allowing DNA to leak inside.

Chemically induced transformation is usually inefficient, and only a small percentage of the cells are ever transformed. Accordingly, the cells must be plated under conditions selective for the transformed cells. Therefore, normally, the DNA used for the transformation should contain a selectable gene, such as one encoding resistance to an antibiotic, or another genetic trick must be used (see chapter 3).

TRANSFORMATION BY PLASMIDS

In contrast to naturally competent cells, cells made permeable to DNA by chemical treatment will take up both single-stranded and double-stranded DNA. Therefore, both linear and double-stranded circular plasmid DNAs can be efficiently introduced into chemically treated cells. This fact has made calcium ion-induced competence very useful for cloning and other applications that require the introduction of plasmid and phage DNAs into cells. Transformation by circular DNAs is generally more efficient, as the DNA is less susceptible to degradation because of the absence of free ends that are susceptible to exonucleases.

TRANSFECTION BY PHAGE DNA

In addition to plasmid DNAs, viral genomic DNAs or RNAs can often be introduced into cells by transformation, thereby initiating a viral infection. This process is called **transfection** rather than transformation, although the principle is the same. To detect transfection, the potentially transfected cells are usually mixed with indicator bacteria that are sensitive to the virus and plated (see the introduction). If the transfection is successful, a plaque forms where the transfected cells have produced phages, which then infect the indicator bacteria.

Some viral infections cannot be initiated merely by transfection with the viral DNA. These viruses cannot transfect cells, because in a natural infection, proteins in the viral head are normally injected along with the DNA, and these proteins are required to initiate the infection. For example, the *E. coli* phage N4 carries a phage-specific RNA polymerase in its head that is injected with the DNA and used to transcribe the early genes (see chapter 7). Transfection with the purified phage DNA does not initiate an infection, because the early genes are not transcribed without this phage-encoded RNA polymerase. Another example of a phage in which the infection cannot be initiated by the nucleic acid alone is phage φ6 (see Box 7.2). This phage has RNA instead of DNA in the phage head and must inject an RNA replicase to initiate the infection, so the cells cannot be transfected by the RNA alone. Such examples of phages that inject required proteins are rare; for most phages, the infection can be initiated by transfection.

TRANSFORMATION OF CELLS WITH CHROMOSOMAL GENES

Transformation with linear DNA is one method used to replace endogenous genes with genes altered *in vitro*. However, most types of bacteria made competent by calcium ion treatment are transformed poorly by chromosomal DNA because the linear pieces of double-stranded DNA entering the cell are degraded by an enzyme called the RecBCD nuclease (or a related enzyme). This nuclease degrades DNA from the ends; therefore, it does not degrade circular plasmid or phage DNAs.

Nevertheless, methods have been devised to transform competent *E. coli* with linear DNA. One way is to use a mutant *E. coli* strain lacking the RecD subunit of the RecBCD nuclease. These *recD* mutants are still capable of recombination, but because they lack the nuclease activity that degrades linear double-stranded DNA, they can be transformed with linear double-stranded DNAs. Other methods, sometimes called **recombineering**, use the recombination systems of phages, such as λ phage, instead of the host recombination functions. As described in chapter 9, recombineering systems have their own strand annealing proteins and will convert double-stranded DNA substrates to single-stranded substrates for recombination. In spite of enzymes like RecBCD, double-stranded DNA can be introduced into cells expressing the λ recombination functions because of the ability of a λ protein named Gam to inhibit RecBCD and related enzymes. The method can also be used with single-stranded DNAs, such as oligonucleotides, which are not degraded by RecBCD.

Electroporation

Another way in which DNA can be introduced into bacterial cells is by **electroporation**. In the electroporation process, the bacteria are mixed with DNA and briefly exposed to a strong electric field using special equipment. It is important that the recipient cells first be washed extensively in buffer with very low ionic strength. The buffer may also contain a nonionic solute, such as glycerol, to prevent osmotic shock. The brief electric fields across the cellular membranes might create artificial pores of H_2O-lined phospholipid head groups. DNA can pass through these temporary hydrophilic pores (see Tieleman, Suggested Reading). Electroporation works with most types of cells, including most bacteria, unlike the methods mentioned above, which are specific for certain species. Also, electroporation is useful with a wider variety of substrates and can be used to introduce linear chromosomal and circular plasmid DNAs and even protein-DNA complexes into cells.

Protoplast Transformation

Disruption of the bacterial cell wall results in production of osmotically sensitive protoplast forms of cells. Protoplasts are highly susceptible to lysis but can be maintained by incubation in a solution of the appropriate high osmolarity to balance the osmolarity of the cytoplasm.

Exposure of protoplasts to DNA in the presence of polyethylene glycol triggers membrane fusion events that can trap the DNA within the cytoplasm, resulting in transformation. The protoplast forms of most bacteria are unable to divide, and the cell wall must be regenerated for transformants to be recovered. This is generally an inefficient process and must be optimized for each organism, which can be tedious. Protoplast transformation is most efficient for circular plasmid DNAs that are stable during the uptake process and can establish independent replication in the transformants. Despite its limitations, protoplast transformation represents a useful tool for introducing plasmid DNA into cells that are not easily transformed by other techniques.

Summary

1. In transformation, DNA is taken up directly by cells. The demonstration that DNA is the transforming principle was the first direct evidence that DNA is the hereditary material. The bacteria from which the DNA was taken are called the donors, and the bacteria to which the DNA has been added are called the recipients. Bacteria that have taken up DNA and are able to maintain the new DNA in a heritable form (for example, by recombination with the genome) are called transformants.

2. Bacteria that are capable of taking up DNA are said to be competent.

3. Some types of bacteria become competent naturally during part of their life cycle. A number of genes whose products form the competence machinery have been identified. Some of these encode proteins related to type II secretion systems, which form type IV pili, or to type IV secretion-conjugation systems.

4. The fate of the DNA during natural transformation is fairly well understood. The double-stranded DNA first binds to the cell surface. Then, one strand of the DNA is degraded by an exonuclease. The surviving single-stranded DNA enters the cell and can invade the chromosome in homologous regions, generating a daughter cell in which the original chromosomal sequence is replaced.

5. Naturally competent cells can be transformed with linear chromosomal DNA but are not as efficiently transformed with monomeric circular plasmid or circular phage DNAs. Transformation by plasmid DNA usually occurs by simultaneous uptake of multiple molecules of the plasmid DNA that can hybridize to each other, recyclize, and be repaired, or by uptake of plasmid multimers that contain multiple tandem copies of the plasmid genome.

6. Some types of bacteria, including *H. influenzae* and *N. gonorrhoeae*, take up DNA of only the same species. Their DNA contains short uptake sequences that are required for import of DNA into the cells. Most other types of bacteria, including *B. subtilis* and *S. pneumoniae*, seem to be capable of taking up any DNA.

7. There are three possible roles for natural competence: a nutritional function allowing competent cells to use DNA as a carbon, energy, and nitrogen source; a recombinational repair function in which cells use DNA from neighboring bacteria to repair damage to their own chromosomes, thus ensuring survival of the species; and an allelic replacement function in which bacteria exchange genetic material among members of their species, increasing diversity and accelerating evolution. Different types of bacteria may maintain transformation for different purposes, and individual types may maintain the process for multiple reasons.

8. Some types of bacteria that do not show natural competence can nevertheless be transformed after some types of chemical treatment or by generation of protoplasts, which lack their cell wall. The standard method for making *E. coli* permeable to DNA involves treatment with calcium ions. Cells made competent by calcium treatment can be transformed with plasmid and phage DNAs, making this method one of the cornerstones of molecular genetics.

9. If the cell is transformed with viral DNA to initiate an infection, the process is called transfection.

10. Brief exposure of cells to an electric field also allows cells to take up DNA, a process called electroporation.

QUESTIONS FOR THOUGHT

1. Why do you think some types of bacteria are capable of developing competence? What is the "real" function of competence?

2. Why do you think different types of competence machinery have evolved in some systems, like *H. pylori*?

3. Why do you think competence is tightly regulated in most systems?

4. How would you determine if the competence genes of *B. subtilis* are turned on by UV irradiation and other types of DNA damage?

5. How would you determine whether antigenic variation in *N. gonorrhoeae* is due to transformation between different bacteria or to recombination within the same bacterium?

SUGGESTED READING

Attaiech L, Charpentier X. 2017. Silently transformable: the many ways bacteria conceal their built-in capacity of genetic exchange. *Curr Genet* **63:**451–455. doi:10.1007/s00294-016-0663-6.

Avery OT, Macleod CM, McCarty M. 1944. Studies on the chemical nature of the substance inducing transformation of pneumococcal types. Induction of transformation by a deoxyribonucleic acid fraction isolated from pneumococcus type III. *J Exp Med* **79:**137–158.

Bongiorni C, Ishikawa S, Stephenson S, Ogasawara N, Perego M. 2005. Synergistic regulation of competence development in *Bacillus subtilis* by two Rap-Phr systems. *J Bacteriol* **187:**4353–4361.

Boonstra M, Vesel N, Kuipers OP. 2018. Fluorescently labeled DNA interacts with competence and recombination proteins and is integrated and expressed following natural transformation of *Bacillus subtilis*. *mBio* **9:**e01161–e01168.

Charpentier X, Kay E, Schneider D, Shuman HA. 2011. Antibiotics and UV radiation induce competence for natural transformation in *Legionella pneumophila*. *J Bacteriol* **193:**1114–1121.

Chen I, Dubnau D. 2004. DNA uptake during bacterial transformation. *Nat Rev Microbiol* **2:**241–249.

Claverys J-P, Martin B, Polard P. 2009. The genetic transformation machinery: composition, localization, and mechanism. *FEMS Microbiol Rev* **33:**643–656.

Cohen SN, Chang ACY, Hsu L. 1972. Nonchromosomal antibiotic resistance in bacteria: genetic transformation of *Escherichia coli* by R-factor DNA. *Proc Natl Acad Sci USA* **69:**2110–2114.

Ellison CK, Dalia TN, Vidal Ceballos A, Wang JC-Y, Biais N,Brun YV, Calia AB. 2018. Retraction of DNA-bound type IV competence pili initiates DNA uptake during natural transformation in *Vibrio cholerae*. *Nat Microbiol* **3:**773–780.

Fontaine L, Wahl A, Fléchard M, Mignolet J, Hols P. 2015. Regulation of competence for natural transformation in streptococci. *Infect Genet Evol* **33:**343–360.

Haijema BJ, van Sinderen D, Winterling K, Kooistra J, Venema G, Hamoen LW. 1996. Regulated expression of the *dinR* and *recA* genes during competence development and SOS induction in *Bacillus subtilis*. *Mol Microbiol* **22:**75–85.

Jakobs M, Meinhardt F. 2015. What renders *Bacilli* genetically competent? A gaze beyond the model organism. *Appl Microbiol Biotechnol* **99:**1557–1570.

Johnston C, Martin B, Fichant G, Polard P, Claverys JP. 2014. Bacterial transformation: distribution, shared mechanisms and divergent control. *Nat Rev Microbiol* **12:**181–196.

Jorth P, Whiteley M. 2012. An evolutionary link between natural transformation and CRISPR adaptive immunity. *mBio* **3:**e00309-12.

Karnholz A, Hoefler C, Odenbreit S, Fischer W, Hofreuter D, Haas R. 2006. Functional and topological characterization of novel components of the comB DNA transformation competence system in *Helicobacter pylori*. *J Bacteriol* **188:**882–893.

Konkol MA, Blair KM, Kearns DB. 2013. Plasmid-encoded ComI inhibits competence in the ancestral 3610 strain of *Bacillus subtilis*. *J Bacteriol* **195:**4085–4093.

Maamar H, Dubnau D. 2005. Bistability in the *Bacillus subtilis* K-state (competence) system requires a positive feedback loop. *Mol Microbiol* **56:**615–624.

Matthey N, Blokesch M. 2016. The DNA-uptake process of naturally competent *Vibrio cholerae*. *Trends Microbiol* **24:**98–110.

Mell JC, Redfield RJ. 2014. Natural competence and the evolution of DNA uptake specificity. *J Bacteriol* **196:**1471–1483.

Mirouz N, Ferret C, Cornilleau C, Carballido-Lopez R. 2018. Antibiotic sensitivity reveals that wall teichoic acids mediate DNA binding during competence in *Bacillus subtilis*. *Nature Commun* **9:**5072 doi:10.1038/s41467-018-07553-8.

Obergfell KP, Seifert HS. 2014. Mobile DNA in the pathogenic *Neisseria*. *Microbiol Spectr* **3:**MDNA3-0015-2014. doi:10.1128/microbiolspec.MDNA3-0015-2014.

Palchevskiy V, Finkel SE. 2006. *Escherichia coli* competence gene homologs are essential for competitive fitness and the use of DNA as a nutrient. *J Bacteriol* **188:**3902–3910.

Potahill M, Lazazzera BA. 2003. The extracellular Phr peptide-Rap phosphatase signaling circuit of *Bacillus subtilis*. *Front Biosci* **8:**d32–d45.

Provvedi R, Chen I, Dubnau D. 2001. NucA is required for DNA cleavage during transformation of *Bacillus subtilis*. *Mol Microbiol* **40:**634–644.

Raymond-Denise A, Guillen N. 1992. Expression of the *Bacillus subtilis dinR* and *recA* genes after DNA damage and during competence. *J Bacteriol* **174:**3171–3176.

Sexton JA, Vogel JP. 2004. Regulation of hypercompetence in *Legionella pneumophila*. *J Bacteriol* **186:**3814–3825.

Slager J, Aprianto R, Veening J-W. 2019. Refining the pneumococcal competence regulon by RNA sequencing. *J Bacteriol* **201:**e00780-18. https://doi.org/10.1128/JB.00780-18.

Tieleman DP. 2004. The molecular basis of electroporation. *BMC Biochem* **5:**10.

Young DM, Parke D, Ornston LN. 2005. Opportunities for genetic investigation afforded by *Acinetobacter baylyi*, a nutritionally versatile bacterial species that is highly competent for natural transformation. *Annu Rev Microbiol* **59:**519–551.

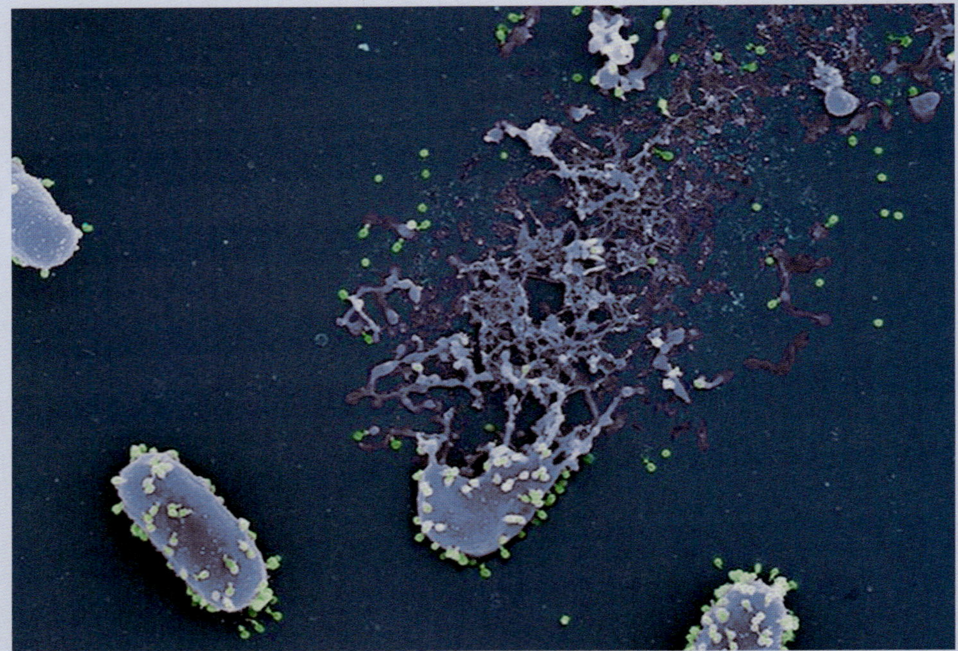

Bacteria (in blue) assaulted and killed by bacteriophages (in green). Adapted from electron microscopic image courtesy of M. Rohde and C. Rohde (Heimholz Centre for Infection Research/Leibniz Institute DSMZ, Brauschweig, Germany), https://www.id-hub.com/2017/08/04/phage-therapy-vivo-important-hosts-immune-response/.

7

Bacteriophages and Transduction

PROBABLY ALL ORGANISMS ON EARTH are parasitized by viruses, and bacteria are no exception. For purely historical reasons, viruses that infect bacteria are usually not called viruses but are called **bacteriophages** (**phages** for short), even though they are also viruses and have lifestyles similar to those of metazoan viruses. Phages and other viruses are also probably very ancient, coexisting with cellular organisms since the earliest life on Earth and probably influencing evolution in substantial ways. As mentioned in the introduction, the name *phage* derives from the Greek verb "to eat," and it describes the eaten-out places, or **plaques**, that are formed on bacterial lawns. The plural of "phage" is also "phage" when referring to a number of phage of the same type, but we add an "s" (phages) when we are discussing more than one type of phage.

Analysis of phage systems has provided major insights into a variety of genetic and biochemical properties of bacterial systems, which have served as paradigms for many basic principles of molecular biology (see Salmond and Fineran, Suggested Reading). Despite the fact that they have been well studied for decades, new insights continue to emerge, in particular, by extending what we have learned from analysis of well-studied model systems to new groups of phages (see Ofir and Sorek, Suggested Reading).

Phages are probably the most numerous biological entities, with an estimated 10^{31} phages on Earth (see Box 7.1). These phages infect an estimated 10^{23} bacteria per second. By their massive predation on cyanobacteria in the oceans alone, they play a major role in the ecology of the planet. They also are extremely diverse biological entities, which has made them a source of interesting and useful functions whose future applications we can only imagine. The smallest phage known is the *Leucomotor* phage, with a genome of 2,435 base pairs (bp), and the current largest is more than 700,000 bp! Most phages have genomes on the order of 40,000 bp.

Like all viruses, phages are so small that they can be seen only under an electron microscope. They contain a nucleic acid genome that is generally packaged within a protein coat in a structure called a capsid. Many also have elaborate tail structures that allow them to penetrate bacterial membranes and cell walls to inject their DNA into the cell (see Figure 7.1A and B). Viruses of metazoans have much simpler shapes because they do not need such elaborate tail structures because they either are engulfed by the cell, in the case of animal viruses, or enter through wounds, in the case of plant viruses. There are some types of phages that

BOX 7.1

Phage Genomics

Because of their relatively small size, phage genomes were the first to be sequenced. As efforts to uncover the vast numbers of microbial species that occupy various environmental niches have blossomed, it has become obvious that the explosion in our knowledge of the bacterial microbiome must be accompanied by recognition of the even larger number of undiscovered bacteriophages that target these newly discovered bacterial species. Furthermore, it is increasingly clear that even for bacterial species that are already well characterized, there are many types of bacteriophages that can infect these bacteria and a wealth of new information to be obtained.

Efforts to investigate what has been referred to as the phage "dark matter" have followed two general approaches. The first utilizes classic methods of selecting particular bacterial target strains and identifying phages isolated from various environmental samples that can form plaques on these strains; this approach has the advantage of immediately pairing a newly discovered phage isolate to a bacterial host and allows growth of the phage and detailed characterization, including determination of complete phage genomes. The limitation, of course, is that only phages capable of infecting the particular strain under laboratory conditions will be identified. Analysis of phage that infect *Mycobacterium* species (subsequently expanded to other members of the *Actinobacteria*) has resulted in a large and growing database of approximately 3,000 new complete phage genomes (available at PhagesDB.org). These studies have revealed new features of phage biology and a snapshot of how genomes of phage that infect a single host can evolve (see Hatfull, Suggested Reading).

The second general approach utilizes metagenomic analysis and bioinformatics to identify new phage sequences. One method is to analyze existing bacterial genomes for the presence of previously unrecognized prophage genomes embedded within the bacterial sequences (see Roux et al. and Hurwitz et al., References). Another more general method is to use metagenomics approaches to identify all possible phage sequences in a particular environmental sample (see Al-Shayeb et al. and Brum and Sullivan, References). This approach is limited by our ability to recognize and properly assemble phage genome sequences, the inherent difficulties in identifying which host matches to the newly discovered phage, and the possible limitations to characterization if the phage and host cannot be cocultured in the laboratory, and the development of new technologies has provided an explosion of new data. Much of the current work has focused on double-stranded DNA viruses, due to technological limitations, but the field is advancing rapidly to include RNA and single-stranded DNA viruses as well. All of this information is transforming our perception of phage diversity and the impact of bacteriophages on bacterial evolution and ecology.

References

Al-Shayeb B, et al. 2020. Clades of huge phages from across Earth's ecosystems. *Nature* **8**:403–407.

Brum JR, Sullivan MB. 2015. Rising to the challenge: accelerated pace of discovery transforms marine virology. *Nat Rev Microbiol* **13**:147–159.

Hurwitz BL, Ponsero A, Thornton J Jr, U'Ren JM. 2018. Phage hunters: computational strategies for finding phages in large-scale 'omics datasets. *Virus Res* **244**:110–115.

Roux S, Hallam SJ, Woyke T, Sullivan MB. 2015. Viral dark matter and virus-host interactions resolved from publicly available microbial genomes. *Elife* **4**:1–20.

do not have tails or even protein capsids. These are usually phages that infect bacteria that lack cell walls, such as mycoplasmas, or are taken up by the cell through its own secretion systems, as is the case for the filamentous phages.

Phages differ greatly in complexity. Smaller phages, such as MS2, usually have no tail, and their heads may consist of as few as two types of proteins. The heads of some of the larger phages, such as T4, have as many as 10 different proteins, each of which can exist in as few as 1 copy to almost 1,000 copies. The tails of phages can also contain many different proteins and have complicated structures, including base plates from which emanate tail fibers that direct binding to specific sites on the surfaces of their bacterial hosts. The tails of phages allow them to infect specific bacterial hosts, and the ability of a particular type of phage to infect a particular set of hosts is called its **host range**.

Phages fall into a relatively small number of families that can be composed of a large number of types that infect different bacterial hosts (Table 7.1). Phages within a family are closely related to each other, and their sequences often bear little relationship to those of phages of other families or to their host bacteria, showing that they have evolved separately from their hosts. Phage genomes also generally share very few genes with their bacterial hosts (see Kristensen et al., Suggested Reading). Comparisons of the genomic sequences of phages within these families have revealed their "mosaic" nature, in which different phages of the same family seem to be assembled from shared "mosaic tiles" composed of individual genes or modules containing

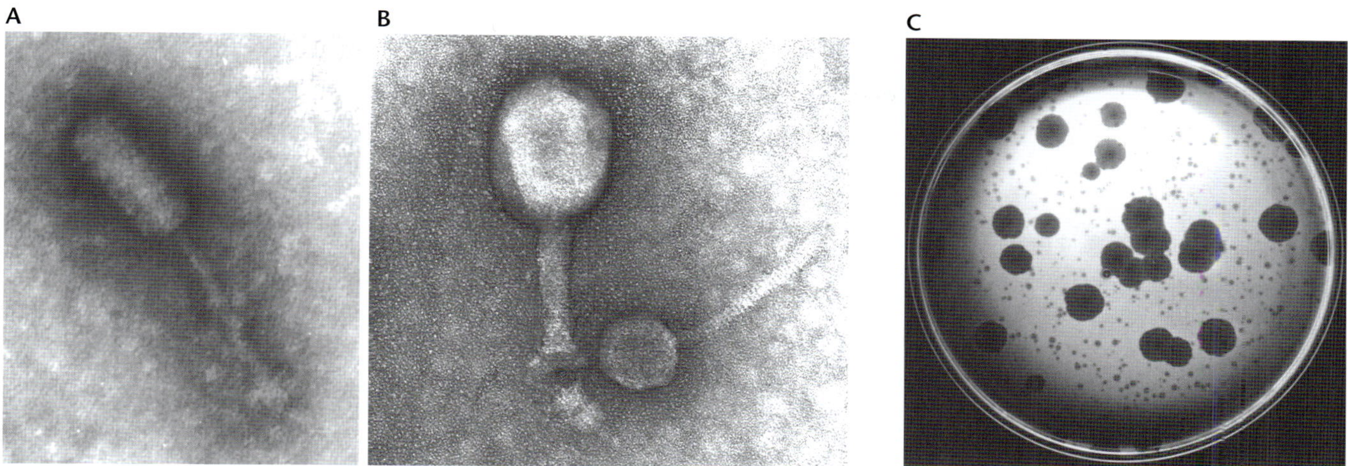

Figure 7.1 Electron micrographs and plaques of some bacteriophages. **(A)** A phage of *Enterococcus*. Electron micrograph provided by Sally Burns, Michigan State University. **(B)** Electron micrograph of phages T4 and λ (left and right, respectively). Electron micrograph provided by Lindsay Black, University of Maryland, Baltimore, and Dwight Anderson, University of Minnesota. **(C)** Plaques of *E. coli* phages M13 (smaller plaques) and T3, a relative of phage T7 (larger plaques). Photograph by Kurt Stepnitz, provided by Michigan State University.

groups of genes for the same function, for example, genes for DNA replication or for formation of the phage head (see Hatfull, Suggested Reading). The assumption is that these genes have been obtained either individually or as an interacting group from different phages of the same family at different evolutionary times. This suggests that phages do not just evolve with their host but also have evolved as a family of phages by exchanging building modules with phages that are from the same family but that now infect different hosts.

Like all viruses, phages are not live organisms capable of living independently and instead are merely a nucleic acid wrapped in a protein and/or membrane coat for protection. This coat is lost during infection, although sometimes one or more different types of proteins that are encapsulated with the nucleic acid also enter the host during infection. The nucleic acid in the capsid carries genes that direct the synthesis of more phage. Depending on the type of phage, either DNA or RNA is carried in the head and is called the **phage genome**. These genome molecules can be very long, because the genome must be long enough to have at least one copy of each of the phage genes. They can also be in more than one segment, at least in the case of some phages with RNA genomes. The length of the DNA or RNA genome therefore reflects the size and complexity of the phage. Long phage genomes, which can be as much as 1,000 times longer than the phage head, must be very tightly packed into the head of the phage, while in filamentous phages, the phage is as long as the genome DNA.

Table 7.1 Some families of phages by morphotype

Family	Example(s)	Characteristic(s)[a]
Siphoviridae	λ, N15, T1, T5	Linear dsDNA, long tail
Podoviridae	T3, T7, P22, N4, φ29	Linear dsDNA, short tail
Myoviridae	T4, P1, P2, Mu, SPO1	Linear dsDNA, large phage, contractile tail
Microviridae	φX174	Circular ssDNA, spherical capsid
Inoviridae	M13, f2, CTXφ	Circular ssDNA, filamentous
Leviviridae	Qβ, R17, MS2	ssRNA, spherical
Cystoviridae	φ6	dsRNA, segmented, envelope
Corticoviridae	PM2	Internal membrane
Plasmaviridae	L2 (*Mycoplasma*)	No capsid, lipid envelope
Lipothrixviridae	TTV-1 (*Archaea*)	Filamentous, tail fibers, dsDNA

[a]ss, single stranded; ds, double stranded.

Because phages are so small, they are usually detected only by the plaques they form on **lawns** of susceptible host bacteria (Figure 7.1C) (see the introductory chapter). Each type of phage makes plaques on only certain host bacteria, which define its host range; host range is determined by the ability to bind to the host cell, deliver its nucleic acid to the host cell cytoplasm, replicate its genome, and form new virions that can be released to infect another cell.

Mutations in the phage DNA can alter the host range of a phage or the conditions under which the phage can form a plaque, which is often how mutations are detected. In this chapter, we discuss what is known about how some representative phages control their gene expression and replicate their genomes, as well as how they exit the cell after new phage have formed. We also discuss how some phages can be used for transfer of genetic material in bacteria in a process called transduction. First, we will review some general features of phage development. We will describe how some phages only carry out a **lytic cycle**, in which the host cell is broken open following phage replication, while other phages have the ability to choose between a lytic mode and a **lysogenic cycle**, in which the phage genome is retained in the cell without production of new virus particles and without killing the host. Most of the discussion will focus on phages that contain DNA as their genetic material, with a brief description of RNA phages.

Lytic Development

Because phages are viruses and are essentially genes wrapped in a protein or membrane coat, they cannot multiply without benefit of a host cell. The virus injects its genes into a cell, and the cell is made to provide some or all of the means to express those genes and make more viruses. Phage undergoing a lytic cycle take over the host cell gene expression machinery and redirect that machinery from bacterial cell growth to production of new viral particles. This occurs through a variety of mechanisms, some of which are described below.

The Lytic Cycle

Figure 7.2 illustrates the basic multiplication process for a typical large DNA phage. To start the infection, a phage adsorbs to an actively growing bacterial cell by binding to a specific receptor on the cell surface. In the next step, the phage injects its entire DNA into the cell, where transcription, usually by the host RNA polymerase, begins almost immediately. However, not all the genes of a phage are transcribed into mRNA when the DNA first enters the cell. The genes that are transcribed first usually have promoters that mimic those of the host cell DNA and can be recognized by the host RNA polymerase. Those transcribed immediately after infection are called the **early genes** of the phage, many of which encode enzymes involved in DNA synthesis, such as DNA polymerase, primase, DNA ligase, and helicase. With the help of these enzymes, the phage DNA begins to replicate, and many copies accumulate in the cell.

Next, mRNA is transcribed from the rest of the phage genes, the **late genes**, which may or may not be intermingled with the early genes in the phage DNA, depending on the phage. These genes often have promoters that are unlike those of the host cell and so are not recognized by the host RNA polymerase alone or are recognized only by a phage-encoded RNA polymerase. Many of these genes encode proteins involved in assembly of the head and tail of the phage and lysis of the cell. After the phage particle is completed, the DNA is packaged into the heads, and the tails are attached. Finally, the host cells break open, or **lyse**, and the new phage are released to infect other sensitive cells. This whole process takes less than 1 hour for many phages, and hundreds of progeny phage can be produced from a single infecting phage.

Most phages use a more complex developmental cycle that proceeds through several intermediate stages in which the expression of different genes is regulated by specific mechanisms. Most of the regulation is achieved by transcribing different classes of genes into mRNA only at certain times; this type of regulation is called **transcriptional regulation** (see chapter 11). Regulation also may occur after the mRNA has been made to determine whether the mRNAs are translated into functional proteins.

Transcriptional Regulation of Phage Gene Expression

Figure 7.3 shows the basic process of phage gene transcriptional regulation, in which one or more of the gene products synthesized during each stage of development turns on the transcription of the genes in the next stage of development. The gene products synthesized during each stage also can be responsible for turning off the transcription of genes expressed in the preceding stage. Genes whose products are responsible for regulating the transcription of other genes are called **regulatory genes**, and this type of regulation is called a **regulatory cascade**, because each step triggers the next step and stops the preceding step. A cascade of gene expression allows step-by-step development of the phage based on information encoded in the phage DNA.

Regulatory genes can usually be identified by mutations. Mutations in most genes affect only the product of the mutated gene. However, mutations in regulatory genes can affect the expression of many other genes. This fact has been used to identify the regulatory genes of many phages. Below, we discuss some of these genes and their functions, selecting our examples either because they are the basis for cloning and other technologies or because of the impact they have had on our understanding of regulatory mechanisms in general.

PHAGES THAT ENCODE THEIR OWN RNA POLYMERASES

Some phages control their developmental cycle by encoding their own RNA polymerases. These RNA polymerases specifically recognize promoters found only on the phage DNA, which allows them to transcribe their own DNA in the presence of much larger amounts of host DNA.

T7: A New RNA Polymerase for the Late Genes

Compared with some of the larger phages, *Escherichia coli* phage T7 has a relatively simple program of gene expression after infection, with only two major classes of genes, the early and late genes. The phage has about 50 genes, many of which are shown on the T7 genome map in Figure 7.4. After infection, expression of the T7 genes proceeds from left to right, with the genes on the extreme

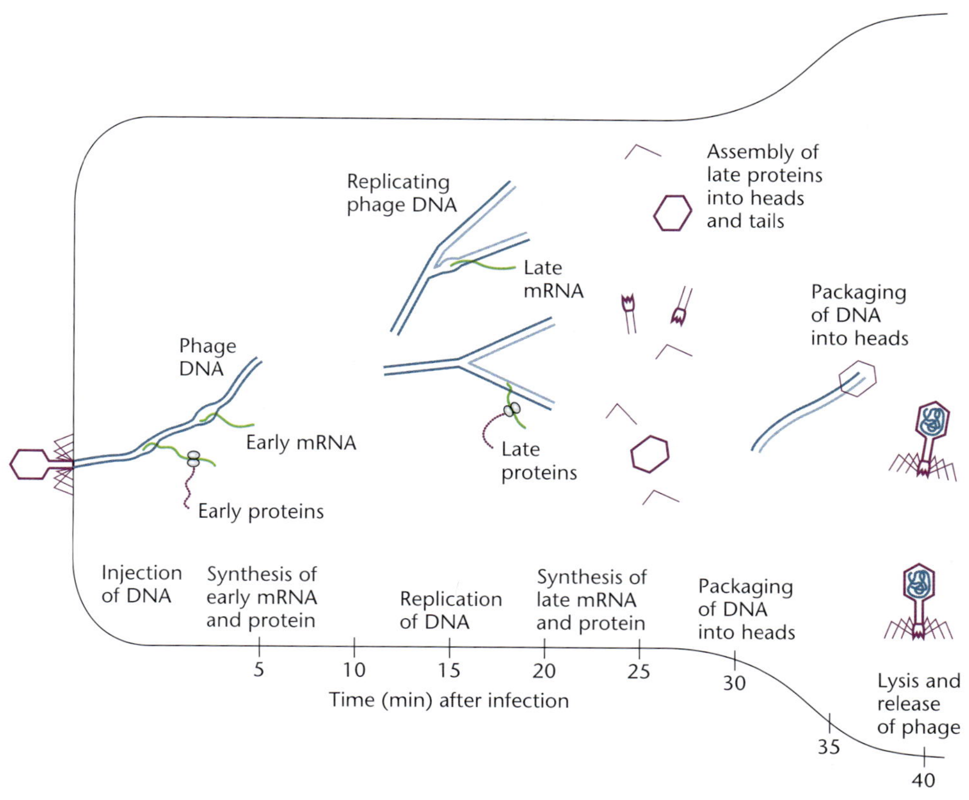

Figure 7.2 A typical bacteriophage multiplication cycle. After the phage injects its DNA, the early genes, most of which encode products involved in DNA replication, are transcribed and translated. Then, DNA replication begins, and the late genes are transcribed and translated to form the head and tail of the phage. The DNA is packaged into the heads, the tails are attached, and the cells lyse, releasing the phage to infect other cells. The phage DNA is shown in blue, mRNAs are shown in green, and proteins are shown in purple.

left of the genetic map, up to and including gene *1.3*, expressed first. These early genes are transcribed using the host RNA polymerase, and no phage-encoded products are required for their transcription. The genes to the right of gene *1.3*—the DNA metabolism and phage assembly genes—are transcribed after a few minutes' delay.

Nonsense and temperature-sensitive mutations (see chapter 3) were used to identify which of the early-gene products are responsible for turning on the late genes.

Figure 7.3 Transcriptional regulation by a regulatory cascade during development of a typical large DNA phage. The blue arrows indicate activation of middle gene expression by early gene products; the orange arrows indicate activation of late gene expression by middle gene products; the red bars indicate repression of early gene expression by middle gene products; the purple bars indicate repression of middle gene expression by late gene products.

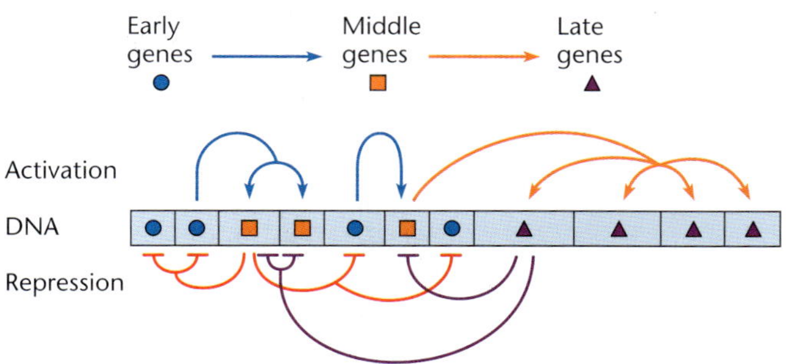

Under nonpermissive conditions, in which functional products were not produced from the mutated genes, nonsense and temperature-sensitive mutations in gene *1* prevented transcription of the late genes, so gene *1* was a candidate for a regulatory gene. Later work showed that the product of gene *1* is an RNA polymerase that recognizes the promoters used to direct transcription of the late genes. The sequences of these late gene promoters differ greatly from those recognized by bacterial RNA polymerases, and these phage late promoters therefore are recognized only by the T7-specific RNA polymerase.

A second level of control of late-gene transcription derives from the fact that binding of phage T7 to the host cell results in initial injection of only part of the phage DNA, corresponding to the left region of the genome and including only a few of the late genes. Transcription of the first of the late genes by the phage-encoded RNA polymerase helps pull the rest of the DNA containing the later genes into the cell, causing sequential late-gene expression based on the order of genes in the genome.

The high specificity of phage T7 RNA polymerase and other phage-encoded RNA polymerases for their own promoters has been exploited in many applications in molecular genetics, including development of high-level expression vectors and synthesis of RNA molecules *in vitro*, and some of these applications are discussed in Box 7.2.

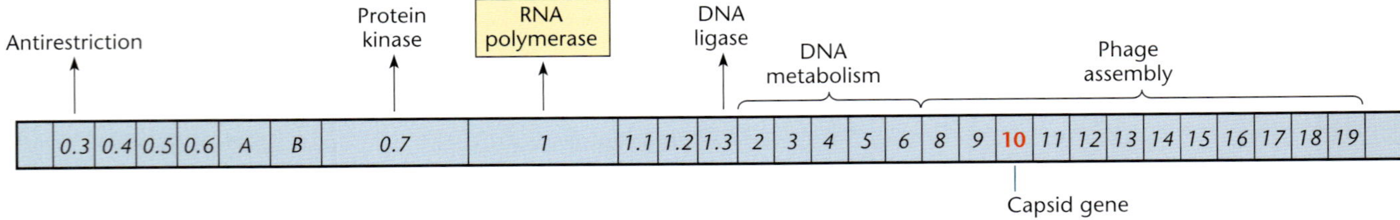

Figure 7.4 Genetic map of phage T7. The genes for the RNA polymerase used for expression vectors and the major capsid protein used for phage display are indicated.

N4: An RNA Polymerase Packaged in the Virion

Like phage T7, *E. coli* phage N4 encodes its own RNA polymerase; in fact, it encodes two forms of RNA polymerase. One of these RNA polymerases is packaged in the phage head, as is the case for many animal viruses (see Choi et al., Suggested Reading). Four molecules of this RNA polymerase are in each phage N4 head, and they are injected into the host cell along with the DNA of the phage. The virion N4 RNA polymerase has a number of interesting properties, including the ability to initiate transcription from unique hairpin-shaped promoters in the DNA. These hairpins can form in supercoiled N4 DNA with the help of the *E. coli* SSB (single-stranded DNA-binding protein [see chapter 1]).

N4 has a more complex transcriptional pattern than T7 in that it has three major classes of genes: early, middle, and late. Early genes are transcribed by the RNA polymerase packaged in the N4 virions. One of the products encoded by genes transcribed by this RNA polymerase is a second RNA polymerase that specifically transcribes the middle genes. This RNA polymerase is unusual in that it can initiate transcription from single-stranded templates, but only with the help of the phage's own SSB protein, which is expressed as a middle gene. Only late in infection is the *E. coli* RNA polymerase used to transcribe the late genes, and recognition of late-gene promoters also requires the phage N4 SSB, which therefore is essential for both middle and late transcription. One of the late genes encodes the virion RNA polymerase that will be encapsidated in the virions and injected with the phage DNA into the next host cell. Note that although N4 and T7 both use host and viral RNA polymerases, N4 uses its own RNA polymerases and the host RNA polymerase in a different order compared to how they are used by T7.

REPROGRAMMING OF HOST RNA POLYMERASE BY SIGMA FACTOR REPLACEMENT

Bacterial host cell RNA polymerases are composed of a core enzyme that carries out transcription elongation, plus a sigma (σ) factor that binds to the core enzyme to form a holoenzyme capable of specific promoter recognition (see chapter 2). Replacement of the σ factor used for transcription of most cellular genes with a new σ factor results in reduction of transcription of the genes recognized by the standard σ factor and increased transcription of genes containing promoters recognized by the new σ factor, often resulting in major shifts in gene expression patterns (see chapter 12). This has been exploited by some phages to redirect host RNA polymerase to transcription of phage genes without the need to synthesize a completely new enzyme.

SPO1: Cascade of Sigma Factors

Bacillus subtilis phage SPO1, like N4, has early, middle, and late classes of genes that are expressed sequentially during the developmental cycle. Early-gene promoters resemble host cell promoters and are recognized by the host cell RNA polymerase holoenzyme containing the σ responsible for transcription of most cellular genes (designated σ^A in *B. subtilis*). One of the early-gene products, called gp28 (the product of gene *28*), is a new σ, which was designated σ^{gp28}. RNA polymerase holoenzyme containing σ^{gp28} transcribes the SPO1 middle genes, the products of which include two polypeptides (gp33 and gp34) that together replace σ^{gp28} and direct transcription of the late genes (Figure 7.5). This sequential replacement of σ factors has been termed a σ cascade, and the SPO1 studies served as a paradigm for the more complex σ cascade used in the *B. subtilis* sporulation process (see chapter 12; see Losick and Pero, Suggested Reading).

MODIFICATION OF RNA POLYMERASE ACTIVITY WITH TRANSCRIPTIONAL REGULATORY PROTEINS

As discussed above, replacement of all or a part of RNA polymerase is an effective strategy to modulate transcriptional activity and direct expression of new sets of genes. A simpler strategy used by many phages relies solely on the host cell RNA polymerase, with the use of transcriptional regulatory proteins. As discussed in chapter 11, many bacterial genes are regulated by transcriptional repressors that inhibit the activity of RNA polymerase and transcriptional activators that increase RNA polymerase activity at specific genes that have binding sites for the regulatory proteins in the promoter region. The pattern of transcriptional regulation can be quite complex and

BOX 7.2

Phage T7-Based Tools

Some of the most widely used expression vectors use the T7 RNA polymerase to express foreign genes in *E. coli*. The pET vectors (for plasmid expression T7) are a family of plasmid expression vectors that use the T7 phage RNA polymerase and T7 gene *10* promoter to express foreign genes in *E. coli*. The promoter that is used to transcribe the head protein gene (gene *10*) of T7 is a very strong promoter, since hundreds of thousands of copies of the T7 head protein must be synthesized within a few minutes after infection, making this one of the strongest known promoters. Downstream of the T7 promoter in the pET vector are a number of restriction sites into which foreign genes can be inserted to allow high-level transcription by the T7 RNA polymerase. A number of variations of pET vectors have been designed. Some of the pET vectors have strong translational initiation regions immediately upstream of sequences for making translational fusions to affinity tags, such as a His tag, which makes the protein easy to detect and purify on nickel columns (see chapter 2 for a discussion of translational fusions). The T7 promoter can also be made inducible by providing the T7 RNA polymerase only when the foreign gene is to be expressed. This is important in cases where the fusion protein is toxic to the cell.

One commonly used strategy for making the T7 RNA polymerase inducible and thereby allowing the synthesis of a toxic gene product from a pET vector is illustrated in the figure. To provide a source of inducible phage T7 RNA polymerase, *E. coli* strains have been constructed in which phage gene *1*, which encodes T7 RNA polymerase, is inserted downstream of the inducible *lac* promoter and integrated into their chromosomes. In these strains, which often have the (DE3) suffix, as in *E. coli* JM109(DE3), the T7 RNA polymerase is synthesized only if an inducer of the *lac* promoter, such as

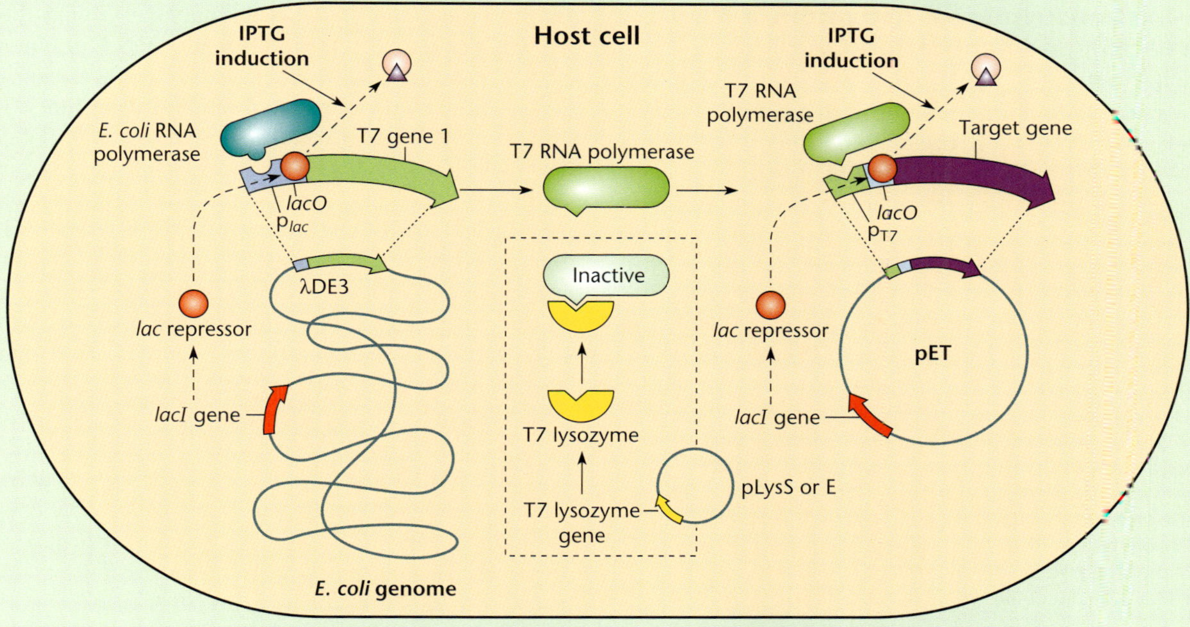

Strategy for regulating the expression of genes cloned into a pET vector. The gene for T7 RNA polymerase (gene *1*) is inserted into the chromosome of *E. coli* (in this case, as part of an integrated phage λ genome) and transcribed from the *lac* promoter; therefore, it is expressed only if the inducer isopropyl-β-D-thiogalactopyranoside (IPTG, purple triangle) is added. The T7 RNA polymerase then transcribes the gene inserted downstream of the T7 promoter on the pET cloning vector. If the protein product of the cloned gene is toxic, it may be necessary to further reduce the transcription of the cloned gene before induction. The T7 lysozyme encoded by a compatible plasmid, pLysS, binds to any residual T7 RNA polymerase made in the absence of induction and inactivates it. Use of pLysE instead of pLysS allows variation in the amount of lysozyme and therefore in the amount of active T7 RNA polymerase. Also, the presence of *lac* operators between the T7 promoter and the cloned gene further reduces transcription of the cloned gene in the absence of IPTG.

(continued)

BOX 7.2 (continued)

Phage T7-Based Tools

isopropyl-β-D-thiogalactopyranoside (IPTG), is added. Growth of the cells without inducer results in very low levels of T7 RNA polymerase synthesis and therefore low levels of synthesis of the inserted foreign gene. When IPTG is added, newly synthesized T7 RNA polymerase then makes large amounts of mRNA for the foreign gene and large amounts of its protein product.

Another application of phage T7 RNA polymerase is in making specific RNAs *in vitro*. RNAs made from a single gene or gene segment are useful as probes for hybridization experiments (**riboprobes**) or as RNAs for analysis of RNA structure and function. This technology is based on the specificity and activity of this class of RNA polymerase. Constructs are made either by cloning or PCR in which the T7 RNA polymerase promoter is inserted immediately upstream of DNA encoding the RNA segment of interest. Purified T7 RNA polymerase is added along with the other ingredients, including the ribonucleoside triphosphates needed for RNA synthesis, and the only RNA that is made is complementary to the transcribed strand of the DNA segment. RNAs can be generated in high concentration for applications such as crystal structure determination. Purified T7 RNA polymerase is available from biochemical supply companies.

can utilize multiple regulatory proteins to generate the appropriate pattern of gene expression.

Phage T4: Transcriptional Activators, a New Sigma Factor, and Replication-Coupled Transcription

Bacteriophage T4 is a large phage, with a complex structure and gene expression cycle. Experiments with this phage, which infects *E. coli* and its relatives, have been so important in the development of molecular genetics that the phage deserves equal status with Mendel's peas in the history of genetics. The function of ribosomes, the existence of mRNA, the nature of the genetic code, the confirmation of codon assignments, and many other basic insights originally came from studies with this phage.

Figure 7.5 Regulation of SP01 gene expression by a cascade of σ factors. Early transcription is carried out by the host RNA polymerase containing the major host σ factor (σ^A in *B. subtilis*). One of the early genes encodes gp28, which is a σ that binds to core RNA polymerase to direct transcription of the middle genes. Two of the middle-gene products, gp33 and gp34, act together as a σ to direct late-gene transcription.

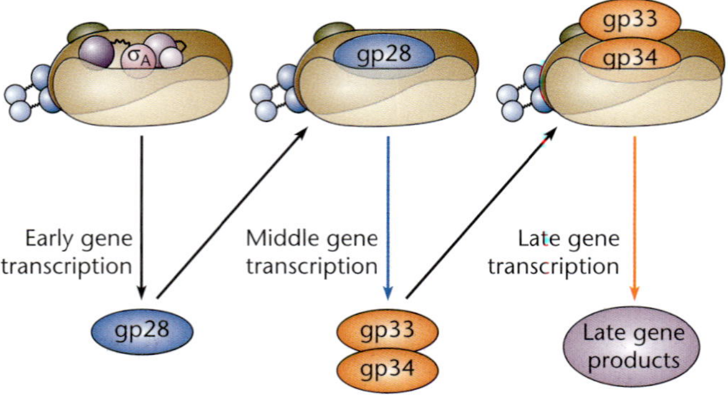

Phage T4 and its relatives are much larger than phages related to T7 and N4, with over 200 genes (the T4 genome map is shown in Figure 7.6A). In contrast to simpler phages that rely solely on the host cell machinery (with the possible exception of specialized forms of RNA polymerase or σ), T4 includes genes that substitute for a variety of host genes, allowing the phage to multiply in different hosts and under many different conditions. For example, T4 encodes enzymes that (i) increase the rate of DNA replication by synthesizing the precursors for DNA replication, (ii) make the unusual nucleoside hydroxymethylcytosine and attach glucose to it, which protects its DNA against restriction endonucleases, (iii) enhance DNA replication by providing its own enzymes for DNA replication (some of which, including the T4 DNA ligase and polynucleotide kinase, are useful enzymes for biotechnology), (iv) increase the rate of protein synthesis by encoding some of its own tRNAs, since its codon usage is somewhat different from that of its host because T4 DNA is very AT rich, and (v) enhance its multiplication when its host is under anaerobic conditions (without oxygen, as occurs in much of its normal environment, the vertebrate intestine). Because T4 has so many genes, the regulation of its gene expression is predictably complex and includes subdivision of the "early" stage genes into "immediate-early" genes, which are expressed as soon as the phage DNA is released into the cytoplasm, and "delayed-early" genes, which are expressed after a few minutes and whose expression relies on products of the immediate-early genes. True "middle" genes are, in turn, expressed after the delayed-early genes, and their expression is followed by that of the late genes, which encode structural proteins for virion production. As each new class of proteins appears, the synthesis of previous classes of proteins is turned off, resulting in the sequential appearance of new sets of proteins.

The regulatory genes responsible for T4 development were identified by isolation of mutants in which large groups of phage genes are no longer expressed. For example, mutations in genes named *motA* and *asiA* prevent the appearance of the middle-gene products. Mutations in genes *33* and *55*, as well as mutations in many of the genes whose products are required for T4 DNA replication, prevent the appearance of the late-gene products. Therefore, *motA*, *asiA*, *33*, and *55*, as well as some genes whose products are required for DNA replication, are predicted to be regulatory genes (see Figure 7.6A). Many genetic and biochemical experiments have been directed toward understanding how these T4 regulatory gene products turn on the synthesis of other proteins.

Immediate-early genes. As noted above, the immediate-early genes of T4 are turned on immediately after the phage DNA enters the cell and are transcribed from promoters recognized by RNA polymerase containing host σ^{70} (see chapter 2). Some of the immediate-early-gene products then allow the synthesis of other early genes after a few minutes' delay to allow accumulation of the newly synthesized regulatory proteins.

Delayed-early genes. The delayed-early genes of T4 are transcribed from the same *E. coli*-type σ^{70} promoters as the immediate-early genes but are regulated by an antitermination mechanism (see chapter 11). Without certain T4 regulatory gene products, the host RNA polymerase, which has initiated at an immediate-early promoter, would stop before it gets to the delayed-early genes. Therefore, the transcription of these genes must await the synthesis of antitermination factors encoded by immediate-early genes. Regulation by antitermination of transcription in the case of phage λ, where the mechanism is better understood, is discussed in chapter 11 and below.

Middle genes. The T4 middle genes are transcribed from their own promoters, which look somewhat different from normal σ^{70} promoters in that their −35 sequence is replaced by a sequence centered at −30 called a Mot box (Figure 7.7). Transcription from these promoters requires the phage-encoded MotA and AsiA proteins, both of which are the products of delayed-early genes. Rather than functioning as σ factors, these two proteins act together as transcriptional activators (see chapter 2 and chapter 11) that remodel σ^{70} on RNA polymerase so that the RNA polymerase complex now recognizes promoters with the Mot box rather than with the consensus σ^{70} −35 sequence. The MotA protein is a sequence-specific DNA-binding protein that, in conjunction with region 4 of σ^{70} (see chapter 2), recognizes the Mot box. AsiA, by binding to σ^{70} on RNA polymerase, releases region 4 of σ^{70} from the β flap of RNA polymerase, allowing region 4 to be repositioned to help MotA bind to the Mot box rather than to the −35 sequence of σ^{70} promoters. Additional stabilization is provided by a direct protein-protein contact between MotA and AsiA (see Yuan and Hochschild, Suggested Reading). Note that whereas most transcriptional activator proteins operate by binding to the DNA upstream of the promoter region and increasing the affinity of RNA polymerase for the promoter (see chapter 2 and chapter 11), MotA and AsiA instead interact with free RNA polymerase holoenzyme to change its promoter recognition properties.

Late genes. The last genes of T4 to be transcribed are the late genes. The principal products of these genes are the head, tail, and tail fiber components of the phage particle itself (see Figure 7.6B). The initiation of transcription of the late genes of T4 is of particular interest because it is coupled to the replication of the phage DNA.

Like the middle genes, the late genes of T4 are transcribed from promoters that are quite different from those of the host σ^{70} promoters (Figure 7.7) (see chapter 2). They have a longer −10 sequence (TATAAATA) rather than the shorter −10 hexameric sequence (TATAAT) of a typical bacterial σ^{70} promoter, and they lack a −35 sequence or any other defined sequence upstream of the −10 sequence. Because of this difference, the host RNA polymerase containing σ^{70} does not normally recognize T4 late promoters. Unlike the σ^{70} remodeling that directs middle-gene transcription, for late transcription the host σ^{70} is displaced from the RNA polymerase altogether, and two new T4-encoded proteins, gp55 and gp33, bind in its place. These two proteins act together to serve as a promoter binding specificity factor, analogous to the role of SPO1 gp33 and gp34 (Figure 7.5). The gp55 portion has a domain similar to σ^{70} region 2 that recognizes the altered −10 sequence of the T4 late promoters, and gp33 contributes some of the functions of σ^{70} region 4 in that it binds to the RNA polymerase β flap, where region 4 of σ^{70} normally binds; however, gp33 does not recognize any specific DNA sequences.

Replication-coupled late-gene transcription. Although the gp33 and gp55 proteins direct RNA polymerase to recognize the late promoters, they cannot efficiently form an open complex and activate late-gene transcription by themselves. Open complex formation requires addition of gp45, which contacts both gp33 and gp55 to the RNA polymerase complex. The gp45 protein is the T4 DNA polymerase accessory protein that acts like the β clamp in *E. coli* DNA replication and wraps around the DNA to form a "sliding clamp" (see chapter 1 for a more detailed description of DNA replication). The gp45 protein then moves with the DNA polymerase and helps to prevent it from falling off the DNA during replication (Figure 7.8). The gp44 and gp62 proteins are subunits of the clamp loader, analogous to the γ complex of *E. coli* (see chapter 1),

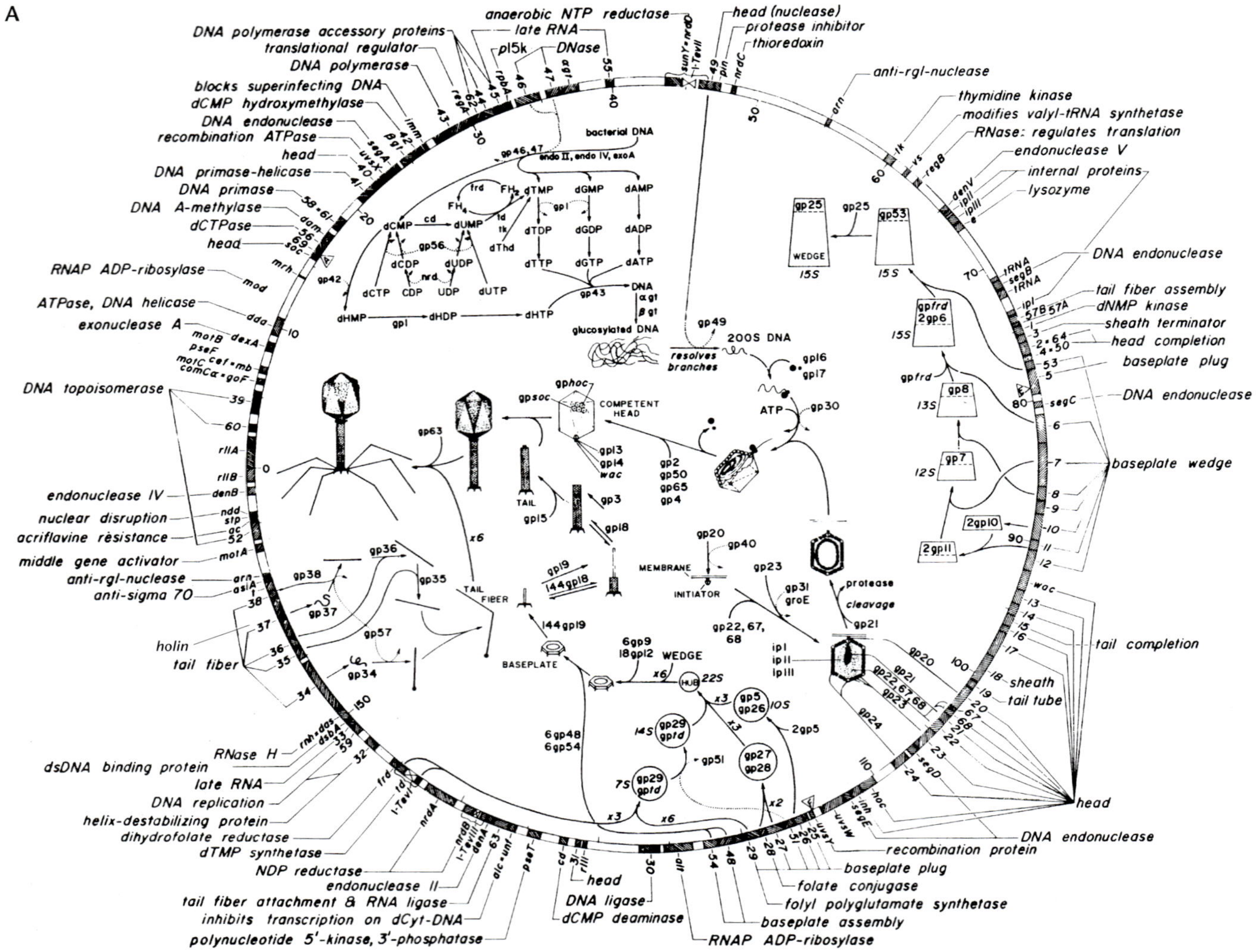

A

B

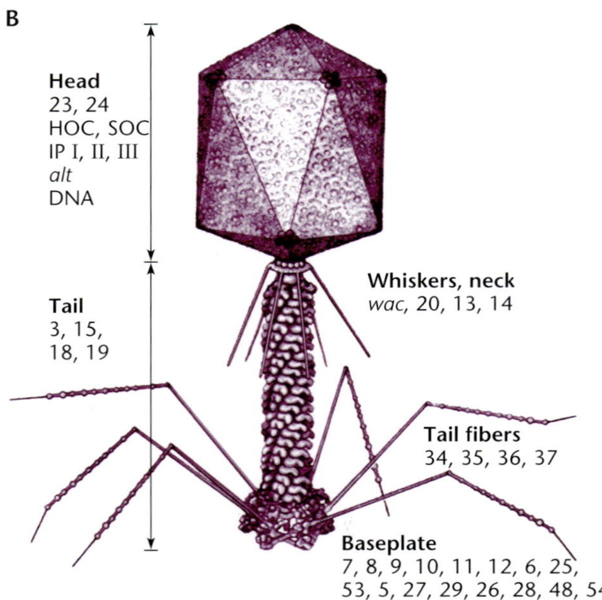

Figure 7.6 **(A)** Genomic map of phage T4. From Karam JD (ed), *Molecular Biology of Bacteriophage T4* (ASM Press, Washington, DC, 1994). **(B)** Structural components of the T4 particle. The features of the particle have been resolved to about 3 nm. The positions of the baseplate and tail fiber proteins are indicated. From Miller ES, Kutter E, Mosig G, et al, *Microbiol Mol Biol Rev* **67:**86–156, 2003.

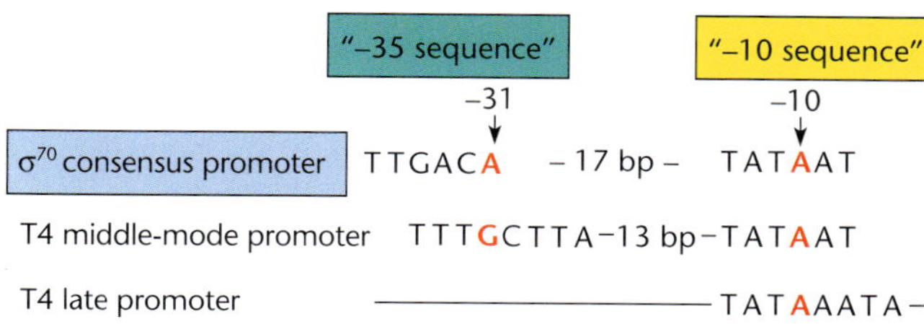

Figure 7.7 Sequence of T4 middle-mode and late promoters. Only the sequences important for recognition by RNA polymerase are shown.

and load the gp45 clamp onto the DNA during replication. Once loaded on the DNA, this sliding clamp interacts with the gp43 DNA polymerase to keep it on the DNA, but it periodically cycles off the replication apparatus. It can then slide around on the DNA by itself and can also bind to gp33 and gp55 on RNA polymerase to activate late transcription.

From the simple depiction in Figure 7.8, it is not obvious how the gp45 ring could contact the C termini of both gp55 and gp33 on the RNA polymerase, but like other such sliding clamps, the gp45 clamp has long protuberances that can extend into the RNA polymerase, as well as other proteins. It is also possible that gp55 extends further toward the back end of the RNA polymerase than is shown in the figure. This ability of sliding clamps to switch their binding from one protein to another is reminiscent of the process that occurs during host DNA replication as found with normal DNA replication and with the recruitment of translesion DNA polymerases (see chapter 1). The requirement of T4 late transcription for continuous T4 DNA replication can also be explained. The gp45 clamps periodically fall off the DNA and are depleted on the DNA unless they are continuously being reloaded at replication forks. The clamp loader can also load the gp45 clamp onto DNA at nicks, partially obviating the need for other replication proteins if the DNA is heavily nicked. Coupling gene expression to the replication of the phage DNA ensures that late gene transcription occurs only if DNA replication is well under way, so that copies of the phage genome are available for packaging into virions generated from late gene products.

ANTITERMINATION MECHANISMS

As noted above, the transition between immediate-early and delayed-early transcription in T4 utilizes transcription antitermination (see chapter 11), which was first discovered in phage λ. In regulation by transcription antitermination, transcription begins at the promoter but then terminates at a site relatively close to the promoter, resulting in generation of a short RNA product. When certain conditions are met, transcription continues past the termination site, generating a longer RNA that results in production of additional protein products. The name "antitermination" is used because the mechanism prevents termination.

Antitermination can occur through a number of mechanisms (see chapter 11). In the case of λ and related phages, specific phage-encoded proteins bind to the nascent RNA as it emerges from RNA polymerase and recruits host proteins that form a complex with RNA polymerase that makes the transcription elongation complex resistant to termination sites it encounters as it moves along the DNA. This is called **processive antitermination** because of the ability to prevent termination at multiple sites.

λ N-mediated antitermination

The circular λ genetic map is shown in Figure 7.9. When the λ DNA first enters the cell, transcription immediately begins from two promoters, p_L and p_R, that face outward from the immunity region (which is involved in regulation of the lysis/lysogeny decision; see "The λ System" below). Transcription from these promoters leads to the synthesis of two RNAs, one leftward on the genome and the other rightward. However, the synthesis of both RNAs terminates at transcription termination sites, $t_L{}^1$ (for leftward transcription) and $t_R{}^1$ (for rightward transcription), leading to synthesis of short RNAs and allowing for the expression of a few important proteins for the lysogeny decision (Figure 7.10A; see Tables 7.2 and 7.3 for lists of key phage λ genes and sites). One of these short RNAs encodes the Cro protein, which is involved in lysogeny control. The other encodes the N protein, the antitermination factor that permits RNA polymerase to bypass the transcription termination sites $t_L{}^1$ and $t_R{}^1$ and continue along the DNA, as shown in Figure 7.10B. This results in generation of longer RNAs that include sequences from the genes downstream of the terminators (*gam red xis int* in the leftward direction, *O P Q* in the rightward direction).

Figure 7.10C and D outline how the N protein antiterminates, showing only rightward transcription. Initially, transcription initiated at the rightward p_R promoter terminates at the transcription terminator designated $t_R{}^1$. Within the sequences transcribed into RNA is the *nutR* (for N utilization rightward) site, which in the nascent transcript binds to the RNA polymerase from which the RNA is emerging. In the meantime, the N protein is being synthesized from the leftward RNA. The N protein interacts with the *nutR* RNA element, which allows interaction of N with RNA polymerase. A set of host-encoded proteins, called the Nus proteins (for N usage substance), help the N protein to bind, and together, they all form an antitermination complex that travels with the RNA polymerase, forming a transcription elongation complex that fails to terminate at $t_R{}^1$ and other transcription termination signals. When this antitermination complex transcribes through $t_R{}^1$, it reaches the genes *O* and *P*, which encode the replication proteins, and replication begins. The gene

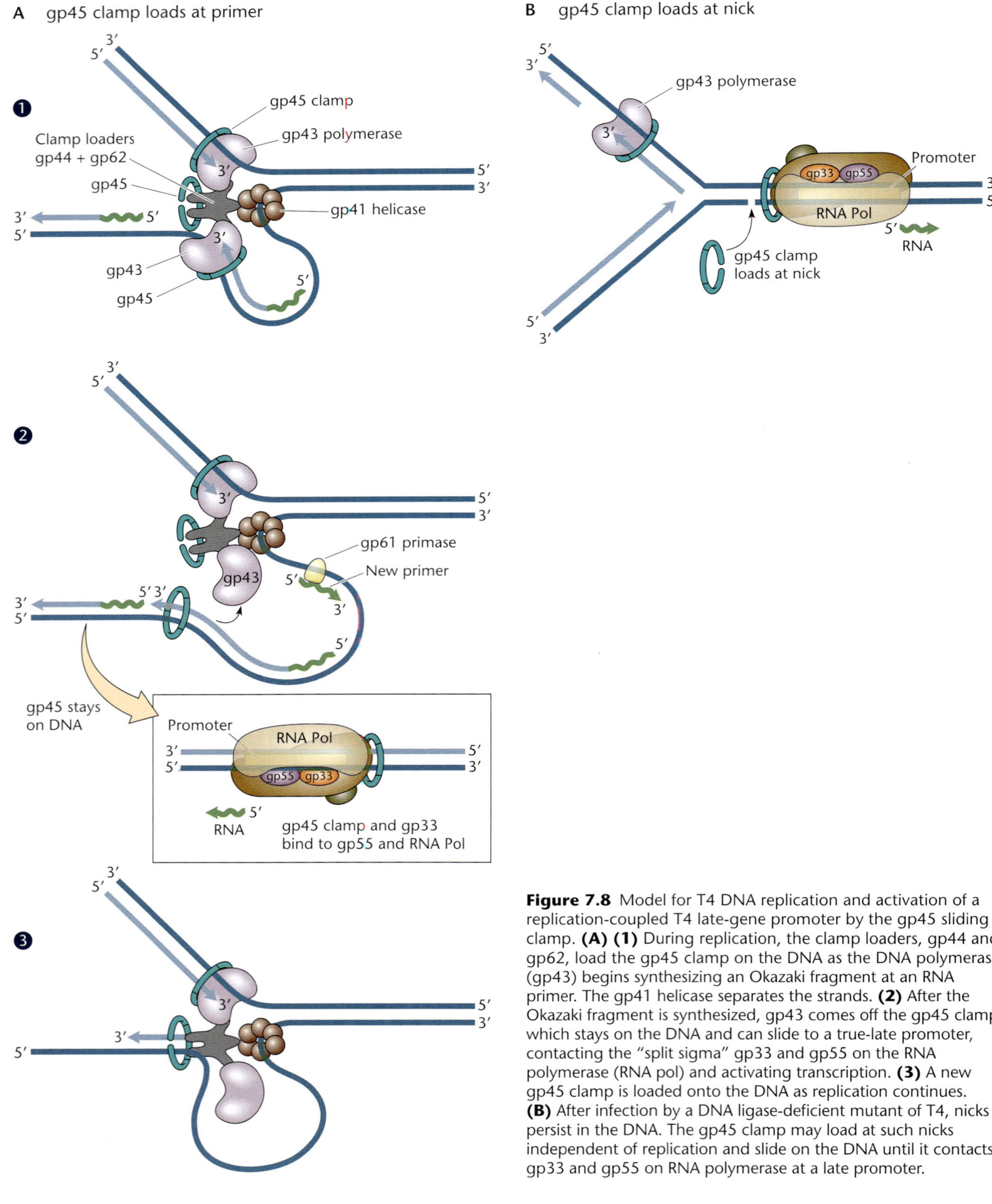

Figure 7.8 Model for T4 DNA replication and activation of a replication-coupled T4 late-gene promoter by the gp45 sliding clamp. **(A) (1)** During replication, the clamp loaders, gp44 and gp62, load the gp45 clamp on the DNA as the DNA polymerase (gp43) begins synthesizing an Okazaki fragment at an RNA primer. The gp41 helicase separates the strands. **(2)** After the Okazaki fragment is synthesized, gp43 comes off the gp45 clamp, which stays on the DNA and can slide to a true-late promoter, contacting the "split sigma" gp33 and gp55 on the RNA polymerase (RNA pol) and activating transcription. **(3)** A new gp45 clamp is loaded onto the DNA as replication continues. **(B)** After infection by a DNA ligase-deficient mutant of T4, nicks persist in the DNA. The gp45 clamp may load at such nicks independent of replication and slide on the DNA until it contacts gp33 and gp55 on RNA polymerase at a late promoter.

for the other antitermination protein, Q, is also transcribed, and this protein turns on the transcription of the late genes (see below).

Similar events occur during leftward transcription. When the *nut* site on the other side, called *nutL* (for N utilization leftward), is transcribed, it allows N protein (with the Nus factors) to bind to the transcription elongation complex and prevent termination at $t_L^{\ 1}$, allowing transcription to continue into other genes on the left side, including *gam* and *red*, which encode λ recombination functions. The RNA polymerase also continues into *int* and *xis*, genes involved in integrating and excising the prophage, although the mRNA from these genes is quickly degraded, as explained below.

Figure 7.11 shows the sequences of the *nut* sites of λ. These sites consist of a sequence, called box B, that forms a "hairpin" secondary structure in the mRNA because it is encoded by a region of the DNA with 2-fold rotational symmetry (see chapter 2). The importance of this structure was demonstrated by identification of mutations that disrupt base-pairing between the two halves of box B, preventing formation of the hairpin in the mRNA; these mutations disrupt N binding and prevent production of λ virions.

The box A and box C sequences in *nut* sites are involved with binding of the host-encoded Nus proteins. Four of these proteins, NusA, -B, -E, and -G, are known to be involved in transcriptional or translational processes in uninfected *E. coli* and have been commandeered by λ to help with antitermination. One of the

Figure 7.9 Genetic map of λ cyclized by pairing at the *cos* sites, shown at the top. The positions and directions of transcription from some promoters are shown in green.

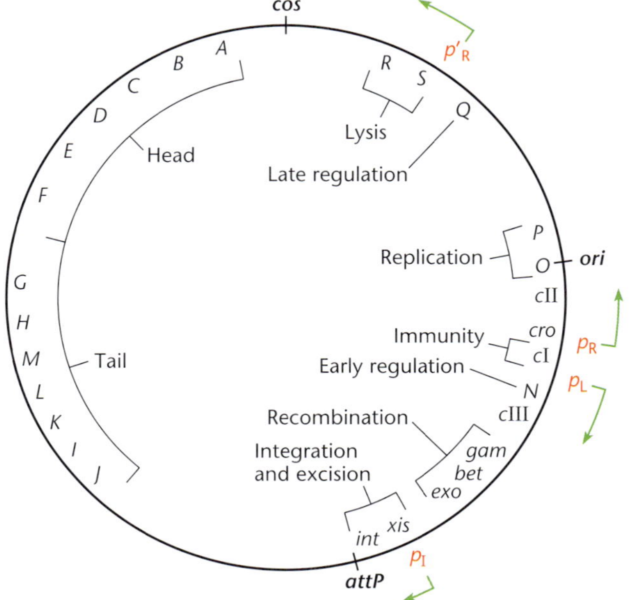

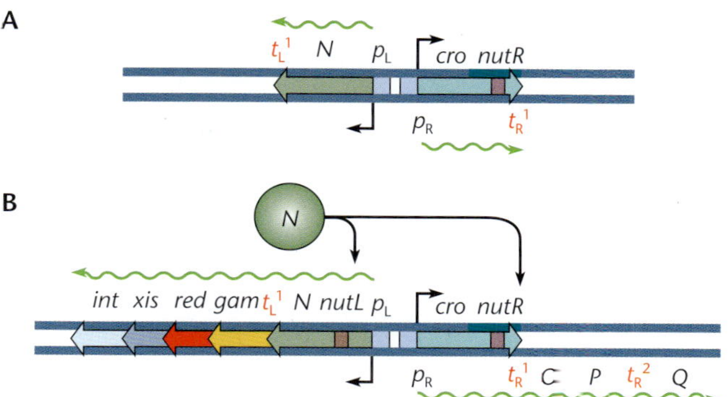

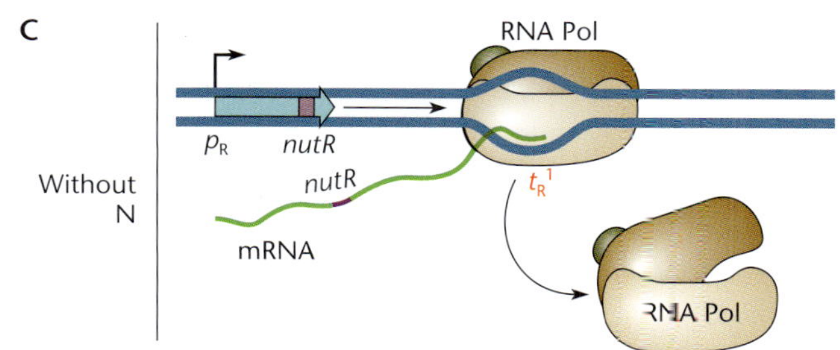

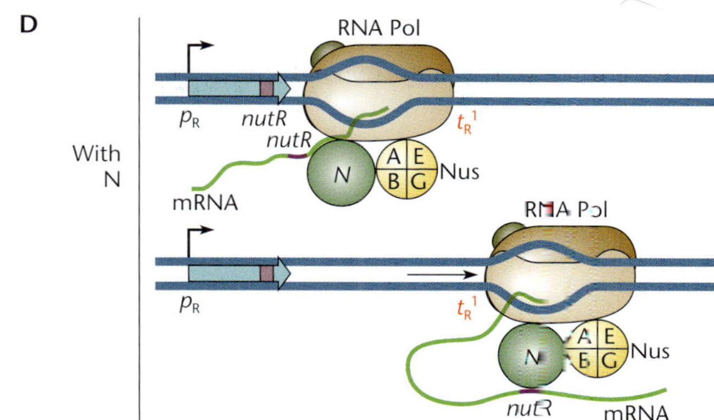

Figure 7.10 Antitermination of transcription in phage λ. **(A)** Before the N protein is synthesized, transcription starts at promoters p_L and p_R and stops at transcription terminators $t_L^{\ 1}$ and $t_R^{\ 1}$. **(B)** The N protein causes transcription to continue past $t_L^{\ 1}$ and $t_R^{\ 1}$ into *gam-red-xis-int* and *O-P-Q*, respectively. **(C and D)** Mechanism of antitermination by N, showing rightward transcription only. **(C)** In the absence of N, transcription initiated at p_R terminates at the terminator $t_R^{\ 1}$. **(D)** If N has been made, it binds to RNA polymerase (RNA Pol) as the polymerase transcribes the *nutR* site. A conformational change occurs in N when it binds to the *nutR* sequence in the RNA, and this change is required before N can bind to the RNA polymerase. The binding of N to RNA polymerase is facilitated by the host Nus proteins NusA, -B, -E, and -G. The antitermination complex, composed of RNA polymerase, N, *nutR*, and NusABEG, then transcribes past transcriptional terminator $t_R^{\ 1}$ plus any other transcriptional terminators downstream. The sites and λ gene products shown here are defined in Tables 7.2 and 7.3.

Table 7.2 Some λ gene products and their functions

Gene product	Function
N	Antitermination protein; loads on RNA polymerase at *nutL* and *nutR* sites and prevents termination at downstream terminators, including t_L^1, t_R^1, and t_R^2
O, P	Initiation of λ DNA replication
Q	Antitermination protein; loads on RNA polymerase at *qut* site and prevents termination at downstream terminators, including t_R'
CI	Repressor, activator; binds to o_L and o_R, preventing transcription from p_L and p_R and activating transcription from p_{RM}
CII	Activator of transcription of cI from p_{RE} and *int* from p_I
CIII	Stabilizes CII by inhibiting a cellular protease
Cro	Repressor of CI synthesis from p_{RM} after infection
Gam	Protein required for rolling-circle replication; inhibits RecBC
Red	Composed of two proteins, β and Exo, involved in λ recombination
Int	Integrase; site-specific recombinase required for recombination between *attP* and *attB* to integrate prophage
Xis	Excisase; directionality factor; works with Int to promote recombination between hybrid sites BOP′ and POB′ to excise prophage

Table 7.3 Some sites involved in phage λ transcription and replication

Site(s)	Function(s)
p_L	Left promoter
p_R, p_R'	Right promoters
o_L	Operator for leftward transcription; binding sites for CI and Cro repressors
o_R	Operator for rightward transcription; binding sites for CI and Cro repressors
t_L^1, t_L^2	Termination sites of leftward transcription
t_R^1, t_R^{234}, t_R'	Termination sites of rightward transcription
nutL	N utilization site for leftward-transcribing RNA Pol; sequence in the mRNA binds to N, allowing it to bind to RNA Pol and resist termination
nutR	N utilization site for rightward-transcribing RNA Pol
qut	Q utilization site for antitermination of transcription from p_R' on DNA); overlaps promoter
p_{RE}	Promoter for repressor establishment; activated by CII
p_{RM}	Promoter for repressor maintenance; activated by CI
p_I	Promoter for *int* transcription; activated by CII
POP′	Phage attachment site (*attP*)
cos	Cohesive ends of λ genome (complementary 12-bp single-stranded ends in linear genome anneal to form a circular genome after infection)

proteins, NusA, binds to the box B sequence after N has bound and helps to stabilize the complex. NusB and NusE bind to the box A sequence, NusG joins the complex, and then all the proteins travel together with RNA polymerase as the antitermination complex. The role of box C is unclear, but it may be required to regulate antitermination in subtle ways not easily detectable in a laboratory situation.

λ Q-mediated Antitermination

As mentioned above, one of the genes that is transcribed by the N-modified transcription complex is gene *Q*, whose product is required for the transcription of the late genes of λ, including the head and tail genes. Thus, λ marches through a regulatory cascade, with one of the earliest gene products (N) directing the synthesis of an initial set of gene products that includes another regulatory gene product (Q), which in turn, directs the transcription of the late genes. Like N, the Q protein of λ is a processive antiterminator protein that binds to RNA polymerase, rendering it resistant to downstream termination sites and allowing transcription from the late promoter p_R' to proceed through terminators into the downstream genes. One of the host Nus proteins, NusA, also travels with the Q-modified transcription complex. Like N, the Q protein loads onto RNA polymerase in response to a sequence located close to the promoter, in this case called *qut* (for Q utilization site) (Figure 7.12 and Table 7.3). However, unlike the *nut* sites, which are functional when they have been transcribed into RNA, the *qut* site is active in the DNA and is not transcribed into RNA.

The details of Q protein antitermination have been studied in some detail (Figure 7.12) (see Roberts et al., Suggested Reading). Transcription of the late genes starts at the late promoter p_R', which is located downstream from the *Q* gene. RNA polymerase transcribes a very short RNA (16 to 17 nucleotides long) from p_R' before it pauses. As the short RNA exits the RNA polymerase, it displaces region 3.2 and region 4 of σ^{70} from the RNA polymerase, leaving only region 2, which normally contacts the −10 region at the promoter (see chapter 2), to contact the DNA. The RNA polymerase pauses 16 bp downstream of the start site of transcription because region 2 recognizes a sequence that is similar to the −10 sequences of the σ^{70} promoters (Figure 7.12 and chapter 2). Because the single-stranded DNA at the transcription bubble is more flexible, it can "scrunch," allowing region 2 to contact the −10-like sequence even though the spacing on the template is less than ideal. The Q protein then loads on the *qut* sequence and contacts the RNA polymerase, stabilizing the interaction between region 4 of σ^{70} and a −35-like sequence that is only 1 bp upstream of the −10-like sequence instead of the usual 17 bp at normal σ^{70} promoters. The displacement of the region 3.2 linker

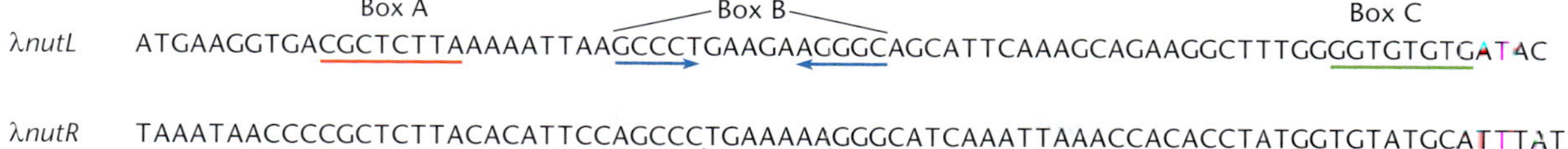

Figure 7.11 Sequences of the *nutL* and *nutR* regions of bacteriophage λ. Box A, box B, and box C are underlined. The 2-fold rotational symmetry in box B that causes a hairpin to form in the mRNA is shown as arrows of opposite polarity in blue.

Figure 7.12 Formation of the Q protein antitermination complex at the p_R' promoter. **(A)** The *qut* site (shown in light blue) overlaps the –35 (turquoise) and –10 sites (yellow) of the σ^{70} promoter. Sequences just downstream, identified as –35-like and –10-like (boxed in light turquoise and light yellow, respectively) function as –35 and –10 sequences to interact with RNA polymerase. **(B)** Transcription initiates as at other promoters, with region 2 of σ contacting the –10 sequence and opening the DNA helix and region 4 contacting the –35 sequence until the nascent 16-base mRNA displaces region 3.2 from the RNA polymerase and region 4 from the DNA and RNA polymerase. **(C)** As the RNA polymerase moves down the template, σ region 2 scans the DNA until it recognizes the –10-like sequence, and the RNA polymerase pauses. This allows σ region 4 to bind to the –35-like sequence with the help of Q (orange circle) bound at the *qut* site. The Q protein then loads on the RNA polymerase and moves with it down the DNA into the late genes, ignoring transcription termination signals. The "scrunching" of the DNA template allows σ region 2 to contact the –10-like element 12 bp downstream of the –10 sequence even though the RNA polymerase has moved 16 bp on the DNA template. Adapted with permission from de Haseth PL, Gott JM, *Mol Microbiol* **75:**543–546, © 2010 Blackwell Publishing Ltd.

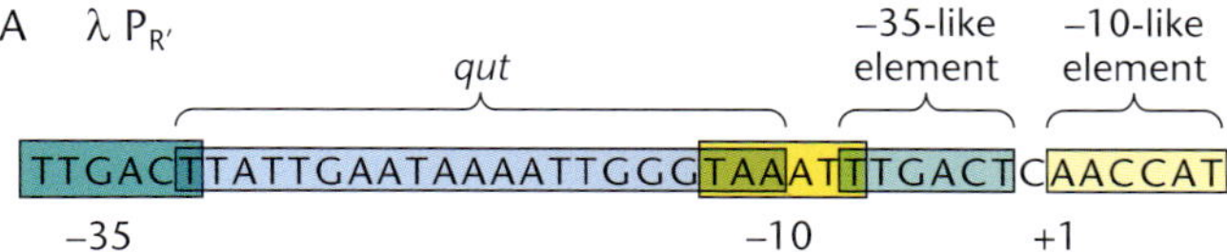

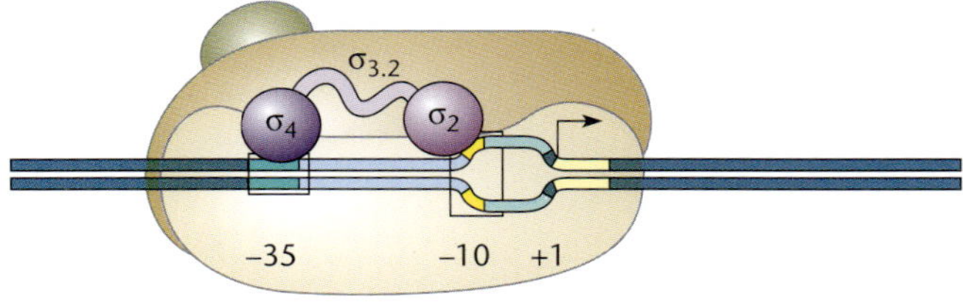

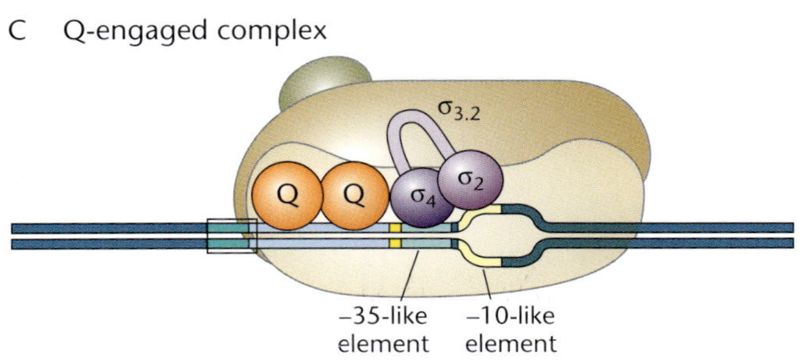

from the RNA polymerase must make σ^{70} flexible enough that region 4 can double back and contact a –35 region that is much closer to the –10 region than it is at normal promoters. The Q protein then loads onto the paused RNA polymerase (with the assistance of the host NusA protein), and the transcript exits through the exit channel and displaces σ^{70} to allow the RNA polymerase to escape from the pause site. Once bound, Q and NusA travel with the RNA polymerase, making it resistant to further transcription termination and pause sites and allowing it to proceed rapidly through the late genes of the phage.

HK022 *put*-Mediated Antitermination

The HK022 phage, which is related to λ, also uses processive antitermination to regulate its gene expression (see King et al., Suggested Reading). However, antitermination in these phages only requires the transcription of *cis*-acting sites on the DNA called *put* sites, transcribed from the early promoters, and not a protein like N. The *put* region of the nascent transcripts binds to the RNA polymerase as they are transcribed and makes the RNA polymerase resistant to termination at sites further downstream. Because they bind to the RNA polymerase as they are made, the *put* RNAs are *cis* acting, like the *nut* sites of λ. In fact, the two sites *putL* and *putR* are located in the HK022 genome approximately where the *nutL* and *nutR* sites are located in λ, just downstream of the p_L promoter and just downstream of the *cro* gene, respectively. However, these RNAs bind to RNA polymerase independently of an antitermination protein and do not require any of the Nus proteins of the host, at least not absolutely, since some antitermination occurs in purified transcription systems with just RNA polymerase and DNA and none of the other factors. Why these phages forego the need for an antitermination protein and simply use the *cis*-acting RNAs for antitermination is unclear but may reflect the fact that antitermination has evolved multiple times using different mechanisms.

Phage Genome Replication and Packaging

Unlike the replication of chromosomal or plasmid DNAs, which must be coordinated with cell division, phage genome replication is governed by one purpose: to make the highest number of copies of the phage genome in the shortest possible time. Phage genome replication can be truly impressive. A single phage DNA or RNA molecule initially entering the cell can replicate to make hundreds or even thousands of copies to be packaged into phage heads in as little as 20 or 30 minutes. This unchecked replication often makes phage genome replication easier to study than replication in other systems. Nevertheless, phage DNA replication shares many of the features of cellular replication found in all types of living organisms, and phage DNA replication has served as a model system to understand DNA replication in bacteria and even in humans. Replication of phage RNA genomes has provided models for replication of eukaryotic RNA viruses.

PHAGES WITH SINGLE-STRANDED CIRCULAR DNA

The genomes of some small phages consist of circular single-stranded DNA. The small *E. coli* phages that fit into this category can be separated into two groups. The representative phage of one group, φX174, has a spherical capsid with spikes sticking out, resembling the ball portion of the medieval weapon known as a "morning star." These phages are called icosahedral phages because their capsids are icosahedrons, like a geodesic dome with mostly six-sided building blocks and an occasional five-sided block. In the other group, represented by M13 and f1 and often referred to as **filamentous phages**, the phages have a single layer of protein covering the extended DNA molecule, making the phages appear as a long filament.

The different shapes of phages of these two families determine how they infect cells and then leave the infected cells. The icosahedral phages enter and leave cells much like other phages. They bind to the cell surface, and only the DNA enters the cell; the newly assembled phage particles exit the cell by lysing the host cell (see "Host Cell Lysis" below). In contrast, M13 and other filamentous phages are called male-specific phages because they specifically adsorb to the pilus encoded by certain plasmids, such as the F plasmid, and so infect only "male," or donor, strains of bacteria (see chapter 5). Unlike most other phages, filamentous phages do not inject their DNA. Instead, the entire phage is drawn into the cell with retraction of the pilus, and the protein coat is removed from the DNA as the phage passes through the inner cytoplasmic membrane of the bacterium. After the phage DNA has replicated, it is again coated with protein as it leaks back out through the cytoplasmic membrane (Figure 7.13). Since these phages do not lyse infected cells and leak out slowly, the cells they infect are "chronically" infected rather than "acutely" infected and are not killed. Nevertheless, the filamentous phages form visible, albeit turbid, plaques, because the infected cells grow more slowly than the surrounding uninfected lawn.

Phage f1 has been used as a model for study of virion structure and assembly. The phage f1 particle has only five proteins, one of which, the major head protein, pVIII, exists in about 2,700 copies and coats the DNA. The other

Figure 7.13 Infection cycle of the single-stranded DNA phage f1. Steps 1 through 8 show the binding of the phage, replication of the DNA, and encapsidation of phage DNA as it is secreted through the membrane pore formed by the pIV secretin to release the phage and infect a new cell. Details are given in the text. ssDNA, single-stranded DNA.

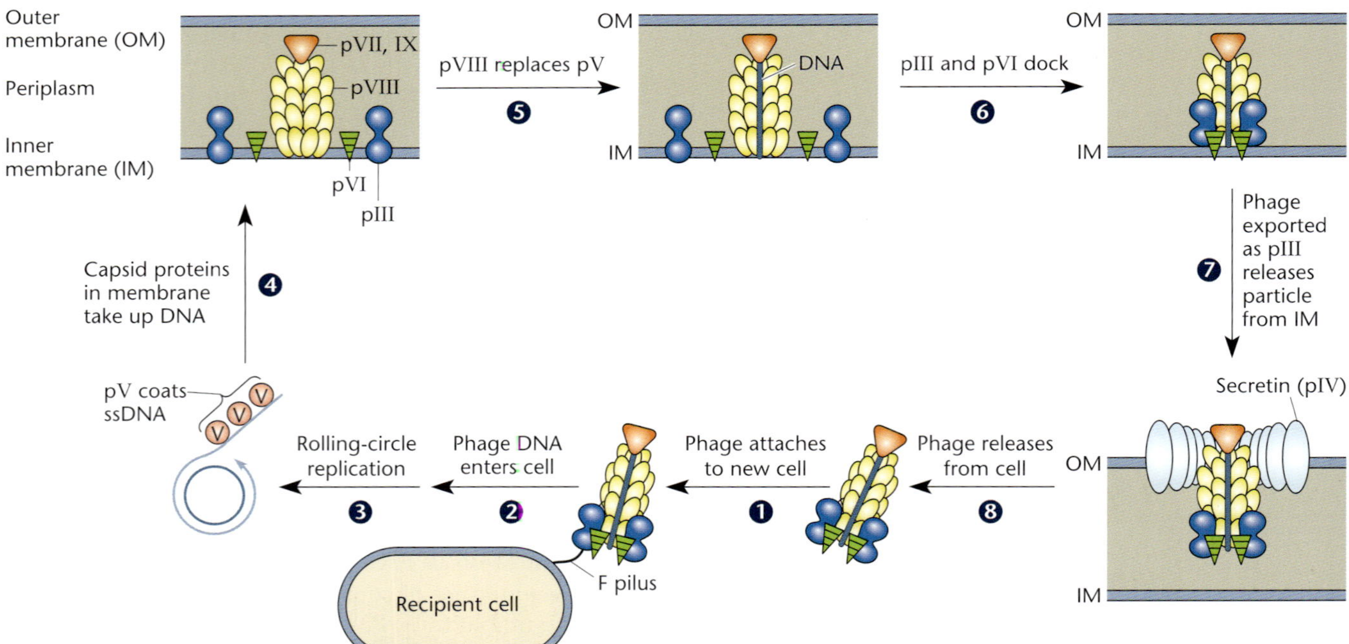

four proteins are on the ends of the phage and exist in only four or five copies per phage particle. Two of these proteins, pVII and pIX, are located on one end of the phage, and the other two, pVI and pIII, are on the other end (Figure 7.14). To start the infection, the pIII protein on one end of the phage first makes contact with the end of the F pilus. The pilus retracts when the phage binds to it, drawing the phage to the cell surface. A different region of the same pIII protein then contacts a host inner membrane protein called TolA, which sticks into the periplasmic space. The phage DNA then enters the cytoplasm, while the major coat protein (pVIII) is stripped off into the host inner membrane.

Unlike uptake of the phage DNA, which must rely exclusively on host proteins, release of the phage particles from infected cells can use phage proteins newly synthesized during the course of the infection (Figure 7.13). After the phage DNA has replicated a few times to produce a double-stranded replication intermediate (see below), it enters the rolling-circle stage of replication. As it rolls off the circle, the newly synthesized single-stranded DNA is coated by another protein, pV. The proteins that make up the phage coat have been synthesized and are waiting in the membrane, and the major head protein, pVIII, replaces the pV protein on the DNA as it enters the membrane. Only single-stranded DNA containing a specific sequence called the *pac* (packaging) site of the phage is packaged into virions. The other phage proteins are then added to the particle. Meanwhile, the phage-encoded secretin protein, pIV, has formed a channel in the outer membrane through which the assembled phage can pass. This channel is related to the channels formed by type II secretion systems to assemble pili on the cell surface and by competence systems to allow DNA into the cell (see chapters 2 and 6).

Replication of Single-Stranded Phage DNA

Studies of the replication of single-stranded phage DNA have contributed much to our understanding of replication in general. It was with these phages that rolling-circle replication was discovered, as well as many proteins required for host DNA replication, including the proteins PriA, PriB, PriC, and DnaT, which are now known to be involved in restarting chromosomal replication forks after they have dissociated upon encountering DNA damage (see chapter 9). Many of the genes for these host proteins were found in searches for host mutants that cannot support the development of the phages and by reconstituting replication systems *in vitro* by adding host DNA replication proteins to phage DNA until replication was achieved. In addition, some types of plasmids use rolling-circle replication (see chapter 4).

Synthesis of the complementary strand to form the first replicative form version of the genome. Replication has been studied in detail for phage M13 because of its importance in molecular biology approaches (see "Phage as a Tool" below and chapter 3). The steps in the replication of phage M13 DNA are outlined in Figure 7.15. The DNA strand of the phage encapsidated in the M13 virion is called the plus strand. Immediately after the single-stranded plus strand DNA enters the cell, a complementary minus strand is synthesized using the plus strand as a template to form a double-stranded DNA called the **replicative form (RF)**. The formation of this first RF is dependent entirely on host functions, because no phage proteins enter the cytoplasm with the phage DNA, and synthesis of phage proteins requires double-stranded DNA as a template for transcription. The synthesis of the complementary strand is primed by an RNA made by the normal host RNA polymerase. Normally, the host RNA polymerase recognizes only double-stranded DNA, but the single-stranded phage DNA forms a hairpin at the origin of replication, making it double-stranded in this small region. Once the RNA primer is synthesized, the host DNA polymerase III and accessory proteins load onto the DNA and synthesize the complementary strand until they have proceeded all of the way around the DNA circle and encounter the RNA primer. The flap endonuclease activity of DNA polymerase I displaces and then removes the RNA primer, its DNA polymerase activity fills the gap with DNA, and the nick is sealed by DNA ligase to leave a double-stranded covalently closed RF, which can then be supercoiled (see chapter 1).

The icosahedral phages, such as φX174, also use the host cellular machinery to generate the initial RF molecules but in this case use a complex primosome related to that used for host chromosomal replication, with the addition of other proteins, PriA, PriB, PriC, and DnaT, that are not required for initiation of chromosomal replication at *oriC* but instead are used to restart chromosomal replication after it collapses at single-stranded interruptions in the chromosome or has been blocked due to encountering damage in the DNA template (see chapter 9). If the replication

Figure 7.14 Schematic representation of the filamentous bacteriophage M13. The single-stranded circular DNA is coated with five viral proteins. The schematic locations of the different proteins are shown. The gpVIII protein is present at about 2,700 copies, while gpIII, gpVI, gpVII, and gpIX are present at about 6 or 7 copies each. All of the coat proteins can be used as platforms for protein display. With the exception of gpIII, the capsid proteins are small, with 33 to 112 amino acids.

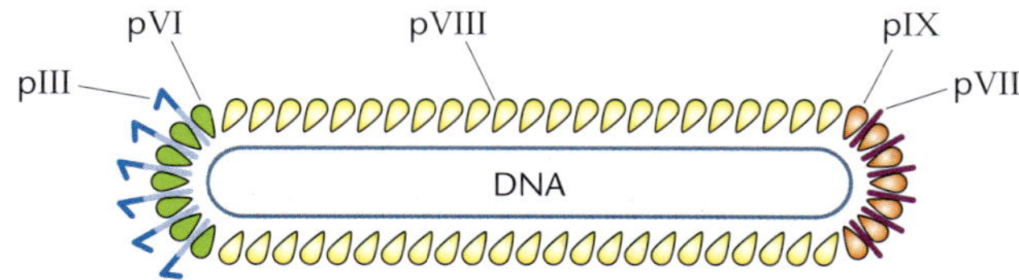

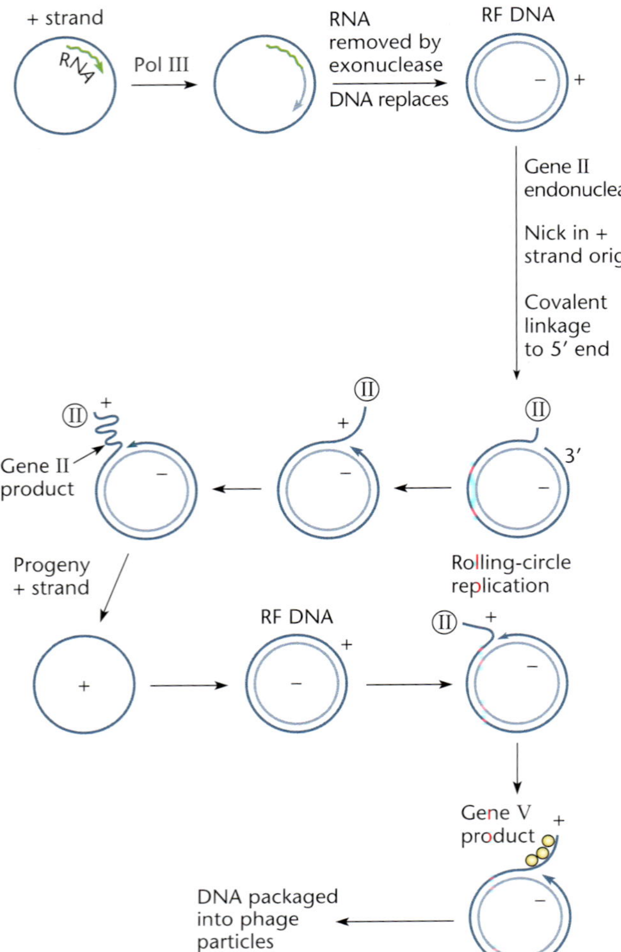

Figure 7.15 Replication of the circular single-stranded DNA phage M13. First, an RNA primer (green) is used to prime the synthesis of the complementary minus strand (in light blue) to form double-stranded replicative form (RF) DNA. The product of gene II, an endonuclease, nicks the plus strand of the RF and remains attached to the 5′ phosphate at the nick. Then, more plus strands (dark blue) are synthesized via rolling-circle replication, and their minus strands are synthesized to make more RFs. Later, the gene V product (yellow circles) binds to the plus strands as they are synthesized, preventing them from being used as templates for more RF synthesis and helping package them into phage heads. PolIII, DNA polymerase III.

fork encounters damage in the template DNA and collapses, the recombination functions can promote the formation of a recombinational intermediate, and the Pri proteins and DnaT are required to load DnaB helicase at such a structure and to reinitiate replication. This is analogous to the recombination-dependent replication (RDR) of phage T4 (see below). The φX174 phage enlists the host functions to initiate its own RF replication, presumably by having an origin of replication that mimics the recombinational intermediate.

Synthesis of more RFs and virion DNA. The steps in replication are probably similar in all single-stranded DNA phages but are best understood with M13. Once the first RF of M13 is synthesized, more RFs are made by semiconservative replication in a process similar to plasmid replication (see chapter 4). This process requires phage proteins that are synthesized from the first RF. The two strands of the RF are replicated separately and by very different mechanisms (Figure 7.15). The plus strand is replicated by rolling-circle replication from a different origin by a process similar to the replication of DNA during transfer of a plasmid by conjugation (see chapter 5). First, a nick is made in the RF, at the origin of plus strand synthesis, by a specific endonuclease, the product of gene II in M13. A host protein called Rep, a helicase, helps unwind the DNA at the nick. The gene II protein remains attached to the 5′ end of the DNA at the nick, and host DNA polymerase III, with its accessory proteins, extends the 3′ end to synthesize more plus strand, displacing the old plus strand. The gene II protein bound to the 5′ end of the old, displaced strand then reseals the ends of the displaced strand by a transesterification reaction in which the phosphate attached to its tyrosine is passed back to the free 3′ end of the old strand. This recyclizes the old plus strand, which can then serve again as the template for minus strand synthesis to create another RF. Such transesterification reactions use little energy and are also used to recyclize plasmids after conjugation (see chapter 5) and in site-specific recombinases and some transposases (see chapter 8).

This process of accumulating RFs continues until the product of phage gene V begins to accumulate. This protein coats the single-stranded plus strand of DNA, probably with the help of the attached gene II product, and prevents the synthesis of more RF. The single-stranded viral DNA is then encapsulated in the head, and the cell is lysed (for icosahedral phages, such as φX174), or it is transferred to the assembling viral particle in the membrane and leaked out of the cell (for filamentous phages, such as f1 and M13), as described above.

PHAGES WITH CIRCULAR DOUBLE-STRANDED DNA GENOMES

The replication of λ DNA has also been studied extensively as an example of replication of a circular double-stranded DNA molecule and has been one of the major model systems for understanding replication in general. The λ DNA is linear in the phage head but cyclizes after it enters the cell through pairing between its cohesive ends, or **cos sites** (Figure 7.9). These ends are single-stranded and complementary to each other for 12 bases and so can join by complementary base pairing, which makes them cohesive, or "sticky." Once the cohesive ends are paired, DNA ligase can join the two ends to form covalently closed circular λ DNA molecules. These circular DNA molecules can replicate in their entirety because there is always DNA upstream to serve as a primer.

Circle-to-Circle, or θ, Replication of λ DNA

Once circular λ DNA molecules have formed in the cell, they replicate by a mechanism similar to the θ replication described for the chromosome in chapter 1 and for plasmids in chapter 4. Replication initiates at the *ori* site in gene *O* (see Figure 7.9) and proceeds in both directions, with both leading- and lagging-strand synthesis in the replication fork (Figure 7.16). When the two replication forks meet somewhere on the other side of the circle, the two daughter molecules separate, and each can serve as the template to make another circular DNA molecule.

Rolling-Circle Replication of λ DNA

After a few circular λ DNA molecules have accumulated in the cell by θ replication, the rolling-circle type of replication ensues. The initiation of λ rolling-circle replication is similar to that of M13 in that one strand of the circular DNA is cut and the free 3′ end serves as a primer to initiate the synthesis of a new strand of DNA that displaces the old strand. DNA complementary to the displaced strand is also synthesized to make a new double-stranded DNA. The λ process differs in that the displaced individual single-stranded molecules are not released when replication around the circle is completed. Rather, the circle keeps rolling, giving rise to long tandem repeats of individual λ DNA molecules linked end to end, called **concatemers** (see Figure 7.16). The production of end-to-end concatemers by rolling-circle replication can be compared to sheets of paper towel on a roll that are torn at the perforations to form individual sheets.

In the final step, the long concatemers are cut at the *cos* sites into λ genome-length pieces as they are packaged into phage heads. Phage λ can package DNA only from concatemers, and at least two λ genomes must be linked end to end in a concatemer, because the packaging system in the λ head recognizes one *cos* site on the concatemeric DNA and takes up DNA until it arrives at the next *cos* site, which it cleaves to complete the packaging. The dependence on concatemeric DNA for λ DNA packaging was important in the discovery of *chi* (χ) sites and their role in RecBCD recombination, as we discuss in chapter 9.

Genetic Requirements for λ DNA Replication

The products of only two λ genes, O and *P*, are required for λ DNA replication. As Figure 7.16 illustrates, both of these proteins are required for loading the host DnaB helicase and initiating DNA replication at the *ori* site. The O protein is thought to bend and open the DNA at this site by binding to repeated sequences, similar to the mechanism by which the DnaA protein bends and opens the DNA at the *oriC* site to initiate chromosome replication (see chapter 1). The P protein binds to the O protein and to the replicative helicase DnaB of the host replication machinery, thus commandeering DnaB for λ DNA replication; it therefore acts like DnaC. Appropriately, the P of the P gene product stands for "pirate."

RNA synthesis by the host RNA polymerase must also occur in the *ori* region for λ replication to initiate. It normally initiates at the p_R promoter and may be required

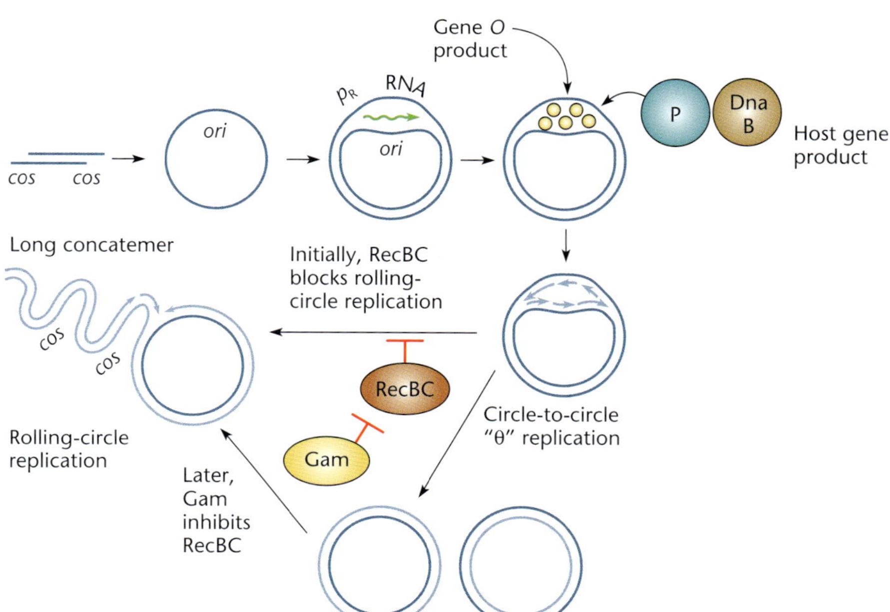

Figure 7.16 Overview of replication of phage λ. See text for details.

to separate the DNA strands at the origin and/or serve as a primer for rightward replication.

A third λ protein, the product of the *gam* gene, is required for the shift to the rolling-circle type of replication, albeit indirectly. The RecBCD nuclease (an enzyme that facilitates recombination of *E. coli*; see chapter 9) inhibits the switch to rolling-circle replication, but Gam inhibits the RecBCD nuclease. Therefore, a *gam* mutant of λ is restricted to the θ mode of replication, and concatemers can form from individual circular λ DNA molecules only by recombination. Because λ requires concatemers for packaging, a *gam* mutant of λ cannot multiply in the absence of a functional recombination system, either its own or the RecBCD pathway of its host.

REPLICATION AND DNA PACKAGING OF DOUBLE-STRANDED DNA LINEAR GENOMES

Whereas many of the smaller double-stranded DNA phages have circular DNA, most of the larger phages have linear DNA, at least as it is packaged into the phage head. This creates special problems for replicating the DNA genome and packaging it into phage heads, as we discuss next.

The Primer Problem

Many of the strategies used by phages with linear DNA genomes to replicate their genomes can be understood in terms of their need to solve the primer problem. The primer problem exists because none of the known DNA polymerases can initiate the synthesis of new DNA and instead can only add to a preexisting primer (see chapters 1 and 4). This property of DNA polymerases presents special problems for replication to the ends of any linear DNA, whether phage or human. When the DNA replication fork gets to the end of a linear template molecule, there is no way to make a DNA primer from which to synthesize the last fragment on the lagging strand, because this would require the DNA polymerase to initiate synthesis of a new strand. RNA polymerases can initiate the synthesis of a new strand of RNA, but even if the primers for the 5′ ends are synthesized as RNA, once the RNA primer is removed, there is no DNA upstream to serve as a primer to replace the RNA primer. Because of this priming problem, a linear DNA molecule would get smaller each time it replicated until essential genes were lost. Eukaryotic chromosomes are linear but have throwaway segments of DNA at their ends, called telomeres, which are dispensable and are enzymatically synthesized from the shortened end by telomerases after each replication, using an RNA in the telomerase enzyme as the template. However, no known phages with linear genomes have a similar kind of telomere, and they must solve this primer problem in other ways. One mechanism used by some bacteria and bacteriophages with linear genomes involves terminal proteins, which can bind the end of the linear genome and initiate a DNA primer (Box 7.3). Still others have hairpin ends, like some linear plasmids, which, in the prophage state as a plasmid, allow them to replicate around the ends and form dimeric circles that can then be resolved by the protein TelN (see "Phage N15" below). Various mechanisms used by linear genomes are discussed in chapter 4 (Box 4.1). A common mechanism used by phages is to circularize the phage genome after infection. Some phages with linear genomes, for example, λ, have complementary *cos* sites on their ends that allow them to form circles after infection, as described above. Other phages have repeated sequences at the ends of their genomic DNA called terminal redundancies, which allow them to form concatemers, as discussed below. In this section, we discuss in more detail how some phages with linear DNA solve the primer problem.

Phage T7: Linear DNA That Forms Concatemers

Some phages, including T7, never cyclize their DNA but form concatemers composed of individual genome-length DNAs linked end to end. The phage DNA can then be cut out of these concatemers so that no information is lost when the phage DNA is packaged.

As shown in Figure 7.17, T7 DNA replication begins at a unique *ori* site and proceeds toward both ends of the molecule, leaving the 3′ ends single stranded because there is no way to prime replication at these ends. However, because T7 has the same sequence at both ends, called a **terminal redundancy**, these single strands are complementary to each other and so can pair, forming a concatemer with the genomes linked end to end. Consequently, the information missing as a result of incomplete replication of the 3′ ends is provided by the complete information at the 5′ end of the other daughter DNA molecule. Individual molecules are then cut out of the concatemers at the unique *pac* sites at the ends of the T7 DNA and packaged into phage heads. It is not clear how the terminal redundancies are re-formed in the mature phage DNAs, but it probably is done by making staggered breaks and then filling them in with DNA polymerase.

Genetic requirements of T7 DNA replication. In contrast to single-stranded DNA phages, which encode only two of their own replication proteins (the products of genes II and V in M13) and otherwise depend on the host replication machinery, and λ, which uses only two phage-encoded proteins (the products of genes *O* and *P*), T7 encodes many of its own replication functions, including DNA polymerase, DNA ligase, DNA helicase, and primase. The phage T7 RNA polymerase is also required to synthesize the initial primer for phage T7 DNA synthesis. In addition to these proteins, the phage encodes a DNA endonuclease and an exonuclease that degrade host DNA to mononucleotides, thereby providing

BOX 7.3

Protein Priming

Some viruses with linear genomes, including the animal adenoviruses and the *Bacillus subtilis* phage φ29, have solved the primer problem by using proteins, rather than RNA, to prime their DNA replication. In the φ29 phage head, a protein is covalently attached to the 5′ end of the virus DNA for protection. After infection, another copy of this protein binds to the attached protein to form a dimer, and this second copy serves as a primer for the synthesis of a new DNA strand, with the first nucleotide attached to a specific serine on the protein. By using the protein as a primer, the virus DNA does not need to form circles or concatemers. The phage DNA polymerase uses this protein to prime its replication by an unusual "sliding back" mechanism. First, the DNA polymerase adds a dAMP to the hydroxyl group of a specific serine on the protein. The incorporation of this dAMP is directed by a T in the template DNA. However, the T used as a template is the second nucleotide from the 3′ end of the template, not the first deoxynucleotide. The DNA polymerase then backs up to recapture the information in the 3′ deoxynucleotide before replication continues. The extra dAMP that is added to the end is removed later. After replication, the protein is transferred to the 5′ end of the newly replicated strand, and replication continues. In this way, no information is lost during replication.

Phage φ29 has also been an important model system with which to study phage maturation, because the phage DNA can be packaged very efficiently into phage heads in a test tube. Interestingly, its packaging motor contains an RNA, six copies of which are joined to form a ring around the entering DNA. This RNA ring binds ATP and might somehow rotate to help pump the DNA into the phage head, although this is difficult to prove, and RNA motors to pump DNA have not yet been identified in other systems.

References

Escarmís C, Guirao D, Salas M. 1989. Replication of recombinant *φ29* DNA molecules in *Bacillus subtilis* protoplasts. *Virology* **169**:152–160.

Lee T-J, Guo P. 2006. Interaction of gp16 with pRNA and DNA for genome packaging by the motor of bacterial virus *φ29*. *J Mol Biol* **356**:589–599.

Meijer WJ, Horcajadas JA, Salas M. 2001. *φ29* family of phages. *Microbiol Mol Biol Rev* **65**:261–287.

another source of deoxynucleotides for phage DNA replication. Analogous host enzymes can substitute for some of these T7-encoded gene products, so they are not absolutely required for T7 DNA replication. For example, the T7 DNA ligase is not required, because the host ligase can act in its place. Nevertheless, T7 DNA replication is a remarkably simple process that requires fewer gene products overall than the replication of bacterial chromosomes and of many other large DNA phages.

Phage T4: Another Phage That Forms Concatemers

Phage T4 also has linear DNA that never cyclizes. It forms concatemers like T7, except that it forms them by recombination rather than by pairing between complementary single-stranded ends. However, T4 and T7 differ greatly in how the DNA replicates and is packaged. Also, befitting its larger size, T4 has many more gene products involved in replication than does T7. As many as 30 T4 gene products participate in replication, as shown in the map in Figure 7.6A. In fact, one of the advantages of studying replication with T4 is that it encodes many of its own replication proteins rather than just using those of its host. It encodes its own DNA polymerase, sliding clamp, clamp-loading proteins, primase, replicative helicase, DNA ligase, etc. Table 7.4 lists some of these functions and their functional equivalents in *E. coli* and in eukaryotes, where known.

Overview of T4 phage DNA replication and packaging. Phage T4 replication occurs in two stages, which are illustrated in Figure 7.18. In the first stage, T4 DNA replicates from a number of well-defined origins around the DNA. This type of replication is analogous to the replication of bacterial chromosomes and leads to the accumulation of single-genome-length molecules. However, these daughter molecules have single-stranded 3′ ends because of the inability of T4 DNA polymerase to completely replicate the end, due to the primer problem discussed above. No information is lost, however, because the sequences at the ends of T4 DNA are repeated, i.e., the DNAs are terminally redundant, as they are in T7. Somewhat later in the infection process, this type of replication ceases and an entirely new type of replication ensues. The single-stranded repeated sequences at the terminally redundant ends of the genome-length molecules can invade the same sequence in other daughter DNAs, forming D loops, which prime replication to form large branched concatemers. This

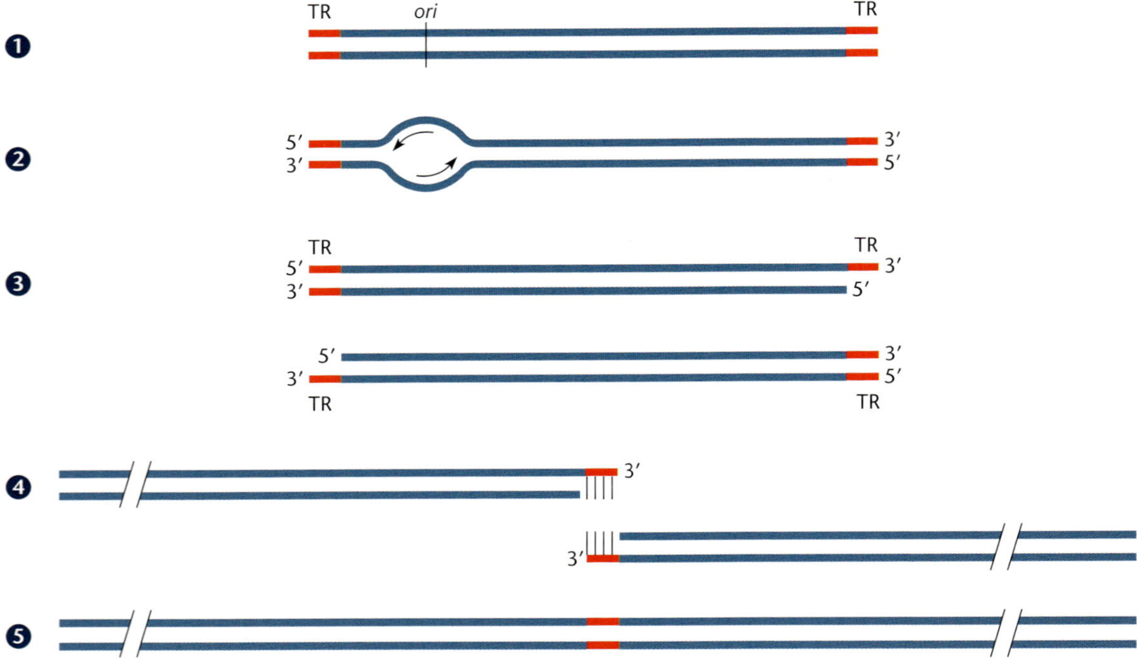

Figure 7.17 Replication of phage T7 DNA. Replication is initiated bidirectionally at the origin (*ori*). The replicated DNAs could pair at their terminally repeated ends (TR) to give long concatemers, as shown.

Table 7.4 T4 gene products involved in replication and their homologs in *E. coli* and eukaryotes

T4 gene product	*E. coli* function	Eukaryotic function[a]
Origin-specific replication		
gp43	PolIII α and ε	DNA Pol α, β, γ
gp45	β sliding clamp	PCNA
gp44, gp62	Clamp loader (γ complex)	RFC
gp41	Replicative helicase (DnaB)	Mcm complex
gp61	Primase (DnaG)	Pol α
gp39, gp52, gp60	GyrAB, topoisomerase IV	Topoisomerase II
gp30	DNA ligase	LIG1, LIG2, LIG3
Rnh	RNase H	RNH1
Recombination-dependent replication		
UvsW	RecG, RuvAB	—[b]
UvsX	RecA	Rad51 (yeast)
UvsY	RecFOR	Rad52 (yeast)
gp46, gp47	RecBCD, SbcCD	Rad50, MreII (yeast)
gp32	SSB	RPA
gp59	PriABC, DnaT, DnaC	—
gp49	RuvC	GEN1 (human), YEN1 (yeast)

[a]PCNA, proliferating-cell nuclear antigen; RPA, replication protein A; RFC, replication factor C; MreII, double-strand break processing.

[b]—, not identified.

recombination-dependent replication (RDR) is analogous to the "replication restart" process discussed in chapter 9 and, while first discovered with this phage, is now known to occur in all organisms. The two stages of T4 DNA replication are discussed in more detail below.

Like T7, T4 DNA is packaged into the phage head from concatemers. Periodic cycles of RDR lead to the synthesis of very large branched concatemers from which individual genome-length DNAs are packaged into phage heads (see Black, Suggested Reading). A pentameric (five-sided) ring around the entrance to the phage head forms a "motor" that sucks DNA into the head at a rate of about 2,000 bp per second, cleaving ATP for energy to drive the motor. When the head is full, the DNA is cut so that the head has somewhat more than a genome length of DNA, accounting for the terminal redundancy, as illustrated in Figure 7.19. This type of packaging is called **headful packaging**, because a length of DNA sufficient to fill the head is taken up. It is like sucking a very long strand of spaghetti into your mouth until your mouth is full and then biting it off. If the packaging motor encounters one of the many branches in the concatemer, the branch is cut by gp49, a Holliday junction resolvase (see chapter 9), which is also involved in recombination.

Because packaging cuts off more than a genome length of DNA, the genomes of the phages that come out of each infected cell are **cyclic permutations** of each other. The

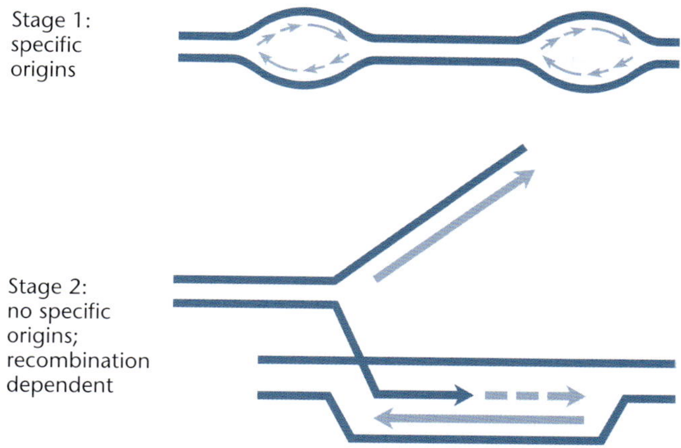

Figure 7.18 Initiation of replication of phage T4 DNA. In stage 1, replication initiates at specific origins using RNA primers. In stage 2, recombinational intermediates furnish the primers for initiation.

mathematical definition of a cyclic permutation is a permutation that shifts all elements of a set by a fixed offset, with the elements shifted off the end inserted back at the beginning. This explains why the genetic map of T4 is circular (Figure 7.6A) even though T4 DNA itself never forms a circle. The way in which different modes of replication and packaging give rise to the various different genetic linkage maps of phages is discussed below (see "Genetic Analysis of Phages" below).

Stage 1 replication of the T4 genome from defined origins. As mentioned above, the first stage of T4 DNA replication, from defined origins, is analogous to chromosome replication from unique origins in cellular organisms, bacteria, and eukaryotes. Consistent with its large size, and also like chromosome replication in eukaryotes, T4 has a number of defined origins around the chromosome that can be used to initiate replication. This is unlike most bacteria and other phages, including T7, which usually use only one unique origin to initiate replication. However, T4 most often uses only one of these origins, *oriE*, to replicate each chromosome.

The first step in initiating replication from a T4 origin is to synthesize RNA primers on the origin, using the host RNA polymerase. These primer RNAs also sometimes double as mRNAs for the synthesis of middle-gene proteins and are made from middle-gene-type promoters. These promoters are first turned on a few minutes after infection and require an RNA polymerase whose σ^{70} has been remodeled by binding the MotA and AsiA proteins (see above). In their role as primers, these short RNAs invade the double-stranded DNA at the origin and hybridize to the strand of the DNA to which they are complementary, displacing the other strand to create a structure called an R loop. The invading RNA can then prime the leading strand of DNA replication from the origin from a replication apparatus made up mostly of T4-encoded proteins. Some of the T4 gene products involved in T4 DNA replication are listed in Table 7.4, and a schematic of how they are involved in replication is shown in Figure 7.8. The gp41 replicative helicase, which plays the role of DnaB in uninfected *E. coli*, is loaded on the DNA. The gp59 helicase-loading protein seems to assist in this but is not absolutely required. Other helicase-loading proteins may assist at particular promoters. Once the gp41 helicase is loaded, replication is under way. The gp41 helicase is associated with the lagging-strand primase (gp61), which primes replication of the lagging strand, similar to the role of the primase DnaG in *E. coli* DNA replication. After replication is under way, the DNA polymerase (gp43) is held on the DNA by a sliding clamp (gp45), which has been loaded on the DNA by the clamp loader, comprising proteins gp44 and gp62. Once synthesis of Okazaki fragments is complete, a T4-encoded DNA ligase (gp30) joins the pieces together, although the host DNA ligase can substitute to some extent for this function. The T4 DNA replication fork moves along the DNA like a trombone,

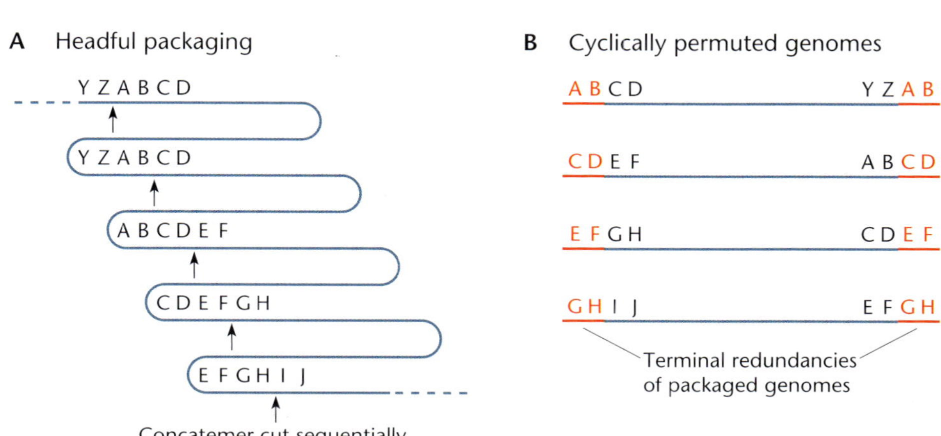

Figure 7.19 T4 DNA headful packaging. Packaging of DNA longer than a single genome equivalent gives rise to repeated terminally redundant ends and cyclically permuted genomes. **(A)** Headfuls of DNA are packaged sequentially from concatemers. The vertical arrows indicate the sites of cleavage during packaging. **(B)** Each packaged genome is a different cyclic permutation with different terminal redundancies (red). Adapted from Kreuzer KN, Michel B, *in* Higgins NP (ed), *The Bacterial Chromosome* (ASM Press, Washington, DC, 2005).

much like the *E. coli* replication fork. In fact, the best evidence for the trombone model has come from this phage. After one or very few copies of the DNA have been made from defined origins, a helicase called UvsW can displace these R loops, suppressing origin-specific replication in favor of the next mode of replication that is dependent on recombination.

Stage 2 recombination-dependent DNA replication of the T4 genome. In the second stage of replication, the leading strand of T4 replication is primed by recombination intermediates rather than by primer RNAs synthesized by RNA polymerase (see Mosig, Suggested Reading). This T4 DNA replication stage is similar to replication restart in uninfected cells (see chapter 10) and uses similar recombination functions, some of which are listed in Table 7.4.

The first step in recombination-dependent replication is the invasion of a complementary double-stranded DNA by a single-stranded 3′ end to form a three-stranded **D loop** (Figure 7.18). This invading single-stranded 3′ end is created during an earlier round of origin-specific replication by the inability of the DNA polymerase to replicate to the end of the molecule and could be extended by the actions of exonucleases, such as gp46 and gp47, on the ends of the molecule. In this respect, gp46 and gp47 are analogous to the RecBCD proteins of *E. coli* (Table 7.4) (see chapter 9). If the cell has been infected by more than one T4 phage particle, the complementary sequence that the free 3′ end invades could be anywhere in the DNA of a coinfecting phage, since T4 DNAs are cyclically permuted (see above). However, if the cell has been infected by a single phage, the newly replicated phage DNAs have the same sequences at their ends, and the single-strand invasion would be into the terminal redundancy of the other daughter DNA. This pairing of the invading strand with the complementary strand in the invaded DNA is promoted by the T4 *uvsX* gene product, which is analogous in function to the RecA protein, the *E. coli* function that promotes single-strand invasion in uninfected cells (see chapter 9). Normally, single-stranded T4 DNA is coated with the T4-encoded single-stranded-DNA-binding protein gp32, and the UvsX protein might need the help of another T4 protein, UvsY, to displace the gp32 protein, much as RecFOR proteins displace the *E. coli* single-stranded-DNA-binding protein SSB in uninfected cells (see chapter 9). Once the D loop has formed, the replicative helicase gp41 is loaded on the DNA by gp59 in a process that is similar to the loading on of the DnaB helicase by the Pri proteins in uninfected cells during replication restarts. The invading 3′ end can then serve as a primer for new leading-strand DNA replication, and the primase gp62 can be loaded on the displaced strand for lagging-strand replication. Later, when replication from defined origins has ceased and there are no double-stranded ends, recombination is initiated by other proteins, gp17 and gp49, that break the DNA and create ends for single-strand invasion, as described above, and the process continues. Repeated rounds of strand invasion and replication lead to very long branched concatemers, which are then packaged into phage heads. This simplified version of RDR ignores some known features of RDR, such as its bidirectionality, as well as the roles of some of the helicases and exonucleases, among other enzymes, known to be required for the process.

RNA PHAGES

Although most studies of phages have focused on those with DNA genomes, phages with RNA genomes also exist. Examples are Qβ, MS2, R17, f2, and φ6. The *E. coli* RNA phages Qβ, MS2, R17, and f2 are similar to each other. All have a single-stranded RNA genome that encodes only four proteins: a replicase, two head proteins, and a single protein lysin (see below). Immediately after the RNA genome enters the cell, it serves as an mRNA and is translated to generate the replicase. This enzyme replicates the RNA, first making complementary minus strands (that do not serve as mRNAs) and then using the minus strands as a template to synthesize more plus strands (that serve as both mRNAs and genomes for newly produced virions). The phage genomic RNA must serve as an mRNA to synthesize the replicase, because no such replicase enzyme exists in *E. coli* with the ability to synthesize RNA from an RNA template. Interestingly, the phage Qβ replicase has four subunits, only one of which is encoded by the phage (see Takeshita and Tomita, Suggested Reading). The other three are components of the host translational machinery: two of the elongation factors for translation, EF-Tu and EF-Ts, and a ribosomal protein, S1. The Qβ replicase also needs another host-encoded protein, Hfq (host factor for Qβ) to replicate the phage genome. This protein, first discovered in the Qβ replicase, is used by the host cell for regulation of gene expression by small RNAs and is discussed in chapter 12. Because the genomes of these RNA phages also function as mRNAs, they have served as a convenient source of a single species of mRNA in studies of translation, including the first sequences of translation initiation regions.

Another RNA phage, φ6, was isolated from the bean pathogen *Pseudomonas syringae* subsp. *phaseolicola*. The RNA genome of this phage is double stranded and exists in three segments in the phage capsid, much like the reoviruses of mammals. These three segments are called the S, M, and L segments, for small, medium, and large. Also like animal viruses, the phage is surrounded by membrane material derived from the host cell, i.e., an envelope, and the phage enters its host cells in much the same way that animal viruses enter their hosts. However, unlike most animal viruses, φ6 is released by lysis.

The replication, transcription, and translation of the double-stranded RNA of a virus such as φ6 present special problems. Not only must the phage replicate its double-stranded RNA, but it also must transcribe it into single-stranded mRNA, since double-stranded RNA cannot be translated. The uninfected host cell contains neither of the enzymes required for these functions, which therefore must be virus encoded and packaged into the phage head so that they enter the cell with the RNA. Otherwise, neither the transcriptase nor the replicase could be made.

Another interesting question with this phage is how three separate RNAs are encapsidated in the phage head. The three segments of the genome are transcribed into plus strand transcripts that are then packaged into a preformed head in sequential order, with the S segment first, the M segment second, and the L segment last. Packaging is initiated at 200-nucleotide *pac* sequences that seem to share little sequence or structural similarity among the three RNAs (see Qiao et al., Suggested Reading). Only after the single plus strand enters the head is its minus strand complement synthesized to make the double-stranded genomic RNA. This expands the prohead into its mature spherical form. The entire process can be performed in the test tube using phage proteins synthesized in *E. coli*, which spontaneously assemble into heads that are able to take up the separate genomic RNAs.

Because the RNA replicases used by phages with RNA genomes have no need for primers, the genomes of RNA viruses can be linear without repeated ends. As we might expect, RNA viruses have higher spontaneous mutation rates during replication, probably because their RNA replicases have no editing functions. This is also true for the replication machineries used for RNA viruses that infect mammalian cells, such as HIV and influenza virus. The high rate of mutation in these viruses plays an important role in their evolution and in the difficulties in generating effective vaccines to protect against them. In contrast, the replicases of coronaviruses, including SARS, MERS, and SARS-CoV-2 (the causative agent of the COVID-19 pandemic), do contain an editing function, which decreases the mutation rate.

Host Cell Lysis

Once the phage genome has replicated, the phage particles assemble and are released from the infected cell to infect other cells. As mentioned above, some phages, such as the small filamentous DNA phages, including M13 and f1, assemble in the membrane and then leak out of the cell, using a modified type II secretion system (see chapter 2). Type II secretion systems were mentioned in chapter 6 because of their relationship to some competence systems for DNA transformation and because they are used to secrete the pilin proteins of type IV pili, and these phages have adapted them to their own use by encoding their own secretin proteins that form a channel through the outer membrane through which the assembled phage can pass.

Phages that leak out of the infected cell without killing it could be considered to cause a chronic infection because the host is not killed and continues to produce phage over multiple rounds of host cell replication. However, most phages cause an acute infection, in that a cell infected by one of these phages suddenly breaks open or lyses, releasing the phage. In addition to their scientific interest, phage lysis systems have potential as antibacterial agents as antibiotics lose their effectiveness due to resistance.

SINGLE-PROTEIN LYSIS

In some of the simplest phages, including the spherical single-stranded DNA phage φX174 and the RNA phage Qβ, a single protein causes host cell lysis (see Chamakura and Young, Suggested Reading). These lysis proteins inhibit enzymes that make precursors of the bacterial cell wall by binding directly to them. When the cell begins to divide, the resultant shortage of cell wall precursors causes the cell to lyse, releasing the phage. While the system adopted by these small phages requires very few phage-encoded gene products, it is inefficient, because the number of phage that are produced from an infected cell depends on the stage in the cell cycle at which the cell was infected. If the cell is infected by a phage shortly before it is to divide, very few phage are produced.

TIMED LYSIS

The larger double-stranded DNA phages have systems to more precisely time the lysis of the host cell so that an optimum number of phage can be produced. If lysis occurs too early, phage will not yet have assembled, and the infection will not be productive. If it occurs too late, the infection process will take longer than necessary, and the phage may miss the opportunity to infect new host cells.

The timing of cell lysis by double-stranded DNA phages is a complicated process that often requires many proteins (see Young, Suggested Reading). Phages of enteric bacteria that are released from the cell by lysis usually encode at least five such proteins. One of these is the endolysin that breaks bonds in the cell wall. The cell wall is a rigid sheet around the cell that gives the cell its structure and integrity and is composed of a sheet of peptidylglycan that must be broken to release phage from the cell. Other proteins required for lysis, at least of enteric bacteria, are the **spanins**, which are composed of a small lipoprotein and a protein that scans the periplasm. Spanins may play a role in separating the outer membrane from the cell wall, which may be required for release of the phage under some conditions. In some systems, the spanins are required for lysis only in the presence of divalent cations, such as Mg^{2+}.

Timing of Lysis by Holins

The endolysins are made early in infection but are not active until later in infection, when they are suddenly activated by the action of **holins**. These are not enzymes but, rather, are membrane proteins that form holes, or pores, in the inner membrane. Still other proteins involved in lysis are **antiholins**, which inhibit holins until it is time for lysis. Understanding the timing of lysis depends upon learning how holins and antiholins work and how they are activated at the right time for lysis.

Some holins form holes in the inner membrane that allow the lysozyme to pass through the inner membrane to reach and degrade the peptidylglycan cell wall in the periplasmic space—hence the name holin. Holins of this type have a number of transmembrane domains (TMDs) and are known to form pores in the membrane large enough for the endolysin to pass through. Others, termed pinholins, form only very small channels that destroy the membrane potential. In fact, it is known that in many systems, destroying the membrane potential causes endolysins to lyse the cell even before lysis normally occurs.

T4 Phage Lysis

The lysis system of the T4 phage may be one of one of those in which the holin channel allows passage of the endolysin; it is illustrated in Figure 7.20A. In this phage, the endolysin, the product of gene *e*, is made in the active form in the cytoplasm relatively early in infection but cannot degrade the cell wall until later, because it cannot pass through the inner membrane. The holin, the product of the gene *t*, gp*t*, is inserted in the membrane but is inactive. At the time for lysis, the *t* gene product becomes active and allows endolysin through the inner membrane, which, with the help of two spanins, causes lysis of the cell. T4 has an antiholin, the product of the *r1* gene, gp*r1*, but this protein is very unstable, so very little accumulates during phage infection. Its job seems to be to delay lysis if there are other T4 phage bound to the outside of the cell, in a process called lysis inhibition. These external phage can inject their DNA into the cell, but the DNA becomes trapped in the periplasmic space rather than entering the cytoplasm. Ordinarily, the puncturing of the membrane by the T4 syringe would activate the holin and lyse the cell. Somehow, this periplasmic DNA stabilizes the gp*r1* antiholin, which then prevents activation of the gp*t* holin by binding to its periplasmic domain, thereby inhibiting cell lysis. Lysis inhibition may ensure that the cells do not lyse when there are already plenty of T4 phage around that have been released from other infected cells nearby, signaling that it may be better to stay in the infected cell and keep multiplying.

λ Phage Lysis

The action of the antiholin of the phage λ in timing lysis is shown in Figure 7.20B (see Young, Suggested Reading). Two copies of the holin, which is named S105 because it is a product of the *S* gene and has 105 amino acids, dimerize to form a pore through which the endolysin can pass. To form a pore, the holin, which has three TMDs, must traverse the membrane three times, so that its N terminus ends up in the periplasm and its C terminus ends up in the cytoplasm, as shown. The antiholin (S107) is 2 amino acids longer than the holin because, even though it is translated from the same open reading frame, its translation starts two codons upstream of the normal AUG codon, adding an extra methionine and lysine at its N terminus. These two extra amino acids prevent the first TMD from entering the membrane, probably because the positively charged lysine cannot enter the membrane against the membrane potential. With only two TMDs in the membrane, S107 cannot form a holin, but it can still dimerize with an S105 holin. However, these hybrid structures are not active as holins and do not form pores. At the appropriate time for lysis, the membrane loses its potential by an unknown mechanism. This may allow the first TMD of the antiholin to enter the membrane and participate with the shorter form in the formation of active pores, which allows access of the endolysin, the product of gene *R*, to the cell wall and, with the help of the spanins Rz and Rz1, causes lysis.

Activation of SAR Endolysins

Some endolysins have long hydrophobic TMDs at their N termini that function as signal sequences and direct the protein through the inner membrane by the *sec* system (see chapter 2). The endolysin of phage P1 is one example. Since these endolysins are transported by the *sec* system, they have no need for porin pores to get through the membrane. Nevertheless, they do also use holins. Such endolysins, called SAR (signal anchor release) endolysins, are inactive after transport to the periplasm because the N-terminal TMD is not cleaved off as they pass through the SecYEG channel and remains anchored in the inner membrane. At the correct time for lysis, the holin releases the trapped domain, activating the endolysin. How the holin does this is not clear, but it might destroy the membrane potential, perhaps by forming pores in the membrane, and without a membrane potential, the N terminus of the endolysin is released. SAR endolysins are often paired with holins called pinholins, which make very small pores that may destroy the membrane potential by allowing the passage of protons but are not large enough to allow the passage of proteins, including the endolysin.

How release of the signal sequence from the membrane activates a SAR endolysin varies but often involves the shuffling, or **isomerization**, of disulfide bonds between cysteine amino acids in the protein. Trapping of the N terminus of the endolysin in the membrane allows an inhibitory disulfide bond to form elsewhere in the protein.

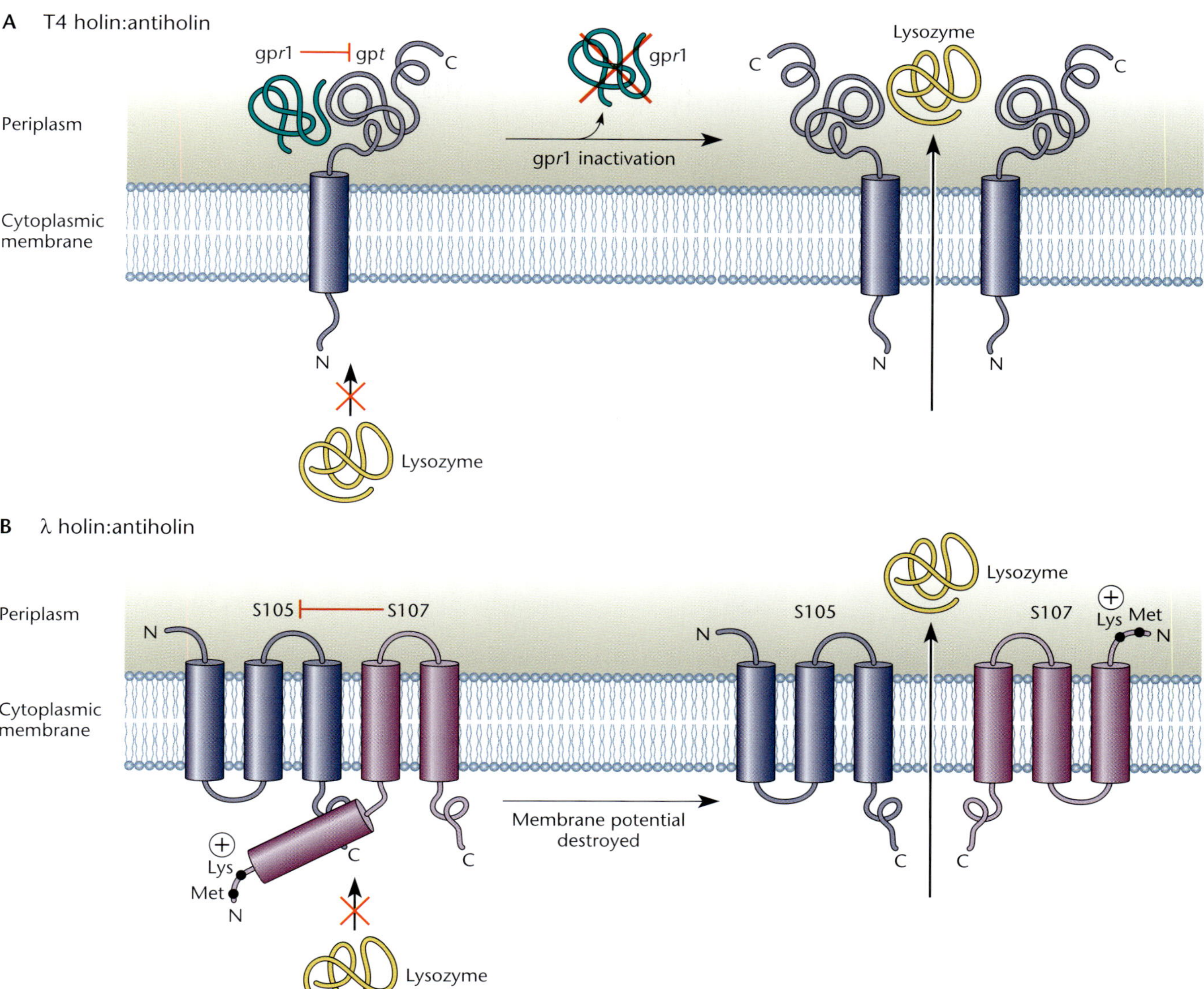

Figure 7.20 Timing of phage lysis by activation of holins. The antiholin keeps the holin inactive until the time of lysis. The holin then becomes active, forming a pore that allows the lysozyme (blue) to traverse the membrane and then to degrade the cell wall and lyse the cell. **(A)** Model for lysis inhibition in T4 phage. If phage are bound to the outside of the cell, the antiholin (gp*r1*) binds to the periplasmic domain of the holin (gp*t*), which prevents it from forming pores. Much later, the antiholin becomes inactive, allowing the holin to form a pore in the membrane through which the lysozyme can pass. **(B)** Model for timing of lysis by λ phage. The antiholin (S107) and holin (S105) differ only in that the antiholin has an extra 2 amino acids, Lys-Met, at its N terminus due to a second upstream translation initiation region (TIR). These extra amino acids prevent the first transmembrane domain (TMD) of S107 from entering the membrane because they add a positive charge that cannot move against the membrane potential. This makes the S107 antiholin inactive as a holin, but it still binds to the S105 holin, interfering with its ability to form pores. At the time of lysis, the membrane loses its potential, allowing TMD1 of S107 to enter the membrane and participate with S105 in the formation of pores through which lysozyme can pass to reach the cell wall. Adapted from Young R, *in* Waldor MK, Friedman DI, Adhya SL (ed), *Phages: Their Role in Bacterial Pathogenesis and Biotechnology* (ASM Press, Washington DC, 2005).

In some cases, one of the cysteines in the inhibitory disulfide bond is in the active center of the endolysin, making it unavailable for the reaction; in other cases, the critical cysteine is not in the active center, but the disulfide bond influences the activity of the endolysin, either by "caging" the active center by surrounding it with other sequences so it is not available or by some other mechanism that affects the conformation of the enzyme. In all these cases, a cysteine residue in the SAR domain trapped in the inner membrane is unpaired because it is not in the periplasm, where disulfide bonds form (see chapter 2). After the SAR domain is released by the holin, Dbs oxidoreductases in the periplasm direct this cysteine to pair with another cysteine to form a new disulfide bond, and formation of this new bond frees the critical cysteine from the inhibitory disulfide bond and activates the enzyme. Not all SAR endolysins are activated by disulfide bond isomerization. In some cases, release of the trapped N-terminal domain is sufficient to cause proper folding and activation of the endolysin.

Lysogenic Development

During lytic development, the phage infects a cell and multiplies, producing more phage that can then infect other cells. However, this is not the only lifestyle phages can employ. Some phages are able to maintain a stable relationship with the host cell in which they neither multiply nor are lost from the cell. Such a phage is called a lysogen-forming or **temperate phage**. Even temperate phages are capable of lytic development, and it is possible that phage currently known as lytic-only may be temperate in an as yet unknown host. In the lysogenic state, the phage DNA either is integrated into the host chromosome or replicates as a plasmid. The phage DNA in the lysogenic state is called a **prophage**, and the bacterium harboring a prophage is a **lysogen** for that phage. Thus, a bacterium harboring the P2 prophage would be a P2 lysogen.

In a lysogen, the prophage acts like any good parasite and does not place too great a burden on its host. The prophage DNA is mostly quiescent; most of the prophage genes that are expressed are those required to maintain the lysogenic state, and most of the other genes are turned off. Often, the only indication that the host cell carries a prophage is that the cell is immune to superinfection by another phage of the same type. The prophage state usually continues until the host cell suffers potentially lethal damage to its chromosomal DNA or is infected by another phage of an unrelated type. Then, like a rat leaving a sinking ship, the phage can be **induced** and enter lytic development, producing more phage. The released phage can then either infect other cells and develop lytically to produce more phage or lysogenize the new bacterial cell.

In addition to prophages that can be induced under some circumstances to make more phage, bacteria carry many **defective prophages** that no longer can form infective phage because they have lost some of their essential genes. These DNA elements are suspected of being defective prophages rather than normal parts of the chromosome because they are not common to all the strains of a species of bacterium and they often carry genes that resemble those from known phage. Defective prophages may be important in evolution because they eventually lose their identity and become part of the normal chromosome; this is one of the ways bacteria can acquire new genes that can evolve to perform new functions. Furthermore, a number of interesting systems have been identified in which the prophage plays an important role in regulation of bacterial genetic systems (see Feiner et al., Suggested Reading).

The λ System

Phage λ is the lysogen-forming phage that has been studied most extensively and the one to which all others are compared. Although lysogeny was suspected as early as the 1920s, the first convincing demonstration that bacterial cells could carry phages in a quiescent state was made with λ in about 1950 (see Lwoff, Suggested Reading). In this experiment, apparently uninfected *E. coli* cells could be made to produce λ virions after irradiation with UV light. Since then, phage λ has played a central role in the development of the science of molecular genetics (see Gottesman, Suggested Reading).

Figure 7.21 gives an overview of the two life cycles of which λ is capable and the fate of the DNA in each cycle, while Figure 7.22 gives a more detailed map of the phage genome for reference. As described above, phage λ DNA is linear in the phage head, and the map shows how it exists in the head. Immediately after the DNA is injected into the cell to initiate the infection, the DNA cyclizes, that is, forms a circular molecule, by pairing between the *cos* sites at the ends (Figure 7.9). This brings the lysis genes (*S* and *R*) and the head and tail genes (*A* to *J*) of the phage together and allows them all to be transcribed from the late promoter p_R', as discussed above. This circular DNA can then either integrate into the host chromosome (lysogenic cycle) or replicate and be packaged into phage heads to form more phage (lytic cycle). Which decision is made depends on the physiological state of the cell, as we discuss below. Later, the integrated DNA in a lysogen can also be excised, replicate, and form more phage (induction).

THE LYSIS-LYSOGENY DECISION

Figure 7.23 illustrates the process of forming a lysogen after λ infection, how the *c*I, *c*II, and *c*III gene products are involved, and the central role of the CII protein.

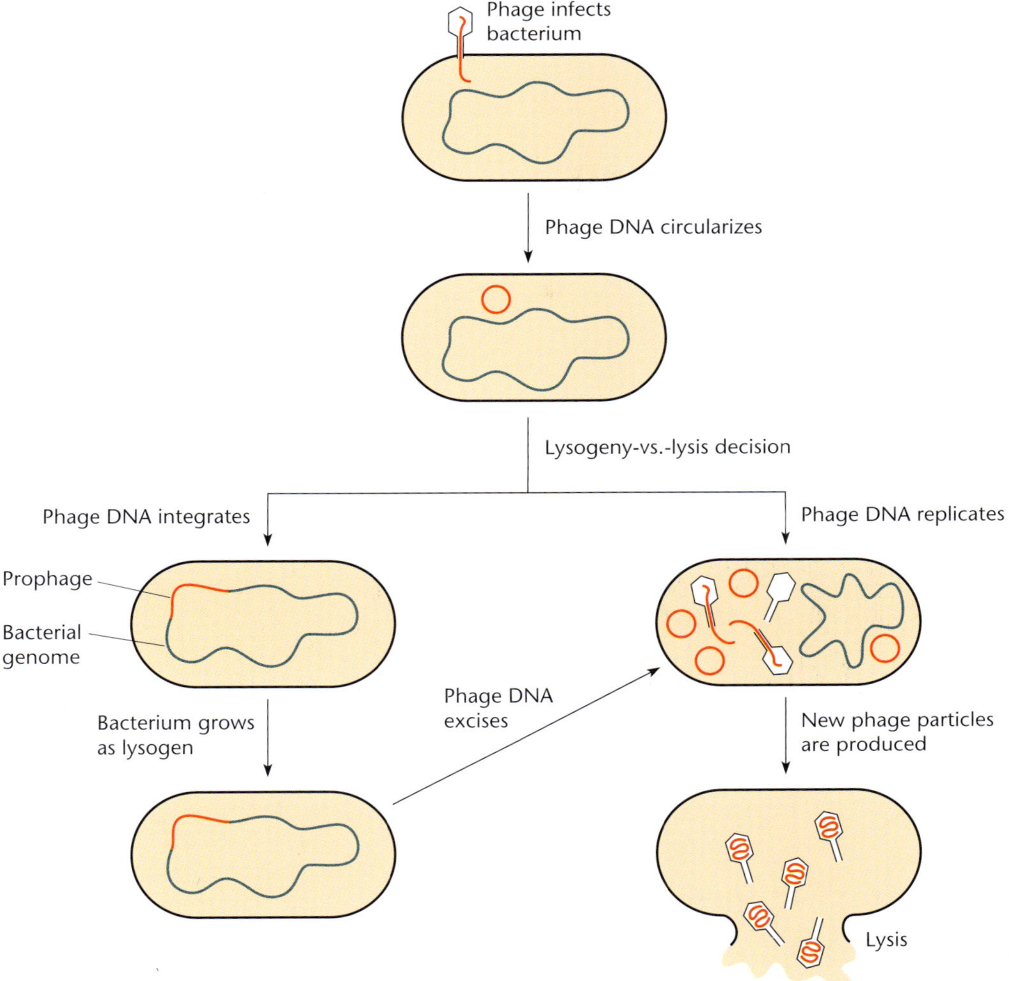

Figure 7.21 Overview of the fate of λ DNA in the lytic and lysogenic pathways. Phage DNA is shown in red; bacterial DNA is shown in blue.

After λ infects a cell, the decision about whether the phage enters the lytic cycle and makes more phage or forms a lysogen depends on the outcome of a competition between the product of the *c*II gene, which acts to form lysogens, and the products of the *cro* gene and of genes in the lytic cycle that replicate the DNA and make more phage particles. Which pathway wins most often depends on the conditions of infection. At a low multiplicity of infection (MOI; see "Multiplicity of Infection" below), the lytic cycle usually wins, and in as many as 99% of the infected cells, the λ DNA replicates and more phage are produced. However, for reasons we explain below, at a high MOI, the CII protein wins more often, and as many as 50% of the infected cells can form lysogens. The richness of the medium also plays a role. One reason is that cells that are growing very fast in rich medium have more RNase III (see chapter 2) than if they are growing more slowly, and more RNase III means more N protein, which favors lytic development (see "λ N-Mediated Antitermination" above; Court et al., Suggested Reading). The reason they have more N protein is that the leftward transcript from p_L contains the *nutL* site just upstream of the translational initiation region (TIR) for the *N* gene (Figure 7.10). The Nus factors and N protein bound to the *nutL* site inhibit N translation from the nearby TIR for the *N* gene, so less N protein is made. There is a cleavage site for RNase III between the *nutL* site and the TIR for gene *N*, and the higher concentrations of RNase III when the cells are growing rapidly cleave the mRNA at a hairpin between the *nutL* site and the TIR for gene *N*, separating the *nutL* site from the *N* gene so more N is made. Note that it seems to make strategic sense to enter the lytic cycle when the MOI is low or when the cells are growing rapidly. If there are many more uninfected cells than phage, as there are if the MOI is low, it offers the opportunity to make many more phage, particularly if

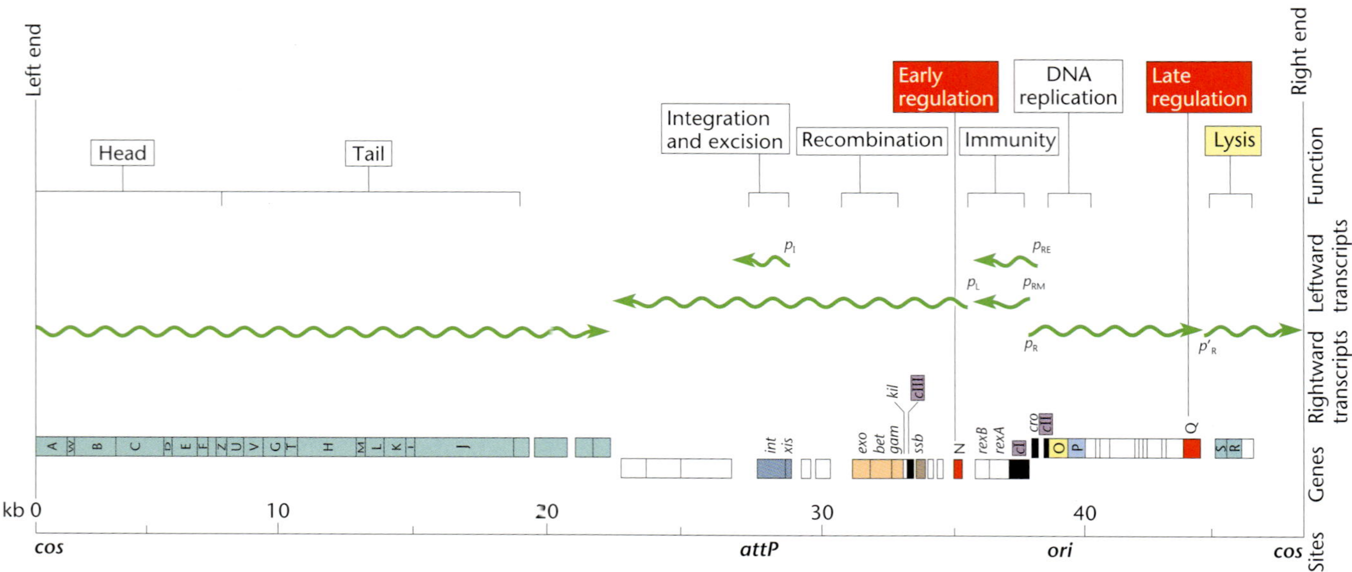

Figure 7.22 Genetic map of phage λ. The locations of key genes and transcripts are shown. The GenBank accession number for the λ genome is NC_001416.

the growth conditions are good. If most of the cells are already infected, as they are if the MOI is high, it seems to make more sense to form a lysogen and wait until conditions improve, especially if growth conditions are poor at the time.

Role of the CII Protein, a Transcriptional Activator

Once the CII protein is made from the p_R promoter, it promotes lysogeny by activating the RNA polymerase to begin transcribing at three promoters, which are otherwise inactive (Figure 7.23). Proteins that enable RNA polymerase to begin transcription at certain promoters are called transcriptional activators (see chapter 11). One of the promoters activated by CII is p_{RE}, which allows transcription of the *c*I gene. The product of this gene, the CI protein, is a transcriptional repressor (see chapter 11) that prevents transcription from the promoters p_R and p_L, which direct transcription of many of the remaining λ genes. The CI repressor is discussed in more detail below. Another promoter activated by the CII protein, p_I, allows transcription of the integrase (*int*) gene and synthesis of the Int protein. The Int enzyme integrates the λ DNA into the bacterial DNA to form the lysogen. Even though the *int* gene is also transcribed from the p_L promoter, the Int protein cannot be made from this transcript, for reasons that are explained below. The third promoter activated by CII is the *pAQ* promoter, which directs the synthesis of an antisense RNA in the Q gene, thereby delaying the synthesis of the Q protein, which would kill the cell by directing late gene expression (see "λ Q-mediated Antitermination" above).

Role of the CIII Protein: A Protease Inhibitor

The role of the *c*III gene product in lysogeny is less direct. CIII inhibits a host-encoded protease, HflB, the product of the *ftsH* gene, that degrades CII. Therefore, in the absence of CIII, the CII protein is rapidly degraded and no lysogens form. Incidentally, this explains why more lysogens form when cells are infected at higher MOIs (see above). More CIII protein is made at higher MOIs, and more CIII protein inhibits more of the HflB protease, leading to accumulation of more CII activator protein and therefore more lysogeny.

PHAGE λ INTEGRATION

As discussed above, as soon as λ DNA enters the cell, it forms a circle by pairing between the complementary single-stranded *cos* sequences at its ends. The Int protein can then promote the integration of the circular λ DNA into the chromosome, as illustrated in Figure 7.24A. Int is a site-specific recombinase, a member of the tyrosine (Y) family of recombinases (see chapter 8), that specifically promotes recombination between the attachment sequence (called *attP*, for attachment phage) on the phage DNA and a site on the bacterial DNA (called *attB*, for attachment bacteria) that lies between the galactose (*gal*) and biotin (*bio*) operons in the chromosome of *E. coli*. This is a nonessential region of the *E. coli* chromosome, so integration of λ at this site causes no observable phenotype. Some phages do integrate their DNA into essential genes of the bacterium, which requires special adaptations. Because the Int-promoted recombination does not occur at the ends of λ DNA but, rather, at the internal *attP*

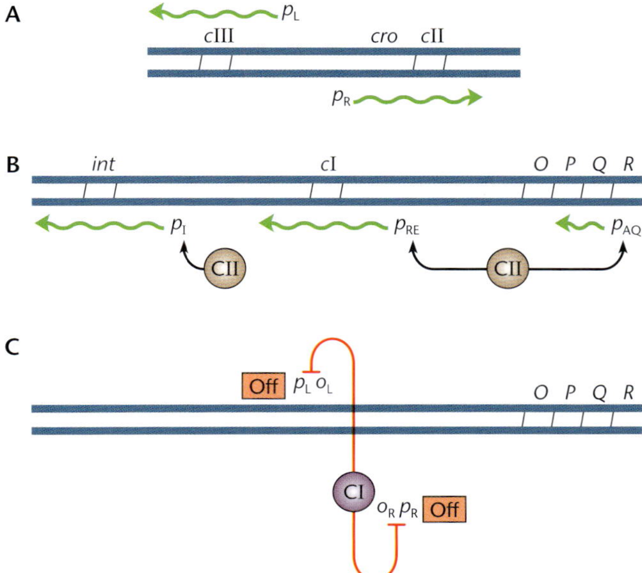

Figure 7.23 Formation of lysogens after λ infection. **(A)** The cII and cIII genes are transcribed from promoters p_R and p_L, respectively. **(B)** CII activates transcription from promoters p_{RE} and p_I, leading to the synthesis of CI repressor and the integrase Int, respectively. It also activates transcription from p_{AQ}, probably inhibiting synthesis of Q. **(C)** The repressor shuts off transcription from p_L and p_R by binding to o_R and o_L. Finally, the Int protein integrates the λ DNA into the chromosome.

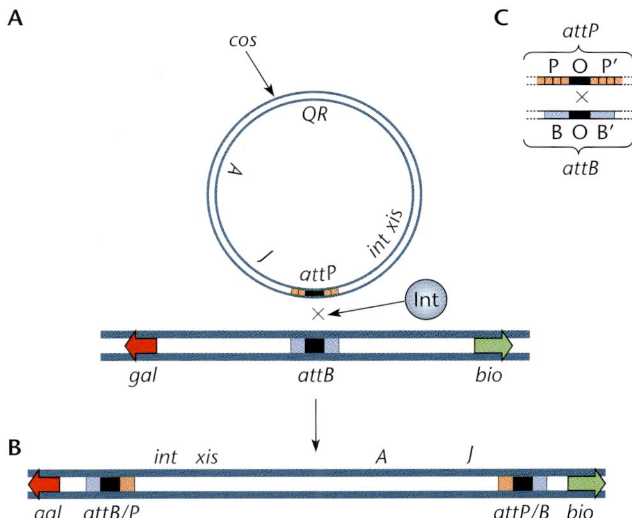

Figure 7.24 Integration of λ DNA into the chromosome of *E. coli.* **(A)** The Int protein promotes recombination between the *attP* sequence (orange) in the λ DNA and the *attB* sequence (light blue) in the chromosome. Panel **(C)** shows the region in more detail, with *attP* sequence POP′ and *attB* sequence BOB′. The common core O sequence of the two sites is shown in black. **(B)** Gene order in the prophage. The locations of the *int*, *xis*, *A*, and *J* genes in the prophage are shown (refer to the λ map in Figure 7.22). The *E. coli gal* (red) and *bio* (green) operons are on either side of the prophage DNA in the chromosome.

site, the prophage map is a cyclic permutation of the map of DNA found in the phage head with different genes at the ends. In the phage head, the λ DNA has the *A* gene at one end and the *R* gene at the other end (Figure 7.22). In contrast, in the prophage, the *int* gene is at one end and the *J* gene is at the other (Figure 7.24B). It was the difference between the phage genetic map and the prophage map that led to this model of integration, which is sometimes called the **Campbell model** after Alan Campbell, who first proposed it.

The recombination promoted by Int is called **site-specific recombination** because it occurs between specific sites, one on the bacterial DNA and another on the phage DNA. This site-specific recombination is a type of nonhomologous recombination, because the sequences of the phage and bacterial *att* sites are mostly dissimilar. They have a common core sequence, O, of only 15 bp, flanked by two dissimilar sequences, B and B′ in *attB* and P and P′ in *attP* (Figure 7.24C). Because the region of homology is so short, this recombination would not occur without the Int protein, which recognizes both *attP* and *attB* and promotes recombination between them. The mechanism of action of Y recombinases and other types of site-specific recombinases is discussed in chapter 8.

MAINTENANCE OF λ LYSOGENY

After the lysogen has formed, the *c*I repressor gene is one of the few λ genes to be transcribed, and its CI repressor product maintains the lysogenic state by binding to two DNA operator sites, o_R and o_L, that are close to promoters p_R and p_L, respectively. Binding of CI to the operator sites prevents transcription from p_R and p_L and therefore transcription of most of the genes of λ, either directly or indirectly. Repressors and operators are discussed further in chapter 11. In the prophage state, the *c*I gene is transcribed from a different promoter, the p_{RM} promoter (for repression maintenance), which is immediately upstream of the *c*I gene, rather than from the p_{RE} promoter used immediately after infection (Figures 7.23 and 7.25). The p_{RM} promoter is not used immediately after infection because its activation requires the CI repressor, as we discuss below.

THE CI REPRESSOR

It is important to know the structure of the CI repressor to understand how it can regulate its own synthesis, as well as that of the other λ genes. Each CI polypeptide consists of two separable parts, or **domains**. It is generally drawn as a dumbbell to illustrate this domain structure (Figure 7.25). One of the domains of the CI polypeptide

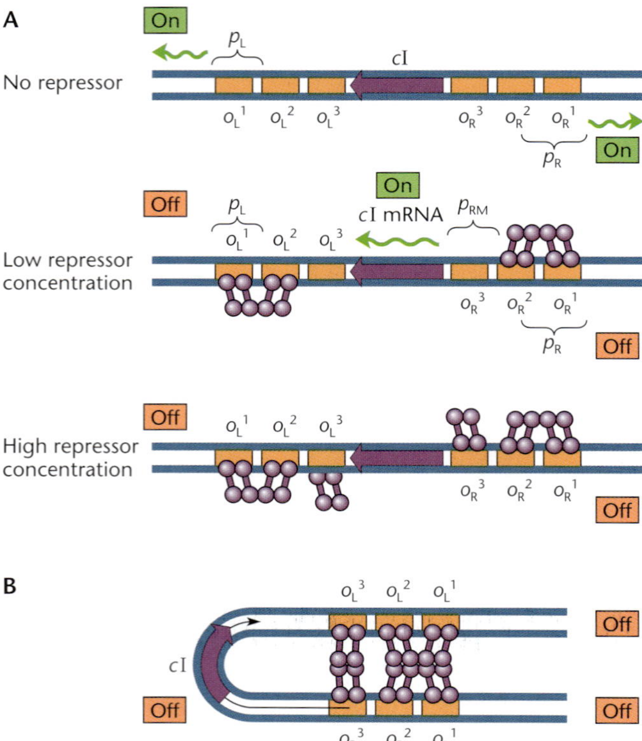

Figure 7.25 Regulation of repressor synthesis in the lysogenic state. The dumbbell shape represents the two domains of the CI repressor. **(A)** The dimeric repressor, shown as two dumbbells, binds cooperatively to o_R^1 and o_R^2 (and o_L^1 and o_L^2), repressing transcription from p_R (and p_L) and activating transcription from p_{RM}. At higher repressor concentrations, it also binds to o_R^3 and o_L^3, repressing transcription from p_{RM}. **(B)** Still higher concentrations cause the formation of tetramers that bend the DNA, further repressing transcription from p_{RM}. The relative affinities of the repressor for the sites is as follows: $o_R^1 > o_R^2 > o_R^3$ and $o_L^1 > o_L^2 > o_L^3$.

promotes the formation of dimers and tetramers by binding to the corresponding domain on other CI polypeptides, shown as two dumbbells binding to each other. For the CI repressor to function, two of these dumbbells must bind to each other through their dimerization domains to form a dimer made up of two copies of the polypeptide. In turn, a tetramer forms when two of these dimer dumbbells bind to each other through their tetramerization regions in the same domain. At very low concentrations of CI polypeptide, the dimers do not form and the repressor is not active. At higher concentrations, the dimers form and the repressor is active. The other domain on each polypeptide specifically binds to an operator sequence on the DNA or, more specifically, to subsequences within these operators, as we discuss below.

Regulation of CI Repressor Synthesis

It is important to maintain the level of the CI repressor in the cell within normal limits, even after a lysogen has formed. If the amount of CI repressor drops below a certain level, transcription of the lytic genes begins and the prophage is induced to produce phage. However, if the amount of repressor increases beyond optimal levels, cellular energy is wasted in making excess repressor, and it might be too difficult to induce the prophage should the need arise. The mechanism of regulation of repressor synthesis in lysogenic cells is well understood and has served as a model for gene regulation in other systems (see Ptashne, Suggested Reading). Figure 7.25 illustrates the regulation of CI repressor synthesis.

The repressor dimer regulates its own synthesis, as well as that of other λ gene products, by binding to the operator sequences, one on the right of the *c*I gene, called o_R, and the other on the left of the *c*I gene, called o_L. The CI protein can be either a repressor or an activator of transcription, depending on how many copies of it are bound to these operators. We first discuss binding to the o_R operator on the right of the repressor gene, since it is the most important operator for regulating repressor synthesis, although both operators have the same structure. The operator o_R can be divided into three repressor-binding sites, o_R^1, o_R^2, and o_R^3. If the concentration of repressor is low, only the o_R^1 site is occupied, which is sufficient to repress transcription from the promoter p_R, which overlaps the operator site (Figure 7.25A). This prevents transcription of the replication genes *O* and *P*, which are immediately downstream of p_R (Figure 7.10). However, as the repressor concentration increases, eventually o_R^2 also becomes occupied, because tetramers can form between the dimers bound at o_R^1 and o_R^2 (Figure 7.25). The formation of a tetramer stabilizes the binding of a CI dimer to o_R^2. This is called **cooperative binding** and is seen with many DNA-binding proteins. Repressor bound at o_R^2 is required to activate transcription from the promoter p_{RM}, which is why p_{RM} is active for transcription of the repressor gene only when there is some repressor in the cell. Since o_R and o_L have the same structure, o_L^1 and o_L^2 are occupied by repressor at the same repressor concentrations as the corresponding sites in o_R. Repressor dimer bound at o_L^1 blocks leftward transcription, and repressor dimers bound at o_L^1 and o_L^2 can bind to those at o_R^1 and o_R^2 to form an octamer, which bends the DNA and helps stabilize their binding and more completely represses transcription from p_L and p_R (Figure 7.25B). Other repressors are known to regulate transcription from a promoter by bending the DNA at the promoter, and we discuss some examples of this in chapter 11.

At very high concentrations, CI repressor dimers also cooperatively bind to o_L^3 and o_R^3, where they also form a tetramer and stabilize each other's binding. The repressor dimer bound at o_R^3 interferes with binding of RNA polymerase to the p_{RM} promoter, which prevents the synthesis of more repressor (Figure 7.25B). This complex regulation allows the prophage to synthesize more repressor when there is less in the cell, and vice versa, so that the cell maintains the levels of repressor within a narrow range. This is a form of **transcriptional autoregulation.** The synthesis of repressor also responds quickly to perturbations in the cell, which explains why λ lysogens are very stable and usually release phage only under unusual circumstances.

IMMUNITY TO SUPERINFECTION

The CI repressor in the cell of a lysogen prevents not only the transcription of the other prophage genes by binding to operators o_L and o_R, but also the transcription of the genes of any other λ phage that infect the lysogenic cell by binding to the operators of that phage. Thus, bacteria lysogenic for λ are immune to λ superinfection. However, λ lysogens can still be lytically infected by any relative of λ phage that has different operator sequences to which the λ CI repressor cannot bind. Any two phages that differ in their operator sequences are said to be **heteroimmune**. If they have the same operator sequences, they can inhibit each other's transcription and are said to be **homoimmune**, no matter how different they are in their other genes. Many hybrid phages have been made between λ and some of its relatives that have λ genes but the repressor gene and operators of another phage, for example, phage 434. Such a phage is then called λ i434 to indicate that it is λ phage with the immunity of phage 434, so it will multiply in a λ lysogen but not a 434 lysogen.

THE Cro PROTEIN

Another protein involved in the regulation of repressor synthesis is Cro. This is one of the first proteins made after infection and inhibits the synthesis of repressor and thereby helps commit the phage to the lytic cycle. Cro does this by binding to the operator sequences, although in reverse order of repressor binding, as illustrated in Figure 7.26. Cro binds first to the $o_R^{\,3}$ site, occluding the p_{RM} promoter, and then to the $o_R^{\,2}$ site, thereby preventing the CI repressor from binding to the $o_R^{\,2}$ site and activating its own synthesis from the p_{RM} promoter. This prevents interference from the lysogenic cycle once the phage is in the lytic cycle.

INDUCTION OF λ

Phage λ remains in the prophage state unless the host cell DNA is severely damaged by irradiation or some types of chemicals that cause extensive damage to the DNA. The prophage is then induced to go through its lytic cycle. Figure 7.27 outlines the process of λ induction. When the cell attempts to repair the damage to its DNA, short pieces of single-stranded DNA accumulate and bind to the RecA protein of the host, which reflects the role RecA

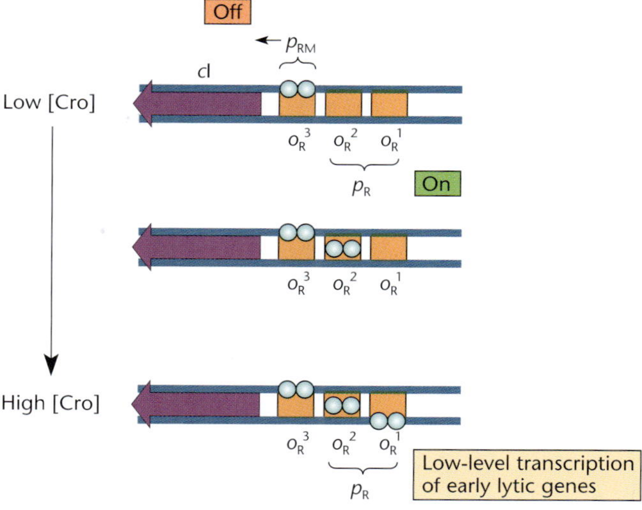

Figure 7.26 Cro prevents repressor binding and synthesis by binding to the operator sites in reverse order from the repressor. By binding to $o_R^{\,3}$, Cro prevents repressor activation of transcription from p_{RM} while allowing transcription from p_R. Eventually, Cro accumulates to the point where it binds to $o_R^{\,1}$ and $o_R^{\,2}$ and blocks transcription of early RNA.

plays in recombination (see chapter 9). This binding activates the **coprotease** activity of RecA, and the activated RecA protein with single-stranded DNA attached then binds to the λ CI repressor, causing the repressor to cleave itself (**autocleavage**). The autocleavage separates the DNA-binding domain in the CI polypeptide from the domain involved in dimer formation. Without the dimerization domain, the CI repressor can no longer form dimers, and the DNA-binding domains can no longer bind to the operators that repress λ lytic gene transcription. As the repressors drop off the operators, transcription initiates from the promoters p_R and p_L, and the lytic cycle begins.

The induction of λ by cleavage of its repressor is yet another case in which studies with λ led to the discovery of more universal phenomena, in this case, SOS induction. The LexA repressor of *E. coli* and other bacteria also cleaves itself when bound by RecA and single-stranded DNA that accumulates following DNA damage. Autocleavage of LexA leads to the induction of many repair genes under its control, including those encoding RecA protein and mutagenic bypass polymerases. Thus, λ relies on the preexisting SOS system of *E. coli* to induce itself after DNA damage. Other lysogenic phages, including the *Vibrio cholerae* phage that makes cholera toxin (CT), "skip the middleman" and use LexA directly to regulate their transcription, so they can be induced following DNA damage by autocleavage of LexA (see below). We discuss

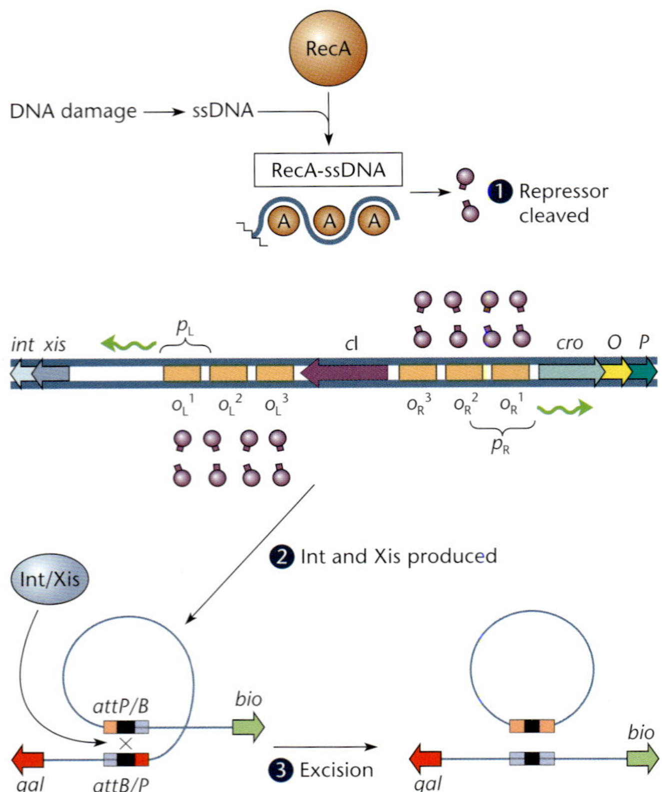

Figure 7.27 Induction of λ. Accumulation of single-stranded DNA (ssDNA) due to damage to the DNA results in activation of the RecA protein, which promotes the autocleavage of the CI repressor protein, separating the dimerization domain of the protein from the DNA-binding domain so that the repressor can no longer form dimers and bind to DNA. Transcription of *int-xis* and *cro*, *O*, and *P* ensues, and the phage DNA is excised from the chromosome and replicates.

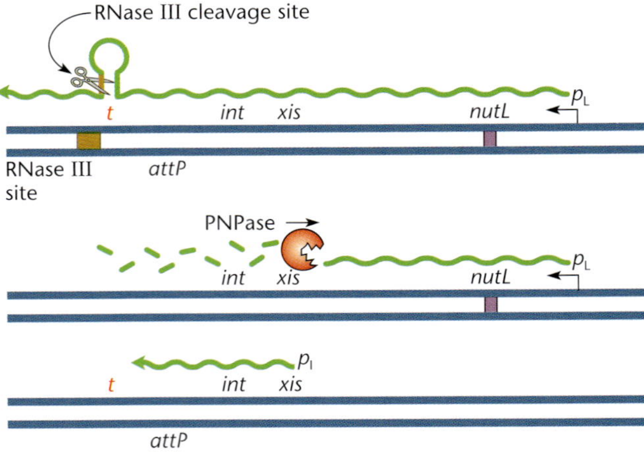

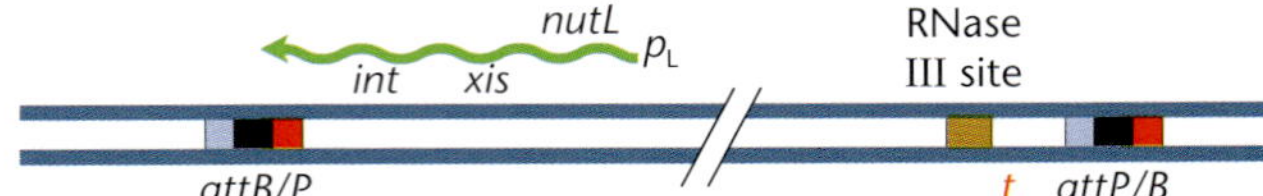

Figure 7.28 Retroregulation. **(A)** After infection, the *xis* and *int* genes cannot be expressed from the p_L promoter, even though transcription extends that far. Because of N, transcription from p_L continues past the terminator, *t*, which results in inclusion of an RNase III cleavage site in the mRNA on the other side of *attP*. The RNA is cleaved and digested back toward *xis* and *int* by PNPase, removing the coding sequences for Int and Xis from the RNA. Xis also cannot be synthesized from transcripts that initiate at p_I because the p_I promoter is within the *xis* gene, and Int cannot excise the DNA in the absence of Xis. The RNA from p_I is stable, however, because the transcript does not contain the *nutL* site and so does not continue through *t* to include the downstream RNase III cleavage site. The gold region indicates the location of the coding information for the RNase III site in the DNA, but RNase III cleaves only the mRNA transcript into which it has been transcribed. **(B)** When the prophage is first induced, however, and before it is excised, the sequence encoding the RNase III cleavage site is separated from the *xis-int* coding sequence, so that the RNA made from p_L is stable and both Int and Xis are made. **(A)** Early after infection; **(B)** early after induction.

SOS induction and mutagenic bypass polymerases in chapters 1 and 10.

EXCISION OF λ

Once the repressor is out of the way, transcription from p_L and p_R can begin in earnest. Some of the genes transcribed from p_L are required to excise the λ DNA from the chromosome. Excision requires site-specific recombination between hybrid *attP-attB* sequences that exist at the junctions between the prophage DNA and the chromosomal DNA (Figure 7.27). These hybrid sequences are different from either *attB* or *attP* and contain sequences from both; therefore, Int alone is not capable of recognizing and promoting recombination between them to excise the prophage. Another protein, called excisase (Xis), is also required to allow the Int protein to recognize these hybrid sequences. Proteins such as Xis are often called **directionality factors** because they promote recombination in only one direction, excision, unlike Int, which is required for both excision and integration. Accordingly, unlike after infection, when only Int is synthesized, after induction, both the Int and Xis proteins are synthesized (see below). In fact, it is necessary that only Int be made after infection, because if both Int and Xis were synthesized after infection, the λ prophage would be excised as soon as it integrates, and lysogens could not form.

Why only Int is made after infection but both Int and Xis are made after induction was a puzzle (Figure 7.28). After infection, Int is synthesized from the p_I promoter, which is in the *xis* gene, so Xis obviously cannot be synthesized from the p_I promoter; however, this does not explain why both Xis and Int are not also synthesized using transcripts that originate from the p_L promoter. After antitermination by N, the transcription from p_L should continue through both *int* and *xis*. To achieve this differential gene expression, λ takes advantage of the fact that the ends of the λ prophage are different from the ends of the DNA in the phage head. After infection, both the *xis* and *int* genes are transcribed from the p_L promoter, but because of N-mediated antitermination, transcription proceeds past *xis* and *int* into sequences on the other side of *attP*, as shown in Figure 7.28. One of these sequences on the other side of *attP* forms a hairpin that is cleaved by RNase III if it appears in an RNA (see chapter 2). Cleavage by RNase III at this site causes the RNA to be degraded by a 3′-5′ exonuclease, PNPase, before either Int or Xis can be translated from the RNA. In contrast, the RNA made from the p_I promoter is stable and can be translated to yield Int because the RNA does not include *nutL*, so it is not antiterminated and instead terminates at a termination site (*t*) before it transcribes the RNase III cleavage site. This system has been termed "retroregulation" because the RNase III cleavage site affects expression of a gene that is located upstream of it.

The situation is very different immediately after induction, when the λ prophage DNA is still integrated into the chromosome. Integration of the phage DNA separates the RNase III cleavage site from promoter p_I, and it is now at the other end of the integrated prophage DNA (Figure 7.28B). Now, the mRNA transcribed from p_L is stable, and both Int and Xis can be synthesized from this mRNA. Regulation through separating genes from promoters during integration is also used by other elements, including integrating conjugative elements (see chapter 5).

While the Int and Xis proteins are excising and cyclizing λ DNA from the chromosome, the *O* and *P* genes are being transcribed from p_R. These proteins promote replication of the excised λ DNA. The late genes, now joined to the early genes by cyclization of the DNA, are transcribed, and phage particles are produced. In about 1 h after the cellular DNA is damaged, depending on the medium and the temperature, the cell lyses, releasing about 100 phage into the medium from a cell that, an hour before, showed few signs of harboring the phage.

SUMMARY OF THE λ LYTIC AND LYSOGENIC CYCLES

The λ lytic and lysogenic cycles, including the induction of prophages, involve so many interacting pathways that it is worth reviewing them. Figure 7.29 and Table 7.5 review the competition for entry into the lysogenic cycle versus the lytic cycle. After infection, when there is no CI repressor in the cell, the *N* and *cro* genes are transcribed. The *N* gene product acts as an antiterminator and allows the transcription of many genes, including *c*II and *c*III, as well as the genes encoding the DNA replication proteins O and P, while the Cro protein inhibits repressor synthesis during the lytic cycle.

Whether the phage enters the lytic or the lysogenic cycle depends on the fate of the CII activator protein, which is determined by the MOI and the metabolic state of the infected cell. At higher MOIs, more CIII protein is made to inhibit the FtsH protease and more CII protein accumulates. More CII means more CI repressor is then made from the p_{RE} promoter, more Int is made from p_I, and less Q is made because of interference from an antisense RNA made from p_{AQ}. Consequently, more of the cells survive and enter the lysogenic phase. In the meantime, the Cro protein is trying to tip the phage in the direction of lytic development. By binding to $o_R{}^3$ it occludes the p_{RM} promoter, so less repressor is made and more of the cells enter the lytic phase. At high enough concentrations, Cro also binds to $o_R{}^2$ and $o_R{}^1$, displacing the CI repressor and allowing transcription of the O and P genes. If the cells are growing in rich medium, more RNase III is made, which separates the *N* gene TIR from the *nutL* site on the same mRNA, allowing more N protein to be made, and more N protein also favors lytic development. Eventually, as replication proceeds, there is too much λ DNA for the repressor to bind to all of it, and transcription of *O* and *P* increases further, followed by yet more DNA replication. The Q protein accumulates, which allows transcription of the head, tail, and lysis genes.

The induction of a prophage is normally caused by damage to the DNA, which causes the CI repressor to cleave itself. Transcription then begins from promoters p_L and p_R, and N antiterminates transcription from these promoters. Now, however, because the prophage DNA ends at the *attP* site, the leftward transcript is stable, and both Int and Xis are made by transcription from the p_L promoter. Int and Xis together promote recombination between the hybrid *attP-attB* sites at the ends of the prophage, and the phage DNA excises as a circle from the chromosome. Lytic development then proceeds much as after infection, although the Cro protein may not play as much of a role because the cell is already committed to lytic development.

Other Lysogenic Systems

Phage λ has served as the paradigm for analysis of phage lysogeny. However, other phage systems provide insight into alternative pathways for entering into a dormant state. A few examples of these are described below.

PHAGE P2

Phage P2 is another lysogen-forming phage of *E. coli*. The phage DNA is linear in the phage head but has cohesive ends, like λ, which cause the DNA to cyclize immediately

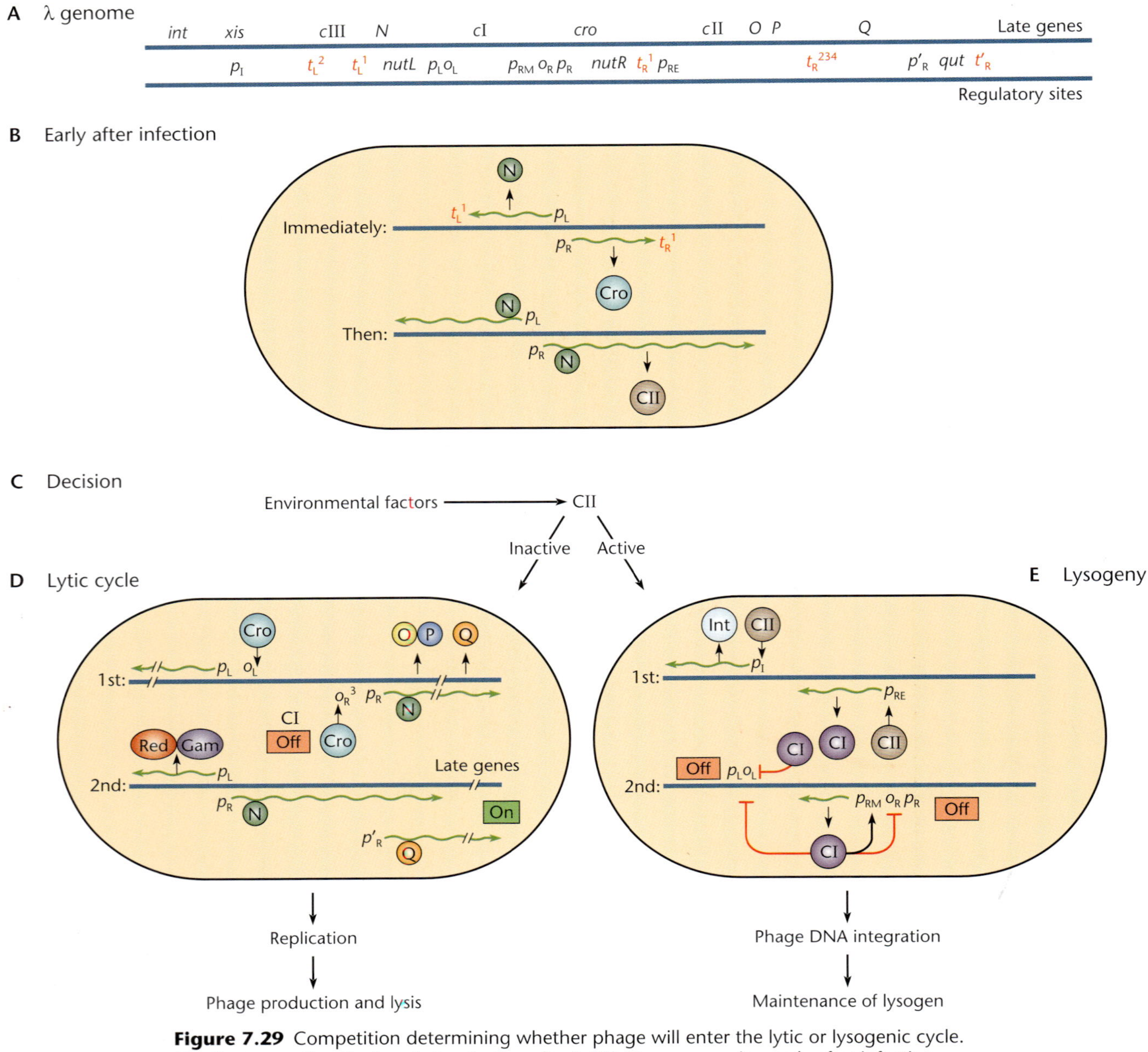

Figure 7.29 Competition determining whether phage will enter the lytic or lysogenic cycle. **(A)** Key genes (top line) and sites (bottom line). **(B)** Gene expression early after infection. **(C)** The abundance of active CII protein determines whether the phage enters the lytic or lysogenic cycle. **(D)** The synthesis of Cro promotes lytic development by repressing the synthesis of CI repressor. Once O and P are synthesized, the replication of λ DNA dilutes out the CI repressor. **(E)** Synthesis of CII promotes lysogeny.

after infection. The phage replicates as a circle, and the DNA is packaged from these circles instead of from concatemers as λ does normally. Also, like λ, the genetic map of P2 phage is linear, because it has a unique *cos* site at which the circles are cut during packaging.

One way in which P2 differs significantly from λ, which almost always integrates into a single site in the *E. coli* chromosome, is that P2 can integrate into many sites in the bacterial DNA, although it uses some sites more than others. Like λ, P2 requires one gene product to integrate and two gene products to be excised. P2 prophage is much more difficult to induce than λ, however. It is not inducible by UV light, and even temperature-sensitive repressor mutations cannot efficiently induce it. The only known way to induce it efficiently is to infect the lysogen with another P2 (or P4) phage (see below).

Table 7.5 Steps leading to lytic growth and lysogeny

Step	Lytic growth	Lysogeny
1	Transcription from p_L and p_R	Same as lytic
2	Synthesis of N and Cro	Same as lytic
3	N allows expression of O and P	N allows expression of CII
4	Cro blocks CI synthesis from p_{RM}	CII activates CI synthesis from p_{RE} and Int synthesis from p_I
5	Low MOI; most CII degraded	High MOI; CII stabilized
6	Fast growth; more RNase III, more N	Slow growth; less RNase III, less N
7	More O, P, Q; lytic growth	More CI; lysogeny

PHAGE P4: A SATELLITE VIRUS

Even viruses can have parasites. Phage P4 is a parasite that depends on phage P2 for its lytic development (see Kahn et al., Suggested Reading). Thus, it is a representative of a group called **satellite viruses**, which need other viruses to multiply. Phage P4 does not encode its own head and tail proteins but, rather, uses those of P2. Thus, P4 can multiply only in a cell that is lysogenic for P2 or that has been simultaneously infected with a P2 phage. When P4 multiplies in bacteria lysogenic for P2, it induces transcription of the head and tail genes of the P2 prophage, which are normally not transcribed in the P2 lysogen.

P4 uses two mechanisms to induce transcription of the late genes of P2. It induces the P2 lysogen because it makes an inhibitor of the P2 repressor protein, which binds to the P2 repressor, inactivating it and inducing P2 to enter the lytic cycle. However, even though the P2 DNA replicates after induction by P4 and all the P2 proteins are made, most of the phage that are made contain P4 DNA genomes rather than P2 genomes. This is because P4 makes a protein called Sid, which causes the P2 proteins to assemble into heads that are smaller than normal, with only one-third the volume of a normal P2 head. These heads are too small to hold P2 DNA but are large enough to hold P4 DNA, which is only about one-third the length of P2 DNA, so that the heads are filled with P4 DNA instead. Because it wears the protein coat of P2, the phage P4 particle looks similar to P2, except that it is smaller to accommodate the shorter DNA. While the DNAs of P2 and P4 have otherwise very different sequences, the *cos* sites at the ends of the DNA are the same, so that the head proteins of P2 can package either DNA.

Phage P4 can also form a lysogen; when it does so, it usually integrates into a unique site on the chromosome. Not only can P4 infection induce a P2 prophage, but P2 infection can also induce a P4 prophage. It does this inadvertently by making a protein called Cox, which induces the P4 prophage. Apparently, P4, which cannot multiply by itself, does not want to be caught sleeping as a prophage if the cell happens to be infected by P2. Again, at least some of the phage that emerge from the infection after P4 is induced have P4 DNA wrapped in a smaller-than-normal P2 coat, even though it was a P2 phage that infected the cell. One phage enters the cell and emerges as a different phage. No matter who initiates the infection, P2 comes out the loser.

Phage P4 also can replicate autonomously as a circle in the prophage state, as does phage P1 (see below). Because of this ability to maintain itself as a circle, phage P4 has been engineered for use as a cloning vector. Phages P2 and P4, as well as their many relatives, have a very broad host range and infect many members of the *Enterobacteriaceae*, including *Salmonella* and *Klebsiella* spp., as well as some *Pseudomonas* spp. They are also related to phage P1, although their lifestyles and strategies for lytic development and lysogeny are very different.

PROPHAGES THAT REPLICATE AS PLASMIDS

Not all prophages integrate into the chromosome of the host to form a lysogen. Some form a prophage that replicates autonomously as a plasmid. These plasmid prophages can be very stable, with efficient and robust copy number control mechanisms and partitioning systems that make them resemble other plasmids (see chapter 4), except that they also encode phage proteins. The prophages can also be either circular or linear plasmids. Other phages are known to sometimes exist as plasmids in the prophage state, including P4 (see above).

Phage P1

P1 is the best-studied phage whose prophage replicates as a circular plasmid. The P1 plasmid prophage is a low-copy-number plasmid that maintains a copy number of only 1 and combines many of the other features of true plasmids, including a partitioning system and even a toxin-antitoxin system used for plasmid maintenance (see Box 4.3). Another interesting aspect of this phage is that it has an invertible segment (see chapter 8). A region of the phage DNA encoding the tail fibers frequently inverts, and the host range of the phage depends on the orientation of this invertible segment; bacteriophage Mu uses a similar DNA inversion to expand its host range. The invertible segment thereby contributes to the very broad host range of P1. It also has a site-specific Y recombinase, Cre, that acts on the *lox* site to resolve plasmid dimers and prevent curing of the prophage. This very active *lox* site forms the end of the linear genetic map. The Cre-*lox* system is the best understood of the Y recombinases and has been put to many uses in molecular genetics (see chapter 8).

Phage N15

Another *E. coli* phage, N15, has a plasmid as its prophage, but this plasmid is linear rather than circular. It has served as a model system for how some types of linear plasmids replicate. The N15 genome has hairpin ends with the 3′ and 5′ ends joined to each other and replicates around the ends from an internal origin to yield a dimeric circle. The dimeric circles are then resolved at a *cis*-acting site, *tos*, by the TelN protein, which forms hairpins at these sites. The replication of N15 prophage and linear plasmids and chromosomes is discussed in Box 4.1.

PHAGE Mu: A TRANSPOSON MASQUERADING AS A PHAGE

Another well-studied phage that forms lysogens is Mu, which integrates randomly into the chromosome. Because it integrates randomly, it often integrates into genes and causes random insertion mutations—hence its name, Mu (for mutator phage). This phage is essentially a transposon wrapped in a phage coat, and it integrates and replicates by transposition. For this reason, the discussion of phage Mu is deferred to chapter 8.

Phi3T: SMALL MOLECULE CONTROL OF LYSOGENY

Most phages monitor conditions within a single host cell to choose between the lytic and lysogenic pathways. However, the *Bacillus* phage Phi3T instead uses levels of a small peptide to monitor conditions in the host population. During a lytic cycle, the phage directs synthesis of a small peptide that is released into the medium and sensed by other nearby cells (see Erez et al., Suggested Reading). When the levels of the peptide are high, due to an increase in the number of phage-infected cells, subsequent phage infections shift to the lysogenic cycle; this makes sense, as a high amount of the peptide signals a low probability of uninfected host cells, which makes a lysogenic pathway a preferable fate, as additional released phage are less likely to find a suitable host. This is similar to other quorum sensing systems (see chapter 12) in which bacterial cells use small molecules to communicate information about the state of other cells in the population.

Genetic Analysis of Phages

Phages are ideal for genetic analysis (see the introductory chapter). They have short generation times and are haploid. Also, phages multiply as clones in plaques, and large numbers can be propagated on plates or in small volumes of liquid media. Different phage mutants can be easily crossed with each other, and the progeny can be readily analyzed. Because of these advantages, phages were central to the development of molecular genetics, and important genetic principles, such as recombination, complementation, suppression, and *cis*- and *trans*-acting mutations, are most easily demonstrated with phages (see Salmond and Fineran, Suggested Reading). In this section, we discuss the general principles of genetic analysis of phages. However, most of the genetic principles presented here are the same for all organisms, including humans. Only the details of how genetic experiments are performed differ from organism to organism.

Infection of Cells

The first step in doing a genetic analysis of phages, or any other virus for that matter, is to infect cells with different mutants of the phage. Phages can infect only cells that are susceptible to them, and they can multiply only in cells that are permissive for their development. To multiply in a cell, not only must the phage adsorb to the cell surface of the bacterium and inject its nucleic acid (either DNA or RNA) but also, a permissive host cell must provide all of the functions needed for multiplication of the phage. Therefore, most phages can infect and multiply in only a very limited number of types of bacteria. The different types of bacterial cells in which a phage can multiply are called its host range. Sometimes, a normally permissive type of cell can become a **nonpermissive host** for the phage as a result of a single mutation or other genetic change in either the host or the phage. Alternatively, a mutant virus or phage may be able to multiply in a particular type of host cell under one set of conditions, for example, at lower temperatures, but not under a different set of conditions, for example, at higher temperatures. As in bacterial genetics, these are called **conditional-lethal mutations** and are very useful in phage genetics. The conditions under which a phage with a conditional-lethal mutation can multiply are **permissive conditions**, while the conditions under which it cannot multiply are **nonpermissive conditions**.

MULTIPLICITY OF INFECTION

Infecting permissive cells with a phage is simple enough in principle. The phage and potential bacterial host need only be mixed with each other, and some bacteria and phage will collide at random, leading to phage infection. However, the percentage of the cells that are infected depends on the concentrations of phage and bacteria. If the phage and bacteria are very concentrated, they collide with each other to initiate an infection more often than if they are more dilute.

The efficiency of infection is affected not only by the concentrations of phage and bacteria, but also by the ratio of phage to bacteria, the **multiplicity of infection (MOI)**. For example, if 2.5×10^9 phage are added to 5×10^8 bacteria, there are $2.5 \times 10^9/5 \times 10^8$, i.e., 5 phage for every cell, and the MOI is 5. If only 2.5×10^8 phage had been added to the same number of bacteria, the MOI would have been 0.5. If the number of phage exceeds the number of cells to infect, the cells are infected at a

high MOI. Conversely, a **low MOI** indicates that the cells outnumber the phage. Whether a high or low MOI is used depends on the nature of the experiment. At a high enough MOI, most of the cells are infected by at least one phage; at a low MOI, many of the cells remain uninfected, but each infected cell is usually infected by only one phage. As described above for the λ life cycle, MOI can have a major effect on the outcome of an infection.

Even at a very high MOI, not all the cells are infected. There are two reasons for this. First, infection by phage is never 100% efficient. The surface of each cell may have only one or very few receptors for the phage, and a phage can infect a cell only if it happens to bind to one of these receptors. There is also the statistical variation in the number of phage that bind to each cell. Because the chance of each phage binding to a cell is random, the number of phage infecting each cell follows a normal distribution. At an MOI of 5, some cells are infected by five phage—the average MOI—but some are infected by six phage, some by four, some by three, and so on. Even at the highest MOIs, some cells by chance receive no phage and so remain uninfected.

The minimum fraction of cells that escape infection due to statistical variation can be calculated by using the Poisson distribution, which can be used to approximate the normal distribution in such situations. In chapter 3, we discuss how Luria and Delbrück used the Poisson distribution to estimate mutation rates. According to the Poisson distribution, the probability of a cell receiving no phage and remaining uninfected (P_0) is at least $e^{-\text{MOI}}$, since the MOI is the average number of phage per cell. If the MOI is 5, then P_0 is equal to e^{-5}, which is equal to 0.0067; i.e., at least 0.67% of the cells remain uninfected. At an MOI of 1, P_0 is equal to e^{1}, or 0.37, so at least 37% of the cells remain uninfected. In other words, at most ~63% of the cells are infected at an MOI of 1. Even this is an overestimation of the fraction of cells infected, since as mentioned, some of the viruses never actually infect a cell.

Phage Crosses

Crosses in phages and other viruses are easy; the cells are infected with different strains of the virus simultaneously. To be certain that many of the cells in a culture are simultaneously infected by both strains of a phage, we must use a high MOI of both phages. Again, the Poisson distribution can be used to calculate the maximum fraction of cells that will be infected by both mutant phages at a given MOI of each. If an MOI of 1 for each mutant phage is used for the infection, then at most 1 – 0.37, or 0.63 (63%), of the cells will be infected with each mutant strain of the phage. Since the chance of being infected with one strain is independent of the chance of being infected with the other strain, at most 0.63 × 0.63, or 0.40 (40%), of the cells will be infected by both phage strains at an MOI of 1. A high MOI of both mutant phages is required to infect most of the bacteria with both strains.

Recombination and Complementation Tests with Phages

As discussed in chapter 3, two basic concepts in classic genetic analysis are recombination and complementation. The types of information derived from these tests are completely different. In recombination, the DNAs of the two parent organisms are assembled in new combinations so that the progeny have DNA sequences from both parents. In complementation, the gene products synthesized from two different DNAs interact in the same cell to produce a phenotype. While the principles of recombination and complementation are the same in bacteria and phages, the ways in which data from recombination tests and complementation tests are obtained and interpreted are different, so we will emphasize the differences here.

RECOMBINATION TESTS

Recombination tests with phages are carried out by infection of cells with both mutants simultaneously at a high MOI. The cells must be permissive for both mutants or the infection must be done under conditions that are permissive for both mutants. The two different phage DNAs then enter the same cell and recombine. Figure 7.30 gives a simplified view of what happens when two DNA molecules from different strains of the same phage recombine. In the example, the two mutant phage strains infecting the cell are almost identical, except that one has a mutation at one end of the DNA and the other has a mutation at the other end. The sequences of the two DNAs therefore differ only at the sites of the two mutations shown as differences in single base pairs at opposite ends of the molecules. Recombination occurs by means of a crossover between the two DNA molecules, and two new molecules are created that are identical in sequence to the original molecules, except that one part now comes from one of the original DNA molecules and the other part comes from the other.

The effect, if any, of the crossover depends on where it occurs. If the crossover occurs between the sites of the two mutations, two new types of recombinant DNA molecules appear: one has neither mutation, and the other has both mutations. The DNAs that are produced, some of which have recombined while others have not, are then packaged into progeny phage particles. Progeny phage that have packaged the DNAs with these new DNA sequences are **recombinant types** because they are unlike either parent (see chapter 3). In the example, progeny

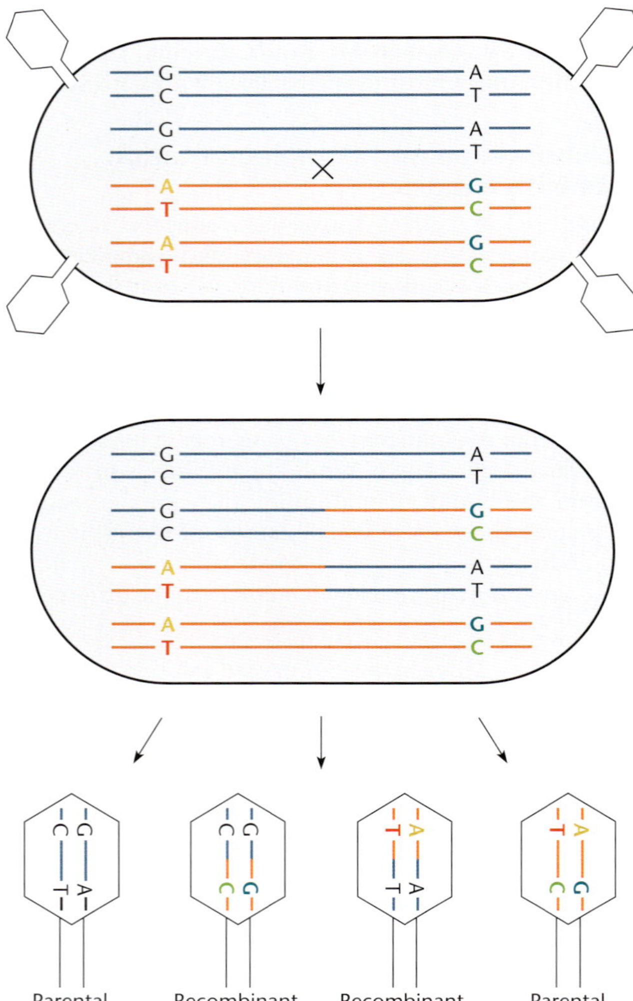

Figure 7.30 Recombination between two phage mutations. The two different mutant parent phages infect the same permissive host cell, and their DNA replicates. Crossovers occur in the region between the two mutations, giving rise to recombinant types that are unlike either parent phage. Only the positions of the mutated base pairs are shown. The DNA of one parent phage is shown in blue, and that of the other is shown in orange.

phage that have packaged a DNA molecule with only one of the mutations are called **parental types** because they are like one of the original phages that infected the cell. The appearance of recombinant types tells us that recombination has occurred.

Recombination Frequency

Unlike in bacterial genetics, where mutations are mapped based on whether they are included in the same piece of donor DNA or on when they are transferred during Hfr crosses, mapping in phages (and most other organisms) is determined by recombination frequencies. The closer together the regions of sequence difference are to each other in the DNA, the less room there is between them for a crossover to occur. Therefore, the frequency of recombinant-type progeny is a measure of how far apart the mutations are in the DNA of the phage. This number is usually expressed as the recombination frequency (see chapter 3). The general equation for recombination frequency used in all organisms from bacteria to fruit flies to humans is as follows: recombination frequency = number of recombinant progeny/total number of progeny. When the recombination frequency is expressed as a percentage, it is called the map unit. For example, the regions of two mutations in the DNA give a recombination frequency of 0.01 if 1 in 100 of the progeny is a recombinant type. The regions of the two mutations are then 0.01×100, or 1, map unit apart.

Different organisms differ greatly in their recombination efficiencies; therefore, map distance is only a relative measure, and a map unit represents different physical lengths of DNA for different organisms. Also, the recombination frequency can indicate the proximity of two mutated regions only when the mutations are not too far apart. If they are far apart, two or more crossovers often occur between them, reducing the apparent recombination frequency. Note that while one crossover between the regions of two mutations creates recombinant types, two crossovers re-create the parental types. In general, odd numbers of crossovers produce recombinant types, and even numbers of crossovers re-create parental types. There is also the technical difficulty of accurately measuring the number of mutant phages of each type that are added. Unless they are added in equal numbers, the apparent recombination frequency will be lowered.

COMPLEMENTATION TESTS

Complementation tests are easy to do with phages and other viruses—in fact, much easier than they are to do with bacteria. As with recombination tests, to perform a complementation test with phages, cells are infected simultaneously with different mutant strains of a particular phage. Now, however, rather than infecting permissive cells or infecting under conditions that are permissive for both mutations, we infect under conditions that are nonpermissive for both mutations. If the two mutations complement each other, both phages will multiply and form plaques. The phage in these plaques will contain both parental types, as well as any recombinant types. If the complementation occurs, the two mutations are probably in different genes (see Table 3.4 for exceptions). Figure 7.31 illustrates complementation tests with phages. Complementation and recombination experiments with phage were crucial in early analyses of gene structure (see Benzer, Suggested Reading) and the genetic code (see Crick et al., Suggested Reading).

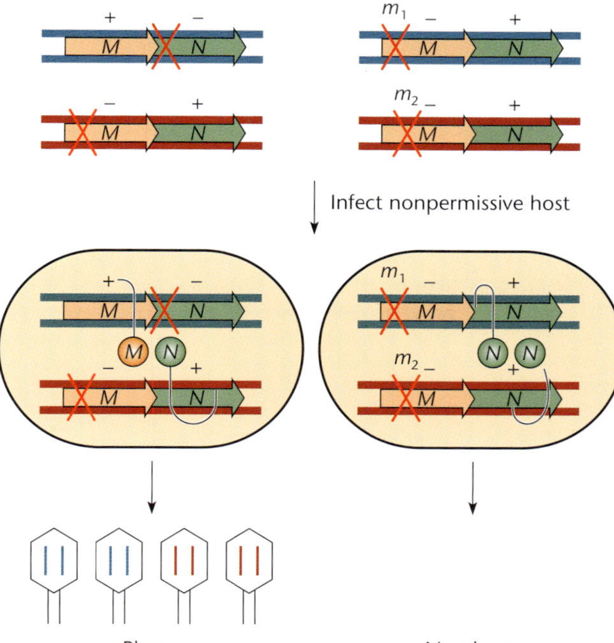

Figure 7.31 Tests of complementation between phage mutations. Phages with different mutations infect the same host cell, in which neither mutant phage can multiply. **(Left)** The mutations, represented by the minus signs, in different genes (*M* and *N*). Each mutant phage synthesizes the gene product that the other one cannot make; complementation occurs, and new phage are produced. **(Right)** Both mutations (minus signs) prevent the synthesis of the *M* gene product. There is no complementation, the mutants cannot help each other multiply, and no phage are produced.

The Genetic-Linkage Map of a Phage

Before the advent of DNA sequencing, the genetic maps of a number of phages were laboriously constructed. A large number of conditional-lethal mutations, both temperature-sensitive and nonsense mutations, as well as other types of mutations, were isolated and assigned to genes using complementation tests. One or more mutations in each gene were then placed on the expanding genetic map by ordering them using recombination frequencies and three-factor crosses (see chapter 3). A picture that shows many of the genes of an organism and how they are ordered with respect to each other is known as the **genetic linkage map** of the organism, so named because it shows the proximity or linkage of the genes to each other. Later, DNA sequencing and other physical methods for mapping DNA, discussed in chapter 1, gave rise to a physical genetic map, which could then be correlated with the genetic linkage map. Nowadays, the ease of sequencing the relatively small DNAs of phages makes it unlikely that the laborious task of constructing a genetic linkage map using phage crosses will ever again be performed for any phage.

FEATURES OF THE GENETIC MAPS OF SOME PHAGES

Using methods such as those described above, physical and genetic linkage maps have been determined for several commonly used phages. Some of these maps appear throughout this chapter, along with the functions of some of the gene products, where known.

The genetic maps of phages have a number of obvious features. One noticeable feature of most phage genetic maps is that genes whose products must physically interact, such as the products involved in head or tail formation, tend to be clustered. In fact, if a gene is not in the same region as other genes whose products are involved in the same function, it is often an indication that the gene has a second, unknown function unrelated to the first. The argument is that the clustering of genes whose products physically interact may allow recombination between closely related phages without disruption of their shared function. Another striking feature of phage genetic maps is that some are linear, while others are circular (compare the T7 map in Figure 7.4 with the T4 map in Figure 7.6A).

That some phage genetic maps are circular came as a surprise in the early days of mapping phage genes. When the genes of the phage were being ordered from left to right by genetic crosses, researchers discovered that they had come full circle and the first gene was now the next one on the right. Also, the form of the genetic map does not necessarily correlate with the linearity or circularity of the phage DNA itself. Some phages with circular DNA in the cell, such as λ during the early stages of replication, have a linear linkage map, while some with linear DNA that never circularizes, such as T4, have a circular map. To understand how the various genetic maps arise, we need to review how some phages replicate their DNA and how it is packaged into phage heads.

Phage λ

Phage λ has a linear genetic map, even though the DNA forms a circle after it enters the cell, because its concatemers are cleaved at unique *cos* sites before being packaged into the phage head. The positions of these *cos* sites determine the end of the linear genetic map. As an illustration, consider a cross between two phages with mutations in the *A* and *R* genes at opposite ends of the phage DNA (see the λ map in Figure 7.22). Even though different parental alleles of the *A* and *R* genes can be next to each other in the concatemers prior to packaging, these alleles are separated when the DNA is cut at the *cos* site during packaging of the DNA. Therefore, the *A* and *R* genes appear to be far apart and essentially unlinked when one measures recombination frequencies in genetic crosses, and the genetic map is linear, with the *A* and *R* genes at opposite ends. For this reason, all types of phages that package DNA from unique *pac* or *cos* sites have linear genetic linkage maps with the ends defined by the position of the *pac* or *cos* site, whether or not the DNA ever cyclizes.

Phage T4

In contrast, phage T4 has a circular genetic map, even though its DNA never forms a circle. The reason for its circular map is that T4 has at most only weak *pac* sites, and the DNA is packaged by a headful mechanism from long concatemers (see Figure 7.19). Consequently, any two T4 phage DNAs in different phage heads from the same infection do not have the same ends but are cyclic permutations of each other. Therefore, genes that are next to each other in the concatemers will still be together in most of the phage heads unless they happen to be on the terminal redundancy, which is only 3% of the genome, and so will appear linked in crosses or when the DNA, which contains a mixture of molecules with different endpoints, is sequenced, producing a circular map.

Phage P22

Phage P22 is a phage of *Salmonella* that has a tail structure different from that of λ and thus is in a different morphological family (Table 7.1), although it replicates by a similar mechanism. However, unlike λ, it has a circular linkage map. The difference is that P22 begins packaging at a unique *pac* site (equivalent to the λ *cos* site) but then packages a few genomes by a processive headful mechanism, like T4, giving rise to a circular genetic map and making it a good transducing phage (see "Generalized Transduction" below).

Phage P1

In the virion, phage P1 has linear DNA that forms a circle by recombination between terminally repeated sequences at its ends after infection. The DNA then replicates as a circle and forms concatemers from which the DNA is packaged by a headful mechanism, which makes it a good transducing phage. However, unlike most phages that package DNA by a headful mechanism, the genetic map of P1 is linear because it has a very active site-specific recombination system called *cre-lox*, which promotes recombination at a particular site in the DNA. Because recombination at this site is so frequent, genetic markers on either side of the site appear to be unlinked, giving rise to a linear map terminating at the *cre-lox* site. This site-specific recombination system in phage development is probably used to resolve dimeric circles in the prophage state, where P1 DNA exists as a circular plasmid.

Phage-Mediated Genetic Transfer

As discussed in the introduction and chapter 3, bacteriophages not only infect and kill cells, but also sometimes transfer bacterial DNA from one cell to another in the process called **transduction**. There are two types of transduction in bacteria: **generalized transduction**, in which essentially any region of the bacterial DNA can be transferred from one bacterium to another, and **specialized transduction**, in which only certain genes, those located close to the attachment site of the prophage in the bacterial genome, can be transferred. The nomenclature of transduction is much like that of transformation and conjugation. Phages capable of transduction are called **transducing phages**. A phage that contains bacterial DNA is called a **transducing particle**. The original bacterial strain in which the transducing particle multiplied and picked up host DNA is called the **donor strain**. The bacterial strain it infects is called the **recipient strain**. Cells that have received DNA from another bacterium by transduction are called **transductants**. Since the analysis of genetic data obtained by transduction is discussed in chapter 3, we restrict ourselves to the mechanism of transduction and the properties a phage must have to be a transducing phage.

Generalized Transduction

Figure 7.32 gives an overview of the process of generalized transduction. While phages are packaging their own DNA, they sometimes mistakenly package the DNA of the bacterial host instead. The resulting phage particles are still capable of infecting other cells, but progeny phages are not produced. What happens to the DNA after it enters the cell depends on the source of the bacterial DNA. If the bacterial DNA is a piece of the bacterial chromosome of the same species, it usually has extensive sequence homology to the chromosome of the recipient cell and may recombine with the host chromosome to form recombinants. If the piece of DNA that was picked up and injected is a plasmid, it may replicate after it enters the cell and thus be maintained. If the incoming DNA contains a transposon, the transposon may hop, or insert itself, into a host plasmid or chromosome, even if the remainder of the DNA contains no sequence in common with the DNA of the bacterium it entered (see chapter 8).

Usually, transductants arise very rarely for a number of reasons. First, mistaken packaging of host DNA is itself generally rare, and transduced DNA must survive in the recipient cell to form a stable transductant. Since each of these steps has a limited probability of success, transduction can usually be detected only by powerful selection techniques.

WHAT MAKES A GENERALIZED TRANSDUCING PHAGE?

Not all phages can transduce. To be a generalized transducing phage, the phage must have a number of characteristics. It must not degrade the host DNA completely after infection, or no host DNA will be available to be

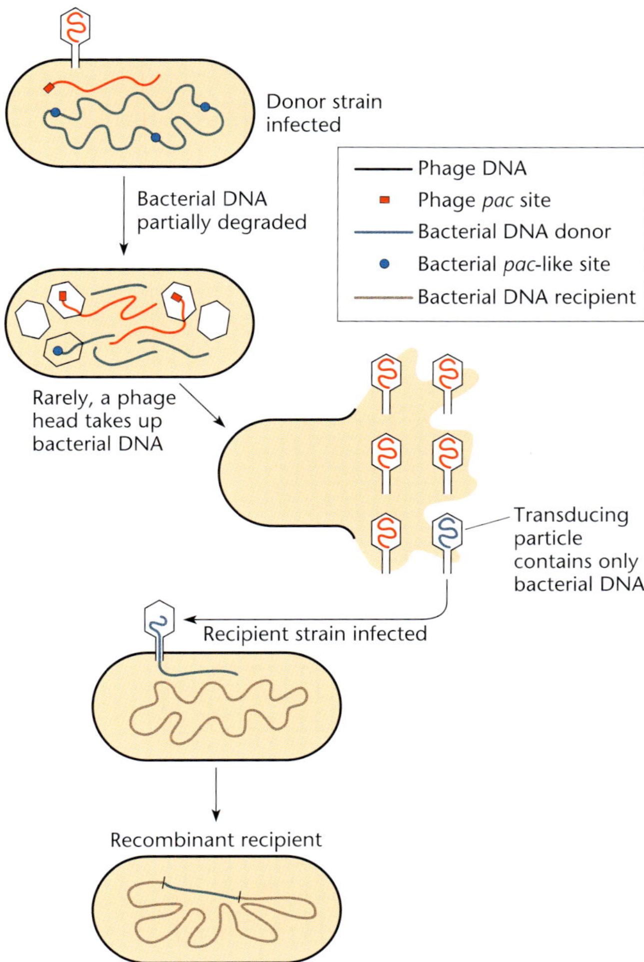

Figure 7.32 Generalized transduction. A phage infects one bacterium, and in the course of packaging DNA into heads, the phage mistakenly packages some bacterial donor DNA (blue) instead of its own DNA (red) into a head because it mistakenly recognizes a *pac*-like site in the bacterial DNA. In the next infection, this transducing particle is different from most of the phage in the population in that it injects the bacterial DNA instead of phage DNA into the recipient bacterium. If the bacterial DNA that was packaged is a plasmid, it might replicate in the recipient cell, or if it is chromosomal DNA, it might recombine with the chromosome and form a recombinant type.

packaged into phage heads when packaging begins. It also must not be too particular about what DNA it packages. As discussed above, many phages package DNA from sites on the DNA called *pac* or *cos* sites. If a DNA lacks such specific sites, it is usually not packaged. The packaging sites of transducing phages must not be so specific that such sequences do not occur by chance in host DNA (Figure 7.32). It also helps if the phage is not too particular about the length of the DNA it packages, so that even relatively nonspecific *pac* sites do not have to be properly spaced.

Some of the most useful transducing phages have a broad host range for adsorption, which allows them to introduce DNA into a variety of bacteria. It is important to realize that to transduce, a phage does not actually have to be capable of multiplying in the recipient host, only of adsorbing to it and injecting its DNA. Transduction from other hosts using a broad-host-range phage sometimes offers a way of introducing plasmids or other DNA elements into hosts for which other means of gene exchange are not available. Even if a type of phage does not have all of the above characteristics, it may be possible to engineer it into a good transducer, and this has been done for some phages (see Yosef et al. and de Jonge et al., Suggested Reading).

Table 7.6 compares two good transducing phages, P1 and P22. Phage P1, which infects enteric bacteria, is a good transducer because it has less *pac* site specificity than most phages and packages DNA via a headful mechanism; therefore, it efficiently packages host DNA. About 1 in 10^6 phage P1 particles transduces a particular marker. It also has a very broad host range for adsorption and can transduce DNA from *E. coli* into a wide variety of other bacteria, including members of the genera *Klebsiella* and *Myxococcus*. It cannot multiply in hosts other than *E. coli*, but it can transfer plasmids, etc., from *E. coli*, on which it can be propagated, into these other hosts.

The *Salmonella enterica* serovar Typhimurium phage P22 is also a very good transducer and, in fact, was the first transducing phage to be discovered (see Zinder and Lederberg, Suggested Reading). Like P1, P22 has *pac* sites that are not too specific and packages DNA by a headful mechanism. From a single *pac*-like site, about 10 headfuls of DNA can be packaged. Because of even this limited *pac* site specificity, however, some regions of *Salmonella* DNA are transduced by P22 at a much higher frequency than others.

As mentioned above, other phages that are not normally transducing phages can be converted into them. For example, T4 normally degrades the host DNA after infection but works well as a transducing phage if its genes for the degradation of host DNA have been mutated. Because phage T4 DNA packaging does not require very specific *pac* sites, it packages any DNA, including the host DNA, with almost equal efficiency. It also packages by a headful mechanism, so even these very nonspecific *pac* sites do not have to be evenly spaced.

In contrast, phage λ does not work well for generalized transduction because it normally packages DNA between two *cos* sites rather than by a headful mechanism. It very infrequently picks up host DNA by mistake, but

Table 7.6 Characteristics of generalized transducing phages P1 and P22

Characteristic	Value in phage: P1	P22
Length (kb) of DNA packaged	100	44
Length (%) of chromosome transduced	2	1
Packaging mechanism	Sequential headful	Sequential headful
Specificity of markers transduced	Almost none	Some markers transduced at low frequency
Packaging of host DNA	Packaged from ends	Packaged from *pac*-like sequences
Transducing particles in lysate (%)	1	2
Transduced DNA recombined into chromosome (%)	1–2	1–2

then it does not cut the DNA properly when the head is filled unless another *cos*-like sequence happens to lie at a genome-length distance along the DNA. Thus, potential transducing particles usually have DNA hanging out of them that must be removed with DNase before the tails can be added. Even with these and other manipulations, λ works poorly as a generalized transducer.

Generalized transducing phages have been isolated for a wide variety of bacteria and have greatly aided genetic analysis of these bacteria. Transduction is particularly useful for moving alleles into different strains of bacteria and making isogenic strains that differ only in a small region of their chromosomes. This makes generalized transduction very useful for strain construction and gene knockouts for functional genomics (see chapter 3). However, if no transducing phage is known for a particular strain of bacterium, finding one can be very time-consuming. Therefore, for such bacteria, transduction needs to be replaced by other methods.

ROLE OF GENERALIZED TRANSDUCTION IN BACTERIAL EVOLUTION

Phages may play an important role in evolution by promoting the horizontal transfer of genes between individual members of a species, as well as between distantly related bacteria. The DNA in phage heads is usually more stable than naked DNA and so may persist longer in the environment. Also, many phages have a broad host range for adsorption. We have cited the example of phage P1, which infects and multiplies in *E. coli* but also injects its DNA into a number of other bacterial species. The host range of P1 is partially affected by the orientation of an invertible DNA segment encoding the tail fibers (see chapter 8). Although incoming DNA from one species does not recombine with the chromosome of a different species if they share no sequences, stable transduction of genes between distantly related bacteria becomes possible when the transduced DNA is a broad-host-range plasmid that can replicate in the recipient strain or contains a broad-host-range transposon that can hop into the DNA of the recipient cell.

Specialized Transduction

In generalized transduction, phages carry or transduce host DNA, instead of their own DNA, and almost any region of the chromosome can be transduced. Some phages that integrate their DNA into the chromosome during lysogeny are also capable of another type of transduction, called **specialized transduction** or **restricted transduction**, because in this type of transduction, only bacterial genes close to the attachment site of the prophage can be transduced. The specialized transducing phage particle carries both bacterial DNA and phage DNA instead of only bacterial DNA, reflecting very different mechanisms for the two types of transduction.

Figure 7.33 illustrates how a λ phage particle capable of specialized transduction of *gal* markers arises. The λ

Figure 7.33 Formation of a λd*gal* transducing particle. A rare mistake in recombination between a site in the prophage DNA (in this case, located between *A* and *J*) and a bacterial site on the left of the prophage in the *gal* operon results in excision of a DNA particle in which some bacterial DNA, including *gal*, has replaced phage DNA.

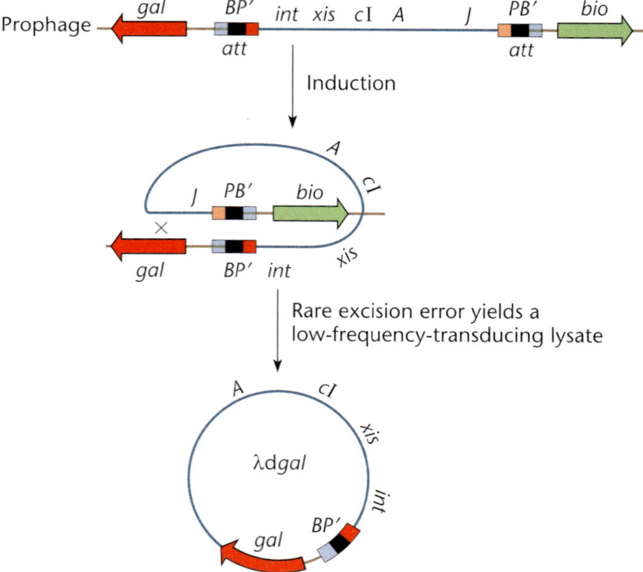

prophage is integrated at its normal *att* site between the closely linked *gal* and *bio* genes in the chromosome. The *gal* gene products degrade galactose for use as a carbon and energy source, while the *bio* gene products make the vitamin biotin. A specialized transducing phage carrying the *gal* genes, called λd*gal*, forms as the result of a rare mistake during the recombination event that results in excision of the prophage DNA. Normally, the prophage DNA is excised from the bacterial DNA by recombination between the hybrid *attP-attB* sites at the junctions between the prophage and host DNA. However, if the recombination occurs between a site internal to the prophage DNA and a neighboring site in the bacterial DNA, in this case in the *gal* operon, the DNA later packaged into the head includes some bacterial sequences from the *gal* operon. The resulting transducing particles are also missing some of the phage genes (*J* in Figure 7.33) and therefore are defective for replication after infection of another host (indicated by the λd*gal* nomenclature, where the "d" stands for defective; see below). Such transducing phages are very rare, because the erroneous recombination that gives rise to them is extremely infrequent, occurring at one-millionth the frequency of normal excision. Furthermore, the recombination must occur by chance between two sites that are approximately a λ genome length apart, with one site in the prophage DNA, or the excised DNA will not contain a *cos* site that allows it to be packaged in a phage λ head. Because of the rarity of these transducing phages, powerful selection techniques are required to detect them. In this case, rare galactose-positive (Gal^+) transductants are selected on plates with galactose as the sole carbon source after transduction of a host with a mutation in its own *gal* genes.

SELECTION OF HFT PARTICLES

In the rare Gal^+ transductants, a λ phage carrying *gal* genes has usually integrated into the chromosome, providing, by complementation, the *gal* gene product that the mutant lacks. If such a Gal^+ lysogen is colony purified and the prophage is induced from it, a high percentage of the resultant phage progeny will carry the *gal* genes. Such a lysogenic strain produces an **HFT lysate** (for high-frequency transduction), because it produces phage that can transduce bacterial genes at a very high frequency.

Normally, not all of the induced phage particles in a *gal*-transducing HFT lysate produced in this way carry the *gal* genes, and the lysate contains a mixture of transducing phage particles and wild-type λ. The reason for this is apparent from Figure 7.33. Because the λ phage head can hold DNA of only a certain length, the transducing particles have of necessity lost some phage genes to make room for the bacterial genes. The properties of the transducing particle are determined by which phage genes are lost. For example, the λd*gal* phage shown in Figure 7.33 lacks essential head and tail genes, beginning with the *J* gene, so it cannot multiply without a wild-type λ **helper phage** to provide the missing head and tail proteins. These phage particles are thus called λd*gal* and usually can be produced only by induction from **dilysogens** that contain the λd*gal* prophage and a wild-type phage integrated next to each other in tandem, as shown in Figure 7.34.

The molecular details of how the HFT lysogens are induced and whether induction requires Int and Xis or just Int depends on how the dilysogen formed. In the example, when the λd*gal* integrated, it could have done so by homologous recombination between λ phage DNA sequences carried by both the λd*gal* DNA and the wild-type λ DNA, creating a structure like that shown in Figure 7.34. Then, the two could be excised together by Int and Xis as shown and then packaged independently from *cos* sites on the concatemers that form during DNA replication. Alternatively, the λd*gal* phage might have integrated next to a preexisting wild-type λ prophage, or the two might have integrated together as a dimeric circle. All of these cases require different combinations of Int and Xis for excision. However they have formed and been induced, the wild-type phage DNA must provide the head and tail genes that the λd*gal* DNA has lost. The two can then multiply together, so that roughly half the phage produced are now λd*gal*; the other half are wild-type λ.

Figure 7.34 Induction of the λd*gal* phage from a dilysogen containing both λd*gal* and a wild-type λ in tandem. Recombination between the hybrid *PB′* and *BP′* sites at the ends excises both phages. The wild-type "helper" phage helps the λd*gal* phage to form phage particles, and both are packaged from repeated *cos* sites in long concatemers. See the text for details.

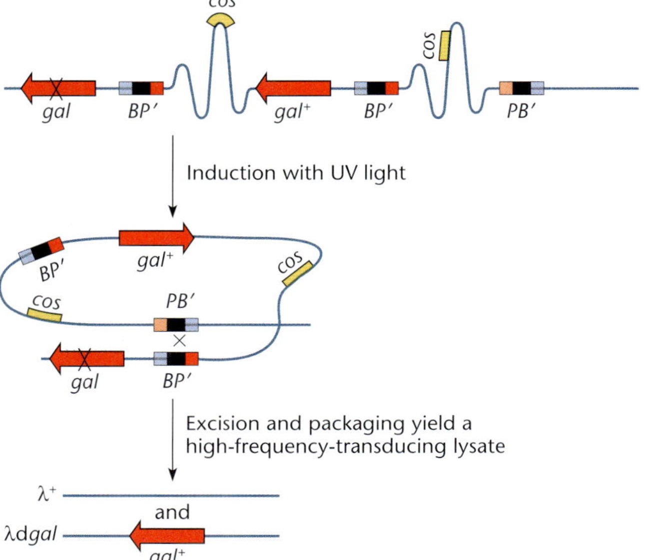

The situation is very different if the HFT transducing phage are created by a mistaken recombination on the other side, replacing the *int* and *xis* genes with *bio* genes. These phage are able to multiply, since the genes on this side of *attP* are not required for multiplication. However, they cannot form a lysogen or be induced without the help of a wild-type phage because the *bio* genes of the host have replaced their *attP* site and their *int* and *xis* genes. Because they can multiply and form plaques, *bio*-transducing phages are called λp*bio*, in which the "p" stands for plaque-forming.

Lysogenic Conversion and Bacterial Pathogenesis

In a surprising number of instances, prophages carry genes for virulence factors or toxins required for virulence by the pathogenic bacteria they lysogenize. This was discovered for diphtheria in 1952, almost as early as lysogeny itself. These genes that confer virulence on their hosts are not found in all the phages of that type, suggesting that they were recently acquired. They also are often expressed from their own promoters and are regulated by bacterial regulators that promote expression when the lysogenic strain is in the eukaryotic host, despite the fact that other prophage genes are usually repressed in the lysogen. Some examples of bacteria that carry prophages with genes that contribute to the diseases they cause are the bacteria that cause diphtheria, bacterial dysentery, scarlet fever, botulism, tetanus, toxic shock syndrome, and cholera. Even λ phage carries genes that confer on its *E. coli* host serum resistance and the ability to survive in macrophages. The process by which a prophage converts a nonpathogenic bacterium to a pathogen is called **lysogenic conversion**.

E. COLI AND DYSENTERY: SHIGA TOXINS

Pathogenic strains of *E. coli* are prime examples of bacteria that are not pathogenic unless they harbor prophages or other DNA elements that carry virulence genes. These bacteria are part of the normal intestinal microbiome. Addition of virulence genes can result in severe diseases, including bacterial dysentery, with symptoms such as bloody diarrhea. The infamous *E. coli* strain O157:H7, which has caused many outbreaks of bacterial dysentery worldwide, is one example of such a lysogenic *E. coli* strain. It harbors more than 18 prophage-like elements, which account for more than 50% of the DNA that is not a normal part of the *E. coli* chromosome.

In one particularly clear example, a group of prophages very closely related to λ can make *E. coli* pathogenic by encoding toxins called Shiga toxins (Stx), so named because they were first discovered in *Shigella dysenteriae*, which is so closely related to *E. coli* that it has been moved into the same genus. Like cholera toxin (CT) and many other toxins, the Shiga toxin is composed of two subunits, A and B. The B subunit helps the A subunit enter an endothelial cell of the host by binding to a specific receptor on the cell surfaces of some tissues. The A subunit is a very specific *N*-glycosylase that removes only a certain adenine base from the 28S rRNA of the host cell ribosome. Removal of this adenine from the 28S rRNA in a ribosome blocks translation by interfering with binding of the translation elongation factor 1α (EF-1α, the eukaryotic equivalent of EF-Tu), to the ribosome. Expression of the toxin is required to convert the disease from just watery diarrhea to hemolytic-uremic syndrome, which is the leading cause of kidney failure in children.

The *stx* genes in prophages are usually in the same place, downstream of p_R' in the phage genome. A prophage genetic map of phage φ361, encoding Stx2, is shown in Figure 7.35A. Note the remarkable similarity between the genetic map of this prophage and the genetic map of its close relative, λ prophage, shown in Figure 7.22.

Because the toxin genes *stx2A* and *stx2B* lie just downstream of the *Q* gene and upstream of the lysis genes *S* and *R*, they are expressed from the phage p_R' promoter only late in induction. However, the toxin genes in φ361 and many of these phages have their own weak promoter, p_{stx}, so they are weakly expressed even in the uninduced lysogen (see Wagner et al., Suggested Reading). Interestingly, the p_{stx} promoter is sometimes regulated by the presence of iron, as is an *stx1* gene found in the chromosome of some *Shigella* strains. Iron deficiency is often used as a sensor of the eukaryotic environment so that virulence genes are turned on only in the eukaryotic intestine (see chapter 12 for a discussion of iron regulation). However, these promoters are very weak, and not enough toxin would be expected to be made from them to account for the illness they cause. Moreover, even if the toxin protein is made at low levels, it is not clear how it would get out of the bacterial cell, which apparently does not harbor a secretion system capable of secreting the toxin into the extracellular environment of the intestine. Therefore, we surmise that the only way sufficient amounts of toxin can be made and get out of the bacterial cell is if the prophage is induced, makes large amounts of toxin from p_R', and then lyses the cell. The phage could be induced by our own immune systems, since neutrophils release H_2O_2, which induces the SOS system in the bacteria and therefore induces the phage. It is likely that some of the bacteria lyse and release the phage, which then infect and lysogenize nonpathogenic *E. coli* strains that are part of the normal bacterial microbiota. If the phage in these normally nonpathogenic bacteria are then induced, they kill the nonpathogenic bacteria that harbored the prophage, but the released Shiga toxin can participate in pathogenesis by the pathogenic strain, thereby multiplying the effects of the toxin and perhaps leading to hemolytic-uremic syndrome and other severe diseases related to the infection. This raises questions about

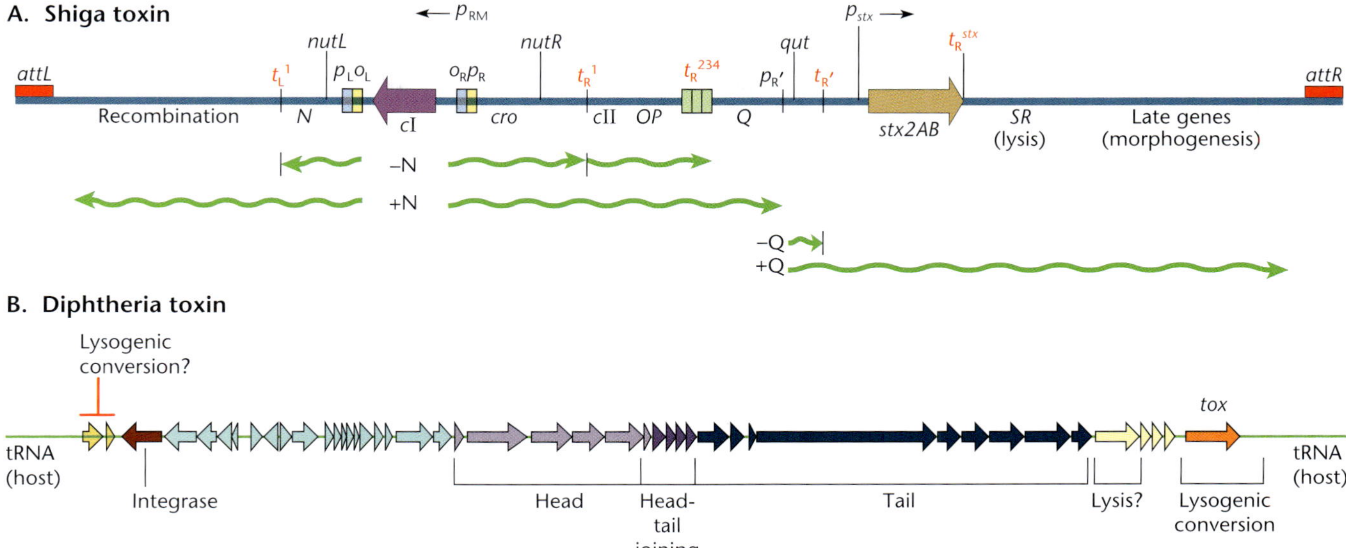

Figure 7.35 Lysogenic conversion. **(A)** Shiga toxins encoded by close relatives of λ. Shown is the prophage genetic map of phage φ361, with the positions of the toxin genes indicated. Modified from Young R, *in* Waldor MK, Friedman DI, Adhya SL (ed), *Phages: Their Role in Bacterial Pathogenesis and Biotechnology* (ASM Press, Washington DC, 2005). **(B)** Genome map of a *Corynebacterium diphtheriae* prophage containing the diphtheria toxin gene (*tox*). Selected genes are annotated. The prophage is bracketed by tRNA genes of the host. Other genes involved in lysogenic conversion may be on the other end of the prophage. Modified from Brussow H, Canchaya C, Hardt WD, *Microbiol Mol Biol Rev* **68:**560–602, 2004.

the effects of some types of antibiotics, such as ciprofloxacin (see chapter 1), that damage DNA and therefore induce phages. They may help to spread these phages and increase, rather than decrease, the severity of the disease. In fact, there is some evidence that the use of ciprofloxacin and similar antibiotics to treat people with bacterial dysentery due to *E. coli* can increase the chance of the disease developing into the more serious hemolytic-uremic syndrome.

DIPHTHERIA TOXIN

Diphtheria is the classic example of a disease caused in part by the product of a gene carried by a phage. As early as 1918, the microbiologist Félix d'Herelle demonstrated plaque formation by phages on the bacterium that causes diphtheria, and in the late 1940s, pathogenic strains of *Corynebacterium diphtheriae* were shown to differ from nonpathogenic strains in that they are lysogenic for a phage called β. These prophages carry a gene, *tox*, that encodes diphtheria toxin (Figure 7.35B), which is largely responsible for the effects of the disease and is the target of the vaccine against it. The diphtheria toxin is another typical AB toxin with two subunits, one that transposes the toxin into the eukaryotic cell, while the other, the toxin, is an enzyme that kills eukaryotic cells by ADP ribosylating (attaching ADP to) the EF-2 translation factor, thereby inactivating it and blocking translation in the cell. The *tox* gene of the β prophage is transcribed only when *C. diphtheriae* infects its eukaryotic host or under low iron concentrations, a condition that mimics this environment. Even though the *tox* gene is on the prophage, it is regulated by the products of chromosomal genes involved in iron regulation, illustrating the close relationship between these bacteria and their phages. The diphtheria toxin has been extensively studied, and the mechanism of action of the diphtheria toxin and the regulation of the *tox* gene are discussed in chapter 12 under "Regulation of Virulence Genes in Pathogenic Bacteria."

CHOLERA TOXIN

Another striking example of toxin genes carried by a phage is CT, encoded by the phage CTXφ of *V. cholerae*, the bacterium that causes cholera (see Waldor and Mekalanos, Suggested Reading). The CTXφ phage is a filamentous phage with single-stranded DNA and is similar to the M13 and fd DNA phages of *E. coli* that are discussed above. CTXφ infects the cell by attaching to a pilus, in this case the toxin-coregulated pilus, and then enters the cell through the TolA, TolQ, and TolR channel. It replicates its DNA through a double-stranded replicative intermediate that can integrate into a specific site in the chromosome to form a prophage.

One striking feature of the CTXφ phage that distinguishes it from most other phages we have discussed is the extent to which it lives in harmony with its host and does not substantially affect the growth rate of the host even when induced (see McLeod et al., Suggested Reading). It also uses many host functions instead of encoding its own. For example, the CTXφ phage does not have its own integrase but uses the host-encoded XerCD recombinase, which resolves chromosomal dimers during replication by recombination at the *dif* site close to the terminus of chromosome replication (see chapter 1). Because of its specificity, the XerCD recombinase integrates the phage into the *dif* site of the chromosome, the site close to the terminus of chromosome replication normally recognized by XerCD. However, even with the prophage integrated, the *dif* site is still usable, and the cells can still resolve chromosome dimers and do not filament.

Like λ, CTXφ has a regulatory protein that can function as both a repressor and an activator, but this repressor-activator is not made in the prophage state, because its promoter is repressed by a host protein, LexA. The function of this protein in the host is to repress repair genes that are part of the SOS system and are induced only following severe DNA damage (see above and chapter 10). The other phage genes (with the exception of those for CT) are also transcribed from the same promoter, so almost no phage genes are transcribed in the prophage state. However, if the DNA is damaged, for example, by UV irradiation, the RecA protein of the host promotes the self-cleavage of the LexA repressor, and the repressor-activator gene and most of the other phage genes are transcribed. Once it is made, the repressor-activator can modulate its own transcription and the transcription of the other phage genes from the same promoter so that only a few phage are ever produced and the growth of the bacterial cell is not seriously affected.

Because the major phage promoter is repressed in the prophage state, the only phage genes transcribed are the CT genes, *ctxA* and *ctxB*, which are transcribed from their own promoter. CT is another AB-type toxin in which the B subunit helps the A subunit toxin enter the eukaryotic cell. The A subunit is an enzyme that ADP ribosylates a mucosal membrane protein, causing cyclic AMP levels to rise and water to be excreted from the cells, leading to watery diarrhea and severe dehydration. The toxin subunits are secreted from the *V. cholerae* cell by a type II secretion system, the same one whose secretin subunit is used to release the phage from the cell. It is also interesting that the CT genes on the prophage are regulated by a transcriptional activator, ToxR, the product of a chromosomal gene, which acts through another transcriptional activator, ToxT, that also regulates the synthesis of the pili that serve as the receptor sites for the phage. This is another example of the close coordination between the phage and host functions involved in pathogenesis. These pili are also important virulence determinants, because they enable the bacteria to adhere to the intestinal mucosa. The regulation by ToxR of many genes involved in pathogenesis, including the *ctx* genes, is a good example of global gene regulation and is discussed in chapter 12.

STAPHYLOCOCCUS AUREUS AND TOXIC SHOCK SYNDROME

Yet another example of a pathogenic bacterium that depends on phages for its virulence is *S. aureus*. Not only are many of the virulence determinants of *S. aureus* carried on prophages, but those that are carried on DNA elements called **pathogenicity islands** often depend on phages for their mobility. Pathogenicity islands are a type of **genetic island** found in pathogenic bacteria and carry genes required for virulence (see chapters 12 and 13). Many types of bacteria harbor genetic islands, often quite large, that are integrated into their chromosomes and encode functions that broaden the physiological capacity of the cells and therefore the number of ecosystems they can inhabit. These genetic islands are not found in all strains of a species and presumably have been acquired fairly recently by horizontal transfer from other bacteria. They are unable to replicate autonomously but usually encode an integrase that allows them to integrate into a specific site in the chromosome. In chapter 8, a specific type of transposon called Tn7 or Tn7-like elements also form genetic islands at their attachment sites in the chromosome. However, in many cases, it remains unclear how a genetic island has moved from one bacterium to another, since most of them cannot transfer in a laboratory setting, and they encode no obvious functions known to promote lateral transfer between bacteria.

S. aureus is a normal inhabitant of the skin of most people, but some strains can cause serious illnesses. One of the most serious illnesses caused by *S. aureus* is toxic shock syndrome. Toxins that elicit toxic shock syndrome, such as TSST-1 (toxic shock syndrome toxin 1), are superantigens that bind to nonvariable regions of antibodies and cause a severe immune response. In those strains of *S. aureus* that cause toxic shock syndrome, the toxin is often encoded on a small pathogenicity island called SaPI (*S. aureus* pathogenicity island). While not technically prophages, because they encode no recognizable phage proteins themselves, these SaPIs could be considered satellite phages, like P4, in that they are induced following infection by a specific type of phage. After they are induced, they replicate, probably with the help of phage proteins, and their DNA is packaged in a shortened version of the phage head, also like P4. The phage can then infect another strain of *S. aureus*, and the SaPI integrates into the chromosome using its own Int protein. This allows the SaPI to move very efficiently between strains of *S. aureus*, increasing the frequency of infection.

A number of different SaPIs exist, and they each have a specific relationship with their own type of helper phage related to λ. For example, the *S. aureus* phage 80α is able to induce and transfer a number of different SaPIs, while a closely related phage, φ11, cannot induce any of these but is able to induce and transfer a different one. The SaPIs normally remain in a dormant state in the chromosome because their transcription is repressed by a repressor, Stl, which binds to a region between two divergent promoters that service most of the genes of the pathogenicity island, simultaneously preventing transcription from both promoters. Phage infection inactivates the Stl repressor, allowing transcription from the promoters and inducing the SaPI.

As with the Shiga toxins, the mobilization of pathogenicity islands by phages raises special questions concerning the use of antibiotics. Some antibiotics, particularly those that inhibit the DNA gyrase, such as the fluoroquinolones, including ciprofloxacin (see chapter 1), cause DNA damage that can induce prophages. If the chromosomes of the cells in which the prophages exist also contain pathogenicity islands, these pathogenicity islands may be induced and encapsulated by the phage and moved into other, previously harmless bacteria.

Host Defenses Against Phage Infection

Bacteria live in a veritable sea of phages, and their survival as species depends upon defending themselves from them. The battle between bacteria and their phages is a continuous "arms race" in which the bacteria develop a defense against the phages and the phages develop a way to circumvent it (see Seed, Suggested Reading). Whether a particular phage defense mechanism works against a particular phage depends on what stage of the arms race the two are in. Also, many defense mechanisms against phage have other purposes in the cell. They may also defend against other incoming DNA elements, such as plasmids, or may play a role in developmental processes, such as killing some cells in a biofilm to aid dispersal from the overgrown biofilm. Many phage defense mechanisms are not a normal part of the chromosome but are encoded in genetic islands analogous to pathogenicity and fitness islands (see Koonin et al., Suggested Reading). Some of these are exchangeable DNA elements, such as prophages and plasmids, so they can be readily exchanged among various bacteria in the population.

Restriction-Modification Systems

Restriction-modification systems are the best-known type of phage defense mechanism. In chapter 13, we discuss the uses of restriction-modification systems in cloning and other manipulations with DNA. These systems degrade DNA by cutting at DNA sequences in which neighboring bases have not been properly modified. This allows the cell to distinguish its own DNA from incoming foreign DNAs, such as phage DNAs. In some cases, it can be the presence rather than the absence of a modified base that distinguishes a foreign DNA from the bacterium's own DNA. For example, T4 and many other T-even phages of *E. coli* and other enteric bacteria have the unusual base hydroxymethylcytosine in their DNA instead of the usual cytosine, and the presence of this unusual base makes them immune to many restriction endonucleases. However, it makes them susceptible to other restriction endonucleases, which only cut DNA in which the cytosines have been modified. In turn, many T-even phages attach glucose to the hydroxymethyl groups on their DNA, which then renders them immune to these restriction systems, and the arms race continues. Incidentally, the specificity of these restriction endonucleases for DNA with modified cytosines suggests that their major purpose in nature is to defend against this ubiquitous family of phages with hydroxymethylcytosine-containing DNA.

Abi Systems

Some phage defense systems kill the cell in response to phage infection. These systems are sometimes called Abi systems for abortive infection because the infection begins normally, but then something happens to the cell that causes it to die and the infection to be aborted (see Labrie et al., Suggested Reading). This has been compared to apoptosis in multicellular eukaryotes, where cells undergo programmed death if they are infected by a virus to prevent spread of the virus through the cells of the organism. There is an obvious advantage in a multicellular organism to such self-inflicted killing of a virus-infected cell, since it blocks multiplication of the virus before it can spread throughout the other cells of the organism. Such suicidal systems seem to make less sense for a single-celled bacterium. By killing themselves, they seem to be acting "altruistically" in helping other bacteria, even though they are doing nothing for themselves. However, even bacteria do not usually live as free single cells but are grouped together in structured multicellular communities, where they are surrounded by others of their kind. Preventing the spread of the phage through their relatives amounts to a kind of sibling selection, where the survival of other bacteria with identical genetic makeup is promoted by the sacrifice of the one infected cell.

The classic example of an Abi phage defense mechanism is the Rex exclusion system, encoded by phage λ genes *rexA* and *rexB*, which are among the few λ genes expressed in a cell that is lysogenic for λ. The mechanism of action of RexA and RexB, and how its activity is regulated, is still under investigation. Some other types of Abi systems are better understood. Two of them are directed against T4 and other T-even phages and stop the development of the infecting phage by cleaving a component of the translation apparatus after infection by the phage, thereby blocking host cell protein synthesis and the multiplication of the infecting phage. One such Abi system is a protease encoded by the defective prophage e14 in most *E. coli* K-12 strains. The protease is activated after infection by the binding of a short region of the newly synthesized major head protein, gp23, to EF-Tu (see chapter 2). The protease will cleave EF-Tu only when it is bound to that region of the phage head protein, so it is not active in an uninfected cell. Cleavage of EF-Tu blocks translation and prevents multiplication of the phage. The function of binding the region of the head protein to EF-Tu may be to delay translation to allow time for the N terminus of the head protein to be taken up by the chaperonin GroEL before translation of the head protein continues. However, this binding signals the Abi system that the cell has been infected by a T-even phage and that it should be sacrificed to prevent spread of the phage. The presence of the e14 defective prophage therefore provides its *E. coli* host with protection from infection by other phages.

Another Abi system is the Prr system, a lysine tRNA-specific RNase encoded by a prophage related to P1. This RNase is normally "masked" by binding to a restriction-modification system. The RNase is activated after T4 infection by a T4-encoded antirestriction peptide that dissociates the restriction-modification system, releasing the RNase, as well as by an increase in cellular dTTP levels required for phage DNA replication. The activated RNase then cleaves the host lysine tRNA, blocking translation and multiplication of the phage. By producing the antirestriction peptide in an attempt to inactivate the restriction-modification system, the phage inadvertently activates the RNase, blocking cellular translation and stopping its own development.

CRISPR/Cas Systems

A very different type of phage defense system is represented by CRISPR/Cas systems. These are adaptive immune systems reminiscent of our own adaptive immune system in that they can establish immunity to a heretofore unknown phage or transmissible plasmid after an initial exposure to it. The acronym CRISPR stands for clustered regularly interspaced short palindromic repeats, and their striking structure caused them to be noticed in bacterial genome sequences long before their role in phage defense was realized. CRISPR loci are composed of directly repeated identical sequences, often about 30 bp long, separated by "spacer" sequences, also often about 30 bp long. The number of repeats varies greatly, with some types of CRISPR loci having only a few repeats and others as many as 500. Almost all archaeal genomes and about a third of bacterial genomes appear to encode CRISPR/Cas systems. Multiple systems can occur in a single genome, and they also can be maintained on mobile elements.

The clue that CRISPR/Cas systems are a type of phage and plasmid defense system came when it was noticed that the spacer sequences often perfectly match a sequence from a phage or plasmid DNA found in that species of bacterium (see Barrangou et al. and Marraffini and Sontheimer, Suggested Reading). Different spacers come from the genomes of different phages or plasmids and from different sequences within the genomes of these phages and plasmids, explaining why the spacer sequences in a CRISPR array are different from each other.

A number of conserved proteins, encoded by *cas* (CRISPR-associated) genes that are adjacent to the locus, play roles in the acquisition of resistance to the phage (the adaptive phase) by the introduction of new spacer sequences or in the attack against incoming DNA, i.e., the interference or immunity phase, which prevents phage infection. As in other host defense systems, some phage have been found to encode enzymes that inhibit the CRISPR/Cas system, providing further evidence for a continuing arms race between the phage and its host. The CRISPR/Cas system has been developed into a set of very useful tools for genetic engineering, which will be discussed in chapter 13 along with more details of the various types of CRISPR/Cas systems.

Small Molecules and Phage Defense

All of the systems described above utilize host-encoded proteins to defend against invading phages. Recent studies have indicated that some bacteria use small molecules to inhibit phage infection. Many bacteria, including *Streptomyces* species, produce an array of small molecules that are known as secondary metabolites, some of which act as antibiotics (see chapters 1 and 2). At least some of these molecules appear to specifically inhibit phage replication at concentrations that do not inhibit bacterial growth (see Kronheim et al., Suggested Reading).

In contrast to the *Streptomyces* small molecules, which act broadly against double-stranded DNA phages, other small molecules may be used to very specifically target certain phage-host interactions. *V. cholerae* uses sensing of a small molecule to monitor cell density and control virulence properties (see chapter 12). A phage that infects *V. cholerae* has the ability to sense the same small molecule, binding of which triggers a switch from lysogenic to lytic development (see Silpe and Bassler, Suggested Reading), and simultaneously induces host cell virulence gene expression. This suggests that the phage has co-opted a host sensing system, with the potential of providing an opportunity for the cell to also control phage behavior.

Phage versus Phage

In conjunction with the phage versus host battle described above, it is also important for phages to compete against other phages in their environment. In addition to the immunity systems described above, which protect a lysogen against infection by another phage, there are also systems in which phages confer a variety of defense mechanisms to their host. These include systems that manipulate cell surface receptors that can allow binding of other phages, as well as phages that deliver new restriction-modification systems to protect the host.

Phages as Tools

Many key tools for molecular biology, including systems for DNA cloning, vectors for gene expression, and powerful enzymes, have been derived from analysis of phage systems. Many of these are discussed in chapter 13, and a few key examples are highlighted here.

Cloning Vectors

Along with the development of plasmids as vehicles for introducing and expressing cloned DNA in bacterial hosts, phages were also heavily used as tools in molecular cloning (see chapter 13). Of particular utility are phage λ and phage M13. Phage λ was especially useful for cloning large pieces of DNA, in particular because of the extensive knowledge of λ genetics and the development of techniques for packaging recombinant DNA into phage particles *in vitro* (see Becker et al., Suggested Reading). Hybrid plasmid-phage vectors, called cosmids, also took advantage of features of both the plasmid (stable intracellular replication) and phage (packaging of DNA into virions).

Because M13 and related phages encapsulate only one of the two DNA strands in their heads, these phages provide a convenient vehicle for cloned DNA to use in other applications that involve single-stranded DNA. Also, because filamentous phages like M13 have no fixed length and the phage particle is as long as its DNA, foreign DNAs of different lengths can be cloned into the phage DNA, producing a molecule that is longer than normal without disrupting the functionality of the phage. These features made M13 extremely important for development of DNA sequencing technologies, but it is less widely used now due to the advent of PCR technologies.

To use a single-stranded DNA phage vector, the double-stranded RF has to be isolated from infected cells, since most restriction endonucleases and DNA ligase require double-stranded DNA. A piece of foreign DNA can then be inserted into the RF by using restriction endonucleases, and the recombinant DNA can be used to **transfect** competent bacterial cells. The term "transfection" refers to the artificial initiation of a viral infection by introduction of viral DNA (see chapter 6). When the RF containing the cloned DNA replicates to form single-stranded progeny DNA, a single strand of the cloned DNA is packaged into the phage particle. The phage plaques obtained when these phage are plated are a convenient source of one strand of the cloned DNA. Which strand of the cloned DNA is represented in the single-stranded DNA depends on the orientation of the cloned DNA in the cloning vector. If it is cloned in one orientation, one of the strands is obtained; if it is cloned in the other orientation, the other strand is obtained.

Some plasmid cloning vectors have also been engineered to contain the *pac* site of a single-stranded DNA phage, which allows it to be packaged in a phage particle. They are called **phasmids** because they are mostly plasmids but have a phage *pac* site. Phasmids are easier than phage RFs to isolate and use for cloning, since they replicate as a double-stranded plasmid. If cells containing such a phasmid are infected with a helper phage, the phasmid is packaged into the helper phage head in a single-stranded form. Which strand is packaged depends on the orientation of the *pac* site in the phasmid. Another form of the cloning vectors, called **phagemids** because they are more phage and less plasmid, also have a plasmid replication origin but encode all of the proteins required to make a phage, so they do not need a helper phage to be packaged. Phasmids and phagemids have many uses in molecular genetics and biotechnology, and we discuss some of these in other chapters.

Phage Display

The relative ease with which phages can be manipulated and propagated has allowed the development of a powerful technology called **phage display**, which was awarded the Nobel Prize in 2018. The principle behind phage display is quite simple. Since most phages are just genes wrapped in a protein coat encoded by those genes, proteins can be expressed (displayed) on the surface of the phage if their coding sequence is translationally fused to the coding sequence for one of the coat proteins encoded by one of the coat protein genes (Figure 7.36). If a phage particle displaying that particular peptide can be separated from most of the other phages that are expressing different peptides, the phage can be propagated to make hundreds of millions of identical copies. This makes it easy to obtain plenty of DNA to determine the sequence of the DNA and, therefore, the peptide displayed by that phage. If the displayed peptide binds to a particular ligand, it is possible to isolate a phage expressing that peptide. This is very useful, because the extreme complexity of protein-protein interactions makes it difficult or impossible to predict what peptide sequences will bind tightly to a particular ligand, such as a hormone or an antibody. With phage display, the ligand is presented with a huge variety of different peptide sequences and asked to choose among them. Many phages have been adapted for phage display, including T7 and M13.

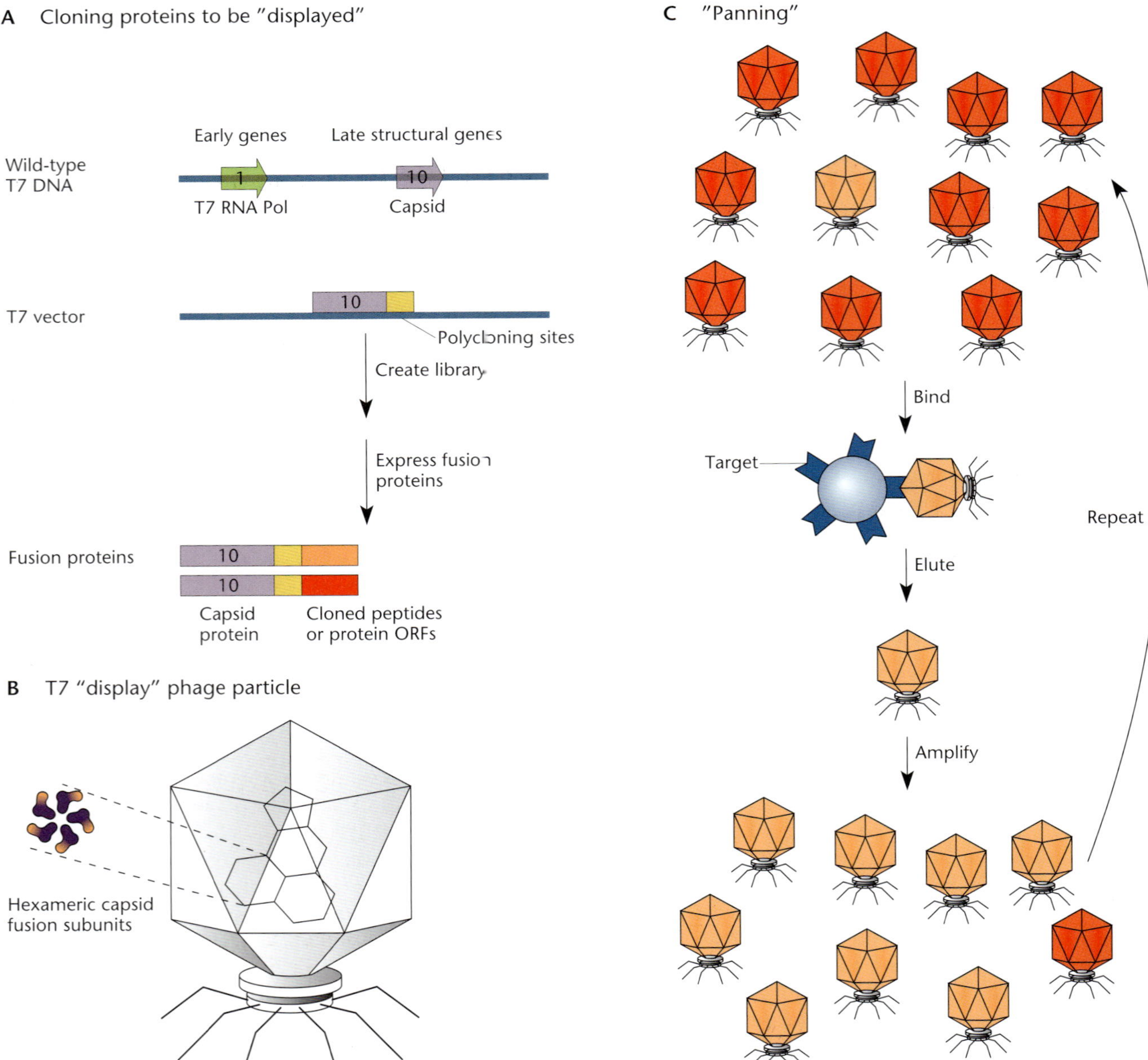

Figure 7.36 Use of phage T7 for phage display. **(A)** A randomized protein-coding sequence is inserted into the polycloning site in the phage DNA to make translational fusions in which random peptides (orange or red) have been fused to the head protein-coding region, gene *10*, of the phage (purple). The DNA is then introduced into cells by transfection or by packaging the DNA into a phage head and using the phage to infect cells. **(B)** When the phage multiply, the progeny of a phage displaying a particular peptide sequence all express that peptide on their surfaces fused to their head proteins. However, since a random collection of peptide-coding sequences has been fused to the gene *10* protein-coding sequence, the descendants of different phage in the population express different fusion peptides on their surfaces. Some of them (shown in orange) express a peptide that happens to bind to the target, while most (shown in red) express peptides that do not bind. When this mixture of phage is exposed to the target and the target is washed, those expressing a peptide that binds are preferentially retained, and more of the others are washed away. The retained phage can then be eluted from the target and propagated. This process is repeated multiple times to further enrich for the phage that specifically bind to the target.

Phage Therapy

The increase in global antibiotic resistance has led to a need for development of new antimicrobial therapies. One possible strategy is the use of phage to target specific disease organisms without causing harm to the general microbial population in and on our bodies. Phage therapy was investigated at some level many years ago, and interest in this concept has increased in recent years (see Abedon et al., Suggested Reading). Further understanding of phage biology and phage-host interactions will be crucial to the success of this approach.

1. Viruses that infect bacteria are called bacteriophages, or phages for short. The plural of phage is also phage if they are all the same type, while phages refers to more than one type.

2. The developmental cycle of a phage that leads to production of new phage particles is called the lytic cycle. Many phages utilize a complex program of gene expression to increase the efficiency of the developmental cycle.

3. The products of phage regulatory genes regulate the expression of other phage genes during development. One or more regulatory genes at each stage of development turns on the genes in the following stage and turns off the genes in the preceding stage, creating a regulatory cascade. In this way, all the information for the stepwise development of the phage can be preprogrammed into the phage DNA.

4. Phage T7 encodes an RNA polymerase that specifically recognizes the promoters for the late genes of the phage. Phage N4 encodes two RNA polymerases, one that is encapsulated in the phage head with the DNA and another that transcribes the middle genes. Only the late genes are transcribed by the host RNA polymerase.

5. Phage SPO1 uses the host RNA polymerase to transcribe its early genes, which include a gene for a new sigma factor, gp28. This sigma factor directs transcription of the middle genes, which include two genes that encode proteins that serve as a "split" sigma factor; these proteins work together to direct RNA polymerase to transcribe the late genes.

6. All the genes of phage T4 are transcribed by the host RNA polymerase, which undergoes many changes in the course of infection. The immediate-early genes are transcribed immediately after infection from σ^{70}-type promoters. The delayed-early genes are transcribed through antitermination of transcription from immediate-early genes. The transcription of the middle genes is from phage-specific promoters and requires transcriptional activators, MotA and AsiA, that remodel the σ^{70} RNA polymerase so it can recognize middle promoters. The late genes are transcribed from a different set of phage-specific promoters that require the two proteins, gp55 and gp33, which act together as a sigma factor to direct RNA polymerase to the late gene promoters. However, they cannot efficiently activate transcription by themselves and must bind to gp45, the T4 DNA replication sliding clamp, which explains the coupling of T4 late transcription to the replication of the T4 DNA.

7. Phage λ regulates its transcription through antitermination proteins N and Q, which allow transcription to continue into the early and late genes, respectively. These proteins bind to the RNA polymerase and allow it to transcribe through transcription termination sites.

8. All phages encode at least some of the proteins required to replicate their nucleic acids. They borrow others from their hosts. Making their own replication proteins may allow them to broaden their host range. Often, these phage-encoded replication proteins are more closely related to those of archaea and eukaryotes than to those of their own host.

9. The requirement of all known DNA polymerases for a primer creates obstacles to the replication of the extreme 5′ ends of linear DNAs. Phages solve this replication primer problem in different ways. Some replicate as circular DNA. Others form long concatemers by rolling-circle replication, by linking single DNA molecules end to end by recombination, or by pairing through complementary ends. They can then package genome-length DNAs from these concatemers by making staggered breaks in the DNA and resynthesizing the ends or package by a headful mechanism, leaving terminally redundant ends.

10. Phage endolysins break the bacterial cell wall, releasing the phages. In some bacteria, these endolysins are made early in infection in an active form but are not active until they pass through the inner membrane. Some holins form pores in the inner membrane that help endolysins pass through the inner membrane. They are inhibited by antiholins until it is time for

(continued)

Summary (continued)

lysis. Other endolysins are transported to the periplasm by the *sec* system and so have no need for holins to transport them through the membrane. However, they remain tethered to the membrane and inactive until it is time for lysis, when they are released by the holin, which probably acts by allowing the passage of protons and destroying the membrane potential.

11. Some phages are capable of lysogeny, in which they persist in the cell as prophages. A bacterium harboring a prophage is called a lysogen. In the lysogen, most of the phage gene products that are produced are involved in maintaining the prophage state. The prophage DNA can either be integrated into the chromosome or replicate autonomously as a plasmid.

12. The *E. coli* phage λ is the prototype of a lysogen-forming phage. It was the first such phage to be discovered, and it is the one to which all others are compared.

13. Whether λ enters the lytic or lysogenic cycles depends on interactions between a set of regulatory proteins and is affected by the MOI and the physiology of the infected cell. Most phage genes are repressed in the lysogenic state. DNA damage can result in induction of the prophage into a lytic cycle and production of phage particles.

14. Bacteriophages are ideal for illustrating the basic principles of classic genetic analysis, including recombination and complementation. To perform recombination tests with phages, cells are infected by two different mutants or strains of phage, and the progeny are allowed to develop. The progeny are then tested for recombinant types. To perform complementation tests with phages, the cells are infected by two different strains under conditions that are nonpermissive for both mutants. If the mutations complement each other, the phages will multiply. Recombination tests can be used to order mutations with respect to each other. Complementation tests can be used to determine whether two mutations are in the same functional unit or gene.

15. The genetic linkage map of a phage shows the relative positions of all of the known genes of the phage with respect to each other. Whether the linkage map of a phage is circular or linear depends on how the DNA of the phage replicates, how it is packaged into phage heads, and whether it contains an active site-specific recombination system.

16. Transduction occurs when a phage accidentally packages bacterial DNA into a head and carries it from one host to another. Generalized transduction is an outcome of a lytic infection, and essentially any region of bacterial DNA can be carried. Specialized transduction occurs after inaccurate excision of a prophage genome and results in transfer only of DNA located very close to the region at which the phage genome was integrated into the bacterial chromosome.

17. Not all types of phages make good generalized transducing phages. To be a good transducing phage, the phage must not degrade the host DNA after infection, must not have very specific *pac* sites, and must accept different-length DNAs or package DNA by a headful mechanism.

18. Some prophages play an important role in bacterial pathogenesis by encoding toxins that damage the mammalian host and increase the severity of the infection. This is called lysogenic conversion, because the prophage converts the bacterial host into a more virulent state.

19. Bacterial cells protect themselves from phages in a number of ways. Restriction-modification systems degrade incoming phage DNA. Abi (aborted-infection) systems kill the infected cell, preventing multiplication of the phage and its spread to other bacteria in the population. CRISPR/Cas systems are a type of adaptive immune system that targets incoming phage DNA for cleavage. CRISPR/Cas systems represent one of many molecular biology tools derived from the study of phage biology.

QUESTIONS FOR THOUGHT

1. Why do you suppose phages regulate their gene expression during development so that genes whose products are involved in DNA replication are transcribed before genes whose products become part of the phage particle? What do you suppose would happen if they did not do this?

2. What do you suppose the advantages are of a phage encoding its own RNA polymerase instead of merely changing the host RNA polymerase? What are the advantages of using the host RNA polymerase?

3. Why do phages often encode their own replicative machinery rather than depending on that of the host?

4. Why is generalized transduction generally a rare event? How is it influenced by phage host range?

5. Is generalized or specialized transduction more likely to have played an important role in bacterial evolution?

6. Why is only one protein, Int, required to integrate the phage DNA into the chromosome while two proteins, Int and Xis, are required to excise it? Why not just make one different Int-like protein that excises the prophage?

SUGGESTED READING

Abedon ST, García P, Mullany P, Aminov R. 2017. Editorial: phage therapy: past, present and future. *Front Microbiol* **8:**981.

Barrangou R, Fremaux C, Deveau H, Richards M, Boyaval P, Moineau S, Romero DA, Horvath P. 2007. CRISPR provides acquired resistance against viruses in prokaryotes. *Science* **315:**1709–1712.

Becker A, Murialdo H, Gold M. 1977. Studies on an in vitro system for the packaging and maturation of phage lambda DNA. *Virology* **78:**277–290.

Benzer S. 1961. On the topography of genetic fine structure. *Proc Natl Acad Sci USA* **47:**403–415.

Black LW. 2015. Old, new, and widely true: the bacteriophage T4 DNA packaging mechanism. *Virology* **479-480:**650–656.

Chamakura K, Young R. 2019. Phage single-gene lysis: finding the weak spot in the bacterial cell wall. *J Biol Chem* **294:**3350–3358.

Choi KH, McPartland J, Kaganman I, Bowman VD, Rothman-Denes LB, Rossmann MG. 2008. Insight into DNA and protein transport in double-stranded DNA viruses: the structure of bacteriophage N4. *J Mol Biol* **378:**726–736.

Court DL, Oppenheim AB, Adhya SL. 2007. A new look at bacteriophage λ genetic networks. *J Bacteriol* **189:**298–304.

Crick FHC, Barnett L, Brenner S, Watts-Tobin RJ. 1961. General nature of the genetic code for proteins. *Nature* **192:**1227–1232.

de Jonge PA, Nobrega FL, Brouns SJJ, Dutilh BE. 2019. Molecular and evolutionary determinants of bacteriophage host range. *Trends Microbiol* **27:**51–63.

Erez Z, Steinberger-Levy I, Shamir M, Doron S, Stokar-Avihail A, Peleg Y, Melamed S, Leavitt A, Savidor A, Albeck S, Amitai G, Sorek R. 2017. Communication between viruses guides lysis-lysogeny decisions. *Nature* **541:**488–493.

Feiner R, Argov T, Rabinovich L, Sigal N, Borovok I, Herskovits AA. 2015. A new perspective on lysogeny: prophages as active regulatory switches of bacteria. *Nat Rev Microbiol* **13:**641–650.

Gottesman M. 1999. Bacteriophage λ: the untold story. *J Mol Biol* **293:**177–180.

Hatfull GF. 2015. Dark matter of the biosphere: the amazing world of bacteriophage diversity. *J Virol* **89:**8107–8110.

Kahn ML, Ziermann R, Dehò G, Ow DW, Sunshine MG, Calendar R. 1991. Bacteriophage P2 and P4. *Methods Enzymol* **204:**264–280.

King RA, Wright A, Miles C, Pendleton CS, Ebelhar A, Lane S, Parthasarathy PT. 2011. Newly discovered antiterminator RNAs in bacteriophage. *J Bacteriol* **193:**5784–5792.

Koonin EVE, Makarova KSK, Wolf YI. 2017. Evolutionary genomics of defense systems in Archaea and Bacteria. *Annu Rev Microbiol* **71:**233–261.

Kristensen DM, Cai X, Mushegian A. 2011. Evolutionarily conserved orthologous families in phages are relatively rare in their prokaryotic hosts. *J Bacteriol* **193:**1806–1814.

Kronheim S, Daniel-Ivad M, Duan Z, Hwang S, Wong AI, Mantel I, Nodwell JR, Maxwell KL. 2018. A chemical defence against phage infection. *Nature* **564:**283–286.

Labrie SJ, Samson JE, Moineau S. 2010. Bacteriophage resistance mechanisms. *Nat Rev Microbiol* **8:**317–327.

Losick R, Pero J. 2018. For want of a template. *Cell* **172:**1146–1152.

Lwoff A. 1953. Lysogeny. *Bacteriol Rev* **17:** 269–337.

Marraffini LA, Sontheimer EJ. 2008. CRISPR interference limits horizontal gene transfer in staphylococci by targeting DNA. *Science* **322:** 1843–1845.

McLeod SM, Kimsey HH, Davis BM, Waldor MK. 2005. CTXφ and *Vibrio cholerae*: exploring a newly recognized type of phage-host cell relationship. *Mol Microbiol* **57:**347–356.

Mosig G. 1998. Recombination and recombination-dependent DNA replication in bacteriophage T4. *Annu Rev Genet* **32:**379–413.

Ofir G, Sorek R. 2018. Contemporary phage biology: from classic models to new insights. *Cell* **172:**1260–1270.

Ptashne M. 2004. *A Genetic Switch: Phage Lambda Revisited*, 3rd ed. Cold Spring Harbor Laboratory Press, Cold Spring Harbor, NY.

Qiao X, Qiao J, Mindich L. 2003. Analysis of specific binding involved in genomic packaging of the double-stranded-RNA bacteriophage φ6. *J Bacteriol* **185:**6409–6414.

Roberts JW, Yarnell W, Bartlett E, Guo J, Marr M, Ko DC, Sun H, Roberts CW. 1998. Antitermination by bacteriophage lambda Q protein. *Cold Spring Harb Symp Quant Biol* **63:**319–325.

Salmond GP, Fineran PC. 2015. A century of the phage: past, present and future. *Nat Rev Microbiol* **13:**777–786.

Seed KD. 2015. Battling phages: how bacteria defend against viral attack. *PLoS Pathog* **11:** e1004847.

Silpe JE, Bassler BL. 2019. A host-produced quorum-sensing autoinducer controls a phage lysis-lysogeny decision. *Cell* **176:**268–280.e13.

Studier FW. 1969. The genetics and physiology of bacteriophage T7. *Virology* **39:**562–574.

Takeshita D, Tomita K. 2012. Molecular basis for RNA polymerization by Qβ replicase. *Nat Struct Mol Biol* **19:**229–237.

Wagner PL, Neely MN, Zhang X, Acheson DWK, Waldor MK, Friedman DI. 2001. Role for a phage promoter in Shiga toxin 2 expression from a pathogenic *Escherichia coli* strain. *J Bacteriol* **183:**2081–2085.

Waldor MK, Mekalanos JJ. 1996. Lysogenic conversion by a filamentous phage encoding cholera toxin. *Science* **272:**1910–1914.

Yosef I, Goren MG, Globus R, Molshanski-Mor S, Qimron U. 2017. Extending the host range of bacteriophage particles for DNA transduction. *Mol Cell* **66:**721–728.e3.

Young R. 2014. Phage lysis: three steps, three choices, one outcome. *J Microbiol* **52:**243–258.

Yuan AH, Hochschild A. 2009. Direct activator/co-activator interaction is essential for bacteriophage T4 middle gene expression. *Mol Microbiol* **74:**1018–1030.

Zinder ND, Lederberg J. 1952. Genetic exchange in *Salmonella*. *J Bacteriol* **64:**679–699.

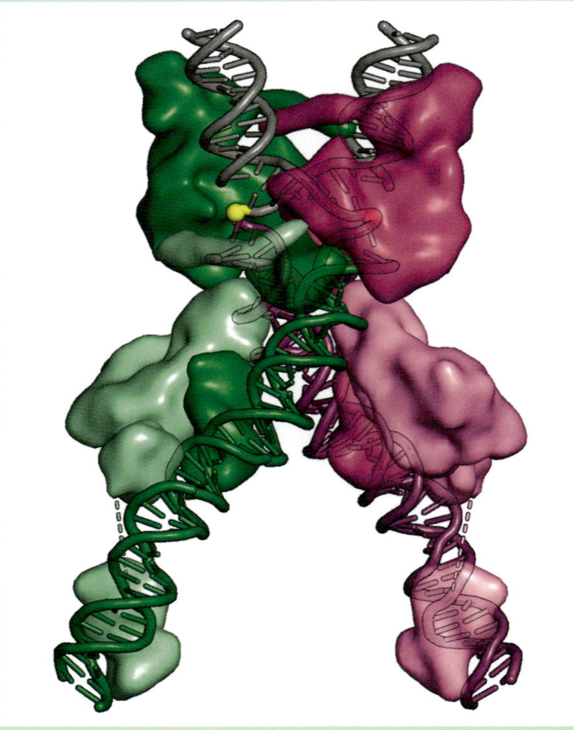

Structure of a tetramer of Mu transposase in complex with the mobile element ends (green and pink DNAs) and target DNA (gray DNA). The target DNA phosphate that has become covalently linked to the element ends is shown as a yellow sphere, and the proteins are shown as smoothed, semi-transparent surfaces. Courtesy Phoebe A. Rice, Univ. of Chicago.

8 Transposition, Site-Specific Recombination, and Families of Recombinases

RECOMBINATION IS THE BREAKING AND REJOINING of DNA in new combinations. In homologous recombination, which accounts for most recombination in the cell, the breaking and rejoining occur only between regions of two DNA molecules that have similar or identical sequences. Homologous recombination requires that the two DNA molecules interact through complementary base pairing, for which they must have the same sequence (see chapter 9). However, other types of recombination, known as **nonhomologous recombination**, also occur in cells. As the name implies, these types of recombination do not depend on extensive homology between the two DNA sequences involved. Some types of nonhomologous recombination are due to the mistaken breaking and rejoining of DNA by enzymes such as topoisomerases (see chapter 1). Other types are not mistakes and have specific functions in the cell. These types depend on specific enzymes that promote recombination between different regions in DNA, which may or may not have sequences in common. This chapter addresses some of these examples of nonhomologous recombination in bacteria and the mechanisms involved, including transposition by DNA transposons, the site-specific recombination that occurs during the integration and excision of prophages and other DNA elements, the inversion of invertible sequences, and the resolution of cointegrates by resolvases and elements that move via an RNA intermediate. Evidence shows that the enzymes that perform these various functions have much in common across all three domains of life.

Transposition

Transposons are elements that can hop, or **transpose**, from one place in DNA to another. Transposable elements were first discovered in corn by Barbara McClintock in the early 1950s and about 20 years later in bacteria by others. Transposons are now known to exist in all organisms on Earth, including humans. In fact, estimates suggest that up to half of the human genome may be transposons. The process by which a transposon moves is called **transposition**, and the enzymes that promote transposition are a type of recombinase called **transposases**. The transposon itself usually encodes its own transposase so that each time it hops, it carries with it the ability to move again. For this reason, transposons have been called "jumping genes." Not all elements that can move are called transposons. In bacteria, the word "transposon" typically refers to DNA transposons that always move as DNA. There are other types of elements that utilize an RNA intermediate. These elements will be described later in this chapter.

Although transposons are common in all domains of life, they are best understood in bacteria, where they play an important role in evolution. While transposons are true parasites of the host, they also can encode beneficial gene products. For example, the spread of antibiotic resistance in bacteria is primarily the result of resistance genes found in transposons, which can be carried between bacteria during transfer of promiscuous plasmids or via transducing phages. As in many fields of study, the growing availability of inexpensive DNA sequencing is providing a fascinating and more comprehensive picture of the diversity of transposons and the role they play in evolution. While a variety of molecular mechanisms are utilized for transposition, and new mechanisms certainly await discovery, a few mechanisms explain the majority of events found in bacteria. These mechanisms are not limited to bacteria but are found in all domains of life. These mechanisms will be the focus of this chapter.

Overview of Transposition

The net result of transposition is that the transposon now resides at a new place in the genome, where it was not originally found. Many transposons are essentially "cut out" of one DNA and inserted into a new position (Figure 8.1), whereas with other transposons a copy is made which is then inserted elsewhere. Regardless of the type of transposon, however, the DNA from which the transposon originated is called the **donor DNA**, and the DNA into which it hops is called the **target** or **recipient DNA**.

In classically defined transposition events, the transposase enzyme catalyzes breakage of the donor DNA at the ends of the transposon and then inserts the transposon into the target DNA. However, the details of the mechanism can vary. Some types of transposons may exist free of other DNA during the act of transposition, but many transposons, before and after they move, remain contiguous with other, flanking DNA molecules. Later in this chapter, we discuss more detailed models for the various types of transposition.

Figure 8.1 Overview of transposition. See the text for details.

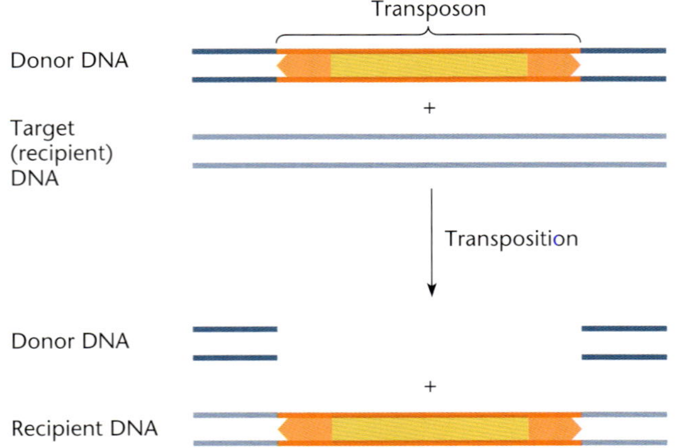

Transposition must be tightly regulated to occur only rarely; otherwise, the cellular DNA would become riddled with transposon copies. Transposons have evolved elaborate mechanisms to regulate transposition so that they move very infrequently and do not often kill the host cell. We mention some of these mechanisms later in this chapter. The frequency of transposition varies from about one in every 10^3 cell divisions to about one in every 10^8 cell divisions, depending on the type of transposon. Thus, the chance of a particular transposon hopping into a gene and inactivating it is sometimes not much higher than the chance that a gene will be inactivated by other types of mutations (see chapter 3).

Structure of Bacterial DNA Transposons

There are many types of bacterial transposons. Some of the smaller elements are about 1,000 bp long and carry only the gene for the transposase that promotes their movement. Larger transposons may also contain one or more other genes that are responsible for regulating transposition or for specialized targeting functions. In other cases, genes directly beneficial to the element or beneficial to the host may also be carried on the element, such as those for resistance to an antibiotic.

Classically defined DNA transposons have *cis*-acting sites at the ends of the element that are recognized by the transposase. These *cis*-acting sites are in opposite orientation and hence are referred to as **inverted repeats** (IRs) (Figure 8.2). As discussed in chapter 1, two regions of DNA are IRs if the sequence of nucleotides on one strand in one region, when read in the 5′-to-3′ direction, is the same or almost the same as the 5′-to-3′ sequence of the opposite strand in the other region. The transposition reaction does not occur until a **synapse** structure is formed, where the transposase proteins bound to each of the IR sequences also bind to one another (Figure 8.2). IRs are one of the features that help to identify transposons in a genome sequence.

Another feature associated with DNA transposons is short **direct repeats** that result from the process of integration. Direct repeats have the same 5′-to-3′ sequence of nucleotides on the same strand. The direct repeats are a by-product of the way the element is joined into the target DNA during transposition. As shown in Figure 8.3, the target DNA originally contains only one copy of the sequence at the place where the transposon inserts. When the 3′ ends of the element are joined to staggered posi-

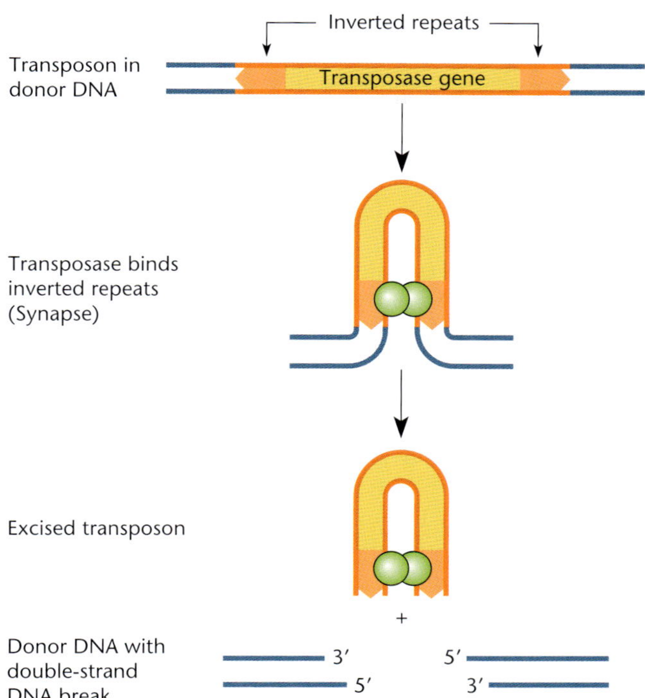

Figure 8.2 Steps in transposon excision. Inverted repeats (IRs) (shown as orange shapes) that are recognized by the element-encoded transposase (green circles) characterize the ends of the transposon. Genes carried by the transposon are not shown but reside in the region shaded light orange. A synapse is formed when the transposase binds the IRs and the ends are paired with the transposase. This signals the transposase to carry out the cleavage and joining events that underlie the process of transposition. A completely excised element is not found for all transposition processes, and not all transposons leave a double-strand break in the donor DNA.

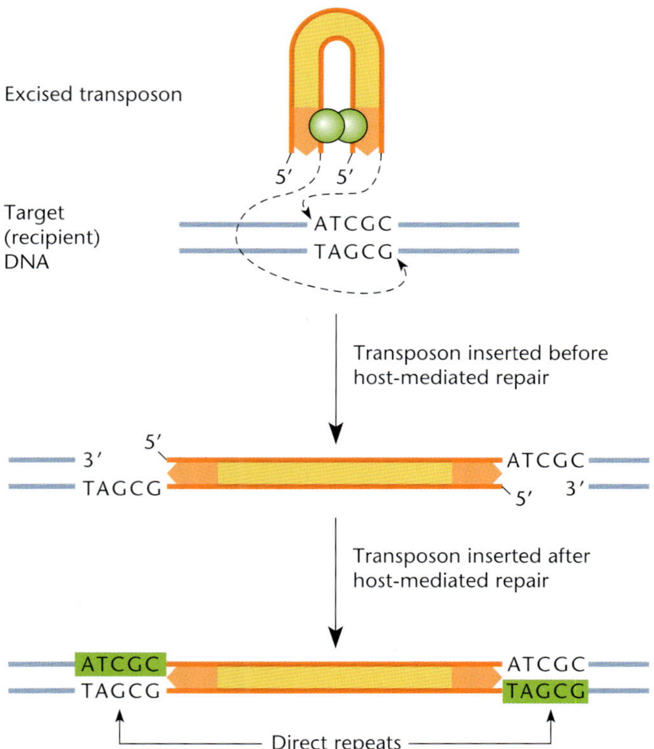

Figure 8.3 Steps in transposon insertion. The transposon inserts into a target DNA using staggered joins to the top and bottom strands (an excised transposon species found with many elements is shown). The process of inserting with staggered joins into the target DNA leaves gaps flanking the transposon that must be filled by a host DNA polymerase. The new DNA is highlighted. Often, a few bases of sequence from the donor DNA also remain bound to the 5′ ends of the element, which must also be processed by host enzymes. The effect of staggered joining events and the subsequent host repair leads to the direct repeats that are indicative of transposition.

tions (i.e., not directly across the DNA backbone) in a target DNA, single-stranded gaps are left behind. These gaps are repaired by replication by DNA polymerase I, which leads to the directly repeated sequence flanking the element. These direct repeat sequences are also referred to as **target site duplications**. DNA polymerase I also removes a few bases that are often carried over from the donor DNA. Most transposons can insert into many places in DNA and have little or only limited target specificity. Thus, the duplicated sequence depends on the sequence at the site in the target DNA into which the transposon inserted. However, even though the duplicated sequences differ, the number of duplicated base pairs is characteristic of each transposon. Some duplicate as few as 2 bp, and others duplicate as many as 9 bp or more.

Types of Bacterial DNA Transposons

In this section, we describe some of the common types of DNA transposons based on basic genetic features.

INSERTION SEQUENCE ELEMENTS

The smallest bacterial transposons are called **insertion sequence (IS) elements**. These transposons are usually only about 750 to 2,000 bp long and encode little more than the transposase enzymes that promote their transposition.

Because IS elements carry no selectable genes, they were discovered only because they generally inactivate a gene if they happen to insert into it. The first IS elements were detected as a type of *gal* mutation that was unlike any other known mutation. This type of mutation resembled deletion mutations in that it was nonleaky; however, unlike deletion mutations, it could revert, albeit at a much lower frequency than base pair changes or frameshifts. Such anomalous *gal* mutations were also very polar and could prevent the transcription of downstream genes (see chapter 2). Later work showed that these mutations resulted from insertion of about 1,000 bp of DNA into a *gal* gene.

Originally, four IS elements were found in *Escherichia coli*: IS*1*, IS*2*, IS*3*, and IS*4*. Most common laboratory strains of *E. coli* K-12 contain approximately six copies of IS*1*, seven copies of IS*2*, and fewer copies of the others. Although almost all bacteria carry IS elements, IS elements tend to have a level of species specificity where the most closely related bacteria have more closely related IS elements. To date, thousands of different IS elements have been found in bacteria. Plasmids also often carry IS elements, which are important in the assembly of the plasmid itself (see below) and in the formation of Hfr strains (see chapter 5). Although the original IS elements were discovered only because they had inserted into a gene, causing a detectable phenotype, IS elements are now more often discovered during genomic sequencing. In addition, DNAs that are introduced into *E. coli* by DNA cloning can also be a target for endogenous IS elements. There are many examples where the IS element sequences inadvertently show up in DNA that was passaged though *E. coli* before being subjected to sequencing. An interesting strategy for getting around this problem involved the construction of an *E. coli* K-12 derivative from which all of the IS elements were removed (see Pósfai, Suggested Reading).

COMPOSITE TRANSPOSONS

Sometimes two IS elements of the same type come together to form a larger transposon, called a **composite transposon**, by bracketing other genes. Figure 8.4 shows the structures of three composite transposons, Tn*5*, Tn*9*, and Tn*10*. Tn*5* consists of genes for kanamycin resistance (Kan^r), bleomycin resistance (Ble^r), and streptomycin resistance (Str^r) bracketed by copies of an IS element called IS*50*. Tn*9* has two copies of IS*1* bracketing a chloramphenicol resistance (Cam^r) gene. In Tn*10*, two copies of IS*10* flank a gene for tetracycline resistance (Tet^r). Some composite transposons, such as Tn*9*, have the bracketing IS elements in the same orientation, whereas others, including Tn*5* and Tn*10*, have them in opposite orientations.

Figure 8.5 compares transposition of an IS element with transposition of a composite transposon. Each IS element can transpose independently as long as the transposase acts on both of its ends. However, because all the ends of the IS elements in a composite transposon are the same (because both IS elements are the same type), a transposase encoded by one of the IS elements can recognize the ends of either IS element. When such a transposase acts on the IRs at the farthest ends of a composite transposon, the two IS elements transpose as a unit, bringing along the genes between them. These two IRs are called the "outside ends" of the two IS elements because they are the farthest from each other.

The two IS elements that form composite transposons are often no longer functional on their own because of mutations in the transposase gene of one of the elements. Thus, only one of the IS elements encodes an active transposase. However, this transposase can act on the outside ends to promote transposition of the composite transposon. Mutations in one or both of the "inside" IRs also enhance transposition of the composite transposon instead of the IS elements as separate units.

Figure 8.4 Structures of some composite transposons. The left (L) and right (R) ends of the elements are indicated. The commonly used Kan^r genes and the Cam^r gene come from Tn*5* and Tn*9*, respectively. The active transposase gene is in one of the two insertion sequence (IS) elements. Note that the IS elements can be in either the same or opposite orientation. Str^r, gene encoding streptomycin resistance; Tet^r, gene encoding tetracycline resistance; Ble^r, gene encoding bleomycin resistance.

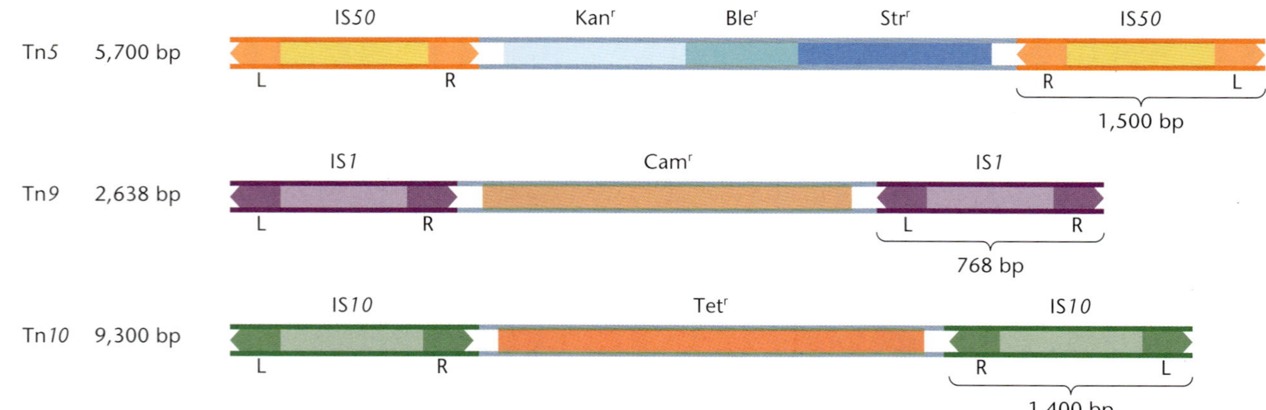

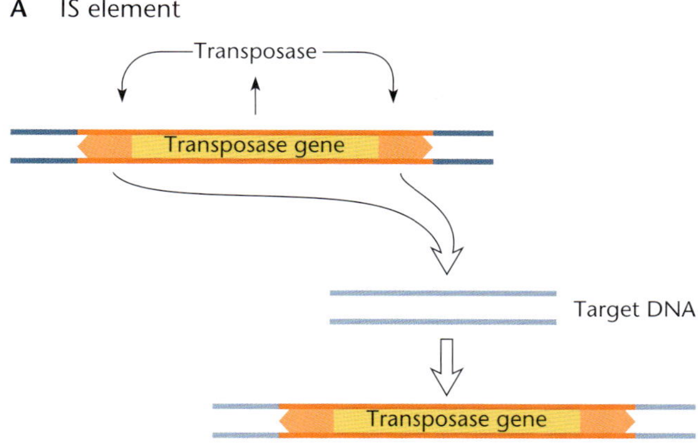

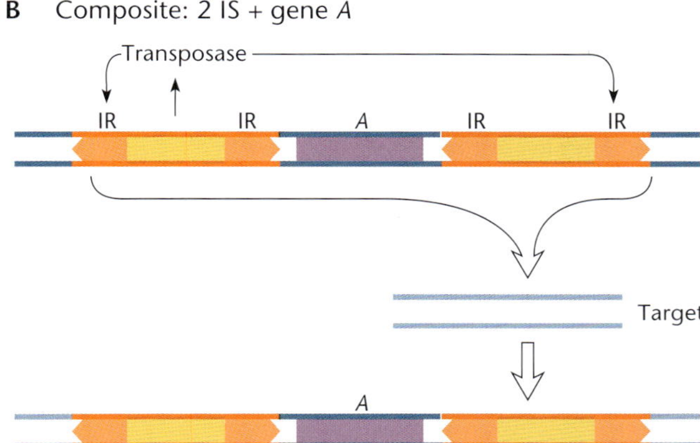

Figure 8.5 Two insertion sequence (IS) elements can transpose any DNA between them. **(A)** Action of the transposase at the ends of an isolated IS element causes it to transpose. **(B)** Two IS elements of the same type are close to each other in the DNA. Action of the transposase on their outside ends causes them to transpose together, carrying along the DNA between them. *A* denotes the DNA between the IS elements (purple), and the arrows indicate the inverted repeat (IR) sequences at the ends of the IS elements. The light blue lines represent the target DNA.

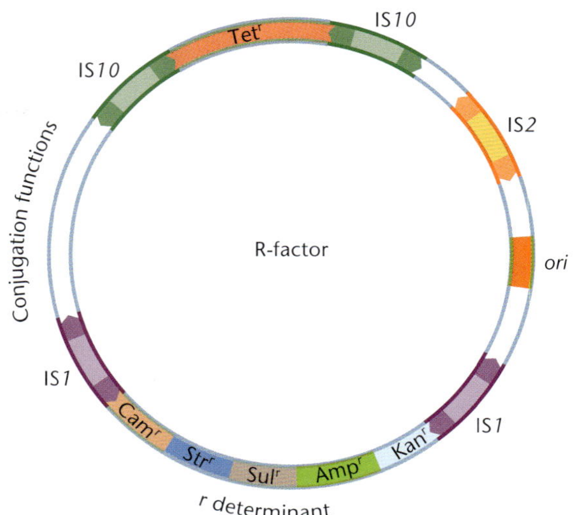

Figure 8.6 R factors, or plasmids containing many resistance genes, may have been assembled, in part, by insertion sequence (IS) elements. The Tet^r gene is bracketed by IS*10* elements (forming the Tn*10* transposon), and the region containing the other resistance genes (the r determinant) is bracketed by IS*1* elements.

Assembly of Plasmids by IS Elements

If two IS elements of the same type happen to insert close to each other on the same DNA, a composite transposon is born. These transposons have not yet evolved a defined structure like the named transposons (e.g., Tn*10*) described above. Nevertheless, the two IS elements can transpose any DNA between them. In this way, "cassettes" of genes bracketed by IS elements can be moved from one DNA molecule to another.

Many plasmids seem to have been assembled from such cassettes. Figure 8.6 shows a naturally occurring plasmid that carries genes for resistance to many antibiotics. Such plasmids are historically called R factors, because they confer resistance to so many different antibiotics (see chapter 4). Notice that many of the resistance genes on the plasmid are bracketed by copies of the same IS elements. IS*10* flanks the tetracycline resistance gene, and IS*1* brackets the genes for resistance to many other antibiotics. Apparently, the plasmid was assembled in nature by insertion of resistance genes onto the plasmid from some other DNA via the bracketing IS elements. In principle, any two transposons of the same type can move other DNA lying between them by a similar mechanism, but because IS elements are the most common transposons and often exist in more than one copy per cell, they probably play the major role in the assembly of plasmids.

NONCOMPOSITE TRANSPOSONS

Composite transposons are not the only ones that carry genes beyond transposition functions such as antibiotic resistance genes. Such genes can also be an integral part of transposons that are sometimes referred to as **noncomposite transposons** (Figure 8.7). There are multiple types of noncomposite elements that carry a variety of different kinds of genes. While in a clinical setting antibiotic or biocide resistance may be common, other types of genes will be found associated with transposons in more natural settings. In some cases, the genes offer an advantage to the bacterial host, while in other cases they more directly benefit the transposon. These genes that are not involved in transposition are often called cargo or passenger genes. Even closely related elements that share almost identical transposition functions can have very different types of cargo genes. One example comes from

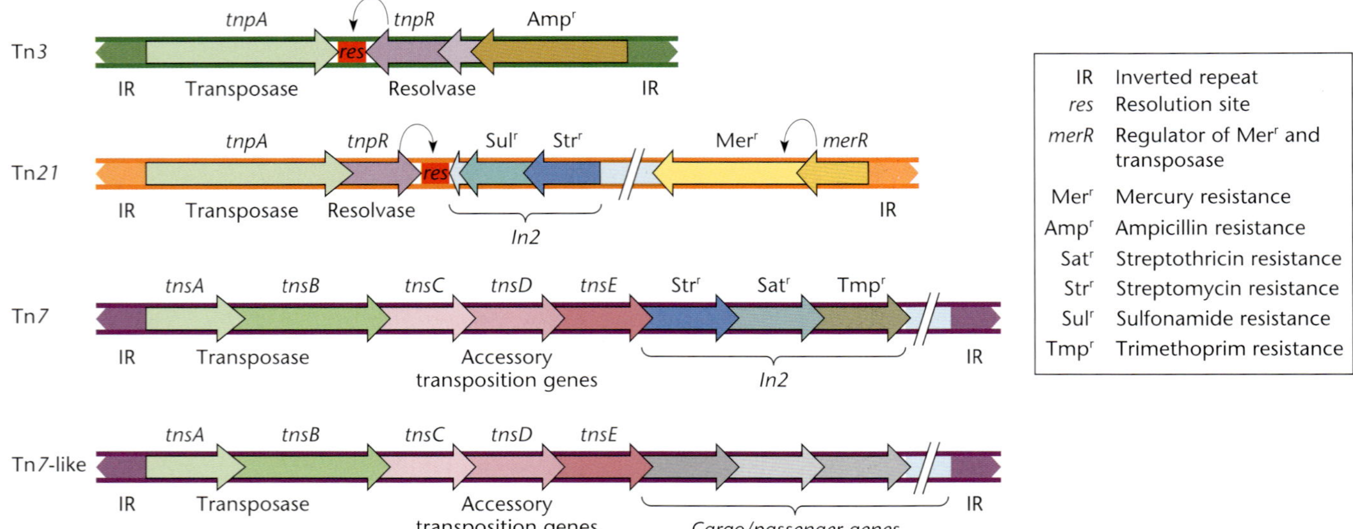

Figure 8.7 Some examples of noncomposite transposons. The positions of the transposase-encoding genes are shown, as well as accessory transposition proteins or resolvase. The open reading frames (ORFs) encoding the proteins are boxed with selected phenotypes shown. The terminal inverted repeat (IR) ends are indicated. The *res* site where resolvase acts is indicated in red. In*2* represents the integron cassettes. The arrows indicate the sites at which the proteins act.

Tn*7* and Tn*7*-like elements where the same set of transposition genes can be found with different sets of genes mobilized between the terminal IRs (see Peters, Suggested Reading) (Figure 8.7).

How genes become associated with noncomposite transposons is not as clear as in the case of the composite elements described above. One type of system that is found in some noncomposite elements is integron cassettes. Integron cassettes utilize site-specific recombination for movement and are described later in the chapter. Unrelated noncomposite transposons can also carry related genes. For example, the same types of integron cassette systems can associate with different families of transposons, such as found with Tn*7* and Tn*21*. New cargo genes are probably integrated rarely into noncomposite elements via a variety of random recombination events, but if the cargo genes provide important new functions benefiting the host or element, they are likely to spread quickly through bacterial populations.

NONAUTONOMOUS ELEMENTS

The ISs and transposons described above can be described as **autonomous elements** because they encode both the transposase enzyme and the *cis*-acting sites where they act as one functional unit. Bacterial genomes can also encode **nonautonomous elements** in which the *cis*-acting sequences required for mobilizing an element are configured around a set of genes that lack the transposition functions. These elements will only be mobilized when another autonomous element also resides in the same cell that can encode the transposase enzyme that acts in *trans* on either the autonomous or nonautonomous element. In chapter 5 we learned about a similar situation with mobilizable plasmids that rely on the conjugation functions encoded in other plasmids for transfer between bacteria. Interestingly, the original discovery of transposons by Barbara McClintock stemmed from the observation that certain crosses between different lineages of corn allowed nonautonomous transposons (*Ds* elements) to be mobilized when recombined into the same background as an autonomous transposon (*Ac* elements). Nonautonomous elements known as miniature inverted-repeat transposable elements (MITEs) can also be found that lack a transposase or any cargo genes.

While nonautonomous elements are sometimes found naturally, synthetic nonautonomous elements are commonly used tools in bacterial genetics and genomics as described below and in other chapters.

Assays of Transposition

To study transposition, we must have assays to monitor the process. As mentioned above, insertion elements were discovered because they generate mutations when they hop into a gene. However, this is usually not a convenient way to assay transposition, because transposition is infrequent, and it is laborious to distinguish insertion mutations from the myriad other mutations that can occur. If the transposon carries a resistance gene, the job of assaying transposition is easier, but how do we know if a transposon has moved from place to place within the cell? In the case of transposons that naturally encode or have been engineered to encode antibiotic resistance, the cells are resistant to the antibiotic no matter where the transposon is inserted in the cellular DNA. Therefore, moving from one place to another makes no difference in the level of resistance of the cell. Obviously, detecting transposition requires special methods.

SUICIDE VECTORS

One way to assay transposition is with **suicide vectors**. Any DNA, including plasmid or phage DNA, that cannot replicate (i.e., is not a replicon) in a particular host can be used as a suicide vector. These DNAs are called suicide vectors because, by entering cells in which they cannot replicate, they essentially kill themselves because they are not passed to progeny cells. To assay transposition with a suicide vector, we use one to introduce a transposon carrying an antibiotic resistance gene into an appropriate host. The way in which the suicide vector itself is introduced into the cells depends on its source. If it is a phage, the cells could be infected with that phage. If it is a plasmid, it could be introduced into the cells through conjugation. However, whatever method is used, it should be very efficient, since transposition is a rare event.

Once in the cell, the suicide vector remains unreplicated and eventually is lost as the cell population grows. The only way the transposon can survive and confer antibiotic resistance to the progeny cells is by moving to another DNA molecule that is capable of autonomous replication in those cells, for example, a replicating plasmid or the chromosome. Therefore, when the cells under study are plated on antibiotic-containing agar and incubated, the appearance of colonies, as a result of the multiplication of antibiotic-resistant bacteria, is evidence for transposition. These cells have been mutagenized by the transposon, since the transposon has moved into a cellular DNA molecule—either the chromosome or a plasmid—causing insertion mutations.

Phage Suicide Vectors

Some derivatives of phage λ have been modified to be used as suicide vectors in *E. coli*. These phages have been rendered incapable of replication in nonsuppressing hosts by the presence of nonsense codons in their replication genes *O* and *P* (see chapter 7). They have also been rendered incapable of integrating into the host DNA by deletion of their attachment region, *attP*. Such a λ phage can be propagated on an *E. coli* strain carrying a nonsense suppressor. However, in a nonsuppressor *E. coli*, it cannot replicate or integrate. Because of the narrow host range of λ, these suicide vectors can normally be used only in strains of *E. coli* K-12. Other bacteriophages have been adapted for delivery into other types of bacteria (see Bardarov et al., Suggested Reading).

Plasmid Suicide Vectors

Plasmid cloning vectors can also be used as suicide vectors, provided that the plasmid cannot replicate in the cells in which transposition is occurring. The plasmid containing the transposon with a gene for antibiotic resistance can be propagated in a host in which it can replicate and is then introduced into a cell in which it cannot replicate. In principle, any plasmid with a conditional-lethal mutation (e.g., a nonsense or temperature-sensitive mutation) in a gene required for plasmid replication can be used as a suicide vector. The plasmid could be propagated in the permissive host or under permissive conditions and then introduced into a nonpermissive host or into the same host under nonpermissive conditions, depending on the type of mutation. Alternatively, a narrow-host-range plasmid could be used (see chapter 4). It can be propagated in a host in which it can replicate and be introduced into a different bacterial species in which it cannot replicate.

Many general methods for assaying transposition are based on promiscuous self-transmissible plasmids because the most efficient way to introduce a plasmid into cells is by conjugation, which can approach 100% under some conditions. If the plasmid containing the transposon contains a *mob* region, it can be mobilized into the recipient cell by using the Tra functions of a self-transmissible plasmid present in the donor cell (see chapter 5). This technique is most highly developed for *Proteobacteria* and especially the *Gammaproteobacteria* (see Box 5.2). Taking advantage of the extreme promiscuity of some self-transmissible plasmids of *Gammaproteobacteria*, the plasmid can be mobilized into almost any *Proteobacteria*. If the mobilizable plasmid has a narrow host range, it might not be able to replicate in the host into which it has been mobilized and will eventually be lost. ColE1-derived plasmids into which a *mob* site has been introduced are often the suicide vectors of choice in such applications because they can replicate only in some enteric bacteria, including *E. coli*, and so can be used as suicide vectors in any distantly related bacteria. The movement of the transposon can then be assayed if it carries a selectable gene, such as for antibiotic resistance, which is expressed in the recipient host.

THE MATING-OUT ASSAY FOR TRANSPOSITION

Transposition can also be assayed by using the "mating-out" assay, which is based on conjugation. In this assay, a transposon in a nontransferable plasmid or the chromosome is not transferred into other cells unless it inserts into a plasmid that is transferable. Figure 8.8 shows a specific example of a mating-out assay with *E. coli*.

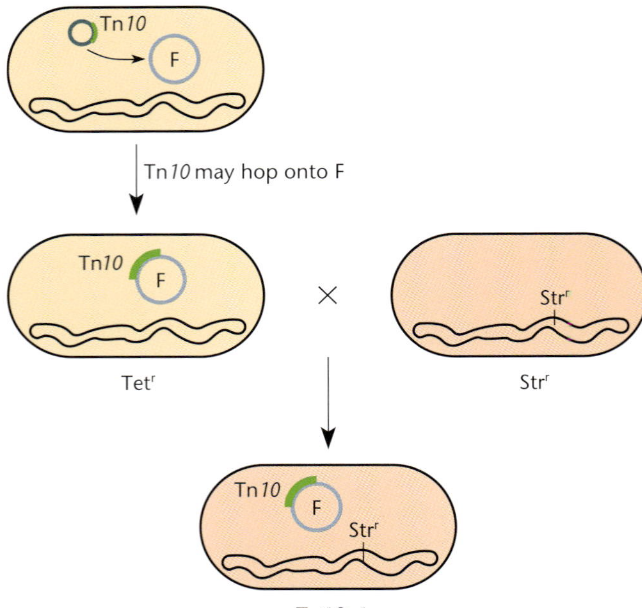

Figure 8.8 Example of a mating-out assay for transposition. See the text for details.

In the example shown, transposon Tn*10*, carrying tetracycline resistance, resides in a small plasmid that is neither self-transmissible nor mobilizable. This small plasmid is introduced by transformation (see chapter 6) into streptomycin-sensitive cells containing F, a larger, self-transmissible plasmid (see chapter 5). While the cells are growing, the transposon may hop from the smaller plasmid into the F plasmid in a few of the cells. Later, when these cells are mixed with streptomycin-resistant recipient cells, any F plasmid into which the transposon hopped will carry the transposon when it transfers to a new cell, thus conferring tetracycline resistance on that cell. Transposition can be detected by plating the mating mixture on agar plates containing tetracycline and by including streptomycin to kill the donor cells (see discussion of counterselection in "Mapping by Gradient of Transfer" in chapter 3).

The appearance of antibiotic-resistant transconjugants in a mating-out assay is not by itself definitive proof of transposition. Some transconjugants could become antibiotic resistant by means other than transposition of the transposon into the larger, self-transmissible plasmid. The smaller plasmid containing the transposon could have been mobilized somehow by the larger plasmid, or the smaller plasmid could have been fused to the larger plasmid by another form of recombination. A few representative transconjugants should be tested by DNA sequencing or other techniques to verify that they contain only the larger plasmid with the transposon inserted into it.

Mechanisms of Transposition

A range of different families of recombinases carry out the process of transposition with DNA elements. Multiple reviews are included in the Suggested Reading section for a more comprehensive understanding of this diversity. In this section, we cover two broad families of enzymes that account for movement of most DNA transposons found in bacteria. These enzyme classes are typically referred to by the single letter codes for the amino acids present in the active site, namely the DDE and the HUH transposases (Table 8.1). The next sections describe models for how these enzymes work and how they function in the movement of DNA elements in bacteria.

DDE Transposons

Many of the transposons found to be prevalent in bacteria are called **DDE transposons**, because their transposases all have two aspartates (D) and one glutamate (E) (see inside front cover) that are essential for their activity. The enzymes are also related both catalytically and structurally to RNase H. The acidic DDE amino acids are not next to each other in the polypeptide, but they come together in the active center when the protein is folded. Their job is to hold (by chelation) two magnesium ions (Mg^{2+}) that participate in the cleavage of the phosphate-sugar backbone of the DNA. The magnesium ions "activate" H_2O, thereby generating a 3′ OH in DNA, which in turn, can join to another DNA end in what is called a **strand transfer reaction**. This is an important property of these enzymes, because, as we will see, this governs how the enzyme has adapted for transposition.

The four major ways that the DDE enzymes allow transposition are based on which particular 3′ OH is joined to which DNA in the reaction (Figure 8.9). This joining event can free both DNA strands from the donor DNA, after which a second reaction is needed to join the excised element to the target DNA. In other cases, the joining event occurs directly to the target DNA, and another set of processing events is used to complete the process of transposition. A way to describe the movement of the element involves indicating whether it is "cut out" or "copied out" of the donor DNA (see Table 8.1 and Figure 8.9). Subsequently, the transposable element can be inserted into the target DNA by a "paste-in" or "copy-in" mechanism dictated by the role of DNA replication in the process.

DETAILS OF THE MECHANISM OF TRANSPOSITION BY Tn*5*: CUT-OUT AND PASTE-IN TRANSPOSITION

The mechanism of transposition of the DDE transposon Tn*5* has been studied extensively and is illustrated in Figure 8.9A (see Reznikoff, Suggested Reading). Transposase bound to each of the ends of the element will come to-

Table 8.1 Characteristics of selected types of bacterial mobile elements

Family	Characteristic			
	Active-site category	Protein-DNA covalent linkage	Target duplication	Example(s)
DDE transposons	DDE	No	Yes	Tn*3*, Tn*5*, Tn*7*, Tn*10*, IS*911*, Mu
HUH transposons	HUH	Yes, to 5′ P	No	IS*91*, IS*200*, IS*605*, IS*608*
Integrating conjugating elements[a]	Tyrosine recombinase	Yes, to 3′ P	No	ICE*Bs*1, SXT/R391, Tn*916*
Group II mobile introns	Catalytic RNA	No	No	L1LtrB

[a]Exceptions exist such as the Tn*GBS1/2* elements, which use DDE-family protein like the DDE transposons (see chapter 5).

gether through dimerization domains in their carboxy termini to form a synapse (see Figure 8.2). The transposase bound to one end of the transposon breaks the DNA at the other end, and vice versa, to leave a reactive 3′ OH at each end of the transposon. These 3′ OH ends attack the phosphodiester bond on the other strand, forming 3′-to-5′ phosphodiester hairpins, as shown in Figure 8.9A. This cuts the transposon out of the donor DNA. When the transposase binds to the target DNA, it breaks the two hairpin ends again, and the 3′ OH ends attack phosphodiester bonds 9 bp apart in the target DNA (on opposite strands), cutting them and allowing the 5′ phosphate ends in the target DNA to join to the 3′ OH ends in the transposon, which inserts the transposon into the target DNA. The 9-bp single-stranded gaps on each side of the transposon are then filled in by DNA polymerase to make the 9-bp repeats in the target DNA characteristic of the Tn*5* transposon.

Figure 8.9 The DDE transpose has been adapted in multiple ways for different families of DNA transposons. In all cases, the DDE transposase (indicated as Tnp or TnsB in green) acts at the position shown with green arrows. Dashed red lines indicate how the end broken by the transposase is joined to another position in the transposition process. While TnsA is part of the heteromeric transposase found in Tn*7*, it is not a member of the DDE family of proteins. TnsA is an endonuclease that breaks DNA at the position indicated to liberate the 3′ ends of the element from the donor DNA. Circled red arrowheads indicate the 3′ OH that is used to initiate DNA replication across the transposon, and the resulting new DNA is indicated in green. For simplicity, the replication involved in making the target site duplication is not shown. Dark and light blue indicate the donor and target DNA, respectively. See the text for details.

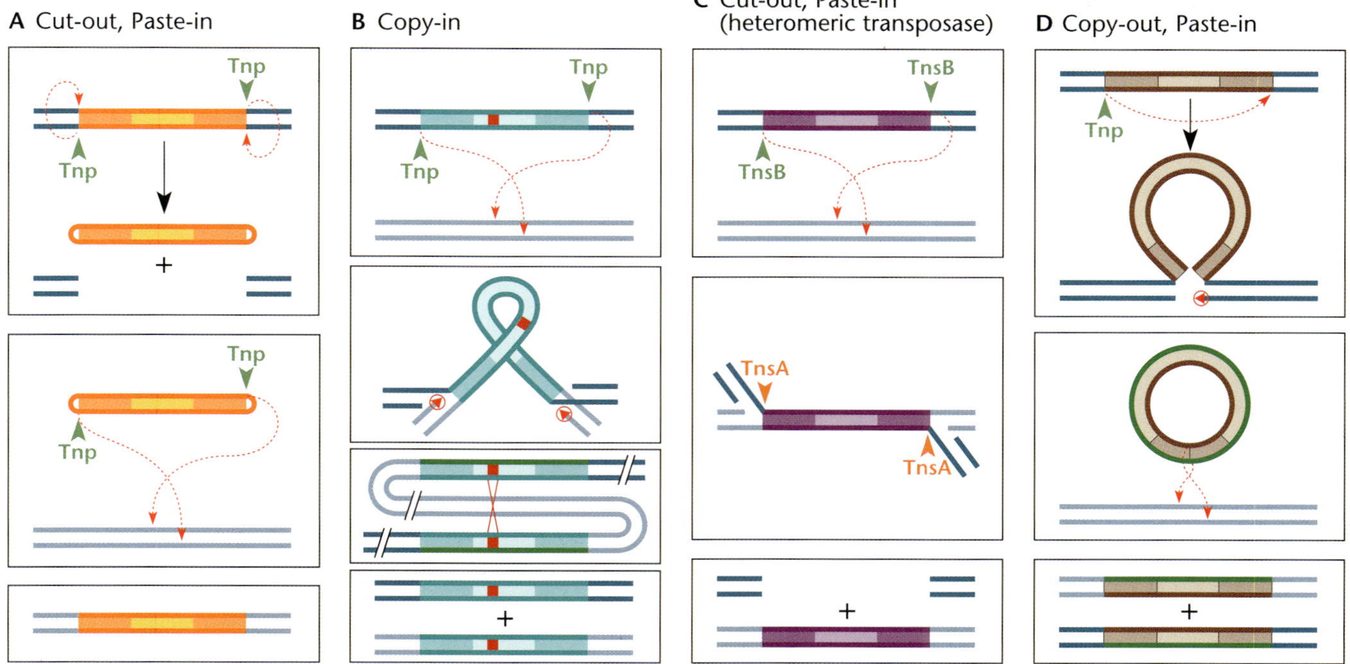

A distinct genetic feature of this process is that a DNA double-strand break is left in the donor DNA. This mechanism is sometimes referred to as **cut-and-paste transposition** or **cut-out and paste-in transposition** because the element is directly moved to a new location. In the cut-and-paste mechanism, there is no increase in the number of transposons by the act of transposition itself. This is difficult to demonstrate genetically, because double-strand break repair (see chapter 9) is extremely efficient in bacteria, and the sister chromosome used for template repair contains the transposon. However, this could be demonstrated by an increase in homologous recombination in the *lac* locus after excision of a cut-and-paste element (see Hagemann and Craig, Suggested Reading). While most repair occurs via recombination, at a very low frequency, the site of excision can be repaired in a way that reestablishes the wild-type allele, something that must be selected for using reversion studies. For a transposon insertion mutation to revert, the transposon must be completely removed from the DNA in a process called **precise excision**. Not a trace of the transposon can remain, including the duplication of the short target sequence, or the gene would probably remain disrupted and nonfunctional.

DETAILS OF THE MECHANISM OF TRANSPOSITION BY Tn*3*: COPY-IN OR REPLICATIVE TRANSPOSITION

Flexibility in how the recombinase acts with DDE transposons can lead to different outcomes. The nonreplicative, cut-and-paste process, found with Tn*5* and Tn*10*, leaves a DNA double-strand break in the donor DNA after transposition. Other transposons utilize a process called replicative transposition in which a copy of the element remains in the donor site. Although replicative transposition and cut-and-paste transposition by DDE transposons seem different, the chemistry is mechanistically the same. Only the fates of the 5′ ends of the element are different in the two cases; unlike in cut-and-paste transposition, the 5′ ends of the element remain unbroken in the process of replicative transposition (see below).

In replicative transposition, the joining event occurs directly to the target DNA with no intermediate step with the transposase; instead, there is a separate series of processing events that involve extensive DNA replication and site-specific recombination (see below) to free the target and donor DNAs (Figure 8.9B). The first detailed model to be developed for transposition attempted to explain all that was known about the genetics of Tn*3* transposition. This model continues to be generally accepted for that type of transposon. In the first step of replicative transposition, the transposase catalyzes the strand transfer of the 3′ OH at the ends of the element to staggered positions in the target DNA, as indicated in Figure 8.10A and B. The inset indicates the actual 3′ OH and 5′ PO_4

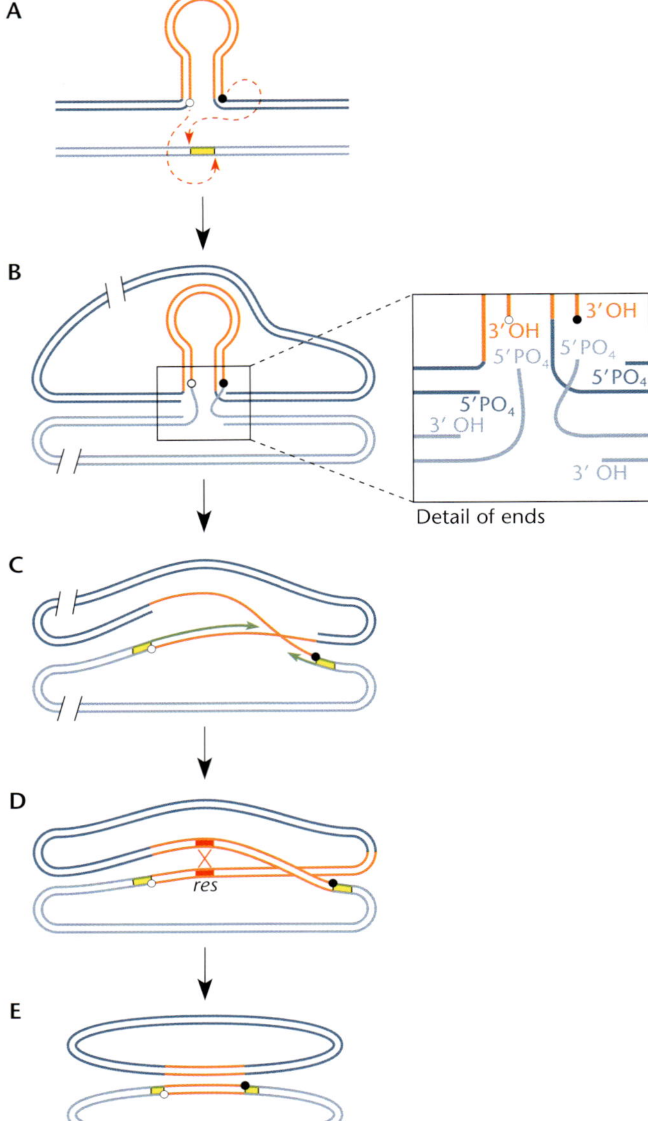

Figure 8.10 Replicative transposition of Tn*3* (orange) and formation and resolution of cointegrates. **(A)** Cleavage at the 3′ ends of the element occurs in a concerted reaction with joins to staggered positions in the target DNA. **(B)** These joining events link the donor and target DNAs. The inset shows details of the boxed region in panel B, with the 3′ OH and 5′ PO_4 that participate in the transactions carried out by the transposase indicated. **(C)** The free 3′ ends are used to prime DNA replication that copies the top and bottom strands of the Tn*3* element (represented in green). **(D)** The cointegrate is resolved by recombination promoted by the resolvase TnpR at the *res* sites (represented by the red rectangles). **(E)** The final product is a Tn*3* element at a new position in a separate target DNA without loss of the original element in the donor DNA. The target site duplication that is formed in the reaction is indicated by a yellow rectangle. The solid and open circles indicate the original 3′ OHs at the transposon ends that were used in the joining reaction to the target DNA.

that participate in the reactions. This reaction was introduced in Figure 8.3; however, in the case of replicative transposition, the 5′ ends of the element always remain attached to the donor DNA. In Figure 8.10, the positions of the two joining events are indicated by open and filled circles. After the 3′ ends of the element are joined at staggered positions to 5′ ends in the target DNA (Figure 8.10B), the 3′ OH ends in the target DNA are then available to initiate DNA replication. In some systems, this is known to require a special replication priming system normally involved in initiating DNA replication for repair (see chapter 9), and this is probably true for all replicative transposons. Replication then proceeds in both directions across the transposon (Figure 8.10C), and ligase is used at the 5′ ends of the donor DNA to complete the **cointegrate**. The new structure where two previously separate molecules are now fused as one contiguous molecule is termed a cointegrate. The last step, **resolution** of the cointegrate (Figure 8.10D), results from recombination between the two *res* sites in the cointegrate promoted by the resolvase of the transposon (see "S Recombinases: Mechanism" below). Resolution of the cointegrate gives rise to two copies of the transposon, one at the former (or donor) site and one at the target site (Figure 8.10E). The host recombination system is not needed to resolve the cointegrate into the original replicons, because the resolvase specifically promotes recombination between the *res* elements in the cointegrates. Although cointegrates can also be resolved by homologous recombination anywhere within the repeated copies of the transposon, the resolvase greatly increases the rate of resolution by actively promoting recombination between the *res* sequences.

This model explains why early genetic studies of the process indicated that cointegrates were obligate intermediates in replicative transposition. After replication has proceeded over the transposon in both directions, the donor DNA and the target DNA are fused to each other, separated by copies of the transposon, as shown (Figure 8.10D). This model also explains why, after transposition, a short target DNA sequence of defined length had been duplicated at each end of the transposon, something that is true for all DDE transposase transposons.

Not all transposons that replicate by this mechanism resolve the cointegrates after they form. An interesting example of this involves the bacteriophage Mu. Phage Mu also provides a nice example of how evolution can mix and match different types of mobile DNA strategies to address the basic needs of a parasitic element (see chapter 7). When bacteriophage Mu replicates itself, it inserts itself into different sites around the chromosome of its bacterial host by a replicative mechanism using its transposase, the MuA protein, similar to the process carried out by TnpA of Tn*3*. However, it does not resolve the cointegrates that form, and soon, the chromosome becomes loaded with Mu genomes. These genomes are then packaged directly from the chromosomal DNA into the phage head. Mu then goes through a typical lysis program to generate virulent particles to infect additional hosts (see chapter 7). A structural model of the Mu transposase after the initial joining event to the target is shown at the start of this chapter.

DETAILS OF THE MECHANISM OF TRANSPOSITION BY Tn*7*: CUT-OUT AND PASTE-IN TRANSPOSITION USING HETEROMERIC TRANSPOSASE

At the most basic level, steps in Tn*7* transposition are most similar to Tn*3* and Mu transposition, in which the transposon ends are joined directly to the target DNA (Figure 8.9C). However, Tn*7* uses a second recombinase, an additional subunit of the transposase, called TnsA, to make nicks at the 5′ ends of the element to prevent the formation of a cointegrate, as is found for Tn*3* transposition. As described below, Tn*7*-like elements use additional accessory proteins to tightly control target site selection.

A dramatic confirmation of the similarity between the cut-and-paste and replicative mechanisms of transposition by DDE transposons came with the demonstration that the cut-and-paste transposon Tn*7* can be converted into a replicative transposon by a single amino acid change in one subunit of the transposase (see May and Craig, Suggested Reading). As defined genetically and biochemically, transposon Tn*7* normally transposes by a cut-and-paste mechanism in which different subunits of the transposase make the breaks in the opposite strands of DNA at the ends of the transposon, leaving a double-stranded DNA break in the donor DNA. Support for the unity of transposition came from studies in which the TnsA active site was inactivated. While TnsA must be part of the transposase for any breaking-and-joining event to occur, it was found that mutating the active site in TnsA allows transposition, but the transposition events occur via a replicative mechanism involving extensive DNA replication, similar to what is found with Tn*3*.

DETAILS OF THE MECHANISM OF TRANSPOSITION BY IS*911*: COPY-OUT AND PASTE-IN TRANSPOSITION

Other DDE transposons use a mechanism of transposition that cannot be described as either a strictly replicative or a cut-and-paste mechanism. For example, the mechanism of transposition of the IS elements IS*2* and IS*3*, as well as IS*911*, has features of both replicative and cut-and-paste transposition (see Chandler et al., Suggested Reading). In the process with IS*911* the two ends of the element on the same strand are joined together to form a conjoined circle (Figure 8.9D). In this "figure-of-eight" structure, the top strand remains contiguous with the donor DNA through the whole process. The molecular steps allowing second-strand synthesis of the element are vaguer and may involve a host DNA replication fork (not shown). The free 3′ OH of the donor DNA likely reestablishes the transposon that remains in the donor site (Figure 8.9D). A free double-stranded circular transposon can then be used in a more standard integration event as indicated (Figure 8.9D). Thus, transposition is replicative because a copy of the transposon appears in the target DNA while the donor DNA retains the transposon. However, a cointegrate between the donor and target DNA does not form, and the transposon is, in a sense, cut out of the donor DNA and pasted into the target DNA.

HUH Transposons

The HUH enzymes possess a characteristic motif consisting of two histidine residues (H) flanking a hydrophobic amino acid (U). They also utilize one or two tyrosine (Y) residues which form a protein-DNA cross-link during their mode of action. Members of the larger family of HUH enzymes were discussed in previous chapters, namely the Rep proteins used for rolling-circle replication with some plasmids (chapter 4) and the relaxase enzymes used in conjugation systems (chapter 5). The best-understood HUH transposons are the IS*200*/*605* family found in bacteria and archaea. Rolling-circle transposons are another group of the HUH transposons that are important in the spread of antibiotic resistance determinants in bacteria, but they are not as well understood.

DETAILS OF THE MECHANISM OF TRANSPOSITION BY IS*608*: SINGLE-STRAND TRANSPOSITION

IS*608*, natively found in *Helicobacter pylori*, is the best-understood member of the IS*200*/*605* elements, owing in large part to its functionality and study in *E. coli* (see Ton-Hoang et al., Suggested Reading). As in other members of the family, transposition occurs downstream of a penta- or tetranucleotide sequence (5′-TTAC in the case of IS*608*). While IS*608* encodes two genes, *tnpA* and *tnpB*, only the TnpA protein is required for transposition, and this HUH endonuclease enzyme is the one that is conserved across this family of elements. The function of TnpB is unknown.

Biases noted in the nature of the donor and target DNA provided the first clues that the element excises and inserts using single-stranded DNA. This was confirmed by reconstitution of the reaction *in vitro* and structural information about the transposase. IS*608* appears to insert only into the lagging-strand template strand during active DNA replication. Interestingly, a number of diverse mobile elements appear to have a requirement for the lagging-strand template for their mobility (see Box 8.1). In contrast to the mechanism used by the DDE transposases, the HUH enzymes form a protein-DNA cross-link during the transposition process. Hairpins that form within the left and right ends of the element when it is in the single-stranded form are recognized by a TnpA dimer that joins the ends, forming a circular single-stranded DNA form of the element and repairing the lagging-strand template donor DNA (Figure 8.11). Helping to explain the preference of the element for DNAs undergoing lagging-strand DNA replication, insertion of the element into a single-stranded DNA substrate also must occur. In the case of IS*608*, a specific DNA sequence, TTAC, is also recognized.

Interestingly, lagging-strand DNA replication is not the only target used by IS*200*/*605* family elements to provide the single-stranded DNA substrate needed for mobilization. While the IS*200*/*605* family member IS*Dra2* found in *Deinococcus radiodurans* shows a bias for lagging-strand DNA templates, this element was also found to be activated when the host is subjected to gamma irradiation (see Pasternak et al., Suggested Reading). *D. radiodurans* is notable for extreme radioresistance, where an incompletely understood mechanism allows a chromosome that has been fragmented by radiation exposure to be reassembled using recombination. In the case of IS*Dra2*, the single-stranded DNA found during fragmentation also appears to provide a substrate for excision and integration of the element, in this case, independent of DNA replication on the lagging-strand template.

ROLLING-CIRCLE TRANSPOSONS

Other HUH transposons, represented by IS*91*, are called **rolling-circle transposons** because they are hypothesized to use a rolling-circle mechanism to transpose themselves into a target DNA (see Garcillán-Barcia et al., Suggested Reading). In addition to the conserved HUH, the transposase used by the IS*91* family has two conserved tyrosine (Y) residues. The details of how they transpose are not completely known, but the enzyme may cut one strand of the DNA close to one end of the transposon (called the *ori* end in analogy to the *ori* sequence of rolling-circle plasmids) and attach the 5′ PO_4 at the cleavage site to one of the tyrosines in the active center of the transposase. The free 3′ OH end could then serve as a primer to replicate over the transposon, ending at the other end of the transposon, called the *ter* end. The displaced old strand of the transposon would be available to enter the target DNA, and its complementary strand synthesized in the target DNA, so that both the donor and target DNAs end up with a copy of the transposon. This is therefore a form of replicative transposition. It is not clear how integration into target DNA occurs in this process. While both tyrosines are required for the pro-

BOX 8.1

Mobile Elements and DNA Replication

A number of elements that utilize a cut-and-paste mechanism of transposition are stimulated to excise from DNA after they are replicated (see Figure 8.13). However, a variety of mobile genetic elements utilize features found during DNA replication to facilitate integration into the host genome. DNA replication on the lagging-strand template has been shown to be particularly attractive for some mobile elements because of the processing events that occur on this strand and greater exposure as single-stranded DNA (see chapter 1). As described in chapter 9, bacteriophage-mediated recombination systems that utilize strand annealing proteins do not have the capacity to survey double-stranded DNA but, instead, act as strand annealing proteins only on single-stranded DNA substrates. Recombination with the λ Beta protein likely shows a bias to recombination on the lagging-strand template because single-stranded DNA is more rarely found on the leading-strand template (panel A of the figure). Multiple elements described in this chapter also favor mobilization on the lagging-strand template during DNA replication. IS*200*/*605* family elements require DNA in the single-strand form to both excise and integrate, biasing transposition with DNA replication forks (excision is shown in panel B). Tn*7* preferentially targets conjugal plasmids, a preference that likely aids dissemination of the element to new bacterial hosts (see Figure 8.14). This behavior is linked to the ability of the transposon-encoded target-selecting protein, TnsE, to recognize 3′ recessed ends and an ability to interact with the major processivity factor of DNA replication, the DnaN sliding clamp protein on lagging-strand templates (panel C). The IS*608* HUH transposase has also been shown to interact with DnaN and structures found in lagging-strand DNA replication templates. While not required in all situations, group II mobile intron integration can be facilitated by using free 3′ OH ends to prime reverse transcription of the element during DNA recombination (shown with a dashed arrow in panel D).

As shown with TnsE-mediated Tn*7* transposition, the ability to target recombination to certain forms of DNA replication

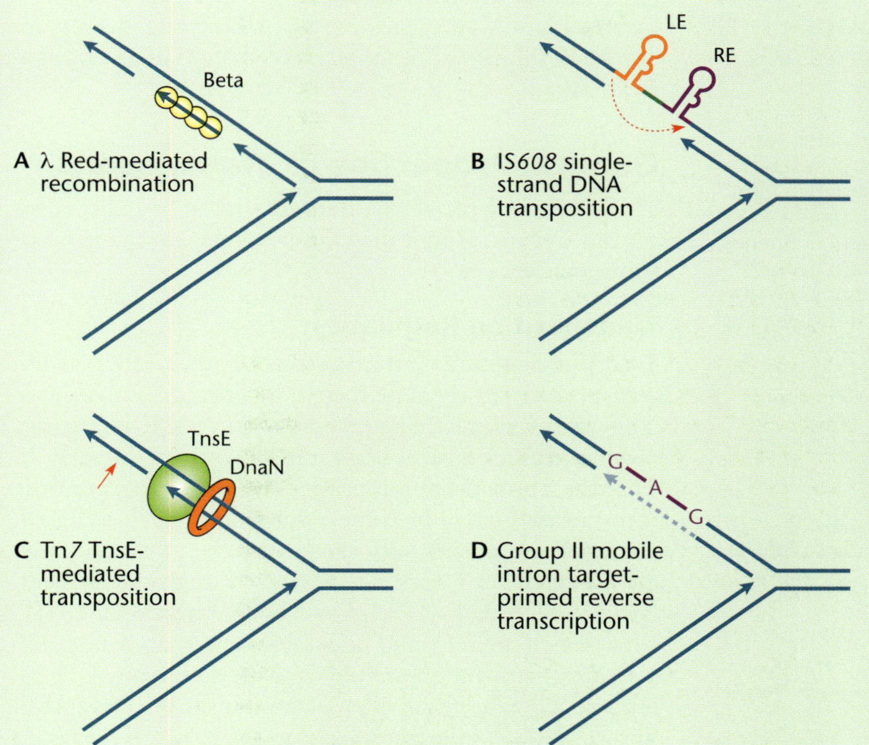

A simplified DNA replication fork where the polymerases, RNA primers, and other aspects are left out for clarity. (A) Model for how the Beta protein (yellow circles) anneals a single-strand DNA onto the lagging-strand template during λ Red recombination. (B) Model for how IS*608* excises out of the lagging-strand template by recognizing hairpin structures marking the left (LE [orange]) and right (RE [purple]) ends of the element. The dashed red line shows how the ends join to form the circular single-strand form of the element. (C) Tn*7* transposition events are recruited to the lagging-strand template by element-encoded TnsE protein that interacts with the DnaN sliding clamp and 3′ recessed ends. The solid red arrow indicates the point of Tn*7* insertion. (D) A group II mobile intron (purple) integrated into the lagging-strand template is subjected to target-primed reverse transcription (dashed arrow line) from one of the Okazaki fragments.

(continued)

BOX 8.1 *(continued)*

Mobile Elements and DNA Replication

may generally favor dispersal of these mobile elements to transmissible plasmids and certain bacteriophage. Targeting certain forms of DNA replication may also favor recombination in bacterial hosts during periods of replication stress.

References

Lavatine L, He S, Caumont-Sarcos A, Guynet C, Marty B, Chandler M, Ton-Hoang B. 2016. Single strand transposition at the host replication fork. *Nucleic Acids Res* **44:**7866–7883.

Li X-T, Costantino N, Lu L-Y, Liu D-P, Watt RM, Cheah KSE, Court DL, Huang J-D. 2003. Identification of factors influencing strand bias in oligonucleotide-mediated recombination in *Escherichia coli. Nucleic Acids Res* **31:**6674–6687.

Parks AR, Li Z, Shi Q, Owens RM, Jin MM, Peters JE. 2009. Transposition into replicating DNA occurs through interaction with the processivity factor. *Cell* **138:**685–695.

Zhong J, Lambowitz AM. 2003. Group II intron mobility using nascent strands at DNA replication forks to prime reverse transcription. *EMBO J* **22:**4555–4565.

cess, the exact roles are unknown; the other examples of rolling-circle replication mentioned above require only one tyrosine in the active center. The free 5′ PO_4 ends created at the two ends of the transposon are presumably shuttled between the two tyrosines to allow replication back over the transposon to create two copies of the transposon.

While the details of how movement occurs with rolling-circle transposons remain to be discovered, it is increasingly clear that a number of antibiotic resistance genes found on integrons are carried on transposons with two conserved tyrosines (Y2 transposons) related to IS*91*. They were first identified because of common sequence elements, which led to them being called **ISCR elements**, for IS common regions. Therefore, the process of rolling-circle transposition is suggested to be another mechanism by which antibiotic genes can recombine and move from one bacterium to another (see Toleman et al., Suggested Reading).

Figure 8.11 Model for single-strand DNA transposition with IS*608*. IS*608* moves as a single-strand circular DNA; it excises and inserts into DNA when DNA is in the single-stranded form. **(A)** Representation of IS*608* in the genome possessing two genes, *tnpA* and *tnpB*, flanked by specialized left and right ends. **(B)** The left and right ends of the elements will form hairpin structures in single-strand DNA that are recognized by the HUH transposase. The transposase joins the two ends of the element to produce the single-stranded DNA circular form of the element (joining event indicated by a dashed red line) and reseals the donor DNA (not shown). **(C)** IS*608* inserts into a single strand adjacent to the sequence TTAC and the second strand produced by host functions.

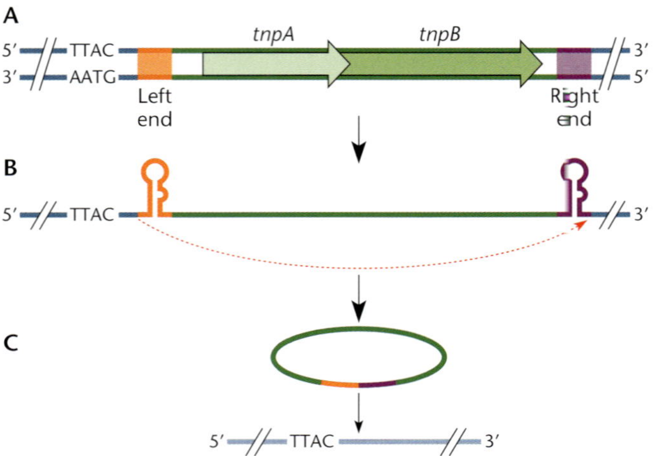

General Properties of Transposons

There are some properties that are shared by many types of transposons, even if they differ in their mechanisms of transposition.

Transposition Regulation

Transposition of most transposons occurs rarely, because transposons self-regulate their transposition activity (see Nagy and Chandler, Suggested Reading). The regulatory mechanisms used by various transposons vary greatly. In Tn*3*, the TnpR protein represses the transcription of the transposase gene. For some transposons, such as Tn*10*, transposition occurs very rarely and then primarily just after a replication fork has passed through the element. Newly replicated *E. coli* DNA is hemimethylated at GATC sites (see chapter 1), and hemimethylated DNA not only activates the promoter of the transposase gene of Tn*10*, but also increases the activity of the transposon ends. Also, the translation of the transposase gene of Tn*10* is repressed by an antisense RNA (see chapter 12). The transposase of Tn*5*, which already carries out only a low frequency of transposition, uses a truncated version of the transposase to further inhibit the active transposase. As illustrated in Figure 8.12, the translation of the truncated transposase is initiated at an internal translational initiation region,

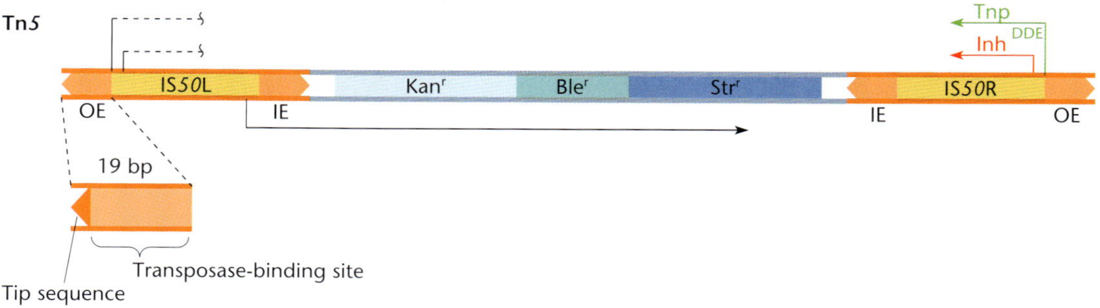

Figure 8.12 Regulation of Tn*5* transposition. Two similar IS*50* elements flank the antibiotic resistance genes. Only IS*50*R encodes the transposase Tnp and the inhibitor, which is an N-terminally truncated version of itself. See the text for details. The inverted repeats with the position of the transposase-binding site at the ends of the elements are indicated as the outer ends (OE) or inner ends (IE). The full-length transposase (Tnp) and a truncated product that is an inhibitor of the transposase (Inh) are indicated.

so it lacks the N terminus but has the C terminus involved in dimer formation. When this defective transposase pairs with the normal transposase, the hybrid transposase is inactive. Most transposons employ similar mechanisms to modulate the level of transposase transcription and/or translation, as well as the level of catalysis. Interestingly, in the case of Tn*7*, transcription and translation of the products required for transposition do not control the frequency or targeting of transposition; instead, the availability of the target site (*attTn7* and conjugal DNA replication) and various host proteins serves as the cue for the highly regulated transposition found with this element (see below).

Regulating transposition with DNA replication is a common theme across all domains of life. For cut-and-paste transposons, one benefit of timing transposition to occur immediately after replication of the element is that the presence of a sister chromosome allows repair of the DNA double-strand break that is left in the donor DNA (Figure 8.13) (see chapter 9). A considerable additional advantage of this type of repair is that the allele in the sister chromosome also has a copy of the element, so that there is a net gain in copy number even though the process of cut-and-paste transposition is itself a conservative process (Figure 8.13).

Target Site Specificity

While the transposition of some elements seems almost totally random, no transposable element inserts completely randomly into target DNA. Most transposable elements show some target specificity, inserting into some sites more often than into others. Even Tn*5* and Mu, which are famous for inserting almost at random, prefer some sites to others, although the preference is generally weak.

A bias for certain DNA sequences is not the only thing that affects where a transposon inserts in the bacterial genome. Transposons are adapted to their host and in a number of cases appear to avoid DNA involved in other cellular processes, such as highly transcribed genes. Transposons can also be attracted to DNA involved in certain cellular processes. As indicated below, Tn*7* preferentially inserts into DNA undergoing active replication in some settings (see Box 8.1). Tn*917* shows a strong bias for the region where DNA replication terminates in many *Firmicutes* (see Shi et al., Suggested Reading). The ability to target certain DNA processes presumably indicates a selective advantage for the element, but on a practical level, these types of biases need to be taken into account when choosing an element as a random insertion mutagen (see below).

Tn*7* is the extreme case of a transposon with target specificity and selectivity. Tn*7* utilizes five proteins for transposition, TnsA, TnsB, TnsC, TnsD, and TnsE, which provide two pathways of transposition. One pathway recognizes a single site found in bacteria, called *attTn7*, and a second pathway recognizes actively conjugating plasmids (Figure 8.14). As described above, TnsA and TnsB make up the transposase that breaks and joins the DNA strands, while the other proteins play roles in regulation and targeting. The TnsD protein identifies *attTn7* as a target site; binding to the site induces a change in the structure of the DNA that helps to recruit the regulatory protein TnsC to establish the target complex. Along with host factors, the TnsD-*attTn7* complex is capable of conveying a signal to the TnsAB transposase to initiate Tn*7* transposition. Insertion into the *attTn7* site may be advantageous, because it appears to be a neutral position in bacteria. The neutrality of the site likely explains why this transposition pathway can occur at a frequency that is about 1,000-fold higher than that of other transposable elements. TnsD likely evolved to recognize this particular sequence because it is found in a highly conserved portion of an essential gene in bacteria. Although an essential gene is recognized, the actual insertion event occurs

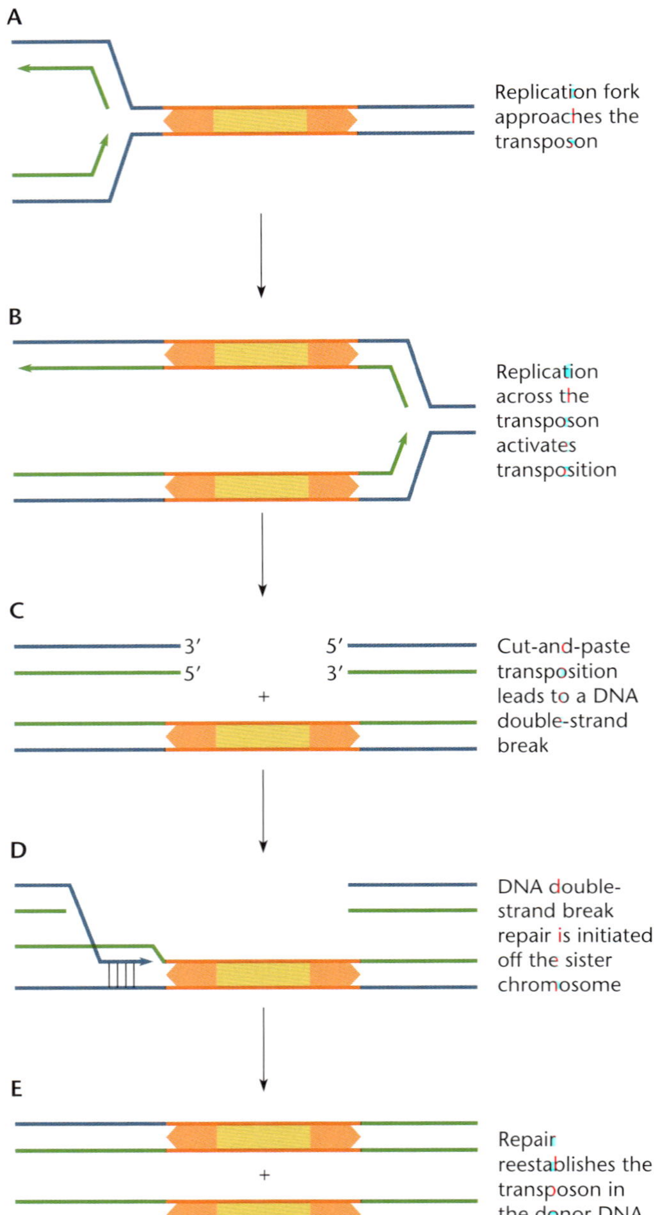
A
Replication fork approaches the transposon
B
Replication across the transposon activates transposition
C
3′
5′
5′
3′
+
Cut-and-paste transposition leads to a DNA double-strand break
D
DNA double-strand break repair is initiated off the sister chromosome
E
+
Repair reestablishes the transposon in the donor DNA

Figure 8.13 Transposition after DNA replication facilitates DNA repair. **(A and B)** For multiple transposons, DNA replication stimulates transposition of the element while producing a copy of the element. **(C and D)** When the element is lost through cut-and-paste transposition, repair of the DNA break can occur by recombination and DNA replication. **(E)** This process reestablishes the element in the donor DNA through the repair process.

3′ to the gene in the transcriptional terminator for the operon, so insertion at that site has no effect on the function of the gene.

Tn7 has a second targeting pathway for transposition that utilizes the Tn7 protein TnsE (Figure 8.14). This

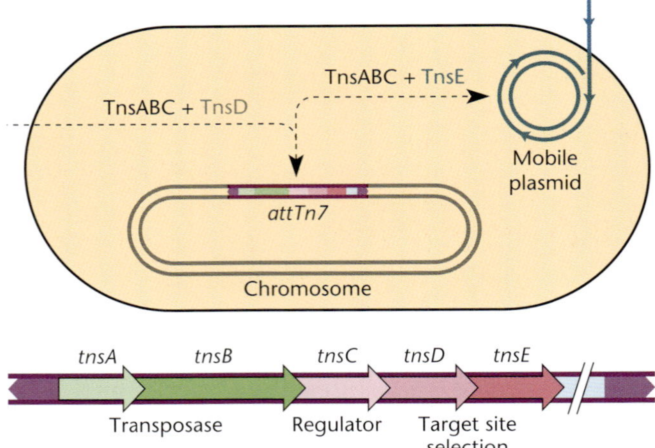
TnsABC + TnsE
TnsABC + TnsD
Mobile plasmid
attTn7
Chromosome
tnsA
tnsB
tnsC
tnsD
tnsE
Transposase
Regulator
Target site selection

Figure 8.14 Transposon Tn*7* uses an element-encoded heteromeric transposase and accessory transposition proteins to control target site selection. A core machinery consisting of a TnsA+TnsB transposase and the TnsC regulator protein is necessary but not sufficient for transposition. The TnsABC + TnsD proteins direct transposition into a single site called its attachment site (*attTn7*) that is conserved in bacterial chromosomes. The TnsABC + TnsE proteins preferentially direct transposition into actively conjugating plasmids in recipient cells. Presumably, transposition with TnsABC + TnsD would allow the element to quickly establish itself in a new bacterial host without the risk of inactivating important host genes, while the TnsABC + TnsE pathway would facilitate transfer of the element between bacterial populations.

pathway is largely cryptic but is activated in the presence of actively conjugating DNA, where transposition events are directed into the conjugal plasmid. Activation involves an interaction with the β sliding-clamp subunit of DNA polymerase III (see Parks et al., Suggested Reading), but how TnsC is recruited to the complex and how transposition shows such a high preference for replication found during conjugation are still being investigated. Having two complementary pathways of transposition appears to be beneficial, given that Tn7 elements are very broadly distributed across diverse bacteria in disparate environments around the world. The existence of two pathways of transposition also allows Tn7-like elements to form genetic islands in bacteria, where horizontally transferred genes accumulate at a specific position in the genome. This is likely to be because Tn7 elements deliver genes to this single position in chromosomes, using the TnsD pathway in a process that is facilitated by the TnsE pathway, which preferentially targets mobile plasmids. The basic mechanism of transposition catalyzed by the heteromeric transposase found with Tn7-like elements has evolved to work with variants of the TnsD protein that allow the element to recognize other specific attachment sites in different Tn7-like elements (see Peters et al., Suggested Reading). Interestingly, research is showing that relatives of the canonical element have evolved as new families with novel mechanisms to facilitate horizontal transfer, similar to the way that canonical Tn*7* evolved the TnsE pathway. One striking example is the recruitment and adaptation of CRISPR/Cas systems as a mechanism to use guide RNAs, not for recognizing invading genetic elements like phage and plasmids for destruction, but instead as a tool to target these mobile DNAs for transposition, thereby facilitating transportation to new bacterial hosts (see chapter 13 and Peters, Suggested Reading).

Effects on Genes Adjacent to the Insertion Site

Most IS and transposon insertions cause polar effects if they insert into a gene transcribed as part of a polycistronic mRNA (see chapter 2). The inserted element contains transcriptional stop signals and may also contain long stretches of sequence that are transcribed but not translated. The latter may cause Rho-dependent transcriptional termination, which prevents transcription of genes located downstream of the insertion site in the target DNA.

In contrast to negative effects on the expression of downstream genes, some insertions may enhance the expression of a gene adjacent to the insertion site. This expression can result from transcription that originates within the transposon. For example, both Tn*5* and Tn*10* contain outward-facing promoters near their termini, and these promoters can initiate transcription into neighboring genes.

Target Immunity

Another feature of some transposons is a property that inhibits transposition into a site in the DNA close to another transposon of the same type. This is called **target immunity**, because DNA sequences close to a transposon in the DNA are relatively immune to insertion of another copy of the same transposon. The immunity can extend over 100,000 bp of DNA, although its reach varies between transposons. Target site immunity is likely to confer many advantages to the host and transposon. If two transposons were to insert close to each other, the resolution of the two copies by the transposon resolvase or homologous recombination between the two copies of the transposon would cause large deletions and might lead to death of the cell. The process of immunity would discourage the insertion of the element into the "sister" element on the duplicated copy chromosome after DNA replication, which would likely destroy one if not both elements.

Target site immunity is limited to only some transposons of the types we have discussed, including the Mu, Tn*3* (Tn*21*), and Tn*7* families of transposons. The mechanism of target site immunity has been addressed at the molecular level with Mu and Tn*7*. Mu and Tn*7* have accessory proteins beyond the transposase that are required for transposition and that play important roles, including for target site immunity. A central paradox in target site immunity stems from the fact that the same proteins that are involved in carrying out transposition are also involved in an interaction that discourages nearby transposition. In the case of Tn*7*, an interaction between TnsB and TnsC discourages TnsC from establishing a transposition complex with TnsD or TnsE anywhere near the element in the genome. In the case of the TnsD-mediated pathway of transposition, target site immunity discourages multiple insertion events from occurring at the *attTn7* site from the same Tn*7* element; this is likely to be essential because of the very high frequency of transposition found with this pathway. In the case of TnsE-mediated transposition, transposition is known to be discouraged over an entire F plasmid when a single Tn*7* insertion resides in the plasmid. This could be important, because conjugal plasmids are preferentially targeted by TnsE-mediated transposition, but multiple Tn*7* insertions could also destabilize these plasmids. The ability of TnsB to bind to the IR sequences is critical for immunity (as is true for the MuA protein in bacteriophage Mu). TnsB binding to these sequences may cause a high concentration of TnsB around the element as it cycles on and off the DNA, which prevents TnsC from forming an active target complex with TnsD or TnsE near the element. An interaction between TnsB and the other subunit of the transposase, TnsA, seems to channel TnsC into an active target complex in regions where no Tn*7* element resides (see Skelding et al., Suggested Reading).

Target immunity is also found with Tn*3* family elements, but the mechanism is enigmatic. Tn*3* does not encode accessory proteins like those found in bacteriophage Mu and Tn*7*, and therefore a distinct mechanism must be involved in this process. Tn*3* family members, like Tn*4430*, use a single transposon-encoded protein for target site immunity (see Lambin et al., Suggested Reading), but it is unclear if host-encoded proteins are involved in the target site immunity process.

Transposon Mutagenesis

One of the most important uses of transposons is in transposon mutagenesis. This is a particularly effective form of mutagenesis, because a gene that has been marked with a transposon is easy to map using special PCR strategies such as arbitrary PCR (Box 8.2) or ligation-mediated PCR and is amenable to high-throughput DNA-sequencing strategies. Furthermore, genes marked with a transposon are also relatively easy to clone by selecting for selectable genes carried on the transposon.

BOX 8.2

Transposons and Genomics

For decades, transposons have been important tools for bacterial genetics. Transposons are a very useful type of insertion mutagen and offer the benefit of providing a selectable marker that is invaluable for genetic mapping. Historically, this has allowed the insertion to be mapped using Hfrs and cotransfer using P1 transduction (see chapters 3 and 5). A set of genetically mapped Tn*10* insertions at approximately 1-minute intervals across the entire *E. coli* genome was an important tool for mapping mutations in the chromosome. Transposon insertions can also be efficiently mapped by DNA sequencing. Early strategies involved cloning the transposon from the genome with some of its flanking DNA to allow determination of the DNA sequence from a plasmid. However, PCR sequencing strategies have all but replaced cloning for this purpose. PCR involves using two short DNA oligonucleotide primers to amplify the intervening DNA sequences in sequential rounds of denaturation, annealing, and amplification with a thermostable DNA polymerase (see chapter 13). The difficulty with using PCR to map transposon insertions stems from the fact that while the sequence of the transposon is known, the flanking

A

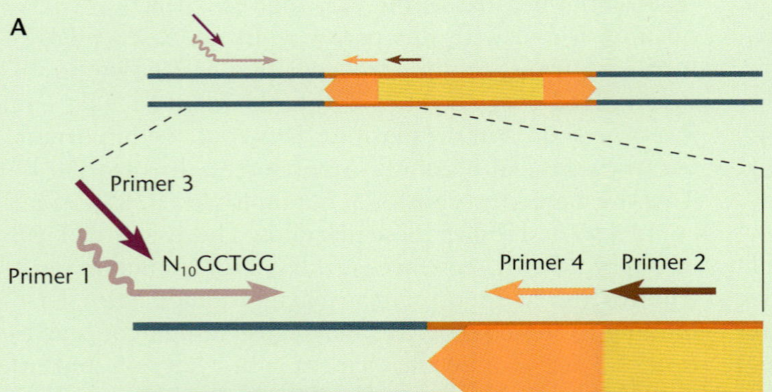

B

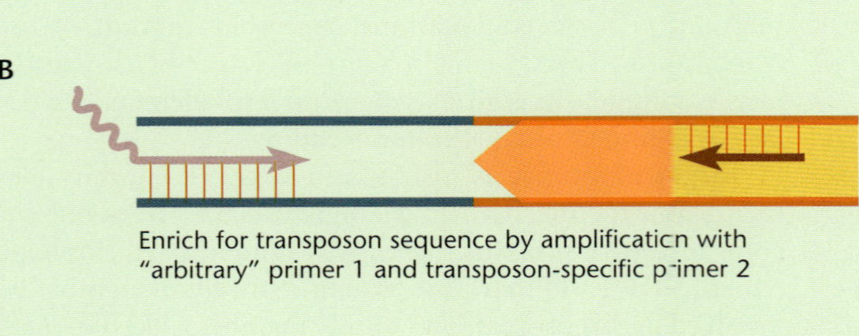

Enrich for transposon sequence by amplification with "arbitrary" primer 1 and transposon-specific primer 2

C

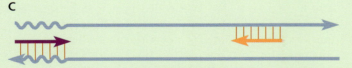

Amplify transposon end sequences and flanking DNA with primer 3 that is specific to the newly added sequence tag and another transposon-specific primer, primer 4

D

Use primer 4 to determine sequence of DNA immediately adjacent to the transposon end

Using arbitrary PCR to map a transposon in genomic DNA. The transposon to be mapped with DNA sequencing (yellow and orange) is shown in chromosomal DNA (dark blue). Four separate primers are used in two PCR amplification reactions as described in the text. Amplified DNA is shown in light blue, and the extension reaction used for sequencing is shown in green.

DNA sequence is not known. There are various techniques for identifying a position where a second primer can anneal for amplification of the transposon end, along with the flanking sequences, but they all involve somehow attaching a known DNA sequence to the flanking DNA sequence.

One efficient way to map insertions is arbitrary PCR, where one of the primers has a series of random nucleotides. In panel A of the figure, a set of oligonucleotides that include 10 random nucleotides (N_{10}) are synthesized; these anneal only at a fixed short sequence, GCTGG, that is found at the very 3′ end of the oligonucleotide. The positions of the four primers used in the experiment are shown. In the first PCR, the arbitrary primer with an extra DNA sequence "tail" (primer 1) is used with a primer that is specific for the position inside the IR of the transposon (primer 2) (panel B). This step enriches for amplified products that contain the transposon end with the DNA sequence found adjacent to the element and also adds the DNA sequence included in the tail of primer 1. The sequence added as a tail found on primer 1 is indicated with a wavy line. In the next step (panel C), a second round of PCR is done with a primer (primer 3) that recognizes the sequence tail (wavy line) and a primer that is specific for the transposon but closer to the outer end of the transposon (primer 4). This product is then subjected to DNA sequencing using primer 4 or a primer that anneals even closer to the end of the transposon (panel D). Part of the value of this procedure is that the template DNA can be whole cells that are added directly to the PCR mixture from a colony or an overnight culture. Another technique involves enzymatically digesting genomic DNA or physically shearing the DNA and directly ligating a known sequence onto the fragments to provide the sequence tag for amplifying the flanking DNA. The arrowheads indicate the 3′ ends of the DNAs.

High-throughput sequencing techniques allow large numbers of insertions to be mapped. They can be used to isolate transposon insertions in every nonessential gene in an organism to greatly facilitate genetic analyses. In addition, tags can be added to allow next-generation sequencing technologies to be used to simultaneously map tens of thousands of insertions in a population of cells containing random insertions. These techniques can be used to identify genes that are likely to be essential because cells containing transposon insertions in these genes do not survive, and therefore transposon insertions should not map to these genes. Also, by mapping insertions in a population of cells with random insertions before and after growth under specialized conditions, one can determine the profile of genes that are required for growth under these conditions. For example, the cell population with random insertions could be constructed and grown first in rich medium. The distribution of transposons in this population of cells could be monitored before and after growth in minimal medium containing only salts and a specific carbon source. Presumably, this would identify all genes required for growth in minimal medium, because individuals in the population that had transposon insertions in genes that were required for growth in minimal medium would be lost from the population by dilution. This procedure has been adapted to many growth conditions and environments (e.g., growth in a biofilm or in a model animal or plant system).

References

Liberati NT, Urbach JM, Miyata S, Lee DG, Drenkard E, Wu G, Villanueva J, Wei T, Ausubel FM. 2006. An ordered, nonredundant library of *Pseudomonas aeruginosa* strain PA14 transposon insertion mutants. *Proc Natl Acad Sci USA* **103**:2833–2838.

Singer M, Baker TA, Schnitzler G, Deischel SM, Goel M, Dove W, Jaacks KJ, Grossman AD, Erickson JW, Gross CA. 1989. A collection of strains containing genetically linked alternating antibiotic resistance elements for genetic mapping of *Escherichia coli*. *Microbiol Rev* **53**:1–24.

van Opijnen T, Bodi KL, Camilli A. 2009. Tn-seq: high-throughput parallel sequencing for fitness and genetic interaction studies in microorganisms. *Nat Methods* **6**:767–772.

Transposon Mutagenesis *In Vivo*

Not all types of transposons are equally useful for mutagenesis. A transposon used for mutagenesis should have the following properties:

1. It should transpose at a fairly high frequency.
2. It should not be very selective in its target sequence.
3. It should carry an easily selectable gene, such as one for resistance to an antibiotic.
4. It should have a broad host range for transposition if it is to be used in several different kinds of bacteria.

Transposon Tn*5* is well suited for random mutagenesis of *Gammaproteobacteria*, because it embodies all of these features. Not only does Tn*5* transpose at a relatively high frequency, but it also has low target specificity and transposes broadly across *Gammaproteobacteria*. Tn*5* also carries a kanamycin resistance gene that is expressed in most of these hosts. In chapter 5 we described a popular method for transposon mutagenesis of *Proteobacteria* other than *E. coli* (see Simon et al., 1983, Suggested Reading) (see Box 5.2). In addition to the broad host range of Tn*5* and the promiscuity of plasmid RP4, this method takes advantage of the narrow host range of ColE1-derived plasmids, which replicate only in *E. coli* and closely related species. Phage Mu is another transposon-like element that can transpose in many types of *Gammaproteobacteria* and that shows little target specificity. Many of the original elements used in *Firmicutes* (such as Tn*917*, Tn*916*, or Tn*10* derivatives) were plagued with targeting biases of one type or another. However, a *mariner*-type element called *Himar1* appears to provide the features elaborated in the four points mentioned above (see Rubin et al., Suggested Reading). Somewhat remarkably, the *Himar1* element was originally obtained from the horn fly, *Haematobia irritans*, but has been adapted with delivery systems and antibiotic resistance genes that work in a variety of bacteria.

Transposon Mutagenesis *In Vitro*

While *in vivo* transposon mutagenesis is a very useful technology, it does have some limitations. One of the limitations is that it is necessary to introduce the transposon on a suicide vector, which may give some residual false-positive results for transposon insertion mutants if the suicide vector is capable of low-level replication. Another limitation is that it is not very efficient and requires powerful positive-selection techniques to isolate the mutants. A third limitation arises if the desired target is a specific plasmid or other smaller DNA sequence. There is no target specificity to the insertion mutants, and so most of the time, the transposon hops into the chromosome; the few transposon insertions in the smaller target DNA must be found among the overwhelming number of mutants in the chromosome. There is also the possibility of multiple transposition events.

In vitro transposon mutagenesis avoids many of these limitations. This technology is made possible by the fact that the transposase enzyme by itself performs the chemical reactions of the "cut-and-paste" transposition reaction. In this procedure, the target DNA is mixed with a donor DNA containing the transposon, and the purified transposase is added, which allows the transposon to insert into the target DNA in the test tube. Multiple transposases have been adapted for this process, and each has its own advantages. For example, mutants of the Tn*5* transposase are available that enhance the transposition frequency, which is necessary because the wild-type Tn*5* transposase is only weakly active *in vitro*. Also, only the sequences at the ends of the IRs of Tn*5* are needed in the donor DNA, and they are only 19 bp long. One disadvantage of Tn*5* is that it is prone to multiple transposition events in a single DNA because the element lacks target site immunity. A mutant form of the Tn*7* regulator protein allows high-frequency transposition with the Tn*7* transposase, resulting in almost undetectable target site selectivity, with the added benefit of target site immunity, which greatly reduces, if not eliminates, multiple insertions (see Biery et al., Suggested Reading). A drawback of Tn*7* is that the *cis*-acting sequences required for optimal transposition with the element are around 100 bp long. In these systems, the transposon that is used has been engineered to lack the transposition genes so that the element does not transpose in subsequent generations or cause genetic instability, such as deletions, once it is in the chromosome. In some of these systems, controllable outward-facing promoters are engineered into the element to allow controllable expression of an adjacent gene or operon. Therefore, in addition to producing loss-of-activity mutations, these elements can also generate mutants with altered expression (see Bordi et al., Suggested Reading).

The target DNA can be either a replicon, such as a plasmid that replicates in the recipient cell, or random linear pieces of the chromosomal DNA of the recipient if it is being introduced into cells that can be transformed with linear DNA (see chapter 6). The linear pieces recombine with the chromosome and replace the chromosomal sequence with the sequence containing the transposon insertion. This offers a method for doing random chromosomal transposon mutagenesis of bacteria for which no workable transposon mutagenesis system is currently available (for example, see Bordi et al., Suggested Reading).

Another variation of this method for performing transposon mutagenesis, which can be applied to mutagenize the DNA of almost any bacterium, and even eukaryotic cells, where transformation works very well, is to use "transpososomes" (see Goryshin et al., Suggested Reading). A transpososome is a complex where the transposase is bound to both ends of the transposon. The benefit of working with a preformed transpososome is that the transposase protein does not have to be made in the cell. This feature is important because the transposase gene on the transposon might not be expressed in a distantly related bacterium and certainly not in a eukaryotic cell. As in other methods of transposon mutagenesis, the transposon should carry a selectable gene that is expressed in the cell to be mutagenized. Transpososomes based on Tn*5* are made by running the transposition reaction *in vitro* in the absence of magnesium ions. Under these conditions, the ends of the transposon in the donor DNA are not cut, but if the transposon has already been cut out of the donor DNA by some other process (for example, with restriction endonucleases), the transposase binds to the ends and remains attached, forming the transpososome. When the transpososome is introduced into the target cells by electroporation (see chapter 6), the transposase attached to the transposon catalyzes the DNA strand exchanges required for transposition of the transposon into the chromosome or other cellular DNA.

The transposase enzyme introduced with the transposon is quickly degraded, preventing further transposition.

Transposon Mutagenesis of Plasmids

One common use of transposon mutagenesis is to identify genes within large clones on a plasmid. If the transposon hops into a gene on the plasmid, it will disrupt the gene. DNA sequencing from the element will identify the gene of interest.

Figure 8.15 illustrates the steps in the selection of plasmids with transposon insertions in *E. coli*. A suicide vector containing the transposon (Tn*5* in the example) is introduced into cells harboring the plasmid. Cells in which the transposon has hopped into cellular DNA (either the plasmid or the chromosome) are then selected by plating them on medium containing the antibiotic to which a transposon gene confers resistance, in this case, kanamycin. Only the cells in which the transposon has hopped into another DNA become resistant to the antibiotic, since the transposons that remain in the suicide vector are lost with the suicide vector. In most antibiotic-resistant bacteria, the transposon will have hopped into the chromosome rather than into the plasmid simply because the chromosome is a larger target. The plasmids in these bacteria are normal. However, if the plasmid being mutagenized is self-transmissible, the plasmids can be isolated from the few bacteria in which the transposon has hopped into the plasmid by mating the plasmid into another *E. coli* strain and selecting for the antibiotic resistance on the transposon. Alternatively, the antibiotic-resistant colonies that have the transposon either in the plasmid or in the chromosome can be pooled and the plasmids can be isolated. This mixture of plasmids, most of which are normal, is then used to transform another strain of *E. coli*, selecting for the antibiotic resistance gene on the transposon (the kanamycin resistance gene in the example). The antibiotic-resistant transformants should contain the plasmid with the transposon inserted somewhere in it. Voilà! In a few simple steps, plasmids with transposon insertion mutations have been isolated. This method can be used to randomly mutagenize a DNA segment cloned in a plasmid or to mutagenize the plasmid itself.

Transposon Mutagenesis of the Bacterial Chromosome

The same methods used to mutagenize a plasmid with a transposon can also be used to mutagenize the chromosome. A gene with a transposon insertion is much easier to map or clone than a gene with another type of mutation (because of the antibiotic resistance marker), making this a popular method for mutagenesis of chromosomal genes.

The major limitation of transposon mutagenesis is that transposon insertions usually inactivate a gene, a lethal event in a haploid bacterium if the gene is essential for growth. Therefore, this method can generally be used only to mutate genes that are nonessential or essential under only some conditions. However, it can still be used to map essential genes by isolating transposon insertion mutations that are not in the gene itself but close to it in the DNA. If the transposon is inserted close enough, it might be used to map or clone the gene. It is also important to remember that insertion of some transposons may increase the expression of genes near the insertion site. As described above, elements that contain controllable outward-facing promoters are also useful for isolating altered expression mutations. Transposon insertions are easily mapped using PCR, and procedures exist for mapping all of the random insertions in a pool of bacteria to identify essential genes or genes that are important or essential in certain environments.

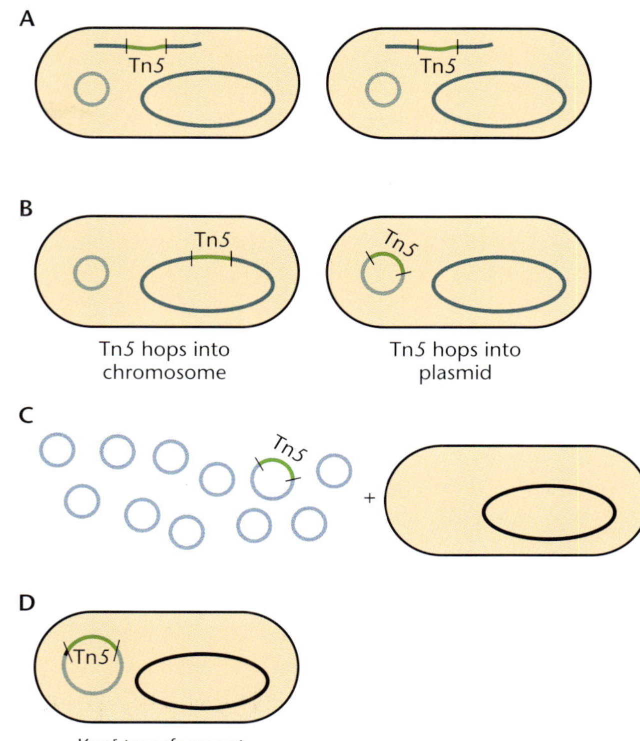

Figure 8.15 Random transposon Tn*5* mutagenesis. Random transposon mutagenesis of a plasmid. **(A)** Transposon Tn*5* is introduced into cells on a suicide vector (indicated as a linear DNA with the green element). **(B)** The culture is incubated, allowing the Tn*5* time to hop, either into the chromosome (dark blue oval) or into a plasmid (light blue circle). Plating on kanamycin-containing medium results in the selection of cells in which a transposition has occurred. **(C)** Plasmid DNA is prepared from Kan^r cells and used to transform a kanamycin-susceptible (Kan^s) recipient. **(D)** Selection for Kan^r allows the identification of cells that have acquired a Tn*5*-carrying plasmid.

Transposon Mutagenesis of All Bacteria

One of the most useful features of transposon mutagenesis is that it can be applied to many types of bacteria, even ones that have not been extensively characterized. Methods have been developed to perform transposon mutagenesis in an increasing number of bacteria. All that is needed is a way of introducing a transposon into the bacterium, provided that the transposon can transpose in that bacterium. The transposon should also carry a gene that can be selected in the target organism. Some of these methods were mentioned above and are outlined in more detail below.

CLONING GENES MUTATED WITH A TRANSPOSON INSERTION

Genes that have been mutated by transposon insertion are usually relatively easy to identify by cloning a DNA segment that includes the easily identified antibiotic resistance gene in the transposon. Since some antibiotic genes, for example, the kanamycin resistance gene in Tn5, are expressed in many types of bacteria, this method can even be used to clone genes from one bacterium into a cloning vector from another. This is particularly desirable because most cloning vectors and recombinant DNA techniques have been designed for *E. coli*. To clone a gene mutated by a transposon from a bacterium distantly related to *E. coli*, the DNA from the mutagenized strain is cut with a restriction endonuclease that does not cut in the transposon, and the resulting DNA fragments are ligated into an *E. coli* plasmid cloning vector cut with the same or a compatible enzyme. The ligation mixture is then introduced into *E. coli* by transformation, and the transformed cells are spread on a plate containing the antibiotic to which the transposon confers resistance. Only cells containing the mutated cloned gene multiply to form a colony.

While DNA sequencing will identify the position of the transposon insertion, in a large unsequenced DNA fragment, it may be useful to be able to clone the subregion of interest that is responsible for a particular phenotype. Transposons can be engineered to make cloning of transposon insertions even more efficient by introducing an origin of replication into the transposon so that the DNA containing the transposon need only cyclize to replicate autonomously in *E. coli*. The use of such a transposon for cloning transposon insertions is illustrated in Figure 8.16. In the example, the transposon carrying a plasmid origin of replication has inserted into the gene to be cloned. The chromosomal DNA is isolated from the mutant cells and cut with a restriction endonuclease that does not cut in the transposon, EcoRI in the example. When the cut DNA is religated, the fragment containing the transposon becomes a circular replicon with the plasmid origin of replication. If the ligation mixture is used

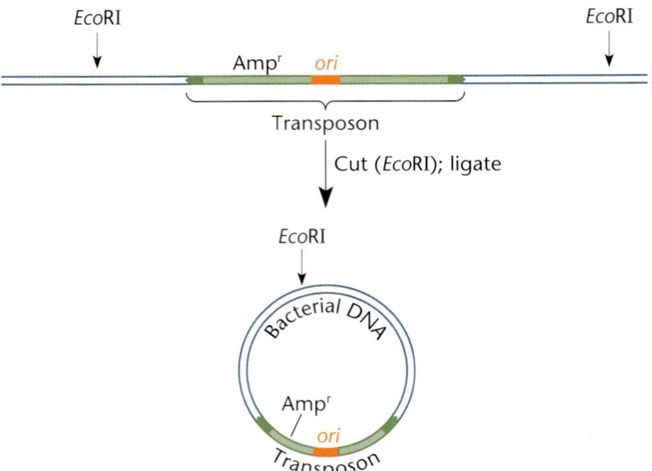

Figure 8.16 Cloning genes mutated by insertion of a transposon. A transposon used for mutagenesis of a chromosome contains a plasmid origin of replication (*ori*), and the chromosome is cut with the restriction endonuclease EcoRI and religated. If the ligation mixture is used to transform *Escherichia coli*, the resulting plasmid in the Amp^r transformants will contain the sequences that flanked the transposon insertion in the chromosome. Chromosomal sequences are shown in blue, and transposon sequences are shown in green.

to transform *E. coli* and the ampicillin resistance gene on the transposon is selected, the chromosomal DNA surrounding the transposon has been cloned. Since the restriction endonuclease cuts outside the transposon, any clones of the transposon cut from the chromosome also include sequences from the gene of interest into which the transposon has inserted.

This method allows the cloning of genes about which nothing is known except the phenotype of mutations that inactivate the gene, and it can be easily adapted to clone genes from any bacterium in which the transposon can hop to create the original chromosomal mutation. Once mutants are obtained with the transposon inserted in the gene of interest, the remaining manipulations are performed in *E. coli*. This can be a particular advantage if the bacterium being studied is difficult to grow or maintain in a laboratory situation.

Using Transposon Mutagenesis To Make Random Gene Fusions

The ability of some transposons to hop randomly into DNA has made them very useful for making random gene fusions to reporter genes (see chapter 2). Fusing a gene to a reporter gene whose product (such as LacZ or green fluorescent protein) is easy to monitor can make regulation of the gene much easier to study or can be used to identify genes subject to a certain type of regulation or those localized to certain cellular compartments. Once a gene subject to a certain type of regulation has been identified in this way, it can be easily cloned and studied using methods such as those described above.

Transposons have been engineered to make either transcriptional or translational fusions. As discussed in chapter 2, in a transcriptional fusion, the coding sequence for a reporter gene is fused to the promoter for the gene of interest so that the two genes are transcribed into mRNA together. In a translational fusion, the open reading frames (ORFs) for the two proteins are fused to each other in the same reading frame so that translation, initiated at the translational initiation region for one protein, continues into the ORF for the reporter protein, making a fusion protein.

Site-Specific Recombination

Another type of nonhomologous recombination, **site-specific recombination**, occurs only between specific sequences or sites in DNA. It is promoted by enzymes called **site-specific recombinases**, which recognize two specific sites in DNA and promote recombination between them. Even though the two sites generally have short sequences in common, the regions of homology are usually too short for normal homologous recombination to occur efficiently. Therefore, efficient recombination between the two sites requires the presence of a specific recombinase enzyme. We have already mentioned some site-specific recombination systems in connection with the resolution of chromosome and ColE1 plasmid dimers by the XerCD site-specific recombinase (see chapter 1). The integrase of phage λ is another example (see chapter 7), as are resolvases, such as TnpR of Tn*3*, that resolve cointegrates formed during replicative transposition. In this section, we discuss some other examples of site-specific recombination in bacteria and phages, which can all be placed into one of two groups, the S and the Y recombinases, based on their mechanism of action.

Integrases

Integrases are a type of site-specific recombinase. They recognize two sequences in DNA and promote recombination between them; therefore, they are no different in principle from the site-specific recombinases that resolve cointegrates. However, integrases act to integrate one DNA into another by promoting recombination between two sites on different DNAs.

PHAGE INTEGRASES

The best-known integrase is the Int enzyme of λ phage, which is responsible for the integration of circular phage DNA into the DNA of the host to form a prophage (see chapter 7). Briefly, the λ phage integrase specifically recognizes the *attP* site in the phage DNA and the *attB* site on the bacterial chromosome and promotes recombination between them. Usually, phage integrases are extremely specific. Only the *attP* and *attB* sites are recognized, so the DNA integrates only at one or at most a few places in the bacterial chromosome. Other integrases seem to be somewhat less specific, including some integrases found with integrative conjugative elements (see chapter 5) and the integrases of integrons (see below), where there seems to be some flexibility in the sequence of the *attB* site. In a reversal of the reaction performed by the integrases, a combination of the integrase (Int) and another enzyme, often called the excisase (Xis), promotes recombination between the hybrid *attP-attB* sites flanking the integrated DNA to excise the integrated DNA, although the integrase is again the enzyme that performs the site-specific recombination.

Because of their specificity, phage integrases have a number of potential uses in molecular genetics. For example, the reaction performed by phage λ Int and Xis has been capitalized on as a cloning technology called the Gateway system (see chapter 13). In this application, a PCR fragment containing the gene of interest is cloned into a plasmid vector, called the entry vector, so that the clone is flanked by one of the hybrid *attB-attP* sites which flank integrated prophage λ DNA in the chromosome. If the vector is mixed with a destination vector containing the other *attP-attB* hybrid site and λ integrase and excisase are added, site-specific recombination between the sites moves the cloned gene into the destination vector, where it becomes flanked by *attB* sites. Once a clone is generated in the entry vector, it can be transferred quickly into a number of different destination vectors. This could be important if, for example, one wished to determine the effect on the solubility of a protein of fusing it to a number of different affinity tags which are encoded on a number of different destination vectors.

INTEGRASES OF TRANSPOSON INTEGRONS

Integrases are also important in the acquisition of cargo/passenger genes by some types of transposons. The first clue was that transposons seemed to have picked up different sets of antibiotic resistance genes so that related members of some families of transposons, such as the Tn*21* family, have different resistance genes inserted in approximately the same place in the transposon. In fact, the first known example of multiple drug resistance in pathogenic bacteria in Japan in the early 1950s was due to transposon Tn*21*, with multiple drug resistance cassettes acquired by integrons. Figure 8.17 shows a more detailed structure of this region, called an **integron**, and how it presumably recruits antibiotic resistance genes. A basic integron consists of an integrase gene (called *intI1* in the figure) next to a site called *attI* (for <u>att</u>achment site <u>i</u>ntegron). A gene in the *attI* site is transcribed from the

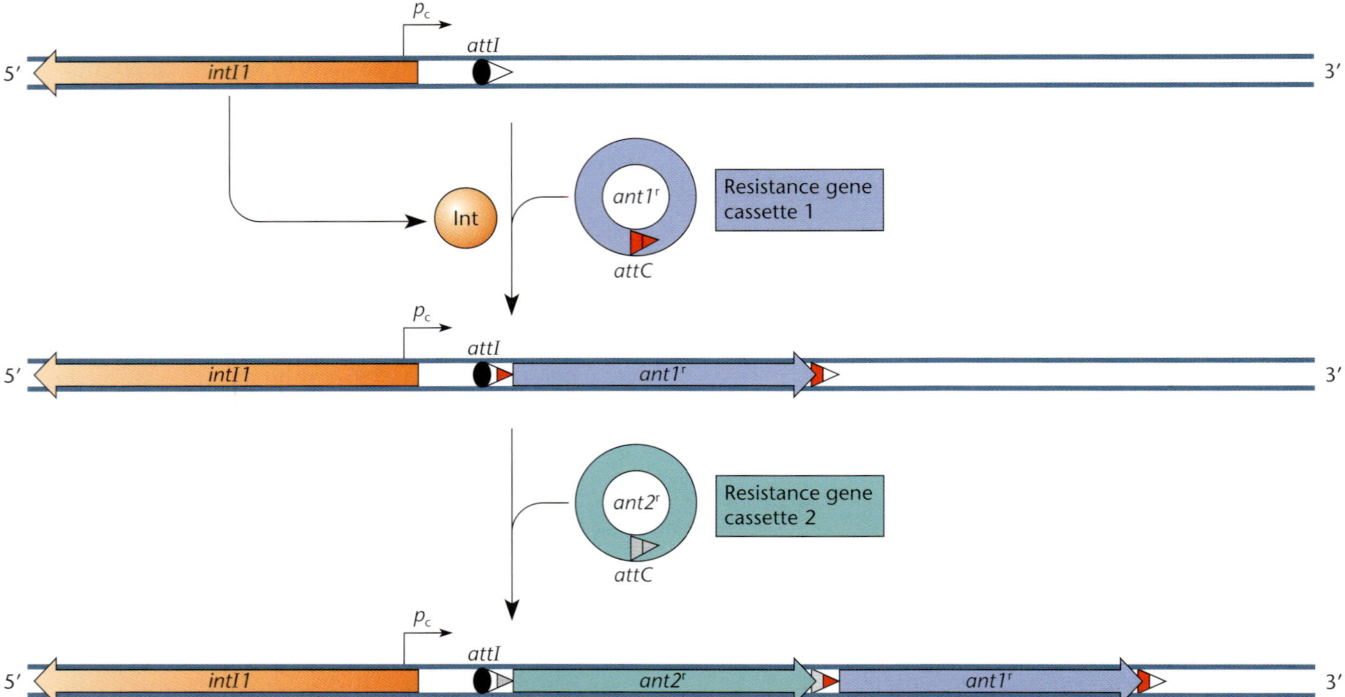

Figure 8.17 Assembly of integrons. The primary transposon carries an integron with a gene, *intI1*, encoding an integrase and a site, *attI*, transcribed by a strong promoter, p_c. A cassette carrying resistance to one antibiotic, *ant1*ʳ, has been excised from elsewhere and is integrated by the integrase by recombination between its *attC* site and the *attI* site on the integron. The antibiotic resistance gene is transcribed from the promoter on the integron. Later, the integrase integrates another cassette carrying a different antibiotic resistance gene, *ant2*ʳ, at the same place. In this way, a number of different antibiotic resistance genes can be assembled by the integron on the transposon. The *attC* sites, represented by triangles, contain conserved regions related to the *attC* sites between the cassettes of superintegrons shown in Figure 8.18. As described in the text, the actual substrate for recombination within the *att* sequences is a single-stranded DNA that has formed a hairpin structure. Modified from Mazel D, *ASM News* **70:** 520–525, 2004.

promoter, p_c. The transposon originally had no antibiotic resistance gene inserted at the *attI* site. Elsewhere in the cell, there was an antibiotic resistance cassette that consisted of an antibiotic resistance gene and a site, *attC* (for attachment site cassette), recognized by the integrase. This cassette was excised and formed a circle, and the integrase then integrated it into the *attI* site on the integron by promoting site-specific recombination between the *attC* site on the cassette and the *attI* site in the integron. At another time, and in another cell, a similar cassette carrying a different resistance gene could integrate next to this one, adding another antibiotic resistance gene to the transposon. The advantages of this system are obvious. By carrying different resistance genes, the transposon allows the cell, and thereby itself, to survive in environments containing various toxic chemicals. An important mechanistic difference from the λ phage integrase that may help account for a more limited role of homology in integron systems is that integration occurs via a single-strand DNA template on the integron cassette that forms a hairpin structure to make the double-stranded DNA substrate used in the reaction (see Bouvier et al., Suggested Reading). Full double-stranded DNA is presumably generated via DNA replication.

Integrons are not only found in transposons. Chromosomal **superintegrons** are found in a number of different types of bacteria (see Rowe-Magnus et al., Suggested Reading). The structure of one of them, from *Vibrio cholerae*, is shown in Figure 8.18. It consists of 179 cassettes that carry ORFs with largely unknown functions separated by partially conserved sequences that might be *attC* sites. Presumably, integrons in transposons will prove to be one example of a larger phenomenon in which useful genes can be recruited as needed from "storage areas" that carry large numbers of such cassettes. While the distribution of integron cassettes is well established, the original source of cassettes remains a mystery.

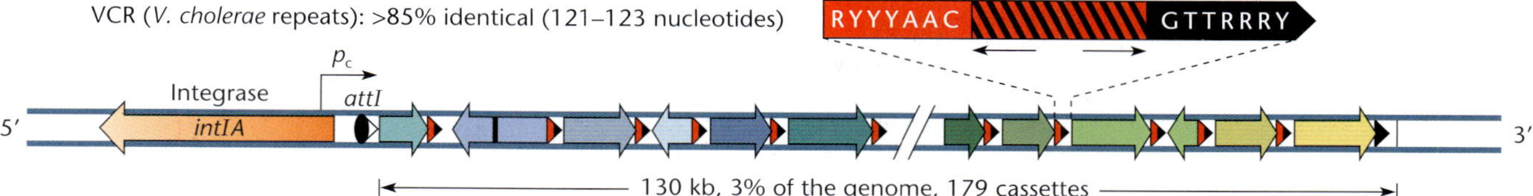

Figure 8.18 Example of a superintegron from *Vibrio cholerae.* More than 100 cassettes encoding resistance to different antibiotics and other functions are associated with partially homologous *attC* sites next to an integrase gene, *intIA*, and an *attI* attachment site. Regions between the cassettes corresponding to possible *attC* sites are shown as arrows. Regions of sequence conservation are shown as triangles. R, purine; Y, pyrimidine. Modified from Mazel D, *ASM News* **70:** 520–525, 2004.

Resolvases

The resolvases of transposons such as the TnpR protein of Tn*3* are another type of site-specific recombinase. In fact, the resolvase of transposon γδ, a close relative of Tn*3*, is one of the best-studied site-specific recombinases and has been crystallized bound to its DNA substrate. These enzymes promote the resolution of cointegrates by recognizing the *res* sequences that occur in one copy in the transposon but in two copies in the same orientation in cointegrates. Recombination between the two *res* sequences in a cointegrate excises the DNA between them, resolving the cointegrate into the donor DNA and the target DNA, both of which contain the transposon.

Other resolvases already mentioned resolve dimers of plasmids. Dimer formation by plasmids reduces their stability, especially if they have a low copy number, because each dimer is treated as one plasmid molecule by the partitioning system and segregated to the same side of the cell (see chapter 4). Because mutations in the resolvase gene can affect the segregation of the plasmid, some of these plasmid resolution systems were originally mistaken for partitioning systems and given the name Par (for partitioning). Examples of resolvases involved in resolving plasmid dimers are the Cre recombinase, which resolves dimers of the P1 prophage plasmid by promoting recombination between repeated *loxP* sites on the dimerized plasmid (see chapter 7), and the XerCD recombinase, which resolves dimers of the ColE1 and pSC101 plasmids by promoting recombination between repeated *cer* and *psi* sites, respectively. The XerCD recombinases also resolve the chromosome dimers between repeated *dif* sites in the dimers during cell division (see chapter 1). Chromosome dimers can form between circular genomes when an uneven number of crossover events occur between two sister chromosomes, something that is common during recombination repair of stalled chromosome replication forks (see chapter 9). The XerCD system is also borrowed by bacteriophages, where it doubles as an integrase to integrate the cholera toxin-producing phage (see chapter 7), illustrating how the ability to promote recombination between specific sequences can be put to many uses.

DNA Invertases

The DNA invertases are like resolvases in that they promote site-specific recombination between two sites on the same DNA. The main difference between the reactions promoted by DNA invertases and those catalyzed by resolvases is that two sites recognized by invertases are in inverse orientation with respect to each other, whereas the sites recognized by resolvases are in direct orientation. As discussed in chapter 3 in the section "Types of Mutations," recombination between two sequences that are in direct orientation deletes the DNA between the two sites, whereas recombination between two sites that are in inverse orientation with respect to each other inverts the intervening DNA.

The sequences acted on by DNA invertases are called **invertible sequences**. These short sequences may carry the gene for the invertase or may be adjacent to it. Therefore, the invertible sequence and its specific invertase form an inversion cassette that sometimes plays an important regulatory role in the cell, some examples of which follow.

PHASE VARIATION IN *SALMONELLA* SPECIES

The classic example of an invertible sequence is the one responsible for **phase variation** in some strains of *Salmonella.* Phase variation was discovered in the 1940s with the observation that some strains of *Salmonella* can change their surface antigens. They do this by shifting from making flagella composed of one flagellin protein, H1, to making flagella composed of a different flagellin protein, H2. The shift can also occur in reverse, i.e., from making H2-type flagella to making H1-type flagella. The flagellar proteins are the strongest antigens on the surfaces of many bacteria, and periodically changing their flagella may help these bacteria escape detection by the host immune system.

Two features of the *Salmonella* phase variation phenomenon suggested that the shift in flagellar type was

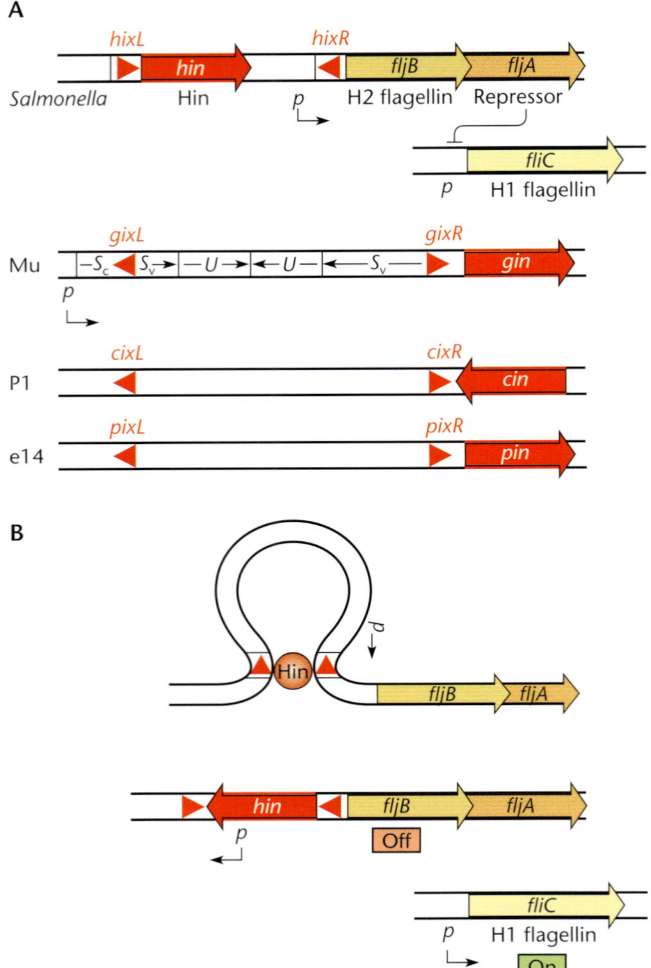

Figure 8.19 Regulation of *Salmonella* phase variation and some other members of the family of Hin invertases. **(A)** Invertible sequences, bordered by triangles, of *Salmonella* and several phages. The recombination sites are designated *hixL*, *hixR*, etc. **(B)** Hin-mediated inversion. In one orientation, the H2 flagellin gene, as well as the repressor gene, is transcribed from the promoter *p*. In the other orientation, neither of these genes is transcribed, and the H1 flagellin is synthesized instead. The invertase Hin is made constitutively from its own promoter.

not due to normal mutations. First, the shift occurs at a frequency of about 10^{-2} to 10^{-3} per cell, much higher than normal mutation rates. Second, both phenotypes are completely reversible—the cells switch back and forth, exhibiting first one type of flagella and then the other.

Figure 8.19 outlines the molecular basis for the *Salmonella* antigen phase variation (see Simon et al., 1980, Suggested Reading). As mentioned above, the two types of flagella are called H1 and H2. A DNA invertase called the Hin invertase causes the phase variation by inverting an invertible sequence upstream of the gene for the H2 flagellin by promoting recombination between two sites, *hixL* and *hixR*. The invertible sequence contains the invertase gene itself and a promoter for two other genes, *fljB*, which encodes the H2 flagellin, and *fljA*, which encodes a repressor of transcription of the *fliC* gene, which encodes the H1 flagellin. When the invertible sequence is in one orientation, the promoter transcribes the H2 gene and the repressor gene, and only the H2-type flagellum is expressed on the cell surface. When the sequence is in the other orientation, neither the H2 gene nor the repressor gene is transcribed, because the promoter is facing away from *fliB* and *fliA*. Now, however, without the repressor, the H1 gene, called *fliC*, can be transcribed; therefore, in this state, only the H1-type flagellum is expressed on the cell surface. The gene that encodes the Hin DNA invertase is also located within the invertible segment, but it has its own promoter and is expressed in either orientation.

OTHER INVERTIBLE SEQUENCES

There are a few other known examples of regulation by invertible sequences in bacteria. For example, fimbria synthesis in some pathogenic strains of *E. coli* is regulated by an invertible sequence. Fimbriae are required for the attachment of the bacteria to the eukaryotic cell surface and may also be important targets of the host immune system. In an interesting case, a symbiont of nematodes, *Photorhabdus*, switches to a pathogenic state in insects by a single inversion that affects the expression of almost 10% of its genes.

Invertible sequences also exist in some phages. An example is the invertible G segment region of phage Mu. Both phage P1 and the defective prophage e14 also have invertible regions (shown in Figure 8.19). These phages use invertible sequences to change their tail fibers. The tail fibers made when the invertible sequences are in one orientation differ from those made when the sequences are in the other orientation, broadening the host range of the phage. In phage Mu, the tail fiber genes expressed when the invertible sequence is in one orientation allow the phage to adsorb to *E. coli* K-12, *Serratia marcescens*, and *Salmonella enterica* serovar Typhi. In the other orientation, the phage is able to adsorb to other strains of *E. coli*, *Citrobacter*, and *Shigella sonnei*.

Not only do these phage invertases perform similar functions, but they are also closely related to each other. Dramatic evidence for their relationship came from the discovery that the Hin invertase inverts the Mu, P1, and e14 invertible sequences, and vice versa (see van de Putte et al., Suggested Reading). Apparently, inversion cassettes can be recruited for many purposes, much like antibiotic resistance cassettes are recruited by integrons.

Y and S Recombinases

As mentioned above, many site-specific recombinases, whether they are integrases, resolvases, or invertases, are closely related to each other. This is not surprising, considering that they all must perform the same basic reactions. First, they must cut a total of four strands of DNA—both strands in two recognition sequences—whether these recognition sequences are on the same DNA (resolvases and invertases) or on different DNAs (integrases). Then, they must join the cut end of each strand to the cut end of the corresponding strand from the other recognition sequence. We can anticipate some of the features that site-specific recombinases must have in order to perform these reactions. First, they must somehow hold the DNA ends after they cut them so that they are not free to flop around and join with the end of any strand they happen to encounter. Second, after the strands are cut, either the DNA or at least part of the recombinase must rotate in a defined way to place the cut ends from different strands in juxtaposition with each other so that the correct rejoining can occur. If they rejoin the ends of the same strands that were cut originally, there will be no recombination, and they will be back to where they started.

All site-specific recombinases appear to fall into two unrelated families, the Y (tyrosine) family and the S (serine) family, based on which of these amino acids, called the catalytic amino acid, plays the crucial role in their active centers. The feature shared by these amino acids is a hydroxyl group in their side chains to which phosphates can be attached. After the DNA is cut, the hydroxyl group on the catalytic amino acid forms a covalent bond with the free phosphate end on the DNA. This protects the end of the strand and holds it while the recombinase moves it into position to be joined to the end of a different cut strand, the essence of recombination. However, when the strands are cut, the end of DNA that is joined to the catalytic amino acid and the structures that form after this cutting differ between the two groups of recombinases. Each of these pathways is outlined in the following sections.

Y Recombinases: Mechanism

The Y recombinases seem to be the most varied group and include recombinases that perform the most complex reactions. Some examples of Y recombinases are listed in Table 8.2, and the structural organizations of some of them are shown in Figure 8.20. Some of them were mentioned in this and previous chapters; they include Cre recombinase, which resolves dimers of the P1 plasmid prophage by recombination between *loxP* sequences, and the XerCD recombinases, which resolve chromosome dimers by recombination between *dif* sites and resolve plasmid ColE1 dimers by recombination between *cer* sites, as well as integrating the cholera toxin-producing phage into one of the *V. cholerae* chromosomes. The λ phage integrase discussed in chapter 7 is also a Y recombinase, as are the integrases of integrons and those of the integrative conjugative elements (see chapter 5). Although the reactions they perform are somewhat different, the terminases that resolve the circular dimerized plasmids generated during replication of linear plasmids, including those in *Borrelia* (see chapter 4), also belong to this family and use a similar mechanism. The Y recombinases are also not limited to eubacteria and include some resolvases found in eukaryotes, such as the Flp recombinase, which inverts a short sequence in the 2-μm circle of yeast (Table 8.2), and they are related to some type I topoisomerases of eukaryotes, suggesting that they might have had a common evolutionary origin.

Table 8.2 Examples of tyrosine (Y) recombinases

Source	Enzyme
Resolvases	
E. coli	Plasmid/phage P1 Cre
	Plasmid F ResD
	Phage N15 telomere resolvase
Borrelia spp.	Telomere resolvase
S. sonnei	Plasmid ColIb-P9 shufflon
Bacteria	XerCD
Saccharomyces spp.	Flp
Integrases	
E. coli	Phage λ integrase
Bacteria	Integrons
Superfamily	
Eukaryotes	Topoisomerases
Viruses, e.g., vaccinia virus	Topoisomerase

Details of the molecular basis of recombination by the Y recombinases are outlined in Figure 8.21, and the structures of the sites recognized by some of them are shown in Figure 8.22. Much of what we know about how Y recombinases work comes from studies of the structure of the relatively simple Cre recombinase, which has been crystallized bound to various forms of its *loxP* DNA substrate. In the absence of evidence to the contrary, we may assume that at least most features of this reaction can be extrapolated to other Y recombinases, even ones that perform more complex reactions. The *loxP* site recognized by the Cre recombinase consists of a short sequence of 8 bp, where the crossover occurs. It also has two almost identical flanking sequences of 11 bp in inverse orientation that are recognized by the recombinase. In the first step, four copies of the Cre resolvase bind to two *loxP* recognition sites (two to each site) and hold them together in a large complex. Then, one strand

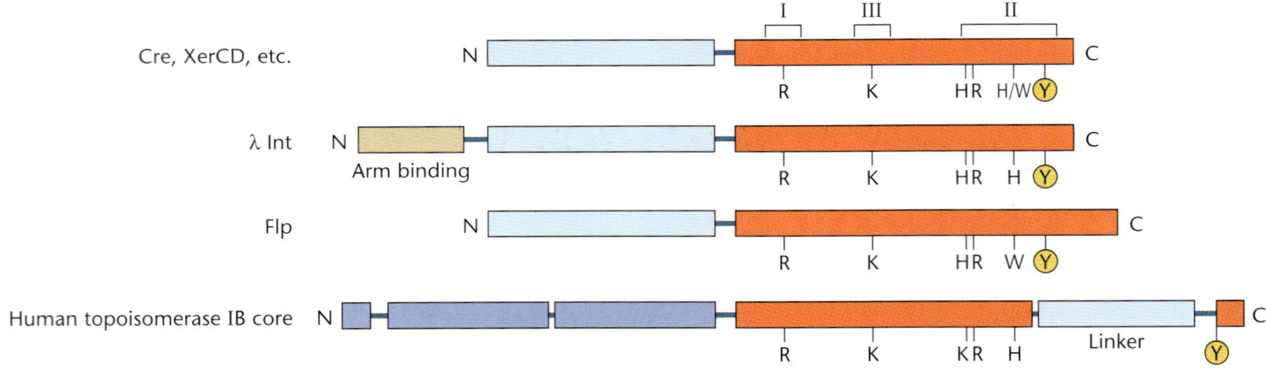

Figure 8.20 Domain structure of tyrosine recombinases (Cre, XerCD, etc.; λ Int; and Flp) and eukaryotic type IB topoisomerases. The conserved C-terminal catalytic domain of the proteins is shown in red. The brackets show the positions of three conserved regions of the catalytic domain: boxes I, II, and III. Residues of the catalytic signature of the family are indicated, and the tyrosine nucleophiles are circled. Other protein regions are shown in different shades of blue to indicate that they are structurally unrelated. Integrases, such as λ Int, have an additional DNA-binding domain at the N terminus to bind the Arm site sequences of the recombination site. In the human type IB topoisomerase core enzyme, residues 215 to 765, the catalytic domain is interrupted by a linker region spanning the region between the active-site histidine and the tyrosine nucleophile. Modified from Hallet B, Vanhooff V, Cornet F, *in* Funnell BE, Phillips GJ (ed), *Plasmid Biology* (ASM Press, Washington, DC, 2004).

of each of the recognition sequences is cut in the crossover region by an attack by the active-site tyrosine, generating 5′ OH ends. As they are cut, the 3′ PO_4 ends are transferred to the side chain of the active-site tyrosine in two of the bound recombinase molecules to form tyrosyl-3′-phosphate bonds. This holds the 3′ phosphate ends and protects them. The free 5′ OH ends then attack the 3′ tyrosyl phosphate bond in the other DNA, rejoining the 5′ OH ends to 3′ phosphate ends on the corresponding strand of the other DNA rather than on their own DNA. This causes two of the strands to cross over and hold the two DNAs together in what is called a Holliday junction. Holliday structures also form during homologous recombination; they are discussed in more detail in chapter 9. The crystal structure reveals that the Holliday junction is held very flat by the resolvase, with the four DNA branches coming out of the complex in the same plane.

What happens next is less clear and differs somewhat in different types of Y recombinases. To achieve recombination, the noncrossover strands must also be cut and joined to the corresponding strands on the other DNA. If the crossover strands are cut again and rejoined to their original strands, no recombination will occur. Therefore, the recombinase has to know which strands to cut in the Holliday junction and which strands to join them to. Holliday junctions can do a number of things, as discussed in chapter 9. They can isomerize so that the crossed strands become the uncrossed strands, and vice versa. They can also migrate so that the position at which the strands are crossed over can move on the DNA, provided that the sequences of the two DNAs are almost the same in the region of migration. One possibility is that part of the resolvase rotates, forcing the isomerization of the Holliday junction so that the correct strands enter the active center of the other two copies of the resolvase to repeat the cutting and rejoining reaction. This seems unlikely, considering that major changes are not seen in the crystal structure of the complex in its various states. The other possibility is that the Holliday junction migrates, rotating

Figure 8.21 Model for the reaction promoted by the Cre tyrosine (Y) recombinase. Four Cre recombinase molecules bind two *loxP* sites, bringing them together. RBE, recombinase-binding element. **(1)** The active-site tyrosines in two of the Cre molecules, indicated by Y, cleave two of the strands in a phosphoryltransferase reaction that forms 3′ phosphotyrosyl bonds and 5′ OH ends (arrows). **(2)** Each 5′ OH end then attacks the opposite 3′ phosphoryltyrosine bond, switching the strands and causing a 3- to 4-nucleotide swap in the complementary region. **(3)** The nucleophilic 5′ OH ends attack the phosphotyrosyl bonds, rejoining the ends to form a Holliday junction that isomerizes. The next steps are essentially a repeat of steps 1 to 3, but the other two Cre molecules cut the other two strands by a phosphotransferase reaction, and the strands are exchanged and rejoined to form the two recombinant DNA molecules. Modified from Hallet B, Vanhooff V, Cornet F, *in* Funnell BE, Phillips GJ (ed), *Plasmid Biology* (ASM Press, Washington, DC, 2004).

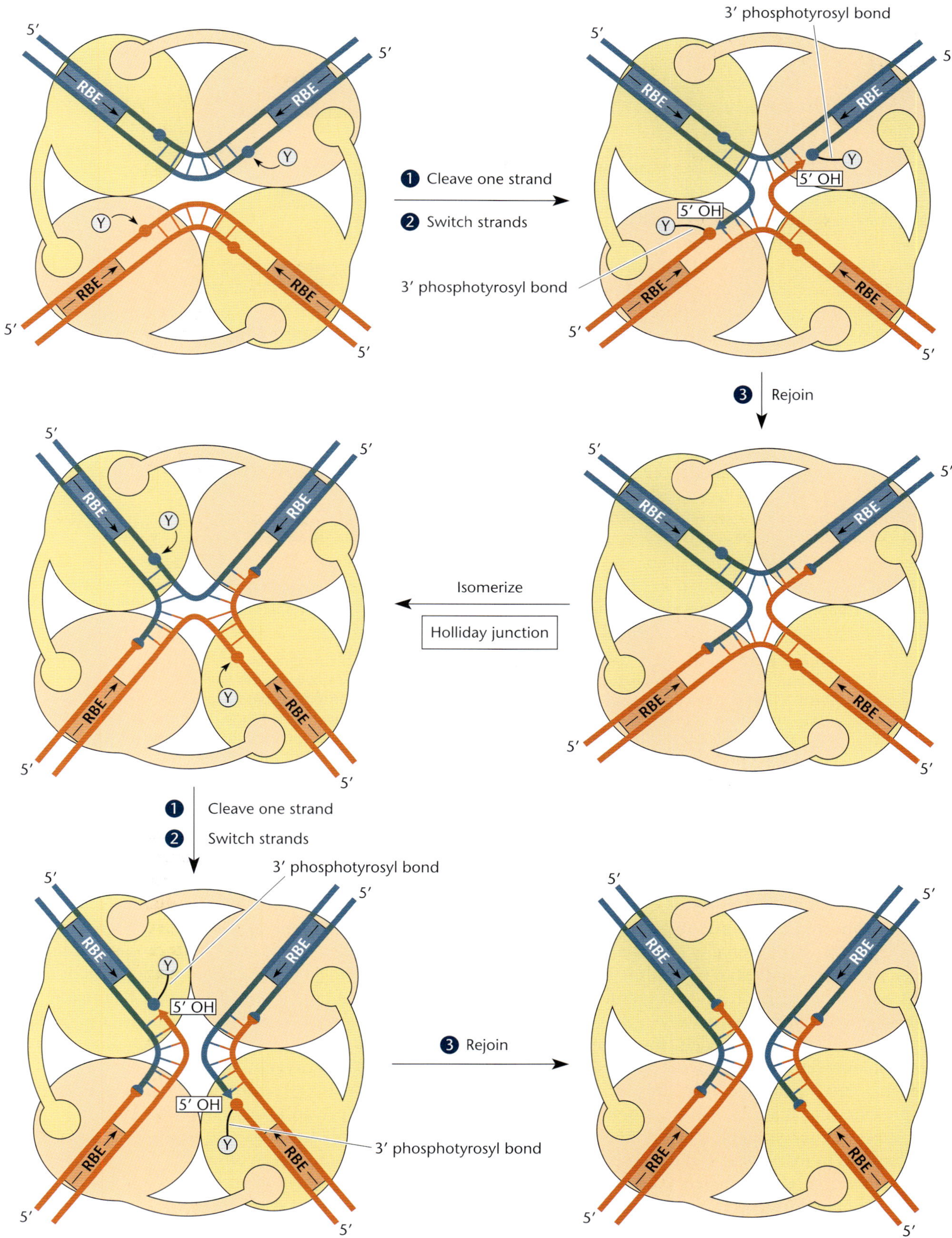

5′
5′
RBE
RBE
Y
Y
RBE
RBE
5′
5′
1 Cleave one strand
2 Switch strands
3′ phosphotyrosyl bond
5′ OH
5′ OH
3′ phosphotyrosyl bond
3 Rejoin
Isomerize
Holliday junction
1 Cleave one strand
2 Switch strands
3′ phosphotyrosyl bond
5′ OH
5′ OH
3′ phosphotyrosyl bond
3 Rejoin

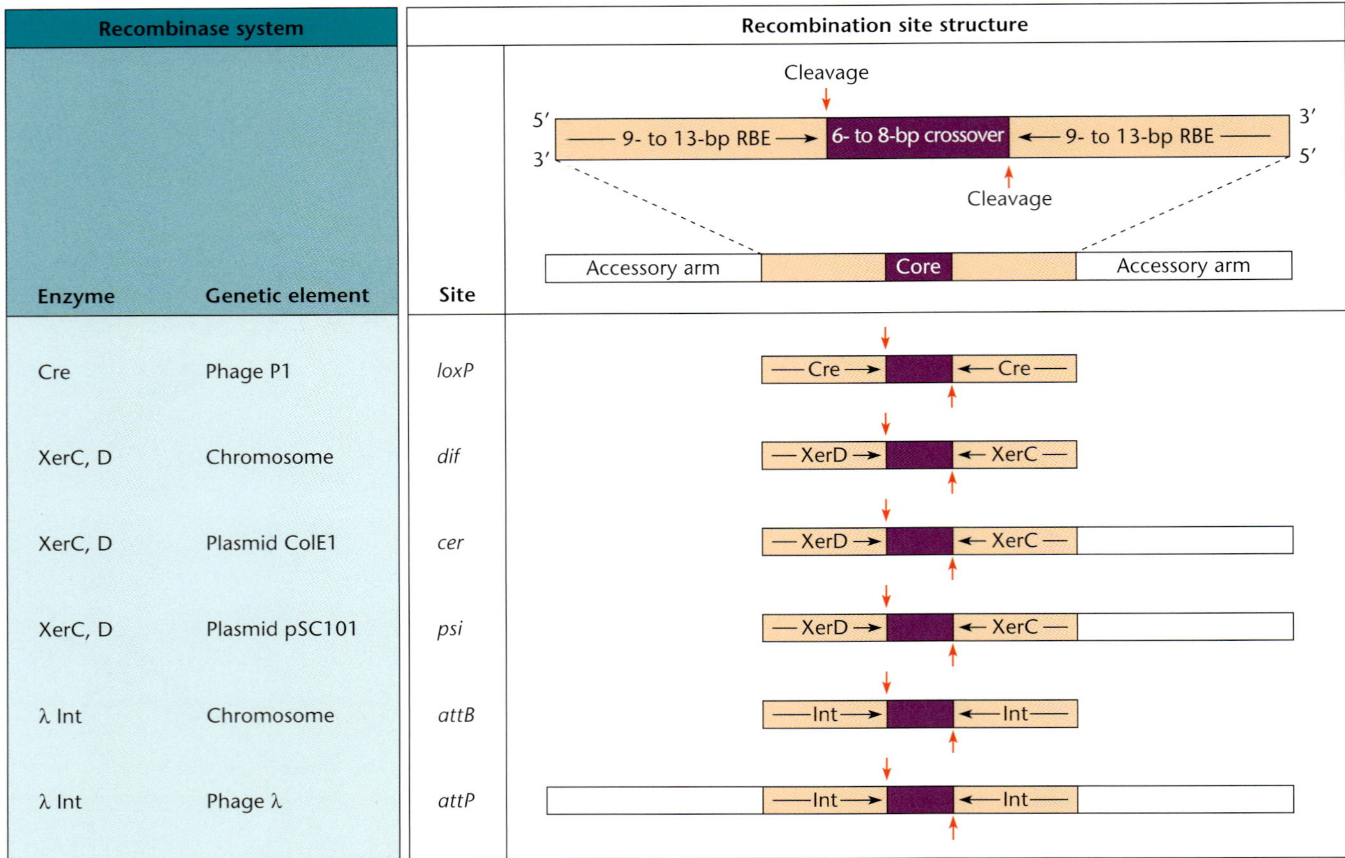

Figure 8.22 Structures of some sites recognized by tyrosine (Y) recombinases. The top part of the figure shows the basic structure, with a core of a 6- to 8-bp crossover region flanked by two 9- to 13-bp sequences required for recombinase binding. The recombinase-binding elements (RBE) are shown in beige and the orientation is indicated with black arrows. Many sites also have flanking accessory protein-binding sites called the accessory arms (white). Proteins bind to these sites and help orient the recombinase and give it specificity. The bottom part of the figure shows the variations on the theme exhibited by some of the sites described in the text. Red arrows indicate the point of cleavage. Modified from Van Duyne GD, *in* Craig NL, Craigie R, Gellert M, Lambowitz AM (ed), *Mobile DNA II* (ASM Press, Washington, DC, 2002).

the DNA so that the correct strands to be cut come in contact with the active centers of the other two copies of the resolvase and can be cut and rejoined. However, it is hard to imagine how the Holliday junction could migrate very far, since that would mean either drastically rotating the DNA arms that emerge from the complex or severely distorting the DNA in the complex.

Further complicating the models, some apparent Y resolvases, including the integrases of the integrating conjugative element, Tn*916*, and the integrases of integrons, do not require extensive homology in the crossover region of the two sites being recombined. The Tn*916* integrative element integrates in many places in the chromosome, suggesting that it can use many different sequences as bacterial *att* sites, although it may prefer some sites over others. Extensive homology in the crossover region should be required for Holliday junction formation and for branch migration. Current research addresses these and other questions about the reactions performed by Y recombinases.

Other proteins besides the Y recombinases themselves are often involved in the recombination reactions. These other proteins bind close to the core region, on what is referred to as the accessory arm regions, and help to stabilize the recombinase-DNA complexes and/or orient the recombinase proteins on the DNA (Figure 8.23). For example, the XerCD recombinase requires two proteins, ArgR and PepA, to promote recombination at *cer* sites in plasmid ColE1 and to resolve dimers (see chapter 4). These proteins also play other, very different roles in the cell: one is the repressor of the arginine biosynthetic operon, and the other is an aminopeptidase. It is not clear why the ColE1 plasmid recruits these particular proteins

Table 8.3 Examples of serine (S) recombinases

Source	Enzyme
Resolvases	
E. coli	Resolvase (TnpR) of Tn*3*
	Resolvase of γδ
	ParA of RP4/RK2
Enterococcus spp.	Resolvase of Tn*917*
Invertases	
S. enterica serovar Typhimurium	Hin flagellar invertase
E. coli phage Mu	Gin tail fiber invertase
Superfamily	
Streptomyces coelicolor	Integrase of phage φC31
B. subtilis	SpoIVCA recombinase
Anabaena spp.	Heterocyst recombinase

for dimer resolution, but they may happen to have structural and/or sequence features that make them easy to adapt to this role. Whatever the reason, such situations are problematic for geneticists trying to deduce the function of a gene product. For example, mutations in the *argR* gene cause constitutive synthesis of the arginine-biosynthetic enzymes but also reduce the stability of the ColE1 plasmid for a completely unrelated reason. The λ integrase also requires that other host proteins, including integration host factor, be bound close to the *attP* site for recombination to occur. This protein is bound at many places in the chromosomal DNA, where it plays multiple roles, making this less surprising.

S Recombinases: Mechanism

The S recombinases are also a large family, comprising many of the plasmid resolution and invertase systems found broadly in bacteria. Some S recombinases are listed in Table 8.3. The TnpR cointegrate resolution systems of Tn*3*-like transposons, including γδ, are members of this family, as are the dimer resolution systems of some plasmids, including the promiscuous plasmids RK2 and RP4 in *Proteobacteria* and the *Enterococcus faecalis* plasmid pAMβ1, to list some plasmids mentioned elsewhere in the book. The integrase of a *Streptomyces* phage, fC131, is an S recombinase, unlike the integrases of most phages, which are Y recombinases. The Hin invertase in *Salmonella* and its relatives that invert tail fiber genes in some phages are also members of this family, as are the excisases mentioned above, which remove the intervening sequences during sporulation in *Bacillus subtilis* and heterocyst formation in *Anabaena*. Some integrative conjugative elements use S integrases rather than Y integrases to integrate into the recipient cell DNA. So far, counterparts of the S recombinases have not been found in eukaryotes, with the possible exception of some in small plasmids in marine diatoms.

The domain structure of some S recombinases is shown in Figure 8.23, and a model of the reaction they perform is shown in Figure 8.24. Again, much of what we know comes from studies with one S recombinase, the resolvase of the transposon γδ, which has been crystallized with its DNA substrate. Superficially, the molecular mechanism of site-specific recombination by S recombinases is similar to that of recombination by Y recombinases. The recognition sites have a short crossover region bracketed by copies of recombinase-binding sites. Four copies of the recombinase bind to two recognition sequences to form a complex in which the recombination occurs. This is where the similarities end, however. Rather than having the active-site tyrosines in two recombinase molecules make a nucleophilic attack on the same phosphodiester bond in the two DNAs, the active-site serines in all four recombinases make nucleophilic attacks a few nucleotides apart to create staggered breaks in both strands, for a total of four breaks. Also, the staggered breaks leave 5′ PO_4 and 3′ OH overhangs. The nucleophilic attacks leave the 5′ phosphate ends joined to the hydroxyl group in the side chain of the active-site serines on all four recombinase molecules to form 5′ phosphorylseryl bonds to the ends of the DNA, rather than 3′ phosphoryltyrosyl bonds, as in the Y recombinases; Figure 8.25 shows these nucleophilic attacks and how they leave 5′ phosphorylseryl bonds. The nucleophilic 3′ OH

Figure 8.23 Domain structure of serine (S) recombinases. The conserved catalytic domain (ca. 120 amino acids) is shown in red. The conserved amino acids that play major roles in catalysis are indicated, and the active-site serines are circled. Modified from Hallet B, Vanhooff V, Cornet F, *in* Funnell BE, Phillips GJ (ed), *Plasmid Biology* (ASM Press, Washington, DC, 2004).

Resolvase/invertase subgroup (e.g., Tn*3* and γδ resolvases and Gin and Hin invertases)

"Large" serine recombinase subgroup (e.g., ϕC31 integrase, Tn*4451* and Tn*5397* transposases, and SpoIVCA)

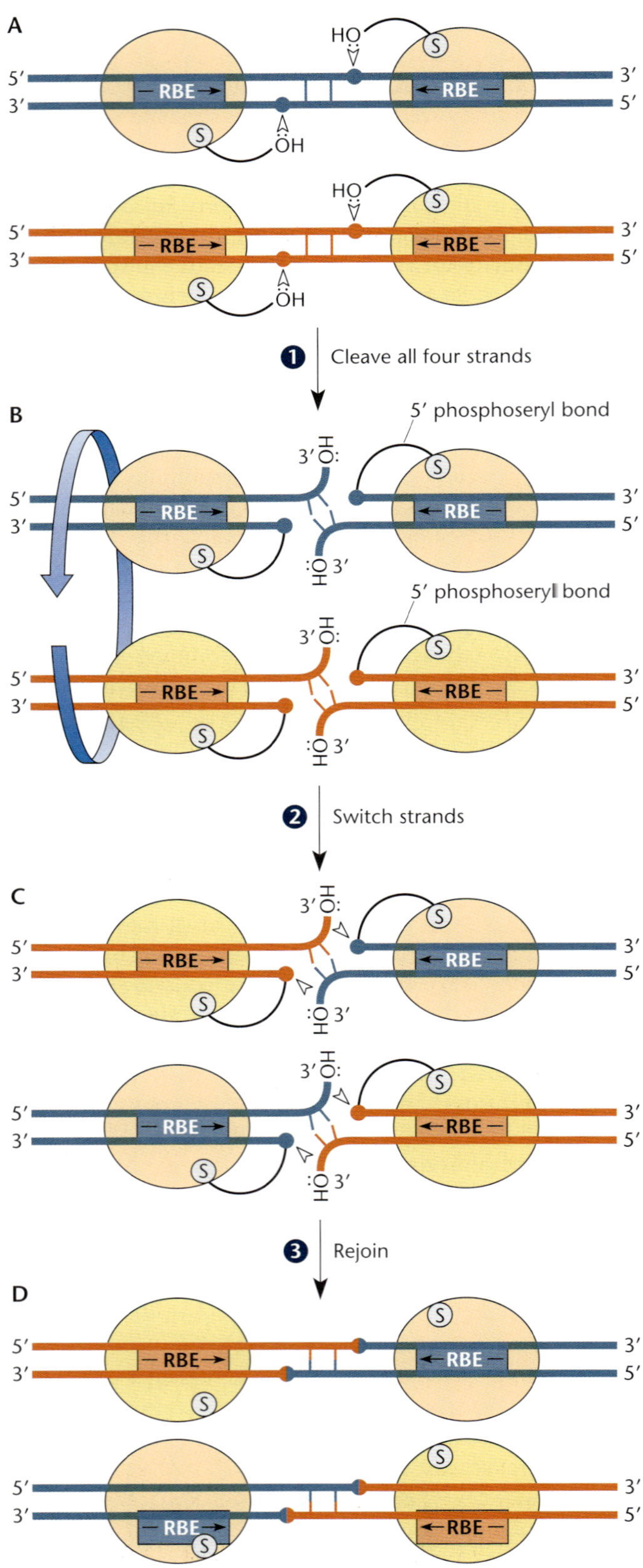

on each cut strand then attacks the phosphorylseryl bond in the corresponding strand of the other recognition sequence. Rejoining of the nicks leaves the recombinant product without the formation of a Holliday junction. A large 180° rotation in the complex appears to explain how the new configuration occurs to allow strand exchange. This was first suggested by the crystal structure, which revealed that the two dimers that are linked to the DNAs interface with each other over a large flat surface (see Li et al., Suggested Reading). This idea has received additional support recently from single-molecule work with a simpler recombinase.

Like the Y recombinases, the S recombinases often use additional proteins bound close to the recognition site to help stabilize the complex during recombination. In some cases, these are extra copies of the recombinase itself.

Group II Mobile Introns: Elements that Move Using an RNA Intermediate

All of the elements described thus far in this chapter move exclusively as DNA. However, there are also elements that are found in bacteria that are transcribed as RNA to produce the mobile form of the element that can insert into a new location in the genome and be reverse transcribed back into DNA. Well-studied elements of this type are group II mobile introns. As the name implies, these elements exist in the coding region of genes but are spliced out like an intron, leaving the protein-coding region intact, a process more typically associated with metazoans. A key feature of this process involves the 2′ OH found on an essential adenosine residue in these elements (see inside cover). After transcription of the gene encoding the element, the group II mobile intron self-splices out of the mRNA, first by a joining event to a conserved guanosine residue at the 5′ end of the element (Figure 8.26A). The free 3′ OH liberated in the reaction then joins to the 3′ end of the element (Figure 8.26B), re-

Figure 8.24 Model for the reaction promoted by the $\gamma\delta$ recombinase. **(A)** Four recombinase molecules bind to two copies of the recombination sites, holding them together, and the active-site serines in all four recombinases attack phosphodiester bonds a few nucleotides apart to leave staggered breaks, with the 5′ phosphate overhangs forming phosphorylserine bonds with the active-site serines. **(B)** Rotation of a domain of the recombinase brings the corresponding ends of the two different recombination sites together. **(C)** The free 3′ OH ends then attack the 5′ phosphorylserine bond on the corresponding strand on the other DNA. **(D)** The nicks are sealed to form the recombinant product. RBE, inverted recombinase-binding elements. Modified from Hallet B, Vanhooff V, Cornet F, *in* Funnell BE, Phillips GJ (ed), *Plasmid Biology* (ASM Press, Washington, DC, 2004).

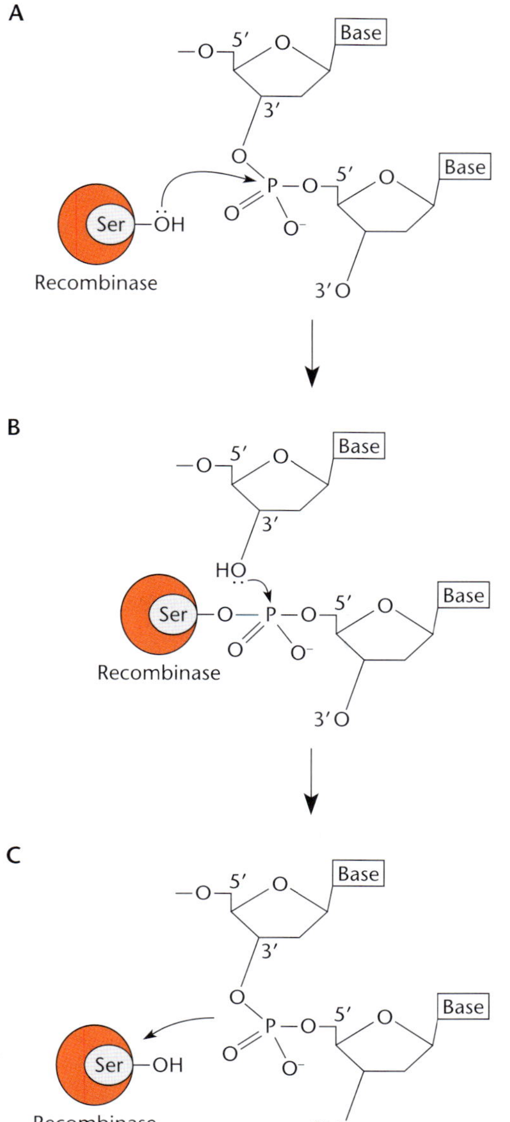

Figure 8.25 How successive attacks by nucleophilic hydroxyl groups of serine (S) recombinases can create a recombinant DNA product. **(A)** A nucleophilic attack by the hydroxyl group on the side chain of serine in the active center of a serine recombinase forms a 5′ phosphorylserine bond and breaks the phosphodiester bond in the DNA. **(B)** The free 3′ OH group can attack another 5′ phosphorylserine bond, breaking the bond. **(C)** This attack results in the re-formation of a phosphodiester bond to a different DNA strand. Modified from Johnson RC, *in* Craig NL, Craigie R, Gellert M, Lambowitz AM (ed), *Mobile DNA II* (ASM Press, Washington, DC, 2002).

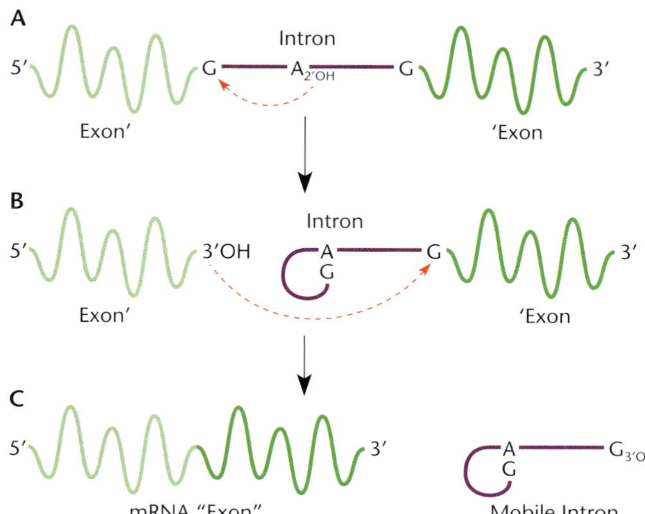

Figure 8.26 Excision of a group II mobile intron from an mRNA. **(A)** After transcription of the mRNA encoding the element, the 2′ OH found on the conserved adenosine joins to a guanosine residue at the 5′ end of the mobile RNA intron element. **(B)** The free 3′ OH liberated in the reaction then joins to the 3′ end of the element. **(C)** The splicing reaction reestablishes the mRNA and leaves the mobile RNA intron element capable of integrating elsewhere in the genome. The 5′ and 3′ portions of the native mRNA are indicated in shades of green, and the mobile element is in purple.

establishing the mRNA and generating a mobile RNA element capable of integrating elsewhere in the genome (Figure 8.26C).

Integration of group II mobile elements occurs at a high frequency into a DNA site that is recognized by homology between this site and the RNA form of the element in a process called **retrohoming**. The RNA element itself is stabilized by an RNA chaperone protein specific to the element and encoded within the element itself called the intron-encoded protein (Figure 8.27A). The intron-encoded protein typically has nuclease activity, producing a nick in the bottom strand of the target DNA (Figure 8.27B). While the 3′ OH of the RNA form of the element allows the RNA to be joined to the top strand, the nick produced in the bottom strand provides a primer for reverse transcription of the element (Figure 8.27C). The final (top strand) RNA form of the element is degraded, and other host repair functions complete the cycle of movement of the group II mobile element (Figure 8.27D). The retrohoming system has been capitalized on as a gene knockout technology called the TargetTron gene knockout system (see Zhong et al., Suggested Reading).

While integration occurs at a high frequency via retrohoming by the strong match between this site and the RNA element, other sites are recognized at a much lower frequency by partial matches in a process known as **retrotransposition**. The retrotransposition process may provide a benefit to the element by allowing dissemination of the element to new positions in the bacterial genome.

As the name implies, group II introns are hypothesized to be the ancestor of spliceosomal introns in metazoans. Group II mobile introns also utilize a similar

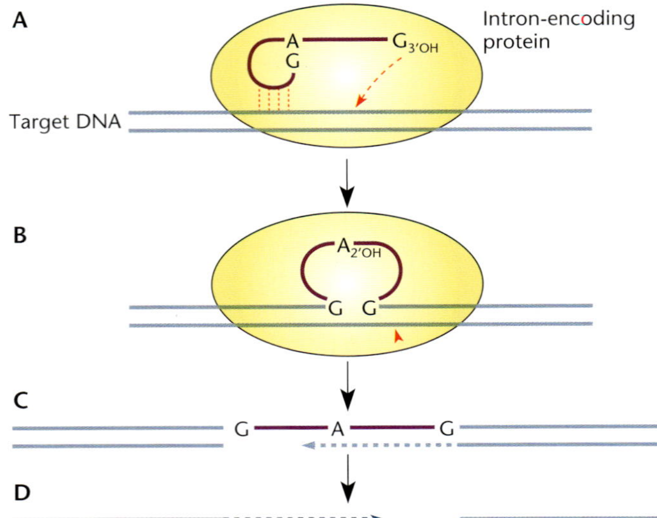

Figure 8.27 Integration of a group II mobile intron into double-stranded DNA by target primed reverse transcription (TPRT). **(A and B)** The group II mobile RNA element (green) with the intron-encoded protein that acts as an RNA chaperone (shown in yellow) inserts into the top strand of double-stranded DNA by reverse splicing, the reverse of the reaction it used to excise from mRNA. Homology between the RNA element and the target DNA provides the specificity for inserting at one position in the host in the retrohoming reaction (shown with a series of thin red dashed lines). **(C)** The endonucease activity of the intron-encoded protein cleaves the bottom strand (red arrowhead) providing a 3′ OH to initiate target-primed reverse transcription (new DNA is shown with dashed blue lines). **(D)** Host processes allow replacement of the RNA top strand with DNA. The important adenine (A) and guanine (G) residues are indicated. See text for details.

template-primed reverse transcription mechanism of mobility as non-long terminal repeat (LTR) retrotransposons that are abundant in primate genomes, including the human genome.

Importance of Transposition and Site-Specific Recombination in Bacterial Adaptation

One of the most important conclusions from studies of transposons and other types of moveable elements in bacteria is the contribution they make to bacterial adaptation. We have seen how integrons can integrate antibiotic resistance genes from large storage areas into transposons that can then transpose into other DNAs. Regardless of how transposons acquire new functions, once encoded within an element, these functions can be mobilized between bacterial populations via conjugative elements, including self-transmissible plasmids and integrating conjugative elements. This presumably allows bacterial genomes to remain small but still have access to many types of genes that increase their ability to adapt to different environments. One way this impacts humans directly is in the acquisition of antibiotic resistance by bacteria. Table 8.4 gives a summary of the moveable elements discussed so far that are known to carry antibiotic resistance genes in various bacteria. If we are to continue to be able to treat bacterial infections effectively, we must have a clear idea of how these moveable elements can contribute to widespread antibiotic resistance and how to combat it.

Table 8.4 Characteristics of genetic elements involved in the spread of antibiotic resistance genes

Genetic element	Characteristics	Role in spread of resistance genes
Self-transmissible plasmid	Circular, autonomously replicating element; carries genes needed for conjugal DNA transfer	Transfers resistance genes; mobilizes other elements that carry resistance genes
Integrating conjugative elements	Integrated element that can excise to form a nonreplicating circular transfer intermediate; carries genes needed for conjugal DNA transfer	Same as self-transmissible plasmid: highly promiscuous, transferring between diverse genera and species
Mobilizable plasmid	Circular, autonomously replicating element; carries site and genes that allow it to use the conjugal apparatus provided by a self-transmissible plasmid	Transfer of resistance genes
Transposon	Moves from one DNA segment to another within the same cell	Carries resistance genes from chromosome to plasmid or vice versa
Gene cassette	Circular, nonreplicating DNA segment containing only ORFs; integrates into integrons	Carries resistance genes
Integron	Integrated DNA segment that contains an integrase, a promoter, and an integration site for gene cassettes	Forms clusters of resistance genes that are transcribed under control of the integron promoter

1. Nonhomologous recombination is the recombination between specific sequences on DNA that occurs even if the sequences are mostly dissimilar.

2. Transposition is the movement of certain DNA sequences, called transposons, from one place in DNA to another. The smallest known bacterial transposons are IS elements, which contain only the genes required for their own transposition. Other transposons carry genes for resistance to substances such as antibiotics and heavy metals. Transposons have played an important role in evolution and are useful for mutagenesis, gene cloning, and random gene fusions.

3. Composite transposons are composed of DNA sequences bracketed by IS elements.

4. Most known transposons are DDE transposons, because their transposases have three amino acids, DDE (aspartate-aspartate-glutamate), which hold magnesium ions that play an essential role in transposition. They are characterized by having IR sequences at their ends and duplicating a short sequence in the target DNA on entry. The same chemistry has been adapted to different classes of transposons that move using distinct types of joining and processing events.

5. HUH transposons form covalent linkages between conserved tyrosine residues in the transposase and DNA as part of recombination. The HUH transposases are part of a larger family of HUH recombinases involved in rolling-circle plasmid replication and in conjugation.

6. IS*200*/*605* family HUH transposase elements like IS*608* move as single-strand circular DNA elements that excise and integrate only when the donor DNA and target DNA are in the single-stranded form. Hairpins that are formed within the ends of the element when in the single-stranded form are one of the features recognized by the HUH transposase. A short (4 to 5 nucleotides) specific sequence is recognized in single-stranded target DNA.

7. In transposon mutagenesis, a transposon insertion can disrupt a gene or activate expression of adjacent genes. Transposons are mapped easily with DNA sequencing and can facilitate cloning in unsequenced genomes.

8. Specially engineered transposons carrying reporter genes can be used to make random gene fusions. Insertion of the transposon into a gene can lead to expression of the reporter gene on the transposon from the promoter or translational initiation region of the disrupted gene, depending on whether the fusion is transcriptional or translational.

9. Genes that have been mutated by insertion of a transposon are often easy to clone in *E. coli* if an antibiotic resistance gene is present on the transposon. Some transposons have been engineered to contain an *E. coli* plasmid origin of replication so that the restriction fragment containing the transposon need not be cloned in another cloning vector but need only be cyclized after ligation to form a replicon in *E. coli*.

10. Site-specific recombinases are enzymes that promote recombination between certain sites on the DNA. Examples of site-specific recombinases are resolvases, integrases, and DNA invertases. The genes for many of these site-specific recombinases have sequences in common and appear to form two families with two distinct common ancestors.

11. Resolvases are site-specific recombinases encoded by replicative transposons that resolve cointegrates by promoting recombination between short *res* sequences in the copies of the transposon in the cointegrate.

12. Integrases promote nonhomologous recombination between specific sequences on a DNA element, such as a phage DNA and the chromosome, integrating the phage DNA into the chromosome to form lysogens. They also integrate antibiotic resistance gene cassettes into transposons. Transposons that encode an integrase that allows them to accept these antibiotic gene cassettes are called integrons. Superintegrons are large DNA elements (50,000 to 100,000 bp) that carry genes, including genes for pathogenicity, that allow the bacterium to occupy unusual ecological niches.

13. DNA invertases promote nonhomologous recombination between short IRs, thereby changing the orientation of the DNA sequence between them. The sequences they invert, invertible sequences, are known to play an important role in changing the host range of phages and the bacterial surface antigens to avoid host immune defenses.

14. Site-specific recombinases can be divided into two types, Y (or tyrosine) recombinases and S (or serine) recombinases. Recombination by both types involves nucleolytic attacks by the hydroxyl group of the side chain of the amino acid on the phosphodiester bond in DNA, forming a phosphoryl bond to the amino acid. They differ in that Y transposons form a 3′ phosphoryltyrosine bond, whereas S transposons form a 5′ phosphorylserine bond; other differences are that Y recombinases cut two strands simultaneously and form a Holliday junction, which can then isomerize, whereas S recombinases cut all four strands, not necessarily in any order, and do not form a Holliday junction. Rather, S recombinases appear to depend on rotation of part of the recombinase to bring the different strands into juxtaposition.

15. Group II mobile introns move as an RNA that is spliced out of an mRNA and reverse spliced into double-stranded DNA. The integrated RNA element is converted back into DNA using target-primed reverse transcription.

QUESTIONS FOR THOUGHT

1. For transposons that transpose replicatively (e.g., Tn*3*), why are there not multiple copies of the transposon around the genome?

2. How do you think that transposon Tn*3* and its relatives have spread throughout the bacterial kingdom?

3. While transposons are parasites of the host, they also mitigate their effects with beneficial functions. List some of the ways they help the host.

4. Where do you suppose the genes that were inserted into integrons in the evolution of transposons came from?

5. If the DNA invertase enzymes are made continuously, why do the invertible sequences invert so infrequently?

SUGGESTED READING

Bardarov S, Kriakov J, Carriere C, Yu S, Vaamonde C, McAdam RA, Bloom BR, Hatfull GF, Jacobs WR Jr. 1997. Conditionally replicating mycobacteriophages: a system for transposon delivery to *Mycobacterium tuberculosis*. *Proc Natl Acad Sci USA* **94:**10961–10966.

Bender J, Kleckner N. 1986. Genetic evidence that Tn*10* transposes by a nonreplicative mechanism. *Cell* **45:**801–815.

Biery MC, Stewart FJ, Stellwagen AE, Raleigh EA, Craig NL. 2000. A simple *in vitro* Tn*7*-based transposition system with low target site selectivity for genome and gene analysis. *Nucleic Acids Res* **28:**1067–1077.

Bordi C, Butcher BG, Shi Q, Hachmann A-B, Peters JE, Helmann JD. 2008. *In vitro* mutagenesis of *Bacillus subtilis* by using a modified Tn*7* transposon with an outward-facing inducible promoter. *Appl Environ Microbiol* **74:**3419–3425.

Bouvier M, Demarre G, Mazel D. 2005. Integron cassette insertion: a recombination process involving a folded single strand substrate. *EMBO J* **24:**4356–4367.

Casadaban MJ, Cohen SN. 1979. Lactose genes fused to exogenous promoters in one step using a Mu-lac bacteriophage: *in vivo* probe for transcriptional control sequences. *Proc Natl Acad Sci USA* **76:**4530–4533.

Chandler M, de la Cruz F, Dyda F, Hickman AB, Moncalian G, Ton-Hoang B. 2013. Breaking and joining single-stranded DNA: the HUH endonuclease superfamily. *Nat Rev Microbiol* **11:**525–538.

Chandler M, Fayet O, Rousseau P, Ton Hoang B, Duval-Valentin G. 2015. Copy-out-paste-in transposition of IS*911*: a major transposition pathway. *Microbiol Spectr* **3:**MDNA3-0031-2014.

Choi W, Harshey RM. 2010. DNA repair by the cryptic endonuclease activity of Mu transposase. *Proc Natl Acad Sci USA* **107:**10014–10019.

Collis CM, Recchia GD, Kim M-J, Stokes HW, Hall RM. 2001. Efficiency of recombination reactions catalyzed by class 1 integron integrase IntI1. *J Bacteriol* **183:**2535–2542.

Comfort NC. 2001. *The Tangled Field: Barbara McClintock's Search for the Patterns of Genetic Control*. Harvard University Press, Boston, MA.

Craig NL. 2002. Tn*7*, p 423–456. *In* Craig NL, Craigie R, Gellert M, Lambowitz AM (ed), *Mobile DNA II*. ASM Press, Washington, DC.

Derbyshire KM, Grindley NDF. 2005. DNA transposons: different proteins and mechanisms but similar rearrangements, p 467–497. *In* Higgins NP (ed), *The Bacterial Chromosome*. ASM Press, Washington, DC.

Foster TJ, Davis MA, Roberts DE, Takeshita K, Kleckner N. 1981. Genetic organization of transposon Tn*10*. *Cell* **23:**201–213.

Garcillán-Barcia M, Bernales I, Mendiola M, De La Cruz F. 2002. IS91 rolling-circle transposition, p 891–904. *In* Craig N, Craigie R, Gellert M, Lambowitz A (ed), *Mobile DNA II*. ASM Press, Washington, DC.

Gill R, Heffron F, Dougan G, Falkow S. 1978. Analysis of sequences transposed by complementation of two classes of transposition-deficient mutants of Tn*3*. *J Bacteriol* **136:**742–756.

Golden JW, Robinson SJ, Haselkorn R. 1985. Rearrangement of nitrogen fixation genes during heterocyst differentiation in the cyanobacterium *Anabaena*. *Nature* **314:**419–423.

Goryshin IY, Jendrisak J, Hoffman LM, Meis R, Reznikoff WS. 2000. Insertional transposon mutagenesis by electroporation of released Tn*5* transposition complexes. *Nat Biotechnol* **18:**97–100.

Groisman EA, Casadaban MJ. 1986. Mini-mu bacteriophage with plasmid replicons for *in vivo* cloning and *lac* gene fusing. *J Bacteriol* **168:**357–364.

Hagemann AT, Craig NL. 1993. Tn*7* transposition creates a hotspot for homologous recombination at the transposon donor site. *Genetics* **133:**9–16.

Hallet B, Vanhooff V, Cornet F. 2004. DNA site-specific resolution systems, p 145–180. *In* Funnell BE, Phillips GJ (ed), *Plasmid Biology*. ASM Press, Washington, DC.

Hughes KY, Maloy SM (ed) 2007. *Methods in Enzymology*, vol. 421. *Advanced Bacterial Genetics: Use of Transposons and Phage for Genomic Engineering*. Elsevier, London, United Kingdom.

Kenyon CJ, Walker GC. 1980. DNA-damaging agents stimulate gene expression at specific loci in *Escherichia coli*. *Proc Natl Acad Sci USA* **77:**2819–2823.

Komano T. 1999. Shufflons: multiple inversion systems and integrons. *Annu Rev Genet* **33:**171–191.

Kunkel B, Losick R, Stragier P. 1990. The *Bacillus subtilis* gene for the development transcription factor sigma K is generated by excision of a dispensable DNA element containing a sporulation recombinase gene. *Genes Dev* **4:**525–535.

Lambin M, Nicolas E, Oger CA, Nguyen N, Prozzi D, Hallet B. 2012. Separate structural and functional domains of Tn*4430* transposase contribute to target immunity. *Mol Microbiol* **83:**805–820.

Li W, Kamtekar S, Xiong Y, Sarkis GJ, Grindley ND, Steitz TA. 2005. Structure of a synaptic gammadelta resolvase tetramer covalently linked to two cleaved DNAs. *Science* **309:**1210–1215.

Martin SS, Pulido E, Chu VC, Lechner TS, Baldwin EP. 2002. The order of strand exchanges in Cre-LoxP recombination and its basis suggested by the crystal structure of a Cre-LoxP Holliday junction complex. *J Mol Biol* **319:**107–127.

May EW, Craig NL. 1996. Switching from cut-and-paste to replicative Tn*7* transposition. *Science* **272:**401–404.

Montaño SP, Pigli YZ, Rice PA. 2012. The Mu transpososome structure sheds light on DDE recombinase evolution. *Nature* **491:**413–417.

Nagy Z, Chandler M. 2004. Regulation of transposition in bacteria. *Res Microbiol* **155:**387–398.

Parks AR, Li Z, Shi Q, Owens RM, Jin MM, Peters JE. 2009. Transposition into replicating DNA occurs through interaction with the processivity factor. *Cell* **138:**685–695.

Parks AR, Peters JE. 2009. Tn7 elements: engendering diversity from chromosomes to episomes. *Plasmid* **61:**1–14.

Pasternak C, Ton-Hoang B, Coste G, Bailone A, Chandler M, Sommer S. 2010. Irradiation-induced *Deinococcus radiodurans* genome fragmentation triggers transposition of a single resident insertion sequence. *PLoS Genet* **6:**e1000799.

Peters JE. 2019. Targeted transposition with Tn7 elements: safe sites, mobile plasmids, CRISPR/Cas and beyond. *Mol Microbiol* **12:**1635-1644. doi: 10.1111/mmi.14383.

Peters JE, Fricker AD, Kapili BJ, Petassi MT. 2014. Heteromeric transposase elements: generators of genomic islands across diverse bacteria. *Mol Microbiol* **93:**1084–1092.

Pósfai G, Plunkett GIII, Fehér T, Frisch D, Keil GM, Umenhoffer K, Kolisnychenko V, Stahl B, Sharma SS, de Arruda M, Burland V, Harcum SW, Blattner FR. 2006. Emergent properties of reduced-genome *Escherichia coli. Science* **312:**1044–1046.

Reznikoff WS. 2003. Tn*5* as a model for understanding DNA transposition. *Mol Microbiol* **47:**1199–1206.

Rowe-Magnus DA, Guerout A-M, Ploncard P, Dychinco B, Davies J, Mazel D. 2001. The evolutionary history of chromosomal super-integrons provides an ancestry for multiresistant integrons. *Proc Natl Acad Sci USA* **98:** 652–657.

Rubin EJ, Akerley BJ, Novik VN, Lampe DJ, Husson RN, Mekalanos JJ. 1999. *In vivo* transposition of mariner-based elements in enteric bacteria and mycobacteria. *Proc Natl Acad Sci USA* **96:**1645–1650.

Shapiro JA. 1979. Molecular model for the transposition and replication of bacteriophage Mu and other transposable elements. *Proc Natl Acad Sci USA* **76:**1933–1937.

Shi Q, Huguet-Tapia JC, Peters JE. 2009. Tn*917* targets the region where DNA replication terminates in *Bacillus subtilis*, highlighting a difference in chromosome processing in the firmicutes. *J Bacteriol* **191:**7623–7627.

Siguier P, Filée J, Chandler M. 2006. Insertion sequences in prokaryotic genomes. *Curr Opin Microbiol* **9:**526–531.

Simon R, Preifer U, Pühler A. 1983. A broad host range mobilization system for *in vivo* genetic engineering: transposon mutagenesis in Gram negative bacteria. *Biotechnology* **1:**784–790.

Simon M, Zieg J, Silverman M, Mandel G, Doolittle R. 1980. Phase variation: evolution of a controlling element. *Science* **209:**1370–1374.

Skelding Z, Queen-Baker J, Craig NL. 2003. Alternative interactions between the Tn7 transposase and the Tn7 target DNA binding protein regulate target immunity and transposition. *EMBO J* **22:**5904–5917.

Toleman MA, Bennett PM, Walsh TR. 2006. IS*CR* elements: novel gene-capturing systems of the 21st century? *Microbiol Mol Biol Rev* **70:**296–316.

Ton-Hoang B, Pasternak C, Siguier P, Guynet C, Hickman AB, Dyda F, Sommer S, Chandler M. 2010. Single-stranded DNA transposition is coupled to host replication. *Cell* **142:** 398–408.

van de Putte P, Plasterk R, Kuijpers A. 1984. A Mu *gin* complementing function and an invertible DNA region in *Escherichia coli* K-12 are situated on the genetic element e14. *J Bacteriol* **158:**517–522.

van Opijnen T, Bodi KL, Camilli A. 2009. Tn-seq: high-throughput parallel sequencing for fitness and genetic interaction studies in microorganisms. *Nat Methods* **6:**767–772.

Zhong J, Karberg M, Lambowitz AM. 2003. Targeted and random bacterial gene disruption using a group II intron (targetron) vector containing a retrotransposition-activated selectable marker. *Nucleic Acids Res* **31:**1656–1664.

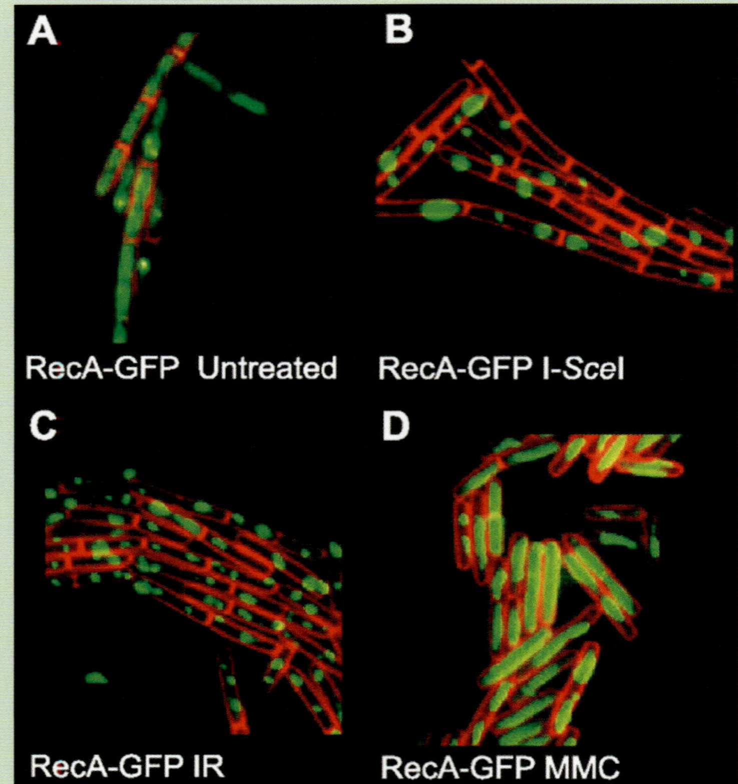

Visualization of RecA localization with RecA-GFP in *B. subtilis* with different mechanisms of inducing DNA damage. (A) Untreated cells with no experimentally induced form of DNA damage. (B) DNA double-strand break formation at a single cellular position with the meganuclease I-SceI. (C) Exposure to 100 Gy of ionizing radiation (IR). (D) Treatment with 1 μg/ml mitomycin C (MMC). Cells were treated with the membrane stain FM4-64, which is shown in red. Experiments were performed as indicated in Fig. 2 of Simmons et al. (Suggested Reading).

Molecular Mechanisms of Homologous Recombination

9

WHEN WE TALK ABOUT RECOMBINATION, we are usually referring to homologous recombination, which is found more generally than the other types of recombination discussed in chapter 8. Homologous recombination can occur between any two DNA sequences that are the same or very similar. Depending on the organism, homology-dependent recombination can occur between regions of homology that are as short as 23 bases, although longer homologies produce more frequent recombination events. Homologous recombination seems to be universal, suggesting that recombination is very important for species survival. It is well advertised that the new combinations of genes obtained through recombination allow the species to adapt more quickly to the environment and speed up the process of evolution. However, it has become increasingly clear that the most important role for homologous recombination is likely to involve facilitating DNA replication of damaged DNA. This is especially clear in bacteria, which invariably have a single origin of DNA replication in their chromosome or in each of their chromosomes (this is also true for most plasmids and bacteriophages), making each DNA replication fork extremely important to maintain. As we will see below, nicks in either of the two template strands cause a DNA replication fork to collapse. A role for homologous recombination in restarting collapsed DNA replication forks was likely the original selective advantage for the evolution of homologous recombination because of the immediate advantage that it provides for an organism.

Because of its importance in genetics, homologous recombination has already been mentioned in previous chapters, for example, in discussions of deletion and inversion mutations and genetic analysis. Determination of recombination frequencies allows us to measure the distance between mutations and thus can be used to map mutations with respect to each other, as we discussed in chapters 3, 6, and 7, among others. Moreover, clever use of recombination can take some of the hard work out of cloning genes and making DNA constructs, and we have already referred to some of these techniques. When we discussed the use of recombination for genetic mapping and many other types of applications in previous chapters, we used a simplified description of recombination: the strands of two DNA molecules were suggested to break at a place where they both have the same sequence of bases, and then the strands of the two DNA molecules join with each other to form a new molecule. In this chapter, we focus on the actual molecular mechanism of recombination—what actually happens to the DNA molecules involved. We will see that the process of homologous recombination is inextricably intertwined with the process of DNA replication with damaged DNA. We also discuss the proteins involved in recombination, mostly in *Escherichia coli*, the bacterium for which recombination is best understood and which has served as the paradigm for recombination in all other organisms.

Homologous Recombination and DNA Replication in Bacteria

One of the most gratifying times in science comes when phenomena that were originally thought to be distinct are discovered to be different manifestations of the same process. Such a discovery is usually followed by rapid progress as the mass of information accumulated on the different phenomena is combined and reinterpreted. This is true of the fields of homologous recombination and DNA replication and the role they jointly play in DNA repair. Because of the connection between the replication, recombination, and repair processes, they are sometimes called the "three R's" in DNA metabolism. Chapter 10 goes into more depth about how recombination is involved in the process of DNA repair. As discussed in chapter 1, all bacteria that have been examined to date appear to have a single origin of chromosomal replication. This can be gleaned from looking at the DNA sequence of a bacterial genome because of many features in the sequence that indicate the direction of DNA replication (see Box 1.1).

While having a single *oriC* offers distinct advantages for regulating DNA replication and coordinating DNA replication with chromosome segregation, it also makes the process vulnerable to problems with the template strands. One such issue arises when there is a nick in the template strand. Even a single nick in the template strand will cause a DNA replication fork to collapse (Figure 9.1). Collapse of one arm of the DNA replication fork at a nick results in a broken end and the inability of the replication fork to continue on the contiguous strand. Presumably, without a way to deal with this event, a considerable amount of chromosomal DNA would need to be degraded, and DNA replication would need to be started again at *oriC*. Bacterial cells likely could not tolerate this kind of loss. Replication forks are believed to collapse nearly every generation, in part, from nicks generated as a by-product of the functioning of many enzymes that act on DNA, as discussed in chapters 1, 8, and 10. Homologous recombination provides a mechanism to restart DNA replication using the broken end of the chromosome (Figure 9.1). Viewing homologous recombination in this light provides an explanation for numerous diverse experimental observations and strongly suggests how homologous recombination occurs during P1 transduction and Hfr crosses and with linear DNA

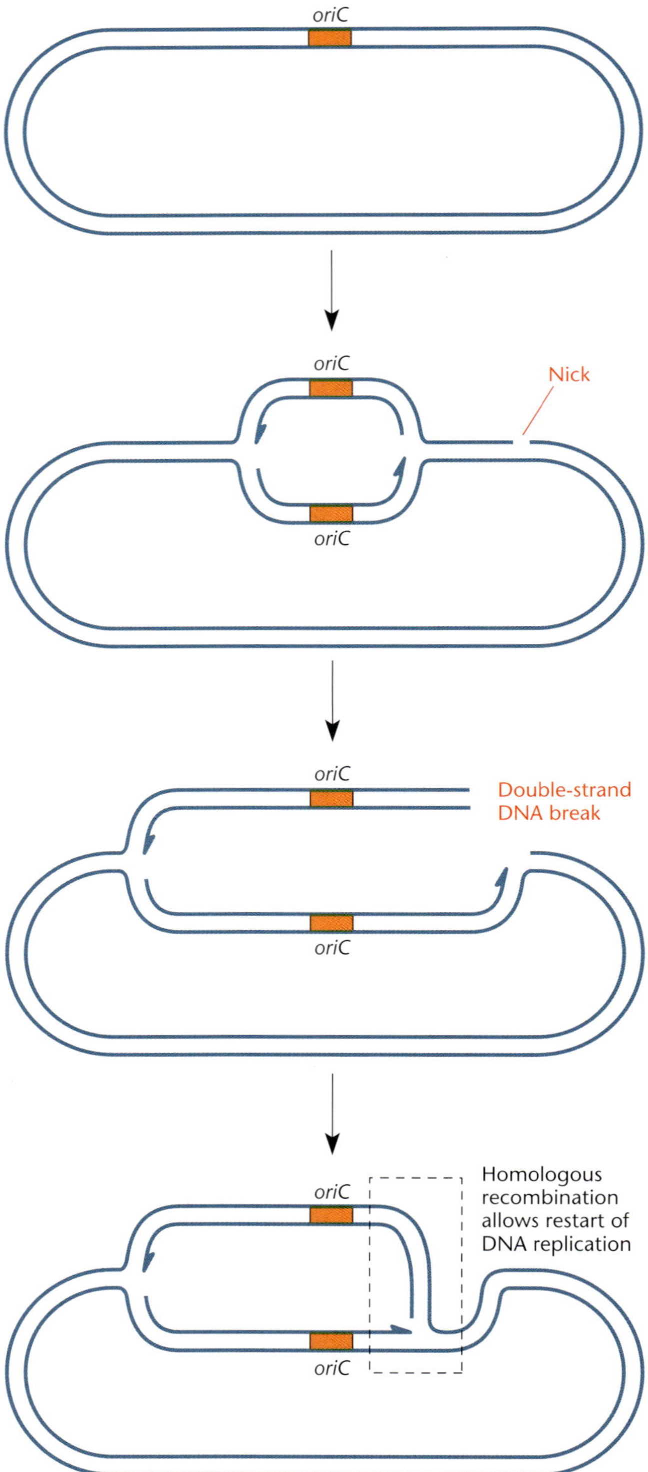

Figure 9.1 Replication forks initiated at *oriC* can collapse when there are nicks in the template strand. In the DNA strand with the nick, a double-strand broken end results, and the replication fork collapses. Recombination with the copy of the chromosome forms a structure from which DNA replication can restart to allow chromosomal replication to continue.

fragments that are transferred into the cell by a process that involves extensive DNA replication (see below).

Early Evidence for the Interdependence of Homologous Recombination and DNA Replication

DNA replication can be initiated in a number of different ways, many of which involve recombination. The process of completing homologous recombination involving DNA double strand breaks itself very likely requires extensive DNA replication. However, it took a long time to recognize the interdependence of these processes, and they were studied in isolation for decades. Early on, it was known that some phages, such as T4, need recombination functions to initiate replication (see chapter 7 and Mosig, Suggested Reading). In these phages, recombination intermediates function to initiate DNA replication later in infection. However, this was thought to be unique to these phages. Normally, initiation of chromosomal replication in bacteria does not require recombination functions. However, after extensive DNA damage due to irradiation or other agents, a new type of initiation, which does require recombination, comes into play. This type of initiation was originally called stable DNA replication because it continued even after protein synthesis stopped (see Kogoma, Suggested Reading). Initiation of DNA replication at the chromosomal *oriC* site normally requires new protein synthesis, and so, in the absence of protein synthesis, replication continues only until all the ongoing rounds of replication are completed and no new rounds are initiated (see chapter 1). However, after extensive DNA damage, such as that caused by UV irradiation, initiation of new rounds of replication occurs even in the absence of protein synthesis. It was shown that recombination is required for stable DNA replication because double-strand DNA breaks formed during irradiation are used to initiate DNA replication through the same process that reestablishes collapsed DNA replication forks (see Kuzminov and Stahl, Suggested Reading).

Homologous recombination allows the broken end that results from collapse of the DNA replication fork to interact with the homologous region that was just replicated (Figure 9.1). However, another process is required to allow a full DNA replication fork to be reestablished. This process is known to involve the Pri proteins, which have a long and interesting history. There are three Pri proteins, PriA, PriB, and PriC, as well as another protein, DnaT, that help reload the replicative helicase DnaB through the use of DnaC. These proteins were first discovered because they are required for the initiation of replication of the DNA of some single-stranded DNA phages (see chapter 7). Mutants lacking the main replication restart protein PriA suffer important growth defects, and mutants lacking PriB and PriC are inviable, so it was assumed that the Pri proteins also play a role in *E. coli* DNA replication. In addition, *priA* mutants are defective in recombination and are hypersensitive to DNA-damaging agents. Later work showed that different combinations of Pri proteins are responsible for acting on different types of substrates that can result when a DNA replication fork is arrested (see Lopper et al., Suggested Reading). The findings that the initiation of DNA replication with the Pri proteins is actually required for homologous recombination and that a role for recombination is essential for restarting DNA replication after replication forks collapse provided some of the evidence for the unity of these processes.

The Molecular Basis for Recombination in *E. coli*

As with many cellular phenomena, much more is known about the molecular basis for recombination in *E. coli* than in any other organism. At least 25 gene products involved in recombination have been identified in *E. coli*, and specific roles have been assigned to many of these proteins (Table 9.1).

chi (χ) Sites and the RecBCD Complex

The first analysis of the genetic requirements for recombination in *E. coli* used Hfr crosses (see Clark and Margulies, Suggested Reading; chapter 3). The *recB* and *recC* genes were among the first *rec* genes found because their products are required for recombination after Hfr crosses. Their products were later shown to also be required for transductional crosses. The products of these genes, together with the product of another gene, *recD*, form a heterotrimer, accordingly named the RecBCD nuclease. The *recD* gene product is not required for recombination after such crosses, so that gene had to be found in other ways. We now know that the RecBCD enzyme is required for the recombination that occurs after conjugation and transduction because the linear pieces of DNA transferred into cells by Hfr crosses or by transduction are natural substrates for the RecBCD enzyme. The RecBCD enzyme processes the ends of these pieces to form single-stranded 3′ ends, which can then invade the chromosomal DNA to form recombinants (see below). Most importantly, RecBCD is of paramount importance in *E. coli* and many other bacteria for repairing the double-strand breaks that are formed after a replication fork collapses at nicks in the template DNA.

HOW RecBCD WORKS

The RecBCD complex is a remarkable enzyme with many activities. It has single-stranded DNA endonuclease and exonuclease activities, as well as DNA helicase and DNA-

Table 9.1 Major genes encoding recombination functions in *E. coli*

Gene	Mutant phenotype	Enzymatic activity	Probable role in recombination
recA	Recombination deficient	Pairing of homologous DNAs	Synapse formation
recBC	Reduced recombination	Exonuclease, ATPase, helicase, χ-specific endonuclease	Initiates recombination by separating strands, degrading DNA up to a χ site, and loading RecA
recD	Rec⁺ χ independent	Stimulates exonuclease	Degrading 3′ ends
recF	Reduced plasmid recombination	Binds ATP and single-stranded DNA	RecA loading at single-strand gaps and overhangs
recJ	Reduced recombination in RecBC⁻	Single-stranded exonuclease	Processing at single-strand gaps
recN	Reduced recombination in RecBC⁻	ATP binding	RecA loading at single-strand gaps
recO	Reduced recombination in RecBC⁻	DNA binding and renaturation	RecA loading at single-strand gaps
recQ	Reduced recombination in RecBC⁻	DNA helicase	Processing at single-strand gaps
recR	Reduced recombination in RecBC⁻	Binds double-stranded DNA	RecA loading at single-strand gaps
recG	Reduced Rec in RuvA⁻ RuvB⁻ RuvC⁻	Branch-specific helicase	Migration of Holliday junctions, dismantling recombination structures
ruvA	Reduced recombination in RecG⁻	Binds to Holliday junctions	Migration of Holliday junctions
ruvB	Reduced recombination in RecG⁻	Holliday junction-specific helicase	Migration of Holliday junctions
ruvC	Reduced recombination in RecG⁻	Holliday junction-specific nuclease	Resolution of Holliday junctions
priA, priB, priC, dnaT	Reduced recombination	Helicase	Restart of replication forks

dependent ATPase activities (see Dillingham and Kowalczykowski, Suggested Reading). To put all these activities in perspective, it is useful to think of the RecBCD complex as a DNA helicase with associated nuclease activities. In actuality, the RecBCD complex has two helicase activities of opposite polarity. When it moves down a duplex DNA, the two helicase motors work in tandem, because the strands are of opposite polarity to one another (Figure 9.2). Having two motors working together contributes to the speed and processivity of the enzyme, especially if there are gaps or other types of damage in the DNA strands. In its role of processing a double-strand broken end to prepare it for recombination, RecBCD is specifically capable of recognizing only broken ends where the top and bottom strands are flush or nearly completely flush. This is something that is important when we consider the two major "pathways" for recombination below. As RecBCD unwinds the DNA, the DNA is cut into small pieces by its 3′-to-5′ exonuclease activity, leaving the 5′-to-3′ strand mostly intact but vulnerable to occasional breaks caused by the same enzyme. This process continues until the RecBCD protein encounters a sequence on the DNA called a *chi* (χ) site (for crossover hotspot instigator), which in *E. coli* has the sequence 5′-GCTGGTGG-3′ but in other types of bacteria has different sequences. These sites were first found because they stimulate recombination in phage λ (see Box 9.1). When RecBCD encounters a χ site, its 3′-to-5′ exonuclease activity is inhibited but its 5′-to-3′ exonuclease activity is stimulated, leading to formation of a free 3′ single-stranded tail, as shown in Figure 9.2. Note that the χ sequence does not have 2-fold rotational symmetry like the sites recognized by many restriction endonucleases. This means that the sequence will be recognized on only one strand of the DNA, and the RecBCD nuclease will pass over the sequence if it occurs on the other strand, making the orientation of the χ site extremely important, as we discuss below.

After a RecBCD enzyme has formed a 3′ single-stranded tail on the DNA, it directs a RecA protein molecule to bind to the DNA next to where it is bound (see below). This is called cooperative binding, because one protein is helping another to bind. In more physicochemical terms, the incoming RecA protein makes contact with both the DNA and with the RecBC protein already on the DNA, which helps to stabilize the binding of RecA to the DNA. This cooperative binding is necessary, because another protein, called single-stranded DNA-binding (SSB), binds single-stranded DNA with greater affinity than RecA. More RecA protein molecules then bind cooperatively to the first RecA protein to form a helical nucleoprotein filament to prepare for the next step in recombination, synapse formation, as discussed in more detail below. Single-stranded DNA is bound by SSB *in vivo*, which prevents RecA binding. The dependence on accessory proteins to load RecA to overcome the SSB barrier provides an important regulator role for recombination.

The discovery of χ sites and their role in recombination took many years and a lot of clever experimentation to figure out how they work (see Box 9.1). A detailed model consistent with much of the available information has emerged from this work. According to this detailed model, the RecD subunit of the RecBCD enzyme does not itself have any exonuclease activity but instead stimulates

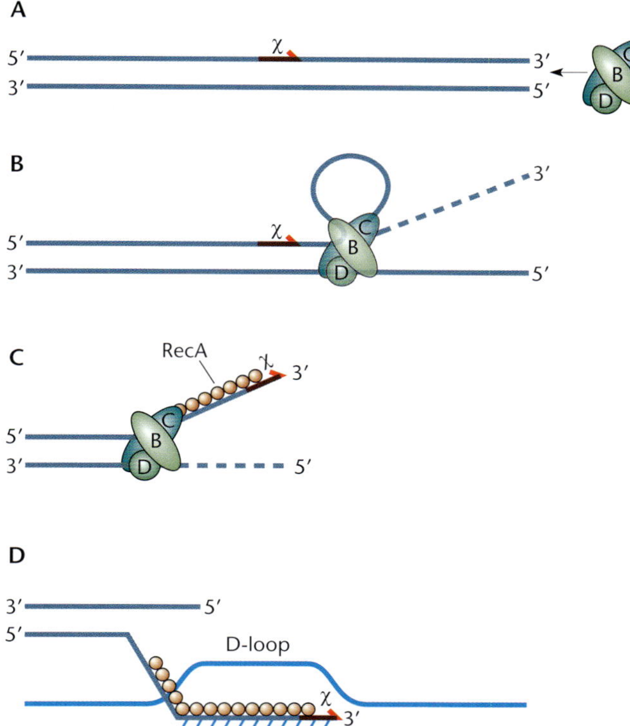

Figure 9.2 Model for promotion of recombination initiation at a χ site by the RecBCD complex. **(A)** RecBCD loads onto a double-stranded end. **(B)** Its helicase activity separates the strands, and its 3′-to-5′ nuclease activity degrades until it encounters a χ site (χ). **(C)** The χ site signals the attenuation of the 3′-to-5′ nuclease and upregulation of the 5′-to-3′ nuclease. As a result of these changes in the activity of the RecBCD complex following interaction with the χ site, a single-stranded 3′ extension is created. Interaction with the χ site also signals the RecBCD complex to load RecA (orange circles) on the single-stranded DNA, forming it into an extended structure. **(D)** This RecA nucleoprotein filament can then invade a complementary double-stranded DNA, forming a D loop or a triple-stranded structure (see Figure 9.6).

the 3′-to-5′ exonuclease activity of the RecB subunit. The χ sites work by inhibiting the 3′-to-5′ helicase activity of the RecD subunit and thereby indirectly inhibiting the 3′-to-5′ exonuclease activity and stimulating the 5′-to-3′ nuclease activity of the remainder of the enzyme. Thus, as the RecBCD enzyme moves along the DNA, opening the strands, it largely degrades the 3′-to-5′ strand until it encounters the sequence 5′-GCTGGTGG-3′ (the χ sequence) on the strand being degraded. The χ site is recognized by RecC. However, this recognition causes a conformational change in the complex that in turn inhibits the 3′-to-5′ strand nuclease activity. Therefore, the RecBCD enzyme continues to move on past the χ site, still degrading the 5′-to-3′ strand but leaving the 3′-to-5′ strand intact. The end result is a single-stranded DNA that has the χ site sequence at its 3′ end, as shown in Figure 9.2C.

The evidence for this detailed model of χ site action is both biochemical and genetic. First, the RecBC enzyme does have some 3′-to-5′ exonuclease activity in the absence of the RecD subunit. Because of this, a linear DNA can be introduced into a RecD⁻ mutant of *E. coli* by transformation (see chapter 6) and is not degraded. Another prediction of the model, which is fulfilled, is that RecD⁻ mutants are proficient for recombination, and this recombination does not require χ sites. This is due to the fact that, since the RecBC enzyme lacking the RecD subunit still has the RecB helicase activity to separate the strands, these single-stranded ends are made and are not degraded even if they do not contain a χ site; they are available to be loaded with RecA and to invade other DNAs and promote recombination. This property of RecD⁻ mutants of not degrading linear DNA but still being proficient for recombination is what historically made RecD⁻ mutants of *E. coli* useful for gene replacements with linear DNA.

This model also explains why recombination is stimulated only on the 5′ side of a χ site. Until the RecBCD enzyme reaches a χ site, the displaced strand is degraded and so is not available for recombination. Only after the RecBCD complex passes completely through a χ site will the strand remain intact to invade another DNA, so that only DNA on the 5′ side of the site survives. This model is also supported by the known enzymatic activities of the RecBCD complex, as well as by electron microscopic visualization of the RecBCD complex acting on DNA. It has also received experimental support from the results of *in vitro* experiments with purified RecBCD enzyme and DNA containing a χ site (see Dixon and Kowalczykowski, Suggested Reading). Recombination on the 5′ side of χ is further stimulated because this is the DNA region where RecBCD actively loads RecA (Figure 9.2C and D).

WHY χ?

The hardest question to answer in biology is often "Why?" To answer this question with certainty, we must know everything about the organism and every situation in which it might find itself, both past and present. Nevertheless, it is tempting to ask why *E. coli* and many other bacteria use such a complicated mechanism involving χ sites for a major pathway of recombination. Adding to the mystery is the fact that they would not need χ sites at all if they were willing to dispense with the RecD subunit of the RecBCD nuclease. As mentioned above, in the absence of RecD, recombination proceeds just fine without χ sites, and the cells are viable. Furthermore, many bacteria have other mechanisms that could repair double-strand DNA breaks that do not involve homologous recombination (see Box 9.2).

One idea is that the self-inflicted dependency on χ sites allows the RecBCD nuclease to play a dual role in recombination and in defending against phages and other foreign

BOX 9.1

Discovery of χ Sites

The discovery on DNA of χ sites that stimulate recombination by the RecBCD nuclease required some interesting genetics and was a triumph of genetic reasoning. Many sites that are subject to single- or double-strand breaks are known to be "hot spots" for recombination. In some cases, such as recombination initiated by homing enzymes (chapter 8), it is clear that breaks at specific sites in DNA can initiate recombination. In general, however, the frequency of recombination seems to correlate well with physical distance on DNA, as though recombination occurs fairly uniformly throughout DNA molecules.

It came as a surprise, therefore, to discover that the major recombination pathway for Hfr crosses in *E. coli*, the RecBCD pathway, occurs through specific sites on the DNA—the χ sites. Like many important discoveries in science, the discovery of χ sites started with an astute observation. This observation was made during studies of host recombination functions using λ phage. The experiments were designed to analyze the recombinant types that formed when the phage Red recombination genes *exo* and *bet* were deleted. Without its own recombination functions, the phage requires the host RecBCD nuclease. In addition, if the phage is also a *gam* mutant, it does not form a plaque unless it can recombine. Therefore, plaque formation by a *gam* mutant is an indication that recombination has taken place.

The reason that *red gam* mutant λ phage cannot multiply to make a plaque without recombination is somewhat complicated. As discussed in chapter 7, λ phage cannot package DNA from genome-length DNA molecules but only from concatemers in which the λ genomes are linked end to end. Normally, the phage makes concatemers by rolling-circle replication. However, if the phage is a *gam* mutant, it cannot switch to the rolling-circle mode of replication because the RecBCD nuclease, which is normally inhibited by Gam, somehow blocks the switch. Therefore, the only way a *gam* mutant λ phage can form concatemers is by recombination between the circular λ DNAs formed via θ replication (Figure 7.16). If the phage is also a *red* mutant (i.e., lacks its own recombination functions), the only way it can form concatemers is by RecBCD recombination, the major host pathway. Therefore, *red gam* mutant λ phage requires RecBCD recombination to form plaques, and the formation of plaques can be used as a measure of RecBCD recombination under these conditions.

The first χ mutations were discovered when large numbers of *red gam* mutant λ phage were plated on RecBCD$^+$ *E. coli*. Very few phage were produced, and the plaques that formed were very tiny. Apparently, very little RecBCD recombination was occurring between the circular phage DNAs produced by θ replication. However, λ mutants that produced much larger plaques sometimes appeared. The circular λ DNAs in these mutants were apparently recombining at a much higher rate. The responsible mutations were named χ mutations because they increase the frequency of crossovers (a crossover in the chromosome is called a chiasma in eukaryotes). Once the mutations were mapped and the DNA was sequenced, χ mutations in λ were found to be base-pair changes that created the sequence 5′-GCTGGTGG-3′ somewhere in the λ DNA. The presence of this sequence appears to stimulate recombination by the RecBCD pathway. Since wild-type λ does not need RecBCD and therefore has no such sequence anywhere in its DNA, recombination by RecBCD is very infrequent unless the χ sequence is created by a mutation.

Further experimentation with χ sites revealed several interesting properties. For example, they stimulate crossovers to only one side of themselves, the 5′ side. Very little stimulation of crossovers occurs on the 3′ side. Also, if only one of the two parental phages contains a χ site, most of the recombinant progeny will not have the χ site, so the χ site itself is preferentially lost during the recombination. These and other properties of χ sites eventually led to the model for χ site action presented earlier in this chapter.

References

Stahl FW, Stahl MM, Malone RE, Crasemann JM. 1980. Directionality and nonreciprocality of chi-stimulated recombination in phage λ. *Genetics* **94:**235–248.

DNAs. Small pieces of foreign DNA, such as a phage DNA entering the cell, are not likely to have a χ site, since 8-bp sequences like χ occur by chance only once in about 65 kb, longer than many phage DNAs. The RecBCD nuclease degrades a DNA until it encounters a χ site, and if it does not encounter a χ site, it degrades the entire DNA. *E. coli* DNA, in contrast, has many more of these sites than would be predicted by chance. In support of the idea that RecBCD evolved to help defend against phages is the extent to which phages go to avoid degradation by the enzyme. Many phages avoid degradation by RecBCD by attaching proteins to the ends of their DNA or by making proteins that inhibit RecBCD in more direct ways, such as the Gam protein of λ (see chapter 7 and below).

Another possible reason for having a complicated system that involves χ sites is that these sites can control the direction of DNA replication following a double-strand break in the chromosome. Ionizing radiation, oxygen radi-

BOX 9.2

Other Types of Double-Strand Break Repair in Bacteria

The RecBCD pathway and the related AddAB (and RecAB) system allow efficient repair of DNA double-strand breaks in bacteria. In many well-studied bacterial species, the RecBCD/AddAB systems process the broken ends to form a 3′ extension of DNA that is actively loaded with RecA for recombination on a "sister" chromosome that results from DNA replication (panel A of the figure; also see Figure 9.2). In other bacteria, the RecF pathway seems to participate in a similar reaction to allow the repair to use a sister chromosome as a template. Recombination-based systems are well represented in eukaryotes, but in many eukaryotes, there are also systems for directly joining breaks where the ends have no homology in a process called nonhomologous end joining (NHEJ) (panel B of the figure). In these systems, the DNA ends are directly joined together. The NHEJ system is inherently risky, because if there is any loss of sequence information before the ends are joined, there will be a permanent loss of genetic information. A second and more serious risk to genome stability comes from the chance of putting the wrong two ends together. One common mechanism of NHEJ in eukaryotes involves an end-binding protein, a heterodimer, Ku70/80, and a ligase that is responsible for working with the end-binding protein to seal the ends of the DNA break. Interestingly, it was shown that many bacteria (but not *E. coli*) have a protein that is homologous to the Ku70/80 proteins. Of additional interest, in many cases, the gene encoding this protein, called Ku, is adjacent to the gene predicted to encode a DNA ligase (e.g., LigD). Subsequent work has shown that the Ku and LigD systems are capable of repairing DNA double-strand breaks in a manner similar to NHEJ in eukaryotes. In addition to a ligase domain, LigD also has nuclease and polymerase domains, which can act as a template-independent terminal transferase on single-stranded or blunt-ended DNAs. The LigD ligase and other ligases not normally involved in sealing nicks during DNA synthesis and repair can work with Ku to allow NHEJ.

Why would some bacteria need an additional mechanism for DNA double-strand break repair beyond the efficient RecBCD/AddAB systems? Intriguingly, many of the types of bacteria that have Ku- and ligase-based systems are found in organisms that spend considerable amounts of time in a quiescent or nonreplicating state. For example, Ku and LigD are found in the spore-forming bacterium *Bacillus subtilis*. Direct end joining may be required when a spore matures and DNA breaks need to be repaired in the absence of a second copy of the chromosome to act as a template for repair. *Mycobacterium tuberculosis*, the causative agent of tuberculosis, also has a Ku-based repair system,

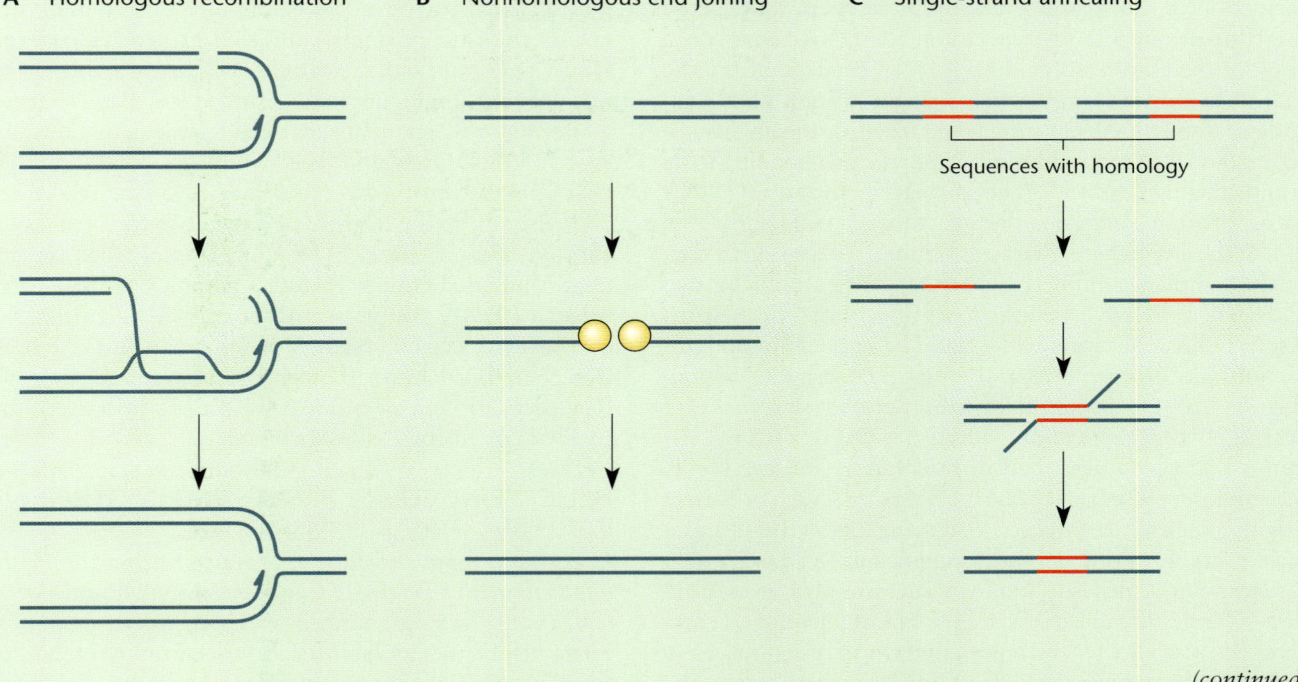

(continued)

BOX 9.2 (continued)

Other Types of Double-Strand Break Repair in Bacteria

which may aid its DNA repair during long periods of inactivity in the latent stage of infection that is characteristic of the disease. Interestingly, eukaryotes seem to regulate their types of DNA repair according to the phase of growth; when cells are not replicating (i.e., in G_1), cells use NHEJ for DNA repair, but when they are in late S/G_2, where a second copy of the chromosome is found, they use homologous recombination for the repair of double-strand breaks. It is also thought that pathogens may benefit from the Ku- and ligase-based NHEJ system. This is because an important part of the defense response of animals involves mechanisms to damage the pathogen's DNA. *M. tuberculosis* actually appears to use three distinct types of double-strand break repair: a form of homology-based repair, as described for the RecBCD/AddAB systems, a Ku- and ligase-based system, and a third system based on single-strand annealing (SSA). SSA is also based on homology, but the homologies are shorter and are found in the same chromosome (panel C of the figure). Repair by SSA is highly error prone and results in deletions between the small regions of homology. Related forms of double-strand break repair involving portions of these repair pathways also exist.

References

Aravind L, Koonin EV. 2001. Prokaryotic homologs of the eukaryotic DNA-end-binding protein Ku, novel domains in the Ku protein and prediction of a prokaryotic double-strand break repair system. *Genome Res* **11**:1365–1374.

Della M, Palmbos PL, Tseng HM, Tonkin LM, Daley JM, Topper LM, Pitcher RS, Tomkinson AE, Wilson TE, Doherty AJ. 2004. Mycobacterial Ku and ligase proteins constitute a two-component NHEJ repair machine. *Science* **306**:683–685.

Gupta R, Barkan D, Redelman-Sidi G, Shuman S, Glickman MS. 2011. Mycobacteria exploit three genetically distinct DNA double-strand break repair pathways. *Mol Microbiol* **79**:316–330.

cals, and other forms of DNA damage (see chapter 10), along with the action of restriction endonucleases, are just some of the ways that double-strand DNA breaks can occur in the chromosome. These breaks can be processed by the RecBCD complex to allow RecA to be loaded for the formation of a displacement loop, or **D loop**, that can be used for priming DNA replication for DNA repair on a copy of the chromosome during DNA replication (Figure 9.3). A D loop is formed when a single strand of DNA invades a duplex DNA of similar sequence, displacing its homologous strand when it binds via hydrogen bonding to its complementary strand. The displaced strand of DNA forms the loop that gives the structure its name. However, unlike the case where a replication fork collapses at a nick in the template strand (Figure 9.1), there would be two ends where RecBCD could load and, with the help of RecA, DNA replication forks could be initiated in both directions (i.e., toward *oriC* and toward the terminus region) (Figure 9.3). However, the distribution of χ sites makes it very likely that replication will be directed toward the terminus region. As described in Box 1.1, χ sites are highly overrepresented in the *E. coli* chromosome, with a distinct origin to terminus polarity, occurring on average about once every 4.5 kb. Even more significant, this enrichment for the polar χ sites is found for χ sites in only one orientation relative to the direction of DNA replication. Therefore, because of the overrepresentation of χ sequences in one orientation between *oriC* and the terminus region, the fates of two RecBCD complexes that load at a double-strand break will likely be very different (Figure 9.3B). The RecBCD complex that degrades DNA toward *oriC* (1 in Figure 9.3B) will quickly encounter a χ site in the correct orientation and, with RecA, aid in the initiation of a DNA replication fork (Figure 9.3C). Therefore, replication in this scenario travels in the same direction as replication events that are normally initiated at *oriC*. However, a RecBCD complex that degrades DNA toward the terminus (labeled 2 in Figure 9.3B) is much less likely to reach a χ site that will attenuate its exonuclease activities. Given that RecBCD may not "see" every χ site, the few sites encountered in this direction may never be recognized, contributing to the degradation of the extra "arm" of the chromosome (Figure 9.3C). While homologous recombination initiated by the RecBCD complex is used for the repair of DNA double-strand breaks in *E. coli*, bacteria can have additional systems for the repair of these breaks (see below). Whatever the purpose or purposes of χ sites, they seem to be a near-universal feature in bacteria with RecBCD or RecBCD-like enzymes.

HELICASE-NUCLEASE PROCESSING IN OTHER BACTERIA: AddAB

E. coli and many, but not all, groups within the *Proteobacteria* utilize RecBCD to process DNA double-strand ends that lack single-strand overhangs. Members of the *Firmicutes*, such as *Bacillus subtilis*, and many bacteria from different phyla contain an enzyme called AddAB (or RexAB) with a function similar to that of RecBCD (see

Chédin et al., Suggested Reading). As described for RecBCD, AddAB loads onto a DNA end that is flush or nearly flush and degrades both DNA strands. AddA resembles RecB, containing both helicase and nuclease domains, and is responsible for unwinding activity of the enzyme and nuclease activity on the 3′-to-5′ strand. AddB is responsible for recognizing χ sites (5′-AGCGG-3′ in *B. subtilis*) but does not contribute helicase activity for the function of the AddAB complex (see Saikrishnan et al., Suggested Reading). AddB encodes the nuclease activity that acts on the 5′-to-3′ strand. Therefore, while the RecBCD enzyme has helicase activities in two separate proteins and only one nuclease activity, AddAB has a single helicase activity with nuclease activities residing in two separate proteins. After detection of a properly oriented χ site, AddAB-mediated cleavage of the 3′-to-5′ strand is essentially eliminated, while cleavage of the 5′-to-3′ strand continues, resulting in a 3′ extended DNA overhang. The AddAB complex remains bound to the χ site containing DNA, which presumably triggers the change in nuclease activity and the ability of the enzyme to coat the DNA substrate with RecA.

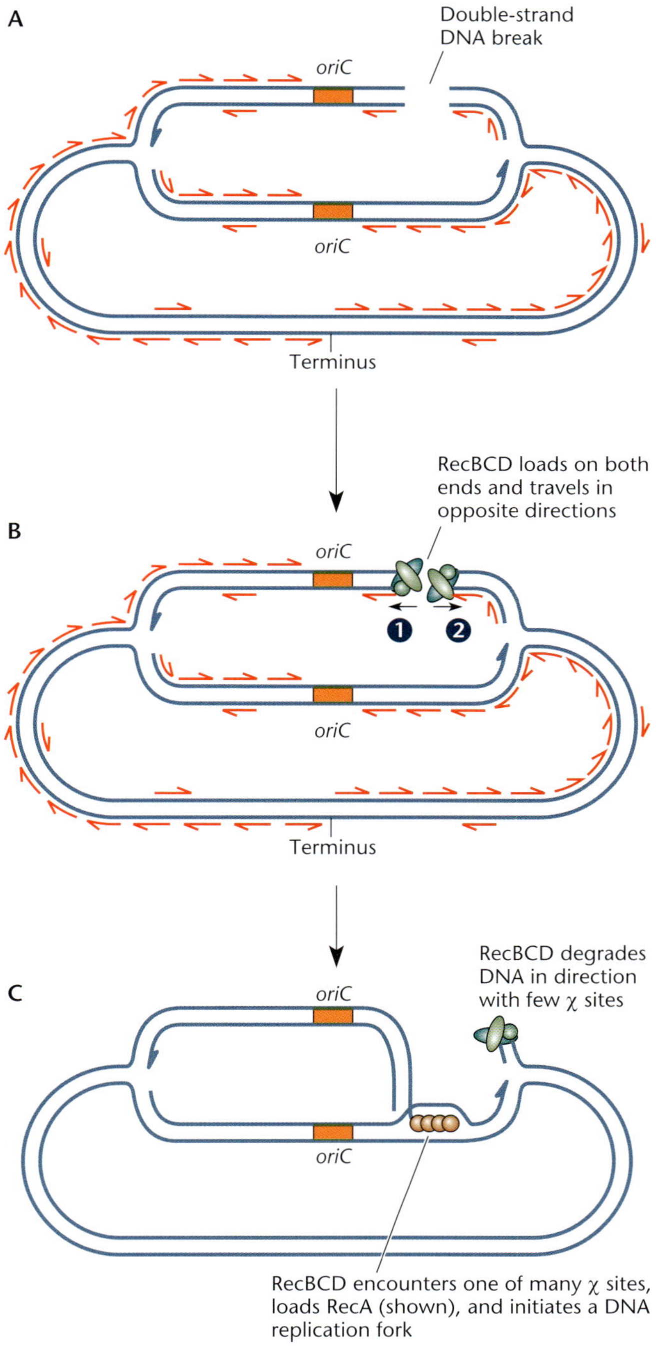

The RecF Pathway

In addition to RecBCD, the other major pathway used in *E. coli* to prepare single-stranded DNA for invading homologous DNA is the RecF pathway (Figure 9.4). This pathway is so named because it requires the products of the *recF* gene, as well as *recO* and *recR*, with important accessory roles for the *recQ* and *recJ* gene products. In *E. coli* and many other bacteria, this pathway is used under different circumstances from the RecBCD (AddAB) pathway, mostly to initiate recombination at single-stranded gaps in DNA or at DNA breaks in the absence of RecBCD and a group of nucleases that destroy 3′ DNA ends. Single-strand gaps may be created during repair of DNA damage or by the movement of the replication fork past a lesion, leaving behind a single-stranded region of DNA (see chapter 1). The RecF proteins can then prepare the single-stranded DNA at the gap to invade a sister DNA. The action of the RecF pathway at single-stranded DNA gaps can also help load repair polymerases in the process

A
Double-strand DNA break
oriC
oriC
Terminus
B
RecBCD loads on both ends and travels in opposite directions
oriC
1 2
oriC
Terminus
C
RecBCD degrades DNA in direction with few χ sites
oriC
oriC
RecBCD encounters one of many χ sites, loads RecA (shown), and initiates a DNA replication fork

Figure 9.3 Model for how χ sites can help RecBCD load RecA to direct DNA replication forks toward the terminus during replication-mediated double-strand break repair. **(A)** A DNA double-strand break occurs during DNA replication. **(B)** Two ends with DNA double-strand breaks are available for processing by the RecBCD complex. One RecBCD complex **(1)** progresses toward the origin, where many correctly oriented polar χ sites are encountered. A second RecBCD complex **(2)** progresses toward the terminus, where there is less chance that it will encounter a χ site that will attenuate its nuclease activity. **(C)** The RecBCD complex progressing toward the origin quickly encounters a χ site, which attenuates its nuclease activity on the top strand and actively loads RecA protein that can produce a structure on the sister chromosome to initiate DNA replication toward the terminus. The RecBCD complex degrading toward the terminus is unlikely to encounter a χ site and continues degrading the arm of the chromosome, possibly providing another important role for removing this structure, which could stop subsequent DNA replication. The terminus region, where DNA replication terminates, and the origin of chromosomal replication (*oriC*) are shown in all panels. The χ sites are indicated in panels A and B by red arrows with half heads.

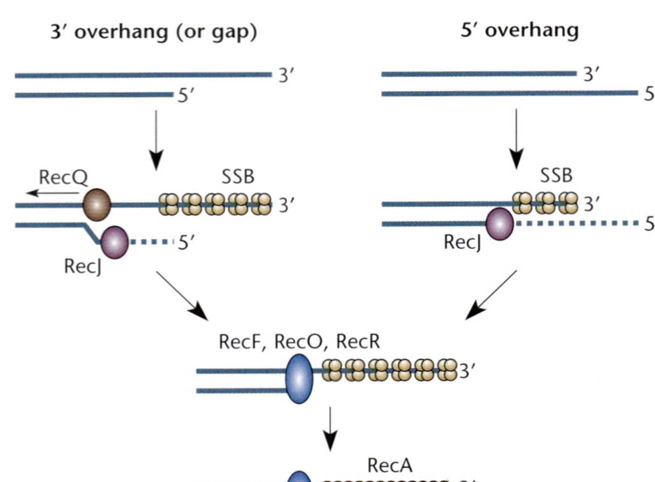

Figure 9.4 Models for recombination initiation by the RecF pathway on substrates with different single-stranded features. The RecF pathway can process substrates with single-strand gaps or with 3′ or 5′ single-strand overhangs. Substrates with a 3′ overhang (or a gap, not shown) are processed by RecQ, a helicase, which can pass the DNA substrate to RecJ, an exonuclease, to enlarge the single-stranded region, thus making it competent for RecA loading and recombination. Substrates with a 5′ overhang can be directly processed by RecJ. SSB protein (tan tetramers of circles) has a high affinity for single-stranded DNA and will coat the single-stranded region produced by RecJ. RecF, RecO, and RecR (RecFOR; indicated as a blue oval) will displace SSB and assemble the RecA (orange circles) nucleoprotein filament. The RecA filament is then competent for invading a sister DNA for recombination (not shown). The dashed line indicates degraded DNA.

of trans-lesion synthesis (see Fujii et al., Suggested Reading; chapter 10).

The RecF pathway does not have a "superstar" like RecBCD that can do it all, so it needs a number of proteins to perform the tasks that the RecBCD complex can perform alone (see Handa et al., Suggested Reading). The RecQ protein is a helicase, like RecBCD, but it lacks a nuclease activity to degrade the strands it displaces. RecQ is important in situations where there are gaps or single-stranded 3′ overhangs. RecQ unwinding produces a 5′-ended single-stranded DNA substrate of the nuclease RecJ to enlarge the gap (see Figure 9.4). In cases where there is a 5′ single-strand DNA extended end, the RecJ protein can directly recognize its substrate. Neither RecJ nor RecQ has the ability to displace SSB and load RecA onto the single-strand substrate for recombination. Instead, in an incompletely understood process, the RecF, RecO, and RecR proteins are loaded, which can, in turn, displace SSB and load RecA (see Morimatsu and Kowalczykowski, Suggested Reading). While the RecF, RecO, and RecR proteins work in a concerted action, they do not form a stable complex. The RecF proteins may also stop RecA from invading the neighboring double-stranded DNA before the synapse with the homologous DNA occurs (see below).

The RecF system is more widespread than the RecBCD/AddAB system across bacteria (see Rocha et al., Suggested Reading). Interestingly, work with *E. coli* suggests that the RecF pathway can process and load RecA at a number of diverse substrates. While *E. coli* strains that are nonfunctional for the RecBCD complex are sick and highly UV sensitive, these phenotypes can be efficiently suppressed if the strain also has mutations in a series of nucleases, *sbcB* and *sbcCD* (suppressor of RecBC) (see Michel and Leach, Suggested Reading). The idea that inactivating a series of nucleases allows *E. coli* to function without RecBCD has led to the speculation that these nucleases curb the role of the RecF pathway in bacteria like *E. coli*. In contrast, in the presence of RecBCD, the Sbc proteins process single-strand extensions so that they can be recognized by RecBCD (Figure 9.5). Therefore, the genetics in *E. coli* is consistent with the idea that the RecF-pathway could be sufficient for loading RecA at gaps and at DNA double-strand breaks but that RecA loading at DNA double-strand breaks can be routed to the RecBCD pathway through the use of additional nucleases.

Synapse Formation and the RecA Protein

Once a single-stranded DNA is created by the RecBCD or RecF pathway, it must find and invade another DNA with the complementary sequence. The joining of two DNAs in this way is called a **synapse**, and the process by which an invading strand can replace one of the two strands in a double-stranded DNA is called strand exchange. This is a remarkable process. Somehow, the single-stranded DNA must find its complementary sequence by scanning possibly all of the double-stranded DNA in the cell, which even in a simple bacterium can be megabases long. How, then, could the single-stranded DNA know when the sequence is complementary, especially if it can only scan the outside of the DNA double helix, and the bases of the double-stranded DNA are on the inside? Not only is synapse formation remarkably fast, but it is also remarkably efficient. Once an incoming single-stranded DNA enters the cell, for example, during an Hfr cross, it finds and recombines with its complementary sequence in the chromosome almost 100% of the time.

Searching for complementary DNA is the job of the RecA protein, whose role in recombination is outlined in Figure 9.6. As the single-stranded DNA is generated by RecBC or RecQ and RecJ, it is coated with RecA, as facilitated by RecBC or RecF, to form a nucleoprotein filament. As mentioned above and shown in Figures 9.2 and 9.4, the RecBCD and RecF proteins help load RecA onto single-stranded DNA in competition with the SSB pro-

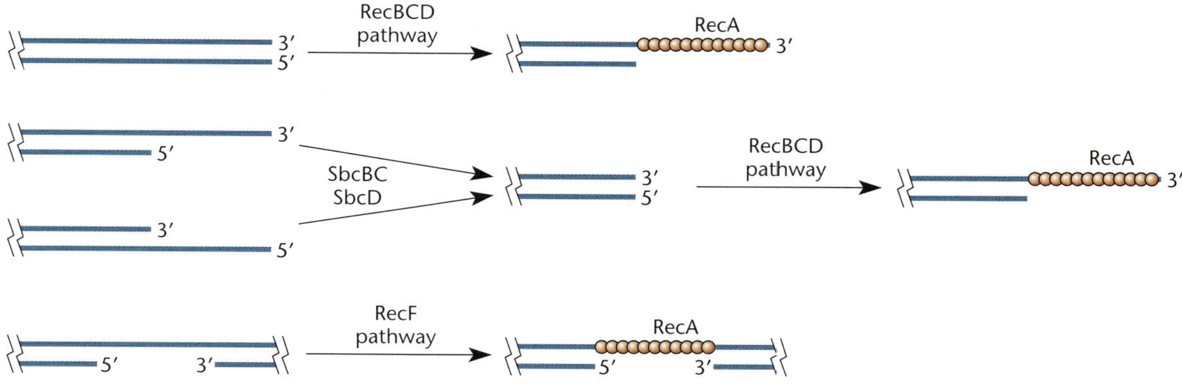

Figure 9.5 Model for how DNA substrates with various types of DNA breaks are processed for RecA loading for eventual recombinational repair in *Escherichia coli*. In *E. coli* and many other bacteria, different substrates are repaired by the RecBCD/AddAB and RecF systems. Genetic and biochemical evidence supports the idea that 3′ and 5′ overhangs are processed by nucleases like SbcB and SbcCD to make them substrates for the RecBCD system for repair (see Figures 9.2 and 9.3) instead of the RecF pathway (see Figure 9.4). This model also suggests why the RecF pathway is more broadly conserved across diverse bacteria.

tein, which is a prerequisite to forming the filament. The DNA in the RecA filament is also helical but is much extended relative to the normal helix in DNA; it takes about twice as many nucleotides to complete a helical turn. The helical nucleoprotein filament then scans the double-stranded DNA in the cell to find a homologous sequence. Some evidence suggests that it might scan double-stranded DNA through its major groove (see chapter 1) and that it can base pair with its complementary sequence in the major groove once it finds it without transiently disrupting the helix. Once the nucleoprotein filament finds its complementary sequence, it pairs with it. There is still some question of what happens next. Either it displaces one strand of the double-stranded DNA to form a D loop, as shown in Figure 9.2, or as some evidence suggests, it actually forms a triple-stranded structure, as shown in Figure 9.6. Other evidence calls into question the formation of triple-stranded structures, and it has been proposed instead that RecA can approach DNA through its minor groove and somehow flip the bases out to test for complementarity. For our present purposes, we often use D loops to schematize strand displacement.

While the details of how a single-stranded DNA-RecA filament and a double-stranded DNA find each other and what kinds of structures are formed remain obscure, a number of observations have been made that may shed light on this process. One observation is that a single-stranded RecA nucleoprotein filament may somehow change, or "activate," double-stranded DNAs merely by binding to them, even if the two DNAs are not comple-

Figure 9.6 Model for synapse formation and strand exchange between two homologous DNAs by RecA protein. The RecA protein (in orange), bound to a single-stranded end, formed as shown in Figures 9.2 and 9.4, and forced into an extended helical structure, pairs with a homologous double-stranded DNA in its major groove to form a three-stranded structure. RecA can drive this three-stranded structure into adjacent double-stranded DNA by a "spooling" mechanism, forming a four-stranded Holliday junction (not shown).

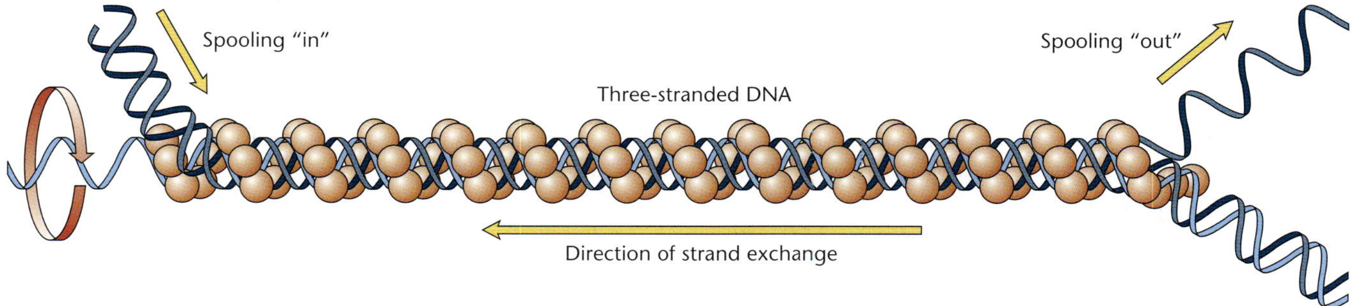

mentary. This activation presumably has something to do with the way in which the RecA nucleoprotein filament scans DNA, looking for its complementary sequence. Once the strands of the double-stranded DNA are activated, even by a noncomplementary RecA nucleoprotein filament, a complementary single-stranded DNA not bound to RecA can invade it and exchange with one of its strands. This process was named *trans* activation because the RecA nucleoprotein filament that activated the DNA is not necessarily the one that invades it (see Mazin and Kowalczykowski, Suggested Reading). It is not clear what happens to a double-stranded DNA when it is activated by a RecA nucleoprotein filament, but the helix may be transiently extended and the strands may be partially separated to allow the single-stranded DNA to search for its complementary sequence. This is an area that needs further investigation.

The RecA protein initially forms a nucleofilament only on single-stranded DNA, either an end or a gap. After the nucleofilament has invaded a double-stranded DNA, the RecA protein can continue to polymerize, extending the nucleofilament into the neighboring double-stranded DNA. The filament grows in the 5′-to-3′ direction in a process called "spooling" (Figure 9.6). It can also apparently depolymerize on the other end, leaving a single-stranded gap. The single-stranded gap can then be repaired by other cellular enzymes.

HOLLIDAY JUNCTIONS

The process by which RecA mediates the invasion of a single-stranded DNA into the homologous sequence in a double-stranded DNA that unwinds the adjacent DNA sequences as it grows creates a special type of junction called a **Holliday junction** (Figure 9.7). Holliday junctions were first proposed in 1964 as part of one of the early models of how homologous recombination occurs. The formation of Holliday junctions is central to many of the roles of recombination in bacterial genetics. Holliday junctions were introduced earlier because of their role during site-specific recombination by Cre and other Y recombinases (see chapter 8). While extension of the filament into double-stranded DNA requires energy in the form of ATP cleavage, in theory, Holliday junctions can migrate spontaneously along the DNA, because as one bond is broken through movement of the junctions, another bond forms on the other side. However, in the cell, specialized enzymes move these junctions, creating what are called **heteroduplexes**. The regions of complementary base pairing between the two DNA molecules are called heteroduplexes because the strands in these regions come from different DNA molecules. The action of RecA and the movement of Holliday junctions both form heteroduplexes; however, filament formation with RecA is unidirectional, while Holliday junctions are presumably free to move in either direction.

Once formed, Holliday junctions can undergo a rearrangement that changes the relationship of the strands to each other. This rearrangement is called an **isomerization** because no bonds are broken. As shown in Figure 9.7, isomerization causes the crossed strands in configuration I to uncross. A second isomerization occurs

Figure 9.7 Holliday junctions can form through the action of RecA. The movement of Holliday junctions creates a heteroduplex region (shown as a region of two shades of blue). Holliday junctions can isomerize between forms I and II. Cutting by specific enzymes and ligating by ligase resolve the Holliday junction. Depending on the position of the nicks introduced by the resolvase, the flanking markers A, B, a, and b recombine or remain in their original conformation. The product DNA molecules contain heteroduplex patches.

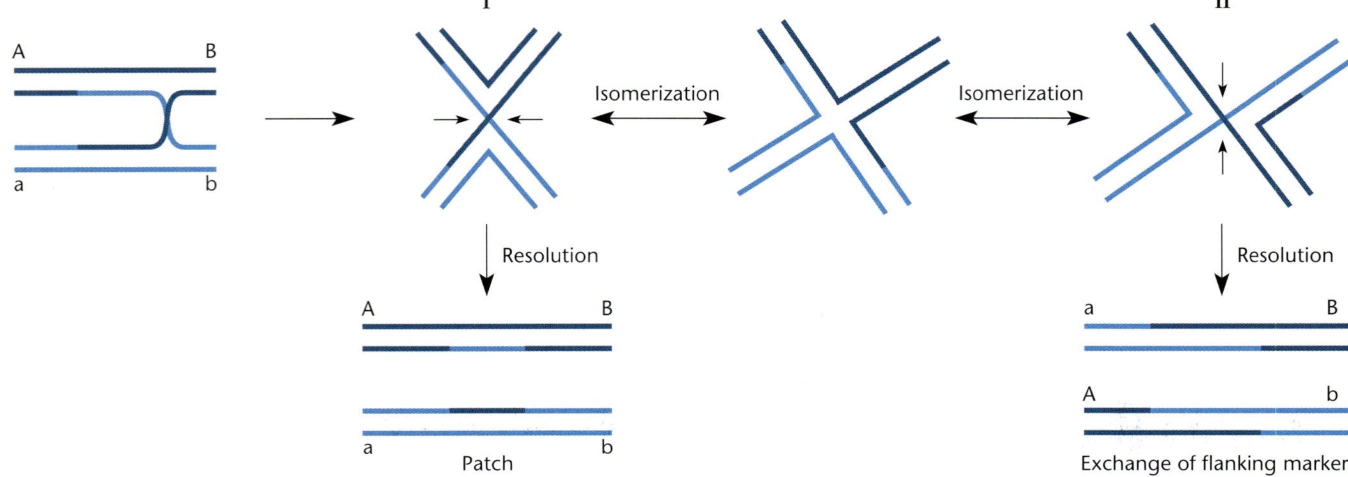

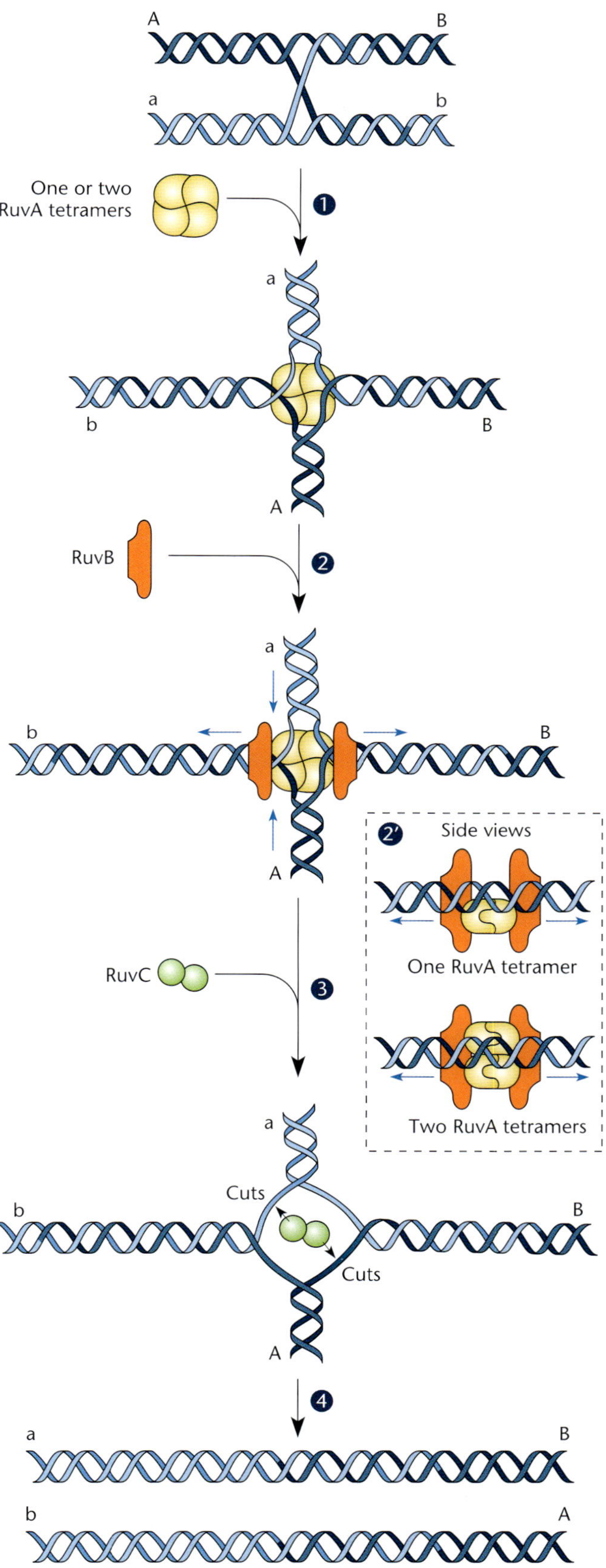

to create configuration II, where the ends of the two double-stranded DNA molecules are in the recombinant configuration with respect to each other. Notice that the strands that were not crossed before are now the crossed strands, and vice versa. It may seem surprising that the strands that have crossed over to hold the two DNA duplexes together can change places so readily, but experiments with models show that the two structures, I and II, are actually equivalent to each other and that the Holliday junction can change from one form to the other without breaking any hydrogen bonds between the bases. Hence, flipping from one configuration to the other requires no energy and should occur quickly, so that each configuration should be present approximately 50% of the time.

The Ruv and RecG Proteins and the Migration and Cutting of Holliday Junctions

As discussed above, Holliday junctions are remarkable structures that can do many things. Once a Holliday junction has formed as the result of the concerted action of RecBCD and RecA or the RecF pathway and RecA, at least two different pathways can resolve the junctions to make recombinant products. One pathway uses the three Ruv proteins, RuvA, RuvB, and RuvC, which are encoded by adjacent genes. Another protein, called RecG, can move Holliday junctions, but its role during homologous recombination is unclear because it lacks its own host-encoded resolvase (see below).

RuvABC

The crystal structures of the Ruv proteins are interesting and give clues to how they function in the migration and resolution of Holliday junctions (Figure 9.8) (see West, Suggested Reading). The RuvA protein is a specific Holliday junction-binding protein whose role is to force the Holliday junction into a certain structure amenable to the subsequent steps of migration and resolution. Four copies of the RuvA protein come together to form a flat structure like a flower with four petals. The Holliday junction lies flat on the flower and thus is forced into a flat (planar) configuration. Binding of the RuvA tetramer also creates

Figure 9.8 Model for the mechanism of action of the Ruv proteins. **(1)** One or two tetramers of the RuvA protein bind to a Holliday junction, holding it in a planar (flat) configuration. Note that at the beginning, the DNA has only one turn of heteroduplex (blue-light blue). **(2)** Two hexamers of the RuvB protein bind to the RuvA complex, each forming a ring around one strand of the DNA. **(2′)** Side view of the complex with one and two tetramers. **(3)** RuvC binds to the complex and cuts two of the strands. **(4)** The Holliday junction has been resolved into a different configuration because of the way the strands were cut. Note that there are now more turns of heteroduplex (blue-light blue).

a short region in the middle of the Holliday junction where the strands are not base paired and the single strands form a sort of square. Another tetramer of RuvA may then bind to the first to form a sort of turtle shell, with the four arms of double-stranded DNA emerging from the "leg holes." The RuvB protein then forms a hexameric (six-member) ring encircling one arm of the DNA, as shown in Figure 9.8. The DNA is then pumped through the RuvB ring, using ATP cleavage to drive the pump, thereby forcing the Holliday junction to migrate.

After the RuvA and RuvB proteins have forced Holliday junctions to migrate, the junctions can be cut (resolved) by the RuvC protein. The RuvC protein cuts only Holliday junctions that are being held by RuvA and RuvB. The RuvC protein is a specialized DNA endonuclease that cuts the two crossed strands of a Holliday junction simultaneously. Like many enzymes that make double-strand breaks in DNA, two identical polypeptides encoded by the *ruvC* gene come together to form a homodimer, which is the active form of the enzyme. Because the enzyme has two copies of the polypeptide, it has two DNA endonuclease active centers, each of which can cut one of the DNA strands to make a double-strand break.

Genetic evidence indicates that RuvC can cut only Holliday junctions that are bound to RuvA and RuvB; mutants with either a *ruvA*, a *ruvB*, or a *ruvC* mutation are indistinguishable in that they are all defective in the resolution of Holliday junctions (see below). To a geneticist, this means that RuvC cannot act to resolve a Holliday junction without RuvA and RuvB being present and bound to the Holliday junction. However, this leads to an apparent conflict with the structural information about RuvA and RuvB discussed above, which indicates that RuvA forms a turtle shell-like structure over the Holliday junction. How could RuvC enter the turtle shell formed by RuvA to cleave the crossed DNA strands inside? One possibility is that a tetramer of RuvA is bound to only one face of the Holliday junction, leaving the other face open for RuvC to bind and cut the crossed strands. Another idea is that the RuvA shell opens up somehow to allow RuvC to enter and cut the crossed strands.

Support for this model of the concerted action of RuvA, RuvB, and RuvC has come from observations of purified Ruv proteins acting on artificially synthesized structures that resemble Holliday junctions (see Parsons et al., Suggested Reading). These junctions are constructed by annealing four synthetic single-stranded DNA chains that have pairwise complementarity to each other (Figure 9.9). These synthetic structures are not completely analogous to a real Holliday junction in that they are not made from two naturally occurring DNAs with the same sequence. Rather, four single strands that are complementary to each other in the regions shown and therefore form a pairwise cross are synthesized. A Holliday junction made in this way is much more stable than a natural Holliday junction because the branch cannot migrate spontaneously all the way to one end or the other. Real Holliday junctions are too unstable for these experiments; they quickly separate into two double-stranded DNA molecules.

Experiments performed with such synthetic Holliday junctions indicated that purified Ruv proteins act sequentially on a synthetic Holliday junction in a manner consistent with the above-mentioned model. First, RuvA protein bound specifically to the synthetic Holliday junctions, and then a combination of RuvA and RuvB caused a disassociation of the synthetic Holliday junctions, simulating branch migration in a natural DNA molecule. The dissociation required the energy in ATP to break the hydrogen bonds holding the Holliday junction together, as predicted. The RuvC protein could also specifically cut the synthetic Holliday junction in two of the four strands.

RuvC has some sequence specificity at the site where it cuts. It always cuts just downstream of two Ts in the DNA preceded by either an A or a T and followed by either a G or a C (as shown in bold in Figure 9.9). At first, it seems impossible that two Ts could be opposite each other in the DNA until we remember that the crossed strands of DNA are not the complementary strands but

Figure 9.9 A synthetic Holliday junction with four complementary strands. The junction cannot migrate but can be disrupted by RuvA and RuvB. It can also be cut by RuvC (red arrows) and other Holliday junction resolvases, such as the gene *49* and gene *3* products of phages T4 and T7, respectively.

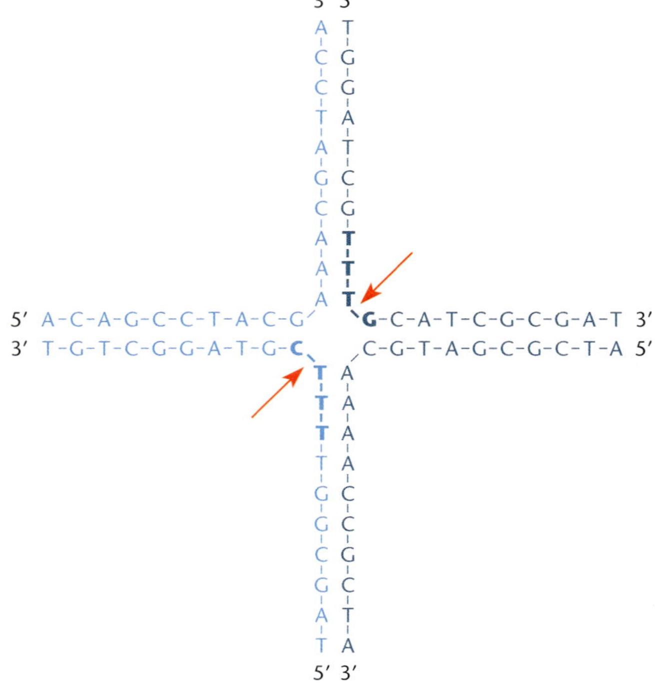

the strands with the same sequence. If one strand has the sequence recognized by RuvC, 5′-(A/T)TT(G/C)-3′, the other strand has the same sequence at that position. Presumably, RuvA and RuvB cause the Holliday junction to migrate until this sequence is encountered and RuvC then can cut the crossed strands. There is a cryptic resolvase protein in *E. coli*, called RusA, encoded by a defective prophage that is not expressed in the wild-type configuration that also has sequence specificity and cuts upstream of two Gs in the DNA. It has been speculated that Holliday junction resolvases may have sequence specificity to distinguish them from nucleases that can cut branches, such as those that arise from single-strand invasion of a double-stranded DNA. Holliday junction resolvases, such as RuvC and RusA, have been shown to cut such branches if the branch point has their recognition sequence. However, such branches cannot migrate, so the crossed-over strand always has the same sequence, which is not apt to be the sequence recognized by the Holliday junction resolvase; therefore, they are seldom cut by RuvC or RusA. Holliday junctions, on the other hand, can migrate, so eventually they migrate to the sequence recognized by the resolvase and are cut.

Firmicutes, like *B. subtilis*, lack RuvC but possess a different protein, called RecU, which acts as the Holliday junction resolvase. The details of how RecU interfaces with the branch migration system are still being worked out (see Khavnekar et al., Suggested Reading).

RecG

Another helicase in *E. coli* that can help junctions to migrate is RecG (Table 9.1). The function of this helicase has been debated for a long time. It has very little effect on recombination when RuvABC is present, which suggested that its role is redundant with respect to RuvABC. The idea was that RecG could play the role of RuvAB and promote the migration of Holliday junctions. Then, another resolvase could play the role of RuvC and resolve the Holliday junction, providing a backup for RuvABC. However, it now seems that the roles of these proteins are not redundant and that they each have their own unique role to play (see Bolt and Lloyd, Suggested Reading). For example, RuvAB and RecG move in opposite directions on single-stranded DNA. The RuvAB helicase moves in the 5′-to-3′ direction on single-stranded DNA, while the RecG helicase moves in the 3′-to-5′ direction. Following DNA repair or another type of processing event (as discussed in detail in chapter 10), replication could restart, in most cases with the help of the PriA proteins and DnaC to load the DnaB helicase back on the DNA. A more recent proposal for the function of RecG is not to facilitate the formation of substrates where the Pri proteins can initiate DNA replication. Instead, it is hypothesized that RecG could unwind these structures to prevent the "overreplication" that would result if DNA replication is initiated at recombination structures. Unregulated replication initiated at these structures would complicate future replication and recombination reactions. The role of RecG is still under active research.

Recombination between Different DNAs in Bacteria

Most of the earlier work on recombination in bacteria concerned recombination between DNAs with different mutations, leading to the formation of recombinant types. As discussed in chapter 3, recombination in bacteria has some differences from recombination in most other organisms. In bacteria, portions of a chromosome are typically introduced from a donor strain that recombines with the recipient chromosome, whereas in other organisms, two long double-stranded DNA molecules of equal size usually recombine. Nevertheless, the requirements for recombination in bacteria are similar to those in other organisms. In fact, accumulating evidence supports the existence of proteins analogous to many of the recombination proteins of bacteria in both eukaryotes and archaea.

How Are Linear DNA Fragments Recombined into the *E. coli* Chromosome?

Throughout chapter 3 and elsewhere in the text, a simplified representation was used to indicate how linear fragments are recombined into the host chromosome (or plasmid or bacteriophage). The requirements for integrating a linear DNA fragment into the chromosome are similar to those required to restart a collapsed replication fork, which is why they use the same functions. The ability of the RecBCD complex to efficiently process double-strand breaks to load the RecA protein to initiate DNA replication immediately suggests a model for how linear fragments could recombine into the chromosome. Figure 9.10 shows how recombination likely initiates DNA replication to copy an allele into the chromosome during DNA replication. The incoming DNA fragment is shown with different shades of blue to emphasize that the stretch of DNA is maintained. Hypothetical alleles that differ between the incoming and existing DNA fragments are indicated with red and black dots. RecBCD processes the ends of the linear fragment to 3′ single-stranded overhangs that are loaded with the RecA protein, which can invade the chromosome. Replication that is initiated by the DNA fragment then continues to the terminus region (clockwise-traveling DNA replication fork) or is met by a replication fork that initiates from *oriC*. After chromosome segregation, each daughter cell will have one of the alleles, the allele from the parent (shown as black dots) or the allele that was recombined into the chromosome (shown as red dots).

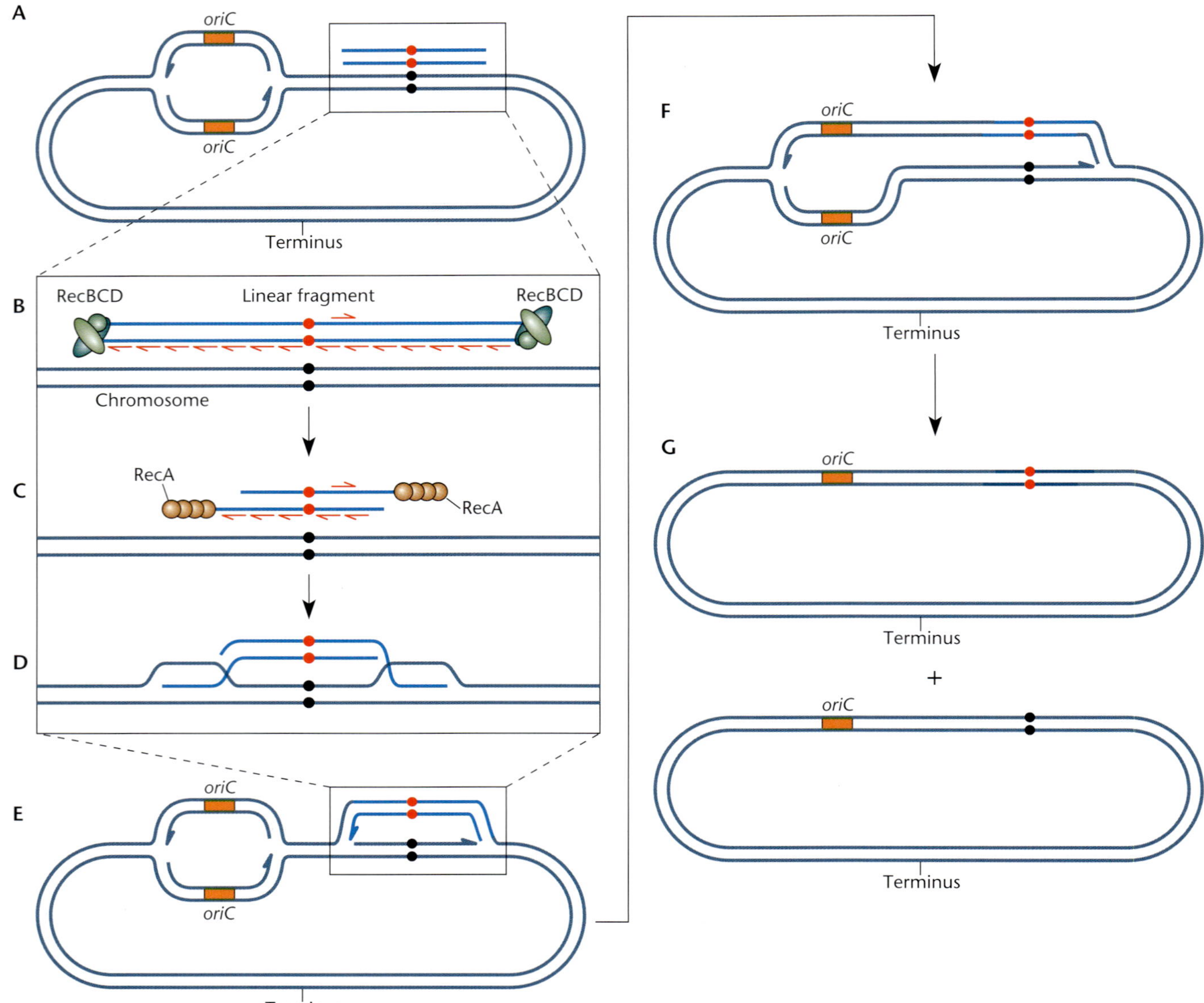

9.10 Model for how linear fragments are recombined into the chromosome using homologous recombination and DNA replication. **(A)** A homologous DNA fragment (light blue) with a different allele than the chromosome (red versus black dots). **(B to D)** Diagram of how the DNAs in the boxed region in panel A are converted to the structure shown in the boxed region in panel E. **(B)** The ends of the linear fragment are processed with RecBCD. The χ sites are indicated by red arrows with half heads. **(C)** The ends are processed into 3′ single-strand extensions that end with a χ site and are coated with the RecA protein. **(D)** The single-stranded DNA coated with RecA invades the homologous DNA sequence and initiates DNA replication in the chromosome. **(E)** Replication that is initiated by the DNA fragment and then continues in opposite directions. **(F)** The replication fork traveling in the clockwise direction progresses to the terminus region, while the counterclockwise replication fork is met by a replication fork that initiates from *oriC*. **(G)** After chromosome segregation, each daughter cell has one of the alleles, the allele from the parent (shown as black dots) or the allele that was recombined into the chromosome (shown as red dots).

Recombination during Natural Transformation

Natural transformation involves the direct uptake of DNA as described in chapter 6. While DNA is transported across the cell membrane, one strand is degraded, making an ideal substrate for recombination, but one that does not fit neatly into the RecBCD and RecF processing pathways described above. As described above, the RecBCD (AddAB) and RecF pathways have mechanisms for displacing SSB (which loads with the highest affinity on single-stranded DNA) and replacing it with RecA as an important process for regulating homologous recombination. A specialized, yet incompletely understood, process is involved in ensuring that the DNA transported into the cell during transformation is protected and used for RecA-mediated recombination. A broadly conserved protein, DprA, is a critical component in this process (see Mortier-Barrière, Suggested Reading). DprA, a specialized SSB (SsbB), and RecA, along with other competence-specific proteins, localize to the cell poles in competent cells, making the single-stranded DNA substrate competent for recombination, but the molecular details of this process remain under investigation.

Phage Recombination Pathways

Many phages encode their own recombination functions, some of which can be important for the multiplication of the phages. As discussed in chapter 7, some phages use recombination to make primers for replication and concatemers for packaging of the DNA into phage particles. Phage recombination systems may also be important for repairing damaged phage DNA and for exchanging DNA between related phages to increase diversity. Phages may encode their own recombination systems to avoid dependence on host systems for these important functions.

Many phage recombination functions are analogous to the recombination proteins of the host bacteria (Table 9.2), and in many cases, the phage recombination proteins were discovered before their host counterparts. As a result, studies of bacterial recombination systems have been heavily influenced by simultaneous studies of phage recombination systems.

Table 9.2 Comparison of phage and host recombination functions

Phage function	Analogous *E. coli* function
T4 UvsX	RecA
T4 gene *49*	RuvC
T7 gene *3*	RuvC
T4 genes *46* and *47*	RecBCD
λ ORF[a] in *nin* region	RecO, RecR, RecF
Rac *recE* gene	RecJ, RecQ
λ *gam*	Inhibits RecBCD
λ *exo*	RecJ
λ *bet*	None
rusA (DLP12 prophage)	RuvC

[a]ORF, open reading frame.

Rec Proteins of Phages T4 and T7

Phages T4 and T7 depend on recombination for the formation of DNA concatemers after infection (see chapter 7). Therefore, recombination functions are essential for the multiplication of these phages. Many of the T4 and T7 Rec proteins are analogous to those of their hosts. For example, the gene *49* protein of T4 and the gene *3* protein of T7 are endonucleases that resolve Holliday junctions and are representative of phage proteins discovered before their host counterparts, in this case RuvC. The gene *46* and *47* products of phage T4 may perform a reaction similar to that of the RecBCD protein of the host, although no evidence indicates the presence of χ-like sites associated with this enzyme. The UvsX protein of T4 and the Bet protein of λ promote strand annealing between single-stranded DNAs. While the strand-annealing ability promotes genetic recombination, these phage proteins lack the strand invasion capacity of RecA found in the host.

The RecE Pathway of the *rac* Prophage

Another classic example of a phage-encoded recombination pathway is the RecE pathway encoded by the *rac* prophage of *E. coli* K-12. The *rac* prophage is integrated at 29 min in the *E. coli* genetic map and is related to λ. This defective prophage cannot be induced to produce infective phage, since it lacks some essential functions for multiplication.

The RecE pathway was discovered by isolating suppressors of *recBCD* mutations, called *sbcA* mutations (for suppressor of *recB* and *recC* deletions), that restored recombination in conjugational crosses. The *sbcA* mutations were later found to activate a normally repressed recombination function of the defective prophage *rac*. Apparently, *sbcA* mutations inactivate a repressor that normally prevents the transcription of the *recE* and *recT* genes. These gene products, or homologous ones from other phages, can be used for recombineering, a powerful technique discussed below.

The Phage λ Red System

Phage λ also encodes recombination functions. The best-characterized is the Red system, which requires the products of the adjacent λ genes *exo* and *bet*. The product of the *exo* gene is an exonuclease that degrades one strand of a double-stranded DNA from the 5′ end to leave a 3′ single-stranded tail. The *bet* gene product is known to help the renaturation of denatured DNA and to bind to the λ exonuclease. Unlike many of the other recombination systems that we have discussed, the λ Red recombination pathway does not require the RecA protein. Instead, the system appears to act at regions of the chromosome when they become available as single-strand DNA during DNA replication through the action of the phage Bet protein (see Box 8.1). The λ Red system was first used for recombineering (see below). Besides the Red system, phage λ encodes another recombination function that can substitute for components of the *E. coli* RecF pathway (see Sawitzke and Stahl, Suggested Reading; Table 9.2). Apparently, phages can carry components of more than one recombination pathway.

Recombineering: Gene Replacements in *E. coli* with Phage λ Recombination Functions

One of the major advantages of using bacteria and other simple organisms for molecular genetic studies is the relative ease of doing gene replacements with some of these organisms (see the discussion of gene replacements in chapter 3). To perform a gene replacement, a piece of the DNA of an organism is manipulated in the test tube to change its sequence in some desired way. The DNA is then reintroduced into the cell, and the recombination systems of the cell cause the altered sequence of the reintroduced DNA to replace the normal sequence of the corresponding DNA in the chromosome. Because it depends on homologous recombination, gene replacement requires that the sequence of the reintroduced DNA has homology with the sequence of the DNA it replaces. However, the homology need not be complete, and minor changes, such as base pair changes, can be introduced into the chromosome in this way as a type of site-specific mutagenesis. Also, the reintroduced DNA need not be homologous over its entire length; homology is needed only at the two fragment ends. This makes it possible to use gene replacement to make large alterations, such as construction of an in-frame deletion to avoid polarity effects, and insertion of an antibiotic resistance gene cassette into the chromosome (see Box 13.2). If the sequences on both sides of the alteration (the flanking sequences) are homologous to sequences in the chromosome, recombination between these flanking sequences and the chromosome will insert the alteration. Historically, methods for gene replacement in *E. coli* have usually relied on the RecBCD-RecA recombination pathway, since this is the major pathway for recombination in *E. coli*. We mention some of these methods in this chapter and chapter 3.

A newer and more useful method for manipulating DNA in *E. coli* is called recombineering. The term "recombineering" is used to describe various applications with bacteriophage recombination proteins that have been optimized for performing site-specific mutagenesis and gene replacements in *E. coli* (Table 9.2). One system in particular, the λ Red system, has many advantages over the RecBCD-RecA pathway for such manipulations. This method makes it possible to use single-stranded DNA oligonucleotides, possibly as short as 30 bases, although those 60 bases long or longer work better. This is important because the synthesis of single-stranded DNAs of these lengths has become routine for making PCR primers, and oligonucleotides with any desired sequence can be purchased for a reasonable cost. Other methods of site-specific mutagenesis for making specific changes in a sequence are more tedious and require a certain amount of technical skill. Probably most importantly, recombineering is very efficient. Minor changes, such as single-amino acid changes in a protein, usually offer no positive selection, and most methods require the screening of thousands of individuals to find one with the replacement.

Figure 9.11 outlines the original procedure for using the λ Red system for gene replacements. Figure 9.11A shows the structure of an *E. coli* strain typically used for this kind of experiment. It carries a defective prophage in which most of the λ genes have been deleted, except the recombination (Red) genes *gam-bet-exo* (Table 9.2; see also Figure 8.2). Figure 9.11B shows the replacement of a sequence in the plasmid by the corresponding region on another plasmid, in which the sequence has been disrupted by introduction of an antibiotic resistance (Ab^r) cassette. This region of the plasmid has been amplified by PCR to produce a double-stranded DNA fragment carrying the antibiotic resistance cassette and some of the flanking sequences. First, the cells are briefly incubated at 42°C, the temperature at which Red genes of the prophage are induced because the mutant repressor is inactive. Then, cells are made competent for electroporation, and the PCR fragment is introduced into the cells by electroporation. The *gam* gene product, Gam, inhibits the RecBCD complex so that the linear DNA fragment is not degraded as soon as it enters the cell. The *exo* gene product, Exo, then processes the fragment for recombination. Exo is an exonuclease that degrades one strand of a double-stranded DNA from the 5′ end, thereby exposing a single-stranded 3′-overhang. The *bet* gene product, Bet, is then responsible for annealing this single-stranded DNA end to a homologous single-stranded region in the host. The cells in which the PCR fragment has recombined with the cellular DNA so that the sequence containing the antibiotic resistance gene has replaced the corresponding sequence in the cellular DNA are then selected on plates containing the antibiotic.

To determine the effect on the cell of inactivating a gene product, it is often useful to delete the entire gene

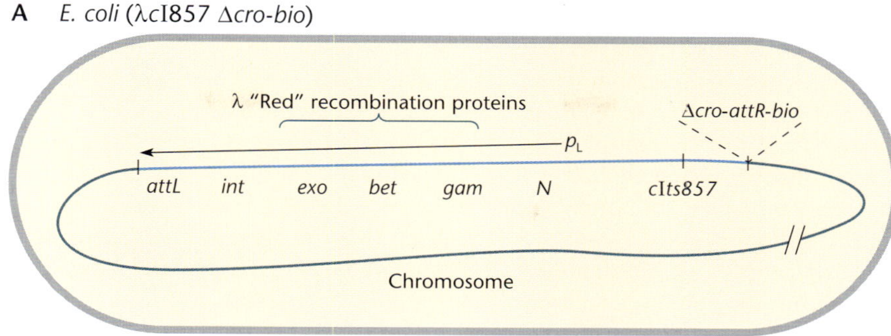

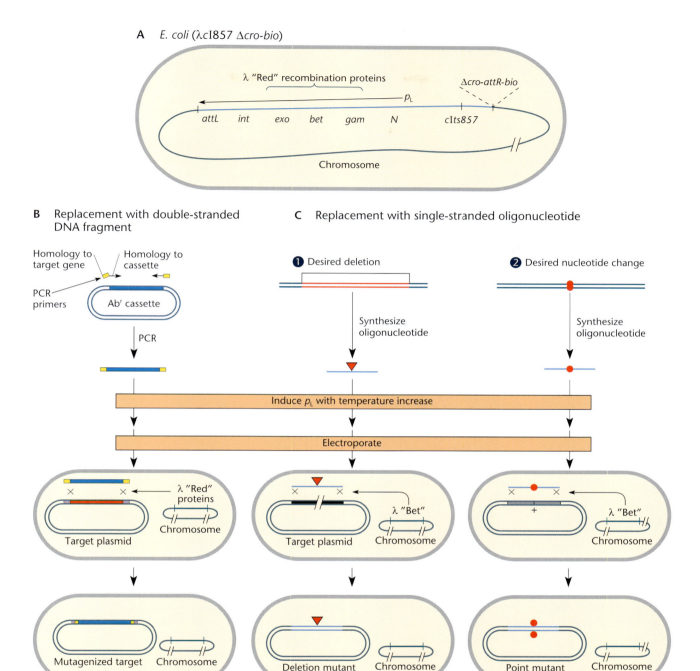

Figure 9.11 Recombineering: *in vivo* DNA modification in *E. coli* using λ phage-encoded proteins. **(A)** A deletion derivative of a λ lysogen with a temperature-sensitive repressor can be used to induce the λ Red recombination functions (light blue). **(B)** Double-stranded DNA cassettes can be amplified using primers with homology to the regions flanking the region to be replaced. Double-stranded DNA can be processed by the λ recombination proteins encoded by the Red genes. Processing by Exo to form single-stranded DNA at the fragment ends is not shown. **(C)** Oligonucleotides (light blue) that contain a deletion or a nucleotide change compared to the target plasmid can be directly synthesized. The region to be deleted is indicated in red, and the missing region is denoted by a red triangle A point mutation is indicated with a red dot. Red recombination with oligonucleotides only requires the Bet protein. Note: in all cases, integration of the DNA substrate occurs by annealing between two single-stranded DNA substrates, as explained in the text.

and replace it with an antibiotic resistance cassette. This can be accomplished by using PCR to amplify the cassette with primers whose 5′ sequences are complementary to sequences flanking the gene to be deleted. Recall that the 5′ sequences on a PCR primer need not be complementary to the sequences being amplified. When this amplified fragment is introduced into the cells, the antibiotic resistance cassette replaces the entire gene.

Introducing an antibiotic resistance cassette into a gene simplifies the task of selecting the gene replacement and inactivating the gene. A variation of this procedure is to include sites recognized by a site-specific recombinase flanking the antibiotic resistance cassette. In this adaptation of the procedure, the recombinase can be expressed later to remove the cassette and leave an in-frame deletion, which is important to reduce the chance of polar effects (see Box 13.2 and Datsenko and Wanner, Suggested Reading).

Sometimes, we want to introduce a small change into the gene for which there is no direct selection, for example, a specific change in one amino acid that we think may play an important role in the protein product of the gene. Variations of the recombineering technique allow construction of recombinants that have a single defined base pair change or some other small change. One such procedure involves using a cassette that has both a gene that can be selected by positive selection and a gene whose product is toxic under some conditions. This allows us to select both for acquisition of the cassette and, later, for its loss. An example of such a system could involve insertion of the recognition site for the I-SceI homing endonuclease, which is long enough to not be found naturally in bacterial genomes (see Box 13.2). The I-SceI recognition site is toxic only when the I-SceI endonuclease is expressed in the same strain. First, a DNA cassette carrying both an antibiotic resistance gene and the cleavage site for the I-SceI homing endonuclease, flanked by sequences for the region of the gene to be replaced, is introduced into the cell by electroporation, selecting for the antibiotic resistance as described above. Then, another DNA fragment (or oligonucleotide), identical to the targeted region of the DNA but carrying the desired base pair change, is introduced. At the same time, a plasmid expressing the I-SceI endonuclease is also introduced by transformation and selected by using a plasmid-borne antibiotic resistance gene. Any DNA retaining the I-SceI recognition site will be destroyed by cleavage by the endonuclease, and only recombinants in which the corresponding region is replaced by a sequence derived from the second DNA fragment will survive. Most of the surviving bacteria, therefore, have the sequence with the base pair change replacing the original sequence in the gene, with no other changes, and this can be verified by DNA sequencing. Various other techniques have been developed to make so-called markerless manipulations of the chromosome.

The utility of this method for making specific changes in a gene increased dramatically when it was discovered that single-stranded DNA can also be used for electroporation (see Ellis et al., Suggested Reading). Single-stranded DNAs with a defined sequence are easily obtainable, as this is how DNA is chemically synthesized for PCR primers, etc. Using single-stranded DNAs also makes it possible to dispense with Gam and Exo, since single-stranded DNA is not degraded by RecBCD and the DNA does not need Exo to make it single stranded. Figure 9.11C shows the replacement if a single-stranded oligonucleotide is used to introduce either an in-frame deletion or a single-base pair change into the target DNA. The procedure is similar, except that only *bet* needs to be expressed from the prophage. The Bet protein promotes pairing between the introduced single-stranded DNA and the chromosomal DNA during DNA replication (see below). Then, replication or repair replaces the normal sequence with the mutant sequence in both strands, as shown.

Interestingly, gene replacement with single-stranded DNA shows a strong strand bias, meaning that a single-stranded oligonucleotide complementary to one strand is more apt to replace the corresponding chromosomal sequence than is an oligonucleotide complementary to the other strand in any particular region. The strand that is preferred in the reaction correlates with the direction of replication in the region. The *E. coli* chromosome replicates bidirectionally from the origin of replication (see chapter 1) so that on one side of the *oriC* region the replication fork moves in one direction, while on the other side it moves in the opposite direction; the sequence to which it binds corresponds to the lagging strand. Presumably, the single-stranded gaps that are produced on the lagging strand at the replication fork are where the Bet protein functions. This is because, unlike RecA, the Bet protein is not able to help a single-stranded DNA invade a completely double-stranded DNA. Bet may need a single-stranded gap in the recipient DNA to pair with before it can "get its foot in the door" and promote the displacement of adjacent double-stranded DNA. Interestingly, there may be an inherent tendency for the lagging strand to be vulnerable to recombination. This stems from the surprising finding that several different species could naturally be manipulated with oligonucleotides without the benefit of discernible phage recombination genes in their genome sequences (see Swingle et al., Suggested Reading). The efficiency of recombineering is also increased in a strain that is deficient in mismatch repair or if a C-C mismatch, which will not be recognized by the mismatch repair system (see chapter 10), is used. Other adaptations to the basic procedure that can increase the efficiency of Red-mediated recombination include using oligonucleotides that are modified in such a way that they are not easily recognized by host enzymes that could degrade them before recombination. All of these improvements greatly increase the frequency of progeny with the desired mutation. The λ Red recombination technique has also been shown to work with multiple oligonucleotides in the same transformation, which allows many sites to be modified in a single reaction (see Isaacs et al., Suggested Reading). Furthermore, specific changes need not always be targeted; instead, random changes can be made using oligonucleotides with randomized nucleotides included at some positions. For example, a collection of oligonucleotides that differed at a certain codon or codons can be utilized. Recombinants from this procedure can then be screened or subjected to genetic selection to identify useful mutations at one or many genetic positions in a bacterial genome.

The high-frequency λ-based method has been adapted to *E. coli* and many of its relatives, such as *Salmonella* and *Yersinia*. The number of bacteria in which it can be used is being expanded by identifying prophages with related recombination functions in other bacteria through genome sequences, such as clinically relevant bacteria like *Mycobacterium tuberculosis* (see van Kessel and Hatfull, Suggested Reading). In some of these, genes for the phage recombination enzymes are more homologous to *recET* genes of the *rac* prophage (see above) than they are to the genes of λ.

Gene Conversion and Other Manifestations of Heteroduplex Formation during Recombination

As mentioned earlier in the chapter, Holliday junctions formed during recombination can migrate. There are consequences to the migration of Holliday junctions when the DNAs involved are not perfectly identical across the region of the heteroduplex (Figure 9.12). This is because mismatches result in a heteroduplex region that can be processed in different ways.

GENE CONVERSION

The first evidence for heteroduplex formation during recombination came from studies of gene conversion in fungi. Understanding this process requires a little knowledge of the sexual cycles of fungi. Some fungi have long been favored organisms for the study of recombination because the spores that are the products of a single meiosis are often contained in the same bag, or ascus (see any general genetics textbook). When two haploid fungal cells mate, they fuse to form a diploid zygote. Then, the homologous chromosomes pair and replicate once to form four chromatids that recombine with each other before they are packaged into spores. Therefore, each ascus contains four spores (or eight in fungi such as *Neurospora crassa*, in which the chromatids replicate once more before spore packaging).

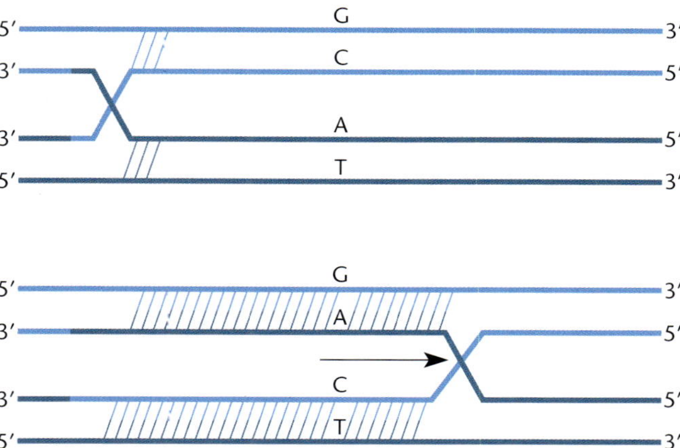

Figure 9.12 Migration of Holliday junctions. By breaking the hydrogen bonds holding the DNAs together in front of the branch and reforming them behind, the junction migrates and extends the regions of pairing (i.e., the heteroduplexes) between the two DNAs. The heteroduplex region is hatched. In the example, two mismatches, GA and CT, form in the heteroduplex regions because one of the DNA molecules has a mutation in the region.

Since both haploid fungi contribute equal numbers of chromosomes to the zygote, their genes should show up in equal numbers in the spores in the ascus. In other words, if the two haploid fungi have different alleles of the same gene, two of the four spores in each ascus should have the allele from one parent, and the other two spores should have the allele from the other parent. This is called a 2:2 segregation. However, the two parental alleles sometimes do not appear in equal numbers in the spores. For example, three of the spores in a particular ascus might have the allele from one parent, while the remaining spore has the allele from the other parent—a 3:1 segregation instead of the expected 2:2 segregation. In this case, an allele of one of the parents appears to have been converted into the allele of the other parent during meiosis—hence the name **gene conversion**.

Gene conversion is caused by repair of mismatches created on heteroduplexes during recombination, and Figure 9.13 shows how such mismatch repair can convert one allele into the other when the DNA molecules of the two parents recombine. In the illustration, the DNAs of the two parents are identical in the region of the recombination except that a mutation has changed a wild-type AT pair into a GC pair in one of the DNA molecules. Hence, the parents have different alleles of the gene. When the two individuals mate to form a diploid zygote and their DNAs recombine during meiosis, one strand of each DNA may pair with the complementary strand of the other DNA in this region. A mismatch will result, with a G opposite a T

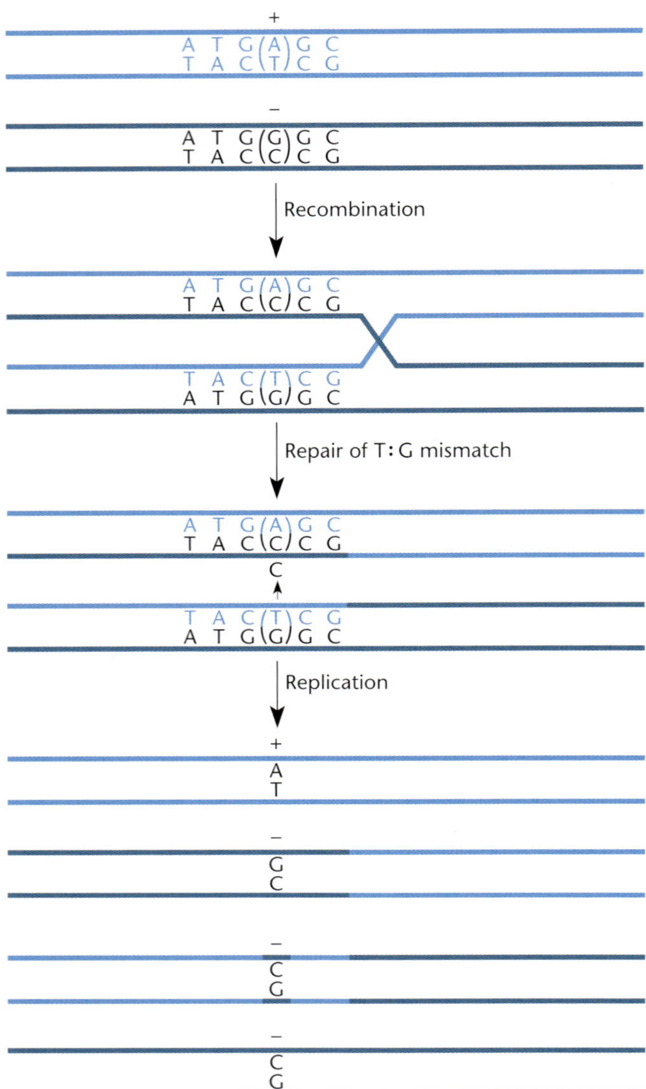

Figure 9.13 Repair of a mismatch in a heteroduplex region formed during recombination can cause gene conversion. A plus sign indicates the wild-type sequence, and a minus sign indicates the mutant sequence. See the text for details.

in one DNA and an A opposite a C in the other DNA (Figure 9.13). If a repair system changes the T opposite the G to a C in one of the DNAs, then after meiosis, three molecules will carry the mutant allele sequence, with GC at this position, but only one DNA will have the wild-type allele sequence, with AT at this position. Hence, one of the two wild-type alleles has been "converted" into the mutant allele.

MANIFESTATIONS OF MISMATCH REPAIR IN HETERODUPLEXES IN PHAGES AND BACTERIA

Gene conversion is more difficult to detect in crosses with bacteria and phages than with fungi, since the products of a single recombination event do not stay together in a bacterial or phage cross. However, the existence of heteroduplexes formed during recombination in bacteria and phages, as well as the repair of mismatches in these structures, is manifested in other ways.

Map Expansion

In prokaryotes, mismatch repair in heteroduplexes can increase the apparent recombination frequency between two closely linked markers, making the two markers seem farther apart than they really are. This manifestation of mismatch repair in heteroduplexes formed during recombination is called map expansion because the genetic map appears to increase in size.

Figure 9.14 shows how mismatch repair can affect the apparent recombination frequency between two markers. In the illustration, the two DNA molecules participating in the recombination have mutations that are very close to each other, so that crossovers between the two mutations to produce wild-type recombinants should be very rare. However, a Holliday junction occurs nearby, and the region of one of the two mutations is included in the heteroduplex, creating mismatches that can be repaired. If the G in the GT mismatch in one of the DNA molecules is repaired to an A, the progeny with the DNA will appear to be a wild-type recombinant. Therefore, even though the potential crossover that caused the formation of heteroduplexes did not occur in the region between the

Figure 9.14 Repair of mismatches can give rise to recombinant types between two mutations. A plus sign indicates the wild-type sequence, and a minus sign indicates the mutant sequence. The positions of the mutations are shown in parentheses. See the text for details.

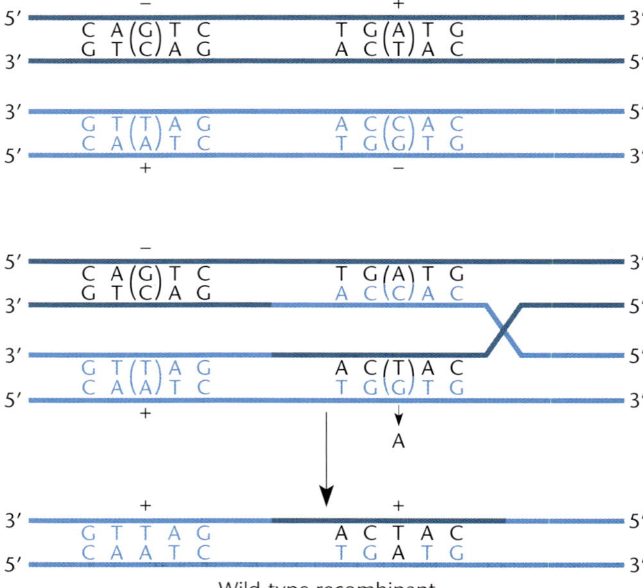

two mutations, apparent wild-type recombinants resulted. The apparent recombination due to mismatch repair might occur even when the Holliday junction is resolved so that the flanking sequences are returned to their original configuration; thus, a true crossover does not result. Therefore, although gene conversion and other manifestations of mismatch repair are generally associated with recombinant DNA molecules, they appear only in DNA molecules with crossovers.

Marker Effects

Mismatch repair of heteroduplexes can also cause marker effects, phenomena in which two different markers at exactly the same locus show different recombination frequencies when crossed with the same nearby marker. For example, two different transversion mutations might change a UAC codon into UAA and UAG codons in different strains. However, when these two strains are crossed with another strain with a third nearby mutation, the recombination frequency between the UAA mutation and the third mutation might appear to be much higher than the recombination frequency between the UAG mutation and the third mutation, even though the UAG and UAA mutations are exactly the same distance on the DNA from the third mutation. Such a difference between the two recombination frequencies can be explained because the UAG and UAA mutations are causing different mismatches to form during recombination, and one of these may be recognized and repaired more readily by the mismatch repair system than the other. In *E. coli* at least, CC mismatches are not repaired by the mismatch repair system. Note that one of the heteroduplexes that the original UAC sequence forms with the UAG mutation at this site contains a CC mismatch, but the heteroduplex formed with the UAA mutation does not.

Marker effects also occur because the lengths of single DNA strands removed and resynthesized by different mismatch repair systems vary (see chapter 10), and the chance that mismatch repair will lead to apparent recombination depends on the lengths of these sequences, or patches. As is apparent in Figure 9.14, a wild-type recombinant arises only if mismatches due to both mutations are not removed on the same repair patch. If the patch that is removed in repairing one mismatch also removes the other mismatch, one of the parental DNA sequences will be restored and no apparent recombination will occur. In *E. coli*, mismatches due to deamination of certain methylated cytosines are repaired by very short patches (VSP repair [see chapter 10]).

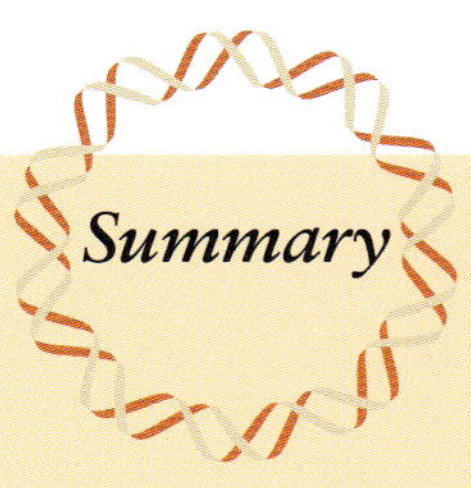

1. Recombination is the joining of DNA strands in new combinations. Homologous recombination occurs only between two DNA molecules that have the same sequence in the region of the crossover.

2. All recombination models involve the formation of Holliday junctions. The Holliday junctions can migrate and isomerize so that the crossed strands uncross and then recross in a different orientation. Holliday junctions can then be resolved by specific DNA endonucleases to produce recombinant DNA.

3. The region over which two strands originating from different DNA molecules are paired following the branch migration of a Holliday junction is called a heteroduplex.

4. Repair of mismatches on the heteroduplex DNA molecules formed as intermediates in recombination can give rise to such phenomena as gene conversion, map expansion, and marker effects.

5. In *E. coli*, the major pathway for recombination during conjugation and transduction is the RecBCD pathway. The RecBCD protein loads on the DNA at a double-strand break and moves along the DNA, looping out a single strand and degrading one strand from the 3′ end. If a sequence in the DNA called the χ sequence is encountered as the protein moves along the DNA, the 3′-to-5′ nuclease activity on RecBCD is inhibited and 5′-to-3′ activity is stimulated, leaving a free 3′ end on which RecBCD loads RecA to form a RecA single-stranded DNA filament that can invade homologous double-stranded DNAs.

6. The RecF pathway is another recombination pathway in *E. coli*. It is required for recombination between single-stranded gaps in DNA and double-stranded DNA. The RecF pathway requires the products of the *recF*, *recO*, *recR*, and eventually *recQ* and *recJ* genes. In *E. coli*, RecFOR act at single-stranded gaps in the DNA but can act also in Hfr crosses only if the *sbcB* and the *sbcC* or *sbcD* genes have been inactivated or mutated. The SbcB and SbcCD enzymes destroy intermediates that would normally allow recombination by the RecF pathway and channel recombination to the RecBCD pathway.

7. The RecA protein promotes synapse formation and strand displacement and is required for recombination by both the

Summary (continued)

RecBCD and RecF pathways. The RecA protein forces a single-stranded DNA into a helical nucleoprotein filament, which can then scan double-stranded DNAs looking for its complementary sequence. If it finds its complementary sequence, it invades it, forming a D loop or a three-strand structure, which can then migrate to form a Holliday junction.

8. Holliday junctions can migrate by at least two separate pathways in *E. coli*, the RuvABC pathway and the RecG pathway. In the first pathway, the RuvA protein binds to Holliday junctions, and then the RuvB protein binds to RuvA and promotes branch migration with the energy derived from cleaving ATP. The RuvC protein is a Holliday junction-specific endonuclease that cleaves Holliday junctions to resolve recombinant products. The RecG protein is also a Holliday junction-specific helicase, but it may be more important to prevent overinitiation of DNA replication.

9. Many phages encode their own recombination systems. Sometimes, phage recombination functions are analogous to host recombination functions. The gene *49* product of the T4 phage and the gene *3* product of T7 resolve Holliday junctions. A phage λ recombination system, the Red system, is encoded by two genes, *exo* and *bet*. The *exo* gene product degrades one strand of a double-stranded DNA to make single-stranded DNA tails. The *bet* gene product promotes annealing of two single-stranded DNAs. These λ genes have become the basis for a very useful way of doing site-specific mutagenesis and gene replacements, sometimes called recombineering.

QUESTIONS FOR THOUGHT

1. Why do you suppose essentially all organisms have recombination systems?

2. Why do you suppose the RecBCD protein promotes recombination through such a complicated process?

3. Why are there overlapping pathways of recombination that can substitute for each other?

4. Why do you think the cell encodes the *sbcB* and *sbcC* gene products that interfere with the RecF pathway?

5. Why do some phages encode their own recombination systems? Why not rely exclusively on the host pathways?

6. Propose a model for how the RecG protein could substitute for RuvABC when it has only a helicase activity and not a resolvase activity.

7. The sequence of χ sites can differ between bacteria. What do you think would happen if you replaced the *recBCD* genes in *E. coli* with those from a species that had a different χ site sequence?

8. Why might it be advantageous to have so many different types of double-strand break repair in bacteria?

SUGGESTED READING

Baharoglu Z, Petranovic M, Flores M-J, Michel B. 2006. RuvAB is essential for replication forks reversal in certain replication mutants. *EMBO J* **25:**596–604.

Bidnenko V, Seigneur M, Penel-Colin M, Bouton MF, Dusko Ehrlich S, Michel B. 1999. *sbcB sbcC* null mutations allow RecF-mediated repair of arrested replication forks in *rep recBC* mutants. *Mol Microbiol* **33:**846–857.

Bolt EL, Lloyd RG. 2002. Substrate specificity of RusA resolvase reveals the DNA structures targeted by RuvAB and RecG *in vivo*. *Mol Cell* **10:**187–198.

Chédin F, Ehrlich SD, Kowalczykowski SC. 2000. The *Bacillus subtilis* AddAB helicase/nuclease is regulated by its cognate *Chi* sequence *in vitro*. *J Mol Biol* **298:**7–20.

Clark AJ, Margulies AD. 1965. Isolation and characterization of recombination-deficient mutants of *Escherichia coli* K12. *Proc Natl Acad Sci USA* **53:**451–459.

Cox MM. 2003. The bacterial RecA protein as a motor protein. *Annu Rev Microbiol* **57:**551–577.

Datsenko KA, Wanner BL. 2000. One-step inactivation of chromosomal genes in *Escherichia coli* K-12 using PCR products. *Proc Natl Acad Sci USA* **97:**6640–6645.

Dillingham MS, Kowalczykowski SC. 2008. RecBCD enzyme and the repair of double-stranded DNA breaks. *Microbiol Mol Biol Rev* **72:**642–671.

Dillingham MS, Spies M, Kowalczykowski SC. 2003. RecBCD enzyme is a bipolar DNA helicase. *Nature* **423:**893–897.

Dixon DA, Kowalczykowski SC. 1993. The recombination hotspot chi is a regulatory sequence that acts by attenuating the nuclease activity of the *E. coli* RecBCD enzyme. *Cell* **73:**87–96.

Ellis HM, Yu D, DiTizio T, Court DL. 2001. High efficiency mutagenesis, repair, and engineering of chromosomal DNA using single-stranded oligonucleotides. *Proc Natl Acad Sci USA* **98:**6742–6746.

Fujii S, Isogawa A, Fuchs RP. 2006. RecFOR proteins are essential for Pol V-mediated translesion synthesis and mutagenesis. *EMBO J* **25:**5754–5763.

Handa N, Morimatsu K, Lovett ST, Kowalczykowski SC. 2009. Reconstitution of initial steps of dsDNA break repair by the RecF pathway of *E. coli*. *Genes Dev* **23:**1234–1245.

Holliday R. 1964. A mechanism for gene conversion in fungi. *Genet Res* **5:**282–304.

Howard-Flanders P, Theriot L. 1966. Mutants of *Escherichia coli* K-12 defective in DNA repair and in genetic recombination. *Genetics* **53:**1137–1150.

Isaacs FJ, Carr PA, Wang HH, Lajoie MJ, Sterling B, Kraal L, Tolonen AC, Gianoulis TA, Goodman DB, Reppas NB, Emig CJ, Bang D, Hwang SJ, Jewett MC, Jacobson JM, Church GM. 2011. Precise manipulation of chromosomes *in vivo* enables genome-wide codon replacement. *Science* **333:**348–353.

Jones M, Wagner R, Radman M. 1987. Mismatch repair and recombination in *E. coli. Cell* **50:**621–626.

Khavnekar S, Dantu SC, Sedelnikova S, Ayora S, Rafferty J, Kale A. 2017. Structural insights into dynamics of RecU-HJ complex formation elucidates key role of NTR and stalk region toward formation of reactive state. *Nucleic Acids Res* **45:**975–986.

Kogoma T. 1997. Stable DNA replication: interplay between DNA replication, homologous recombination, and transcription. *Microbiol Mol Biol Rev* **61:**212–238.

Kuzminov A, Stahl FW. 1999. Double-strand end repair via the RecBC pathway in *Escherichia coli* primes DNA replication. *Genes Dev* **13:**345–356.

Lloyd RG. 1991. Conjugational recombination in resolvase-deficient *ruvC* mutants of *Escherichia coli* K-12 depends on *recG*. *J Bacteriol* **173:**5414–5418.

Lopper M, Boonsombat R, Sandler SJ, Keck JL. 2007. A hand-off mechanism for primosome assembly in replication restart. *Mol Cell* **26:**781–793.

Lovett ST. 2005. Filling the gaps in replication restart pathways. *Mol Cell* **17:**751–752.

Mazin AV, Kowalczykowski SC. 1999. A novel property of the RecA nucleoprotein filament: activation of double-stranded DNA for strand exchange *in trans*. *Genes Dev* **13:**2005–2016.

Meselson MS, Radding CM. 1975. A general model for genetic recombination. *Proc Natl Acad Sci USA* **72:**358–361.

Michel B, Leach D. 2012. Homologous recombination-enzymes and pathways. *Ecosal Plus* **5:**1–46.

Morimatsu K, Kowalczykowski SC. 2003. RecFOR proteins load RecA protein onto gapped DNA to accelerate DNA strand exchange: a universal step of recombinational repair. *Mol Cell* **11:**1337–1347.

Mortier-Barrière I, Velten M, Dupaigne P, Mirouze N, Piétrement O, McGovern S, Fichant G, Martin B, Noirot P, Le Cam E, Polard P, Claverys JP. 2007. A key presynaptic role in transformation for a widespread bacterial protein: DprA conveys incoming ssDNA to RecA. *Cell* **130:**824–836.

Mosig G. 1987. The essential role of recombination in phage T4 growth. *Annu Rev Genet* **21:**347–371.

Parsons CA, Tsaneva I, Lloyd RG, West SC. 1992. Interaction of *Escherichia coli* RuvA and RuvB proteins with synthetic Holliday junctions. *Proc Natl Acad Sci USA* **89:**5452–5456.

Radding CM. 1991. Helical interactions in homologous pairing and strand exchange driven by RecA protein. *J Biol Chem* **266:**5355–5358.

Renzette N, Gumlaw N, Sandler SJ. 2007. DinI and RecX modulate RecA-DNA structures in *Escherichia coli* K-12. *Mol Microbiol* **63:**103–115.

Rocha EP, Cornet E, Michel B. 2005. Comparative and evolutionary analysis of the bacterial homologous recombination systems. *PLoS Genet* **1:**e15.

Saikrishnan K, Yeeles JT, Gilhooly NS, Krajewski WW, Dillingham MS, Wigley DB. 2012. Insights into *chi* recognition from the structure of an AddAB-type helicase-nuclease complex. *EMBO J* **31:**1568–1578.

Sawitzke JA, Stahl FW. 1992. Phage λ has an analog of *Escherichia coli recO*, *recR* and *recF* genes. *Genetics* **130:**7–16.

Sharples GJ. 2001. The X philes: structure-specific endonucleases that resolve Holliday junctions. *Mol Microbiol* **39:**823–834.

Simmons LA, Goranov AI, Kobayashi H, Davies BW, Yuan DS, Grossman AD, Walker GC. 2009. Comparison of responses to double-strand breaks between *Escherichia coli* and *Bacillus subtilis* reveals different requirements for SOS induction. *J Bacteriol* **191:**1152–1161.

Singleton MR, Dillingham MS, Gaudier M, Kowalczykowski SC, Wigley DB. 2004. Crystal structure of RecBCD enzyme reveals a machine for processing DNA breaks. *Nature* **432:**187–193.

Swingle B, Markel E, Costantino N, Bubunenko MG, Cartinhour S, Court DL. 2010. Oligonucleotide recombination in Gram-negative bacteria. *Mol Microbiol* **75:**138–148.

Szostak JW, Orr-Weaver TL, Rothstein RJ, Stahl FW. 1983. The double-strand-break repair model for recombination. *Cell* **33:**25–35.

Taylor AF, Smith GR. 2003. RecBCD enzyme is a DNA helicase with fast and slow motors of opposite polarity. *Nature* **423:**889–893.

van Kessel JC, Hatfull GF. 2007. Recombineering in *Mycobacterium tuberculosis*. *Nat Methods* **4:**147–152.

Wang HH, Isaacs FJ, Carr PA, Sun ZZ, Xu G, Forest CR, Church GM. 2009. Programming cells by multiplex genome engineering and accelerated evolution. *Nature* **460:**894–898.

West SC. 1998. RuvA gets X-rayed on Holliday. *Cell* **94:**699–701.

Yeeles JTP, Dillingham MS. 2010. The processing of double-stranded DNA breaks for recombinational repair by helicase-nuclease complexes. *DNA Repair (Amst)* **9:**276–285.

Zhang J, Mahdi AA, Briggs GS, Lloyd RG. 2010. Promoting and avoiding recombination: contrasting activities of the *Escherichia coli* RuvABC Holliday junction resolvase and RecG DNA translocase. *Genetics* **185:**23–37.

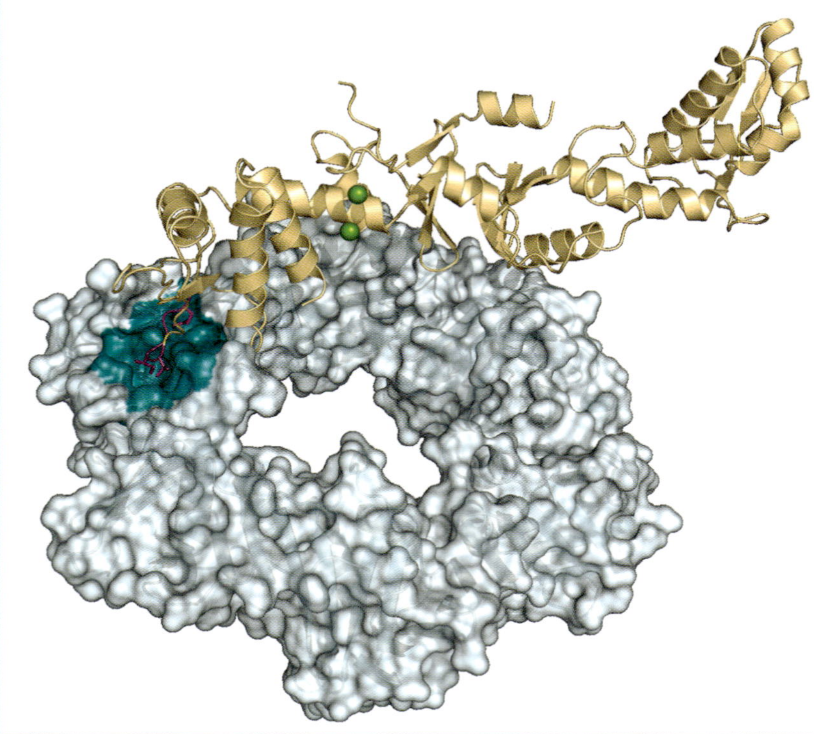

Model of the interaction between the endonuclease domain of *Bacillus subtilis* MutL and the *B. subtilis* clamp, based on the crystal structure of the endonuclease domain of BsMutL (Pillon et al., *Mol Cell* 39:145–151, 2010) and the structure of Bsclamp fused to the regulatory domain of BsMutL (Almawi et al., *Nucleic Acids Res* 47:4831–4842, 2019). Courtesy of Alba Guarné, McGill University.

DNA Repair and Mutagenesis

10

THE CONTINUITY OF SPECIES from one generation to the next is a tribute to the stability of DNA. If DNA were not so stable and were not reproduced so faithfully, there could be no species. Before DNA was known to be the hereditary material and its structure was determined, a lot of speculation centered on what types of materials would be stable enough to ensure the reliable transfer of genetic information over so many generations (see, for example, Schrodinger, Suggested Reading, in the introduction). Therefore, the discovery that the hereditary material is DNA—a chemical polymer no more stable than many other chemical polymers—came as a surprise.

Evolution has provided the selection for a DNA replication apparatus that minimizes mistakes (see chapter 1). However, mistakes during replication are not the only threats to DNA. Since DNA is a chemical, it is constantly damaged by chemical reactions. Many environmental factors can damage the molecule. Heat can speed up spontaneous chemical reactions, leading, for example, to the deamination of bases. Chemicals can react with DNA, adding groups to the bases or sugars, breaking the bonds of the DNA, or fusing parts of the molecule to each other. Irradiation at certain wavelengths can also chemically damage DNA, which can absorb the energy of the photons. Once the molecule is energized, bonds may be broken or parts may be erroneously fused. DNA damage can be very deleterious to cells because DNA polymerase may not be able to replicate over the damaged area, preventing the cells' multiplication. Even if the damage does not block replication, replicating the damaged region can cause mutations, some of which will be deleterious or even lethal. Obviously, cells need mechanisms for DNA damage repair.

To describe DNA damage and its repair, we need first to define a few terms. Chemical damage in DNA is called a **lesion**. Chemical compounds or treatments that cause lesions in DNA can kill cells and can also increase the frequency of mutations in the DNA. Such treatments that generate mutations are called **mutagenic treatments** or **mutagens**. Some mutagens, known as ***in vitro* mutagens**, can be used to damage DNA in the test tube, which then produces mutations when the DNA is introduced into cells. Other mutagens damage DNA only in the cell, for example, by interfering with base pairing during replication. These are called ***in vivo* mutagens**.

In this chapter, we discuss the types of DNA damage, how each type of damage to DNA might cause mutations, how bacterial cells repair the damage to their DNA, and more generally, how bacteria tolerate DNA damage even in cases when it is not immediately repaired. Many of these mechanisms seem to be universal, as related systems are found in other organisms, including humans.

Evidence for DNA Repair

Before discussing specific types of DNA damage, we should make some general comments on the outward manifestations of DNA damage and its repair. The first question is how we even know a cell has the means to repair a particular type of damage to its DNA. One way is to measure killing by exposure to chemicals or by irradiation. The chemical agents and radiation that damage DNA also often damage other cellular constituents, including RNA and proteins. Nevertheless, cells exposed to these agents usually die as a result of chemical damage to the DNA. The other components of the cell can usually be resynthesized and/or exist in many copies, so that even if some molecules are damaged, more of the same type of molecule will be there to substitute for the damaged ones. However, a single chemical change in the enormously long chromosomal DNA of a cell can prevent the replication of that molecule and subsequently cause cell death unless the damage is repaired.

To measure cell killing—and thereby demonstrate that a particular type of cell has DNA repair systems—we can compare the survival of the cells exposed intermittently to small doses of a DNA-damaging agent with that of cells that receive the same amount of treatment continuously. If the cells have DNA repair systems, more cells survive the short intervals of treatment because some of the damage is repaired between treatments. In contrast, if DNA is not repaired during the rest periods, it should make no difference whether the treatment occurs at intervals or continuously. The cells accumulate the same amount of damage regardless of the different treatments, and the same fraction of cells should survive both regimens, unless the cell has the ability to repair the damage.

Another indication of repair systems comes from the shape of the **killing curves**. A killing curve is a plot of the number of surviving cells versus the extent of treatment by an agent that damages DNA. The extent of treatment can refer to the length of time the cells are irradiated or exposed to a chemical that damages DNA or to the intensity of irradiation or the concentration of the damaging chemical.

The two curves in Figure 10.1 contrast the shapes of killing curves for cells with and without DNA repair systems. In the curve for cells without a repair system for the DNA-damaging treatment, the fraction of surviving cells drops exponentially, since the probability that each cell will be killed by a lethal "hit" to its DNA is the same during each time interval. This exponential decline gives rise to a straight line when plotted on a semilog scale, as shown (Figure 10.1).

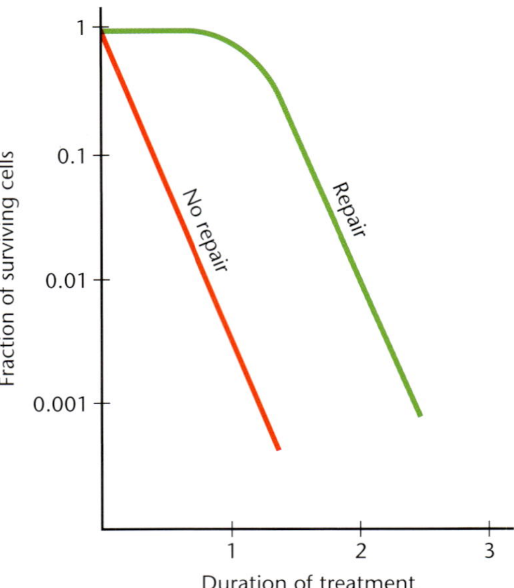

Figure 10.1 Survival of cells as a function of the time or extent of treatment with a DNA-damaging agent. The fraction of surviving cells is plotted against the duration of treatment. The shoulder on the survival curve indicates the presence of a repair mechanism.

The other curve shows what happens if the cell has DNA repair systems. Rather than dropping exponentially with increasing treatment, this curve extends horizontally first, forming a "shoulder." The shoulder appears because repair mechanisms repair lower levels of damage, allowing many of the cells to survive. Only with higher treatment levels, when the damage becomes so extensive that the repair systems can no longer cope with it, will the number of surviving cells drop exponentially with increasing levels of treatment.

Among the survivors of DNA-damaging agents, there may be many more mutants than before. However, it is very important to distinguish DNA damage from mutagenesis. In particular types of cells, some types of DNA damage are mutagenic, while others are not, independent of their effects on cell survival. Recall from chapter 3 that mutations are permanent heritable changes in the sequence of nucleotides in DNA. The damage to DNA caused by a chemical or by irradiation is not by itself a mutation because it is not heritable. A mutation might occur because the damage is not repaired and the replication apparatus must proceed over the damage, making mistakes because complementary base pairing does not occur properly at the site of the damage. Alternatively, mistakes might be made during attempts to repair the damage, causing changes in the sequence of nucleotides at the site of the damage. In the following sections, we discuss some types of DNA damage, how they might cause mutations, and the repair systems that can fix them. Most of what we describe is best known for *Escherichia coli*, for which these systems are best understood, but the universality of many of them is clear, and humans share many of the same systems.

Specific Repair Pathways

Different agents damage DNA in different ways, and different repair pathways operate to repair the various forms of damage. Some of these pathways repair only a certain type of damage, whereas others are less specific and repair many types. We first discuss examples of damage repaired by specific repair pathways.

Deamination of Bases

One of the most common types of damage to DNA is the deamination of bases. Some of the amino groups in adenine, cytosine, and guanine are particularly vulnerable and can be removed spontaneously or by many chemical agents (Figure 10.2). When adenine is deaminated, it becomes **hypoxanthine**. When guanine is deaminated, it becomes **xanthine**. When cytosine is deaminated, it becomes **uracil**.

Deamination of DNA bases is mutagenic because it results in base mispairing. As shown in Figure 10.2, hypoxanthine derived from adenine pairs with the base cytosine instead of thymine, and uracil derived from the deamination of cytosine pairs with adenine instead of guanine.

The type of mutation caused by deamination depends on which base is altered. For example, the hypoxanthine that results from deamination of adenine pairs with cytosine during replication, incorporating C instead of T at that position. In a subsequent replication, the C pairs with the correct G, causing an AT-to-GC transition in the DNA. Similarly, uracil resulting from the deamination of a cytosine pairs with an adenine during replication, causing a GC-to-AT transition type of mutation.

DEAMINATING AGENTS

Although deamination often occurs spontaneously, especially at higher temperatures, some types of chemicals react with DNA and remove amino groups from the bases. Treatment of cells or DNA with these chemicals, known as **deaminating agents**, can greatly increase the rate of mutation. Which deaminating agents are mutagenic in a particular situation depends on the properties of the chemical.

Hydroxylamine

Hydroxylamine specifically removes the amino group of cytosine and consequently causes only GC-to-AT transitions in the DNA. However, hydroxylamine, an *in vitro* mutagen, cannot enter cells, so it can be used only to mutagenize purified DNA or viruses. Mutagenesis by hydroxylamine is particularly effective when the treated DNA is introduced into cells deficient in repair by the uracil-*N*-glycosylase enzyme for reasons discussed below.

Nitrous Acid

Nitrous acid not only deaminates cytosines, but also removes the amino groups of adenine and guanine (Figure 10.2). It also causes other types of damage. Because it is less specific, nitrous acid can cause both GC-to-AT and AT-to-GC transitions, as well as deletions. Nitrous acid can enter some types of cells and so can be used as a mutagen both *in vivo* and *in vitro*.

REPAIR OF DEAMINATED BASES

Given that base deamination is potentially mutagenic, it is not surprising that special enzymes have evolved to remove deaminated bases from DNA. These enzymes, **DNA glycosylases**, break the glycosyl bond between the damaged base and the sugar in the nucleotide. These enzymes are highly specific, and a unique DNA glycosylase exists for each type of deaminated base and removes only that particular base. Specific DNA glycosylases remove bases damaged in other ways; these types of glycosylases are discussed in later sections. There are at least a dozen specific *N*-glycosylases that remove damaged bases in *E. coli*, and they all work by basically the same mechanism.

Figure 10.3 illustrates the removal of damaged bases from DNA by DNA glycosylases. There are two types of glycosylases: those that just remove the base and others, called **AP lyases**, that both remove the base and cut the DNA backbone on the 3′ side of the damaged base. If just the base has been removed by the specific glycosylase, nucleases called **AP endonucleases** cut the sugar-phosphate backbone of the DNA on the 5′ side of the missing base, leaving a 3′ OH group. These enzymes can cut either next to the spot from which a pyrimidine (C or T) has been removed (an apyrimidinic site) or next to where a purine (A or G) has been removed (an apurinic site). After the cut is made and processed, the free 3′ OH end is used as a primer by the repair DNA polymerase (DNA polymerase I [Pol I] in *E. coli*) to synthesize more DNA, while the 5′ exonuclease activity associated with the DNA polymerase displaces and degrades the strand ahead of the DNA polymerase. In this way, the entire region of the DNA strand around the deaminated base is resynthesized, allowing a normal base to be inserted in place of the damaged one.

VERY-SHORT-PATCH REPAIR OF DEAMINATED 5-METHYLCYTOSINE

Most organisms have 5-methylcytosine bases instead of cytosines at specific sites in their DNA. These bases are cytosines with a methyl group at the 5 position on the pyrimidine ring instead of the usual hydrogen (Figure 10.2B). Specific enzymes called **methyltransferases**

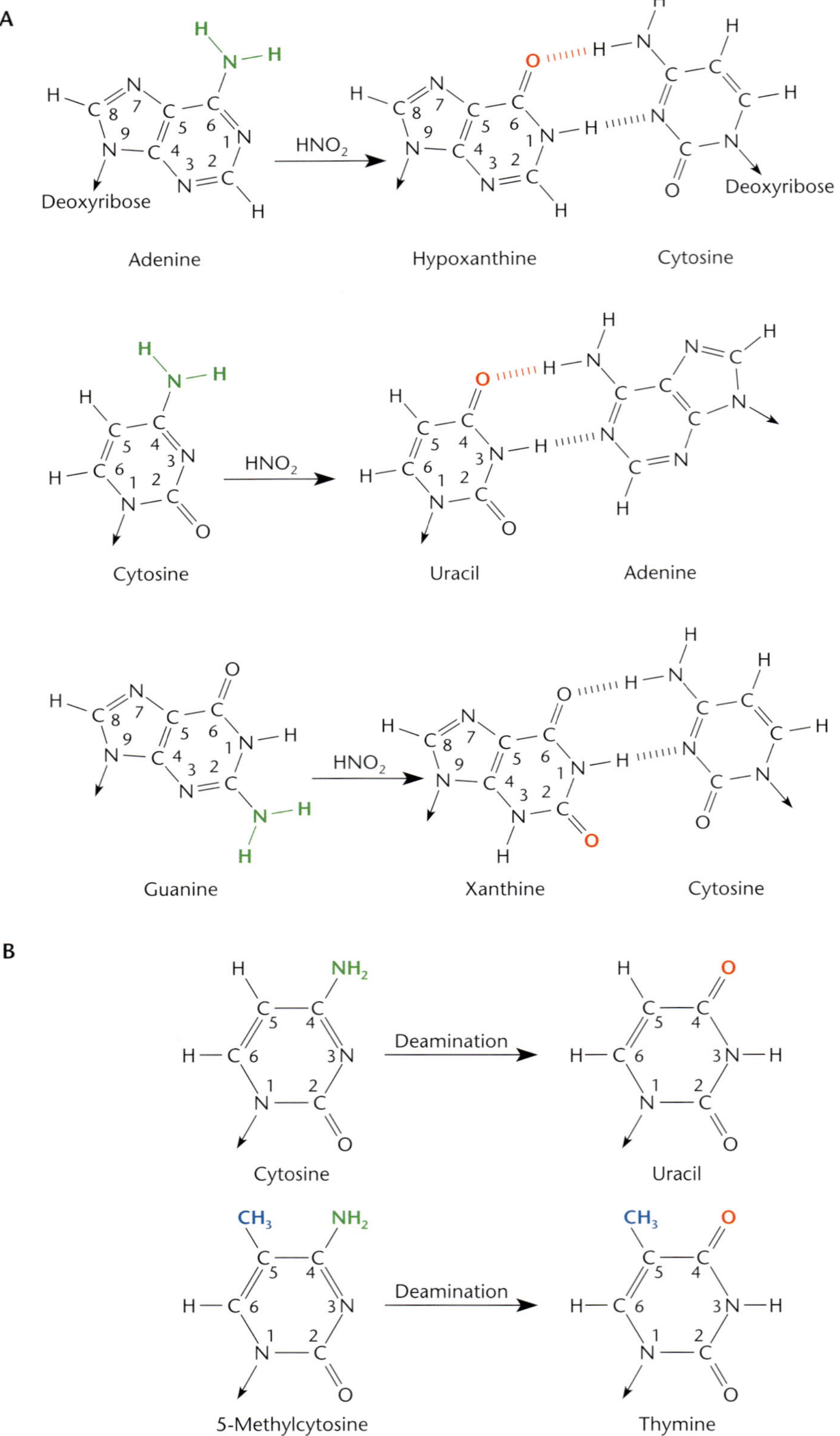

Figure 10.2 (A) Modified bases created by deaminating agents, such as nitrous acid (HNO_2). Some deaminated bases pair with the wrong base, causing mutations. **(B)** Spontaneous deaminations. Deamination of 5-methylcytosine produces a thymine that is not removable by the uracil-*N*-glycosylase.

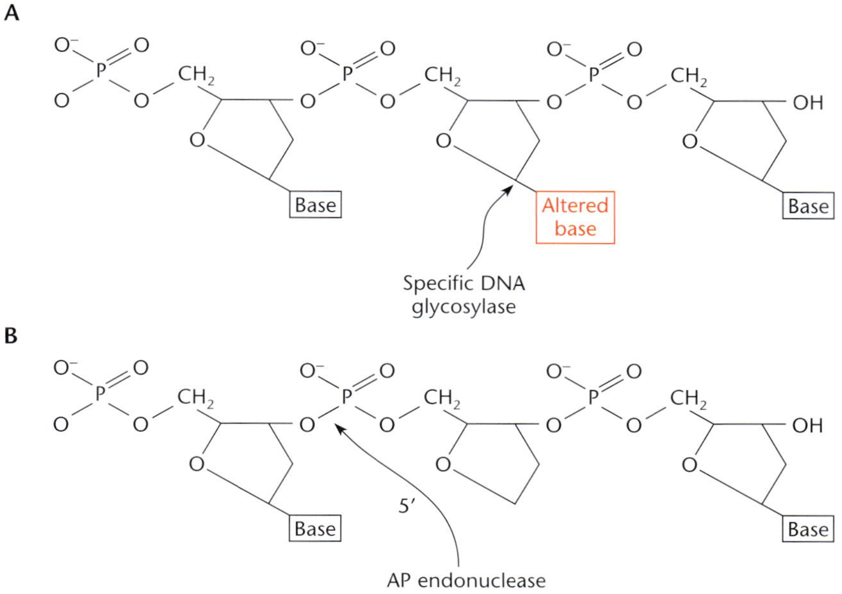

Figure 10.3 Repair of altered bases by DNA glycosylases. **(A)** The specific DNA glycosylase removes the altered base. **(B)** An apurinic, or AP, endonuclease cuts the DNA backbone on the 5′ side of the AP site. The strand is degraded and resynthesized, and the correct base is restored (not shown).

transfer the methyl group to this position after the DNA is synthesized, using *S*-adenosylmethionine as the methyl donor. The functions of these 5-methylcytosines are often obscure, but we know that they sometimes help protect DNA against cutting by restriction endonucleases in bacteria and play additional roles in eukaryotes.

The sites of 5-methylcytosine in DNA are often hot spots for mutagenesis, because deamination of 5-methylcytosine yields thymine rather than uracil (Figure 10.2B), and thymine in DNA is not recognized by the uracil-*N*-glycosylase, since it is a normal base in DNA. These thymines in DNA are located opposite guanines and so could in principle be repaired by the mismatch repair system (see below). However, in a GT mismatch created by a replication mistake, the mistakenly incorporated base can be identified because it is in a newly replicated strand (see below), whereas the GT mismatches created by the deamination of 5-methylcytosine are generally not found in newly synthesized DNA. Repairing the wrong strand causes a GC-to-AT transition in the DNA.

In *E. coli* K-12 and some other enteric bacteria, most of the 5-methylcytosine in the DNA occurs in the second C of the sequence 5′-CCWGG-3′/3′-GGWCC-5′, where the middle base pair (W) is generally either AT or TA. The second C in this sequence is methylated by an enzyme called DNA cytosine methylase (Dcm) to give C^mCAGG/GGTC^{m}C, where C^m indicates the position of the 5-methylcytosine. The mutation potential of this modification likely provided the selection for a special repair mechanism for deaminated 5-methylcytosines that occur in this sequence in *E. coli* K-12. This repair system specifically removes the thymine whenever it appears as a TG mismatch in this sequence.

Because a small region, or "patch," of the DNA strand containing the T is removed and resynthesized during the repair process, the mechanism is called **very-short-patch (VSP) repair**. In VSP repair, the Vsr endonuclease, the product of the *vsr* gene, binds to a TG mismatch in the C^mCWGG/GGWTC sequence and makes a break next to the T. The T is then removed, and the strand is resynthesized by the repair DNA polymerase (DNA Pol I; see chapter 1), which inserts the correct C.

The VSP repair system is very specialized and usually repairs TG mismatches only in the sequence shown above. Therefore, this repair system would have only limited usefulness if methylation did not occur in the sequence. The *vsr* gene is immediately downstream of the gene for the Dcm methylase, ensuring that cells that inherit the gene to methylate the C in C^mCWGG/GGWC^{m}C also usually inherit the ability to repair the mismatch correctly if it is deaminated. While only enterics like *E. coli* have been shown to have this particular repair system, many other organisms have 5-methylcytosine in their DNA, and we expect that similar repair systems will be found in these organisms.

Damage Due to Reactive Oxygen

Although molecular oxygen (O_2) is not very damaging to DNA and most other macromolecules, more reactive forms of oxygen (collectively called reactive oxygen species) are very damaging. These more reactive forms of oxygen include hydrogen peroxide, superoxide radicals, and hydroxyl free radicals, which are produced as byproducts by flavoenzymes in the presence of molecular oxygen (see Korshunov and Imlay, Suggested Reading). The molecular oxygen strips electrons from the flavin, converting itself into these more reactive forms. Of these, hydroxyl free radicals are the most damaging to DNA. Hydrogen peroxide, which has many normal functions in cells, is also produced but has no unpaired electrons and is not particularly damaging to DNA; however, it can be converted into hydroxyl free radicals in the presence of iron (Fe^{2+}) atoms. Reactive forms of oxygen can also arise as a result of environmental factors, including UV irradiation and chemicals, such as the herbicide paraquat. Interestingly, the killing effect of most common antibiotics appears to work, in part, through the formation of reactive oxygen species (see Dwyer et al., Suggested Reading). Among the mechanisms implicated in this process is that by blocking translation, these antibiotics may change the relative concentrations of proteins in the respiratory chains, causing them to produce more reactive oxygen.

Because the reactive forms of oxygen appear in cells growing in the presence of oxygen, all aerobic organisms must contend with the resulting DNA damage and have evolved elaborate mechanisms to remove these chemicals from the cellular environment. In bacteria, some of these systems are induced by the presence of the reactive forms of oxygen (Box 10.1), and the induced genes encode enzymes such as superoxide dismutases, catalases, and peroxide reductases that help destroy or scavenge the reactive forms. The same systems that induce enzymes to scavenge reactive forms of oxygen also induce the synthesis of proteins involved in exporting antibiotics or preventing the uptake of antibiotics, consistent with the idea that the lethal effects of antibiotics can occur with the participation of oxidative damage, so the two are linked. Other genes induced by reactive oxygen species encode repair enzymes that help repair the oxidative DNA damage. The accumulation of this type of damage stemming from reactive oxygen species may be responsible for the increase in cancer rates with age and for many age-related degenerative diseases (Box 10.1).

8-oxoG

One of the most mutagenic lesions in DNA caused by reactive oxygen is the oxidized base **7,8-dihydro-8-oxoguanine** (8-oxoG, or GO) (Figure 10.4). This base appears frequently in DNA because of damage caused by internally produced free radicals of oxygen, and unless repair systems deal with the damage, DNA Pol III often mispairs it with adenine, causing spontaneous GC-to-AT transition mutations. Because of the mutagenic potential of 8-oxoG, *E. coli* has evolved many mechanisms for avoiding the resultant mutations, and we discuss these below.

MutM, MutY, and MutT

The *mut* genes of an organism were so named because they were identified by their ability to reduce the normal rates of spontaneous mutagenesis (see "Isolation of *mut* Mutants" below). Organisms with a mutation that inactivates the product of a *mut* gene will suffer higher than normal rates of spontaneous mutagenesis. We have already discussed some *mut* genes of *E. coli*, identified because of mutations that increase the spontaneous mutation rate, which include the genes of the mismatch repair system and the *dnaQ* (*mutD*) gene encoding the editing function ε (see chapter 1). Other *mut* genes of *E. coli* are *mutM*, *mutY*, and *mutT*. The products of these *mut* genes are exclusively devoted to preventing mutations due to 8-oxoG. The generally high rate of spontaneous mutagenesis during aerobic growth in *mutM*, *mutT*, and *mutY* mutants is testimony to the fact that internal oxidation of DNA is an important source of spontaneous mutations and that 8-oxoG, in particular, is a very mutagenic form of damage to DNA (Box 10.1).

The discovery that these three *mut* genes are dedicated to relieving the mutagenic effects of 8-oxoG in DNA, as well as the role played by each of them, was the result of some clever genetic experiments (see Michaels et al., Suggested Reading). First, there was the evidence that the functions of the three *mut* genes are additive to reduce spontaneous mutations, since the rate of spontaneous mutations is higher if two or all three of the *mut* genes are mutated than if only one of them is mutated. There was also evidence that mutations in each of the *mut* genes increased the frequency of some types of spontaneous mutations but not others. Below, we discuss the roles of the products of each of these *mut* genes and then discuss how the genetic evidence is consistent with each of these roles.

MutM

The MutM enzyme is an *N*-glycosylase that specifically removes the 8-oxoG base from the deoxyribose sugar in DNA (Figure 10.4). This repair pathway functions like other *N*-glycosylase repair pathways discussed in this book, except that the depurinated strand is cut by the AP endonuclease activity of MutM itself, degraded by an exonuclease, and resynthesized by DNA Pol I (Figure 10.3). The MutM protein is present in larger amounts in cells that have accumulated reactive oxygen, because the *mutM* gene is part of a regulon induced in response to oxidative stress. We discuss regulons in more detail in chapter 12.

MutY

The MutY enzyme is also a specific *N*-glycosylase. However, rather than removing 8-oxoG directly, the MutY *N*-glycosylase specifically removes adenine bases that have been mistakenly incorporated opposite an 8-oxoG in DNA (Figure 10.4). Repair synthesis by DNA Pol I then usually introduces the correct C to prevent a mutation, as with other *N*-glycosylase-initiated repair pathways.

BOX 10.1

The Role of Reactive Oxygen Species in Cancer and Degenerative Diseases

To respire, all aerobic organisms, including humans, must take up molecular oxygen (O_2). At normal temperatures, molecular oxygen reacts with very few molecules. However, some of it is converted into more reactive forms, such as superoxide radicals (O_2^-), hydrogen peroxide (H_2O_2), and hydroxyl radicals. Hydrogen peroxide is also produced deliberately in the liver to help detoxify recalcitrant molecules and by lysozomes to kill invading bacteria. Iron in the cell can catalyze the conversion of hydrogen peroxide to hydroxyl radicals ($^{\cdot}OH$), which may be the form in which oxygen is most damaging to DNA (see the text).

As described in the text, many enzymes have evolved to help reduce this damage. In *E. coli*, the major mechanisms for signaling the presence of reactive oxygen species is through OxyR. Interestingly, it is the direct effect of reactive oxygen species on the OxyR protein that causes a specific disulfide bond to form between cysteine residues 199 and 208 within the protein, which changes OxyR from a repressor to a transcriptional activator of the target genes (see Lee et al., 2004, References). While the actual protein involved in signaling oxidative stress is completely different in humans, we employ the same strategy of using disulfide bond formation to detect reactive oxygen species (see Guo et al., References). In this case, human cells use one of the major DNA damage-signaling proteins, ATM. ATM becomes activated to signal DNA repair when a disulfide bond forms a covalent link between dimers of the protein at residue Cys 2991 in the presence of reactive oxygen species.

Accumulation of DNA damage due to reactive oxygen has been linked to many degenerative diseases, such as cancer, arthritis, cataracts, and cardiovascular disease. Overexpression of catalase in the energy-producing mitochondria of mice leads to increases in the life span (see Schriner et al., References). This increase in life span may result from partial mitigation of the age-associated decline in the mitochondria, along with other health benefits, such as a decrease in the incidence of insulin resistance, a precondition associated with type 2 diabetes (see Lee et al., 2010, References).

The importance of internally generated reactive oxygen in cancer has received dramatic confirmation (see Chmiel et al., References). The authors report that a genetic disease characterized by increased rates of colon cancer is due to mutations in the human repair gene *MYH*, which is analogous to the *mutY* gene of *E. coli*. Siblings who have inherited this predisposition to cancer, called familial adenomatous polyposis, are heterozygous for different mutant alleles of the *MYH* gene. The *mutY* gene product of *E. coli* is a specific *N*-glycosylase that removes adenine bases that have mistakenly paired with 8-oxoG in the DNA (see the text), and the human enzyme is known to have a similar activity. Also, mice that have had their *Ogg-1* and *Myh* genes inactivated (so-called knockout mice) are much more prone to lung and ovarian tumors, as well as lymphomas. The Ogg-1 gene product of mice is functionally analogous to MutM of *E. coli*. Furthermore, the primary types of mutations in these knockout mice are GC-to-TA transversions, like they are in *E. coli* (see the text and Xie et al., References).

Obviously, any mechanism for reducing the levels of these active forms of oxygen should increase longevity and reduce the frequency of many degenerative diseases. Fruits and vegetables produce antioxidants, including ascorbic acid (vitamin C); tocopherol (vitamin E); carotenes, such as beta-carotene (in carrots and many other vegetables), lutein (in green leafy vegetables), and lycopene (in tomatoes and other fruits); and multiple forms of vitamin A, that destroy these molecules and thereby protect the DNA in their seeds and the photosynthetic apparatus in their leaves from damage due to oxygen free radicals produced by UV irradiation. Evidence suggests that consumption of the many fruits and vegetables that contain these compounds reduces the rates of cancer and degenerative diseases.

References

Chmiel NH, Livingston AL, David SS. 2003. Insight into the functional consequences of inherited variants of the *hMYH* adenine glycosylase associated with colorectal cancer: complementation assays with *hMYH* variants and pre-steady-state kinetics of the corresponding mutated *E. coli* enzymes. *J Mol Biol* **327**:431–443.

Guo Z, Kozlov S, Lavin MF, Person MD, Paull TT. 2010. ATM activation by oxidative stress. *Science* **330**:517–521.

Lee HY, Choi CS, Birkenfeld AL, Alves TC, Jornayvaz FR, Jurczak MJ, Zhang D, Woo DK, Shadel GS, Ladiges W, Rabinovitch PS, Santos JH, Petersen KF, Samuel VT, Shulman GI. 2010. Targeted expression of catalase to mitochondria prevents age-associated reductions in mitochondrial function and insulin resistance. *Cell Metab* **12**:668–674.

Lee C, Lee SM, Mukhopadhyay P, Kim SJ, Lee SC, Ahn WS, Yu MH, Storz G, Ryu SE. 2004. Redox regulation of OxyR requires specific disulfide bond formation involving a rapid kinetic reaction path. *Nat Struct Mol Biol* **11**:1179–1185.

Schriner SE, Linford NJ, Martin GM, Treuting P, Ogburn CE, Emond M, Coskun PE, Ladiges W, Wolf N, Van Remmen H, Wallace DC, Rabinovitch PS. 2005. Extension of murine life span by overexpression of catalase targeted to mitochondria. *Science* **308**:1909–1911.

Xie Y, Yang H, Cunanan C, Okamoto K, Shibata D, Pan J, Barnes DE, Lindahl T, McIlhatton M, Fishel R, Miller JH. 2004. Deficiencies in mouse *Myh* and *Ogg1* result in tumor predisposition and G to T mutations in codon 12 of the K-ras oncogene in lung tumors. *Cancer Res* **64**:3096–3102.

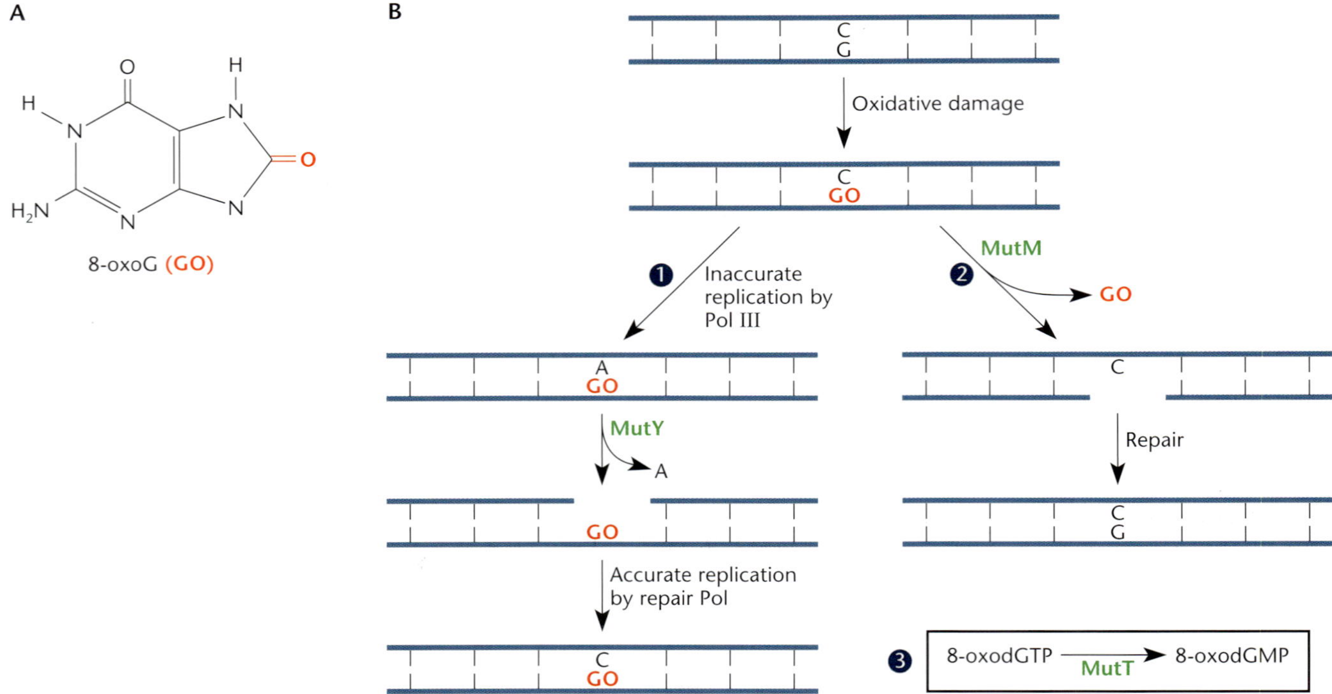

Figure 10.4 **(A)** Structure of 8-oxoG. **(B)** Mechanisms for avoiding mutagenesis due to 8-oxoG (GO). In pathway 1, an A mistakenly incorporated opposite 8-oxoG is removed by a specific glycosylase (MutY), and the strand is degraded and resynthesized with the correct C. In pathway 2, the 8-oxoG is itself removed by a specific glycosylase (MutM), and the strand is degraded and resynthesized with a normal G. In the third pathway, the 8-oxoG is prevented from entering the DNA by a specific phosphatase (MutT) that degrades the triphosphate 8-oxodGTP to the monophosphate 8-oxodGMP. Modified from Michaels ML, Miller JH, *J Bacteriol* **174:**6321–6325, 1992.

The MutY enzyme also apparently recognizes a mismatch that results from accidental incorporation of an A opposite a normal G and removes the A. However, its major role in avoiding mutagenesis in the cell seems to be to prevent mutations due to 8-oxoG. As evidence, mutations that cause the overproduction of MutM completely suppress the mutator phenotype of *mutY* mutants. The interpretation of this result is as follows. If a significant proportion of all spontaneous mutations in a *mutY* mutant resulted from misincorporation of A's opposite normal G's, excess MutM should have no effect on the mutation rate, because removal of 8-oxoG should not affect this type of mispairing. However, the fact that excess MutM almost completely suppresses the extra mutagenesis in *mutY* mutants suggests that very few of the extra spontaneous mutations in a *mutY* mutant are due to A-G mispairs and that most are due to A–8-oxoG mispairs.

MutT

The MutT enzyme operates by a very different mechanism (Figure 10.4): it prevents 8-oxoG from entering the DNA in the first place. The reactive forms of oxygen can oxidize not only guanine in DNA to 8-oxoG, but also the base in dGTP to form 8-oxodGTP. Without MutT, 8-oxodGTP is incorporated into DNA by DNA Pol III, which cannot distinguish 8-oxodGTP from normal dGTP. The MutT enzyme is a phosphatase that specifically degrades 8-oxodGTP to 8-oxodGMP so that it cannot be used in DNA synthesis.

GENETICS OF 8-oxoG MUTAGENESIS

There are a number of ways in which the genetic evidence obtained with *mutM*, *mutT*, and *mutY* mutants is consistent with these functions for the products of the genes. First, these activities explain why the effects of mutations in the genes are additive. If *mutT* is mutated, more 8-oxodGTP will be present in the cell to be incorporated into DNA, increasing the spontaneous-mutation rate. If MutM does not remove these 8-oxoGs from the DNA, spontaneous-mutation rates will increase even further. If MutY does not remove some of the A residues that mistakenly pair with the 8-oxoGs, the spontaneous-mutation rate will be higher yet.

Mutations in the *mutM*, *mutY*, and *mutT* genes also increase the frequency of only some types of mutations, which again can be explained by the activities of these enzymes. For example, only the frequency of GC-to-TA transversion mutations is increased in *mutM* and *mutY* mutants. This is meaningful because, in general, transversions are less common than transition mutations (see chapter 3). The fact that mutations in both genes increase the frequency of the same type of relatively uncommon mutation first suggested that they function in the same pathway and also makes sense considering the functions of the gene products. If MutM does not remove 8-oxoG from DNA, mispairing of the 8-oxoG with A can lead to GC-to-TA transversions. Moreover, GC-to-TA transversions will occur if MutY does not remove the mispaired A residues opposite the 8-oxoGs in the DNA. In contrast, while *mutT* mutations can increase the frequency of relatively uncommon GC-to-TA transversion mutations, they can also increase the frequency of TA-to-GC transversions. This is possible because an 8-oxodGTP molecule, which owes its existence to the lack of MutT to degrade it, may cause mutations in two ways. It may incorrectly enter the DNA by pairing with an A and then, once in the DNA, correctly pair with a C to result in an AT-to-CG transversion, or it can enter the DNA correctly as a G by pairing with a C but then, once in the DNA, pair incorrectly with an A to result in a GC-TA transversion.

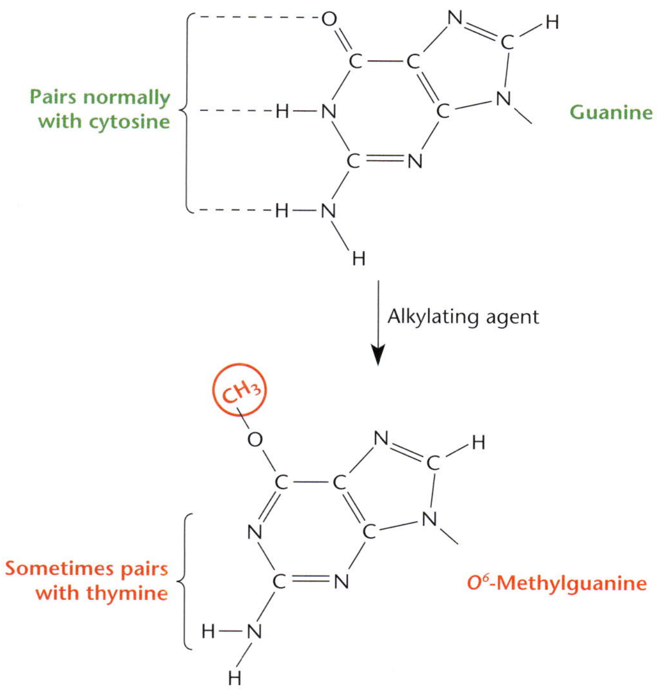

Figure 10.5 Alkylation of guanine to produce O^6-methylguanine. The altered base sometimes pairs with thymine, causing mutations.

Damage Due to Alkylating Agents

Alkylation is another common type of damage to DNA. Both the bases and the phosphates in DNA can be alkylated. The responsible chemicals, known as **alkylating agents**, usually add alkyl groups (CH_3, CH_3CH_2, etc.) to the bases or phosphates in DNA, although any electrophilic reagent that reacts with DNA could be considered an alkylating agent. For example, the anticancer drug *cis*-diamminedichloroplatinum (cisplatin) is an alkylating agent that reacts with guanines in the DNA. Other examples of alkylating agents are ethyl methanesulfonate (EMS), nitrogen mustard gas, methyl methanesulfonate, and *N*-methyl-*N*′-nitro-*N*-nitrosoguanidine (nitrosoguanidine, NTG, or MNNG). Some of these alkylate DNA directly, whereas others react with cellular constituents, such as glutathione, that are supposed to inactivate them but instead convert them into alkylating agents for DNA and worsen their effects. Not all alkylating agents are artificially synthesized; some are produced normally in cells or in the environment. For example, methylchloride, produced in large quantities by marine algae, is a DNA-alkylating agent, as are *S*-adenosylmethionine and methylurea, which are produced as normal cellular metabolites. Given the endogenous sources of alkylating agents, it makes sense that repair systems to deal with alkylation damage to DNA evolved.

Many reactive groups of the bases can be attacked by alkylating agents. The most reactive are N^7 of guanine and N^3 of adenine. These nitrogens can be alkylated by EMS or methyl methanesulfonate to yield methylated or ethylated bases, such as N_7-methylguanine and N_3-methyladenine, respectively. Alkylation of the bases at these positions can severely alter their pairing with other bases, causing major distortions in the helix.

Some alkylating agents, such as nitrosoguanidine, can also attack other atoms in the rings, including the O^6 of guanine and the O^4 of thymine. The addition of a methyl group to these atoms makes O^6-methylguanine (Figure 10.5) and O^4-methylthymine, respectively. Altered bases with an alkyl group at these positions are particularly mutagenic because the helix is not significantly distorted, so the lesions are not recognized by the more general repair systems discussed below. However, the altered base often mispairs, producing a mutation. In this section, we discuss the repair systems specific to these types of alkylated bases.

SPECIFIC *N*-GLYCOSYLASES

Some types of alkylated bases can be removed by specific *N*-glycosylases. The repair pathways involving these enzymes work in the same way as other *N*-glycosylase repair pathways in that the alkylated base is first removed by the specific *N*-glycosylase, and then the apurinic, or AP, DNA strand is cut by an AP endonuclease. Exonucleases degrade the cut strand, which is then resynthesized by DNA Pol I. In *E. coli*, two *N*-glycosylases that remove methylated and ethylated bases from the DNA have been identified. One of these, TagA (for three methyladenine glycosylase A) removes the base 3-methyladenine and some related methylated and ethylated bases. Another, AlkA, is less specific. It not only removes 3-methyladenine from the DNA, but also removes many other alkylated bases, including 3-methylguanine and 7-methylguanine. This enzyme, which is encoded by the *alkA* gene, is induced as part of the adaptive response (see below).

METHYLTRANSFERASES

Other repair systems for alkylated bases act by repairing the damaged base rather than removing it and resynthesizing the DNA. These proteins, called methyltransferases, directly remove the alkyl group from the base by transferring the methyl or other alkyl group from the altered base in the DNA to themselves. Therefore, they are not true enzymes, because they do not catalyze the reaction but rather are consumed during it. Once they have transferred a methyl or other alkyl group to themselves, they become inactive and are eventually degraded. The two major methyltransferases in *E. coli* are Ada and Ogt, sometimes called alkyltransferases I and II, respectively. Both of these proteins repair bases damaged from alkylation of the O^6 carbon of guanine and the O^4 carbon of thymine. Ogt plays the major role when the cells are growing actively, but when the cells reach stationary phase and stop growing, or if the cell is exposed to an external methylating agent, Ada is induced as part of the adaptive response and then becomes the major methyltransferase (see below). The fact that the cell is willing to sacrifice an entire protein molecule to repair a single O^6-methylguanine or O^4-methylthymine lesion is a tribute to the mutagenic potential of these lesions.

AlkB AND AidB

Two other enzymes that repair damage induced by alkylating agents are AlkB and AidB. The enzyme AlkB basically oxidizes the methyl groups on 1-methyladenine and 3-methylcytosine to formaldehyde, releasing them and restoring the normal base. More precisely, it is an α-ketoglutarate-dependent dioxygenase that couples the decarboxylation of α-ketoglutarate to the hydroxylation of the methyl group on 1-methyladenine or 3-methylcytosine to release formaldehyde (see Trewick et al., Suggested Reading). Its cofactor, α-ketoglutarate, is an intermediate in the tricarboxylic acid cycle with many uses in nitrogen assimilation and amino acid biosynthesis and therefore is always available in large quantities. The function of AidB is not proven; nevertheless, the structure of the protein suggests that it is not involved in the direct repair of DNA but instead may have a role in deactivating alkylating agents that would otherwise damage DNA (see Bowles et al., Suggested Reading).

THE ADAPTIVE RESPONSE

The genes whose products repair alkylation damage in *E. coli* are part of the **adaptive response**, which includes the specific *N*-glycosylases and methyltransferases. The products of these genes are normally synthesized in small amounts, but they are produced in much greater amounts if the cells are exposed to an alkylating agent. The name "adaptive response" comes from early evidence suggesting that *E. coli* "adapted" to damage caused by alkylating agents. If *E. coli* cells are briefly treated with an alkylating agent, such as nitrosoguanidine, they will be better able to survive subsequent treatments with this and other alkylating agents. We now know that the cell adapts to the alkylating agents by inducing the expression of a number of genes whose products are involved in repairing alkylation damage to DNA. The adaptive-response genes seem to be the most important for conferring resistance to alkylating agents that transfer methyl (CH_3) groups to DNA. Resistance to alkylating agents that transfer longer groups, such as ethyl (CH_3CH_2) groups, to DNA seems to be due mostly to excision repair (see below).

Regulation of the Adaptive Response

Treatment of *E. coli* cells with an alkylating agent causes the levels of some of the proteins involved in repairing alkylation damage to increase from a few to many thousands of copies. The genes induced as part of the adaptive response include *ada*, *aidB*, *alkA*, and *alkB*, discussed above. The regulation is achieved through the state of methylation of one of the alkylation-repairing proteins, the Ada protein (Figure 10.6). The *ada* gene is part of an operon with *alkB*, while the *aidB* and *alkA* genes are transcribed separately. The Ada protein can regulate the transcription of its own gene, as well as the other genes under its control, because in addition to its role in repairing alkylation damage to DNA, it is a transcriptional activator. However, the Ada protein becomes a transcriptional activator only if the alkylation damage is quite extensive. It can discern the level of damage in the DNA by having two amino acids to which methyl groups can be transferred, cysteine-321 (the 321st amino acid from its N terminus) and cysteine-38 (the 38th amino acid from its N terminus). After mild alkylation damage has occurred, most of the methylation is confined to the bases; these methyl groups can be transferred from either O^6-methylguanine or O^4-methylthymine to cysteine-321, close to the C terminus. This removes the methyl groups from the damaged DNA bases but inactivates the Ada protein as far as transfer of more methyl groups from the bases is concerned. At higher levels of damage, some of the phosphates in the DNA background also become methylated in the form of phosphomethyltriesters, and these methyl groups can be transferred only to cysteine-38, close to the N terminus. The presence of a methyl group on cysteine-38 converts Ada

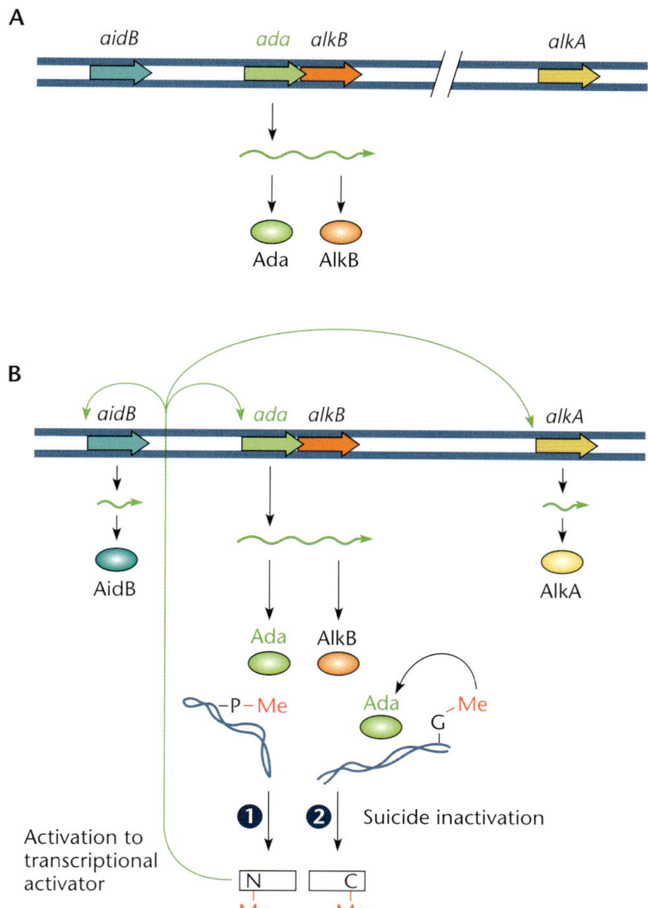

Figure 10.6 **(A)** The adaptive response. **(B)** Regulation of the adaptive response. Only a few copies of the Ada protein normally exist in the cell. After damage due to alkylation, the Ada protein, a methyltransferase, transfers alkyl groups from methylated DNA phosphates to an amino acid in the N terminus (N) of itself, converting itself into a transcriptional activator (pathway 1), or from a methylated base to an amino acid in its C terminus (C), inactivating itself (pathway 2). See the text for details.

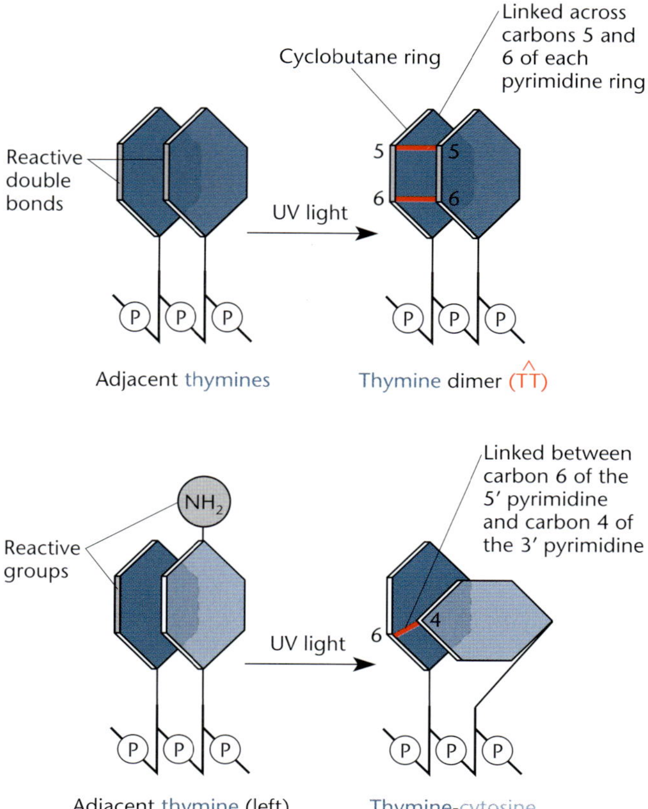

Figure 10.7 Two common types of pyrimidine dimers caused by UV irradiation. In the top diagram, two adjacent thymines are linked through the 5- and 6-carbons of their rings to form a cyclobutane ring. In the bottom diagram, a 6-4 dimer is formed between the 6-carbon of a cytosine and the 4-carbon of a thymine 3′.

into a transcriptional activator; modification at cysteine-38 alleviates a repulsive DNA-protein charge and therefore allows the protein to bind to a specific DNA sequence in the promoter region of its target genes, permitting the modified protein to activate transcription of its own gene, as well as the other genes under its control (see He, Suggested Reading). Transcriptional activators are discussed in detail in chapter 11.

Damage Due to UV Irradiation

One of the major sources of natural damage to DNA is UV irradiation due to sun exposure. Every organism that is exposed to sunlight must have mechanisms to repair UV damage to its DNA. The conjugated-ring structure of the bases in DNA causes them to strongly absorb light in the UV wavelengths. The absorbed photons energize the bases, causing their double bonds to react with other nearby atoms and hence to form additional chemical bonds. These chemical bonds result in abnormal linkages between the bases in the DNA and other bases or between bases and the sugars of the nucleotides.

One common type of UV irradiation damage is the **pyrimidine dimer**, in which the rings of two adjacent pyrimidines become fused (Figure 10.7). In one of the two possible dimers, the carbon atoms at positions 5 and 6 of two adjacent pyrimidines are joined to form a **cyclobutane ring**. In the other type of dimer, the carbon at position 6 of one pyrimidine is joined to the carbon at position 4 of an adjacent pyrimidine to form a **6-4 lesion**.

PHOTOREACTIVATION OF CYCLOBUTANE DIMERS

A special type of repair system called **photoreactivation** provides an efficient way to repair cyclobutane-type pyrimidine dimers due to UV irradiation. The photoreactivation repair systems separate the fused bases of the cyclobutane pyrimidine dimers rather than replacing them. This mechanism is named photoreactivation because this type of repair occurs only in the presence of visible light (and so used to be called light repair).

In fact, photoreactivation was the first DNA repair system to be discovered. In the 1940s, it was observed that the bacterium *Streptomyces griseus* was more likely to survive UV irradiation in the light than in the dark. Photoreactivation is now known to exist in most organisms on Earth that are exposed to UV irradiation, with the important exception of placental mammals, such as humans. Humans (and other placental mammals) instead appear to rely more heavily on nucleotide excision repair (see below).

The mechanism of action of photoreactivation is shown in Figure 10.8. The enzyme responsible for the repair is called **photolyase**. This enzyme, which contains a reduced flavin adenine dinucleotide group that absorbs light with wavelengths between 350 and 500 nm, binds to the fused bases. Absorption of light then gives photolyase the energy it needs to separate the fused bases.

Figure 10.8 Photoreactivation. The photoreactivating enzyme (photolyase) binds to cyclobutane pyrimidine dimers (red), even in the dark. Absorption of light by the photolyase causes it to cleave the bond between the two pyrimidines, restoring the bases to their original form.

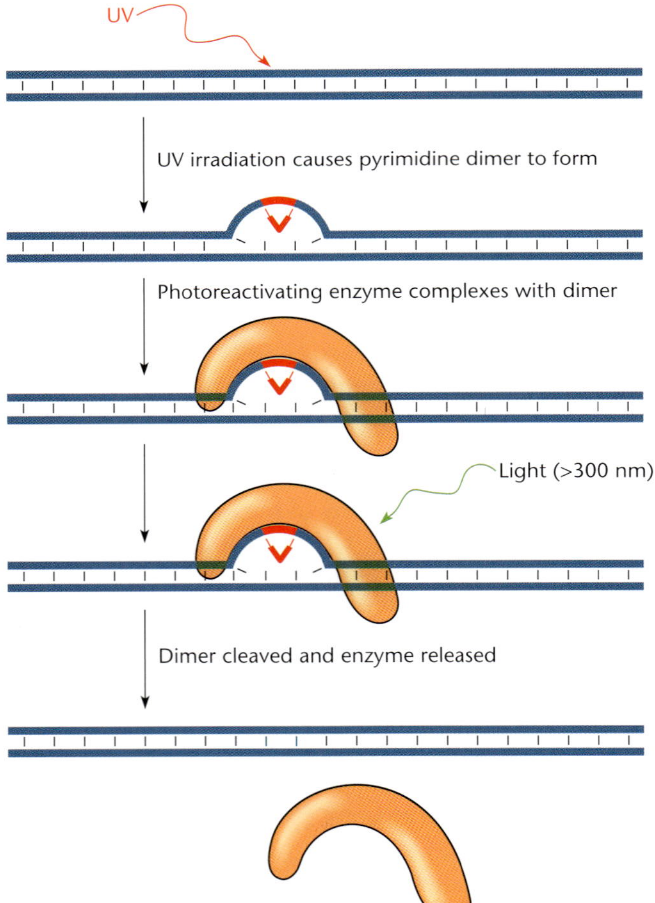
UV
UV irradiation causes pyrimidine dimer to form
Photoreactivating enzyme complexes with dimer
Light (>300 nm)
Dimer cleaved and enzyme released

N-GLYCOSYLASES SPECIFIC TO PYRIMIDINE DIMERS

There are also specific *N*-glycosylases that recognize and remove pyrimidine dimers. This repair mechanism is similar to the mechanisms for deaminated and alkylated bases discussed above and involves AP endonucleases or lyases and the removal and resynthesis of strands of DNA containing the dimers.

General Repair Mechanisms

As mentioned above, not all repair mechanisms in cells are specific for a certain type of damage to DNA. Some types of repair systems can repair many different types of damage. Rather than recognizing the damage itself, these repair systems recognize distortions in the DNA structure caused by improper base pairing and repair them, independent of the type of damage that caused the distortion. There are a number of factors that can cause slight distortions in DNA, including 8-oxoG, incorporation of base analogs, frameshifts that occur by polymerase slippage, some types of alkylation damage, and mismatches. In general, DNA damage that causes only a minor distortion of the helix is repaired by the methyl-directed mismatch repair system, whereas other pathways, including nucleotide excision (see below), repair damage that causes more significant distortions.

Base Analogs

Base analogs are chemicals that resemble the normal bases in DNA. Because they resemble the normal bases, these analogs are sometimes converted into a deoxynucleoside triphosphate and enter the DNA. Incorporation of a base analog can be mutagenic, because the analog often pairs with the wrong base, leading to base pair changes in the DNA. Figure 10.9 shows two base analogs, 2-aminopurine (2-AP) and 5-bromouracil (5-BU). 2-AP resembles adenine, except that it has the amino group at the 2 position rather than at the 6 position. The other base analog pictured, 5-BU, resembles thymine, except for a bromine atom instead of a methyl group at the 5 position.

Figure 10.10 shows how mispairing by a base analog might cause a mutation. In the illustration, the base analog 2-AP has entered a cell and has been converted into the nucleoside triphosphate. The deoxyribose 2-AP triphosphate is then incorporated during synthesis of DNA, sometimes pairing with cytosine in error. Which type of mutation occurs depends on when the 2-AP mistakenly

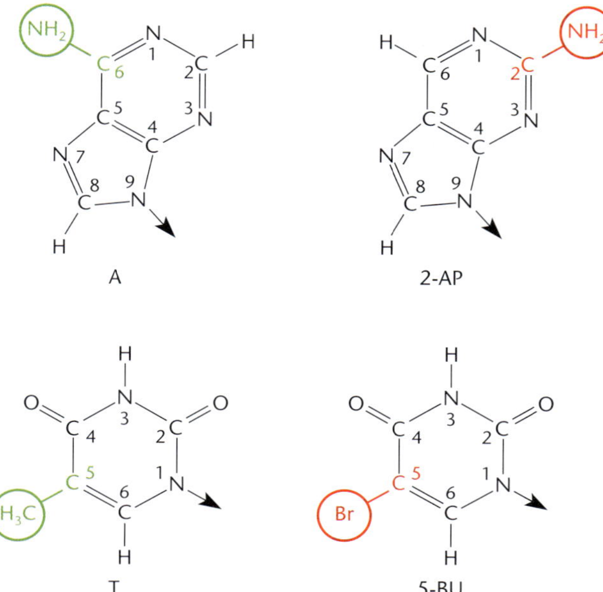

Figure 10.9 Base analogs 2-aminopurine (2-AP) and 5-bromouracil (5-BU). The amino groups that are at different positions in adenine (A) and 2-AP are indicated, as are the methyl group in thymine (T) and the bromine group in 5-BU.

pairs with cytosine. If the 2-AP enters the DNA by mistakenly pairing with C instead of T, in subsequent replications it usually pairs correctly with T. This causes a GC-to-AT transition mutation. However, if it is incorporated properly by pairing with T but pairs mistakenly with C in subsequent replications, an AT-to-GC transition mutation is the result. Similar arguments can be made for 5-BU, which sometimes mistakenly pairs with cytosine instead of adenine.

Frameshift Mutagens

Another type of damage repaired by the methyl-directed mismatch repair system is those caused by the incorporation of frameshift mutagens, which are usually planar molecules of the acridine dye family. These chemicals are mutagenic because they intercalate between bases in the same strand of the DNA, likely by increasing the distance between the bases and preventing them from aligning properly with bases on the other strand. The frameshift mutagens include acridine dyes, such as 9-aminoacridine, proflavine, and ethidium bromide, as well as some aflatoxins made by fungi.

Figure 10.11 illustrates a model for mutagenesis by a frameshift mutagen. In the model, intercalation of the dye forces two of the bases apart, causing the two strands to slip with respect to each other. One base is thus paired with the base next to the one with which it previously paired. This slippage is most likely to occur

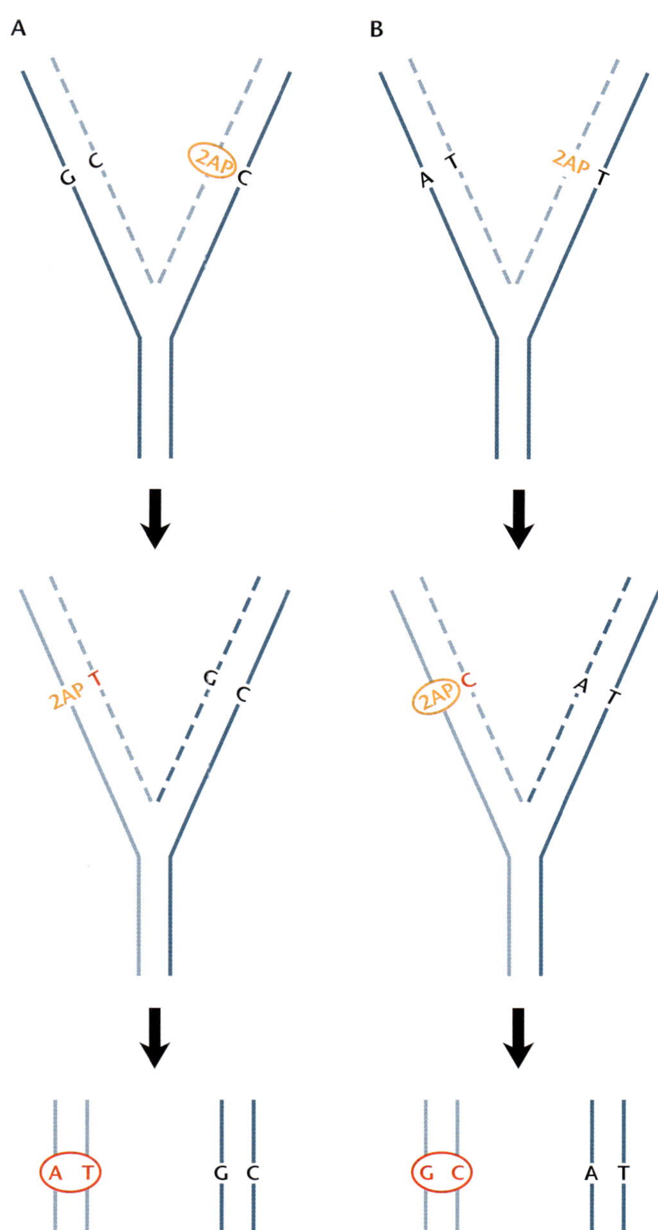

Figure 10.10 Mutagenesis by incorporation of the adenine analog 2-AP into DNA. The 2-AP is first converted to the deoxyribose nucleoside triphosphate and is then inserted into the DNA. **(A)** The analog incorrectly pairs with a C during its incorporation into the DNA strand. **(B)** The analog is incorporated correctly opposite a T but mispairs with C during subsequent replication. The mutation is circled in both panels.

where a base pair in the DNA is repeated, for example, in a string of AT or GC base pairs. Whether a deletion or addition of a base pair occurs depends on which strand slips, as shown in Figure 10.11. If the dye is intercalated into the template DNA prior to replication, the newly synthesized strand might slip and incorporate an extra

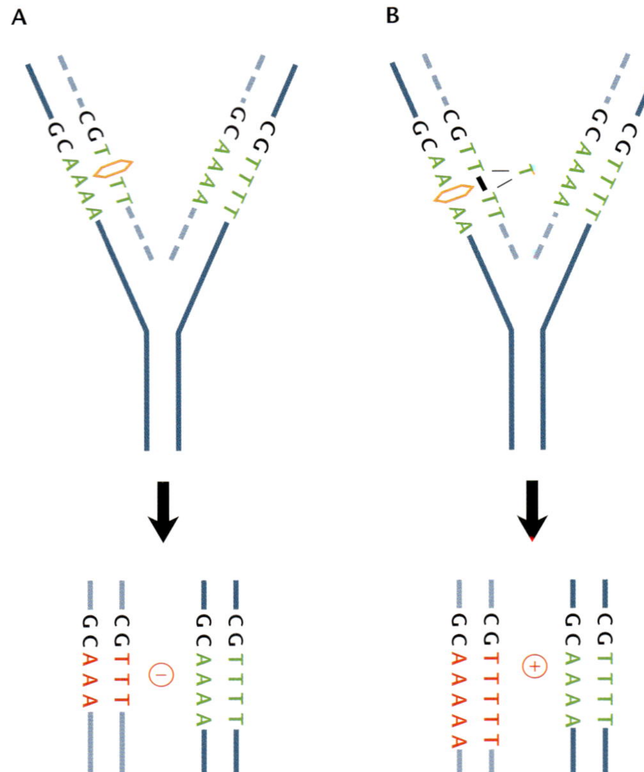

Figure 10.11 Mutagenesis by a frameshift mutagen. Intercalation of a planar acridine dye molecule between two bases in a repeated sequence in DNA forces the bases apart and can lead to slippage. **(A)** The dye inserts itself into the new strand, resulting in deletion of a base pair (–). **(B)** The dye comes into the old strand, adding a base pair (+).

nucleotide. However, if it is incorporated into the newly synthesized strand, the strand might slip backward, leaving out a base pair in subsequent replication. A variation on this idea is that slippage basically occurs at the same rate but that asymmetric intercalation of the compound stabilizes the structure with the extra base on one strand during replication.

Mismatch Repair

One of the major pathways for avoiding mutations in *E. coli* is the **mismatch repair system**. Mismatch repair systems are most important immediately following DNA replication (see Liao et al., Suggested Reading). This is because in spite of the vigilance of the editing functions, sometimes the wrong base pair is inserted into DNA (see chapter 1). Mismatch repair allows the cell another chance to prevent a permanent mistake or mutation. This system recognizes the mismatch and removes it, as well as DNA in the same strand around the mismatch, leaving a gap in the DNA that is filled in by the action of DNA polymerase, which inserts the correct nucleotide.

The mismatch repair system is very effective at removing mismatches from DNA. However, by itself, it would not significantly lower the rate of spontaneous mutagenesis unless it repaired the correct strand of DNA at the mismatch. In the example shown in Figure 10.12, a T was mistakenly incorporated opposite a G. If the mismatch repair system changes the T to the correct C in the newly replicated DNA, a GC base pair will be restored at this position, and no change in the sequence or mutation will occur. However, if it repairs the G in the old DNA in the mismatch to an A, the mismatch will have been removed and replaced by an AT base pair with correct pairing, but a GC-to-AT change would have occurred in the DNA sequence at the site of the mismatch, creating a mutation. To prevent mutations, the mismatch repair system must have some way of distinguishing the newly synthesized strand from the old strand so that it can replace the base in the right DNA strand.

Different organisms seem to use different mechanisms to distinguish the new and old strands during the process of mismatch repair. Some organisms, including *E. coli*, use specific changes in the state of methylation of the DNA strands to allow the mismatch repair system to distinguish the new strand from the old strand after replication. In *E. coli* and some other members of the *Gammaproteobacteria*, the A residues in the symmetric sequence GATC/CTAG are methylated at the 6′ position of the larger of the two rings of the adenine base. These methyl groups are added to the bases by the enzyme deoxyadenosine methylase (Dam methylase) (see chapter 1), but this occurs only after the nucleotides have been incorporated into the DNA. Since DNA replicates by a semiconservative mechanism, the A in the GATC/CTAG sequence in the newly synthesized strand remains temporarily unmethylated after replication of a region containing the sequence. The DNA at this site is said to be hemimethylated if the bases on only one strand are methylated. Figure 10.12 shows that a hemimethylated GATC/CTAG sequence indicates to the mismatch repair system which strand is newly synthesized and should be repaired. For this reason, the repair system is called the methyl-directed mismatch repair system. The use of Dam-mediated hemimethylation to direct the mismatch repair system to the newly synthesized strand seems to be restricted to certain groups in the *Gammaproteobacteria*, since most bacteria and eukaryotes do not have a methylation-sensing protein like that found in *E. coli* or the Dam methylase (see below). Nevertheless, nearly all organisms do possess a mismatch repair system, indicating that other mechanisms exist to distinguish the new from the old strand immediately following DNA replication.

The methyl-directed mismatch repair system in *E. coli* requires the products of the *mutS*, *mutL*, and *mutH* genes. Like the *mut* genes whose products are involved in avoiding mutagenesis due to 8-oxoG, these *mut* genes were

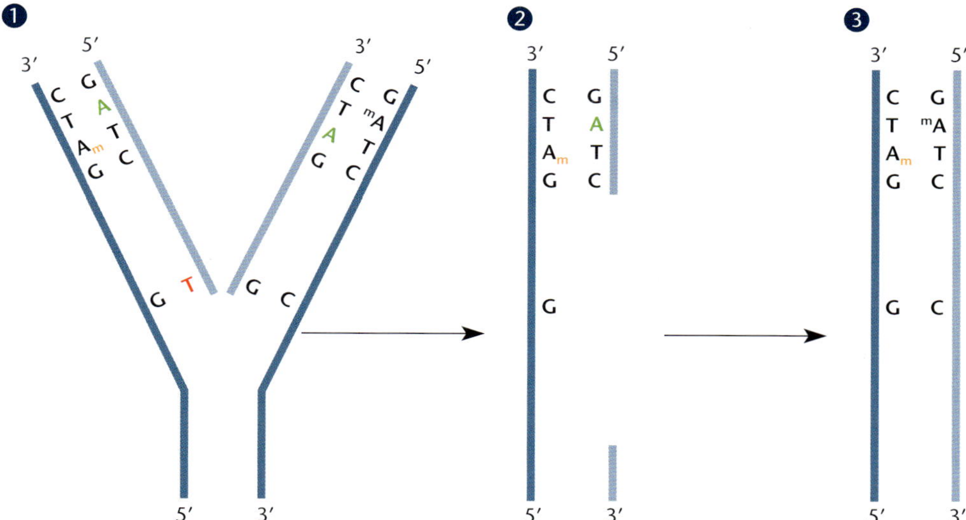

Figure 10.12 The methyl-directed mismatch repair system. The newly replicated DNA contains a GT mismatch **(1)**. The newly synthesized strand is recognized because it is not methylated at the nearby GATC sequence, and the T in the mismatch is removed, along with neighboring sequences **(2)**. The sequence is resynthesized, replacing the T with the correct C. The neighboring GATC sequence is then methylated by the Dam methylase **(3)**. The newly synthesized strand is shown in light blue.

found in a search for mutations that increase spontaneous-mutation rates (see below). It is somewhat mysterious how the mismatch repair system can use the state of methylation of the GATC/CTAG sequence to direct itself to the newly synthesized strand, even though the nearest GATC/CTAG sequence is probably some distance from the alteration. Figure 10.13 presents a model that has been used to explain this mechanism. First, a dimer of the MutS protein binds to the alteration in the DNA that is causing a minor distortion in the helix (marked with an X in the figure). A nearby GATC sequence is still unmethylated on the newly synthesized strand. Then, two copies of MutL bind to MutS, and a copy of MutH binds to the MutS-MutL complex. This binding activates the MutH nuclease to cut the nearest hemimethylated GATC sequence in the newly synthesized unmethylated strand, and exonucleases degrade the DNA past the site of the original mismatch. DNA Pol III then fills the gap, and ligase seals the break. This model is supported by experiments showing that the mismatch repair system preferentially repairs mismatches on hemimethylated DNA by correcting the sequence on the unmethylated strand of DNA to match the sequence on the methylated strand (see below and Pukkila et al., Suggested Reading). How, then, does the cell know in what direction to degrade the DNA? The DNA is degraded in either the 3′-to-5′ or 5′-to-3′ direction, depending on which side of the nearest GATC sequence the mismatch has occurred. According to the model, *E. coli* solves this problem by using four different exonucleases, two of which, ExoVII and RecJ, can degrade only in the 5′-to-3′ direction and two of which, ExoI and ExoX, can degrade only in the 3′-to-5′ direction. Also, these exonucleases can degrade only single-stranded DNA, so a helicase is needed to separate the strands. This is the job of the UvrD helicase, which is a general helicase also used by the excision repair system (see below). After the strand containing the mismatch has been degraded, DNA Pol III resynthesizes a new strand, using the other strand as the template and removing the cause of the distortion.

While this model more or less explains the genetic and biochemical evidence concerning the mismatch repair system (see below), evidence is accumulating that it may not be the whole story. For one thing, it is somewhat surprising that DNA Pol III, the replication DNA polymerase, is used for this repair. Most other repair reactions use DNA Pol I, the normal repair enzyme. How does DNA Pol III load itself on the DNA at the gap created by the exonucleases? It normally loads on DNA only at the origin of replication or with the help of Pri proteins at recombinational intermediates at blocked replication forks. Suspicions are also raised by the fact that other organisms seem to be able to identify the new strand for mismatch repair without the help of Dam methylation. Mismatch repair systems are universal, as is the ability to distinguish the newly synthesized strand from the old strand, even though most organisms, even most types of bacteria, do not have a Dam methylase and so could not use hemimethylation to identify the new strand. There is also evidence that the mismatch repair system is in much closer contact with the replicating polymerase than is indicated by the model. MutS may bind to the β clamp that holds the replicating DNA polymerase on the DNA, and MSH, the eukaryotic equivalent of MutS (Box 10.2), binds to the proliferating-cell nuclear antigen protein, which is the eukaryotic equivalent of the β clamp protein in bacteria. The MutL protein in *Bacillus subtilis* interacts with the β clamp, and this interaction regulates an endonuclease activity in the protein, providing nuclease activity similar to that provided by MutH in *E. coli* (see Pillon et al., Suggested Reading). It will be interesting to see how new evidence obtained with the bacterial systems furthers our understanding of this most interesting and important of DNA repair mechanisms, especially considering its role in preventing cancer in humans (Box 10.2).

GENETIC EVIDENCE FOR METHYL-DIRECTED MISMATCH REPAIR

Models like the one presented in Figure 10.12 are based on biochemical and genetic evidence. The results of these experiments support the conclusion that the state of methylation of GATC sequences in the DNA helps direct the mismatch repair system to the newly synthesized strand of DNA. Some of this evidence is briefly reviewed in this section.

Isolation of *mut* Mutants

The *mut* genes of *E. coli* were discovered because mutations in these genes increase spontaneous-mutation rates. The resulting phenotype is often referred to as the "mutator phenotype." Other mutations with this phenotype, including the *mutT*, *mutY*, and *mutM* mutations in the genes for reducing mutations due to oxygen damage to DNA and the *mutD* mutations in the gene for the editing function of DNA polymerase, are discussed in other sections. Mutations in the *uvrD*, *vsr*, and *dam* genes of

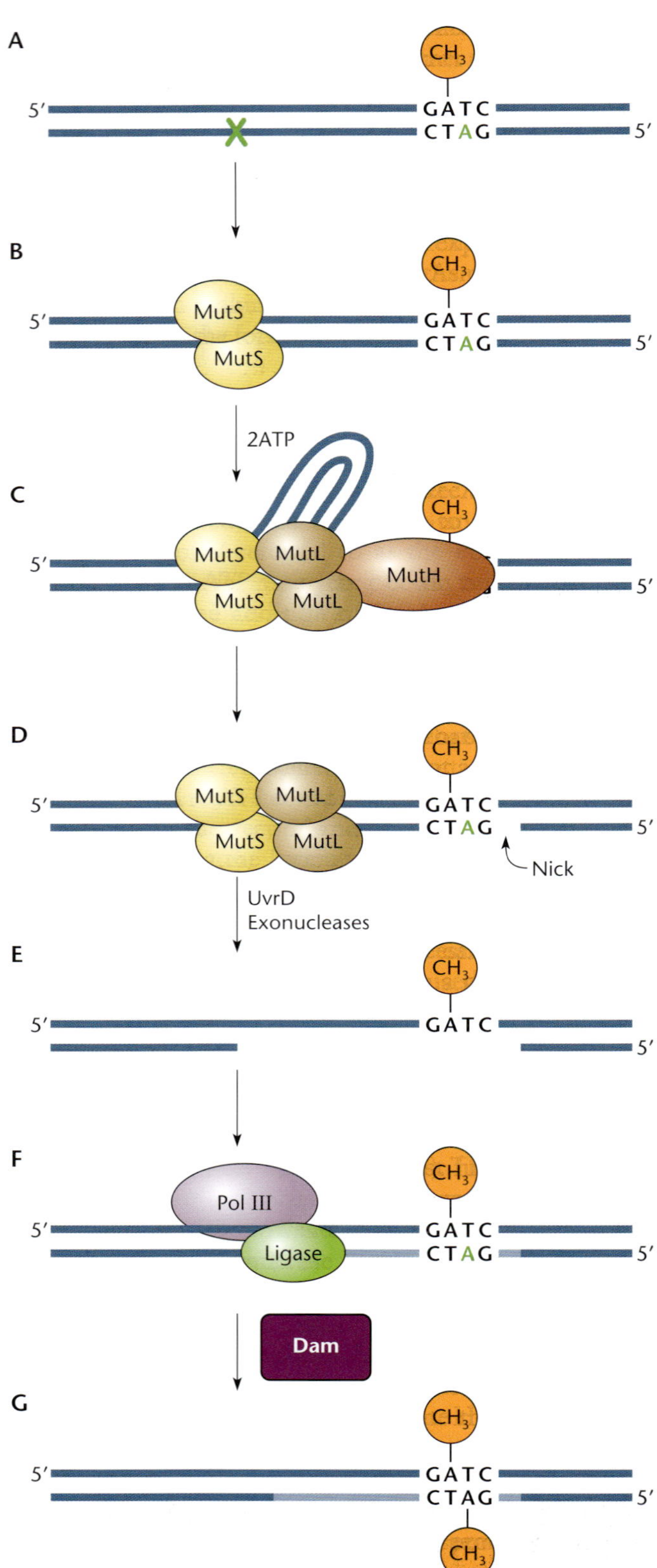

Figure 10.13 MutSLH DNA repair in *E. coli.* **(A)** One arm of a replication fork is shown at the top of the figure, with methylated and unmethylated GATC sequences and a replication mistake (X) generating a base-base or deletion/insertion mismatch. **(B)** The mismatch is bound by MutS. **(C)** In an ATP-dependent reaction, a ternary complex is formed with MutS, MutL, and MutH proteins. **(D)** Incision by activated MutH occurs in the newly synthesized strand at the unmethylated GATC sequence. **(E)** The nick is extended into a gap by excision in either the 3′-to-5′ direction or the 5′-to-3′ direction. Only the 5′-to-3′ direction is depicted. The gap is formed by the actions of exonucleases, including exonuclease I (ExoI), ExoVII, ExoX, and RecJ, and the direction of excision is determined by the UvrD helicase. **(F)** Resynthesis is accomplished by the DNA Pol III holoenzyme, and the nick is sealed by DNA ligase. **(G)** Subsequent methylation by Dam completes the process. Modified from Marinus MG, *in* Higgins NP (ed), *The Bacterial Chromosome* (ASM Press, Washington, DC, 2005).

BOX 10.2

DNA Repair and Cancer

Cancer is a multistep process that is initiated by mutations in oncogenes and other tumor-suppressing genes. It is not surprising, therefore, that DNA repair systems form an important line of defense against cancer. All organisms probably have mismatch repair systems that help reduce mutagenesis. However, unlike bacteria, which have only one or two different MutS proteins and one MutL protein, humans have at least five MutS analogs and four MutL homologs (see Kang et al., References). Attention was focused on the role of the mismatch repair system in cancer with the discovery that people with a mutation in a mismatch repair gene are much more likely to develop some types of cancer, including cancers of the colon, ovary, uterus, and kidney. One such genetic predisposition, called Lynch syndrome, results from mutations in a gene called *hMSH2* (for human *mutS* homolog 2) and leads to an increase in the incidence of certain types of colon cancer. This gene was first suspected to be involved in mismatch repair because people who inherited the mutant gene showed a higher frequency of short insertions and deletions that should normally be repaired by the mismatch repair system (see the text). People with this hereditary condition were found to have inherited a mutant form of a gene homologous to the *mutS* gene of *E. coli.*

Another reason why the mismatch repair system is relevant to cancer research in humans is the role it plays in making cells sensitive to cancer therapeutic agents. Apparently, much of the toxicity of some antitumor agents, such as cisplatin and other alkylating agents, is due to the mismatch repair system. When tumor cells become resistant to the drug, it is often because they have acquired a mutation in a human *mut* gene (see Karran, References). Interestingly, tumor cells that are resistant to cisplatin can be resensitized to the drug by suppressing a Y polymerase (such as the *E. coli* translesion Pol IV and V) in human Rev1 cells (see Doles et al., References). This indicates that translesion polymerases may be a target to make existing cancer therapies more effective.

Many other examples exist where processes involved in various forms of DNA repair discovered in bacteria are being found to play important roles in DNA repair in humans, and when defective, they lead to cancer. Mutations found in genes encoding the proteins BRCA1 and BRCA2, which are involved in repair involving homologous recombination, are associated with increases in the incidence of breast and ovarian cancer. The condition xeroderma pigmentosum stems from mutations in the nucleotide excision repair system in humans, resulting in extreme UV light sensitivity and skin cancer predisposition. Glioblastoma, a severe type of brain cancer, is associated with downregulation of the MGMT enzyme that repairs damage due to alkylating agents. These and numerous other examples indicate how findings from highly tractable bacterial systems will continue to help elucidate the link between DNA repair and cancer and cancer treatment.

References

Doles J, Oliver TG, Cameron ER, Hsu G, Jacks T, Walker GC, Hemann MT. 2010. Suppression of Rev3, the catalytic subunit of Polζ, sensitizes drug-resistant lung tumors to chemotherapy. *Proc Natl Acad Sci USA* **107:**20786–20791.

Fishel R, Lescoe MK, Rao MRS, Copeland NG, Jenkins NA, Garber J, Kane M, Kolodner R. 1993. The human mutator gene homolog MSH2 and its association with hereditary nonpolyposis colon cancer. *Cell* **75:**1027–1038.

Gradia S, Subramanian D, Wilson T, Acharya S, Makhov A, Griffith J, Fishel R. 1999. hMSH2-hMSH6 forms a hydrolysis-independent sliding clamp on mismatched DNA. *Mol Cell* **3:**255–261.

Kang J, Huang S, Blaser MJ. 2005. Structural and functional divergence of MutS2 from bacterial MutS1 and eukaryotic MSH4-MSH5 homologs. *J Bacteriol* **187:**3528–3537.

Karran P. 2001. Mechanisms of tolerance to DNA damaging therapeutic drugs. *Carcinogenesis* **22:**1931–1937.

E. coli also increase the rates of at least some spontaneous mutations, although these genes were found in other ways and so were not named *mut*.

A common method for detecting mutants with abnormally high mutation rates is colony papillation. This scheme is based on the fact that all of the descendants of a bacterium with a particular mutation are mutants of the same type. Colonies grow from the middle out, so if a mutation occurs in a growing colony, the mutant descendants of the original mutant stay together to form a sector, or papilla, as the colony grows. A growing colony contains many papillae composed of mutant bacteria of various types. A *mut* mutation increases the frequency of papillation for many types of mutants, so the phenotype used in a colony papillation test is purely a matter of convenience.

Lac$^+$ revertants of *lac* mutations in *E. coli* are an obvious choice for papillation tests, because the revertants form conspicuous blue papillae on 5-bromo-4-chloro-3-indolyl-β-D-galactopyranoside (X-Gal) plates. Figure 10.14 shows a colony from a strain with a *mutY* mutation along with wild-type colonies in a papillation test using reversion of a *lac* mutation as an indicator of mutator activity. If the bacteria forming a colony are *mut* mutants with a higher-than-normal spontaneous-mutation

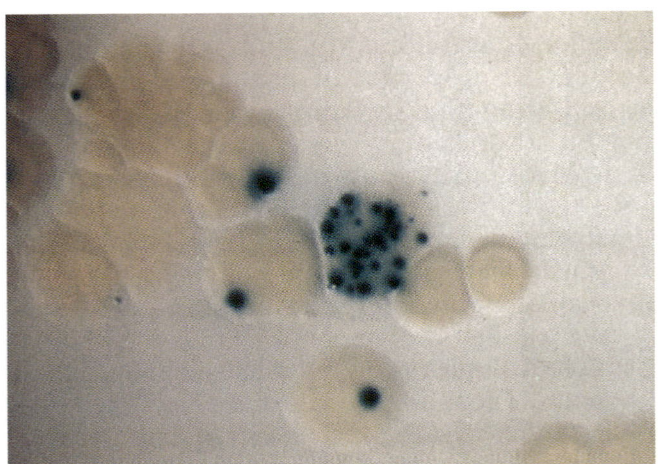

Figure 10.14 Colonies due to *mut* mutants have more papillae. A *lacZ* mutant was plated on medium containing X-Gal (5-bromo-4-chloro-3-indolyl-β-D-galactopyranoside) and a low amount of lactose. On this medium, revertants of the *lacZ* mutation produce blue papillae capable of overgrowing the white Lac$^-$ cells in a colony of bacteria. The *mut* mutants produce more blue papillae than normal owing to increased spontaneous-mutation frequencies. Reprinted with permission from Nghiem Y, Cabrera M, Cupples CG, Miller JH, *Proc Natl Acad Sci USA* **85:**2709–2713, 1988.

frequency, the colony will have more blue papillae than normal. This is how many of the *mut* genes we have discussed were detected.

Once a number of *mut* genes were detected, they were classified into complementation groups and arbitrarily assigned letters (see chapter 3). The mutations were also combined in all of the available permutations to see how they interacted. The products of three of the *mut* genes, *mutS*, *mutL*, and *mutH*, were predicted to participate in the same repair pathway because of the observation that double mutants did not exhibit higher spontaneous-mutation rates than cells with mutations in each of the single genes alone. In other words, the effect of inactivating two of these *mut* genes by mutation is not additive. Generally, mutations that affect steps of the same pathway do not have additive effects.

Among other experiments, evidence supporting the role of methylation in directing the mismatch repair system came from genetic studies of 2-AP sensitivity in *E. coli* (see Glickman and Radman, Suggested Reading). These experiments were based on the observation that a *dam* mutation makes *E. coli* particularly sensitive to killing by 2-AP. This is expected from the model, since as mentioned above, cells incorporate 2-AP indiscriminately into their DNA because they cannot distinguish it from the normal base adenine. The cellular mismatch repair system repairs DNA containing 2-AP because the incorporated 2-AP causes a slight distortion in the helix. Since the 2-AP is incorporated into the newly synthesized strand during the replication of the DNA, the strand containing the 2-AP is normally transiently unmethylated, so that the 2-AP-containing strand is repaired, removing the 2-AP. In a *dam* mutant, however, neither strand is methylated, so the mismatch repair system cannot tell which strand was newly synthesized and may try to repair both strands. The rationale was that if two 2-APs are incorporated close enough to each other, the mismatch repair system may try to simultaneously remove the two mismatches by repairing the opposite strands. However, in this scenario, cutting two strands at sites opposite each other could cause extensive double-strand breaks in the DNA, which may kill the cell.

One prediction based on this proposed mechanism for the sensitivity of *dam* mutants to 2-AP is that the toxicity of 2-AP in *dam* mutants should be reduced if the cells also have a *mutL*, *mutS*, or *mutH* mutation. Without the products of the mismatch repair enzymes, the DNA is not cut on either strand, much less on both strands simultaneously. The cells may suffer higher rates of mutation, but at least they will survive. This prediction was fulfilled. Double mutants that have both a *dam* mutation and a mutation in one of the three *mut* genes were much less sensitive to 2-AP than were mutants with a *dam* mutation alone. Furthermore, *mutL*, *mutH*, and *mutS* mutations can be isolated as suppressors of *dam* mutations on media containing 2-AP. In other words, if large numbers of *dam* mutant *E. coli* cells are plated on medium containing 2-AP, the bacteria that survive are often double mutants with the original *dam* mutation and a spontaneous *mutL*, *mutS*, or *mutH* mutation.

ROLE OF THE MISMATCH REPAIR SYSTEM IN PREVENTING HOMEOLOGOUS AND ECTOPIC RECOMBINATION

As discussed in chapter 3, some DNA rearrangements, such as deletions and inversions, are caused by recombination between similar sequences in different places in the DNA. This is called **ectopic recombination**, or "out-of-place" recombination. Many sequences at which ectopic recombination occurs are similar but not identical. Also, recombination between DNAs from different species often occurs between sequences that are similar but not identical. In general, recombination between similar but not identical sequences is called **homeologous recombination**. Recall that recombination between identical sequences is called homologous recombination.

The mismatch repair system helps reduce ectopic and other types of homeologous recombination. As evidence, recombination between similar but unrelated bacteria, such as *E. coli* and *Salmonella enterica* serovar Typhimurium, is greatly enhanced if the recipient cell has a *mutL*, *mutH*, or *mutS* mutation. Also, the frequency of deletions and other types of DNA rearrangements is enhanced among bacteria with a *mutL*, *mutH*, or *mutS* mutation, since these rearrangements often depend on recombination between similar but not identical sequences. The frequency of mutations engineered using recombineering is also enhanced by *mutS* mutations (see chapter 9), since this also depends on homeologous recombination between the DNA introduced into the cell by electroporation and the endogenous DNA. It is incompletely understood how the mismatch repair system actually inhibits homeologous and ectopic recombination. However, mismatches recognized by MutSL will recruit UvrD, which can unwind homeologous structures to reject incomplete matches, providing part of the explanation for this process (see Tham et al., Suggested Reading).

Nucleotide Excision Repair

One of the most important general repair systems in cells is **nucleotide excision repair**, so named because the entire damaged nucleotides are cut out of the DNA and replaced. This type of repair is very efficient and seems to be common to most types of organisms. It is also relatively nonspecific and is responsible for repairing many different types of damage. Because of its efficiency and relative lack of specificity, the nucleotide excision repair system is very important to the ability of the cell to survive damage to its DNA.

The nucleotide excision repair system is relatively nonspecific because, like the mismatch repair system, it recognizes general changes in the normal DNA helix that result from damage, rather than the chemical structure of the damage itself. This makes it capable of recognizing and repairing damage to DNA as diverse as most types of alkylation and almost all the types of damage caused by UV irradiation, including cyclobutane dimers, 6-4 lesions, and base-sugar cross-links. Nucleotide excision repair can collaborate with transcription in multiple ways to allow the RNA polymerase elongation complex to help survey the chromosome for DNA damage. Nucleotide excision repair also collaborates with recombinational repair to remove cross-links formed between the two strands by some chemical agents, such as psoralens, cisplatin, and mitomycin. However, because nucleotide excision repair recognizes only major distortions in the helix, it does not repair lesions such as base mismatches, O^6-methylguanine, O^4-methylthymine, 8-oxoG, or base analogs, all of which result in only minor distortions, and therefore these lesions must be repaired by other repair systems.

MECHANISM OF NUCLEOTIDE EXCISION REPAIR

Because nucleotide excision repair is such an important line of defense against some types of DNA-damaging agents, including UV irradiation, mutations in the genes whose products are required for this type of repair can make cells much more sensitive to these agents. In fact, mutants defective in excision repair were identified because they are killed by much lower doses of irradiation than is the wild type. Table 10.1 lists the *E. coli* genes whose products are required for nucleotide excision repair. Comparative genomic analysis has found UvrA, UvrB, and UvrC orthologs across bacterial species, as well as in some members of the archaea. The products of some of these genes, such as *uvrA*, *uvrB*, and *uvrC*, are involved only in excision repair, while the products of others, including the *polA* and *uvrD* genes, are also required for other types of repair.

Table 10.1 Genes involved in the UvrABC endonuclease repair pathway

Gene	Function of gene product
uvrA	DNA-binding protein
uvrB	Loaded by UvrA to form a DNA complex; nicks DNA 3′ of lesion
uvrC	Binds to UvrB-DNA complex; nicks DNA 5′ of lesion
uvrD	Helicase II; helps remove damage-containing oligonucleotide
polA	Pol I; fills in single-strand gap
lig	Ligase; seals single-strand nick

Nucleotide excision repair is recruited to DNA damage in two distinct ways, one that is coupled to transcription (see below) and another that is not. In the transcription-independent pathway distortions produced by lesions anywhere in the genome can be recognized as illustrated in Figure 10.15. The products of the *uvrA*, *uvrB*, and *uvrC* genes interact to form what is called the **UvrABC endonuclease**. The function of these gene products is to make a nick close to the damaged nucleotide, allowing it to be excised. In more detail, two copies each of the UvrA and UvrB proteins form a complex that binds nonspecifically to DNA, even if it is not damaged. This complex then migrates on and off the DNA in a search in three dimensions until it identifies DNA damage (in the figure, because of a thymine dimer). The critical feature of nucleotide excision repair that allows recognition of such a wide variety of lesions stems from the ability of UvrA to bind the regions flanking the lesion (see Jaciuk et al., Suggested Reading). This allows UvrA to recognize both a deformity in the DNA helix and a change in its rigidity in the helix before signaling the activity of the greater complex. Once the lesion is identified, UvrA facilitates UvrB interaction with the lesion in a verification step where upon confirming DNA damage, UvrA is replaced with the UvrC nuclease in the complex. UvrC has two distinct nuclease domains, one which cuts 4 or 5 phosphodiester bonds 3′ of the damage and the other cuts 8 phosphodiester bonds 5′ of the damage. Once the DNA is cut, the UvrB recruits the UvrD helicase to remove the

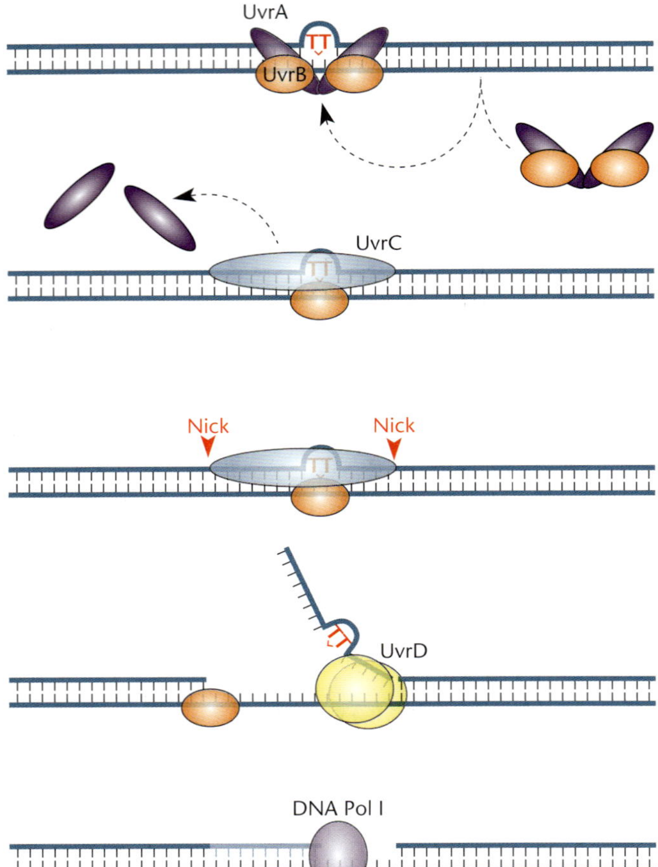

Figure 10.15 Model for nucleotide excision repair by the UvrABC endonuclease. See the text for details. The final ligation step is not shown.

oligonucleotide containing the damage. UvrB is also responsible for recruiting DNA Pol I to resynthesize the strand that was removed, using the complementary strand as a template. *E. coli* and some other bacterial species have an additional protein, Cho, a UvrC homolog that can also carry out the 3′ function in nucleotide excision repair, and with UvrC this may expand the types of lesions that can be excised by nucleotide excision repair in these species (see Moolenaar et al., Suggested Reading).

TRANSCRIPTION-COUPLED REPAIR

Damage in some regions of the DNA presents a more immediate problem for the cell than does damage in other regions. For example, pyrimidine dimers in transcribed regions of DNA block not only replication of the DNA, but also transcription of mRNA when RNA polymerase stalls at the damage. It makes sense for the cell to first repair the damage that occurs in transcribed genes so that these genes can be transcribed and translated into proteins. RNA polymerase that stalls during transcription can also stall translation because transcription and translation are coupled (see chapter 2). Stalled transcription/translation complexes would consequently stall DNA replication if not removed, thereby corrupting three of the most important processes in the cell. These risks likely explain why there are special mechanisms to deal with RNA polymerase that has stalled at a DNA lesion and to recruit DNA repair.

Mfd-dependent transcription coupled repair

The *mfd* gene (for mutation frequency decline) was discovered in *E. coli* more than 50 years ago by Evelyn Witkin by isolation of mutations that prevent a decrease in mutations if protein synthesis is inhibited immediately following DNA damage (see Witkin, Suggested Reading). Similar systems exist in eukaryotes, where they are called Rad26 in yeast and CSB in humans. One of the early pieces of evidence suggesting the existence of transcription-coupled repair was the fact that DNA damage that occurs within transcribed regions of the DNA and in the transcribed strand is repaired preferentially by nucleotide excision repair. Also, mutations occur more frequently when damage occurs in the nontranscribed strand of DNA, as would be expected if damage in this strand were not repaired by the relatively mistake-free excision repair system. It was some time before the Mfd protein was linked to these phenomena.

RNA polymerase stalled at a DNA lesion creates two potential problems for the cell. One is that the stalled RNA polymerase can block access of the nucleotide excision repair system to the damage and thereby prevent its repair. Another is that the stalled RNA polymerase can block the passage of replication forks. The Mfd protein overcomes these potential problems by binding to the DNA behind the stalled RNA polymerase and translocating (moving) itself forward, pushing the RNA polymerase ahead of it (see Park et al., Suggested Reading) (Figure 10.16). The action of Mfd can have one of two effects. As shown in Figure 10.16, if the damage is still there, the RNA polymerase cannot move forward, and the forward movement of Mfd will disrupt the RNA-DNA transcription bubble that was otherwise holding the RNA polymerase on the DNA. This causes the RNA polymerase to be released, getting it out of the way of repair and replication. If the RNA polymerase can move forward, the block is relieved and the RNA polymerase can continue to make the RNA (not shown in Figure 10.16). Probably of greater significance, the Mfd protein is also capable of recruiting the nucleotide excision repair system to allow repair. Mfd has a region that is homologous to the UvrB protein. This region of Mfd accounts for its ability to recruit the nucleotide excision repair system; deletion of the domain results in a protein

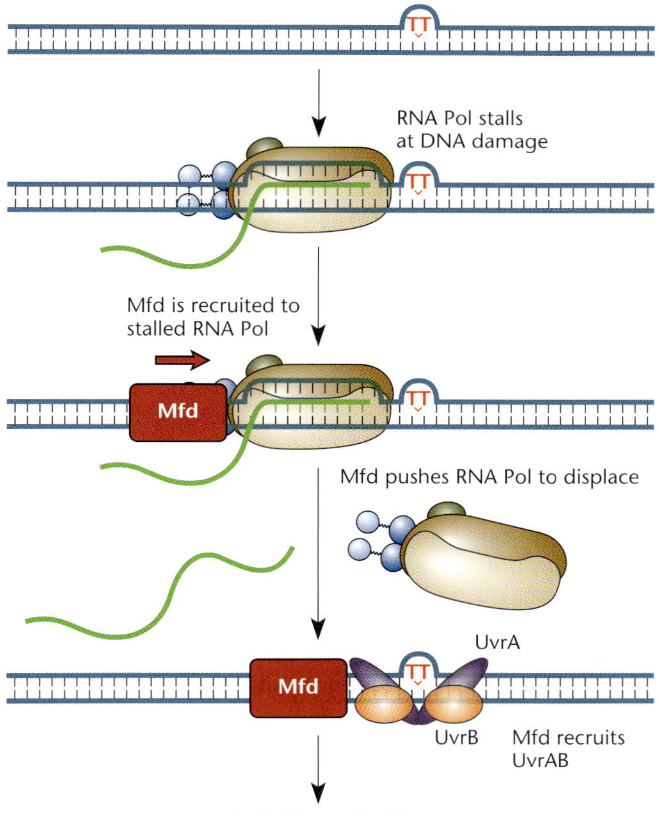
TT
RNA Pol stalls
at DNA damage
Mfd is recruited to
stalled RNA Pol
Mfd
Mfd pushes RNA Pol to displace
UvrA
Mfd
UvrB
Mfd recruits
UvrAB
As in Figure 10.15

Figure 10.16 Model for transcription-coupled nucleotide excision repair. Mfd-dependent transcription-coupled repair is modeled to involve the Mfd protein translocating the elongation complex in the direction of transcription, thereby dislodging the enzyme and recruiting the nucleotide excision repair proteins as explained in the text. Following nucleotide excision repair, transcription can resume (not shown).

that is capable of displacing RNA polymerase but that is unable to carry out transcription-coupled repair (see Manelyte et al., Suggested Reading). The Mfd protein also likely functions to help with clearing backtracked RNA polymerase from transcripts (a function also carried out by the Gre factors) (see chapter 2).

INDUCTION OF NUCLEOTIDE EXCISION REPAIR

Although the genes of the excision repair system are almost always expressed at low levels, *uvrA*, *uvrB*, *cho*, and *uvrD* are expressed at much higher levels after DNA has been damaged. This is a survival mechanism that ensures that larger amounts of the repair proteins are synthesized when they are needed. The *uvr* genes, along with many other proteins involved in DNA repair, are induced by DNA damage. The *uvr* genes are induced following exposure to DNA-damaging agents or irradiation and are members of the *din* genes (for damage inducible), which include *recF*, *recA*, *umuC*, *umuD*, and many others. Many *din* genes, including *uvrA*, *uvrB*, *cho*, and *uvrD*, are induced because they are part of the SOS regulon (see below).

DNA Damage Tolerance Mechanisms

In all of the repair systems discussed above, the cell removes the damage from the DNA, often using the information in the complementary strand to restore the correct DNA sequence. In each of these systems, one might imagine that the goal is to repair the damage before the replication apparatus arrives on the scene and tries to replicate over the damaged sites. However, what happens if the damage is not repaired before a replication fork arrives? In all the instances where the damage is not repaired, the cell has no choice but to replicate over the damage. Mechanisms that allow the cell to tolerate DNA damage without repairing it at that moment are called **damage tolerance mechanisms**. Genetically, mutants in these systems show increased sensitivity to DNA damage with or without a change in the frequency of mutations. While some of these systems were presented in chapters 1 and 9, we now have enough background to go into more detail on the varied ways that DNA recombination allows bacteria to tolerate DNA damage to facilitate DNA repair and replication.

Homologous Recombination and DNA Replication

Homologous recombination is a significant mechanism for damage tolerance. In chapter 9, we learned how homologous recombination is important for reestablishing a replication fork that collapses at a nick in one of the template strands. Homologous recombination also plays an important role in allowing a DNA replication fork to bypass damage in DNA. After a replication fork has moved on, the damage remains in the chromosome, where it can be repaired. Such damage could involve various types of alterations to the bases so that they cannot base pair properly or alterations that make them too bulky move through the DNA polymerase involved in normal DNA replication. There are also structures that can form during DNA replication that provoke endogenous nucleases to make double-strand breaks in the chromosome. These breaks will be fixed using recombination to initiate replication using the same pathway that repairs collapsed DNA replication forks.

LAGGING-STRAND DAMAGE

The consequences of encountering damage in the template strand undergoing lagging-strand replication (the strand running in the 5′-to-3′ direction) are expected to be very different from the consequences of encountering it in the leading strand (running in the 3′-to-5′ direction).

If damage is on the lagging strand, the replication fork presumably has the opportunity to pass right over the damage, provided that the DnaB helicase, which encircles the lagging strand (see Figure 1.10), can proceed over the damage. While DNA damage is likely to stall DNA replication by DNA Pol III (in the example in Figure 10.17, the block is due to a cyclobutane thymine dimer), DNA replication on this template strand is naturally discontinuous, and replication regularly initiates at a new primer made by DnaG as a normal part of lagging-strand DNA replication (see Figure 1.13A). As indicated in Figure 10.17, this leaves a gap in the template strand that must be repaired. The gap structure that results is recognized by the RecF pathway (see Figure 9.5). The RecF pathway can direct loading of RecA at the gap, which would allow the damaged strand to invade the other daughter chromosome formed by replication (Figure 10.17). In this model, invasion into the undamaged chromosome provides a template to allow nucleotide excision repair to recognize the damage and for DNA polymerase to repair the damaged strand.

LEADING-STRAND DAMAGE

The consequences of encountering damage on the leading strand have historically been assumed to be more severe than those for encountering damage on the lagging-strand template. However, *in vitro* evidence with the proteins from *E. coli* suggests multiple types of repair that could help deal with damage to the leading strand. As described in chapter 1, it has been shown that the DnaG primase can prime DNA replication past DNA damage on the leading-strand template when a block is found on the template strand (see Figure 1.13B). This process is dependent on the PriC pathway for reinitiated DNA replication (see Heller and Marians, Suggested Reading). Presumably, the gap that was left behind in this scenario would be repaired using the same series of events as modeled in Figure 10.17 for DNA damage on the DNA template replicated by lagging-strand replication.

Replication that is stalled on the DNA template replicated by leading-strand replication may have an additional

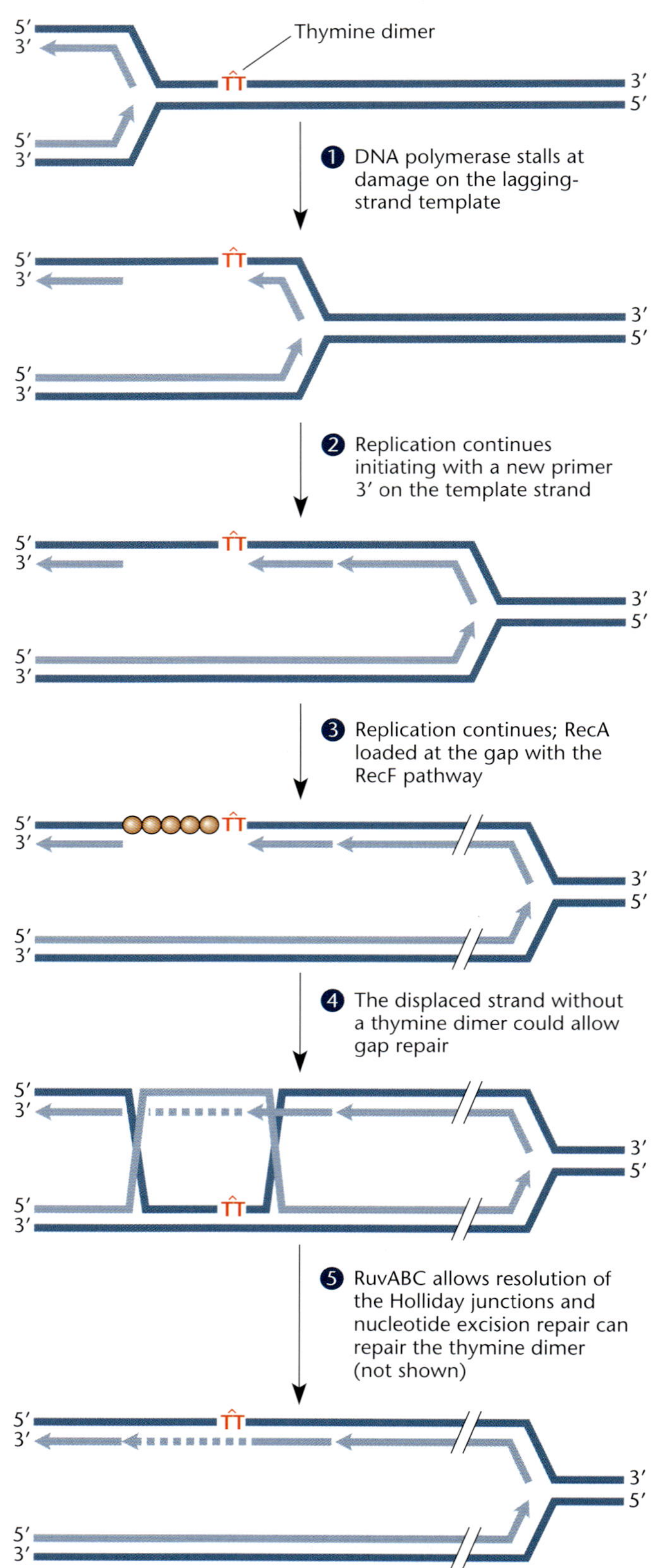
Thymine dimer
5′
3′
TT
1 DNA polymerase stalls at damage on the lagging-strand template
2 Replication continues initiating with a new primer 3′ on the template strand
3 Replication continues; RecA loaded at the gap with the RecF pathway
4 The displaced strand without a thymine dimer could allow gap repair
5 RuvABC allows resolution of the Holliday junctions and nucleotide excision repair can repair the thymine dimer (not shown)

Figure 10.17 Model for recombination-mediated bypass of DNA damage in the DNA strand undergoing lagging-strand replication. DNA Pol III stalls at the thymine dimer on the lagging strand DNA template **(1)**. While replication on both template strands can continue, a gap is left behind **(2)**. The RecF pathway loads RecA onto the single-stranded DNA at the gap **(3)**, and the RecA nucleoprotein filament invades the sister DNA **(4)**. The gap is now opposite a good strand and can be filled in, leaving two Holliday junctions, which are resolved by RuvABC **(5)**. The DNA in which the thymine dimer remains depends on how the Holliday junctions are resolved.

method for facilitating DNA repair. In this model for repair of the thymine dimer, the replication fork can regress, as shown in Figure 10.18. The model starts out the same as the one discussed above, in that single-stranded DNA forms when replication of the leading strand is blocked at the damage. The newly synthesized strands might then pair with each other in a strand exchange reaction promoted by the RecFOR and RecA proteins. This creates a branch that can migrate backward (fork regression) to form a Holliday junction in what is sometimes called a "chicken foot" structure, in which the two new strands are hybridized to each other. The shorter new strand that results from blocked replication of the leading strand could then furnish a 3′ hydroxyl primer for a repair polymerase using an undefined DNA polymerase to extend the shorter strand until the two strands are the same length. On the right side of the figure, some arbitrary base pairs are shown to illustrate how this allows replication past the site of the thymine dimer. The chicken foot Holliday junction could then migrate forward, possibly with the help of RecG or some other helicase, past the site of the damage. We then have a bona fide replication fork on the other side of the damage, with a free 3′ hydroxyl group to prime replication of the leading strand. The DnaB helicase and other replication proteins could then be reloaded, perhaps by the PriA-PriB-DnaT-DnaC pathway, and replication is again under way.

These are just some scenarios to explain the genetic evidence for how the recombination functions can help the cell to tolerate damage to the DNA template. Final proof of any model requires more detailed studies of what happens when the extremely complex replication fork encounters damage on the DNA.

BREAKING THE CHROMOSOME TO REPAIR THE CHROMOSOME

Interestingly, there appear to be scenarios where repair enzymes actually break the chromosome, possibly as a tool to facilitate DNA repair. An impressive early example of evidence that regressed forks can occur in living cells and that they could be recognized by a nuclease was based on the observation that double-strand breaks occur in the chromosome in strains in which the Rep helicase was inactivated (Seigneur et al., Suggested Reading). The DnaB helicase travels on the 5′-to-3′ template strand and unwinds DNA ahead of the replication fork, and the Rep helicase interacts with DnaB but travels on the 3′-to-5′ template strand to help displace proteins ahead of DNA replication (see chapter 1). In a *rep* strain background, the RecBCD complex is essential; in a *rep* strain with temperature-sensitive (Ts) mutations in *recB* and *recC*, cell death occurs at the nonpermissive temperature through the accumulation of double-strand DNA breaks that are not repaired. In the *rep recB*(Ts) *recC*(Ts) strain, mutations in *ruvA* and *ruvB* restore viability at the otherwise nonpermissive temperature.

In other experiments, it was found that in a *dnaB*(Ts) mutant, and even in a wild-type background, replication forks likely reverse using RuvAB branch migration and become targets for various processing enzymes (Figure 10.19). Replication forks stall at various impediments in the chromosome, which is exacerbated in *rep* and *dnaB* helicase mutants (Figure 10.19, step 1). A replication fork may reverse through the action of RuvAB (possibly aided by RecG) (step 2). The Holliday junction RuvC resolvase can cleave the Holliday junction formed by RuvAB (step 3). This resolution of the Holliday junction forms a double-strand break in the chromosome (step 4). Processing of the broken end by RecBCD allows RecA-mediated strand invasion (step 5). The PriA primosome can then restart DNA replication, allowing the replication fork to be restored (step 6). Alternatively, the extruded DNA formed as the replication fork reverses could be recognized by RecBCD (step 7), and the PriA primosome could then restart DNA replication, allowing the replication fork to be restored (step 8).

In *E. coli*, self-inflicted double-strand breaks are also found through the action of the SbcCD nuclease. The gene encoding the SbcCD nuclease was originally identified based on its ability to suppress mutations in *recB* and *recC* (see chapter 9). The SbcCD nuclease cleaves preferentially at large palindromes that are capable of forming hairpin structures. These structures appear to form only following DNA replication and may occur in the transient single-stranded DNA that occurs following separation of the strands. What benefit could the cell derive from self-cleavage of the chromosome? One possibility is that the palindromes themselves lead to genome instability, and by forming breaks at these structures, the enzyme may facilitate their removal. Interestingly, in the absence of SbcBC, large experimentally introduced palindromes can lead to chromosomal inversions, which can be highly detrimental to the cell (see Darmon et al., Suggested Reading).

REPAIR OF INTERSTRAND CROSS-LINKS IN DNA

There are other situations where recombination functions may also collaborate with other repair functions to repair damage to DNA. One example might be in the repair of chemical cross-links in the DNA (Figure 10.20). Many chemicals, such as light-activated psoralens, mitomycin, cisplatin, and EMS, can form **interstrand cross-links** in the DNA, in which two bases in the opposite strands of the DNA are covalently joined to each other (hence the prefix "inter," or "between"). Interstrand cross-links present special problems and cannot be repaired by either nucleotide excision repair or recombination repair alone. Cutting both strands of the DNA with the UvrABC

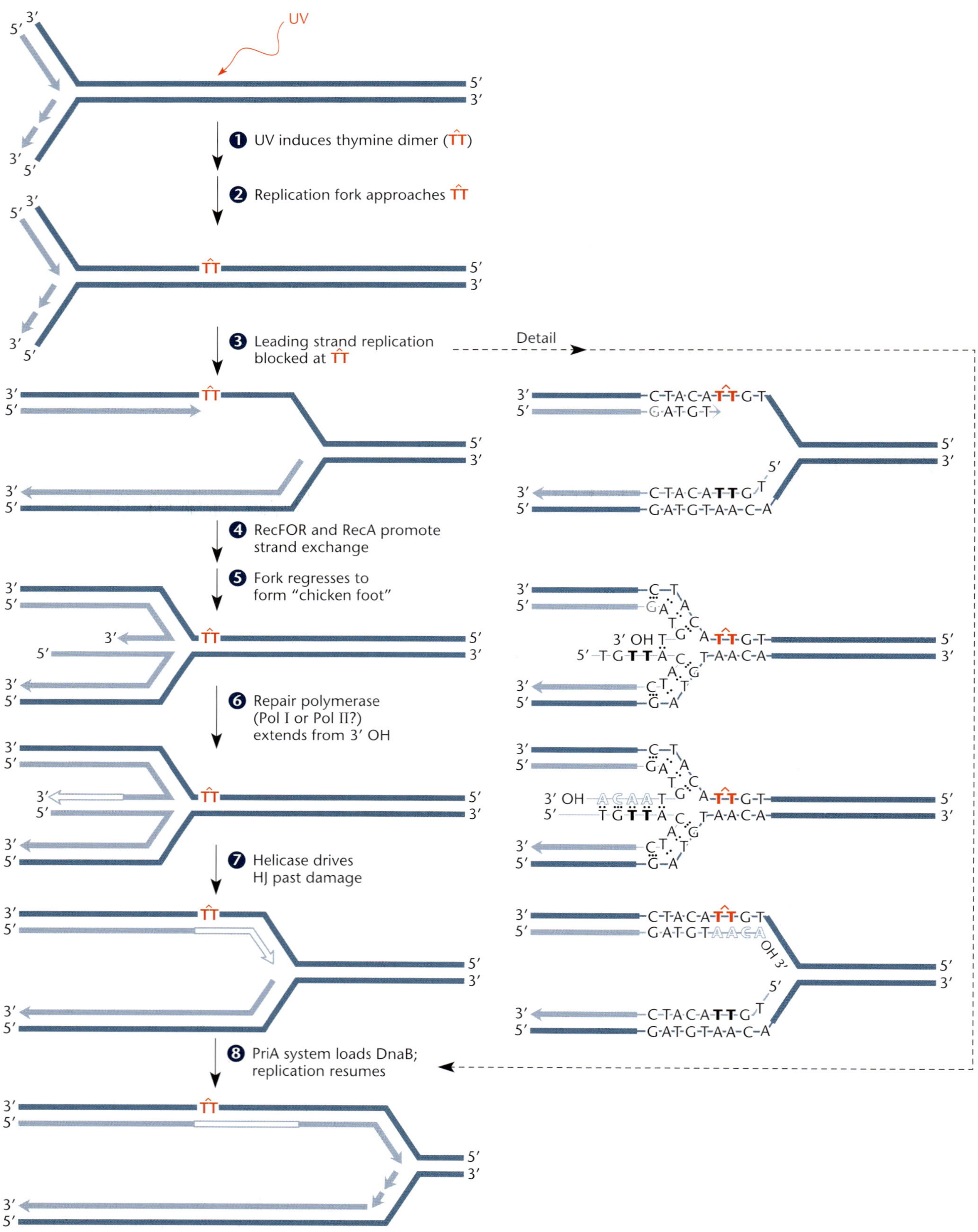
UV
1 UV induces thymine dimer (T̂T)
2 Replication fork approaches T̂T
3 Leading strand replication blocked at T̂T
Detail
4 RecFOR and RecA promote strand exchange
5 Fork regresses to form "chicken foot"
6 Repair polymerase (Pol I or Pol II?) extends from 3′ OH
7 Helicase drives HJ past damage
8 PriA system loads DnaB; replication resumes
C-T-A-C-A-T-T-G-T
G-A-T-G-T
C-T-A-C-A-T-T-G
G-A-T-G-T-A-A-C-A
3′ OH
5′
OH 3′

endonuclease would cause double-strand breakage that would be difficult to repair because recombination functions could not separate the strands.

Although DNA cross-links cannot be repaired by either excision or recombination repair alone, they can be repaired by a combination of recombination repair and nucleotide excision repair, as shown in Figure 10.20. In the first step, the UvrABC endonuclease makes nicks in one strand on either side of the interstrand cross-link, as though it were repairing any other type of damage. This leaves a gap opposite the DNA damage, as shown in the figure. In the second step, recombination repair replaces the gap with an undamaged strand from the other daughter DNA in the cell (see above). The DNA damage is then confined to only one strand and is opposite an undamaged strand; therefore, it can be repaired by the nucleotide excision repair pathway. Like the other recombination-based strategies for repair, this repair is possible only when the DNA has already replicated and there are already multiple copies of the DNA in the cell.

SOS-Inducible Repair

As mentioned above, DNA damage leads to the induction of genes whose products are required for DNA repair. In this way, the cell is better able to repair the damage and survive. The first indication that repair systems for UV damage are inducible came from the work of Jean Weigle on the reactivation of UV-irradiated λ phage (see Weigle, Suggested Reading). He irradiated phage λ and its hosts and tested the survival of the phage by their ability to form plaques when they were plated on *E. coli*. He observed that more irradiated phage survived when plated on *E. coli* cells that had themselves been irradiated than when plated on unirradiated *E. coli* cells. Because the phage was restored to viability, or reactivated, by being plated on preirradiated *E. coli* cells, this phenomenon was named **Weigle reactivation**, or **W reactivation**. Apparently, in the irradiated cells, a repair system that could repair the damage to the λ DNA when the DNA entered the cell was being induced.

THE SOS RESPONSE

Earlier in this chapter, we mentioned a class of genes induced after DNA damage, called the *din* genes, which includes genes that encode products that are part of the excision repair and recombination repair pathways. The products of other *din* genes help the cell survive DNA damage in other ways. For example, some *din* gene products transiently delay cell division until the damage can be repaired, and some allow the cell to replicate past the DNA damage (see below).

Many *din* genes are regulated by the **SOS response**, so named because the mechanism rescues cells that have suffered severe DNA damage. Genes under this type of control are called **SOS genes**. Originally, classical genetic analysis uncovered some 30 SOS genes. More recent microarray analyses have found a few more, not all of which may be directly regulated by the SOS system. For example, some of the new genes are on cryptic prophages that are induced by DNA damage, indirectly increasing the expression of the gene (see below and Courcelle et al., Suggested Reading).

Figure 10.21 illustrates how genes under SOS control are induced. The SOS genes are normally repressed by a protein called the LexA repressor, which has a dumbbell shape and binds to sequences called the **SOS box** upstream of the SOS genes and prevents their transcription. The SOS box is the operator sequence that binds the LexA repressor by analogy to other operator sites, such as *lacO* and the operator sites that bind the λ repressor (see chapters 7 and 11). Any gene directly regulated by LexA would therefore have one of these SOS boxes close to its promoter. Quite often, genes in the same regulon have a common upstream sequence that binds the regulatory protein, and these are often referred to as boxes, with the name of the regulon. In fact, this is often how the genes of a regulon are first identified, by the presence of one of these boxes close to their promoter. Examples of regulons and their boxes are discussed in chapter 12.

The LexA repressor remains bound to the SOS boxes upstream of the promoters for the genes, repressing their transcription, until the DNA is extensively damaged by UV irradiation or other DNA-damaging agents. This stimulates the LexA repressor to cleave itself, an action known as **autocleavage**, thereby inactivating itself and allowing transcription of the SOS genes. This is reminiscent of the cleavage of the λ repressor, which also cleaves itself following DNA damage during induction of phage λ, as described in chapter 7. The two mechanisms are remarkably similar, as discussed below.

Figure 10.18 Fork regression model for recombination-mediated replicative bypass of a thymine dimer in DNA when the damage is on the leading strand **(1 and 2)**. The leading-strand replicating polymerase could stall, but the lagging-strand replicating polymerase keeps going, making good double-stranded DNA opposite the damage **(3)**. The fork could then back up (regress) to form a Holliday junction (HJ)-like "chicken foot" in which the newly synthesized strands have paired with each other, perhaps with the help of the RecFOR pathway and RecA **(4 and 5)**. The free 3′ end due to truncated synthesis of the leading strand could then serve as a primer for the synthesis of DNA past the original site of the damage **(6)**. This is illustrated on the right. The junction could then migrate back over the damage **(7)**, and the replication fork machinery could be reloaded by the PriA pathway **(8)**. The site of the thymine dimer is in boldface.

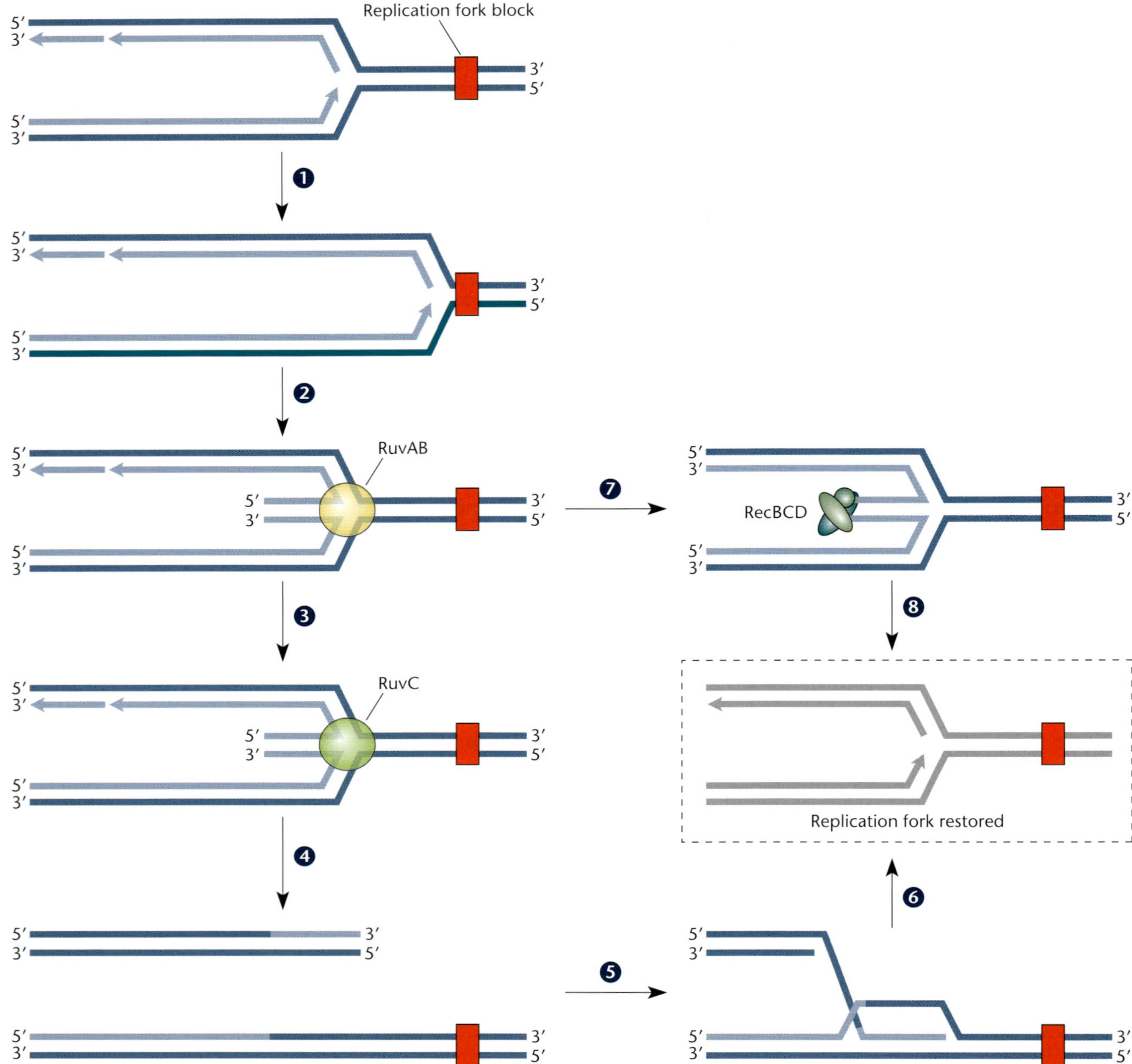

Figure 10.19 Models for how regressed replication forks can be repaired by multiple mechanisms. **(1)** DNA replication can stall due to various features (indicated by a red rectangle). **(2 and 3)** Annealing of the newly synthesized strands allows the replication fork to reverse or regress (Figure 10.18), providing a Holliday junction structure that can be recognized by RuvAB. **(4)** Processing by RuvC, in collaboration with RuvAB, results in a break in the chromosome. **(5 and 6)** Processing of the break by RecBCD and loading of RecA (not shown) allow strand invasion and the initiation of replication with a sister chromosome. **(7)** Alternatively, the arm of the regressed replication fork that is formed by annealing of the newly synthesized strands can be processed by RecBCD. **(8)** This structure can be converted via the exonuclease activity of RecBCD without the formation of a double-strand break in the chromosome, which can, in turn, be restarted by the Pri proteins.

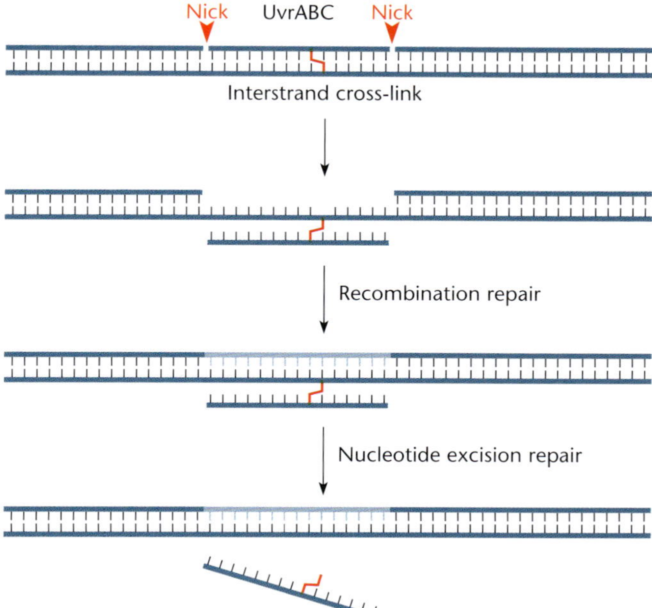

Figure 10.20 Repair of a DNA interstrand cross-link through the combined action of nucleotide excision repair and recombination repair. See the text for details.

The reason why the LexA repressor no longer binds to DNA after it is cleaved is also well understood. Each LexA polypeptide has two separable domains. One, the **dimerization domain**, binds to another LexA polypeptide to form a dimer (hence its name), and the other, the **DNA-binding domain**, binds to the DNA operators upstream of the SOS genes and blocks their transcription. The LexA protein binds to DNA only if it is in the dimer state. The point of cleavage of the LexA polypeptide is between the two domains, and autocleavage separates the DNA-binding domain from the dimerization domain. The DNA-binding domain cannot dimerize and so by itself cannot bind to DNA and block transcription of the SOS genes.

Figure 10.21 also illustrates the answer to the next question: Why does the LexA repressor cleave itself only after DNA damage? After DNA damage, single-stranded DNA is found in the cell due to blockage of replication forks at the damaged sites. This single-stranded DNA is recognized and bound by the RecA protein to form RecA nucleoprotein filaments, which then bind to LexA and cause it to autocleave. This feature of the model—that LexA cleaves itself in response to RecA binding rather than being cleaved by RecA—is supported by experimental evidence. In work with the protein purified from bacteria, certain conditions were identified under which, even in the absence of RecA, LexA eventually cleaves itself. This result indicates that RecA acts as a **coprotease** to facilitate LexA autocleavage, rather than doing the cleaving itself like a standard protease.

The RecA protein thus plays a central role in the induction of the SOS response. It senses the single-stranded DNA that accumulates in the cell as a by-product of attempts to repair damage to the DNA and then causes LexA to cleave itself, relieving the repression of the SOS genes and allowing them to be expressed. We have already discussed another activity of RecA, the recombination function involved in synapse formation; below, we discuss yet another in **translesion synthesis** (**TLS**).

Another level of control in this system comes from the relative affinity of LexA for the SOS boxes. While the SOS boxes recognized by LexA are very similar, they are not identical. These differences result in the need for different amounts of LexA to repress transcription from different SOS genes. These differences in the operators allow a tunable effect like a rheostat, where the SOS boxes that are bound weakly cause the gene they control to be turned on with a small decrease in LexA (i.e., these genes are turned on almost immediately after SOS induction). In contrast, genes that are controlled by SOS boxes that are bound very tightly by LexA remain off unless the level of LexA falls very low (i.e., these genes are turned on only after prolonged DNA damage). As shown in Figure 10.21A, some SOS boxes, like those controlling *recA*, allow some leaky expression even without SOS induction to provide a basal level of activity in the cell (Figure 10.21A). Some genes, like the *polB* gene encoding DNA Pol II, turn on relatively quickly even under lower levels of DNA damage, while others, like *umuDC*, require prolonged exposure to DNA damage for extensive cleavage of LexA to allow release of LexA repression.

The RecA protein therefore plays two roles, one in recombination and one in inducing the SOS response. We can speculate why the RecA protein might play these two different roles. It must bind to single-stranded DNA to promote synapse formation during recombination, so it can easily serve as a sensor of single-stranded DNA in the cell. Also, even though it is itself encoded by an SOS gene that is induced following DNA damage, it is always present in large enough amounts to bind to all of the LexA repressor and to quickly promote autocleavage of all of the repressor after DNA damage.

As mentioned above, the regulation of the SOS response through cleavage of the LexA repressor is strikingly similar to the induction of λ through autocleavage of the CI repressor (see chapter 7). Like the LexA repressor, the λ repressor must be a dimer to bind to the operator sequences in DNA and repress λ transcription. Each λ repressor polypeptide consists of an N-terminal DNA-binding domain and a C-terminal dimerization domain. Autocleavage of the λ repressor, stimulated by the activated RecA nucleoprotein filaments, also separates the DNA-binding domain and the dimerization

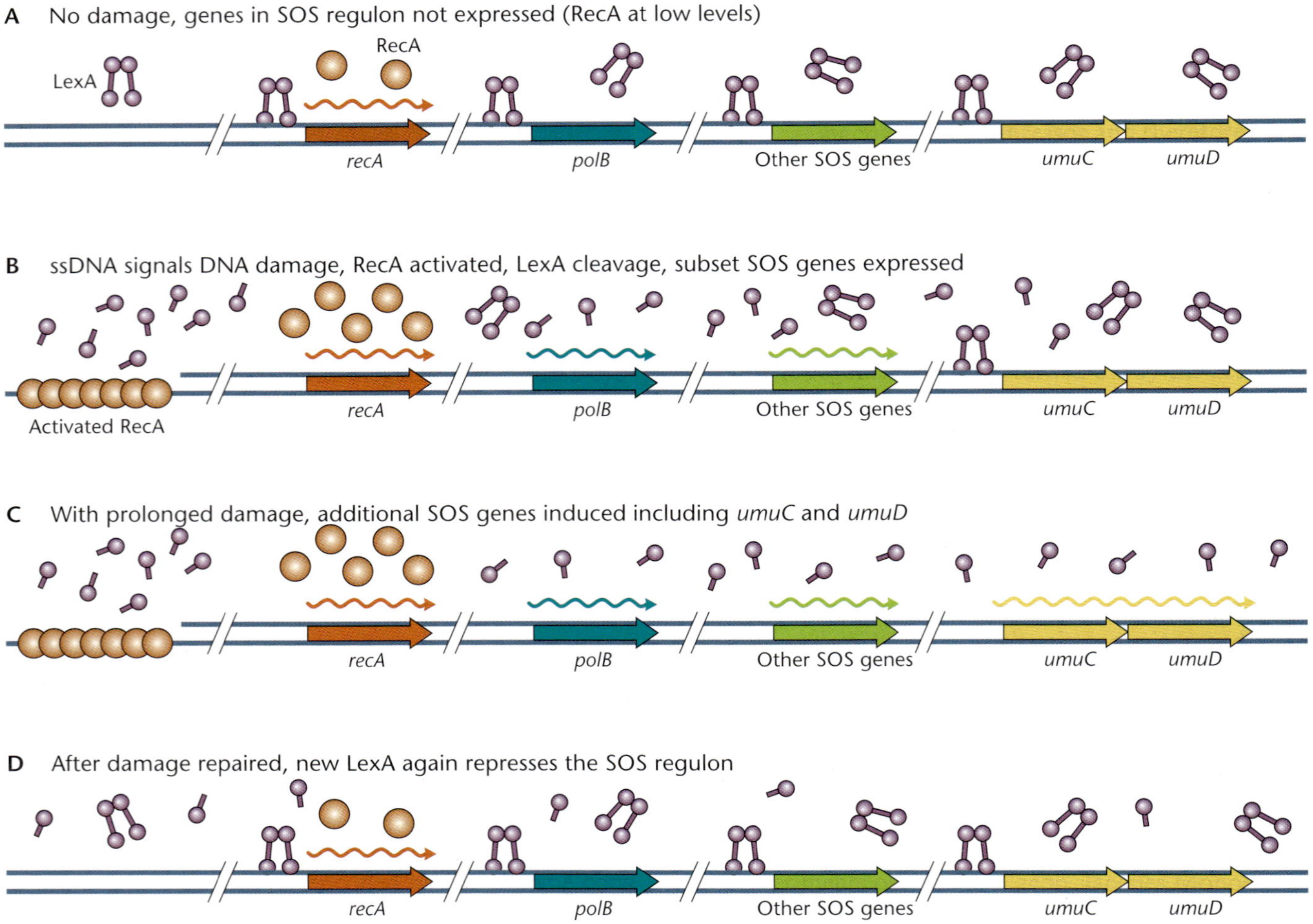

A No damage, genes in SOS regulon not expressed (RecA at low levels)

B ssDNA signals DNA damage, RecA activated, LexA cleavage, subset SOS genes expressed

C With prolonged damage, additional SOS genes induced including *umuC* and *umuD*

D After damage repaired, new LexA again represses the SOS regulon

Figure 10.21 Regulation of the SOS response regulon in *Escherichia coli.* About 30 genes around the *E. coli* chromosome are normally repressed by the binding of a LexA dimer (barbell structure) to their operators; only a few of these genes are shown. **(A)** Some SOS genes, like *recA*, are expressed at low levels at all times. **(B)** After DNA damage, the single-stranded DNA that accumulates in the cell when replication forks stall binds to RecA (orange circles), forming a RecA nucleoprotein filament, which binds to LexA, causing LexA to cleave itself. The cleaved repressor can no longer bind to the operators of the genes, and the genes are induced. Some genes, like *polB* and others, are released from repression even with small reductions in LexA levels allowing induction early in the SOS response. **(C)** Other genes, like *umuDC*, are expressed only after prolonged DNA damage when the levels of LexA are much lower. **(D)** Following repair of DNA damage when replication resumes and there is no longer pervasive single-stranded DNA, newly produced LexA protein is capable of again repressing the genes in the SOS regulon.

domain, preventing the binding of the λ repressor to the operators and allowing transcription of λ lytic genes. The sequences of amino acids around the sites at which the LexA and λ repressors are cleaved are also similar. The evolution of the phage λ repressor, therefore, appears to have taken advantage of the endogenous SOS regulatory system of the host to induce its genes following DNA damage by mimicking LexA, thereby allowing λ to escape a doomed host cell. The activated RecA coprotease also promotes the autocleavage of the UmuD protein involved in SOS mutagenesis (see below).

Once DNA repair has had time to act, there will no longer be an excess of single-stranded DNA to activate the RecA protein. Newly produced full-length LexA (and other proteins that are coaxed to autocleave by activated RecA) will again reach sufficient concentrations, allowing the cell to leave the SOS-induced state and thereby shutting off the response.

GENETICS OF SOS-INDUCIBLE MUTAGENESIS

It is well known that many types of DNA damage, including UV irradiation, are mutagenic and increase the number of mutations in the surviving cells. This is true of all organisms, from bacteria to humans (Box 10.2). The finding that the surviving UV-irradiated cells accumulate mutations implies that one or more of the repair mechanisms used to repair damage are mistake prone. A system that involves TLS allows the replication fork to proceed over damaged DNA so that the molecule can be replicated and the cell can survive. In addition to not repairing the lesion, the polymerases that are responsible for TLS typically copy DNA with a lower fidelity. While accumulating mutations is not ideal, for this particular situation it is preferable to death of the cell.

The first indications that a mistake-prone pathway for UV damage repair is inducible in *E. coli* came from the same studies that showed that repair itself is inducible (see Weigle, Suggested Reading). In addition to measuring the survival of UV-irradiated λ plated on UV-irradiated *E. coli*, Weigle counted the clear-plaque mutants among the surviving phage. Recall from chapter 7 that lysogens form in the centers of wild-type phage λ plaques, making the plaques cloudy, but mutants that cannot lysogenize form clear plaques that can be easily identified. There were more clear-plaque mutants among the surviving phage if the bacterial hosts had been UV irradiated prior to infection than if they had not been irradiated. Therefore, the increased mutagenesis was due to induction of a mutagenic repair system after DNA damage. This inducible mutagenesis was named **Weigle mutagenesis**, or just **W mutagenesis**. Later studies showed that the inducible mutagenesis results from induction of one or more of the SOS genes; it does not occur without RecA and cleavage of the LexA repressor (see Steinborn, Suggested Reading). Thus, the inducible mutagenesis Weigle observed is now often called **SOS mutagenesis**.

Determining Which Repair Pathway Is Mutagenic

Although Weigle's experiments showed that one of the inducible UV damage repair systems in *E. coli* is mistake prone and causes mutations, they did not indicate which system was responsible. A genetic approach was used to answer this question (see Kato et al., Suggested Reading). To detect UV-induced mutations, the experimenters used the reversion of a *his* mutation. Their basic approach was to make double mutants with both a *his* mutation and a mutation in one or more of the genes of each of the repair pathways. The repair-deficient mutants were then irradiated with UV light and plated on medium containing limiting amounts of histidine so that only His^+ revertants could multiply to form a colony. Under these conditions, each reversion to his^+ results in only one colony, making it possible to measure directly the number of his^+ reversions that have occurred. This number, divided by the total number of surviving bacteria, gives an estimation of the susceptibility of the cells to mutagenesis by UV light.

The results of these experiments led to the conclusion that recombination repair of blocked replication forks does not seem to be mistake prone. While *recB* and *recF* mutations reduced the survival of the cells exposed to UV light, the number of his^+ reversions per surviving cell was no different from that of cells lacking mutations in their *rec* genes. Also, nucleotide excision repair does not seem to be mistake prone. Addition of a *uvrB* mutation to the *recB* and *recF* mutations made the cells even more susceptible to killing by UV light but also did not reduce the number of his^+ reversions per surviving cell.

However, *recA* mutations did seem to prevent UV mutagenesis. While these mutations made the cells extremely sensitive to killing by UV light, the few survivors did not contain additional mutations. We now know that two genes, *umuC* and *umuD*, must be induced for mutagenic repair and that *recA* mutations prevent their induction. Thus, the UmuC and UmuD proteins act in the opposite way to most repair systems. Rather than repairing the damage before mistakes in the form of mutations are made, the UmuC and UmuD proteins actually make the mistakes themselves. If they were not present, the cells that survive UV irradiation and some other types of DNA damage would have fewer mutations. The payoff, however, is that presumably, more cells survive. Besides its role in inducing the SOS genes, the RecA protein is directly involved in SOS mutagenesis (see below).

Isolation of *umuC* and *umuD* Mutants

Once it was established that most of the mutagenesis after UV irradiation can be attributed to the products of SOS-inducible genes and that these gene products are not the known ones involved in recombination and nucleotide excision repair, the next step was to identify the unknown genes. Mutations that inactivate a *mut* gene or another repair pathway gene increase the rate of spontaneous mutations because, normally, the gene products repair damage before it can cause mistakes in replication. However, as noted above, mutations that inactivate gene products of a mutagenic or mistake-prone repair pathway have the opposite effect, decreasing the rates of at least some induced mutations. Because the repair pathway itself is mutagenic, mutations should be less frequent if the repair pathway does not exist than if it does exist. Cells with a mutation in a gene of the mistake-prone repair pathway may be less likely to survive DNA damage, but there should be a lower percentage of newly generated mutants among the survivors.

Mutations that reduce the frequency of mutants after UV irradiation were found to fall in two genes, named *umuC* and *umuD* (for ultraviolet-induced mutations C and D). The first *umuC* and *umuD* mutants were isolated in two different laboratories by essentially the same method—reversion of a *his* mutation that makes cell growth require histidine to measure mutation rates—but we describe only the one used by Kato and Shinoura (see Suggested Reading). The basic strategy was to treat the *his* mutant with a mutagen that induces DNA damage similar to that caused by UV irradiation and to identify mutant bacteria in which fewer *his*+ revertants occurred. These mutants would have a second mutation that inactivated the mutagenic repair pathway and reduced the rate of reversion of the *his* mutation.

Figure 10.22 illustrates the selection in more detail. The *his* mutant of *E. coli* was heavily mutagenized so that some of the bacteria would have mutations in the putative mutagenic repair genes. Individual colonies of the bacteria were then patched with a loop onto plates with medium containing histidine, and the plates were incubated until patches due to bacterial growth first appeared. Each plate was then replicated onto another plate containing 4-nitroquinoline-1-oxide (4NQO), which causes DNA damage similar to that caused by UV irradiation. After the patches developed, this plate was replicated onto a third plate containing limiting amounts of histidine. Most bacteria formed patches with a few regions of heavier growth due to His+ revertants, indicating that they were being mutagenized by the 4NQO. However, a few bacteria formed patches with very few areas of heavier growth and therefore had fewer His+ revertants. The bacteria in these patches were candidates for descendants of mutants that could not be mutagenized by 4NQO or, presumably, by UV irradiation, since, as mentioned above, the types of damage caused by the two mutagens are similar.

Figure 10.22 Detection of a mutant defective in mutagenic repair. Colonies of mutagenized *his* bacteria are picked individually from a plate and patched onto a new plate containing histidine. This plate (plate 1) is then replicated onto a plate containing 4-nitroquinoline-1-oxide (4NQO) (plate 2) to induce DNA damage similar to that induced by UV irradiation. The 4NQO-containing plate is then replicated onto another plate with limiting amounts of histidine (plate 3). After incubation, mottling of a patch caused by many His+ revertants indicates that the bacterium that made the colony on the original plate was capable of mutagenic repair. The colony circled in red on the original plate may have arisen from a mutant deficient in mutagenic repair, because it produces fewer His+ revertants when replicated onto plate 3.

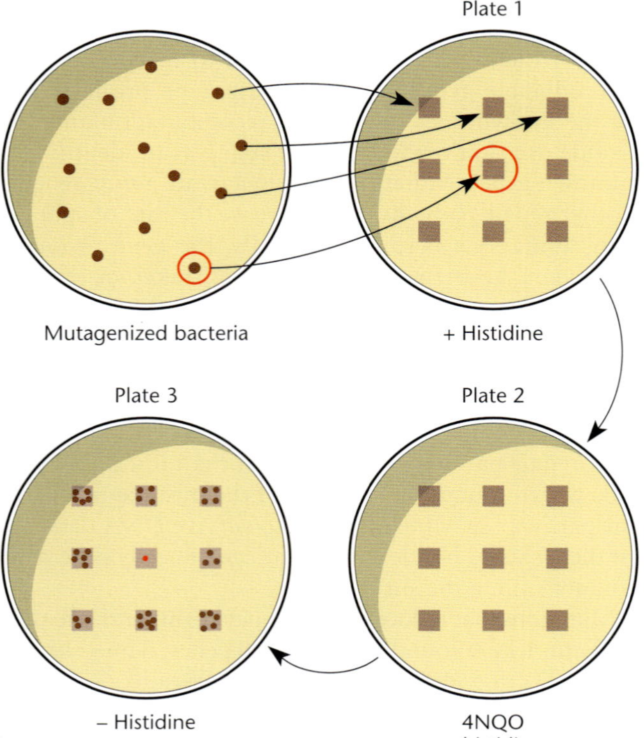

The next step was to map the mutations in some of the mutants. Some of the mutations that prevented mutagenesis by UV irradiation mapped to either *recA* or *lexA*. These mutations could have been anticipated, because mutations in these genes could prevent the induction of all the SOS genes, including those for mutagenic repair. The *recA* mutations presumably inactivate the coprotease activity of the RecA protein, preventing it from causing autocleavage of the LexA repressor and thereby preventing induction of the SOS genes. The *lexA* mutations in all probability change the LexA repressor protein so that it cannot be cleaved, presumably because one of the amino acids around the site of cleavage has been altered. This is a special type of *lexA* mutation called an Ind− mutation. If the LexA protein is not cleaved following UV irradiation of the cells, the SOS genes, including the genes for mutagenic repair, will not be induced.

Some of the mutations that prevented UV mutagenesis mapped to a previously unknown locus distant from *recA* or *lexA* on the *E. coli* map. Complementation tests between mutations at this locus revealed two genes at the site required for UV mutagenesis, named *umuC* and *umuD*. Later experiments also showed that these genes are transcribed into the same mRNA and so form an operon, *umuDC*, in which the *umuD* gene is transcribed first. Experiments with *lacZ* fusions revealed that the *umuDC* operon is inducible by UV light and is an SOS operon, since it is under the control of the LexA repressor (see Bagg et al., Suggested Reading).

Experiments Showing that Only *umuC* and *umuD* Must Be Induced for SOS Mutagenesis

The fact that *umuC* and *umuD* are inducible by DNA damage and are required for SOS mutagenesis does not mean that they are the only genes that must be induced for this pathway. Other genes that must be induced for SOS mutagenesis might have been missed in the mutant selections. Some investigators sought an answer to this question (see Sommer et al., Suggested Reading). Their experiments used a *lexA*(Ind−) mutant, which, as mentioned above, should permanently repress all the SOS genes. They also mutated the operator site of the *umuDC* operon so that these genes would be expressed constitutively and would no longer be under the control of the LexA repressor. Under these conditions, the only SOS gene products that should be present are those of the *umuC* and *umuD* genes, since all other such SOS genes are permanently repressed by the LexA(Ind⁻) repressor. In addition, a shortened form of the *umuD* gene was used. This altered gene synthesizes only the carboxyl-terminal UmuD′ fragment that is the active form for SOS mutagenesis (see below). With the altered *umuD* gene, the RecA coprotease is not required for UmuD to be autocleaved to the active form.

The experiments showed that UV irradiation is mutagenic if UmuC and UmuD′ are expressed constitutively, even if the cells are *lexA*(Ind⁻) mutants, indicating that *umuC* and *umuD* are the only genes that need to be induced by UV irradiation for SOS mutagenesis. However, this result does not entirely eliminate the possibility that other SOS gene products are involved. As discussed below, the RecA protein is also directly required for UV mutagenesis. The *recA* gene is induced to higher levels following UV irradiation but apparently is present in large enough amounts for UV mutagenesis even without induction. The GroEL and GroES proteins are also required for UV mutagenesis, presumably because they help fold mutagenic repair proteins, but the *groEL* and *groES* genes are not under the control of the LexA repressor and so are expressed even in the *lexA*(Ind⁻) mutants.

Experiments Show that RecA Has a Role in UV Mutagenesis in Addition to Its Role as a Coprotease

Similar experiments were performed to show that RecA has a required role in UV mutagenesis in addition to its role in promoting the autocleavage of LexA and UmuD. A sufficient explanation for why mutations in the *recA* gene can prevent UV mutagenesis was that they prevent the induction of all the SOS genes, including *umuC* and *umuD*, and that they also prevent the autocleavage of UmuD to UmuD′, which is required for TLS (see below). However, this did not eliminate the possibility that RecA plays another role in TLS besides its role as a LexA and UmuD coprotease. The mutants that express UmuC and the cleaved form of UmuD′ constitutively could also be used to answer this question. If the coprotease activity of RecA alone is required for TLS, *recA* mutations should not affect UV mutagenesis in this genetic background. However, it was found that *recA* mutations still prevented UV mutagenesis, even if UmuC and UmuD′ were made constitutively, indicating that RecA also participates directly in mutagenesis. This inspired new models in which UmuCD-based mutagenic polymerase needed a RecA nucleoprotein filament for a separate role beyond inducing the autocleavage of UmuD (Figure 10.23; also see below).

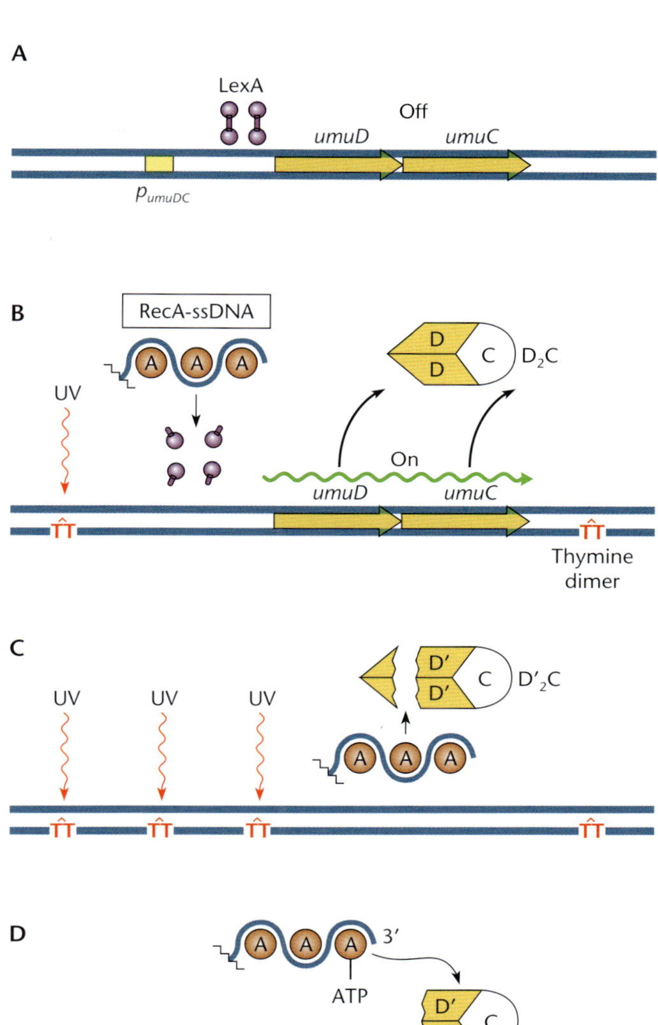

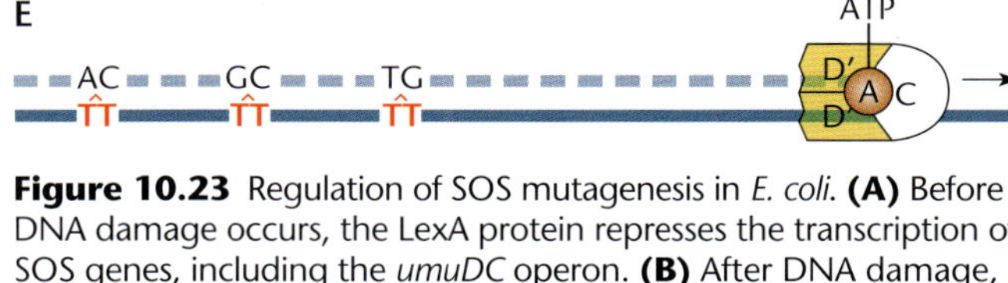

Figure 10.23 Regulation of SOS mutagenesis in *E. coli*. **(A)** Before DNA damage occurs, the LexA protein represses the transcription of SOS genes, including the *umuDC* operon. **(B)** After DNA damage, the RecA protein binds to the single-stranded DNA (ssDNA), which accumulates, forming RecA nucleoprotein filaments. These filaments bind to LexA, promoting its autocleavage and inducing the SOS genes. One of the operons induced late in SOS induction is the *umuDC* operon. The UmuC and UmuD proteins form a heterotrimer composed of two copies of UmuD and one copy of UmuC ($UmuD_2C$). **(C)** More damage causes more RecA nucleoprotein filaments to accumulate, eventually promoting the autocleavage of UmuD to form $UmuD'_2C$. **(D)** A RecA monomer that is bound to ATP can be delivered to the $UmcD'_2C$ complex from the very 3′ end of a RecA nucleoprotein filament, allowing the formation of the Pol V mutasome. **(E)** The Pol V mutasome can replicate right over the damage, often making mistakes in the process; some wrong bases are shown mistakenly incorporated opposite thymine dimers.

Mechanism of TLS by the Pol V Mutasome

As expected of a process that will increase the frequency of mutation, regulation is of paramount importance. As described throughout this book, multiple systems have evolved to ensure the fidelity of DNA replication, and it would not make sense to have translesion DNA polymerases unless they could be kept in check until they were absolutely needed. Regulation of the UmuC and UmuD proteins from *E. coli* that combine to form DNA polymerase V (**DNA Pol V**) provides an excellent example of how a highly error-prone polymerase is regulated. As is often the case, the findings have implications far beyond the UmuC and UmuD proteins and DNA repair in *E. coli* (Box 10.2). DNA Pol V that, in contrast to the normal replicative DNA polymerase, can replicate right over some types of damage to the DNA, including the thymine cyclobutane dimers and cytosine-thymine 6-4 dimers induced by UV irradiation, is said to be capable of TLS. It can also replicate over abasic sites in which the base has been removed by a glycosylase (see above); obviously, therefore, it does not require correct base pairing before it can move on. This suggested an explanation of why DNA Pol V could be mutagenic. Because thymine dimers and other types of damaged bases cannot pair properly, DNA Pol V must incorporate bases almost at random opposite the damaged bases, causing mutations.

UmuC is almost undetectable during normal growth, but with SOS induction, it can go as high as 200 molecules per cell. While at first pass, this might seem to explain why Pol V is only active with SOS induction, in actuality, the regulation of UV mutagenesis is more complicated than this, presumably because it is in the best interest of the cell not to induce this mutagenic system unless the damage cannot be repaired in other ways. Figure 10.23 illustrates why SOS mutagenesis occurs only after extensive damage to the DNA. The first factor that limits SOS mutagenesis only to situations where severe DNA damage is found involves the amount of LexA needed to regulate the *umuDC* operon; even a small amount of LexA will prevent transcription of these genes, and it is estimated that prolonged exposure to RecA-inducing conditions (>30 minutes) is required for the genes to be expressed. Therefore, these proteins will not even be available until very late in SOS induction (Figure 10.21C). Once expressed, the newly synthesized UmuC and UmuD proteins come together to form a heterotrimer complex with two copies of the UmuD protein and one copy of the UmuC protein ($UmuD_2C$). This complex is inactive as DNA polymerase, although it might bind to DNA Pol III and temporarily arrest replication to create a "checkpoint" (see Sutton et al., Suggested Reading), while $UmuD_2$ appears to play an additional regulatory role (see below). However, as the single-stranded DNA-RecA nucleoprotein filaments accumulate, they also cause UmuD to cleave itself (to form UmuD′), in much the same way that they cause LexA and the λ repressor to cleave themselves. The autocleavage of UmuD to UmuD′ requires a higher concentration of RecA nucleoprotein filaments than does the autocleavage of LexA; therefore, rather than happening immediately, it occurs only if the damage is so extensive that it cannot be immediately repaired by other SOS functions (Figure 10.23C). Once UmuD is cleaved to UmuD′, the UmuC in the $UmuD'_2C$ complex is ready for activation using a separate additional property of activated RecA that is distinct from changes in expression through LexA or induced cleavage of UmuD. As described below, the final step in making the complex capable of TLS appears to involve the formation of an $UmuD'_2C$-ATP-RecA complex.

Figure 10.23 shows an extended model for how the $UmuD'_2C$ complex performs TLS and causes mutations. This model integrates the earlier observations and a more recent finding that activated RecA bound to ATP must be integrated with $UmuD'_2C$ to make an error-prone DNA Pol V referred to as the Pol V mutasome, $UmuD'_2C$-RecA-ATP (see Jiang et al., Suggested Reading). The required addition of RecA-ATP is carried out with a nucleoprotein filament separate from the DNA that is actually subject to replication. In the earlier biochemical work, it was impossible to know if RecA was acting in *trans* or in *cis* to the DNA that was to be replicated. It was shown biochemically that a RecA monomer that was bound to ATP could be delivered to the $UmcD'_2C$ complex from the very 3′ end of a RecA nucleoprotein filament, allowing the formation of the Pol V mutasome (Figure 10.23D and E). Mechanisms must also be in place for reestablishing high-fidelity DNA replication. Pol V is generally considered to be less processive than Pol III, limiting the extent of replication on a given DNA template. In addition, DNA Pol V is also specifically targeted for proteolysis after DNA damage is repaired and the SOS response subsides (see Gonzalez et al., Suggested Reading).

There are still questions to be answered about how TLS functions. The Pol V mutasome must function with the β clamp processivity factor, which is very likely coopted from DNA Pol III. However, the regulation that allows the exchange of polymerase with the β clamp is still the subject of active investigation. There are five separate DNA polymerases in *E. coli* (additional TLS polymerases are also encoded on some plasmids). Given that every polymerase must work with a β clamp, there must be a molecular mechanism that helps coordinate these transfers at the DNA replication fork. The idea of polymerase switching is relevant to our own human genome; currently, it is believed that there are at least 15 different DNA polymerases that are also subject to similar processivity clamp exchanges.

Other Specialized Polymerases and Their Regulation

The UmuC polymerase is a member of a large group of polymerases called the **Y family of polymerases**. Structural studies have shown that the active site of the Y family of polymerases is more open than those of the replicative polymerases and has fewer contacts between the polymerase and the DNA template, allowing the Y family of polymerases to replicate over many types of lesions in the DNA. They also lack editing functions that would cause them to stall at mismatches. Most types of bacteria, archaea, and eukaryotes are known to have Y DNA polymerases related to UmuC (Box 10.2). In addition, some naturally occurring plasmids carry analogs of the *umuC* and *umuD* genes. The best studied of these genes are the *mucA* and *mucB* genes of plasmid R46. The products of these plasmid genes can substitute for UmuC and UmuD′ in mutagenic repair of *E. coli*, and *Salmonella* spp. carrying the R46 derivative plasmid pKM101 are more sensitive to mutagenesis by many types of DNA-damaging agents. Because of the sensitivity that the *mucA* and *mucB* genes impose on a bacterial strain with the pKM101 plasmid, it has been introduced into the *Salmonella* strains used for the Ames test for chemical mutagens (Box 10.3).

DNA Pol II

In addition to DNA Pol V, *E. coli* has two other DNA polymerases that are induced as part of the SOS response, DNA Pol II and DNA Pol IV. DNA Pol II is present at 50 or more copies per cell during normal growth, but it is induced about 10-fold upon SOS induction. DNA Pol II is capable of doing some TLS. However, unlike Pol V, but typical of B polymerases, DNA Pol II has exonuclease activity for proofreading and is processive for DNA replication.

The crystal structure of DNA Pol II has shed light on how the polymerase can replicate both normal DNA and damaged DNA (see Wang and Yang, Suggested Reading). The structure revealed small protein cavities outside the active site of the enzyme that may allow the template DNA to loop out of the enzyme to facilitate primer extension past DNA damage or possibly allow it to skip over DNA damage that would otherwise halt another DNA polymerase. DNA Pol II is expressed early during SOS induction. Early synthesis of the polymerase after DNA damage may be beneficial, because the enzyme could allow replication past lesions that stall DNA Pol III while still reducing the risk of mutagenesis that goes along with increased levels of the other SOS-induced DNA polymerases. For example, if DNA Pol II is induced early and can replicate past the lesion that stalled DNA Pol III, it will prevent single-stranded DNA from lingering long enough to allow SOS induction of the error-prone polymerases.

DNA Pol IV (DInB)

DNA Pol IV (also known as DinB) is a Y family DNA polymerase, and its participation in TLS has been known

BOX 10.3

The Ames Test

It is now well established that cancer is initiated by mutations in genes, including oncogenes and tumor-suppressing genes. Therefore, chemicals that cause mutations are often carcinogenic for humans. Many new chemicals are being used as food additives or otherwise come into contact with humans, and each of these chemicals must be tested for its carcinogenic potential. However, such testing in animal models is expensive and time-consuming. Because many carcinogenic chemicals damage DNA, they are mutagenic for bacteria, as well as humans. Therefore, bacteria can be used in initial tests to determine if chemicals are apt to be carcinogenic. The most widely used of these tests is the Ames test, developed by Bruce Ames and his collaborators. This test uses revertants of *his* mutations of *Salmonella* spp. to detect mutations. The chemical is placed on a plate lacking histidine on which has been spread a His$^-$ mutant of *Salmonella*. If the chemical can revert the *his* mutation, a ring of His$^+$ revertant colonies will appear around the chemical on the plate. A number of different *his* mutations must be used, because different mutagens cause different types of mutations, and they all have preferred sites of mutagenesis (hot spots). Also, the test is made more sensitive by eliminating the nonmutagenic nucleotide excision repair system with a *uvrA* mutation and introducing a plasmid containing the *mucA* and *mucB* genes. These genes are analogs of *umuC* and *umuD*, so they increase mutagenesis (see the text). Some chemicals are not mutagenic themselves but can be converted into mutagens by enzymes in the mammalian liver. To detect these precursors of mutagens, we can add a liver extract from rats to the plates and spot the chemical on the extract. If the extract converts the chemical into a mutagen, His$^+$ revertants will appear on the plate.

Reference

McCann J, Ames BN. 1976. Detection of carcinogens as mutagens in the *Salmonella*/microsome test: assay of 300 chemicals: discussion. *Proc Natl Acad Sci USA* **73:**950–954.

for some time. Unlike Pol V, it causes spontaneous mutations when it is overproduced, even without DNA damage. Many of the mutations found with Pol IV are frameshifts, but only –1 frameshifts. While its levels increase 10-fold following DNA damage, it is always present in the cell at a fairly high level of about 150 to 250 copies per cell. DinB is largely responsible for the spontaneous mutations that occur during special types of long-term selection experiments, when the cells are not growing. The mutation potential (and TLS activity) of DinB appears to be regulated by its interaction with other important proteins that have been discussed in this chapter, $UmuD_2$ and RecA (see Godoy et al., Suggested Reading). The three proteins interact physically and functionally, and structural modeling suggests that the DinB interaction with $UmuD_2$ and RecA helps enclose the relatively open active site of DinB, thereby preventing the template bulging responsible for –1 frameshifts. While the finding that a subunit of DNA Pol V (i.e., UmuD) can function with DNA Pol IV (i.e., DinB) blurs the historical nomenclature designation, it also reveals an exciting and unexpected regulatory option for cells for DNA replication. Presumably, in the earliest stages of SOS induction (<30 minutes), DinB interacts with $UmuD_2$ and RecA to allow replication that can bypass some lesions without the formation of –1 frameshifts. However, in the presence of higher levels of DNA damage and prolonged SOS induction (>30 minutes), the mutagenic TLS activity of DinB could allow replication over DNA with different bulky adducts or (controversially) could actually provoke mutations that might help the population overcome stressful conditions. DinB is the only Y polymerase that is conserved across all three domains of life. It is therefore of additional interest that amino acids in the proposed region of interaction between UmuD, RecA, and DinB covary throughout evolution, suggesting that this regulatory activity for mutation potential could be conserved.

Eukaryotes have a large number of different mutagenic polymerases related to UmuC and Pol IV (Box 10.2). Each type of mutagenic DNA polymerase may be required to replicate over a particular type of damage to DNA and thereby may play a role in avoiding mutations due to a particular type of DNA damage. Mutations in some of these genes have been implicated in increased cancer risk. In the absence of the specialized Y-type polymerase for a particular type of lesion in DNA, another one will take over, causing the mutations that lead to increased risk of cancer. Interestingly, these same types of DNA polymerases seem to make some types of cancer resistant to chemotherapy, such as treatment with cisplatin, presumably because the DNA damage caused by the compound is responsible for cancer cell death, and the polymerases protect cancer cells from that damage.

Summary of Repair Pathways in *E. coli*

Table 10.2 shows the repair pathways that have been discussed and some of the genes whose products participate in each pathway. Some pathways, such as photoreactivation and most base excision pathways, have evolved to repair specific types of damage to DNA. Some, such as VSP repair, mend damage only in certain sequences. Others, such as mismatch repair and nucleotide excision repair, are much more general and repair essentially any damage to DNA, provided that it causes a distortion in the DNA structure. Overlap in the systems involved in repairing DNA lesions likely contributes to the efficiency and speed of repair.

The separation of repair functions into different pathways is in some cases artificial. Some repair genes are inducible, and the repair enzymes themselves can play a role in their induction, as well as in the induction of genes in other pathways. For example, the RecBCD nuclease is involved in recombination repair but can also help make the single-stranded DNA that activates the RecA coprotease activity after DNA damage to induce SOS functions. The RecA protein is required for both recombination repair and induction of the SOS functions, and it is directly involved in SOS mutagenesis. The proposed role of UmuD in the Pol V mutasome and with DinB/Pol IV provides another example of such an artificial grouping. Needless to say, the overlap of the functions of the repair gene products in different repair pathways has complicated the assignment of roles in these pathways.

Bacteriophage Repair Pathways

The DNA genomes of bacteriophages are also subject to DNA damage, either when the DNA is in the phage particle or when it is replicating in a host cell. Not surprisingly, many phages encode their own DNA repair enzymes. In fact, the discovery of some phage repair pathways preceded and anticipated the discovery of the corresponding bacterial repair pathways. By encoding their own repair pathways, phages avoid dependence on host pathways, and repair proceeds more efficiently than it might with the host pathways alone.

The repair pathways of phage T4 are perhaps the best understood. This large phage encodes at least seven different repair enzymes that help repair DNA damage due to UV irradiation, and some of these enzymes are also involved in recombination (see chapter 9). Table 10.3 lists the functions of some T4 gene products and their homologous bacterial enzymes. The phage also uses some of the corresponding host enzymes to repair damage to its DNA.

One of the most important functions for repairing UV damage in phage T4 is the product of the *denV* gene,

Table 10.2 Genetic pathways for damage repair and tolerance

Repair mechanism	Genetic locus	Function
Methyl-directed mismatch repair	*dam*	DNA adenine methylase
	mutS	Mismatch recognition
	mutH	Endonuclease that cuts at hemimethylated sites
	mutL	Interacts with MutS and MutH
	uvrD (*mutU*)	Helicase
VSP repair	*dcm*	DNA cytosine methylase
	vsr	Endonuclease that cuts at 5′ side of T in TG mismatch
GO (guanine oxidations)	*mutM*	Glycosylase that acts on GO
	mutY	Glycosylase that removes A from A-GO mismatch
	mutT	8-OxodGTP phosphatase
Alkylation/adaptive response	*ada*	Alkyltransferase and transcriptional activator
	alkA	Glycosylase for alkylpurines
	alkB	α-Ketoglutarate-dependent dioxygenase
Nucleotide excision	*uvrA*	Component of UvrABC
	uvrB	Component of UvrABC
	uvrC	Component of UvrABC
	uvrD	Helicase
	polA	Repair synthesis
Base excision	*xthA*	AP endonuclease
	nfo	AP endonuclease
Photoreactivation	*phr*	Photolyase
Recombination repair	*recA*	Strand exchange
	recBCD	Recombination repair double-strand breaks
	recFOR	Recombination repair at ssDNA gaps and overhangs
	ssb	Single-stranded DNA-binding protein
SOS system	*recA*	Coprotease
	lexA	Repressor
	umuDC	TLS (Pol V)
	dinB	Mutagenic polymerase (Pol IV)
	polB	Replication restarts (Pol II)

which is an AP lyase (see above) that has both *N*-glycosylase and DNA endonuclease activities (see Dodson and Lloyd, Suggested Reading). The *N*-glycosylase activity specifically breaks the bond holding one of the pyrimidines to its sugar in pyrimidine cyclobutane dimers of the *cis-syn* type (Figure 10.7). The endonuclease activity then cuts the DNA just 3′ of the pyrimidine dimer, and the dimer is removed by the exonuclease activity of the host cell DNA Pol I. As mentioned above, AP lyases thus work very differently from the UvrABC endonuclease of *E. coli* and, in a sense, combine both the *N*-glycosylase and AP endonuclease activities of some other repair pathways. The endonuclease activity of the DenV protein also functions independently of the *N*-glycosylase activity and cuts apurinic sites in DNA, as well as heteroduplex loops caused by short insertions or deletions. Because it cuts next to pyrimidine dimers in DNA, the purified DenV endonuclease of T4 is often used to determine the persistence of pyrimidine dimers after UV irradiation. The bacterium *Micrococcus luteus* has an enzyme similar to DenV. It will be interesting in future work to see how repair enzymes have been adapted across the diversity of bacteria and what yet undiscovered additional mechanisms have evolved to ensure the accurate maintenance of bacterial chromosomes.

Table 10.3 Bacteriophage T4 repair enzymes

Repair enzyme	Host analog
DenV	UV endonuclease of *M. luteus*
UvsX	RecA
UvsY	RecOR
UvsW	RecG
gp46/gp47 exonuclease	RecBCD recombination repair
gp49 resolvase	RuvABC recombination repair
gp59	PriA

1. All free-living organisms on Earth probably have mechanisms to repair damage to their DNA. Some of these repair systems are specific to certain types of damage, while others are more general and repair any damage that makes a significant distortion in the DNA helix.

2. Specific DNA glycosylases remove some types of damaged bases from DNA. Specific DNA glycosylases that remove uracil, hypoxanthine, some types of alkylated bases, 8-oxoG, and any A mistakenly incorporated opposite 8-oxoG in DNA are known. After the damaged base is removed by the specific glycosylase, the DNA is cut by an AP endonuclease, and the strand is degraded and resynthesized to restore the correct base.

3. The positions of 5-methylcytosine in DNA are particularly susceptible to mutagenesis, because deamination of 5-methylcytosine produces thymine instead of uracil, and thymine is not removed by the uracil-*N*-glycosylase. *E. coli* has a special repair system, called VSP repair, that recognizes the thymine in the thymine-guanine mismatch at the site of 5-methylcytosine and removes it, preventing mutagenesis. The products of the *mutS* and *mutL* genes of the mismatch repair system also make this pathway more efficient, perhaps by helping attract the Vsr endonuclease to the mismatch.

4. The photoreactivation system uses an enzyme called photolyase to specifically separate the bases of one type of pyrimidine dimer created during UV irradiation. The photolyase binds to pyrimidine dimers in the dark but requires visible light to separate the fused bases.

5. Methyltransferase enzymes remove the methyl group from certain alkylated bases and phosphates and transfer it to themselves. Others are dioxygenases that oxidize the methyl group, converting it into formaldehyde and removing it. In *E. coli*, some of these alkylation defense proteins are inducible as part of the adaptive response. Methylation of the Ada protein converts it into a transcriptional activator for its own gene, as well as for some other repair genes involved in repairing alkylation damage.

6. The methyl-directed mismatch repair system recognizes mismatches in DNA and removes and resynthesizes one of the two strands, restoring the correct pairing. The products of the *mutL*, *mutS*, and *mutH* genes participate in this pathway in *E. coli*. The Dam methylase product of the *dam* gene helps identify the strand to be degraded. The region including the mismatch in the newly synthesized strand is degraded and resynthesized because the A in neighboring GATC sequences on that strand has not yet been methylated by the Dam enzyme.

7. The nucleotide excision repair pathway encoded by the *uvr* genes of *E. coli* removes many types of DNA damage that cause gross distortions in the DNA helix. The UvrABC endonuclease cuts on both sides of the DNA damage, and the entire oligonucleotide including the damage is removed and resynthesized. Transcription-coupled repair targets nucleotide excision repair to actively transcribed genes, playing an important role in identifying targets for repair.

8. Recombination repair is important for the repair of double-strand DNA breaks and gaps. Recombination repair does not actually repair damage, as described in this chapter, but instead helps the cell tolerate it. Lesions on either strand will stop DNA polymerases, resulting in gaps in the template strands as replication is initiated ahead of the damage, leaving a gap opposite the damaged base. Recombination with the other strand can put a good strand opposite the damage, and the replication can continue. This type of repair in *E. coli* requires the recombination functions RecBCD, RecFOR, RecA, RecG, and RuvABC, as well as PriA, PriB, PriC, DnaC, and DnaT, to restart replication forks.

9. A combination of excision and recombination repair may remove interstrand cross-links in DNA. The excision repair system cuts one strand, and the single-strand break is enlarged by exonucleases to leave a gap opposite the damage. Recombination repair can then transfer a good strand to a position opposite the damage, and excision repair can remove the damage, since it is now confined to one strand. Interstrand cross-links can be repaired only if there are two or more copies of that region of the chromosome in the cell.

10. The SOS regulon includes many genes that are induced following DNA damage. The genes are normally repressed by the LexA repressor, which cleaves itself (autocleavage) after extensive DNA damage. The autocleavage is triggered by single-stranded DNA-RecA nucleoprotein filaments that accumulate following DNA damage.

11. SOS mutagenesis is due to the products of the genes *umuC* and *umuD*, which are induced following DNA damage. After induction, these proteins form a heterotrimer, UmuD_2C, which is not active. The UmuD and UmuC proteins can form a mutagenic polymerase called a Pol V mutasome with the aid of RecA. A RecA single-stranded DNA nucleoprotein filament promotes the autocleavage of UmuD to UmuD′, and the nucleoprotein filament also delivers RecA bound to ATP to activate the mutasome.

12. The Pol V mutasome can replicate over abasic sites and the two forms of pyrimidine dimers formed by UV irradiation, as

Summary (continued)

well as some other types of DNA damage, inserting bases with lower fidelity opposite the damage and causing mutations. This polymerase is a member of a large family of translesion DNA polymerases called the Y polymerases, which are found in all the kingdoms of life, and mutations in the genes for these polymerases are known to be the cause of some types of genetic susceptibility to cancer.

13. In *E. coli*, three polymerases are induced during the SOS response at different times. These different polymerases may provide multiple options to deal with various types of DNA damage. The regulation of this system likely allows the cell to balance mutagenic repair to enable replication of the chromosome while minimizing the accumulation of mutations.

QUESTIONS FOR THOUGHT

1. Some types of organisms, for example, yeasts, do not have methylated bases in their DNA and so would not be expected to have a methyl-directed mismatch repair system. Can you think of any other ways besides methylation that a mismatch repair system could be directed to the newly synthesized strand during replication?

2. How would you determine if a bacterium isolated from the gut of a marine organism at the bottom of the ocean has a photoreactivation-based DNA repair system?

3. Why do you suppose so many pathways exist to repair some types of damage in DNA?

4. Why do you think the SOS mutagenesis pathway exists if it contributes so little to survival after DNA damage?

SUGGESTED READING

Bagg A, Kenyon CJ, Walker GC. 1981. Inducibility of a gene product required for UV and chemical mutagenesis in *Escherichia coli*. *Proc Natl Acad Sci USA* **78**:5749–5753.

Bowles T, Metz AH, O'Quin J, Wawrzak Z, Eichman BF. 2008. Structure and DNA binding of alkylation response protein AidB. *Proc Natl Acad Sci USA* **105**:15299–15304.

Cohen SE, Lewis CA, Mooney RA, Kohanski MA, Collins JJ, Landick R, Walker GC. 2010. Roles for the transcription elongation factor NusA in both DNA repair and damage tolerance pathways in *Escherichia coli*. *Proc Natl Acad Sci USA* **107**:15517–15522.

Courcelle J, Khodursky A, Peter B, Brown PO, Hanawalt PC. 2001. Comparative gene expression profiles following UV exposure in wild-type and SOS-deficient *Escherichia coli*. *Genetics* **158**:41–64.

Darmon E, Eykelenboom JK, Lincker F, Jones LH, White M, Okely E, Blackwood JK, Leach DR. 2010. *E. coli* SbcCD and RecA control chromosomal rearrangement induced by an interrupted palindrome. *Mol Cell* **39**:59–70.

Deaconescu AM, Chambers AL, Smith AJ, Nickels BE, Hochschild A, Savery NJ, Darst SA. 2006. Structural basis for bacterial transcription-coupled DNA repair. *Cell* **124**:507–520.

Dodson ML, Lloyd RS. 1989. Structure-function studies of the T4 endonuclease V repair enzyme. *Mutat Res* **218**:49–65.

Dwyer DJ, Belenky PA, Yang JH, MacDonald IC, Martell JD, Takahashi N, Chan CTY, Lobritz MA, Braff D, Schwarz EG, Ye JD, Pati M, Vercruysse M, Ralifo PS, Allison KR, Khalil AS, Ting AY, Walker GC, Collins JJ. 2014. Antibiotics induce redox-related physiological alterations as part of their lethality. *Proc Natl Acad Sci USA* **111**:E2100–E2109.

Epshtein V, Kamarthapu V, McGary K, Svetlov V, Ueberheide B, Proshkin S, Mironov A, Nudler E. 2014. UvrD facilitates DNA repair by pulling RNA polymerase backwards. *Nature* **505**:372–377.

Friedberg EC, Lehmann AR, Fuchs RPP. 2005. Trading places: how do DNA polymerases switch during translesion DNA synthesis? *Mol Cell* **18**:499–505.

Friedberg EC, Walker GC, Siede W, Wood RD, Schultz RA, Ellenberger T. 2006. *DNA Repair and Mutagenesis*, 2nd ed. ASM Press, Washington, DC.

Glickman BW, Radman M. 1980. *Escherichia coli* mutator mutants deficient in methylation-instructed DNA mismatch correction. *Proc Natl Acad Sci USA* **77**:1063–1067.

Godoy VG, Jarosz DF, Simon SM, Abyzov A, Ilyin V, Walker GC. 2007. UmuD and RecA directly modulate the mutagenic potential of the Y family DNA polymerase DinB. *Mol Cell* **28**:1058–1070.

Gonzalez M, Rasulova F, Maurizi MR, Woodgate R. 2000. Subunit-specific degradation of the UmuD/D′ heterodimer by the ClpXP protease: the role of *trans* recognition in UmuD′ stability. *EMBO J* **19**:5251–5258.

He C, Hus JC, Sun LJ, Zhou P, Norman DP, Dötsch V, Wei H, Gross JD, Lane WS, Wagner G, Verdine GL. 2005. A methylation-dependent electrostatic switch controls DNA repair and transcriptional activation by *E. coli ada*. *Mol Cell* **20**:117–129.

Heller RC, Marians KJ. 2005. The disposition of nascent strands at stalled replication forks dictates the pathway of replisome loading during restart. *Mol Cell* **17**:733–743.

Jaciuk M, Nowak E, Skowronek K, Tańska A, Nowotny M. 2011. Structure of UvrA nucleotide excision repair protein in complex with modified DNA. *Nat Struct Mol Biol* **18**:191–197.

Jiang Q, Karata K, Woodgate R, Cox MM, Goodman MF. 2009. The active form of DNA polymerase V is UmuD′$_2$C-RecA-ATP. *Nature* **460**:359–363.

Kamarthapu V, Nudler E. 2015. Rethinking transcription coupled DNA repair. *Curr Opin Microbiol* **24:**15–20.

Kato T, Rothman RH, Clark AJ. 1977. Analysis of the role of recombination and repair in mutagenesis of *Escherichia coli* by UV irradiation. *Genetics* **87:**1–18.

Kato T, Shinoura Y. 1977. Isolation and characterization of mutants of *Escherichia coli* deficient in induction of mutations by ultraviolet light. *Mol Gen Genet* **156:**121–131.

Kisker C, Kuper J, Van Houten B. 2013. Prokaryotic nucleotide excision repair. *Cold Spring Harb Perspect Biol* **5:**a012591.

Korshunov S, Imlay JA. 2010. Two sources of endogenous hydrogen peroxide in *Escherichia coli*. *Mol Microbiol* **75:**1389–1401.

Landini P, Volkert MR. 2000. Regulatory responses of the adaptive response to alkylation damage: a simple regulon with complex regulatory features. *J Bacteriol* **182:**6543–6549.

Liao Y, Schroeder JW, Gao B, Simmons LA, Biteen JS. 2015. Single-molecule motions and interactions in live cells reveal target search dynamics in mismatch repair. *Proc Natl Acad Sci USA* **112:**E6898–E6906.

Manelyte L, Kim YI, Smith AJ, Smith RM, Savery NJ. 2010. Regulation and rate enhancement during transcription-coupled DNA repair. *Mol Cell* **40:**714–724.

Michaels ML, Cruz C, Grollman AP, Miller JH. 1992. Evidence that MutY and MutM combine to prevent mutations by an oxidatively damaged form of guanine in DNA. *Proc Natl Acad Sci USA* **89:**7022–7025.

Moolenaar GF, van Rossum-Fikkert S, van Kesteren M, Goosen N. 2002. Cho, a second endonuclease involved in *Escherichia coli* nucleotide excision repair. *Proc Natl Acad Sci USA* **99:**1467–1472.

Nghiem Y, Cabrera M, Cupples CG, Miller JH. 1988. The *mutY* gene: a mutator locus in *Escherichia coli* that generates G.C—-T.A transversions. *Proc Natl Acad Sci USA* **85:**2709–2713.

Nohmi T, Battista JR, Dodson LA, Walker GC. 1988. RecA-mediated cleavage activates UmuD for mutagenesis: mechanistic relationship between transcriptional derepression and posttranslational activation. *Proc Natl Acad Sci USA* **85:**1816–1820.

Pandya GA, Yang IY, Grollman AP, Moriya M. 2000. *Escherichia coli* responses to a single DNA adduct. *J Bacteriol* **182:**6598–6604.

Park JS, Marr MT, Roberts JW. 2002. *E. coli* transcription repair coupling factor (Mfd protein) rescues arrested complexes by promoting forward translocation. *Cell* **109:**757–767.

Pillon MC, Babu VMP, Randall JR, Cai J, Simmons LA, Sutton MD, Guarné A. 2015. The sliding clamp tethers the endonuclease domain of MutL to DNA. *Nucleic Acids Res* **43:**10746–10759.

Pukkila PJ, Peterson J, Herman G, Modrich P, Meselson M. 1983. Effects of high levels of DNA adenine methylation on methyl-directed mismatch repair in *Escherichia coli*. *Genetics* **104:**571–582.

Robertson AB, Matson SW. 2012. Reconstitution of the very short patch repair pathway from *Escherichia coli*. *J Biol Chem* **287:**32953–32966.

Rupp WD, Wilde CEI III, Reno DL, Howard-Flanders P. 1971. Exchanges between DNA strands in ultraviolet-irradiated *Escherichia coli*. *J Mol Biol* **61:**25–44.

Seigneur M, Bidnenko V, Ehrlich SD, Michel B. 1998. RuvAB acts at arrested replication forks. *Cell* **95:**419–430.

Sommer S, Knezevic J, Bailone A, Devoret R. 1993. Induction of only one SOS operon, *umuDC*, is required for SOS mutagenesis in *Escherichia coli*. *Mol Gen Genet* **239:**137–144.

Steinborn G. 1979. Uvm mutants of *Escherichia coli* K12 deficient in UV mutagenesis. II. Further evidence for a novel function in error-prone repair. *Mol Gen Genet* **175:**203–208.

Sutton MD, Farrow MF, Burton BM, Walker GC. 2001. Genetic interactions between the *Escherichia coli umuDC* gene products and the beta processivity clamp of the replicative DNA polymerase. *J Bacteriol* **183:**2897–2909.

Tang M, Shen X, Frank EG, O'Donnell M, Woodgate R, Goodman MF. 1999. $UmuD'_2C$ is an error-prone DNA polymerase, *Escherichia coli* pol V. *Proc Natl Acad Sci USA* **96:**8919–8924.

Teo I, Sedgwick B, Kilpatrick MW, McCarthy TV, Lindahl T. 1986. The intracellular signal for induction of resistance to alkylating agents in *E. coli*. *Cell* **45:**315–324.

Tham K-C, Hermans N, Winterwerp HHK, Cox MM, Wyman C, Kanaar R, Lebbink JHG. 2013. Mismatch repair inhibits homeologous recombination via coordinated directional unwinding of trapped DNA structures. *Mol Cell* **51:**326–337.

Trewick SC, Henshaw TF, Hausinger RP, Lindahl T, Sedgwick B. 2002. Oxidative demethylation by *Escherichia coli* AlkB directly reverts DNA base damage. *Nature* **419:**174–178.

Wang F, Yang W. 2009. Structural insight into translesion synthesis by DNA Pol II. *Cell* **139:** 1279–1289.

Weigle JJ. 1953. Induction of mutation in a bacterial virus. *Proc Natl Acad Sci USA* **39:**628–636.

Witkin EM. 1956. Time, temperature, and protein synthesis: a study of ultraviolet-induced mutation in bacteria. *Cold Spring Harb Symp Quant Biol* **21:**123–140.

Worth L Jr, Clark S, Radman M, Modrich P. 1994. Mismatch repair proteins MutS and MutL inhibit RecA-catalyzed strand transfer between diverged DNAs. *Proc Natl Acad Sci USA* **91:** 3238–3241.

A 0.5% ara

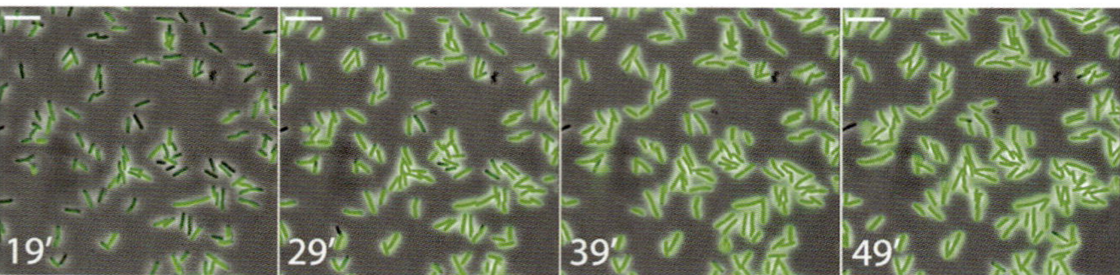

B 0.05% ara

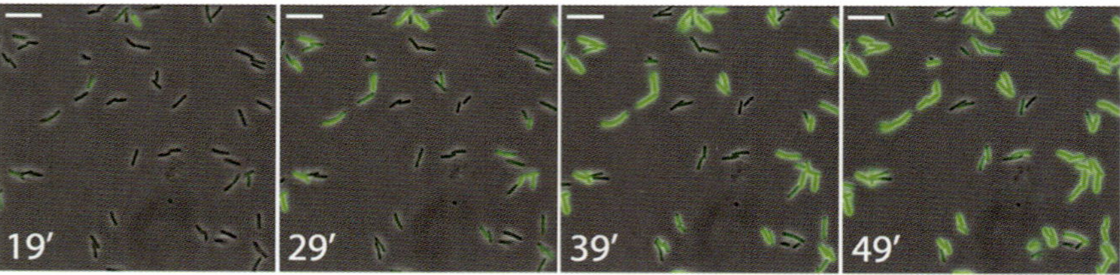

C 0.01% ara

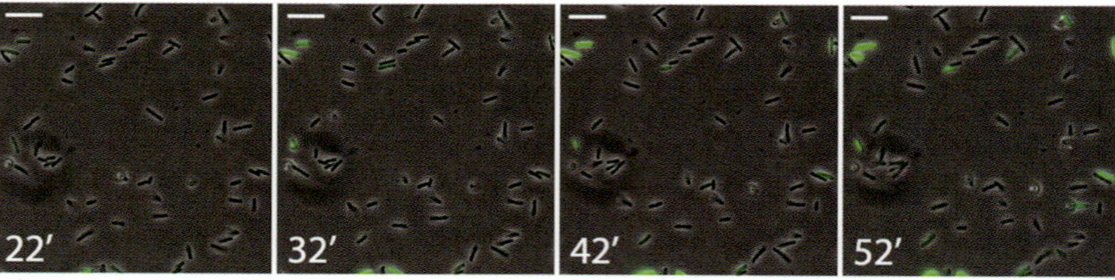

E. coli cells containing the *araBAD* regulatory region fused to a fluorescent reporter show changes in fluorescence during growth in varying amounts of the inducer arabinose. At low concentration of arabinose, nearly all of the cells are non-fluorescent, indicating very low *ara* operon expression in all but a few cells (panel C), and the proportion of fluorescent cells increases as the concentration of arabinose increases (panels B and A). Individual cells turn on expression during the incubation period (compare left to right for each concentration), rather than a general increase in fluorescence of every cell (see Fritz et al., Suggested Reading).

Regulation of Gene Expression: Genes and Operons

11

THE DNA OF A CELL contains thousands to hundreds of thousands of genes, depending on whether the organism is a relatively simple single-celled bacterium or a complex multicellular eukaryote, like a human. All of the features of the organism are due, either directly or indirectly, to the products of these genes. However, all the cells of a multicellular organism do not always look or act the same, despite the fact that they share essentially the same genes. Even the cells of a single-celled bacterium can look or act differently depending on the conditions under which the cells find themselves, because the genes of a cell are expressed at different levels. The process by which the expression of genes is turned on and off at different times and under different conditions is called the **regulation of gene expression**.

Cells regulate the expression of their genes for many reasons. A cell may express only the genes that it needs in a particular environment so that it does not waste energy making RNAs and proteins that are not needed at that time, or the cell may turn off genes whose products might interfere with other processes going on in the cell at the time. Cells also regulate their genes as part of developmental processes, such as sporulation. Genes can be expressed independently, as monocistronic units, or their expression can be coordinated through cotranscription in a polycistronic unit, or operon (see chapter 2). Groups of genes and operons can also be coordinately regulated via global regulatory mechanisms.

As described in chapter 2, the expression of a gene moves through multiple stages, each of which offers an opportunity for regulation. First, the information in the nucleotide sequence in the DNA is copied into RNA. Even if RNA is the final gene product, as is the case for rRNAs and tRNAs, the transcript may require further processing or modification for activity. If the final product of the gene is a protein, the mRNA synthesized from the gene might have to be processed before it can be translated into protein. Protein synthesis provides another opportunity for regulatory input. The protein might have to be further processed or transported to its final location to be active. Even after the gene product is synthesized in its final form, its activity might be modulated under certain environmental conditions by posttranslational modification or association with other cellular partners. The rate of degradation of the RNA or protein can also dramatically affect the levels of the final active product.

By far the most common type of regulation occurs at the first stage, when RNA is made. Genes that are regulated at this level are said to be **transcriptionally**

regulated. This form of gene regulation is the most efficient, since synthesizing mRNA that will not be translated seems wasteful. This type of regulation can occur either at the transcription initiation step or during transcription elongation, to modulate the amount of full-length mRNA that is made. However, not all genes are transcriptionally regulated, at least not exclusively. Examples in which the expression of a gene is regulated after RNA synthesis abound.

Any regulation that occurs after the gene is transcribed into mRNA is called **posttranscriptional regulation**. There are many types of posttranscriptional regulation; the most common is **translational regulation**. If a gene is translationally regulated, the mRNA may be continuously transcribed from the gene, but its translation is sometimes inhibited. Despite the apparent inefficiency of producing an mRNA that is not translated, it has advantages for certain classes of genes and is particularly useful when a rapid response is favored. Examples of a variety of mechanisms that can be used to regulate individual genes or operons are discussed in this chapter. Expression of multiple genes and operons can be coordinated in more complicated global regulatory systems, which are discussed in chapter 12.

Transcriptional Regulation in Bacteria

Thanks to the relative ease of doing genetics with bacteria, transcriptional regulation in these organisms is better understood than in other systems and has served as a framework for understanding transcriptional regulation in eukaryotic organisms. There are important differences between the mechanisms of transcriptional regulation in bacteria and eukaryotes, many of which are related to the presence of a nuclear membrane in eukaryotes and the complexity of the transcriptional machinery. Archaea are prokaryotes, like bacteria, but use a complex transcriptional apparatus that more closely resembles that found in eukaryotes. Despite these differences, many of the strategies used are similar throughout the biological world, and many general principles have been uncovered through studies of bacterial transcriptional regulation.

As discussed in chapter 2, most transcriptional regulation occurs at the level of transcription initiation at the promoter. Transcriptional regulation occurs primarily through proteins called **transcriptional regulators**, which usually bind to DNA, often with helix-turn-helix (HTH) motifs (Box 11.1). Regulation of transcription initiation can be either negative or positive, which is defined by whether the regulatory protein turns gene expression off or on (Table 11.1). If regulation is negative, expression is controlled by a **repressor** that binds to a repressor-binding site or **operator sequence** in the DNA and prevents initiation of transcription by RNA polymerase. If the regulation is positive, initiation of transcription is controlled by an **activator** that is required for efficient initiation by RNA polymerase at the promoter. The activity of a regulatory protein can also be changed in response to a physiological signal, such as the binding of a small molecule. A signal that increases gene expression (either by inactivating a repressor or by activating an activator) is called an **inducer**, while a signal that decreases gene expression (by activating a repressor or inactivating an activator) is called a **corepressor**. Inducers are often used for catabolic systems, where expression of genes involved in degradation of a substrate is induced only when the substrate is available (an **inducible** system). Expression of genes encoding biosynthetic pathways is often repressed by the end product of the pathway, so that the cell expresses these genes only when the cell needs more of the product of the pathway (a **repressible** system). Activators, repressors, inducers, and corepressors can be used in all four combinations, as shown in Table 11.1. Examples of each of these types of systems are described below.

Some regulatory proteins can be either repressors or activators, depending on where they bind to the promoter region. If the binding site overlaps the binding site for RNA polymerase to the promoter, the protein might sterically inhibit (i.e., physically get in the way of) the binding of RNA polymerase to the promoter and repress transcription. However, if the regulatory protein binds further upstream, it might make contacts with RNA polymerase that stabilize the binding of RNA polymerase to the promoter and activate transcription. Binding of a regulatory protein can have different effects on different promoters, depending on the characteristics of the promoter and its interactions with RNA polymerase.

While transcription initiation is a common level of gene regulation, transcription also can be regulated after RNA polymerase leaves the promoter. This can occur by premature termination of transcription, or **transcription attenuation**. Changes in the processivity of RNA polymerase can also be used to regulate transcription elongation, for example, by affecting the probability of termination at Rho-dependent terminators. As is the case for regulation of transcription initiation, transcription elongation and termination effects can be mediated by regulatory proteins and other factors that interact with the nascent RNA transcript. These factors can act either negatively (to repress gene expression) or positively (to activate gene expression). One of the most important initial steps in investigating the mechanism of gene regulation is to determine whether regulation is positive or negative (or even involves a combination of mechanisms).

BOX 11.1

The Helix-Turn-Helix Motif of DNA-Binding Proteins

Proteins that bind to DNA, including repressors and activators, often share similar structural motifs determined by the interaction between the protein and the DNA helix. One such motif is the helix-turn-helix (HTH) motif. A region of approximately 7 to 9 amino acids forms an α-helical structure called helix 1. This region is separated by about 4 amino acids from another α-helical region of 7 to 9 amino acids called helix 2. The two helices are at approximately right angles to each other—hence the name HTH. When the protein binds to DNA, helix 2 lies in the major groove of the DNA double helix, while helix 1 lies crosswise to the DNA, as shown in the figure. Because they lie in the major groove of the DNA double helix, the amino acids in helix 2 can contact and form hydrogen bonds with specific bases in the DNA. Thus, a DNA-binding protein containing an HTH motif recognizes and binds to specific regions on the DNA. Many DNA-binding proteins exist as dimers and bind to inverted-repeat DNA sequences. In such cases, the two polypeptides in the dimer are arranged head to tail so that the amino acids in helix 2 of each polypeptide can make contact with the same bases in the inverted repeats.

In the absence of structural information, the existence of an HTH motif in a protein can often be predicted from the amino acid sequence, since some sequences of amino acids cause the polypeptide to assume an α-helical form, and the bent region between the two helices usually contains a glycine. The presence of an HTH motif in a protein helps to identify it as a DNA-binding protein. Proteins with this domain can act as either repressors (such as LacI and GalR) or activators (such as CAP), depending on the location of the binding site in the target gene and whether it acts to prevent or facilitate binding of RNA polymerase to the DNA.

A variant on the HTH domain is the winged HTH domain, in which the "winged turn" is 10 amino acids or more in length, longer than the 3 or 4 amino acids of the "turns" of other HTH domains (see Kenney, References).

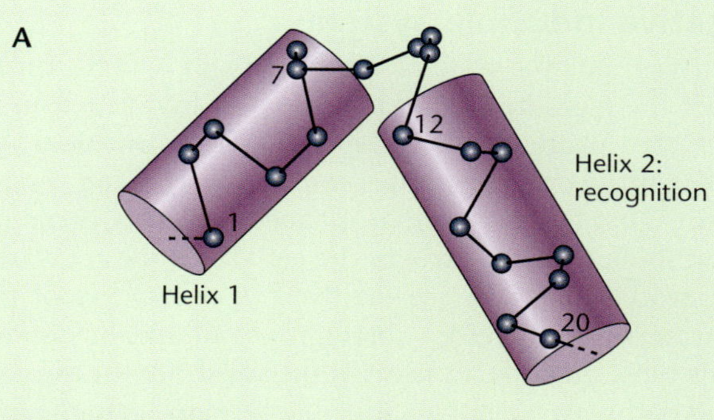

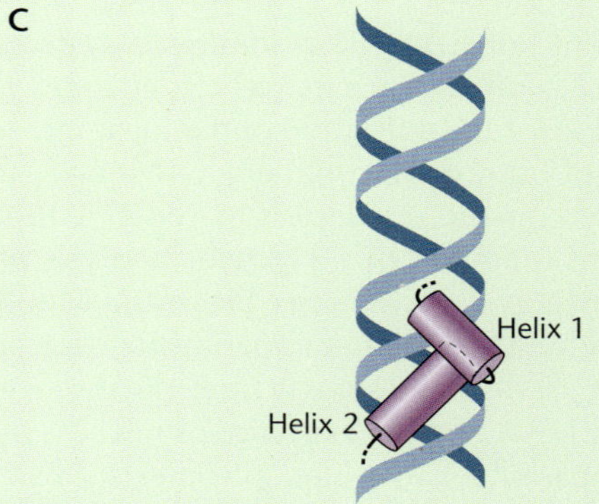

The HTH motif of DNA-binding proteins. (A) The structure of a helix-turn-helix (HTH) domain. **(B)** Number of amino acids in the HTH domain of the catabolite activator protein (CAP) protein. **(C)** Interactions of helix 1 and helix 2 with double-stranded DNA.

References

Kenney LJ. 2002. Structure/function relationships in OmpR and other winged-helix transcription factors. *Curr Opin Microbiol* **5**:135–141.

Steitz TA, Ohlendorf DH, McKay DB, Anderson WF, Matthews BW. 1982. Structural similarity in the DNA-binding domains of catabolite gene activator and *cro* repressor proteins. *Proc Natl Acad Sci USA* **79**:3097–3100.

Genetic Evidence for Negative and Positive Regulation

One of the easiest ways to see the differences between negatively and positively regulated genes or operons is to look at their behavior in genetic tests. One difference is in the effect of mutations that inactivate the regulatory gene. In **negative regulation**, a mutation that inactivates the regulatory gene allows transcription of the target genes, even in the absence of whatever physiological signal normally turns on expression. If the regulation is positive, mutations that inactivate the regulatory gene prevent transcription of the target genes, even in the presence of the regulatory signal, giving an **uninducible** phenotype. A mutant in which the target genes are always transcribed, even in the absence of the inducing signal, is called a **constitutive mutant**. Constitutive mutations

Table 11.1 Transcriptional regulatory systems

Effector molecule	Regulatory protein	
	Negative (repressor)	**Positive (activator)**
Inducer	Negative inducible (*E. coli lac* operon; LacI)	Positive inducible (*E. coli ara* operon; AraC)
Corepressor	Negative repressible (*E. coli trp* operon; TrpR)	Positive repressible (*E. coli fab* operon; FadR)

are much more common with negatively regulated systems than with positively regulated systems because any mutation that inactivates the repressor will result in the constitutive phenotype. With positively regulated systems, a constitutive phenotype can be caused only by mutational changes that do not inactivate the activator protein but instead alter it so that it can activate transcription without the inducing signal. Such changes tend to be rare and may not be possible for a particular activator.

Complementation tests reveal another important difference between negatively and positively regulated systems. Constitutive mutations of a negatively regulated system are often recessive to the wild type (i.e., the phenotype of the mutant is masked if the cell contains a second wild-type copy of the regulatory gene [see chapter 3 for genetic definitions]). This is because any normal repressor protein in the cell encoded by a wild-type copy of the gene binds to the operator and blocks transcription, even if the repressor encoded by the mutant copy of the gene in the same cell is inactive. In contrast, constitutive mutations in a positively regulated operon are often dominant to the wild type. A mutant activator protein that is active without the inducing signal might activate transcription even in the presence of a wild-type activator protein. In the following sections, we describe some examples of transcriptional regulatory systems and how genetic evidence contributed to our understanding of these systems.

Negative Regulation of Transcription Initiation

By definition, a negative regulatory system uses a repressor to turn off gene expression. A repressor can prevent initiation of transcription in a number of ways. The most common way is that the operator sequence overlaps with the promoter sequence so that binding of the repressor prevents binding of the RNA polymerase to the promoter. Alternatively, or in addition, the repressor might bend the promoter DNA so that RNA polymerase can no longer bind. The repressor might also hold the RNA polymerase on the promoter so that it cannot leave as it attempts to make RNA (see Rojo, Suggested Reading). The activity of the repressor can in turn be controlled by a physiological signal, often a small molecule. If the signal inactivates the repressor and therefore turns on expression of the system, then the system is inducible. If the signal activates the repressing activity of the repressor and turns off gene expression, the system is repressible.

Negative Inducible Systems

Negative inducible systems are systems in which expression of the regulated genes is turned off by the action of a repressor protein. The effector acts as an **inducer** because it turns on gene expression by inactivating the repressor. This type of regulatory mechanism is common for catabolic operons, where the cell needs the regulated gene products, which are involved in utilization of the substrate, only when the substrate is available. In systems of this type, the inducer is often the substrate for the regulated pathway. Negative inducible systems were the first type of regulatory system to be characterized and serve as a general paradigm for gene regulation. We will use the *Escherichia coli lac* and *gal* operons as examples.

THE *E. COLI lac* OPERON

The classic example of a negative inducible system is provided by the *E. coli lac* operon, which encodes the enzymes responsible for the utilization of the sugar lactose. The experiments of François Jacob, Jacques Monod, and their collaborators on the regulation of the *E. coli lac* genes are excellent examples of the genetic analysis of a biological phenomenon in bacteria (see Jacob and Monod, Suggested Reading). Although these experiments were performed in the late 1950s, only shortly after the discovery of the structure of DNA and the existence of mRNA, they still stand as the paradigm to which all other studies of gene regulation are compared. It is also important to note that while we refer to the structure of the *lac* operon throughout this section, Jacob and Monod had no such information available to guide their work.

Mutations of the *lac* Operon

When Jacob and Monod began their classic work, it was known that the enzymes of lactose metabolism are inducible in that they are expressed only when the sugar lactose is present in the medium. If no lactose is present, the enzymes are not made. From the standpoint of the cell, this is a sensible strategy, since there is no point in making the enzymes for lactose utilization unless lactose is available for use as a carbon and energy source.

To understand the regulation of the lactose genes, Jacob and Monod performed a genetic analysis (see chapter 3). First, they isolated many mutants in which lactose metabolism and regulation were affected. These mutants fell into two fundamentally different groups. Some mutants were unable to grow with lactose as the sole carbon and energy source and so were called Lac⁻ mutants. Other mutants made the lactose-metabolizing enzymes whether or not lactose was present in the medium and so were called constitutive mutants. Note that the fact that many constitutive mutants were isolated provided the first clue that regulation is in fact negative.

Complementation Tests with *lac* Mutations

Jacob and Monod needed to know which of the mutations affected *trans*-acting gene products—either protein or RNA—involved in the regulation and how many different genes these mutations represented. They also wished to know if any of the mutations were *cis* acting (affecting sites on the DNA involved in regulation).

To answer these questions, they first determined whether a particular *lac* mutation is dominant or recessive, which required that the organisms be diploid, with two copies of the genes being tested. Bacteria are normally haploid, with only one copy of each of their genes, but are "partial diploids" for any genes carried on an introduced F′ plasmid. Recall that an F′ plasmid is a plasmid into which some of the bacterial chromosomal genes have been inserted (see chapter 5). They introduced an F′ plasmid carrying the wild-type *lac* region into a strain with the *lac* mutation in the chromosome (Figure 11.1). If the partial diploid bacteria are Lac⁺ and can multiply to form colonies on minimal plates with lactose as the sole carbon and energy source, the *lac* mutation is recessive. If the partial diploid cells are Lac⁻ and cannot form colonies on lactose minimal plates, the *lac* mutation is dominant. Jacob and Monod discovered that most *lac* mutations are recessive to the wild type and so presumably inactivate genes whose products are required for lactose utilization.

The question of how many genes are represented by recessive *lac* mutations could be answered by performing pairwise complementation tests between different *lac* mutations, rather than between a *lac* mutation and the wild-type gene as described above. F′ plasmids carrying the *lac* region with one *lac* mutation were introduced into a mutant strain with another *lac* mutation in the chromosome (Figure 11.1). In this kind of experiment, if the partial diploid cells are Lac⁺, the two recessive mutations can complement each other and are members of different complementation groups, indicating that they represent different genes. If the partial diploid cells are Lac⁻, the two mutations cannot complement each other and are members of the same complementation group or are in the same gene. Jacob and Monod found that most of the *lac* mutations sorted into two different complementation groups, which they named *lacZ* and *lacY*. We now know of a third gene, *lacA*, which was not discovered in their original selections because its product is not required for growth on lactose or its regulation.

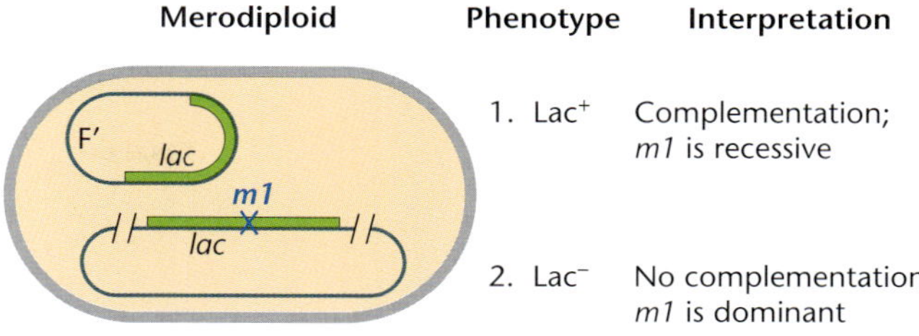

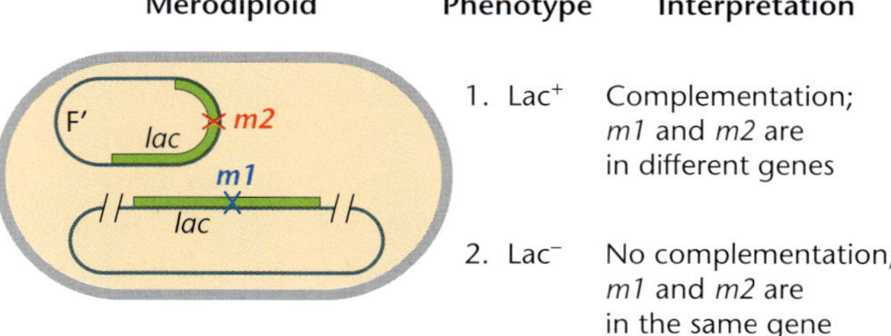

Figure 11.1 Complementation of *lac* mutations. One mutation (*m1*) is in the chromosome, and the other (*m2*) is in an F′ plasmid. If the two mutations complement each other, the cells will be Lac⁺ and will grow with lactose as the sole carbon and energy source. The mutations will not complement each other if they are in the same gene or if one affects a regulatory site or is polar.

cis-*Acting* lac *mutations.* Not all Lac⁻ mutants have *lac* mutations that affect diffusible gene products and can be complemented. Immediately adjacent to the *lacZ* mutations are other *lac* mutations that are much rarer and have radically different properties. These mutations cannot be complemented to allow the expression of the *lac* genes on the same DNA, even in the presence of wild-type copies of the *lac* genes. Mutations that cannot be complemented are *cis* acting and presumably affect a site on DNA rather than a diffusible gene product like an RNA or protein (see chapter 3).

To show that a *lac* mutation is *cis* acting, i.e., affects only the expression of genes on the same DNA where it occurs, we could introduce an F′ plasmid containing the potential *cis*-acting *lac* mutation into cells containing either a *lacZ* or *lacY* mutation in the chromosome (a *lacZ* mutation is shown in Figure 11.2). Any *trans*-acting gene products encoded by the *lacZ* (or *lacY*) gene on the F′ plasmid can complement the chromosomal *lacZ* (or *lacY*) mutation. However, if a Lac⁻ phenotype resulted, the *lac* mutation in the F′ plasmid must prevent expression of the LacZ (and LacY) proteins from the F′ plasmid. The mutation in the F′ plasmid is therefore *cis* acting.

As discussed below, Jacob and Monod named one type of the *cis*-acting *lac* mutations "*lacp* mutations" and

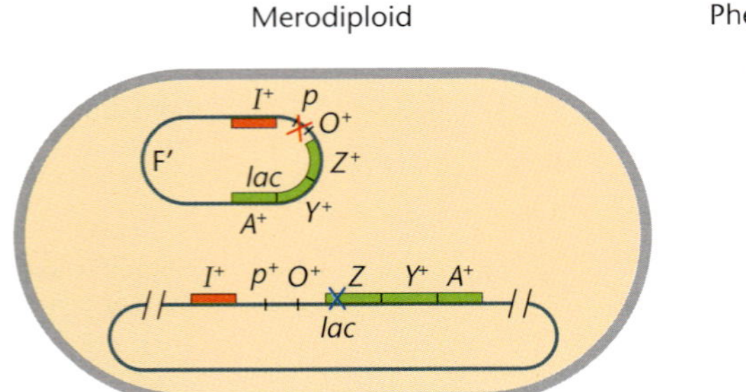

Figure 11.2 The p_{lac} mutations cannot be complemented and are *cis* acting. A p_{lac} mutation in the F′ plasmid prevents the expression of any of the other *lac* genes on F′, so a *lac* mutation in the chromosome will not be complemented. Partial diploid cells will be Lac−. Mutated *lac* regions are shown with an X.

hypothesized that they affect the binding of RNA polymerase to the beginning of the gene, i.e., are mutations in the promoter region p_{lac} (see chapter 2). We now know that along with promoter mutations, they also identified strong polar mutations in the beginning of the *lacZ* gene that also prevent the transcription of the downstream *lacY* gene (see chapter 2 and below).

Lac− mutants with dominant mutations. Some Lac− mutants have mutations that affect diffusible gene products but are dominant rather than recessive. A dominant *lac* mutation makes the cell Lac− and unable to use lactose even if there is another good copy of the *lac* operon in the cell, either in the chromosome or in the F′ plasmid. These dominant *lac* mutations are called *lacI*^{s} mutations, for "superrepressor mutations." As shown below, these mutants have mutations that change the repressor so that it can no longer bind the inducer.

Constitutive* lac *mutations. As mentioned above, some *lac* mutations do not make the cells Lac−, but instead make them constitutive, so that they express the *lacZ* and *lacY* genes even in the absence of the inducer lactose. In complementation tests between constitutive mutations, partial diploids are made in which either the chromosome or the F′ plasmid, or both, carry a constitutive mutation. The partial diploid cells are then tested to determine whether they express the *lac* genes constitutively or only in the presence of the inducer. If the partial diploid cells express the *lac* gene in the absence of the inducer, the constitutive mutation is dominant. However, if the partial diploid cells express the *lac* genes only in the presence of the inducer, the mutations are recessive. Using this test, Jacob and Monod found that some of the constitutive mutations, which could be recessive or dominant, affect a *trans*-acting gene product, either protein or RNA. Complementation between the recessive constitutive mutations revealed that they are all in the same gene, which these investigators named *lacI* (Figure 11.3A).

cis*-acting* lacO^c *mutations.* A rarer constitutive mutation in this and related regulatory systems is *cis* acting, allowing constitutive expression of the *lacZ* and *lacY* genes from the DNA that has the mutation, even in the presence of a wild-type copy of the *lac* DNA. Jacob and Monod named these *cis*-acting constitutive mutations *lac*O^c mutations, for *lac* operator-constitutive mutations. Figure 11.3B shows the partial diploid cells used in these complementation tests.

trans*-acting dominant constitutive mutations.* Some *lacI* mutations, called *lacI*$^{-d}$ mutations, are *trans* dominant, making the cell constitutive for expression of the *lac* operon even in the presence of a good copy of the *lacI* gene. These *lacI*$^{-d}$ mutations are possible because the LacI polypeptides form a homotetramer (a complex of four copies of the LacI protein). A mixture of normal and defective subunits can be nonfunctional, causing the constitutive LacI− phenotype. Hence, the *lacI*$^{-d}$ mutations are *trans* dominant, because one defective subunit in the tetramer results in an inactive complex. Table 11.2 summarizes the behavior of the various *lac* mutations in complementation tests.

The Jacob and Monod Operon Model

On the basis of this genetic analysis, Jacob and Monod proposed their **operon model** for *lac* gene regulation (Figure 11.4). The *lac* operon includes the genes *lacZ* and *lacY* and the third gene, *lacA*, which was identified later. These genes, known as the **structural genes** of the operon, encode the enzymes required for lactose utilization. The *lacZ* gene product is a β-galactosidase that cleaves lactose to form glucose and galactose, which can then be used by other pathways. The *lacY* gene product is a permease that allows lactose into the cell. The *lacA* gene product is a transacetylase whose function is still unclear.

The operon model explained why the structural genes are expressed only in the presence of lactose. The product of the *lacI* gene is a repressor protein. In the absence of lactose, this repressor binds to the operator sequence (*lacO*) close to the promoter and thereby prevents binding of RNA polymerase to the promoter and blocks the transcription of the structural genes. In contrast, when lactose is available, the inducer binds to the repressor and changes its conformation so that it can no longer bind to

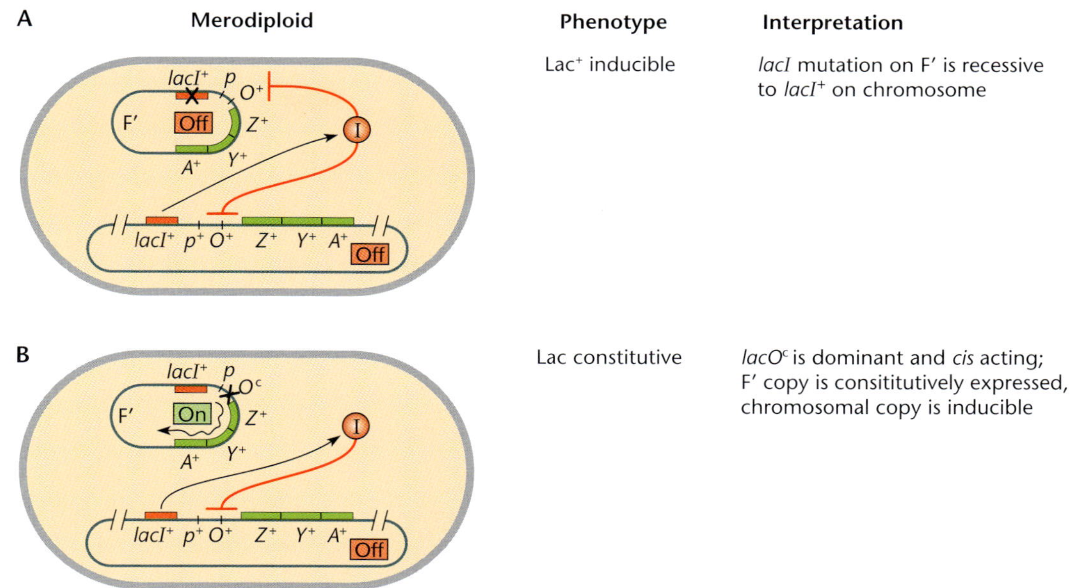

Figure 11.3 Complementation with two types of constitutive mutations. **(A)** The *lacI* mutation can be complemented, so other genes on the F′ plasmid are inducible in the presence of a wild-type copy of the *lacI* gene. **(B)** In contrast, *lacO*ᶜ mutations cannot be complemented by a wild-type *lacO* region in the chromosome and so are *cis* acting. Mutated regions are shown with an X.

the operator sequence. RNA polymerase can then bind to p_{lac} and transcribe the *lacZ*, *lacY*, and *lacA* genes. The LacI repressor is very effective at blocking transcription of the structural genes of the operon. Transcription is about 1,000 times more active in the absence of repressor than in its presence.

It is worth emphasizing how the Jacob and Monod operon model explains the behavior of mutations that affect the regulation of the *lac* enzymes. Mutants with *lacZ* and *lacY* mutations are Lac⁻ because they do not make an active β-galactosidase or permease, respectively, both of which are required for lactose utilization. These mutations are clearly *trans* acting, because they are recessive and can be complemented. An active β-galactosidase or permease made from another DNA in the same cell can provide the missing enzyme and allow lactose utilization.

Table 11.2 *lac* operon mutations

Mutation	Function affected	Phenotype
lacZ	β-Galactosidase	Lac⁻, recessive, *trans* acting
lacI	Operator binding	Constitutive, recessive, *trans* acting
*lacI*ˢ	Inducer binding	Uninducible, dominant, *trans* acting
lacI⁻ᵈ	Operator binding	Constitutive, dominant, *trans* acting
*lacI*ᑫ	Quantity of LacI	Inducible,[a] dominant, *trans* acting
*lacO*ᶜ	Repressor binding	Constitutive, dominant, *cis* acting
lacP	RNA polymerase binding	Uninducible, recessive, *cis* acting

[a]Requires higher concentration of inducer.

The behavior of p_{lac} mutations is also explained by the model. Jacob and Monod proposed that p_{lac} mutations change the sequence on the DNA to which the RNA polymerase binds. This explains why p_{lac} mutations are *cis* acting; if the site on DNA at which RNA polymerase initiates transcription is changed by a mutation so that it no longer binds RNA polymerase, the *lacZ*, *lacY*, and *lacA* genes on that DNA molecule are not transcribed into mRNA, even in the presence of a good copy of the p_{lac} region elsewhere in the cell.

Their model also explains the behavior of the two constitutive mutations: *lacI* (*trans* acting, recessive) and *lacO*ᶜ (*cis* acting, dominant). The *lacI* mutations affect a *trans*-acting function because they inactivate the repressor protein that binds to the operator and prevents transcription. The LacI repressor made from a functional copy of the *lacI* gene anywhere in the cell can bind to the operator sequence and block transcription in *trans*, masking the effect of the *lacI* mutation. However, the *lacO*ᶜ mutations change the sequence on DNA to which the LacI repressor binds to block transcription. The LacI repressor cannot bind to this altered *lacO* sequence, even in the absence of lactose. Therefore, RNA polymerase is free to bind to the promoter and transcribe the structural

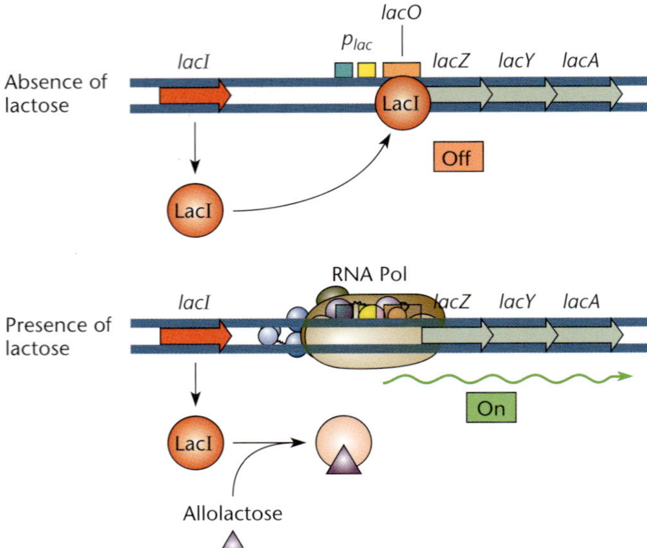

Figure 11.4 The Jacob and Monod model for negative regulation of the *lac* operon. In the absence of the inducer, lactose, the LacI repressor binds to the operator region, preventing transcription of the other genes of the operon by RNA polymerase (RNA Pol). In the presence of lactose, the repressor can no longer bind to the operator, allowing transcription of the *lacZ*, *lacY*, and *lacA* genes. It was later determined that the true inducer is allolactose, a metabolite of lactose, rather than lactose.

genes. The *lacO*^{c} mutations are *cis* acting and dominant because they allow constitutive expression of the *lacZ*, *lacY*, and *lacA* genes from the same DNA, even in the presence of a good copy of the *lac* operon elsewhere in the cell.

The existence of superrepressor *lacI*^{s} mutations is also explained by Jacob and Monod's model. These are mutations that change the repressor molecule so that it can no longer bind the inducer. The mutated repressor binds to the operator even in the presence of inducer, making the cells permanently repressed and phenotypically Lac$^-$. The fact that this type of mutation is dominant over the wild type is also explained. The mutated repressor proteins repress the transcription of any *lac* operon in the same cell, even in the presence of inducer and functional LacI protein, so they make the cell Lac$^-$ even in the presence of a good *lacI* gene, either in the chromosome or on an F′ plasmid.

The *lac* genes provide a good example of what is meant by "operon." An operon includes all the genes that are transcribed into the same mRNA plus any adjacent *cis*-acting sites that are involved in the transcription or regulation of transcription of the genes. The *lac* operon of *E. coli* consists of the three structural genes, *lacZ*, *lacY*, and *lacA*, which are transcribed into the same mRNA, as well as the *lac* promoter from which these genes are transcribed. It also includes the *lac* operator, since it is a *cis*-acting regulatory sequence involved in regulating the transcription of the structural genes. However, the *lac* operon does not include the gene for the repressor, *lacI*. The *lacI* gene is adjacent to the *lacZ*, *lacY*, and *lacA* genes and regulates their transcription, but it is not transcribed onto the same mRNA as the structural genes. Moreover, its product is *trans* acting rather than *cis* acting.

Refinements to the Regulation of the *lac* Operon

The operon model of Jacob and Monod has survived the passage of time. In 1965, it earned them a Nobel Prize, which they shared with André Lwoff. Because of its elegant simplicity, the operon model for the regulation of the *lac* genes of *E. coli* serves as the paradigm for understanding gene regulation in other organisms. The *lac* genes and *cis*-acting sites also have many uses in the molecular genetics of all organisms. They have been introduced into many other types of organisms, where they are used to study many aspects of gene regulation and developmental and cell biology.

While still largely intact, the operon model has undergone a few refinements over the years. As mentioned above, Jacob and Monod did not know of the existence of the *lacA* gene and thought that *lacY* encoded the transacetylase rather than the permease, which was unknown at the time. Also, most of the mutations that Jacob and Monod defined as p_{lac} were not, in fact, promoter mutations but, instead, were strong polar mutations in *lacZ* that prevent the transcription of all three structural genes (*lacZ*, *lacY*, and *lacA*). Later studies also revealed that the true inducer that binds to the LacI repressor is allolactose, a metabolite of lactose, rather than lactose. In most experiments, an analog of allolactose called isopropyl-β-D-thiogalactopyranoside (IPTG) is used as the inducer because it is not metabolized by the cells and therefore is maintained at a higher concentration.

The most significant refinement of the Jacob and Monod operon model came from the discovery that the LacI repressor can bind to not just one but three operators, called o_1, o_2, and o_3 (Figure 11.5). The operator closest to the promoter, o_1, seems to be the most important for repressing the transcription of *lac* expression and acts by sterically interfering with the binding of RNA polymerase to the promoter, because the sequence is centered at position +11 relative to the transcription start point (+1) and therefore overlaps with the position occupied by RNA polymerase during transcription initiation. Deletion of either o_2 or o_3 has little effect on repression. However, deleting both o_2 and o_3 diminishes repression as much as 50-fold.

Why does the *lac* operon have more than one operator, especially since the o_2 and o_3 operators are so far from the RNA polymerase binding site that it seems unlikely that they could block binding of the RNA polymerase to the

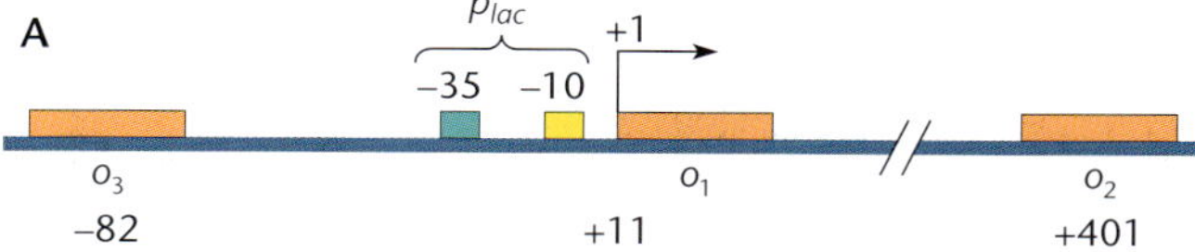

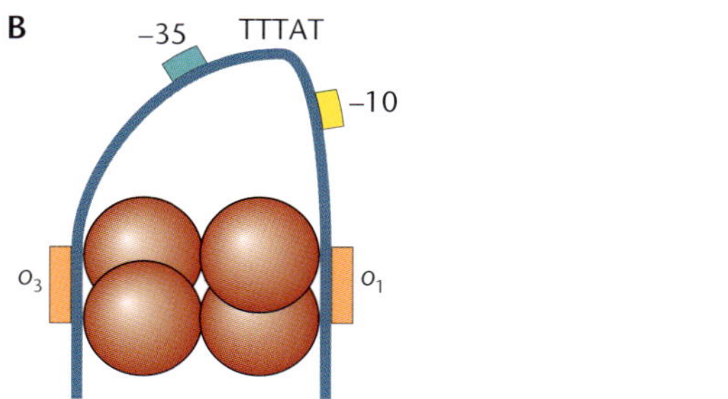

Figure 11.5 Locations of the three operators in the *lac* operon **(A)** and a model for how binding of the tetrameric LacI repressor to two of these operators may help repress the operon **(B)**. Repressor (red spheres) bound to o_1 and o_3, or o_1 and o_2 (not shown), can bend the DNA in the promoter region and stabilize binding of LacI to the DNA to help prevent binding of RNA polymerase to the promoter. The AT-rich region between the –35 and –10 sequences of the promoter may facilitate bending. Modified from Choy H, Adhya S, *in* Neidhardt FC, et al (ed), Escherichia coli *and* Salmonella: *Cellular and Molecular Biology*, 2nd ed. (ASM Press, Washington, DC, 1996).

promoter? The answer is that the presence of more than one operator allows one LacI repressor tetramer to bind to two operators simultaneously, bending the DNA (and the promoter, in the case of o_1 and o_3) between them to form a DNA loop (Figure 11.5B). This DNA loop stabilizes the interaction of LacI with the DNA. As discussed below, many other operons, including the *gal* and *ara* operons, also contain multiple operators that promote DNA looping in the promoter region. In the *lac* system, formation of the DNA loop assists in repression but is not required for regulation to occur. New technologies, including single-molecule studies that allow visualization of interactions of LacI and RNA polymerase on an individual template DNA molecule, are likely to provide new insights into the molecular mechanisms that underlie transcriptional regulation in this system (see Sanchez et al., Suggested Reading).

Catabolite Regulation of the *lac* Operon

In addition to being under the control of its own specific repressor, the *lac* operon is regulated in response to the availability of other carbon sources through **catabolite repression**. Catabolic pathways are used to break down substrates to yield carbon or energy. The catabolite repression system ensures that the genes for lactose utilization are not expressed if a better carbon and energy source, such as glucose, is available. The name "catabolite repression" is a misnomer, at least in *E. coli*, since the expression of operons under catabolite control in *E. coli* requires a transcriptional activator, the catabolite activator protein (CAP, historically called Crp, for catabolite repression protein), and the small-molecule effector cyclic AMP (cAMP). Many operons are under the control of CAP-cAMP, and we defer a detailed discussion of the mechanism of **catabolite regulation** until chapter 12, since it is a type of global regulation.

Structure of the *lac* Control Region

Figure 11.6 illustrates the structure of the *lac* control region in detail, showing the nucleotide sequences of the *lac* promoter and two of the operators (o_1 and o_3), as well as the region to which CAP binds. The *lac* promoter is a typical σ^{70} bacterial promoter with characteristic –10 and –35 regions (see chapter 2). One of the operators to which the LacI repressor binds (o_1) overlaps the initiation site (+1 in Figure 11.6A) for transcription of *lacZ*, *lacY*, and *lacA*. The other *lac* operator sequences are positioned upstream and downstream of the promoter, and the sequences of all three operator sites are shown in Figure 11.6B. Each symmetrical half of an operator binds a LacI monomer. Simultaneous binding of the LacI tetramer to two operator sites (e.g., o_1 and o_3, as shown in Figure 11.5B) allows each of the four subunits of the tetramer to interact with an operator half-site. The CAP-binding site, which enhances binding of RNA polymerase in the absence of glucose, is just upstream of the promoter, as shown.

Locations of *lacI* Mutations in the Three-Dimensional Structure of the LacI Repressor

When the *lacI* gene was sequenced, the exact amino acid changes in the *lac* repressor due to some of the mutations described above (and listed in Table 11.2) could be identified, which provided information about the regions of the LacI protein that are involved in the various functions of the repressor. For example, the *lacI*^{s} mutations should be largely confined to the site on the protein where the inducer binds, allowing that site to be identified. Likewise, *lacI*$^{-d}$ mutations should not inactivate the regions involved in dimer or tetramer formation, since the mutant proteins must form mixed dimers or tetramers with the wild-type polypeptide to be dominant. We would expect that constitutive *lacI*$^{-}$ mutations would be scattered around the protein, since they could affect almost any activity of the protein, including its ability to bind to DNA and its ability to fold into its final conformation. However, it was found that all of the various mutation types are often scattered around the gene and are not always concentrated in a certain region. Many years later, when there was a three-dimensional structure of the protein, the distribution of the mutations was easier to understand, since amino acids that are not in the same region

A The *lac* operon regulatory region

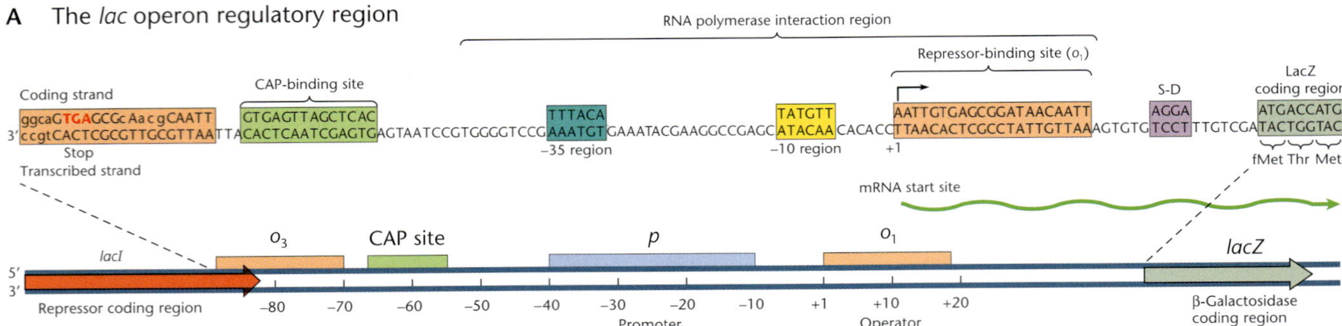

B The *lac* operator sequences

o_1 5′ AATTGTGAGCGGATAACAATT 3′

o_2 5′ AAaTGTGAGCGAGTAACAAcc 3′

o_3 5′ ggcaGTGAGCGcAacgCAATT 3′

Symmetrical operator halves

Figure 11.6 (A) DNA sequence of the promoter and operator regions of the *lac* operon. The entire region is only 100 bp long. The positions of the o_1 and o_3 operator sequences are shown. o_2 is centered at +401, in the LacZ coding region. **(B)** Alignment of the three natural *lac* operator sequences. Nucleotides of o_3 and o_2 that do not match o_1 are shown in lowercase.

in the linear polypeptide might be close to each other in the folded protein.

The LacI protein was crystallized, and its three-dimensional structure was determined by X-ray diffraction (see Lewis et al., Suggested Reading). Nuclear magnetic resonance spectroscopy was also used to determine its interaction with the *lac* operators and how this structure changes when the inducer IPTG is bound. Figure 11.7 shows the structure of the LacI repressor and the regions affected by amino acid changes due to some types of mutations. For example, the amino acid changes due to *lacI*^{s} mutations are often in the inducer-binding pocket (see Pace et al., Suggested Reading). The spacing between the DNA-binding N-terminal domains changes when inducer is bound, preventing its binding to the operator. An interaction in which inducer binding in one region of the protein can promote changes in another region is called an **allosteric interaction**. Some *lacI*^{s} mutations also change amino acids in the allosteric signaling region, the region that signals that inducer is bound to the DNA-binding domains in the inducer-binding pocket. This region doubles as the dimerization domain, which helps hold two monomers together, and changing the orientation of the two monomers may be part of the signal. The way in which some other repressors interact with their operators and how this interaction changes with the binding of inducers are discussed later in this chapter in the sections on the *gal* and *trp* repressors (see "The *E. coli gal* Operon" and "The *E. coli trp* Operon").

Experimental Uses of the *lac* Operon

The *lac* genes and regulatory regions have found many uses in molecular genetics (see Shuman and Silhavy, Suggested Reading). For example, the *lacZ* gene is probably

Figure 11.7 Three-dimensional structure of the LacI protein, showing regions devoted to its various functions and the sites of some types of mutational changes mentioned in the text. From Lewis M, Chang G, Horton NC, Kercher MA, Pace HC, Schumacher MA, Brennan RG, Lu P, *Science* **271:**1247–1254, 1996. Reprinted with permission from AAAS.

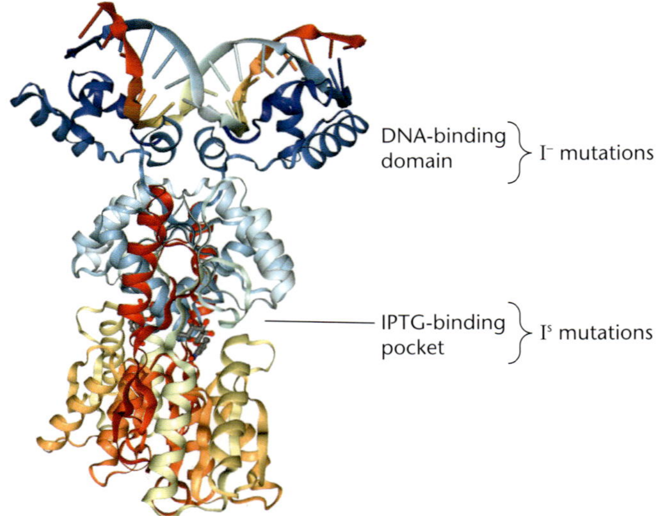

the most widely used reporter gene (see chapter 2) and has been introduced into a wide variety of different organisms, ranging from bacteria to fruit flies to human cells. It is so popular because its product, β-galactosidase, is easily detected by colorimetric assays using substrates such as X-Gal (5-bromo-4-chloro-3-indolyl-β-D-galactopyranoside) and ONPG (*o*-nitrophenyl-β-D-galactopyranoside). Also, the N-terminal portion of the protein is nonessential for activity, making it easier to make translational fusions. However, the *lacZ* polypeptide product is very large, which can be a disadvantage in some types of expression systems.

The *lac* promoter or its derivatives are used in many expression vectors (see chapter 4). This promoter offers many advantages in these expression vectors. It is fairly strong, allowing high levels of transcription of a cloned gene in *E. coli*. It is also inducible, which makes it possible to clone genes whose products are toxic to the cell. The cells can be grown in the absence of the inducer IPTG so that the cloned gene is not transcribed. When the cells reach a high density, the inducer IPTG is added and the cloned gene is expressed. Even if the protein is toxic and kills the cell when made in such large amounts, enough of the protein is usually synthesized before the cell dies.

There are many derivatives of the *lac* promoter in use. These derivatives retain some of the desirable properties of the wild-type *lac* promoter but have additional features. For example, the mutated *lac* promoter *lacUV5* is no longer sensitive to catabolite regulation and so is active even if glucose is present in the medium (see chapter 12). A hybrid *trp-lac* promoter called the *tac* promoter has also been widely used. The *tac* promoter has the advantages that it is stronger than the *lac* promoter and is insensitive to catabolite regulation but still retains its inducibility by IPTG.

Because it binds so tightly to its operator sequences, the LacI repressor protein also has many uses. One current use is to locate regions of DNA in the cell in bacterial cell biology. In one application, the LacI protein is translationally fused to green fluorescent protein, which can be detected in the cell by fluorescence microscopy. If this fusion protein is expressed in a cell in which multiple copies of the operator sequence have been introduced into a region of the chromosome, for example, close to the origin of replication, the fusion protein binds in multiple copies to the origin region of the chromosome. The location of the origin region of the chromosome can then be tracked in the cell as it goes through its cell cycle by monitoring the fluorescence given off by the fusion protein. The same system can be used to experimentally stall DNA replication forks to analyze mechanisms for chromosome organization (see Possoz et al., Suggested Reading).

THE *E. COLI gal* OPERON

The *gal* operon of *E. coli*, which is involved in the utilization of the sugar galactose, is another classic example of

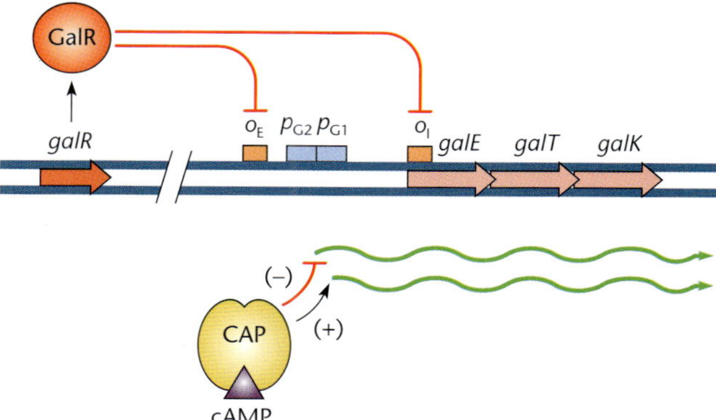

Figure 11.8 Structure of the galactose operon of *E. coli*. The *galE*, *galT*, and *galK* genes are transcribed from two promoters, p_{G1} and p_{G2}. Catabolite activator protein (CAP) with cyclic AMP (cAMP) bound turns on p_{G1} and turns off p_{G2}, as shown. There are also two operators, o_E and o_I. The repressor genes are some distance away, as indicated by the broken line. Only the *galR* repressor gene is shown.

negative regulation. Figure 11.8 shows the organization of the genes in this operon. Transcription initiates from two promoters, p_{G1} and p_{G2}. The products of three structural genes, *galE*, *galT*, and *galK*, are required for the conversion of galactose into glucose, which can then enter the glycolysis pathway. As in the *lac* operon, expression of the *gal* operon is induced during growth in the presence of galactose, which prevents binding of the repressor (a negative inducible system).

The specific reactions catalyzed by each of the enzymes of the *gal* pathway are shown in Figure 11.9. The *galK* gene product is a kinase that phosphorylates galactose to make galactose-1-phosphate. The product of the *galT* gene is a transferase that transfers the galactose-1-phosphate to UDP-glucose, displacing the glucose to make UDP-galactose. The released glucose can then be used as a carbon and energy source. The *galE* gene product is an epimerase that converts UDP-galactose to UDP-glucose to continue the cycle. It is not clear why *E. coli* cells use this seemingly convoluted pathway to convert galactose to glucose so that the latter can be used as a carbon and energy source. However, many organisms, including both plants and animals, use this pathway.

Unlike the genes for lactose utilization, not all of the genes for galactose utilization are closely linked in the *E. coli* chromosome. The *galU* gene, whose product synthesizes UDP-glucose, is located in a different region of the chromosome. Also, the genes for the galactose permeases, which are responsible for transporting galactose into the cell, are not part of the *gal* operon. Another difference from the *lac* operon is that not only are there two repressor genes, *galR* and *galS*, but they are nowhere near the operon, unlike the *lacI* gene, which is adjacent

Galactose + ATP —GalK→ Galactose-1-P + ADP

Galactose-1-P + UDP-glucose —GalT→ UDP-galactose + glucose-1-P

UDP-galactose —GalE→ UDP-glucose

UTP + glucose-1-P —GalU→ UDP-glucose

Figure 11.9 Pathway for galactose utilization in *E. coli*.

to the *lac* operon. This scattering of the genes for galactose metabolism reflects the fact that galactose not only serves as a carbon and energy source, but also plays other roles. For example, the UDP-galactose synthesized by the *gal* operon donates galactose to make polysaccharides for lipopolysaccharide and capsular synthesis.

Two *gal* Repressors: GalR and GalS

The two repressors that control the *gal* operon are GalR and GalS, encoded by the *galR* and *galS* genes, respectively. GalR was discovered first because mutations in *galR* cause constitutive expression of the *gal* operon. However, it was apparent that the *gal* operon is also subject to other levels of regulation. If the GalR repressor were solely responsible for regulating the *gal* operon, mutations that inactivate the *galR* gene should result in the same level of *gal* expression whether galactose is present or not. However, some regulation of the *gal* operon could be observed, even in *galR* mutants. When galactose was added to the medium in which *galR* mutant cells were growing, more of the enzymes of the *gal* operon were made than if the cells were growing in the absence of galactose. The product of another gene, *galS*, was found to be responsible for the residual regulation. Double mutants with mutations that inactivate both *galR* and *galS* are fully constitutive.

The GalS and GalR repressor proteins are closely related, and they both bind the inducer galactose. However, they play somewhat different roles. The GalR repressor is responsible for most of the repression of the *gal* operon in the absence of galactose. The GalS repressor plays only a minor role in regulating the *gal* operon but is the only repressor for the genes of the galactose transport system, which transports galactose into the cell. The reason for this two-tier regulation is unclear but is likely to be related to the diverse roles of galactose in the cell.

Two *gal* Operators

There are also two operators in the *gal* operon. One is upstream of the promoters, and the other is internal to the first gene, *galE* (Figure 11.8). The two operators are named o_E and o_I for operator external to the *galE* gene and operator internal to the *galE* gene, respectively. Both of the operator sequences are palindromic, similar in organization to the *lac* operator sequences.

The first mutant with an o_I mutation was isolated as part of a collection of constitutive mutants of the *gal* operon (see Irani et al., Suggested Reading). These mutants are easier to isolate in strains with superrepressor *galR*^{s} mutations than in wild-type *E. coli*. The *galR*^{s} mutations are analogous to *lacI*^{s} mutations. The superrepressor mutation makes a strain Gal$^-$ and uninducible because galactose cannot bind to the mutated repressor. Therefore, *E. coli* with a *galR*^{s} mutation cannot multiply to form colonies on plates containing only galactose as the carbon and energy source. However, a constitutive mutation that inactivates the GalR^s repressor or changes the operator sequence prevents the mutant repressor from binding to the operator and allows the cells to use the galactose and multiply to form a colony. Thus, if bacteria with a *galR*^{s} mutation are plated on medium with galactose as the sole carbon and energy source, only constitutive mutants multiply to form a colony. However, most of the constitutive mutants isolated in this way have mutations in the *galR* gene that inactivate the GalR^s repressor rather than operator mutations, since the operator is by far the smaller target. Many *galR* mutants would have to be screened before a single operator mutant was found. Therefore, to make this method practicable for isolating constitutive mutants with operator mutations, the frequency of *galR* mutants must be decreased until it is not too much higher than that of constitutive mutants with mutations in the operator sequences.

One way to reduce the frequency of *galR* mutants is to use a strain that is diploid for (has two copies of) the *galR*^{s} gene. Then, even if one *galR*^{s} gene is inactivated by a mutation, the other *galR*^{s} gene continues to make the GalR^s protein, making the cell phenotypically Gal$^-$. Two independent mutations, one in each *galR*^{s} gene, are required to make the cell constitutive. Since the frequency of two independent mutations is the product of the frequency of each of the single mutations, the presence of two independent *galR* mutations should be very rare, probably no more frequent than single operator mutations, making cells with operator mutations a significant fraction of the total constitutive mutants and easier to identify. Moreover, constitutive mutants with operator mutations can be distinguished from the double mutants with mutations in both *galR*^{s} genes by the locations where they map. Operator mutations map in the *gal* operon, unlike mutations in the *galR*^{s} genes, which map elsewhere in the genome.

Accordingly, a partial diploid was constructed that contains one copy of the *galR*^{s} gene in the normal position and another copy in a specialized transducing λ phage integrated at the λ attachment site (see chapter 7). When this strain was plated on medium containing galactose as the sole carbon and energy source, a few Gal⁺ colonies arose due to constitutive mutants. The mutations in two of these constitutive mutants mapped in the region of the *gal* operon and so were presumed to be operator mutations. When the DNA of the two mutants was sequenced, it was discovered that one mutation had changed a base pair in the known operator region (o_E) just upstream of the promoters, as expected. However, the other operator mutation had changed a base pair downstream in the *galE* gene, suggesting that a sequence in that gene also functions as an operator. This operator was named o_I. Furthermore, the o_I mutation occurred in a sequence that was similar to the 15-bp known o_E operator. Moreover, the mutation in the *galE* gene was *cis* acting for constitutive expression of the *gal* operon, one of the criteria for an operator mutation.

As was observed for the *lac* operon, both the o_E and o_I sites are important for regulation. However, unlike *lac*, in which at least two functional operators are required for full repression, but o_1 alone provides significant levels of repression, repression of the *gal* operon requires both operators. Figure 11.10 shows a model for how the two operators cooperate to block transcription in the absence of galactose. Repressor molecules bound to the two operators interact with each other to bend the DNA of the promoter that is located between the two operators. The bent promoter cannot bind RNA polymerase, because the –35 and –10 regions that need to contact the σ subunit are misaligned, so there is no initiation of transcription at the promoter. Repressor bound to only one of the two operators might still interfere with transcription to some extent, but the promoter would not be bent, and the repression is therefore much less severe.

Further studies showed that the position of the two operator sites relative to each other is critical for regulation. In addition, a basic DNA-binding protein called HU binds between the two operators and introduces a 180° turn in the DNA that helps to position the operators so that GalR dimers bound at each of the two operator sites can interact with each other to form a stable DNA loop (see Geanacopoulos et al., Suggested Reading). The overall complex is called the repressosome (Figure 11.10). An additional requirement is that the regulation normally occurs only on supercoiled DNA; the supercoiling of the DNA is presumably required for accurate placement of the GalR dimers to form the DNA loop required for repression.

Two *gal* Promoters and Catabolite Regulation of the *gal* Operon

As mentioned, the *gal* operon has two promoters called p_{G1} and p_{G2} (Figure 11.8). One of these promoters is like the *lac* promoter in that it is poorly expressed if a better carbon source, such as glucose, is available. However, the other promoter is active even in the presence of glucose and continues to promote expression of the Gal enzymes so that other cellular constituents containing galactose can continue to be made. Catabolite regulation is discussed in more detail in chapter 12. Although only one of the promoters is subject to catabolite regulation, both are repressed by GalR in the absence of galactose.

A

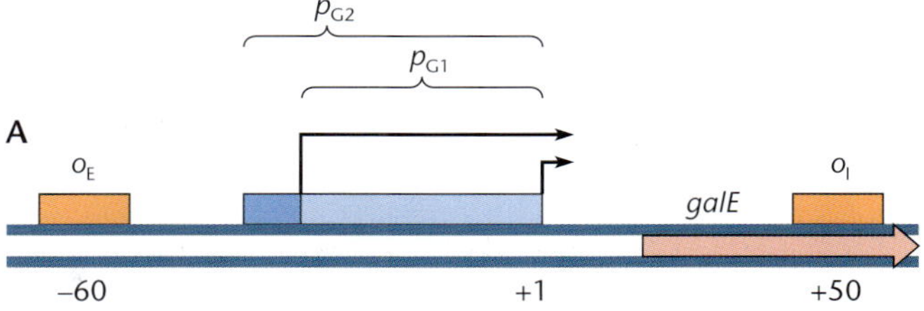

B

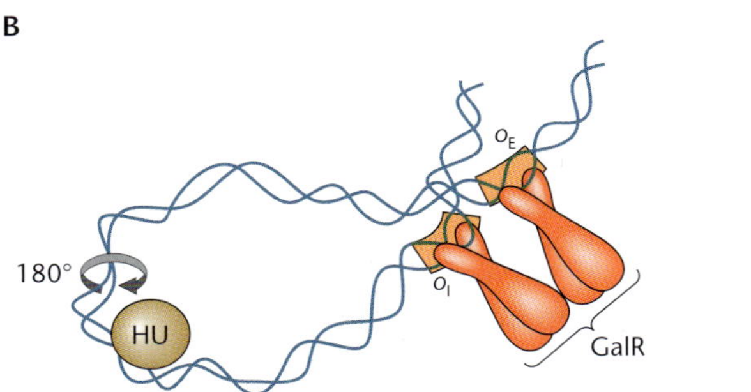

Figure 11.10 Formation of the *gal* operon repressosome. **(A)** Structure of the *gal* operon regulatory region. **(B)** Two dimers of the *galR* repressor gene product bind to each other and to the operators o_E and o_I to bend the DNA between the operators. A DNA-binding protein, HU, introduces a 180° twist in the DNA (shown by the arrow) so that the repressor can bind to the two operators. Bending of the DNA in the promoter regions inactivates the promoters.

Negative Repressible Systems

The enzymes encoded by the *lac* and *gal* operons are involved in the catabolism of specific sugars as sources of carbon and energy. Consequently, these operons are called **catabolic operons** or **degradative operons**. As described for the *lac* and *gal* systems, expression is induced only when the substrate of the enzymes encoded in the regulated operons is available to the cell, i.e., the appropriate sugar (lactose or galactose) acts as an inducer that inactivates the repressor. Not all operons are involved in degrading compounds, however. The enzymes encoded by some operons synthesize compounds needed by the cell, such as nucleotides, amino acids, and vitamins. These operons are called **biosynthetic operons**.

The regulation of a biosynthetic operon is essentially opposite to that of a degradative operon. The enzymes of a biosynthetic pathway should not be synthesized in the presence of the end product of the pathway, since the product is already available and energy should not be wasted in synthesizing more. However, the basic mechanisms by which degradative and biosynthetic operons are regulated are often similar. Like catabolic operons, biosynthetic operons can also be regulated negatively by repressors. If the genes of a biosynthetic operon are constitutively expressed when the regulatory gene is inactivated by mutation, even if the end product of the pathway is present in the medium, the biosynthetic operon is negatively regulated, and systems of this type are referred to as **negative repressible**.

The terminology used to describe the negative regulation of biosynthetic operons differs somewhat from that used for catabolic operons, despite shared principles. In a negative repressible system, the effector that binds to the repressor and allows it to bind to the operators is called the **corepressor** because it functions together with the repressor to promote repression. A repressor that negatively regulates a biosynthetic operon is not active in the absence of the corepressor and in this state is called the **aporepressor**. However, once the corepressor is bound, the protein is able to bind to the operator and so is now called the repressor.

THE *E. COLI trp* OPERON

The tryptophan (*trp*) operon of *E. coli* is the classic example of a negative repressible system (see Yanofsky and Crawford, Suggested Reading). The enzymes encoded by the *trp* operon (Figure 11.11) are responsible for synthesizing the amino acid L-tryptophan, which is a constituent of most proteins and so must be synthesized if none is available in the medium. The products of five structural genes in the operon are required to make tryptophan from chorismic acid. These genes are transcribed from a single promoter, p_{trp}. The *trp* operon is negatively regulated by the TrpR repressor protein, whose gene (*trpR*), like the *gal* repressor genes, is unlinked to the rest of the operon. This may reflect the fact that TrpR regulates not only the *trp* operon, but also the *aroH* operon, which encodes enzymes involved in synthesis of chorismic acid, a precursor for other pathways in addition to tryptophan biosynthesis.

Figure 11.12 shows the model for the regulation of the *trp* operon by the TrpR repressor. Binding of the TrpR repressor to the operator prevents transcription from the p_{trp} promoter. However, the TrpR repressor can bind to the operator only if the corepressor tryptophan is present in the cell. Tryptophan binds to the TrpR aporepressor protein and changes its conformation so that it can bind to the operator.

The TrpR repressor has been crystallized, and its structure has been determined in both the aporepressor form (when it cannot bind to DNA) and the repressor form with tryptophan bound (when it can bind to the operator). These structures have led to a satisfying explanation of how tryptophan binding changes the TrpR protein so that it can bind to the operator (Figure 11.13). The TrpR repressor is a dimer, and each copy of the *trpR* polypeptide has α-helical structures (shown as cylinders in the figure). Helices D and E form the HTH DNA-binding domain (Box 11.1). Helix D corresponds to helix 1, the nonspecific DNA-binding helix, and helix E corresponds to helix 2, the DNA sequence-specific recognition helix. In the aporepressor state, the conformations of the two HTH domains in the dimer do not allow proper interactions with successive major grooves appearing on one side of the DNA helix. Binding of the tryptophan corepressor alters the HTH conformations, allowing the repressor to bind to the operator.

Isolation of *trpR* Mutants

As is the case for other negatively regulated operons, constitutive mutations of the *trp* operon are quite common, and most map in *trpR*, inactivating the product of the

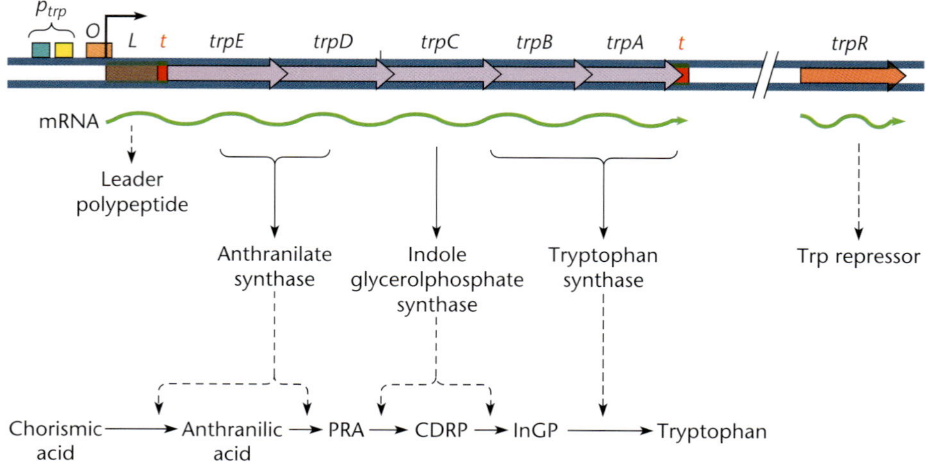

Figure 11.11 Structure of the tryptophan biosynthetic (*trp*) operon of *E. coli*. The structural genes *trpEDCBA* are transcribed from the promoter p_{trp}. Upstream of the structural genes is a short coding sequence for the leader peptide called *trpL*. The *trpR* repressor gene is unlinked, as shown by the broken line. PRA, phosphoribosyl anthranilate; CDRP, 1-(*o*-carboxyphenylamino)-1-deoxyribulose-5-phosphate; InGP, indole 3-glycerol phosphate.

A In the absence of tryptophan

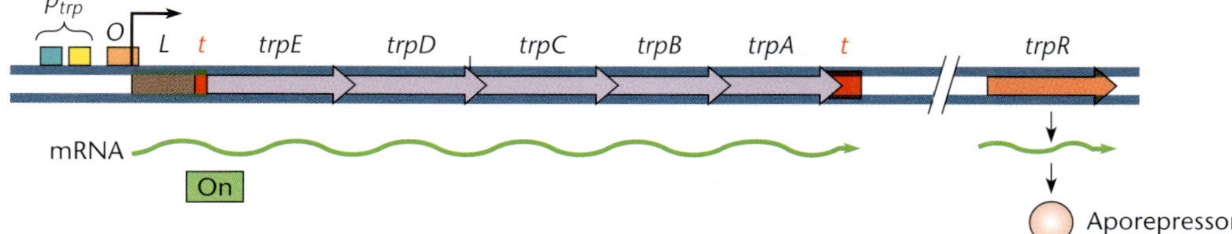

B In the presence of tryptophan

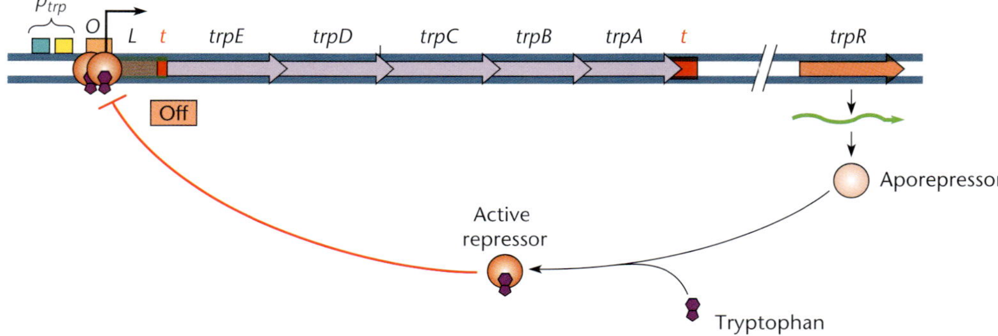

Figure 11.12 Negative regulation of the *trp* operon by the TrpR repressor. Binding of the corepressor tryptophan to the aporepressor converts it to the repressor conformation, which is active in DNA binding.

gene or preventing binding of the corepressor. Mutants with constitutive expression of the *trp* operon can be obtained by selecting for mutants resistant to the tryptophan analog 5-methyltryptophan in the absence of tryptophan. This tryptophan analog binds to the TrpR repressor and acts as a corepressor. However, 5-methyltryptophan cannot be used in place of tryptophan in protein synthesis. Therefore, in the presence of the analog, the *trp* operon is not induced even in the absence of tryptophan, and the cells will starve for tryptophan. Only constitutive mutants that continue to express the genes of the *trp* operon in the presence of 5-methyltryptophan can multiply to form colonies on plates with this analog but without tryptophan.

Other Types of Regulation of the *E. coli trp* Operon

The *trp* operon is also subject to a second level of regulation called **transcription attenuation**. This type of regulation is discussed below. Also, as in many biosynthetic pathways, the first enzyme of the *trp* pathway is subject to **feedback inhibition** by the end product of the pathway, tryptophan. We also return to feedback inhibition below.

Molecular Mechanisms of Transcriptional Repression

Repression of transcription initiation can occur by a variety of mechanisms. The simplest mechanism is steric hindrance, where binding of the repressor protein to the DNA physically blocks access of RNA polymerase to the promoter site. The *E. coli* LacI protein provides an example of this mechanism. A second possibility is that binding of the repressor protein to the DNA results in a change in DNA structure, like a DNA loop, that in turn prevents binding of RNA polymerase to the promoter, as is the case for the *E. coli gal* operon. More complicated DNA structural changes also can occur, some of which are discussed in the context of global regulators (see chapter 12). Another possibility is that a repressor does not directly inhibit transcription but, instead, prevents the positive activity of an activator that is required for efficient transcription. This mechanism, which is termed antiactivation, is discussed below in the context of transcriptional activation.

Positive Regulation of Transcription Initiation

The first part of this chapter covers the classic examples of negative regulation by repressors. However, many operons are regulated positively by activators. An operon under the control of an activator protein is transcribed only in the presence of that protein. As we saw in the discussion of negative regulation, the activity of the activator can be activated by an inducer (e.g., in the arabinose operon) or inactivated by a corepressor (e.g., in the fatty acid biosynthetic operon). Activators often work by increasing

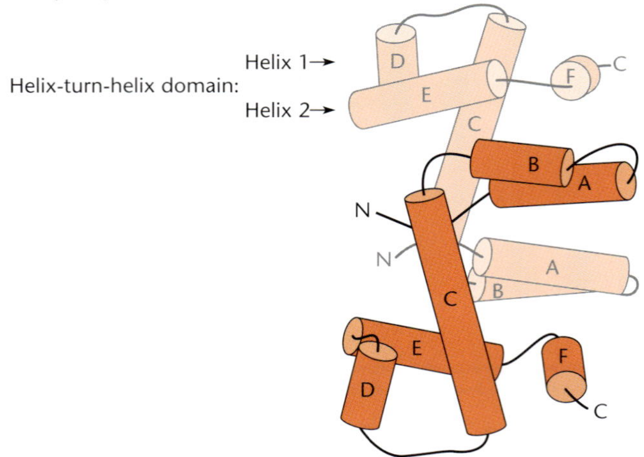

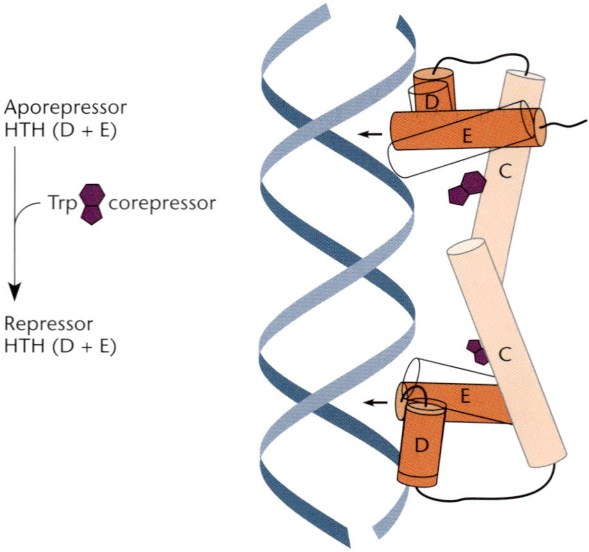

Figure 11.13 Structure of the TrpR repressor and an illustration of how tryptophan binding allows it to convert from the aporepressor to the repressor that binds to the operator. **(A)** The helices (shown as cylinders) of the aporepressor dimer in the inactive state with no tryptophan bound. **(B)** The shift of the E cylinders shows the repositioning of the helix-turn-helix (HTH) domains of the active repressor with tryptophan bound.

the tightness of binding of the RNA polymerase to the promoter, by allowing it to open the strands of DNA at the promoter, or by rotating and bending the promoter DNA to bring the recognition sites together.

Positive Inducible Systems

Positive regulatory systems utilize an activator to turn on gene expression, often by aiding in the recruitment of RNA polymerase. However, in positive inducible systems, the activator protein is unable to bind to DNA to increase transcription in the absence of its effector molecule, which serves as an inducer, or can bind but cannot activate transcription without the effector. As is the case with other inducible systems, positive inducible systems are often involved in catabolism of the inducer or a related molecule, so that expression of the regulated genes is necessary only when the substrate of the pathway, often the inducer itself, is present.

THE *E. COLI ara* OPERON

The *E. coli ara* operon was the first example of **positive regulation** in bacteria to be discovered (although it actually shows many levels of regulation [see below]). The *ara* operon is responsible for conversion of the five-carbon sugar L-arabinose into D-xylulose-5-phosphate, which can be used by other pathways. *E. coli* can also utilize D-arabinose, an isomer of L-arabinose, but the enzymes for D-arabinose utilization are encoded by a different operon, which lies elsewhere in the chromosome.

Figure 11.14A illustrates the structure of the *ara* operon. Three structural genes in the operon, *araB*, *araA*, and *araD*, are transcribed from a single promoter, p_{BAD}. Upstream of the promoter is the activator region, *araI*, where the activator protein AraC binds to activate transcription in the presence of L-arabinose, and the CAP site, at which CAP binds to mediate catabolite regulation (see chapter 12). There are also two operators, *ara*O_1 and *ara*O_2, at which the AraC protein binds to repress transcription (see below). The *araC* gene, which encodes the regulatory protein, is also shown. This gene is transcribed from the promoter p_C in the opposite direction from *araBAD*, as shown by the arrows in the figure. As described below, the AraC protein is a positive activator of transcription. As such, it is a member of a large family of activator proteins (Box 11.2).

Genetic Evidence for Positive Regulation of the *ara* Operon

Early genetic evidence indicated that the *lac* and *ara* operons are regulated by very different mechanisms (see Englesberg et al. and Schleif, Suggested Reading). One observation was that loss of the regulatory proteins results in very different phenotypes. For example, deletions and nonsense mutations in the *araC* gene—mutations that presumably inactivate the protein product of the gene—lead to an uninducible phenotype in which the genes of the *ara* operon are not expressed, even in the presence of the inducer arabinose. Recall that deletion or nonsense mutations in the regulatory gene of a negatively regulated operon such as *lac* result in a constitutive phenotype, not an uninducible phenotype. Another difference between *ara* and negatively regulated operons is in the frequency of constitutive mutants. Mutants that con-

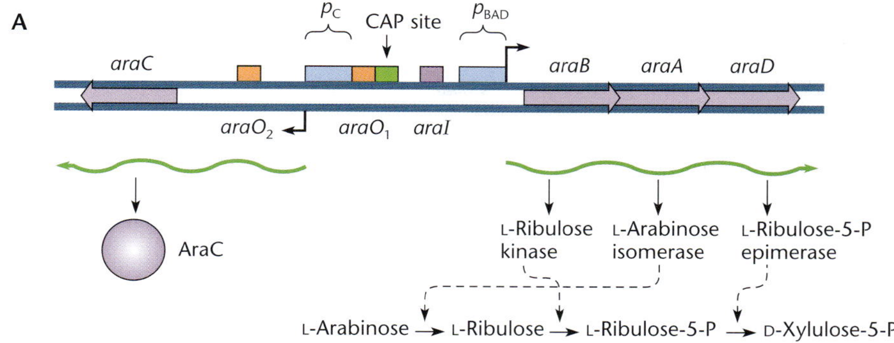

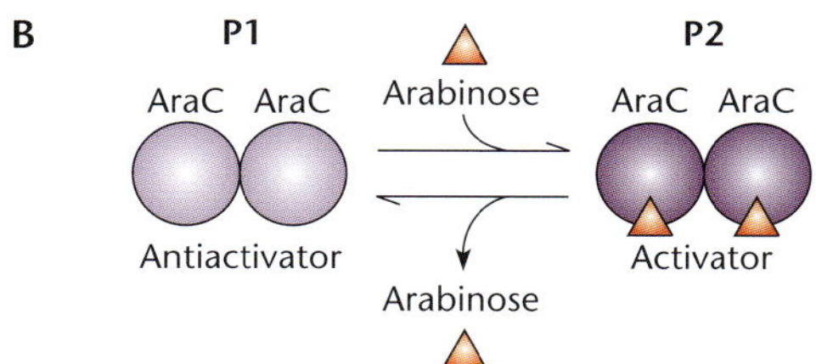

Figure 11.14 (A) Structure and function of the L-arabinose operon of *E. coli.* **(B)** Binding of the inducer L-arabinose converts the AraC protein from an antiactivator P1 form to an activator P2 form. See the text for details.

stitutively express a negatively regulated operon are relatively common, because any mutation that inactivates the repressor gene causes constitutive expression. However, mutants that constitutively express *ara* are very rare, which suggests that mutations that result in the constitutive phenotype do not merely inactivate AraC.

Isolating Constitutive Mutations of the *ara* Operon

Because constitutive mutations of the *ara* operon are so rare, special tricks are required to isolate them. One method for isolating rare constitutive mutations in *araC* uses the anti-inducer D-fucose. This anti-inducer binds to the AraC protein and prevents it from binding L-arabinose, thereby preventing induction of the operon. As a consequence, wild-type *E. coli* cannot multiply to form colonies on agar plates containing D-fucose with L-arabinose as the sole carbon and energy source because the cells are unable to induce *ara* expression. Only mutants that constitutively express the genes of the *ara* operon can form colonies under these conditions.

Another selection for constitutive mutations in the L-arabinose operon cleverly plays off the operon responsible for the utilization of its isomer, D-arabinose. The enzymes produced by the L-arabinose and D-arabinose operons cannot use the intermediate of the other operon, with one exception. The product of the *araB* gene (the ribulose kinase enzyme of the L-arabinose operon pathway, which phosphorylates L-ribulose as the second step of the pathway) can also phosphorylate D-ribulose, so that the L-arabinose kinase can substitute for that of the D-arabinose operon. Nevertheless, *E. coli* mutants that lack the D-arabinose kinase cannot multiply to form colonies on plates containing only D-arabinose as a carbon and energy source, because D-arabinose is not an inducer of the L-*ara* operon. Only constitutive mutants of the L-*ara* operon can grow if D-arabinose kinase-deficient mutants are plated on agar plates containing D-arabinose as the sole carbon and energy source.

A Model for Positive Regulation of the *ara* Operon

The contrast in phenotypes between mutations that inactivate the *lacI* and *araC* genes led to an early model for the regulation of the *ara* operon. According to this early model, the AraC protein can exist in two states, called P1 and P2 (Figure 11.14B). In the absence of the inducer, L-arabinose, the AraC protein is in the P1 state and inactive. If L-arabinose is present, it binds to AraC and changes the protein conformation to the P2 state. In this state, AraC binds to the DNA at the site called *araI* (Figure 11.14A) in the promoter region and activates transcription of the *araB*, *araA*, and *araD* genes.

This model explained some, but not all, of the behavior of the *araC* mutations. It explained why mutations in *araC* that cause the constitutive phenotype are rare but do occur at a very low frequency. According to this model, these mutations, called *araC*^{c} mutations, change AraC so that it is permanently in the P2 state, even in the absence of L-arabinose, and thus the operon is always transcribed. Such mutations would be expected to be very rare because only a few amino acid changes in the AraC protein could

BOX 11.2

Families of Regulators

The techniques of comparative genomics have made it possible to identify repressor and activator genes in a wide variety of bacteria. These transcriptional regulatory proteins belong to a limited number of known families, based on sequence and structural conservation, even though they regulate operons with very different functions and respond to different effectors. Transcriptional regulators can be assigned to a family based on sequence and structural homology, the organization of their motifs, and whether they use the same motif to bind to DNA. Many of them use an HTH motif or a winged HTH motif (Box 11.1), while others use a looped-hinge helix or a zinc finger motif, among others. There are at least 15 different families of transcriptional regulators. Some of these families are quite large, with many members. Some families consist only of repressors, others consist only of activators, and some consist of both repressors and activators. Most of the families are named after the first member of the family to be studied. For example, the LacI family includes GalR and consists mostly of repressors that regulate operons involved in carbon source utilization and generally also respond to CAP and cAMP. These repressors usually function as homotetramers and have an HTH DNA-binding motif to bind to DNA at their N termini, an effector-binding motif in the middle, and dimerization and tetramerization motifs at their C termini.

The TetR family of repressors, named after the repressor that regulates the tetracycline resistance gene in Tn*10*, is even larger and includes LuxR, which regulates light emission in chemiluminescent bacteria (see chapter 12). These family members are found in all types of bacteria, and they repress operons involved in a variety of functions, including antibiotic resistance and synthesis, osmotic regulation, efflux pumps for multidrug resistance, and virulence genes in pathogenic bacteria. The TetR repressor itself is widely used to regulate gene expression in eukaryotic cells because it binds very tightly to its operators and because tetracycline readily diffuses into eukaryotic cells. TetR has an HTH motif that binds to DNA, but only when the repressor has not bound tetracycline, and partially unwinds the operator DNA through its major groove. Presumably, this can change the structure of the DNA at the promoter. Interestingly, a member of this family that regulates a multidrug efflux pump has a very broad effector-binding pocket, allowing it to bind a variety of antibiotics.

The AraC family of activators is also very large. These activators seem to fall into at least two subfamilies, those that regulate carbon source utilization, such as AraC, and function as dimers, and those that respond to stress responses, such as SoxS, and function as monomers. A signature of this family is that the HTH motif that binds to DNA is in the C-terminal part of the protein, not in the N terminus like many other activators.

Other large classes are the LysR activators and the NtrC activators. The NtrC activators are particularly interesting in that they activate transcription only from σ^N promoters (the nitrogen sigma factor [see chapter 12]). These activators are organized into distinguishable domains (see Box 12.3), the structure of which depends on whether the activity of the protein is modulated by binding of an effector molecule or by phosphorylation of the activator by a partner kinase protein (see chapter 12). The domain of the activator protein that either binds the inducer or is phosphorylated is located at the N terminus, and the DNA-binding domain is located at the C terminus. The middle region of the polypeptide contains a region that interacts with RNA polymerase and has an ATPase activity required for activation.

Experiments with hybrid activators, made by fusing the C-terminal DNA-binding domain of one activator protein to the N-terminal inducer-binding domain of another activator protein from the same family, provide a clear demonstration that

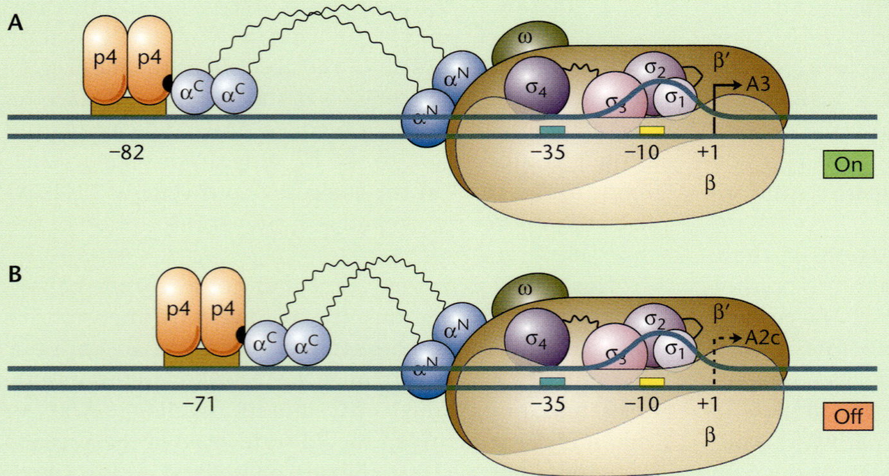

Figure 1 Figures 1–4 adapted from Dove SL, Hochschild A, *in* Higgins NP (ed), *The Bacterial Chromosome* (ASM Press, Washington, DC, 2005), with permission.

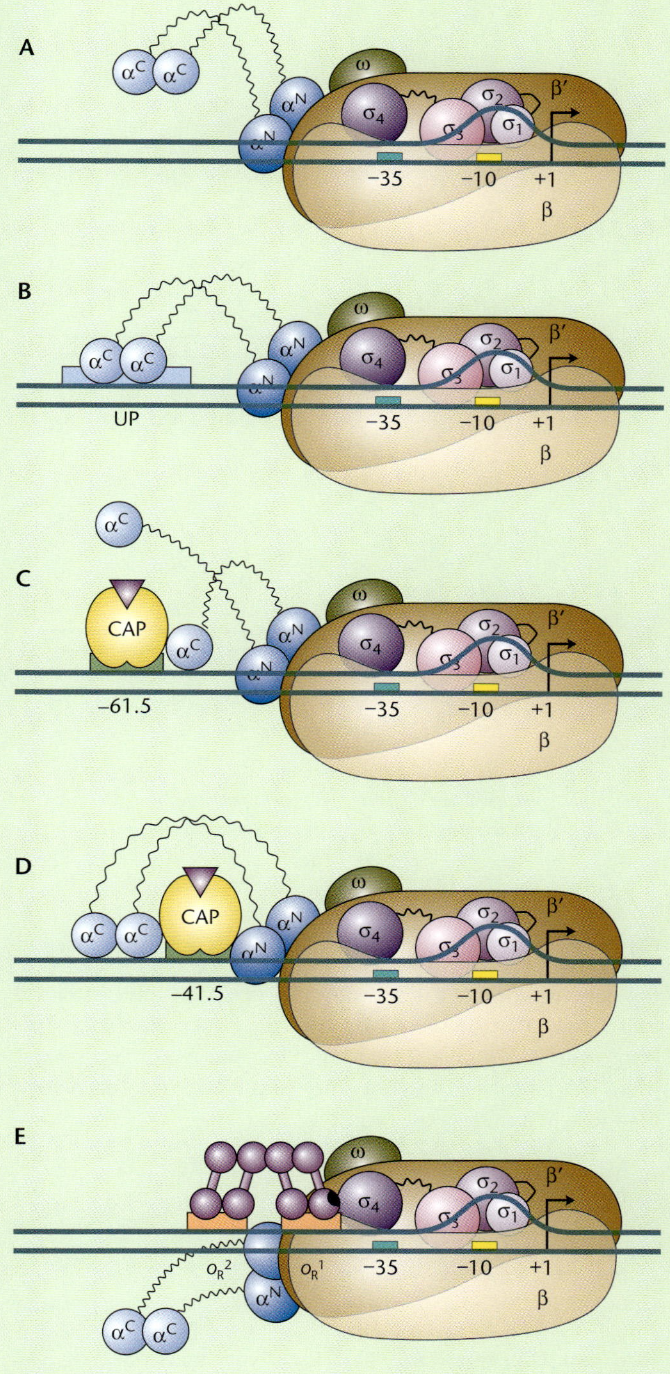

Figure 2

members of a family of regulators all use the same basic strategy to activate transcription of their respective operons (see Parsek et al., References). Sometimes, such hybrid activators can still activate the transcription of an operon, but which operon is activated by the hybrid activator depends on the source of the C-terminal DNA-binding domain, which determines target site specificity. On the other hand, the identity of the effector molecule that induces the operon depends on the source of the N-terminal inducer-binding domain. Artificial combination of these domains from different activators leads to a situation where an operon is induced by the inducer of a different operon. It is intriguing to think that all activator proteins may have evolved from a single precursor protein through simple changes in its effector-binding and DNA-binding regions, yet they continue to activate RNA polymerase by the same basic mechanism.

Regulatory proteins use almost every conceivable mechanism to regulate transcription. Repressors act on essentially every step required for initiation of transcription, and many affect more than one step. Some repressors act, at least in part, by preventing the binding of RNA polymerase to the promoter, either by getting in the way (steric hindrance) or by bending the DNA at the promoter. They can also act by preventing RNA polymerase from separating the strands of DNA at the promoter (open-complex formation) or even by hindering the ability of RNA polymerase to move out of the promoter and begin making RNA (promoter escape). In one particularly illustrative case, shown in Figure 1, the repressor/activator protein p4 of a *Bacillus subtilis* phage represses an already strong promoter by making it so strong that the RNA polymerase cannot escape it and begin transcription. Figure 1A shows how binding of p4 protein to a sequence at −82 relative to the start site of the A3 promoter activates transcription from that promoter (shown as On in the figure), while binding of p4 protein to a sequence at −71 relative to the start site of another promoter, A2c, inhibits transcription from that promoter (shown as Off in the figure). In both cases, the p4 binding site is upstream of the promoter and should therefore not interfere with binding of RNA polymerase. The crucial difference between these promoters is that the A3 promoter has weak affinity for RNA polymerase and therefore benefits from activation by p4, while the A2C promoter has strong affinity for RNA polymerase, so the extra affinity provided by the p4 protein locks RNA polymerase in place and prevents promoter escape (see chapter 2).

Activators can also act at any of the steps of transcription initiation. Many of them "recruit" the RNA polymerase to the promoter by binding both to the DNA close to the promoter and to the RNA polymerase, thereby stabilizing the binding of the RNA polymerase to the promoter. Figure 2 gives some examples of how activators and sequences around a promoter can recruit RNA polymerase to promoters. Figure 2A shows the RNA polymerase binding to the −35 and −10 regions of a σ^{70} promoter. Figure 2B shows how binding of the αCTD domains (the C-terminal domains of the α subunits of RNA polymerase) to a sequence upstream of the promoter called an UP element can stabilize the binding. Figure 2C and D show how CAP bound at different sites upstream of a promoter can contact different regions of the RNA polymerase and stabilize its binding. In Figure 2C, CAP is bound further upstream and contacts one

(continued)

Families of Regulators

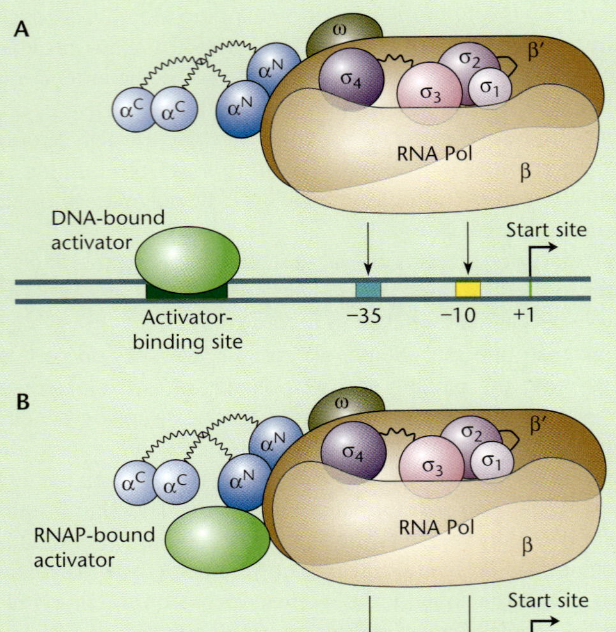

Figure 3

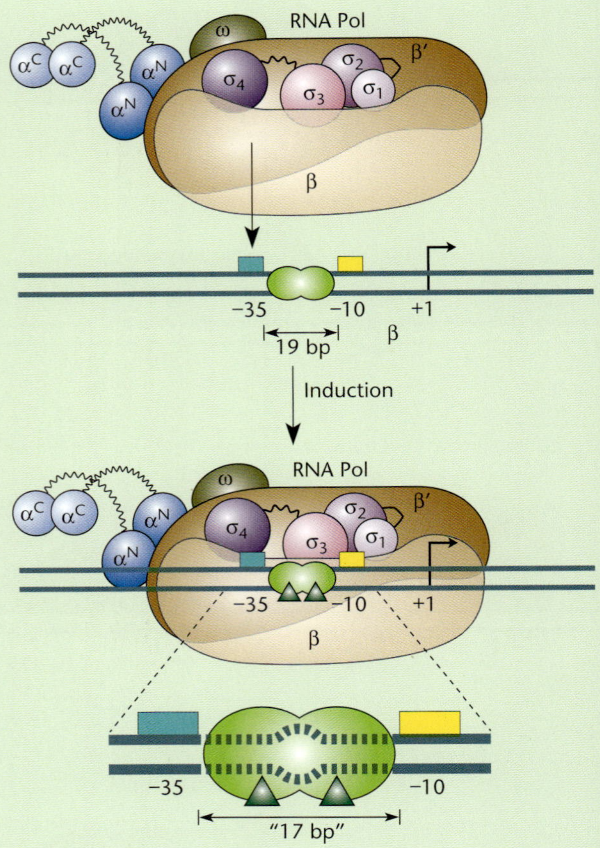

Figure 4

αCTD domain. In Figure 2D, it is bound closer to the start site and can contact one of the αNTDs (N-terminal domains of the α subunit), as well as the αCTDs. Figure 2E shows how the CI protein of phage λ (shown as dumbbells) can activate transcription from the p_{RM} promoter by binding cooperatively to the operators o_R^1 and o_R^2 (see chapter 7), from where they can contact both σ^{70} and the αCTDs. Unlike most activators, which bind to DNA (Figure 3A), activators like SoxS can bind to the RNA polymerase before it binds to the promoter (Figure 3B). Only after the SoxS activator has bound to it can the RNA polymerase bind to the promoter. Others, such as the NtrC-type activators, do not recruit the RNA polymerase but allow an RNA polymerase already bound at the promoter to open the DNA to form an open complex (see chapter 12). Activators in the MerR family activate transcription by changing the conformation of the promoter DNA to increase binding of RNA polymerase (see Huffman and Brennan, References). Normally the spacing between the −35 sequence and the −10 sequence in a σ^{70}-type promoter is 17 bp, but this promoter has a spacer region that is 19 bp (Figure 4), which rotates the two elements by about 70°, making it difficult for the σ^{70} subunit of the RNA polymerase to contact both of them (see chapter 2). The activator binds to the promoter in the absence of inducer but fails to activate under these conditions. When the inducer binds, the conformation of the activator changes, and this bends and twists the DNA between the two promoter elements. This brings the −35 and −10 elements closer together and rotates them so that their orientation and spacing more closely resemble those in a normal σ^{70} promoter.

References

Dove SL, Hochschild A. 2005. How transcription initiation can be regulated in bacteria, p 297–310. *In* Higgins NP (ed), *The Bacterial Chromosome.* ASM Press, Washington, DC.

Egan SM. 2002. Growing repertoire of AraC/XylS activators. *J Bacteriol* **184:**5529–5532.

Huffman JL, Brennan RG. 2002. Prokaryotic transcription regulators: more than just the helix-turn-helix motif. *Curr Opin Struct Biol* **12:**98–106.

Parsek MR, McFall SM, Shinabarger DL, Chakrabarty AM. 1994. Interaction of two LysR-type regulatory proteins CatR and ClcR with heterologous promoters: functional and evolutionary implications. *Proc Natl Acad Sci USA* **91:**12393–12397.

Ramos JL, Martínez-Bueno M, Molina-Henares AJ, Terán W, Watanabe K, Zhang X, Gallegos MT, Brennan R, Tobes R. 2005. The TetR family of transcriptional repressors. *Microbiol Mol Biol Rev* **69:**326–356.

specifically change the conformation of the AraC protein to the P2 state.

AraC Is Not Just an Activator

If AraC acts solely as an activator, partial diploid cells that have both an *araC*ᶜ allele and the wild-type allele would be expected to constitutively express the *araB*, *araA*, and *araD* genes. That would make *araC*ᶜ mutations dominant over the wild type, since the mutant AraC in the P2 state should activate transcription of *araBAD*, even in the presence of wild-type AraC protein in the P1 state. This prediction was tested with complementation. An F′ plasmid carrying the wild-type *ara* operon was introduced into cells with an *araC*ᶜ mutation in the chromosome. Figure 11.15 illustrates that the partial diploid cells were inducible, not constitutive, indicating that *araC*ᶜ mutations were recessive rather than dominant. This observation was contrary to the prediction of the model. Therefore, the model had to be changed.

Figure 11.16 illustrates a more detailed model to explain why *araC*ᶜ mutations are recessive. In this model, transcriptional activation requires that the P2 form of AraC, which exists in the presence of arabinose, binds at both the *araI*$_1$ and *araI*$_2$ sites (which together make up the *araI* site in the original model [Figure 11.14]). The P1 form of the AraC protein that exists in the absence of arabinose is not simply inactive but takes on a new role as an **antiactivator** (Figure 11.16A). The P1 state is called an antiactivator rather than a repressor because it does not repress transcription like a classical repressor but, instead, acts to prevent activation by the P2 state of the protein. In the P1 state, the AraC protein preferentially binds to the operators *araO*$_2$ and *araI*$_1$. Binding of AraC to both *araO*$_2$ and *araI*$_1$ bends the DNA between the two sites in a manner similar to the DNA looping caused by the GalR repressor.

Figure 11.15 Recessiveness of *araC*ᶜ mutations. The presence of a wild-type copy of the *araC* gene prevents activation of transcription of the operon by AraCᶜ. The P1 form of AraC binds to operators of both the chromosomal and F′ copies, preventing activation of *araBAD* expression by the AraC in the P2 state from the AraCᶜ allele.

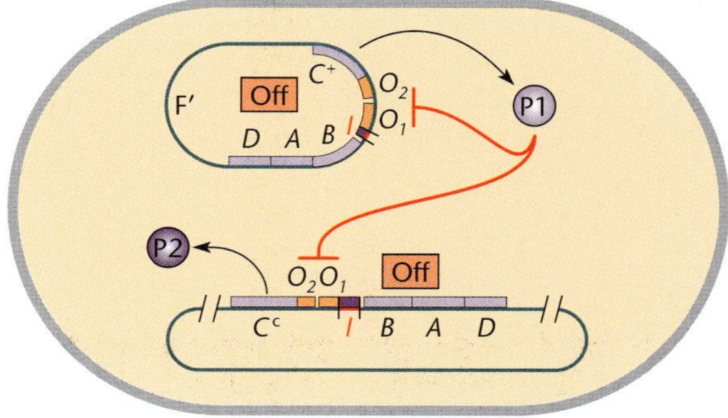

Because AraC in the P1 form preferentially binds to *araO*$_2$, it cannot bind to *araI*$_2$ and activate transcription from the p_{BAD} promoter. In the presence of L-arabinose, however, the AraC protein changes to the P2 form and now preferentially binds to *araI*$_1$ and *araI*$_2$, activating transcription of the operon by RNA polymerase (Figure 11.16B).

This model explains why the *araC*ᶜ mutations are recessive to the wild type in complementation tests, because AraCᶜ in the P2 form can no longer bind to *araI*$_1$

Figure 11.16 A model to explain how AraC can be a positive activator of the *ara* operon in the presence of L-arabinose and an antiactivator in the absence of L-arabinose, as well as how AraC can negatively autoregulate transcription of its own gene (see Box 11.2, Figure 1). **(A)** In the absence of arabinose, AraC molecules in the P1 state (light purple spheres) preferentially bind to *araI*$_1$ and *araO*$_2$, preventing any AraC in the P2 state from binding to *araI*$_2$. No transcription occurs, because AraC must bind to *araI*$_2$ to activate transcription from p_{BAD}. Bending of the DNA between the two sites may also inhibit transcription of the *araC* gene itself by inhibiting transcription from the p_C promoter. The bend in the DNA is also facilitated by the binding by the CAP protein to the CAP-binding site (green rectangle) (see chapter 12). **(B)** In the presence of arabinose, AraC shifts to the P2 state (dark purple spheres) and preferentially binds to *araI*$_1$ and *araI*$_2$. AraC bound to *araI*$_2$ activates transcription from p_{BAD}. **(C)** If the AraC concentration becomes very high, it will also bind to *araO*$_1$, thereby repressing transcription from its own promoter, p_C.

A Absence of L-arabinose

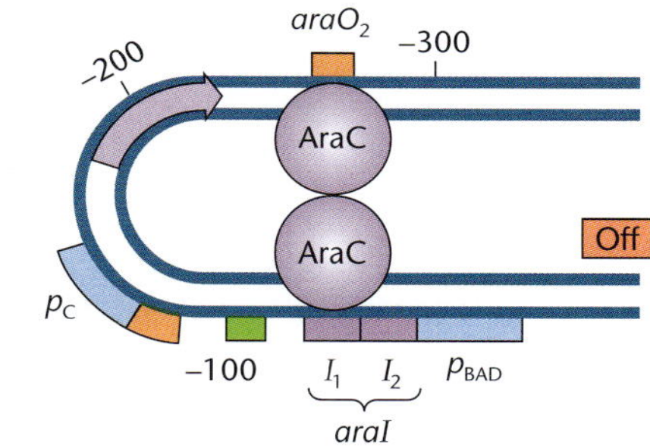

B Presence of L-arabinose

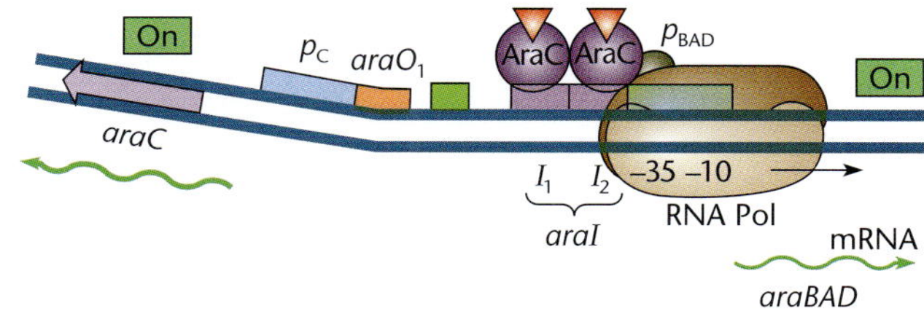

C Excess of AraC

and *araI*$_2$ to activate transcription as long as wild-type AraC in the P1 state is already bound to *ara*O$_2$ and *araI*$_1$.

Face-of-the-Helix Dependence

The *ara* operon provides a classic example of a regulatory system that exhibits **face-of-the-helix dependence**. In systems of this type, the binding site for a regulatory protein on the regulated gene is constrained by how a regulatory protein bound to that site will be positioned relative to some other site (for example, the binding site for another regulatory protein or RNA polymerase). This effect is identified by construction of mutants containing small insertions or deletions. In the case of the *ara* operon, repression by AraC in the absence of arabinose requires that AraC bound to *araI*$_1$ is able to interact with AraC bound to *ara*O$_2$ to form the DNA loop. If 5 bp of DNA is inserted between the *araI*$_1$ and *ara*O$_2$ sites, AraC bound to one of the sites will be unable to interact with AraC bound to the other site, because addition of 5 bp of DNA (half of a turn of the DNA helix [see chapter 1]) positions the second AraC not only further away in linear distance, but also on the opposite side of the DNA helix (Figure 11.17). Addition of 10 bp of DNA, which corresponds to a full turn of the DNA helix, extends the linear distance between the binding sites, but more importantly, it repositions the second AraC so that it can now interact with the first AraC. Therefore, addition of 5 bp of DNA blocks repression, but addition of 10 bp does not. This pattern of response to insertion (or deletion) of partial or full turns of the DNA helix is the hallmark of face-of-the-helix dependence and is interpreted to indicate a requirement for an interaction between factors bound at the two sites on the same face of the DNA helix. In the case of the *ara* operon, the key interaction is between the two molecules of AraC that must interact to form the DNA loop.

Autoregulation of *araC*

The AraC protein not only regulates the transcription of the *ara* operon, but also negatively autoregulates its own transcription. Like TrpR, the AraC protein represses its own synthesis, so less AraC protein is synthesized in the absence of arabinose than in its presence. However, if the concentration of AraC becomes too high, its synthesis will again be repressed.

Figure 11.16 also shows a model for the **transcriptional autoregulation** of AraC synthesis. In the absence of arabinose, the interaction of two AraC monomers bound at *ara*O$_2$ and *araI*$_1$ bends the DNA in the region of the *araC* promoter p_C, thereby inhibiting transcription from the promoter (Figure 11.16A). In the presence of arabinose, the AraC protein is no longer bound to *ara*O$_2$, so the p_C promoter is no longer bent and transcription from p_C occurs. However, if the AraC concentration becomes too high, the excess AraC protein binds to the operator *ara*O$_1$, preventing further transcription of *araC* from the p_C promoter (Figure 11.16C). A similar autoregulation of the λ *c*I gene by its product, the CI protein, was described in chapter 7.

Figure 11.17 Face-of-the-helix dependence. **(A)** Molecules of AraC in the P1 state are bound to *araI*$_1$ and *araO*$_2$ on the same face of the DNA and can interact with each other to form a DNA loop. Insertion or deletion of 10 bp (a full turn of the DNA helix) between the operator sites is tolerated, as the protein molecules remain on the same side of the helix. **(B)** Insertion or deletion of 5 bp between the operator sites (half of a turn of the DNA helix) positions the protein on opposite sides of the helix, so the proteins are no longer able to interact with each other to form a DNA loop.

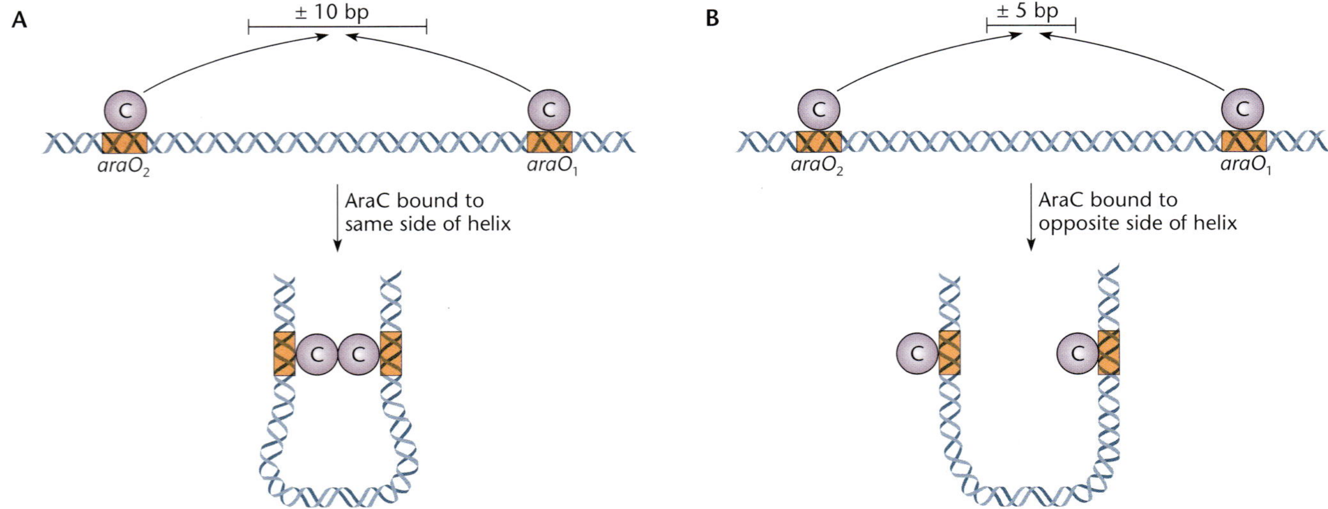

Catabolite Regulation of the *ara* Operon

The *ara* operon is also regulated in response to the availability of other carbon sources, such as glucose (see chapter 12). CAP, which regulates the transcription of genes subject to catabolite regulation, is a transcriptional activator, like the AraC protein; CAP activates gene expression only when the cell lacks a better carbon source, which is signaled by an increase in cAMP. By binding to the CAP-binding site shown in Figure 11.14A, CAP may help to open the DNA loop created when AraC binds to $araO_2$ and $araI_1$. Opening the loop may prevent binding of AraC to $araO_2$ and $araI_1$ and facilitate the binding of AraC to $araI_1$ and $araI_2$ and the activation of transcription from p_{BAD}. Thus, the absence of glucose or another carbon source better than arabinose enhances the transcription of the *ara* operon.

Uses of the *ara* Operon

Besides its historic importance in the pioneering studies of positive regulation, the *ara* operon has many uses in biotechnology. The p_{BAD} promoter, from which the genes of the *ara* operon are transcribed, is often used instead of the *lac* promoter in expression vectors because it is more tightly regulated than the *lac* promoter. Because of its combination of positive and negative regulation by the AraC activator, very little transcription occurs from the promoter unless L-arabinose is present in the medium. The p_{BAD} promoter is also more tightly regulated by catabolite regulation than the *lac* promoter, making it more suitable for expression of large amounts of a toxic gene product by first growing the cells in medium containing glucose and then washing out the glucose and adding L-arabinose. A widely used series of *E. coli* expression vectors use the p_{BAD} promoter and have other desirable features (see Guzman et al., Suggested Reading). They have a variety of antibiotic resistance genes for selection, and some have origins of replication that allow them to coexist with the more standard cloning vectors that have the ColE1 origin of replication (see chapter 4).

Positive Repressible Systems

The activity of an activator protein can be controlled either by the action of an inducer that activates it (such as arabinose for AraC in the *ara* operon) or by an effector that inactivates it and therefore is formally a corepressor. As was discussed above in the context of negative repressible systems, expression of genes involved in biosynthesis of a cellular component is often turned off in the presence of high concentrations of the product of the pathway, for example, in the *E. coli trp* operon, where the TrpR repressor requires tryptophan for its DNA-binding activity. In the case of a positively regulated system that is repressible, the role of the effector molecule is to inactivate the activator, which has the same genetic effect as a corepressor that activates a repressor (as tryptophan does for TrpR). The common feature is that the effector molecule turns off gene expression, regardless of the mechanism.

THE *E. COLI fab* OPERON

The *E. coli fab* operon encodes genes involved in fatty acid biosynthesis. The FadR protein serves as a transcriptional activator, analogous to AraC or CAP. As in those systems, mutations that disrupt function of the FadR regulatory protein result in reduced expression of the *fabHDG* operon. However, in this case, FadR binds to its target site upstream of the *fab* operon promoter in the absence of an effector molecule, so the default state of the system in the absence of the effector is for gene expression to be on. Fatty acyl-CoAs, the end product produced by the *fab* operon gene products, bind to FadR and prevent it from binding to the DNA (Figure 11.18A). The products of the pathway signal the cell that the enzymes involved in producing them are no longer required. Since the role of the effector is to turn gene expression off, this represents a repressible system. The regulatory protein is an activator, which turns gene expression on, so this system is an example of a positive repressible system.

FadR was initially discovered not because of its role as an activator of *fab* gene expression but rather because it also acts as a repressor of fatty acid degradation (*fad*) genes. In this case, binding of FadR to DNA, which occurs when fatty acid levels are low, prevents binding of RNA polymerase to the *fad* operon promoter because the binding site overlaps the promoter site (Figure 11.18B). Repression is relieved in the presence of the fatty acyl-CoA inducer, which binds to FadR and inhibits DNA binding. The *fad* operon therefore exhibits a negative inducible pattern of regulation, analogous to that of the *lac* operon. These studies demonstrated that FadR has a dual role in the cell, simultaneously regulating fatty acid biosynthesis and degradation (see Cronan and Subrahmanyam, Suggested Reading). In both roles, binding of FadR to the DNA is modulated by the same effector, but its effect is determined by the location of the FadR-binding site relative to the promoter. This is similar to the dual role of the *Bacillus subtilis* CcpA and *E. coli* Cra regulatory proteins (see chapter 12).

Molecular Mechanisms of Transcriptional Activation

Like transcriptional repression, activation can utilize a variety of mechanisms (see Browning and Busby, Suggested Reading, and Box 11.2). Promoters that are dependent upon the action of an activator protein generally have relatively low affinity for binding of RNA polymerase. This results in low levels of transcription in the absence of the activator, a condition that is necessary for

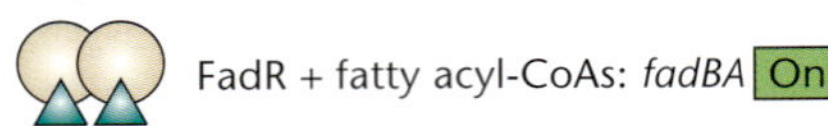

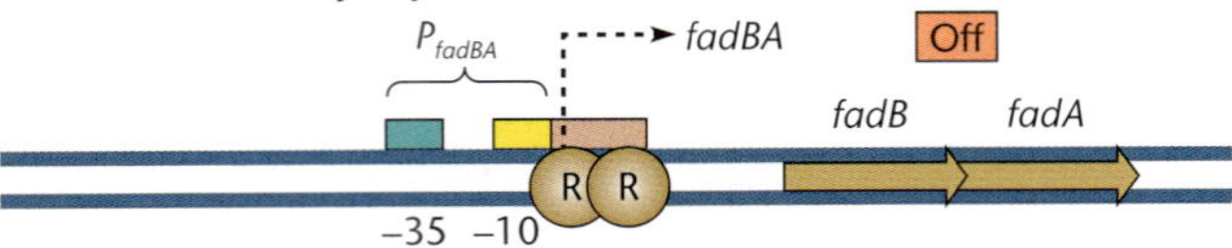

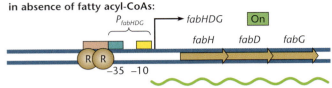

Figure 11.18 Regulation of fatty acid biosynthesis and degradation pathways. **(A)** Transcription of fatty acid biosynthesis genes such as the *fabHDG* operon is activated by binding of FadR upstream of the promoter to increase binding of RNA polymerase. Binding of FadR to the DNA is inhibited by fatty acyl-CoAs. This is therefore a positive repressible system because FadR is an activator, and fatty acyl-CoAs inhibit activation, therefore resulting in loss of expression when fatty acyl-CoA levels are high. **(B)** Transcription of fatty acid degradation genes including the *fadBA* operon is repressed by binding of FadR to the operator site, which blocks binding of RNA polymerase to the promoter. In this case, FadR acts as a repressor, and fatty acyl-CoAs inhibit binding of FadR to the DNA, thereby inducing *fadBA* expression. This is therefore a negative inducible system. In both cases, binding of the effector to FadR inhibits binding of the protein to the DNA; the different outcomes are due to different positioning of the FadR binding site relative to the promoter.

transcription to be dependent upon the activator. The poor affinity for RNA polymerase is usually reflected by weak similarity to the consensus recognition sequence for RNA polymerase holoenzyme containing σ^{70} (see chapter 2). In many cases, the promoter contains a very poor −35 region, which results in loss of contact between region 4 of σ^{70} and the promoter DNA. Many transcriptional activators replace this protein-DNA contact between RNA polymerase and the promoter with a protein-protein contact between the activator and RNA polymerase by positioning the activator on the DNA immediately upstream of the RNA polymerase-binding site.

A large class of activators function by interacting with the carboxyl-terminal domains of the RNA polymerase α subunits (αCTDs). As described in chapter 2, the αCTDs are connected to the rest of the RNA polymerase by a flexible linker. The αCTD can interact directly with the DNA upstream of the −35 region of the promoter if the DNA contains an UP element (see chapter 2 and Box 11.2). UP elements are found in promoters of rRNA operons and contribute to the very high transcriptional activity of these promoters (see chapter 12). Binding of certain transcriptional activators (including CAP [see chapter 12]) upstream of the −35 region allows a protein-protein interaction between the αCTD and the activator to mimic the protein-DNA interaction between αCTD and an UP element and activates transcription by increasing the affinity of RNA polymerase for the promoter. Both UP elements and binding sites for activators that interact with αCTD generally exhibit face-of-the-helix dependence for their position relative to the −35 region of the promoter. Addition of 10 bp (a full turn of the DNA helix) between the −35 region and the activator-binding site (or UP element) is often tolerated, because the flexible linker allows αCTD to reach further upstream on the DNA. However, addition of 5 bp (half a turn of the DNA helix) usually disrupts activation, because αCTD is unable to reach around to the other side of the DNA helix.

Transcriptional activators can also interact with other parts of RNA polymerase to facilitate binding to the promoter. An example of this is provided by the phage λ CI protein, which is a central regulator of the phage λ gene expression program (see chapter 7). The CI protein acts primarily as a repressor but also activates transcription of its own gene when levels of the protein are very low. Transcriptional activation requires binding of CI immediately upstream of the −35 region of the promoter, where the protein interacts with the σ subunit. Activators that interact with σ do not exhibit face-of-the-helix dependence of their binding site relative to the binding site for σ, because, unlike αCTD, σ is unable to "stretch" to reach a binding site that is further upstream on the DNA. For activators of this type, the binding site must be very close to the position of RNA polymerase, and insertions of even 1 to 2 nucleotides are not tolerated.

Activators that function to increase the affinity of RNA polymerase for a promoter can in some cases act as repressors if the promoter already exhibits high affinity for RNA polymerase. This occurs because RNA polymerase must release from the promoter DNA to move into the elongation phase of transcription (see chapter 2). This represents another example of how a regulatory protein can act as either an activator or a repressor, depending on either the position of its binding site relative to the promoter or features of the promoter itself (Box 11.2).

INTERACTIONS OF ACTIVATORS AND REPRESSORS

Individual genes or operons are often subject to regulation by multiple regulatory proteins that act together to modulate expression of the target genes in response to multiple physiological signals. In some cases, different regulatory proteins cooperate to allow activation, where either protein alone is insufficient. For example, binding of the activator that interacts with RNA polymerase may be dependent on binding of a second activator that either allows the first activator to recognize a poor target site on the DNA or shifts the position of the activator to allow interaction with RNA polymerase. Other systems coordinate multiple signals by using antiactivation. As described above for negative regulation, repressor proteins usually bind to the promoter region at a position that overlaps with the RNA polymerase-binding site and prevent transcription simply by preventing access of RNA polymerase to the promoter. In contrast, antiactivators bind to the DNA upstream of the promoter, at a site that overlaps the position where a required activator protein must bind. The presence of the repressor protein prevents binding of the activator and therefore inhibits transcription by preventing activation. These types of complexities illustrate how cells can use a fixed set of regulatory proteins in a variety of ways to integrate physiological signals of a variety of types. Additional examples of complex regulatory interactions are discussed in chapter 12.

Regulation by Transcription Attenuation

In the above examples, the transcription of a gene or operon is regulated through effects on the initiation of RNA synthesis. However, transcription can also be regulated after RNA polymerase leaves the promoter. One way is by **transcription attenuation**. Unlike repressors and activators, which turn transcription from the promoter on or off, the attenuation mechanism works by allowing transcription to begin constitutively at the promoter but then terminates it in the **leader region** of the transcript before the RNA polymerase reaches the first structural gene of the operon (Figure 11.19A). The termination event occurs only if the gene products of the operon are not needed (see Merino and Yanofsky, Suggested Reading). The classic example of regulation by transcriptional attenuation is the *trp* operon of *E. coli*, which uses translation of a short coding sequence to regulate transcription termination. The *B. subtilis trp* operon and the *E. coli bgl* operon use RNA-binding proteins, and in riboswitch systems, the leader RNA directly monitors a regulatory signal. In all of these cases, physiological effects cause changes in the secondary structure of the leader RNA, which in turn affects termination of transcription. The processivity of RNA polymerase can also be modified by proteins that associate with RNA polymerase, as is the case for the phage λ N and *E. coli* RfaH proteins. Representative examples of these types of systems are discussed below.

Modulation of RNA Structure

Many transcription attenuation mechanisms operate by effects on the leader RNA secondary structure. Factor-independent transcriptional terminators are composed of a short helix immediately adjacent to a stretch of U residues in the newly synthesized (or nascent) RNA. As described in chapter 2, pausing of RNA polymerase during synthesis of the U residues allows the nascent RNA that is emerging from RNA polymerase to fold into the helix, which triggers RNA polymerase to terminate transcription. Termination can be prevented by folding of the nascent RNA into a different RNA structure that captures sequences that would otherwise form the terminator helix (Figure 11.19B). This alternate structure is called an **antiterminator**, because it competes with formation of the structure that acts as a factor-independent terminator. The choice between termination and antitermination is often mediated by interaction of regulatory molecules with the RNA, which determines which of the competing structures will form.

REGULATION OF THE *E. COLI trp* OPERON BY LEADER PEPTIDE TRANSLATION

The archetype of transcription attenuation control is the *trp* operon of *E. coli*. As discussed above, the *trp* operon is an example of a negative repressible system, in which binding of the TrpR repressor protein, with tryptophan as the corepressor, inhibits transcription initiation. However, early genetic evidence suggested that this is not the only type of regulation for the *trp* operon. If the *trp* operon were regulated solely by the TrpR repressor, the levels of the *trp* operon enzymes in a *trpR* mutant would be the same in the absence and the presence of tryptophan. However, even in a mutant that completely lacks the TrpR protein, or a mutant in which the operator site to which TrpR binds is absent, the expression of these enzymes is higher in the absence of tryptophan than in its presence, indicating that the *trp* operon is subject to a regulatory mechanism in addition to that provided by the TrpR repressor and that this second level also responds to tryptophan availability.

Early evidence suggested that $tRNA^{Trp}$ (the tRNA that is responsible for insertion of tryptophan at UGG tryptophan codons) plays a role in the regulation of the *trp* operon (see Morse and Morse, Suggested Reading). Mutations in the structural gene for $tRNA^{Trp}$ and in the gene encoding

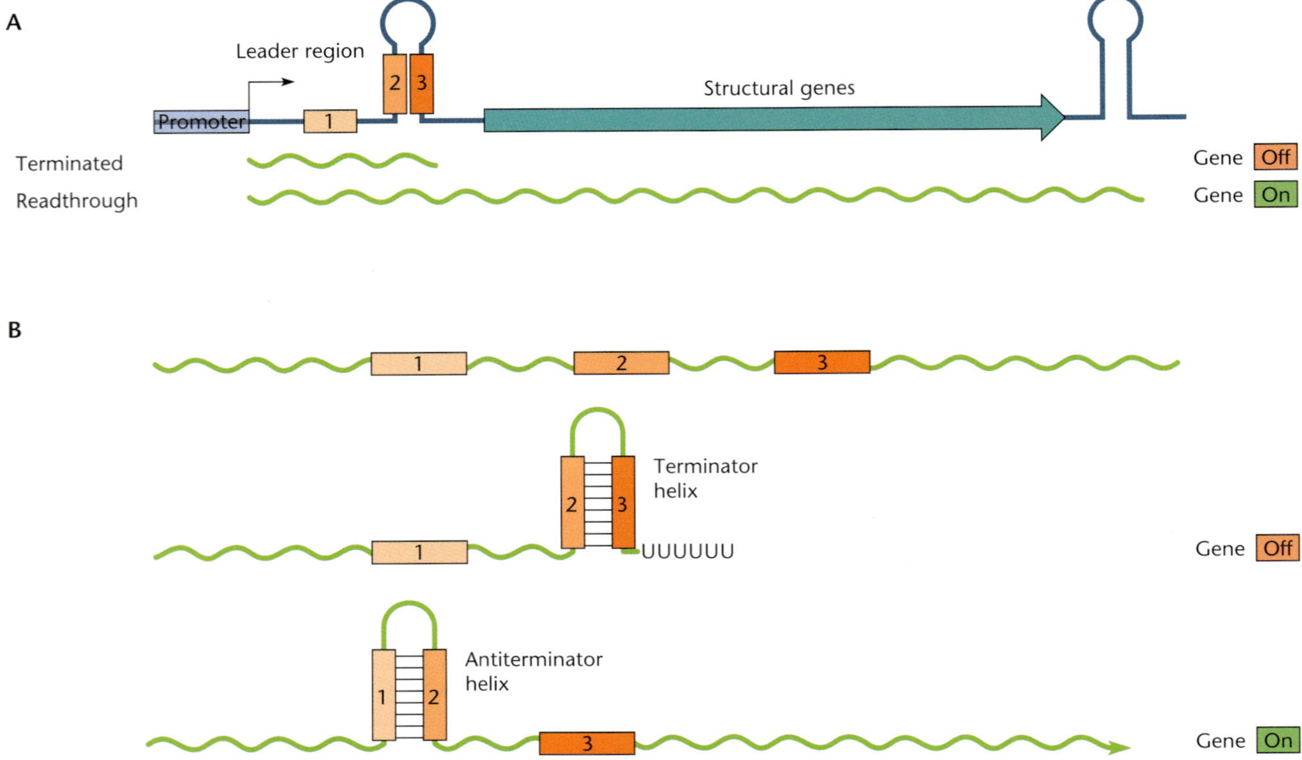

Figure 11.19 Transcription attenuation. **(A)** The presence of a transcription terminator (2:3 helix) in the leader region of a gene can lead to premature termination of transcription and synthesis of only a short terminated RNA that does not include the downstream coding sequence. Readthrough of the termination site is required for expression of the downstream gene. **(B)** Alternate folding of the leader RNA into either the 2:3 terminator helix or the 1:2 antiterminator helix determines whether the short terminated RNA or the full-length RNA is made.

the aminoacyl-tRNA synthetase (aaRS) responsible for transferring tryptophan to $tRNA^{Trp}$, as well as mutations in genes whose products are responsible for modifying $tRNA^{Trp}$, increase the expression of the operon. All these mutations presumably lower the amount of tryptophanyl-$tRNA^{Trp}$ in the cell, suggesting that this other regulatory mechanism senses the amount of tryptophan bound to $tRNA^{Trp}$.

Other evidence suggested that the region targeted by this other type of regulation is the leader region, or *trpL* (Figures 11.11 and 11.20). Deletions in this region, which lies between the promoter and *trpE*, the first structural gene of the operon, eliminate the regulation, so that double mutants, with both a deletion mutation of the leader region and a *trpR* mutation, are completely constitutive for expression of the *trp* operon. Deletions of the leader region are also *cis* acting and affect only the expression of the *trp* operon on the same DNA. Later evidence indicated that transcription terminated in this leader region in the presence of tryptophan because, under those conditions, the cells had abundant levels of aminoacylated $tRNA^{Trp}$ (tryptophanyl-$tRNA^{Trp}$). Because the regulation seemed to be able to stop, or attenuate, transcription that had already initiated at the promoter, it was called attenuation of transcription; an analogous type of regulation was observed in the *Salmonella his* (histidine biosynthesis) operon.

Model for Regulation of the *E. coli trp* Operon by Attenuation

Figures 11.20 and 11.21 illustrate the current model for regulation of the *trp* operon by attenuation (see Oxender et al., Suggested Reading). According to this model, the percentage of the $tRNA^{Trp}$ that is aminoacylated (i.e., has tryptophan attached) determines which of several alternative secondary-structure helices will form in the leader RNA. Recall from chapter 2 that formation of an RNA helix results from complementary pairing between the bases in RNA transcribed from inverted-repeat sequences in the DNA.

The *trpL* region, which contains two consecutive Trp codons, monitors tryptophan levels by sensing availability of tryptophanyl-$tRNA^{Trp}$. If levels of tryptophan are low, the levels of tryptophanyl-$tRNA^{Trp}$ will also be low.

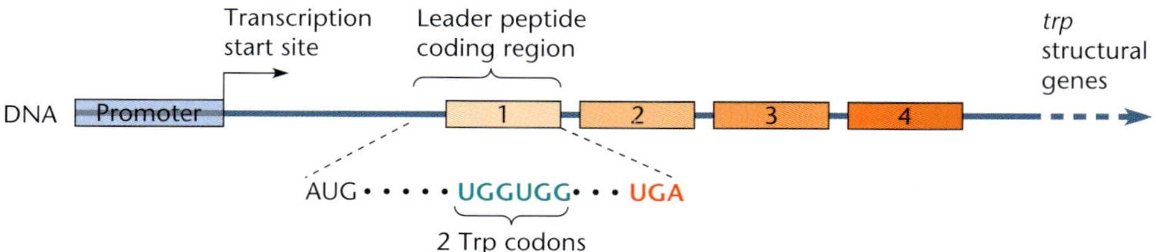

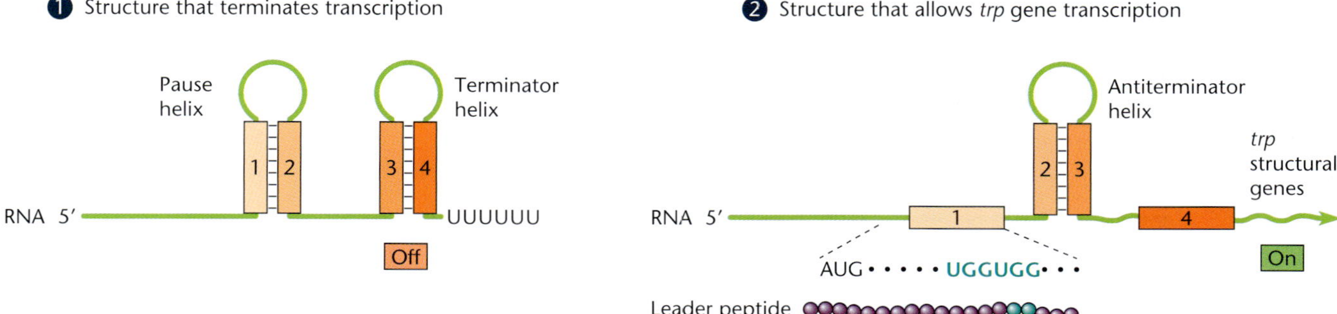

Figure 11.20 Structure of the leader region of the *trp* operon. **(A)** Key features involved in regulation by attenuation. UGGUGG (in turquoise) indicates the two Trp codons in the leader region. **(B)** Alternative pairing of region 2 with either region 1 (to form the pause helix) or region 3 (to form the antiterminator helix) determines whether the 3:4 helix transcription terminator will form; the 1:2 helix causes RNA polymerase to pause, which allows initiation of translation of the leader peptide coding region.

When a ribosome encounters one of the UGG Trp codons, it temporarily stalls, unable to insert a tryptophan. This ribosome stalled in the *trpL* region therefore communicates that the tryptophan concentration is low and that transcription should continue (Figure 11.21). This mechanism is dependent on the fact that transcription and translation are coupled in bacteria, and a ribosome can initiate translation of a transcript before synthesis of the transcript is complete (see chapter 2); this allows the translational machinery to coordinate with the transcriptional machinery. The position of the stalled ribosome affects the structure of the mRNA to which it is bound and determines which sequences are available for pairing to form RNA helices that affect the activity of the RNA polymerase that is engaged in transcription of the leader RNA that the ribosome is engaged in translating.

Figures 11.20 and 11.21 show how the helices operate in attenuation. Four different regions in the *trpL* leader RNA, regions 1, 2, 3, and 4, can form three different helices (1:2, 2:3, and 3:4) by alternative pairing of region 2 with either region 1 or region 3 (1:2 versus 2:3) and alternative pairing of region 3 with either region 2 or region 4 (2:3 versus 3:4), as shown in Figure 11.20. The formation of helix 3:4 causes RNA polymerase to terminate transcription, because this helix is part of a factor-independent transcription termination signal (see chapter 2); note that termination requires both the helix and the adjacent run of U residues in the transcripts, so only the 3:4 helix promotes termination because only this helix has the adjacent U residues. Whether helix 3:4 forms is determined by the dynamic relationship between ribosomal translation of the Trp codons (UGGUGG) in the *trpL* region and the progress of RNA polymerase through the *trpL* region (illustrated in Figure 11.21). After RNA polymerase initiates transcription at the promoter, it moves through the *trpL* region to a site located just after region 2, where it pauses. The helix formed by leader RNA regions 1 and 2 (1:2) is an important part of the signal that causes RNA polymerase to pause immediately after synthesis of region 2. The pause is short, but it ensures that a ribosome has time to load onto the leader RNA before the RNA polymerase proceeds to region 3. The moving ribosome may help to release the paused RNA polymerase by catching up and colliding with it.

The progress of ribosome translation through the Trp codons of *trpL* determines whether helix 3:4 will form, causing termination, or whether region 2 will pair instead with region 3 to form the 2:3 helix. The 2:3 helix

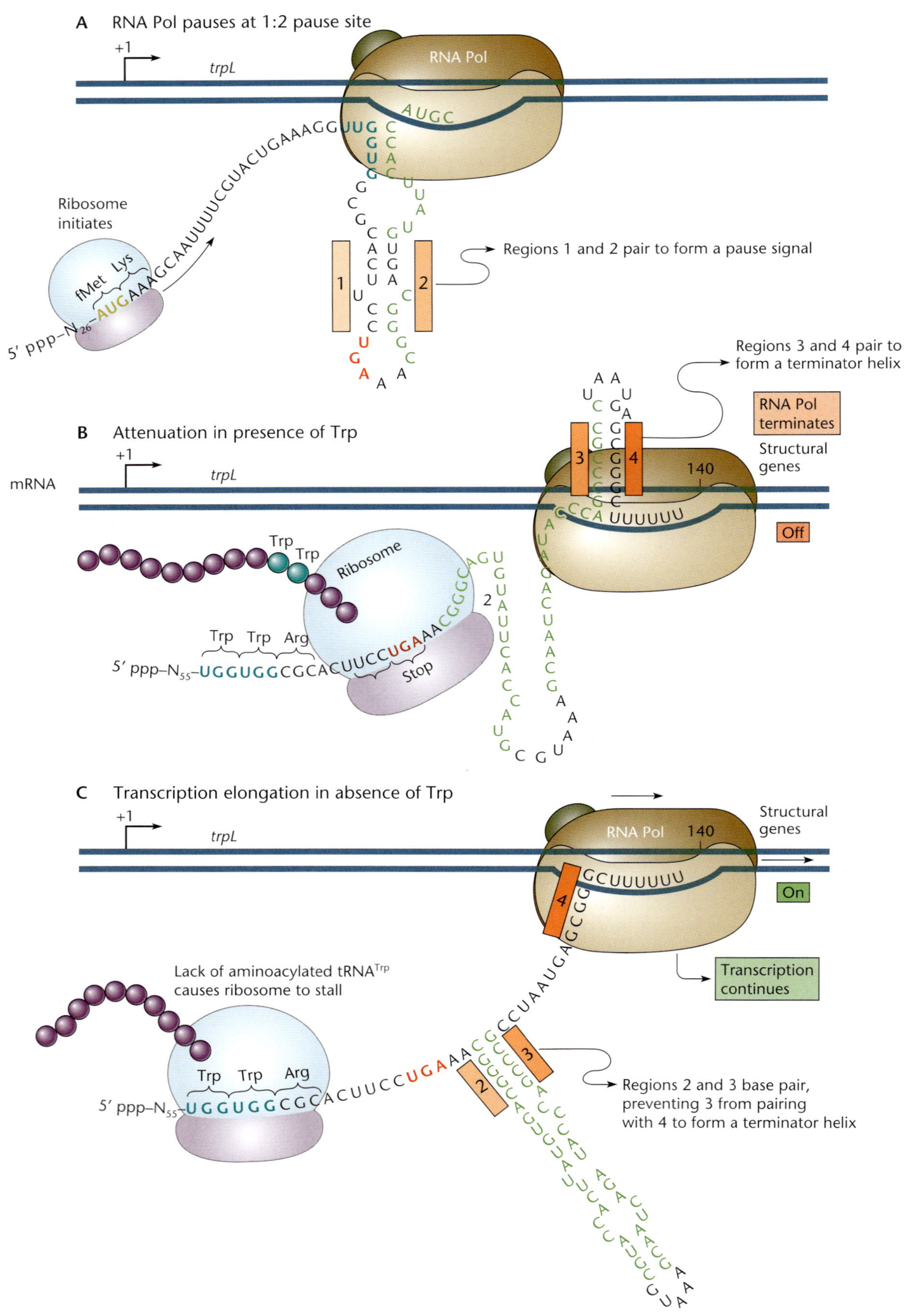
A
RNA Pol pauses at 1:2 pause site
+1
trpL
RNA Pol
Ribosome initiates
fMet
Lys
5′ ppp–N26–AUGAAAGCAAUUUUCGUACUGAAAGGUUG
Regions 1 and 2 pair to form a pause signal
1
2
Regions 3 and 4 pair to form a terminator helix
RNA Pol terminates
B
Attenuation in presence of Trp
+1
mRNA
trpL
3
4
140
Structural genes
Off
Trp
Trp
Ribosome
Trp Trp Arg
5′ ppp–N55–UGGUGGCGCACUUCCUGAAACGGGCAGU
Stop
2
C
Transcription elongation in absence of Trp
+1
trpL
RNA Pol
140
Structural genes
GCUUUUUU
On
4
Transcription continues
Lack of aminoacylated tRNATrp causes ribosome to stall
Trp Trp Arg
5′ ppp–N55–UGGUGGCGCACUUCCUGAAA
2
3
Regions 2 and 3 base pair, preventing 3 from pairing with 4 to form a terminator helix

sequesters region 3, preventing formation of the 3:4 helix; helix 2:3 therefore acts as an antiterminator, because its formation prevents termination. If the ribosome stalls at the Trp codons because of low tryptophan concentrations and therefore low availability of aminoacylated $tRNA^{Trp}$, region 3 will pair with region 2 (Figure 11.21C). If the ribosome does not stall at the Trp codons, it will continue until it reaches the UGA stop codon at the end of *trpL*. By remaining at the stop codon while region 4 is synthesized, the ribosome prevents formation of helix 2:3. Therefore, helix 3:4 can form and terminate transcription (Figure 11.21B).

Genetic Evidence for the *trp* Attenuation Model

The existence and *in vivo* function of helix 2:3 were supported by the phenotypes produced by mutations within the sequences predicted to form the helix. For example, a mutation that changes nucleotides predicted to base pair within helix 2:3 should destabilize the helix. Mutations of this type result in termination of transcription in the *trpL* region even in the absence of tryptophan, consistent with the model that formation of helix 2:3 normally prevents formation of helix 3:4.

The idea that translation of the leader peptide from the *trpL* region is essential to regulation is supported by the phenotype of a mutation that changes the AUG start codon of the leader peptide to AUA, which prevents initiation of translation. In this mutant, termination occurs even in the absence of tryptophan, consistent with the prediction that translation is necessary for antitermination. The observation of efficient termination in the absence of translation suggests that the RNA polymerase paused at helix 1:2 will eventually move on, even without a translating ribosome to nudge it, and will eventually transcribe the 3:4 region. Without a ribosome stalled at the *trp* codons, however, helix 1:2 will persist, preventing the formation of helix 2:3. If helix 2:3 does not form, helix 3:4 will form and transcription will terminate.

One final prediction of the model is that stalling translation at codons other than those for tryptophan in *trpL* should also cause increased transcription through the termination site. Changing the tryptophan codons in the *trpL* region to codons for other amino acids results in increased readthrough in response to starvation for those amino acids, as predicted by the model. Several other amino acid-related operons in *E. coli* have similar leader peptide coding regions, with the codons appropriate for those operons (e.g., phenylalanine codons in the leader region of a phenylalanine-related operon). This mechanism is easily adaptable to new types of genes during evolution simply by changing the identities of the codons within the leader peptide coding region.

REGULATION OF THE *B. SUBTILIS trp* OPERON BY AN RNA-BINDING PROTEIN

Comparing the regulation of the same operon in different types of bacteria often reveals that the same regulatory response can be achieved in different ways. The *trp* operon of *B. subtilis* consists of seven genes whose products are required to make tryptophan from chorismic acid. Interestingly, although *B. subtilis* uses different mechanisms than *E. coli* to regulate its *trp* operon, the result is the same: the operon can respond both to the amount of free tryptophan in the cell and to the aminoacylation state of $tRNA^{Trp}$.

Like the *E. coli trp* operon, the *B. subtilis trp* operon has a leader region that includes a factor-independent transcriptional terminator (Figure 11.22). However, rather than depending on pausing by the ribosome at tryptophan codons to sense limiting tryptophan and alter the secondary structure of the mRNA, the *B. subtilis trp* operon uses an RNA-binding repressor protein called TRAP (for *trp* RNA-binding attenuation protein) (see Babitzke and Gollnick, Suggested Reading). This protein forms a complex with 11 subunits, each of which can bind tryptophan. The subunits are arranged in a wheel, and each subunit binds to a repeated 3-base (triplet) sequence (either GAG or UAG) in the leader mRNA, but binding occurs only if the TRAP subunit is bound to a tryptophan molecule. The optimal spacing between the triplets is 2 or 3 bp. This causes the leader RNA to wrap around TRAP, which sequesters sequences needed for formation of an antiterminator helix. If the antiterminator helix does not form,

Figure 11.21 Details of regulation by transcription attenuation in the *trp* operon of *Escherichia coli*. **(A)** RNA polymerase pauses after transcribing regions 1 and 2. This allows time for a ribosome to load onto the mRNA and begin translating the leader peptide coding sequence, eventually reaching the RNA polymerase and bumping it off the pause site. **(B)** In the presence of tryptophan, aminoacylated $tRNA^{Trp}$ is available; the ribosome translates through the Trp codons and prevents the formation of the 2:3 helix, thereby allowing the formation of helix 3:4, which is part of a transcription terminator, and transcription terminates, which prevents expression of the downstream *trp* operon structural genes. **(C)** In the absence of tryptophan, aminoacylated $tRNA^{Trp}$ is not available, the ribosome stalls at the Trp codons, and helix 2:3 forms, preventing the formation of helix 3:4 and allowing transcription to continue through the terminator site, which allows expression of the downstream *trp* operon structural genes.

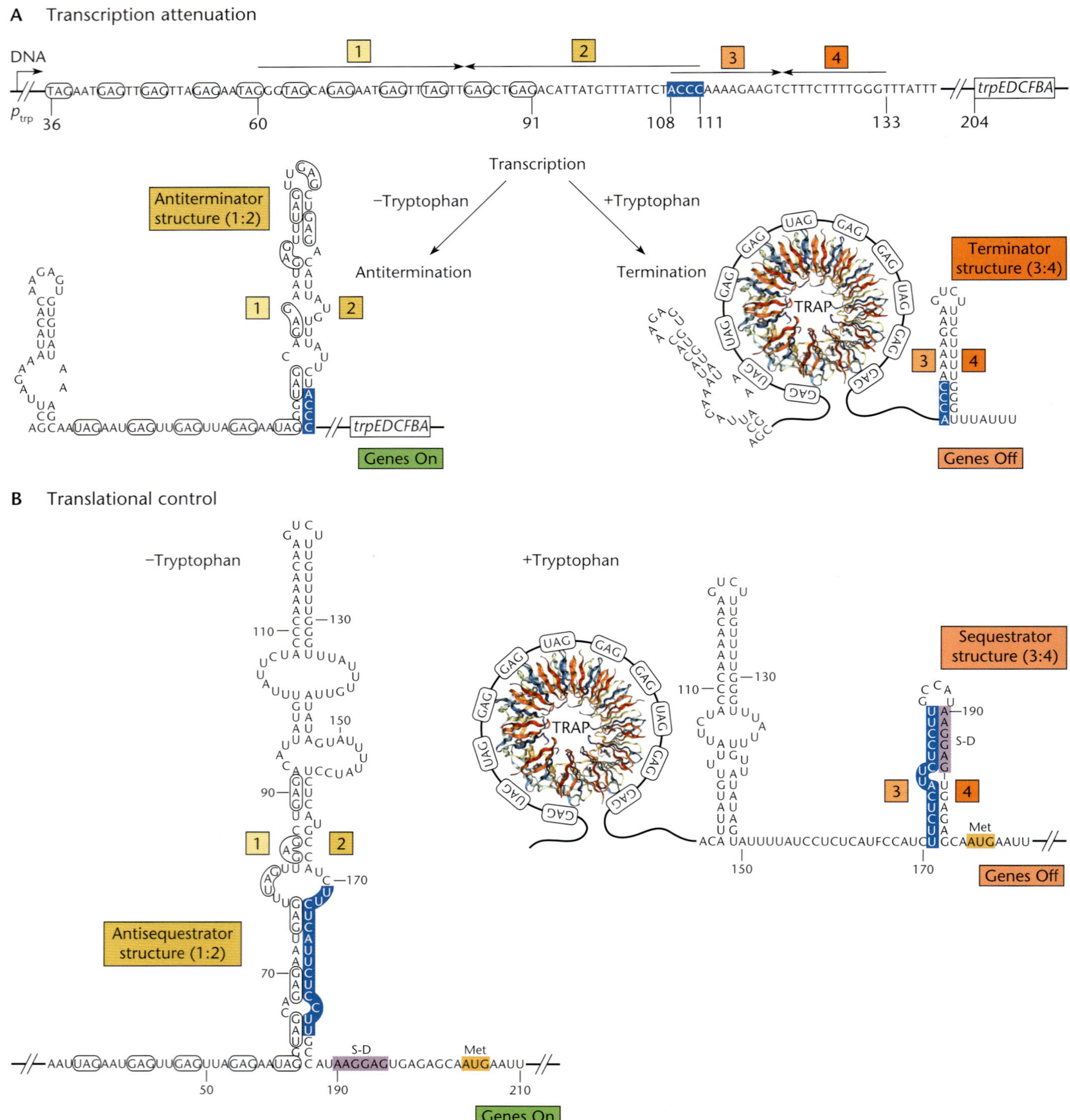
A Transcription attenuation
DNA
1
2
3
4
P_{trp}
36
60
91
108
111
133
204
trpEDCFBA
Transcription
−Tryptophan
+Tryptophan
Antitermination
Termination
Antiterminator structure (1:2)
Terminator structure (3:4)
TRAP
Genes On
Genes Off
B Translational control
−Tryptophan
+Tryptophan
Sequestrator structure (3:4)
Antisequestrator structure (1:2)
S-D
Met
Genes On
Genes Off

sequences needed for formation of a less stable downstream terminator helix are available, and transcription termination occurs. Therefore, transcription termination prevents expression of the operon only if tryptophan is present in the medium. The fact that there are 11 sites on TRAP for binding tryptophan may allow the regulation to be finely tuned in response to intermediate levels of tryptophan when only some of the 11 sites are occupied. TRAP is formally a repressor (like LacI or TrpR), because mutations that inactivate the gene that encodes TRAP result in constitutive *trp* operon expression. Like the TrpR repressor, TRAP requires tryptophan as a corepressor.

TRAP, when bound to tryptophan, can also directly block translation of the first gene of the *trp* operon by binding to similar 3-base repeats just upstream of the Shine-Dalgarno (S-D) sequence of the *trpE* gene, as well as other tryptophan-related genes. TRAP binding close to the S-D sequence prevents the ribosome from binding and blocks translation of the gene (see Figure 11.22B and below). Therefore, tryptophan in the medium can inhibit both the transcription and the translation of the genes of the *trp* operon. Yet another level of regulation that controls the activity of the TRAP RNA-binding protein in response to the aminoacylation state of $tRNA^{Trp}$ utilizes the T box riboswitch mechanism, which is discussed below.

REGULATION OF THE *E. COLI bgl* OPERON BY AN RNA-BINDING PROTEIN

TRAP acts as a repressor by destabilizing an antiterminator helix and promoting termination. An RNA-binding protein can also act as an activator of gene expression by stabilizing an antiterminator helix and preventing termination. This is the strategy used by the *bgl* operon of *E. coli*. The *bgl* operon encodes proteins required for degradation of β-glucoside sugars for use as carbon and energy sources (see Fux et al., Suggested Reading). The BglG RNA-binding protein binds to and stabilizes an antiterminator helix that shares sequences with a terminator helix in the leader region of the *bgl* operon (Figure 11.23). If the BglG protein stabilizes the antiterminator helix, the terminator helix does not form, and transcription continues into the operon, which results in transcription of the genes for β-glucoside uptake and degradation.

As with other catabolic pathways, the *bgl* operon is induced only when the β-glucoside substrates are present in the medium. However, the β-glucoside inducer does not bind directly to the BglG protein. Instead, the activity of BglG is coupled to the transport of β-glucosides into the cell by the sugar phosphotransferase system (PTS). The BglG protein is normally phosphorylated by the BglF protein, which is part of the transport system responsible for import of β-glucosides. When β-glucosides are transported by BglF, BglF transfers a phosphate group onto the sugar and does not phosphorylate BglG. If no β-glucoside transport is occurring, BglF instead phosphorylates the BglG protein. Phosphorylation of BglG prevents BglG monomers from associating with each other to form dimers; since dimers are the active form of the BglG protein that can bind to the antiterminator helix, phosphorylation inactivates BglG and results in transcription attenuation. This ensures that the genes for β-glucoside utilization are turned on only if β-glucosides are being transported into the cell. Additional interactions with the sugar transport system have also been demonstrated that suggest that BglG is captured by binding to BglF in the membrane and is released by sugar transport, which facilitates its activity in transcription antitermination (see Raveh et al., Suggested Reading).

The BglG protein is an activator, as inactivation of the *bglG* gene results in loss of *bgl* operon expression, even in the presence of the β-glucoside inducer. BglF is formally a repressor, since it inactivates the BglG activator. Inactivation of the *bglF* gene results in constitutive *bgl* operon expression, because BglG is not phosphorylated and is therefore always active. However, growth on β-glucoside

Figure 11.22 TRAP regulation of the *trp* operon in *Bacillus subtilis*. **(A)** Model for transcription attenuation of the *trp* operon. When tryptophan is limiting (–Tryptophan), TRAP is not activated and is unable to bind the RNA. During transcription, formation of the antiterminator (1:2 helix) prevents formation of the terminator (3:4 helix), which results in transcription of the downstream *trp* operon structural genes. When tryptophan is in excess (+Tryptophan), TRAP is activated for RNA binding. Tryptophan-activated TRAP binds to the (G/U)AG repeats and promotes termination by preventing formation of the 1:2 antiterminator structure. The overlap between the antiterminator and terminator structures is shown in blue. **(B)** Translational control of *trpE* by TRAP. Under tryptophan-limiting conditions, TRAP is not activated and is unable to bind to the *trp* leader transcript. In this case, the *trp* leader RNA adopts a structure such that the *trpE* Shine-Dalgarno (S-D) sequence is single stranded and available for translation. When tryptophan levels are high, TRAP is activated and binds to the (G/U)AG repeats. As a consequence, a helix that sequesters the *trpE* S-D forms, which prevents ribosome binding and translation. The overlap between the two alternative structures is shown in blue. Numbering is from the start of transcription. Modified from Babitzke P, Gollnick P, *J Bacteriol* **183:**5795–5802, 2001.

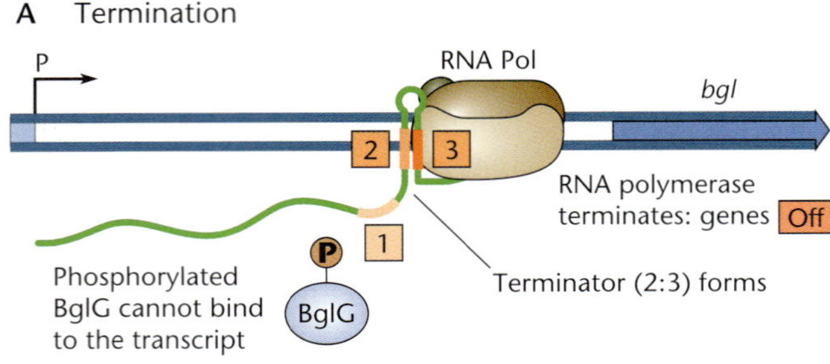

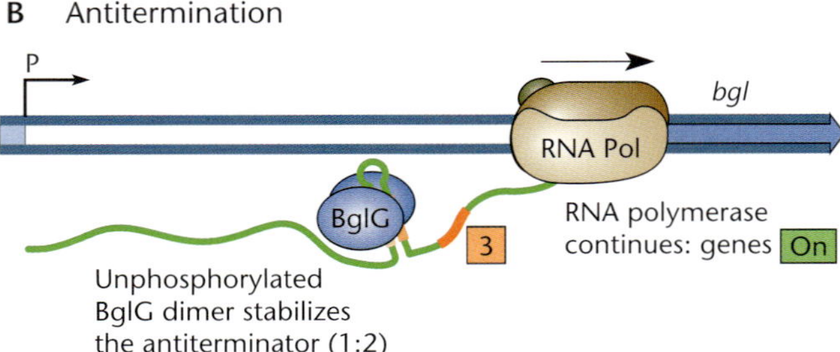

Figure 11.23 Regulation of the *bgl* operon by protein-mediated antitermination. **(A)** The leader region of the *bgl* operon contains a transcriptional terminator. When the levels of β-glucoside sugars are low, the sugar phosphotransferase protein BglF phosphorylates BglG protein; phosphorylated BglG is unable to dimerize and does not bind to the leader RNA, allowing the terminator helix (2:3) to form, resulting in termination of transcription. **(B)** When β-glucoside sugars are present, they are transported by BglF, which is therefore not available to phosphorylate BglG. Unphosphorylated BglG binds to the leader RNA and stabilizes an antiterminator structure (1:2) that prevents formation of the terminator, allowing RNA polymerase to synthesize the full-length mRNA.

sugars cannot occur in a *bglF* mutant because transport of the sugars across the membrane requires the active BglF transporter.

RIBOSWITCH RNAs DIRECTLY SENSE METABOLIC SIGNALS

A **riboswitch** is an RNA element, usually found in leader RNAs, that controls expression of the downstream coding sequences by sensing a regulatory signal directly, without a requirement for translation (as in the *E. coli trp* operon) or a regulatory protein (as in the *B. subtilis trp* and *E. coli bgl* systems). These RNA elements usually fold into a complex three-dimensional structure that specifically recognizes the cognate ligand and not related molecules. Binding of the ligand results in a conformational change in the RNA, and this conformational change is responsible for the effect on gene expression. Regulation can occur at the level of transcription attenuation, by folding of the RNA into alternate terminator and antiterminator helices, or at the level of translation initiation, by folding of the RNA into a structure that sequesters the ribosome-binding site (see below). Riboswitches have been identified that respond to a variety of ligands, including tRNA and small molecules, such as nucleotides, cofactors, amino acids, and metal ions.

The T Box Mechanism: tRNA-Sensing Riboswitches

The first example of a signal molecule acting directly on the mRNA to regulate gene expression was uncovered in analysis of the regulation of the transcription of the genes for the aminoacyl-tRNA synthetase (aaRS) enzymes in *B. subtilis*. Depriving the cell of an amino acid causes accumulation of the uncharged form of the tRNA for that amino acid. The bacteria respond by synthesizing higher levels of the aaRS for that tRNA, which allows more efficient attachment of the limiting amino acids to their cognate tRNAs.

The expression of the aaRS genes in *B. subtilis* is regulated through transcription attenuation. If the tRNA that is normally aminoacylated by that aaRS is mostly unaminoacylated (with no amino acid attached), transcription terminates less often in the leader sequence, more of the full-length mRNA is synthesized, and more aaRS is made. If most of the tRNA for that amino acid has the amino acid attached (i.e., is aminoacylated), transcription of the aaRS gene is more likely to terminate in the leader region, which prevents synthesis of more of the aaRS. Therefore, whether transcription terminates in the leader sequence of each gene is determined by the relative levels of the unaminoacylated and aminoacylated cognate tRNA for that aaRS.

The leader RNAs of genes that are regulated by this mechanism contain a series of conserved sequence and structural features, which include a terminator helix and a competing antiterminator helix (see Grundy and Henkin, Suggested Reading). An additional crucial element is a large helix that contains a small internal unpaired region called the Specifier Loop (Figure 11.24). This unpaired region includes a triplet sequence that matches a codon that corresponds to the amino acid specificity of the regulated aaRS (e.g., the tyrosyl-tRNA synthetase gene leader RNA contains a UAC tyrosine codon, and a tryptophanyl-tRNA synthetase gene leader RNA contains a UGG tryptophan codon). A simple genetic experiment proved that this triplet codon is responsible for the specific response of each gene to the aminoacylation status of its cognate amino acid. Mutation of the UAC Tyr codon in the *tyrS* gene to a UUC Phe codon resulted in loss of response to tyrosine availability and a switch to a response to limitation of phenylalanine. Mutation of the triplet to a UAG nonsense codon resulted in termination under all conditions (i.e., an uninducible phenotype) and could be suppressed by introduction into the cell of a tRNA in

A Low tRNA aminoacylation

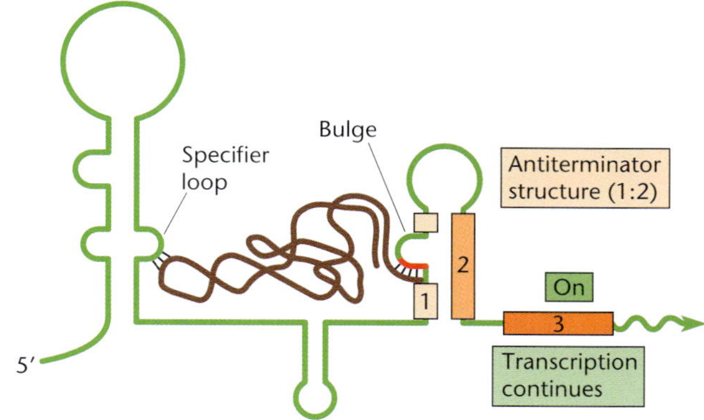

B High tRNA aminoacylation

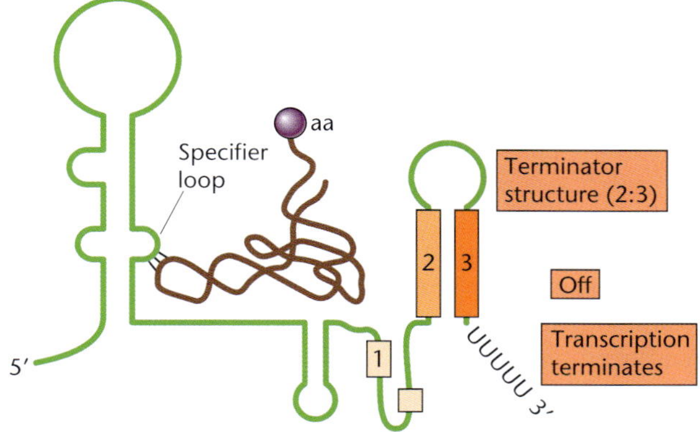

Figure 11.24 The tRNA-responsive T box riboswitch system. The leader RNAs for amino acid-related genes contain competing terminator (2:3) and antiterminator (1:2) helices. An upstream helix includes the specifier loop, which contains a triplet sequence corresponding to a codon of the same amino acid class as the regulated gene. A specific tRNA binds to the specifier loop through codon-anticodon base pairing. **(A)** If the tRNA is not aminoacylated, the 3′ end of the tRNA makes a second interaction with a bulge in the antiterminator helix, which stabilizes the antiterminator and prevents termination, allowing expression of the downstream gene. **(B)** If the tRNA is aminoacylated, it is unable to pair with the bulge. The antiterminator helix is not stabilized, and the terminator helix forms, which results in transcription termination.

which the anticodon was mutated to match the UAG codon (a nonsense suppressor tRNA) (see chapter 3); the ability of a tRNA mutant to suppress the effect of a leader RNA mutation suggested that tRNA is the effector molecule recognized by this system.

These and other experiments resulted in a model in which binding of a specific unaminoacylated tRNA to the leader RNA stabilizes the antiterminator, which sequesters sequences that would otherwise participate in formation of the terminator helix. The presence of the amino acid on the 3′ end of the aminoacylated tRNA prevents the interaction with the antiterminator, which allows the system to differentiate between aminoacylated and unaminoacylated tRNA. The triplet sequence, which was designated the Specifier Sequence, determines which tRNA a particular leader RNA will monitor.

In vitro experiments have confirmed many aspects of this model, including the fact that binding of tRNA to the leader RNA can affect its secondary structure and stabilize the antiterminator hairpin (see Yousef et al., Suggested Reading). Antitermination occurs when the appropriate unaminoacylated tRNA is added to a purified system that has only the RNA polymerase and the DNA template, confirming that no other factors are required. Many different aaRS genes (as well as genes involved in amino acid biosynthesis, transport, and gene regulation) can be regulated by the same mechanism, but each gene responds only to levels of the cognate amino acid, monitored via the aminoacylation status of the corresponding tRNA.

As described above, the *B. subtilis trp* operon is regulated in response to tryptophan by the TRAP RNA-binding protein. However, like the *E. coli trp* operon, the *B. subtilis trp* operon also responds to the aminoacylation of $tRNA^{Trp}$. Another regulatory gene that contains a series of 3-base repeats characteristic of a TRAP-binding site in its leader RNA was identified in the *B. subtilis* genome. The leader sequence of this gene also contained a T box riboswitch with a UGG Trp specifier sequence, suggesting that its expression would be induced only when $tRNA^{Trp}$ is poorly aminoacylated. Transcription of this gene was shown to be induced when aminoacylation of $tRNA^{Trp}$ was low, and overexpression of the regulatory gene resulted in constitutive expression of the *trp* operon. The protein product was shown to bind directly to TRAP, even in the presence of tryptophan, blocking the ability of the TRAP-Trp complex to bind to the *trp* operon leader RNA, and the protein was named anti-TRAP because it interferes with TRAP activity. The *trp* operon of *B. subtilis* therefore behaves much like the *trp* operon of *E. coli* in that expression is induced by both low tryptophan and low $tRNA^{Trp}$ aminoacylation, although the molecular mechanisms used are very different.

Metabolite-Binding Riboswitches

Riboswitch regulation occurs most commonly by binding of small molecules directly to the leader RNA. Many riboswitches bind metabolites that are the end products of biosynthetic pathways and regulate genes that are involved with the pathways or with uptake of the metabolites into the cell. Since binding of the ligand turns off gene expression, these systems are repressible. As with the tRNA-binding T box riboswitch, binding of the ligand affects the secondary structure of the leader RNA, and the resulting structural change affects expression of the downstream genes. Examples of small molecules that are

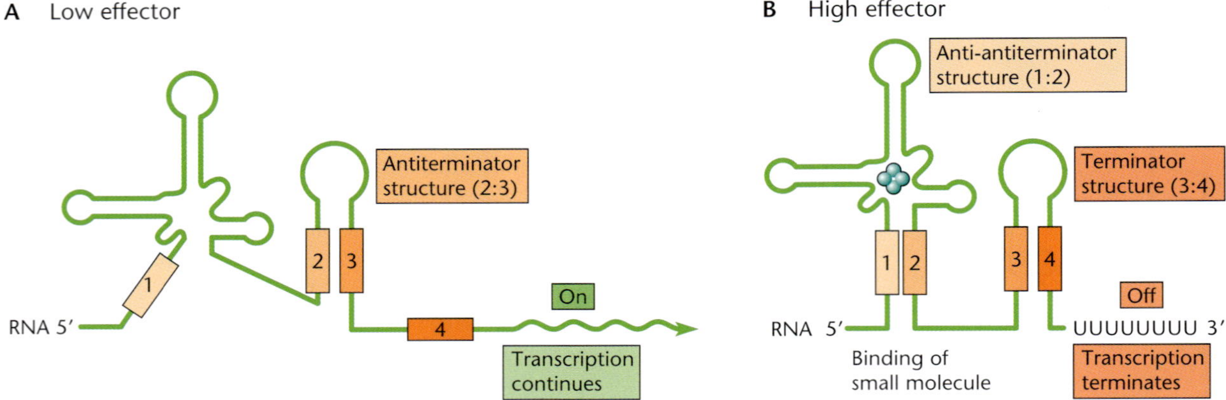

Figure 11.25 Metabolite-binding riboswitch regulation of transcription attenuation. Regions 1, 2, 3, and 4 represent RNA sequences that can form alternative secondary structures. **(A)** Limiting effector. The antiterminator structure (2:3) forms, allowing transcription to continue. **(B)** High levels of effector. Binding of the effector molecule causes the anti-antiterminator structure (1:2) to form. This sequesters sequences (2) that would otherwise participate in formation of the antiterminator (2:3). Consequently, region 3 is available to pair with region 4 to form the terminator helix (3:4), and transcription terminates.

used in riboswitch regulation are amino acids (lysine and glycine), vitamins (B_{12} and thiamine pyrophosphate [B_1]), nucleic acid bases (guanine and adenine), cofactors (flavin mononucleotide and *S*-adenosylmethionine [SAM]), and metal ions (magnesium). The regulation frequently occurs at the level of transcription attenuation, as with the *B. subtilis* aaRS genes, but also can occur through translation by blocking the translation initiation region (TIR) on the mRNA (see below). A few examples have been found in eukaryotes, where binding of the ligand affects mRNA stability or RNA splicing.

One well-studied example of a metabolite-binding riboswitch is the S box riboswitch that regulates genes involved in methionine metabolism in *B. subtilis* and related organisms by transcription attenuation (see McDaniel et al., Suggested Reading). These genes contain leader RNA sequences with similar structural features that include a terminator and a competing antiterminator. If methionine is limiting, the leader RNA forms the more stable antiterminator helix, and the genes for methionine biosynthesis are expressed; high methionine levels result in stabilization of the terminator helix and transcription attenuation (Figure 11.25). However, methionine does not act directly as the signal molecule. In addition to its role as a building block in protein synthesis, methionine is also converted into SAM, which is the donor of methyl groups in many biochemical reactions in the cell. We mentioned some of these reactions in earlier chapters, including the methylation of DNA by restriction endonucleases, but there are many more, some of which are essential to the survival of the cell. The cell therefore determines its need for methionine by monitoring the level of SAM, and SAM is the effector that directly binds to the S box riboswitches upstream of operons involved in biosynthesis of methionine and SAM.

Binding of SAM to the S box riboswitch stabilizes a structural element in the RNA that includes sequences that would otherwise participate in formation of the stable antiterminator helix. The SAM-binding domain therefore serves as an anti-antiterminator, as it prevents formation of the antiterminator. If the anti-antiterminator is not stabilized by binding of SAM, the antiterminator can form, resulting in transcription attenuation only when SAM levels are high. Like other metabolite-binding riboswitches, the S box RNA binds very specifically to SAM and does not recognize related compounds, including *S*-adenosylhomocysteine, which is the product that remains after SAM donates its methyl group in methyltransferase reactions. This specificity is very important, as it ensures that the genes for biosynthesis of methionine (and therefore SAM) are repressed only when SAM is abundant. As with the tRNA-responsive T box riboswitch, SAM-dependent termination can occur in a purified transcription system consisting only of RNA polymerase and template DNA, demonstrating that no other factors are required. The crystal structures of a number of riboswitches bound to their cognate ligand have been determined, and they provide insight into how each RNA specifically recognizes the appropriate metabolite.

A rarer class of metabolite-binding riboswitches show the opposite response to the ligand and are inducible. In these RNAs, the RNA folds into the terminator helix in the absence of ligand, and ligand binding stabilizes the antiterminator rather than an anti-antiterminator. The genes regulated by riboswitches of this type are generally involved in utilization of the ligand in a biosynthetic pathway or in export of the compound from the cell. This illustrates the versatility of riboswitches, which can shift their regulatory response by simple changes in the relative stability of the terminator and antiterminator helices.

Changes in Processivity of RNA Polymerase

As described in the previous sections, the RNA structure can have a major effect on transcription termination, because of the dependence of factor-independent termination on formation of a helix in the nascent RNA immediately upstream of the run of U's in the RNA that form the U-A RNA-DNA hybrid within the transcription elongation complex (TEC) (see chapter 2). However, factors other than RNA structure can also affect transcription termination at both factor-independent and Rho-dependent terminators. They include proteins that interact with RNA polymerase and affect its **processivity**, which means the efficiency with which it moves from nucleotide to nucleotide along the DNA template. Transcription termination requires that RNA polymerase pause at the termination site. In the case of factor-independent terminators, pausing allows the RNA to fold into the terminator helix before the RNA polymerase moves past the termination site. For Rho-dependent termination, pausing provides an opportunity for the Rho factor (which binds to the newly synthesized RNA and migrates in a 5′-to-3′ direction behind RNA polymerase) to catch up to the RNA polymerase and promote termination.

Factors that affect RNA polymerase processivity can target specific genes if those genes include sequences responsible for recruitment of the factors to the TEC. These factors often remain associated with the TEC and can therefore promote readthrough of termination sites encountered by the TEC over long distances. This phenomenon is called **processive antitermination** and results in generation of long transcripts. Processive antitermination is important both in regulatory systems, as described here and in chapter 7, and in transcription of the long rRNA operons (see chapter 12).

PHAGE λ N AND Q PROTEINS

As described in chapter 7, phage λ shifts its gene expression into the lytic cycle by using the phage-encoded N protein to direct the TEC to ignore transcription termination sites, resulting in generation of longer transcripts that include genes required for phage DNA replication. N is an RNA-binding protein that binds to two specific sites, one on each of the two major transcripts that originate from the p_L and p_R promoters. These sites are called *nut* sites, for N utilization, and they are composed of a short helix. Binding of N to the nascent RNA transcript as it emerges from RNA polymerase results in recruitment of additional host cell-encoded factors, termed *nus* factors (for N usage), and the resulting complex has high processivity and ignores subsequent termination sites that it encounters. This results in expression of a new set of genes.

A similar antitermination event is responsible for transcription of genes required for the late stage of phage λ development. One of the proteins produced as a result of the activity of N is the Q protein, which promotes antitermination of transcripts that initiate at a different set of promoters. Unlike N, which binds to the newly synthesized RNA transcript, Q binds to the DNA at a specific site near the promoters for the genes on which it acts. Binding of Q modifies the TEC in a manner similar to the effect of N and results in processive antitermination and late gene expression.

The HK022 phage and some other phages related to λ also use progressive antitermination to regulate their gene expression. However, antitermination in these phages only requires the transcription of *cis*-acting sites called *put* sites, transcribed from the early promoters, and not a protein like N. The small *put* RNAs, which are part of the transcripts generated from the early promoters, bind to the RNA polymerase as they are transcribed and make it resistant to termination at sites further downstream. Because they bind to the RNA polymerase as they are made, the *put* RNAs are *cis* acting, like the *nut* sites of λ. In fact, the two sites *putL* and *putR* are located in the HK022 genome approximately where the *nutL* and *nutR* sites are located in λ, adjacent to p_L and downstream of *cro* for p_R-dependent transcription. However, these RNAs bind to RNA polymerase independent of an antitermination protein and do not require the Nus proteins of the host, at least not absolutely, since some antitermination occurs in purified transcription systems with just RNA polymerase and DNA and none of the other factors.

BACTERIAL PROCESSIVE ANTITERMINATION SYSTEMS

Bacteria also use processive antitermination to increase expression of their genes. As noted above, transcription of very long operons (including rRNA operons) poses a special problem for the cell because of the high probability that RNA polymerase will terminate at some point during transcription. This is especially important in operons that are not efficiently translated or are not translated at all (such as rRNA operons), because of the presence of sites at which Rho protein may bind to promote termination before the end of the operon. The cell has developed special antitermination systems to modify the TEC so that transcription becomes more processive and therefore less likely to terminate. These systems, which are especially important for ribosome biosynthesis (see chapter 12), likely represent the evolutionary basis for the presence of the host-encoded Nus factors that phage λ has borrowed for its own purposes.

The RfaH System

Transcription of several long operons in *E. coli*, including the genes for lipopolysaccharide production, certain genes involved in virulence, and plasmid-borne *tra* genes, relies on the RfaH protein. This protein acts in a manner similar to that described for the λ Q protein in that it binds to the template DNA at specific sites (called *ops* sites) at which RNA polymerase pauses during transcription of the operon. RfaH interacts with the nontemplate strand of the DNA in the transcription bubble, which is displaced to the outside of the TEC when RNA polymerase is transcribing the template strand of the same region. The protein remains associated with RNA polymerase once the complex leaves the *ops* site and promotes high processivity and resistance to Rho-dependent termination. As with rRNA operons, the presence of systems like this in the host organism is likely to have provided the tools for an invading virus like λ to develop a related regulatory mechanism.

Regulation of mRNA Degradation

Synthesis of the mRNA is obviously a crucial step in gene expression. However, how efficiently an mRNA can be translated, and how many protein products can be generated per mRNA molecule, is dependent not only on the synthesis rate of the transcript (regulated by transcription initiation or elongation, as described above), but also on the length of time that mRNA persists in the cell. RNA stability is quantified by measurements of the persistence of an mRNA in the cell. The standard approach is to measure the **half-life** of the transcript, which is the time it takes for the amount of a particular transcript to be reduced to half of the amount observed at the start of the experiment. Most bacterial mRNAs have a relatively short half-life of 1 to 5 min; this contrasts with much greater mRNA stability in eukaryotic organisms, where transcripts can persist for hours (see chapter 2). Increasing the half-life of an individual transcript can result in higher gene expression, while decreasing the half-life results in lower gene expression.

RNAs are destroyed by endoribonucleases that cleave internally in the RNA chain and by exoribonucleases that degrade from the end(s) of the molecule (see chapter 2). The susceptibility of a particular RNA to degradation is dependent on the features of that RNA (for example, whether it is highly structured or single stranded) and can be modified by changes in the structure or by changes in the accessibility of sites at which the ribonucleases (RNases) can act. This can be exploited as a mechanism for regulation of gene expression. Analysis of regulation at the level of mRNA degradation is complicated by the fact that mRNAs that are efficiently translated (and are therefore densely covered with elongating ribosomes) are often more stable than mRNAs that are not efficiently translated. This means that regulatory mechanisms that directly affect translation (see below) may result in indirect effects on mRNA stability. Nevertheless, there are several systems in which it is clear that mRNA degradation is specifically controlled by a regulatory mechanism.

Protein-Dependent Effects on RNA Stability

RNA degradation usually requires binding of RNases to the transcript. The positioning of an RNA-binding protein on a transcript can affect the ability of an RNase to access its target site and therefore can affect the stability of the mRNA. The RNA-binding protein may hide a site at which the RNase would otherwise act, resulting in increased stability of the transcript. An RNA-binding protein can also recruit an RNase to a transcript that it otherwise fails to recognize, decreasing the stability of the transcript.

The simplest case of regulation of mRNA stability by an RNA-binding protein is provided by the *rne* gene, which encodes RNase E, one of the major endoribonucleases in *E. coli* (see chapter 2). The *rne* gene includes a leader region that causes this transcript to be highly sensitive to cleavage by RNase E. When RNase E levels are high, the transcript is rapidly degraded, which prevents synthesis of additional RNase E protein (Figure 11.26). Low intracellular levels of RNase E allow the transcript to persist and to be translated, which results in an increase in the amount of the enzyme (see Schuck et al., Suggested Reading). This is another example of autoregulation, where a gene product controls the expression of its own gene. Autoregulation allows tight control of the amount of a gene product in the cell by direct measurement of the gene product itself. This is especially important in genes like *rne*, because the RNase E enzyme is essential for cell survival, but too much of the enzyme could lead to inappropriate degradation of mRNAs that the cell needs.

The importance of the leader region as the target for regulation was established in this case by transplanting the *rne* leader region upstream of a different mRNA. This resulted in RNase E-dependent destabilization of this new mRNA, which demonstrated that the leader region alone was sufficient to confer the regulatory response. The role of RNase E as the enzyme responsible for degradation was more difficult to establish, because the *rne* gene is essential. It was therefore not possible to inactivate the gene to test the effect on gene expression. Instead, a temperature-sensitive conditional mutation (see chapter 3) in *rne* that resulted in reduced RNase E activity at high growth temperatures was used to allow comparison of expression levels in the presence of high versus low RNase E activity.

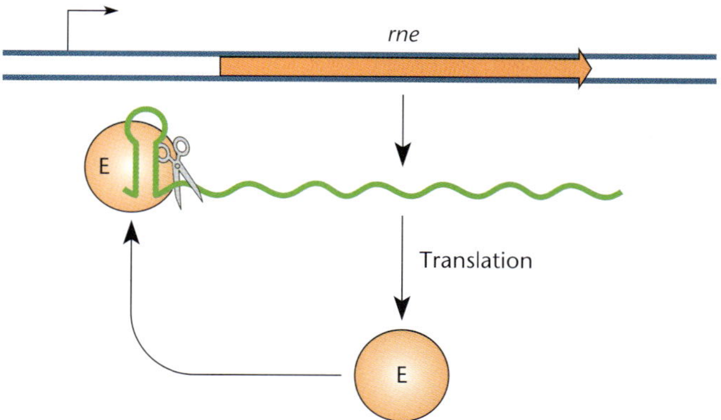

Figure 11.26 Regulation by mRNA degradation. The *E. coli rne* gene, which encodes the endoribonuclease RNase E, contains an RNase E binding site at the 5′ end of the transcript. When RNase E levels are low, the transcript is stable and is translated to give more RNase E protein. When RNase E levels are high, RNase E binds to the transcript and triggers degradation of the mRNA, resulting in reduced synthesis. This is an example of autogenous regulation.

This illustrates some of the complexities that are encountered during analysis of the regulation of cellular processes that are essential for survival.

RNA-Dependent Effects on RNA Stability

The stability of an mRNA can be affected by the action of a regulatory RNA. In most cases, this is mediated by a *trans*-acting small regulatory RNA (sRNA) that binds to the target mRNA and changes its stability. A *cis*-acting regulatory element in the target mRNA itself can also affect mRNA stability. In both cases, the regulatory RNA acts by altering the susceptibility of the target mRNA to RNases, thereby changing its stability and the level of gene expression.

REGULATION OF mRNA DEGRADATION BY sRNAs

The role of sRNAs in regulation of gene expression has become increasingly evident in the last 10 years (see Storz et al., Suggested Reading). Most (but not all) sRNAs regulate multiple targets by base pairing between the sRNA and the mRNA target; as many of these sRNAs are part of global regulatory systems, they are discussed in more detail in chapter 12. The RNA-binding protein Hfq often facilitates the interaction between an sRNA and its target. Regulation usually occurs either at the level of translation initiation (by sequestration of the TIR [see below]) or at the level of mRNA degradation. As noted above, translational repression can have secondary effects on transcript stability, but there are several cases in which there appears to be a primary effect on the susceptibility of an mRNA to specific RNases.

sRNAs can affect the stability of their target mRNAs by a variety of mechanisms. In some cases, the sRNA (probably in conjunction with the Hfq protein) facilitates the recruitment of an RNase to the complex, which results in simultaneous degradation of both the sRNA and its target; this appears to be the case for the *E. coli* RyhB sRNA, which is involved in iron regulation (see chapter 12). Degradation usually involves RNase E or RNase III (see chapter 2), both of which are endoribonucleases and presumably trigger attack by additional RNases. In some cases, a specific endonucleolytic cleavage can increase mRNA stability by removing a segment of the mRNA that otherwise leads to rapid degradation. Stabilization can also occur simply by formation of the sRNA-mRNA complex if the complex blocks the access of an endoribonuclease or the processivity of an exoribonuclease, resulting in protection of the mRNA. Analysis of the complexities of how different sRNAs act on their targets, and the mechanisms of their effects on gene expression, is a very active field of research.

THE *glmS* RIBOZYME

The *B. subtilis glmS* gene is an example of a gene in which expression is controlled by modulation of mRNA stability without an additional protein or sRNA partner. This gene encodes the enzyme required for biosynthesis of glucosamine-6-phosphate (GlcN6P), a cell wall component. Like metabolite-binding riboswitches, the leader region of the *glmS* gene binds GlcN6P, and the binding results in reduced synthesis of the enzyme. However, in this case, binding of the signal molecule does not result in a change in the leader RNA structure, but instead activates the RNA to cleave itself at a specific position. The *glmS* leader RNA therefore acts as a metabolite-induced ribozyme, an RNA enzyme (see chapter 2). Cleavage of the RNA results in removal of the 5′ end of the transcript, and the mRNA now contains a 5′-hydroxyl end instead of the normal 5′ phosphate. The presence of the 5′ hydroxyl appears to target the mRNA for degradation by RNase J1, which is a 5′-3′ exoribonuclease (see chapter 2). Because the ribozyme activity is activated by GlcN6P, the end product of the pathway, the mRNA is stable when the supply of GlcN6P is low, and degradation of the mRNA (and repression of *glmS* gene expression) occurs when the cell has an adequate supply of GlcN6P (see Collins et al., Suggested Reading).

Regulation of Translation

Translation of an mRNA to generate a protein product requires that the 30S ribosomal subunit be able to access the TIR, positioning the initiator methionyl tRNA (fMet-tRNA^{fmet}) at the AUG start codon, and that the resulting 70S translation elongation complex can move processively down the mRNA (see chapter 2). Most known examples of translational regulation operate at the level of translation initiation, as that step is highly sensitive to the structure of the target mRNA (see Geissmann et al., Suggested Reading), whereas translation elongation is usually (but not always) highly processive and relatively insensitive to the mRNA structure. Examples of both levels of regulation are described below. It is interesting that several of the regulatory mechanisms described above that affect transcription attenuation or mRNA stability can also be modified in simple ways to operate instead at the level of translation initiation.

Regulation of Translation Initiation

The first step of translation involves binding of the 30S initiation complex, which includes the 30S ribosomal subunit, the initiator fMet-tRNA^{fmet}, and initiation factors, to the TIR, which is composed of the S-D sequence and the initiator codon (usually AUG), as described in chapter 2. Anything that inhibits access of the initiation complex to the TIR, such as secondary structure in the mRNA that sequesters the TIR or physical blocking of the TIR by binding of an RNA-binding protein or sRNA to the mRNA, inhibits translation initiation. This has been exploited by a number of regulatory mechanisms. Changes in the composition of the 30S initiation complex can also affect its ability to recognize TIRs with specific characteristics, resulting in effects on the efficiency of translation.

RNA THERMOSENSORS: REGULATION BY MELTING SECONDARY STRUCTURE IN THE mRNA

An mRNA leader region can affect the expression of a gene directly through the effects of temperature on its secondary structure. Base pairing between complementary sequences on the mRNA can cause secondary structures to form in the RNA in the form of helices and more complicated structures (see chapter 2). Secondary structures are less stable at higher temperatures because the base pairing that holds them together dissociates at these temperatures. One way in which temperature can regulate the expression of a gene is if the secondary structures that have formed block access of the ribosome to the TIR of the mRNA, for example, if they include the S-D sequence and/or the initiator codon. At lower temperatures, the structure is stable, and translation initiation is inhibited. When the temperature rises, these secondary structures can melt, exposing the TIR so that the ribosomes can bind and initiate translation of the mRNA. RNAs of this type are called RNA **thermosensors**, as they have the ability to directly sense a change in temperature.

The *E. coli rpoH* Heat Shock Thermosensor

One of the best-characterized RNA thermosensors is in the *E. coli rpoH* gene, which encodes the heat shock sigma factor, σ^H. The RNA polymerase holoenzyme containing σ^H instead of the normal σ^{70} recognizes a new set of promoters that are responsible for transcription of genes that help the cell to respond to an abrupt increase in temperature called a heat shock. The cell always contains a certain amount of *rpoH* mRNA, but this mRNA is usually inactive for translation because of a secondary structure that sequesters the TIR (Figure 11.27). An abrupt increase in temperature causes the

Figure 11.27 Regulation of the *E. coli rpoH* gene by an RNA thermosensor. Transcription of the *rpoH* gene under normal growth conditions yields an mRNA in which the Shine-Dalgarno (S-D) sequence for translation of the *rpoH* coding region is sequestered into a helix. Exposure of the cells to heat shock results in melting of the helix and release of the S-D sequence. The mRNA is active for ribosome binding, and σ^H is synthesized.

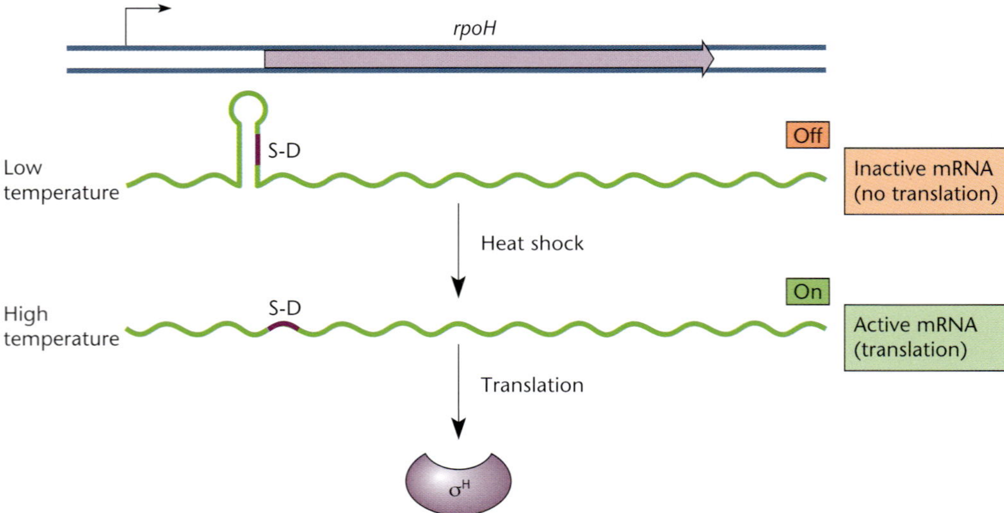

secondary structure in the mRNA to melt and allows translation of the transcript, which results in a rapid increase in the amount of σ^H in the cell. This allows the cell to respond very quickly to the temperature increase, because the *rpoH* mRNA is already present in the cell before the heat shock occurs. The heat shock response is a form of global regulation and is discussed in more detail in chapter 12.

RNA Thermosensors That Control Virulence

Some pathogenic bacteria use RNA thermosensors to regulate their virulence genes. Our body temperature, and that of most other warm-blooded hosts, is much higher than those of the outside environments usually inhabited by bacteria. Pathogenic bacteria often use temperature as one of the clues that they are in a mammalian host, and a rise in temperature tells them that it is time to turn on the virulence genes that allow them to survive and multiply in the host. Placement of an RNA thermosensor in the leader region of a gene that regulates virulence allows virulence gene expression to be repressed when the temperature is low and rapidly induced when the organism enters the host. As with the heat shock sigma factor discussed above, the response is rapid because the mRNA for the regulatory protein is already present in the cell prior to entry into the host, and melting of the secondary structure to allow protein synthesis to proceed is nearly instantaneous. One example of this type of temperature regulation is in the expression of the *lcrF* gene in *Yersinia pestis*, the bacterium that causes bubonic plague. The product of this gene is the transcriptional activator that turns on virulence genes in mammals. During growth at low temperature (e.g., in the flea, which is an intermediate host for this organism), an RNA thermosensor in the *lcrF* mRNA blocks the S-D sequence, preventing translation; this secondary structure melts at 37°C (the body temperature of the mammalian host), allowing translation of the mRNA and synthesis of the LcrF transcriptional regulator. This allows the cell to quickly activate transcription of the genes involved in virulence, so that the organism can multiply and kill the mammalian host.

RIBOSWITCH REGULATION OF TRANSLATION INITIATION

Many riboswitches operate at the level of transcription attenuation, as described above. However, the same types of ligand-dependent RNA rearrangements can also regulate gene expression at the level of translation initiation. The key difference is that in riboswitches of this type, the transcription terminator helix is replaced by a sequestrator helix that sequesters the S-D sequence (by pairing of the S-D with a complementary anti-S-D [ASD] sequence), and the antiterminator helix is replaced by an anti-sequestrator helix that sequesters the ASD sequence.

A crucial difference between these two classes of riboswitches is that in transcriptional riboswitches, binding of the signal molecule determines whether the full-length transcript is synthesized, whereas in translational riboswitches, the full-length transcript is always made, but binding of the ligand determines whether the mRNA will be translated. A second important distinction is that for riboswitches that operate at the level of transcription attenuation, the ligand must bind during transcription of the leader RNA, before RNA polymerase reaches the termination site. In contrast, for translational riboswitches, the ligand can bind either during transcription or after the mRNA is completely synthesized. A single class of riboswitches (e.g., the S box SAM-binding riboswitch described above) is often found in both transcriptional and translational forms, often in different groups of organisms. Riboswitches that function at the level of transcription attenuation tend to predominate in *B. subtilis* and related organisms, while translational control is found more frequently in the *Gammaproteobacteria*, for reasons that are unknown but that may reflect evolutionary history.

TRANSLATIONAL REGULATION BY RNA-BINDING PROTEINS

Binding of an RNA-binding protein to a TIR can sequester the TIR and block binding of the 30S initiation complex. As noted above, the TRAP transcription attenuation protein can also regulate *trp* gene expression at the translational level by binding to a site that overlaps the TIR of target genes (Figure 11.22B). Other examples include many operons for the genes that encode ribosomal proteins. For each operon, a single ribosomal protein encoded in the operon acts as the regulator for the entire operon, usually at the level of translation. Binding of the regulatory protein to the mRNA, at a site that overlaps the TIR for the first gene in the operon, represses translation of the entire mRNA. The effect on the first gene is straightforward, because the presence of the protein prevents binding of the 30S initiation complex. This is a form of autoregulation, as the protein represses its own expression, and is straightforward for a monocistronic operon. However, for a polycistronic operon, the normal expectation is that the other genes in the operon have their own independent TIR. Repression of the downstream genes is likely to occur by **translational coupling** (see chapter 2). A probable mechanism is that the TIRs of the downstream genes are sequestered in a secondary structure by pairing of the TIR with a complementary region in the 3′ part of the coding sequence of the upstream gene. If the upstream gene is being translated (because its TIR is not blocked by the ribosomal protein that acts as the repressor), the mRNA region that would otherwise sequester the TIR is occupied by translating ribosomes, and the TIR for the downstream gene remains accessible. If the upstream gene is not translated (because its TIR is blocked by the regulatory protein), the TIR for the downstream gene will be sequestered by base pairing, and its translation will also be inhibited. Regulation of ribosome biosynthesis is discussed in more detail in chapter 12.

TRANSLATIONAL REGULATION BY sRNAs

sRNAs can regulate target genes in a variety of ways and commonly operate by base pairing with a complementary region of the target mRNA (see Storz et al., Suggested Reading). Binding of the sRNA to its target can affect degradation of the target mRNA, as described above. In addition, if the sRNA binds to a site that overlaps the TIR, binding of the sRNA can inhibit translation of the mRNA. This is similar to the effect of an RNA-binding protein in that the major effect is that the TIR is made unavailable for binding of the 30S translation initiation complex. In many cases, interaction between the sRNA and the mRNA is dependent on the Hfq protein, which facilitates the pairing of the two RNAs. A number of sRNAs have been shown to function in this way, including the *E. coli* DsrA sRNA, which inhibits translation of multiple genes, including the *hns* gene, which encodes the H-NS global regulatory protein (see chapter 12). Interestingly, in addition to its role as a repressor of the *hns* gene, DsrA can also activate expression of another global regulator, the RpoS stationary-phase sigma factor, which is encoded by the *rpoS* gene. The *rpoS* mRNA contains a structure that sequesters its TIR, so that the transcript is inactive unless that structure is disrupted. Binding of the DsrA sRNA to the region of the mRNA that would otherwise pair with the TIR releases the TIR, which is now available for binding of the 30S initiation complex. Repression of *hns* and activation of *rpoS* utilize different regions of the DsrA sRNA, so DsrA has the ability to regulate multiple sets of targets. Since both RpoS and H-NS are global regulators (see chapter 12), DsrA impacts a large number of genes in the cell.

TRANSLATIONAL AUTOREGULATION OF INITIATION FACTOR IF3

All of the translational mechanisms described so far involve physical blocking of the TIR either by a protein or an sRNA or by secondary structure within the leader RNA. An alternative mechanism involves changes in the translation initiation complex itself, which can affect the ability of the complex to recognize specific mRNAs. This type of mechanism is used for the regulation of the *infC* gene, which encodes translation initiation factor IF3, a normal component of the 30S translation initiation complex.

As mentioned in chapter 2, the initiation codon for translation is usually AUG but can also be GUG or, more rarely, UUG. An AUU initiation codon is used only to initiate the translation for the *infC* gene. Use of an AUU initiation codon only for *infC* is conserved in many organisms, suggesting that it is an important feature of the gene. The AUU codon plays an important role in regulation of the *infC* gene so that the cell produces only as much IF3 as it needs. IF3 is responsible for directing the 30S translation initiation complex to authentic TIRs and for positioning the fMet-tRNA^{fMet} accurately at the initiation codon. Translation initiation complexes that contain IF3 will therefore discriminate against the *infC* mRNA because the mRNA has an unfavorable AUU initiation codon. If translation initiation complexes are saturated with IF3, the *infC* mRNA will not be translated and no additional IF3 will be synthesized. If there is insufficient IF3 in the cell, translation initiation complexes that lack IF3 will be present, and they will fail to discriminate against the AUU codon, which allows IF3 synthesis to increase. This regulatory mechanism takes advantage of the function of the regulated gene product in monitoring the levels of that product. We will see a similar principle in the regulation of the gene that encodes the RF-2 translation termination factor (see "Regulation of Translation Termination" below).

Translational Regulation in the Exit Channel of the Ribosome

After the details of translation were worked out in the early 1960s, it was assumed that ribosomes translate an mRNA independent of its sequence, using the sequence of nucleotides in the mRNA only to direct the insertion of amino acids into the growing polypeptide. External features, such as the secondary structure of the mRNA and codon usage, were recognized to influence the initiation, termination, and rate of translation, but the ribosome itself did not appear to discriminate between different coding sequences. It came as a surprise to discover that ribosomes could interact differently with different newly synthesized polypeptides as they emerge through the exit channel and that some polypeptides could cause translation to pause or stop.

As peptide bonds form in the peptidyl transfer center of the ribosome, the growing peptide enters the exit channel in the large 50S subunit of the ribosome and emerges on the other end about 30 amino acids later. This exit channel is constructed mostly of the 23S rRNA, but some proteins, including L4 and L22, help form a constriction that narrows the channel, which the growing polypeptide enters some 9 amino acids from the peptidyl transfer center. The growing polypeptide is exposed both before it enters the narrower part of the channel and after its N terminus emerges from the ribosome, providing opportunities for external factors to influence the movement of the polypeptide through the channel and regulate the translation of the polypeptide. Specific polypeptide sequences called **stalling sequences** in growing polypeptides are capable of contacting specific regions in the walls of the exit channel, causing translation to arrest in the peptidyl transfer center unless certain conditions are met.

Two general types of such regulation have been described (see Ito et al., Suggested Reading). In both types, the specific stalling sequence can cause translation to arrest when it enters the narrow part of the exit channel. In some situations, the binding of a small effector to the stalling sequence is necessary for the arrest to occur, and in others, the arrest always occurs unless it is relieved by entry of the N terminus of the polypeptide into some cellular structure, such as the membrane, forcing coordinate translation of the polypeptide with its insertion into the structure. In general, it is the translation of an upstream leader polypeptide that is arrested, rather than the translation of the gene itself. The translational arrest of this upstream leader polypeptide then regulates the expression of the downstream gene, either translationally, through translational coupling, or transcriptionally, through attenuation.

Examples of the first type, in which a small-molecule effector is required for regulation of a gene by translational arrest, are genes whose products confer resistance to the antibiotic erythromycin or chloramphenicol, both of which block translation by binding to the ribosome (see chapter 2). This regulation is illustrated for an erythromycin resistance gene in Figure 11.28A. In this case, the mRNA is constitutively synthesized, and only the *ermCL* leader peptide coding sequence is translated in the absence of erythromycin. In the presence of sublethal concentrations of erythromycin, the antibiotic binds to the ribosome translating the *ermCL* coding sequence and the ErmCL leader peptide as the peptide enters the narrow part of the exit channel, causing translation to arrest, even though this concentration of antibiotic would not normally inhibit translation. Stalling of the ribosome affects the structure of the RNA and prevents the formation of a helix that would otherwise sequester the TIR of the downstream *ermC* gene, which encodes an enzyme that methylates a specific base in 23S rRNA. This methylation prevents binding of erythromycin to the ribosome and confers resistance to the antibiotic. This mechanism allows rapid induction of the *ermC* resistance gene when the cell is first exposed to low concentrations of the antibiotic, so that the cell is prepared if higher concentrations of the antibiotic accumulate.

Another example of an effector-dependent translational arrest is the regulation of the tryptophan utilization *tna* operon of some enteric bacteria, including *E. coli*. In this case, binding of tryptophan to the stalling peptide in the ribosome arrests translation of the leader peptide. The stalled ribosome masks a Rho utilization site (*rut* site), thereby preventing Rho-dependent termination and allowing transcription of the downstream *tna* operon. In this way, the genes for use of tryptophan as a carbon, nitrogen, and energy source are turned on only if tryptophan is present. Note that in this case, regulation of *tna* operon expression occurs at the level of transcription attenuation; however, sensing of tryptophan is mediated by the ribosome during leader RNA translation.

Other known examples of regulation in the ribosome exit channel are proteins involved in transporting other proteins into and through the inner membrane (see chapter 2). In one well-studied example in *E. coli*, translational arrest during synthesis of a leader polypeptide, SecM, after the N terminus of the protein has exited the ribosome, prevents translation of the downstream *secA* gene, whose product is required for insertion of some proteins into the SecYEG translocon in the inner membrane (see Figure 11.28B and chapter 2). If the N-terminal region of the SecM protein is inserted into the SecYEG translocon, the arrest is relieved and the downstream *secA* gene is translated. It is not clear how the binding of SecM to the SecYEG translocon relieves binding of the stalling peptide to the channel. The SecM protein itself has no apparent function and is quickly degraded when it enters the periplasm. The translational arrest occurs only at very low temperatures or if the SecYEG channels are disrupted in some way. This mechanism may reduce the synthesis of SecA protein when the other translocation systems are inadequate and/or may have the effect of localizing the SecA protein to the SecYEG translocon in the membrane, where it performs its function.

Recent work has begun to clarify how translation arrest might occur. Specific amino acids in the stalling peptide contact specific bases in the 23S rRNA and L4 and L22 proteins in the channel. It has been proposed that binding of the stalling peptide to the channel, with or without the help of a small-molecule effector (depending on the system), causes a distortion in the orientation of the peptidyl tRNA in the P site of the ribosome and prevents formation of the next peptide bond. These effects depend on which amino acid is attached to the aminoacyl-tRNA in the A site, and sometimes which amino acid in the growing peptide it has to bond to. The molecular basis for these effects is not yet fully understood.

Regulation of Translation Termination

As described above for transcription attenuation systems, it is possible to regulate expression of a gene by positioning a signal that should stop the gene expression machinery early in an mRNA and providing a regulatory mechanism that permits bypassing of that signal under certain circumstances to allow expression of the gene. The *E. coli prfB* gene, which encodes ribosome release factor 2 (RF2), provides an example of how that type of event can operate at the translational level.

RF2 recognizes UGA and UAA nonsense codons to terminate translation (see chapter 2). The *prfB* coding sequence is very unusual in that it contains a UGA codon early in the coding sequence. Furthermore, the downstream part of the coding sequence is "out of frame"

A Erythromycin resistance
Transcription start site
ermCL leader peptide coding region
ermC gene
DNA
Promoter
1
2
3
4
AUG • • • • • AUCAGC • • • UGA
Alternative pairing structures for *ermCL* RNA
1 Without erythromycin
S-D
Sequestrator structure (3:4)
1
2
3
4
AUG
ermCL
ermC
RNA 5′ AUG
Off
2 With erythromycin
Erythromycin
Antisequestrator structure (2:3)
2
3
ermCL
ermC
S-D
RNA 5′ AUG
AUC 1 AGC
4
AUG
On
B SecM
30S
50S
Arrest codons
mRNA
secM
S-D
secA
Periplasm
SecYEG
Cytoplasmic membrane
Cytoplasm
SecA
SecA
SecA
Arrest peptide
secM mRNA
S-D
secA
mRNA released

with the portion that is upstream of the UGA, so synthesis of full-length RF2 requires that the ribosome slip back on the mRNA by 1 nucleotide into the −1 frame. When the translating ribosome reaches the UGA codon in the mRNA, it terminates and is released from the mRNA only if RF2 levels are high enough to allow efficient recognition of the UGA codon. If RF2 levels are low, the translating ribosome pauses at the UGA codon, which results in a high frequency of the frameshift event that allows translation to continue in the −1 frame and synthesis of more RF2 protein. This is an example of **programmed frameshifting** (see Box 2.5), and like the *infC* system described above, it provides an example of translational autoregulation that exploits the biological function of the gene product (in this case, translation termination) as a key feature of the regulatory mechanism.

Posttranslational Regulation

We usually consider gene expression to be complete once a polypeptide has been released by the translating ribosome. However, other steps may be necessary to determine how much active protein product is in the cell. As described in chapter 2, the protein may need to fold correctly, to form higher-order complexes, or to be modified to be fully active. The level of protein product in the cell also depends on how long each protein molecule persists in the cell. Finally, the activity of the protein can be affected by binding of small molecules, as in feedback inhibition of a biosynthetic enzyme by the end product of the pathway in which it participates. Each of these represents an opportunity for posttranslational regulation.

Posttranslational Protein Modification

Posttranslational regulation can occur by reversible modification of specific sites on a protein. These modifications, which in bacteria most commonly include phosphorylation, methylation, and acetylation, can change the activity of the protein. Therefore, changing the activity of the enzymes responsible for the modification can control the activity of the modified proteins.

One example of an important posttranslational modification is the adenoribosylation (addition of AMP) to the *E. coli* glutamine synthetase enzyme, which synthesizes glutamine from glutamate. At high concentrations of glutamine, this enzyme is adenoribosylated, which temporarily inhibits its activity until glutamine levels drop. The AMP groups are then removed, and the activity is restored (see chapter 12). Phosphorylation is another common modification that can change protein activity. This is used to control the activity of a large set of regulatory proteins through partnership of a phosphorylation enzyme (a "sensor kinase") and a DNA-binding protein (a "response regulator") in what are termed **two-component regulatory systems**, which are responsible for a wide range of global regulatory responses (see chapter 12). Other common posttranslational modifications include the methylations that are used in bacterial chemotaxis systems to allow the cell to monitor gradients of molecules in its external environment and to respond by moving toward attractants (compounds they want to use) and away from toxic compounds. Protein acetylation has also emerged recently as a mechanism to regulate enzyme activity in response to cellular metabolism.

A common theme of these modification systems is that the modification is reversible, allowing the cell to continually sense whether conditions have changed. The cellular response is therefore dependent on the relative activities of the enzymes responsible for addition and removal of the modification, providing opportunities for tight control of the activity of the target protein.

Regulation of Protein Turnover

We discussed earlier that the level of an mRNA is determined not only by its synthesis rate, but also by its rate of degradation. This also applies to protein levels. A protein that is very stable can accumulate to high levels even if its synthesis rate is relatively low, while a protein that is very unstable will not accumulate to high levels even if its synthesis rate is high. The stability of a protein (measured by its half-life, which is the time it takes for the amount of protein present at an initial time point to be reduced by half) is determined by intrinsic properties of the protein (i.e., its susceptibility to cellular proteases [see chapter 2]) but can also be changed in response to changes in environmental conditions, such as heat shock (see chapter 12); this can occur because the stress condition is

Figure 11.28 Regulation by translational arrest in the ribosome. **(A)** Regulation of the *ermC* gene by erythromycin. In the absence of erythromycin, the ribosome translates the *ermCL* region without stalling, and helix 1:2 forms, which allows the 3:4 sequestrator to form; this prevents binding of a new ribosome to the *ermC* TIR, and the *ermC* coding sequence is not translated. Binding of erythromycin to the stalling peptide in the exit channel causes the ribosome to stall on the mRNA, which prevents formation of the 1:2 helix. This allows formation of the 2:3 helix and releases region 4 to allow binding of the ribosome to the TIR of the *ermC* gene. **(B)** Arrest of translation of the upstream coding sequence for SecM prevents translation of the downstream gene for SecA. If SecM enters the SecYEG channel in the membrane, the *secA* TIR is exposed and translation of SecA occurs. See chapter 2 for details of protein translocation through the membrane.

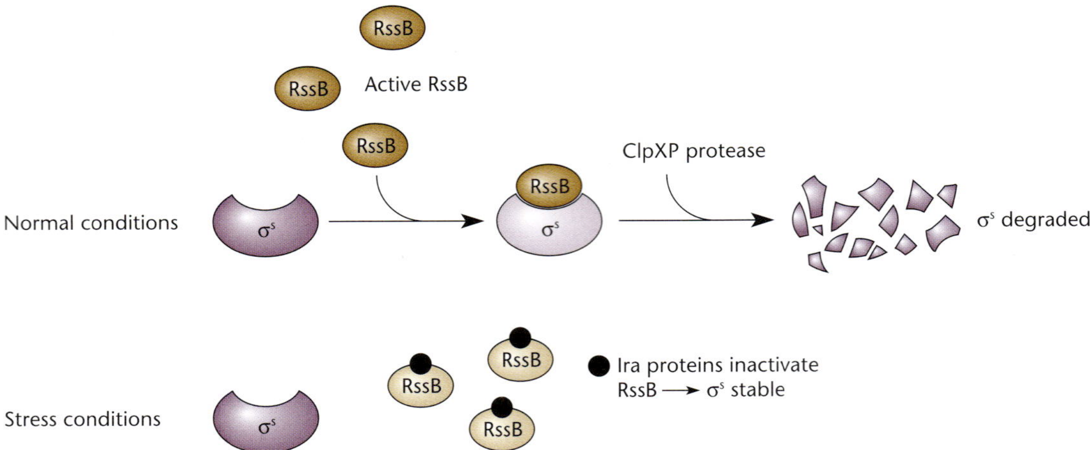

Figure 11.29 Regulated proteolysis of σ^S by adaptors and antiadaptors. Under normal growth conditions, both σ^S and the RssB adaptor protein are present in the cell. RssB is active and binds to σ^S and delivers it to the ClpXP protease for degradation. When the cell enters stationary phase or is subjected to certain stressful conditions, the Ira antiadaptor proteins are synthesized. The Ira proteins bind to RssB and inactivate it. This prevents degradation of σ^S, which is now able to direct the transcription of genes required for response to the stressful conditions. Adapted from Wilson DN, *Mol Cell* **41:**247–248, 2011.

damaging to cellular proteins or can be targeted to specific proteins in a process called **regulated proteolysis**. Individual proteins can be targeted for destruction by proteins called **adaptors**, whose activity can in turn be controlled by other proteins, called **antiadaptors** because they prevent the activity of the adaptors; adaptors result in decreased stability, and antiadaptors result in increased stability of the protein that would otherwise be recognized by the partner adaptor. Adaptors and antiadaptors are usually highly specific in their interactions with each other, the proteins they regulate, and specific proteases.

REGULATION OF THE RpoS SIGMA FACTOR BY ADAPTORS AND ANTIADAPTORS

One of the best-characterized systems of regulated proteolysis is the *E. coli* RpoS protein, which is a sigma factor (σ^S) responsible for transcription of stress response genes during stationary phase and in response to certain stress conditions (see chapter 12). Some of the σ^S protein is synthesized under nonstress conditions. However, in the absence of stress, the σ^S protein is rapidly degraded by the ClpXP protease (see chapter 2). Rapid degradation of σ^S requires the RssB protein, which acts as an adaptor protein to specifically bind σ^S and deliver it to ClpXP for destruction (Figure 11.29). This results in low levels of σ^S under conditions when transcription of the stress response genes is not required. When the cell encounters starvation or stress conditions, the RssB protein is inactivated by one of a set of antiadaptor proteins, the Ira (inhibitor of RssB activity) proteins, each of which is induced in response to a different stressful condition (e.g., starvation or low phosphate). Binding of an Ira protein to RssB prevents binding of RssB to σ^S and therefore results in increased stability of σ^S and transcription of the stress response genes (see Bougdour et al., Suggested Reading). This system allows the cell to be poised for response to a variety of environmental stresses by synthesis of σ^S before it is needed. Activity of σ^S is maintained at a low level by regulated proteolysis until the cell experiences a stress condition for which σ^S directs the response.

Feedback Inhibition of Enzyme Activity

Biosynthetic pathways are not regulated solely through transcriptional and translational regulation of their operons and by covalent modifications or degradation of their enzymes; they are also often regulated by **feedback inhibition** of the enzymes once they are made. In feedback inhibition, the end product of a pathway binds to the first enzyme of the pathway and inhibits its activity. This blocks the activity of the pathway even though the enzymes are present in the cell. Feedback inhibition is common to many types of biosynthetic pathways and is a more sensitive and rapid mechanism for modulating the amount of the end product than are transcriptional regulation and translational regulation, which respond more slowly to changes in the concentration of the end product of the pathway. Feedback inhibition is also easily reversible, because the enzymes remain in the cell, ready to resume activity if the level of the end product of the pathway drops as the cell utilizes the compound.

FEEDBACK INHIBITION OF THE *trp* OPERON

In addition to the transcriptional regulation described earlier, the tryptophan biosynthetic pathway of *E. coli* is also subject to feedback inhibition. Tryptophan binds to the first enzyme of the tryptophan synthesis pathway, anthranilate synthetase, and inhibits its activity, thereby preventing the synthesis of more tryptophan. The tryptophan analog 5-methyltryptophan has been used to study this process. At high concentrations, 5-methyltryptophan binds to anthranilate synthetase in place of tryptophan and inhibits the activity of the enzyme, starving the cells for tryptophan. Only mutants defective in feedback inhibition because of a missense mutation in the *trpE* gene that prevents the binding of tryptophan (and 5-methyltryptophan) to the anthranilate synthetase enzyme can multiply to form a colony in the absence of tryptophan.

A similar method is described above for isolating constitutive mutants with mutations of the *trp* operon, but selection of constitutive mutants requires lower concentrations of 5-methyltryptophan. If the concentration of this analog is high enough, even constitutive mutants will be starved for tryptophan, because binding of the analog inactivates all of the anthranilate synthase that is synthesized.

Why Are There So Many Mechanisms of Gene Regulation?

It is evident from the discussion in this chapter (and chapter 12) that gene expression can be regulated in a wide variety of ways and by a wide variety of mechanisms. It is relatively easy to understand why a catabolic system is induced by the substrate of the regulated pathway and a biosynthetic system is repressed by the end product of the pathway. But why isn't all regulation at the level of transcription initiation? It would appear to be wasteful to regulate gene expression at the level of transcript degradation or translation, since the cell is producing mRNAs that it may never use. Posttranslational regulation seems even more wasteful, as the cell produces both mRNAs and proteins it does not use. In some cases, the level of gene expression at which regulation occurs is dictated by the function of the gene product. For example, the gene encoding IF3 or RF2 exploits the function of its product in its regulatory mechanism. Another advantage of posttranscriptional regulation is that the cell always contains an adequate supply of the mRNA for the regulated gene, and the decision to utilize that mRNA to synthesize the protein product can be made very rapidly in response to subtle physiological changes. This is especially important in stress responses, such as the heat shock response, which in *E. coli* utilizes an RNA thermosensor to directly and rapidly sense an increase in temperature, enabling the cell to immediately begin synthesis of gene products that protect it from this stressful condition. Posttranslational regulation allows an even faster response to stressful conditions.

It is not always obvious why one set of genes in one set of organisms uses one mechanism while related genes in other organisms use a different mechanism. It is important to note that these mechanisms evolved in response to selective pressures faced by an ancestral cell, and we may not know what those selective pressures were. Nevertheless, the bacterial cells we see today must find that these regulatory solutions serve their current needs, and it is likely that regulatory patterns will shift as organisms face new sets of selective pressures.

1. Regulation of gene expression can occur at any stage in the expression of a gene. If the amount of mRNA synthesized from the gene differs under different conditions, the gene is transcriptionally regulated. If the regulation occurs after the mRNA is made, the gene is posttranscriptionally regulated. A gene is translationally regulated if the mRNA is made but not always translated at the same rate. The mRNA or protein product of a gene can also be stabilized or degraded, or the protein product can be modified.

2. In bacteria, more than one gene is sometimes transcribed into the same mRNA. Such a cluster of genes, along with their adjacent *cis*-acting regulatory sites, is called an operon.

3. The regulation of operon transcription can be negative, positive, or a combination of the two. If a protein blocks the transcription of the operon, the operon is negatively regulated and the regulatory protein is a repressor. If a protein is required for transcription of an operon, the operon is positively regulated and the regulatory protein is an activator.

4. If an operon is negatively regulated, mutations that inactivate the regulatory gene product result in constitutive mutants in which the operon genes are always expressed. If the operon is positively regulated, mutations that inactivate the regulatory protein cause permanent loss of expression of the operon. In general, because most mutations are inactivating mutations,

(continued)

Summary (continued)

constitutive mutations are much more common with negatively regulated operons than with positively regulated operons.

5. Sometimes the same protein can be both a repressor and an activator in different situations, which complicates the genetic analysis of the regulation.

6. The regulation of transcription of bacterial operons is often achieved through small molecules called effectors, which bind to the repressor or activator protein, changing its conformation. If the presence of the effector causes the operon to be transcribed, it is called an inducer; if its presence blocks transcription of the operon, it is called a corepressor. The substrates of catabolic operons are usually inducers, whereas the end products of biosynthetic pathways are usually corepressors.

7. The regions on DNA to which repressors bind are called operators. Some repressors act by physically interfering with the binding of the RNA polymerase to the promoter (preventing closed-complex formation). Others allow repressor binding but prevent opening of the DNA at the promoter (preventing open-complex formation). Yet others prevent the RNA polymerase from escaping the promoter to begin RNA synthesis (preventing promoter clearance). Some repressors act by binding to two operators on either side of the promoter simultaneously, bending the DNA between them and inactivating the promoter.

8. The regions to which activator proteins bind are called activator sequences. Some activator proteins recruit RNA polymerase to the promoter by binding both to a region on the DNA close to the promoter and to an exposed region of the RNA polymerase, thereby stabilizing the binding of the RNA polymerase to the promoter. Others interact with RNA polymerase already at the promoter and allow it to form an open complex. Still others remodel the promoter by changing the conformation of the DNA, thereby optimizing the spacing and orientation of the –10 and –35 regions.

9. Binding sites for repressor proteins usually overlap with the RNA polymerase-binding site or are downstream. Binding sites for activators are usually upstream of the promoter so that the activator does not interfere with access of RNA polymerase to the promoter. Some regulatory proteins can bind to different regions on DNA, depending on the location of the binding site, and can act as both repressors and activators.

10. Some operons are transcriptionally regulated by a mechanism called attenuation. In operons regulated by attenuation, transcription begins on the operon but then terminates after a short leader sequence has been transcribed if the enzymes encoded by the operon are not needed.

11. Attenuation is sometimes determined by whether certain codons in the leader sequence are translated. Pausing of the ribosome at these codons can cause secondary-structure changes in the leader RNA, leading to termination of transcription by RNA polymerase before it reaches the first gene of the operon. In other operons, attenuation is mediated by RNA-binding proteins that either stabilize or destabilize leader RNA structural elements that determine whether transcription will terminate.

12. Leader RNA structural changes can affect both transcription attenuation and translation initiation. Riboswitches are leader RNA elements that directly sense a physiological signal. Binding of the signal changes the leader RNA secondary structure, affecting transcription attenuation or translation of the downstream gene. RNA-binding proteins can also affect translation initiation by sequestration of the TIR.

13. Gene expression can be regulated by affecting either the synthesis or the degradation of the mRNA and protein products. The half-life of the mRNA or protein product is the time required for the amount of the product to be reduced to half of the amount that was present at a given time point. Changes in half-life can have major effects on the amount of product present in the cell.

14. The activity of a protein can be regulated through reversible regulation of the activities of the enzymes of the pathway. This reversible regulation can occur by feedback inhibition, which results from binding of the end product of the biosynthetic pathway to the first enzyme of the pathway, or it can occur by reversible covalent modification of the protein.

QUESTIONS FOR THOUGHT

1. Why do you suppose both negative and positive mechanisms of transcriptional regulation are used to regulate bacterial operons?

2. Why are regulatory protein genes sometimes autoregulated?

3. Why do you suppose the genes for the biosynthesis of most amino acids, such as tryptophan, isoleucine-valine, and histidine, are arranged together in operons?

4. What advantages or disadvantages are there to regulation by attenuation? Would it not be less wasteful to regulate all operons through initiation of RNA synthesis at the promoter by repressors or activators?

5. Why does the stability of an mRNA transcript or protein product matter? Why is the synthesis rate not the only important parameter?

SUGGESTED READING

Babitzke P, Gollnick P. 2001. Posttranscription initiation control of tryptophan metabolism in *Bacillus subtilis* by the *trp* RNA-binding attenuation protein (TRAP), anti-TRAP, and RNA structure. *J Bacteriol* **183:**5795–5802.

Bougdour A, Cunning C, Baptiste PJ, Elliott T, Gottesman S. 2008. Multiple pathways for regulation of sigmaS (RpoS) stability in *Escherichia coli* via the action of multiple anti-adaptors. *Mol Microbiol* **68:**298–313.

Browning DF, Busby SJ. 2004. The regulation of bacterial transcription initiation. *Nat Rev Microbiol* **2:**57–65.

Collins JA, Irnov I, Baker S, Winkler WC. 2007. Mechanism of mRNA destabilization by the *glmS* ribozyme. *Genes Dev* **21:**3356–3368.

Cronan JE Jr, Subrahmanyam S. 1998. FadR, transcriptional co-ordination of metabolic expediency. *Mol Microbiol* **29:**937–943.

Englesberg E, Squires C, Meronk F Jr. 1969. The L-arabinose operon in *Escherichia coli* B/r: a genetic demonstration of two functional states of the product of a regulator gene. *Proc Natl Acad Sci USA* **62:**1100–1107.

Fux L, Nussbaum-Shochat A, Lopian L, Amster-Choder O. 2004. Modulation of monomer conformation of the BglG transcriptional antiterminator from *Escherichia coli*. *J Bacteriol* **186:**6775–6781.

Geanacopoulos M, Vasmatzis G, Zhurkin VB, Adhya S. 2001. Gal repressosome contains an antiparallel DNA loop. *Nat Struct Biol* **8:**432–436.

Geissmann T, Marzi S, Romby P. 2009. The role of mRNA structure in translational control in bacteria. *RNA Biol* **6:**153–160.

Grundy FJ, Henkin TM. 1993. tRNA as a positive regulator of transcription antitermination in *B. subtilis*. *Cell* **74:**475–482.

Guzman LM, Belin D, Carson MJ, Beckwith J. 1995. Tight regulation, modulation, and high-level expression by vectors containing the arabinose P_{BAD} promoter. *J Bacteriol* **177:**4121–4130.

Irani MH, Orosz L, Adhya S. 1983. A control element within a structural gene: the *gal* operon of *Escherichia coli*. *Cell* **32:**783–788.

Ito K, Chiba S, Pogliano K. 2010. Divergent stalling sequences sense and control cellular physiology. *Biochem Biophys Res Commun* **393:**1–5.

Jacob F, Monod J. 1961. Genetic regulatory mechanisms in the synthesis of proteins. *J Mol Biol* **3:**318–356.

Lewis M, Chang G, Horton NC, Kercher MA, Pace HC, Schumacher MA, Brennan RG, Lu P. 1996. Crystal structure of the lactose operon repressor and its complexes with DNA and inducer. *Science* **271:**1247–1254.

McDaniel BA, Grundy FJ, Henkin TM. 2005. A tertiary structural element in S box leader RNAs is required for *S*-adenosylmethionine-directed transcription termination. *Mol Microbiol* **57:**1008–1021.

Merino E, Yanofsky C. 2005. Transcription attenuation: a highly conserved regulatory strategy used by bacteria. *Trends Genet* **21:**260–264.

Morse DE, Morse ANC. 1976. Dual-control of the tryptophan operon is mediated by both tryptophanyl-tRNA synthetase and the repressor. *J Mol Biol* **103:**209–226.

Oxender DL, Zurawski G, Yanofsky C. 1979. Attenuation in the *Escherichia coli* tryptophan operon: role of RNA secondary structure involving the tryptophan codon region. *Proc Natl Acad Sci USA* **76:**5524–5528.

Pace HC, Kercher MA, Lu P, Markiewicz P, Miller JH, Chang G, Lewis M. 1997. Lac repressor genetic map in real space. *Trends Biochem Sci* **22:**334–339.

Possoz C, Filipe SR, Grainge I, Sherratt DJ. 2006. Tracking of controlled *Escherichia coli* replication fork stalling and restart at repressor-bound DNA *in vivo*. *EMBO J* **25:**2596–2604.

Raveh H, Lopian L, Nussbaum-Shochat A, Wright A, Amster-Choder O. 2009. Modulation of transcription antitermination in the *bgl* operon of *Escherichia coli* by the PTS. *Proc Natl Acad Sci USA* **106:**13523–13528.

Rojo F. 1999. Repression of transcription initiation in bacteria. *J Bacteriol* **181:**2987–2991.

Sanchez A, Osborne ML, Friedman LJ, Kondev J, Gelles J. 2011. Mechanism of transcriptional repression at a bacterial promoter by analysis of single molecules. *EMBO J* **30:**3940–3946.

Schleif R. 2000. Regulation of the L-arabinose operon of *Escherichia coli*. *Trends Genet* **16:**559–565.

Schuck A, Diwa A, Belasco JG. 2009. RNase E autoregulates its synthesis in *Escherichia coli* by binding directly to a stem-loop in the *rne* 5′ untranslated region. *Mol Microbiol* **72:**470–478.

Shuman HA, Silhavy TJ. 2003. The art and design of genetic screens: *Escherichia coli*. *Nat Rev Genet* **4:**419–431.

Storz G, Vogel J, Wassarman KM. 2011. Regulation by small RNAs in bacteria: expanding frontiers. *Mol Cell* **43:**880–891.

Yanofsky C, Crawford IP. 1987. The tryptophan operon, p 1453–1472. *In* Neidhardt FC, Ingraham JL, Low KB, Magasanik B, Schaechter M, Umbarger HE (ed), Escherichia coli *and* Salmonella typhimurium*: Cellular and Molecular Biology*, vol 2. American Society for Microbiology, Washington, DC.

Yousef MR, Grundy FJ, Henkin TM. 2005. Structural transitions induced by the interaction between tRNA$^{(Gly)}$ and the *Bacillus subtilis glyQS* T box leader RNA. *J Mol Biol* **349:**273–287.

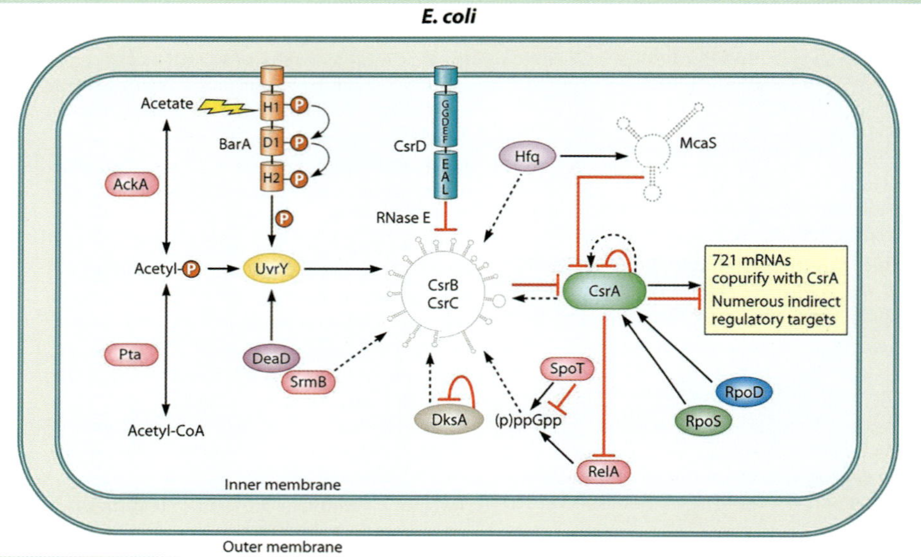

The Csr global regulatory network in *E. coli.* Solid lines indicate regulatory interactions for which the molecular mechanism is well studied, and dashed lines indicate regulatory effects that are not yet fully understood. Black arrows indicate positive effects; red T bars indicate negative effects. Modified from Vakulskas et al., 2015 (see Suggested Reading).

12 Global Regulation: Regulons and Stimulons

BACTERIA MUST BE ABLE TO ADAPT to a wide range of environmental conditions to survive. Nutrients are limiting in most natural environments, so bacteria must be able to recognize the availability of nutrients and protect themselves against starvation until an adequate food source becomes available. Different environments may vary greatly in the amount of water or in the concentration of solutes, so bacteria must be able to adjust to desiccation and differences in osmolarity, as well as other stresses. Temperature fluctuations are also a problem for bacteria. Unlike humans and other warm-blooded animals, bacteria cannot maintain their own cell temperature and must be able to function over wide ranges of temperature. Pathogenic bacteria must be able to sense that they have entered the host and adapt to the new conditions in this environment.

Survival alone is not enough for a species to prevail, however. The species also must compete effectively with other organisms in its environment. Competing effectively might mean being able to use scarce nutrients efficiently or rapidly taking advantage of plentiful ones to achieve higher growth rates and thereby become a higher percentage of the total population of organisms in the environment. Moreover, different compounds may be available for use as carbon and energy sources. The bacterium may need to choose the carbon and energy source it can use most efficiently and ignore the rest if it is to compete most effectively and so that it does not waste energy making extra enzymes.

Not only do conditions vary in the environment to which the bacterium is exposed, but the changes also can be abrupt. The bacterium may have to adjust the rate of synthesis of its cellular constituents quickly in response to a change in growth conditions. For example, different carbon and energy sources allow different rates of bacterial growth. Different growth rates require different rates of synthesis of cellular macromolecules, such as DNA, RNA, and proteins, which in turn require different concentrations of the components of the cellular macromolecular synthesis machinery, such as ribosomes, tRNA, and RNA polymerase. Moreover, the relative rates of synthesis of the different cellular components must be coordinated so that the cell does not accumulate more of some component than it needs.

Many groups of bacteria also have the ability to undergo complicated developmental cycles, such as sporulation, in response to nutrient limitation. These cycles often require long periods of time without cell division, so the decision to embark on this pathway has major repercussions on the ability of the organism to increase its population size. Pathways of this type often also require coordination of expression of large numbers of genes for success.

Adjusting to major changes in the environment requires regulatory systems that simultaneously modulate the expression of numerous genes and operons. These systems are called **global regulatory mechanisms**. Often in global regulation, a single regulatory protein (or RNA) controls a large number of genes and operons, which are then said to be members of the same **regulon**. Most genes are part of at least one regulon, and some regulons are very large. Individual genes or operons can also be members of multiple regulons, which allows a response to multiple input signals. Regulons often overlap in their responses to changing conditions. The collection of regulons that respond to the same set of environmental conditions is called a **stimulon**. Large-scale genomic analyses can be used to identify most of the genes of a regulon or stimulon. Such studies reveal that only seven regulators control almost half of all the genes of *Escherichia coli*.

Table 12.1 lists some well-characterized global regulatory mechanisms in *E. coli*. If the genes are under the control of a single regulatory gene (and thus are members of the same regulon), the regulatory gene is also listed. Some examples of regulons are discussed in previous chapters. For example, all the genes under the control of the TrpR repressor, including the *trpR* gene itself, are part of the TrpR regulon (see chapter 11). The Ada regulon comprises the adaptive-response genes, including those encoding the methyltransferases that repair alkylation damage to DNA; all of these genes are under the control of the Ada protein. Similarly, the SOS genes that are induced after UV irradiation and some other types of DNA-damaging treatments are all under the control of the same protein, the LexA repressor, and so are part of the LexA regulon (see chapter 10).

In this chapter, we discuss how some global regulatory mechanisms operate on the molecular level and describe some of the genetic experiments that have contributed to this knowledge. In many cases, the molecular basis of the type of global regulatory mechanism may involve a complex interaction among several cellular signals or regulators. Ongoing studies of the molecular basis of global regulatory mechanisms represent one of the most active areas of research involving bacterial molecular genetics.

Carbon Catabolite Regulation

One of the best-characterized global regulatory systems in bacteria coordinates the expression of genes involved in carbon and energy source utilization. All cells must have access to high-energy, carbon-containing compounds, which they degrade to generate ATP for energy and smaller molecules needed as building blocks for cellular constituents. Smaller molecules resulting from the metabolic breakdown of larger molecules are called **catabolites**.

In nutrient-rich environments, bacterial cells may be growing in the presence of several different carbon and energy sources, some of which can be used more efficiently than others. Energy must be expended to synthesize the enzymes needed to metabolize the different carbon sources, and the utilization of some carbon compounds requires more enzymes, and yields less energy, than does the utilization of others. By making only the enzymes for utilization of the carbon and energy source that yields the highest return, the cell gets the most catabolites and energy, in the form of ATP, for the energy it expends. The mechanism for ensuring that the cell preferentially uses the best carbon and energy source available is called **catabolite regulation**. Operons that are subject to catabolite regulation are said to be **catabolite sensitive**. Historically, this regulatory response has been referred to as "catabolite repression" based on the fact that cells growing in better carbon sources, such as glucose, seem to repress the expression of operons for the utilization of poorer carbon sources. However, as we shall see, the name "catabolite repression" is often a misnomer, because in at least some of the regulatory systems in *E. coli*, the genes under catabolite control are activated when poorer carbon sources are the only ones available. Catabolite regulation is also sometimes called the **glucose effect** because glucose, which yields the highest return of ATP per unit of expended energy, usually strongly represses operons for other carbon sources. To use glucose, the cell need only convert it to glucose-6-phosphate, which can enter the glycolytic pathway. Thus, glucose is the preferred carbon and energy source for most (but not all) types of bacteria.

Figure 12.1 illustrates what happens when *E. coli* cells are growing in a mixture of glucose and galactose. The cells first use the glucose, and only after it is depleted do they begin to use the galactose. When the glucose is gone, the cells stop growing briefly while they synthesize the enzymes for galactose utilization. Following the appropriate regulatory changes in the cell, growth resumes, but at a slightly lower rate. This growth pattern is called **diauxie** and is commonly observed during growth in a mixture of carbon sources.

Carbon Catabolite Regulation in *E. coli*: Catabolite Activator Protein (CAP) and cAMP

Most bacteria and single-cell eukaryotes are known to have systems for catabolite regulation. The best understood is the **cyclic AMP (cAMP)**-dependent system of *E. coli* and other enteric bacteria. cAMP is similar to AMP

Table 12.1 A sampling of *E. coli* global regulatory systems

System	Response	Regulatory gene(s) (protein[s])	Category of mechanism	Some genes, operons, regulons, and stimulons
Nutrient limitation				
Carbon	Catabolite regulation	*crp* (CAP, also called CRP)	DNA-binding activator or repressor	*lac*, *ara*, *gal*, *mal*, and numerous other C source operons
	Control of fermentative versus oxidative metabolism	*cra* (Cra)	DNA-binding activator or repressor	Enzymes of glycolysis, Krebs cycle
Nitrogen	Response to ammonia limitation	*rpoN*	Sigma factor (σ^N)	*glnA* (GS) and operons for amino acid degradation
		ntrBC (NtrBC)	Two-component system	
Phosphorus	Starvation for inorganic orthophosphate (P_i)	*phoBR* (PhoBR)	Two-component system	>38 genes, including *phoA* (bacterial alkaline phosphatase) and *pst* operon (P_i uptake)
Growth limitation				
Stringent response	Response to lack of sufficient aminoacylated-tRNAs for protein synthesis; perturbation of carbon metabolism	*relA* (RelA), *spoT* (SpoT)	(p)ppGpp metabolism	rRNA, tRNA, ribosomal proteins, amino acid biosynthesis operons
Stationary phase	Switch to maintenance metabolism and stress protection	*rpoS* (RpoS)	Sigma factor (σ^S)	Many genes with σ^S promoters; complex effects on many operons
Oxygen	Response to anaerobic environment	*fnr* (FNR)	CAP family of DNA-binding proteins	>31 transcripts, including *narGHJI* (nitrate reductase)
	Response to presence of oxygen	*arcAB* (ArcAB)	Two-component system	>20 genes, including *cob* (cobalamin synthesis)
Stress				
Osmoregulation	Response to abrupt osmotic upshift	*kdpDE* (KdpD/KdpE)	Two-component system	*kdpFABC* (K^+ uptake system)
	Adjustment to osmotic environment	*envZ*/*ompR* (EnvZ/OmpR)	Two-component system	OmpC and OmpF outer membrane proteins
		micF	sRNA	*ompF* (porin)
Oxygen stress	Protection against reactive oxygen species	*soxS* (SoxS)	AraC family of DNA-binding proteins	Regulon, including *sodA* (superoxide dismutase) and *micF* (sRNA regulator of *ompF*)
		oxyR (OxyR)	LysR family of DNA-binding proteins	Regulon, including *katG* (catalase)
Heat shock	Tolerance of abrupt temperature increase	*rpoH* (RpoH)	Sigma factor (σ^H)	Stimulon; Hsps (heat shock proteins), including *dnaK*, *dnaJ*, and *grpE* (chaperones) and *lon*, *clpP*, *clpX*, and *hflB* (proteases)
Envelope stress	Misfolded Omp proteins	*rpoE* (RpoE)	Sigma factor (σ^E)	>10 genes, including *rpoH* (σ^H) and *degP* (encoding a periplasmic protease)
	Misfolded pilus	*cpxAR* (CpxAR)	Two-component system	Overlap with RpoE regulon
pH shock	Tolerance of acidic environment	Many	Many	Complex stimulon

(a constituent of an RNA chain; see chapter 2), with a single phosphate group on the ribose sugar; however, the phosphate is attached to both the 5′-hydroxyl and the 3′-hydroxyl groups of the sugar, thereby making a circle out of the phosphate and sugar (Figure 12.2A). Only *E. coli* and other closely related enteric bacteria seem to use this cAMP-dependent system. Some bacteria have an entirely different catabolite regulation system that does not involve cAMP (see below). Even *E. coli* has a second, cAMP-independent system for catabolite regulation, which is discussed in Box 12.1.

REGULATION OF cAMP SYNTHESIS

Catabolite regulation in *E. coli* is achieved through fluctuation in the levels of cAMP, which vary inversely with the availability of readily metabolizable carbon sources,

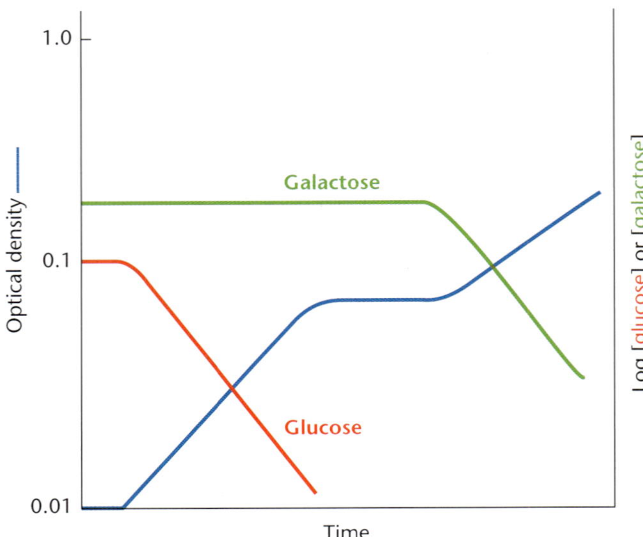

Figure 12.1 Diauxic growth of *E. coli* in a mixture of glucose and galactose. The concentrations of the sugars in the medium are shown as red and green lines. The optical density, a measure of cell growth, is shown as a blue line. The cells first deplete all the glucose and then show a short lag while they induce the *gal* operon (plateau in optical density, shown in blue). They then grow more slowly on the galactose.

such as glucose. In other words, cellular concentrations of cAMP are higher when levels of easily metabolized carbon sources are lower; this occurs when the bacteria are growing in a relatively poor source of carbon, such as lactose or maltose. The synthesis of cAMP is controlled through the regulation of the activity of adenylate cyclase. This enzyme, which makes cAMP from ATP, is more active when glucose is low and less active when glucose is high. The adenylate cyclase enzyme is associated with the inner membrane and is the product of the *cya* gene.

Figure 12.2 outlines the current picture of the regulation of adenylate cyclase activity. An important factor in the regulation is the phosphoenolpyruvate (PEP)-dependent sugar phosphotransferase system (PTS), which is responsible for transporting certain sugars, including glucose, into the cell. We mentioned the PTS in connection with the regulation of the *bgl* operon in chapter 11. One of the protein components of the PTS, named IIA^{Glc}, can exist in either an unphosphorylated (IIA^{Glc}) or a phosphorylated (IIA^{Glc}~P) form. The IIA^{Glc}~P form activates adenylate cyclase to make cAMP. When levels of glucose or another sugar that IIA^{Glc} transports are high in the growth medium, most of the IIA^{Glc} is in the unphosphorylated form. As a result, little of the IIA^{Glc}~P form is available to activate the adenylate cyclase, and cAMP levels drop.

The ratio of IIA^{Glc}~P to IIA^{Glc} is determined largely by the ratio of PEP to pyruvate in the cell. When a rapidly

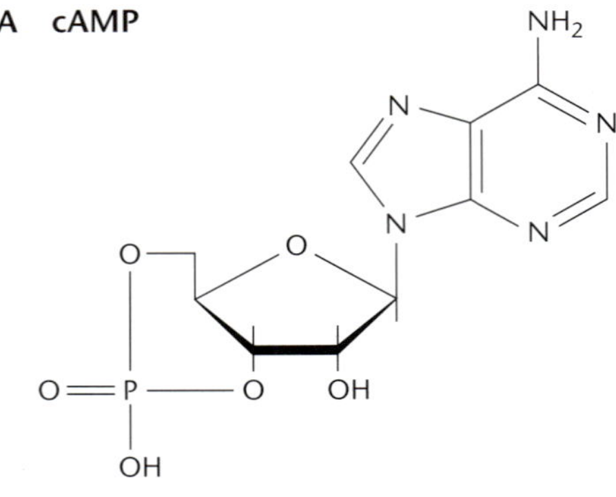

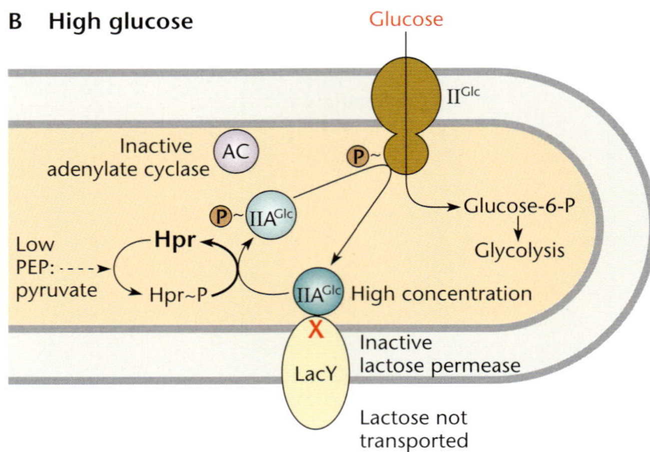

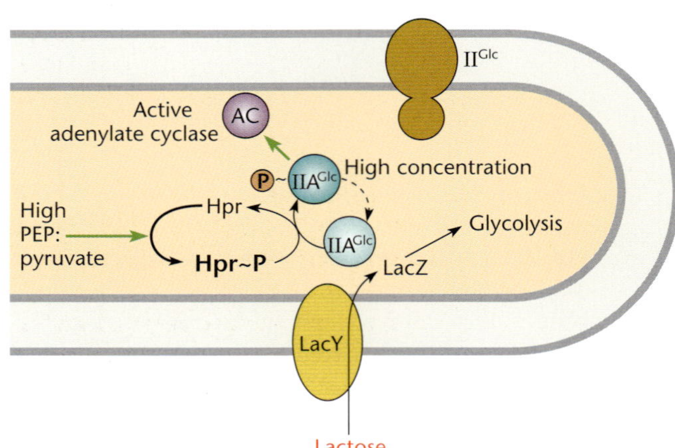

Figure 12.2 Exogenous glucose inhibits both cAMP synthesis and the uptake of other sugars, such as lactose. **(A)** Structure of cAMP. **(B)** In the presence of glucose, the ratio of IIA^{Glc} to IIA^{Glc}~P is high, as glucose is phosphorylated when it is transported by the glucose transporter, II^{Glc}. Unphosphorylated IIA^{Glc} inhibits the lactose permease (LacY), resulting in "inducer exclusion." **(C)** In the absence of glucose, the IIA^{Glc}~P concentration is high, and it activates adenylate cyclase. Also, lactose transport is permitted.

BOX 12.1

cAMP-Independent Carbon Catabolite Regulation in *E. coli*

Not all catabolite regulation in bacteria is mediated by cAMP. In fact, most *Firmicutes*, including *B. subtilis*, do not even have cAMP and use a mechanism for catabolite regulation that is very different from the CAP-cAMP system. Even in *E. coli* and other enteric bacteria, there is a mechanism of catabolite repression that does not depend on cAMP. This mechanism involves the Cra protein, named for its function as a catabolite repressor/activator (originally called FruR, for fructose repressor). The Cra protein is encoded by the *cra* gene and is a DNA-binding protein similar to LacI and GalR. Cra was discovered during the identification of mutations that suppress *ptsH* mutations of *E. coli* and *Salmonella enterica* serovar Typhimurium. The *ptsH* gene encodes the Hpr protein, which is involved in phosphorylation of many sugars that can be transported by the PTS system, including glucose (Figure 12.2). Therefore, *ptsH* mutants cannot use these sugars, because phosphorylation is the first step in the glycolytic pathway. The *cra* mutations suppress *ptsH* mutations and allow growth on PTS sugars by allowing the constitutive expression of the fructose catabolic operon, which includes a gene that encodes a protein that can substitute for Hpr. The *cra* mutants were found to be pleiotropic in that they are unable to synthesize glucose from many substrates, including acetate, pyruvate, alanine, and citrate. They also demonstrate elevated expression of genes involved in glycolytic pathways.

The pleiotropic phenotype of *cra* mutants suggested that Cra functions as a global regulatory protein, activating the transcription of some genes and repressing the synthesis of others. As was noted for other regulatory proteins, including *B. subtilis* CcpA (Figure 12.7), whether Cra activates or represses transcription depends on where it binds relative to the promoter of the regulated gene. If its binding site is upstream of the promoter, it activates transcription of the operon; if its binding site overlaps or is downstream of the promoter, it represses transcription. In either case, the DNA-binding activity of Cra depends on the presence of a signal molecule. In the absence of the signal molecules, Cra is active in DNA binding. Cra is inactivated by binding of fructose-1-phosphate (F1P) or fructose-1,6-bisphosphate (FBP), which are present at high concentrations during growth in the presence of sugars, such as glucose. The effect of this on the transcription of a particular operon depends on whether Cra functions as a repressor or an activator of that operon. If it functions as a repressor, the transcription of the operon increases when levels of glucose (and therefore F1P or FBP) are high; if it functions as an activator, the transcription of the operon decreases when levels of F1P or FBP are high. In general, the Cra protein represses operons whose products are involved in central pathways for sugar catabolism, such as the Embden-Meyerhof and Entner-Doudoroff pathways, so the transcription of these genes increases when glucose and other good carbon sources are available. In contrast, it usually activates operons whose products are involved in synthesizing glucose from pyruvate and other metabolites (gluconeogenesis), so it does not activate the transcription of these operons if glucose is present. The activity of Cra allows another layer of regulation of carbon metabolism independent of the activity of the sugar transport system, which is what is monitored (indirectly) by the CAP-cAMP system. Cra instead monitors internal pools of sugar metabolites.

Reference

Saier MH Jr, Ramseier TM. 1996. The catabolite repressor/activator (Cra) protein of enteric bacteria. *J Bacteriol* **178:**3411–3417.

metabolizable substrate, such as glucose, is present in the medium, the PEP/pyruvate ratio is low; when only poorer carbon sources are available, the PEP/pyruvate ratio is high. The PEP transfers its phosphate to another protein, called Hpr (for histidine protein; histidine is the amino acid in the protein to which the phosphate is transferred to generate Hpr~P), and becomes pyruvate. The phosphate from Hpr~P is then transferred to IIA^{Glc} to make IIA^{Glc}~P. Therefore, the higher the PEP/pyruvate ratio, the higher the Hpr~P/Hpr ratio and the higher the IIA^{Glc}~P/ IIA^{Glc} ratio. High IIA^{Glc}~P results in high cAMP, which serves as the signal that only poorer carbon sources are available. The transfer of phosphate from PEP to HPr to IIA^{Glc} is called a **phosphorylation cascade** because phosphates are transferred from one molecule to another, much like water is transferred down a cascade of waterfalls. We give other examples of phosphorylation cascades later in this chapter.

The unphosphorylated form of IIA^{Glc} (which is present when glucose is high) also inhibits other sugar-specific permeases that transport sugars, such as lactose (Figure 12.2). Therefore, less of these other sugars enters the cell if glucose or another, better carbon source is available, and less inducer is present to induce transcription of their respective operons (see chapter 11). This effect is called **inducer exclusion**. For systems like the *lac* and *gal* operons that require both activation when glucose is low and induction by their specific sugar substrate (lactose or galactose, respectively), it is often difficult to distinguish the effects of inducer exclusion on operon induction from the effects of cAMP on the promoter (see Inada et al., Suggested Reading, and below).

CATABOLITE ACTIVATOR PROTEIN

The mechanism by which cAMP turns on catabolite-sensitive operons in *E. coli* is well understood and has served as a model for transcriptional activation (see chapter 11). The cAMP binds to the protein product of the *crp* gene, which is an activator of transcription of catabolite-sensitive operons. This activator protein goes by two names, CAP (for catabolite activator protein) and CRP (for cAMP receptor protein). We will use the term "CAP," as it reflects the molecular mechanism by which the protein acts on transcription. The activator CAP with cAMP bound (CAP-cAMP) functions like other activator proteins discussed in chapter 11 in that it interacts with RNA polymerase to activate transcription from promoters for operons under its control, including *lac*, *gal*, *ara*, and *mal*. These operons are all members of the **CAP regulon** or the catabolite-sensitive regulon (Table 12.1). However, the mechanism of CAP-cAMP regulation varies. CAP also can function not only as an activator, but also as a repressor, depending on where it binds relative to the promoter (see below).

REGULATION BY CAP-cAMP

The mechanism by which CAP activates transcription varies from promoter to promoter. Some of these mechanisms are shown in Figure 12.3. Upstream of the promoter is a short sequence called the **CAP-binding site**, which is similar in sequence in all catabolite-sensitive operons and so can be easily identified. CAP is active as a DNA-binding protein only when it is bound to cAMP, so the site is occupied only when cAMP levels are high. CAP functions like many other activators to make contact with the RNA polymerase at the promoter and to stimulate one or more of the steps in the initiation of transcription (see Browning and Busby, Suggested Reading). Transcriptional activators are discussed in chapter 11. CAP can contact different regions of the RNA polymerase and stimulate different steps in initiation, depending on where it is bound relative to the promoter. This is illustrated in Figure 12.3B. At class I CAP-dependent promoters, such as the *lac* promoter, a dimer of CAP in complex with cAMP binds upstream of the promoter and contacts the C-terminal end of the α subunit of RNA polymerase (α C-terminal domain [αCTD]) (see chapters 2 and 11). This contact strengthens the binding of RNA polymerase to the promoter (to form the closed complex). In class II CAP-dependent promoters, such as the *gal* promoter p_{G1}, the CAP dimer-binding site slightly overlaps that of RNA polymerase, and CAP contacts a region in the N terminus of the α subunit (α N-terminal domain [αNTD]). In this position, CAP stimulates the opening of the DNA at the promoter (open complex formation). There are even promoters in which more than one CAP dimer binds to stimulate both RNA polymerase binding and open complex formation. CAP also can bend the DNA when it binds to the CAP-binding site, which can affect access to other regulatory proteins and to RNA polymerase.

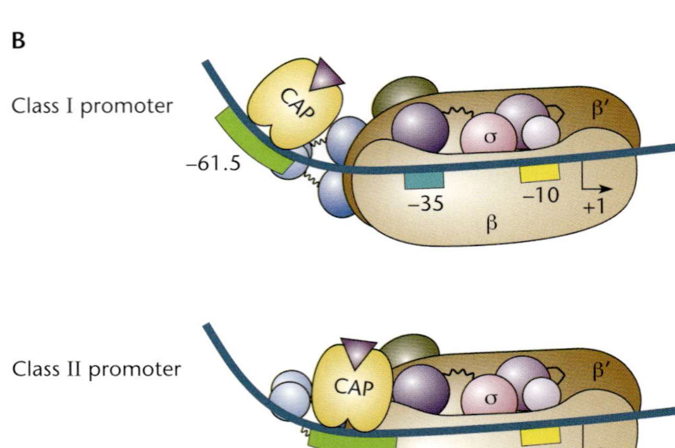

Figure 12.3 Model for CAP activation at class I and class II CAP-dependent promoters. **(A)** Sequence of the CAP-binding site upstream of the class I *lac* promoter. RNA Pol, RNA polymerase. **(B)** Binding and location of interactions of CAP-cAMP with the C-terminal domain (CTD) and N-terminal domain (NTD) of the α subunit with class I and class II promoters, respectively. The purple triangle represents cAMP bound to CAP. Modified from Savery NJ, Lloyd GS, Busby SJW, Thomas MS, Ebright RH, Gourse RL, *J Bacteriol* **184:**2273–2280, 2002.

The position of the CAP-binding sequence relative to the promoter can vary widely. In the *ara* operon, the CAP-binding site is further upstream, with the AraC-binding site between it and the promoter (Figure 12.4). Nevertheless, CAP can still make contact with the αCTD of RNA polymerase, which can reach up along the DNA, as shown. The ability of the αCTD to interact with CAP at this site

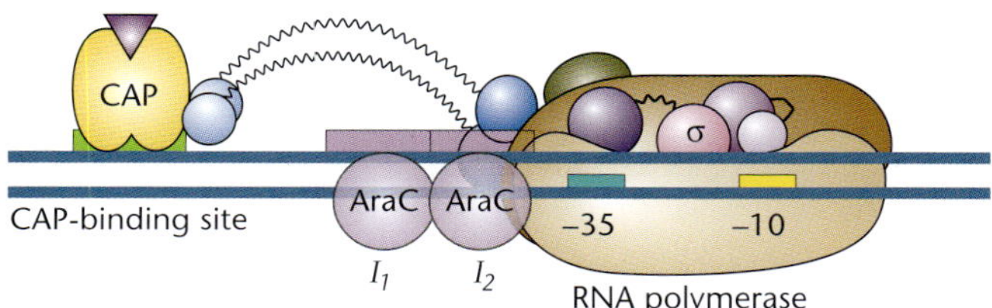

Figure 12.4 Summary of the RNA polymerase-promoter and activator-promoter interactions at the p_{BAD} promoter of the *ara* operon. The σ^{70} subunit of RNA polymerase contacts the –35 and –10 hexamers. Occupancy of the I_1 and I_2 half-sites by AraC activates transcription with the aid of catabolite activator protein (CAP-cAMP), utilizing the α subunit-activator interactions, as shown. The binding sites of σ^{70} and AraC overlap by 4 bp at p_{BAD} Modified from Dhiman A, Schleif R, *J Bacteriol* **182:**5076–5081, 2000.

requires that the CAP-binding site is on the same face of the DNA helix as the promoter, so that the αCTD can reach upstream without reaching around to the other side of the DNA helix. This is an example of face-of-the-helix dependence (see chapter 11). CAP can also stimulate transcription by interacting with another activator or can stimulate transcription by preventing the binding of a repressor (see chapter 11).

Certain operons in the CAP regulon, such as *gal*, are less sensitive to catabolite repression than others. As discussed in chapter 11, some transcription of the *gal* operon is maintained when glucose is present in the medium because it has two promoters, p_{G1} and p_{G2} (see Figure 11.8). Transcription from p_{G2} does not require CAP-cAMP for its activation, and therefore the p_{G2} promoter permits some expression of the *gal* operon even in the presence of glucose. This low level of expression is necessary to allow the synthesis of cell wall components that contain galactose, since the UDP-galactose synthesized by the operon serves as the donor of galactose in biosynthetic reactions. However, the level of expression of the *gal* operon from p_{G2} is not high enough for the cells to grow well on galactose as a carbon and energy source.

RELATIONSHIP OF CATABOLITE REGULATION TO INDUCTION

An important point about CAP-dependent regulation of catabolite-sensitive operons is that it occurs in addition to any other regulation to which the operon is subject. Two conditions must be met before catabolite-sensitive operons can be transcribed: better carbon sources, such as glucose, must be absent, and the inducer of the operon must be present. Take the example of the *lac* operon (Figure 12.5). If a carbon source better than lactose is available, cAMP levels are low, and CAP-cAMP does not bind upstream of the *lac* promoter to activate transcription. Also, the *lac* transport system is inhibited, excluding the inducer from the cell. However, even at high cAMP levels, the *lac* operon is not transcribed unless the inducer, allolactose (a metabolite of lactose), is also present. In the absence of inducer, the LacI repressor is bound to the operator and prevents the RNA polymerase from binding to the promoter and transcribing the operon (see chapter 11).

GENETIC ANALYSIS OF CATABOLITE REGULATION IN *E. COLI*

The above model for the regulation of catabolite-sensitive operons is supported by both genetic and biochemical analyses of catabolite regulation in *E. coli*. These analyses have involved the isolation of mutants defective in the global regulation of all catabolite-sensitive operons, as well as mutants defective in the catabolite regulation of specific operons.

Isolation of *cya* and *crp* Mutations

According to the model presented above, mutations that inactivate the *cya* and *crp* genes for adenylate cyclase and CAP, respectively, should prevent transcription of all the catabolite-sensitive operons. In these mutants, there is no CAP with cAMP attached to bind to the promoters. In other words, *cya* and *crp* mutants should be Lac^-, Gal^-, Ara^-, Mal^-, and so on. In genetic terms, *cya* and *crp* mutations are **pleiotropic**, because they cause many phenotypes, i.e., the inability to use many different sugars as carbon and energy sources.

The fact that *cya* and *crp* mutations should prevent cells from using several sugars was used in the first isolations of *cya* and *crp* mutants (see Schwartz and Beckwith, Suggested Reading). The identification of these mutants was based on the fact that colonies of bacteria turn tetrazolium salts red as they multiply, provided that the pH remains high. However, bacteria that are fermenting a carbon source give off organic acids, such as lactic acid, that lower the pH, preventing the conversion to red. As a consequence, wild-type *E. coli* cells growing on a fermentable carbon source form white colonies on tetrazolium-containing plates, whereas mutant bacteria that cannot use the fermentable carbon source utilize a different carbon source in the medium and so form red colonies. Some of these red-colony-forming mutants might have *cya* or *crp* mutations, although most would have mutations that inactivate a gene within the operon for the utilization of the fermentable carbon source. Thus, without a way to increase the frequency of *cya* and *crp* mutants among the red-colony-forming mutants, many red-colony-forming mutants would have to be tested to find any with mutations in either *cya* or *crp*.

For these experiments, the investigators reasoned that they could increase the frequency of *cya* and *crp* mutants

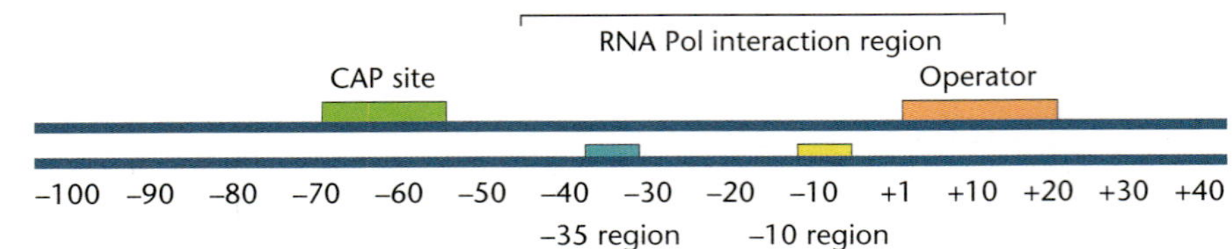

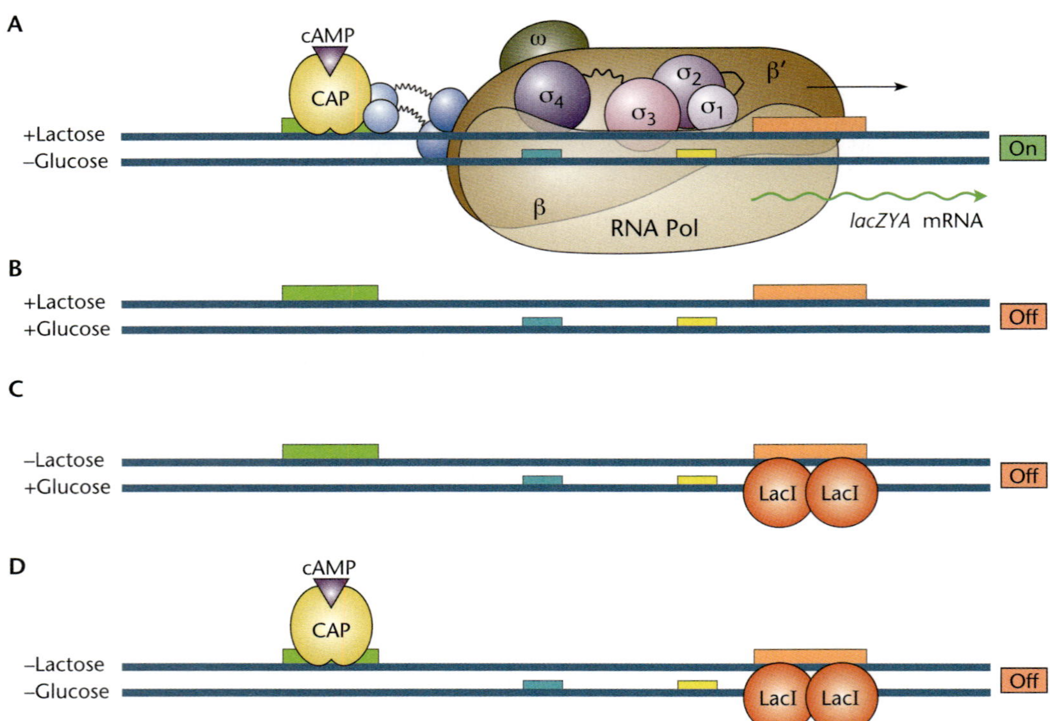

Figure 12.5 Regulation of the *lac* operon by both glucose and the inducer lactose. **(A)** The operon is expressed to maximum levels only in the absence of glucose and the presence of lactose, which is converted into the inducer allolactose. **(B and C)** The operon is off in the presence of glucose whether or not lactose is present, because the catabolite activator protein-cyclic AMP (CAP-cAMP) complex is not bound to the CAP site. **(D)** The operon is also off if lactose is not present, even if glucose is also not present, because the LacI repressor is bound to the operator and prevents binding of RNA polymerase. The relative positions of the CAP-binding site, operator, and promoter are shown. The entire regulatory region covers about 100 bp of DNA.

by plating heavily mutagenized bacteria on tetrazolium agar containing two different fermentable sugars, for example, lactose plus galactose. Failure to utilize either of the two sugars would require either two mutations, one in each sugar utilization operon, or a single mutation in *cya* or *crp*. Since mutants with single mutations should be much more frequent than mutants with two independent mutations, the *cya* and *crp* mutants should be a much larger fraction of the total red-colony-forming mutants growing on two carbon and energy sources. Indeed, when the red-colony-forming mutants that could not use either of the two sugars provided were tested, most of them were found to be deficient in adenylate cyclase activity or to lack the protein, now named CAP, that was later shown to be required for the activation of the *lac* and *gal* promoters.

Promoter Mutations That Affect Activation by CAP-cAMP

Genetic experiments with the *lac* promoter (p_{lac}) also contributed to the models of CAP activation. Three classes of mutations have been isolated in the *lac* promoter. Those belonging to class I change the CAP-binding site so that CAP can no longer bind to it. The *lac* promoter mutation L8 is an example (Figure 12.6). By preventing the binding of CAP-cAMP upstream of the promoter, this mutation weakens the *lac* promoter. As a result, the

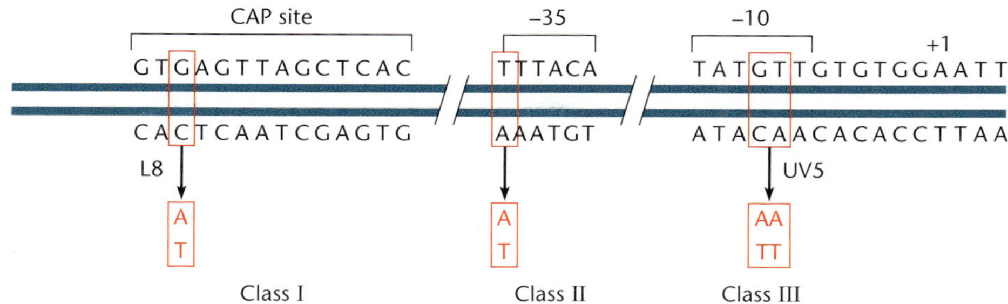

Figure 12.6 Mutations in the *lac* regulatory region that affect activation by catabolite activator protein (CAP). The class I mutation L8 changes the CAP-binding site so that CAP can no longer bind and the promoter cannot be turned on even in the absence of glucose (high cyclic AMP [cAMP]). The class II mutation changes the first position of the conserved −35 region of the promoter and causes reduced RNA polymerase binding. The class III mutation UV5 changes 2 bp in the −10 sequence of the promoter so that the promoter no longer requires activation by CAP and the operon can be induced even in the presence of glucose (low cAMP). The changes in the sequence in each mutation appear in red.

lac operon is expressed poorly, as measured by β-galactosidase activity, even when cells are growing in lactose without glucose and cAMP levels are high. The low level of expression of the *lac* operon in the mutant is less strongly affected by the carbon source and is not reduced much more if glucose is added and cAMP levels drop, presumably because the remaining expression occurs without binding of CAP to the promoter and therefore fails to respond to cAMP levels.

Other promoter mutations, called class II mutations, change the −35 region of the RNA polymerase-binding site so that the promoter is less active even when cAMP levels are high. However, with this type of mutation, the residual expression of the *lac* operon is still sensitive to catabolite repression. Consequently, the amount of β-galactosidase synthesized when cells grow in the presence of lactose plus glucose, when cAMP levels are low, is less than the amount synthesized when the cells are growing in the presence of a poorer carbon source plus lactose, when cAMP levels are high.

A third, very useful mutated *lac* promoter type, termed class III, was found by isolating Lac$^+$ revertants of class I mutations, such as L8, or of *cya* or *crp* mutations. One such mutation is called p_{lac}*UV5*. This mutant promoter is stronger than the wild-type *lac* promoter and no longer requires CAP-cAMP for activation. As shown in Figure 12.6, the p_{lac}*UV5* mutation changes 2 base pairs (bp) within the −10 region of the *lac* promoter so that the sequence reads TATAAT instead of TATGTT. This mutant −10 sequence perfectly matches the sequence of a consensus σ^{70} promoter (see chapter 2), which increases its affinity for RNA polymerase and results in loss of dependence on activation by CAP-cAMP. Some expression vectors use the p_{lac}*UV5* promoter rather than the wild-type *lac* promoter, so the promoter can be induced (by lactose or a related compound like isopropyl β-D-1-thiogalactopyranoside [IPTG], a non-metabolizable analog of lactose) even if the bacteria are growing in glucose-containing medium.

Carbon Catabolite Regulation in *Bacillus subtilis*: CcpA and Hpr

The use of cAMP as a signal for carbon catabolite regulation is not universal. *Bacillus subtilis* and its relatives use a completely different mechanism. Unlike *E. coli*, this bacterium does not produce cAMP and so depends exclusively on cAMP-independent pathways to regulate its carbon source utilization pathways. In this system, catabolite regulation involves a transcriptional repressor (in contrast to the use of a transcriptional activator in *E. coli*). The repressor protein, called CcpA (for catabolite control protein A), is a helix-turn-helix DNA-binding protein and is a member of the LacI/GalR family of regulators (see chapter 11). The CcpA repressor binds to operator sites called *cre* (for catabolite repressor) sites in the promoters of many catabolite-sensitive genes and represses the transcription of genes involved in carbon source utilization. Approximately 100 genes in *B. subtilis* are known to be under the control of CcpA.

CcpA also acts as an activator of transcription of genes that encode functions needed by the cell when carbon source availability is high. During growth in the presence of high glucose levels, *B. subtilis* produces large amounts of acetate, which is excreted from the cell. This results in a drop in the extracellular pH. To avoid too great a drop in pH, which is toxic, the cells can shift to production of a pH neutral compound called acetoin. The genes for production of both acetate and acetoin contain *cre* sites and are dependent on CcpA for transcriptional activation. The major difference between the genes CcpA activates and those that it represses is that the *cre* sites for the activated genes are positioned upstream of the promoter, while the genes that are repressed by CcpA have

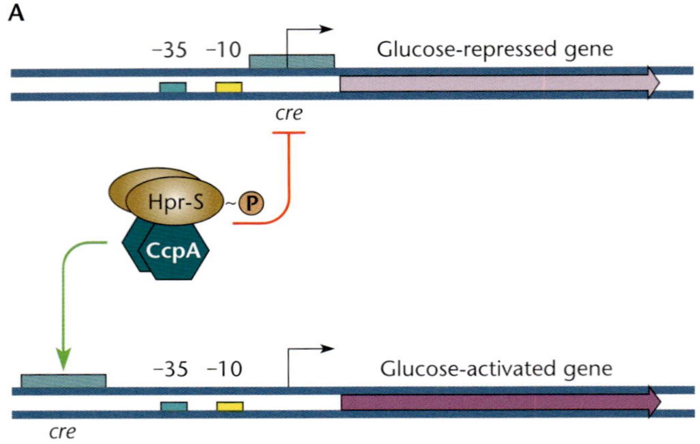

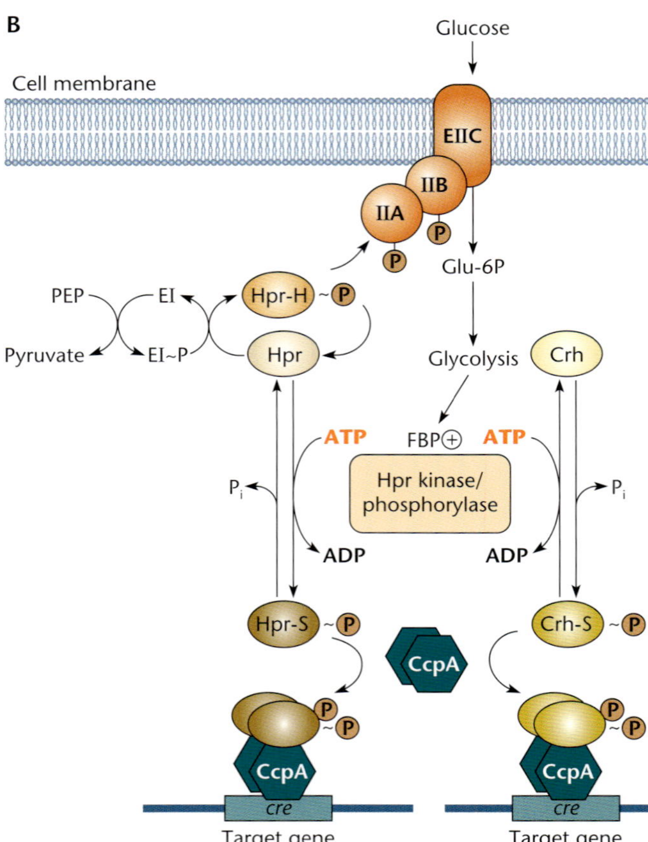

Figure 12.7 Carbon catabolite regulation in *B. subtilis*. **(A)** The CcpA regulatory protein represses genes for carbon source utilization pathways, which have *cre* sites within or downstream of the promoter, and activates genes for carbon excretion pathways, which have *cre* sites upstream of the promoter. **(B)** Binding of CcpA to *cre* sites requires a protein-protein interaction with the Hpr protein phosphorylated on a specific serine residue (Hpr-S~P) by Hpr kinase, which is activated by the glycolytic intermediate fructose-1,6-bisphosphate (FBP). Crh-S~P can replace Hpr-S~P for this interaction. Modified from Lorca G, Chung Y-J, Barbote R, Weyler W, Schilling C, Saier M Jr, *J Bacteriol* **187:**7826–7839, 2005.

cre sites that overlap the promoter, so that binding of CcpA to the *cre* site inhibits binding of RNA polymerase (Figure 12.7A).

How is CcpA activity controlled in response to carbon source availability? The DNA-binding activity of CcpA depends on the Hpr protein, which we discussed above in reference to the regulation of adenylate cyclase in *E. coli*. The phosphorylation state of Hpr also plays a crucial regulatory role in *B. subtilis*, but the mechanism is different. If cells are growing in a high-energy carbon source, such as glucose, high levels of intermediates in the glycolytic pathway, including fructose-1,6-bisphosphate (FBP), accumulate. In *B. subtilis*, FBP causes phosphorylation of Hpr on a specific serine residue in the protein. Phosphorylation of Hpr to generate Hpr-S~P utilizes another protein, called Hpr kinase (HprK), the activity of which is stimulated by FBP. Hpr-S~P binds to CcpA, and the CcpA-Hpr-S~P complex binds to the *cre* sites (Figure 12.7B). CcpA activity is therefore controlled by protein-protein interactions rather than binding of a small molecule (like cAMP in *E. coli*). HprK is also responsible for dephosphorylating Hpr-S~P when FBP is low.

Interestingly, the same Hpr protein that serves as a co-regulator with CcpA when it is phosphorylated on the serine residue also serves as the phosphate donor in the PTS for sugar transport in a role similar to that described for the *E. coli* system. Sugar transport involves phosphorylation of a histidine residue in Hpr (Hpr-H~P), and the phosphate is then donated to the PTS transport protein IIA^{Glc}. Phosphorylation of Hpr at the serine can inhibit phosphorylation at the histidine and therefore inhibits the transport of sugars that use the PTS system. This allows the close coordination of sugar transport and the regulation of catabolite-sensitive operons. A second Hpr-like protein, called Crh, can act in parallel with Hpr to activate CcpA binding to target gene *cre* sites (Figure 12.7B).

Regulation of Nitrogen Assimilation

Nitrogen is a component of many biological molecules, including nucleotides, amino acids, and vitamins. Thus, all organisms must have a source of nitrogen atoms for growth to occur. For most bacteria, possible sources include ammonia (NH_3) and nitrate (NO_3^-), as well as nitrogen-containing organic molecules, such as amino acids and the bases in nucleosides. Some bacteria can even use atmospheric nitrogen (N_2) as a nitrogen source in a process known as nitrogen fixation. "Fixing" atmospheric nitrogen is a crucial step in the nitrogen cycle on Earth that few organisms can do, and the nitrogen cycle is one of the many cycles on the planet that are carried out by bacteria and are required for the existence of humans (Box 12.2).

Whatever the source of nitrogen, all biosynthetic reactions that utilize it ultimately involve either the incorpo-

BOX 12.2

Nitrogen Fixation

Some bacteria can use atmospheric nitrogen (N_2) as a nitrogen source by converting it to NH_3 in a process called nitrogen fixation, which appears to be unique to bacteria. However, N_2 is a very inconvenient source of nitrogen. The very stable bond holding the two nitrogen atoms together must be broken, and 16 moles (mol) of ATP must be cleaved to cleave 1 mol of dinitrogen. Bacteria that can fix nitrogen include members of the cyanobacteria and members of the genera *Klebsiella*, *Azotobacter*, *Rhizobium*, and *Azorhizobium*. These organisms play an important role in nitrogen cycles on Earth.

Some types of nitrogen-fixing bacteria, including members of the genera *Rhizobium* and *Azorhizobium*, are symbionts that fix N_2 in nodules on the roots or stems of plants and allow the plants to live in nitrogen-deficient soil. In return, the plant furnishes nutrients and an oxygen-free atmosphere in which the bacterium can fix N_2. This symbiosis therefore benefits both the bacterium and the plant. An active area of biotechnology is the use of N_2-fixing bacteria as a source of natural fertilizers.

The fixing of N_2 requires the products of many genes, called the *nif* genes. In free-living nitrogen-fixing bacteria, such as *Klebsiella* spp., there are about 20 *nif* genes arranged in eight adjacent operons. Some of the *nif* genes encode the nitrogenase enzymes directly responsible for fixing N_2. Others encode proteins involved in assembling the nitrogenase enzyme and in regulating the genes. Plant-symbiotic bacteria also require many other genes whose products produce the nodules on the plant (*nod* genes) and allow the bacterium to live and fix nitrogen in the nodules (*fix* genes).

Because nitrogen fixation requires a large investment of energy, the genes involved in N_2 fixation are part of the Ntr regulon and are under the control of the NtrC activator protein. In *Klebsiella pneumoniae*, in which the regulation of the *nif* genes has been studied most extensively, the phosphorylated form of NtrC (NtrC~P) does not directly activate all eight operons involved in N_2 fixation. Instead, NtrC~P activates the transcription of another activator gene, *nifA*, whose product is directly required for the activation of the eight *nif* operons. The nitrogenase enzymes are very sensitive to oxygen, and in the presence of oxygen, the *nif* operons are negatively regulated by the product of the *nifL* gene. The NifL protein is able to sense oxygen because it is a flavoprotein with a bound flavin adenine dinucleotide group, which is oxidized in the presence of oxygen. The NifL protein then forms a stable complex with NifA and inactivates it so that the *nif* genes are not transcribed.

References

Martinez-Argudo I, Little R, Shearer N, Johnson P, Dixon R. 2004. The NifL-NifA system: a multidomain transcriptional regulatory complex that integrates environmental signals. *J Bacteriol* **186:**601–610.

Masson-Boivin C, Sachs JL. 2018. Symbiotic nitrogen fixation by rhizobia: the roots of a success story. *Curr Opin Plant Biol* **44:**7–15.

Oldroyd GE, Murray JD, Poole PS, Downie JA. 2011. The rules of engagement in the legume-rhizobial symbiosis. *Annu Rev Genet* **45:**119–144.

ration of nitrogen in the form of NH_3 or the transfer of nitrogen in the form of an NH_2 group from glutamate and glutamine, which in turn are synthesized by directly adding NH_3 to α-ketoglutarate and glutamate, respectively. Thus, because NH_3 is directly or indirectly the source of nitrogen in biosynthetic reactions, most other forms of nitrogen must be reduced to NH_3 before they can be used in these reactions. This process is called **assimilatory reduction** of the nitrogen-containing compounds, because the nitrogen-containing compound converted into NH_3 is introduced, or assimilated, into biological molecules. In another type of reduction, **dissimilatory reduction**, oxidized nitrogen-containing compounds, such as NO_3^-, are reduced when they serve as electron acceptors in anaerobic respiration (in the absence of oxygen). However, these compounds are generally not reduced all the way to NH_3 in this process, and the nitrogen is not assimilated into biological molecules and may be released as atmospheric N_2. Here, we discuss only the assimilatory uses of nitrogen-containing compounds, as they provide an excellent example of global regulation. The genes whose products are required for anaerobic respiration in *E. coli* are members of a different regulon, the FNR regulon, which is turned on only in the absence of oxygen, when other, less efficient electron acceptors are required (Table 12.1).

Pathways for Nitrogen Assimilation

Enteric bacteria use different pathways to assimilate nitrogen depending on whether NH_3 concentrations are low or high (Figure 12.8), and the regulatory pathways responsible for sensing NH_3 abundance will be described below. When NH_3 concentrations are low, for example, when the nitrogen sources are amino acids which must be degraded to release their NH_3, an enzyme named **glutamine synthetase**, the product of the *glnA* gene, adds the NH_3 directly to glutamate to make glutamine. About 75% of this glutamine is then converted to glutamate by

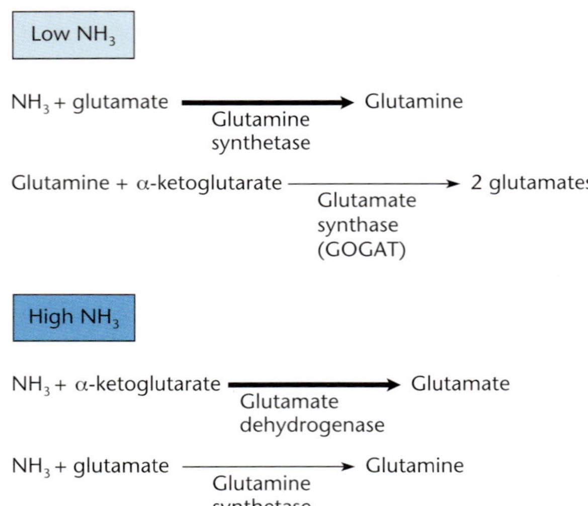

Figure 12.8 Pathways for nitrogen assimilation in *E. coli* and other enteric bacteria. When NH_3 concentrations are low, the glutamine synthetase enzyme adds NH_3 directly to glutamate to make glutamine. Glutamate synthase (GOGAT) can then convert the glutamine plus α-ketoglutarate into two glutamates, which can reenter the cycle. In the presence of high NH_3 concentrations, the NH_3 is added directly to α-ketoglutarate by glutamate dehydrogenase to make glutamate, which can be subsequently converted to glutamine by glutamine synthetase.

another enzyme, **glutamate synthase**, sometimes called GOGAT, which removes an $-NH_2$ group from glutamine and adds it to α-ketoglutarate to make two glutamates. These glutamates can in turn be converted into glutamine by glutamine synthetase. Because the NH_3 must all be routed by glutamine synthetase to glutamine when NH_3 concentrations are low, the cell needs high amounts of the glutamine synthetase enzyme under these conditions. This pathway requires a lot of energy, but it is necessary if nitrogen availability is limited. The significance of this is addressed later.

If NH_3 concentrations are high because the medium contains NH_3 (usually in the form of NH_4OH) but carbon sources are limited, the nitrogen is assimilated through a very different pathway. This pathway requires less energy but is possible only if NH_3 concentrations are high. In this case, the enzyme **glutamate dehydrogenase** adds the NH_3 directly to α-ketoglutarate to make glutamate. Some of the glutamate is subsequently converted into glutamine by glutamine synthetase. Much less glutamine is required for protein synthesis and biosynthetic reactions than for assimilation of limiting nitrogen from the medium. Therefore, cells need much less glutamine synthetase when growing in high concentrations of NH_3 than when growing in low concentrations.

Regulation of Nitrogen Assimilation Pathways in *E. coli* by the Ntr System

The operons for nitrogen utilization in *E. coli* are part of the **Ntr system**, for nitrogen regulated. Ntr regulation ensures that the cell does not waste energy making enzymes for the use of nitrogen sources such as amino acids or nitrate when NH_3 is available. Transport systems for alternative nitrogen sources are also part of this regulon. In this section, we discuss what is known about how the Ntr global regulatory system works. As usual, geneticists led the way by identifying the genes whose products are involved in the regulation, so that a role could eventually be assigned to each one. The Ntr regulatory systems in a variety of Gram-negative bacteria, including *Escherichia*, *Salmonella*, *Klebsiella*, and *Rhizobium*, are similar, but with important exceptions, some of which are pointed out here.

REGULATION OF THE *glnA-ntrB-ntrC* OPERON BY A SIGNAL TRANSDUCTION PATHWAY

Since cells need more glutamine synthetase when growing at low NH_3 concentrations than when growing at high NH_3 concentrations, the expression of the *glnA* gene, which encodes glutamine synthetase, must be regulated according to the nitrogen source that is available (see Leigh and Dodsworth 2007, Suggested Reading). This gene is part of an operon that includes three genes, *glnA*, *ntrB*, and *ntrC*. The products of the *ntrB* and *ntrC* genes are involved in regulating the operon. (These proteins are also called NR_{II} and NR_I, respectively, but we use the Ntr names in this chapter.) Because the *ntrB* and *ntrC* genes are part of the same operon as *glnA*, their genes are autoregulated, and their products are also synthesized at higher levels when NH_3 concentrations are low.

Figure 12.9 illustrates the regulation of the *glnA-ntrB-ntrC* operon and other Ntr genes. In addition to NtrB and NtrC, the proteins GlnD and P_{II} participate in the regulation of the operon. These four proteins form a **signal transduction pathway** in which information about nitrogen source availability is passed (or transduced) from one protein to another until it gets to its final destination, the transcriptional regulator NtrC, which activates genes in the Ntr regulon.

The availability of nitrogen is sensed through the level of glutamine in the cell. If the cell is growing in a nitrogen-rich environment, the levels of glutamine are high, whereas if the cell is growing under limiting nitrogen conditions, the levels of glutamine are low. How the levels of glutamine affect the regulation of the Ntr genes involved in using alternative nitrogen sources is probably best explained by working backward from the last protein in the signal transduction pathway, NtrC. The NtrC protein can be phosphorylated to form NtrC~P, the form in which it is a transcriptional activator that

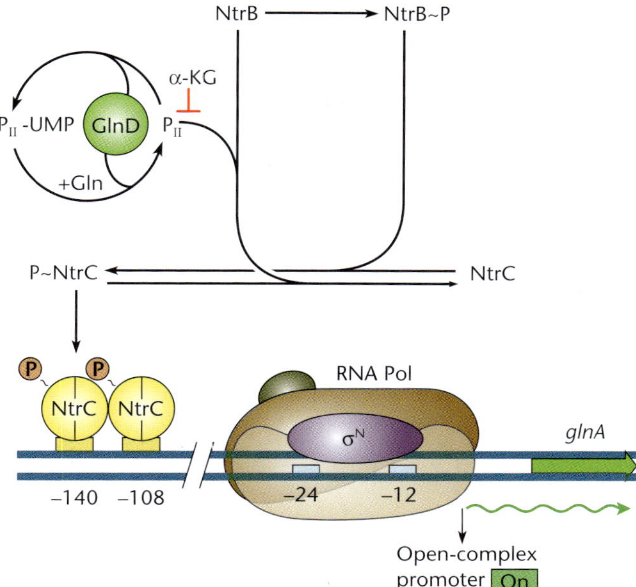

Figure 12.9 Regulation of nitrogen assimilation genes by a signal transduction pathway in response to NH_3 levels. At low NH_3 concentrations, the reactions shown in blue predominate. High glutamine (Gln) causes low levels of uridylylation of P_{II}, which in turn affects the activity of the NtrB sensor kinase. The effect of α-ketoglutarate (α-KG) on P_{II} is indirect. Phosphorylated NtrC activates transcription by RNA polymerase containing σ^N. See the text for details.

activates transcription of the Ntr genes, which are turned on under limiting nitrogen (Figure 12.9). The penultimate protein in the regulatory pathway is NtrB. Together, NtrB and NtrC form a **two-component regulatory system** (Box 12.3) in which the NtrB protein is the **sensor kinase** and NtrC is the **response regulator**. NtrB is a protein kinase that can phosphorylate itself on a specific histidine residue to form NtrB~P; NtrB~P can then transfer this phosphate to NtrC to form NtrC~P, which is the active form that can activate transcription of the Ntr genes. The autophosphorylation activity of NtrB occurs only if nitrogen is limiting and depends on the state of modification of another regulatory protein, called P_{II}. The P_{II} protein is modified, not by phosphorylation, but by having UMP attached to it (to form P_{II}~UMP) in a process called uridylylation. If nitrogen is limiting, making the glutamine level low, most of this protein exists as P_{II}~UMP. However, if nitrogen is in excess and the glutamine level is high, the glutamine stimulates an enzyme called GlnD to remove the UMP from P_{II}. The unmodified P_{II} protein binds to NtrB and inhibits its autokinase activity so that it cannot phosphorylate itself to form NtrB~P. If it cannot phosphorylate itself, it cannot transfer a phosphate to NtrC, and most of the NtrC remains in the unphosphorylated state that is unable to activate transcription of the Ntr genes involved in using alternate nitrogen sources.

Regulation of Other Ntr Operons

The operons other than *glnA-ntrB-ntrC* that are activated by NtrC~P depend on the type of bacteria and the other nitrogen sources they can use. In general, operons under the control of NtrC~P are those involved in using poorer nitrogen sources. For example, genes for the uptake of the amino acids glutamine in *E. coli* and histidine and arginine in *Salmonella enterica* serovar Typhimurium are under the control of NtrC~P. An operon for the utilization of nitrate as a nitrogen source in *Klebsiella pneumoniae* is activated by NtrC~P, but neither *E. coli* nor *S. enterica* serovar Typhimurium has a nitrate utilization operon.

In some bacteria, the Ntr genes are not regulated directly by NtrC~P but are under the control of another gene product whose transcription is activated by NtrC~P. For example, operons for amino acid degradative pathways in *Klebsiella aerogenes* do not require direct activation by NtrC~P. However, they are indirectly under the control of NtrC~P because transcription of the gene for their transcriptional activator, *nac*, is activated by NtrC~P. The nitrogen fixation genes of *K. pneumoniae* are similarly under the indirect control of NtrC~P because NtrC~P activates transcription of the gene for their activator protein, *nifA* (Box 12.2). As a result, genes for nitrogen fixation are expressed only when cells are limited for nitrogen and NtrC~P levels are high.

TRANSCRIPTION OF THE *glnA-ntrB-ntrC* OPERON BY THE NITROGEN SIGMA FACTOR, σ^N

The most important promoter that directs transcription of the *glnA-ntrB-ntrC* operon is the p_2 promoter, activity of which requires activation by NtrC~P. The p_2 promoter and other NtrC~P-dependent promoters are unusual in terms of the RNA polymerase holoenzyme that recognizes them. Most promoters are recognized by the RNA polymerase holoenzyme with σ^{70} attached, but the Ntr-type promoters are recognized by a holoenzyme containing a special σ factor (see chapter 2), designated σ^{54} or σ^N (Box 12.4). As shown in Figure 12.10, promoters recognized by the σ^N holoenzyme look very different from promoters recognized by the σ^{70} holoenzyme. Unlike the typical σ^{70} promoter, which has RNA polymerase-binding sequences centered at bp –35 and –10 relative to the transcription start site (see chapter 2), the σ^N promoters have very different binding sequences centered at bp –24 and –12. Because promoters for the genes involved in Ntr regulation are recognized by RNA polymerase with σ^N, this sigma factor was named the nitrogen sigma factor and the gene was named *rpoN* (Table 12.1). However, σ^N-type promoters have been found in many operons unrelated

BOX 12.3

Signal Transduction Systems in Bacteria

Many regulatory mechanisms require that the cell sense changes in the external environment and change the expression of its genes or the activities of its proteins accordingly. The sensors can be of many types. Some are serine-threonine-tyrosine kinases-phosphatases (STYK in Figure 1A), such as the Rsb proteins that activate the stress sigma factor, σ^B, of *B. subtilis* and many other bacteria. Others already discussed are adenyl cyclases (ACyc, Figure 1B), which make cAMP in enteric bacteria, such as *E. coli*, in response to nutritional conditions, such as a relatively poor carbon source. An interesting type of signaling molecule is cyclic diGMP (c-di-GMP, Figure 1C), which consists of two guanosine monophosphate (GMP) molecules linked to each other's 3′ carbons through their 5′ phosphates to form a sort of circle of phosphate-ribose sugar groups. Specific enzymes called diguanylate cyclases make this small-molecule effector, and specific phosphodiesterases destroy it. These enzymes were discovered primarily through genomic analysis because the cyclases have the domain GGDEF and the phosphodiesterases have the domain EAL (see front endsheet for amino acid assignments). Many signaling proteins in these pathways have both a diguanylate cyclase and a diguanylate phosphodiesterase domain. They are widespread, having been found in many types of bacteria, and play diverse roles in the attachment of bacteria to surfaces, in the formation of biofilms, in the regulation of photosynthesis, and in motility (see Povolotsky and Hengge, and Jenal et al., References).

Some of the most common and widely studied sensor systems are the so-called two-component signal transduction systems (Figure 1D), which consist of a sensor kinase (SK) that autophosphorylates (transfers phosphates to itself) on a histidine residue and a response regulator (RR) that accepts the phosphate from the sensor kinase onto an aspartate residue and then performs a specific action in the cell. Removal of the phosphate from the response regulator is another important step, to allow the system to be "reset" to monitor another signal; this can occur spontaneously, through the action of a specific phosphatase, or through the phosphatase activity of the sensor kinase.

Two-component systems have been found in all bacteria and some plants but, at least in this form, are absent from animals. Bacteria with large genomes can have hundreds of these systems. As the name implies, they usually consist of two proteins, but in some cases, the sensor kinase and the response regulator activities are domains of the same protein. In only a few cases is the stimulus to which the sensor kinase responds known (see Gao and Stock, References). The output responses of the systems also vary (see Galperin, References). To name just a few, the response regulator is quite often a DNA-binding transcriptional regulator with a helix-turn-helix domain or an antiterminator protein, but among the varied possibilities is to destabilize a protein by targeting it for proteolysis. The cellular functions that enlist two-component signal transduction systems also vary widely, including involvement in motility in response to chemical attractants, i.e., the methylated chemotaxis proteins (MCP in Figure 1E), the induction of pathogenesis operons after entry into a suitable host, and the activation of extracellular stress responses.

The way in which these two-component sensor kinase and response regulator systems operate in general is illustrated

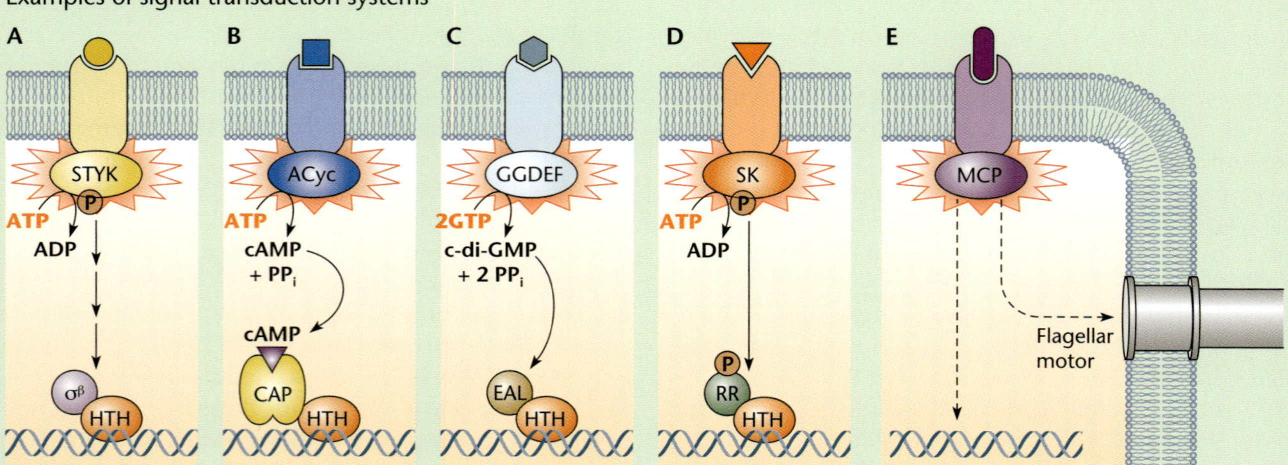

Figure 1 Modified from Dhiman A, Schleif R, *J Bacteriol* **182:**5076–5081, 2000.

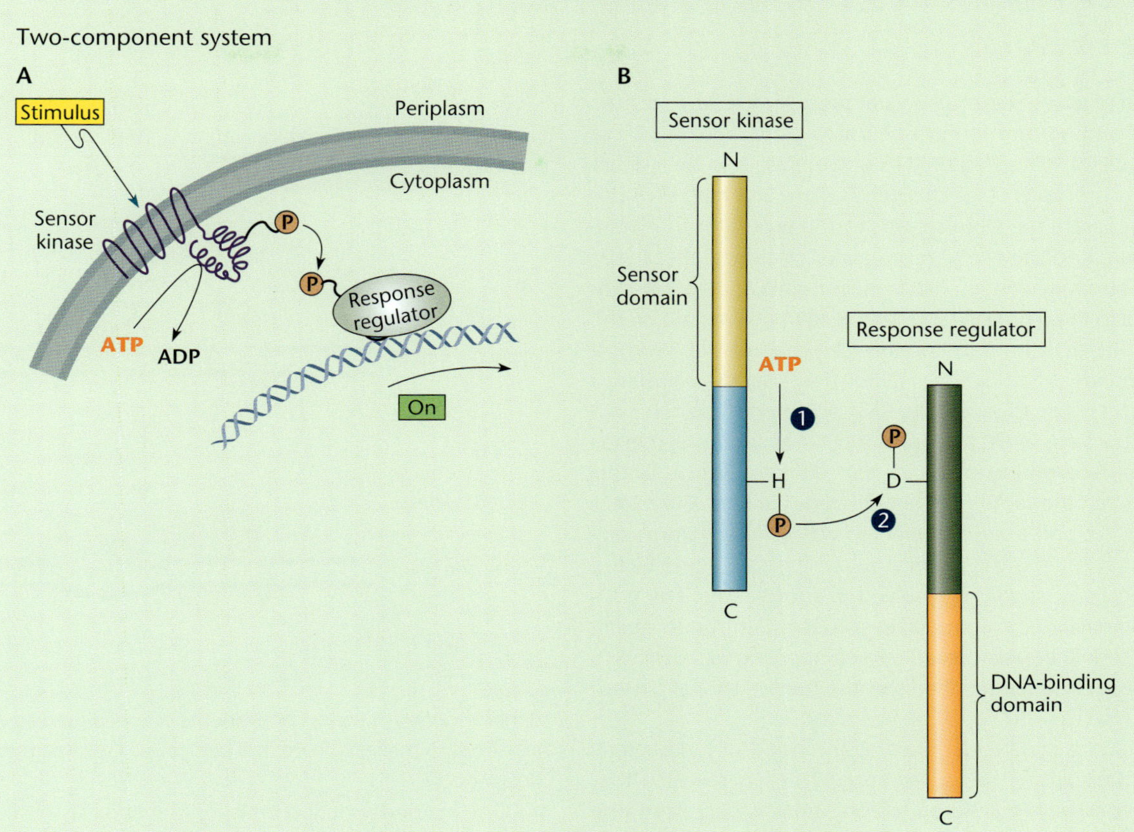

Figure 2

in Figure 2. Panel A shows that sensor kinases are often integral membrane proteins responsive to external signals. Panel B shows the functions of the protein domains. The CTD of a sensor kinase has the conserved histidine that is phosphorylated (step 1). The response regulators are similar in their N-terminal regions, which includes the phosphorylated aspartate (step 2). The remainder of the protein differs depending on its function, although different subfamilies of response regulators show regions of high homology in other parts of the protein, including the helix-turn-helix motif of many transcriptional regulators.

References

Galperin MY. 2006. Structural classification of bacterial response regulators: diversity of output domains and domain combinations. *J Bacteriol* **188:**4169–4182.

Gao R, Stock AM. 2009. Biological insights from structures of two-component proteins. *Annu Rev Microbiol* **63:**133–154.

Jenal U, Reinders A, Lori C. 2017. Cyclic di-GMP: second messenger extraordinaire. *Nat Rev Microbiol* **15:**271–284.

Povolotsky TL, Hengge R. 2012. 'Life-style' control networks in *Escherichia coli*: signaling by the second messenger c-di-GMP. *J Biotechnol* **160:**10–16.

to nitrogen utilization in other organisms, including in the flagellar genes of *Caulobacter* spp. and some promoters of the toluene-biodegradative operons of the *Pseudomonas putida* Tol plasmid. Interestingly, all of the known σ^N-type promoters require activation by an activator protein.

The Transcription Activator NtrC

The polypeptide chains of NtrC-type activators have the basic arrangement shown in Figure 12.11A. A DNA-binding domain that recognizes a sequence present upstream of σ^N-type promoters is located at the carboxyl-terminal end of the NtrC polypeptide. A regulatory domain that is phosphorylated (or in some cases, binds a regulatory factor) is present at the amino-terminal end. The region of the polypeptide responsible for transcriptional activation is in the middle. This region has an ATP-binding domain and an ATPase activity that cleaves ATP to ADP. The NTD masks the middle domain for activation unless the NTD has been phosphorylated (or has bound its inducer, in some members of the NtrC family).

BOX 12.4

Sigma Factors

Sigma (σ) factors seem to be unique to bacteria and their phages and are not found in eukaryotes or archaea. These proteins cycle onto and off of RNA polymerase and help direct it to specific promoters (see chapter 2). They also help RNA polymerase separate the DNA strands at the promoter to initiate transcription, and they help in promoter clearance after initiation. They also may contain the contact points of activator proteins that help these proteins stabilize the RNA polymerase on the promoter and activate transcription (see chapter 11). Promoters are often identified by the σ factor they use; for example, a σ^{70} promoter is one that uses the RNA polymerase holoenzyme with σ^{70} attached, while a σ^{H} promoter uses RNA polymerase holoenzyme containing σ^{H}. A caveat: different sigma factors are often given the same name in different bacteria. For example, σ^{E} in *E. coli* and *B. subtilis* refers to very different sigma factors; the σ^{E} in *E. coli* is the extracytoplasmic stress sigma factor, while the σ^{E} in *B. subtilis* is involved in sporulation. The transcriptional activity of a sigma factor is dependent on its ability to bind to core RNA polymerase, and competition for binding to the core RNA polymerase influences the transcriptional pattern in the cell (see Feklístov et al., References).

Sigma factors can be identified bioinformatically in bacteria based on their sequence conservation. This has revealed that the number of different sigma factors varies widely from one bacterial type to another. The bacterium with the least known so far, *Helicobacter pylori*, has only 3 different types, while the current record holder, *Streptomyces coelicolor*, has 63. In general, bacteria that are free living have more sigma factors than do obligate parasites, probably reflecting the greater environmental challenges faced by free-living bacteria.

There are two major classes of sigma factors in bacteria: the σ^{70} class, which comprises most of the sigma factors discussed in this book, including σ^{S}, σ^{H}, σ^{B}, and σ^{E}, and another class, σ^{N}, which seems to form a class by itself (see below). While all sigma factors in the σ^{70} class have some sequence and functional homology, there is no sequence similarity between members of this class and members of the σ^{N} class. The two classes also seem to differ fundamentally in their mechanisms of action (see below). Members of the σ^{70} class are found in all bacteria and play many diverse roles, some of which are discussed in previous chapters. The σ^{N}-type promoters are also widely distributed across bacteria, but they are not universal. This sigma factor was originally named the "nitrogen sigma factor," σ^{N}, because the promoters it uses were first found in the genes for Ntr regulation that are turned on during nitrogen-limited growth in *E. coli* (see the text). However, it is now known that there is no common theme for σ^{N}-expressed genes. In some soil bacteria, this sigma factor is used to express biodegradative genes, for example, to degrade toluene, and in other species, including some pathogens, it is used by some of the flagellar genes to make components of type III secretion systems, as well as to make alginate in *P. aeruginosa* (see Kazmierczak et al., References).

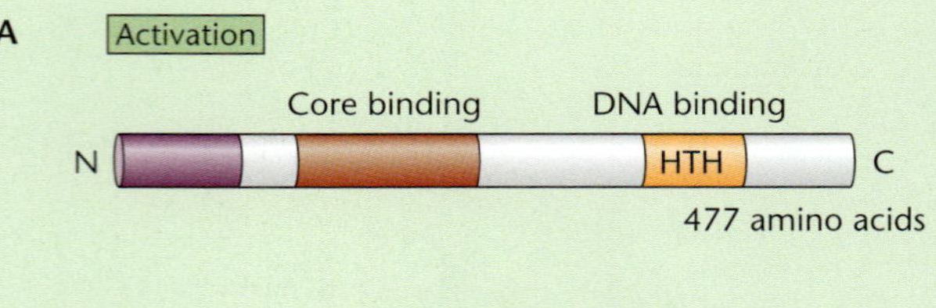

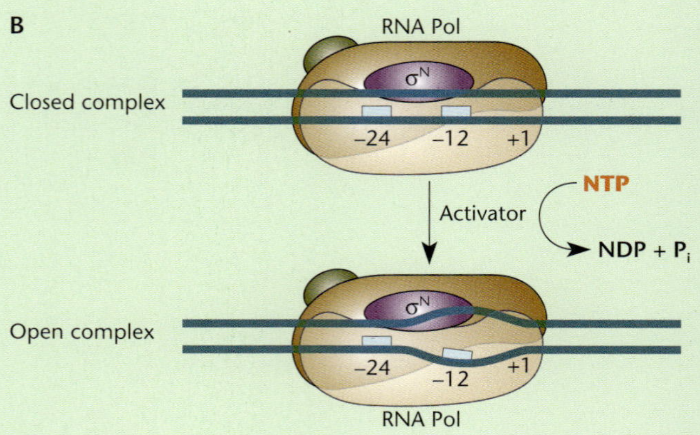

One major distinction between the two classes of sigma factors is in the way their promoters are activated. For example, most promoters that use a sigma factor of the σ^{70} family can initiate transcription without the help of an activator protein. If a σ^{70} promoter requires an activator, it generally binds adjacent to the promoter and helps to recruit RNA polymerase to the promoter (see chapter 11). However, all σ^{N} promoters studied thus far absolutely require a specialized activator protein with ATPase activity, which binds to an upstream activator sequence that can be hundreds of base pairs upstream of the promoter. In some ways, this makes σ^{N} promoters more like the RNA polymerase II promoters of eukaryotes. They also differ from σ^{70} promoters in their mechanism of activation. The activation of a σ^{N} promoter is illustrated in the figure (modified from Buck M, Gallegos MT, Studholme DJ, Guo Y, Gralla JD, *J Bacteriol* **182**:4129–4136, 2000). Panel A shows the functionally important domains of σ^{N} that allow binding to core RNA polymerase and DNA. The NTD allows σ^{N} to respond to activators. Panel B shows that the σ^{N}-RNA polymerase forms a stable but closed complex with the promoter, even in the absence of the activator bound to the upstream activator sequence (not shown in figure). The activator has a latent ATPase activity, which becomes activated by phosphorylation that is often passed down from a sensor kinase, either directly or through a phosphorylation cascade, with NtrC being the prototype (see the text). Phosphorylation of the N terminus of the activator alters its activity or its affinity for upstream activator sequences. Multimerization and formation of a DNA-bound complex activates the ATPase activity in its central domain. Once activated, the ATPase can cause σ^N to undergo a conformational change that stimulates open-complex formation by the σ^N-RNA polymerase and allows initiation at the promoter.

References

Feklístov A, Sharon BD, Darst SA, Gross CA. 2014. Bacterial sigma factors: a historical, structural, and genomic perspective. *Annu Rev Microbiol* **68**:357–376.

Kazmierczak MJ, Wiedmann M, Boor KJ. 2005. Alternative sigma factors and their roles in bacterial virulence. *Microbiol Mol Biol Rev* **69**:527–543.

Österberg S, del Peso-Santos T, Shingler V. 2011. Regulation of alternative sigma factor use. *Annu Rev Microbiol* **65**:37–55.

Paget MSB, Helmann JD. 2003. The σ^{70} family of sigma factors. *Genome Biol* **4**:203–215.

Wigneshweraraj S, Bose D, Burrows PC, Joly N, Schumacher J, Rappas M, Pape T, Zhang X, Stockley P, Severinov K, Buck M. 2008. Modus operandi of the bacterial RNA polymerase containing the σ^{54} promoter-specificity factor. *Mol Microbiol* **68**:538–546.

NtrC-activated promoters, including p_2 of the *glnA-ntrB-ntrC* operon, are unusual in that NtrC binds to an **upstream activator sequence (UAS)**, which lies more than 100 bp upstream of the promoter. For most positively regulated promoters, the activator protein-binding sequences are adjacent to the site at which RNA polymerase binds (see chapter 11). Activation at a distance, such as occurs with the NtrC-activated promoters, is much more common in eukaryotes, where many examples are known.

Figure 12.11B shows a model of how NtrC~P activates transcription from p_2 and other Ntr-activated promoters. The RNA polymerase holoenzyme containing σ^N can bind to the promoter even when nitrogen is not limiting and NtrC is not phosphorylated. NtrC also can bind to the UAS even if it is not phosphorylated, but no transcriptional activation occurs unless NtrC is phosphorylated. When nitrogen is limiting and NtrC becomes phosphorylated, oligomers of NtrC~P bound at the UAS activate transcription from the promoter. Activation of the promoter requires contact between NtrC~P bound at the UAS and RNA polymerase bound at the promoter, which must involve bending of the DNA. For some NtrC-type activators, DNA bending is facilitated by an additional accessory protein (such as IHF, the integration host factor first identified by its role in phage λ integration into the chromosome [see chapter 7]).

Figure 12.10 Sequence comparison of promoters recognized by the RNA polymerase holoenzyme carrying the normal sigma factor (σ^{70}), the nitrogen sigma factor (σ^N), and the heat shock sigma factor (σ^H). Instead of consensus sequences centered at bp −10 and −35 with respect to the RNA start site, the σ^N promoter has consensus sequences at bp −12 and −24. The σ^H promoter has consensus sequences centered at approximately bp −10 and −35, but they are different from the consensus sequences of the σ^{70} promoter. N indicates that any base pair can be present at this position. +1 is the start site of transcription.

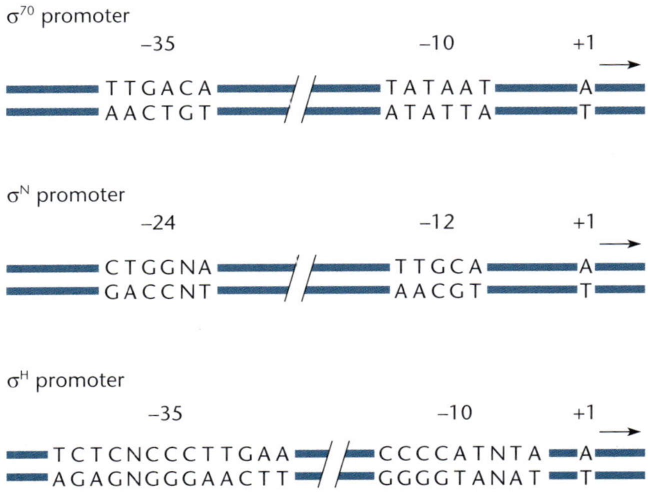

At most promoters transcribed by RNA polymerase containing σ^{70}, the limiting factor for transcription initiation is binding of RNA polymerase to the promoter, and open complex formation occurs spontaneously (see chapter 2). In contrast, transcription at a σ^N promoter requires cleavage of ATP by NtrC~P for open complex formation. The fact that the limiting step for transcription is open complex formation rather than binding is consistent with the observation that binding of both RNA polymerase containing σ^N and NtrC to the DNA can occur in the absence of activation. This may allow the system to be poised for an immediate response as soon as NtrC becomes phosphorylated.

ADENYLYLATION OF GLUTAMINE SYNTHETASE

Regulating the transcription of the *glnA* gene is not the only way that the activity of glutamine synthetase is controlled in the cell. The activity is also modulated by the adenylylation of (transfer of AMP to) a specific tyrosine in the glutamine synthetase enzyme by an adenylyltransferase enzyme when NH_3 concentrations are high. As mentioned in chapter 11, posttranslational modification of proteins can have a major effect on protein activity, and adenylylation of glutamine synthetase represents an example of this. The adenylylated form of the glutamine

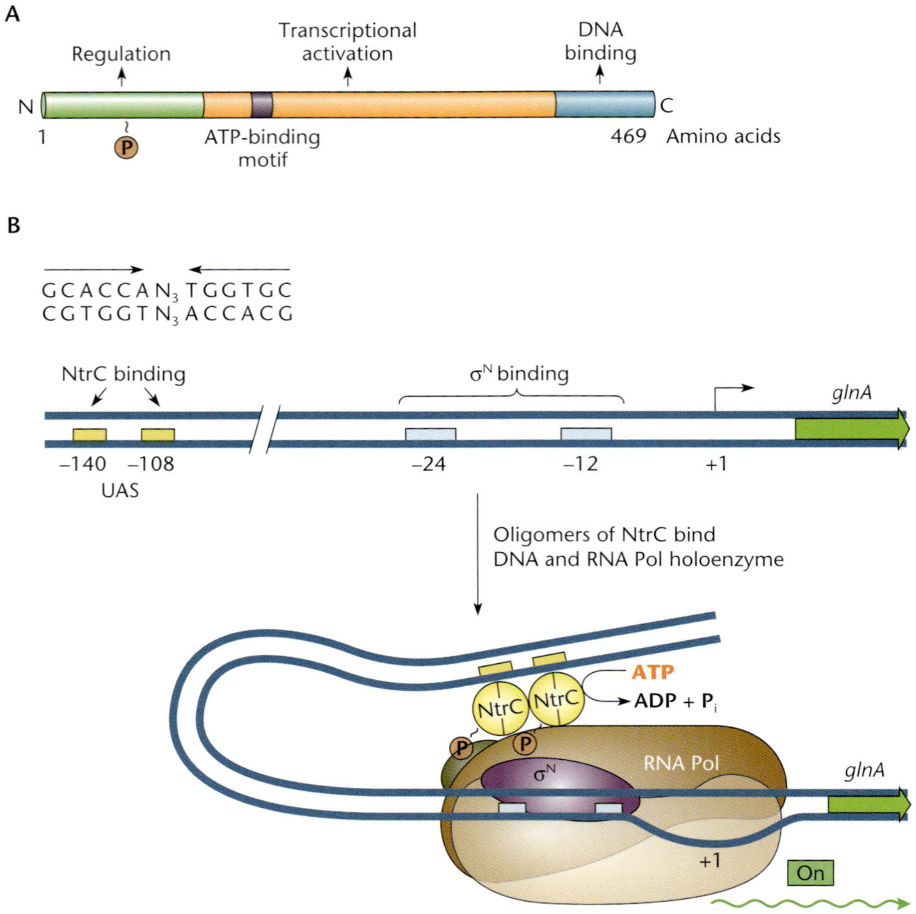

Figure 12.11 Model for the activation of the p_2 promoter by phosphorylated NtrC protein. **(A)** Functions of the various domains of NtrC. P denotes a phosphate. **(B)** Oligomers (shown as two dimers) of phosphorylated NtrC bind to the inverted repeats in the upstream activator sequence (UAS). The DNA is bent between the UAS and the promoter, allowing contact between NtrC and the RNA polymerase containing σ^N bound at the promoter more than 100 bp downstream. Cleavage of ATP due to the NtrC ATPase activity is required for the RNA polymerase to form open complexes at the promoter.

synthetase enzyme is less active and is also much more susceptible to feedback inhibition by glutamine than is the unadenylylated form (see chapter 11 for an explanation of feedback inhibition). This makes sense, considering that glutamine synthetase plays different roles when NH_3 concentrations are high and when they are low. When NH_3 concentrations are high, the primary role of glutamine synthetase is to make glutamine for protein synthesis, which requires less enzyme; the residual glutamine synthetase activity is feedback inhibited by glutamine to ensure that the cells do not accumulate too much glutamine. When NH_3 concentrations are low, more enzyme is required. In this situation, the enzyme should not be feedback inhibited, because its major role is to assimilate nitrogen.

The state of adenylylation of the glutamine synthetase enzyme is also regulated by GlnD and the state of the P_{II} protein. When the cells are in low NH_3 concentrations, so that the P_{II} protein has UMP attached (Figure 12.9), the adenylyltransferase removes AMP from glutamine synthetase. When the cells are in high NH_3 concentrations, so that the P_{II} protein does not have UMP attached, the P_{II} protein binds to and stimulates the adenylyltransferase to add more AMP to the glutamine synthetase.

COORDINATION OF CATABOLITE REPRESSION, THE Ntr SYSTEM, AND THE REGULATION OF AMINO ACID-DEGRADATIVE OPERONS

Not only must bacteria sometimes use one or more of the 20 amino acids as a nitrogen source, but they also must sometimes use amino acids as carbon and energy sources. The types of amino acids that different bacteria can use vary. For example, *E. coli* can use almost any amino acid as a nitrogen source except tryptophan, histidine, and the branched-chain amino acids, such as valine. However, it can use only alanine, tryptophan, aspartate, asparagine, proline, and serine as carbon sources. *Salmonella* also can use many amino acids as nitrogen sources but can use only alanine, cysteine, proline, and serine as carbon sources. Like the sugar utilization operons, the amino acid utilization operons not only have their own specific regulatory genes, so that they are transcribed only in the presence of their own inducer (usually the amino acid that the gene products can utilize), but often are also part of larger regulons. As discussed in this section, they are often under Ntr regulation and so are not induced in the presence of their inducer while a better nitrogen source, such as NH_3, is in the medium. In some bacteria, these operons are also under the control of the CAP catabolite regulation system and are not expressed in the presence of better carbon sources, such as glucose.

In addition to their potential as nitrogen, carbon, and energy sources, amino acids are necessary for other purposes. Most obviously, an amino acid is needed for protein synthesis if it is in short supply relative to the other amino acids. However, other functions include the use of proline in osmoregulation. Therefore, the use of amino acids as carbon and nitrogen sources can present strategic problems for the cell. The way in which all these potentially conflicting regulatory needs are resolved is often complicated (see Commichau et al., Suggested Reading).

Regulation of Nitrogen Assimilation in *B. subtilis*

As described above, nitrogen regulation in enteric bacteria relies heavily on posttranslational protein modification (e.g., phosphorylation of NtrC, uridylylation of P_{II}, and adenylylation of glutamine synthetase) and a dedicated sigma factor (σ^N). In contrast, regulation of the same pathway in *B. subtilis* depends primarily on protein-protein interactions, using a regulatory protein called TnrA, for *trans*-acting nitrogen regulation.

THE TnrA PROTEIN REGULATES NITROGEN METABOLISM IN *B. SUBTILIS*

When glutamine is high, TnrA acts as a transcriptional repressor of the glutamine synthetase gene and also forms a complex with existing glutamine synthetase and glutamine to inhibit TnrA activity (see Sonenshein, Suggested Reading). When glutamine levels are low, the inhibitory activity of TnrA is blocked, and TnrA shifts its role to serve as an activator of transcription of genes for utilization of secondary nitrogen sources. The expression of the gene for GOGAT (see above), which is involved in glutamate biosynthesis, is repressed by TnrA when glutamine is high but also is subject to regulation by the GltC transcriptional activator. GltC-dependent activation of GOGAT synthesis is increased if the cell has large amounts of α-ketoglutarate (the substrate for glutamate synthesis) and is decreased by large amounts of glutamate (the product of GOGAT activity). This allows the cell to produce the enzyme only when glutamate biosynthesis should occur.

The regulatory activity of TnrA is not modulated by posttranslational modification but, instead, is determined by formation of a complex with glutamine synthetase enzyme in the presence of glutamine. Binding of TnrA to glutamine synthetase and glutamine occurs only when glutamine is abundant and results in activation of TnrA as a repressor of *glnA* gene expression. Glutamine synthetase therefore regulates expression of its own gene (an example of **transcriptional autorepression**), although repression requires both high levels of glutamine synthetase enzyme and the end product of the pathway.

THE CodY GLOBAL REGULATOR

The CodY protein is a global regulator that is conserved in many *Firmicutes*. It serves as a general sensor of cell physiology and acts primarily as a repressor of genes whose products help the cell to adjust to nutrient limitation (see Sonenshein, Suggested Reading). The DNA-binding activity of CodY is activated by either high GTP levels or high concentrations of the branched-chain amino acids (leucine, isoleucine, and valine). These serve as useful signal molecules of metabolic activity, as their synthesis is dependent on the availability of nitrogen, carbon, and even phosphorus, so that high levels indicate that the cell is not starved for any of these crucial nutrients. High GTP levels also signal efficient energy production. CodY represses a number of genes involved in nitrogen metabolism, including the gene for GOGAT, thereby inhibiting synthesis of glutamate, and also represses genes for enzymes involved in the uptake and utilization of other amino acids that can be used to generate glutamate and glutamine. Interactions with CcpA, the central regulator of carbon metabolism in Gram-positive bacteria that responds to the FBP levels (see above), further allow the cell to integrate information about general metabolism to control a large variety of pathways.

Regulation of Ribosome Components and tRNA Synthesis

To compete effectively in the environment, cells must make the most efficient use possible of the available energy. We have already talked about the competitive advantage of using the carbon source that allows the most efficient energy production first. However, a competitive edge can also be gained by economizing within the cell. One of the major ways in which cells conserve energy is by regulating the synthesis of their ribosomes and tRNAs so that they make only enough to meet their needs. More than half of the RNA made by the cell at any one time is rRNA and tRNA. Moreover, each ribosome is composed of about 50 different proteins, and there are about as many different tRNAs. Synthesis of the translational machinery therefore represents a major investment of cellular resources and requires careful regulation, not only to ensure the appropriate total quantity of the components, but also to maintain all of the components in the correct relative amounts so that the machinery can operate at maximum efficiency.

The numbers of ribosomes and tRNA molecules needed by the cell vary greatly, depending on the growth rate. Fast-growing cells require many ribosomes and tRNA molecules to maintain the high rates of protein synthesis required for rapid doubling of cell mass. Cells growing more slowly, either because they are using a relatively poor carbon and energy source or because some nutrient is limiting, need fewer ribosomes and tRNAs. As a consequence, a rapidly growing *E. coli* cell contains as many as 70,000 ribosomes, but a slow-growing cell has fewer than 20,000. As with most of the global regulatory systems we have discussed, the regulation of ribosome and tRNA synthesis is much better understood for *E. coli* than for any other organism. In this section, we confine our discussion to *E. coli*, with occasional references to other bacteria when information is available.

Ribosomal Protein Gene Regulation

Ribosomes are composed of both proteins and RNAs (see chapter 2). A special nomenclature is used for the components of the ribosome. The ribosomal proteins are designated by the letter L or S, to indicate whether they are from the large (50S) or small (30S) subunit of the ribosome, respectively, followed by a number for the particular protein. Thus, protein L11 is protein number 11 from the large 50S subunit of the ribosome, whereas protein S12 is protein number 12 from the small 30S subunit. The gene names begin with *rp*, for ribosomal protein, followed by a lowercase *l* or *s* to indicate whether the protein product resides in the large or small subunit. Another capital letter designates the specific gene. For example, the gene *rplK* is ribosomal protein gene K encoding the L11 protein; note that K is the 11th letter of the alphabet. Similarly, *rpsL* encodes the S12 protein; L is the 12th letter of the alphabet.

MAPPING OF RIBOSOMAL PROTEIN GENES

A total of 54 different genes encode the 54 polypeptides that comprise the *E. coli* ribosome, and mapping these genes was a major undertaking. Some ribosomal protein genes were mapped by mapping mutations that caused resistance to antibiotics, such as streptomycin, which binds to ribosomal protein S12, blocking the translation of other genes (see chapter 2). More complex techniques involving specialized transducing phages and DNA cloning (see chapters 7 and 13) were needed to map the other ribosomal protein genes. Often, clones containing these genes were identified because they direct synthesis of a particular ribosomal protein in coupled *in vitro* transcription-translation systems. As these genes are highly conserved, identification in newly sequenced genomes is usually based on similarity to the related gene in *E. coli*.

The mapping results revealed some intriguing aspects of the organization of the ribosomal protein genes in the chromosome of *E. coli*. Rather than being randomly scattered around the chromosome, most of the 54 genes are organized into large clusters of operons. Some of these operons also contain genes for other components of macromolecular synthesis, including subunits of RNA polymerase and genes for proteins of the DNA replication apparatus. Clustering of this type is found in many bacterial genomes.

Several hypotheses have been proposed to explain why genes involved in macromolecular synthesis are clustered in the *E. coli* genome. First, the products of these genes must all be synthesized in large amounts to meet the cellular requirements. Some clusters are near the origin of replication, *oriC*, and cells growing at high growth rates have more than one copy of the genes near this site (see chapter 1), which allows higher rates of synthesis of the gene products. Other possible reasons have to do with the structure of the bacterial nucleoid (see chapter 1). Clustered genes are probably on the same loop of the nucleoid. A loop for the macromolecular synthesis genes might be relatively large and extend out from the core of the nucleoid to allow RNA polymerase and ribosomes to gain easier access to the genes. A third possible explanation is that the genes for macromolecular synthesis must be coordinately regulated with the growth rate, and their assembly in clusters of operons may facilitate their coordinate regulation.

REGULATION OF THE SYNTHESIS OF RIBOSOMAL PROTEINS

The regulation of the synthesis of ribosomal proteins and rRNA presents an interesting case of coordinate regulation. Even though the ribosomal proteins and rRNAs are synthesized independently and only later are assembled into mature ribosomes, there is never an excess of either free ribosomal proteins or free rRNA in the cell, suggesting that their synthesis is somehow coordinated.

The regulation of ribosomal protein synthesis is best understood in *E. coli* (see Burgos et al., Suggested Reading). However, it is likely that many aspects of this regulation are conserved in other bacteria. The synthesis of many of the ribosomal proteins is translationally autoregulated (see chapter 11). The ribosomal proteins bind to translation initiation regions (TIRs) in their own mRNA and repress their own translation. However, rather than having each ribosomal gene of the operon translationally regulate itself independently, the protein product of only one of the genes of the operon is responsible for repressing the translation of all the ribosomal proteins encoded by the operon. Basically, this protein translationally represses the first gene in the operon, and the translation of the other protein-coding regions on the same mRNA is also translationally regulated, because their translation is coupled to the translation of the first

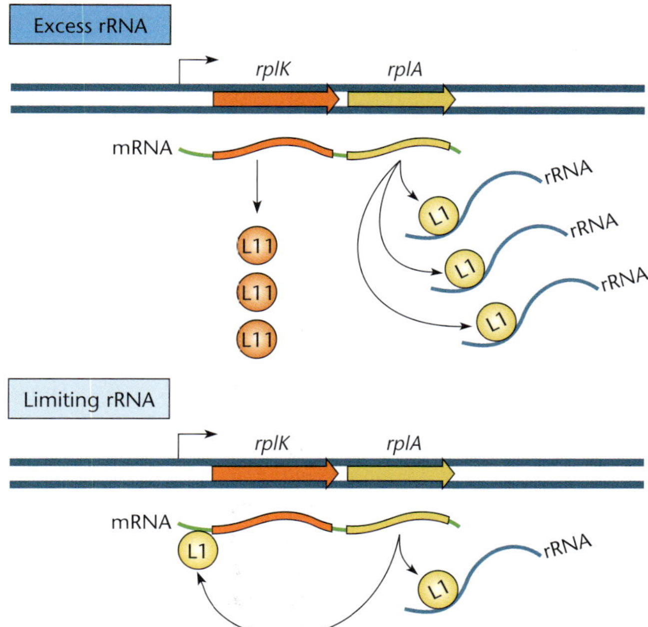

Figure 12.12 Translational autoregulation of ribosomal protein gene expression, as illustrated by the *rplK-rplA* operon. When rRNA is in excess, the regulatory ribosomal protein (L1 in the example) binds to rRNA to assemble more ribosomes, and the *rplK-rplA* mRNA is efficiently translated. When free rRNA is limiting, L1 binds instead to a target site on the mRNA that overlaps the TIR for *rplK*. This directly inhibits *rplK* translation by blocking binding of the ribosome and indirectly inhibits *rplA* translation by translational coupling (because the TIR for *rplA*) is available for ribosome binding only when the upstream *rplK* coding sequence is being translated.

protein (see chapter 2 for an explanation of translational coupling). The designated regulatory protein can also bind to free rRNA in the cell, which, as discussed below, is what coordinates the synthesis of the ribosomal proteins with the synthesis of rRNA.

Figure 12.12 illustrates this regulatory mechanism with the relatively simple *rplK-rplA* operon, which encodes L11 and L1. The L1 protein serves as the regulator of the operon. The normal role of the L1 protein is to bind to free rRNA to assemble new ribosomes. L1 can also bind to the TIR for the *rplK* gene on the mRNA for the *rplK-rplA* operon. Binding of L1 to the *rplK* TIR directly inhibits translation of the *rplK* coding region by blocking binding of the ribosome for translation initiation and also inhibits translation of its own gene, because the *rplA* TIR is sequestered in the mRNA and is unavailable for translation unless *rplK* is actively translated, which releases the *rplA* TIR (see chapter 2). However, if the cell contains free rRNA, the L1 protein preferentially binds to it instead of binding to the *rplK* TIR, and synthesis of L11 (and L1) resumes. When there is no longer any free rRNA in the cell—because it has all been assembled into the ribosomes—the free L1 protein begins to accumulate, and it can bind to the TIR for the *rplK* gene and again repress the translation of the *rplK-rplA* mRNA. Note that this is a form of feedback repression, but in this case, the protein product of a gene directly represses its own synthesis.

A similar pattern is observed in other ribosomal protein gene operons. Their designated regulatory protein accumulates when rRNA is not freely available, and the free regulatory ribosomal protein binds to the TIR for an early gene in the operon and represses the synthesis of all the ribosomal proteins encoded in the operon. The synthesis resumes when more free rRNA accumulates. In this way, the cell ensures that there will be neither an excess of free ribosomal proteins nor an excess of free rRNA, and regulation of rRNA synthesis in response to growth conditions (see below) results in coordinate regulation of ribosomal protein synthesis.

The protein in each operon that is designated as the regulator is usually a protein that normally binds to rRNA early during the assembly of the ribosome and so already has an RNA-binding activity. In at least some cases, the RNA structural element to which the designated protein binds in the mRNA is related to the binding site for that protein in the rRNA. Note that whereas translational regulation is not the only mechanism used for ribosomal protein gene regulation, some form of autogenous regulation and response to rRNA levels is common.

Regulation of rRNA and tRNA Synthesis

If the translation of the ribosomal proteins is coupled to the synthesis of the rRNAs under different growth conditions, then how is the synthesis of rRNA regulated? As discussed in chapter 2, the 16S, 23S, and 5S rRNAs are synthesized together as a long precursor RNA, often with tRNA sequences positioned between the rRNA sequences. After synthesis, the long precursor RNA is processed into the individual rRNAs and tRNAs. Every ribosome contains one copy of each of the three types of rRNAs, and synthesizing all three rRNAs as part of the same precursor RNA ensures that all three are made in equal amounts.

REGULATION OF rRNA TRANSCRIPTION

Each cell has tens of thousands of ribosomes, requiring the synthesis of large amounts of rRNA. To meet this need, bacteria have evolved many ways to increase the output of their rRNA genes. For example, many bacteria have more than one copy of the genes for rRNA. For example, *E. coli* has 7 copies of the rRNA operons, and *B. subtilis* has 10 copies. Many bacteria also have very strong promoters for their rRNA genes, with high affinity for RNA polymerase. The RNA polymerase molecules initiate transcription and start down the operon, one immediately after another, so that under conditions when rRNA synthesis is occurring at the highest level, as many as 50 RNA polymerase molecules can be transcribing each rRNA operon simultaneously. The rRNA promoters are so strong because their sequences at the −10 and −35 elements closely match the preferred recognition sites for RNA polymerase holoenzyme containing σ^{70}, and they also have a sequence called the UP element upstream of the promoter (see Figure 2.10B). This sequence enhances initiation of transcription from the promoter by interacting with the α-CTD region of RNA polymerase to increase the affinity of the promoter for RNA polymerase. Protein activators such as CAP also interact with α-CTD and use protein-protein contacts with RNA polymerase in the place of the DNA-protein contacts between the UP element sequence and α-CTD in rRNA operons (see chapter 11).

An additional mechanism for increasing transcription from rRNA operon promoters utilizes the FIS protein. FIS is a DNA-binding protein that participates in a number of regulatory systems, as well as in certain site-specific recombination systems, including integration of the phage λ prophage during lysogeny. FIS is more abundant in rapidly growing cells, and its levels drop as cells enter stationary phase. Binding of FIS upstream of rRNA operon promoters increases their transcriptional activity, and since FIS is more abundant during rapid growth, this increases rRNA synthesis under these conditions.

GROWTH RATE REGULATION OF rRNA AND tRNA TRANSCRIPTION

As mentioned above, cells growing quickly in rich medium have many more ribosomes and a higher concentration of tRNA than do cells growing slowly in poor medium. This regulation of rRNA and tRNA synthesis is called **growth rate regulation**.

rRNA (and tRNA) operon promoters are unusual in that the rate of initiation of transcription is very sensitive to the concentration of the initiating nucleotide (see Barker and Gourse, Suggested Reading). Like most transcription, rRNA synthesis begins with ATP or GTP, so one or the other of these initiates transcription of each of the rRNA operons. The rRNA promoters are unusual in that they have a high affinity for RNA polymerase but form very short-lived open complexes (see chapter 2 for a discussion of closed and open complexes and transcription initiation). The high affinity for RNA polymerase allows them to form closed complexes at a high rate and compete very effectively for RNA polymerase. They then quickly form the open complex. However, because these open complexes are unstable, the transcription complex rapidly returns to the closed complex state unless the initiating nucleotide immediately enters through the secondary channel and transcription initiates. High concentrations of the initiating nucleotide increase the probability of initiation before the open complex reverts to the closed complex state. When growth rates are high, the concentrations of ATP and GTP increase, leading to more frequent initiations at the rRNA promoters and therefore more synthesis of rRNAs. When growth rates are low, the concentrations of ATP and GTP are usually lower, which results in less efficient formation of the open complex at these promoters and lower transcription of the rRNA and tRNA genes. Since ribosomal protein synthesis is coupled to rRNA availability, changes in rRNA abundance lead to changes in ribosome production.

In addition to the response of rRNA operon promoters to the steady-state growth rate via monitoring of the availability of GTP or ATP, there is also a response to sudden changes in growth conditions, e.g., starvation for amino acids. This is called the stringent response and is discussed below.

ANTITERMINATION OF rRNA OPERONS

Another mechanism that promotes a high rate of rRNA gene transcription is the presence of antitermination sequences that are positioned just downstream of the promoter and in the spacer region between the 16S and 23S coding sequences. These antitermination sequences reduce pausing by RNA polymerase and prevent termination at sites that can promote binding of the ρ factor to cause ρ-dependent transcription termination (see Figure 2.15). Since rRNA is not translated, transcription of these long operons is particularly susceptible to ρ-dependent termination. The antitermination sequences therefore allow the synthesis of more full-length rRNAs and also allow them to be completed in a shorter time because RNA polymerase moves more rapidly along the DNA (see Condon et al., Suggested Reading).

Stringent Response

In addition to the growth rate regulation discussed above, the rRNA and tRNA genes in *E. coli* are subject to another type of regulation called the **stringent response**, which causes rRNA and tRNA synthesis to cease when cells are starved. Both growth rate regulation and the stringent response involve the use of nucleotides as signals of the nutritional status of the cell.

Protein synthesis requires all 20 amino acids, which must be made by the cell or furnished in the medium. A cell is said to be starved for an amino acid when the amino acid is missing from the environment and the cell cannot make a sufficient amount. A translating ribosome therefore will stall when it encounters a codon for the missing amino acid because the corresponding aminoacyl-tRNA is not available for insertion into the growing polypeptide chain (see chapter 2).

In principle, rRNA synthesis can continue in cells starved for an amino acid, since RNA does not contain amino acids. However, in *E. coli* and many other types of cells, the synthesis of rRNA and tRNA ceases when an amino acid is lacking. This reduction of the synthesis of rRNA and tRNA after amino acid starvation is called the stringent response. Because ribosomal protein gene expression is coupled to rRNA availability, turning off rRNA synthesis also turns off ribosomal protein synthesis. The stringent response saves energy and cellular resources and conserves any remaining stores of the amino acid that is limiting for growth; there is no point in making ribosomes and tRNA if one of the amino acids is not available for protein synthesis.

SYNTHESIS OF ppGpp DURING THE STRINGENT RESPONSE

In *E. coli* and many other organisms, the reduction of rRNA and tRNA synthesis results from the accumulation of an unusual nucleotide, **guanosine tetraphosphate (ppGpp)**. This nucleotide is first made as **guanosine pentaphosphate (pppGpp)** by transferring two phosphates from ATP to the 3′ hydroxyl of GTP but is then quickly converted to ppGpp by a phosphatase. These nucleotides were originally called "magic spot" I and II (MSI and MSII) because they show up as distinct spots during some types of chromatography (see Cashel and Gallant, Suggested Reading).

Figure 12.13 shows a model for how amino acid starvation stimulates the synthesis of ppGpp. The nucleotide is generated by an enzyme called RelA (for relaxed control gene A), which is bound to the ribosome. When *E. coli* cells are starved for an amino acid (lysine in the example), the tRNAs for that amino acid (e.g., $tRNA^{Lys}$) are uncharged. The low level of $tRNA^{Lys}$ with lysine attached ($Lys\text{-}tRNA^{Lys}$) results in failure of elongation factor Tu (EF-Tu) to bring $Lys\text{-}tRNA^{Lys}$ into the aminoacyl (A) site of the ribosome (see chapter 2). EF-Tu has very low affinity for uncharged $tRNA^{Lys}$, and therefore, when a ribosome moving along an mRNA encounters a codon for that amino acid (the codon AAA in the example), the ribosome stalls. If the ribosome stalls long enough, an uncharged tRNA ($tRNA^{Lys}$ in the example) may eventually enter the A site of the ribosome even though it is not bound to EF-Tu. Uncharged tRNA entering the A site stimulates the RelA protein, which is transiently associated with the ribosome, to synthesize pppGpp, which is then converted to ppGpp.

The intracellular levels of ppGpp during amino acid starvation are also regulated by the SpoT (for magic spot)

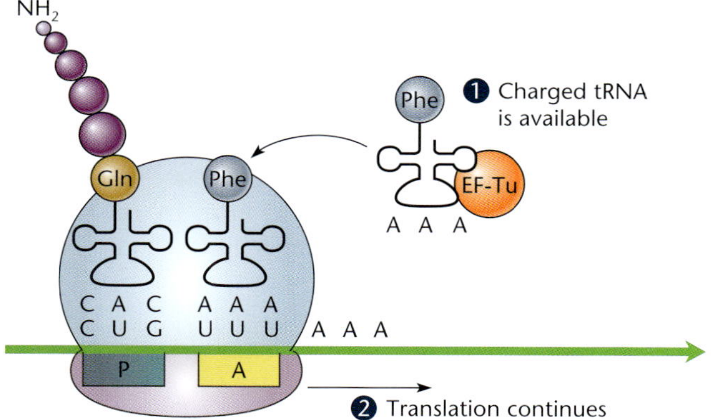

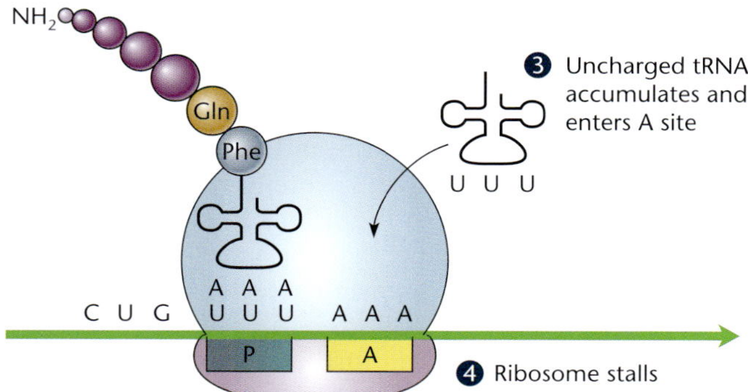

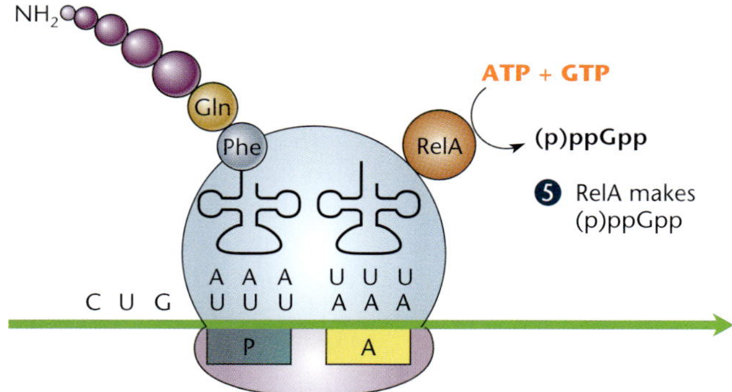

Figure 12.13 Model for synthesis of ppGpp after amino acid starvation. Cells are starved for the amino acid lysine. Translation of a phenylalanine codon occurs normally (steps 1 and 2). The $tRNA^{Lys}$ has no lysine attached, and elongation factor Tu (EF-Tu) cannot bind to a tRNA that is not aminoacylated. A ribosome moving along the mRNA stops when it arrives at a codon for lysine (AAA) because it has no aminoacylated $tRNA^{Lys}$ to translate the codon. This results in a ribosome with a peptidyl-tRNA in the P site and an empty A site (see chapter 2). If the ribosome remains stalled at the codon long enough, an unaminoacylated $tRNA^{Lys}$ (anticodon UUU in the example) binds to the A site of the ribosome even though EF-Tu is not bound (steps 3 and 4). This binding causes RelA to synthesize (p)ppGpp (step 5).

protein, which is the product of the *spoT* gene. The SpoT protein has pppGpp synthesis activity, like RelA, as well as a hydrolase activity that degrades ppGpp. The ppGpp degradation activity of SpoT is inhibited after amino acid starvation, leading to greater accumulation of ppGpp. Therefore, after amino acid starvation, the cellular concentration of ppGpp is determined both by the activation of the pppGpp synthesis activity of RelA and by the inhibition of the ppGpp-degrading activity of SpoT. The *spoT* gene product also can synthesize pppGpp in response to other starvation conditions. While *E. coli* has separate RelA and SpoT proteins, many other types of bacteria, including *B. subtilis*, have only a single protein with both of these activities. This class of enzyme is designated Rsh (RelA-SpoT homolog) and performs both roles.

Mutants in *relA* (or *rsh*) lack a normal stringent response. These mutants do not accumulate ppGpp after amino acid starvation and do not shut off rRNA and tRNA synthesis. Because rRNA synthesis and tRNA synthesis are not stringently coupled with protein synthesis in *relA* mutants, strains with *relA* mutations are called **relaxed strains**, which is the origin of the gene name. These mutants have a difficult time recovering after amino acid starvation because the cells continue to produce ribosomes after starvation, resulting in further depletion of cellular resources that made recovery less efficient once nutrients were restored.

SYNTHESIS OF ppGpp BY SpoT

In organisms like *E. coli* that have RelA and SpoT proteins, both have ppGpp synthesis activities. However, the two proteins respond to different physiological signals. RelA is specific for a response to amino acid limitation via its association with translating ribosomes. In contrast, SpoT does not associate with ribosomes and instead monitors perturbations in carbon metabolism through a direct interaction with acyl carrier protein (ACP). ACP plays a crucial role in fatty acid biosynthesis by binding to short-chain fatty acids and accepting additional fatty acid substrates to build longer fatty acid chains. ACP bound to long-chain fatty acids does not bind to SpoT with high affinity. The complex of ACP bound to short-chain fatty acids, which accumulates when carbon metabolism is not functioning efficiently, binds to SpoT, and this interaction simultaneously stimulates the ppGpp synthesis activity and represses the ppGpp hydrolysis activity, resulting in accumulation of ppGpp in the absence of amino acid starvation (Figure 12.14) (see Dalebroux et al., Suggested Reading).

ROLE OF ppGpp IN GROWTH RATE REGULATION, AFTER STRESS, AND IN STATIONARY PHASE

ppGpp levels increase when *E. coli* cells are growing more slowly in poorer medium as well as when cells run out of nutrients and begin to reach stationary phase. Many other stress conditions can also cause ppGpp levels to increase. Under these conditions, the rates of rRNA and tRNA synthesis are reduced, but other genes whose products are required in stationary phase or for stress responses are turned on. This suggests that ppGpp serves as a general "alarmone," signaling that major changes in gene expression will soon be required. Mutations in both *relA* and *spoT* are required to completely block ppGpp synthesis, and a *relA spoT* double mutant completely lacks ppGpp (and is sometimes referred to as a ppGpp^0 strain).

Because *relA spoT* double mutants lack ppGpp, many experiments comparing these mutants with the wild type have been done to determine the effect of ppGpp on cells. However, in spite of many years of such experimentation, it is not completely clear how ppGpp affects transcription, and it is likely that more than one mechanism is involved. The ppGpp nucleotide seems to have both positive and negative effects on gene transcription. For example, in addition to the effect on rRNA and tRNA synthesis, *relA spoT* mutants do not grow in minimal medium without added amino acids, i.e., they are effectively auxotrophic for some amino acids. In *E. coli*, *relA spoT* mutants require nine different amino acids, suggesting that ppGpp serves as an inducer for the transcription of the operons to make these amino acids when they are lacking in the medium. Also, *relA spoT* double mutants show phenotypes similar to those of mutants that lack σ^S (the stationary-phase sigma factor), which responds to many types of stress in the cell (see below). The activities of other alternate "stress" sigma factors, including σ^H (the heat shock sigma factor) and σ^E (the extracytoplasmic stress sigma factor), are also enhanced by ppGpp (see below).

One factor that complicates the interpretation of these experiments and necessitates the use of careful controls is the possibility of indirect effects due to competition for RNA polymerase. Genes that seem to be positively regulated by ppGpp might be those with promoters that do not bind RNA polymerase as tightly as the promoters for rRNA and tRNA genes; they would therefore require higher concentrations of free RNA polymerase to be active. During normal rapid growth, at least half of all the RNA polymerase in the cell is devoted to transcription of rRNA and tRNA genes. If the initiation of synthesis of the rRNAs and tRNAs is blocked by ppGpp, more RNA polymerase becomes available to transcribe other genes. However, *in vitro* experiments have indicated that competition for RNA polymerase is not the sole explanation for the apparent positive effect of ppGpp on the transcription of some genes (see below).

DksA: A PARTNER IN ppGpp ACTION

In *E. coli*, the effect of ppGpp on RNA polymerase is enhanced by a protein named DksA. It was first found be-

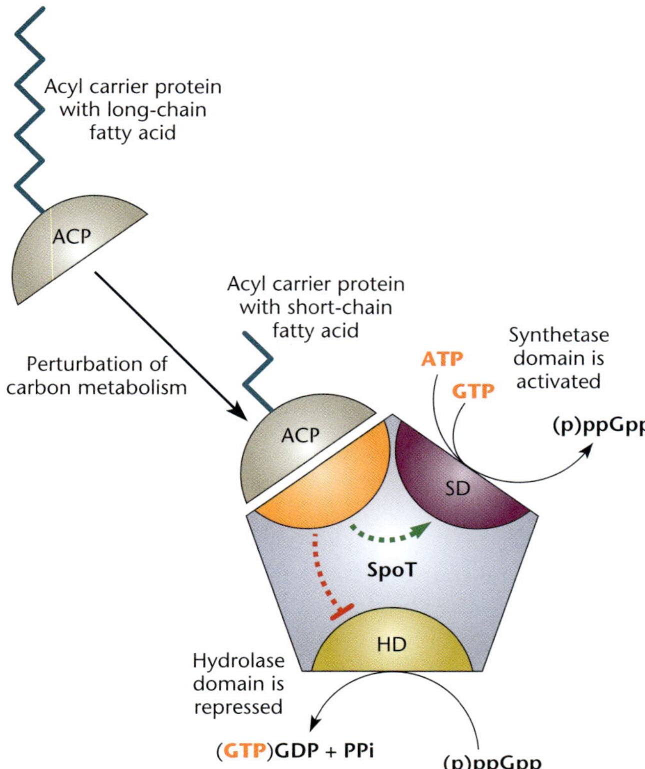

Figure 12.14 Regulation of SpoT activity. SpoT has both (p)ppGpp synthetase and hydrolase activity. Perturbation of carbon metabolism results in accumulation acyl carrier protein (ACP) associated with short-chain fatty acids, and binding of this form of ACP to SpoT activates the (p)ppGpp synthetase activity and represses the hydrolase activity, resulting in induction of the stringent response. Modified from Dalebroux et al. 2010 (see Suggested Reading). SD, synthetase domain; HD, hydrolase domain.

cause overexpression of the *dksA* gene suppresses growth defects of *dnaK* mutants (the name DksA means *dnaK* suppressor A). It is not clear, even in retrospect, why excess DksA would have this effect. Recall from chapter 2 that DnaK is the bacterial Hsp70 protein chaperone that helps to fold some proteins as they emerge from the ribosome and also plays the role of the cellular thermometer that induces heat shock (see below). The *dnaK* null mutants are sick because DnaK normally binds to the heat shock sigma, σ^H, and targets it for degradation. In the absence of DnaK, σ^H accumulates, and the heat shock genes are expressed constitutively, slowing cell growth. Apparently, excess DksA relieves the toxicity of the constitutive induction of the heat shock response, but how it does this is not clear.

The role of DksA in ppGpp-mediated regulation was not discovered until later, when it was noticed that the phenotypes of deletion mutants of the *dksA* gene (Δ*dksA*) are similar to the phenotypes of *relA spoT* double mutants that do not make ppGpp. The Δ*dksA* mutants are effectively auxotrophic for some (but not all) of the same amino acids; they also show increased rates of rRNA transcription, and they lack stringent control and growth rate regulation of rRNA synthesis. *In vitro* experiments that investigated the effect of adding ppGpp to RNA polymerase, with or without purified DksA, showed that DksA markedly enhances the ability of ppGpp to inhibit transcription from rRNA promoters. It also greatly increases the dependence of initiation from rRNA promoters on the concentration of the initiating nucleotide (ATP or GTP) and seems to further shorten the half-life of the open complexes on these promoters. DksA also enhances the ppGpp stimulation of transcription from the promoters for some amino acid biosynthetic operons, indicating that this effect of ppGpp is not all due to competition for RNA polymerase.

Clues to how the DksA protein has these effects came from the observation that DksA bears a remarkable structural similarity to the GreA protein. Recall from chapter 2 that GreA is a transcription factor that inserts into the secondary channel in RNA polymerase through which deoxynucleotide triphosphates enter (see Figure 2.13). It has a long extended coiled-coil probe that extends all the way into the channel to the active center and has two acidic amino acids on the end of the probe (aspartate and glutamate) that may bind the Mg^{2+} in the active center, which plays a role in the polymerization reaction. From this position, GreA can trigger cleavage of backtracked RNAs and release stalled transcription complexes. DksA also has an extended coiled-coil structure of about the same length, with two conserved acidic amino acids in the end (in this case, both aspartates), although there is little amino acid sequence similarity otherwise. This striking similarity in structure suggests that DksA may also enter RNA polymerase through the secondary channel, like GreA. Recent studies have demonstrated that ppGpp binds to two sites on RNA polymerase, one of which is at the interface between DksA and RNA polymerase (see Ross et al., Suggested Reading), a location that is consistent with the importance of DksA for the transcriptional response to ppGpp.

STRINGENT RESPONSE IN OTHER BACTERIA

Homologs of RelA and/or SpoT are found in nearly all bacterial genomes (see Dalebroux et al., Suggested Reading). However, it is not clear that all bacteria respond similarly to ppGpp accumulation. There is no apparent homolog of DksA in many organisms, including *B. subtilis*. Furthermore, it appears that the crucial parameter that regulates rRNA operon transcription in *B. subtilis* is the GTP concentration, rather than ppGpp itself. Recall that ppGpp is synthesized from GTP. Therefore, activation of ppGpp synthesis results in a rapid drop in the GTP concentration. As in *E. coli*, the promoters for rRNA operons in *B. subtilis* are highly sensitive to the concentration of the initiating nucleoside triphosphate, and in this organism, nearly all of the rRNA and tRNA promoters use GTP as the initiating nucleotide. It therefore may be the case in *B. subtilis* that the major effect of the stringent response (at least for rRNA transcription) is due to the reduction in GTP pools rather than a direct effect of the interaction of ppGpp with RNA polymerase.

Stress Responses in Bacteria

Many of the global regulons in bacteria are designed to deal with stress. In order to survive, all organisms must be able to deal with abrupt changes in the conditions in which they find themselves. The osmolarity, temperature, or pH of their surroundings might abruptly increase or decrease or they might be suddenly deprived of nutrients required for growth and have to enter a dormant state. If they have invaded a eukaryotic host, they might suddenly be exposed to reactive forms of oxygen or nitric oxide as part of the host defense. Not only must they be able to respond quickly to such changes, but also, their response must be flexible enough to deal with a variety of different stresses, or even more than one stress at a time. Furthermore, they must be able to sense when the stress condition has ended so that the cell can return to a normal state. As expected, bacteria and other organisms have evolved complicated interactive pathways to deal with such changes, and this is a subject of active current research. In this section, we discuss what has been learned about some major pathways and how they interact.

Heat Shock Regulation

The regulation of gene expression following a heat shock is one of the most extensively studied and highly conserved global regulatory responses in bacteria and other organisms. One of the major challenges facing cells is to survive abrupt changes in temperature (see Schumann, Suggested Reading). To adjust to abrupt temperature increases, cells induce at least 30 different genes encoding proteins called the **heat shock proteins (HSPs)**. The concentrations of these proteins quickly increase in the cell after a temperature upshift and then slowly decline, a phenomenon known as the **heat shock response**. Besides being induced by abrupt increases in temperature, the heat shock genes are often induced by other types of stress that damage proteins, such as the presence of ethanol and other organic solvents in the medium. Therefore, the heat shock response is more of a general stress response rather than a specific response to an abrupt increase in temperature, although there is a component of the regulatory mechanism that responds specifically to heat.

Unlike most shared cellular processes, the heat shock response was observed in cells of eukaryotes long before it was seen in bacteria. Some of the HSPs are remarkably similar in all organisms and presumably play similar roles in protecting all cells against heat shock. Some of the mechanisms of regulation of the heat shock response may also be similar in organisms ranging from bacteria to mammals.

HEAT SHOCK REGULATION IN *E. COLI*

The molecular basis for the heat shock response was first understood in *E. coli* and is illustrated in Figure 12.15. In this bacterium, about 30 genes encoding 30 different HSPs are turned on following a heat shock. The functions of many of these HSPs are known (Table 12.1). Most of the HSPs play roles during the normal growth of the cell and so are always present at low concentrations, but after a heat shock, their rate of synthesis increases markedly and then slowly declines to normal levels.

Some HSPs, including GroEL, DnaK, DnaJ, and GrpE, are chaperones that direct the folding of newly synthesized proteins (see chapter 2). The names of these proteins do not reflect their functions but, rather, reflect how they were originally discovered. For example, DnaK and DnaJ were found because they affect the assembly of a protein complex required for phage λ DNA replication, but they are not themselves involved directly in replication. Chaperones help the cell survive a heat shock by binding to proteins denatured by the sudden rise in temperature and either helping them to refold properly or targeting them for destruction. As mentioned above, the chaperones are among the most highly conserved proteins across all three domains of life. Other HSPs, including Lon and Clp, appear to degrade and not refold denatured proteins. These HSPs therefore may be involved in managing proteins that are so badly damaged that they are irreparable and so are best degraded before they poison the cell.

Knowing that many HSPs are involved in helping proteins to fold properly or in destroying denatured proteins helps to explain the transient nature of the heat shock response. Immediately after the temperature increases, the concentrations of salts and other cellular components that were adjusted for growth at lower temperatures are not appropriate for protein stability at the higher temperature; this leads to massive protein unfolding. Later, after the temperature has been elevated for some time, the internal conditions have had time to adjust, so the level of denatured proteins is lower and the increased number of chaperones and other HSPs is no longer necessary. Hence, the synthesis and activation of the HSPs decline.

Genetic Analysis of Heat Shock in *E. coli*

As with other regulatory systems, the analysis of the heat shock response was greatly aided by the discovery of mutants with defective regulatory genes. The first such

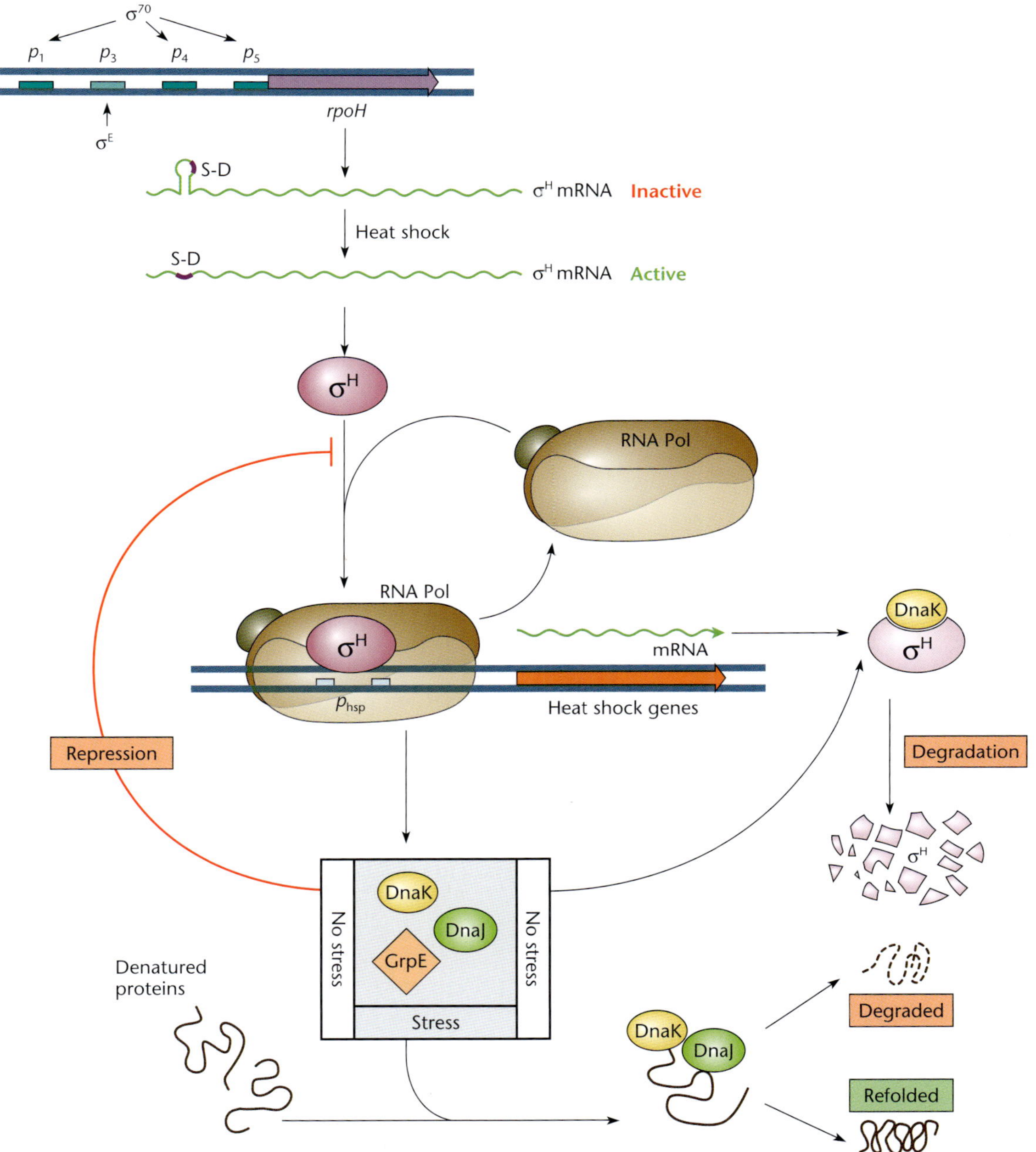

Figure 12.15 Induction of the heat shock response in *E. coli*. The *rpoH* mRNA is an RNA thermosensor, which is inactive for translation at low temperature and is unfolded and translated at high temperature. This results in rapid synthesis of σ^H after heat shock. In the absence of heat shock, σ^H is quickly inactivated by binding to DnaK, which both makes it less active and targets it for degradation. After an abrupt increase in temperature, many other proteins are denatured, and DnaK, with the help of GrpE and DnaJ (see chapter 2), binds to them to help them refold or be degraded. Binding of DnaK to other protein substrates frees σ^H, stabilizing it and making it more active for transcription initiation at heat shock gene promoters (p_{hsp}). When the cell adjusts to the higher temperature and DnaK accumulates to the point where some is again available to bind to σ^H, the activity of σ^H in the cell drops, and the transcription of the heat shock genes returns to basal levels. Modified from Mager WH, de Kruijff AJJ, *Microbiol Rev* **59:**506–531, 1995.

mutant was found in a collection of temperature-sensitive mutants and was shown to be unable to induce HSP production after a shift to high temperature (see Zhou et al., Suggested Reading). This mutant, which failed to make the regulatory gene product, made it possible to clone the regulatory gene by complementation. A library of wild-type *E. coli* DNA was introduced into the mutant strain, and clones that permitted the cells to survive at high temperatures were isolated. When the sequence of the cloned gene for the regulatory HSP was compared with those of genes encoding other proteins, the gene was found to encode a new type of sigma factor that was only 32 kilodaltons (kDa) in mass; therefore, it was named σ^{32}, heat shock sigma factor, or σ^{H}. The RNA polymerase holoenzyme with σ^{H} attached recognizes promoters for the heat shock genes that are different from the promoters recognized by the normal σ^{70} factor and the nitrogen sigma factor, σ^{N} (Figure 12.10). The gene for the heat shock sigma factor was named *rpoH* (for RNA polymerase subunit heat shock).

Regulation of σ^{H} Synthesis

Normally, very few copies of σ^{H} exist in the cell. However, immediately after an increase in temperature from 30 to 42°C, the amount of σ^{H} in the cell increases 15-fold. This increase in concentration leads to a significant rise in the rate of transcription of the heat shock genes, since they are transcribed from σ^{H}-type promoters. Understanding how the heat shock genes are turned on after a heat shock requires an understanding of how this increase in the amount of σ^{H} occurs.

An abrupt increase in temperature might increase the amount of σ^{H} through several mechanisms. A crucial observation is that the amount of *rpoH* mRNA does not increase substantially after heat shock. This indicates that the increase in σ^{H} protein must occur at a posttranscriptional level. In fact, immediately after the temperature upshift, the translation rate of the *rpoH* mRNA increases 10-fold (i.e., there is translational regulation), and the half-life of the σ^{H} protein increases markedly (i.e., there is posttranslational regulation [see chapter 11]).

rpoH mRNA: An RNA Thermosensor

As described in chapter 11, the RNA structure that sequesters the TIR of an mRNA can prevent binding of the 30S ribosomal subunit and therefore can prevent translation initiation. The RNA structure can be stabilized by binding of an RNA-binding protein or small molecule. Another possibility is that the RNA structure is intrinsically stable at low temperature and unstable at high temperature because the hydrogen bonds that hold the structure together have a higher probability of melting at higher temperature; an RNA element of this type is called an RNA thermosensor and allows translation initiation to be sensitive to temperature. This is the case for the *rpoH* mRNA, which is poorly translated during growth at low temperature but is efficiently translated after heat shock (see Nagai et al., Suggested Reading). The *rpoH* mRNA, therefore, preexists in the cell under normal growth conditions (because it is transcribed primarily from promoters recognized by holoenzyme containing σ^{70}) but is inactive, and exposure to heat shock allows the mRNA to be translated. This type of mechanism is advantageous for a stress response, as the increase in σ^{H} protein is very rapid, allowing the cell to quickly induce synthesis of the HSPs. An additional promoter recognized by holoenzyme containing another σ factor, σ^{E} (see below), results in higher levels of *rpoH* mRNA under heat shock conditions, resulting in further induction of the heat shock response.

DnaK: The *E. coli* Cellular Thermometer

In addition to the posttranscriptional regulation of σ^{H} synthesis, there is also posttranslational regulation of σ^{H} levels due to the protein chaperone DnaK. Like *rpoH* mRNA, DnaK senses the change in temperature and promotes the heat shock response. However, unlike *rpoH* mRNA, which directly senses the temperature increase because of its effect on mRNA structure, DnaK senses temperature indirectly by measuring the cellular consequences of the temperature increase. It can play this role because it normally binds to nascent proteins in the process of being synthesized and helps them to fold properly. Under heat shock conditions, the DnaK chaperone can also bind to denatured proteins and help them refold.

Figure 12.15 shows a model for how the ability of the DnaK chaperone to bind unfolded proteins indirectly regulates the synthesis of the heat shock proteins. One of the proteins to which DnaK binds is σ^{H}. By binding to σ^{H}, the DnaK protein regulates the transcription of the heat shock genes in two ways. First, the binding of DnaK inhibits the activity of σ^{H} so that the σ^{H}-DnaK complex is less active in transcription, which lowers the transcription of the heat shock genes. Second, it affects the stability of σ^{H} because the σ^{H} protein with DnaK bound is more susceptible to a cellular protease called FtsH than is free σ^{H} (see Meyer and Baker, Suggested Reading). The σ^{H} bound to DnaK is rapidly degraded, so that any σ^{H} that is produced prior to heat shock is unable to direct the transcription of the heat shock genes.

How, then, does a sudden increase in temperature result in increased σ^{H}-dependent transcription? The answer lies in the chaperone role of DnaK. DnaK binds to denatured proteins to help them to refold properly. After heat shock, many denatured proteins appear in the cell, and most of the DnaK protein binds to these unfolded proteins. This leaves less DnaK available to bind to σ^H. The σ^H protein is therefore more stable and accumulates in the cell. It is also more active, increasing the transcription of the heat shock genes, including the *dnaK* gene itself. Synthesis of the σ^H protein also rises, due to the activation of translation of the *rpoH* transcript, resulting in rapid upregulation of the heat shock regulon.

Returning to Normal: Turning Off the Response

This model explains how the concentration of the HSPs increases sharply after an increase in temperature. Another key aspect of the heat shock response is that after the cell adjusts to the higher temperature, the synthesis of HSPs slowly declines. When enough DnaK has accumulated to bind to all the unfolded proteins and internal conditions have adjusted so that the proteins are more stable at the higher temperature, extra DnaK once again becomes available to bind to and inactivate σ^H, leading to the observed drop in the rate of synthesis of the HSPs. Furthermore, if the temperature returns to normal, the *rpoH* mRNA again folds into the repressed state, and additional synthesis of σ^H is inhibited. This reduction in HSP synthesis is crucial to cell survival, as uncontrolled synthesis of HSPs is toxic to the cell because of excessive proteolysis.

Another alternative sigma factor, σ^E, is responsible for transcription of some heat shock genes at high temperature, including the *rpoH* gene encoding σ^H. Because σ^E is activated by damage to the outer membrane of the cell by heat and other agents, it is discussed below in connection with extracytoplasmic stress responses.

HEAT SHOCK REGULATION IN OTHER BACTERIA

Once heat shock regulation in *E. coli* was fairly well understood, it was of interest to see whether other bacteria use the same mechanism. Surprisingly, many other bacteria, including *B. subtilis*, in which it has been studied in the greatest detail outside of *E. coli*, use a very different mechanism. Rather than using a heat shock sigma factor analogous to σ^H, *B. subtilis* and many other types of bacteria use the normal σ and a repressor protein named HrcA to repress transcription from heat shock genes during growth at lower temperatures. The HrcA repressor binds to an operator sequence called CIRCE, which is highly conserved among bacteria as diverse as *B. subtilis* and cyanobacteria, suggesting that this type of regulation may be very ancient. Bacteria that use HrcA do use a chaperone as a cellular thermometer, but rather than using DnaK, they use the chaperonin GroEL (see chapter 2). GroEL may be required to fold HrcA; when the temperature increases abruptly and GroEL is recruited to help other proteins fold, HrcA may remain misfolded. This would then cause derepression of the heat shock genes HrcA would otherwise repress. A return to normal temperature results in a drop in the level of other denatured proteins; this releases GroEL to return to folding of HrcA, so that the heat shock genes can again be repressed.

General Stress Response in Enteric Bacteria

In addition to the heat shock sigma factor σ^H, which responds to an abrupt increase in temperature, *E. coli* has another sigma factor, called the stationary-phase sigma factor (σ^S), which is used to transcribe genes that are involved in the general stress response (Box 12.4). The gene for this sigma factor, *rpoS*, is turned on following many different types of stress, including nutritional deprivation, oxidative damage, and acidic conditions, and σ^S is active during stationary phase (as suggested by its name). σ^S is closely related to the normal vegetative sigma factor, σ^{70}, which transcribes most genes in *E. coli* and recognizes very similar promoters. The only difference is that the −10 sequence recognized by σ^S may be somewhat extended, and in fact, some promoters may be recognized by both σ^S and σ^{70}.

More than 10% of the total number of genes in *E. coli* are affected, either positively or negatively, by the absence of σ^S (see Landini et al., Suggested Reading). Of these, some are affected under all the conditions tested in early stationary phase, while others are transcribed under only some conditions, such as low pH or high osmolarity in the medium. Many of these are regulatory genes that are activated only under a certain set of conditions. In addition to genes whose products are obviously involved in stress responses, many genes involved in central energy metabolism, such as glycolysis, are also affected. The products of some of these genes probably play roles in switching the cell from aerobic metabolism to anaerobic metabolism. Others are transporters that may help to remove toxic compounds from the cell or to scavenge for rare nutrients. The picture thus arises of σ^S as the master regulator at the top of a large regulatory pyramid, turning on expression of a number of genes for more specialized activators that then respond to more individualized stress conditions.

With so much of the fate of *E. coli* in its hands, σ^S must be able to respond quickly to a number of different environmental signals. Transcriptional regulation is efficient but rather slow. While different stress conditions do affect *rpoS* transcription, the levels of *rpoS* mRNA remain high throughout the exponential phase, indicating that most of the regulation occurs posttranscriptionally, either in the ability of the *rpoS* mRNA to be translated or in the stability of σ^S itself; we therefore concentrate on these.

One way in which translation of *rpoS* mRNA is regulated is through the action of a small RNA (sRNA) called DsrA, whose synthesis increases at low temperatures. Most sRNAs act as negative regulators by reducing the translation or levels of their target mRNAs (see Box 12.5). In contrast, DsrA stimulates *rpoS* mRNA translation by binding to the 5′ untranslated region of the *rpoS* mRNA with the help of a protein, Hfq (Figure 12.16). The *rpoS* mRNA normally contains a structural element that sequesters the translation initiation region (similar to the case for the *rpoH* mRNA described above). Binding of DsrA to a region that would otherwise pair with the translation initiation region opens up the mRNA secondary structure and exposes the region for translation initiation. Another sRNA, RprA, can substitute for DsrA under some conditions, suggesting redundancy in the regulation.

DsrA also acts as a regulator of other genes, including the *hns* gene, which encodes the H-NS global regulatory protein. H-NS plays a major role in repressing the expression of a variety of genes involved in growth and metabolism in *E. coli*, often by generating DNA loops that prevent transcription of its target genes. DsrA represses translation of the *hns* mRNA under the same conditions under which *rpoS* translation is enhanced, further amplifying the shift in gene expression under stress conditions. It is interesting to note that repression of *hns* mRNA translation utilizes a different region of the DsrA sRNA than that used to bind to and activate translation of the *rpoS* mRNA, so that DsrA acts as a compound sRNA with two different domains, each of which has different mRNA targets (Figure 12.16).

Levels of σ^S are also regulated through effects on the stability of the protein after it is made. While cells are growing exponentially, σ^S is being continuously made, but very little of it accumulates, since it has a half-life of only 1 to 2 minutes because it is rapidly degraded by the ClpXP protease (see chapters 2 and 11). However, when the cells run out of energy or are subjected to some other stress, the half-life of σ^S increases, leading to its rapid accumulation. The ClpXP protease consists of two proteins, a barrel-shaped chaperone made up of six copies of the ClpX protein that unfolds proteins, and the ClpP protease, which then degrades the unfolded protein (see chapter 2). However, ClpXP can degrade σ^S only if σ^S is bound to another protein, RssB. The RssB protein is therefore an adaptor protein that targets σ^S for degradation during normal growth. Exposure of cells to stress results in induction of the synthesis of the Ira antiadaptor proteins, which inactivate RssB and therefore allow σ^S to accumulate (see chapter 11). This is an example of regulated proteolysis, and in this case, even small changes

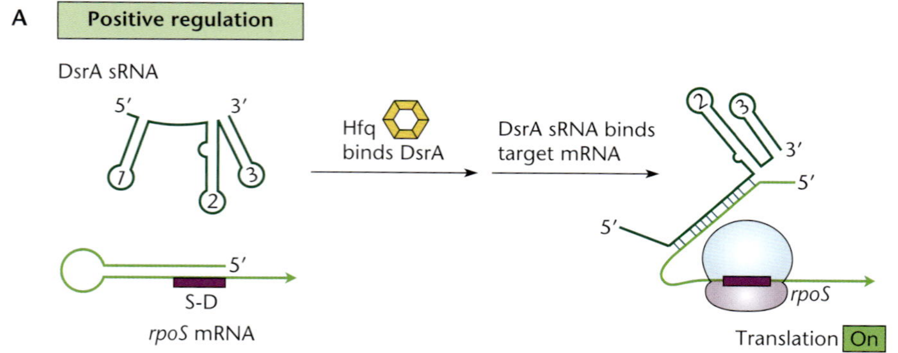

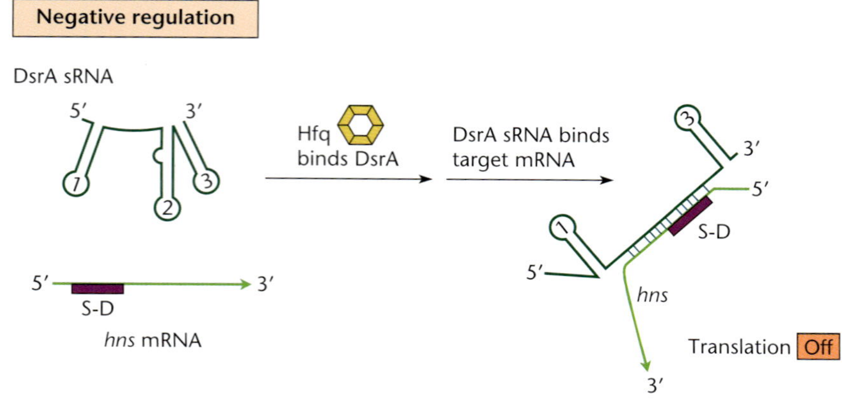

Figure 12.16 Repression and activation by the DsrA sRNA. **(A)** Domain 1 of the DsrA sRNA can bind to the 5′ region of the *rpoS* mRNA. This mRNA normally contains a structure that sequesters the Shine-Dalgarno (S-D) sequence and prevents *rpoS* translation. Binding of DsrA to the region of the RNA that would otherwise pair with the S-D sequence releases the S-D sequence and allows synthesis of σ^S. Binding requires the assistance of the Hfq protein. **(B)** Domain 2 of DsrA binds to the *hns* mRNA (also with the help of Hfq) to repress translation by direct sequestration of the S-D sequence.

BOX 12.5

Regulatory RNAs

It is increasingly obvious that small regulatory RNAs (sRNAs) play a significant role in regulation in bacterial systems (see Storz et al. and Kavita et al., References). This mode of regulation is sometimes called "riboregulation" and can occur at many different levels. The majority of regulatory RNAs that have been identified function by base pairing with the RNA they regulate. This type of sRNA can be encoded on the strand of the DNA opposite that from which the target RNA is derived; these sRNAs are therefore *cis* encoded, because they come from the same region of the DNA, and their sequences are perfect matches to those of their RNA targets. sRNAs of this type can inhibit translation if they are complementary to the TIR of the target mRNA. Other *cis*-encoded sRNAs regulate plasmid replication, as discussed in chapter 4.

Most sRNAs are not encoded in the same DNA region as their targets. These sRNAs are referred to as *trans* encoded, and they usually are not perfectly complementary to their targets. In addition, they often have multiple target genes. Some of these sRNAs are mentioned in the text; they include the DsrA, MicF, and RyhB RNAs, which regulate *rpoS*, *ompF*, and the genes for iron-containing proteins, respectively. Their synthesis is usually highly regulated, and they then affect the expression of the genes they control. By binding to their RNA targets, they can regulate gene expression in many ways. For example, they can inhibit translation by binding close to the TIR of an mRNA and blocking access by the ribosome to the TIR (e.g., MicF repression of *ompF*), or they can stimulate translation by binding close to the TIR and melting a secondary structure that includes the TIR (DsrA activation of *rpoS*). They can also target the mRNA for degradation by a cellular RNase (RyhB). For these *trans*-encoded sRNAs, the regions of complementarity are short and interrupted by mismatches. Because of such short interactions, a single antisense RNA may be able to regulate more than one target gene, with different regions of the sRNA base pairing with the various target sequences (e.g., DsrA regulation of *rpoS* and *hns*).

Many *trans*-encoded sRNAs require a protein, Hfq, to bind to their target RNAs (see Updegrove et al., References). This protein was first found as a host-encoded protein in the phage Qβ replicase that is required to replicate the genomic RNA of this small phage (see chapter 7)—hence its name, Hfq (for host factor Qβ). This protein is found in many bacteria and is homologous to the Sm RNA-binding proteins involved in RNA splicing in eukaryotes. Six polypeptide products of the *hfq* gene form a ring (a hexameric ring). The protein helps the sRNA bind to a specific region of the target RNA, even though there is very little complementary base pairing to hold them together. Hfq may also act as an RNA chaperone to hold the sRNA and mRNA in the proper conformation to allow them to interact with each other.

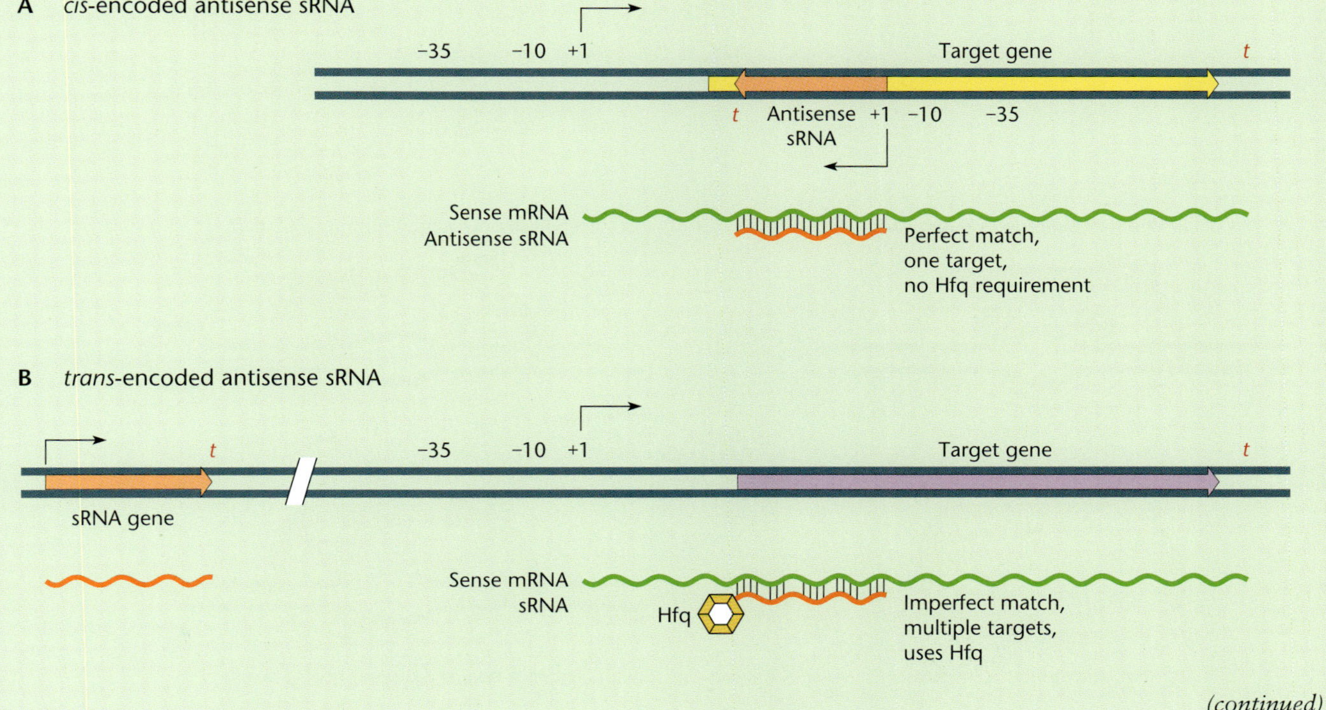

(continued)

BOX 12.5 (continued)

Regulatory RNAs

A very different class of sRNAs do not base pair with their regulatory targets. Instead, these sRNAs bind to proteins and regulate their activity. One example is provided by the CsrB family of sRNAs, which exhibit a repeated structure with multiple small helical domains. These domains mimic the binding site of an RNA-binding repressor protein called CsrA, which was first identified because of its role in the production of glycogen, which is made by *E. coli* as a carbon storage compound (see Romeo et al., References). The genes that CsrA controls are normally repressed by binding of CsrA to the target mRNA, which stabilizes a helical domain that sequesters the TIR of the mRNA. When the CsrB sRNAs are produced, they bind multiple copies of CsrA protein, titrating the protein so that the target mRNAs can be translated. Similar systems have now been uncovered in many bacteria (including *V. cholerae* [see the text]), where they control a variety of systems, including virulence and biofilm formation.

Another unusual sRNA is the 6S RNA, which binds to RNA polymerase containing σ^{70} in stationary phase and inhibits its activity. As described in the text, 6S RNA is a structural mimic of open-complex DNA, and inhibition of RNA polymerase containing σ^{70} enhances the transition to transcription by RNA polymerase containing σ^{S} during stationary phase (see Cavanagh and Wassarman, References).

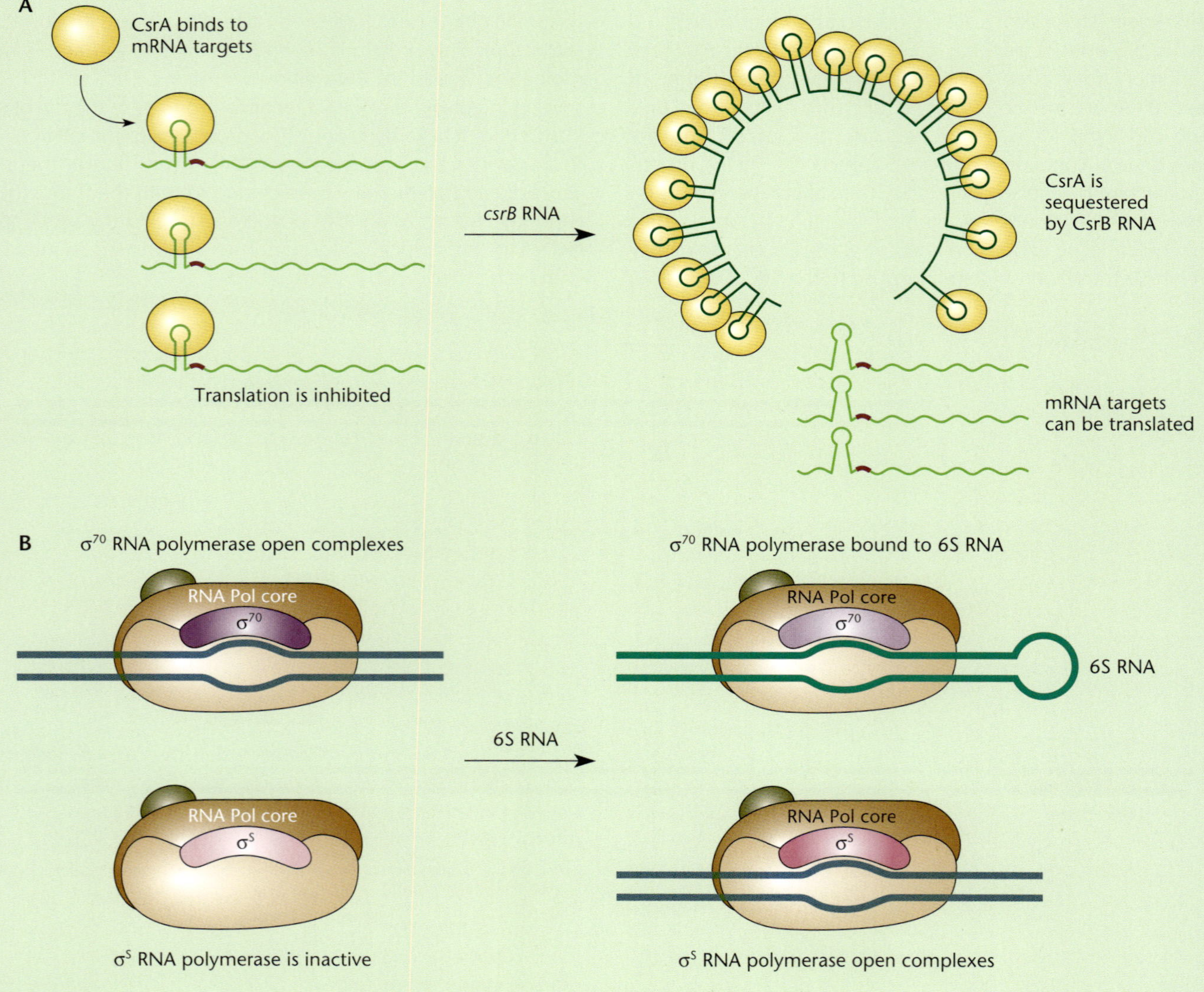

The first sRNAs were found by accident, for example, because they inhibited the synthesis of a gene product when they were overproduced from a multicopy plasmid. They rarely are found in classical genetic analyses, perhaps because these genes are small and therefore are small targets for mutagenesis or because they often have redundant functions, so that inactivating mutations have no obvious phenotypes. Now, however, new sRNAs can be found by analysis of the genomic sequences of bacteria and by direct sequencing of RNAs isolated from cells (see Barquist and Vogel, References). With such methods, over 100 sRNAs have been found in *E. coli*. The sRNAs are often encoded in intergenic regions (the regions between genes) and can sometimes be recognized because they have consensus promoters and transcription terminators but lack an obvious open reading frame encoding a protein between them. Alternatively, they might be recognized as conserved sequences in intergenic regions by a comparison of the genomes of closely related bacteria or by special types of genomics analyses, including tiling microarrays and high-throughput sequencing of cDNAs generated from total RNA isolated from the cell. Another approach uses the Hfq protein to pull sRNAs out of the RNA pool from cells, since many of them bind to Hfq. Any approach that relies on isolation of the RNA pool from the cell is limited by the fact that only sRNAs that are expressed under the conditions under which the cells were grown will be present in the pool. These approaches are currently being applied to a growing number of bacterial genomes, and the interesting task is to identify the targets of these sRNAs and their regulatory roles.

References

Barquist L, Vogel J. 2015. Accelerating discovery and functional analysis of small RNAs with new technologies. *Annu Rev Genet* **49:**367–394.

Cavanagh AT, Wassarman KM. 2014. 6S RNA, a global regulator of transcription in *Escherichia coli*, *Bacillus subtilis*, and beyond. *Annu Rev Microbiol* **68:**45–60.

Kavita K, de Mets F, Gottesman S. 2018. New aspects of RNA-based regulation by Hfq and its partner sRNAs. *Curr Opin Microbiol* **42:**53–61.

Romeo T, Vakulskas CA, Babitzke P. 2013. Post-transcriptional regulation on a global scale: form and function of Csr/Rsm systems. *Environ Microbiol* **15:**313–324.

Storz G, Vogel J, Wassarman KM. 2011. Regulation by small RNAs in bacteria: expanding frontiers. *Mol Cell* **43:**880–891.

Updegrove TB, Zhang A, Storz G. 2016. Hfq: the flexible RNA matchmaker. *Curr Opin Microbiol* **30:**133–138.

in σ^S levels can have a dramatic effect on transcription because σ^S will redirect RNA polymerase to recognize new sets of promoters.

Yet another posttranslational level of regulation of σ^S-dependent transcription is exerted by a different type of sRNA that, unlike DsrA, does not act by base pairing with its target (Box 12.5). The 6S RNA is a structural mimic of promoter DNA recognized by RNA polymerase containing σ^{70}, but it does not bind to RNA polymerase containing σ^S. 6S RNA accumulates as cells enter stationary phase and binds to and inactivates the normal RNA polymerase containing σ^{70} (see Cavanagh and Wassarman, Suggested Reading). This then allows transcription by RNA polymerase containing σ^S to proceed more efficiently, as scarce resources are not being used in transcription of normal promoters. The combination of all of these effects results in an efficient reprogramming of transcription under stressful conditions, which enables the cell to respond most effectively to these conditions.

General Stress Response in *Firmicutes*

Many enteric bacteria are known to have a general stress response based on σ^S and probably similar to that of *E. coli*. However, *Firmicutes*, including *B. subtilis*, use another sigma factor, σ^B, to transcribe stress-induced genes. This sigma factor is not a homolog of σ^S in sequence and is activated using a very different mechanism. Its mode of activation has features in common with the activation of σ^E in *E. coli* and σ^F and σ^G in *B. subtilis* sporulation (see below) in that it depends on inactivation of an **anti-sigma factor**, a partner protein that otherwise binds to the sigma factor and prevents it from interacting with RNA polymerase to direct transcription. The signaling pathway that activates σ^B is fairly long and complicated, using a complex phosphorelay system involving serine-threonine kinases and phosphatases that is more reminiscent of eukaryotes than it is of bacteria. Also, there seems to be one pathway to sense energy deficiency when the cells run out of a carbon and energy source and a different pathway to sense an environmental stress, such as heat shock or a pH change.

One interesting aspect of the system is the question of how environmental signals are communicated to the first step of the pathway (see Kim et al., Suggested Reading). A total of five periplasmic proteins called the Rsb proteins are involved in the induction of the stress response as part of a large complex (the stressosome) that senses external stresses and induces the stress response by activating the signal transduction system that activates σ^B. All five Rsb proteins have similar carboxyl termini, but one of them, RsbS, is shorter, consisting of only the shared carboxyl terminus. These proteins are all serine or threonine kinases, and phosphate groups are added to or removed from them in response to external stresses. Each of the longer Rsb proteins (RsbRA, RsbRB, RsbRC, and RsbRD) responds to a different but overlapping external stress, and they all phosphorylate RsbS in response to their particular stress. Phosphorylation of RsbS triggers a cascade of events that involve phosphorylation or dephosphorylation of a series of regulatory proteins. The final outcome of this regulatory cascade is the release of the σ^B sigma factor from an anti-sigma factor that holds it in an inactive state under normal growth conditions. RNA polymerase containing σ^B then directs the transcription of genes whose products help the cell to recover from all of the various stress conditions to which the Rsb proteins respond.

Extracytoplasmic (Envelope) Stress Responses

The membranes of bacteria are the first line of defense against external stresses. They are also particularly sensitive to abrupt changes in osmolarity and damaging agents, such as hydrophobic toxins, heat shock, and pH changes. Not surprisingly, many stress responses are dedicated to preserving the integrity of the bacterial membranes. Responses of this type are often referred to as extracytoplasmic stress responses because they respond to changes outside the cytoplasm (see Grabowicz and Silhavy, Suggested Reading).

REGULATION OF PORIN SYNTHESIS

One of the challenges often faced by bacteria is a change in osmolarity due to changing solute concentrations outside the cell. The osmotic pressure is normally higher inside the cell than outside it. This pressure would cause water to enter the cell and the bacterium to swell, but the rigid cell wall can help mitigate this by keeping the cell from expanding. However, even the cell wall is not invincible, and bacteria must keep the difference in osmotic pressure inside and outside the cell from becoming too great. The ability to monitor osmolarity can be important to bacteria for a second reason: bacteria also sometimes sense changes in their external environment by detecting changes in osmolarity. In fact, one way in which pathogenic bacteria sense that they are inside a host, and induce their virulence genes, is by the much higher osmolarity inside the host (see "Regulation of Virulence Genes in Pathogenic Bacteria" below). The systems by which bacterial cells sense these changes in osmolarity and adapt are global regulatory mechanisms, and many genes are involved.

Much is known about how bacteria respond to media with different osmolarities. One way they regulate the differences in osmotic pressure across the membrane is by excreting or accumulating K^+ ions and other solutes, such as proline and glycine betaine. Enteric bacteria, such as *E. coli*, have the additional problem of maintaining osmotic pressure in the periplasm, as well as in the cytoplasm. They achieve this in part by synthesizing oligosaccharides in the periplasmic space to balance solutes in the external environment.

One of the major mechanisms by which enteric bacteria balance osmotic pressure across the outer membrane is by synthesizing pores to let solutes into and out of the periplasmic space. These pores are composed of outer membrane proteins called **porins**. To form pores, three of the polypeptide products of these genes come together (trimerize) in the outer membrane to form what are called β barrels with central channels that selectively allow hydrophilic molecules through the very hydrophobic outer membrane.

The two major porin proteins in *E. coli* are OmpC and OmpF (for outer membrane protein). Pores composed of OmpC are smaller than those composed of OmpF, and the sizes of the pores can determine which solutes can pass through the pores and thus confer protection under some conditions. For example, the smaller pores, composed of OmpC, may prevent the passage of some toxins, such as the bile salts in the intestine. The larger pores, composed of OmpF, may allow more rapid passage of solutes and so confer an advantage in dilute aqueous environments. Accordingly, *E. coli* cells growing in a medium of high osmolarity, such as the human intestine, have more OmpC than OmpF, whereas *E. coli* cells growing in a medium of low osmolarity, such as dilute aqueous solutions, have less OmpC than OmpF.

Environmental factors other than osmolarity can alter the ratio of OmpC to OmpF. This ratio increases at higher temperatures or pH or when the cell is under oxidative stress due to the accumulation of reactive forms of oxygen. The ratio also increases during growth in the presence of organic solvents, such as ethanol, or some antibiotics and other toxins. Presumably, the smaller size of OmpC pores limits the passage of many toxic chemicals into the cell. Many of these abrupt changes in porin proteins occur when the *E. coli* bacterium leaves the external environment and passes through the stomach into the intestine of a warm-blooded vertebrate host, its normal habitat. It then must synthesize mostly OmpC-containing pores to keep out toxic materials, such as bile salts, as mentioned above. Other conditions cause a decrease in both OmpC and OmpF concentrations. To respond to all of these other changes, the *ompC* and *ompF* genes are in a number of different regulons, which respond to different external stresses. We first discuss one of these pathways in *E. coli*, the regulation of *ompC* and *ompF* expression by EnvZ and OmpR. This system has served as a model for two-component signal transduction systems (Box 12.3) that allow the cell to sense the external environment and adjust its gene expression accordingly; therefore, we discuss this subject in some detail.

GENETIC ANALYSIS OF PORIN REGULATION

As in the genetic analysis of any regulatory system, the first step in studying the osmotic regulation of porin synthesis in *E. coli* was to identify the genes whose products are involved in the regulation. The isolation of mutants defective in the regulation of porin synthesis was greatly aided by the fact that some of the porin proteins also serve as receptors for phages and bacteriocins, so that mutants that lack a particular porin are resistant to a given phage or bacteriocin. This offers an easy selection for mutants defective in porin synthesis, as only mutants that lack a certain porin in the outer membrane are able to form colonies in the presence of the corresponding phage or bacteriocin.

Using such selections, investigators isolated mutants that had reduced amounts of the porin protein OmpF in their outer membranes. These mutants were found to have mutations in *ompF*, which is the structural gene for OmpF, and an operon containing two genes, *envZ* and *ompR*. Mutations in the *ompF* locus can completely block OmpF synthesis, whereas mutations in *envZ* or *ompR* only partially prevent its synthesis and encode proteins that are required for optimal transcription of the *ompF* gene (see Hall and Silhavy, Suggested Reading).

EnvZ and OmpR: A Sensor Kinase and Response Regulator Partnership

The *envZ* and *ompR* genes were cloned and sequenced, and similarities in amino acid sequence between EnvZ and OmpR and other sensor kinase and response regulator pairs, including NtrB and NtrC, suggested that these proteins are also a sensor kinase and response regulator pair of proteins. Like many (but not all) sensor proteins, EnvZ is an inner membrane protein, with its NTD in the periplasm and its CTD in the cytoplasm (Box 12.3). The NTD of EnvZ senses an unknown signal in the periplasm that reflects the osmolarity and transfers this information to the cytoplasmic domain, resulting in autophosphorylation of EnvZ protein when osmolarity is high. EnvZ~P transfers its phosphate to the OmpR protein, which acts as a transcriptional regulator that regulates transcription of the porin genes. Like the NtrB protein, the EnvZ protein is autophosphorylated, and its phosphate is transferred to OmpR. High levels of OmpR~P are required for transcription of *ompC*. However, it was observed that null mutations in *envZ* and *ompR* result in very low transcription of *ompF*, suggesting that some amount of OmpR~P is required for *ompF* transcription.

The Affinity Model for Regulation of *ompC* and *ompF*

The phenotypes of the *envZ* and *ompR* mutants can be explained by a model that takes into account the concept that phosphorylation of OmpR is a reversible event and that the crucial parameter is the concentration of OmpR~P in the cell. Like many sensor kinases, the EnvZ protein is known to have both phosphotransferase and phosphatase activities, allowing it to both donate a phosphoryl group to and remove one from OmpR. Whether the phosphorylated or unphosphorylated form of OmpR predominates depends on whether the phosphotransferase or phosphatase activity of EnvZ is most active. Under conditions of high osmolarity, the phosphotransferase activity predominates and the levels of OmpR~P are high. Under conditions of low osmolarity, the phosphatase activity predominates, and most of the OmpR is unphosphorylated.

To explain how *envZ* null mutations can prevent the optimal transcription of both genes, we must propose that OmpR~P is required to activate transcription of both the *ompC* and *ompF* genes. However, how could higher levels of OmpR~P (under high-osmolarity conditions) favor the transcription of *ompC* while lower levels of OmpR~P (under low-osmolarity conditions) favor the transcription of *ompF* even though both require OmpR to be phosphorylated? One model is based on the existence of multiple binding sites for OmpR~P upstream of the promoters for the *ompF* and *ompC* genes (see Yoshida et al., Suggested Reading). Some of these sites bind OmpR~P more tightly than others; in other words, some are high-affinity sites, whereas others are low-affinity sites. By binding to these sites, the OmpR~P protein can either activate or repress transcription from the promoters, depending on the position of the binding site relative to the promoter (see chapters 2 and 11). The presence of low-affinity sites at an activating position upstream of the *ompC* promoter would result in transcription only when OmpR~P levels are high, which is what was observed. The dependence of *ompF* transcription on some amount of OmpR~P further suggests that the *ompF* gene should contain high-affinity sites at an activating position so that activation of the promoter occurs when levels of OmpR~P are low. However, the observation that *ompF* is not transcribed when levels of OmpR~P are high suggests the additional presence of low-affinity binding sites at a repressive position in the *ompF* promoter that would result in inhibition of transcription when OmpR~P levels are high, which fits the observed pattern of expression. Further studies have indicated that regulation of this system is even more complicated, as we will see below, but this model illustrates how a combination of binding sites with different affinities for a regulatory protein can be combined with differential positioning of the binding sites relative to the promoter to give different regulatory outcomes.

REGULATION OF *ompF* BY THE MicF sRNA

As mentioned above, the OmpC/OmpF ratio increases not only when the osmolarity increases, but also when the temperature or pH increases or when toxic chemicals, including active forms of oxygen, nitric oxide, or organic solvents are in the medium. In general, under conditions where levels of nutrients and toxins are high, such as in the vertebrate intestine, OmpC levels are high. Since OmpC forms narrower channels, fewer toxins can get in. Fewer nutrients can get in, as well, but since their concentration in the intestine is high, this is not a problem. If nutrient levels are low, such as in water outside the vertebrate host, OmpF levels are higher, because allowing the available nutrients to get into the cell becomes more important than keeping toxins out. The rationale for regulating the porins by temperature or pH is less obvious. One possibility is that the bacterium uses temperature and pH as signals to indicate that the water in which it lives has just been drunk by a vertebrate and the bacterium is about to pass into the vertebrate intestine. The porins would then need to be regulated to combat the onslaught of toxins that will be faced by the bacterium, including oxidative bursts by macrophages and bile salts.

Much of the regulation of the OmpF porin in response to these other forms of stress is through the MicF sRNA. The mechanisms of action of regulatory sRNAs, including DsrA, the sRNA that regulates the translation of σ^S in *E. coli* (see above), are discussed in Box 12.5. The MicF RNA was one of the first sRNAs to be discovered, and it was found by chance because its gene is adjacent to the *ompC* gene but is transcribed in the opposite direction (i.e., *ompC* and *micF* are divergently transcribed). When the region of the chromosome containing the *ompC* gene was cloned into a high-copy-number plasmid and introduced into *E. coli* cells, the synthesis of OmpF was inhibited. At first, it was assumed that the OmpC protein was somehow inhibiting the synthesis of the OmpF protein. However, it was observed that the OmpC coding region was not required for inhibition, which instead required an sRNA encoded just upstream of the *ompC* gene. The sRNA was named MicF for (multicopy inhibitor of OmpF). Part of MicF is complementary to the 5′ region of the mRNA for *ompF*, including the *ompF* TIR, and pairing of MicF with the *ompF* mRNA inhibits its translation. High concentrations of MicF sRNA therefore result in lower levels of OmpF synthesis. As is the case for many sRNAs that are only partially complementary to their mRNA targets, repression requires the Hfq RNA chaperone (Box 12.5).

The cellular levels of MicF sRNA increase under certain conditions because the promoter for the *micF* gene contains binding sites for many transcriptional activators. The activators seem to work independently, and each activates transcription from the *micF* promoter under its own particular set of conditions. For example, the transcriptional activator SoxS activates transcription from the *micF* promoter when the cell is under oxidative stress. Another activator, MarA, activates the transcription of *micF* when weak acids or some antibiotics are present. A third activator, Rob, induces transcription of *micF* in the presence of cationic peptide antibiotics. OmpR~P, which binds upstream of the *ompC* promoter, can also activate *micF* transcription; the dependence of this activation on high levels of OmpR~P (conditions under which *ompF* transcription is repressed) allows MicF to rapidly block translation of any *ompF* mRNA that already is present in the cell, resulting in faster shutoff of OmpF synthesis.

The MicF sRNA inhibits the translation of the mRNA for OmpF, but not OmpC. Translation of the mRNA for OmpC is inhibited by its own sRNA, named MicC by analogy to the MicF sRNA. The MicC sRNA was discovered because it has some sequence complementarity to the 5′ region of *ompC* mRNA, and overexpression of the *micC* gene was shown to inhibit OmpC synthesis. Interestingly, the *micC* gene is also adjacent to a gene for a porin, in this case OmpN, which is very poorly expressed in *E. coli* under laboratory conditions, but might be expected to form pores with sizes similar to those of OmpF. The regulation of porin synthesis by osmolarity and other environmental factors is obviously central to cell survival, which is why it is so complicated and involves so many interacting systems and regulatory molecules.

REGULATION OF THE ENVELOPE STRESS RESPONSE BY THE CpxA-CpxR TWO-COMPONENT SYSTEM

Another way that *E. coli* senses stress to the outer membrane is via the two-component system CpxA-CpxR (Figure 12.17). This two-component system works like many two-component systems (Box 12.3) in that CpxA is a sensor kinase that phosphorylates itself in response to a signal and then transfers the phosphate to the response regulator CpxR, a transcriptional activator that activates the transcription of more than 100 genes under its control (see Raivio, Suggested Reading). The CpxA sensor kinase phosphorylates itself when it senses that proteins to be secreted, such as pili or curli fibers that play a role in attaching the bacterium to surfaces such as eukaryotic cells, are piling up in the periplasm as a result of some defect in transport due to damage to the outer membrane (see Ruiz and Silhavy, Suggested Reading). The accumulation of these proteins in the periplasm is

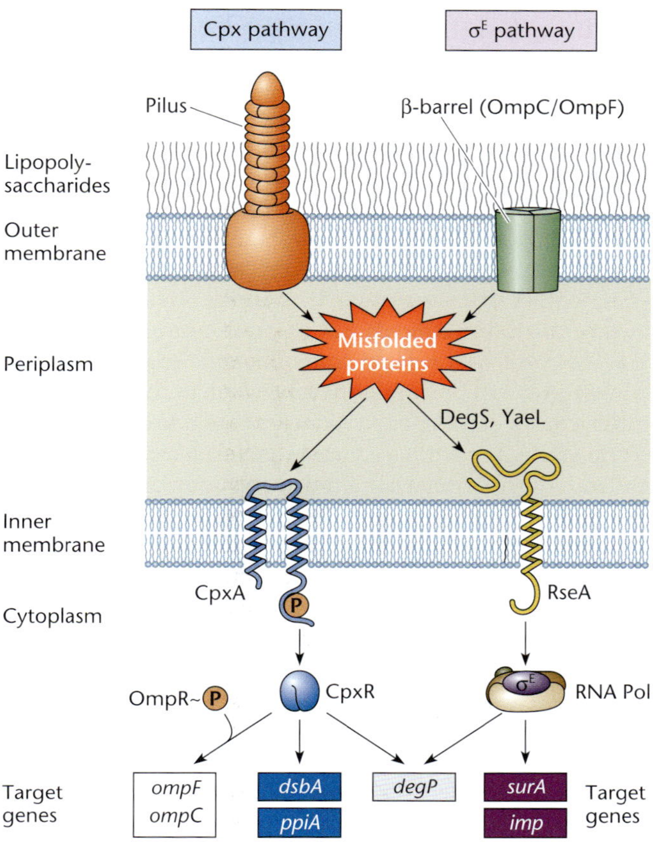

Figure 12.17 Two envelope stress responses in *E. coli* respond to different stress signals in the periplasm. The Cpx pathway responds to the accumulation of pilin subunits in the periplasm, while the σ^E pathway responds to the accumulation of OmpA. DsbA is a periplasmic chaperone and oxidoreductase that makes disulfide bonds in exported proteins (see chapter 2), PpiA is a peptidyl-prolyl *cis-trans* isomerase, DegP is a protease and chaperone, SurA is a chaperone, and Imp is an usher protein for outer membrane proteins. Modified from Digiuseppe PA, Silhavy TJ, *ASM News* **70:**71–79, 2004.

toxic, and many of the genes regulated by CpxR encode proteases and chaperones in the periplasm that help fold and degrade these proteins. The synthesis of the proteins that make up pili and curli fibers is also inhibited by phosphorylated CpxR, presumably so that their intermediates will not accumulate in the periplasm.

THE EXTRACYTOPLASMIC SIGMA FACTOR σ^E

In *E. coli*, extracytoplasmic stress is also sensed by another system that uses a specialized sigma factor to transcribe stress response genes (see Sineva et al., Suggested Reading). This alternative sigma factor is called the extracytoplasmic function (ECF) sigma factor, σ^E, because it is used mostly to express genes that function in the periplasm, such as proteases that degrade defective proteins in the periplasm and chaperones that help fold proteins as they pass through the periplasm. It was first discovered because one of the promoters that transcribe the *rpoH* gene for the heat shock sigma factor at high temperatures is recognized by this sigma factor. Sigma factors in this family are generally referred to as σ^{ECF}, and some bacteria have multiple σ^{ECF} family members.

The activation of σ^E in *E. coli* is similar to the activation of the σ^B general stress response sigma factor in *B. subtilis* in that it involves inactivation of an anti-sigma factor that holds σ^E in an inactive state in the absence of stress conditions. In this case, the activity of the σ^E anti-sigma factor is controlled by proteolytic degradation in response to envelope stress. This aspect of the induction of the envelope stress response is quite well understood and is illustrated in Figure 12.17. The anti-sigma factor RseA is an inner-membrane-spanning protein with domains in both the periplasm and the cytoplasm. The cytoplasmic domain binds σ^E, inactivating it and sequestering it to the membrane. When the outer membrane is damaged, Omp proteins, including OmpC, accumulate in the periplasm because they cannot be assembled into porins in the damaged outer membrane. The carboxyl-terminal domain of the Omp proteins that have accumulated in the periplasm binds to the carboxyl-terminal domain of a protease called DegS in the periplasm. The carboxyl terminus of DegS may normally inhibit its protease activity, and binding of the carboxyl terminus of an Omp protein to DegS activates the proteolytic activity of DegS, which then cleaves off the periplasmic domain of RseA. A second protease, named YaeL, then degrades the transmembrane domain of RseA, which releases σ^E from RseA and allows σ^E to bind RNA polymerase and transcribe genes in the envelope stress response.

There is experimental evidence for each of these steps. Mutations that inactivate the *rpoE* gene for σ^E are lethal at any temperature or in the absence of conditions that trigger envelope stress, showing that the σ^E protein is essential for viability under all conditions. It is not clear why σ^E is essential, even in the absence of envelope stress conditions, but some of the periplasmic proteins under its control, such as protein chaperones in the periplasm that help insert other proteins into the outer membrane, are likely to be essential under all conditions. This suggests that some level of σ^E-dependent transcription occurs under nonstress conditions, although it increases under stress conditions. Mutations that inactivate *rseA* cause constitutive induction of the σ^E response, as expected, since there is no anti-sigma factor to inhibit σ^E even in the absence of stress. Mutations that inactivate DegS and YaeL are lethal, but double mutants with both an *rseA* mutation and either a *degS* or *yaeL* mutation are viable. In genetic terms, *rseA* mutations are suppressors of *degS* and *yaeL* mutations. This shows that the only essential role for DegS and YaeL is to degrade RseA and, by extension, induce the σ^E stress response. The fact that inactivation of RseA is required for viability even under nonstress conditions indicates that some level of σ^E-dependent transcription is essential. The role of the carboxyl terminus of DegS in inhibiting its own protease activity was found when the region of the *degS* gene encoding the carboxyl terminus was deleted, which led to constitutive activation of σ^E. Finally, the role of the carboxyl termini of the Omp proteins in activating the protease activity of DegS was discovered when it was shown that overproducing the carboxyl termini of these proteins was sufficient to induce σ^E but did not increase the induction further if the carboxyl terminus of DegS had been deleted.

As described for other stress responses, it is important that activation of σ^E is reversible. When the stress responsible for activation of DegS has been resolved by both removal of the stress and destruction or repair of misfolded proteins in the periplasm, newly synthesized RseA will not be cleaved by DegS and YaeL, and it will sequester and inactivate σ^E, resulting in turnoff of the transcription of σ^E-dependent genes.

The combination of the detection of environmental signals by the EnvZ-OmpR two-component system, the detection of defects in protein assembly into the outer membrane and/or periplasmic damage caused by toxins by the CpxA-CpxR two-component system, and the detection of accumulation of damaged or misplaced (notably Omp) proteins in the periplasm by the DegS protease, which in turn activates σ^E, allows the cell to detect a variety of effects that require modification of the outer cell surface. The resulting changes in gene expression include modification of the ratios of different porins (e.g., OmpF versus OmpC) that modulate the rate of transport into and out of the cell. Synthesis of chaperones and proteases that act specifically in the periplasm to facilitate outer membrane assembly is also induced (Figure 12.17), which allows the cell to repair the damage and restore outer membrane and periplasmic functions. The complexity and interconnection of these responses reflect the importance of the outer cell surface to cell integrity and function. The presence of multiple σ^{ECF} family members in some organisms further illustrates the importance of this mode of regulation.

Iron Regulation in *E. coli*

Iron is an important nutrient, both for bacteria and for humans (see Chandrangsu et al., Suggested Reading). Many enzymes use iron as a catalyst in their active centers, transcriptional regulators (such as FNR, which regulates genes for anaerobic metabolism), use it as a sensor of oxygen levels, and hemes use it as an oxygen carrier. However, too much iron can also be very damaging to cells and requires a stress response. Iron catalyzes the conversion of hydrogen peroxide and other reactive forms of oxygen to hydroxyl free radicals, the most mutagenic form of oxygen. It is therefore essential to maintain an adequate supply of iron without accumulation of excess levels.

Iron exists in two states in the environment, the ferric (Fe^{3+}) state, with a valence of three, and the ferrous (Fe^{2+}) state, with a valence of two. Iron in the ferric state forms largely insoluble compounds that are not easily used by bacteria and other organisms. Because oxygen quickly converts iron in the ferrous state to iron in the ferric state, most iron in aerobic environments exists in the insoluble ferric state. Accordingly, many bacteria secrete proteins called siderophores that bind ferric ions and transport them into the cell, where they can be converted to ferrous ions in the reducing atmosphere of the cytoplasm. These siderophores are made and secreted only if iron is limiting, to avoid the damaging effects of too much iron in the cell. Iron limitation is especially important for bacterial pathogens, as discussed below.

There are three basic mechanisms in *E. coli* and most other bacteria for regulating genes involved in iron metabolism. The Fur system uses a DNA-binding repressor protein that regulates transcription initiation. One of the genes regulated by Fur encodes an sRNA called RyhB, which represses gene expression primarily by triggering degradation of target mRNAs. Finally, the aconitase enzyme of the tricarboxylic acid (TCA) cycle is an iron-binding protein that doubles as a translational repressor of mRNAs that contain a site to which the enzyme can bind. We discuss the Fur repressor first.

The Fur Regulon

The Fur repressor is a classic repressor with a helix-turn-helix DNA-binding domain in its amino terminus (see Box 11.1), a dimerization domain in its carboxyl terminus, and an effector-binding pocket in the middle. Figure 12.18 illustrates regulation by Fur, which is much like the regulation of the *trp* operon by the TrpR repressor (see chapter 11). When ferrous iron (Fe^{2+}) is in excess inside the cell, it acts as a corepressor by binding to the Fur aporepressor, changing its conformation to the repressor form, which can bind to an operator sequence called a Fur box. By binding to the Fur box, which overlaps the −10 sequence of the promoters it regulates, the Fur protein blocks access of RNA polymerase to the promoters and represses transcription of all of the genes and operons of the Fur regulon. If ferrous ions are in short supply, the repressor is in the aporepressor state and cannot bind to the operator sequences. The transcription of the genes under Fur control, including genes for the siderophores and iron transporters in the membrane (called

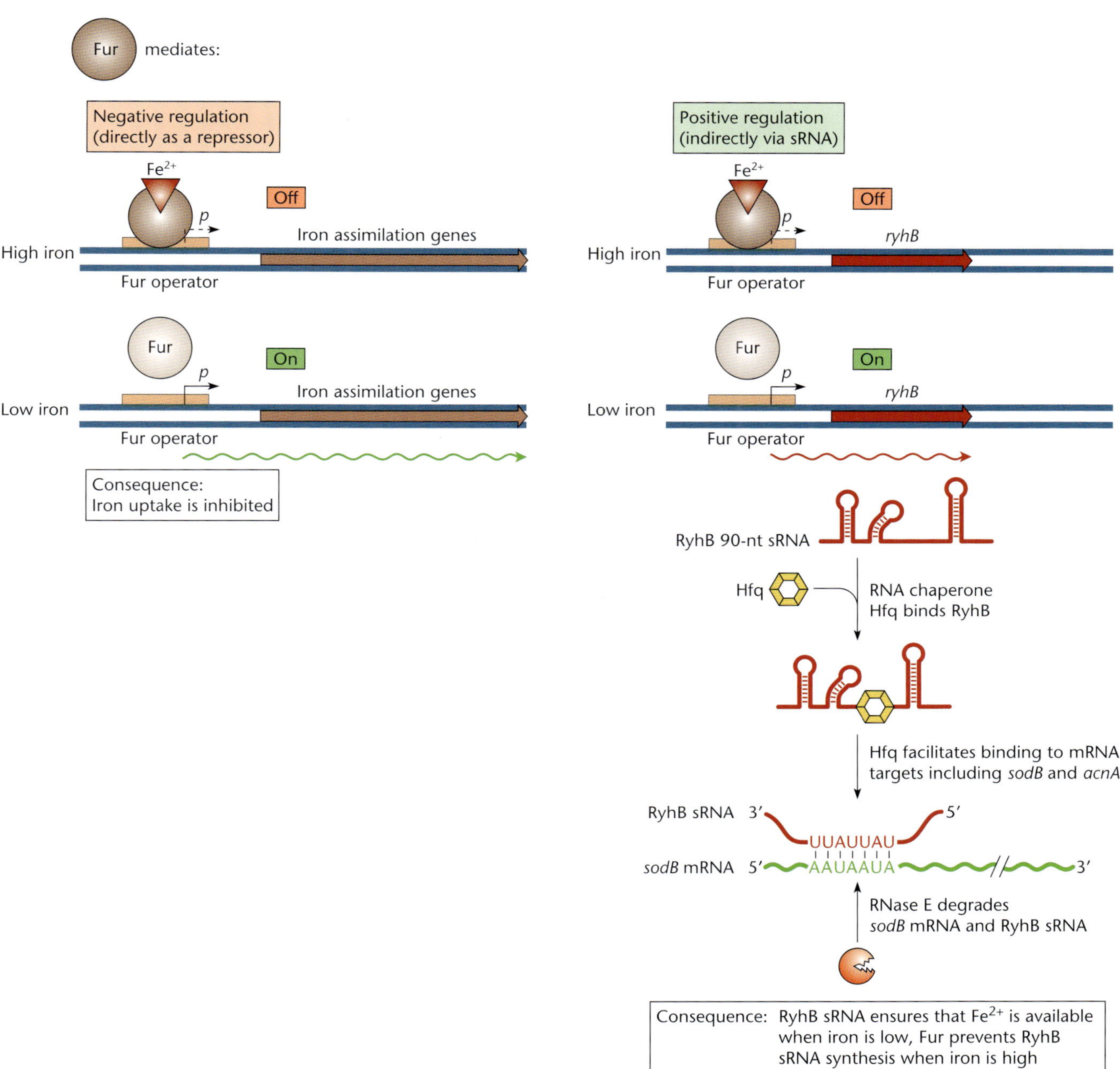

Figure 12.18 Regulation of operons in the Fur regulon. **(Left)** Negative regulation. Fur is a repressor that binds to the Fur operator in the presence of iron, blocking transcription of operons under its control, many of which encode proteins involved in iron uptake. **(Right)** Indirect positive regulation by Fur. Fur in the presence of iron represses synthesis of the RyhB sRNA. RyhB is made when iron is limiting, and it base pairs (with the help of the Hfq protein) to the 5′ ends of SodB mRNA and the mRNAs of other genes that encode proteins that utilize iron as a cofactor. Binding of RyhB creates a double-stranded RNA region that is the target for cleavage by a cellular RNase called RNase E, which degrades both the sRNA and its mRNA target. This prevents expression of the genes when iron levels are low but allows their expression when iron levels are high, when Fur represses RyhB synthesis. Turning off genes for nonessential iron-binding proteins when iron is limiting makes more iron available for essential iron-containing proteins.

iron assimilation genes in Figure 12.18), is therefore high when ferrous iron is low and is repressed when ferrous iron is high. The Fur regulation system is highly conserved among *Proteobacteria*. Some *Firmicutes* have a similar system based on repressors related to the DtxR repressor of the *Actinobacteria* member *Corynebacterium diphtheriae* (see below), which has little sequence similarity to Fur but is structurally similar and probably works in similar ways.

The RyhB sRNA

While many genes are turned off when iron is in excess, other genes are turned on. They include the ferritin-like iron storage proteins and many other proteins that contain iron, including aconitase A (AcnA), an iron-containing enzyme of the TCA cycle, and an iron-containing superoxide dismutase (SodB) that destroys peroxides in the cell before they can be converted into hydroxyl radicals by the iron and damage DNA and other cellular constituents. In the presence of excess iron, the concentration of these proteins increases rather than decreases. However, very little of these proteins is made in a Fur$^-$ mutant, and their levels do not increase in the presence of iron, suggesting that Fur activates their transcription. Comparisons of the levels of mRNA for all genes of *E. coli* in the presence and absence of iron revealed that the transcription of 53 genes increases when iron is in excess, while that of 48 genes decreases. Accordingly, it was proposed that Fur could act either as a repressor or as an activator, like many other transcriptional regulators, repressing some genes in the presence of iron and activating others. However, unlike some regulatory proteins that can act directly as either a repressor or activator depending on where they bind to the DNA relative to the site of RNA polymerase binding (see chapter 11), activation by Fur does not occur by a direct effect on the target genes but instead occurs indirectly via an sRNA called RyhB. This sRNA is expressed at high levels when iron levels are low and inhibits the expression of iron-responsive genes under its control. In the presence of excess iron, Fur represses the synthesis of RyhB, and RyhB repression of the genes it regulates is relieved, causing an increase in expression of these iron-responsive genes. This helps the cell by ensuring that enzymes that are required only when iron levels are high are synthesized only under those conditions.

The way in which the RyhB sRNA regulates the genes under its control is shown in Figure 12.18. As is the case for many sRNAs, the RyhB sRNA sequence is partially complementary to sequences close to the 5′ end of the mRNA of genes under its control. This complementarity allows RyhB to pair with the mRNA, with the help of the Hfq protein, in a manner similar to that described for DsrA and MicF (Box 12.5 and Figure 12.16). However, rather than blocking translation of the mRNAs, like most other sRNAs discussed so far, RyhB binding creates a partially double-stranded RNA structure that is a substrate of the RNase E enzyme, one of the major RNases in *E. coli* (see chapter 2). RNase E cleaves both the mRNA and the sRNA, resulting in degradation of both RNAs (Figure 12.18). Prevention of the synthesis of many iron-containing proteins by the RyhB sRNA when iron is limiting reserves the available iron for the most essential iron-containing enzymes in *E. coli*, including ribonucleotide reductase, which is required to make deoxynucleotides and hence DNA. When iron levels are high, the inhibition of RyhB synthesis by Fur allows these iron-containing enzymes to be produced.

The Aconitase Translational Repressor

The more we learn about cells, the more we discover about how many functions are coordinated. These discoveries are often serendipitous, as was the case when aconitase was discovered not only to function in metabolism, but also to play a role in iron regulation in bacteria and eukaryotes. This dual role was first discovered in eukaryotes during studies of proteins called iron-responsive proteins (IRPs). When iron is limiting, IRPs bind to the mRNAs for proteins involved in iron metabolism. They either can inhibit the translation of an mRNA by binding to the 5′ end of the transcript and preventing access of the ribosome to the initiator codon or they can increase expression by binding to the 3′ end of the mRNA and stabilizing the mRNA. In general, IRPs inhibit the translation of proteins such as ferritins that are needed only when iron is high and stimulate the synthesis of proteins such as iron transport proteins or transferrins that are needed when iron is limiting. The sequences to which they bind are highly conserved in evolution and are called iron-responsive elements.

It came as a complete surprise when IRPs were purified, partially sequenced, and discovered to be aconitases. Aconitases are enzymes that function in the TCA cycle to convert citrate into isocitrate. The TCA cycle produces essential carbon-containing compounds (including α-ketoglutarate) used in many biochemical reactions and generates reducing power in the form of reduced NADH to feed electrons into the electron transport system to make ATP. Aconitases contain iron in the form of an iron-sulfur cluster, often written $[4Fe\text{-}4S]^{2+}$. Many iron-containing enzymes have the iron in this prosthetic group, and it usually makes them sensitive to oxygen.

Bacteria and mitochondria have aconitases related to the cytoplasmic aconitases of eukaryotes. *E. coli* has two aconitases, which differ in their regulation and their sensitivity to oxygen. Aconitase A (AcnA) is induced following stress and in the stationary phase and is an iron-containing protein whose expression is repressed by RyhB (and therefore is induced when iron levels are high). The other aconitase, aconitase B (AcnB), is the major aconitase synthesized during exponential growth and is more sensitive to oxygen. AcnB also regulates the translation and stability of mRNAs involved in iron metabolism in response to iron deficiency, much like the aconitases from eukaryotes. A similar dual role has been found for the aconitases of other bacteria, including *B. subtilis* and some pathogenic bacteria. Also, some pathogenic bacteria may use their aconitases to sense the availability of iron as part of the signal that they are inside a eukaryotic host and to adjust their physiology accordingly.

It is not clear why aconitase plays the dual role of sensing iron levels and performing an essential step in the TCA cycle. One possibility is related to the extreme toxicity and mutagenic properties of hydroxyl radicals (see Box 10.1). These are produced from hydrogen peroxide in the presence of high iron concentrations. If the cellular iron concentration is high, the TCA cycle may run at full capacity to increase the reducing power in the cell and to reduce the amount of such dangerous reactive oxygen species. If the cellular iron concentration is low, the TCA cycle can run at a lower rate, just fast enough to produce essential intermediates and electron donors for the electron transport system.

Regulation of Virulence Genes in Pathogenic Bacteria

Many of the stress responses and other types of regulons discussed above are directly relevant to the ability of bacterial pathogens to survive in a eukaryotic host. The pathogen must recognize that it has entered a new environment, adapt its metabolism to allow it to exploit this new environment, and express virulence genes that allow the organism to survive in the host and cause disease. The virulence genes of pathogenic bacteria represent a type of global regulon. Most pathogenic bacteria express their virulence genes only in the eukaryotic host and respond to specific conditions inside the host to turn on the expression of these genes. Virulence genes can be identified because mutations that inactivate them render the bacterium nonpathogenic but do not affect its growth outside the host. Note that many of the basic regulatory responses of bacterial cells discussed above, such as catabolite regulation, the stringent response, and a variety of stress responses, are also very important for a bacterial pathogen to cause disease (see Eisenreich et al., Suggested Reading). Some pathogens occupy multiple sites in the host during infection and must shift their program of gene expression accordingly. Furthermore, interactions among bacteria can be crucial for disease, for example, in sensing the presence of other members of the same or different species or for formation of complex community structures, such as biofilms, which can have a major impact, not only on the disease process, but also on the susceptibility of the bacterial cells to antibiotics. In this section, we focus on examples of the regulation of genes that affect bacterial pathogenesis.

Diphtheria

Diphtheria is caused by the bacterium *Corynebacterium diphtheriae*, a member of the *Actinobacteria* that colonizes the human throat. It is spread from human to human through aerosols created by coughing or sneezing. The colonization of the throat by itself results in few symptoms. However, strains of *C. diphtheriae* that harbor a prophage named β (see chapter 7) produce diphtheria toxin, which is responsible for most of the disease symptoms. The toxin is excreted from the bacteria in the throat and enters the bloodstream, where it does its damage.

DIPHTHERIA TOXIN

Diphtheria toxin is a member of a large group of A-B toxins, so named because they have two subunits, A and B. In most A-B toxins, the A subunit is an enzyme that damages host cells, and the B subunit helps the A subunit enter the host cell by binding to specific cell receptors. The two parts of the diphtheria toxin are first synthesized from the *tox* gene as a single polypeptide chain, which is cleaved into the A and B subunits as it is excreted from the bacterium. These two subunits are held together by a disulfide bond until they are translocated into the host cell, where the disulfide bond is reduced and broken, releasing the individual A subunit into the cell.

The action of the diphtheria toxin A subunit on eukaryotic cells is well understood. The A subunit enzyme specifically ADP-ribosylates (adds ADP-ribose to) a modified histidine amino acid of the translation elongation factor EF-2 (equivalent to EF-G in bacteria [see chapter 2]). The ADP-ribosylation of the translation factor blocks translation and kills the cell. The opportunistic pathogen *Pseudomonas aeruginosa* makes a toxin that is identical in action to the diphtheria toxin, although it has a somewhat different sequence.

Regulation of the *tox* Gene of *C. diphtheriae*

Iron limitation presents a problem for bacteria in general and for pathogenic bacteria in particular. All of the iron in the human body is tied up in other molecules, such as transferrins and hemoglobin. Thus, to multiply in a eukaryotic host, a pathogenic bacterium must extract the iron from the transferrins and other proteins to which it is bound and transport it into its own cell. For this purpose, *C. diphtheriae* and many other pathogenic bacteria synthesize very efficient siderophores, much like those of free-living bacteria (see above). These small siderophores are excreted from the bacterial cells into the host, where they bind Fe^{2+} more tightly than do other molecules and so can extract Fe^{2+} from them. The siderophore-Fe^{2+} complexes

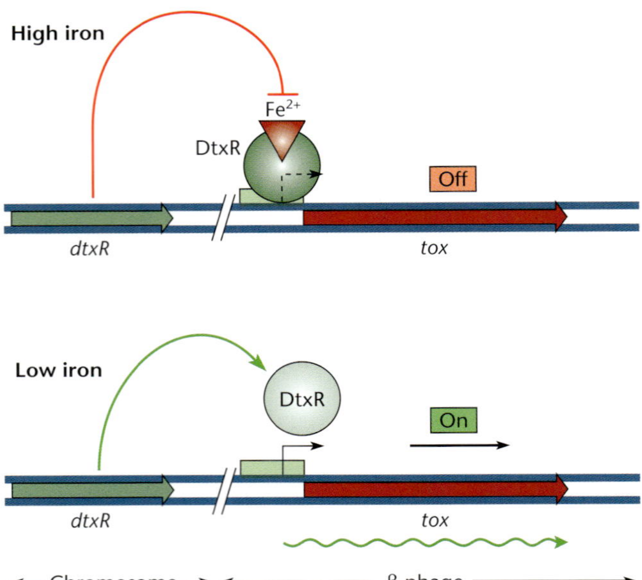

Figure 12.19 Regulation of the *C. diphtheriae tox* gene of prophage β. The DtxR repressor protein, which is encoded by the chromosomal *dtxR* gene, binds to the operator for the *tox* gene, which is encoded on the β prophage. DtxR repressor activity requires ferrous ions (Fe^{2+}).

are transported back into the bacterial cell. As in free-living bacteria, the genes for making the siderophores and a high-efficiency transport system for iron are expressed only when iron is limiting.

Iron limitation is often used as a signal that the bacterium is in the eukaryotic host environment and the virulence genes should be turned on. In *C. diphtheriae*, the virulence genes, including the *tox* gene and iron uptake genes, are under the control of the same global regulator, DtxR (for diphtheria toxin regulator). The DtxR protein of *C. diphtheriae* is a repressor that functions similarly to the Fur repressor protein of Gram-negative bacteria, including *E. coli* (Figure 12.19). Like Fur, the DtxR protein requires iron as a corepressor to bind to the operators of genes under its control. Interestingly, even though the *tox* gene encoding the toxin of *C. diphtheriae* is carried on the lysogenic β phage (see chapter 7), it is regulated by the DtxR repressor, which is encoded on the chromosome (Figure 12.19). Most other genes controlled by DtxR are chromosomal genes. This is just one of many examples of the contribution of lysogenic phages to the pathogenicity of bacteria.

Cholera and Quorum Sensing

Cholera is another well-studied example of the global regulation of virulence genes. *Vibrio cholerae*, the causative agent, is a proteobacterium that is spread through water contaminated with human feces. The disease continues to be a major health problem worldwide, with periodic outbreaks, especially in countries with poor sanitation. When ingested by a human, *V. cholerae* colonizes the small intestine, where it synthesizes cholera toxin, which acts on the mucosal cells to cause a severe form of diarrhea. Other virulence determinants are the flagellum, which allows the bacterium to move in the mucosal layer of the small intestine, and pili called toxin-coregulated pili (TCP) that allow it to stick to the mucosal surface. Quorum sensing, which allows the bacterium to sense when other *V. cholerae* cells are nearby (see below), and **biofilm formation**, in which the bacteria band together and surround themselves with an impregnable layer of polymers to keep from being washed out of the intestine and to resist host defense systems, also play important roles in the disease process.

CHOLERA TOXIN

The mechanism of action of cholera toxin has been the subject of intense investigation, in part because of what it reveals about the normal action of eukaryotic cells. The cholera toxin is composed of two subunits, CtxA and CtxB, which are exported from the bacterial cell by a type II secretion system (see chapter 2). The two subunits are secreted through the inner membrane by the SecYEG channel and then assemble in the periplasm before being released to the outside of the cell through a large structure called a secretin in the outer membrane (see chapter 2). Once outside the cell, the CtxB subunit helps the CtxA subunit enter the eukaryotic cell. Like diphtheria toxin, the CtxA subunit of cholera toxin is an ADP-ribosylating enzyme. However, rather than ADP-ribosylating a translation elongation factor, CtxA ADP-ribosylates a mucosal cell membrane protein called Gs, which is part of a signal transduction pathway that regulates the activity of the adenylate cyclase enzyme that makes cAMP. The ADP-ribosylation of Gs causes cAMP levels to rise in host cells and alters the activities of transport systems for sodium and chloride ions. This results in loss of sodium and chloride ions from the cells, and the change in osmotic pressure releases water from the cells, resulting in severe diarrhea and dehydration. This facilitates further spread of the disease to new hosts. The treatment is to aggressively administer water orally or intravenously until the condition of the patient improves.

Regulation of the Synthesis of Cholera Toxin and Other Virulence Determinants

The *ctxA* and *ctxB* genes encoding the cholera toxin are part of a large regulon containing as many as 20 genes. In addition to the *ctx* genes, the genes of this regulon include those that encode pili, colonization factors, and outer membrane proteins (e.g., OmpT and OmpU) related to osmoregulation. Although some of these genes, including the *ctx* genes, are carried on a prophage (see chapter 7), others, including the pilin genes, are carried on the bacterial chromosome. The transcription of the genes of this regulon is activated only under conditions of high osmolarity and in the presence of certain amino acids, conditions that may mimic those in the small intestine. Here, we describe the cascade of genes involved in the regulation of transcription of *ctx* and other virulence genes.

ToxR-ToxS. The cholera virulence regulon was first found to be under the control (either directly or indirectly) of the activator protein ToxR, the product of the *toxR* gene. The ToxR protein combines in single-polypeptide elements that are normally part of two different proteins of the two-component sensor and response regulator type of system (Box 12.3). The ToxR polypeptide spans the inner membrane so that the carboxyl-terminal part of the protein is in the periplasm, where it can sense the external environment. The amino-terminal part is in the cytoplasm, where it contains an OmpR-like DNA-binding domain that can activate the transcription of genes under its control.

While the ToxR protein resembles other response regulators in some respects, it is unlike most response regulator proteins in that it is not activated by phosphorylation. Also, it is not known to bind to any small-molecule effectors. A clue to its mechanism of activation could be found in another protein, ToxS. The *toxS* gene is immediately downstream of *toxR* in the same operon, and the gene was discovered because mutations that inactivate it also prevent expression of the genes of the ToxR regulon. Like ToxR, the ToxS protein is anchored in the inner membrane, but a large domain protrudes into the periplasm.

One model for how ToxS might activate ToxR came from experiments designed to investigate the membrane topology of ToxR. The purpose of these experiments was to determine if part of ToxR spans the inner membrane and extends into the periplasm. A method for determining the membrane topology of proteins in enteric bacteria uses translational fusions of various regions of the protein to PhoA (see chapter 2 for a general discussion of reporter gene fusions). The PhoA protein is an alkaline phosphatase enzyme that cleaves XP (5-bromo-4-chloro-3-indolylphosphate), which is like X-Gal (5-bromo-4-chloro-3-indolyl-β-D-galactopyranoside), the indicator for β-galactosidase activity, except that the dye is fused to phosphate instead of galactose. If PhoA cleaves the phosphate off XP, the colonies turn blue. However, PhoA must be in the periplasm to be active, probably because it must form dimers that are held together by disulfide bonds between cysteines in its subunits, and these disulfide bonds form only in the periplasm and not in the cytoplasm. Therefore, if a region of a protein is fused to PhoA, the alkaline phosphatase will be active and the strain carrying that fusion will form blue colonies on XP plates only if that region of the protein is in the periplasm. To use this method to determine which parts of ToxR are in the periplasm, fusion proteins composed of the PhoA reporter fused to various portions of ToxR were generated. Bacteria containing fusions to the amino-terminal portion of ToxR did not form blue colonies on XP plates, while strains containing fusions to the carboxyl-terminal portion did, suggesting that the carboxyl-terminal portion of the ToxR protein is in the periplasm but the amino-terminal portion is in the cytoplasm. A surprising result was that some of the ToxR-PhoA fusions functioned like wild-type ToxR in activation of transcription, even when *toxS* was inactivated by a mutation. One explanation for this result is that the PhoA part of these fusions was driving the dimerization of the ToxR portion in the periplasm and that dimerization was required for ToxR to activate transcription. This suggested that the normal function of ToxS is to dimerize ToxR, since ToxS could be dispensed with if PhoA promoted the dimerization.

Other evidence suggested that dimerization, while important, may not be all that is needed to activate the ToxR protein. These results suggested that some feature related to the membrane anchoring of ToxR may also be required. If dimerization were sufficient, attaching other dimerization domains, such as the ones from the CI repressor protein of λ phage (see chapter 7), to the cytoplasmic domain of the ToxR protein should cause ToxR to dimerize in the cytoplasm and activate transcription. However, fusion proteins composed of ToxR and other dimerization domains are still inactive unless the ToxR protein retains its transmembrane domain. This suggests that the ToxR protein must be at least partly in the membrane to be active and that ToxS may play an additional role in stabilizing ToxR. Anchoring ToxR in the membrane may allow ToxR to be activated directly by external signals, and ToxS may help in this activation. It is known that membrane-damaging agents, such as bile, can activate ToxR, consistent with the idea that ToxR senses properties of the membrane.

ToxT and TcpP. The regulation of virulence genes in *V. cholerae* is much more complicated than a single activator, ToxR, turning on virulence genes. A general outline of the various regulatory pathways used to turn on the genes required for *V. cholerae* pathogenicity is given in Figure 12.20. As mentioned above, *V. cholerae* has many virulence genes besides the toxin genes that are also part of the ToxR regulon, since *toxR* mutations prevent their expression. However, while ToxR can activate the transcription of the toxin genes directly, the ToxR protein does not directly control most of the other ToxR regulon genes. Instead, another transcriptional activator, the ToxT protein, which is a member of the AraC family of regulators (see chapter 11), controls these genes. ToxR activates the transcription of the *toxT* gene, and ToxT then activates

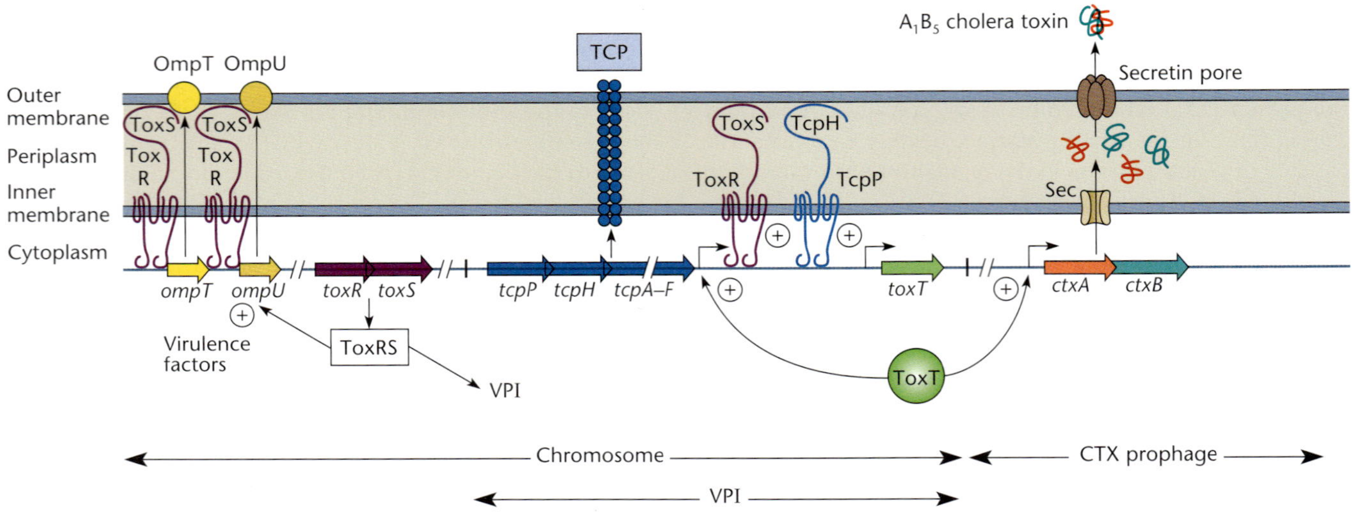

Figure 12.20 Regulatory cascade for *V. cholerae* virulence factors. The ToxR-ToxS proteins directly regulate *omp* virulence factors and the *toxT* regulatory gene located on a *V. cholerae* pathogenicity island (VPI). The VPI-encoded TcpP-TcpH proteins also regulate *toxT* transcription. ToxT activates the Ctx prophage-borne *ctxAB* toxin genes and the toxin-coregulated pili (TCP) genes. ToxT also positively regulates its own expression from the promoter for transcription of the *tcpA-F* operon.

the transcription of the other genes. The ToxT-activated genes are therefore part of the ToxR regulon, as their expression is dependent on ToxR, and mutations in *toxR* result in loss of expression because of the loss of *toxT* expression. ToxT also activates the expression of its own gene, so that an initial signal from ToxR can be amplified to generate high levels of ToxT synthesis. This is an example of an **autoinduction loop**, where a regulatory gene activates its own expression and can result in a very strong regulatory response.

Activation of the *toxT* gene also requires yet another activator, TcpP. The activity of the TcpP activator requires the activity of another protein, TcpH. Both TcpP and TcpH are inner membrane proteins, but it is not clear how TcpH regulates TcpP activity. Transcription of the *tcpP-tcpH* operon responds to environmental cues, but we do not understand how these two genes, together with the *toxR-toxS* genes, transduce environmental signals that the bacterium is in the intestine of its host into activation of ToxT. This type of serial activation, where one regulator activates another regulator, which activates yet another regulator, is called a **regulatory cascade**. Obviously, much more needs to be done before we can begin to understand this important model system of bacterial pathogenicity.

Interestingly, the *toxT*, *tcpP*, and *tcpH* genes are all located on a DNA element in *V. cholerae* called VPI (*V. cholerae* pathogenicity island). A **pathogenicity island** is a segment of DNA that contains a cluster of genes involved in virulence, and this set of genes has usually been acquired by horizontal gene transfer from another organism (see chapter 13). The cholera toxin is encoded by a different horizontally acquired DNA element, a lysogenic phage (see chapter 7). These genes are regulated by ToxR, encoded by a chromosomal gene. This is yet another example of virulence traits in pathogenic bacteria that are encoded on exchangeable DNA elements but interact with the products of chromosomal genes, as well as another example of pathogenic bacteria that are derived from their free-living relatives by acquisition of virulence genes encoded on interchangeable DNA elements.

QUORUM SENSING

As mentioned above, **quorum sensing** is an important contributor to virulence in *V. cholerae*. For a long time, it was thought that single-celled bacteria, such as *V. cholerae*, live as isolated cells, with no way of telling if other members of the same species are nearby. However, with the discovery of quorum sensing, it became apparent that many bacteria have ways of communicating with each other. The basic observation for a phenomenon that uses quorum sensing is illustrated in Figure 12.21. Cells growing at low density exhibit very low expression of the regulated genes, and expression turns on only when the cells reach a threshold density, at which point expression increases rapidly. The bacteria monitor cell density by releasing small molecules that can be taken up by other bacteria, usually members of the same species. When the concentration of bacteria in a particular location is high, the concentration of these secreted small molecules also becomes high, signaling that the bacterium is in the pres-

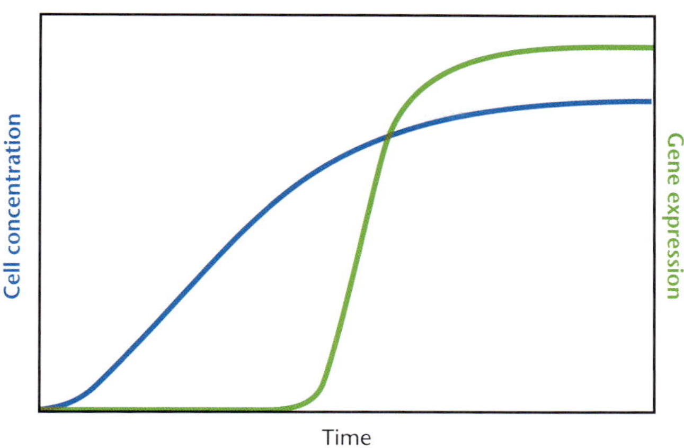

Figure 12.21 Quorum sensing. In systems regulated by quorum sensing, expression of the regulated genes (green line) remains very low until cell density (blue line) reaches a threshold level. At this point, the quorum sensing signal accumulates to a high enough level to induce expression of target genes, and the expression level rapidly increases.

ence of many other bacteria of the same type. In response, the bacteria induce certain types of genes to adapt them for a more communal existence, such as in a biofilm. We have discussed cell-cell signaling in reference to competence gene regulation in *Firmicutes*, which use small peptides as their signals (see chapter 6).

Quorum sensing was first discovered in the marine bacterium *Photobacterium fischeri* (*Vibrio fischeri*), which is a facultative symbiont that can live either free in the ocean or in the light organs of some fishes and squids, where they give off light due to chemiluminescence. The light is due to induction of the *lux* genes, which are turned on only when a bacterium is close to a high number of other bacteria of the same type. The squid and fish with which these types of bacteria form a symbiosis live in the ocean, where there is little light, or at the edge of the water, where available light changes from day to night. By emitting light when they are concentrated in the light organ, the bacteria help the marine organisms find each other and hide from predators. The host organism provides an environment that the bacteria find hospitable and controls bacterial numbers to control light emission. Recent work has concentrated on a free-living marine bacterium, *Vibrio harveyi*, which also gives off light when the bacteria are concentrated in certain regions of the ocean, but it is not known to be a symbiont of any marine animal. A likely explanation is that *V. harveyi* also forms such a symbiosis, but with an unknown marine organism partner. This explanation is supported by the fact that some of the genes induced along with *lux* form a type III secretion system. In other types of bacteria, such systems are used to inject proteins into eukaryotic cells and are important in avoiding host defenses and establishing infections or symbioses (see chapter 2).

Extensive research has revealed that the chemiluminescence of *V. harveyi* is induced because the bacteria give off two small molecules, called autoinducers AI-1 and AI-2. AI-1 is a homoserine lactone, and AI-2 is a furanosyl borate diester; the structures of these compounds are shown in Figure 12.22. Autoinducer AI-2 is made by the product of the *luxS* gene, which is found in many bacteria. Because AI-2 is made by so many different types of bacteria, it has been proposed that it is a universal autoinducer allowing different types of bacteria to communicate with each other, for example, in the formation of biofilms.

Both AI-1 and AI-2 act through two-component sensor kinase and response regulator cascades that determine the state of phosphorylation of LuxO, a transcriptional regulator (Figure 12.22A). LuxO is a member of the NtrC family of activators and, like NtrC, is active only if it is phosphorylated. It also activates transcription only from σ^N-type promoters, like the other members of this family of regulators. Genes under the control of the LuxO activator include four noncoding sRNAs. These sRNAs, with the help of the Hfq protein (see Box 12.5), inhibit the synthesis of an activator called LuxR by binding to and destabilizing its mRNA. If LuxR is not made, the *lux* operon is not transcribed, and the cells do not give off light. When the cell population density is low, the concentrations of AI-1 and AI-2 are low, and the kinase activity of the sensor kinases is high, which leads to the phosphorylation of LuxO, so that it is active as a transcriptional activator. The four sRNAs are made, LuxR is not made, the *lux* operon is not transcribed, and the cells do not give off light. At a high concentration of cells, and therefore at a high concentration of the autoinducers, the binding of the autoinducers to the sensor kinases activates the phosphatase activity of the sensor kinases, which results in removal of phosphate from the response regulator protein in the next step of the pathway until eventually LuxO loses its phosphate and is unphosphorylated. If LuxO is not phosphorylated, the sRNAs are not made; this allows synthesis of the LuxR activator, the *lux* operon is transcribed, and the cells give off light.

Quorum Sensing in *V. cholerae*

Analysis of the nonpathogenic *V. harveyi* provided crucial insights into mechanisms of virulence gene regulation in the pathogenic *V. cholerae*, which was also found to have quorum-sensing systems, as do many types of pathogenic bacteria, including plant pathogens such as *Agrobacterium*. While it lacks the AI-1 system, *V. cholerae* has the AI-2 system and two other systems with unknown autoinducers named CA-1 and VarS-VarA, which also function through sensor kinases and response regulators to finally determine the state of phosphorylation of LuxO.

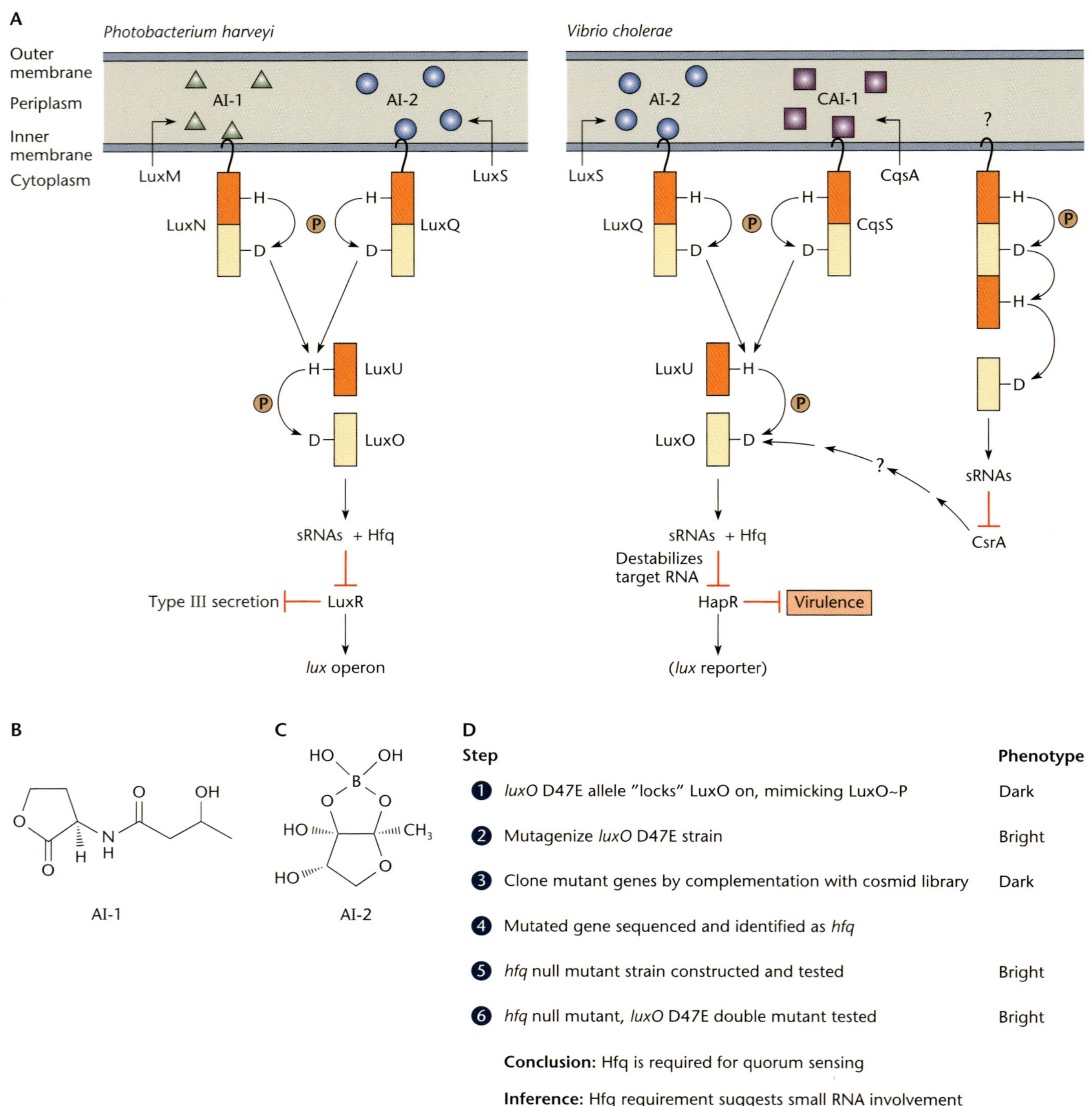

Figure 12.22 Quorum sensing in *Photobacterium harveyi* and *Vibrio cholerae.* **(A)** *V. cholerae* shares the AI-2 quorum-sensing system with *P. harveyi* and contains two additional phosphorelay systems to regulate virulence. AI-1 (triangles), AI-2 (circles), and CAI-1 (squares) are small molecules. LuxS synthesizes AI-2, while LuxM and Cqs synthesize AI-1 and CAI-1, respectively. LuxN, LuxQ, LuxS, LuxU, and LuxO form a phosphorelay. The flow of phosphate depicts a low-cell-density state. High cell density would reverse the flow. **(B)** Structure of AI-1. **(C)** Structure of AI-2. **(D)** Steps in identifying quorum-sensing regulatory small RNAs (sRNAs).

Use of a quorum sensing mechanism by an intestinal pathogen like *V. cholerae* is advantageous to the organism, as it allows it to delay toxin synthesis until bacterial numbers are high enough to allow a coordinated effort to cause damage to the host, which in the case of this system results in release of intestinal contents and dissemination of the pathogen to new hosts.

A comparison of quorum sensing in *V. cholerae* and *V. harveyi* is presented in Figure 12.22. A total of seven sRNAs are used by the pathways for quorum-sensing signaling in *V. cholerae* (see Lenz et al., Suggested Reading). Not only does *V. cholerae* have the four sRNAs whose transcription is activated by LuxO in *V. harveyi*, but it also has three others that are used in the VarS-VarA pathway. These other sRNAs bind to and inhibit the activity of a regulatory protein named CsrA (Box 12.5), which in turn affects the state of phosphorylation of LuxO by an unknown mechanism. Instead of an activator, LuxR, which activates the transcription of the *lux* gene and other genes appropriate for the free-living *V. harveyi*, *V. cholerae* has HapR, which differentially regulates *lux*, virulence genes, genes involved in biofilm formation, etc., that are important for pathogenesis.

Whooping Cough

Another well-studied disease used to illustrate global regulation of virulence genes is whooping cough, caused by *Bordetella pertussis*. Whooping cough is mainly a childhood disease and is characterized by uncontrolled coughing, hence the name. The bacteria colonize the human throat and are spread through aerosols that result from the coughing. Effective vaccines have been developed, but the disease continues to kill thousands of children worldwide, mainly in areas where the vaccines are not available. Because outbreaks occur sporadically even in populations with high vaccination rates, an additional vaccination in later years of life has become routine.

Despite their very different symptoms, the diseases caused by *V. cholerae* and *B. pertussis* have similar molecular mechanisms. *B. pertussis* makes a complex A-B toxin (pertussis toxin) that is in some ways similar to the cholera toxin. The pertussis toxin has six subunits, although only two of them are identical. One of the subunits (S1) is the active portion of the toxin, while the others are involved in adhesion to the mucosal surface of the throat. The toxin is first secreted through the outer membrane by the SecYEG translocase and is then exported with a type IV secretion system (see chapter 2). Once outside the bacterial cell, the B domains of the toxin bind to receptors on ciliated epithelial cells and transfer the A domains into the cells, where they ADP-ribosylate the G protein in a signal transduction pathway involved in deactivating the adenylate cyclase, which leads to elevated cAMP levels. However, rather than causing a loss of water from the cells, as is the case for cholera toxin, the elevated cAMP levels in throat epithelial cells cause an increase in mucus production that triggers the uncontrolled coughing characteristic of whooping cough.

In addition to pertussis toxin, *B. pertussis* synthesizes a number of other toxins and other virulence proteins. They include a bacterially encoded adenylate cyclase enzyme that enters host cells and presumably directly increases intracellular cAMP levels by synthesizing cAMP. This observation supports the importance of increased cAMP levels to the pathogenesis of the bacterium, although the contribution of this adenylate cyclase to the symptoms is unknown. Other known toxins include one that causes necrotic lesions on the skin of mice and a cytotoxin that is a peptidoglycan fragment that kills ciliated cells of the throat. Other virulence factors are involved in the adhesion of the bacterium to the mucosal layer and fimbriae and factors required to survive nutrient deprivation, allow motility, etc.

REGULATION OF *B. PERTUSSIS* VIRULENCE GENES

Like the virulence genes of *C. diphtheriae* and *V. cholerae*, many of the virulence genes of *B. pertussis* are expressed only when the bacterium enters the eukaryotic host at the same time that other genes that are expressed in the free-living state are repressed. The regulation of the virulence genes of *B. pertussis* is achieved by a sensor kinase and response regulator pair of proteins encoded by linked genes, *bvgA* and *bvgS* (for *Bordetella virulence* genes); *bvgS* encodes the sensor kinase, and *bvgA* encodes the transcriptional regulator.

The BvgS-BvgA system is similar to many other sensor kinase and response regulator pairs in that the BvgS protein is a transmembrane protein, with its N terminus in the periplasm and its C terminus in the cytoplasm, allowing it to communicate information from the external environment across the membrane to the inside of the cell (Box 12.3). It also exists as a dimer in the inner membrane. Furthermore, like many other sensor kinase proteins that work in two-component systems, the BvgS protein autophosphorylates in response to a signal from the external environment and donates this phosphate to the BvgA protein, which then regulates transcription of the virulence genes.

Attempts have been made to determine the signals to which BvgS responds to phosphorylate itself. In laboratory cultures, the signal transduction pathway necessary for transcription of the pertussis toxin gene and other virulence genes is normally constitutively active but is repressed by high nicotinamide and magnesium levels. Another factor that is likely to be important for regulation of pathogenesis is that expression of the virulence genes is highest at 37°C, the temperature of the human body.

Once phosphorylated, the BvgA response regulator both activates the transcription of genes required in the host and represses genes that allow it to survive outside the host. This is particularly clear for *Bordetella bronchiseptica*, a close relative of *B. pertussis* that can infect other mammals besides humans and can survive for longer times outside the host. This more versatile relative has a BvgS-BvgA system very closely related to that of *B. pertussis*; under conditions that mimic those inside its hosts, the phosphorylated BvgA protein represses a number of genes whose products are required in the free-living state.

One difference between BvgS and many other sensor kinases is that it has more than one amino acid that can accept phosphates when it receives its signal. As with other sensor kinases, one histidine in the cytoplasmic N terminus of the protein phosphorylates itself in response to conditions that mimic entrance into the host bronchial tubes. This phosphate group can then be transferred to an aspartate closer to the C terminus of the same BvgS polypeptide and then transferred again to another histidine that is even closer to the C terminus before it is finally transferred to an aspartate in the BvgA response regulator. Thus, in successive steps, the phosphate is transferred closer and closer to the C terminus of the sensor kinase and therefore closer to the BvgA response regulator in the cytoplasm; this serial transfer of phosphates from site to site is called a **phosphorelay**. Sensor kinases like BvgS, which transfer phosphates within themselves in a phosphorelay, have been called multidomain sensors because they transfer phosphates in a phosphorelay within the same polypeptide as opposed to phosphorelays from one protein to another (Box 12.3).

There is speculation about why *Bordetella* uses a multidomain sensor to signal that it is in a mammalian throat rather than just having a single site of phosphorylation in the sensor kinase or a multiprotein phosphorelay, as in many other signal transduction systems. This speculation centers on research indicating that changes in gene expression occur in more than one stage. When *Bordetella* encounters conditions like those in the host, the virulence genes are not just turned on, but rather, they proceed through an intermediate stage. The three stages have been named Bvg$^-$ for outside the host, Bvg$^+$ for inside the host, and Bvg^i for an intermediate state when the bacterium has just entered the host and has not yet established an infection and is not in a state where it can be spread to another host in an aerosol through coughing. Genes expressed in the Bvg$^-$ state include those that allow it to survive in a free-living state, such as genes for carbon source utilization and growth at low temperatures. Genes expressed in the Bvg^i state are those that promote attachment to the epithelial cells in the throat, and those expressed in the Bvg$^+$ state are the toxin genes and others that should be expressed only when the bacterium has already established an infection and is ready to spread to other hosts. The multidomain sensor of BvgS may facilitate sensing of multiple signals to modulate the final concentration of BvgA~P.

Developmental Regulation: Sporulation in *B. subtilis*

As mentioned in the introductory chapter, many bacteria undergo complex developmental cycles. In their development, some bacteria perform many functions reminiscent of eukaryotes: they undergo regulatory cascades; their cells communicate with each other, differentiate, and form complex multicellular structures; different cells in these multicellular structures often perform different distinct functions, which require compartmentalization and cell-cell communication; and the cells use phosphorelays to respond to changes in communication with other cells and with the external environment. Because of the relative ease of molecular genetic analysis with some bacteria, some of these developmental processes have been extensively investigated as potential model systems for even more complex developmental processes in eukaryotes. We will use an example of this type of bacterial developmental system to illustrate how a variety of regulatory mechanisms can be integrated to control a complex multistage process.

The best-understood bacterial developmental system is sporulation in *B. subtilis* (see Higgins and Dworkin 2012, Suggested Reading). When starved, *B. subtilis* cells undergo genetically programmed developmental changes. They first attempt to obtain nutrients from neighboring organisms by producing antibiotics and extracellular degradative enzymes. They even cannibalize their siblings. If starvation conditions persist, the cells sporulate, producing endospores that are metabolically dormant and highly resistant to environmental stresses, including heat, UV light, acid, and organic solvents. Exposure of the spore to nutrients results in germination and a return to active metabolism and vegetative growth.

The process of sporulation starts with an asymmetric division that produces two cell types with different morphological fates. The larger cell, which is called the mother cell, engulfs the smaller forespore and provides nutrients and gene products important for spore development. Eventually, the mother cell lyses, releasing the endospore. The sporulation process takes about 8 hours under normal laboratory conditions, so this is represents a major commitment that the cell makes only if other options for survival fail.

Many of the changes that occur in the sporulating cell can be visualized by electron microscopy (Figure 12.23). The figure also describes the stages of spore development and shows some of the proteins that have been identified as key regulators of specific stages of development. We describe below some of the regulators and how they were identified.

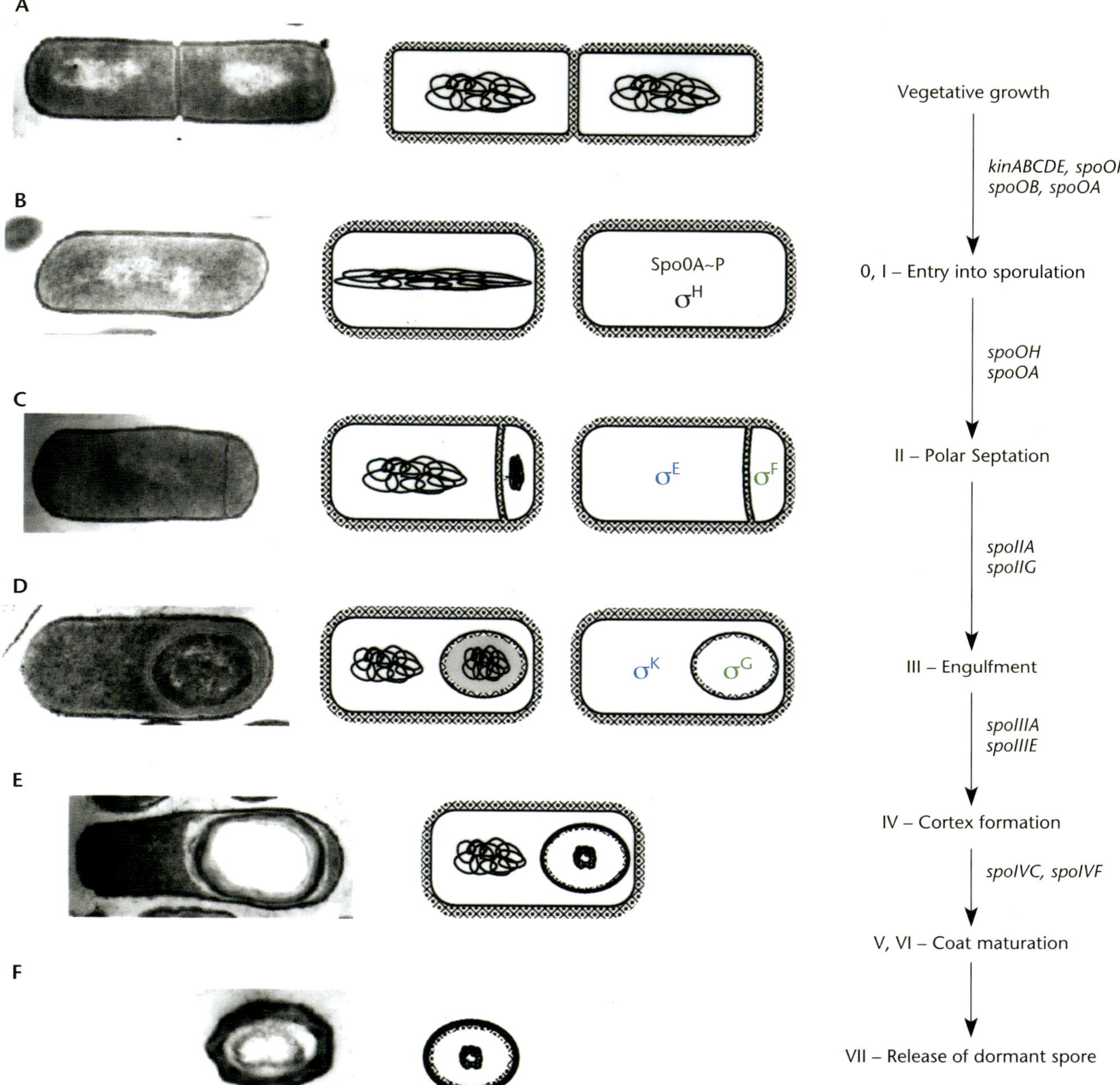

Figure 12.23 Stages of sporulation. The left side of each panel shows an electron micrograph of the stage of sporulation. The middle section shows, in cartoon form, the disposition of the chromosomes and the time and site of action of the principal regulatory proteins that govern sporulation gene expression. The right section shows the names of each stage and key genes required at each step. Electron micrographs reprinted by permission from Springer Nature: from p. 21–33, chapter 2 (Driks) in Russo VEA, et al (ed), *Development: Genetics, Epigenetics and Environmental Regulation,* ©1999. Line drawings modified from Levin PA, Losick R, *in* Brun YV, Shimkets LJ (ed), *Prokaryotic Development* (ASM Press, Washington, DC, 2000).

Identification of Genes That Regulate Sporulation

Isolation of mutants was crucial to the process of identifying the important regulators of sporulation. Many mutants were isolated on the basis of a phenotype referred to as Spo$^-$ (for sporulation minus). Such mutants could be identified as nonsporulating colonies because plate-grown cultures of the wild type develop a dark brown spore-associated pigment, whereas the nonsporulating colonies remain unpigmented.

Spo$^-$ mutants were phenotypically characterized by electron microscopy and then grouped according to the stage at which development was arrested (Figure 12.23). The names of *B. subtilis* sporulation genes reflect three aspects of the genetic analysis of these genes. They do not fall into the standard Demerec convention for gene names used today, but they do provide a handy system for remembering the stage of development that is affected by mutations in the genes. The roman numerals refer to the results of phenotypic categorization of the mutant strains, with the numbers 0 through V indicating the stages of sporulation at which mutants were found to be blocked. For example, all loci designated *spoII* resulted in cells that could not progress from stage II to stage III. The gene names also contain one or two letters. The first letter designates the different loci in which mutations were found to cause similar phenotypes; for example, mutations in *spoIIA* and *spoIIG* both result in a block at stage II, but these two loci are not genetically linked to each other. Each such locus was defined by the set of mutations that caused the same morphological block and that were genetically closely linked to each other, indicating that the mutations were likely to be in the same gene or operon. The second letter in the names indicates the individual open reading frames that were found when DNA sequencing revealed that a locus contained several open reading frames, so that *spoIIAA* and *spoIIAB*, both of which are in the *spoIIA* locus, encode two different polypeptides.

Regulation of Sporulation Initiation

Much of what we understand about the mechanism of sporulation initiation is based on studies of a class of sporulation-minus mutants that failed to begin the sporulation process. These were designated *spo0* because they were blocked at stage 0. Many of these mutants have pleiotropic phenotypes, meaning that they are altered in several characteristics. Besides being unable to sporulate, they fail to produce the antibiotics or degradative enzymes that are characteristically produced by starving cultures, and they do not develop competence for transformation (see chapter 6).

Two of the *spo0* genes, *spo0A* and *spo0H*, encode transcriptional regulators. The *spo0A* gene encodes a response regulator of a two-component system (Box 12.3). The Spo0A protein is responsible for regulating a number of cellular responses to starvation. The product of *spo0H* is a sigma factor (σ^H; note that this *B. subtilis* σ^H is distinct from *E. coli* σ^H, which is involved in heat shock). Many of the genes that are targets for Spo0A regulation are transcribed by the σ^H-containing RNA polymerase holoenzyme.

Like many response regulators, Spo0A must be phosphorylated in order to carry out its regulatory functions. In contrast to most response regulators that respond directly to a single sensor kinase (Box 12.3), phosphorylation of Spo0A utilizes a phosphorelay system (Figure 12.24) that includes another two of the *spo0* gene products, Spo0F

Figure 12.24 The phosphorelay activation of the transcription factor Spo0A. The phosphorelay is initiated by at least five histidine kinases, which autophosphorylate on a histidine residue in response to different signals. Kinases A and B phosphorylate to high levels and initiate sporulation, and kinases C, D, and E phosphorylate only to low levels for competence, biofilm formation, antibiotic synthesis, and degradative enzymes. The phosphate is transferred to Spo0F, from Spo0F to Spo0B, and finally from Spo0B to Spo0A to generate Spo0A~P. The phosphorelay is also controlled by dephosphorylation. RapA and RapB are phosphatases for SpoF~P, and Spo0E is a phosphatase for Spo0A~P. RapA and RapB are inhibited by the PhrA and competence-stimulating factor (PhrC) pentapeptides, respectively. The levels of the Spo0E phosphatase are controlled by degradation by the FtsH protease. The output of the system is dependent on the final levels of Spo0A~P.

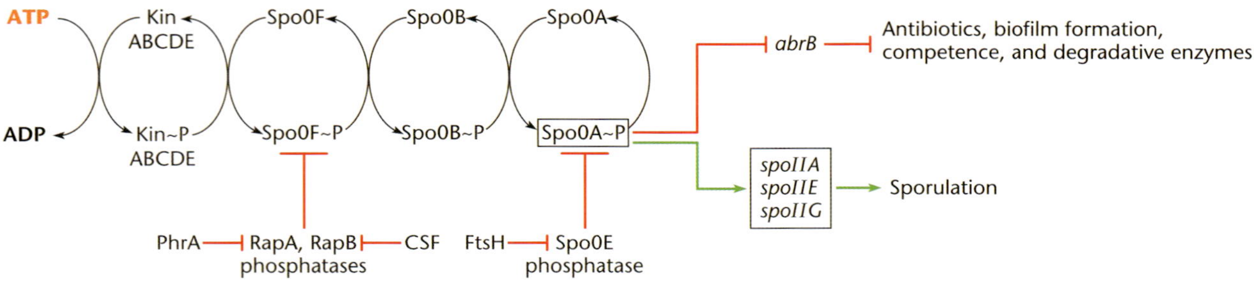

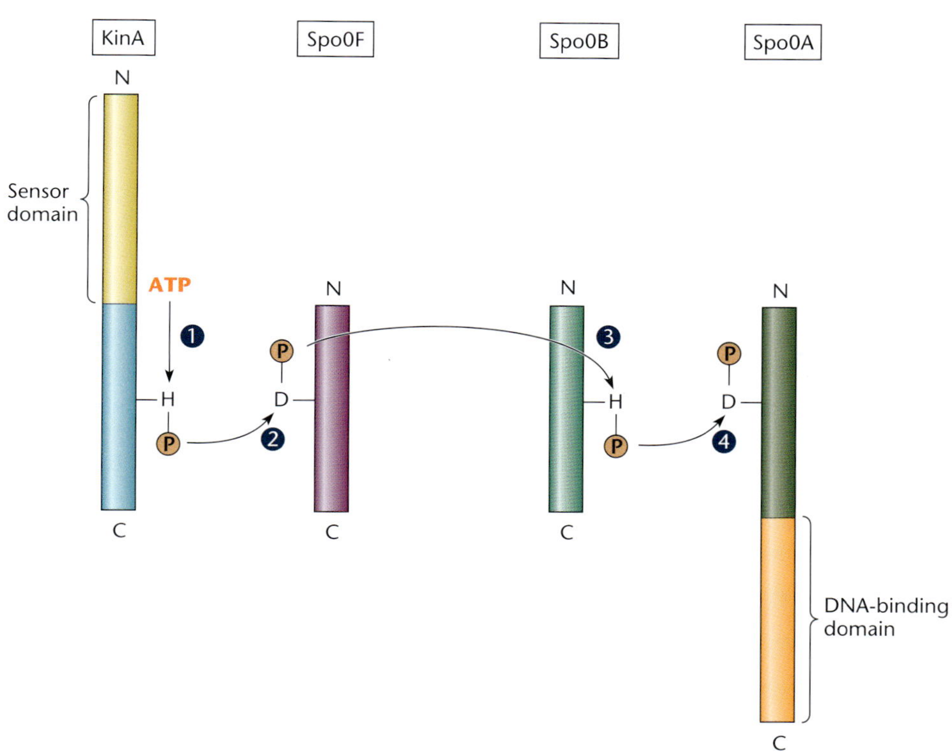

Figure 12.25 Phosphate transfer through the sporulation phosphorelay. Unlike most signal transduction pathways, in this system the histidine kinase and the phosphorylated aspartate domains are found on separate polypeptides. KinA is shown as an example of the five kinases. The kinase autophosphorylates on a histidine residue and then the phosphoryl group is transferred to an aspartate on Spo0F, then to a histidine on Spo0B, and then to an aspartate on Spo0A.

and Spo0B. The phosphorelay also involves at least five protein kinases that each phosphorylate Spo0F under certain conditions. Spo0B is a phosphotransferase enzyme that removes phosphoryl groups from Spo0F~P and transfers them to Spo0A. At each step, phosphates are transferred between specific histidine and aspartate residues on the phosphorelay proteins (Figure 12.25). Spo0A~P then regulates its target genes by binding to their promoter regions, activating some and repressing others. Note that the Spo0A phosphorelay is much more complicated than that described for *B. pertussis* BvgA, because the sporulation system uses multiple kinases and then transfers the phosphate between different proteins before it finally reaches the Spo0A transcriptional regulator.

The regulatory effect of Spo0A on a given target gene depends on the amount of Spo0A~P in the cell, reminiscent of OmpR~P regulation of porin genes and *B. pertussis* BvgA~P regulation of virulence genes. At low levels, Spo0A~P positively regulates genes involved in the synthesis of antibiotics and degradative enzymes, as well as competence and biofilm formation. These represent activities used by the cell to try to overcome the nutritional limitation that triggered the process that ultimately leads to sporulation if a new source of nutrients is not obtained. This positive regulation results in large part from what is actually a "double-negative" series of events, in which the direct effect of Spo0A~P action is repression of a gene called *abrB*, which itself encodes a repressor that acts on the antibiotic and degradative-enzyme genes (Figure 12.24). At higher levels, Spo0A~P directly activates transcription of several sporulation operons, including *spoIIA*, *spoIIE*, and *spoIIG*. Activation of these genes leads to commitment to the sporulation process. As described for differential regulation by OmpR~P, repression of *abrB* requires only low concentrations of Spo0A~P, whereas activation of *spoIIA*, *spoIIE*, and *spoIIG* occurs only when Spo0A~P levels are high, because the binding sites for SpoOA~P on these target promoters have lower affinity for Spo0A~P.

REGULATION OF THE SpoOA PHOSPHORELAY SYSTEM

Numerous genes participate in regulating the amount of Spo0A~P produced in a cell. Several of them encode the kinases mentioned above, which phosphorylate Spo0F and therefore increase SpoOA~P levels. Two of these kinases, KinA and KinB, phosphorylate Spo0F and consequently Spo0A to high levels in response to severe extended starvation and direct the cell to initiate sporulation; the others, KinC, KinD, and KinE, phosphorylate Spo0F (and therefore Spo0A) only to low levels and direct the cell to degradative enzyme synthesis, competence, and biofilm formation. The signals that activate these kinases are unknown. Other signals, such as DNA damage, could also activate KinA and KinB unless there is intervention by a checkpoint protein, Sda. This protein binds to the kinases and inhibits them, preventing sporulation in response to DNA damage (see Ruvolo et al., Suggested Reading). *B. subtilis* also encodes an addiction module that kills some cells in the population in response to starvation, allowing the killed cells to be cannibalized by other cells to delay or prevent their sporulation. Clearly, the cell sporulates only if it is absolutely necessary.

Genetic analysis of two of the *spo0* loci, *spo0E* and *spo0L*, revealed that their gene products function as negative regulators of sporulation. This conclusion was based on the observation that null mutations in these genes resulted in increased sporulation, whereas overexpression of the genes resulted in decreased sporulation. Biochemical analysis revealed that Spo0E acts as a phosphatase that dephosphorylates Spo0A~P, and Spo0L (and the related Spo0P) is a phosphatase that acts on Spo0F~P, thereby draining phosphate out of the phosphorelay and diminishing Spo0A~P levels. Spo0L and Spo0P were renamed RapA and RapB, respectively, to reflect their roles as response regulator aspartyl-phosphate phosphatases.

Inhibition of the phosphorelay by the RapA, RapB, and Spo0E phosphatases is shown in Figure 12.24. Since these phosphatases function to reduce the accumulation of Spo0A~P, their activities must be inhibited under conditions that promote sporulation and antibiotic synthesis. Levels of Spo0E are controlled posttranslationally by the FtsH protease, which targets Spo0E for degradation. For RapA and RapB, the known regulatory signals are small peptides named PhrA and competence-stimulating factor that are produced by *B. subtilis* cells as quorum sensor molecules. As discussed above, the quorum sensors of enteric bacteria are typically homoserine lactones or other small molecules, while those of *Firmicutes* are more typically peptides. These and related signals allow the cell to monitor a variety of signals, including starvation, cell density, metabolic states, cell cycle events, and DNA damage, and use that information to modulate Spo0A~P levels and therefore to determine the developmental pathway to be followed.

Compartmentalized Regulation of Sporulation Genes

During the early stages of sporulation, levels of Spo0A~P and other regulators are likely to be shifting rapidly as conditions shift and key physiological signals are assessed. The key step in driving the sporulation process forward is the formation of the asymmetric septum, which separates the cell into two compartments with different fates. Only the smaller forespore compartment will become a mature spore, while the larger mother cell's activities are directed to assisting that process. The mother cell and the forespore are genetically identical, but certain proteins must be made specifically in the developing spore, and others (such as those that form the sturdy spore coat) must be made in the surrounding mother cell cytoplasm. Thus, the set of genes transcribed from the mother cell DNA must differ from the set transcribed from the forespore DNA. This compartmentalization of gene expression requires a complex set of regulatory mechanisms that will be discussed below.

Table 12.2 *B. subtilis* sporulation regulators

Stage of mutant arrest	Gene	Function
0	*spo0A*	Transcription regulator
	spo0B	Phosphorelay component
	spo0F	Phosphorelay component
	spo0E	Phosphatase
	spo0L	Phosphatase
	spo0H	σ^H
II	*spoIIAA*	Anti-anti-σ^F
	spoIIAB	Anti-σ^F
	spoIIAC	σ^F
	spoIIE	Phosphatase
	spoIIGA	Protease
	spoIIGB	Pro-σ^E
III	*spoIIIG*	σ^G
IV	*spoIVCB-spoIIIC (sigK)*[b]	Pro-σ^K
	spoIVF (operon)[c]	Regulator of σ^K

[a]Loss-of-function mutations in *spoIIAB* do not produce a Spo⁻ phenotype; rather, they cause lysis.

[b]In *B. subtilis* and some other bacilli, two gene fragments undergo recombination to produce the *sigK* gene, which encodes the σ^K coding region.

[c]Subsequent work has further defined two genes, *spoIVFA* and *spoIVFB* (see the text).

The Role of Sigma Factors in Sporulation Regulation

The entire collection of sporulation genes can be sorted into a handful of classes on the basis of transcription by a specific sigma factor. The sporulation sigma factors replace the principal vegetative-cell sigma factor A (σ^A) in RNA polymerase holoenzyme, possibly by outcompeting σ^A for binding to RNA polymerase core. The σ^A of *B. subtilis* plays a role similar to that of σ^{70} of *E. coli* (see chapter 2). As shown in Table 12.2, there are five distinct sigma factors associated with sporulation, sigma H (σ^H), sigma E (σ^E), sigma F (σ^F), sigma G (σ^G), and sigma K (σ^K). Each of the sigma factors is active at a specific time during sporulation. σ^H is active before septation (in the predivisional cell), σ^E and σ^K are sequentially active in the mother cell, and σ^F and σ^G are sequentially active in the forespore.

TEMPORAL PATTERNS OF REGULATION

Measurements of the times of expression of the sporulation genes indicated that many of the genes undergo dramatic increases in expression at specific times after the sporulation process started. Use of gene fusions allowed large-scale comparisons of the complete set of sporulation genes. The most commonly used reporter genes were *lacZ* and *gus* from *E. coli* (see chapter 2). The product of the *lacZ* gene, β-galactosidase, and the product of the *gus* gene, β-glucuronidase, can be assayed by adding "artificial" substrates (such as *o*-nitrophenyl-β-D-galactopyranoside [ONPG] or methylumbelliferyl-β-glucuronide [MUG]) to samples of the test culture at various times after induction of sporulation by a nutritional downshift. The appearance of β-galactosidase or β-glucuronidase activity indicated the onset of gene expression. In addition, direct measurements of mRNAs of various sporulation genes correlated well with the results of *lacZ* and *gus* fusion experiments. Therefore, the use of such fusions became widespread because of the relative ease and convenience of fusion assays.

A significant outcome of comprehensive fusion experiments was the extensive assessment and comparison of the times of expression of many sporulation genes. Moreover, the timing of reporter gene expression could be correlated with the timing of morphological changes that were visible as sporulation progressed.

TRANSCRIPTION FACTOR DEPENDENCE PATTERNS OF EXPRESSION

Fusions with *lacZ* were also used to determine whether the expression of one gene depended on the activity of a second gene. If the expression of one gene depends on a second gene, the second gene may encode a direct or indirect regulator of the first. The use of *spo* mutations in combination with *spo-lacZ* fusions allowed the testing of many regulatory dependencies. For example, expression of a *spoIIA*::*lacZ* fusion was dependent on *spo0* loci (*spo0A*, *spo0B*, *spo0F*) but not on any of the "later" loci.

The discovery that several of the *spo* genes encode sigma factors led to analysis of which sporulation genes are transcribed by each RNA polymerase containing each sigma factor. This could be tested directly by *in vitro* transcription studies with RNA polymerase containing specific sigma factors. In some cases, it was possible to infer the sigma factor dependence of many of the other sporulation genes on the basis of conserved sequences near the transcription start sites, which were shown to be the promoter sites for recognition by the sigma factors. For some transcription factors, such as Spo0A, chromatin immunoprecipitation followed by microarray analysis has been used to determine binding sites throughout the genome, identifying genes likely to be under direct control of the transcription factor.

CELLULAR LOCALIZATION

Several methods have been used to determine the cellular locations of expression of sporulation genes. For example, expression of β-galactosidase in the forespore can be distinguished from that in the mother cell on the basis that the forespore is more resistant to lysozyme. Therefore, a mild treatment with lysozyme results in release of β-galactosidase from the mother cell compartment but not from the forespore. Immunoelectron microscopy has been useful for visualizing the expression of β-galactosidase, and more recently, the use of fusions to green fluorescent protein has allowed the direct visualization of the cellular locations of numerous sporulation proteins.

From studies like these, it could be seen that all of the genes turned on after septation were expressed in only one compartment. The genes transcribed by RNA polymerase with σ^F and σ^G were expressed only in the forespore compartment, and the genes transcribed by RNA polymerase with σ^E and σ^K were expressed only in the mother cell.

Intercompartmental Regulation during Development

When the observations of timing, dependence relationships, and localization of sporulation gene expression are combined, a complex pattern of regulation that includes both spatial and temporal regulation is revealed. This pattern is dependent on sequential activation of sigma factors and signaling between the developing compartments.

Figure 12.26 shows that after septation, gene expression in the forespore depends at first on σ^F and later on σ^G. σ^F is encoded by the *spoIIA* locus, which is transcribed by RNA polymerase containing σ^H prior to septation; however, σ^F becomes active only in the forespore after septation. An early σ^F-dependent transcript, *gpr*, encodes a protease that is important during spore germination (and therefore must be packaged within the mature spore). Another σ^F-transcribed operon is *spoIIIG*, which encodes the late forespore sigma factor σ^G. Transcription of *spoIIIG* differs from that of *gpr* in that it occurs later and, although confined to the forespore, requires functioning of the *spoIIG* locus (Table 12.2), which encodes the mother cell-specific sigma factor σ^E. This dependence of transcription of a gene in the forespore on a sigma factor active only in the mother cell provides an indicator of interdependence of events in the two compartments, which will be discussed below. Once σ^G is produced in the forespore, it directs transcription of a set of genes including the *ssp* genes, which encode spore-specific proteins that condense the nucleoid. Transcription of the *ssp* genes requires not only *spoIIIG*, which encodes σ^G, but also *spoIIAC* (which encodes σ^F) and *spoIIG* (which encodes the mother cell σ^E, a product of which is involved in σ^G production; see below).

Gene expression in the mother cell also reveals intercompartmental regulation. Figure 12.26 shows that one of the genes transcribed relatively early in the mother cell by σ^E RNA polymerase is *gerM*, which encodes a protein required for germination. Later, σ^E RNA polymerase transcribes the genes necessary for production of σ^K (*spoIVCB-spoIIIC*). RNA polymerase with σ^K then transcribes *cotA*,

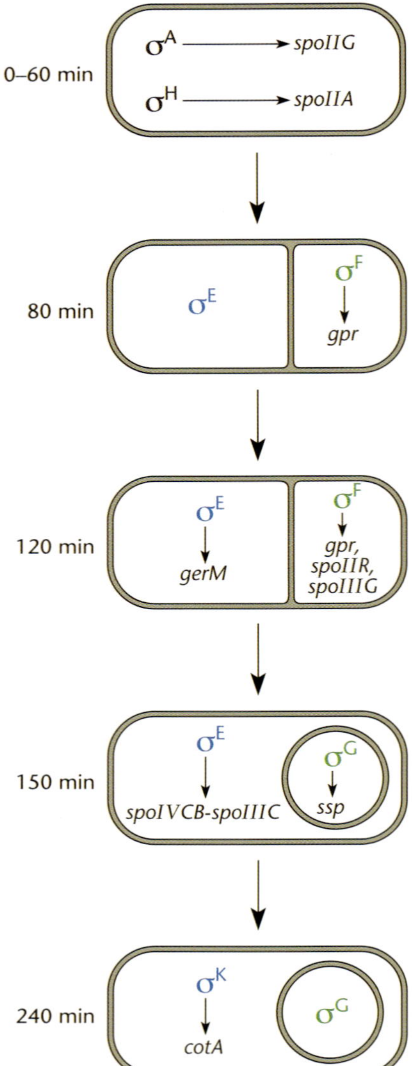

Figure 12.26 Compartmentalization of sigma factors and temporal regulation of transcription within compartments. The genes for σ^E and σ^F are transcribed before polar septation. σ^F is activated in the forespore compartment and is required for transcription of the gene for σ^G, which succeeds it. σ^E is activated in the mother cell and is required for transcription of the genes for its successor, σ^K. Each sigma factor directs synthesis of genes required at the appropriate time in the appropriate compartment.

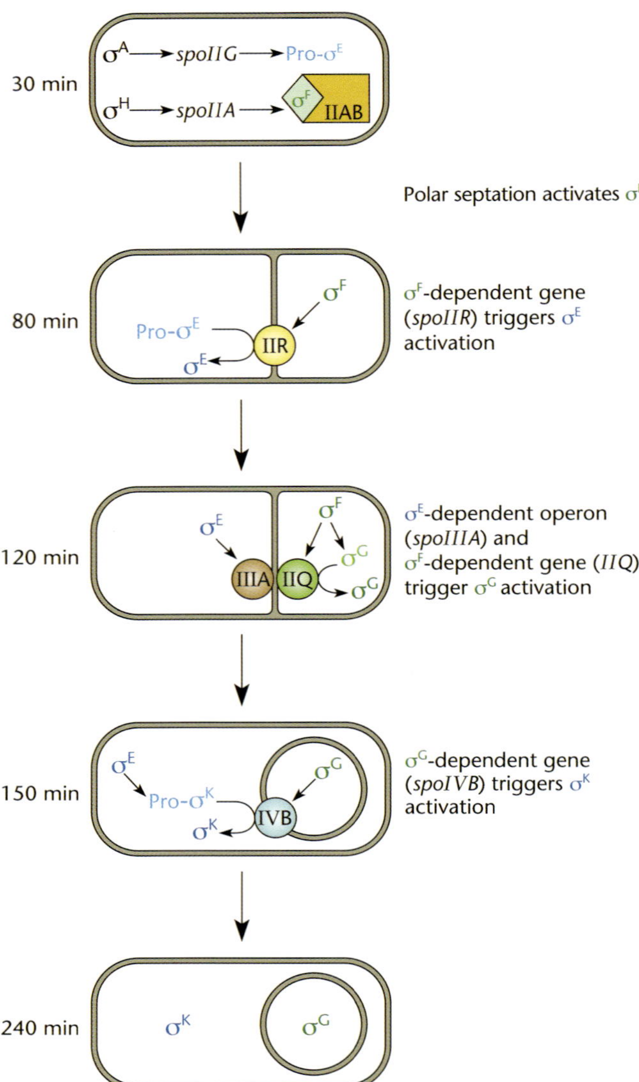

Figure 12.27 Sequential and compartmentalized activation of the *B. subtilis* sporulation sigma factors. A series of signals allows communication between the two developing compartments, as described in the text, at approximately the times shown after induction of sporulation. The forespore sigma factors σ^F and σ^G are controlled by protein-protein interactions with an inhibitory anti-sigma factor, and the mother cell sigma factors σ^E and σ^K are controlled by proteolysis of an inactive precursor to release the active mature protein.

one of a set of *cot* genes that encode proteins incorporated into the spore coat. Transcription of *cotA* also requires the activity of the late forespore sigma factor σ^G.

Activation of the sigma factors alternates between the two developing compartments. As shown in Figure 12.27, each successive activation step requires intercompartmental communication. The critical information is whether morphogenesis and/or gene expression in the other compartment has progressed beyond a "checkpoint." The way in which the two compartments communicate their status to each other is a fascinating area of current research and is discussed in the next sections.

TEMPORAL REGULATION AND COMPARTMENTALIZATION OF σ^E AND σ^F

Both the *spoIIGB* and *spoIIAC* genes, encoding σ^E and σ^F, respectively (Table 12.2), are transcribed in the developing cell before the sporulation septum divides off the forespore compartment (Figure 12.26). However, neither sigma factor starts to transcribe its target genes until after the septum forms, and then, as mentioned above, each sigma factor becomes active in only one compartment: σ^F in the forespore and σ^E in the mother cell. Before septation, the sigma factors are held in inactive states, with a different inhibitory mechanism acting on each sigma factor. For σ^F, the active protein must be released from a complex that contains an inhibitory anti-sigma factor, SpoIIAB. For σ^E, the active form of the protein must be proteolytically released from an inactive precursor, Pro-σ^E. Once the sporulation septum forms, σ^F becomes active in the forespore. Subsequently, σ^E becomes active in the mother cell.

Regulation of σ^F

Activation of σ^F in the forespore requires the interplay of a set of proteins. Two of these are binding partners named SpoIIAA and SpoIIAB (Figure 12.28). SpoIIAB is an anti-sigma factor that binds to and inactivates σ^F. SpoIIAA is an anti-anti-sigma factor that nullifies the anti-sigma factor activity of SpoIIAB; it does this because of its own ability to bind SpoIIAB. Regulation of σ^F utilizes a "partner switching" mechanism in which SpoIIAB switches between an inactive state (bound to SpoIIAA) and an active state (bound to σ^F).

A cycle of phosphorylation and dephosphorylation of SpoIIAA modulates the binding of SpoIIAA to SpoIIAB. Only unphosphorylated SpoIIAA can bind to SpoIIAB. Before septation, SpoIIAA is in a phosphorylated state and so does not bind to SpoIIAB in the preseptational cell. Unbound by SpoIIAA, SpoIIAB is free to bind to and inactivate σ^F. After septation, SpoIIAA is in the unphosphorylated state in the forespore; hence, it binds SpoIIAB, releasing σ^F. Dephosphorylation of SpoIIAA occurs only in the forespore and not in the mother cell, so that σ^F in the mother cell remains bound to SpoIIAB and is therefore inactive.

The enzymes that phosphorylate and dephosphorylate SpoIIAA are SpoIIAB and SpoIIE, respectively. SpoIIAB therefore acts not only as an anti-sigma factor for σ^F but also as a negative regulator of the activity of its own anti-anti-sigma factor. These opposing phosphorylation and dephosphorylation activities, before and after septation, determine the balance between the two forms of SpoIIAA. Before septation and in the mother cell after septation, SpoIIAB phosphorylation of SpoIIAA predominates, resulting in the inability of SpoIIA to bind to SpoIIAB, resulting in inactive σ^F. Once the spore septum has formed, the SpoIIE phosphatase, which is localized to the septum itself, is activated, but only in the forespore. SpoIIE dephosphorylates SpoIIAA~P, and the resulting SpoIIAA can bind to and inactivate SpoIIAB, releasing σ^F for binding to RNA polymerase core.

Regulation of σ^E

Mother cell transcription depends on σ^E, which is encoded by the *spoIIGB* gene of the *spoIIG* operon. The primary product of translation of the *spoIIGB* mRNA is an inactive precursor protein, named Pro-σ^E, which is processed to form the active sigma factor. Pro-σ^E must be cleaved by a specific protease to remove the amino-terminal region of the precursor, releasing active σ^E. The protease that cleaves Pro-σ^E is the product of the first gene in the *spoIIG* operon, *spoIIGA* (Figure 12.29A). Although the *spoIIG* operon is expressed in the predivisional cell, the *spoIIGA* product does not process SpoIIGB immediately but waits about an hour, until after septation has occurred. Then, notification from the forespore that development is proceeding and that σ^F has become active comes via the

Figure 12.28 Model for the regulation of σ^F activity. SpoIIAB holds σ^F in an inactive state (red outline) in both the predivisional sporangium and the mother cell. SpoIIAB phosphorylates SpoIIAA. SpoIIAA~P cannot bind SpoIIAB. The SpoIIE phosphatase controls the generation of dephosphorylated SpoIIAA, which complexes with SpoIIAB and prevents its interaction with σ^F. These reactions allow the release of active σ^F (green outline).

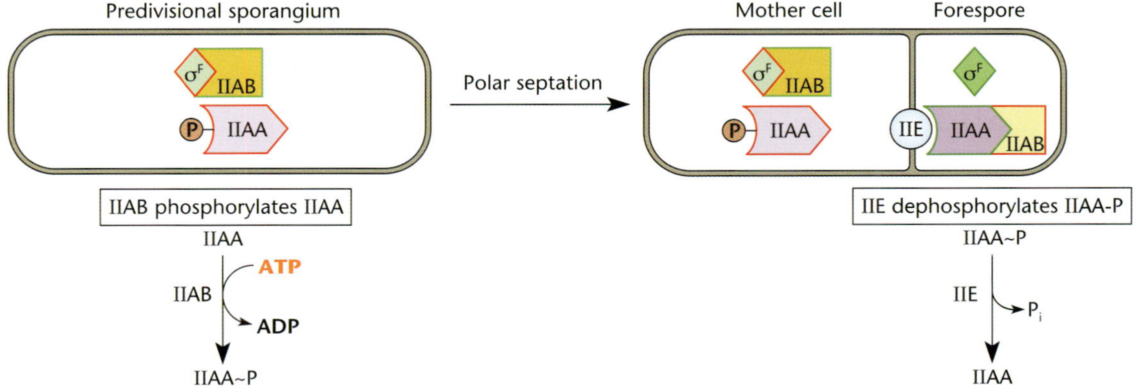

A

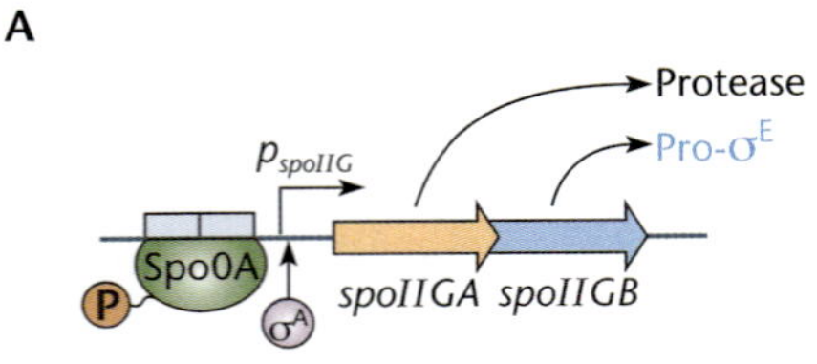

B

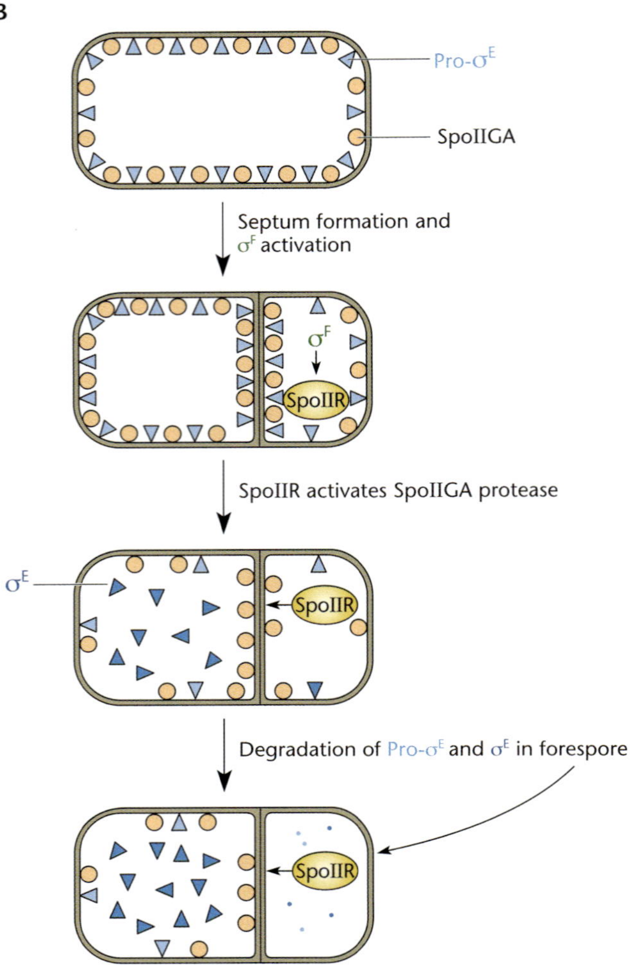

Figure 12.29 Model for activation of σ^E in the mother cell compartment. **(A)** The *spoIIG* operon is transcribed by σ^A RNA polymerase and requires activation by Spo0A~P. **(B)** Pro-σ^E (light blue triangles) and SpoIIGA (gold circles) are associated with the cytoplasmic membrane in the sporulating cell. After septum formation, both proteins are associated with all cell membranes. Then SpoIIR is expressed in the forespore under the control of σ^F, and SpoIIR activates the SpoIIGA protease, which cleaves Pro-σ^E to form active σ^E (dark blue triangles), which is released in the cytoplasm of the forespore. Finally, any Pro-σ^E or σ^E is degraded in the forespore.

messenger protein SpoIIR, which is the product of a σ^F-transcribed gene.

An important clue to the explanation for the time delay in Pro-σ^E processing was the observation that mutants that lacked σ^F activity failed to process Pro-σ^E to σ^E. This suggested that a σ^F-transcribed gene is required for the processing mechanism. A genetic search for such a gene identified the *spoIIR* gene, mutations in which resulted in synthesis of Pro-σ^E that was never processed to active σ^E. It appears that the SpoIIR protein is secreted into the spaces between the septal membranes, where it activates SpoIIGA, which is embedded in the cell membrane, to process Pro-σ^E. One important question is why SpoIIGA, which is synthesized prior to septation and is therefore present in both compartments, fails to cleave Pro-σ^E in the forespore as soon as SpoIIR protein is synthesized there. Restriction of σ^E accumulation to the mother cell appears to be due to a combination of forespore-specific degradation of Pro-σ^E (so that the precursor is no longer available in that compartment) and continued expression of Pro-σ^E in the mother cell (but not the forespore) due to compartment-specific activity of Spo0A~P after septation (see Kroos, Suggested Reading) (Figure 12.29).

σ^G, A SECOND FORESPORE-SPECIFIC SIGMA FACTOR

As described above, formation of the polar septum is the key trigger for activation of σ^F in the forespore. Similarly, engulfment of the forespore by the mother cell membrane, which is dependent on σ^E-dependent transcription in the mother cell, is essential for activation of the second forespore sigma, σ^G. The σ^G-encoding gene, *spoIIIG*, is transcribed in the forespore by σ^F RNA polymerase, which causes synthesis of σ^G to be dependent on σ^F. Its transcription lags behind that of other σ^F-transcribed genes, evidently because it requires a signal from the mother cell sigma factor, σ^E. The evidence for the existence of such a signal is indirect—the *spoIIIG* gene is not transcribed in a *spoIIG* (σ^E) mutant—but the signal has not yet been identified.

Transcription of *spoIIIG* also depends on σ^F-dependent expression of SpoIIQ in the forespore, but the reason for this dependence is unknown. Another aspect of σ^G regulation involves an anti-sigma factor, called CsfB or Gin by different groups, that may prevent premature σ^G activity in the forespore. However, σ^G fails to become active in the forespore unless the products of the *spoIIIA* operon are expressed in the mother cell and SpoIIQ is expressed in the forespore (Figure 12.27). The SpoIIIA and SpoIIQ proteins form channels connecting the mother cell and forespore. The channels are proposed to act as feeding tubes through which the mother cell nurtures the forespore by providing molecules needed for biosynthesis (see Camp and Losick, Suggested Reading). Without the channels, genes

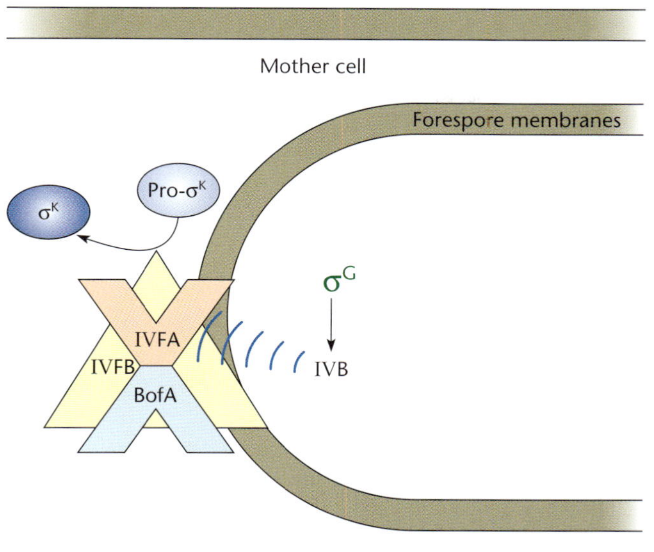

Figure 12.30 Model for regulation of Pro-σ^K processing. Proteolytic cleavage of inactive Pro-σ^K (light blue oval) to active σ^K (dark blue oval) requires activation of the SpoIVFB intramembrane protease by SpoIVFA and BofA, which is dependent on the σ^G-transcribed SpoIVB in the forespore.

under σ^G control in the forespore fail to be expressed, and the forespore collapses. A second anti-sigma factor, named Fin, is transcribed by RNA polymerase containing σ^F and specifically inhibits σ^F activity; this inhibitor helps to facilitate the transition from σ^F- to σ^G-dependent transcription in the forespore.

σ^K, A MOTHER CELL SIGMA

The last sigma to be made, σ^K, is synthesized only in the mother cell. σ^E RNA polymerase transcribes the *sigK* gene, which localizes the transcript to the mother cell. Like σ^E, σ^K is cleaved from a precursor protein, Pro-σ^K. *In B. subtilis, sigK* is generated by a recombination event that joins the *spoIVCB* and *spoIIIC* gene fragments (Table 12.2). Also, like σ^E processing, σ^K processing depends on a signal from the forespore. In this case, the signal is expression of the *spoIVB* gene under the control of the forespore sigma factor σ^G (Figure 12.27).

The SpoIVB protein is thought to be secreted across the innermost membrane surrounding the forespore and to communicate with the Pro-σ^K-processing factors across the membrane, thus activating the SpoIVFB protease that cleaves Pro-σ^K (Figure 12.30). SpoIVB activation of the σ^K-specific protease does not occur directly but, rather, occurs by deactivation of proteins SpoIVFA and BofA, which inhibit the protease. SpoIVFB is an intramembrane-cleaving metalloprotease, meaning that active-site residues are in transmembrane helices. This type of protease cleaves a membrane-associated substrate and is broadly conserved from bacteria to humans. Interestingly, SpoIVFB depends on ATP in order to cleave Pro-σ^K, perhaps sensing the energy level in the mother cell, and the SpoIVFA-SpoIVFB-BofA complex is associated with the SpoIIIA-SpoIIQ channels discussed above that are necessary for σ^G activation in the forespore.

Other Sporulation Systems

Formation of endospores has been found only in the *Firmicutes* and has been characterized most completely in *B. subtilis*, which does not cause disease. However, spore formation underlies the danger of *Bacillus anthracis* as a bioterrorism threat, as the spore form is very stable, difficult to destroy, and easy to disseminate. Sporulation also contributes to pathogenesis in organisms such as *Clostridioides* (formerly *Clostridium*) *difficile*, which is a major problem in hospital settings in part because of the high resistance of the spore state, which increases the probability of spread from patient to patient. *Clostridium botulinum* spores are responsible for a deadly foodborne illness due to improper canning or storage of food contaminated with spores from the soil; the high heat resistance of the spores, and production of a heat-stable toxin within the contaminated food, increases the risk of botulism poisoning.

An unusual member of the *Clostridia*, called *Acetonema longum*, presents special problems to the sporulation process. This organism, while genetically a member of the *Firmicutes*, has an outer structure that more closely resembles that of the enteric bacteria, with a thin cell wall and both an inner and outer cell membrane. This requires inversion and remodeling of the mother cell inner membrane during spore formation in a process even more complicated than that described for *B. subtilis* (see Tocheva et al., Suggested Reading).

Other organisms, such as *Streptomyces* and *Myxococcus*, form spore-like cells as part of complex developmental processes. Like endospore formation, sporulation in these systems is a response to nutritional limitation, and the resulting spores are metabolically inert, allowing survival of inhospitable environments and reinitiation of vegetative growth once the spores are in a new environment. However, the spores formed by these organisms are formed from individual cells, not within another cell as is the case for *Bacillus* endospores, and the resulting spores are not as resistant to environmental stresses as are true endospores. Spore formation in these systems also requires a temporary stop to normal growth, and therefore occurs only when the cell is no longer able to grow in its current environment. These other systems also utilize complex signaling pathways that integrate multiple types of information and serve as important models for understanding developmental processes.

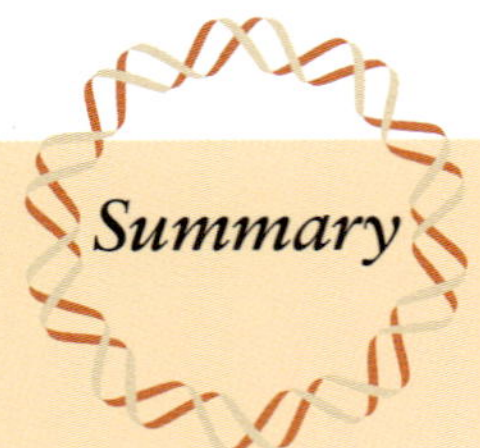

1. The coordinated regulation of a large number of genes is called global regulation. Operons that are regulated by the same regulatory protein are part of the same regulon.

2. In catabolite regulation, the operons for the use of alternate carbon sources are not induced when a better carbon and energy source, such as glucose, is present. In *E. coli* and other enteric bacteria, this catabolite regulation is achieved, in part, by cAMP, which is made by adenylate cyclase, the product of the *cya* gene. When the bacteria are growing in a low-energy carbon source, such as lactose or galactose, the adenylate cyclase is activated and cAMP levels are high. When the bacteria are growing in a high-energy carbon source, such as glucose, cAMP levels are low. The cAMP acts through a protein called CAP (also called CRP), the product of the *crp* gene. CAP is a transcriptional activator, which, with cAMP bound, activates the transcription of catabolite-sensitive operons, such as *lac* and *gal*.

3. *Firmicutes*, such as *B. subtilis*, use a regulatory protein called CcpA to repress genes for carbon source utilization pathways and activate genes for carbon excretion pathways when glucose is high. CcpA activity is controlled by protein-protein interactions with a protein called Hpr that is part of the sugar transport system.

4. Bacterial cells induce different genes depending on the nitrogen sources available. Genes that are regulated through the nitrogen source are called Ntr genes. Most bacteria, including *E. coli*, prefer NH_3 as a nitrogen source and do not transcribe genes for using other nitrogen sources when growing in NH_3. Glutamine concentrations are low when NH_3 concentrations are low. A signal transduction pathway is then activated, culminating in the phosphorylation of NtrC. This signal transduction pathway begins with the GlnD protein, a uridylyltransferase, which is the sensor of the glutamine concentration in the cell. At low concentrations of glutamine, GlnD transfers UMP to the P_{II} protein, inactivating it. However, at high concentrations of glutamine, the GlnD protein removes UMP from P_{II}. The P_{II} protein without UMP attached can bind to NtrB, somehow preventing the transfer of phosphate to NtrC and causing the removal of phosphates from NtrC. The phosphorylated NtrC protein activates the transcription of the *glnA* gene, the gene for glutamine synthetase, as well as the *ntrB* gene and its own gene, *ntrC*, since they are part of the same operon as *glnA*. It also activates the transcription of operons for using other nitrogen sources.

5. The NtrB and NtrC proteins form a sensor kinase and response regulator pair and are related to other sensor kinase and response regulator pairs in bacteria.

6. NtrC-regulated promoters of *E. coli* and other enterics require a special sigma factor called σ^N. Transcription by RNA polymerase containing σ^N requires activation by phosphorylated NtrC.

7. The genes encoding the ribosomal proteins, rRNAs, and tRNAs are part of a large regulon in bacteria, with hundreds of genes that are coordinately regulated. A large proportion of the cellular energy goes into making the rRNAs, tRNAs, and ribosomal proteins; therefore, regulating the expression of these genes saves the cell considerable energy.

8. The synthesis of ribosomal proteins is coordinated by coupling the translation of the ribosomal protein mRNAs to the amount of free rRNA that is not yet in a ribosome. The ribosomal protein genes are organized into operons, and one ribosomal protein of each operon plays the role of translational repressor. The same protein also binds to free rRNA, so that when there is excess rRNA in the cell, all of the repressor protein binds to the free rRNA, and none is available to repress translation.

9. In *E. coli*, the synthesis of rRNA and tRNA following amino acid starvation is inhibited by ppGpp, synthesized by an enzyme called RelA that is found associated with the ribosome. All types of bacteria contain ppGpp, so the regulation may be universal, but the mechanism of action of ppGpp varies. In some organisms, including *E. coli*, a protein named DksA is important in mediating the response of RNA polymerase to ppGpp.

10. Cells contain fewer ribosomes when they are growing more slowly in poorer media. This is called growth rate control and may be due to the lower concentration of the initiating ribonucleosides, ATP and GTP, in slower-growing cells. RNA polymerase forms short-lived open complexes on the promoters for the rRNA genes, and these may have to be stabilized by immediate initiation of transcription with high concentrations of ATP and GTP.

11. Bacteria induce a set of proteins called the heat shock proteins in response to an abrupt increase in temperature. Some of the heat shock proteins are chaperones, which assist in the refolding of denatured proteins; others are proteases, which degrade denatured proteins. The heat shock response is common to all organisms, and some of the heat shock proteins have been highly conserved throughout evolution.

12. In *E. coli*, the promoters of the heat shock genes are recognized by RNA polymerase holoenzyme with an alternative sigma factor called the heat shock sigma factor, or σ^H. The amount of this sigma factor markedly increases following heat shock, leading to increased transcription of the heat shock genes. The increase in σ^H following heat shock involves an RNA thermosensor in the mRNA encoding σ^H and DnaK, a chaperone that is one of the heat shock proteins. The DnaK protein normally binds to σ^H, targeting the sigma factor for degradation. Immediately after a heat shock, DnaK binds to other denatured proteins, making less DnaK available to bind to σ^H so that the sigma factor is stabilized and more of it accumulates.

Summary (continued)

13. In addition to the heat shock sigma factor, bacteria have other stress sigma factors that are activated by a wide variety of different stresses.

14. Bacteria have ways of detecting stress to their membranes, including osmotic stress and damage to the outer membrane. These are called extracytoplasmic stresses.

15. One of the ways that bacteria adjust to changes in the osmolarity of the medium is by changing the ratio of their porin proteins, which form pores in the outer membrane through which solutes can pass to equalize the osmotic pressure on both sides of the membrane. The major porins of *E. coli* are OmpC and OmpF, which make pores of different sizes, thereby allowing the passage of different-size solutes. The relative amounts of OmpC and OmpF change in response to changes in the osmolarity of the medium. The *ompC* and *ompF* genes in *E. coli* are regulated by a sensor kinase and response regulator pair of proteins, EnvZ and OmpR. The EnvZ protein is an inner membrane protein with both kinase and phosphatase activities that, in response to a change in osmolarity, can transfer a phosphoryl group to or remove one from OmpR, a transcriptional activator. The state of phosphorylation of OmpR affects the relative rates of transcription of the *ompC* and *ompF* genes.

16. The ratio of OmpF to OmpC porin proteins is also affected by an antisense RNA named MicF. A region of the MicF RNA can base pair with the TIR of the OmpF mRNA and block access by ribosomes, thereby inhibiting OmpF translation. The *micF* gene is regulated by a number of transcriptional regulatory proteins, including SoxS, which induces the oxidative stress regulon.

17. Some bacteria detect damage to their outer membrane by detecting the accumulation of outer membrane proteins in the periplasm. The two systems in *E. coli* are Cpx and σ^E, which respond to the accumulation in the periplasm of pilin subunits and Omp proteins, respectively.

18. The virulence genes of pathogenic bacteria can also be members of global regulons and are normally expressed only when the bacterium is in its host.

19. The diphtheria toxin gene, *tox*, encoded by a prophage of *C. diphtheriae*, is turned on only when iron is limiting, a condition mimicking that in the host. The *tox* gene is regulated by a chromosomally encoded repressor protein, DtxR, which is similar to the Fur protein involved in regulating the genes of iron availability pathways in *E. coli* and other enteric bacteria.

20. The toxin genes of *V. cholerae* are carried on a prophage and are regulated by a regulatory cascade that begins with a transcriptional activator, ToxR. The ToxR protein traverses the inner membrane and is activated by a second protein, ToxS. ToxR and ToxS act in concert with another gene pair, TcpP-TcpH, to activate the transcription of *toxT*, whose gene product in turn activates the transcription of virulence genes.

21. The virulence genes of *B. pertussis* are regulated by a sensor kinase and response regulator pair of proteins, BvgS and BvgA. The regulation goes through multiple stages as the bacterium enters its host.

22. Many sRNAs play important roles in gene regulation in bacteria. Most sRNAs function by pairing with complementary sequences in mRNA, often with the help of the Hfq RNA-binding protein, and block or enhance translation or mRNA degradation by RNases. Other sRNAs function by direct interaction with RNA polymerase or by titration of a regulatory protein.

23. *B. subtilis* sporulation is the best-understood bacterial developmental system. It involves a regulatory cascade of sigma factors and communication between cellular compartments involving signal proteins, proteases, and channels.

QUESTIONS FOR THOUGHT

1. Why do you think genes for the utilization of amino acids as a nitrogen source are not under Ntr regulation in *Salmonella* spp. but are under Ntr regulation in *Klebsiella* spp.?

2. Why are the corresponding sensor kinase and response regulator genes of the various two-component systems so similar to each other?

3. Why do some organisms use different enzymes for ppGpp synthesis and degradation while others use a single enzyme with both activities?

4. Why do bacteria use small molecules or peptides to sense other bacteria in their environment? What benefit might this have to the bacteria?

5. Why are *B. subtilis* cells so averse to sporulating that they sporulate only after prolonged starvation? What does this say about the purpose of sporulation?

SUGGESTED READING

Barker MM, Gourse RL. 2001. Regulation of rRNA transcription correlates with nucleoside triphosphate sensing. *J Bacteriol* **183:**6315–6323.

Browning DF, Busby SJ. 2004. The regulation of bacterial transcription initiation. *Nat Rev Microbiol* **2:**57–65.

Burgos HL, O'Connor K, Sanchez-Vazquez P, Gourse RL. 2017. Roles of transcriptional and translational control mechanisms in regulation of ribosomal protein synthesis in *Escherichia coli*. *J Bacteriol* **199:** e00407-17.

Camp AH, Losick R. 2009. A feeding tube model for activation of a cell-specific transcription factor during sporulation in *Bacillus subtilis*. *Genes Dev* **23:**1014–1024.

Cashel M, Gallant J. 1969. Two compounds implicated in the function of the RC gene of *Escherichia coli*. *Nature* **221:**838–841.

Cavanagh AT, Wassarman KM. 2014. 6S RNA, a global regulator of transcription in *Escherichia coli*, *Bacillus subtilis*, and beyond. *Annu Rev Microbiol* **68:**45–60.

Chandrangsu P, Rensing C, Helmann JD. 2017. Metal homeostasis and resistance in bacteria. *Nat Rev Microbiol* **15:**338–350.

Commichau FM, Forchhammer K, Stülke J. 2006. Regulatory links between carbon and nitrogen metabolism. *Curr Opin Microbiol* **9:**167–172.

Condon C, Squires C, Squires CL. 1995. Control of rRNA transcription in *Escherichia coli*. *Microbiol Rev* **59:**623–645.

Dalebroux ZD, Svensson SL, Gaynor EC, Swanson MS. 2010. ppGpp conjures bacterial virulence. *Microbiol Mol Biol Rev* **74:**171–199.

Eisenreich W, Dandekar T, Heesemann J, Goebel W. 2010. Carbon metabolism of intracellular bacterial pathogens and possible links to virulence. *Nat Rev Microbiol* **8:**401–412.

Grabowicz M, Silhavy TJ. 2017. Envelope stress responses: an interconnected safety net. *Trends Biochem Sci* **42:**232–242.

Hall MN, Silhavy TJ. 1981. Genetic analysis of the *ompB* locus in *Escherichia coli* K-12. *J Mol Biol* **151:**1–15.

Higgins D, Dworkin J. 2012. Recent progress in *Bacillus subtilis* sporulation. *FEMS Microbiol Rev* **36:**131–148.

Inada T, Kimata K, Aiba H. 1996. Mechanism responsible for glucose-lactose diauxie in *Escherichia coli*: challenge to the cAMP model. *Genes Cells* **1:**293–301.

Kim T-J, Gaidenko TA, Price CW. 2004. A multicomponent protein complex mediates environmental stress signaling in *Bacillus subtilis*. *J Mol Biol* **341:**135–150.

Krásný L, Gourse RL. 2004. An alternative strategy for bacterial ribosome synthesis: *Bacillus subtilis* rRNA transcription regulation. *EMBO J* **23:**4473–4483.

Kroos L. 2007. The *Bacillus* and *Myxococcus* developmental networks and their transcriptional regulators. *Annu Rev Genet* **41:**13–39.

Landini P, Egli T, Wolf J, Lacour S. 2014. σ^S, a major player in the response to environmental stresses in *Escherichia coli*: role, regulation and mechanisms of promoter recognition. *Environ Microbiol Rep* **6:**1–13.

Leigh JA, Dodsworth JA. 2007. Nitrogen regulation in bacteria and archaea. *Annu Rev Microbiol* **61:**349–377.

Lenz DH, Miller MB, Zhu J, Kulkarni RV, Bassler BL. 2005. CsrA and three redundant small RNAs regulate quorum sensing in *Vibrio cholerae*. *Mol Microbiol* **58:**1186–1202.

Magasanik B. 1982. Genetic control of nitrogen assimilation in bacteria. *Annu Rev Genet* **16:** 135–168.

Meyer AS, Baker TA. 2011. Proteolysis in the *Escherichia coli* heat shock response: a player at many levels. *Curr Opin Microbiol* **14:**194–199.

Nagai H, Yuzawa H, Yura T. 1991. Interplay of two *cis*-acting mRNA regions in translational control of σ^{32} synthesis during the heat shock response of *Escherichia coli*. *Proc Natl Acad Sci USA* **88:**10515–10519.

Raivio TL. 2014. Everything old is new again: an update on current research on the Cpx envelope stress response. *Biochim Biophys Acta* **1843:**1529–1541.

Ross W, Sanchez-Vazquez P, Chen AY, Lee JH, Burgos HL, Gourse RL. 2016. ppGpp binding to a site at the RNAP-DksA interface accounts for its dramatic effects on transcription initiation during the stringent response. *Mol Cell* **62:**811–823.

Ruiz N, Silhavy TJ. 2005. Sensing external stress: watchdogs of the *Escherichia coli* cell envelope. *Curr Opin Microbiol* **8:**122–126.

Ruvolo MV, Mach KE, Burkholder WF. 2006. Proteolysis of the replication checkpoint protein Sda is necessary for the efficient initiation of sporulation after transient replication stress in *Bacillus subtilis*. *Mol Microbiol* **60:**1490–1508.

Schumann W. 2016. Regulation of bacterial heat shock stimulons. *Cell Stress Chaperones* **21:**959–968.

Schwartz D, Beckwith JR. 1970. Mutants missing a factor necessary for the expression of catabolite-sensitive operons in *E. coli*, p 417–422. *In* Beckwith JR, Zipser D (ed), *The Lactose Operon*. Cold Spring Harbor Laboratory Press, Cold Spring Harbor, NY.

Sineva E, Savkina M, Ades SE. 2017. Themes and variations in gene regulation by extracytoplasmic function (ECF) sigma factors. *Curr Opin Microbiol* **36:**128–137.

Sonenshein AL. 2007. Control of key metabolic intersections in *Bacillus subtilis*. *Nat Rev Microbiol* **5:**917–927.

Tocheva EI, Matson EG, Morris DM, Moussavi F, Leadbetter JR, Jensen GJ. 2011. Peptidoglycan remodeling and conversion of an inner membrane into an outer membrane during sporulation. *Cell* **146:**799–812.

Vakulskas CA, Potts AH, Babitzke P, Ahmer BMM, Romeo T. 2015. Regulation of bacterial virulence by Csr (Rsm) systems. *Microbiol Mol Biol Rev* **79:**193–224.

Wray LV Jr, Zalieckas JM, Fisher SH. 2001. *Bacillus subtilis* glutamine synthetase controls gene expression through a protein-protein interaction with transcription factor TnrA. *Cell* **107:**427–435.

Yoshida T, Qin L, Egger LA, Inouye M. 2006. Transcription regulation of *ompF* and *ompC* by a single transcription factor, OmpR. *J Biol Chem* **281:**17114–17123.

Zhou YN, Kusukawa N, Erickson JW, Gross CA, Yura T. 1988. Isolation and characterization of *Escherichia coli* mutants that lack the heat shock sigma factor σ^{32}. *J Bacteriol* **170:**3640–3649.

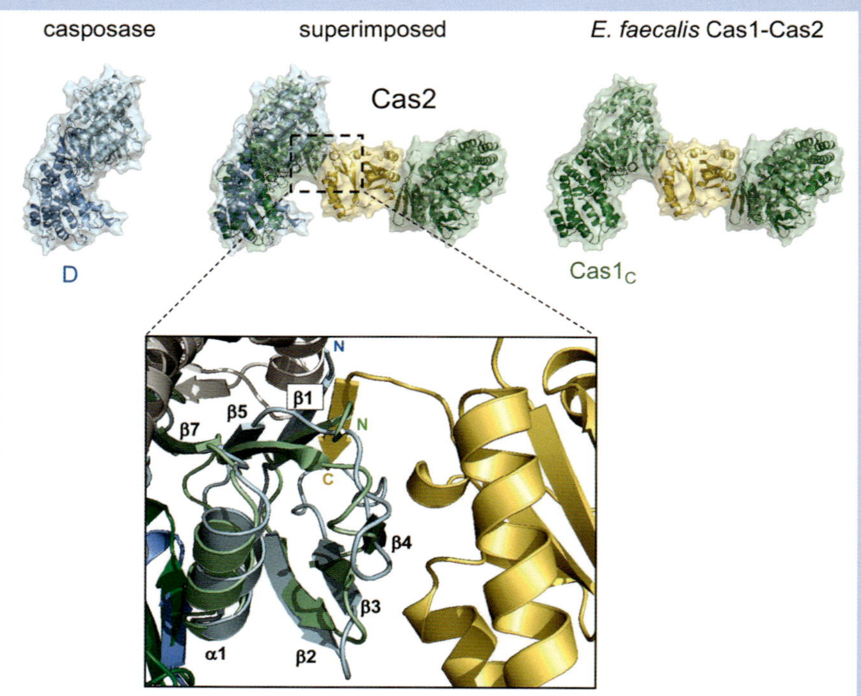

Casposase structure and the mechanistic link between DNA transposition and spacer acquisition by CRISPR-Cas. (See Hickman et al., Suggested Reading.) The work is made available under the Creative Commons CC0 public domain dedication.

13 Genomes and Genomic Analysis

DNA SEQUENCING HAS REVOLUTIONIZED BIOLOGY, including our understanding of bacteria and archaea. The pace with which DNA sequencing has progressed is remarkable. The first bacterial viruses were sequenced in the 1970s, but the larger genomes of the first free-living bacterium (*Haemophilus influenzae*) and archaeon (*Methanococcus jannaschii*) were not completed until 1995. The first genome sequences of laboratory strains of *Escherichia coli* and *Bacillus subtilis* were completed in the next couple of years. Fast forward to the publication of this textbook; a researcher now can readily download hundreds of thousands of complete bacterial genomes, including thousands of genomes for each of the well-studied species of medical, agricultural, or industrial importance. This level of information has allowed for a comprehensive view of some of the underlying rules for how genes are organized and how genomes evolve. Full genome sequences have also facilitated development of a number of technologies for the study of bacteria and archaea in what is sometimes referred to as **genomics**. The term "genomics" is generally used to refer to the mechanisms of study that are enabled by having complete genomes.

While not a focus of this textbook, the study of microbial communities has also been completely revolutionized as DNA sequencing has become less expensive. **Metagenomics** generally refers to the types of studies used on communities of organisms using DNA sequence information. Metagenomics is particularly important for microbial communities, because a very small percentage of microorganisms that are present in any particular environment can be cultured. Therefore, culture-dependent techniques provide only a very small picture of what is happening in most microbial communities. Of particular interest are microbes that only grow inside a host or are dependent on being grown with other microbes, thereby making it impossible to obtain a pure culture for DNA sequencing.

The Bacterial Genome

Examination of genome sequences is revealing much about the nature of bacterial genomes. For example, bacterial genomes are densely packed with coding information. In bacteria, the major types of coding information include genes encoding proteins or rRNAs or tRNAs along with regulatory elements such as small RNAs that are discussed in earlier chapters. In bacterial and archaeal genomes, there can be 500 to 1,000 genes in every megabase of sequence. In contrast, the human genome is estimated to have 12 to 15 genes per megabase primarily because of introns and other noncoding DNA.

Another outcome of comparing genome sequences has been the observation of the high degree of conservation in genetic linkage, called **synteny**. All bacteria share some level of synteny over stretches of chromosome sequence, but this synteny is very obvious within a genus. Major changes that break up synteny are of two types, inversions and insertions. Inversion within bacterial genomes appears to be constrained by the nature of the genomes themselves; many sequence features in the chromosome must be maintained in a gradient that progresses from the origin to the terminus of replication (see Box 1.1), and natural inversions occur predominantly in a way that does not cause changes in this gradient across the genome (see Box 3.1). The syntenic regions of the chromosome are also interspersed with unique insertions of DNA sequence that were originally acquired by **horizontal transfer**, i.e., by DNA transfer from other types of bacteria rather than solely by vertical inheritance from their ancestors. These regions are typically composed of mobile elements such as integrated bacteriophages called prophages (see chapter 7), insertion sequence elements and transposons (see chapter 8), and integrative conjugative elements (see chapter 5).

Other collections of genes that break up the synteny in closely related organisms but show no obvious mechanism for mobilization are referred to as **genomic islands**. Genomic islands generally encode information that provides an advantageous mechanism or allows adaptation to a specific environment. Certain genomic islands for which specific functions are known are sometimes referred to as pathogenicity islands (see chapter 12) or fitness islands. Genomic islands can be of particular interest to a molecular geneticist or bioinformatician studying bacteria and archaea because they can offer clues into the gene products that allowed a strain of bacteria to adapt to a given environment. This is particularly appreciated with pathogenic bacteria because this information can suggest ways to better understand and combat bacterial diseases, especially emergent diseases. For example, genome sequencing has revealed why *E. coli* strains can be either "intestinal friends" or "intestinal foes." Since the significance of *E. coli* as a model organism is discussed in the text, it is important to clarify what distinguishes the disease-causing *E. coli* strains, such as the O157:H7 strain that is the causative agent of deadly infections worldwide, from the harmless *E. coli* laboratory strains, such as K-12. *E. coli* K-12 and *E. coli* O157:H7 share 4.1 Mb of DNA sequence homology, but scattered throughout the genome of O157:H7 are long DNA regions that encode virulence characteristics. Much of this additional DNA is the result of horizontal transfer.

E. coli provides an example of what is sometimes called an "open" genome because of the large and obvious impact of horizontal gene transfer and the accumulation of genomic islands. The rampant exchange of genetic information between bacteria has even brought into question the very idea of whether bacteria can have true species that represent a linear ancestry if genetic exchange between species is so pervasive. However, if we dig a little deeper, we can tell that there is a core set of genes that is common to a particular species. The genes shared among all members of a species are called the **core genome**, and these genes are very similar within a species. Figure 13.1 shows the relationship between three hypothetical strains from a species with an open genome. Those genes that are shared among all three strains are considered the core genome. **Pan genome** is a term that applies to any gene that is found in any strain from the same species, in other words, the total of all of the genes in all of the sequenced strains from the species (Figure 13.1). As more and more genomes from a given species with an open genome are sequenced, we can expect the pan genome to continue to grow. In contrast, the size of the core genome of a given species will decrease as more

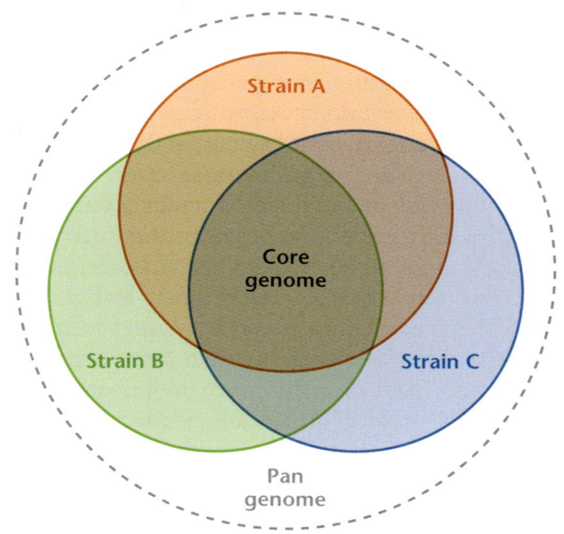

Figure 13.1 Bacterial strains from within the same species can be significantly different at the genetic level. Bacteria that show evidence of extensive horizontal transfer in their genome are referred to as having open genomes. In the example, three strains from the same species are compared with a Venn diagram that indicates the number of genes shared across some or all of the sequenced strains. Genes shared across all of the sequenced strains (three in our example) are the core genome for the species. Genes found to occur in the species in any of the strains form the pan genome.

genomes are sequenced and examples are identified that lack genes that were previously considered to be universal to that species, and this number will presumably reach a plateau that includes all of the crucial genes. Mobile elements such as plasmids, transposons, and bacteriophage (as discussed in chapters 4, 5, 7, and 8) will be much more common in the pan genome than in the core genome. The entirety of mobile elements in a given host is sometimes referred to as the **mobilome**.

Earlier chapters often referred to information gathered from genome sequences. In the introduction we learned about the incredible diversity of bacteria and archaea, information that comes from comparing features of genomes through the process of molecular phylogeny, which examines changes in highly conserved features found in all bacteria and archaea as a tool to determine how they are related. In chapter 1 we learned about DNA sequence information in prokaryotic genomes found between and within the coding features of the genome that allow it to be efficiently replicated, repaired, compacted, and segregated. Bacterial genomes are highly organized with many features that show a distinct polarity from the origin of DNA replication (*oriC*) to the region where DNA replication forks meet, typically approximating the *dif* site where dimer chromosomes resolve (see Box 1.1). Understanding the details of these polar informational aspects of bacterial chromosomes and plasmids, and new yet-to-be-discovered features, requires that full genome sequences are known.

As expected, even with the relatively simple genomes of bacteria and archaea, determining the functional coding properties within the string of A, T, C, and G nucleotides is not trivial. Computational systems have been developed for this task that allow for **genome annotation**, by which the coding regions for proteins and functional RNAs are identified and predications about their function are made based on work with other bacteria and archaea, often with the model systems discussed in detail in this textbook (Box 13.1). While gene function can often be predicted, genomics tools are required to confirm these functions in newly sequenced genomes, something that is often complicated by the operon structure found in bacteria and archaea (Box 13.2). In many chapters we also learned about regulatory features that often can be predicted using techniques in comparative genomics but also must be confirmed using transcript and proteome analysis, which are enabled by knowing the full genome sequence.

In the remainder of this chapter we discuss additional mechanisms that contribute to how bacterial genomes evolve and the tools molecular geneticists use to study bacteria and archaea. As we will see, many of the tools are also used throughout molecular biology in the study of all domains of life, with important applications in the fields of medicine, agriculture, and engineering, among many others.

DNA Sequencing

A variety of technologies have been developed for DNA sequencing over the years. One important tool in determining DNA sequences has been DNA polymerases derived from a variety of bacteria, but other chemistries have also been important. As discussed in chapter 1, DNA polymerases all extend a primer polynucleotide chain by attaching the 5′ phosphate of an incoming deoxynucleotide triphosphate to the 3′ end of the growing primer chain. They can synthesize DNA only by extending primers that are hybridized to a template DNA, and the choice of which deoxynucleotide to add at each step is determined by complementary base pairing between the incoming deoxynucleotide and the template DNA, leading to synthesis of DNA that is a complementary copy of the template.

In the original **Sanger sequencing** method, chain-terminating dideoxynucleotides are used, in which the absence of the 3′ OH prevents further extension of the DNA chain (see Figure 1.2). Four reactions are run, each containing three of the deoxynucleotides; for the fourth nucleotide, a mixture of the normal deoxynucleotide and the corresponding dideoxynucleotide was used. The nucleotides are incubated with the template DNA being sequenced, as well as a primer complementary to a sequence on the DNA close to the region being sequenced, and a DNA polymerase. The four reactions correspond to the use of a dideoxynucleotide for G, A, T, or C. Each time the dideoxynucleotide is incorporated in place of the normal deoxynucleotide, the chain terminates, so that the reaction containing dideoxy-G will stop only at positions at which a G should be incorporated, the reaction containing dideoxy-A will stop only at positions at which an A should be incorporated, etc. If the four reactions are then run next to each other on a polyacrylamide gel, a "sequencing ladder" is obtained. Which lane has the next "rung" or "step" in the ladder indicates what nucleotide is next in the template DNA that is being sequenced. An advance in this strategy involves the ability to have each of the four dideoxynucleotides labeled with a different fluorescent molecule, each emitting light at a different frequency so that each fragment of DNA is marked with a certain color at the terminal nucleotide. This technology allows all of the "rungs" on the ladder to be read in one lane, an advance that also allows the samples to be run in narrow tubing instead of large sequencing gels, which facilitates greater automation in the DNA sequencing process.

BOX 13.1

Annotation and Comparative Genomics

Genome sequencing is one way to begin the study of a bacterium. However, this information is most useful in the context of other information about the bacterium. In this book, we show how the methods of genetic analysis and genomic analysis complement each other to permit a more complete understanding of how a bacterium functions. Figure 1 summarizes many of the types of genomic and genetic experiments available. Here, we briefly describe some of these experiments, many of which are more fully discussed and illustrated in previous chapters. For an example of a journal publication that describes a comprehensive annotation analysis of *E. coli* K-12, see Riley et al. (References).

Genome Sequence

Genome-sequencing methodologies are discussed in the text.

Annotation and Comparative Genomics

For analysis of a new genome sequence, the use of bioinformatics resources, as described below, can give us a profile of the similarities of DNA sequences and gene products to those of other organisms.

Functional Annotation

Genome sequence information (e.g., Figure 2) is accumulating faster than we can understand it, but tools for analyzing genome sequences are also rapidly increasing in number and sophistication. Examples of tools available for public use are provided below.

RNA-Encoding Sequences

rRNA-encoding sequences are extremely highly conserved and so are easily recognized (see Chan and Lowe, References). Methods for identifying small, noncoding, regulatory RNAs are only now being developed (see chapter 12).

Protein-Encoding Sequences

For finding genes in prokaryotes (i.e., distinguishing coding from noncoding DNA), an especially useful tool is **GLIMMER** (for **Gene Locator and Interpolated Markov Modeler**; see Delcher et al., References). A **Markov model** is a statistical tool useful in a situation in which a system, in this case a protein sequence, undergoes a series of changes in its state (i.e., amino acid substitutions), and a change from one state to the next is independent of the history of the state. A type of Markov model that is especially useful in genome annotation is the **hidden Markov model** (**HMM**). An HMM uses previous data sets to weight an analysis; in other words, an HMM is able to "train itself" if it is "given" a set of about 50 related sequences. Once trained, the HMM places more value on states that are conserved, e.g., common amino acid substitutions. This allows HMMs to be highly sensitive. In addition, an HMM can be further trained to recognize codons or DNA sequences characteristic of a particular organism; this type of model is called an **interpolated Markov model**.

An HMM is able to consider all possible combinations of factors, such as gaps, matches, and mismatches, that could affect the alignment of a set of sequences. Thus, an HMM can pick out amino acid positions that are or are not conserved. The basis for using HMMs is the HMMER statistical tool (see Potter et al., References).

For protein-encoding sequences, the most common sequence analysis tool for predicting function is called **BLAST (Basic Local Alignment Search Tool)**. The U.S. National Center for Biotechnology Information (NCBI) has a publicly available website with step-by-step tutorials (https://blast.ncbi.nlm.nih.gov/Blast.cgi). To use BLAST, a query of a nucleotide or amino acid sequence is submitted for comparison to the publicly available databases. Searches sometimes ask that a sequence be submitted in **FASTA** format, which means as an uninterrupted sequence, using the standard nucleotide and single-letter amino acid codes.

The BLAST algorithm can translate a sequence in all six possible reading frames (Figure 2.44). Moreover, the BLAST search can be performed in several ways: translated query versus protein database (blastx), protein query versus translated database (tblastn), or translated query versus translated database (tblastx). Numerous additional variations of BLAST are available at the NCBI BLAST website, including protein-protein BLAST (blastp) and position-specific iterated BLAST (**PSI-BLAST**).

After a query is submitted, the BLAST algorithm calculates the statistical significance of any matches found. The significance of a similarity is expressed as an **E value**. The E (for expected) value is a term that indicates the significance of an alignment found between two sequences. An E value is the number of database hits of similar quality that you would expect to find by chance. One sequence would be the query sequence, and the other would be a related sequence found in a database, for example, by a BLAST search. The lower the E value, the closer the similarity found. Generally, an E value greater than 0.01 to 0.05 is considered to be insignificant.

Related gene sequences can be categorized as **homologs**, which are genes or sequences that share common ancestry. Homologs can be classified as **orthologs**, which are genes that are similar in sequence and have a common ancestor but are found in different species with (in some cases) similar functions, and **paralogs**, which are genes that arose by duplication within a given species and may have similar functions. In addition, proteins are categorized into "families," in which the individual members share certain features, as discussed below.

It is important to note that the matches that result from some BLAST searches, such as blastp, are matches to protein domains rather than to genes per se. A protein domain is gen-

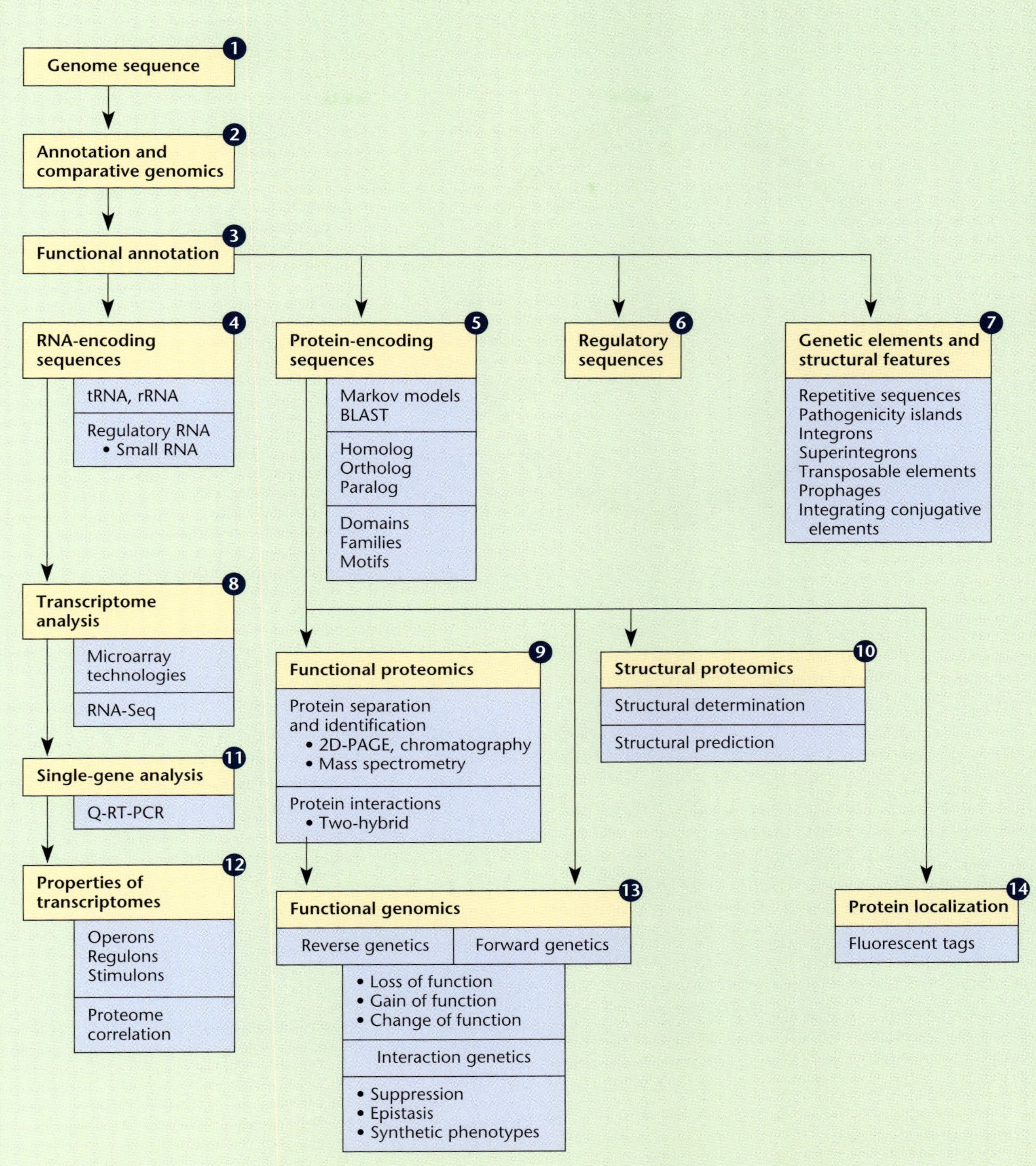

Figure 1

erally an independently folding element of a protein; thus, proteins are mosaics of domains. Examples are the σ factor domains in Figure 2.6.

The regions of sequence conservation among proteins that are found by an HMM analysis can be used to categorize proteins into families and so can provide information about the function of a protein. The term "family" is used in many contexts and can refer to many types of categories. In a scheme that broadly defines families, proteins are divided into three types of families. One type of family contains sequences

(continued)

Annotation and Comparative Genomics

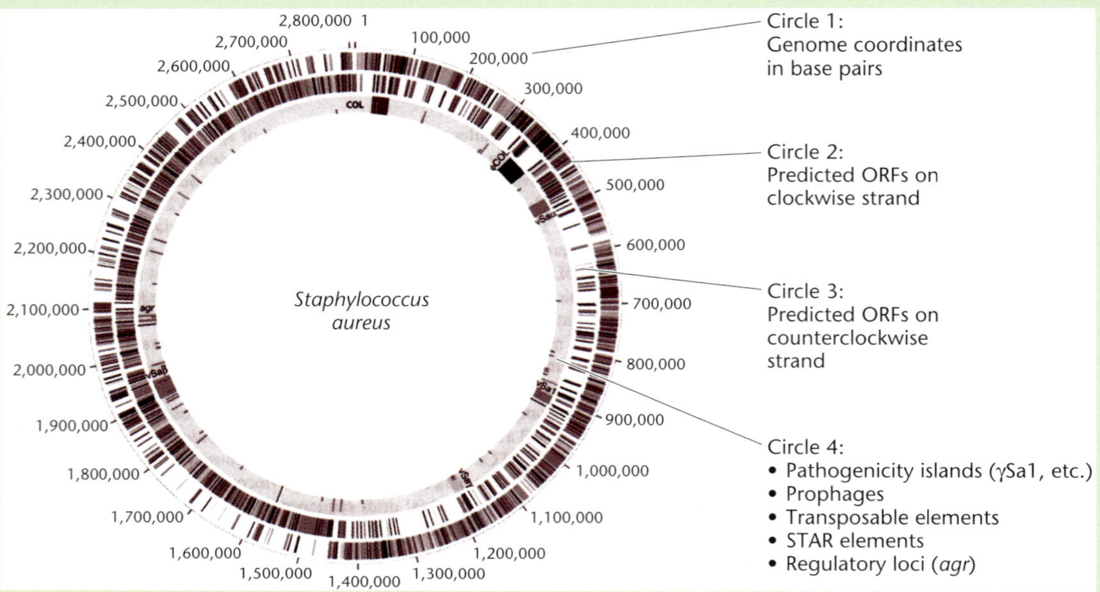

Figure 2 Modified from Gill SR, Fouts DE, Archer GL, et al, *J Bacteriol* **187:**2426–2438, 2005.

based on one or more shared protein domains. A given domain may be found in more than one functional type of protein. A second type of family is enzyme families, in which the gene products all perform the same biological function. Enzyme names are assigned by the Enzyme Commission (https://enzyme.expasy.org/). A very large collection of metabolic pathways has been compiled at the website https://www.genome.jp/kegg/. This site shows so-called **Kegg maps**, which provide a preliminary suggested pathway in which an enzyme might function; experimental support for the enzyme classification is important. A third family type is the "superfamily," which contains two or more proteins that are related by sequence but have not necessarily been tested for biochemical function. Thus, the importance of experimental study of protein function (see below) cannot be overstated.

A COG is a **cluster of orthologous genes**. The NCBI has defined approximately 25 COGs by comparing protein-encoding regions of dozens of complete genomes of the major phylogenetic lineages; each COG is defined by proteins from at least three lineages. NCBI COGs are grouped into the following four major categories:

1. Information storage and processing, containing translation, ribosomal structure, and biogenesis functions; transcription functions; and DNA replication, recombination, and repair functions
2. Cellular processes and signaling, containing functions for cell division and partitioning; posttranslational modification, protein turnover, and chaperones; cell envelope biogenesis and the outer membrane; cell motility and secretion; inorganic ion transport and metabolism; and signal transduction mechanisms
3. Metabolism, containing functions for energy production and conversion; carbohydrate transport and metabolism; amino acid transport and metabolism; nucleotide transport and metabolism; coenzyme metabolism; lipid metabolism; and secondary-metabolite biosynthesis, transport, and catabolism
4. "Poorly characterized," containing "general function prediction only" and "function unknown"

The Pfam families are more likely to describe domains than full-length proteins, for example, indicating evidence for an ATP-binding domain in a protein (see El-Gebali et al., References).

The TIGRFAM families are a versatile resource for protein classifications and include superfamilies, which comprise all proteins that have amino acid homology but that may differ in biological function; equivalogs, which include proteins that are conserved in function; and subfamilies, which contain proteins incompletely evaluated for function (see Finn et al., References).

Motifs are conserved patterns of amino acids, often the amino acids making up the active site of a protein. Thus, a motif can indicate that a protein has biochemical activity similar to that of other proteins with a related motif.

Regulatory Sequences

Regulatory sequences are usually determined experimentally, but bioinformatics techniques are becoming more effective.

Genetic Elements and Structural Features

Chapters 4, 5, 7, and 8 describe the features of DNA elements, such as prophages and transposons. Structural features, such as repetitive sequences, are discussed throughout the book.

Transcriptome Analysis

RNA profiling can be carried out through a variety of techniques, including high-density DNA microarrays (see Rhodius and LaRossa, References) and RNA-Seq (see Wang et al., References). The technologies for sequencing and adaptations for microarray strategies are evolving very rapidly, and many tools and services are available at genomics centers at universities or commercially. The basic principle of microarray technologies is to have known segments of a genome positioned at known positions on a solid substrate (a "gene chip") and then isolating RNAs from cells grown under two different conditions, or from wild-type versus mutant strains, and using different fluorescent dyes to differentially label the cDNAs generated from those RNAs by reverse transcription. The two cDNA preparations are hybridized separately to the genomic arrays, and the intensities of the hybridization signals for the two cDNA preparations at each genomic position are compared to identify genes for which expression increases or decreases as a result of the change in growth conditions or introduction of a mutation in one of the strains.

A more recent approach to identifying the set of RNAs present in cells under a specific condition involves high-throughput sequencing of cDNA generated from the total pool of cellular RNA. This approach, commonly called RNA-Seq, allows quantitative measurements of even low-abundance RNAs that might have been missed in traditional microarray experiments. The improvements in DNA sequencing technologies have made RNA-Seq increasingly accessible as a tool.

Proteome Analysis

The techniques of **proteomics** can be used to identify proteins to analyze the levels of proteins in cells, to determine relative changes in protein levels of regulons or stimulons, to evaluate protein-protein interactions, and to study subcellular localization of proteins. Two-dimensional polyacrylamide gel protein electrophoresis (**2D-PAGE**) can separate many of the proteins of the cell into individual spots on a membrane, and the spots can be further characterized. Because sample complexity limits the ability to identify individual proteins, nongel protein separation techniques, such as liquid chromatography, are often used. This method separates the protein samples into subsamples that are less complex.

Mass spectrometry (MS) is an important tool of proteomics. It is used both to identify proteins and to quantitate protein expression. Protein MS involves the ionization of proteins and peptides and subsequent measurements of mass-to-charge (m/z) ratios. Once proteins have been obtained from a biological sample, they are fragmented into peptides, often by trypsin digestion, because it produces small peptides that are suitable for the first step in MS, which is ionization. Two popular ionization methods are matrix-assisted laser desorption ionization (**MALDI**) and electrospray ionization (**ESI**) (see Kolker et al., References).

In **tandem mass spectrometry** (**MS-MS**), after a protein has been subjected to MS, a single peptide is isolated and shunted through a "collision" chamber in which nitrogen or argon gas breaks the peptide into subfragments. These fragments are then further analyzed so that the sequence of the peptide can be deduced. Computerized comparison to databases can identify the sequence of the peptide and, if the genome sequence is known, the protein and gene from which the peptide came.

Structural Proteomics

Three-dimensional protein structures are determined from the crystal structure or from nuclear magnetic resonance spectroscopy. The development of predictive algorithms has allowed some structures to be predicted from the sequence alone; predicting protein structures from amino acid sequences is a very challenging but active research area.

Functional Genomics

Reverse genetics can be used to determine the function of a gene whose sequence is known. **Forward genetics** can be used to identify and sequence a gene whose function is known. Often, this requires a large repertoire of genetic techniques rather than merely knocking out the gene. **Interaction genetics** seeks to elucidate the significant interrelationships and subtle interactions of genes and gene products. Methods for all of these aspects of genetic analysis are discussed throughout the book.

Protein Localization

Gene fusion techniques using, for example, fluorescent probes can often locate the gene product within the cell.

References

Chan PP, Lowe TM. 2019. tRNAscan-SE: searching for tRNA genes in genomic sequences. *Methods Mol Biol* **1962:**1–14.

Delcher AL, Bratke KA, Powers EC, Salzberg SL. 2007. Identifying bacterial genes and endosymbiont DNA with GLIMMER. *Bioinformatics* **23:**673–679.

El-Gebali S, Mistry J, Bateman A, Eddy SR, Luciani A, Potter SC, Qureshi M, Richardson LJ, Salazar GA, Smart A, Sonnhammer ELL, Hirsh L, Paladin L, Piovesan D, Tosatto SCE, Finn RD. 2019. The Pfam protein families database in 2019. *Nucleic Acids Res* **47**(D1):D427–D432.

Finn RD, et al. 2017. InterPro in 2017-beyond protein family and domain annotations. *Nucleic Acids Res* **45**(D1):D190–D199.

Gill SR, et al. 2005. Insights on evolution of virulence and resistance from the complete genome analysis of an early methicillin-resistant

(continued)

BOX 13.1 *(continued)*

Annotation and Comparative Genomics

Staphylococcus aureus strain and a biofilm-producing methicillin-resistant *Staphylococcus epidermidis* strain. *J Bacteriol* **187:**2426–2438.

Kolker E, Higdon R, Hogan JM. 2006. Protein identification and expression analysis using mass spectrometry. *Trends Microbiol* **14:**229–235.

Potter SC, Luciani A, Eddy SR, Park Y, Lopez R, Finn RD. 2018. HMMER Web server: 2018 update. *Nucleic Acids Res* **46**(W1):W200–W204.

Rhodius VA, LaRossa RA. 2003. Uses and pitfalls of microarrays for studying transcriptional regulation. *Curr Opin Microbiol* **6:**114–119.

Riley M, Abe T, Arnaud MB, Berlyn MKB, Blattner FR, Chaudhuri RR, Glasner JD, Horiuchi T, Keseler IM, Kosuge T, Mori H, Perna NT, Plunkett G III, Rudd KE, Serres MH, Thomas GH, Thomson NR, Wishart D, Wanner BL. 2006. *Escherichia coli* K-12: a cooperatively developed annotation snapshot: 2005. *Nucleic Acids Res* **34:**1–9.

Typas A, Nichols RJ, Siegele DA, Shales M, Collins SR, Lim B, Braberg H, Yamamoto N, Takeuchi R, Wanner BL, Mori H, Weissman JS, Krogan NJ, Gross CA. 2008. High-throughput, quantitative analyses of genetic interactions in *E. coli*. *Nat Methods* **5:**781–787.

Wang Z, Gerstein M, Snyder M. 2009. RNA-Seq: a revolutionary tool for transcriptomics. *Nat Rev Genet* **10:**57–63.

BOX 13.2

Special Problems in Genetic Analysis of Operons

Alleles of Operon Genes

Mutant alleles of structural genes may alter the function of a gene to eliminate, change, or increase its activity. A mutation that eliminates gene function creates a null allele. The term "loss-of-function allele" generally is interchangeable with "null allele." One mechanism for studying gene function is to isolate a transposon insertion in the gene. This is sometimes referred to as a "knockout" mutation. Caution is warranted with transposon insertions because, while this implies a null mutation, additional experimentation would need to be done, since some activity could remain, depending on where the transposon was inserted in the coding sequence. Although many valuable collections of mutants have been obtained by transposon mutagenesis, it should really be considered a method of identifying a candidate for follow-up experiments, because there is no assurance that all of these are null mutations. Further complicating the issue, such mutations can be polar and prevent expression of other genes downstream in the same operon. In some cases, they might even increase the expression of a downstream gene or, alternatively, provide promoter activity and express genes upstream that are transcribed in the direction opposite that of the gene in which the insertion occurred. Exciting strategies involving high-density transposon mutagenesis combined with high-throughput DNA-sequencing techniques (see Box 8.2) are getting around some of the issues with classical transposon mutagenesis experiments when looking over the entire genome, but the analysis of individual insertions will still involve the same safeguards.

Using Reverse Genetics To Construct Null, Nonpolar Alleles

When the sequence of the DNA of a bacterium is known, it is often possible to make null mutations of genes using systematic methods that avoid many of the complications discussed above. Figure 1 illustrates one such method for *E. coli* (see Baba et al., References). This procedure is designed to delete the entire gene but leave the translation initiation region at the beginning of the gene and the terminator codon at the end of the gene plus some sequences upstream of the terminator codon in case the coding sequence for the gene being deleted includes the S-D sequence for the downstream gene. Overlap of the terminator codon for one gene with the translational initiation region for the next gene in an operon often occurs and can cause translational coupling (see chapter 2). In this method, upstream and downstream PCR primers are designed to amplify an antibiotic resistance gene from a plasmid (Figure 1B). However, additional sequences are included on the 5′ ends of the primers (H1 and H2 in Figure 1B). They are not needed for the amplification part of PCR, but instead, these synthetic ends are added to the final product from PCR, which allows recombination with the chromosome (not shown). These sequences provide homology to the N-terminal and C-terminal coding information in "gene *B*" (Figure 1A). This fragment is then introduced into an *E. coli* strain that expresses the Red functions of phage λ to promote recombineering between the amplified fragment and the gene in the chromosome (see chapter 9). Recombination between the sequences flanking the antibiotic resistance gene in the amplified fragment and the corresponding sequences in the gene in the chromosome deletes most of the gene, replacing it with the antibiotic resistance gene (Figure 1B). As an additional feature, the antibiotic resistance gene can be removed later if the antibiotic resistance gene on the plasmid is flanked by sequences for a site-specific recombinase, for example, the Flp recombinase of yeast. When another plasmid expressing this recombinase is introduced into the *E. coli* strain, the antibiotic resistance gene is excised, leaving behind

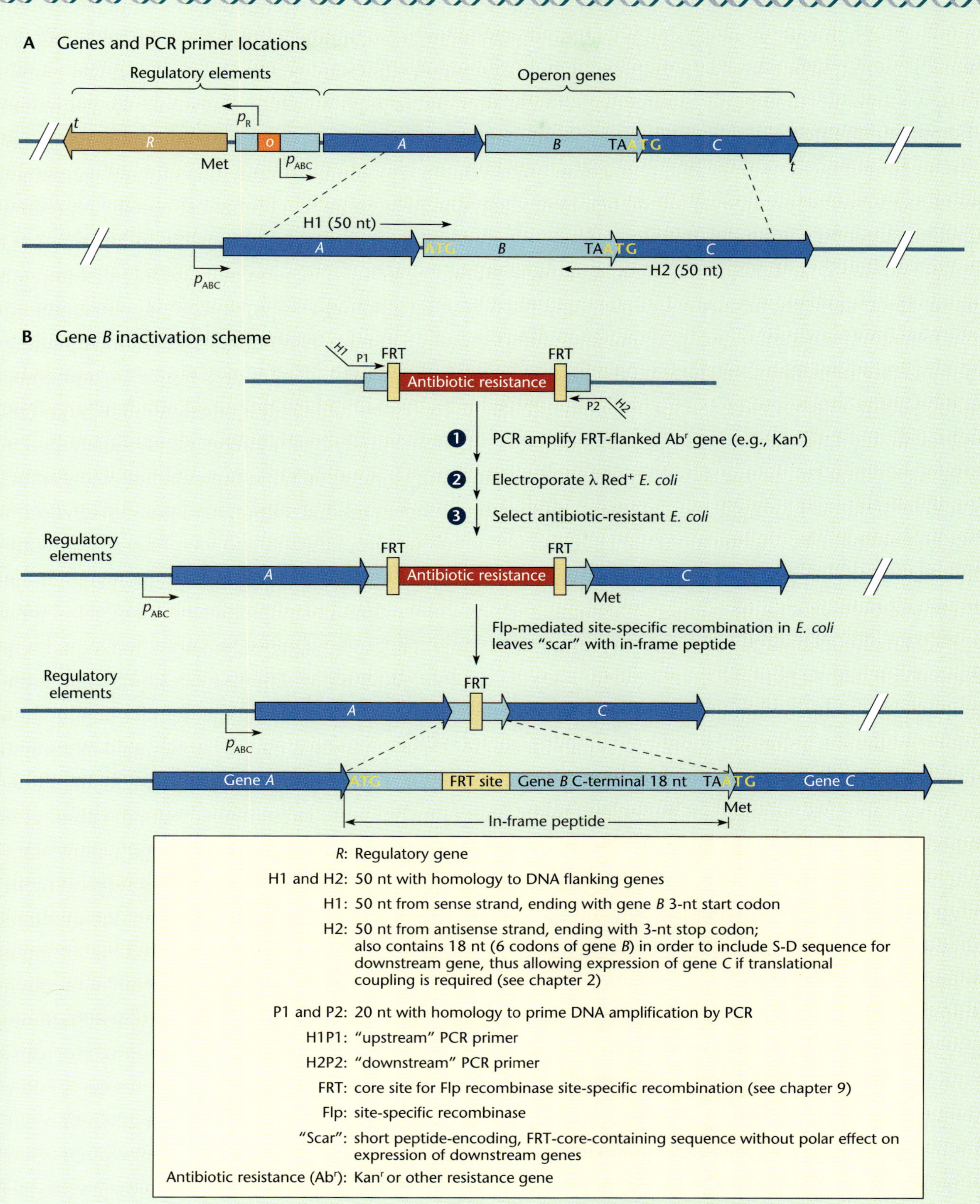

Figure 1

(continued)

BOX 13.2 (continued)

Special Problems in Genetic Analysis of Operons

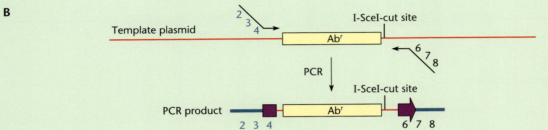

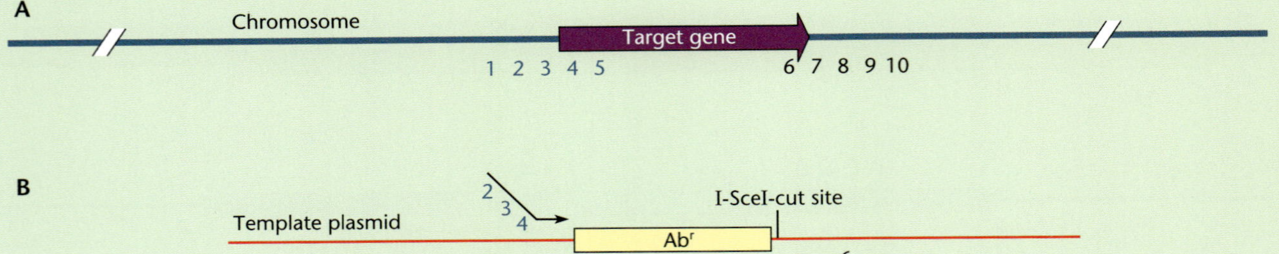

B
Template plasmid
2 3 4
I-SceI-cut site
Ab^r
6 7 8
PCR
I-SceI-cut site
PCR product
Ab^r
2 3 4
6 7 8

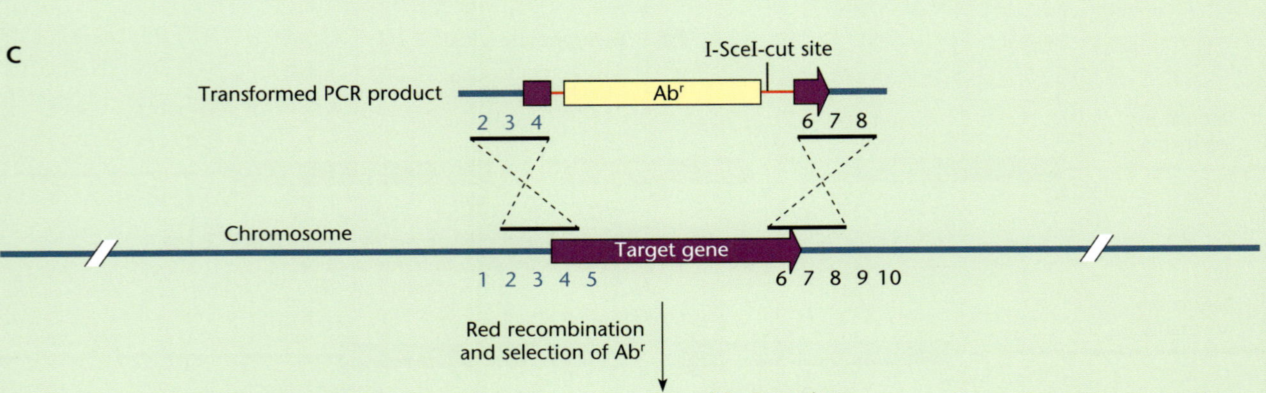

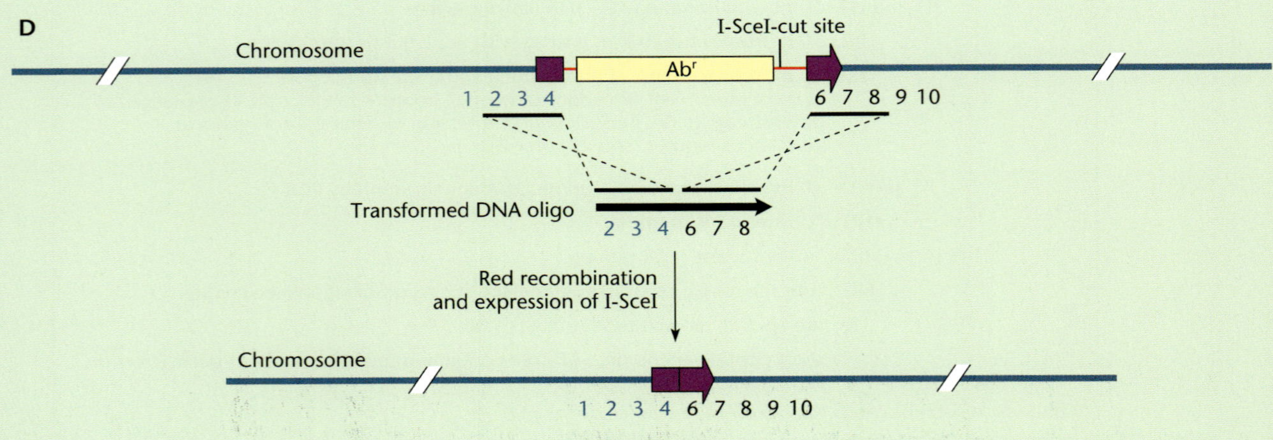

Figure 2

a "scar." Encoded in this scar is a short polypeptide that contains too few sequences from the original gene to be active but whose expression prevents polarity or effects of translational coupling on the expression of the downstream gene. The Keio collection has gene knockouts for all of the nonessential genes in the MG1655 genome and is available through the National Institute of Genetics of Japan or the *E. coli* Genetic Stock Center in the United States.

The technique shown in Figure 1 is efficient, but the "scar" that is left behind can complicate procedures if it is necessary to knock out multiple genes with the Flp recombinase. Other techniques allow a "scarless" method that depends on having both a gene that can be selected, e.g., for antibiotic resistance (Ab^r), and an element that can be counterselected. The homing endonuclease I-SceI provides an especially good tool for counterselection for multiple reasons. For one, the cut site is large enough (30 bp) that no bacterial genome studied to date has it by chance but small enough that it can easily be included with the positive-selection marker. Another big advantage comes from providing a double-strand break that allows loading of the Red recombination machinery. In the procedure, primers are designed to amplify a positive selection marker and the I-SceI cut site sequence from a template. Like the procedure in Figure 1, additional sequences are included on the 5′ ends of the primers that have homology regions encoding the N-terminal and C-terminal regions of the protein (Figure 2A and B). The positions along the template DNA are indicated with numbers (1 to 10) to emphasize the regions that are deleted in the procedure. The PCR product is introduced by transformation into cells induced for the λ Red recombination system (Figure 2B). In the final step, a DNA oligonucleotide with the desired deletion is designed. It is introduced into the cell at the same time that the λ Red recombination system is induced and the I-SceI endonuclease is expressed. The efficient cleavage of the chromosome at the I-SceI cut site provides a strong counterselection for the desired recombination event (Figure 2D). Successful recombinants will not be sensitive to the positive-selection marker, but sequencing is used to confirm the modification. CRISPR technologies with Cas9 and a synthetic guide RNA can now alleviate the need for the intermediate step shown in Figure 2B and C, because the double-strand break can be made in basically any gene at protospacers with the correct PAM (see the text and Jiang et al., References). Even single base pair changes can be made if the change in the protospacer allows the guide RNA to recognize the original allele but not the new recombinant allele.

Alleles of Regulatory Genes and Elements

The methods discussed above can also be applied to assess a gene to determine if it has a regulatory function. For example, a gene that is adjacent to an operon in a genome sequence, e.g., *lacI* or *araC*, is a candidate for encoding a regulator, especially if it contains a helix-turn-helix motif (see chapter 11). Determination of the null phenotype of the gene can show whether the gene product is a positive or a negative regulator. Generation of other allele types can add to an understanding of how the regulator functions. However, it is still very difficult to predict which amino acid changes will elicit a particular phenotype, for example, a superrepressor phenotype, even if the regulator falls into one of the families discussed in chapter 11. Imagine how naive our view of regulation would be if some regulatory genes had not been intensively studied by using selectional genetics to identify the amino acids important for the various phenotypes.

References

Baba T, Ara T, Hasegawa M, Takai Y, Okumura Y, Baba M, Datsenko KA, Tomita M, Wanner BL, Mori H. 2006. Construction of *Escherichia coli* K-12 in-frame, single-gene knockout mutants: the Keio collection. *Mol Syst Biol* **2**:2006.0008.

Herring CD, Glasner JD, Blattner FR. 2003. Gene replacement without selection: regulated suppression of amber mutations in *Escherichia coli. Gene* **311**:153–163.

Jiang W, Bikard D, Cox D, Zhang F, Marraffini LA. 2013. RNA-guided editing of bacterial genomes using CRISPR-Cas systems. *Nat Biotechnol* **31**:233–239.

Advanced Genome-Sequencing Techniques

DNA sequencing with the Sanger sequencing method, which takes advantage of terminating a strand of DNA using a dideoxynucleotide, continues to be widely used. However, newer high-throughput techniques were developed for ever-larger sequencing projects involving many thousands of much shorter templates being read at the same time. Initial genome sequencing efforts involved sequencing individual fragments of the genome cloned into plasmids or phage. For larger, genome-size DNA molecules, this involved sequencing a huge collection of random clones of segments of the genome. By collecting many times more DNA sequence than the actual size of the genome, enough overlapping DNA sequences are collected to reconstruct the virtual genome using computers. Other strategies allow researchers to fill in the gaps where information was missing to produce a continuous data set that represented the entire genome of the bacterium. The first genome sequence from a free-living organism, *H. influenzae*, published in 1995 was sequenced using this strategy.

Newer technologies have entirely removed the cloning step, so that DNA is sequenced directly as thousands of randomly sheared pieces of the original. While these sequencing runs are short (typically <500 bp), the sheer number of random sequences collected allows enough overlapping pieces of information to assemble the sequence and put entire genomes together or to address other biological questions. The strategy of sequencing thousands of reads at one time is referred to as **massive parallel sequencing** and involves a variety of technologies with some commonalities. If all of the pieces of sequence information can be assembled to give the entire chromosome, it is considered to have a **complete closed genome sequence**. More typically, all of the pieces of sequence cannot be assembled together as one unit, but instead, the investigator is left with a number of large stretches of sequence referred to as **contigs**, which is still useful if almost all of the genome is represented in these contigs. Repetitive sequences from mobile elements can complicate assembling a given genome. Having extensive amounts of similar mobile genetic elements and/or having an insufficient number of sequence reads are common reasons for failing to get a single closed genome sequence with a given strain of bacteria.

Popular DNA-sequencing strategies involve fragmenting the DNA substrate, cleaning up the ends, and ligating short synthetic DNAs to both ends of the fragments (Figure 13.2). The DNAs are attached to a special solid substrate, where they are amplified in a modified polymerase chain reaction (see "Polymerase Chain Reaction" below), called bridge-PCR or bridge amplification, where the primers used in the amplification procedure are affixed to the same substrate. This procedure yields spots of clusters or wells, each with the same DNA. Nucleotides tagged with fluorescent molecules flow over these collections of DNA fragments to allow incorporation of the next complementary base extending from the primer. A sensitive detection device is able to read the fluorescent signal from a fluorophore attached to the nucleotide in the terminal position in each spot that corresponds to one of the four individual bases. The fluorophore is then removed and the terminal nucleotide activated to allow extension with the next base as the nucleotide flow is repeated. While the read length of each individual DNA is short, many thousands of fragments can be read at the same time, allowing the collection of a massive amount of DNA sequence information. As mentioned above, software that can identify overlap between the sequences allows the short reads to be compiled into longer strings of sequence information.

An interesting adaptation to the first steps in the process shown in Figure 13.2 is called **tagmentation**, where

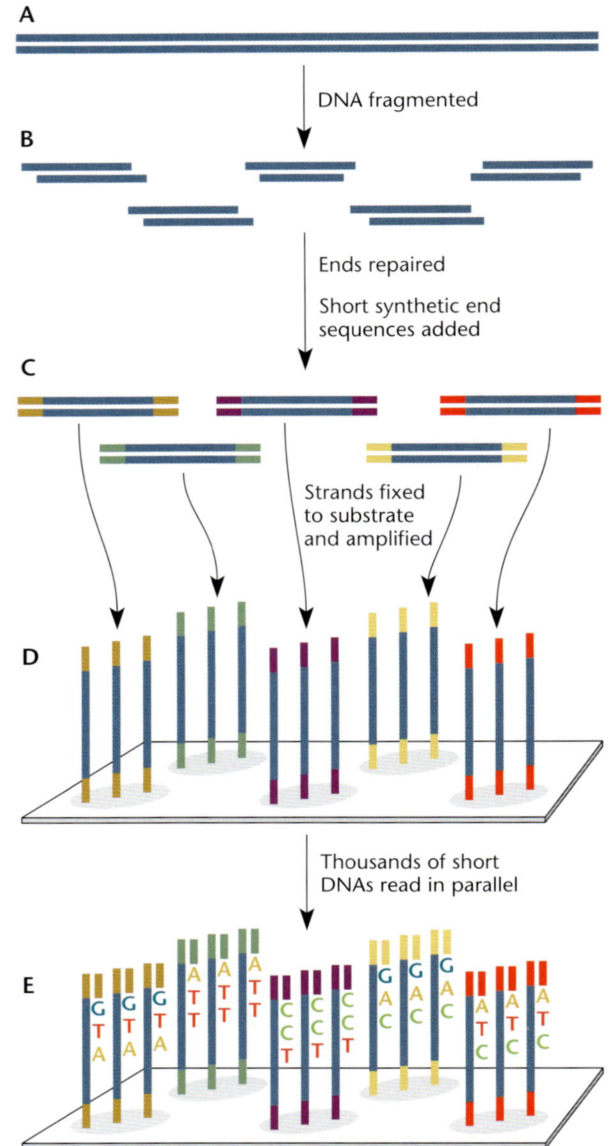

Figure 13.2 Popular DNA-sequencing strategies involve fragmenting the DNA substrate **(A and B)** and processing the ends to allow short synthetic DNAs to be ligated to both ends of the fragments **(C)**. The DNAs are fixed to a solid substrate, where they are amplified in a way that yields discernible spots of clusters or wells, each with the same DNA (the amplification is not shown for simplicity) **(D)**. Special nucleotides flow over the spots, allowing the complementary base to integrate, extending the substrates with the matching base **(E)**. A sensitive detection device is able to read the fluorescent signal from each spot that corresponds to individual bases. As described in the text, while the read length of each individual DNA is short, many thousands of fragments can be read at the same time, which allows the collection of a massive amount of DNA sequence information. Software that can identify overlap between the sequences allows the short reads to be compiled into longer strings of sequence information.

a special bacterial transposase is used in the preparation of DNAs for sequencing. This combines what is equivalent to the fragmentation and ligation steps explained above. As described in more detail in chapter 8, during transposition, each of the *cis*-acting ends of a transposon directly joins to an insertion site in a largely sequence-nonspecific way. In the tagmentation procedure, instead of using a separate ligation step, a synthetic transposon end sequence recognized by the transposase can directly join the short synthetic DNAs to the DNA to be sequenced, allowing these to be joined to the template in a process that also produces the small fragments needed for the sequencing procedure.

Exciting techniques involving sensitive detection devices that allow long individual strands of DNA to be read are now becoming more common. One of the important advantages of longer reads is that they are easier to assemble computationally, especially if the genome to be sequenced has a lot of repetitive features that could otherwise complicate the assembly process from shorter sequencing reads. As with the other newer sequencing technologies, it is also done in a massively parallel fashion, allowing large amounts of sequence information to be collected. Given the importance of collecting quality DNA sequence information at the lowest possible price, other technologies are certain to be developed in the future.

Technologies that reduce the cost and time needed to do DNA sequencing allow its broader application beyond genome sequencing. Massive parallel sequencing can be used in other types of experiments. For example, the profile of all RNA transcripts of an organism under a particular growth condition can be determined by converting all of the RNA transcripts to DNA that can be sequenced in a process called **RNA-Seq**. This technique provides detailed information about all of the genes that are expressed in the cell under certain growth conditions. Massive parallel sequencing can also determine protein occupancy across a genome in a procedure called **ChIP-Seq**, for chromatin immunoprecipitation and sequencing. In the first step, the protein of interest is chemically cross-linked to the host chromosome in living cells. The genome is fragmented, and the protein is selectively pulled from cell extracts using an antibody that recognizes the protein. Chromosomal fragments to which the protein is bound can be recovered by reversing the chemical cross-link, freeing the thousands of previously bound DNA fragments, which are then sequenced using the techniques described above. ChIP-Seq provides information about all sites at which the protein was bound at the time the cells were exposed to the cross-linking agent. Tn-Seq, described in chapter 8, is another application of DNA sequencing in which identification of the insertion sites of transposons located at thousands of different positions in a population of bacteria can be used as a technique in systems biology to identify essential genes or genes that are required under certain conditions, because cells with insertions in those genes will be unable to survive under those conditions (see Box 8.2).

Polymerase Chain Reaction

The uses of bacterial and archaeal DNA polymerases extend beyond determining genome sequences. One of the most useful technical applications involving DNA polymerases is the **polymerase chain reaction** (PCR). This technology makes it possible to selectively amplify regions of DNA out of much longer DNAs. It is called PCR because each newly synthesized DNA serves as the template for more DNA synthesis in a sort of chain reaction until large amounts of DNA have been amplified from a single DNA molecule. The power of this method is that it can be used to detect and amplify sequences from just a few molecules of DNA from any biological specimen, for example, a drop of blood or a single hair; this has made it very useful in criminal investigations to identify the perpetrators of crimes on the basis of DNA typing. However, for our purposes here, it also has many other applications, including the physical mapping of DNA, gene cloning, mutagenesis, and DNA sequencing.

The principles behind the use of PCR to amplify a region of DNA are outlined in Figure 13.3. PCR takes advantage of the same properties of DNA polymerases that are important in other applications, i.e., their ability to make a complementary copy of a DNA template starting from the 3′ hydroxyl of a primer DNA. PCR uses two primers complementary to sequences on either side of the region to be amplified. As described in chapter 1, normal cellular DNA replication uses an RNA primer for initiating DNA replication. In PCR, DNA primers are used that also become part of the final product that is amplified. These primers are also designed so they anneal to opposite DNA strands, allowing the replication events extending from the primers to converge as shown in Figure 13.3.

In the process of PCR amplification, the reaction is cycled through different temperatures to allow multiple rounds of DNA amplification. The DNA is first denatured by increasing the temperature to 95°C to separate the strands. The temperature is then lowered to a point at which the primers bind specifically to the complementary sequences on the DNA. Finally, the reaction is brought to a third temperature at which the DNA polymerase can incorporate nucleotides to extend the sequence from the 3′ ends of the primers. Each primer primes the synthesis of DNA over the region to be amplified (which is delineated by the two primer-binding sites) and continues polymerizing past the end of the region. If the two strands are again separated by heating and the temperature is again lowered, a new primer can then hybridize to this

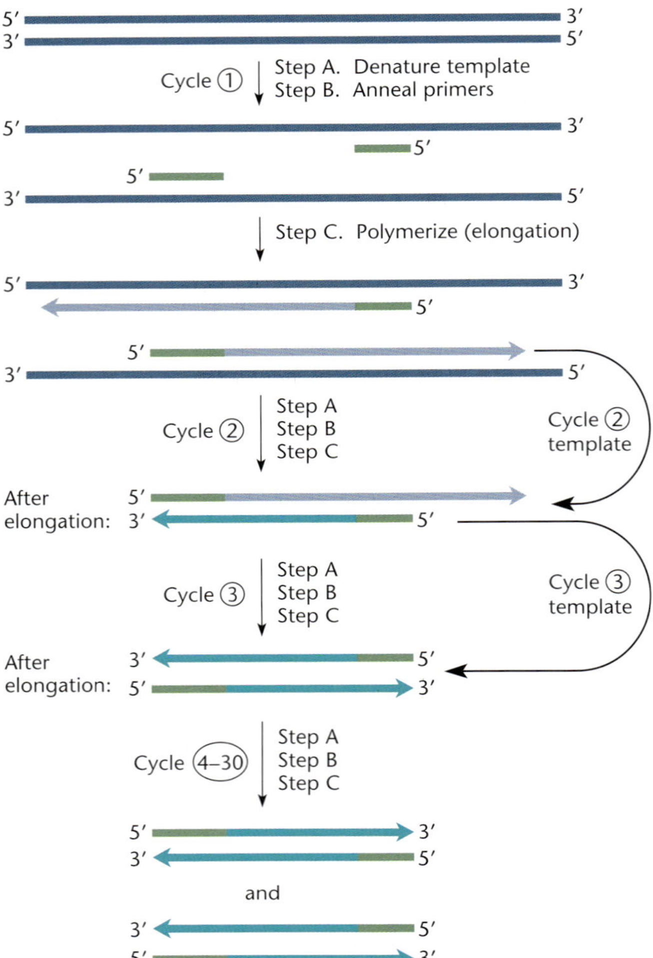

Figure 13.3 Steps in PCR. In the first cycle, the template is denatured by heating. Primers are added that hybridize to the separated strands for the synthesis of the complementary strand. The strands of the DNA are separated by the next heating cycle, and the process is repeated. The DNA polymerase survives the heating steps because it is from a thermophilic bacterium. The DNA sequence is amplified approximately a billion-fold.

newly synthesized strand and prime replication back over the region. Now, however, the DNA polymerase runs off the end of the template DNA when it reaches the 5′ end of the first primer sequence, leading to the synthesis of a shorter piece of DNA with the primer sequences on both ends. If this DNA is then heated again to separate the strands, this shorter piece can bind another primer molecule, which can then prime the synthesis of the complementary strand of the shorter DNA, and so on. This process of heating and cooling can be repeated 30 or 40 times until large numbers of copies of the DNA region between the primer binding sites have accumulated, beginning from one or very few longer DNA molecules that contain the region.

In principle, any DNA polymerase could be used to perform PCR. However, most DNA polymerases would be inactivated by the high temperatures required to separate the strands of double-stranded DNA, making it necessary to add fresh DNA polymerase after each heating step. This is where the DNA polymerases from thermophilic bacteria, such as *Thermus aquaticus*, come to the rescue. These bacteria normally live at very high temperatures, so their DNA polymerase, called the *Taq* polymerase in the case of *T. aquaticus*, can survive the high temperatures needed to separate the strands of DNA, obviating the need to add new DNA polymerase at each step. We can just mix the primers, a tiny amount of biological material containing DNA with the region to be amplified, and the *Taq* polymerase, set a thermocycler machine to heat and cool over and over again, and come back some time later. At the end of this process, large amounts of the region of the amplified DNA have been synthesized, and we can detect the amplified DNA on a gel or use it in other procedures. As described above for massive parallel sequencing, PCR can even be carried out on DNA that has been attached to a surface, resulting in many copies of each DNA fragment for sequencing.

PCR MUTAGENESIS

In addition to accurate amplification of a DNA segment, PCR can be used either to make specific changes in a DNA sequence or to randomly mutagenize a region of DNA. Site-specific mutagenesis allows the investigator to make a desired change at a particular site in the sequence of DNA rather than relying on more traditional methods of mutagenesis that more or less randomly cause mutations and do not target them to a particular site (see chapter 3). Methods for site-specific mutagenesis rely on synthetic DNA primers that are mostly complementary to the sequence of the DNA being mutagenized, except for the desired mutational change. When this primer is hybridized to the DNA and used to prime the synthesis of new DNA by the DNA polymerase, the synthesized DNA has the same sequence as the template, except for the change in the attached primer. This method of site-specific mutagenesis can be used to make only minor changes, such as single-base-pair changes, in the DNA sequence, because if the sequence of the primer is altered too much, it no longer hybridizes to the template DNA.

PCR can be used to make random changes in a sequence because the *Taq* polymerase makes many mistakes because it lacks an editing function; the presence of manganese ions increases the error rate even higher. In fact, the mistake level during normal amplification by *Taq* polymerase is so high that new forms of thermally stable DNA polymerase that are more accurate than the *Taq* polymerase are now typically used for amplifying

DNA fragments that will be used for cloning, to avoid introduction of sequence changes. These newer polymerases are also derived from organisms that live at high temperatures, but they possess a proofreading activity. Regardless, clones made from PCR fragments are normally subjected to DNA sequencing to be certain that no unwanted mutations have been introduced.

Barriers to Horizontal Transfer: Genome Gatekeepers and the Molecular Biologist's Toolkit

One of the things we have learned from bacterial genomes is that mobile elements such as bacteriophages, transposons, and integrating and independently replicating plasmids make up a significant part of the gene content in most free-living bacteria and archaea. While many important adaptive functions can come from these mobile elements, they are still selfish elements that are more likely to kill or slow the growth of the host than they are to provide an important function. A good way to think about mobile elements in bacteria is the same way one thinks about mutations; although mutations can be beneficial and are essential for evolution, in actuality almost all mutations reduce or abolish the function of a gene, not enhance it. The same can be said for mobile elements with bacteria and archaea. As expected, bacteria and archaea have evolved many mechanisms to deter the horizontal transfer of mobile elements into their genome.

The study of the wide variety of mechanisms that bacteria and archaea use to protect themselves from horizontal transfer is interesting on many levels. For one, each mechanism of protection developed by the host will inevitably lead to an advantage to mobile elements that can defeat this protective mechanism. The massive number of microorganisms and mobile elements that are found in the environment and the billions of years they have been mounting this competition reveal some of the greatest examples of molecular ingenuity that can be found in nature (see Koonin et al. 2017b, Suggested Reading). This molecular ingenuity resulted in the development of systems for modifying and manipulating DNA in the test tube or in living organisms. Enzymes such as restriction endonucleases that were discovered in the 1960s and 1970s help to protect the bacterial or archaeal host genome from invading DNAs but also provided tools that have allowed for the age of molecular biology. More recently, enzyme systems such as CRISPR/Cas complexes were identified as advanced systems of acquired immunity in bacteria and archaea that provide important protection from bacteriophages and other mobile elements. CRISPR/Cas systems have now been adapted as one of the most important tools in molecular genetics and genomics in our time and are discussed later in this chapter.

Restriction Endonucleases

Among the most useful enzymes that alter DNA are the restriction endonucleases. These are enzymes that recognize specific sequences in DNA and cut the DNA at a position that is contiguous with this sequence. The cell that produces these enzymes also usually produces enzymes with methylating activities that modify DNA at the recognition sequence, making it immune to cutting by the endonuclease activity. The production of the methylase therefore protects the cell that produces the restriction endonuclease from cleaving its own genome. These enzymes are made exclusively by bacteria, and the major role for many of them may be to defend against incoming bacteriophages and other DNA elements, generally by cleaving unmodified DNA. Restriction endonucleases are important barriers that help protect bacteria from rampant horizontal transfer of incoming DNA but are only the first line of defense for the bacterial genomes. These relatively simple systems are sometimes compared to the innate immune systems found in many multicellular organisms, such as humans, where general features are recognized as an attempt to distinguish between self and non-self. Restriction endonucleases have also been suggested to persist because they act as "selfish" DNAs. Given that the effect of the endonuclease lasts longer than the methylation activity (which is lost by subsequent rounds of DNA replication if the methylating enzyme is not continuously produced), there is selection against losing the information that encodes the restriction-modification system by gene deletions; loss of the protective methylating activity can leave the whole genome sensitive to cleavage by the restriction enzyme, so that cells that have lost the genes and no longer produce methylase will die. Therefore, restriction systems can serve as so-called toxin-antitoxin or plasmid addiction systems that help prevent the loss of plasmids by killing the cell if it is cured of the plasmid (see Box 4.3).

RESTRICTION ENDONUCLEASES THAT LIMIT GENE EXCHANGE IN *E. COLI*

Restriction endonucleases are classified into four groups, types I, II, III, and IV (Table 13.1). The enzymes in these groups differ mostly in the relationship between their methylating and cleaving activities. All four types of systems are important for protecting the host from horizontally transferred DNA sequences. Unfortunately, this sometimes includes DNA that an investigator wants to introduce into a bacterial host in the laboratory. For example, the native *E. coli* K-12 strains of bacteria used in most laboratories have a type I *Eco*K1 system. The naming system is based on where the enzymes were first identified (in this case *E. coli*, *Eco*) and then an independent identifier. The EcoK1 system has three components, one for DNA cleavage (HsdR), one for transferring a methyl group (HsdM), and

Table 13.1 Types of restriction endonucleases

Type	Attributes
Type I	Cuts DNA at random locations far from the recognition sequence
Type II	Cuts within or close to the recognition sequence
Type III	Cuts between two recognition sites found in opposite orientation
Type IV	Typically recognizes and cuts a modified recognition sequence

one for recognition of the sequence that is specific to the particular system (HsdS) for cleavage or methylation. In the type I system, "self" DNA is recognized because the recognition sequence is methylated at one position, protecting it from the cleavage component of the system (HsdR). DNA that lacks this methylation at the recognized sequence and enters a cell with a functional HsdRMS system is cleaved. DNAs that are produced in types of bacteria that lack the *Eco*K1 system are degraded when introduced into a strain of *E. coli* K-12 bacteria that is *hsdRMS*$^{+}$, making molecular biology experiments difficult if not impossible. Common versions of *E. coli* K-12 strains that have been engineered for cloning experiments (e.g., DH5alpha, JM109, and XL1-Blue) have an *hsdR* mutation in the *hsdRMS* system; as a result, DNA will not be restricted (cleaved) in this mutant background, but it will be modified such that if DNAs are subsequently isolated from this strain and moved into other *E. coli* K-12 strains, the DNAs will not be restricted.

Type IV restriction systems work in what could be considered the opposite way. Instead of cleaving DNA that lacks methylation at a specific sequence, these systems cut at specific sequences only when that sequence is methylated. Type IV systems in standard *E. coli* K-12 strains are encoded in the *mcrA*, *mcrBC*, and *mrr* genes. Mutants of these systems should be considered if DNA with an unusual type of methylation is being introduced into *E. coli* K-12 and a restriction barrier seems to be implicated because of failure to successfully introduce the new DNA.

OTHER DEFENSE SYSTEMS USING NUCLEASES AND METYHLATION

Even with the restriction endonucleases described above, the CRISPR/Cas systems described later in this chapter, and other systems described in this textbook, there is reason to believe we are only beginning to understand the number of systems prokaryotes use to defend their genome from mobile elements. These systems seem to accumulate in bacteria and archaea in regions sometimes referred to as defense islands (see Makarova et al. 2011, Suggested Reading). Approaches for computationally identifying new systems within defense islands that can be tested in the laboratory have successfully identified multiple intriguing new antiphage and antiplasmid systems (see Doron et al., Suggested Reading). It is not unreasonable to think that many new and potentially useful molecular tools derived from these defense systems await discovery in the thousands of new genomes that are regularly becoming available.

TYPE II RESTRICTION ENZYMES AND CLONING

The type II enzymes are most familiar to molecular biologists, because they have proven to be extremely useful for gene cloning for many years. Hundreds of type II enzymes are known, and many of them can be purchased from molecular biology supply companies. One thing that makes them so useful is that the methylating activity can be separated from the cleaving activity. Second, they each recognize their own specific sequence in DNA and then cut the DNA at or close to the recognition sequence (Figure 13.4). This is different from type I enzymes like EcoK1, which are not useful for cloning because they cut at multiple unpredictable positions that are distant from the site that is recognized (Figure 13.4). Third, the sequences recognized by many of the type II enzymes are palindromic and, by making staggered breaks in these sequences, they leave complementary (sticky) ends that can be used for DNA cloning (see below). The recognition sequences are often 4, 6, or 8 bp long, and the sequences recognized by some restriction endonucleases are shown in Table 13.2.

Using Restriction Endonucleases To Generate Recombinant DNAs

One of the properties of some type II restriction endonucleases that make them so useful is that the sequences they recognize read the same in the 5′-to-3′ direction on both strands. Such a sequence is said to have 2-fold rotational symmetry or to be a **palindrome** because it reads the same if you rotate it through 180° and read the other strand. The term "palindrome" in English refers to words or phrases in which the letters read the same in both directions, as in "madam, I'm Adam." Because the sequence reads the same on both DNA strands, the restriction endonuclease binds to the identical sequence on both strands and then cuts the two strands at the same place in the sequence. For example, the restriction endonuclease HindIII (so called because it was the third restriction endonuclease found in *H. influenzae* strain D) recognizes the 6-bp sequence 5′-AAGCTT-3′/3′-TTCGAA-5′ and cuts between the two A's on each strand (Table 13.2, Figure 13.4), which is in the same place in the sequences of the two strands read in the 5′-to-3′ direction. Such a break is called a "staggered break" because the breaks in the two strands are not exactly opposite each other in the DNA, which has the effect of leaving short single-stranded ends on both ends of the broken DNA. Because

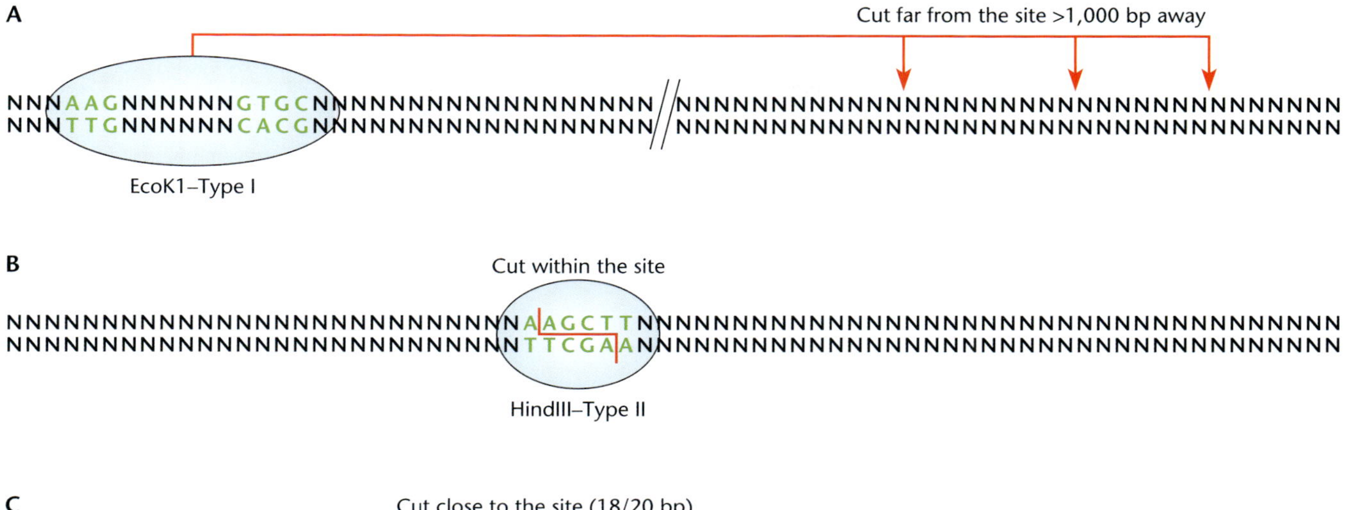

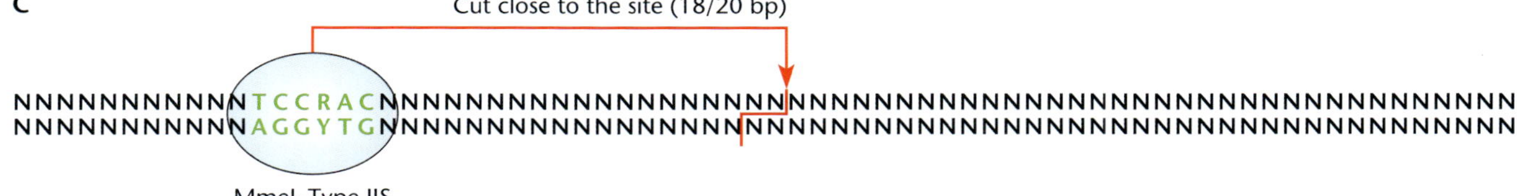

Figure 13.4 Multiple types of restriction endonucleases exist where the DNA sequence that is recognized is related in different ways to where cleavage occurs. **(A)** EcoK1, a type I restriction endonuclease, is naturally encoded in *E. coli* K-12 strains. In type I restriction endonuclease systems, cleavage occurs at multiple locations displaced from the recognition site in an unpredictable fashion, which makes them of little value for cloning applications. **(B)** Type II restriction enzymes like HindIII cleave within a palindromic site at staggered positions and are useful for DNA cloning experiments. **(C)** A special subtype of type II enzyme, called type IIS, can be useful in specific applications because the site of cleavage occurs at a specific distance offset from the site recognized in DNA (see text).

the original sequence that was cut had a 2-fold rotational symmetry, the two single-stranded ends are complementary to each other and so can pair with each other. More importantly, each of the single-stranded ends can pair with the single-stranded end of any other DNA that was cut with the same restriction endonuclease, because they all have single-stranded ends with the same sequence. These single-stranded ends are called sticky ends because they can pair with ("stick to") any other single-stranded ends with the complementary sequence (Figure 13.5). Other restriction endonucleases might leave the same sticky ends even if they recognize a somewhat different sequence. Such restriction endonucleases are said to be compatible. For example, PstI and NsiI have different recognition sequences (Table 13.2), but the sticky ends that result from their cleavage are complementary and therefore are compatible for cloning. Pairing of two complementary sticky ends with each other forms a double-stranded DNA with staggered nicks in the two strands, which can then be sealed by DNA ligase (see below). The new DNA that has been generated in this way is called **recombinant DNA** because two DNA molecules have been recombined into new sequence combinations.

Table 13.2 Recognition sequences of some restriction endonucleases[a]

Enzyme	Recognition sequence
Sau3A	*GATC/CTAG*
BamHI	G*GATCC/CCTAG*G
BsaI	GGTCTCN*/CCAGAGNNNN*
EcoRI	G*AATTC/CTTAA*G
HindIII	A*AGCTT/TTCGA*A
NotI	GC*GGCCGC/CGCCGG*CG
NsiI	ATGCA*T/T*ACGTA
PstI	CTGCA*G/G*ACGTC
SmaI	CCC*GGG/GGG*CCC

[a]Asterisks indicate where the endonucleases cut within or adjacent to the recognition sequence.

Cloning and Cloning Vectors

Ligation of two DNA molecules results in formation of only one molecule of a recombinant DNA. In order for it to be useful, many copies of the recombinant DNA molecule

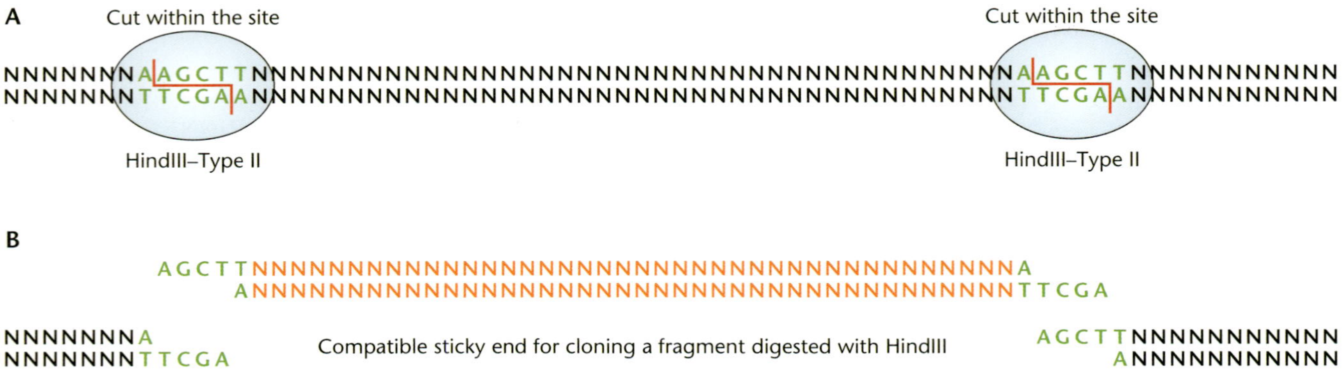

A

Cut within the site

Cut within the site

HindIII–Type II

HindIII–Type II

B

Compatible sticky end for cloning a fragment digested with HindIII

Figure 13.5 Recombinant DNAs can be joined using compatible ends formed by digestion with restriction endonucleases. Cleavage of substrate DNAs with the same restriction endonuclease will leave the same single-strand overhang that can be used for specifically joining these ends, and the resulting breaks in the DNA backbone can be fixed using a DNA ligase. The single-strand extended ends that are produced by digestion with restriction endonucleases are also referred to as sticky ends because of the utility they have for molecular cloning.

are needed. This is the function of cloning vectors. A **cloning vector** is a DNA that has its own origin of replication and is capable of independent replication in the cell. DNAs that have an *oriV* sequence that makes them capable of independent replication in the cell are called replicons. Plasmids are an example of replicons. The process of cloning a piece of DNA into a circular plasmid with its own *oriV* sequence is illustrated in Figures 13.6 and 13.7. Once it has been joined to another piece of DNA and introduced into a cell, the cloning vector replicates itself, along with the piece of DNA to which it has been joined, making many exact copies of the original DNA molecule. These exact replicas of the piece of DNA are called **DNA clones** by analogy to the genetic replicas of an organism that are made when an organism replicates itself asexually. Phages also are capable of independent replication in their bacterial hosts, and some of these have been modified to serve as convenient cloning vectors. Some examples of cloning vectors are discussed in previous chapters.

In the early days of using restriction enzymes for cloning, the investigator used restriction enzyme recognition sites that were already found by chance in a given DNA sequence to be cloned. The use of PCR provides more flexible options for choosing the region of DNA that is cloned. Figure 13.6 shows a hypothetical region of DNA with one monocistronic and one bicistronic operon and the associated mRNAs. In the example, a cloning vector is shown with a regulatable promoter, so that the native promoter for the *yfgB* gene is not needed. The use of PCR primers that hybridize downstream of the promoter but upstream of the translation signals of the gene allows expression of the cloned gene to be controlled at the level of transcription. While, typically, the entire oligonucleotide primer sequence is identical to the region to be amplified, synthetic tails can be added to the primer. In the example shown in Figure 13.6, the extra sequence includes the recognition sequence for the HindIII restriction enzyme. The amplified DNA (or amplicon) resulting from this PCR reaction will include the open reading frame with the requisite translational signals flanked by sites recognized by the HindIII restriction enzyme. As explained in chapter 4, cloning vectors often will have a multicloning site that includes the sequences recognized by many different restriction endonuclease enzymes all in a small region that is positioned downstream of a regulatable promoter. For example, the promoter from the *lacZYA* operon can be regulated with lactose, or more commonly, the lactose analog IPTG (see chapter 11). Another option would be to design the oligonucleotide primers such that both transcriptional and translational signals are encoded in the vector and not carried over from the native gene; in this case, the translational signals in the vector must be fused in-frame with the amplified open reading frame to allow translation of the gene.

The cloning vector and the amplicon produced by PCR are both digested with HindIII and purified (Figure 13.7). The fragments are ligated together, and the resulting DNAs are introduced into *E. coli* by transformation (see chapter 6). Cells in which the recombinant DNA has been introduced are identified by selecting for the antibiotic resistance that is also encoded on the plasmid. This simple procedure allows the expression vector with the *yfgB* gene to be captured or cloned for producing this product. One common problem with this approach is that the plasmid can recircularize without acquiring the desired insert DNA; this can be reduced by using a plasmid vector digested with two different restriction enzymes (e.g., HindIII and EcoRI) and PCR primers that introduce matching

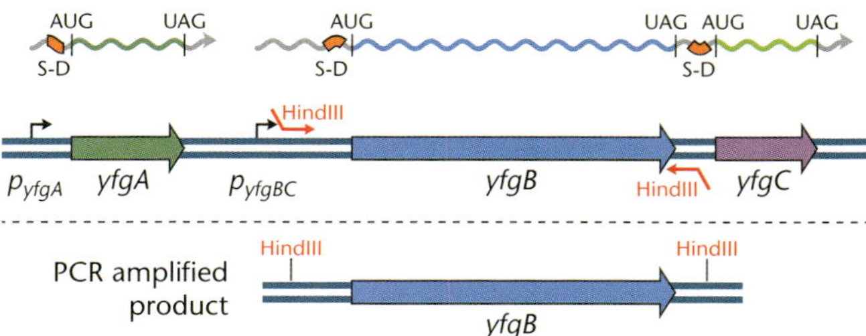

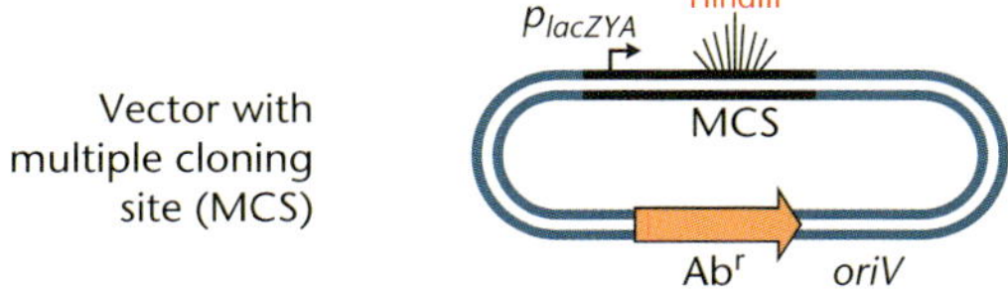

Figure 13.6 A single gene from a region of the genome can be cloned using PCR to allow controlled expression in a cloning vector. **(A)** The diagram shows two hypothetical operons with individual promoters and translation signals and the resulting mRNA. PCR primers can be used to affix sites recognized by a restriction endonuclease, such as HindIII, for cloning. The resulting PCR product can subsequently be digested to produce specific sites found in a compatible cloning vector. **(B)** Cloning vectors can have multiple cloning sites (MCS) where many DNA sites for commercially available restriction endonuclease are found. These vectors include an origin of replication (*oriV*) that allows the vector to replicate apart from the chromosome in bacteria and a selectable marker such as resistance to an antibiotic (Ab^r). Cloning vectors referred to as expression vectors can also have a regulatable promoter for controlling expression of the gene products such as the lactose operon promoter, p_{lacZYA}.

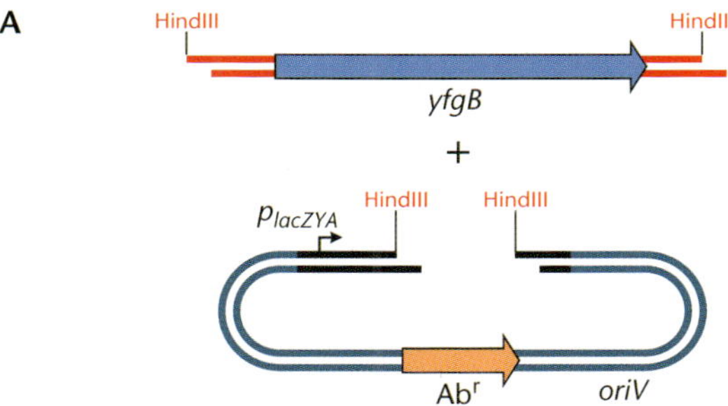

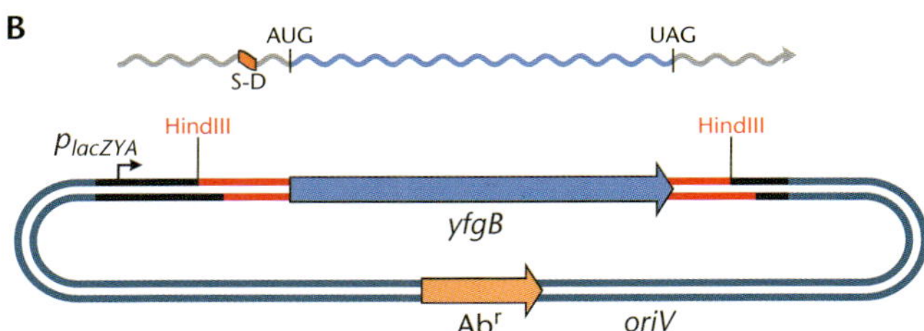

Figure 13.7 DNA products produced by PCR and cleaved with restriction endonucleases can be cloned into expression vectors. **(A)** DNA substrates produced as described in Figure 13.6 have compatible sticky ends from HindIII digestion that allow them to be joined together by DNA ligase. **(B)** The resulting product can be introduced into and selected for maintenance in an appropriate strain of *E. coli*. Expression of the cloned gene can be controlled by the concentration of lactose in the media or a synthetic inducing substrate, IPTG.

restriction sites upstream and downstream of the coding sequence. There are many more options available today for cloning genes, as described later in the chapter.

DNA Libraries

A **DNA library** is a collection of DNA clones that includes all, or at least almost all, the DNA sequences of an organism. One way to make a DNA library is with restriction endonucleases. The entire DNA of an organism is cut with restriction endonucleases, and the pieces are ligated into a cloning vector cut with a compatible enzyme. The mixture is then introduced into cells by transformation; if a phage is used for cloning, the process of introducing the DNA into hosts is called transfection and the resulting genetically distinct clones are isolated as plaques (see chapter 7). If the collection is large enough, every DNA sequence of the organism is represented somewhere in the pooled clones, and the library is said to be complete. The trick then is to find the clone you want out of all the clones in the library; some methods to do this are mentioned in previous chapters.

Techniques for Nontraditional Cloning and Assembly

Traditional cloning approaches as described above that use restriction enzymes to capture DNA fragments into cloning vectors are still commonly used. However, over the years, a variety of new techniques have been developed using molecular systems derived from bacteria. Molecular biologists now have a broad arsenal of techniques to efficiently clone one or assemble multiple DNA fragments. Each of these procedures comes with its own set of advantages and disadvantages and is described in this section (Table 13.3).

TA CLONING OF PCR-AMPLIFIED FRAGMENTS

As described above, PCR is useful for adding sequences at the ends of an amplified DNA fragment, such as restriction sites for cloning. Although the primers used for PCR amplification must be complementary to the sequence being amplified at the 3′ end, they need not be complementary at the 5′ end. Therefore, the 5′ end of the primer sequence can include, for example, the recognition sequence for a specific restriction endonuclease, making it easier to clone the PCR-amplified fragment

Table 13.3 Methods of DNA cloning and assembly

Method	Advantage(s)	Disadvantage(s)
Traditional cloning	Lower cost	Lower efficiency Single DNA fragment Ligation step required Limited by restriction sites
TA cloning	Medium cost	Medium efficiency Single DNA fragment Ligation step required Very limited vector options
Topo cloning	Higher efficiency No ligation step	Higher cost Single DNA fragment Very limited vector options
Gateway cloning	Higher efficiency No ligation step Easy exchange with other vectors	Higher cost Single DNA fragment Leaves recombination sites
USER cloning	Higher efficiency No ligation step Multiple DNA fragments Seamless	Higher cost for primers and reagent
Golden Gate assembly	Higher efficiency Multiple DNA fragments Seamless	Medium cost Ligation step Somewhat limited vector options
Gibson/HiFi assembly	Higher efficiency Multiple DNA fragments Seamless Flexible options for vectors	Higher cost (but also saves costs at other steps)

(Figures 13.6 and 13.7). PCR fragments can also be cloned as blunt-ended fragments into plasmid cloning vectors. A derivative of these procedures stems from a behavior of *Taq* DNA polymerase, which usually adds a single deoxy adenosine (A) to the end of amplified DNA fragments because the *Taq* polymerase also has a terminal transferase activity (Figure 13.8A). Plasmid vectors can be purchased or produced in the lab that have a complementary deoxythymidine (T) to facilitate cloning of these fragments in a process that is sometimes referred to as TA cloning (Figure 13.8B).

Topo I CLONING

One problem with the above-mentioned cloning methods is that they all involve ligation steps. Even under optimal conditions, ligation is somewhat inefficient, which limits the numbers of clones that can be obtained. This is a bigger problem in some applications, such as making libraries for genomic sequencing. A more efficient system that does not rely on ligation is called Topo cloning because it relies on the type I topoisomerase (Topo I) of vaccinia virus (see Shuman, Suggested Reading). Like other type I topoisomerases, this enzyme makes a single-strand break in one strand of DNA and holds the broken ends while the other strand passes through the break. The break is then rejoined, thereby introducing or removing supercoils in the DNA one at a time (see chapter 1). However, unlike most topoisomerases, Topo I of vaccinia virus has strong sequence specificity and generates breaks only next to the 5-bp sequence shown in Figure 13.8C. Topo I makes a break on the 3′ side of the final T in the sequence, and one of the tyrosines in the enzyme remains attached to the phosphate bond to form a 3′ phosphoribosyltyrosine bond, much like Y recombinases (see Figure 8.20). Normally, the DNA would then rotate around the other strand, and the topoisomerase would rejoin the strands of DNA to remove a supercoil. However, when prepared under certain conditions, the topoisomerase can remain covalently bound to the end of the linear substrate at a T overhang as a useful reagent for molecular cloning (see below). Note that this resembles the overhang left after a restriction endonuclease makes a staggered cut in the DNA. If another DNA with the complementary overhang pairs with this DNA, the topoisomerase will join the two ends, thereby allowing insertion of a DNA fragment when the pairing and joining events occur at both ends of the fragment.

A plasmid cloning vector that has two recognition sites for the topoisomerase is required for this process. This vector is cut on the 3′ side of the Topo I recognition sites, for

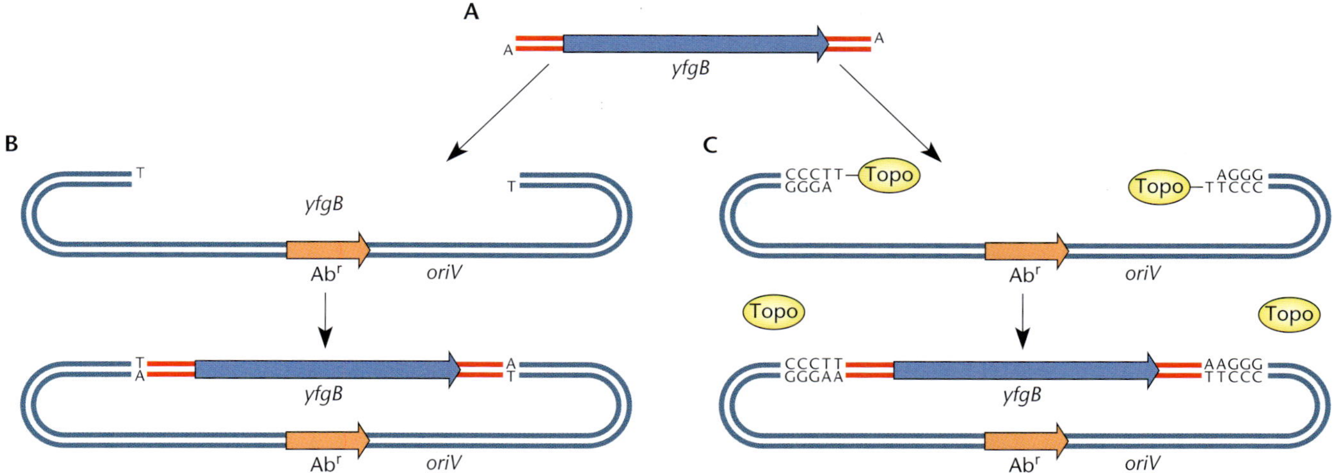

Figure 13.8 TA cloning and Topo TA cloning. **(A)** TA and Topo TA cloning take advantage of the observation that amplification with *Taq* DNA polymerase results in an extra A nucleotide overhang on the 3′ ends of the amplified fragment. **(B)** In TA cloning, a vector is used that has an overhanging T on the 5′ ends of the vector, providing a sticky end to facilitate ligation-mediated cloning. **(C)** In Topo cloning, a vector with the recognition site for vaccinia virus topoisomerase is treated so that the resulting DNA has the topoisomerase enzyme covalently linked to the 5′ overhang T ends as described in the text. The commercially available vector reagent can be used to directly clone DNA amplicons produced by PCR with *Taq* DNA polymerase without the use of ligase.

example, with a restriction endonuclease, so that the desired overhang sequences remain after the topoisomerase acts. The topoisomerase is then added; it cuts the DNA, leaves the desired overhangs, and remains attached to the DNA. Once the activated vector has been prepared, it can then be mixed with any DNA fragment with the deoxy adenosine (A) overhang, and the bound topoisomerases insert the fragment into the vector with high efficiency. One particularly useful application depends on the fact that the *Taq* polymerase that is commonly used in PCR amplifications naturally leaves the single base, A, as a 5′ overhang in the amplified fragment as noted above (Figure 13.8A). If the activated vector has been constructed so that it has a single base, T, as a 5′ overhang, any PCR-amplified fragment can be effectively cloned into it (Figure 13.8C). Topo cloning has the disadvantage that preparing the activated vector is time-consuming and technically difficult. Therefore, in most cases, the Topo cloning kit is purchased from a vendor with the vector already prepared.

CLONING WITH λ SITE-SPECIFIC RECOMBINATION: GATEWAY CLONING

Gateway cloning takes advantage of the site-specific recombination systems used by bacteriophage λ for integrating and excising from its attachment site in the bacterial chromosome (see chapter 7; Hartley et al., Suggested Reading). Recall that recombination occurs efficiently between the *attP* site on the phage and an *attB* site in the bacterial chromosome using the phage integrase (Int) and the host integration host factor (IHF). Excision occurs with these same proteins plus the excisionase (Xis) product. The Gateway system takes advantage of using these proteins, which can be purchased from biotechnology suppliers, in an *in vitro* reaction. In this system there are actually two *attB* sites, one on either end of the linear fragment to be cloned (Figure 13.9). The system works because there are compatible sets of *attP* sites in the vector. Compatibility between the two sets of *attB*/*attP* sites stems from a requirement of the site-specific reaction where the sequences in the crossover region have to be identical for the process to work. Therefore, two forms of *attB* and two forms of *attP* are used that have different sequences in the crossover regions (i.e., *attB1* and *attB2* and a compatible set of *attP1* and *attP2*; Figure 13.9). In the first step of the procedure, an insert is prepared that includes the *attB* sites, which were provided in 5′ tails on the PCR primers during amplification. The vector that is included in the system includes an antibiotic resistance marker and has the complementary *attP1* and *attP2* sites to allow the efficient and directional replacement of a vector fragment in the presence of purified Int and IHF proteins in the test tube.

The overall success rate is further improved because the region of the vector that will be replaced encodes the toxin component, CcdB, from one of the F′ plasmid toxin-antitoxin systems (see Table 5.1). The plasmid encoding the CcdB toxin cannot be maintained unless the host

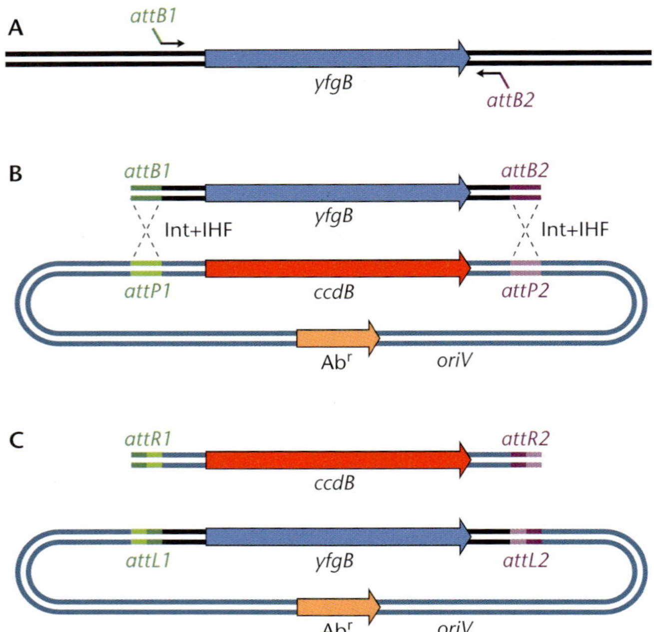

Figure 13.9 Cloning with λ Int and host integration host factor (IHF). The site-specific recombinase used for integration and excision with bacteriophage λ has been adapted for cloning in the Gateway system. **(A)** A DNA substrate to be cloned is PCR amplified using primers that add compatible λ attachment sites derived from the chromosomal site found in the *E. coli* chromosome at the ends of the DNA fragments for expression of the hypothetical gene *yfgB*. The *attB* sites from the *E. coli* chromosome are derivatives (i.e., *attB1* and *attB2*) with different crossover regions so they only recognize specific λ bacteriophage-derived sites (i.e., *attP1* and *attP2*). **(B)** A vector with compatible bacteriophage λ *att* sites will give the desired product when incubated with λ integrase and *E. coli* IHF as described in the text. The *ccdB* gene product in the vector allows for negative selection because it produces a toxic product if the host cell does not also encode the antitoxin protein CcdA. **(C)** The product that results from the steps described above is called an entry clone because it can be reacted with compatible vectors, called destination vectors, with matching *attB* and *attP* sites. Destination vectors allow the reading frame to be subcloned to include a variety of expression systems with any number of different fusions for purification or other procedures. Vectors have origins of replication (*oriV*) and a selectable antibiotic resistance determinant (Ab^r) for maintenance in an *E. coli* cloning strain.

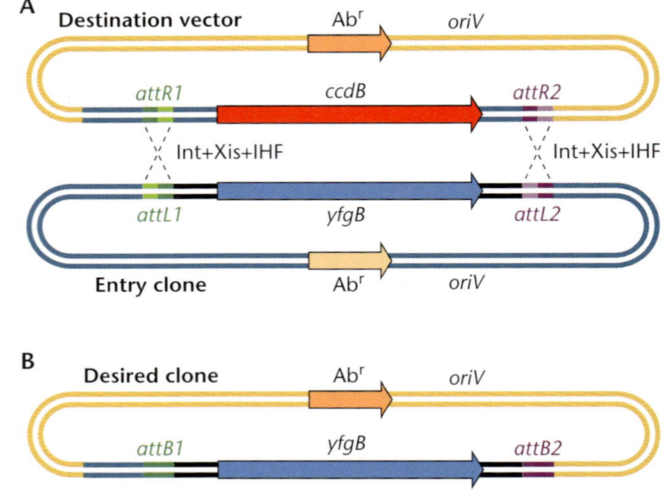

Figure 13.10 Subcloning with λ Int and Xis and host integration host factor (IHF) in the Gateway system. The directionality of λ integration and excision is controlled by the proteins used, allowing for an *in vitro* system for subcloning into compatible vectors. **(A)** Once a gene product of interest (*yfgB*) has been cloned into an entry clone, it can be subcloned into compatible destination vectors encoding a number of different expression systems for use in different kinds of cells. As described in Figure 13.9 and the text, compatible *attB1*/*attP1* and *attP2*/*attB2* sites with matching crossover regions result in *attR1*/*attL1* and *attR2*/*attl2* sites that are also compatible. **(B)** The *attR* and *attL* sites will react with λ Int and Xis and host IHF to form *attB1*/*attP1* and *attP2*/*attB2* sites. Products of the Int, Xis, and IHF reaction will result in an expression clone that can be directly selected in *E. coli*. The side product of the reaction will encode the *ccdB* gene product. The *ccdB* gene product in the vector allows for negative selection because it produces a toxic product if the host cell does not also encode the antitoxin protein CcdA. Vectors have origins of replication (*oriV*) and a selectable antibiotic resistance determinant (Ab^r) for maintenance in an *E. coli* cloning strain.

produces the CcdA antitoxin product. Therefore, only the successful product that contains the *yfgB* gene should be recovered because only these products will have the *ccdB* coding gene removed from the vector. The new product produced in this step of the procedure is also known as the "entry clone," because it can be used with other partner vectors to swap the DNA fragment to other "destination" vectors that differ by the type of expression system they use or different type of tags that can be fused to the gene of interest (Figure 13.10). This type of procedure is useful in part because a large variety of destination vectors exist. For example, if you are trying to isolate a protein, this system would allow you to easily swap the insert that encodes your protein into a variety of different vectors to facilitate the purification process. Compatible vector options allow the investigator to affix various affinity tags for purification or special protein fusions that increase the solubility of the target protein. Alternatively, other destination vectors may allow expression of your target protein in various types of cells, including eukaryotic cells.

LIGATION-INDEPENDENT CLONING USING LONG OVERHANGS: URACIL-*N*-GLYCOSYLASE-MEDIATED AND T4 DNA POLYMERASE-MEDIATED CLONING

Most of the mechanisms described in this chapter involve the use of DNA ligase to seal the final nicks in the DNA fragments before transformation into *E. coli*. However, if the complementary overhangs are sufficiently long, they will remain annealed following transformation into bacteria such that the host DNA ligase can perform the task *in vivo*. Two mechanisms are described below that allow the ends of PCR-amplified DNA fragments to be treated enzymatically to allow single-stranded overhangs of greater than 6 bases, which will allow DNAs to be joined together *in vivo* without the need for ligation *in vitro*.

A clever mechanism for producing long single-stranded complementary substrates at the ends of virtually any two DNAs to be joined involves the use of bacterial repair enzymes. As long as the sequences to be joined encode a deoxythymine, a procedure utilizing uracil-*N*-glycosylase (UNG) enzyme and AP endonuclease (endonuclease IV) or endonuclease VIII can be used (see Bitinaite et al., Suggested Reading). Recall that UNG is an important repair enzyme that removes uracil nucleotides that are mistakenly incorporated into DNA (see chapter 10). The UNG enzyme does not cleave the DNA backbone, and therefore an AP endonuclease is also needed in the repair system. A combination of UNG and endonuclease VIII has been called uracil-specific excision reagent (USER) and is available commercially. In the cloning procedure, the primers that are used to amplify DNA fragments to be cloned are ordered from the vendor with one deoxyuracil replacing a thymine at a position 6 to 10 base pairs from the 5′ end of the primer (Figure 13.11). Amplification with the deoxyuracil-containing primer and a polymerase that can integrate an A opposite the U (such as *Taq* polymerase) results in a double-stranded DNA that can be treated with the UNG and AP endonuclease or endonuclease VIII to produce a DNA end sequence that now contains a long single-stranded overhang. In addition to *Taq* polymerase, commercially available proofreading polymerases have also been engineered to efficiently read and amplify templates containing uracil (PfuTurbo Cx, Phusion U, and Q5U DNA polymerase). The same procedure is used to produce the cloning vector with the complementary single-stranded region. Alternatively, multiple products can be assembled at the same time by ensuring that compatible single-stranded regions are produced at each end to be joined (see below). Because the overhangs produced in this procedure are rather long and can stably persist, an *in vitro* ligation step is not required. The stability of the ends allows the products to remain annealed during transformation into the host, and the host ligase will join the fragments after the DNA has been introduced into the cell. Therefore, a UNG and AP endonuclease-mediated procedure circumvents the need for both restriction enzymes and a DNA ligase.

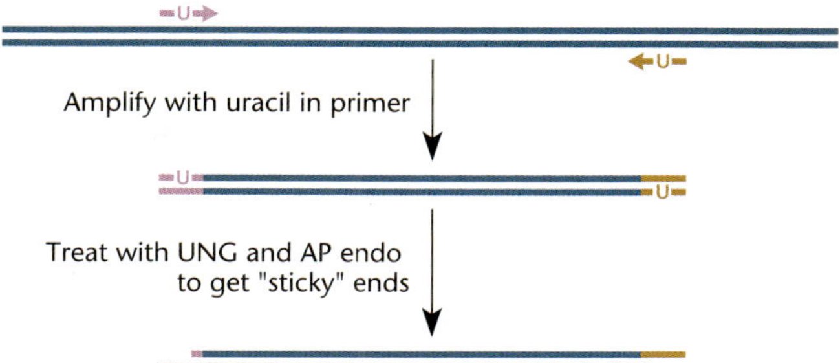

Figure 13.11 Cloning with products treated with uracil-*N*-glycosylase (UNG) and AP endonuclease (AP endo) with the USER system. Long single-strand extensions useful for ligase-free cloning can be produced with UNG and AP endo. In the procedure, PCR amplification is carried out with primers containing a U nucleotide instead of a T at a position 6 to 10 bases from the 5′ end of the primer. The resulting PCR product is treated with a combination of UNG and AP endo, which removes the U, allowing a 3′ single-stranded extension at the ends of the DNA fragment. Cloning can be done without ligase because single-stranded extensions are sufficiently long to allow the construct to be held together during transformation into *E. coli*, where the substrates will be ligated *in vivo*. The only requirement for designing the ends of the fragments to be joined is that a T can be located in the amplification primers; therefore, it is also amenable to joining multiple DNA fragments in a single reaction.

Another technique developed for processing the ends of linear fragments to allow ligation-independent cloning (LIC) involves the use of T4 DNA polymerase (see Aslanidis and de Jong, Suggested Reading). This mechanism takes advantage of the exonuclease function associated with T4 DNA polymerase to produce a 10- to 15-base overhang. In a standard version of the procedure, the insert to be cloned is treated with T4 DNA polymerase with only one of the deoxynucleoside triphosphates. T4 DNA polymerase will chew back from the 3′ end of the DNA using its exonuclease function but will not be able to polymerize forward at the positions coinciding with the three missing bases—only with the one base that is included. For example, if only dGTP was included and the end of the top strand had the sequence 5′-CTACTGTCCATACTCA-3′, the 3′ exonuclease activity of T4 DNA polymerase would remove everything from the 3′ end up to the G base (i.e., the G would be added by the polymerase activity favored over removal by the exonuclease activity of the T4 DNA polymerase enzyme). The cloning vector would be produced with the complementary sequence to the overhang in an analogous manner. Variations on the technique will also allow the assembly without sequence constraints to allow scar-less cloning (see Li and Elledge, Suggested Reading).

GOLDEN GATE CLONING

Golden Gate cloning provides a mechanism to clone multiple DNA fragments at one time (currently, up to 25 fragments in a single reaction). In this procedure, compatible end sequences are generated by the use of a special group of restriction endonucleases, called type IIS enzymes, that cleave at a position that is slightly offset from the sequence that is recognized (Figure 13.4C). A common enzyme used in this procedure is BsaI. The BsaI recognition site is 5′-GGTCTC-3′, but instead of cutting the DNA within this sequence, as found with most type II enzymes, it cleaves one base pair 3′ to the recognition sequence on the top strand and five bases 5′ to the recognition sequence on the bottom strand (Figure 13.12A). The Golden Gate procedure therefore allows the investigator to design a cleavage site with a 4-base overhang in any DNA sequence. As indicated in Figure 13.12, the BsaI recognition site is introduced into the 5′ end of the primers used in the procedure. However, unlike the more traditional cloning procedures, as described in Figure 13.7, in the Golden Gate procedure, the recognition site for the enzyme will be removed, leaving only the native sequence used for amplification. All of the multiple DNA fragments that the investigator wishes to be joined need only to have compatible end sequences after the synthetic end sequences with the BsaI recognition site are removed (Figure 13.13). The same procedure involving the addition of the BsaI recognition site is carried out with the cloning vector. An additional benefit is that digestion with the BsaI restriction endonuclease and the ligation reaction can be done in the same tube, because the BsaI recognition site is removed in the procedure. If a standard type II restriction enzyme is used in a similar scenario, the ligation products would be recut since the recognition site would be re-created upon ligation.

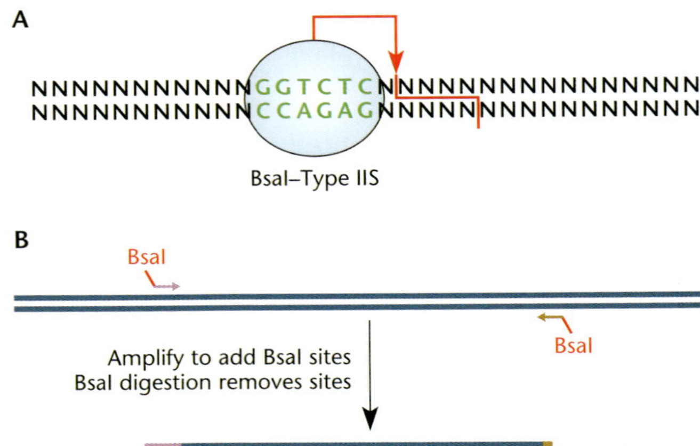

Figure 13.12 Cloning with products treated with type IIS restriction endonucleases with the Golden Gate system. **(A)** Type IIS restriction endonucleases like BsaI cleave the DNA at a position offset from the site that is recognized by the enzyme. These restriction enzymes are useful because the site that is recognized can be removed in the procedure. They also can be used on any DNA sequence producing ends with single-stranded extensions that are useful for assembly-style cloning. **(B)** In the procedure, the recognition site for BsaI is worked into the primer used for PCR amplification. By designing the BsaI recognition site at the end of the primer in the correct orientation, treatment results in removal of the site and a long single-strand extension, results that are useful for assembly of multiple DNA fragments (see Figure 13.13).

GIBSON ASSEMBLY

Another form of assembly cloning was first used in synthetic biology for the assembly of complete genomes in a procedure sometimes called "Gibson assembly" (see Gibson et al. 2008, Suggested Reading; see also Box 13.3, page 560). In this procedure, three distinct enzyme activities are used to assemble two or multiple DNA fragments that are commercially available under this and other names, such as HiFi DNA assembly (see Gibson et al. 2009, Suggested Reading). Like the UNG-mediated and Golden Gate cloning procedures described above, compatible end sequences are planned in advance (see Figure 13.13). However, Gibson assembly and related techniques do not require modified primers, they do not require addition of restriction enzymes sites, and the junctions between fragments are seamless (Figure 13.14). In this procedure, a 5′ exonuclease activity in the reaction mixture generates the single-stranded ends on all of the linear DNA fragments. The complementarity that was planned in advance allows these extended regions to anneal. A DNA polymerase activity that is also in the reaction mixture fills in any gaps that remain following annealing of the ends. Finally, a third activity in the reaction mix, a DNA ligase, allows the fragments to be joined together. The enzymatic activities in the reaction mix are thermostable and the reaction is isothermal, meaning that in a standard version of the procedure, the whole process is done at 50°C.

Systems like the Gibson cloning system or any number of other enzyme systems using similar steps provide an efficient way of making large DNA molecules from blocks of DNA sequences that also can be ordered as smaller synthetic fragments from biotechnology vendors. These kinds of procedures are especially ideal in cases where it may be difficult to obtain the strain of interest or if the organism is pathogenic or requires highly specialized growth conditions, making it difficult to obtain chromosomal DNA for insert preparation.

IN VIVO CLONING AND ASSEMBLY TECHNIQUES

The techniques described in this section involve enzymatic reactions carried out *in vitro* with purified proteins. However, cloning and assembly techniques can also utilize a host cell that has an efficient form of recombination to join DNA fragments *in vivo*. Useful variations

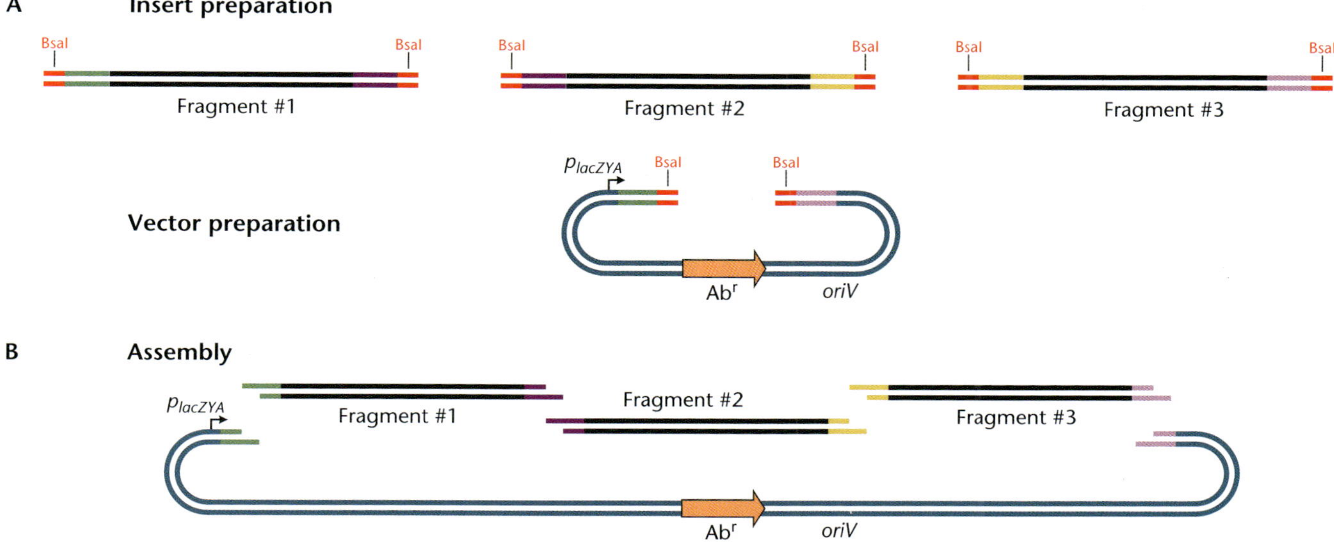

Figure 13.13 Multiple DNA fragments can be joined when they have compatible ends by using Golden Gate assembly and other assembly technologies. **(A)** In the procedure, type IIS restriction endonuclease recognition sites (shown in red), like those for BsaI, are situated in the primers such that they can be subsequently removed with BsaI digestion to leave 4-base single-stranded extensions (see Figure 13.12). **(B)** In the diagram, the green, purple, yellow, and pink regions are complementary and can anneal during the assembly step, allowing the fragments and the vector to assemble in the correct order and orientation. An additional benefit of the procedure is that BsaI digestion and ligation can be carried out in a single tube, as described in the text.

typically involve systems where only small regions of homology are needed for joining by recombination and/or with organisms where transforming DNA can be made very efficient through natural or artificial means (see chapter 6). For example, in chapter 9, *in vivo* recombination techniques were described using λ Red recombination for efficiently recombining linear DNAs into the chromosome or plasmids (Figure 9.11), but λ Red recombination can also be used to join linear DNAs that are electroporated into the host in another type of plasmid-based cloning (see Sharan et al., Suggested Reading). The naturally efficient recombination system of yeast is providing an invaluable tool for the assembly of large DNAs transformed into the cell to assemble larger than megabase-size circular molecules (Box 13.3), while the efficient natural transformation system of *B. subtilis* can assemble and stably maintain chromosomal insertions of multiple megabases (see Juhas and Ajioka, Suggested Reading).

CRISPR/Cas Systems

One of the most unexpected and exciting discoveries in molecular biology over the last several years is the CRISPR/Cas systems. CRISPR/Cas systems are examples of adaptive immune systems in bacteria and archaea. They are adaptive in the sense that, like our own adaptive immune system, they are capable of maintaining a memory of a previous infection. In the case of bacteria and archaea, this is typically an infection from a bacteriophage, but one that for some reason was not successful in killing the cell. By maintaining memory of the previous encounter with the virus, the strain possessing this system can protect itself from this virus if it infects the cell again at a later time. CRISPR/Cas systems are common, found in about 90% of archaea and in about half of bacteria.

The name CRISPR is unfortunate because it bears no relation to the actual function of the system. CRISPR stands for cluster of regularly interspaced short palindromic repeats, which refers to a set of repeated sequences that contain a palindrome (Figure 13.15). When they were first noticed in bacterial genomes, it was not appreciated that one of the most important parts of the system was actually the DNA sequences found between the repeats, which were called **spacers** (Figure 13.15). The repeats and spacers are typically referred to as a **CRISPR array**. One of the first clues to the function of CRISPR/Cas systems was recognition that the spacers in the array were not random sequences but, instead, were segments of DNAs found in mobile elements, especially bacteriophages (see Bolotin et al., Suggested Reading). Other investigators recognized how these spacer sequences related to the fully functioning system, including the proteins en-

BOX 13.3

Synthesizing and Cloning Complete Bacterial Genomes

Investigators in many fields will find themselves using the cloning and assembly techniques described in this chapter to make genetic constructs, some of which are very large. However, an ambitious goal taken over the course of a number of years was not to make a construct to move into a given bacterial strain, but instead to make a new entire bacterial strain. The goals of this endeavor were multifold. On the application side, the ability to have complete control over every aspect of the genetics of bacterial strains used in industrial or therapeutic applications could potentially have significant commercial and humanitarian value. However, there are also numerous basic science questions that can be addressed with such an ability; probably greatest of all is knowing what can constitute a functional minimal genome for a free-living lifeform. Previous work on the idea of a minimal genome has involved assessing which genes can be deleted or inactivated one by one without losing viability of the strain. However, a more direct way to address the question of a minimal genome is to build the genome from the beginning.

A series of experiments aimed at creating a bacterial strain with an entirely synthetic genome was undertaken by the lab of J. Craig Venter. The entire project involved bacteria from the genus *Mycoplasma* because of their small genome size and a simple membrane configuration that makes genome transplantation more straightforward. An important initial step was establishing a system for replacing the endogenous genome of one *Mycoplasma* species with that from another species (see Lartigue et al., References). A significant challenge with genome transplantation in bacteria was ensuring that there were no breaks in the circular chromosome, because even a single break would not allow the transferred chromosome to replicate.

A second important technical step was developing the methods needed to construct an extremely large circular chromosomal DNA molecule from the assembly of smaller DNA blocks or oligonucleotides. This challenge led to the development of the "Gibson assembly" technique that is described in this chapter (see Figure 13.14). A significant complication with the initial work was ensuring that the sequence of the 582,970-base pair *Mycoplasma genitalium* genome was free of any errors by checking subassemblies of the genome along the way as well as the final construct (see Gibson et al. 2008, References). Early steps in the project involved working with partially overlapping double-stranded DNA fragments of 5 to 7 kb that could be ordered commercially and assembled into larger DNA blocks. DNA blocks of 24 and then 72 kb were replicated using an F plasmid derivative vector (bacterial artificial chromosome) for replication in *E. coli* (see chapter 4). Larger assemblies and the final entire *M. genitalium* genome had to be assembled and cloned with yeast recombination.

Combining the technologies described above allowed the group to make what was by one measure the first synthetic organism, a derivative of *Mycoplasma mycoides* with a 1.08-Mb genome (see Gibson et al. 2010, References). A switch was made from experimenting with very slowly growing *M. genitalium* species to *M. mycoides*, which grew much more quickly. Even though the genome of *M. mycoides* was twice as large as that of *M. genitalium*, advances had been made for assembling much larger DNAs more easily by working in yeast with its highly efficient homologous recombination system that allows many fragments to be assembled *in vivo*. Therefore, genome size was less of a limiting factor than growth rate in making progress with the work.

In 2016, the investigators finally reached their goal of designing and building a minimal genome (see Hutchison et al. 2016, References). Building from what was learned from the earlier work, assembly of multiple DNA fragments was relatively efficient. In an additional change from the earlier work, instead of assembling commercially sourced double-stranded DNA blocks, in this work, approximately 50 oligonucleotides (Figure, red) were assembled into fragments of 1.4 kb *in vitro* (Figure, blue). The 1.4-kb fragments were subsequently assembled into 7-kb cassettes (Figure, purple) that were maintained in *E. coli* and confirmed by sequencing to be free from errors. The 7-kb blocks were directly assembled in yeast with *in vivo* recombination and maintained as eight separate segments which could be confirmed with DNA sequencing (Figure, orange). These eight pieces were subsequently amplified using rolling-circle replication, and then a final complete genome was assembled again in yeast (Figure, green circle).

Once the technological aspects of cloning large DNA fragments were solved, the most significant hurdles in synthesizing a functioning minimal genome involved determining what genes were really essential for robust growth and how to construct new operons in a way that would not interfere with expression of adjacent operons. Success hinged on the ability to test multiple permutations of changes in the efficient recombination system established in yeast. Unexpectedly, many more genes were needed than were predicted to be essential from earlier experiments done in multiple labs. One reason for the mismatch between what was predicted to be essential based on earlier work versus what was really needed in the minimal genome was often explained by synthetic lethality. Synthetic lethality is the genetic property where gene A and gene B are not found essential when inactivated on their own, but inactivating both genes A and B together is not tolerated (see chapter 3). In the work from the Venter lab, many *Mycoplasma* genes could be deleted in isolation in the parent strain, but they were still required when attempting to produce the min-

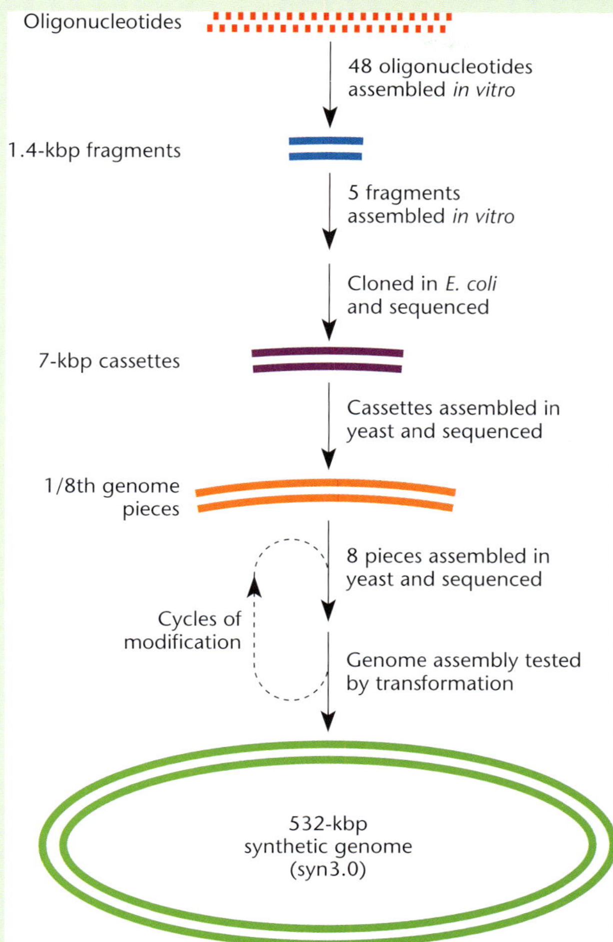

imal genome from scratch. Subsequently, multiple synthetically lethal gene pairs could be established, and once identified, both genes were included in the final minimal genome strain. Remarkably, of the 473 genes that were needed in the minimal *M. mycoides* 532-kb genome, the function of 149 of these genes was unclear, suggesting that molecular microbiologists still have much work to do before we have a comprehensive understanding of the molecular foundations of life.

The idea of what constitutes a minimal genome is itself complicated. What is considered a minimal genome also needs to take into account the environment where the strain is grown. For example, genes required for making all 20 amino acids and various cofactors will be essential in the laboratory if the strain is grown in minimal medium containing only salts and a carbon source. However, these same genes may be nonessential when the strain is grown in a complete medium such as standard laboratory LB medium. Additionally, growth in the laboratory may not require genes needed for stress conditions, such as changes in osmolarity, exposure to DNA-damaging agents, or changes in nutrient availability, that could otherwise be essential to maintain a bacterial strain in the environment. It is probably worth pointing out that even in the context of the experiments from the Venter laboratory, the final organism with its minimal genome displayed a much slower growth rate than the parent strain and produced cells with highly irregular cell sizes, suggesting that basic growth features had been compromised. It will be exciting to see if the basic principles of genome assembly and genome transplantation can be applied to more relevant free-living bacteria to broaden the application potential of these exciting methods.

References

Gibson DG, Benders GA, Andrews-Pfannkoch C, Denisova EA, Baden-Tillson H, Zaveri J, Stockwell TB, Brownley A, Thomas DW, Algire MA, Merryman C, Young L, Noskov VN, Glass JI, Venter JC, Hutchison CA III, Smith HO. 2008. Complete chemical synthesis, assembly, and cloning of a *Mycoplasma genitalium* genome. *Science* **319:**1215–1220.

Gibson DG, et al. 2010. Creation of a bacterial cell controlled by a chemically synthesized genome. *Science* **329:**52–56.

Hutchison CA, et al. 2016. Design and synthesis of a minimal bacterial genome. *Science* 351:aad6253.

Lartigue C, Glass JI, Alperovich N, Pieper R, Parmar PP, Hutchison CA III, Smith HO, Venter JC. 2007. Genome transplantation in bacteria: changing one species to another. *Science* **317:**632–638.

coded adjacent to the CRISPR array, which were called CRISPR-associated proteins, or Cas proteins (see Barrangou et al., Suggested Reading).

SPACER ACQUISITION, crRNA PROCESSING, AND INTERFERENCE

CRISPR/Cas systems fall into a number of diverse types, which was confusing during early work with these systems (see below). However, all of the CRISPR/Cas systems work through the same three major steps: (i) spacer acquisition, (ii) pre-CRISPR RNA (pre-crRNA) processing, and (iii) interference (Figure 13.16). We will use the process of acquisition of immunity to a bacteriophage as an example. However, CRISPR/Cas systems will also defend against DNAs introduced by plasmid transfer or natural transformation (see Marraffini and Sontheimer, Suggested Reading). In the first step of the process, the cell must be infected but somehow survive, possibly because the infecting virus is defective in some way or is slowed down sufficiently by other host defense systems,

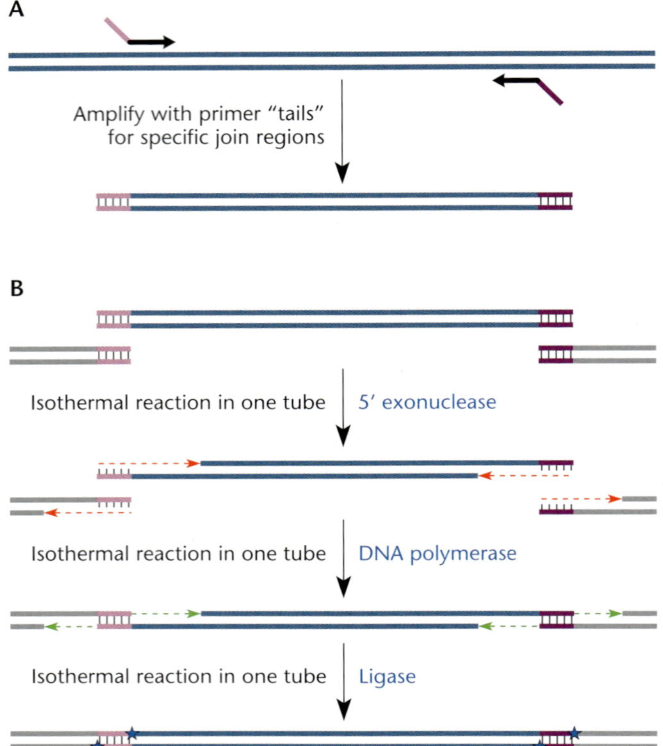

Figure 13.14 Gibson assembly can be used to join multiple DNA fragments in a single isothermal reaction. **(A)** Amplification primers are designed to amplify the DNA fragments to be assembled with 15 to 80 base pairs of overlap between the fragments. The pink and purple regions indicate matching DNA sequence. **(B)** Three separate enzymatic activities are utilized in the procedure. A 5′ endonuclease activity produces 3′ single-stranded extensions at the ends of the DNA fragments (red dashed arrow-headed line). The complementary ends anneal in the isothermal reaction, and a DNA polymerase included in the reaction extends the DNA to fill the gaps (green dashed arrow-headed line). Ligase will seal the ends of the fragments (blue stars) to produce closed circular DNA substrates that can be introduced into a standard *E. coli* cloning strain by transformation.

such as restriction endonucleases. Dedicated Cas proteins that are part of the system collect a small segment of DNA from the invading virus that becomes a spacer when it is incorporated into the CRISPR array in a process called **spacer acquisition** or **adaptation** (Figure 13.16). In a second step, called **pre-crRNA processing**, the CRISPR array is transcribed from a promoter associated with the system to form a pre-crRNA that is processed by one or more additional Cas proteins into individual RNAs called **guide RNAs** because they will later guide the homology-based process to recognize the bacteriophage from which they were obtained. In the final step of the process, called **interference**, the guide RNA assembles with one or more Cas proteins to form an **effector complex** that can recognize the viral genome if the cell is infected again by the same bacteriophage. The effector complex matches the guide RNA to the bacteriophage DNA by complementary base pairing and uses a nuclease activity to cut and incapacitate the viral DNA; with other types of systems, binding of the effector complex to the target is followed by recruitment of a dedicated Cas nuclease that degrades the viral genome and prevents it from killing the host. The process is called interference because it interferes with the normal virus function of infecting the bacterial host.

CLASSES AND TYPES OF CRISPR/Cas SYSTEMS

Six types of CRISPR/Cas systems have been found in bacteria and archaea, but one feature that is shared among all of the systems that behave as described above is the ability to acquire spacers that recognize invading DNAs such as bacteriophages (Figure 13.16). The process of spacer acquisition is carried out by two or three proteins, Cas1, Cas2, and occasionally, Cas4. Apparently, once this part of the system had evolved, the processes of pre-crRNA processing and interference evolved independently to give the various types of CRISPR/Cas systems (see Koonin and Makarova, Suggested Reading).

The six types of CRISPR/Cas systems are grouped into two large classes (see Makarova et al. 2020, Suggested Reading). The class 2 systems are architecturally simpler, with processing of the crRNA and interference carried out by one large protein or effector complex (Figure 13.17). In the class 1 systems, the functions utilized for crRNA processing and interference are in individual proteins that form a complex called Cascade, standing for CRISPR-associated complex for antiviral defense. The organization of CRISPR/Cas systems into six types is based on signature proteins specific to the individual systems as well as by examining the phylogenetic relationships between the systems (see Makarova et al. 2015, Suggested Reading). In the case of the class 1 systems, the most common types, type I and type III, are grouped based on the signature endonucleases, Cas3 and Cas10, respectively, that degrade the phage or other invading DNA as part of the system (Figure 13.17). Type IV systems resemble type I systems but lack the spacer acquisition system; the function of these systems is incompletely understood but may be involved in competition between similar plasmids. In the class 2 systems, there is a single protein that carries out all of the functions other than spacer acquisition. The best-understood system is the type II system that uses the Cas9 protein (Figure 13.17). The other two class 2 systems are the type V and type VI systems, which use the Cas12 and Cas13 signature proteins. Gene organization and phylogeny were also used to establish subtypes of the systems that are designated with letters. For example, type I systems fall into seven subtypes, subtypes I-A to I-F.

Basic features of Class 1 CRISPR/Cas systems

A Bacterial chromosome

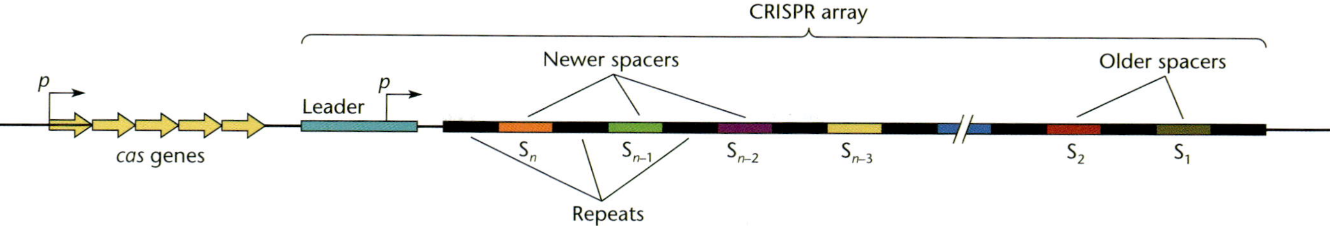

CRISPR (clustered regularly interspaced short palindromic repeats)
CRISPR array: one to hundreds of repeats

Repeat: 21–47 bp (CAGTTCCCCGCGCCAGCGGGGATAAACCG)

Spacer sample: ~30 bp $(\text{CTTTCGCAGACGCGCGGCGATACGCTCACGCA})_n$

Leader: Several hundred noncoding base pairs

cas genes (cascade): *cas* complex for CRISPR-associated antivirus defense; acquisition of new spacers; immunity, e.g., processing pre-crRNAs

B Phage or plasmid DNA

ps_n

PAM

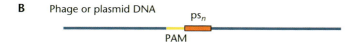

PAM: protospacer-adjacent motif (example) (CT/AT)

ps_n: protospacer $(\text{CTTTCGCAGACGCGCGGCGATACGCTCACGCA})_n$

Figure 13.15 Features of CRISPR/Cas systems. **(A)** Typical class 1 CRISPR array with CRISPR-associated (*cas*) genes. A number of repeated sequences are interspaced with spacers s_n to s_1 of about the same length but with different sequences. A leader region array contains a promoter (p) from which the CRISPR array is transcribed and CRISPR-associated *cas* genes whose products are involved in taking up sequences and targeting protospacer sequences. Spacers are always added adjacent to the leader region; hence, the oldest spacers will be most distal from the leader region. **(B)** A protospacer (ps_n) from a phage or plasmid with exactly the same sequence as one of the spacers (s_n) in the CRISPR.

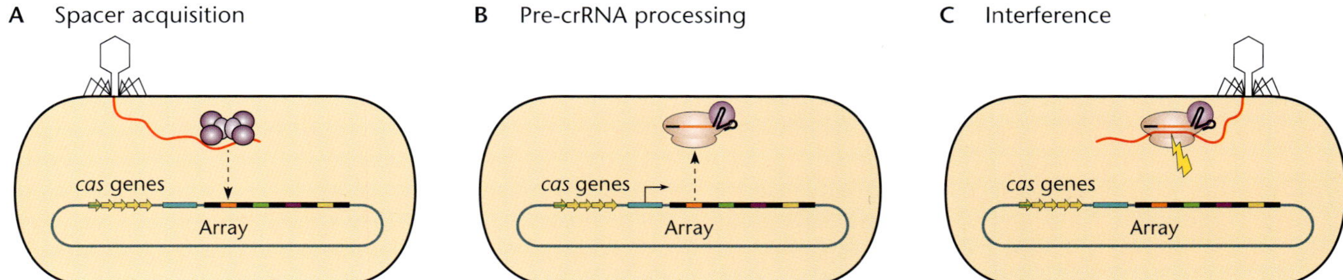

Figure 13.16 Three major processes are involved in the functioning of CRISPR/Cas systems. CRISPR/Cas systems are remarkably diverse, but all function with three major steps. In spacer acquisition (also known as adaptation), short stretches of DNA sequence called spacers are harvested from mobile elements, such as a bacteriophage that infected but for some reason was unable to kill the cell. A subset of the proteins harvest the spacers that are inserted immediately adjacent to the leader region found upstream of the array. The *cas* genes are typically adjacent to the array. During pre-CRISPR RNA (pre-crRNA) processing, the CRISPR array is transcribed from a promoter found in the leader region. One or more Cas proteins will process the long pre-crRNA into an individual guide or crRNA that forms a complex with one or more Cas proteins, depending on the system. In the interference stage, the guide RNA complex surveys the cell for a match in the DNA sequence. When a match to the guide RNA has been identified, nuclease activity is either engaged or recruited through a separate protein to cleave or degrade the bacteriophage DNA.

Figure 13.17 Simplified representation of five of the six major types of CRISPR/Cas systems. The five major types of CRISPR/Cas systems all have related Cas1 and Cas2 proteins involved in spacer acquisition (shown in purple). In the class 1 systems, multiple Cas proteins and in some subtypes the small subunit (SS) are involved in pre-CRISPR RNA (pre-crRNA) processing and interference. In the class 2 systems, a single effector protein has the pre-crRNA processing and interference functions. Colors indicate CRISPR/Cas function: guide RNA processing (blue); guide RNA binding and protospacer-adjacent motif (PAM) recognition (red); and nuclease activity (yellow). Class 1, type IV systems omitted for simplicity. See text for details.

A Spacer acquisition

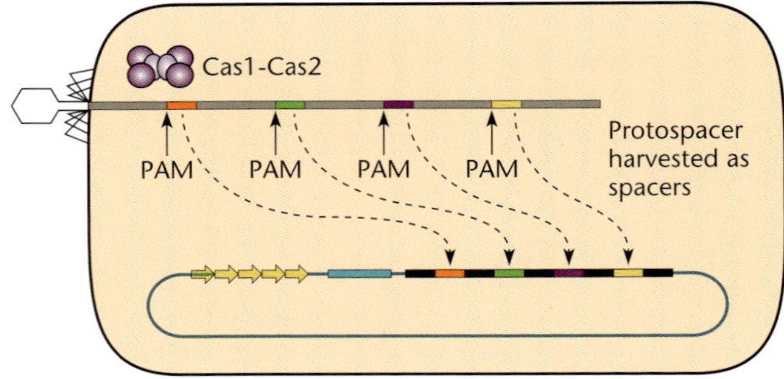

B Interference

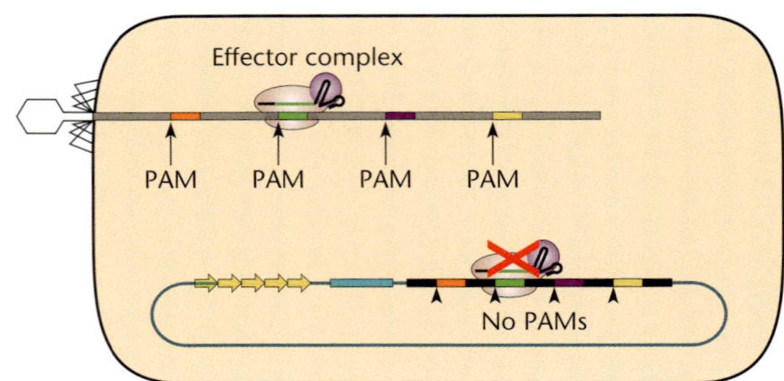

Figure 13.18 Protospacer-adjacent motifs (PAMs) play an important role in controlling the spacer acquisition and interference stages of CRISPR/Cas function. PAMs allow CRISPR/Cas systems to recognize a match to the guide RNA in an invading mobile element in the interference phase but to ignore the spacer in the CRISPR array that also matches guide RNA. (A) The spacer acquisition system only harvests fragments of sequence as spacers (protospacers) found next to a PAM. (B) The short PAM sequence is not found in the repeat of the CRISPR array, which allows the effector complex to discriminate between guide RNA matches in the cell. This is because the system will only activate interference nuclease activity with protospacers that contain the PAM and not the spacer in the array that lacks the PAM.

PROTOSPACER-ADJACENT MOTIFS (PAMs)

One of the initial major questions concerning CRISPR/Cas systems was how guide RNAs could target invading bacteriophage and other mobile elements for destruction without recognizing the spacer sequence found in the CRISPR array, which has the same sequence as the target. A clue to how this could happen came from the observation that not all stretches of sequences could be recognized and integrated as spacers in a CRISPR array. Those sequences that could end up in the array, called **protospacers**, always have a specific short sequence motif next to them (see Mojica et al., Suggested Reading). This sequence motif adjacent to the protospacer is referred to as a **protospacer-adjacent motif**, or PAM (Figure 13.18). The Cas1-Cas2 spacer acquisition system will only collect DNA fragments that are found adjacent to its specific PAM sequence (Figure 13.18). In contrast, the PAM sequence is specifically missing in the CRISPR array. This same PAM sequence is also used by the effector complex such that only spacers with the adjacent PAM will trigger cleavage or degradation in the interference stage. This system allows for an effector complex to recognize the difference between a protospacer to be recognized as part of an invading DNA that is to be degraded versus the same sequence found as a spacer in the CRISPR array because the spacer in the array lacks the PAM (Figure 13.18). The PAM differs between the various types of CRISPR/Cas systems. Knowing the PAM is important when harnessing CRISPR/Cas systems as a tool to target DNA breaks or degradation or with related applications because engineered guide RNAs will only recognize protospacers with the correct PAM.

SPACER ACQUISITION WITH Cas1-Cas2

As noted above, spacer acquisition is the unifying mechanistic process found across the CRISPR/Cas systems. The Cas1 and Cas 2 proteins are required at minimum, although an accessory protein, Cas4, is also utilized in many systems. Protospacer sequences that are acquired from a bacteriophage or other mobile element are integrated at the position closest to the leader region of the CRISPR array, into the first CRISPR repeat (Figure 13.19). Interestingly, the process of integrating the DNA frag-

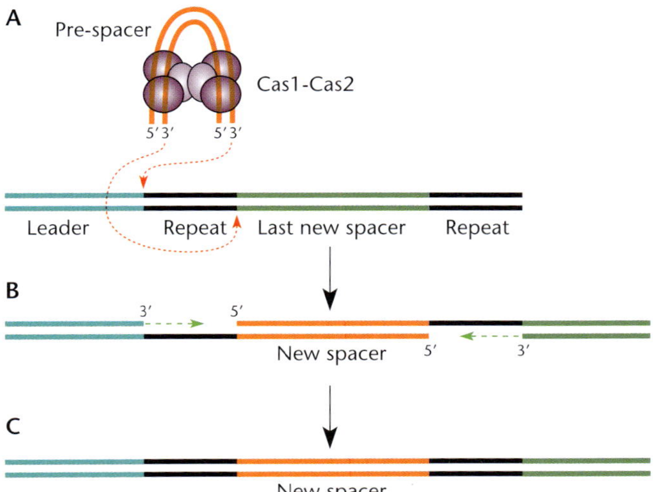

Figure 13.19 Model for how a new spacer is integrated into the CRISPR array at the repeat closest to the leader region with Cas1 and Cas2. **(A)** Joining events during the integration process occur at the 5′ ends of the repeat closest to the leader region. **(B)** This process results in large gaps flanking the newly integrated spacer. The Cas1-Cas2 proteins are removed and a DNA polymerase activity fills in the missing DNA strand (red dashed line). **(C)** This process allows the new spacer to reside in the position closest to the leader region.

ment that will form the spacer (called a prespacer prior to integration) into the array occurs in a process reminiscent of the process used for the integration of DNA transposons (see Figure 8.3; see Xiao et al., Suggested Reading). In both transposition and prespacer integration, the incoming DNA ends are joined at staggered positions in the target DNA (Figure 13.19A). As a result of this process, gaps are formed at both ends of the prespacer following integration that must be repaired by host DNA polymerase enzymes (Figure 13.19B). In the case of transposition, this gap-filling process results in the target site duplication that is a characteristic of integration of DNA transposons (see Figure 8.3). In the case of spacer acquisition by Cas1 and Cas2, this process forms the characteristic repeats in the CRISPR array (Figure 13.19C). The similarities between the processes of spacer acquisition and transposition are a clue to the origin of the CRISPR/Cas systems. The discovery of a sister phylogenetic group of proteins related to Cas1 endonucleases provides another important clue that the spacer acquisition system was actually itself acquired from transposons, in which current group members are called casposons because they move as transposons and are related to the Cas1 used for spacer acquisition (see Krupovic et al., Suggested Reading). It is interesting to note that our own human immune system, which utilizes the RAG1 and RAG2 recombinases, was also originally derived from an ancient transposon similar to the *Transib* elements found today (see Kapitonov and Jurka, Suggested Reading).

TYPE I CRISPR/Cas SYSTEMS

Type I CRISPR/Cas systems are the most common type of system found in both bacteria and archaea, comprising about half of the CRISPR/Cas systems found in each of these domains (see Makarova et al. 2015, Suggested Reading). The signature feature that relates all of the type I systems is the nuclease that is recruited for degrading the invading DNA, Cas3 (Figure 13.20). The various subtypes within this group can have one or more Cas3 nucleases, and in some cases, it is fused to proteins involved in spacer acquisition. In class 1 systems, the proteins that process the long pre-crRNA into guide RNAs (crRNAs) in effector complexes are made up of multiple proteins. The 5′ region and 3′ stem loop of the guide RNA are encoded in regions of the repeat that flank an individual spacer (Figure 13.20A and B). Cas6 is responsible for cleaving the pre-crRNA to form the guide RNA, and Cas6 is the first protein to enter the growing Cascade complex. Cas7 subunits are normally present in six copies in the effector complex and coat the length of the guide RNA. The 5′ end of the guide RNA is recognized by Cas5, which also associates with Cas8 (sometimes called the large subunit) (Figure 13.20C). Cas8 is responsible for recognizing the PAM in target DNAs; this is essential for ensuring that only matches to the guide RNA in foreign DNAs (protospacers), and not the spacer in the CRISPR array, are recognized for degradation by Cas3. In some of the type I systems, an additional protein (called small subunit or SS) helps to stabilize the R-loop when one strand of DNA is displaced during the recognition process (Figure 13.20C). Among the multiple differences between the class 1 and 2 systems is that the PAMs are found on different sides of the protospacers in the two classes.

With the type 1 systems, Cas1-Cas2 act independently from the effector complex to acquire and integrate spacers into the CRISPR array. The first spacer that is acquired from an invading bacteriophage or other element occurs by a process called **naive spacer acquisition**. How spacers are preferentially harvested from mobile elements and rarely from the much larger chromosome is still incompletely understood. One intriguing hypothesis is that the host RecBCD nuclease (see chapter 9) could be involved in such a process. Participation of RecBCD would help to explain how foreign DNAs could be preferred, because foreign DNAs would lack the *Chi* sequences found in the host and would be subject to extensive degradation (see Levy et al., Suggested Reading). In this RecBCD-based model, the fragments produced as the foreign DNA sequence is degraded would be the substrates to be integrated as spacers; however, the model

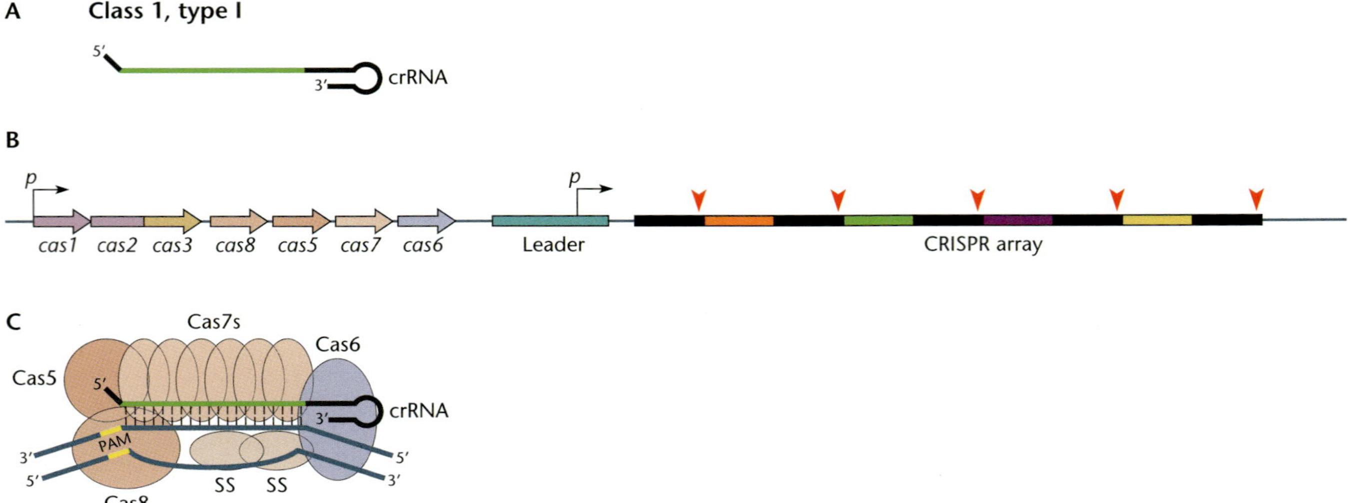

Figure 13.20 Schematic representation of the Cas genes and CRISPR array and a guide RNA complex from an idealized class 1, type I system. **(A)** CRISPR RNA (crRNA) is shown with the region encoded in the spacer shown in green and the 5′ and 3′ features encoded in the repeat region (black). **(B)** Representation of the *cas* genes and CRISPR array showing the upstream leader region with promoter. Red arrowheads indicate the position where the pre-crRNA is cleaved in the RNA encoded in the array. Promoters used to transcribe the components are shown as bent black arrows. **(C)** Representation of an idealized class 1 guide RNA complex (cascade) showing the R-loop formed with recognition of a protospacer with the appropriate protospacer-adjacent motif (PAM) (yellow) in DNA (blue). In some systems, an additional protein called small subunit (SS) stabilizes the displaced strand in the system.

remains controversial because it is not clear that the product of RecBCD could be directly used by the Cas1-Cas2 integration machinery.

The process of naive spacer acquisition occurs at a low frequency, but there is a separate process called **primed spacer acquisition** that is much more efficient (see Datsenko et al., Suggested Reading). This process occurs as a collaboration between Cas1, Cas2, and Cas3 along with the cascade complex (see Dillard et al., Suggested Reading). In the primed acquisition process, there is already a spacer in the CRISPR array that exactly or closely matches sequences in the invading bacteriophage. This recognition is used as a cue to collect many additional spacers from the same bacteriophage genome. The primed spacer acquisition process probably provides two very important functions in type I systems. For one, it allows immunity of the CRISPR/Cas system to increase because there will be multiple guide RNA complexes that recognize different sequences in the same bacteriophage. A second very important benefit to primed spacer acquisition is the ability to greatly reduce so-called escape mutants from bypassing the system. While type I CRISPR/Cas systems are very effective at interfering with killing by bacteriophage, the whole CRISPR/Cas system can be compromised if one of the infecting phage genomes has a mutation that no longer allows the guide RNA to recognize its match in the phage. This phage would then overtake the population of bacteria because the bacteria would no longer be immune via the spacer. However, the primed acquisition system can utilize the partial match to collect additional spacers from the phage, greatly reducing the sensitivity of the system to escape mutants.

TYPE II Cas9 CRISPR/Cas SYSTEMS

The type II Cas9 CRISPR/Cas systems are the most well known because of their utility for guide RNA programmable DNA nuclease activity (see Jinek et al., Suggested Reading). They are the third most common type of system found in bacteria but are not known to be natively found in archaea. As with all CRISPR/Cas systems, spacer acquisition is carried out by Cas1 and Cas2. However, in the type II systems, Cas1-Cas2 spacer acquisition occurs in coordination with the large Cas9 effector protein, which is also responsible for pre-crRNA processing and interference. The guide RNAs in this system require an additional *trans*-acting RNA, tracrRNA, that is part of the guide RNA effector complex (Figure 13.21). To simplify the system for genome modification, an engineered system was designed in which the crRNA is fused to the tracrRNA to form a so-called synthetic guide RNA (sgRNA; Figure 13.21C). Two nuclease activities within Cas9 are involved in separately nicking the top and bottom strands during the interference stage of CRISPR/Cas function.

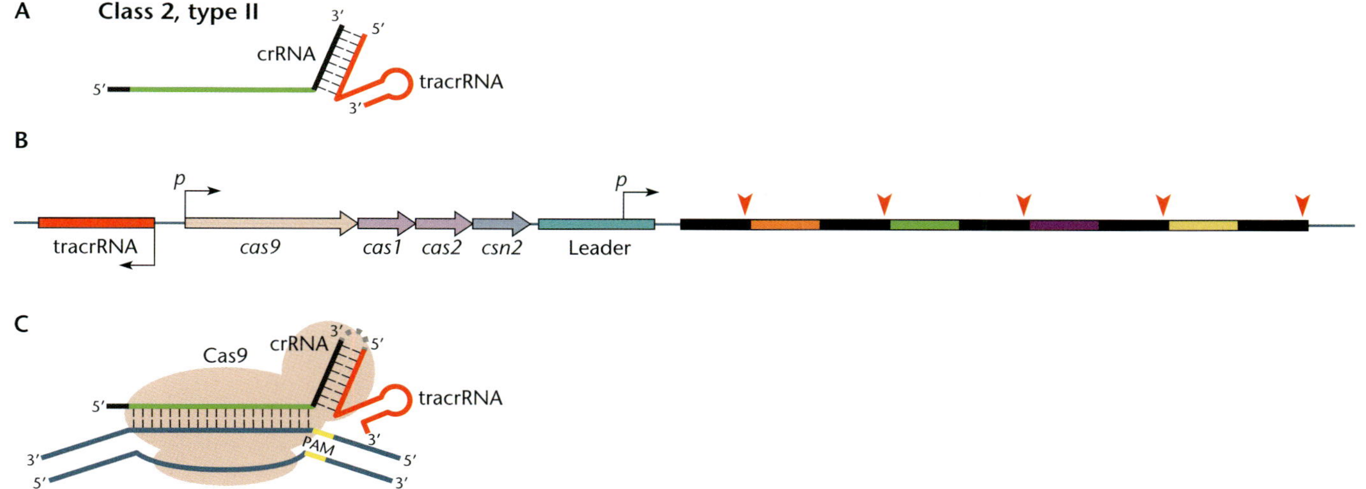

Figure 13.21 Schematic representation of the *cas/csn* genes, *trans*-acting RNA (tracrRNA), and CRISPR array and a guide RNA complex from the class 2, type II system. **(A)** CRISPR RNA (crRNA) is shown with the region encoded in the first spacer shown in green and the 3′ and 5′ features encoded in the repeat region (black). The tracrRNA in the system anneals to the crRNA (red). **(B)** Representation of the *cas/csn* genes and CRISPR array showing the upstream leader region with promoter. Red arrows indicate the positions where the pre-crRNA (not shown) is cleaved in the RNA encoded in the array by an RNA-processing enzyme from the host. Promoters used to transcribe the components are shown as bent black arrows. **(C)** Representation of a class 2, type II Cas9 guide RNA complex showing the R-loop formed with recognition of a protospacer with the appropriate protospacer-adjacent motif (PAM) (yellow) in DNA (blue). In some engineered systems, a synthetic guide RNA (sgRNA) is produced, where the guide RNA is fused to tracrRNA for simplification with expressing the components (linkage between the guide RNA and tracrRNA shown with a dashed gray line).

It would be difficult to overstate the multitude of ways that the type II Cas9 systems have revolutionized genome manipulation across eukaryotes (see Cong et al., Suggested Reading). In these systems, a programmed DNA break allows the investigator to replace the endogenous copy of the gene with a DNA fragment that is introduced into the cells by transformation. Such a procedure allows a markerless replacement at the locus recognized by the guide RNA in the system. Other techniques used in eukaryotes may simply involve inactivating the target gene when repair occurs, through a process that favors deletion of the target gene. Type II Cas9 systems have shown considerable utility in bacteria as well, although in most bacteria, including *E. coli*, a separate Red recombination system is often needed to facilitate double-strand break repair (see Figure 9.11; see Jiang et al., Suggested Reading). In numerous variations on the system, the Cas9 protein has been modified in different ways to allow other guide-RNA-directed processes. For example, a nuclease-defective version of Cas9 (dCas9) can also allow silencing of genes in bacteria and eukaryotes in a process referred to as CRISPRi (see Qi et al., Suggested Reading). This process can be useful as a complement to genome-wide screens such as Tn-Seq (see chapter 8), with the capacity to turn the system on and off or titrate silencing with versions that are also adaptable to diverse bacteria (see Peters et al. 2018, Suggested Reading). Cas9 has also been fused to a wide variety of other protein domains to target specific enzymatic activities to a position recognized by the guide RNA, and many more applications for this important technology are certain to be engineered in the future.

OTHER TYPES OF CRISPR/Cas SYSTEMS

Type III CRISPR/Cas systems are the second most common type of system found in bacteria and archaea. The signature protein for the type III CRISPR/Cas systems is the Cas10 nuclease. One fascinating finding with the type III systems is an adaptation in one subtype that allows spacers to be harvested from RNA instead of DNA (see Silas et al., Suggested Reading). Normally, in the other systems, Cas1-Cas2 harvest spacers only from DNA, but in this case, the Cas1 protein is fused to a reverse transcriptase, allowing an RNA to be converted to DNA and captured as a spacer in the array. Of further interest, examining the phylogeny of the reverse transcriptase domain suggests that this activity was repurposed from the enzyme found in group II introns. The ability to harvest spacers from RNA suggests that a variety of applications may be possible for tracking gene expression based on these systems, where the spacers collected in the CRISPR array could provide a chronological history of mRNAs expressed in the cell (see Schmidt et al. Suggested Reading).

The class 2 type V and type VI systems, which utilize the single effector proteins Cas12 and Cas13, respectively, are relatively rare in prokaryotes and have not been studied to the same degree as Cas9. Interestingly, the nuclease activity associated with the Cas12 and Cas13 proteins appears to have been independently repurposed from other proteins. The nuclease activity of the Cas12 effector protein appears to have been derived from a type of insertion sequence (see chapter 8), while the ribonuclease activity associated with the Cas13 effector protein in type VI systems appears to have been repurposed from the toxin component of a toxin/antitoxin addiction system as described in plasmids (see Box 4.3). Cas12 and Cas13 show promise in sensitive nucleic acid sequence detection applications. In these procedures, enzymatic cleavage of the Cas12 and Cas13 proteins can be coupled to specific reporters with fluorophores that could allow sensitive detection applications with pathogens and DNA sequence variants (see Gootenberg et al., Suggested Reading).

DEFENSE AND COUNTERDEFENSE WITH CRISPR/Cas SYSTEMS

One fascinating aspect of CRISPR/Cas biology is the evolved response to these systems in mobile elements. CRISPR/Cas systems provide a potent mechanism to protect the host from invading DNAs, but evolution has also provided mechanisms to circumvent CRISPR/Cas systems. For example, an active area of research involves diverse anti-CRISPR/Cas systems found in mobile elements (see Bondy-Denomy et al., Suggested Reading). These systems can compromise various steps in the CRISPR/Cas protection pathway. One interesting class of anti-CRISPR/Cas proteins takes the form of nucleic acid mimics, literally, proteins that show properties that are similar enough to DNA or RNA that they can engage and inactivate Cas proteins by mimicking a target DNA (see Shin et al., Suggested Reading). This type of defense is also known to occur as a mechanism to protect bacteriophage from restriction endonucleases with proteins also identified in bacteriophage (see Walkinshaw et al., Suggested Reading).

There are also multiple examples where CRISPR/Cas systems have been acquired by mobile elements as a mechanism to gain some advantage over the host (see Faure et al., Suggested Reading). Full type I CRISPR/Cas systems have been identified in bacteriophage from *Vibrio cholerae* that allow the phage to defeat DNA-mediated host defense systems that would otherwise limit their spread through the bacterial population during outbreaks of cholera (see Seed et al., Suggested Reading). Also of interest are multiple cases where Tn7-like transposons have acquired and adapted CRISPR/Cas systems not as mechanisms of defense from other mobile elements, but instead, as targeting mechanisms where the guide RNA is used to target transposition into other mobile elements using the system (see Peters et al. 2017, Suggested Reading).

Final Thoughts

Bacteria comprise the most broadly diverse domain of life, and along with archaea, are providing a fascinating reservoir of genetic information. It is safe to assume that we are just scratching the surface of what we can learn from the numerous and growing number of bacterial and archaeal genome sequences. An exciting future lies ahead for students of molecular genetics in bacteria and archaea, not just for a better understanding of our natural world, but also in mining this information to develop tools to help solve current and future problems faced on our planet.

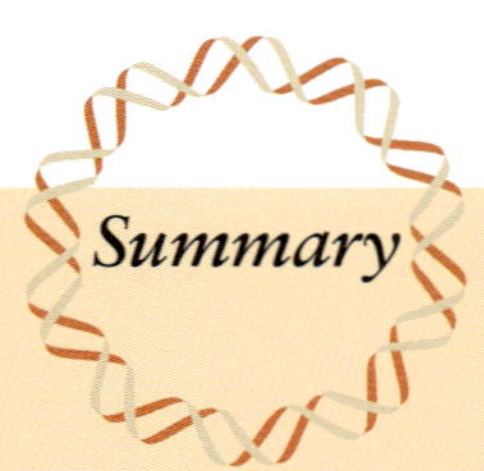

1. The first bacterial genomes were sequenced in the mid-1990s. Approximately 25 years later, hundreds of thousands of bacterial genomes are available for download from public databases.

2. Comparisons across bacterial genomes reveal recognizable genetic linkages, especially in organisms from the same genus. In addition to deletions, gross changes between genomes from the same species result from the addition of blocks of genes via the horizontal exchange of DNA. New blocks of genes are referred to as genomic islands, and more specialized names such as pathogenicity islands or defense islands are used if the function of the gene sets is known.

Summary (continued)

3. Most bacteria have open genomes with evidence for extensive gene addition by horizontal transfer. Genes shared with all members of the species make up the core genome. The compilation of all genes ever found in genomes from members of a species make up the pan genome.

4. Early DNA sequencing technologies involved the use of dideoxynucleotides that cannot be extended, terminating DNA polymerase extension at an A, T, C, or G, thereby making a product that can be resolved by size to reveal the sequence. Newer technologies involve sequencing short stretches of DNA in a massively parallel fashion. This information is used to predict the full genome by assembling the sequence information, a process facilitated by having information from so many small fragments. The newest technologies allow sequencing at the rate of DNA polymerization and allow much longer read lengths of sequence.

5. The polymerase chain reaction (PCR) has enabled numerous technologies that allow a small amount of DNA to be amplified by cycles of denaturation of DNA strands at a high temperature, annealing DNA primers at a lower temperature, and then extending from the primers with a thermostable DNA polymerase. PCR also plays an important role in multiple new cloning and gene assembly technologies.

6. Bacteria and archaea have numerous mechanisms for reducing the impact of horizontal transfer of genetic information from bacteriophage, plasmids, transposons, and other integrating mobile genetic elements. Restriction endonucleases and CRISPR/Cas systems are two broad families of systems that have revolutionized the fields of molecular biology and molecular genetics.

7. There are four types of restriction endonucleases. Type II enzymes that cut within the DNA sequence recognized by the enzyme have defined much of molecular biology we know today, especially with the ease of amplifying of DNAs by PCR.

8. TA cloning and Topo (topoisomerase) cloning both take advantage of the property of *Taq* and some other DNA polymerase to leave a 3′ A overhang at the ends of amplified DNA products.

9. Bacteriophage λ integration and excision have been adapted to a directional *in vitro* cloning system also referred to as Gateway cloning.

10. Two major systems have been developed for producing 3′ extensions in virtually any DNA amplified by PCR. Uracil-*N*-glycosylase combined with AP endonuclease and digestion with type IIs enzymes that cut at a set distance from their recognition site have allowed useful assembly technologies that can join multiple DNA fragments.

11. Gibson or HiFi DNA assembly technologies involve three separate enzymatic activities (5′ exonuclease, DNA polymerase, and DNA ligase) in one reaction mix that can assemble multiple DNA fragments in a single isothermal reaction.

12. CRISPR/Cas systems are adaptive immune systems found in archaea and bacteria. Although there are multiple distinct types that evolved independently, they basically all use the same three steps: (i) spacer acquisition, (ii) pre-crRNA processing, and (iii) interference.

13. Spacer acquisition (also known as adaptation) is the process in which the short DNA sequences known as spacers are harvested from an invading element. The information encoded in the spacers is subsequently used to recognize the element if it invades the cell again.

14. Information in the CRISPR array containing a collection of spacers is used to identify invading genetic elements. The CRISPR array is read as a long pre-crRNA transcript that is subsequently processed into individual guide RNAs.

15. Interference is the process whereby spacer matches in an invading element (called protospacer) are used to recognize the element, which is then inactivated by nuclease activities in the Cas protein or proteins. The spacer in the host CRISPR array is not targeted for degradation/cleavage because it lacks a specific motif called the protospacer-adjacent motif (PAM).

16. Class 1 CRISPR/Cas systems are the most common systems in bacteria and archaea and use multiple proteins and a dedicated nuclease in the interference step, either Cas3 (type I systems) or Cas10 (type III systems).

17. Class 2 CRISPR/Cas systems use a single protein for guide RNA formation and interference. The most common of these systems (type II) uses the Cas9 protein and a dedicated tracrRNA.

18. In many cases, protein domains/activities that are found in the diverse CRISPR/Cas systems have been repurposed from mobile elements, such as transposons.

19. Mobile genetic elements have evolved anti-CRISPR/Cas systems and anti-restriction endonuclease systems to circumvent host defense systems.

20. CRISPR/Cas systems have been captured by bacteriophage for overcoming host defense and by transposons as a mechanism of targeting transposition via guide RNAs.

QUESTIONS FOR THOUGHT

1. What types of lifestyles might bacteria have that would lead to a closed genome versus an open genome?

2. Given all of the mechanisms bacteria and archaea have for limiting horizontal transfer of mobile genetic elements, why do you think these mechanisms remain so common?

3. Darwinian evolution holds that natural selection acts with random mutations to produce the most fit variants in forming new species. How might CRISPR/Cas systems be considered a violation of the idea of Darwinian evolution?

4. CRISPR/Cas systems are found in 90% of archaea but only about half of bacterial species. Why might this be the case?

5. A case has been shown where quorum sensing has been found to specifically activate CRISPR/Cas function at high cell density, but not at low cell density. Why might this be advantageous?

6. Genes and gene systems acquired through horizontal transfer provide a common way that new functions evolve in bacteria, a process that is comparatively rare in mammals. Why might this be the case?

SUGGESTED READING

Aslanidis C, de Jong PJ. 1990. Ligation-independent cloning of PCR products (LIC-PCR). *Nucleic Acids Res* **18:**6069–6074.

Barrangou R, et al. 2007. CRISPR provides acquired resistance against viruses in prokaryotes. *Science* 315:1709–1712.

Bitinaite J, Rubino M, Varma KH, Schildkraut I, Vaisvila R, Vaiskunaite R. 2007. User friendly DNA engineering and cloning method by uracil excision. *Nucleic Acids Res* **35:**1992–2002.

Bolotin A, Quinquis B, Sorokin A, Ehrlich SD. 2005. Clustered regularly interspaced short palindrome repeats (CRISPRs) have spacers of extrachromosomal origin. *Microbiology* **151:**2551–2561.

Bondy-Denomy J, Garcia B, Strum S, Du M, Rollins MF, Hidalgo-Reyes Y, Wiedenheft B, Maxwell KL, Davidson AR. 2015. Multiple mechanisms for CRISPR-Cas inhibition by anti-CRISPR proteins. *Nature* **526:**136–139.

Borges AL, Davidson AR, Bondy-Denomy J. 2017. The discovery, mechanisms, and evolutionary impact of anti-CRISPRs. *Annu Rev Virol* **4:**37–59.

Cong L, Ran FA, Cox D, Lin S, Barretto R, Habib N, Hsu PD, Wu X, Jiang W, Marraffini LA, Zhang F. 2013. Multiplex genome engineering using CRISPR/Cas systems. *Science* **339:**819–823.

Datsenko KA, Pougach K, Tikhonov A, Wanner BL, Severinov K, Semenova E. 2012. Molecular memory of prior infections activates the CRISPR/Cas adaptive bacterial immunity system. *Nat Commun* **3:**945.

Dillard KE, Brown MW, Johnson NV, Xiao Y, Dolan A, Hernandez E, Dahlhauser SD, Kim Y, Myler LR, Anslyn EV, Ke A, Finkelstein IJ. 2018. Assembly and translocation of a CRISPR-Cas primed acquisition complex. *Cell* **175:**934–946.E15.

Doron S, Melamed S, Ofir G, Leavitt A, Lopatina A, Keren M, Amitai G, Sorek R. 2018. Systematic discovery of antiphage defense systems in the microbial pangenome. *Science* **359:**eaar4120.

Faure G, Shmakov SA, Yan WX, Cheng DR, Scott DA, Peters JE, Makarova KS, Koonin EV. 2019. CRISPR-Cas in mobile genetic elements: counter-defence and beyond. *Nat Rev Microbiol* **17:**513–525.

Gibson DG, Benders GA, Andrews-Pfannkoch C, Denisova EA, Baden-Tillson H, Zaveri J, Stockwell TB, Brownley A, Thomas DW, Algire MA, Merryman C, Young L, Noskov VN, Glass JI, Venter JC, Hutchison CA III, Smith HO. 2008. Complete chemical synthesis, assembly, and cloning of a *Mycoplasma genitalium* genome. *Science* **319:**1215–1220.

Gibson DG, Young L, Chuang R-Y, Venter JC, Hutchison CA III, Smith HO. 2009. Enzymatic assembly of DNA molecules up to several hundred kilobases. *Nat Methods* **6:**343–345.

Gootenberg JS, Abudayyeh OO, Kellner MJ, Joung J, Collins JJ, Zhang F. 2018. Multiplexed and portable nucleic acid detection platform with Cas13, Cas12a, and Csm6. *Science* 360:439–444.

Hartley JL, Temple GF, Brasch MA. 2000. DNA cloning using *in vitro* site-specific recombination. *Genome Res* **10:**1788–1795.

Høyland-Kroghsbo NM, Paczkowski J, Mukherjee S, Broniewski J, Westra E, Bondy-Denomy J, Bassler BL. 2017. Quorum sensing controls the *Pseudomonas aeruginosa* CRISPR-Cas adaptive immune system. *Proc Natl Acad Sci USA* **114:**131–135.

Jiang W, Bikard D, Cox D, Zhang F, Marraffini LA. 2013. RNA-guided editing of bacterial genomes using CRISPR-Cas systems. *Nat Biotechnol* **31:**233–239.

Jinek M, Chylinski K, Fonfara I, Hauer M, Doudna JA, Charpentier E. 2012. A programmable dual-RNA-guided DNA endonuclease in adaptive bacterial immunity. *Science* **337:**816–821.

Juhas M, Ajioka JW. 2017. High molecular weight DNA assembly *in vivo* for synthetic biology applications. *Crit Rev Biotechnol* **37:**277–286.

Kapitonov VV, Jurka J. 2005. RAG1 core and V(D)J recombination signal sequences were derived from *Transib* transposons. *PLoS Biol* **3:**e181.

Koonin EVE, Krupovic M. 2015. Evolution of adaptive immunity from transposable elements combined with innate immune systems. *Nat Rev Genet* 16:184–192.

Koonin EVE, Makarova KSK. 2017. Mobile genetic elements and evolution of CRISPR-Cas systems: all the way there and back. *Genome Biol Evol* **9:**2812–2825.

Koonin EVE, Makarova KSK, Zhang F. 2017a. Diversity, classification and evolution of CRISPR-Cas systems. *Curr Opin Microbiol* **37:**67–78.

Koonin EVE, Makarova KSK, Wolf YI. 2017b. Evolutionary genomics of defense systems in archaea and bacteria. *Annu Rev Microbiol* **71:**233–261.

Krupovic M, Béguin P, Koonin EVE. 2017. Casposons: mobile genetic elements that gave rise to the CRISPR-Cas adaptation machinery. *Curr Opin Microbiol* **38:**36–43.

Levy A, Goren MG, Yosef I, Auster O, Manor M, Amitai G, Edgar R, Qimron U, Sorek R. 2015. CRISPR adaptation biases explain preference for acquisition of foreign DNA. *Nature* **520:**505–510.

Li MZ, Elledge SJ. 2012. SLIC: a method for sequence- and ligation-independent cloning, p 51–59. *In* Peccoud J (ed), *Gene Synthesis: Methods and Protocols.* Humana Press, Totowa, NJ.

Makarova KSK, Wolf YIY, Snir S, Koonin EVE. 2011. Defense islands in bacterial and archaeal genomes and prediction of novel defense systems. *J Bacteriol* **193:**6039–6056.

Makarova KSK, et al. 2015. An updated evolutionary classification of CRISPR-Cas systems. *Nat Rev Microbiol* **13:**722–736.

Makarova KSK, et al. 2020. Evolutionary classification of CRISPR-Cas systems: a burst of class 2 and derived variants. *Nature Rev Microbiol* **18:**67–83.

Marraffini LA, Sontheimer EJ. 2008. CRISPR interference limits horizontal gene transfer in staphylococci by targeting DNA. *Science* **322:**1843–1845.

Meeske AJ, Nakandakari-Higa S, Marraffini LA. 2019. Cas13-induced cellular dormancy prevents the rise of CRISPR-resistant bacteriophage. *Nature* **570:**241–245.

Mojica FJM, Díez-Villaseñor C, García-Martínez J, Almendros C. 2009. Short motif sequences determine the targets of the prokaryotic CRISPR defence system. *Microbiology* **155**:733–740.

Peters JE, Makarova KSK, Shmakov S, Koonin EVE. 2017. Recruitment of CRISPR-Cas systems by Tn7-like transposons. *Proc Natl Acad Sci USA* **114**:E7358–E7366.

Peters JM, Koo BM, Patino R, Heussler GE, Hearne CC, Qu J, Inclan YF, Hawkins JS, Lu CHS, Silvis MR, Harden MM, Osadnik H, Peters JE, Engel JN, Dutton RJ, Grossman AD, Gross CA, Rosenberg OS. 2019. Enabling genetic analysis of diverse bacteria with Mobile-CRISPRi. *Nat Microbiol* **4**:244–250.

Qi LS, Larson MH, Gilbert LA, Doudna JA, Weissman JS, Arkin AP, Lim WA. 2013. Repurposing CRISPR as an RNA-guided platform for sequence-specific control of gene expression. *Cell* **152**:1173–1183.

Schmidt F, Cherepkova MY, Platt RJ. 2018. Transcriptional recording by CRISPR spacer acquisition from RNA. *Nature* **562**:380–385.

Seed KD, Lazinski DW, Calderwood SB, Camilli A. 2013. A bacteriophage encodes its own CRISPR/Cas adaptive response to evade host innate immunity. *Nature* **494**:489–491.

Sharan SK, Thomason LC, Kuznetsov SG, Court DL. 2009. Recombineering: a homologous recombination-based method of genetic engineering. *Nat Protoc* **4**:206–223.

Shin J, et al. 2017. Disabling Cas9 by an anti-CRISPR DNA mimic. *Sci Advances* **3**(7):e1701620.

Shuman S. 1994. Novel approach to molecular cloning and polynucleotide synthesis using vaccinia DNA topoisomerase. *J Biol Chem* **269**:32678–32684.

Silas S, Makarova KS, Shmakov S, Páez-Espino D, Mohr G, Liu Y, Davison M, Roux S, Krishnamurthy SR, Fu BXH, Hansen LL, Wang D, Sullivan MB, Millard A, Clokie MR, Bhaya D, Lambowitz AM, Kyrpides NC, Koonin EV, Fire AZ. 2017. On the origin of reverse transcriptase-using CRISPR-cas systems and their hyperdiverse, enigmatic spacer repertoires. *MBio* **8**:e00897-17.

Walkinshaw MD, Taylor P, Sturrock SS, Atanasiu C, Berge T, Henderson RM, Edwardson JM, Dryden DT. 2002. Structure of Ocr from bacteriophage T7, a protein that mimics B-form DNA. *Mol Cell* **9**:187–194.

Xiao Y, Ng S, Nam KH, Ke A. 2017. How type II CRISPR-Cas establish immunity through Cas1-Cas2-mediated spacer integration. *Nature* **550**:137–141.

Zhang Y, Cheng TC, Huang G, Lu Q, Surleac MD, Mandell JD, Pontarotti P, Petrescu AJ, Xu A, Xiong Y, Schatz DG. 2019. Transposon molecular domestication and the evolution of the RAG recombinase. *Nature* **569**:79–84.

Glossary

A (aminoacyl) site The site on the ribosome to which the incoming aminoacylated tRNA binds.

aaRS *See* Aminoacyl-tRNA synthetase.

Abortive initiation A step in transcription in which RNA polymerase initiates transcription but stops after synthesizing short RNAs, which are released. RNA polymerase then reinitiates transcription from the promoter.

Accretion-mobilization A process by which integrating conjugating elements gain and lose features needed for conjugal transfer by merging with other elements over time.

Activator A protein that increases expression of a gene or operon.

Activator site A sequence in DNA upstream of the promoter to which the activator protein binds.

Adaptation A change that allows an organism to be better suited for a particular environment.

Adaptive response Activation of transcription of genes of the Ada regulon, which is involved in the repair of some types of alkylation damage to DNA.

Adaptor A protein that controls the activity of a cellular protease. *See* Regulated proteolysis.

Adenine One of the two purine (two-ringed) bases in DNA and RNA.

Alkylating agent A chemical that reacts with DNA, forming a carbon bond to one of the atoms in the DNA.

Allele One of the forms of a gene, e.g., the gene with a particular mutation. The term can refer to either the wild-type form or a mutant form.

Allele-specific suppressor A second-site mutation that alleviates the effect of only one type of mutation in a gene.

Allelism test A complementation test to determine if two mutations are in the same gene, i.e., if they create different alleles of the same gene.

Allosteric interaction *See* Allosterism.

Allosterism A change in the conformation of a domain of a protein as a result of a change in a different domain, e.g., when binding of the allolactose inducer to the inducer-binding pocket of the LacI repressor changes the angle of the DNA-binding domain.

Amber The codon UAG. Mutations to this codon in the reading frame for a protein are called amber mutations. Mutations that suppress mutations to this codon are called amber suppressors.

Amino group The NH_2 chemical group.

Amino terminus *See* N terminus.

Aminoacyl-tRNA synthetase (aaRS) The enzyme that attaches a specific amino acid to a specific class of tRNAs.

Antiactivator A regulatory protein that acts as a repressor by preventing activation by an activator.

Antiadaptor A protein that controls the activity of an adaptor and therefore regulates the activity of a cellular protease. *See* Adaptor, Regulated proteolysis.

Antibiotic Generally, a substance—often a natural microbial product or its semisynthetic derivative—that kills bacteria (i.e., is bacteriocidal) or inhibits the growth of bacteria (i.e., is bacteriostatic). Some antibiotics are entirely chemically synthesized.

Antibiotic resistance gene cassette A fragment of DNA, usually bracketed by restriction sites for ease of cloning, that contains a gene whose product confers resistance to an antibiotic for easy selection.

Anticodon The 3-base sequence in a tRNA that pairs with the codon in mRNA by complementary base pairing. *See* Codon.

Antiholin A phage protein that binds to a holin and prevents transport of endolysin through the hole in the membrane formed by the holin. *See* Holin.

Antiparallel A configuration in which, moving in one direction along a double-stranded DNA or RNA, the phosphates in one strand are attached 3′ to 5′ to the sugars while the phosphates in the other strand are attached 5′ to 3′.

Antisense RNA RNA that contains a sequence complementary to a sequence in an mRNA.

Antiterminator A sequence in RNA that prevents downstream transcription termination. It can be either an RNA structure that forms in the RNA after it is transcribed or a protein or protein complex that binds to the transcription elongation complex. The RNA structures often prevent termination by preventing the formation of the helix of a factor-independent transcriptional terminator. *See* Transcription attenuation.

AP endonuclease A DNA-cutting enzyme that cuts on the 5′ side of a deoxynucleotide that has lost its base, usually due to a DNA glycosylase, i.e., an apurinic or apyrimidinic site. This cutting allows the DNA strand to be degraded and resynthesized, replacing the apurinic or apyrimidinic site with a normal nucleotide.

AP lyase A DNA-cutting enzyme, usually associated with an *N*-glycosylase, that cuts on the 3′ side of the apurinic or apyrimidinic site created by the *N*-glycosylase activity of the enzyme.

Aporepressor A protein that can be converted into a repressor by undergoing a conformational change if a small molecule called the corepressor is bound to it.

Apparent revertant In a genetic analysis, a mutant that seems to have overcome the effect of a mutation. Without further evidence, it could either be a true revertant or be suppressed.

Archaea One of the three divisions of organisms, which, like bacteria, exists as single-celled organisms. While they share some of the features of both eukaryotes and bacteria, archaea are convincingly placed within the same lineage as eukaryotes.

Assimilatory reduction Addition of electrons to nitrogen-containing compounds to reduce them to NH_3 for incorporation into cellular constituents.

Attenuation Regulation of an operon by premature termination of transcription under conditions where less of the gene product(s) is needed.

Autocleavage The process by which a protein cuts itself.

Autoinduction loop A system in which an activator activates its own expression, resulting in amplification of the signal and a very strong regulatory response.

Autonomous elements Mobile genetic elements that possess all of the features that allow them to move without a dependence on another element.

Autophosphorylation A process by which a protein transfers a PO_4 group to itself, independent of the source of the PO_4 group.

Autoregulation The process through which a gene product controls the level of its own synthesis.

Autotransporter *See* Type V secretion system.

Auxotrophic mutant A mutant that cannot make a growth substance that the normal or wild-type organism can make.

Backbone The chain of alternating phosphates and ribose sugars (in RNA molecules) or deoxyribose sugars (in DNA molecules) that provides the structure in which bases are attached in long polymers of RNA or DNA.

Backtrack Movement of a complex backward along its template, e.g., backward movement of RNA polymerase along the DNA, resulting in a displacement of the 3′ end of the RNA from the active site.

Bacteria Members of the domain of organisms characterized by a relatively simple cell structure free of many cellular organelles, the presence of 16S and 23S rRNAs with specific sequence conservation patterns that differ from those found in the archaea, and a simple core RNA polymerase, among other features.

Bacterial artificial chromosome (BAC) cloning vector An *Escherichia coli* plasmid cloning vector that can accept very large clones of DNA (>300 kb) because it is derived from the F plasmid. It is often used to make DNA libraries of large fragments in genome-sequencing projects to complement small insert libraries. BACs are also used as a platform for producing genetic constructs in *E. coli* that are subsequently shuttled into other bacterial and nonbacterial hosts.

Bacterial lawn The layer of bacteria on an agar plate that forms when many bacteria are plated at a sufficiently high density that individual bacterial colonies are not found.

Bacteriophage A virus that infects bacteria.

Base Carbon-, nitrogen-, and hydrogen-containing chemical compounds with structures composed of one or two rings that are constituents of the DNA or RNA molecule.

Base analog A chemical that resembles one of the bases and can mistakenly incorporate into DNA or RNA during synthesis.

Base pair (bp) Each set of opposing bases in the two strands of double-stranded DNA or RNA that are held together by hydrogen bonds and thereby help hold the two strands together. Also used as a unit of length.

Base pair change A mutation in which one type of base pair in DNA (e.g., an AT pair) is changed into a different base pair (e.g., a GC pair).

Binding A process by which molecules are physically joined to each other by noncovalent interactions.

Biofilm formation A developmental process in which groups of bacterial cells form a structured assembly on a surface.

Bioinformatics A repertoire of computer technologies that allow prediction of open reading frames and regulatory sites, pIs and molecular weights of proteins, posttranslational modifications of proteins, subcellular localization of proteins, the level of expression of genes, and the functions of gene products, and establish the interrelatedness of genes and proteins.

Biopanning The process by which a phage expressing a specific peptide on its surface is separated from other phage not expressing the peptide by its ability to bind to a molecule fixed to a solid matrix.

Biosynthesis The synthesis of chemical compounds by living organisms.

Biosynthetic Processes involved in generation of more complicated molecules from simpler building blocks.

Biosynthetic operon An operon composed of genes whose products are involved in synthesizing compounds, such as amino acids or vitamins, rather than degrading them.

Bistability A state in which a population of cells is found in two subpopulations with different physiological properties and these subpopulations are maintained over time. Cells can switch between states, but do so rarely, so that the balance of the two different states is maintained.

BLAST (Basic Local Alignment Search Tool) A genome annotation tool that uses bioinformatics to find similar regions in DNA sequences. It can compare DNA or amino acid sequences and identify nucleotide or protein sequences in a database that are similar to a query sequence.

Broad host range The property that allows a phage, plasmid, or other DNA element to enter and/or replicate in a wide variety of bacterial species.

C terminus The end of a polypeptide chain with the free carboxyl (COOH) group.

cAMP *See* Cyclic AMP.

Campbell model The model in which the λ phage DNA forms a circle and then integrates into the chromosome by recombination between a site normally internal to the λ phage DNA and a site on the chromosome, which creates a prophage genetic map that is a cyclic permutation of the phage genetic map. It was named after the person who first proposed it.

Canonical sequence The type representative to which all others are compared.

CAP *See* Catabolite activator protein.

CAP regulon All of the operons that are regulated, either positively or negatively, by the CAP protein.

CAP-binding site The sequence on DNA to which the CAP protein binds.

Capsid The protein and/or membrane coat that surrounds the genomic nucleic acid (DNA or RNA) of a virus.

Carboxyl group The chemical group COOH.

Carboxyl terminus *See* C terminus.

Catabolic Processes involved in breaking down more complicated molecules into simpler products.

Catabolic operon An operon composed of genes whose products degrade organic compounds. *See* Degradative operon.

Catabolite A small molecule produced by the degradation of larger carbon-containing organic compounds, such as sugars.

Catabolite activator protein (CAP) The DNA- and cAMP-binding protein that regulates catabolite-sensitive operons in enteric bacteria by binding to their promoter regions. Also called catabolite repressor protein (Crp).

Catabolite regulation The reduced expression of some operons in the presence of high cellular levels of catabolites due to growth on an efficiently utilized carbon source.

Catabolite repression *See* Catabolite regulation.

Catabolite-sensitive operons Operons whose expression is regulated by the cellular levels of catabolites.

Catenanes Structures formed when two or more circular DNA molecules are joined like links in a chain.

Cell division cycle The events occurring between the time a cell is created by division of its mother cell and the time it divides.

Cell divisions The total number of times in a growing culture that individual cells have grown and divided. Sometimes called cell generations.

Central dogma The tenet that information in DNA is copied into mRNA by the process of transcription and that that information is translated to generate protein.

Change-of-function mutation A mutation that changes the activity of a protein rather than inactivating all or part of it, e.g., a mutation that makes a regulatory protein respond to a different inducer.

Channel gating Blocking a membrane channel unless the substrate is being transported. This prevents other molecules from leaking into or out of the cell through the membrane channel.

Chaperone A protein that binds to other proteins or RNAs and helps them to fold correctly or prevents them from folding prematurely.

Chaperone-usher secretion *See* Type V secretion system.

Chaperonin A protein that forms double back-to-back chambers, which alternate in taking up denatured proteins and refolding them. It is represented by the GroEL (Hsp60) protein in *Escherichia coli*.

ChIP-Seq Chromosome immune precipitation with DNA sequencing is a method for determining all of the positions where a protein normally binds across the genome. The protein of interest is chemically crosslinked to DNA in living cells, and the DNA is subsequently isolated and sheared. Antibodies or affinity tags are used to physically isolate the DNAs associated with the protein of interest, and these sequences are determined.

***chi* (χ) site** The sequence 5′-GCTGGTGG-3′ in *Escherichia coli* DNA. It is recognized by the *E. coli* RecBCD complex, thereby signaling changes in the complex, upregulating 5′-to-3′ nuclease activity, inhibiting 3′-to-5′ nuclease activity, and activating its ability to load RecA.

Chromosome In a bacterial cell, the DNA molecule that contains most of the genes required for cellular growth and maintenance; usually the largest DNA molecule in the cell and the one that contains a characteristic *oriC* sequence.

CI repressor The phage λ-encoded protein that binds to the phage operator sequences close to the p_R and p_L promoters and prevents transcription of most of the genes of the phage.

***cis*-acting site** A functional region on a DNA molecule that does not encode a gene product and affects only the DNA molecule that it resides in or near (e.g., a promoter or an origin of replication).

Clamp loader A protein complex that places the ring-like sliding clamp accessory protein of the replicative DNA polymerase around the DNA strands.

Classical genetics The study of genetic phenomena by using only intact living organisms.

Clonal A situation in which all members of a population are the descendants of a single organism or replicating DNA molecule, as in colonies growing from a single bacterium on an agar plate or DNA molecules all derived from replication of the same single DNA molecule.

Clone A collection of DNA molecules or organisms that are all identical to each other because they result from replication or multiplication of the same original DNA or organism.

Cloning vector An autonomously replicating DNA (replicon), usually a phage or plasmid, into which other DNA molecules that are not capable of replicating themselves can be introduced so that the non-self-replicating DNA can be cloned.

Closed complex The complex that forms when RNA polymerase first binds to a promoter, before the strands of DNA at the promoter separate.

Cluster of orthologous genes (COG) A compilation of presumptive genes from a diversity of organisms (representing the major phylogenetic lineages) that are proposed to be functionally analogous, based on sequence similarity.

Cochaperone A smaller protein that helps chaperones fold proteins or cycle their adenine nucleotide. It is represented by DnaJ and GrpE in *Escherichia coli*.

Cochaperonin A smaller protein that helps chaperonins fold proteins by forming the cap on the chamber once the protein has been taken up. It is represented by the GroES protein (Hsp10) in *Escherichia coli*.

Coding strand The strand of DNA in the part of a gene coding for a protein that has the same sequence as the RNA transcribed from the gene. *See* Nontemplate strand.

Codon A 3-base sequence in mRNA that stipulates one of the amino acids or serves as a translation termination signal.

COG *See* Cluster of orthologous genes.

Cognate aaRS (cognate aminoacyl-tRNA synthetase) The enzyme that attaches the correct amino acid to a specific tRNA.

Cointegrate A DNA molecule after a replicative transposition event from a donor DNA into a target DNA in which the donor and target DNAs are joined, separated by copies of the transposon.

Cold-sensitive mutant A mutant that cannot live and/or multiply in the lower temperature ranges at which the normal or wild-type organism can live and/or multiply.

Colony A small lump or pile made up of millions of cells on an agar plate that were derived from a single cell and therefore are (essentially) genetically identical.

Competence pheromones Small peptides given off by bacterial cells that induce competence in neighboring cells when the cells are at high concentrations.

Competent The state during which cells are capable of taking up DNA.

Complementary base pair A pair of nucleotides that can be held together by hydrogen bonds between their bases, e.g., dGMP and dCMP or dAMP and dTMP.

Complementation Restoration of a wild-type phenotype when two alleles (either wild-type and mutant alleles, or two different mutations that cause the same phenotype) are combined in the cell. For wild-type and mutant alleles, this suggests that the phenotype was caused by the mutant allele. For two different mutations, complementation indicates that the mutations are probably in different genes that affect the same process.

Complete closed genome sequence A dataset that has all of the DNA sequence information collected and assembled to indicate the natural state of the circular or linear chromosome(s) of an organism.

Composite transposon A transposon made up of two almost identical insertion (IS) elements plus the DNA between them that move as a unit.

Concatemer Two or more almost identical DNA molecules linked tail to head.

Condensins Proteins that bind chromosomal DNA in two different places, folding it into large loops and thereby making it more condensed. They are represented in model bacteria by the Smc protein in *Bacillus subtilis* and by the MukB protein in *Escherichia coli*.

Conditional-lethal mutation A mutation that inactivates an essential cellular component but only under a certain set of circumstances, for example, a temperature-sensitive mutation that inactivates an essential gene product only at relatively high temperatures or a nonsense mutation that inactivates an essential gene product only in the absence of a nonsense suppressor.

Congression Cotransformation as the result of uptake of more than one fragment of transforming DNA.

Conjugation The transfer of DNA from one bacterial cell to another by the transfer functions of a self-transmissible DNA element, such as a plasmid or integrating conjugative element.

Conjugative transposon A somewhat outdated way of referring to a type of integrating conjugative element. *See* Integrating conjugative element.

Consensus sequence A nucleotide sequence in DNA or RNA, or an amino acid sequence in a protein, in which each position in the sequence has the nucleotide or amino acid that has been found most often at that position in molecules with the same function and similar sequences.

Constitutive mutant A mutant in which a regulatory system is active regardless of the presence or absence of the regulatory signal.

Contigs A series of datasets of DNA sequence information from an organism that are collected and assembled into a series of contiguous units but not the entire natural state of the circular or linear chromosome(s) of an organism.

Cooperative binding Process in which the binding of one protein molecule to a site (e.g., on DNA) greatly enhances the binding of another protein molecule of the same type to an adjacent site. The proteins bound at adjacent sites interact through their multimerization domains, which stabilizes the binding.

Coprotease A protein that binds to another protein and thereby activates the second protein's autocleavage or other protease activity, although it itself is not a protease.

Copy A molecule of a particular type identical to another in the same cell. The term often refers to a gene that has been moved somewhere else so that it now exists in more than one place in the genome.

Copy number The ratio of the number of plasmids of a particular type in the cell to the number of copies of the chromosome.

Copy number effect An effect on the cell due to having more than the one copy of the gene, leading to increased levels of the gene product.

Core enzyme The catalytic part of an enzyme complex that lacks one or more additional subunits that modify its properties. *See* Core RNA polymerase.

Core genome The genes that are found in all sequenced representatives within a species. Generally all of the housekeeping functions and genes that define the species and not transiently associating genes from mobile genetic elements.

Core polymerase The part of a DNA or RNA polymerase that actually performs the polymerization reaction and functions independently of accessory and regulatory proteins that cycle on and off the protein.

Core RNA polymerase RNA polymerase composed of two α subunits and the β, β', and ω subunits but lacking the σ subunit. This form of RNA polymerase is catalytically active but does not recognize promoter sequences for accurate initiation of transcription.

Corepressor A small molecule that binds to an aporepressor protein and converts it into a repressor.

***cos* site** The sequence of deoxynucleotides at the ends of λ DNA in the phage head. A staggered cut in this sequence at the time the phage DNA is packaged from concatemers gives rise to complementary or cohesive ends that can base pair with each other to form circular DNA on infection of another host cell.

Cotranscribed Two or more contiguous genes transcribed by a single RNA polymerase molecule from a single promoter.

Cotransducible Two genetic markers that are close enough together on the DNA that they can be carried in the same phage particle during transduction.

Cotransduction frequency The percentage of transductants selected for being recombinant for one genetic marker that have also become recombinant for another genetic marker. It is a measure of how far apart the markers are on DNA.

Cotransformable Being close enough together to be carried on the same piece of DNA during transformation; the term is used to describe a pair of genetic markers that meet this criterion.

Cotransformation frequency The frequency of transformants recombinant for the selected marker that are also recombinant for an unselected marker. It is a measure of how far apart the markers are on DNA.

Cotranslational translocation A type of translocation in which a protein is inserted into the membrane by the signal recognition particle system as it is being translated. This is required if the protein is to be inserted in the inner membrane and is highly hydrophobic.

Counterselect To apply a treatment (e.g., addition of an antibiotic) that prevents multiplication of the donor in a genetic cross.

Coupling hypothesis A model for the regulation of replication of iteron plasmids in which two or more plasmids are joined by binding to the same Rep protein through their iteron sequences.

Coupling protein A protein that is part of the Mpf system of self-transmissible plasmids. The coupling protein binds to the relaxase of the Dtr (DNA transfer) system to communicate that contact has been made with a recipient cell.

Covalent A bond that holds two atoms together by sharing their electron orbits.

Covalently closed circular Having no breaks or discontinuities in either strand (of a double-stranded circular DNA).

CRISPR array The memory component of diverse genetic systems primarily used for defense from bacteriophage and other mobile elements. Initially identified as a clustered regularly interspaced short palindromic repeats of DNA sequence. The repeat DNA sequences flank spacer DNA sequences that are complementary to the mobile elements recognized by the systems typically for degradation. RNA transcripts from CRISPR arrays are processed by one or multiple CRISPR-associated (Cas) proteins typically to function as acquired immune systems in bacteria and archaea.

Crossover The site of joining of two DNA molecules during recombination.

C-terminal amino acid The amino acid on one end of a polypeptide chain that has a free carboxyl (COOH) group unattached to the amino group of another amino acid.

Cut-and-paste transposition A mechanism of transposition in which the entire transposon is excised from one place in the DNA and inserted into another place.

Cut-out and paste-in transposition *See* Cut-and-paste transposition.

Cyclic AMP (cAMP) Adenosine monophosphate with the phosphate attached to both the 3′ and 5′ carbons of the ribose sugar.

Cyclically permuted Lacking unique ends. The mathematical definition of a cyclic permutation is a permutation that shifts all elements of a set by a fixed offset with the elements shifted off the end inserted back at the beginning. In a cyclically permuted genome, there are no unique ends. If the genome of such a phage is drawn as a circle, each genome starts somewhere on the circle and extends around the circle until it returns to the same place or just past it, so that the individual genomes have different endpoints but contain all of the genes in the same order.

Cyclobutane ring A ring structure of four carbons held together by single bonds that is present in some types of pyrimidine dimers in DNA.

Cytokinesis The process of one cell dividing into two.

Cytoplasmic domain A region of the polypeptide chain of a transmembrane protein that is in the interior, or cytoplasm, of the cell.

Cytosine One of the pyrimidine (one-ringed) bases in DNA and RNA.

D loop Displacement loop. The three-stranded structure that forms when a single strand of DNA invades a double-stranded DNA, displacing one of the strands.

Dam methylase *See* Deoxyadenosine methylase.

Damage tolerance mechanism A way of dealing with damage to DNA that does not involve repairing the damage, for example, replication restart or translesion synthesis.

Daughter cell One of the two cells arising from division of a mother cell.

Daughter DNA One of the two DNAs arising from replication of another DNA.

DDE transposon A transposon containing the DDE (aspartate-aspartate-glutamate) motif in its transposase. These amino acids coordinate the magnesium ions required in the active center for transposase activity.

Deaminating agent A chemical that reacts with DNA, causing the removal of amino (NH_2) groups from the bases in the DNA.

Deamination The process of removing amino (NH_2) groups from a molecule. In mutagenesis, it is the removal of amino groups from the bases in DNA.

Decatenation The process performed by type II topoisomerases of passing DNA strands through each other to resolve catenanes.

Decoding site The site on the ribosome responsible for pairing between the mRNA codon and the tRNA anticodon.

Defective prophage A DNA element in the bacterial chromosome that contains phage-like DNA sequences and presumably was once capable of being induced to form phage virions but has lost genes essential for lytic development.

Degradative operon An operon whose genes encode enzymes required for the breakdown of molecules into smaller molecules with the concomitant release of energy and/or compounds needed for other pathways. *See* Catabolic operon.

Deletion mutation A mutation in which one or a number of contiguous base pairs have been removed from the DNA.

Deoxyadenosine An adenine base attached to a deoxyribose sugar.

Deoxyadenosine methylase (Dam methylase) An enzyme that attaches a CH_3 (methyl) group to the adenine base in DNA. It is typified by the Dam methylase in *Escherichia coli* that methylates the A in the sequence GATC.

Deoxycytidine A cytosine base attached to a deoxyribose sugar.

Deoxyguanosine A guanine base attached to a deoxyribose sugar.

Deoxynucleoside A base (usually A, G, T, or C) attached to a deoxyribose sugar.

Deoxynucleotide A base (usually A, G, T, or C) attached to a deoxyribose sugar and a phosphate group.

Deoxythymidine The thymine base attached to deoxyribose sugar.

Diauxie A two-stage growth pattern that occurs when cells are grown on a mixture of two carbon sources where the preferred carbon source is used first.

Dilysogen A lysogen containing two copies of the prophage, usually joined tail to head in tandem.

Dimer A protein or other molecule made up of two polypeptides or other subunits.

Dimerization domain The region of a polypeptide that binds to another polypeptide of the same type to form a dimer.

Diploid The state of a cell containing two copies of each of its genes, which are not derived from replication of the same DNA. *See* Haploid.

Direct repeat A short sequence of deoxynucleotides in DNA closely followed by an identical or almost identical sequence on the same strand and read in the same direction.

Directed-change hypothesis The discredited hypothesis that mutations in DNA occur preferentially when they benefit the organism or help it adapt to a new environment.

Directional cloning Cloning a piece of DNA into a cloning vector in such a way that it can be inserted in only one orientation, for example, by using restriction endonucleases that leave incompatible overhang sequences at each end.

Directionality factor A protein that binds to a site-specific recombinase, such as an integrase, and changes its specificity so it will promote recombination in only one direction, e.g., excision.

Dissimilatory reduction The reduction of nitrogen-containing compounds, such as nitrate, that occurs when they are used as terminal electron acceptors in anaerobic respiration. The reduced nitrogen-containing compounds are not necessarily incorporated into cellular molecules.

Disulfide bonds Covalent bonds between two sulfur atoms, such as those between the side chain sulfur atoms in two cysteine amino acids in a polypeptide.

Disulfide oxidoreductases Enzymes in the periplasmic space that can form or break disulfide bonds between cysteines by reducing or oxidizing the bonds. They contain the motif CXXC, where X can be any amino acid, and exchange cysteine bonds in the protein with the cysteines in the Dsb protein.

Division time The time taken by a type of newborn bacterial cell to grow and divide again in a particular growth environment.

DNA clones Identical copies of a single DNA molecule that are usually made by cloning the DNA in a cloning vector and propagating the vector.

DNA glycosylase An enzyme that removes bases from DNA by cleaving the bond between the base and the deoxyribose sugar.

DNA library A collection of individual clones of the DNA of an organism that together represent all the DNA sequences of that organism.

DNA polymerase V A translesion DNA polymerase from *E. coli* produced from the UmuC protein and a cleaved version of UmuD, UmuD′. Activation requires RecA and multiple levels.

DNA polymerase accessory proteins Proteins that travel with the DNA polymerase during replication.

DNA polymerase III holoenzyme The replicative DNA polymerase in *Escherichia coli*, including all the accessory proteins, sliding clamp, editing functions, etc.

DnaA box The sequence 5′-TTATCCACA-3′ in DNA to which the DnaA protein binds with the greatest affinity. The DnaA protein is required for the initiation of chromosome replication in *Escherichia coli*.

DNA-binding domain An independently folding region of a protein that interacts either nonspecifically with DNA or with a specific DNA sequence in DNA.

Domain A region of a polypeptide with a particular function or localization.

Dominant mutation A mutation that exhibits its phenotype even in a diploid organism containing a wild-type allele of the gene.

Donor bacterium The strain of bacterium used as a source of DNA to transfer into another strain in a genetic cross, whether by transformation, transduction, or conjugation. *See* Donor strain.

Donor DNA A region of DNA that is transferred into another DNA molecule (the recipient). In transposition, it is the DNA in which the transposon originally resides before it transposes to the target DNA.

Donor strain The bacterial strain that is the source of the transferred DNA in a bacterial cross. For example, in a transductional cross, the donor strain is the strain in which the phage was previously propagated; in conjugation, it is the strain harboring the transmissible plasmid.

Downstream Lying in the 3′ direction from a given point on RNA or in the 3′ direction on the coding strand of a DNA region.

Dtr component The DNA transfer component of a plasmid transmission system; the *tra* or *mob* genes of the plasmid involved in preparing the plasmid DNA for transfer.

Duplication junction The point at which an ectopic crossover occurred, resulting in a tandem-duplication mutation in DNA.

Duplication A mutation that causes a region of DNA to be repeated elsewhere in the chromosome. *See* Tandem duplication.

E (exit) site The site on the ribosome at which the tRNA binds after it has contributed its amino acid to the growing polypeptide and just before it exits the ribosome. It may help maintain the correct reading frame.

E value A measure of the number of similarities or local alignment scores that are reported in a sequence search based on comparing a query sequence with a database, identifying matching sequences, and calculating the probability that a particular match could have occurred by chance. For example, the score of a query sequence and the same sequence in a database would be very close to 0. For genome annotation, only E values less than $1/e^5$ are usually considered evidence of a reliable match.

Early gene A gene expressed early during a developmental process, for example, during bacterial sporulation or phage infection.

Eclipse phase A step during natural transformation during which DNA cannot be isolated from the recipient cells in a state that is active for transformation of a new recipient.

Ectopic recombination "Out-of-place" recombination; homologous recombination occurring between two sequences, usually but not always nonidentical, in different regions of the DNA. It is often responsible for deletions, inversions, and other types of DNA rearrangements and is sometimes called "unequal crossing over" when it occurs between two DNA molecules. *See* Homeologous recombination.

Effector A small molecule that binds to a protein and changes its properties.

EF-G *See* Translation elongation factor G.

EF-Tu *See* Translation elongation factor Tu.

8-OxoG A damaged DNA base commonly caused by reactive forms of oxygen in which an oxygen atom has been added to the 8 position of the small ring of the base guanine. Abbreviation for 7,8-dihydro-8-oxoguanine; also abbreviated GO.

Electroporation The introduction of nucleic acids, proteins, or nucleoprotein complexes into cells through exposure of the cells to a strong electric field.

Electrospray ionization (ESI) A method used for preparation of samples for mass spectrometry that produces singly and multiply charged ions from a peptide so that multiple peaks are seen in a mass spectrometric analysis. The sample is introduced into an electric field in a liquid solution. Ions are formed when the solution is sprayed from a fine needle into the electric field. As solvent evaporates, intact peptides are left with different numbers of charges, depending on the sequence of the peptide.

Elongation factor G (EF-G) *See* Translation elongation factor G.

Elongation factor Tu (EF-Tu) *See* Translation elongation factor Tu.

Endonuclease An enzyme that can cut phosphodiester bonds between nucleotides internal to a polynucleotide.

Endoribonuclease An endonuclease that cleaves an RNA chain.

Endosymbiosis A collaboration between two organisms where one lives inside the other.

Enol One of two forms of a chemical equilibrium in a keto-enol shift where they differ only in the position of the hydrogen.

Enrichment The process of increasing the frequency of a particular type of mutant in a population, often by using an antibiotic, such as ampicillin, that kills cells only if they are growing.

ESI *See* Electrospray ionization.

Essential genes Genes whose products are required for maintenance and/or growth of the cell under all known conditions.

E site See E (exit) site.

Eubacteria Another term for bacteria, proposed by Carl Woese after the discovery of archaea to distinguish between these two different domains of one-celled organisms. The organisms are now usually simply referred to as bacteria.

Eukaryotes Members of the kingdom of organisms whose cells contain a nucleus, usually surrounded by a nuclear membrane, and many other cellular organelles, including a Golgi apparatus and an endoplasmic reticulum. They have 18S and 28S rRNAs rather than the 16S and 23S rRNAs of bacteria.

Exit site *See* E (exit) site.

Exonuclease A nuclease enzyme that can remove nucleotides only from the end of a polynucleotide.

Exoribonuclease An exonuclease that digests RNA from the end of the molecule.

Exported proteins Proteins that leave the cytoplasm after they are made and end up outside the cell.

F plasmid Fertility plasmid, the first plasmid discovered that is capable of doing conjugation. The F plasmid was utilized in early genetic experiments with *Escherichia coli* indicating that bacteria can exchange genes and evolved using the same underlying rules as other types of organisms.

Face-of-the-helix dependence A situation in which the position of a binding site for a regulatory protein on the DNA must be on the same side of the DNA helix as the position for binding of another factor (e.g., another protein or RNA polymerase).

Factor-dependent transcription termination site A DNA sequence that causes transcription termination only in the presence of a particular protein, such as the ρ protein of *Escherichia coli.*

Factor-independent transcription termination site A DNA sequence that causes transcription termination by RNA polymerase alone, in the absence of other proteins. In bacteria, it is characterized by a GC-rich region with an inverted repeat followed by a string of A's on the template strand.

Feedback inhibition Inhibition of synthesis of the product of a pathway by the end product of the pathway; often results from binding of the end product of the pathway to the first enzyme of the pathway, thereby inhibiting the activity of the enzyme.

Filamentous phage A type of phage with a long, floppy appearance. The nucleic acid genome of these phages is merely coated with protein, making the phage particle length proportional to the length of the genome and giving the floppy appearance. In contrast, the nucleic acids of most phages are encapsulated in a rigid, almost spherical, icosahedral head.

Fimbriae Another name for pili, except for conjugative sex pili encoded by self-transmissible plasmids, which are always called pili and never fimbriae.

Firmicutes One of the phylogenetic groups of bacteria; it includes *Bacillus subtilis*. Members of this phylum lack an outer membrane and typically stain Gram positive.

5′ end The end of a nucleic acid strand (DNA or RNA) at which the 5′ carbon of the ribose sugar is not attached through a phosphate to another nucleotide.

5′ phosphate end In a polynucleotide, a 5′ end that has a phosphate attached to the 5′ carbon of the ribose sugar of the last nucleotide.

5′-to-3′ direction The direction on a polynucleotide (RNA or DNA) from the 5′ end to the 3′ end.

5′ untranslated region The untranslated sequence of nucleotides that extends from the 5′ end of an mRNA to the first initiation codon for a polypeptide encoded by the mRNA.

Flanking sequences The sequences that lie on either side of a gene or other DNA element.

Flap endonuclease A special kind of endonuclease activity where a phosphodiester bond is cleaved, liberating a DNA or RNA segment after it is displaced from the reverse complement strand; for example, in the process of RNA primer removal by DNA polymerase 1 during DNA replication.

fMet-tRNA^{fMet} The special tRNA in prokaryotes that has formylmethionine attached to its 3′ end and is used to initiate translation at prokaryotic translational initiation regions. It binds to translation initiation factor IF2 and responds to the initiator codons AUG and GUG and, more rarely, to other codons in a translational initiation region. *See* Initiator tRNA.

Forward genetics The classical genetic approach in which genes are first identified by the phenotypes of mutations in the genes.

Frameshift mutation Any mutation that adds or removes one or very few (but not a multiple of 3) base pairs from DNA, whether or not it occurs in the coding region for a protein.

Fusion protein A protein created when coding regions from different genes are fused to each other in frame so that one part of the protein is encoded by sequences from one gene and another part is encoded by sequences from a different gene. *See* Translational fusion.

Gain-of-function mutation A mutation that creates a new activity for the gene product or causes the expression of a gene that is quiescent in the wild type.

Gene A region on DNA encoding a particular polypeptide chain or functional RNA, such as an rRNA, tRNA, or small noncoding RNA.

Gene cassette A piece of DNA containing a selectable gene that can be easily cloned into another gene or cloning vector.

Gene conversion Nonreciprocal apparent recombination associated with mismatch repair on heteroduplexes that are formed between two DNA molecules during recombination. The name comes from genetic experiments with fungi in which the alleles of the two parents were not always present in equal numbers in an ascus, as though an allele of one parent had been "converted" into the allele of the other parent.

Gene replacement A molecular genetic technique in which a cloned gene is altered in the test tube and then reintroduced into the organism, allowing selection for organisms in which the altered gene has replaced the corresponding normal gene in the organism by homologous recombination.

Generalized recombination *See* Homologous recombination.

Generalized transduction The transfer, via phage transduction, of essentially any region of the bacterial DNA from one bacterium to another. The transducing phage particle contains only bacterial DNA.

Generation time The time it takes for the cells in an exponentially growing culture to double in number. *See* Division time.

Genetic code The assignment of each mRNA nucleotide triplet to an amino acid (or a translation termination signal). *See* Codon.

Genetic island A DNA element in the chromosome that is not an obvious prophage, transposon, or plasmid but that shows evidence of having been recently horizontally transferred into the chromosome based on features of the DNA and absence of the sequences in related species at that position.

Genetic linkage map An ordering of the genes of an organism solely on the basis of recombination frequencies between mutations in the genes in genetic crosses.

Genetic marker A difference in sequence of the DNAs of two strains of an organism in a particular region that causes the two strains to exhibit different phenotypes that can be used for genetic mapping of the region of sequence difference.

Genetics The science of studying organisms on the basis of their genetic material.

Genome annotation The process where the DNA sequence of A, T, C, and G from an organism is interpreted into the predicted gene products they encode.

Genomic islands Stretches of DNA sequence that were acquired by horizontal gene transfer from other organisms and not found in all members of the same species. These may or may not have the features to allow subsequent transfer to other bacteria.

Genomics The process of using the entire DNA sequence of an organism to study its physiology and relationship to other organisms. The term also refers to investigations enabled by large sets of information from RNA or proteins and compiled sequences for environments (metagenomics).

Genotype The sequence of nucleotides in the DNA of an organism, usually discussed in terms of the alleles of its genes.

GLIMMER (Gene Locator and Interpolated Markov Modeler) A software program that is used to find coding regions in bacterial genomes.

Global regulatory mechanism A regulatory mechanism that affects many genes and operons scattered around the genome.

Glucose effect The regulation of genes involved in carbon source utilization based on whether glucose is present in the medium. *See* Catabolite regulation.

Glutamate dehydrogenase An enzyme that adds ammonia directly to α-ketoglutarate to make glutamate. It is responsible for assimilation of nitrogen in high ammonia concentrations.

Glutamate synthase An enzyme that transfers amino groups from glutamine to α-ketoglutarate to make glutamate. Also called GOGAT.

Glutamine synthetase An enzyme that adds ammonia to glutamate to make glutamine. Responsible for the assimilation of nitrogen in low ammonia concentrations.

Gradient of transfer In a conjugational cross, the decrease in the transfer of chromosomal markers the farther they are in one direction from the origin of transfer of an integrated plasmid.

Gram-negative bacteria Bacteria characterized by an outer membrane and a thin peptidoglycan cell wall that stains poorly with a staining procedure invented by the Danish physician Hans Christian Gram in the 19th century. The term is generally used to describe bacteria in the phylum *Proteobacteria*.

Gram-positive bacteria Bacteria characterized by having no outer membrane and a thick peptidoglycan layer that stains well with the Gram stain. This term is generally used to describe bacteria in the phylum *Firmicutes*, even though not all members of the phylum stain Gram positive.

Growth rate regulation of ribosomal synthesis The regulation of ribosomal synthesis that ensures that cells growing more slowly have fewer ribosomes. It is proposed to be at least partially due to the levels of the initiating nucleotides GTP and ATP, which affect the stability of open complexes on the promoters for rRNAs.

Guanine One of the two purine (two-ringed) bases in DNA and RNA.

Guanosine pentaphosphate (pppGpp) The nucleoside guanosine with two phosphates attached to the 3′ carbon and three phosphates attached to the 5′ carbon of the ribose sugar. It is quickly converted to ppGpp, which is responsible for the stringent response.

Guanosine tetraphosphate (ppGpp) The nucleoside guanosine with two phosphates attached to each of the 3′ and 5′ carbons of the ribose sugar. It is responsible for the stringent response and other stress responses. *See* Stringent response.

Guide RNA The RNA component that directs a protein::RNA complex to a site of action using Watson-Crick base pairing in CRISPR/Cas systems found in bacteria and archaea.

Half-life The time required for the amount of an exponentially decaying substance to decrease by one-half.

Haploid The state of a cell containing only one copy or allele of each of its chromosomal genes. *See* Diploid.

Headful packaging A mechanism of encapsulation of DNA in a virus head in which the concatemeric DNA is cut after uptake of a length of DNA sufficient to fill the head, rather than at *pac* or *cos* sites.

Heat shock protein (HSP) One of a group of highly evolutionarily conserved proteins whose rate of synthesis markedly increases after an abrupt increase in temperature or certain other stresses on the cell.

Heat shock response The cellular changes that occur in the cell after an abrupt rise in temperature.

Helicases Enzymes that unwind double-stranded nucleic acids.

Helper phage A wild-type phage that furnishes gene products that a deleted form of the phage or other DNA element cannot make, thereby allowing the DNA element to replicate and be packaged into a phage particle.

Heterodimer A protein made of two polypeptide chains that are different in primary sequence because they are encoded by different genes. *See* Heteromultimer, Homodimer.

Heteroduplex A double-stranded DNA region in which the two strands come from different DNA molecules and so can have somewhat different sequences, leading to mismatches.

Heteroimmune phages Related phages capable of lysogeny that carry different immunity regions and therefore cannot repress each other's transcription, so they can multiply on cells lysogenic for the other phage. *See* Homoimmune phages.

Heteromultimer A protein made of more than one polypeptide chain (usually more than two) that are different in primary sequence because they are encoded by different genes. *See* Heterodimer.

Hfr strain A bacterial strain that contains a self-transmissible plasmid integrated into its chromosome and thus can transfer its chromosome by conjugation.

HFT lysate transduction The lysate of a lysogenic phage containing a significant percentage of a transducing phage with bacterial DNA substituted for some of the phage DNA.

Hidden Markov model (HMM) Generally, a mathematical tool used to predict an unknown or "hidden" event based on observable features. In the field of bioinformatics, the hidden Markov model is used for a variety of tasks to predict gene function, align biological sequences, or predict biological structures.

High MOI (multiplicity of infection) A state of a virus or phage infection in which the number of viruses greatly exceeds the number of cells being infected, so that most cells are infected by more than one virus.

Holin A phage protein that forms a channel in the inner membrane, allowing access of the endolysin to the cell wall or destroying the membrane potential (proton motive force) and causing cell lysis.

Holliday junction An intermediate in homologous recombination in which one strand from each of two DNAs crosses over and is rejoined to the corresponding strand on the opposite DNA.

Holoenzyme An enzyme complex (e.g., RNA polymerase or DNA polymerase) attached to all of its accessory proteins that allows complete function.

Homeologous recombination Homologous recombination in which the deoxynucleotide sequences of two participating regions are somewhat different from each other, usually because they are in different regions of the DNA or because the DNAs come from different species. *See* Ectopic recombination.

Homodimer A protein made up of two polypeptide chains that are identical in primary sequence, usually because they are encoded by the same gene. *See* Heterodimer, Homomultimer.

Homoimmune phages Two related phages that have the same immunity region so that they repress each other's transcription; hence, one cannot multiply on a lysogen of the other. *See* Heteroimmune phages.

Homologous recombination A type of recombination that depends on the two DNAs having identical or at least very similar sequences in the regions being recombined because complementary base pairing between strands of the two DNAs must occur as an intermediate state in the recombination process.

Homologs Two or more nucleotide or protein sequences that are similar because they derive from a common ancestor.

Homomultimer A protein made up of more than one polypeptide (usually more than two) that are identical in primary sequence, usually because they are encoded by the same gene. *See* Homodimer.

Horizontal transfer Transfer of DNA between individuals in the population rather than by inheritance from ancestors. *See* Vertical transfer.

Host range All of the types of host cells in which a DNA element, plasmid, phage, etc., can multiply.

Hot spot A position in DNA that is particularly prone to a type of event, such as mutagenesis by a particular mutagen or insertion of a transposable element.

Hsp70 A highly evolutionarily conserved heat shock-induced protein chaperone of 70 kDa, represented by DnaK in bacteria.

Hypoxanthine A purine base derived from the deamination of adenine.

***In vitro* mutagen** A mutagen that reacts only with purified DNA or with viruses or phage. It cannot be used to mutagenize living cells, either because it cannot get in or because it is too reactive and is destroyed before it reaches the DNA.

***In vivo* mutagen** A mutagen that can enter living intact cells and mutagenize the DNA.

Incompatibility (Inc) group A set of plasmids that interfere with each other's replication and/or partitioning and so cannot be stably maintained together in the descendants of the same bacterium.

Induced mutations Mutations that are caused by deliberately irradiating cells or treating cells or DNA with a mutagen, such as a chemical.

Inducer A small molecule that can increase the expression of a gene or operon.

Inducer exclusion The process by which the inducer of an operon, such as a sugar, is kept out of the cell by inhibiting its transport through the membrane. Often, a more efficiently utilized sugar, such as glucose, inhibits the transport of other, less efficiently used sugars, such as lactose.

Inducible Able to have expression increased by an inducer.

Induction In gene regulation, the turning on of the expression of the genes of an operon. In phages, the initiation of lytic development of a prophage.

Informational suppressors Mutations in the gene expression machinery that result in suppression of a particular class of mutations. Nonsense suppressor tRNAs, which insert an amino acid at the position of a nonsense codon in an mRNA, are the most common class. *See* Nonsense suppressor tRNA.

In-frame deletion A deletion mutation in an open reading frame that removes a multiple of 3 bp and so does not cause a frameshift. These deletions are particularly useful because they cannot be polar and can remove a specific domain without removing the rest of the protein.

Initial transcription complex *See* Initiation complex.

Initiation codon The 3-base sequence in an mRNA that specifies the first amino acid to be inserted in the synthesis of a polypeptide chain. In prokaryotes, it is the 3-base sequence (usually AUG or GUG) within a translational initiation region at which formylmethionine is inserted to begin translation. In eukaryotes, the AUG closest to the 5′ end of the mRNA is usually the initiation codon, and methionine is inserted to begin translation.

Initiation complex In transcription, the complex that forms after the first ribonucleoside triphosphate enters into RNA polymerase and pairs with the DNA nucleotide at the position corresponding to the transcription start site.

Initiation factors Proteins (IF1, IF2, and IF3 in bacteria) that assist in initiation of translation.

Initiator tRNA A dedicated tRNA that is charged with formylmethionine and binds to the initiation codon during translation initiation. *See* tRNA^{fMet}.

Injectisome *See* Type III secretion system.

Inner membrane protein A protein that resides, at least in part, in the cytoplasmic (inner) membrane of Gram-negative bacteria.

Insertion element *See* Insertion sequence (IS) element.

Insertion mutation A change in a DNA sequence due to the incorporation of another DNA sequence, such as a transposon or antibiotic resistance cassette, into the sequence.

Insertion sequence (IS) element A small transposon in bacteria that carries only the gene or genes encoding the products needed to promote transposition.

Insertional inactivation Inactivation of the product of a gene by an insertion mutation, usually denoting inactivation of the product of a gene on a plasmid cloning vector by cloning a fragment of DNA into the gene.

Integrase A type of site-specific recombinase that promotes recombination between two defined sequences in DNA, causing the integration of one DNA into another DNA (e.g., the integration of a phage DNA into the chromosome).

Integrating conjugative element A mobile element that can excise from the genome and transfer between bacteria using element-encoded functions. Depending on the element, integration can occur at a specific site or almost randomly. Previously called a conjugative transposon.

Integron An integrase and an *att* site for integration of gene cassettes, often for antibiotic resistance. A promoter downstream of the integrase is situated to allow transcription of cassette genes inserted into the *att* site. They are typically, but not exclusively, found in transposons and plasmids.

Intein A parasitic DNA that encodes a polypeptide sequence that, when inserted into the gene for another polypeptide, introduces a polypeptide sequence into the other polypeptide that must be removed (spliced out) before the other polypeptide can be active. Inteins are usually self-splicing and are removed at the protein level.

Interaction genetics Characterization of a system by analysis of how genes and gene products affect each other.

Interference The step when an invading mobile element, such as a bacteriophage, is recognized and subsequently cleaved or degraded by a CRISPR/Cas system.

Interpolated Markov model One type of hidden Markov model that is useful for locating genes in a particular organism because it has been trained on known sequences of that organism.

Interstrand cross-links Covalent chemical bonds between the two complementary strands of DNA in a double-stranded DNA.

Intragenic complementation Complementation between two mutations in the same gene. It is rare and allele specific; it usually occurs only if the protein product of the gene is a homodimer or homomultimer.

Intragenic suppressor A suppressor mutation that occurs in the same gene as the mutation it is suppressing.

Intron A parasitic DNA that, when inserted into a gene for a protein, introduces polynucleotide sequences into the mRNA that must be removed (spliced out) before the mRNA can be translated into a functional protein. Introns are removed at the RNA level.

Inversion junctions The points where the recombination events that inverted a sequence occurred.

Inversion mutation A change in DNA sequence as a result of flipping a region within a longer DNA so that it lies in reverse orientation. It is usually due to homologous recombination between inverted repeats in the same DNA molecule. Inversions due to site-specific recombinases or invertases are not considered inversion mutations.

Inverted repeat Two nearby sequences in DNA that are the same or almost the same when read in the 5′-to-3′ direction on opposite strands.

Invertible sequence A sequence in DNA bracketed by inverted repeats that depends upon a site-specific recombinase protein for its inversion rather than the generalized homologous recombination system.

IS element *See* Insertion sequence (IS) element.

IS*CR* element A type of Y2 transposon.

Isogenic Strains of an organism that are almost identical genetically except for one small region or gene.

Isolation of mutants The process of obtaining a pure culture of a particular type of mutant from among a myriad of other types of mutants and the wild type.

Isomerization Changing of the spatial conformation of a molecule without breaking any bonds. In DNA recombination, it refers to the rotating of the DNAs in a Holliday junction, thus changing the strands that are crossed without breaking any hydrogen or other types of bonds. In transcription, it refers to an open-complex formation.

Iteron sequences Short DNA sequences, often repeated many times in the origin regions of some types of plasmids, that bind the Rep protein required for replication of the plasmid and play a role in regulating the copy number.

KEGG maps An automated reconstruction of the metabolic pathways of an organism. See http://www.genome.jp/kegg/.

Killing curves A graph showing the relationship between exposure to a specific condition and the proportion of cells in the population killed.

Kinase An enzyme that transfers a phosphate group from ATP to another molecule.

Knockout mutation A mutation that presumably eliminates the function of a gene, i.e., is presumably a null mutation.

Lagging strand During DNA replication, the newly synthesized strand that is made from the template strand that is produced in the opposite direction from the overall movement of the replication fork, i.e., in the 3′-to-5′ direction overall.

Late gene A gene that is expressed only relatively late in the course of a developmental process, e.g., a late gene of a phage.

Lawn *See* Bacterial lawn.

Leader region An RNA sequence close to the 5′ end of an mRNA that may be translated but does not encode a functional polypeptide. Also called leader sequence.

Leading strand During DNA replication, the newly synthesized strand that is made from the template strand that is produced in the same direction as the overall direction of movement of the replication fork, i.e., in the 5′-to-3′ direction.

Leaky mutation A mutation that does not totally inactivate the product of the gene.

Lep protease One of the enzymes that cleaves the signal sequence off secreted proteins as they pass through the SecYEG channel, and probably also the Tat pathway.

Lesion Any change in a DNA or RNA molecule as a result of chemical alteration of a base, sugar, or phosphate.

Ligase An enzyme that joins two molecules.

Linked A genetic term referring to the fact that two markers are close enough on the DNA that they are separated by recombination less often than if they sorted randomly.

Low MOI (multiplicity of infection) The state of a virus or phage infection in which the number of cells almost equals or exceeds the number of viruses, so that most cells remain uninfected or are infected by at most one or very few viruses.

Lyse To break open cells and release their cytoplasm into the medium.

Lysogen A strain of bacterium that harbors a prophage.

Lysogenic conversion A property of a bacterial cell caused by the presence of a particular prophage.

Lysogenic cycle The series of events following infection by a bacteriophage and culminating in the formation of a stable prophage.

Lysogenic phage A phage that is known to be capable of entering a prophage state in some host.

Lytic cycle The series of events following infection by a bacteriophage or induction of a prophage and culminating in lysis of the bacterium and the release of new phage into the medium.

Macrodomain A region of the chromosome that more readily genetically recombines within the region than with loci outside the region.

Major groove In double-stranded DNA, the larger of the two tracks between the two strands of DNA as they twist around each other as a helix.

MALDI *See* Matrix-assisted laser desorption ionization.

Male strain A bacterium or strain harboring a self-transmissible plasmid or other conjugative element capable of producing a sex pilus.

Marker rescue Acquisition of a genetic marker by the genome of an organism or virus through recombination with a cloned DNA fragment containing the marker.

Markov model A statistical tool that can be applied to a system that is represented by discrete states. For example, when used for protein annotation, a discrete state could be 1 of the 23 amino acids at a position in the protein.

Mass spectrometry (MS) An analytical method that measures ion abundances based on their mass-to-charge (*m/z*) ratios. First, gas phase ions are produced from the compound of interest (*see* Matrix-assisted laser desorption ionization and Electrospray ionization). Then, the ions are separated on the basis of their *m/z* ratios. Finally, the ions at different *m/z* ratios are detected and counted.

Massive parallel sequencing A DNA sequencing process involving reading many short stretches of DNA at the same time.

Matrix-assisted laser desorption ionization (MALDI) A procedure that is used to measure peptide mass. A singly charged ion is produced from a peptide, resulting in one peak on a mass spectrometric analysis. The technique can be used to directly analyze complex peptide mixtures if a complete genome sequence is available.

Merodiploid A bacterial cell that is mostly haploid but is diploid for some region of the genome due to some chromosomal genes being carried on a prophage or plasmid. *See* Partial diploid.

Messenger RNA (mRNA) An RNA transcript that includes the coding sequences for at least one polypeptide.

Metagenomics A type of study where DNA sequence information is collected from a number of organisms in a community without first isolating the individuals as a pure culture.

Methionine aminopeptidase An enzyme that removes the N-terminal methionine from newly synthesized polypeptides.

Methyl-directed mismatch repair The repair system that recognizes mismatches in newly replicated DNA and specifically removes and resynthesizes the new strand. In some enteric bacteria, the new strand is distinguishable from the old strand because it is the strand that is not methylated in the nearest hemimethylated GATC sequences.

Methyltransferase In DNA repair, an enzyme that removes a CH_3 (methyl) or CH_3CH_2 (ethyl) group from a base in DNA by catalyzing transfer of the group from the DNA to the protein itself. In DNA modification, any one of a number of enzymes that catalyze the transfer of a methyl group to a specific base on a DNA sequence recognized by the enzyme.

Minor groove In double-stranded DNA, the smaller of the two tracks between the two strands of DNA as they twist around each other as a helix.

–10 sequence In a bacterial σ^{70}-type promoter, a short sequence that is centered 10 bp upstream of the transcription start site. The canonical or consensus sequence is TATAAT/ATATTA.

–35 sequence In a bacterial σ^{70}-type promoter, a short sequence that is centered 35 bp upstream of the transcription start site. The canonical or consensus sequence is TTGACA/AACTGT.

mismatch Improper pairing of the normal bases in DNA, e.g., an A opposite a C.

Mismatch repair system A pathway for removing mismatches in DNA by degrading a strand containing the mismatched base and replacing it by synthesizing a new strand containing the correctly paired base.

Missense mutation A base pair change mutation in a region of DNA encoding a polypeptide that changes an amino acid in the polypeptide.

***mob* genes** The genes on a mobilizable DNA element that allow it to be mobilized by a self-transmissible element, such as a self-transmissible plasmid. They often encode Dtr (DNA transfer) functions and a coupling protein that allows the element to communicate with the mating-pair formation (Mpf) system of the self-transmissible element.

***mob* region** A region in DNA carrying an origin of transfer (*oriT* sequence) and often genes whose products allow the plasmid or other DNA element to be mobilized by self-transmissible elements.

Mobilizable plasmid A plasmid that possesses only a subset of the features needed for it to be mobilized by conjugation and therefore requires that a separate plasmid reside in the cell to provide the remaining *trans*-acting features.

Mobilization The process by which a mobilizable DNA element, incapable of self-transmission, is transferred into other cells by the conjugation functions of a self-transmissible element.

Mobilome All of the genetic elements present in an organism or community that are capable of moving to new genetic loci within or between individuals.

MOI *See* Multiplicity of infection.

Molecular genetic techniques Methods for manipulating DNA in the test tube and reintroducing the DNA into cells.

Monocistronic mRNA An mRNA that encodes a single polypeptide.

Motif A conserved nucleotide or amino acid sequence that is relatively short and suggests similarity of function.

Mpf component A component made up of *tra* gene products of a self-transmissible plasmid involved in making the surface structures (pilus, etc.) that contact another cell and transfer the DNA during conjugation, as well as the coupling protein that communicates with the Dtr component. The term "Mpf" is derived from "mating pair formation."

mRNA *See* Messenger RNA.

Multicopy suppression A process that relieves the effects of a mutation in a different gene when it is expressed in higher than normal amounts from a multicopy plasmid or other vector. It is often a complication of cloning by complementation.

Multimer A protein or other molecule that consists of more than one polypeptide chain or other subunit (usually more than two).

Multiple cloning site A region of a cloning vector that contains the sequences cut by many different type II restriction endonucleases.

Multiplicity of infection (MOI) The ratio of phages or viruses to cells that initiates an infection.

Mutagen A chemical or type of irradiation that causes mutations by damaging DNA.

Mutagenic repair A pathway for repairing damage to DNA that sometimes changes the sequence of deoxynucleotides as a consequence.

Mutagenic treatments or chemicals Treatments or chemicals that cause mutations by damaging DNA.

Mutant An organism that differs from the normal, or wild type, as a result of a change in the sequence (mutation) of its DNA.

Mutant allele The mutated gene of a mutant organism that makes it different from the wild-type gene.

Mutant phenotype A characteristic that makes a mutant organism different from the wild type.

Mutation Any heritable change in the sequence of deoxynucleotides in DNA.

Mutation rate The probability of occurrence of a mutation causing a particular phenotype each time a newborn cell grows and divides.

N terminus The end of a polypeptide chain with the N-terminal amino acid.

Naive spacer acquisition The first spacer that is acquired from a particular invading genetic element and integrated into the CRISPR array.

Narrow host range A range of hosts (that a DNA element can enter and/or replicate in) that includes only a few closely related types of cells. The term typically refers to *Escherichia coli* and its close relatives.

Naturally competent Able to take up DNA at a certain stage in the bacterial growth cycle without chemical or other treatments. A characteristic of some types of bacteria. Also called naturally transformable.

Naturally transformable *See* Naturally competent.

Negative regulation A type of regulation in which a protein or RNA molecule, in its active form, inhibits a process, such as the transcription of an operon or translation of an mRNA.

Negative repressible A regulatory system in which the regulatory protein turns gene expression off in the presence of a small signal molecule.

Negatively supercoiled A state of a DNA molecule in which the two strands of the double helix are wrapped around each other less than about once every 10.5 bp.

Nested deletion A set of deletion mutations with one common end point and with the other end point extending varying distances into a gene or region of DNA.

Nicks Broken phosphate-deoxyribose bonds in the phosphodiester backbone of double-stranded DNA.

Nonautonomous element A mobile genetic element that possesses only a subset of the features needed for it to be mobilized and therefore requires a separate element to provide the missing features in *trans*.

Noncoding strand The strand of DNA that is used as a template for RNA synthesis. *See* Template strand, Transcribed strand.

Noncomposite transposon A transposon in which the transposase genes and the inverted-repeat ends are included in the minimum transposable element and are not part of autonomous insertion sequence (IS) elements. *See* Composite transposon.

Noncovalent change Any change in a molecule that does not involve the making or breaking of a chemical covalent bond due to shared electron orbits in the molecule.

Nonhomologous recombination The breaking and rejoining of two DNAs into new combinations, which does not necessarily depend on the two DNAs having similar sequences in the region of recombination.

Nonpermissive conditions Conditions under which a mutant organism or virus cannot multiply but the wild type can multiply.

Nonpermissive host A host organism in which a mutant phage or virus cannot multiply but the wild type can multiply.

Nonpermissive temperature A temperature at which the wild-type organism or virus, but not the mutant organism or virus, can multiply.

Nonselective conditions Conditions or media in which both the mutant and wild-type strains of an organism or virus can multiply.

Nonsense codon A codon that does not stipulate an amino acid but, rather, triggers the termination of translation. In most organisms, the codons UAG, UGA, and UAA are nonsense codons.

Nonsense mutation In a region of DNA encoding a protein, a base pair change mutation that causes one of the nonsense codons to be encountered in frame when the mRNA is translated.

Nonsense suppressor A suppressor mutation that allows an amino acid to be inserted at some frequency for one or more of the nonsense codons during the translation of mRNAs.

Nonsense suppressor tRNA A tRNA that, usually as a result of a mutation, allows the tRNA to pair with one or more of the nonsense codons in mRNA during translation and therefore causes an amino acid to be inserted for the nonsense codon. The mutation usually changes the anticodon on the tRNA.

Nontemplate strand The strand of DNA that is complementary to the template strand and that is not used as a template for RNA synthesis; it has the same base sequence as the RNA product. *See* Coding strand.

N-terminal amino acid The amino acid on the end of a polypeptide chain whose amino (NH_2) group is not attached to another amino acid in the chain through a peptide bond.

Ntr system A global regulatory system that regulates a number of operons in response to the nitrogen sources that are available.

Nuclease An enzyme that cuts the phosphodiester bonds in DNA or RNA polymers.

Nucleoid A compact, highly folded structure formed by the chromosomal DNA in the bacterial cell and in which the DNA appears as a number of independent supercoiled loops held together by a core.

Nucleoid occlusion proteins Proteins that prevent the formation of the division septum in a region of the cell still occupied by the nucleoid.

Nucleotide excision repair A system for the repair of DNA damage in which the entire damaged nucleotide is removed rather than just the damaged base. A cut is made on either side of the damage on the same strand, and the damaged strand is removed and resynthesized.

Null mutation A mutation in a gene that totally abolishes the function of the gene product.

Ochre The nonsense codon UAA. An ochre mutation is a mutation to this codon in the reading frame for a protein. An ochre suppressor suppresses ochre mutations.

Okazaki fragments The short pieces of DNA that are synthesized in the opposite direction of movement of the replication fork during replication from a lagging-strand template.

Oligopeptide A short polypeptide only a few amino acids long.

Opal The nonsense codon UGA. Mutations to this codon in the reading frame of a protein are called opal mutations. Opal suppressors suppress opal mutations. Sometimes called umber.

Open complex The complex of RNA polymerase and DNA at a promoter in which the strands of the DNA have been separated after isomerization.

Open reading frame (ORF) A sequence of DNA, read 3 nucleotides at a time, that is not interrupted by any nonsense codons.

Operator Usually a sequence on DNA to which a repressor protein binds to block transcription. More generally, any sequence in DNA or RNA to which a negative regulator binds.

Operon A DNA region encompassing genes that are transcribed into the same mRNA, as well as any adjacent *cis*-acting regulatory sequences.

Operon model The model proposed by Jacob and Monod for the regulation of the *lac* operon, in which transcription of the structural genes of the operon is prevented by binding of the LacI repressor to the operator region, thereby preventing access of RNA polymerase to the promoter. The inducer lactose binds to LacI and changes its conformation so that it can no longer bind to the operator, and as a result, the structural genes are transcribed.

ORF *See* Open reading frame.

oriC A sequence of DNA consisting of the site in the bacterial chromosome at which initiation of a round of replication normally occurs and all of the surrounding *cis*-acting sequences required for initiation.

Origin of replication The site on a DNA, plasmid, phage, chromosome, etc., at which replication initiates, including all of the surrounding *cis*-acting sequences required for initiation.

Orthologs Genes in different species that are derived from a common ancestor. They may differ in function, but they usually have identical functions.

Outer membrane protein A protein that resides, at least in part, in the outer membrane of Gram-negative bacteria.

P (peptidyl) site The site on the ribosome to which the peptidyl tRNA, which contains the growing peptide chain, is bound.

Palindrome A DNA sequence that reads the same in the 5′-to-3′ direction on the top strand and bottom strand.

PAM (Protospacer Adjacent Motif) A short DNA sequence that must be present for a spacer to be acquired and inserted into a CRISPR array (i.e., allowing it to be a protospacer). The PAM must also be present for a protospacer to be recognized by the guide RNA complex.

Pan genome All of the genes that are ever found in any of the sequenced representatives from a species.

Par function A site or gene product that is required for the proper partitioning of a plasmid.

Paralogs Genes that have resulted from duplication of a gene. They generally have similar functions but may have distinct functions.

Parent One of the two strains of an organism participating in a genetic cross.

Parental types Progeny of a genetic cross that are genetically identical to one or the other of the parents.

Partial diploid A bacterium that has two copies of part of its genome, usually because a plasmid or prophage in the bacterium contains some bacterial DNA. Also called a merodiploid.

Partitioning An active process by which at least one copy of a replicon (plasmid, chromosome, etc.) is distributed into each daughter cell at the time of cell division.

Pathogenicity island (PAI) A DNA element integrated into the chromosome of a pathogenic bacterium that carries genes whose products are required for pathogenicity and that, based on its base composition and codon usage, shows evidence of having been acquired fairly recently by horizontal transfer. It may or may not carry genes for its own integration. Pathogenicity islands form a subset of a more general class of integrated elements called genomic islands.

PCR *See* Polymerase chain reaction.

Peptide bond A covalent bond between the amino (NH_2) group of one amino acid and the carboxyl (COOH) group of another.

Peptide deformylase An enzyme that removes the formyl group from the amino-terminal formylmethionine of newly synthesized polypeptides.

Peptidyltransferase The ribozyme of the 23S rRNA (28S rRNA in eukaryotes) that forms a bond between the carboxyl group of the growing polypeptide and the amino group of the incoming amino acid.

Periplasm The space between the inner and outer membranes in Gram-negative bacteria.

Periplasmic domain A region of a membrane protein located in the periplasm of the cell.

Periplasmic protein A protein located in the periplasm.

Permissive conditions Conditions under which a mutant organism or virus can multiply.

Permissive temperature A temperature at which both a temperature-sensitive mutant (or cold-sensitive mutant) and the wild type can multiply.

Persister A cell that survives a normally lethal treatment (such as exposure to an antibiotic) even though it lacks a specific genetic system allowing it to survive the treatment upon exposure.

Pfams Protein family and domain databases that are useful for categorizing predicted genes or proteins, primarily based on a compilation of protein domains.

Phage Short for bacteriophage.

Phage display A technology that allows the purification and amplification of a phage particle expressing a particular polypeptide on its surface based on the ability of the polypeptide to bind to another protein or chemical compound. The DNA of the phage can then be sequenced to determine the sequence of a polypeptide that binds to the other protein or compound.

Phage genome The nucleic acid (DNA or RNA) that is packaged into the phage particle and contains all the genes of the phage.

Phagemid A cloning vector that contains mostly phage sequences and that can replicate as a phage and be packaged in a phage head but can also replicate as a plasmid.

Phase variation The reversible change of one or more cellular phenotypes—for example, of the cell surface antigens of a bacterium—at a frequency higher than normal mutation frequencies. It can be due to an invertible sequence, etc.

Phasmid A hybrid cloning vector containing mostly plasmid but some phage sequences, including a *pac* site, so it can be packaged in a phage head but does not contain all of the genes to make a phage particle and can usually only replicate as a plasmid.

Phenotype Any identifiable characteristic of a cell or organism that can be altered by mutation.

Phenotypic lag The delay between the time a mutation occurs in the DNA and the time the resulting change in the phenotype of the organism becomes apparent.

Phosphate The chemical group PO_4.

Phosphorylation cascade An interacting set of proteins with kinase, phosphotransferase, and phosphatase activities that transfer signals in the cell by the successive phosphorylation and dephosphorylation of the proteins.

Photolyase An enzyme that uses the energy of visible light to split pyrimidine butane dimers in DNA, restoring the original pyrimidines.

Photoreactivation The process by which cells exposed to visible light after DNA damage achieve greater survival rates than cells kept in the dark. It is due to restoration of pyrimidine dimers to the individual pyrimidines by photolyase.

Physical map A map of DNA showing the actual distance in deoxynucleotides between identifiable sites, such as restriction endonuclease cleavage sites.

Pilin A protein that makes up the structure of pili. *See* Pilus.

Pilus A protrusion or filament composed of protein attached to the surface of a bacterial cell. *See* Sex pilus, Fimbriae.

Pilus-specific phages A phage that infects only cells carrying a particular self-transmissible plasmid. The plasmid contains the genes for the pilus used by the phage as its adsorption site.

Plaques Clear spots in a bacterial lawn as a result of phage killing and lysing the bacteria as the bacterial lawn is forming and the phage is multiplying.

Plasmid Any DNA molecule in cells that replicates independently of the chromosome and regulates its own replication so that the number of copies of the DNA molecule remains relatively constant.

Pleiotropic mutation A mutation that causes many phenotypic changes in the cell.

Point mutation A mutation that maps to a single position in the DNA, usually the change in or addition or deletion of a single base pair.

Poisson distribution A mathematical distribution that can be used to calculate probabilities in certain situations. It can be used to approximate a binomial distribution when the probability of success in a single trial is low but the number of trials is large. It is named after the mathematician who first derived it.

Polarity A condition in which a mutation in one gene reduces the expression of a downstream gene that is cotranscribed into the same mRNA. It can be due either to premature termination of transcription before RNA polymerase reaches the second gene or to dependence of the second gene on translational coupling.

Polyadenylation Addition of multiple adenosine residues to the 3′ end of an RNA to generate a poly(A) tail.

Polycistronic mRNA An mRNA that contains more than one translational initiation region so that more than one polypeptide can be translated from the mRNA.

Polymerase chain reaction (PCR) A technique involving a succession of heating and cooling steps that uses the DNA polymerase from a thermophilic bacterium and two primers to make many copies of a given region of DNA occurring between sequences complementary to the primers.

Polymerase switching The process by which a different type of DNA polymerase replaces the DNA polymerase already found at the primed DNA template.

Polymerization A reaction in which small molecules are joined in a chain to make a longer molecule.

Polymerizing The act of joining small molecules to form a chain.

Polypeptide A long chain of amino acids held together by peptide bonds. Polypeptides are the product of a single gene.

Porin A protein that forms channels in the outer membrane of Gram-negative bacteria by forming a β-barrel in the outer membrane.

Positive regulation A type of regulation in which the gene is expressed only if the active form of a regulatory protein (or RNA) is present.

Positive selection The process of determining conditions under which only a strain with the desired mutation or a particular recombinant type can multiply. Usually just called selection.

Positively supercoiled A DNA molecule in which the two strands of the double helix are wrapped around each other more than about once every 10.5 bp, as predicted by the Watson-Crick structure.

Posttranscriptional regulation Regulation of the expression of a gene that occurs after the mRNA has been synthesized from the gene, for example, in the rate of translation of the mRNA.

Posttranslational regulation Regulation of the expression of a gene that occurs after the protein has been synthesized, e.g., by regulated proteolysis or feedback inhibition.

ppGpp *See* Guanosine tetraphosphate, Stringent response.

Precise excision Removal of a transposon or other foreign DNA element from a DNA in such a way that the original DNA sequence is restored.

Pre-crRNA processing The process by which the RNA transcript from a CRISPR array is cleaved into individual CRISPR RNAs (crRNA) and assembled into the larger guide RNA complex.

Precursors The smaller molecules that are polymerized to form a polymer.

Presynthetic Occurring before actual DNA or RNA synthesis.

Primary structure The sequence of nucleotides in an RNA or of amino acids in a polypeptide.

Primase An enzyme that synthesizes short RNAs to prime the synthesis of DNA chains.

Prime factor A self-transmissible plasmid carrying a region of the bacterial chromosome.

Primed spacer acquisition A process by which an existing spacer in the CRISPR array facilitates the acquisition of additional spacers from the same genetic element.

Primer A single-stranded DNA or RNA polymer that can hybridize to a single-stranded template DNA or RNA and provide a free 3′ hydroxyl end to which DNA polymerase can add deoxynucleotides to synthesize a chain of DNA complementary to the template DNA or RNA.

Primosome A complex of proteins involved in making primers for the initiation of synthesis of DNA strands.

Processive antitermination The process by which RNA polymerase is altered by binding either a protein or a newly transcribed RNA so that it becomes insensitive to downstream transcription termination signals.

Processivity The efficiency with which a biosynthetic complex (e.g., RNA polymerase) moves along its template to carry out its activity without stopping or falling off.

Programmed frameshifting A system in which gene expression (or regulation) depends on a specific shift in the reading frame by the translating ribosome.

Prokaryote A somewhat outdated name for bacteria and archaea whose cells do not contain a nuclear membrane and visible nucleus or many of the other organelles characteristic of the cells of higher organisms. The name erroneously suggests that present-day eukaryotes descended from present-day bacteria and archaea.

Prolyl isomerase An enzyme, often associated with chaperones, that can catalyze the conversion of one isomer of proline to the other isomer. Proline is the only amino acid that has more than one isomer because the carbon in the carboxyl group is not free to rotate.

Promiscuous plasmid A self-transmissible plasmid that can transfer itself into many types of bacteria, which need not be closely related to each other.

Promoter A region on DNA to which RNA polymerase holoenzyme binds in order to initiate transcription.

Promoter escape The step during transcription when RNA polymerase releases the promoter site and enters the elongation phase.

Promoter recognition The first step in transcription, when RNA polymerase holoenzyme binds to the promoter sequence on the DNA.

Prophage The state of phage DNA in a lysogen in which the phage DNA is integrated into the chromosome of the bacterium or replicates as a plasmid.

Protein export The transport of proteins through the cellular membranes to the outside of the cell.

Protein secretion system A cellular structure composed of a number of different proteins that allows the transfer of certain selected proteins through one or both membranes to the outside of the cell.

Proteobacteria One of the largest phylogenetic groups of bacteria, which includes *Escherichia coli.* This phylum within the division *Bacteria* is divided into six classes, *Alpha-*, *Beta-*, *Gamma-*, *Delta-*, *Epsilon-*, and *Zetaproteobacteria*.

Proteome The complete set of proteins expressed in an organism under a particular growth condition.

Proteomics Global analysis of protein expression patterns and protein interactions. It includes techniques such as mass spectrometry, phage display, and two-hybrid analysis.

Protospacer A DNA sequence that is capable of being acquired as a spacer and recognized by a guide RNA complex with the appropriate spacer from a CRISPR array.

Prototroph A strain that can make all of the growth substances made by the original isolate. *See* Auxotrophic mutant.

Pseudopilus A structure that resembles a pilus. *See* Pilus.

PSI-BLAST A program that can find a set of related sequences based on the presence of common sequence patterns. Reiterative sequence alignments are performed with the goal of defining as large a potential family of functionally related proteins as possible.

Purines Bases found in DNA and RNA possessing two rings, adenine and guanine.

Pyrimidines Bases found in DNA (thymine and cytosine) and RNA (uracil) possessing one ring.

Pyrimidine dimer A type of DNA damage in which two adjacent pyrimidine bases are joined by covalent chemical bonds, requiring repair.

Quaternary structure The complete three-dimensional structure of a protein complex, including all the polypeptide chains making up the complex and how they are wrapped around each other.

Quorum sensing A phenomenon in which populations of cells sense cell density by measuring the accumulation of a usually small signal molecule.

Random-mutation hypothesis The generally accepted hypothesis explaining the adaptation of organisms to their environment. It states that mutations occur randomly, free of influence from their consequences, but that mutant organisms preferentially survive and reproduce themselves if the mutations inadvertently confer advantages under the conditions experienced by the organisms.

Reading frame of translation Any sequence of nucleotides in RNA or DNA read three at a time in succession, as during translation of an mRNA.

Recessive mutation In complementation tests, a mutation that does not exhibit its phenotype in the presence of a wild-type allele of the gene.

Recipient The bacterium that receives a mobile plasmid during the process of conjugation, or donor DNA during transformation of transduction.

Recipient bacterium The bacterial strain that receives DNA in a genetic cross, whether transformation, transduction, or conjugation.

Recipient DNA In transposition, the DNA into which the transposon inserts.

Recipient strain *See* Recipient, Recipient bacterium.

Reciprocal cross A genetic cross in which the alleles of the donor and recipient strains are reversed relative to an earlier cross. An example would be a transduction in which the phage was grown on the strain that had the alleles of what was previously the recipient strain and used to transduce a strain with the alleles of what was previously the donor strain. In bacterial crosses, generally what was previously the selected marker becomes an unselected marker.

Recombinant DNA A DNA molecule derived from the sequences of different DNAs joined to each other in a test tube.

Recombinant types In a genetic cross, progeny that are genetically unlike either parent in the cross because they have DNA sequences that are the result of recombination between the parental DNAs.

Recombinase An enzyme that specifically recognizes two sequences in DNA and joins the strands to cause a crossover within the sequences.

Recombination The rejoining of DNA into new combinations.

Recombination-deficient mutant A mutant strain in which DNA shows a reduced capacity for recombination due to a mutation in a *rec* gene whose product is involved in recombination.

Recombineering A technique that uses phage recombination functions to promote recombination between introduced DNA and cellular DNA. It can be used for site-specific mutagenesis, etc., without the need for cloning of the cellular DNA.

Redundancy A feature of the genetic code that results in multiple codons that encode the same amino acid.

Regional mutagenesis Any technique of mutagenesis in which mutations are restricted to a small region of the DNA or genome.

Regulated proteolysis A regulatory mechanism in which degradation of a protein occurs at different rates under different conditions.

Regulation of gene expression Modulation of the rate of synthesis of the active product of a gene so that the active gene product can be synthesized at different rates, depending on the state in which the organism finds itself.

Regulatory cascade A strategy for regulating the expression of genes in a stepwise manner (e.g., during developmental processes) in which the products of genes expressed during one stage of development turn on the expression of genes for the next stage of development and often turn off genes from the previous stage.

Regulatory gene A gene whose product regulates the expression of other genes as well as, sometimes, its own expression.

Regulon The set of operons that are regulated by the product of the same regulatory gene.

Relaxase The protein of a self-transmissible or mobilizable plasmid that makes a cut at the *oriT nic* site, remains attached to the 5′ end at the cut, is secreted into the recipient cell, and rejoins the cut ends in the recipient cell.

Relaxed plasmid A plasmid that has a high copy number because the copy number is not tightly controlled.

Relaxed strain A bacterial strain that continues to make rRNA and other stable RNAs even if starved for an amino acid. These strains usually have a mutation in the *relA* gene that inactivates the RelA enzyme, so they do not synthesize ppGpp in response to amino acid starvation.

Relaxosome The complex of proteins, including the relaxase, which is bound to the *oriT* sequence of a self-transmissible or mobilizable plasmid in the donor cell.

Release factors Nonsense codon-specific proteins that are required, along with EF-G and RRF, for the termination of polypeptide synthesis and the release of the newly synthesized polypeptide from the ribosome when the ribosome encounters an in-frame nonsense codon in the mRNA.

Replica plating A technique in which bacteria grown on one plate are transferred to a fuzzy cloth and then are transferred from the fuzzy cloth onto another plate so that the bacteria on the first plate are transferred to the corresponding position(s) on the second plate.

Replication fork The region in a replicating double-stranded DNA molecule where the two strands are actively separating to allow synthesis of the complementary strands.

Replicative form (RF) The double-stranded DNA or RNA that forms by synthesis of the complementary minus strand after infection by a phage or virus that has a single-stranded genome.

Replicative transposition A type of transposition in which the breaks that form at each end of the transposon are directly joined to the target DNA and replication is integral to completing transposition. The free 3′ ends at the extremities of the transposon are used as primers to synthesize over the transposon, giving rise to a cointegrate that can be resolved, leaving a copy of the element at the original position.

Replicon A DNA molecule capable of autonomous replication because it contains an origin of replication that functions in the cell in which it is located.

Replisome The collection of proteins that act together to produce a second copy of the genome during DNA replication.

Reporter gene A gene whose product is stable and easy to assay and so is convenient for detecting and quantifying the expression of genes to which it is fused. *See* Transcriptional fusion, Translational fusion.

Repressible Able to have expression reduced by a corepressor.

Repressor A protein or RNA that negatively regulates transcription or translation so that synthesis of the gene product is reduced when it is active.

Resolution A process by which two DNAs are separated at a specific site by a resolvase enzyme.

Resolvase A type of site-specific recombinase that breaks and rejoins DNA in *res* sequences in the two copies of the transposon in a cointegrate, thereby resolving the cointegrate into separate DNAs, each with one copy of the transposon.

Response regulator *See* Response regulator protein.

Response regulator protein A protein that is part of a two-component regulatory system. The response regulator receives a signal (usually in the form of a phosphoryl group) from another protein (the sensor kinase) and performs a regulatory function (e.g., activates transcription).

Retrohoming The process by which an element inserts itself into the same site in a different DNA that lacks it by first making an RNA copy of itself and then making a DNA copy of this RNA with a reverse transcriptase, while it inserts the DNA copy into the target DNA by a sort of reverse splicing.

Retrotransposition *See* Retrotransposon.

Retrotransposon A class of transposable element that produces a copy of itself by transcription, where the RNA copy is later converted into DNA for integration into a new site.

Reverse genetics The process in which the function of the product of a gene is determined by first altering the sequence in DNA in the test tube, using molecular biology techniques, and then reintroducing the DNA into a cell to see what effect the mutation has on the organism. Contrast this with forward genetics, in which the mutation is first recognized because of the phenotype it causes.

Reversion Restoration of a mutated sequence in DNA to the wild-type sequence.

Reversion rate The probability that a mutated sequence in DNA will change back to the wild-type sequence each time the organism multiplies.

Revert *See* Reversion.

Revertant An organism in which the mutated sequence in its DNA has been restored to the wild-type sequence.

Reverted *See* Reversion.

RF *See* Replicative form.

Ribonuclease (RNase) An enzyme that cleaves RNA.

Ribonucleoside triphosphate (rNTP) A base (A, U, G, or C) attached to a ribose sugar with three phosphate groups attached in tandem to the 5′ carbon of the sugar.

Ribonucleotide reductase An enzyme that catalyzes the reduction of nucleoside diphosphates to deoxynucleoside diphosphates by removing the hydroxy group at the 3′ carbon of the ribonucleoside diphosphate and replacing it with a hydrogen.

Riboprobe A hybridization probe made of RNA rather than DNA.

Ribosomal proteins The proteins that, in addition to the rRNAs, make up the structure of the ribosome.

Ribosomal RNA (rRNA) Any one of the three RNAs (16S, 23S, and 5S in bacteria) that make up the structure of the ribosome.

Ribosome The cellular structure, made up of about 50 different proteins and three different RNAs, that is the site of protein synthesis.

Ribosome cycle The association and dissociation of the 30S and 50S ribosomes during initiation and termination of translation.

Ribosome release factor (RRF) A factor that acts with release factors and EF-G to release the polypeptide from the peptidyl-tRNA during termination of translation.

Ribosome-binding site (RBS) Site at which the 30S ribosomal subunit binds to initiate translation; includes S-D sequence and initiation codon. *See* Translational initiation region (TIR).

Riboswitch A regulatory RNA element that directly senses a regulatory signal, often a small molecule, and the resulting change in RNA structure modulates the function of the RNA, for example, expression of the downstream coding sequences.

Ribozyme An RNA that has enzymatic activity.

RNA modification Any covalent change to RNA, such as methylation of a base, that does not involve the breaking and joining of phosphate-phosphate or phosphate-ribose bonds in the backbone of the RNA.

RNA polymerase An enzyme that polymerizes ribonucleoside triphosphates to make RNA chains by using a DNA or RNA template.

RNA polymerase core enzyme The $\alpha_2\beta\beta'\omega$ complex of RNA polymerase without a σ factor attached.

RNA polymerase holoenzyme The $\alpha_2\beta\beta'\omega$ complex of RNA polymerase with a σ factor attached.

RNA processing Covalent changes to RNA that involve the breaking and/or joining of phosphate-phosphate or phosphate-ribose bonds in the backbone of the RNA.

RNA-Seq A procedure used to identify and enumerate the RNAs in a cell by reverse transcribing them into DNA and subjecting them to DNA sequencing.

RNase *See* Ribonuclease.

Rolling-circle (RC) replication A type of replication of circular DNAs in which a single-stranded nick is made in one strand of the DNA and the 3′ hydroxyl end is used as a primer to replicate around the circle, displacing the old strand.

Rolling-circle transposon *See* Y2 transposon.

Round of replication The cycle of replication of a circular DNA in which a complete copy of the DNA is made.

rRNA *See* Ribosomal RNA.

Sacculus The cell wall surrounding the bacterial cell, i.e., the murein layer.

Sanger sequencing A process of DNA sequencing that takes advantage of nucleotides that can be integrated into a growing DNA strand by DNA polymerase but that cannot be further extended because they lack a 3′OH.

Satellite virus A naturally occurring virus that depends on another virus for its multiplication.

Saturation genetics A mutant search that is so extensive that, presumably, all of the genes whose products participate in a biological process are represented by mutations.

Screening The process (usually streamlined) of testing a large number of organisms for a particular mutant type.

S-D sequence *See* Shine-Dalgarno sequence.

Sec system The general system encoded by the *sec* genes of bacteria for transporting proteins across the cytoplasmic membrane; it consists of the targeting factors SecA and SecB and includes the components of the SecYEG channel in the inner membrane.

Secondary structure A structure of a polynucleotide or polypeptide chain that results from noncovalent pairing between nucleotides or amino acids in the chain.

Secretion The process of transporting proteins through the membranes.

Segregate *See* Segregation.

Segregation The process by which newly replicated DNAs or genetic alleles are separated into daughter cells or spores.

Selected marker A difference in DNA sequence between two strains in a bacterial or phage cross that is used to select recombinants. The cells are plated under conditions in which only recombinants that have received the donor sequence or allele can multiply.

Selection A procedure in which bacteria or viruses are placed under conditions in which only the wild type or the desired mutant or recombinant can multiply, allowing the isolation of even very rare mutants and recombinants.

Selective conditions Conditions under which only the wild type or the desired mutant can multiply.

Selective media Media that have been designed to allow multiplication of only the desired mutant or the wild type. Such media often lack one or more nutrients or contain a substance that is toxic.

Selective plate An agar plate made with selective medium.

Self-transmissible Carrying all of the genes for its transfer into other bacteria; the term is used to describe DNA elements, including plasmids and transposons.

Self-transmissible plasmid A plasmid that encodes all the gene products needed to transfer itself to other bacteria through conjugation.

Semiconservation replication A type of DNA replication in which the daughter DNAs are composed of one old strand and one newly synthesized strand.

Sensor kinase In two-component systems, a protein that transfers the γ phosphate of ATP to itself in response to a certain environmental or cellular signal and then transfers the phosphate to a response regulator protein that performs some cellular function. *See* Sensor protein.

Sensor kinase protein *See* Sensor kinase.

Sensor protein The protein in a two-component system that detects changes in the environment and communicates that information to the response regulator, usually by transferring a phosphoryl group. *See* Sensor kinase.

Serial dilution A procedure in which an aliquot of a solution is diluted in one tube and then an aliquot of the solution in this tube is diluted in a second tube, and so forth. The total dilution is the product of each of the individual dilutions.

7,8-Dihydro-8-oxoguanine *See* 8-OxoG.

Sex pilus A rod-like structure that forms on the surface of a bacterium containing a self-transmissible plasmid and facilitates transfer of the plasmid or other DNA into another bacterium.

Shine-Dalgarno (S-D) sequence A short sequence, usually about 10 nucleotides upstream of the initiation codon in a bacterial translational initiation region, that is complementary to a sequence in the 3′ end of the 16S rRNA; it helps to position the ribosome for initiation of translation. It is named after the individuals who discovered it.

Shuttle vector A plasmid cloning vector that contains two origins of replication that function in different types of cells so that the plasmid can replicate in both types of cells.

Siblings In microbial genetics, two cells or viruses that arose from the multiplication of the same mutant cell or virus.

Sigma (σ) factor A subunit of RNA polymerase holoenzyme that directs promoter recognition. *See* RNA polymerase holoenzyme.

Signal recognition particle (SRP) A universally evolutionarily conserved particle, composed of RNA and protein, that in bacteria binds the first transmembrane domain of a protein destined for the inner membrane as the protein emerges from the ribosome and directs it to the SecYEG translocon.

Signal sequence A sequence, composed of mostly hydrophobic amino acids, that is located at the N terminus of some membrane and exported proteins and that targets the protein for transport into or through the cytoplasmic membrane. In bacterial proteins transported by the Sec system or the Tat system, the signal sequence is removed as the protein passes into or through the membrane; in proteins targeted by the signal recognition particle to the cytoplasmic membrane, it is usually the first transmembrane domain and is not removed.

Signal transduction pathway A set of proteins that pass a signal from one to the other by direct contact. They do this by chemically altering each other by proteolysis, by transferring a chemical group such as a phosphoryl or methyl group, or by simply binding to each other.

Silent mutation A change in the DNA sequence of a gene encoding a protein that does not change the amino acid sequence of the protein, usually because it changes the last base in a codon, thereby changing it to another codon but one that encodes the same amino acid.

Single mutation A mutation due to a single event that changed the DNA sequence, independent of how many base pairs were changed by the event.

Site-specific mutagenesis One of many methods for mutagenizing DNA in such a way that the change is localized to a predetermined base pair in the DNA.

Site-specific recombinases Enzymes that recognize two specific sites on DNA and promote recombination between them.

Site-specific recombination Recombination that occurs only between defined sequences in DNA. It is usually performed by site-specific recombinases.

6-4 lesion A type of damage to DNA in which the carbon at the 6 position of a pyrimidine is covalently bound to the carbon at the 4 position of an adjacent pyrimidine.

SMC proteins *See* Condensins.

Sortase An enzyme in Gram-positive bacteria that cuts a protein to be displayed on the cell surface at its sorting signal and attaches it through a new peptide bond to a peptide cross bridge in the cell wall.

SOS box The operator sequence to which the LexA repressor binds. It is found close to all of the promoters of SOS genes.

SOS gene A gene that is a member of the LexA regulon, so that its transcription is normally repressed by LexA repressor.

SOS mutagenesis The increased mutagenesis that occurs after induction of the SOS response and that is primarily due to the induction of the *umuC* and *umuD* genes and the subsequent autocleavage of UmuD, which, with UmuC, forms a translesion DNA polymerase that is mistake prone. *See* Weigle mutagenesis.

SOS response Induction of transcription of the SOS genes in response to DNA damage. It is due to stimulation of autocleavage of LexA repressor by the RecA single-stranded DNA nucleoprotein coprotease.

Spacer DNA sequence within a CRISPR array that is complementary to genetic elements that are recognized for destruction by the CRISPR/Cas system.

Spacer acquisition The process of spacers being added to the CRISPR array. Also known as adaptation.

Spanin A protein encoded by some phages that infect Gram-negative bacteria that somehow releases the outer membrane from the cell wall, promoting lysis.

Specialized transduction A type of transduction in which only some selected regions of the chromosome can be transduced from one strain to another. It is usually due to a phage that integrates into only one site in the bacterial chromosome and sometimes mistakenly picks up neighboring chromosomal sequences when it excises.

Spontaneous mutations Mutations that occur in organisms without deliberate attempts to induce them by irradiation or chemical treatment.

SRP *See* Signal recognition particle.

Stalling sequence A polypeptide sequence that interacts with the exit channel of the ribosome and causes translation to pause.

Stimulon The collection of all of the operons that are turned on by a particular environmental condition, independent of whether they are part of the same regulon. *See* Regulon.

Stop codon A codon in an mRNA that signals translation termination. *See* Nonsense codon.

Strain A group of organisms that are identical to each other but differ genetically from other organisms of the same species. A strain is a subdivision of a species.

Strand passage A reaction performed by topoisomerases in which one or two strands of a DNA are cut and the ends of the cut DNA are held by the enzyme to prevent rotation while other strands of the same or different DNAs are passed through the cuts.

Strand transfer reaction The process by which a strand of DNA is broken and joined to another strand of DNA.

Stringent plasmid A plasmid that exists in only one or very few copies per cell because plasmid replication is tightly controlled.

Stringent response Cessation of synthesis of rRNA and other stable RNAs in the cell when the cells are starved for an amino acid. In *Escherichia coli*, it is due to the accumulation of ppGpp synthesized by the RelA protein on the ribosome.

Structural gene One of the genes in an operon for a pathway that encodes one of the enzymes of the pathway.

Sugar A simple carbohydrate with the general formula $(CH_2O)_n$, as found in nature; n is 3 to 9.

Suicide vector A cloning vector, usually plasmid or phage DNA that cannot replicate in the cells into which it is being introduced, so it is degraded or diluted out by subsequent cell divisions.

Supercoiling A condition in which the two strands of the DNA double helix are wrapped around each other either more or less often than predicted from the natural distances dictated by the unstressed helical structure of DNA, i.e., more or less than about 10.5 bp per turn.

Superintegrons Large arrays of single gene cassettes, often without promoters, that are flanked by recombination signals and can be rearranged by a cognate recombinase. They often encode drug resistance or pathogenicity determinants.

Suppression Alleviation of the effects of a mutation by a second mutation elsewhere in the DNA.

Suppressor mutation A mutation elsewhere in the DNA that alleviates the effects of another mutation.

Surface exclusion A process that prevents a recipient bacterium from receiving a second plasmid of the same type by conjugation.

Synapse In recombination, a structure in which two DNAs are held together by pairing between their strands via recombination proteins.

Synteny Conservation of gene order or genetic linkage in the genomes of different types of organisms.

Synthetic genomics An area of investigation enabled by techniques that create very large fragments of DNA or even complete genomes with minimal or no dependence on existing DNA template sequences.

Synthetic lethality screen A selection system set up to isolate mutants with mutations in genes whose products are required for viability only in the absence of another gene product. Often, the other gene is set up so that it is transcribed only from an inducible promoter. The mutations being sought are lethal only in the absence of inducer, when the other gene is not being expressed.

Tagmentation A process by which specific DNA sequences are randomly joined to DNA to provide binding sites for subsequent DNA amplification by PCR.

Tandem duplication A type of mutation that causes a DNA sequence to be followed immediately by the same sequence in the same orientation.

Tandem mass spectrometry (MS-MS) Two mass spectrometric analyses run sequentially so that the first analysis allows selection of a specific peptide ion and the second analysis includes fragmentation of the selected peptide, analyzing the masses of the pieces and thereby determining partial peptide sequences. In the fragmentation step, the bonds that break are almost exclusively along the peptide backbone, and therefore, the ion species detected in the second analysis mostly represent peptide ions.

Target In transposition, the DNA into which a transposon inserts.

Target immunity The process by which transposons are inhibited in their ability to transpose into a DNA region that already contains the same transposon.

Target site duplications A hallmark of the process of DNA transposition resulting from the staggered joining events that occur during the process that results in a sequence being duplicated on either side of the element following host repair.

Tautomer A (usually temporary) form of a molecule in which the electrons are distributed among the atoms differently than the normal configuration.

TEC *See* Transcription elongation complex.

Temperate phage A phage that is known to be capable of lysogeny.

Temperature-sensitive mutant A mutant that cannot grow in the temperature range in which the wild type can multiply, usually indicating that a lower temperature is needed to permit growth.

Template strand The strand of DNA that is used to make a copy. In the case of DNA replication, an RNA primer is most often used to initiate polymerization of the DNA copy. In the case of transcription, the term refers to a region from which RNA is synthesized that serves as the template for RNA synthesis; the sequence of this DNA strand is complementary to the sequence of the RNA. *See* Transcribed strand, Noncoding strand.

Terminal redundancy Containing direct repeats at both ends of a DNA (usually a phage genome), that is, the sequences at both ends are the same in the direct orientation.

Terminate End a process, such as transcription or translation.

Termination of transcription The process by which the RNA polymerase leaves the DNA and the RNA chain is released.

Termination of translation The process by which the ribosome leaves the mRNA and the polypeptide is released when a nonsense codon in the mRNA is encountered in frame.

Termination codons *See* nonsense codon.

Terminator A sequence in the DNA and/or structure in the RNA that causes RNA polymerase to stop transcription and to release both the DNA template and the RNA transcript.

Tertiary structure The three-dimensional structure of a polypeptide or RNA.

Theta replication A type of replication of circular DNA in which the replication apparatus initiates at an origin of replication and proceeds in one or both directions around the circle with leading and lagging strands of replication. The expanded molecule in an intermediate state of replication resembles the Greek letter theta (θ).

3′ End The terminus of a polynucleotide chain (DNA or RNA) ending in the nucleotide that is not joined at the 3′ carbon of its deoxyribose or ribose to the 5′ phosphate of another nucleotide.

3′ Hydroxyl end In a polynucleotide, a 3′ end that has a hydroxyl group on the 3′ carbon of the ribose sugar of the last nucleotide without a phosphate group attached.

3′ Untranslated region (3′ UTR) In an mRNA, the sequences downstream or 3′ of the nonsense codon of an open reading frame that encodes a protein.

Thymine One of the pyrimidine (one-ringed) bases in DNA and some tRNAs.

TIR *See* Translational initiation region.

tmRNA *See* Transfer-messenger RNA.

Topoisomerase An enzyme that can alter the topology of a DNA molecule by cutting one or both strands of DNA, passing other DNA strands through the cuts while holding the cut ends so that they are not free to rotate, and then resealing the cuts.

Topoisomerase IV A type II topoisomerase of *Escherichia coli* that is responsible for decatenation of daughter chromosomes after replication.

Topology The relationship of the strands of DNA to each other in space.

tra Genes encoding products allowing self-transmissible DNA elements to transfer themselves into other bacteria.

***trans* acting** A function or mutation that affects DNA in the cell other than the DNA site from which it is derived.

***trans*-acting function** A gene product that can act on DNAs in the cell other than the one from which it was made.

Transconjugant A recipient cell that has received and integrated DNA from another cell by conjugation.

Transcribe To make an RNA that is a complementary copy of a strand of DNA.

Transcribed strand In a region of a double-stranded DNA that is transcribed into RNA, the strand of DNA that is used as a template and so is complementary to the RNA. *See* Template strand.

Transcript A complementary RNA made from a region of DNA.

Transcription antitermination The process by which RNA polymerase can be made to proceed through one or more termination sites. This can be accomplished through changes in the secondary structure of the mRNA, creating a new secondary structure that masks a transcription terminator secondary structure, or through changes to the transcription elongation complex. *See* Processive antitermination.

Transcription attenuation Regulation of gene expression by initiating transcription but then prematurely terminating transcription unless certain conditions are met.

Transcription bubble The ~17-bp region in DNA during transcription in which the two strands of DNA have been separated by the RNA polymerase and within which the newly synthesized RNA forms a short RNA-DNA duplex with the transcribed strand of DNA.

Transcription elongation The step of transcription during which the RNA polymerase core moves along the DNA and synthesizes RNA. *See* Transcription elongation complex.

Transcription elongation complex (TEC) The RNA polymerase complex that is active in transcription elongation.

Transcription start site The nucleotide in the coding strand of DNA in a promoter that corresponds to the first nucleotide polymerized into RNA from the promoter.

Transcription termination *See* Termination of transcription.

Transcriptional activator A protein that is required for transcription of an operon. The protein makes contact with the RNA polymerase and allows the RNA polymerase to initiate transcription from the promoter of the operon.

Transcriptional autoregulation The process by which a protein regulates the transcription of its own gene, by being either a repressor or an activator of its own transcription.

Transcriptional autorepression The process by which a protein represses the transcription of its own gene.

Transcriptional fusion Introduction of a gene downstream of the promoter for another gene or genes so that it is transcribed from the promoter for the other gene(s) into the same mRNA but is translated as a separate polypeptide from its own translational initiation region.

Transcriptional regulation Regulation in which the amount of product of a gene that is synthesized under certain conditions is determined by how much mRNA is made from the gene.

Transcriptional regulator Any protein that regulates the transcription of genes, e.g., a repressor, activator, or antitermination protein.

Transcriptome The complete set of transcripts expressed in an organism at a given time under a given condition. The transcripts actually detected depend on their abundance under the experimental conditions used.

Transducing particle A phage whose head contains bacterial DNA, as well as, or instead of, its own DNA.

Transducing phage A type of phage that sometimes packages bacterial DNA during infection and introduces it into other bacteria during infection of those bacteria.

Transductant A bacterium that has received DNA from another bacterium by transduction.

Transduction A process in which DNA other than phage DNA is introduced into a bacterium via infection by a phage containing the DNA.

Transfection Initiation of a virus infection by introducing virus DNA or RNA into a cell by transformation or electroporation, rather than by infection by viral particles.

Transfer-messenger RNA (tmRNA) A small RNA found in bacteria that is a hybrid between a tRNA and an mRNA. It can be aminoacylated with alanine like a tRNA and enter the A site of the ribosome if the A site is unoccupied, e.g., if the ribosome has reached the 3′ end of an mRNA without encountering a nonsense codon. A short reading frame on the tmRNA is then translated, fusing a short peptide sequence to the C terminus of the truncated protein, which targets the protein for degradation by a protease.

Transfer RNA (tRNA) The small, stable RNAs in cells to which specific amino acids are attached by aminoacyl tRNA synthetases. The tRNA with the amino acid attached enters the ribosome and base pairs through its anticodon sequence with a 3-nucleotide codon sequence in the mRNA to insert the correct amino acid into the growing polypeptide chain.

Transformant A cell that has received DNA by transformation.

Transformation Introduction of DNA into cells by mixing the DNA and the cells.

Transformylase The enzyme that transfers a formyl (CHO) group onto the amino group of methionine to make formylmethionine.

Transient diploids Cells in the temporary state of diploidy that exists after a DNA that cannot replicate in that type of cell enters the cell and before it is lost or degraded.

Transition mutation A type of base pair change mutation in which the purine base has been changed into the other purine base and the pyrimidine base has been changed into the other pyrimidine base (e.g., AT to GC or GC to AT).

Translated region A region of an mRNA that encodes a protein.

Translation The process by which the information in an mRNA is used to dictate the amino acid sequence of a protein product.

Translation elongation factor G (EF-G) The protein required to move the peptidyl-tRNA from the A site to the P site on the ribosome with the concomitant cleavage of GTP to GDP after the peptide bond has formed.

Translation elongation factor Tu (EF-Tu) The protein that binds to aminoacylated tRNA and accompanies it into the A site of the ribosome. It then cycles off the ribosome with the concomitant cleavage of GTP to GDP, leaving the aminoacylated tRNA behind.

Translation termination *See* Termination of translation.

Translational coupling A gene arrangement in which the translation of one protein-coding sequence on a polycistronic mRNA is required for the translation of the second, downstream coding sequence. Often, translation of the upstream coding sequencing is required to remove secondary structure in the mRNA that blocks the translational initiation region for the downstream coding sequence.

Translational fusion The fusion of parts of the coding regions of two genes so that translation initiated at the translational initiation region for one polypeptide on the mRNA will continue into the coding region for the second polypeptide in the correct reading frame for the second polypeptide. A polypeptide containing amino acid sequences from the two genes that were joined to each other will be synthesized. *See* Fusion protein.

Translational initiation region (TIR) The initiation codon, the Shine-Dalgarno sequence, and any other surrounding sequences in mRNA that are recognized by the ribosome as a place to begin translation. *See* Ribosome-binding site (RBS).

Translational regulation Variation, under different conditions, in the amount of synthesis of a polypeptide due to variation in the rate at which the polypeptide is translated from the mRNA.

Translationally autoregulated Able to affect the rate of translation of its own coding sequence on its mRNA. Usually in such cases, the protein binds to its own translational initiation region or that of an upstream gene to which it is translationally coupled; hence, the protein represses its own translation.

Translationally coupled A condition where coding regions are present on the same transcript, where translation of the downstream coding region is dependent on translation of the upstream coding sequence. *See* Translational coupling.

Translesion synthesis (TLS) Synthesis of DNA over a template region containing a damaged base or bases that are incapable of proper base pairing.

Translocase The evolutionarily highly conserved channel in the cytoplasmic membrane through which proteins are exported. In bacteria, it is represented by the SecYEG membrane channel.

Transmembrane domain The region in a polypeptide between a region that is exposed to one surface of a membrane and a region that is exposed to the other surface. This region must traverse and be embedded in the membrane. Usually, transmembrane domains have a stretch of at least 20 mostly hydrophobic amino acids that is long enough to extend from one face of a bilipid membrane to the other.

Transmembrane protein A membrane protein that has surfaces exposed on both sides of the membrane.

Transposase An enzyme encoded by a transposon that breaks the DNA at both ends of the transposon and joins these ends to a target DNA during transposition.

Transposition Movement of a transposon from one place in DNA to another.

Transposon A DNA sequence that can move from one place in DNA to another, using a specialized recombinase called a transposase. It should be distinguished from homing DNA elements, which usually move only into the same sequence in another DNA and depend on homologous recombination, or DNA elements that insert into other DNAs, using recombinases called integrases.

Transposon mutagenesis A technique in which a transposon is used to make random insertion mutations in DNA. The transposon is usually introduced into the cell in a suicide vector, so it must transpose into another DNA in the cell to become established.

Transversion mutation A type of base pair change mutation in which the purine in the base pair is changed into the pyrimidine, and vice versa, e.g., GC to TA or GC to CG.

Trigger factor A chaperone in *Escherichia coli* that is closely associated with the exit pore of the ribosome and that helps proteins fold as they emerge from the ribosome. It can partially substitute for DnaK.

Triparental mating A conjugational mating for introducing mobilizable plasmids into cells in which three strains of bacteria are mixed. One strain contains a self-transmissible plasmid, which transfers itself into the second strain, containing a mobilizable plasmid, which in turn is mobilized into the third strain.

Triple-stranded structure Three strands of DNA held together in a triple-stranded structure that has been hypothesized to form when a RecA nucleoprotein filament invades a double-stranded DNA.

tRNA *See* Transfer RNA.

$tRNA^{fMet}$ The tRNA to which formylmethionine is attached and that pairs with the initiator codon in a translational initiation region to initiate translation of a polypeptide in bacteria. *See* Initiator tRNA.

True revertant A strain in which a mutation was changed back to the original wild-type sequence.

Two-component regulatory system A pair of proteins, one of which, the sensor kinase, undergoes a change in response to a change in the environment and communicates this change, usually in the form of a phosphate, to another protein, the response regulator, which then causes the appropriate cellular response. Different two-component systems are often highly similar to each other, which allows them to be identified in sequenced bacterial genomes. Also referred to as two-component signal transduction.

Two-dimensional polyacrylamide gel electrophoresis (2D-PAGE) A separation technique in which proteins are applied to a pI (isoelectric point) strip and separated by charge by using isoelectric focusing; this strip is then attached to another slab acrylamide gel containing sodium dodecyl sulfate so that the proteins move at right angles to the first gel and are separated by size.

Type I secretion system A protein secretion system in Gram-negative bacteria based on a specific ATP-binding cassette (ABC) transporter and the TolC channel, a multiuse channel. Proteins secreted by type I systems recognize a signal sequence in the carboxyl terminus of the protein that is not cleaved off during transport.

Type II secretion system A protein secretion system of Gram-negative bacteria that uses either the SecYEG channel or the Tat channel to transport proteins through the inner membrane and then uses a specific secretin β-channel to transport the protein through the outer membrane. It makes a complicated structure called a pseudopilus, which may push the protein through the inner membrane channel and through the secretin channel to the outside of the cell.

Type III secretion system A multicomponent protein secretion system of pathogenic Gram-negative bacteria that forms a syringe-like structure, sometimes called an injectisome, that injects effector proteins directly through both bacterial membranes into eukaryotic cells.

Type IV secretion system A protein secretion system of Gram-negative bacteria that can inject proteins directly through both bacterial membranes into other cells, although some seem to use the SecYEG channel to transport the protein through the inner membrane. Plasmid conjugation systems, also found in Gram-positive bacteria, are essentially type IV secretion systems.

Type V secretion system A group of secretion systems that includes the autotransporters, the two-partner secretion systems, and the chaperone-usher systems. These secretion systems all form a dedicated β-barrel in the outer membrane that exports only one or a select group of proteins, and they also all use the SecYEG channel to transport the exported protein through the inner membrane.

Type VI secretion system A secretion system encoded by a highly conserved cluster of at least 12 genes with as many as 9 other genes, which might encode specific effectors.

UAS *See* Upstream activator sequence.

Uninducible A regulatory system that cannot be expressed even in the presence of the inducing signal.

Unselected marker A difference between the DNA sequences of two bacteria or phages involved in a genetic cross that can be used for genetic mapping. Mapping information can be obtained by testing recombinants that have been selected for being recombinant for one marker, the selected marker, to determine if they have the sequence of the donor or the recipient for another marker, an unselected marker.

UP element A region of DNA upstream of a promoter to which the α subunit of RNA polymerase binds during transcription initiation to enhance promoter recognition.

Upstream Lying in the 5′ direction from a given point on RNA or in the 5′ direction on the coding strand of a DNA region from which an RNA is made.

Upstream activator sequence (UAS) A DNA sequence upstream of a promoter that increases transcription from the promoter by binding an activator protein. It is usually associated with NtrC family activators and σ^{54} promoters. It can be many hundreds of base pairs upstream from the promoter. Also called upstream activator site.

Uptake sequences Short DNA sequences that allow DNA containing the sequence to be bound and taken up by some types of bacteria during natural transformation.

Uracil (U) One of the pyrimidine (one-ringed) bases; naturally found in RNA.

Uracil-*N*-glycosylase An enzyme that removes the uracil base from DNA by cleaving the bond between the base and the deoxyribose sugar.

UvrABC endonuclease A complex of three proteins that cuts on both sides of any DNA lesion that causes a significant distortion of the helix as a first step in excision repair of the damage. Also called UvrABC excinuclease.

Variants Usually, different strains of the same type of bacteria as they are isolated from nature.

Vertical transfer Transfer of DNA from an organism to its progeny solely through reproduction. *See* Horizontal transfer.

Very-short-patch (VSP) repair A type of repair in enteric bacteria that removes the mismatched T along with a very short stretch of DNA in the sequence CT(A/T)GG(T/A)CC and replaces it with a C. It is due to deamination of 5-methylcytosine at this position in these bacteria.

W mutagenesis *See* Weigle mutagenesis.

W reactivation *See* Weigle reactivation.

Weigle mutagenesis Another name for SOS mutagenesis. It refers to the increase in the number of phage mutations if phage infect cells that have been preirradiated with UV. It is due to SOS induction of the *umuCD* genes, as well as *recA*. It is named after Jean Weigle, who first observed it.

Weigle reactivation The increased ability of phages to survive UV irradiation damage to their DNA if the cells they infect have been previously exposed to UV irradiation. It is due to SOS induction of repair functions. It is named after Jean Weigle, who first observed it.

Wild type The normal type. Literally, the term refers to the organism as it was first isolated from nature. In a genetic experiment, it is the strain from which mutants are derived.

Wild-type allele The form of a gene as it exists in the wild-type organism.

Wild-type phenotype The particular outward trait characteristic of the wild type that is different in the mutant.

Wobble The ability of the base of the first nucleotide (read 5′ to 3′) in the anticodon of a tRNA to pair with more than one base in the third nucleotide (read 5′ to 3′) of a codon in the mRNA. The term also generally applies to the ability of G and U residues in RNA to pair with each other.

Xanthine A purine base that results from deamination of guanine.

Y2 transposon A transposon with two Ys (tyrosines) in its active center, sometimes called a rolling-circle transposon because the mechanism of transposition resembles rolling-circle replication of phages and plasmids.

Y family of polymerases A large group of DNA polymerases, represented by DinB (Pol IV), and UmuC (Pol V) in *Escherichia coli*, that are capable of translesion synthesis, perhaps because they have a more open active center and lack editing functions.

YidC An inner membrane protein of unknown function that cooperates with the SecYEG channel in inserting inner membrane proteins into the inner membrane. Some proteins are inserted by YidC alone.

Zero frame In the coding region of a gene, the sequence of nucleotides, taken three at a time, in which the polypeptide encoded by the gene is translated.

Index

D

E

F

G

H

M

Q

R

S

T

U

V

W

X

Y

Z